CHEMISTRY

About the cover

The cover art illustrates how chemistry embraces not only everything on Earth but also the entire universe. From the gigantic structures of billions of galaxies scattered across billions of light years to the level of atoms and molecules measured in billionths of meters, there is chemistry. Water, one of the most prevalent molecules in the universe and one that is essential for life on Earth, is illustrated at the molecular level by space-filling models on the left of the cover. We begin our study of chemistry with water. Interactions of water with other molecules are part of the chemistry of life and are represented on the right by an electron-density model of a mixture of water and an alcohol. Interactions like these multiplied many times over with a great variety of even larger molecules lead to the organized structures that make up living cells shown at the right. And further organization leads to more complex multicellular organisms, including you. Thus, we come full circle, because it is humans like you who have organized all these concepts into the discipline called "chemistry."

			13 III	14 IV	15 V	16 VI	17 VII	18 VIII
				+1	−3	−2	−1	
				1 H Hydrogen 1.0079 $1s^1$				1s 2 He Helium 4.00 $1s^2$
			5 B Boron 10.81 $2s^22p^1$	6 C Carbon 12.01 $2s^22p^2$	7 N Nitrogen 14.01 $2s^22p^3$	8 O Oxygen 16.00 $2s^22p^4$	9 F Fluorine 19.00 $2s^22p^5$	10 Ne Neon 20.18 $2s^22p^6$
10	11	12	13 Al Aluminum 26.98 $3s^23p^1$	14 Si Silicon 28.09 $3s^23p^2$	15 P Phosphorus 30.97 $3s^23p^3$	16 S Sulfur 32.06 $3s^23p^4$	17 Cl Chlorine 35.45 $3s^23p^5$	18 Ar Argon 39.95 $3s^23p^6$
28 Ni Nickel 58.71 $3d^84s^2$	29 Cu Copper 63.54 $3d^{10}4s^1$	30 Zn Zinc 65.37 $3d^{10}4s^2$	31 Ga Gallium 69.72 $3d^{10}4s^24p^1$	32 Ge Germanium 72.59 $3d^{10}4s^24p^2$	33 As Arsenic 74.92 $3d^{10}4s^24p^3$	34 Se Selenium 78.96 $3d^{10}4s^24p^4$	35 Br Bromine 79.91 $3d^{10}4s^24p^5$	36 Kr Krypton 83.80 $3d^{10}4s^24p^6$
46 Pd Palladium 106.42 $4d^{10}$	47 Ag Silver 107.87 $4d^{10}5s^1$	48 Cd Cadmium 112.40 $4d^{10}5s^2$	49 In Indium 114.82 $4d^{10}5s^25p^1$	50 Sn Tin 118.69 $4d^{10}5s^25p^2$	51 Sb Antimony 121.75 $4d^{10}5s^25p^3$	52 Te Tellurium 127.60 $4d^{10}5s^25p^4$	53 I Iodine 126.90 $4d^{10}5s^25p^5$	54 Xe Xenon 131.29 $4d^{10}5s^25p^6$
78 Pt Platinum 195.08 $4f^{14}5d^96s^1$	79 Au Gold 196.97 $4f^{14}5d^{10}6s^1$	80 Hg Mercury 200.59 $4f^{14}5d^{10}6s^2$	81 Tl Thallium 204.37 $4f^{14}5d^{10}6s^26p^1$	82 Pb Lead 207.19 $4f^{14}5d^{10}6s^26p^2$	83 Bi Bismuth 208.98 $4f^{14}5d^{10}6s^26p^3$	84 Po Polonium (209) $4f^{14}5d^{10}6s^26p^4$	85 At Astatine (210) $4f^{14}5d^{10}6s^26p^5$	86 Rn Radon (222) $4f^{14}5d^{10}6s^26p^6$
110 Ds Darmstadtium (271) $5f^{14}6d^87s^2$	111	112	113					

62 Sm Samarium 150.35 $4f^66s^2$	63 Eu Europium 151.96 $4f^76s^2$	64 Gd Gadolinium 157.25 $4f^75d^16s^2$	65 Tb Terbium 158.92 $4f^96s^2$	66 Dy Dysprosium 162.50 $4f^{10}6s^2$	67 Ho Holmium 164.93 $4f^{11}6s^2$	68 Er Erbium 167.26 $4f^{12}6s^2$	69 Tm Thulium 168.93 $4f^{13}6s^2$	70 Yb Ytterbium 173.04 $4f^{14}6s^2$
94 Pu Plutonium (244) $5f^67s^2$	95 Am Americium (243) $5f^77s^2$	96 Cm Curium (247) $5f^76d^17s^2$	97 Bk Berkelium (247) $5f^97s^2$	98 Cf Californium (251) $5f^{10}7s^2$	99 Es Einsteinium (254) $5f^{11}7s^2$	100 Fm Fermium (257) $5f^{12}7s^2$	101 Md Mendelevium (258) $5f^{13}7s^2$	102 No Nobelium (255) $5f^{14}7s^2$

CHEMISTRY

*A Project of the
American Chemical Society*

W. H. Freeman and Company
New York

Publisher: Susan Finnemore Brennan
Acquisitions Editor: Clancy Marshall
Marketing Manager: Mark Santee
Media Editors: Victoria Anderson, Amanda McCorquodale, Charles Van Wagner
Project Editor: PreMediaONE, A Black Dot Group Company, Vivien Weiss
Photo Editor: Patricia Marx
Photo Researcher: Nigel Assam
Text Designer: Circa 86
Illustrations: Network Graphics
Illustration Coordination/Text Composition: PreMediaONE,
    A Black Dot Group Company
Production Coordinator: Susan Wein
Printing and Binding: Quebecor World
Illustrations created by Network Graphics. Spartan images provided by I. D.
    Eubanks. Some illustrations also provided by American Chemical Society.

Library of Congress Cataloging-in-Publication Data

Bell, Jerry A.
Chemistry : a project of the American Chemical Society/by Jerry Bell.

    p. cm.

Includes index.
ISBN 0-7167-3126-6 (pbk.)
1.   Chemistry. I. American Chemical Society. II. Title.
QD31.3.B45 2004
540—dc22
    2004040421

W. H. Freeman and Company
41 Madison Avenue, New York, NY 10010
Houndmills, Basingstoke RG21 6XS, England
www.whfreeman.com

2001–2002 Field Testers

Priscilla Bell, Whittier College, Whittier, CA

Kent A. Chambers, Hardin-Simmons University, Abilene, TX

Michelle M. Dose, Hardin-Simmons University, Abilene, TX

Timothy T. Ehler, Buena Vista College, Storm Lake, IA

Laura Eisen, The George Washington University at Mount Vernon College, Washington D.C.

Amina K. El-Ashmawy, Collin County Community College, Plano, TX

Mike Falcetta, Roberts Wesleyan College, Rochester NY

Jeanne Gulnick, Saint Joseph's College, Standish, Maine

Martin B. Jones, Adams State College, Alamosa, CO

Glenn L. Keldsen, Purdue University North Central, Westville, IN

Jonathan Mitschele, Saint Joseph's College, Standish, Maine

Chuck F. Reeg, Whittier College, Whittier, CA

R. Neil Rudolph, Adams State College, Alamosa, CO

Reginald B. Shiflett, Meredith College, Raleigh, NC

Thomas J. Whitfield, Community College of Rhode Island, Warwick, RI

Marie Wolff, Joliet Junior College, Joliet, IL

These faculty and their students made substantial contributions to improving earlier draft versions of this textbook. The Editorial/Writing Team acknowledges their past and continuing efforts with gratitude and appreciation.

Brief Contents

Contents

Introduction

Everything you hear, see, smell, taste, and touch involves chemistry and chemicals (matter). And hearing, seeing, smelling, tasting, and touching all involve intricate series of chemical reactions and interactions in your body. With such an enormous range of topics, chemistry offers you fascinating opportunities to explore and to study. At the same time, the sheer breadth of these possibilities may make chemistry seem a daunting subject to study. Aware of both the fascination and the challenge of studying chemistry, the American Chemical Society chose a team of chemists to consider what concepts would help you open the doors to opportunities that require a knowledge of chemistry without being overwhelming. The team also took up the challenge to develop effective approaches to learning and teaching chemistry. The result of the team's efforts is this textbook, *Chemistry,* and its complementary materials, including project-based laboratory experiments, your molecular model kit, the ***Web Companion,*** and the ***Personal Tutor.***

Learning chemistry, even with a limited range of concepts and content, requires a good deal of effort from both you and your instructors. To facilitate your efforts, we have written ***Chemistry*** in a conversational tone designed to be accessible and engaging. But you cannot learn chemistry only by reading about it, just as you cannot learn how to write a short story or how to find fossils simply by reading about how others do it. Learning how others do something you want to do is important, but you must also practice doing it yourself. Chemists and other scientists learn about the world through experimenting. They then try, often in collaborative efforts, to develop models of the world at the molecular level that explain their results and allow them to predict the outcomes of other possible experiments. We have tried to incorporate this same approach in this textbook.

Throughout ***Chemistry,*** we present activities and thought-provoking questions that are intended to promote active small-group and whole-class participation. To encourage your participation and collaborative learning efforts, four features appear often in each chapter:

 Intro.1 INVESTIGATE THIS

An *Investigate This* usually involves short experiments that introduce the chemical concepts explored in the subsequent paragraphs. The investigations are designed to be carried out in small groups or in the whole-class setting.

Intro.2 CONSIDER THIS

A *Consider This* follows each *Investigate This* and usually asks you to discuss and develop hypotheses or explanations for what you have observed. At other places, a *Consider This* will ask you to think about and discuss the consequences of what has just been presented or to anticipate what is to come. The intent in all cases is to involve the class in a discussion.

Intro.3 WORKED EXAMPLE

Each *Worked Example* guides you through the reasoning involved in solving a problem. Thinking about *how* to solve a problem is often more important and more challenging than actually carrying out the solution procedure, so we place an emphasis on this thinking. Almost all *Worked Examples* include the following components, after the statement of the problem:

Necessary information:
What do you need to know, including the information from the problem statement, to solve the problem?

Strategy:
How do you put the information together to solve the problem? What concepts are involved and how are they to be used? With an appropriate strategy (there is often more than one) in hand, the problem is essentially solved.

Implementation:
Carry out the strategy using the needed information to solve the problem. Calculations, if necessary, are done at this stage.

Does the answer make sense?
Once you get an answer to a problem, you should always check to be sure it makes sense. You should also check to be sure you have carried out any numerical calculations correctly; but making sense of the answer is a distinct task (and can sometimes flag possible numerical problems). Is the answer about the size you would expect (based on other experiences, for example)? Does it have the expected direction (sign or change from some baseline)? And so on . . . .

Intro.4 CHECK THIS

At least one *Check This* follows each *Worked Example* and presents a similar problem or problems so that you can practice the strategy presented in the *Worked Example*. *Check This* problems also appear in other places, where you are asked to practice some technique or answer questions based on what has just been presented in the text. ***Chemistry*** is designed to be used with paper, pencil, calculator, and model kit at hand, so you can try each *Check This* as you come to it.

In addition to these features within each chapter, there is an *Outcomes Review* section near the end of each chapter, a pause for *Reflection and Projection* at the end of every few sections, and *End-of-Chapter Problems* you can use to test your problem-solving skills. Use the *Outcomes Review* to remind yourself of the important ideas from the chapter and the *Reflection and Projection* pause to think back on the preceding sections and consider where they are heading. Use the *End-of-Chapter Problems* to check your understanding of all these ideas. Some of these problems will give you more practice with the kinds of problems you meet in the *Worked Example* and *Check This* activities throughout the chapter. Other *End-of-Chapter Problems* are included to stretch your thinking and engage you in problem-solving strategies that are combinations of strategies introduced in the chapter or that extend a bit beyond them. Most scientists work cooperatively, and we encourage you to try working on these problems collaboratively as well. Often a group can come up with more and better solutions than an individual working in isolation. The numerical answers to *Check This* and *End-of-Chapter Problems* are given in Appendix A, so you can check your solutions.

Throughout ***Chemistry,*** you will find an emphasis on understanding and reasoning and on models of all kinds: physical, computer, and mathematical models and analogies. We use models, because it is difficult to observe individual atoms or molecules as they undergo the changes and interactions that lead to the events we can easily observe in nature or in the laboratory. As we try to understand the physical and chemical properties of atoms and molecules and how they cause observable effects, we will use three levels of description, which are exemplified by this page from the ***Web Companion:***

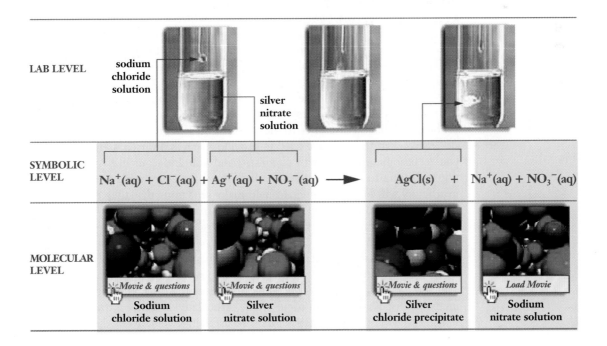

LAB LEVEL	sodium chloride solution	silver nitrate solution		
SYMBOLIC LEVEL	$Na^+(aq) + Cl^-(aq)$	$+ Ag^+(aq) + NO_3^-(aq)$	$\longrightarrow$	$AgCl(s) + Na^+(aq) + NO_3^-(aq)$
MOLECULAR LEVEL	*Movie & questions* Sodium chloride solution	*Movie & questions* Silver nitrate solution	*Movie & questions* Silver chloride precipitate	*Load Movie* Sodium nitrate solution

- *Lab Level:* These are the observations you make on macroscopic systems, such as that shown in these frames from a movie of a reaction between two solutions in a test tube or in your *Investigate This* activities.
- *Symbolic Level:* Intermediate between and connecting the Lab and Molecular Levels is the Symbolic Level of description. This is the descriptive level that you probably associate with chemistry. This level is essential because it combines a great deal of information in a succinct format. It is also the most abstract level of representation, since all the symbols need to be interpreted to make sense of the description. The usual approach in *Chemistry* will be first to try to understand systems at the Lab and Molecular Levels and only then proceed to the Symbolic Level.
- *Molecular Level:* These are our models of what is going on among the particles (atoms, ions, and molecules) that gives rise to the effects observed in the laboratory. These models are animated in the *Web Companion* and are usually shown as less complicated still figures in the text. You will also often use models you build yourself with your molecular model kit.

Web Companion

> **WEB Chapter X,**
> **Section X.5.3-4**
> ① ② ③ ④
>
> A brief description of what you will find on these pages is given here.

The *Web Companion,* from which the preceding figure is taken, is designed to afford you opportunities to use interactive animations, movies, and other resources that provide a visual (usually moving) means to examine many of the concepts included in the written text. When a *Web Companion* page (or pages) is available for some concept, a marginal icon and a description appear beside the text. The mouse, mousepad, and hand icon remind you that there is a computer-based complement to the accompanying written text. The rectangle contains the chapter and section reference in the *Companion* and the dangling circles are a reminder of how you can navigate within a section of the *Companion*. As indicated, each reference will be accompanied by a brief description of what to expect when you access the *Companion*.

To access the *Web Companion,* visit www.whfreeman.com/acsgenchem and select "Web Companion." Find the appropriate location in the *Companion* by selecting the chapter and subsection referenced. The *Companion* may also be available on your institutional computers and, if so, your instructors will tell you how to access it. There are also *Consider This* and *Check This* problems, as well as *End-of-Chapter Problems,* based on the *Web Companion*. These are denoted by this icon, 🖱, a smaller version of the mouse, mousepad, and hand, and the problem will have a reference to the chapter and section you will need to access.

As you can see, we have tried to provide a rich and varied menu of ways for you to become engaged with chemistry and you and your class will probably not have time to delve deeply into all of them for every topic you study. Your instructors will be your guides to help make selections that they think best for what your course is intended to accomplish. To help them in this task, we have prepared an on-line *Faculty Resource and Organizational Guide* (the *FROG*) that includes details and results for each *Investigate This* activity as well as alternatives to the ones in the text, further leading questions for discussions that result when you tackle the *Consider This* problems, possible solutions for all the *Check This* and *End-of-Chapter Problems,* and estimates of the time required for development and discussion of most topics in the text. Many of the ideas and suggestions in the *FROG* come from instructors who used draft versions of these materials.

We plan the *FROG* to be updated regularly. When your instructors use a strategy, an activity, or an example that is not in the text but helps you understand a concept better, we urge you to encourage them to submit their ideas for inclusion as the *FROG* is updated, so other students might benefit as well.

A very large percentage of students who take the general chemistry course in a college or university have already had at least one year of a high school chemistry course. We assume that you are in this category and have probably been exposed to a good deal of the nomenclature and methods that are part of the study of chemistry. We take advantage of this background to move quickly into an examination of the properties of water that depends on some familiarity with the properties of atoms and molecules. You probably have also done some of the algebraic and arithmetic calculations that are a part of essentially all beginning chemistry courses. We take advantage of this experience as well, by providing a review of only necessary concepts and then using them to try to answer questions based on our initial studies of water. The brief reviews we provide in the text may not, however, be enough to make you comfortable with the problems we pose, so we have provided a *Personal Tutor*.

The primary purpose of the *Personal Tutor* is to give you more guidance and practice with problems and computations in the areas that seem to give students trouble. Using the *Personal Tutor* is much like visiting your instructor during office hours or going to a human tutor. Marginal boxes like this one will alert you to a section in the *Personal Tutor* that might be helpful for the topic under discussion. Before using the *Tutor* for the first time, visit www.whfreeman.com/acsgenchemhome, select "Personal Tutor," and take the diagnostic exam. When you have completed the exam, you will get feedback on what sections of the *Personal Tutor* would be helpful for you to study. You may need to take advantage of it frequently or you may need its assistance few times or not at all during the course. The questions in the *Tutor* can be a good review when preparing for tests, even if you are confident that you know how to solve the problems or do the computations. We urge you to take advantage of this resource in whatever ways it can be helpful.

We have outlined above how we designed this textbook and its complementary materials to provide you, your classmates, and your instructors the resources to learn and teach chemistry actively and interactively. Now let us return to the first task the American Chemical Society team considered, what concepts to include in *Chemistry*. Several concepts or "big ideas" recur in one form or another throughout the book. Here are brief statements of these concepts:

- Attractions between positive and negative centers hold matter together and are responsible for chemical reactions.
- The lower its energy, the more stable the system.
- During change, energy is conserved: $\Delta E_{net} = 0$.
- The properties of elements repeat periodically as the atomic number of their atoms increases.
- Electrons in atoms and molecules act like matter waves with quantized energies; the more spread out a matter wave, the lower (more favorable) its energy.

Personal Tutor
A brief description directs you to the section you might find helpful for this part of the text.

- Change occurs in the direction that increases the number of distinguishable arrangements of particles and/or energy quanta. Entropy, S, is a measure of this number, and in all spontaneous processes, net entropy increases, $\Delta S_{net} > 0$.
- Reactions are at equilibrium when $\Delta S_{net} = 0$ for the change from reactants to products.
- When a reaction at equilibrium is disturbed, the system reacts to minimize the disturbance. This is Le Chatelier's principle. Reactions at equilibrium are quantitatively described by a temperature dependent equilibrium constant ratio.
- Electric current can produce reduction–oxidation chemical reactions. Reduction–oxidation chemical reactions can produce an electric current.
- The rate of a chemical reaction depends on the concentrations of species and the temperature of the system. These are a result of the reaction pathway.

Some of the concepts in this list may look familiar and others probably do not. Our goal in structuring **_Chemistry_** to emphasize active and collaborative learning has been to provide the means for you to understand these concepts. The understanding you gain will allow you to apply the concepts not only to the problems we and your instructors provide but also to the problems and systems you meet in other courses, and most important, to interesting and intriguing systems you meet in the world outside the classroom. It has been an enjoyable challenge to write **_Chemistry_** and to develop the complementary materials. We hope it is an equally enjoyable challenge to use them to learn chemistry.

CHEMISTRY

O plunge your hands in water
Plunge them up to the wrist;
Stare, stare in the basin
And wonder what you've missed.

As I Walked Out One Evening
W. H. Auden (1907–1973)

Water in all its forms is essential in many ways for life on Earth and is also inviting for recreation and relaxation. Molecular-level illustrations of the three phases of water are shown in the circles and you can see both solid and liquid water in the photograph. Gaseous water is invisible, but the clouds (fog) of tiny water droplets you see are evidence that water molecules leave the surface of the warm water as a gas and then condense as they move into the cold air above the pool.

Water: A Natural Wonder

The quotation from W. H. Auden describes something we do many times every day, but few of us stop to wonder what we have missed. Water is by far our most familiar chemical compound, and we spend much of our lives in contact with water as a liquid, solid, or gas. The example in the picture on the facing page shows people enjoying a plunge in the pool formed by a warm thermal spring that is surrounded by new-fallen snow. Because it is so familiar, we seldom pause to consider what a unique and remarkable substance water is.

About 70% of the human body mass is water, and everything we drink and much of what we eat is mostly water. Living systems are fundamentally **aqueous** systems, and one emphasis of this textbook is the chemistry that is applicable to living systems. We will often use this chemistry as a jumping-off point to develop a systematic understanding of chemistry that applies to organisms as well as to chemistry in the laboratory and in industrial processes. While we will concentrate on the important chemistry that occurs in aqueous media, we will apply this understanding to other conditions as well.

We will start with the bulk properties of water and then go on to see how the structure of the water molecule helps to explain these properties. Along the way, as we develop concepts to explain how water acts, we will apply these concepts to other molecules and substances. You are probably familiar with several of the concepts in this chapter, so this will be a review of what you have already learned. Many of these ideas will be developed in more detail in later chapters, so this will also be a look ahead to where we are going. In most cases, we will begin with experiments and observations and use these to develop molecular-level explanations that will account for the results. In all cases, we will emphasize reasoning from evidence and always look for further evidence to support our explanations.

Personal Tutor Try the online diagnostic exam to check your background knowledge and find out how you might best use the **Tutor.** Also, skim Sections 1.1–1.5 of the textbook and try several of the Check This and Consider This problems as well as End-of-Chapter problems to find out what concepts you may need to spend more time on.

1.1. Phases of Matter

1.1 CONSIDER THIS

What do you already know about water?

Use words and/or drawings to make a list of all the properties of water that you know from experience or have learned about in previous courses. Which properties do you think are essential for life on Earth? Why are they essential? Compare your list and analysis with those of your classmates. What properties of water did you learn about from others in your class?

Aqueous is from Latin, *aqua* = water, which also shows up in "aquarium," "aquatic," and so on.

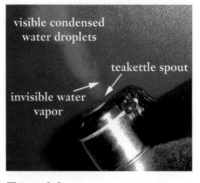

Figure 1.1.

Water vapor and a cloud of liquid water droplets from a teakettle. Invisible water vapor leaving the spout expands into the cooler surrounding air and condenses to form a visible cloud of tiny water droplets (water dust) a few centimeters from the spout.

Your list and analysis from Consider This 1.1 probably confirmed that you already know a lot about the properties of water at both the macroscopic and molecular levels. Did your list include the fact that water is the only common substance found in all three **phases of matter,** solid, liquid, and gas, on the surface of the earth under normal conditions? The photograph that opens this chapter exemplifies this observation. There you can see the liquid water of the thermal spring pool and the solid water of the snow covering the shrubs, trees, and mountains in the background. You cannot see any gaseous water, because gaseous water, which we usually call water vapor, is invisible. But you have evidence that water vapor is leaving the warm water surface, because you can see the mist or fog of tiny liquid water droplets (water dust) formed as the vapor condenses (liquefies) in the cold air above the pool. You can also observe this effect in the water vapor leaving the spout of a teakettle of boiling water, as shown in Figure 1.1.

"Steam" is another name for gaseous water near its boiling point. The mist of visible condensed water droplets is often incorrectly called steam, so we usually will not use the word *steam.*

Density The amount of matter that occupies a given volume is its **density.** Mass is used to measure how much matter is present, so the density of a substance is expressed as the mass of the substance per unit volume in units of $g \cdot mL^{-1}$ or $kg \cdot L^{-1}$. Density is one of the physical properties we can use to identify and characterize a pure substance.

Solids A **solid** has a definite volume and shape that does not depend upon the size and shape of its container. The density of solid water is 0.917 $g \cdot mL^{-1}$ (at 0 °C). The densities of other pure solids range from about 0.5 to 20 $g \cdot mL^{-1}$. Figure 1.2(a) shows a molecular-level model of solid water (ice). The molecules in ice form an orderly array, with the molecules attracted to one another, close to each other, and fixed in place with respect to one another. If Figure 1.2(a) represents a "snapshot" of the molecules taken at a particular instant, another snapshot taken later would show the same molecules in the same locations.

(a) Molecules in a solid occupy fixed positions in an orderly closely packed array.

(b) Molecules in a liquid can move about from place to place but stay close together.

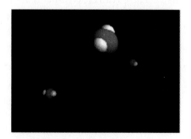

(c) Molecules in a gas are far apart and move almost independently of one another.

Figure 1.2.

Molecular-level representations of solid, liquid, and gaseous water. These are space-filling representations of water molecules, which are discussed in Section 1.6.

Liquids A **liquid** has a definite volume but takes the shape of its container; it is fluid (pourable). The density of a pure liquid substance is nearly the same as the density of the same substance in the solid phase. For example, at the melting point of water (the temperature at which solid ice becomes liquid water), the density of liquid water is 1.000 $g \cdot mL^{-1}$. To explain the density,

Figure 1.2(b) shows the molecules in liquid water attracted to one another and close together, as they are in the solid. To explain the fluidity of liquids, our model assumes the molecules in a liquid can move about, so they are shown in a disorderly jumble in the figure. A snapshot taken at a later time would show a different jumble than the one we see in Figure 1.2(b) and would probably show some different water molecules.

Note that the density of liquid water is greater than the density of solid water. Did your list in Consider This 1.1 include the fact that solid water (ice) floats on liquid water? Water is the only common substance that exhibits this property. For all other substances, the density of the solid phase is higher than the density of the liquid phase, so the solids sink in the liquids, as shown in Figure 1.3 for solid and liquid paraffin compared to ice and water.

Web Companion

Chapter 1, Section 1.1 ①

Interactive molecular-level ②
animations of the phases ③
and phase changes of water ④

Figure 1.3.

Comparison of the densities of solid and liquid paraffin and water. Paraffin, on the left, is a waxy substance whose molecules contain only carbon and hydrogen atoms.

1.2 CONSIDER THIS

How do the volumes of liquid and solid water compare?

(a) If you freeze 1 L of liquid water, will the volume of the solid ice be larger, smaller, or the same as the liquid? Explain the reasoning for your answer.

(b) 👆 Is there evidence in the *Web Companion*, Chapter 1, Section 1.1.3, to support your answer in part (a)? What is the evidence and how does it support your answer?

Gases A **gas** takes the volume and shape of its container. The density of a pure gaseous sample varies with the pressure and temperature of the gas because the volume of the sample depends strongly on these variables. To compare gas densities, we choose a standard set of conditions and measure or calculate the density of all gases under these conditions. A common choice for this comparison is the gas at one atmosphere pressure and 0 °C, which are called **standard temperature and pressure (STP)** conditions. The calculated density of gaseous water at STP is 8.03×10^{-4} g·mL^{-1}. Our representation of gaseous water [water vapor; Figure 1.2(c)] shows the molecules far apart and moving essentially independent of one another. Since there are many fewer molecules in a given volume of gas than the same volume of liquid, the density of a pure gaseous substance is much less than the density of the same substance in the liquid or solid phase. Because the motions of the molecules are independent of one another, the molecules move about to occupy as much volume as is available to them.

The densities of solids and liquids also vary with temperature, but the variation is much smaller than for gases. The variation of solid and liquid densities with pressure is usually negligible.

Phase changes To change a solid to liquid or a liquid to gas requires an input of energy to break the attractions that hold the molecules in place in the solid or together in the liquid. The **melting point (mp)** and **boiling point (bp)** are the temperatures at which, respectively, the solid to liquid and liquid to gas phase changes occur for a pure substance at a pressure of one atmosphere. Melting and boiling point temperatures are characteristic of a substance. The melting and boiling points of water, 0 °C and 100 °C, are familiar examples.

1.3 CHECK THIS

Molecular level representation of boiling water

(a) When water boils at one atmosphere pressure and 100 °C, you observe lots of bubbles forming and rising to the surface of the liquid. Make a sketch representing the gas in a bubble and the liquid surrounding it. Use circles or other geometric shapes to represent molecules and label them to indicate what molecule each represents.

(b) 🐾 Compare your representation in part (a) with the one that is animated in the *Web Companion*, Chapter 1, Section 1.1.5–6.

> We will discuss these properties in more detail in Sections 1.10, 1.11, and 1.12.

The change of a liquid to a gas, **vaporization,** can occur at temperatures below the liquid boiling point, but the pressure of the gas that is formed is less than one atmosphere. Did you note in Consider This 1.1 that vaporization (which we also call evaporation) of water from your skin helps cool you down when you exercise? Water absorbs more heat per gram than any other substance when it changes from liquid to gas. Water also requires a great deal of heat just to warm it up. Since we are about 70% water, this capacity to absorb a lot of heat without warming too much helps us maintain a constant body temperature, even on hot days.

Vaporization of water from the warm thermal spring shown in the chapter-opening photograph explains where the water molecules come from that condense to form the mist above the pool. **Condensation,** the formation of a liquid from its gas (vapor) is the reverse of vaporization, and **freezing,** the formation of a solid from its liquid, is the reverse of melting. We often refer to both liquids and solids as **condensed phases** because their molecules are more tightly packed together than in the gas phase. Energy is released to the surroundings when a gas condenses or a liquid freezes. This is why you can be badly burned by hot water vapor (steam) if it condenses to water on your skin.

> In localized hot spots, like volcanoes or deep-sea vents, molten rock from deep beneath the surface actually reaches the surface of the earth.

Because the range of temperatures at the surface of the earth is from somewhat below to somewhat above 0 °C, water exists on the earth in all three phases. Water is unique; all other common substances exist naturally in only a single phase. For example, the major gases in the atmosphere, nitrogen (bp = -196 °C) and oxygen (bp = 183 °C), are never found naturally as liquids (or solids). Iron (mp = 1535 °C) and silicon dioxide (quartz, mp = 1610 °C) are never found naturally as liquids (or gases) at the surface of the earth.

Reflection and Projection

Because water is so familiar, perhaps you had never considered how strange its properties are compared to the properties of the large numbers of other substances you also see around you. Water occurs naturally as a solid, liquid, and gas at the surface of the earth. Solid water floats on liquid water. A large quantity of thermal energy (heat) is required to raise the temperature of water and to vaporize it. We have shown how the physical phases of water can be visualized at the molecular level to help explain some of the observed properties of solids, liquids, and gases. However, these models are applicable to other substances as well as water and do not explain what is unique about water. We need to have a more detailed picture or model of water molecules and their interactions in order to explain the properties of water that we can see, feel, and measure. Since molecules are made of atoms, we will begin with a brief reminder about atomic properties and then go on to consider water and other molecules. Briefly review those concepts in the next four sections that are already familiar and spend more time working through those that are new or less familiar.

1.2. Atomic Models

It is common knowledge, even among people who have never studied chemistry, that an atom of oxygen and two atoms of hydrogen are somehow connected to form a water molecule, H_2O (pronounced "aitch-two-oh"). On the other hand, it often comes as a surprise to learn that the connections result from the electrical nature of matter.

 1.4 INVESTIGATE THIS

What electrical effects can you observe?

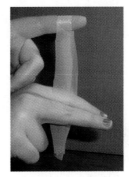

Drape a 1×15-cm strip of thin plastic sheet from a plastic bag over one finger, so that equal lengths hang down on each side. Place one finger of your other hand *between* the two halves of the plastic strip. Quickly slide your hand down the full length of the free ends of the strip, with the plastic loosely pinched between your fingers and thumb. Immediately release the free ends. Observe what happens. Try rubbing the strips again. Observe what happens.

1.5 CONSIDER THIS

Do electrical effects explain your Investigate This 1.4 results?

What, if anything, does electricity have to do with your observations in Investigate This 1.4? Have you ever observed other effects similar to what you observed when you rubbed the plastic strip? What were they? What might they have to do with electricity?

Electrical nature of matter The electrical nature of matter was recognized for centuries before scientists developed a model to explain it. You have probably experienced an electric shock after scuffing across a carpet or sliding across an automobile seat, and then touching some metal object. The phenomenon of objects acquiring electrical charge is illustrated by rubbing a balloon on cloth or your hair. Initially, as Figure 1.4(a) illustrates, the balloon and the cloth are each electrically neutral. The number of positive and negative particles in each is balanced and they just cancel one another out.

Figure 1.4(b) illustrates how negative charges are rubbed off the cloth and onto the balloon. This produces a negative electric charge on the surface of the balloon and leaves behind a net positive charge on the cloth. The charged balloon attracts electrically neutral objects, such as the small bits of paper in Figure 1.4(c), by inducing a shift in the charge distribution in the paper, which has no net electrical charge. The surface of the paper nearest the balloon develops a charge whose sign is opposite to that of the charge on the balloon. **Electrostatic attraction** occurs between electrical charges of opposite sign, and the paper sticks to the balloon. Conversely, electrical charges of the same sign repel one another.

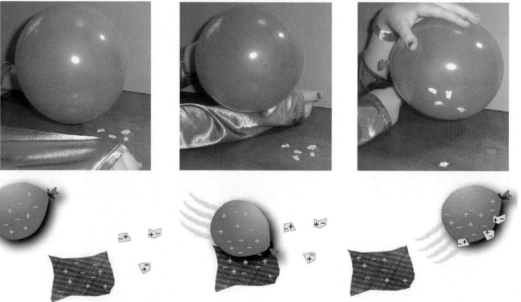

Figure 1.4.

Electrostatic attraction between an electrically charged balloon and bits of paper.

(a) Balloon, cloth, and bits of paper are all electrically neutral; the positive and negative charges cancel.

(b) Rubbing transfers some negative electrical particles from the cloth onto the balloon.

(c) Balloon induces charge separation in bits of paper, and they are attracted to the balloon.

1.6 CONSIDER THIS

Can an electrostatic model explain the Investigate This 1.4 results?

Make drawings, modeled after those in Figure 1.4, that show how the electrostatic model can explain your observations in Investigate This 1.4 when two strips of thin plastic hanging next to one another are rubbed simultaneously. Compare

continued

your explanations and sketches with those of other students. Discuss any differences and try to reach agreement on an explanation.

Atoms and elements Our electrostatic model for matter assumes that matter is composed of negative and positive particles. Atoms are made of these positive and negative particles, so we need to consider how the particles are combined in the structure of atoms. Our model for atoms has evolved during the two hundred years since John Dalton (English minister and scientist, 1766–1844) proposed an atomic model based on the assumption that all matter is composed of indivisible atoms. Scientists have since learned that every **atom** is made up of a positively charged **nucleus** (plural: *nuclei*) surrounded by negatively charged **electrons,** e^-. The nucleus consists of particles called **protons,** p, and **neutrons,** n. Protons and neutrons have nearly identical mass, which is almost 2000 times the mass of an electron. Protons are positively charged and neutrons are electrically neutral, as their name implies. The value of the charge on a proton or electron is called a **unit charge** and is written as $1+$ (said as *one plus*) for protons and $1-$ (said as *one minus*) for electrons. Atoms are electrically neutral; they have equal numbers of protons and electrons, so the positive and negative charges balance one another. Numerical values for the mass and charge of these subatomic particles are given in Table 3.1, Chapter 3, Section 3.1.

Elements are pure substances that are composed of all the same kind of atom. The identity of an element is defined by the number of protons in the nuclei of its atoms. Every atom of the same element has the same number of protons, and therefore the same nuclear charge and the same number of electrons to balance the nuclear charge. For example, the heaviest element that occurs on the earth in quantities large enough to mine is uranium, with 92 protons. The lightest element is hydrogen, with 1 proton. At the time of this writing, 116 elements are known; 84 of them occur naturally in measurable quantities. The other 32 exist on the earth only because they have been made in nuclear reactors and particle accelerators or result from nuclear decay of other elements.

The inside front cover and end paper of the book list names, symbols, and nuclear charges for all the named elements. The listing is in the form of a **periodic table** of the elements. The periodic table arranges all the known elements into rows (called **periods**) and columns (called **groups**). Elements with similar chemical properties are placed in the same group. The periodic table is the chemist's principal tool for organizing information about the elements. Dmitri Mendeleev (Russian chemist, 1834–1907), devised one of the first periodic tables in 1869. We will frequently use the periodic table (or parts of it) to illustrate trends in atomic properties. The development and structure of the periodic table are central themes of Chapter 4.

Electron-shell atomic model The current model for the atom is based on the observation that atomic nuclei have radii on the order of 10^{-15} m, while atoms themselves have radii on the order of 10^{-10} m, some 100,000 times larger than the radius of the nucleus. To put these values in perspective, if a nucleus were the size of a pea, the entire atom would be about the size of the baseball stadium in Figure 1.5. For any atom, the region outside the nucleus is the

Atom is derived from the Greek *atomos,* which means indivisible or not able to be cut.

Processes that result in nuclear transformations are discussed in Chapter 3.

 Web Companion

Chapter 1, Section 1.2 — ①

Explore the shell model and an interactive periodic table to investigate atomic properties.

②
③
④

domain of its electrons. The nucleus contains essentially the total mass of the atom, but the size of an atom is determined entirely by the distribution of its electrons.

nucleus

Figure 1.5.

Relative size of a nucleus and an atom. Imagine a pea, the nucleus, at second base. The atom would be about the size of the entire stadium.

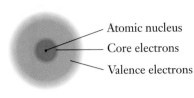

Atomic nucleus

Core electrons

Valence electrons

Figure 1.6.

Electron-shell model of an atom.

Valence (Latin, *valentia* = strength or power) was originally used to characterize the combining properties of atoms.

A simple atomic model, called the **electron-shell model,** illustrated in Figure 1.6, is sufficient for understanding much of chemistry. Negatively charged electrons are held in atoms by their attraction to the positively charged nucleus, but the locations of electrons are only approximately known. Spherical shells arranged concentrically around the nucleus represent the volumes of space where electrons are most likely to be found. Two categories of electrons are distinguished in Figure 1.6. **Core electrons** are nearest the nucleus and are so strongly attracted to the nucleus that they never interact with other atoms. The nucleus plus the core electrons are called the **atomic core.** The atomic core has a positive charge, the **core charge,** equal to the charge on the nucleus minus the number of core electrons. **Valence electrons** are farther from the nucleus and more weakly attracted than the core electrons, because the attraction is mainly from the core charge, which is less than the full nuclear charge. Valence electrons are responsible for interactions of atoms with one another.

Figure 1.7 illustrates the shell model, as applied to the first 20 elements of the periodic table. The field for each element includes the element name, its one- or two-letter symbol, the units of positive charge in its atomic nucleus (which is called the **atomic number**), and the units of negative charge (symbolized as e^-) associated with its core and valence electrons. The sum of the unit charges for core electrons and valence electrons must equal the units of positive charge for the nucleus. For the first 20 elements, the number of core electrons is the same for all the elements in any period (row). The first period has no core electrons, the second period has two, the third has 10, and the fourth (up to element 20) has 18.

The number of valence electrons increases across a period. The beginning of each period corresponds to a new electron shell, the **valence shell** for that period. The valence shell for the first element in the period contains one electron, and all the electrons added across the previous period are now core electrons in the new element. Then electrons are added one at a time to the new valence shell as each new element is added across the period. The result is

	I	II	III	IV	V	VI	VII	VIII
	1 valence e^-	2 valence e^-	3 valence e^-	4 valence e^-	5 valence e^-	6 valence e^-	7 valence e^-	8 valence e^-
1	Hydrogen **H**			**Scale** $\longleftrightarrow$ 200 pm				Helium **He**
No core electrons	Nucleus: 1+ Core e^-: 0– Valence e^-: 1–							Nucleus: 2+ Core e^-: 0– Valence e^-: 2–
2	Lithium **Li**	Beryllium **Be**	Boron **B**	Carbon **C**	Nitrogen **N**	Oxygen **O**	Fluorine **F**	Neon **Ne**
2 core electrons	Nucleus: 3+ Core e^-: 2– Valence e^-: 1–	Nucleus: 4+ Core e^-: 2– Valence e^-: 2–	Nucleus: 5+ Core e^-: 2– Valence e^-: 3–	Nucleus: 6+ Core e^-: 2– Valence e^-: 4–	Nucleus: 7+ Core e^-: 2– Valence e^-: 5–	Nucleus: 8+ Core e^-: 2– Valence e^-: 6–	Nucleus: 9+ Core e^-: 2– Valence e^-: 7–	Nucleus: 10+ Core e^-: 2– Valence e^-: 8–
3	Sodium **Na**	Magnesium **Mg**	Aluminum **Al**	Silicon **Si**	Phosphorus **P**	Sulfur **S**	Chlorine **Cl**	Argon **Ar**
10 core electrons	Nucleus: 11+ Core e^-: 10– Valence e^-: 1–	Nucleus: 12+ Core e^-: 10– Valence e^-: 2–	Nucleus: 13+ Core e^-: 10– Valence e^-: 3–	Nucleus: 14+ Core e^-: 10– Valence e^-: 4–	Nucleus: 15+ Core e^-: 10– Valence e^-: 5–	Nucleus: 16+ Core e^-: 10– Valence e^-: 6–	Nucleus: 17+ Core e^-: 10– Valence e^-: 7–	Nucleus: 18+ Core e^-: 10– Valence e^-: 8–
4	Potassium **K**	Calcium **Ca**						
18 core electrons	Nucleus: 19+ Core e^-: 18– Valence e^-: 1–	Nucleus: 20+ Core e^-: 18– Valence e^-: 2–						

Figure 1.7.

Periodic variation of valence electrons and atomic size for the first 20 elements.

that *the number of valence electrons is the same for each element in a group* (a column) of the periodic table. (Helium in the eighth column is an exception with only two valence electrons; all the other eighth-column elements—noble gases—have eight valence electrons.)

The periodic table in Figure 1.7 has the atoms arranged in 8 groups rather than the 18 shown on the inside front cover. The arrangement of 8 groups is often used when discussion is limited to the first 20 elements.

1.7 CHECK THIS

Number of valence and core electrons

(a) Use the periodic table on the inside front cover and the data in Figure 1.7 to determine the number of valence electrons in (i) bromine, (ii) strontium, and (iii) selenium atoms. Explain how you get your answers.

(b) What is the core charge for each of the atoms in part (a)? How many core electrons does each atom have? Explain how you get your answers.

(c) How many electron shells surround the nucleus of a phosphorus atom? How many electrons are in each shell? Are your numbers consistent with those in the electron shell model in the *Web Companion*, Chapter 1, Section 1.2? Explain your answers.

The models in Figure 1.7 are scaled to be proportional to the relative sizes of the atoms. Compare atomic sizes within a vertical group, and you will see that the sizes of atoms increase with increasing atomic number. Not surprisingly, a core of 10 electrons is larger than a core containing 2 electrons, and a core of 18 electrons is larger than a core of 10 electrons. Therefore, the valence shell diameters of the elements increase in size from the top to the bottom (smaller to larger atomic numbers) of a group. However, across a period from left to right, the sizes of atoms generally *decrease* despite the fact that the numbers of electrons *increase*. Remember that the nuclear charge is also increasing across a period; the increased charge attracts all the electrons more strongly and draws them all closer to the nucleus.

1.8 WORKED EXAMPLE

Trends in atomic size

Predict the relative sizes of calcium, strontium, and barium atoms.

Necessary information: We need the periodic table on the inside front cover to find the relationship of these three elements to one another.

Strategy: Determine whether these elements are in the same period (row) or the same group (column) and use the trends noted in the previous paragraph to make our prediction.

Implementation: Calcium, strontium, and barium are in the second column of the periodic table and increase in atomic number in this same order. Since atoms of elements in the same group increase in size with increasing atomic number, we predict that the sizes of these atoms are in the order: calcium < strontium < barium.

Does the answer make sense?   The periodic table in Figure 1.7 shows the second column elements increasing in size from top to bottom; it seems reasonable to expect the trend to continue, as we have predicted.

1.9 CHECK THIS

Trends in atomic sizes

Predict the relative sizes of arsenic, selenium, and bromine atoms. Explain the basis of your prediction.

1.3. Molecular Models

A **molecule** is a collection of atoms that are all strongly attached to one another and move and act together as a single entity. Atoms of essentially all 110 elements are able to associate with each other to form molecules. Molecules may contain as few as two atoms, or as many as your imagination permits. Biologi-cally important molecules often contain thousands of atoms and some contain millions. Molecules may contain only one kind of atom (O_2, oxygen gas, for example) or more than one kind (H_2O, for example). In living systems, the great majority of molecules contain two or more of the elements carbon, hydrogen, oxygen, nitrogen, sulfur, and phosphorus.

Molecules are difficult to study individually, so much of what we know about their properties we infer from observations of the **macroscopic** or **bulk properties** of collections of molecules. We then invent physical and mathe-matical models of molecules to help interpret these experimental observations. Once suitable **molecular models** have been developed, they can be used to explain new observations and to suggest avenues for further study. Our goal is to help you become comfortable working in both directions—from observa-tions to molecular-level explanations, and from molecular models to predicted properties.

> "Macroscopic" describes properties that we can observe directly with our senses. "Bulk properties" are also associated with enough of a substance for the properties to be directly observable.

Figure 1.8 shows six models or representations of the water molecule. Some may already be familiar to you. You may even have included one or more of these representations in your list from Consider This 1.1 and you should rely on what you have learned previously as you use them. Others of the representations are probably unfamiliar. We will introduce them in this chapter and continue to use them throughout the book.

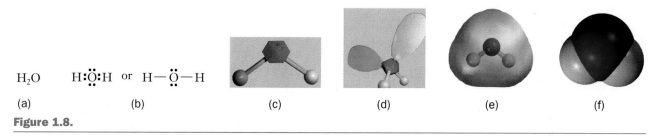

H_2O

(a)

H:Ö:H or H—Ö—H

(b)

(c)

(d)

(e)

(f)

Figure 1.8.

Six different models of the water molecule.

You may not think of the **molecular formula** for water, H_2O, Figure 1.8(a), as a molecular model at all. However, the formula tells you how many atoms of hydrogen, H, and oxygen, O, combine to produce a water molecule. A **compound** is a substance whose molecules are all the same; they all have the same molecular formula. The composition of a compound, such as water or sucrose (table sugar), is the same, no matter where you find it, because all its molecules have the same number of atoms of each element.

The molecular formula H_2O is called a **line formula** because it is written on a single line of text. Line formulas often provide no explicit information about the way the atoms are connected to each other, although you can often figure out the linkages by examining the sequence of atomic symbols in the formula. Figures 1.8(b) through (f) each show the connectivity within the water molecule. Figure 1.8(b) presents two representations of what is called an **electron-dot model,** which shows the distribution of valence electrons in the molecule. The remaining representations all show that the water molecule has a bent structure.

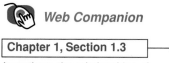

Web Companion

Chapter 1, Section 1.3 — ①

Investigate the relationships of these molecular models to one another.

②
③
④

Representation (c) is called a **ball-and-stick model.** It shows connectivity within the molecule as well as the bent structure. The representations (d) through (f) provide basic shape information as well as other details about the properties of water molecules. Each of these models is useful within the context of specific discussions of the properties of water. We will explain them as they are used later in the chapter.

Molecular bonding and structure Each pair of atoms in a molecule is linked by a **covalent bond,** which results from pairs of valence electrons being shared between the two atoms. Figure 1.9(a) is a representation of the formation of a covalent bond between two hydrogen atoms to produce a hydrogen molecule, H_2. As the electron shells of the two atoms overlap, a new distribution of the electrons forms. In this distribution, the electrons are shared by and attracted to both nuclei. Figure 1.9(b) represents this H—H bond formation with electron-dot structures for the atoms and molecule. Figure 1.9(c) extends this representation to show the shared electron-pair bonds in water. A more detailed discussion of molecular bonding, which explains how a pair of electrons makes up a covalent bond, is the subject of Chapter 5.

Figure 1.9.

Representations of shared pairs of electrons forming covalent bonds. (a) and (b) show the formation of H_2 and (c) shows the formation of H_2O. Valence electrons are represented by dots, and shared electron pairs are shown in red.

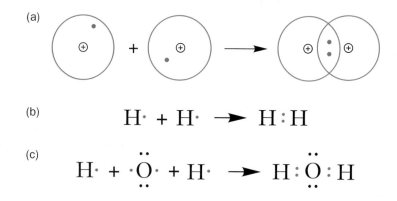

Figure 1.8(c) and Figure 1.10 show a photograph of a ball-and-stick model for the water molecule made with a molecular-model kit like yours. The red ball in the figures represents an oxygen atomic core, the nucleus plus core electrons, and the two aqua balls represent the two hydrogen atomic cores, protons. Each of the sticks that connect the balls represents a pair of valence electrons, forming a covalent bond. Ball-and-stick models show the **molecular shape,** or geometry, which is *the arrangement of atomic nuclei relative to each other in a molecule.* **Bond lengths** (the distances between pairs of covalently bonded atomic nuclei) and **bond angles** (the angle formed by the two bonds of three linked atoms) define the shape of a molecule. The numerical values in Figure 1.10 are the results of many experiments, which show that a water molecule has two identical H—O bonds, each of which is 94 picometers, pm (1 pm = 10^{-12} m) long. The H—O—H bond angle is 104.5° so that water is described as being bent, or as a V-shaped molecule.

Ball-and-stick models provide useful information about molecular shape, but provide no information about valence electrons—beyond the fact that each stick represents a pair of bonding electrons. Before you can build a ball-and-stick model, you need to know how many atoms of each type combine to form

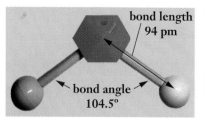

bond length 94 pm

bond angle 104.5°

Figure 1.10.

Geometry of the water molecule.

the molecules. Then you must consider how atoms use valence electrons to form molecules. We need a model that helps visualize all the valence electrons.

1.4. Valence Electrons in Molecular Models: Lewis Structures

Use the partial periodic table, Figure 1.7, to verify that an oxygen atom has six valence electrons and two core electrons, and that a hydrogen atom has one valence electron. The two oxygen core electrons do not participate in forming chemical bonds. In a water molecule, only the eight valence electrons (six from the oxygen atom and one each from the two hydrogen atoms) have a role in binding the three atoms together and in determining the shape of the molecule. The eight valence electrons in water are represented as four electron pairs, as shown by the electron-dot models in Figures 1.8(b), 1.9(c), and 1.11.

Note carefully that the meanings for the letter symbols for elements vary with context. In electron-dot diagrams like Figure 1.11, O represents the *atomic core*, the nucleus plus *core* electrons. The valence electrons are explicitly indicated using dots (or strokes). When atoms of oxygen are specified, as in the periodic table of Figure 1.7, the symbol O represents the nucleus plus all the electrons. The O symbol is sometimes used to represent only the oxygen nucleus, as you will see in Chapter 3. The inconsistency in the way chemists use atomic symbols should not create confusion if you remain attentive to the context of the discussion.

Lewis structures Electron-dot models are also called **Lewis structures** to honor G. N. Lewis (American chemist, 1875–1946), who first proposed the electron-pair, covalent bond. Lewis structures are often written with the covalent bonds shown as a pair of dots [the red dots in Figure 1.11(a)] to reinforce the idea that the bond involves a pair of electrons. A Lewis structure may also be written with the covalent bonds shown as a stroke (line) between the bonded atomic cores [the red lines in Figure 1.11(b)]. The strokes are quicker to write, and make it easier to see the relationship between a Lewis structure and a ball-and-stick model. Valence electrons that are written adjacent to only one element symbol (the black dots in Figure 1.11) are associated exclusively with that element and are called **nonbonding electrons.**

When we write Lewis structures, we keep in mind that second-period elements (carbon through fluorine) in molecules typically have eight associated valence electrons. The "rule of eight," or **octet rule,** is useful for figuring out appropriate Lewis structures for molecules that contain these second-period elements. Also, a hydrogen atom always has two valence electrons associated with it in molecules, and the pair of electrons is always written between H and the atomic core to which it is bound.

The first step in writing a Lewis structure is to determine the total number of valence electrons in the atoms that make up the molecule. Then these electrons are distributed among the atomic cores so that each second-period element is surrounded by an octet (four pairs) of electrons. For the purposes of counting these octets, valence electrons written between a pair of element symbols are considered to be associated with both of the bound atoms. *Electrons shared between a pair of atoms count toward satisfying the octet rule for both atoms.*

Ball-and-stick model kits use balls drilled with several patterns of holes to represent atomic centers. Thus, molecules of various geometries can be made. In your model kit, the "balls" (except for hydrogen) are polyhedra, but the kits are still referred to as "ball-and-stick." Model kits usually use red for oxygen, white or aqua for hydrogen, blue for nitrogen, and black for carbon.

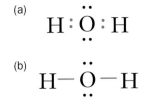

Figure 1.11.

Two representations of the electron-dot model (Lewis structure) of water. Electron pairs in covalent bonds may be represented as (a) dots or (b) strokes connecting the atomic symbols.

1.10 WORKED EXAMPLE

The Lewis structure for methane

Write the Lewis structure for the methane molecule, CH_4.

Necessary information: We need the number of valence electrons for carbon and hydrogen atoms from the periodic table.

Strategy: Determine the total number of valence electrons in the molecule, arrange them to surround the second period element (carbon) with four electron pairs, and then add the hydrogen atomic cores.

Implementation: From Figure 1.7, we find that the C atom has four valence electrons and each of the four H atoms has one; the total number of valence electrons in CH_4 is eight. C is the only second-period element in the molecule, so we surround it with four pairs of electrons:

$$:\overset{..}{\underset{..}{C}}:$$

Each H atomic core can share one pair of electrons with another atomic core, so we use each of the four electron pairs on the C to form an electron-pair bond with one of the H atomic cores:

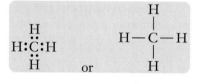

or

Does the answer make sense? All the valence electrons in the methane molecule are accounted for. There are two-electron bonds between all four H atomic cores and the C atomic core and eight electrons surround the carbon, as our Lewis structure rules require. The structure with bonding electron pairs shown as strokes reminds us that there are no nonbonding electrons in this molecule; all the valence electrons are bonding electrons.

> In this chapter, we use names of compounds without telling you the origin of the names. For now, please focus on the connections, shapes, and polarities of the molecules, not their names.

1.11 CHECK THIS

Writing Lewis structures

(a) Write the Lewis structure for the ammonia molecule, NH_3.
(b) Write the Lewis structure for the hydrogen fluoride molecule, HF.

Connectivity You can write Lewis structures if you know or can figure out the connectivity (which atoms are linked to which) in the molecule. Figuring out the connectivity is simplified somewhat by the fact that hydrogen can have only two valence electrons. Thus hydrogen is covalently bonded to only one other atom. Second-period elements have characteristic numbers of covalent bonds and nonbonding electrons in their molecules, as shown in Table 1.1.

Table 1.1	**Bonding and nonbonding electron pairs for second-period elements in molecules.**

Element	C	N	O	F
covalent bonds	4	3	2	1
nonbonding pairs	0	1	2	3

Source: Dickerson and Geis, *Chemistry, Matter, and the Universe*
(W. A. Benjamin, Inc., Menlo Park, CA, 1976), p. 165.

Out of every 100,000 atoms
in your body, there are

60,560	H
25,670	O
10,680	C
2,440	N
650	all others

The prevalence of H and O
is not surprising. Approximately
70% of your body mass is water.
Source: Dickerson and Geis, *Chemistry,
Matter, and the Universe* (W. A.
Benjamin, Inc., Menlo Park, CA, 1976),
p. 165.

We are particularly interested in the second-period elements because many of the molecules you will meet in this book are **biomolecules,** the molecules essential for life. Biomolecules are composed mainly of hydrogen, carbon, nitrogen, and oxygen. Lewis structures can help you understand and interpret the properties of these molecules. Writing Lewis structures for molecules containing two or more second-period elements is only slightly more complicated than writing the structures for molecules containing one second-period element.

1.12 WORKED EXAMPLE

Lewis structure of methanol

Write the Lewis structure for methanol, CH_4O.

Necessary information: We need the number of valence electrons for carbon, oxygen, and hydrogen atoms from the periodic table and the number of covalent bonds to carbon and oxygen from Table 1.1.

Strategy: Determine the possible connectivity of atoms and the total number of valence electrons in the molecule. Place two valence electrons between each pair of atomic cores and distribute any that are left over to give every second-period element eight electrons surrounding its atom symbol.

Implementation: Hydrogen atoms can bond to only one other atom, so we must connect the C and O atomic cores together and distribute the H atomic cores around them in accordance with Table 1.1:

$$\begin{array}{c} H \\ H \ C \ O \\ H \ H \end{array}$$

From Figure 1.7, we find that a C atom has four valence electrons, an O atom has six, and each of the four H atoms has one, so the total number of valence electrons in CH_4O is 14. We distribute these to make two-electron bonds between each pair of atomic cores:

$$\begin{array}{c} H \\ H\!:\!\ddot{C}\!:\!O \\ \ddot{H} \ \ddot{H} \end{array}$$

continued

Only 10 valence electrons are accounted for in this structure and there are only four electrons around the O atomic core. Table 1.1 shows that an O atomic core usually has two nonbonding electron pairs in addition to the two bonding pairs:

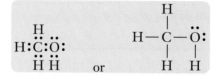

All 14 valence electrons are accounted for, so this is the Lewis structure.

Does the answer make sense? There are two-electron bonds between all four H atomic cores and the C and O atomic cores, a two-electron bond between the two C and O atomic cores, and the C and O are each surrounded by eight electrons, as our Lewis structure rules require. The structure with bonding electron pairs shown as strokes reminds us that there are both nonbonding and bonding electrons in this molecule. The line formula for methanol is often written as CH_3OH to make the connectivity and structure more apparent.

1.13 CHECK THIS

Lewis structure of ethane

Write the Lewis structure for ethane, C_2H_6. Check your result to see that it is consistent with the information in Table 1.1. Try writing a line formula for ethane that more clearly shows its connectivity.

Reflection and Projection

The preceding sections focused primarily on building physical and conceptual models to represent atoms and molecules. The discussion centered on a few second-period elements and hydrogen, all of which form covalent bonds with each other. The physical models are based on the electron-shell description for atoms, which takes into account the size and charge of the nucleus, core electrons that do not take part in chemical interactions, and valence electrons that interact with electrons from other atoms.

Lewis structures for molecules represent all the valence electrons, but the connectivity must be known before the structure can be written, unless the molecule is so simple that the atoms can be assembled in only one way. Constraints on how molecules can be assembled are imposed by (a) the octet rule for second-period elements, (b) hydrogen's limit of two electrons, and (c) the usual number of covalent bonds that each second-period element forms (four for carbon, three for nitrogen, and two for oxygen). Ball-and-stick molecular models show the three-dimensional geometry of the nuclei in a molecule, but do not always represent all the valence electrons. The role of the valence electrons in producing the molecular geometry is the topic of the next section.

1.5. Arranging Electron Pairs in Three Dimensions

The Lewis structure of H_2O in Figure 1.11, of CH_4 in Worked Example 1.10, and of those you wrote for NH_3 and HF in Check This 1.11 show four pairs of valence electrons arranged around the second period elements, but the representations are written on a sheet of paper, which has only two dimensions. Two-dimensional Lewis structures cannot unambiguously reveal molecular geometry in three dimensions. Though the H—O—H diagram shows that water has a central oxygen atom to which two hydrogen atoms are connected, it does not show the 104.5° H—O—H angle. Accurate representations of three-dimensional geometries of molecules reflect the spatial arrangements of electron pairs around the atomic cores. Knowing these geometries is important because the arrangement of electrons and the three-dimensional structure of molecules are responsible for most of their physical and chemical properties. The next step in building a useful three-dimensional model is to figure out how electron pair arrangements produce observed molecular geometries.

To begin, consider that each of the four valence electron pairs around oxygen is strongly attracted to the 6+ atomic core charge (the 8+ positive nucleus and the two core electrons, 2− negative charge). As we will discuss in Chapter 5, the four pairs of electrons around oxygen occupy four separate regions of space. The strength of the attraction between positive and negative charges increases as the distance between the charges decreases. Therefore, the most favorable attraction occurs when each valence-electron pair is located as close as possible to the positive charge. Finally, we have to consider the shape of the region occupied by the electron pairs. For simplicity, think of each of the four regions of space occupied by valence-electron pairs as roughly spherical, like four tennis balls or round balloons.

We often say that core electrons "shield" valence electrons from the full nuclear charge. This shielding is not 100% effective, so the valence electrons are attracted by a charge that is somewhat larger than just the core charge.

1.14 INVESTIGATE THIS

What arrangement of balloons around a point is stable?

For this investigation, use four 8- or 10-inch round balloons. Inflate all four balloons to the same size and tie off their stems so they remain inflated. Attach two of the balloons together by wrapping their stems together. Attach a third and then a fourth balloon in the same way. Try to make different geometries of the four attached balloons. For the most stable arrangement you can find, describe and record the geometry of the centers of the balloons with respect to their mutual point of attachment.

1.15 CONSIDER THIS

What is the geometry of four valence electron pairs?

Imagine lines connecting the center of each of the balloons in Investigate This 1.14 to the mutual point of attachment. For the most stable arrangement, try drawing this set of four lines on a piece of paper. Find an atom center in your molecular model kit that has holes in the correct geometric arrangement and place connectors in the holes to help visualize the lines you are trying to draw.

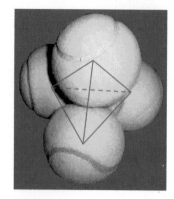

Figure 1.12.

Tetrahedral arrangement of four balls. The red lines represent connections between the centers of the balls that form a regular tetrahedron.

Figure 1.13.

Ball-and-stick tetrahedral array of sticks.

Figure 1.14.

Bonding and nonbonding electrons in the molecular model for water. The sticks represent bonding electron pairs and the paddles nonbonding electron pairs.

Tetrahedral arrangement In Investigate This 1.14 you probably found an arrangement of four balloons that places their centers all the same distance from a central point, as shown for a set of tennis balls in Figure 1.12. Here, three balls are arranged in a triangle, and the fourth ball is placed on top of the triangle. This arrangement is called **tetrahedral** because lines joining the centers of the balls form a regular **tetrahedron,** the four-sided geometric figure shown in red in the figure. The tetrahedral arrangement puts all the balls as close as possible to the central point (the mutual point of attachment for the four balloons in Investigate This 1.14). Four valence electron pairs always form a tetrahedral arrangement around second-period elements because that arrangement places the valence electron pairs nearest the nucleus.

To visualize the angles between lines from the center of a tetrahedron to its corners, use your ball-and-stick model kit that has some polyhedra with four holes drilled tetrahedrally. The black polyhedra (carbon) are tetrahedral, and nitrogen (blue) and oxygen (red) are drilled the same way. Put a stick of the same length in each of the four holes, as in Figure 1.13. The angle formed by any two of these sticks is called the **tetrahedral angle,** which is 109.5°. If the arrangement of valence-electron pairs around the oxygen atom in water were exactly tetrahedral, the H—O—H bond angle would be 109.5°. The value of 104.5° (see Figure 1.10), obtained experimentally, is close to this predicted angle. The tetrahedral arrangement is very symmetric. Any two sticks can be chosen for the H—O covalent bonds, and the same V-shaped molecular model results.

Modified ball-and-stick model Your model kit has parts that you can use to visualize all valence electrons in a molecule—both bonding and nonbonding electrons. Figure 1.14 shows a ball-and-stick model for water made with sticks and "paddles." In this model, each stick represents a pair of electrons in a covalent bond. Each paddle represents a pair of nonbonding electrons. Think of the paddles as cross-sections of volumes of space (like your balloons in Investigate This 1.14) within which electron pairs are likely to be found. Note the relationship between the model in Figure 1.14 and the Lewis structure for water, Figure 1.11(b).

1.16 CONSIDER THIS

How is a Lewis structure translated into a molecular model?

(a) From their Lewis structures, construct models of water, methane, ammonia, and hydrogen fluoride. Show nonbonding as well as bonding electrons in your models.

(b) How would you describe the shape of each molecule you have modeled? First describe the arrangement of the electron pairs around each second-period element. Then, using water as a guide, describe the shape of the molecule. Remember that shape refers to the arrangement of the nuclei with respect to one another. Nonbonding electrons are not considered when determining shape. Discuss your descriptions with your classmates. Develop consistent descriptions that everyone understands and agrees upon.

Models such as the one pictured in Figure 1.14 have limitations that must be taken into account when we use them to visualize molecular properties. For example, although this type of model helps keep track of the number of valence electrons and the shape of the molecule, volumes in space occupied by valence electrons and electron densities within those volumes are not well represented. We will introduce other models for these purposes in the next section.

1.6. Polarity of the Water Molecule

Several kinds of experiments show that the molecules of some substances, **polar molecules,** have their centers of positive and negative charge separated from each other. Polar molecules are **electric dipoles,** that is, they have a positive end and a negative end, as illustrated in Figure 1.15(a). A lowercase Greek delta, δ, is used to symbolize "a small amount." In Figure 1.15(a), this means that there are partial charges, not full + or −, on the ends of the molecule. This charge separation is similar to the induced charge separation shown on the bits of paper in Figure 1.4(c) except that it is a permanent part of the molecule's properties and does not depend on the presence of an external electric charge. In representing molecular polarity, we generally use red to denote parts of molecules or systems that are more negative and blue to represent parts that are more positive. Regions where the positive and negative charges cancel are in green. The text that accompanies Figure 1.17 explains how these regions are determined.

(a) Charge separation makes a polar molecule a permanent electric dipole.

(b) Coincidence of charges leaves a nonpolar molecule with no permanent electric dipole.

Figure 1.15.

Schematic illustrations of (a) polar and (b) nonpolar molecules.

Nonpolar molecules are molecules with no electric dipole. In a nonpolar molecule, the centers of positive charge (from the nuclei) and negative charge (from the electrons) coincide, as illustrated in Figure 1.15(b). In a polar molecule, the arrangement of electrons is skewed so that the centers of positive and negative charge do not coincide.

1.17 CONSIDER THIS

How might an electric charge affect polar molecules?

Imagine that you have a liquid containing polar molecules like those in Figure 1.15(a). If the liquid is sandwiched between two electrically charged metal plates, one negative and the other positive, how would you expect the molecules to be oriented? Draw a sketch to illustrate your answer.

Effect of electric charge on polar molecules One kind of experiment that is used to test for the polarity of molecules in a liquid is illustrated in Figure 1.16. Two metal plates are placed in the liquid and then one plate is given a negative charge and the other a positive charge. Initially, as shown in Figure 1.16(a), when the plates are not charged, the molecules of liquid are randomly oriented. When the plates are charged, as seen in Figure 1.16(b), polar molecules orient themselves with their negative ends toward the positive plate and positive ends toward the negative plate. The molecules are continuously jostling each other in the liquid, so they do not line up exactly. The electric circuit that is used to charge the plates responds differently when polar and nonpolar liquids are tested, making it possible to distinguish polar and nonpolar molecules from one another.

Figure 1.16.

Orientation of the molecules of a polar liquid in response to an electric charge.

(a) Polar molecules in a liquid between two metal plates are randomly oriented.

(b) When the plates are charged, the polar molecules are attracted and are oriented.

Water molecules are polar Experiments such as the one illustrated in Figure 1.16 show that water molecules are polar. A charge-density model of a water molecule is shown in Figure 1.17(a). The surface is drawn to enclose the volume in space where electrons are most likely to be found. These color-coded models are generated from calculations that show net electrical charge on various parts of the molecule. As in Figure 1.15, regions where negative charge from the electrons and positive charge from the nuclei balance one another are colored green. Regions where negative charge from the electrons is not balanced by positive charge (where the electron density is high) have colors at the red end of the spectrum. Regions where positive charge from the nuclei is not balanced by negative charge (where the electron density is low) have colors at the blue end of the spectrum. The extremes, red and blue, respectively, represent the regions with the highest amounts of negative and positive charge.

For comparison to the charge-density (computer-generated) model, Figure 1.17(b) shows a space-filling model of water, a physical model molded of plastic with the "atoms" identified by the same colors used for ball-and-stick models. **Space-filling models** are approximate representations of the space (volume) occupied by the electrons around each atomic core in molecules, and you can see that the volume and shape of the two models are similar. Space-filling models often help us visualize the interactions of molecules, as in Figure 1.2. But we still need representations like Figure 1.17(a) to visualize the polarity of molecules.

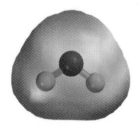 

(a) charge-density model   (b) space-filling model

Figure 1.17.

Charge-density and space-filling models of the water molecule. The computed surface in (a) encloses the volume occupied by the electrons in the molecule. The internal ball-and-stick representation is included to show where the nuclei are. Color coding is explained in the text.

The significant feature of Figure 1.17(a) is that the side of the molecule where the oxygen atom is located is more negative than the side of the molecule where the hydrogen atoms are located. This unsymmetrical distribution of electric charge makes the molecule an electric dipole. Every polar molecule can be represented using a dipole arrow, as shown in Figure 1.18. The point of the dipole arrow represents the negatively charged end of the molecule, while the "+" tail of the arrow represents the positively charged end. Quantitative measurements of the effect of electrical charge on polar molecules, as in Figure 1.16, are used to calculate their dipole moments. The **dipole moment** of a molecule is a measure of magnitude of the charge difference between the two ends of the dipole: The larger the dipole moment, the larger the effective charge difference.

Bond polarity Charge distribution is an important factor in understanding chemical reactions, and is used extensively for explanations throughout the text. Charge-density models, like the one in Figure 1.17(a), are useful for visualizing charge distribution in molecules. They are not always available, however, so it is important to be able to estimate charge distributions using simpler models. Unsymmetrical charge distribution results from two factors: the presence of nonbonding electrons on some atoms in the molecule and the unequal sharing of valence electrons in covalent bonds. It isn't surprising that nonbonding electrons produce an unsymmetrical charge distribution. For example, the nonbonding electrons represented in Figure 1.14 for the oxygen atom in water are on one side of the molecule. They are largely responsible for the region of high negative charge on that side of the molecule in the charge-density model [Figure 1.17(a)].

In addition, bonding electrons are not always shared equally between the two bonded atomic cores. For example, the oxygen atom in water attracts the electrons in the covalent bonds more strongly than do the hydrogen atoms. Thus, the oxygen end of the bonds is slightly negative and the hydrogen ends are slightly positive. The bonds each act like little dipoles, **bond dipoles,** as shown by the dipole arrows in Figure 1.19.

Individual bond dipoles are not experimentally measurable, but we can use bond dipoles, together with the effect of nonbonding electrons, to figure out the direction of the molecular dipole, which can be measured. In Figure 1.19, one bond dipole points up and to the right and the other up and to the left. The two bonds in water are identical, so their bond dipoles should also be the same. The net effect of the dipoles pointing right and left is to cancel each other because they are the same magnitude and point in opposite directions. But both dipoles are pointing up, so they add together (reinforce each other) in the vertical direction. The nonbonding pairs of electrons (not shown in Figure 1.19) are also at the top of model and their negative charge adds to the effect of the bond dipoles. The net effect is the molecular dipole shown in Figure 1.18.

Electronegativity The positive and negative ends of bond dipoles result from differences in electronegativity (*EN*) of the two atoms. The concept of electronegativity originated with Linus Pauling (American chemist, 1901–1994). The **electronegativity** of an element is a measure of its attraction for the electrons it shares with another element in a covalent bond.

The units and numerical values for dipole moments will not concern us except as a relative measure of polarity. For reference, the dipole moment of water is 1.87 Debye (a unit of charge times length).

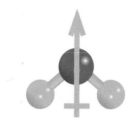

Figure 1.18.

The electric dipole in the water molecule.

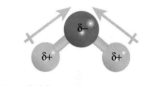

Figure 1.19.

Unequal sharing of electron charge and bond dipoles in the water molecule.

The higher the electronegativity, the greater the attraction for shared electrons. Electronegativity values for the first 20 elements are shown in Figure 1.20. Electronegativity values increase from element to element across a period. Within groups, the lighter elements are the most electronegative. These trends converge at fluorine, which has the highest electronegativity of all the elements. Note that hydrogen is placed in group IV because its electronegativity is intermediate between the extremes at either end of the table.

	I	II	III	IV	V	VI	VII	VIII
1				Hydrogen H 2.2				Helium He —
2	Lithium Li 1.0	Beryllium Be 1.6	Boron B 2.0	Carbon C 2.6	Nitrogen N 3.0	Oxygen O 3.4	Fluorine F 4.0	Neon Ne —
3	Sodium Na 0.9	Magnesium Mg 1.3	Aluminum Al 1.6	Silicon Si 1.9	Phosphorus P 2.2	Sulfur S 2.6	Chlorine Cl 3.2	Argon Ar —
4	Potassium K 0.8	Calcium Ca 1.3						

Figure 1.20.

Periodic variation of electronegativity for the first 20 elements. Relative magnitudes for electronegativity values are indicated using the red–green–blue color-coding scheme that was used earlier to indicate charge density.

Web Companion

Chapter 1, Section 1.6 — ①
②
Use interactive animations to investigate electronegativity and polarity. ③
④

As a rule of thumb, a significant bond dipole exists for a covalent bond between the atoms of two elements that differ by 0.5 or more *EN* units. For example, a bond between oxygen (*EN* = 3.4) and hydrogen (*EN* = 2.2), which differ by 1.2 *EN* units, has a large bond dipole. In contrast, a bond between carbon (*EN* = 2.6) and hydrogen (*EN* = 2.2), which differ by 0.4 *EN* units, has a small, almost negligible, bond dipole.

Of the elements commonly encountered in living systems, oxygen and nitrogen are the most electronegative, while carbon, hydrogen, phosphorus, and sulfur have intermediate values. Polar bonds formed between N or O and any of these others make a large contribution to molecular dipole moments. Molecules (or parts of molecules) containing these elements are usually polar. Nonpolar bonds formed between any two members of the intermediate electronegativity group make only a small contribution to molecular dipole moments. Molecules composed of hydrogen and carbon (for example, the molecules in oil and gasoline) are nonpolar.

1.18 CHECK THIS

Predicting bond dipoles

(a) Predict the direction of the bond dipoles in each of the following bonds and indicate whether the bond dipole is relatively small or relatively large. Explain the reasoning for your answers.

  (i) H—F **(ii)** N—H **(iii)** H—S **(iv)** C—Cl **(v)** N—F **(vi)** Li—H

(b) Do the charge-density models in the *Web Companion*, Chapter 1, Section 1.6.1–2, support your answers in part (a)? Explain why or why not.

1.19 WORKED EXAMPLE

Predicting molecular dipoles

Write the Lewis structure for nitrogen trifluoride, NF_3. On a three-dimensional drawing or model of the molecule, show the direction of the bond dipoles and the molecular dipole.

Necessary information: We need the number of valence electrons in N and F atoms and the electronegativity of these elements. We also need to know that N forms three and F forms one covalent, two-electron bond to other atoms.

Strategy: Write a Lewis structure, as we have done previously. Then make a ball-and-stick model representing the geometry of this structure, determine the direction of the bond dipoles from the relative electronegativities, and use the geometry and direction of the dipoles and nonbonding electrons to predict the direction of the molecular dipole.

Implementation: There are 26 valence electrons in NF_3: 5 from the N and 7 each from the Fs. N bonds with three other atoms and F with only one, so the Fs must all be bonded to the N:

Six electrons are used for the bonds. We can use the remaining 20 electrons to surround all of the atom symbols with 8 electrons to give the Lewis structure:

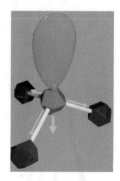

The tetrahedral array of three covalent bonds between N and F and the nonbonding electron pair on N are shown in the picture, at the right, of a ball-and-stick model of NF_3. The nonbonding electron pairs on the Fs are not shown, since they are not relevant to the molecular shape (but do contribute to the net dipole moment). The molecule can be described as a triangular pyramid with the Fs at the corners of the base and the N at the apex.

The electronegativities of N and F are 3.0 and 4.0, respectively, so there is a substantial N—F bond dipole with the negative end toward F, the more electronegative element. Thin arrows in the picture show the bond dipoles. The three bond dipoles point from the apex toward the corners of the base of the pyramid. They are symmetrically arranged around the pyramid, so there is no net dipole in a direction parallel with the base of the pyramid. All three bond dipoles point down from the apex toward the base. The nonbonding electron pairs on the Fs also add a negative region at the base of the pyramid. The nonbonding electron pair on the nitrogen adds a negative region at the apex of the pyramid, but cannot completely counteract the effects of the bond dipoles and Fs nonbonding electrons. There is a net molecular dipole from the apex perpendicular to the base shown by the thick arrow in the picture.

continued

Does the answer make sense? The Lewis structure obeys our rules and accounts for all the valence electrons, so it seems to be correct. The only way we can test the molecular dipole prediction is to look up NF_3 in a reference handbook; the molecule has a dipole moment of 0.24 Debye, which shows that it does have a permanent molecular dipole. The direction of the dipole is harder to determine, but experiments prove that it is in the direction we have shown.

1.20 CHECK THIS

Predicting molecular dipoles

(a) On a three-dimensional drawing or model of the ammonia molecule, NH_3, show the direction of the bond dipoles and the resulting molecular dipole. Refer to Check This 1.11 for the Lewis structure.
(b) Do the same for the hydrogen fluoride molecule, HF.

1.21 CONSIDER THIS

How does molecular geometry affect molecular dipoles?

(a) Carbon dioxide, CO_2, is a linear molecule with the carbon atom midway between the two oxygen atoms. On a Lewis structure of the molecule show partial charge separations, bond dipole arrows, and overall molecular dipole arrow. Would you classify carbon dioxide as polar or nonpolar? What is the basis for your classification?
(b) 👆 How is the polarity of the BH_3 molecule, *Web Companion*, Chapter 1, Section 1.6.3–4, similar to CO_2? Make a sketch of a likely charge-density model for CO_2.

Reflection and Projection

The three-dimensional shape of a molecule is determined by the arrangement of valence electron pairs that puts them as close as possible to the center of each atom. In molecules, four pairs of valence electrons (the octet rule) arranged tetrahedrally about the atomic core surround each second-period element (carbon through fluorine). However, only the arrangement of the nuclei is considered in describing the shape of a molecule. Nonbonding electron pairs are ignored when describing molecular shape; ball-and-stick molecular models are useful for visualizing molecular shape.

Nonbonding electrons in a molecule often cause unsymmetrical charge distribution in the molecule. In addition, unequal sharing of electrons between atoms that differ substantially in electronegativity causes small charge separations in covalent bonds between them and produces bond dipoles. If the nonbonding electrons and/or the bond dipoles in a molecule do not cancel out, the molecule as a whole has a molecular electric dipole and is said to be polar.

One result of this polarity is that polar molecules tend to orient their positive ends toward negative charges and their negative ends toward positive charges. The chemical and physical properties of a substance composed of polar molecules are most unusual when a hydrogen atom is associated with the center of partial positive charge and a highly electronegative atom (nitrogen, oxygen, or fluorine) is associated with the center of partial negative charge. Figures 1.17, 1.18, and 1.19 show that water is one such substance, and the consequences are considered in the next sections.

The previous paragraph reminds us that we must not lose sight of why we are spending so much time developing these molecular models. Our primary interest in this here, and in many of the chapters that follow, is the observable properties of water and other substances. We must show how the models answer questions such as: *"Why is water a liquid at ordinary temperatures?" "Why is the solid less dense than the liquid?" "Why is so much energy needed to change the temperature of water?" "Why is so much energy needed to vaporize water?"* And in later chapters, we will be interested in further questions such as: *"What kinds of compounds does water dissolve?" "What chemical reactions occur in aqueous solutions?" "What chemical reactions involve water itself—either as a reactant or as a product?"* Explanations for all of these phenomena are based on models for water and associated molecules and ions. The models we develop in this chapter and extend in later chapters are essential for visualizing the processes and interpreting bulk phenomena in terms of molecular-level explanations. And, as your study of the linkage between macroscopic properties and the molecular model for water progresses, you will see that what you learn about water extends to other substances as well.

1.7. Why Is Water Liquid at Room Temperature?

In order for a substance to exist as a liquid, the attractions between its molecules have to be strong enough to prevent them from breaking away from one another and going into the gas phase. As energy is added to a liquid, its temperature goes up, the molecules move faster, and finally they are moving fast enough to break away from one another to become a gas. If the pressure exerted by this gas is equal to the pressure of the atmosphere, we say that the liquid is boiling. The boiling point of a liquid is a measure of the strength of the attractions between the molecules in the liquid: the higher the boiling point, the stronger the intermolecular attractions that have to be overcome to produce the gas. We will try to discover how our molecular model accounts for these interactions and leads to the observed behaviors of different compounds.

"Room temperature" indicates a temperature of about 25 °C.

1.22 INVESTIGATE THIS

How does boiling point vary with the number of molecular electrons?

Figure 1.21 shows the boiling points of CH_4 (methane), SiH_4 (silane), GeH_4 (germane), and SnH_4 (stannane) plotted as a function of the total number of electrons in each molecule. These hydrides (compounds of the element with

continued

hydrogen) of group IV elements all have the same symmetrical, tetrahedral structure you observed for methane in Consider This 1.16 and all are nonpolar. Describe the relationship between the boiling point and the number of electrons. What conclusions can you draw about the relative attractions between the molecules in these compounds? Discuss in your group and with the class how you might explain any trends you see.

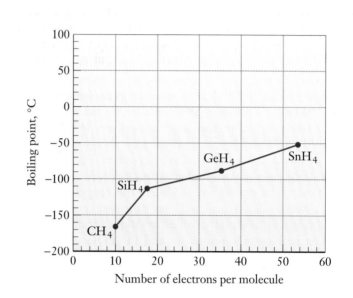

Figure 1.21.

Boiling point *vs*. electrons per molecule for group IV hydrides. The points are connected only to highlight the trend in the data.

Induced-dipole attractions

The boiling points of nonpolar compounds generally increase as the number of electrons in the molecules increases. This is the trend shown for the group IV hydrides in Figure 1.21. Nonpolar molecules have negligible dipole moments. Nonetheless, many of these compounds are liquids or solids at room temperature. Examine the charge-density models of methane, CH_4, ethane, CH_3CH_3, and hexane, $CH_3(CH_2)_4CH_3$, in Figure 1.22. Recall from Section 1.6, that green is used for an electrically neutral surface. Notice that the entire surface of these nonpolar molecules is green.

The electron distributions in Figure 1.22 represent averages of the likely locations for finding the electrons. Electrons are continuously in motion and, at any instant, electron distribution can be skewed one way or another. The skewing leads to a fleeting polarity of the molecule. Polarity in one molecule

Figure 1.22.

Charge-density models of methane, ethane, and hexane. For comparison, Figure 1.17(a) is a charge-density model of the water molecule computed in the same way.

 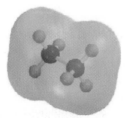 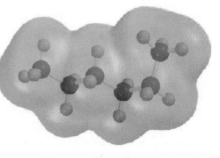

methane, CH_4 ethane, CH_3CH_3 hexane, $CH_3(CH_2)_4CH_3$

induces polarity in a neighboring molecule, as illustrated in Figure 1.23. The two polarized molecules attract one another. These attractions are called **induced-dipole attractions.** Induced-dipole attraction is a molecular-scale effect similar to the observable effect illustrated in Figure 1.4(c), where a charged balloon induces polarity in a neutral object and attracts it. Induced-dipole attractions are short-lived, but as there are many electrons in molecules, there is plenty of opportunity for these attractions to occur. Since the skewing of the electron distribution can occur anywhere in the molecule, these attractions are often called **dispersion forces** or **London dispersion forces** to honor Fritz London (German physicist, 1900–1954) who first proposed these induced-dipole attractions. Dispersion forces exist across the entire surface of a molecule, so dispersion forces among large molecules produce substantial attraction.

Web Companion

| Chapter 1, Section 1.7 | ①
②
③
④

Use interactive animations to investigate intermolecular attractions.

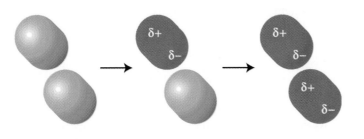

Figure 1.23.

Induced-dipole attraction between nonpolar molecules.
The plus sign represents the center of positive charge in the molecule. Displacement of the electron distribution in one molecule creates a transient dipole that induces a dipole in the second molecule.

1.23 CONSIDER THIS

Do boiling points correlate with the number of molecular electrons?

(a) This graph shows boiling points for a related series of non-polar **hydrocarbons,** compounds containing only carbon and hydrogen atoms. The series starts at methane, CH_4, and adds one "CH_2" after another. The boiling points are plotted as a function of the number of electrons in these molecules. Propose a molecular level interpretation for the observed behavior of the boiling points.

(b) Does your interpretation in part (a) also apply to the data in Figure 1.21? Explain why or why not.

(c) 🖑 Do the examples in the *Web Companion*, Chapter 1, Section 1.7.1–2, follow the same pattern of boiling points as the group IV hydrides in Figure 1.21 and the carbon-hydrogen compounds in part (a)? Does the animation apply to these other cases as well? Explain why or why not.

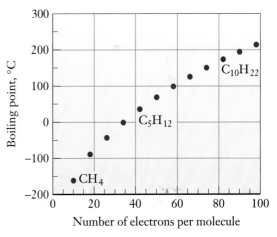

Attractions among polar molecules Dispersion forces increase as the size of molecules increase because more electrons are available to form instantaneous dipoles. All molecules have dispersion forces that attract them to one another. Many molecules also have permanent electric dipoles, and we might expect that attractions between the molecular dipoles would increase the attraction of these molecules for one another. In **dipole-dipole attractions,** the positive end of one dipolar molecule and the negative end of another attract one another.

1.24 CONSIDER THIS

How can dipole-dipole attractions be visualized?

As they jostle about in the liquid, dipolar molecules favor arrangements that maximize their dipole-dipole attractions, such as

and

Sketch favorable arrangements for dipole-dipole attractions among *four* dipoles. Compare your arrangements with those made by other students. How many different arrangements did members of the class find?

You found, in Consider This 1.24, that dipole-dipole attractions do not require one particular alignment of the molecules. There are many orientations for dipole-dipole attractions between a polar molecule and its neighbors in the liquid state. As they are tumbling about in the liquid, polar molecules must often be in position to attract one another. Do the boiling points of polar molecules reflect these additional attractions?

1.25 CONSIDER THIS

How does polarity affect boiling point?

Figure 1.24 adds the boiling points of the group V, VI, and VII hydrides to the data for the group IV hydrides from Figure 1.21. The structures of the hydrides in each group are the same:

Group	Molecules	Structure
IV	CH_4, SiH_4, GeH_4, SnH_4	tetrahedral like CH_4
V	NH_3, PH_3, AsH_3, SbH_3	pyramid like NH_3
VI	H_2O, H_2S, H_2Se, H_2Te	bent like H_2O
VII	HF, HCl, HBr, HI	linear like HF

(a) Do the polarities of the hydrides increase or decrease as you go from left to right in Figure 1.24? Explain your reasoning. What trend in boiling points would you expect for this trend in polarities?

(b) Do the dispersion forces (induced dipole attractions) increase or decrease as you go from left to right in Figure 1.24? Explain your reasoning. What trend in boiling points would you expect for this trend in dispersion forces?

(c) What conclusions can you draw about the relative strengths of polarity and dispersion force attractions between the molecules in these compounds? Explain your reasoning. Discuss in your group and with the class how you might explain any deviations from trends you see.

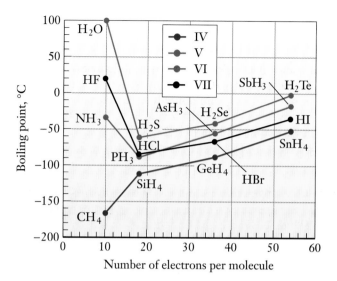

Figure 1.24.

Boiling point *vs.* electrons per molecule for groups IV through VII hydrides. The points are connected only to highlight the trends in the data.

There are two striking things to note in Figure 1.24. One is that all the third and higher period (row) hydrides increase in boiling point as the number of electrons per molecule increases. This is the direction we would expect, if dispersion forces are the main contributors to the interactions among these molecules in their liquids. All the valence electron pairs in the group IV hydrides are bonding pairs, while the group V, VI, and VII hydrides have one or more nonbonded electron pairs. These nonbonded electrons are more free to respond to induced dipoles from neighboring molecules, which helps to explain why the third and higher period group V, VI, and VII hydrides have somewhat higher boiling points than the comparable group IV hydrides.

The other feature that stands out in Figure 1.24 is how far out of line the boiling points of ammonia (NH_3), water (H_2O), and hydrogen fluoride (HF) are compared to all the other compounds. The boiling points of these second-period compounds are 150 to 250 °C higher than the boiling point of methane (CH_4), even though all four of these molecules have 10 electrons. For comparison, note that the spread in boiling points is only about 50 °C for the hydrides with 18, 36, and 54 electrons/molecule.

Because NH_3, H_2O, and HF are quite polar molecules, perhaps their high boiling points are a result of dipole-dipole attractions. However, HCl and HBr are also quite polar, since the electronegativities of chlorine and bromine are more than 0.5 units higher than the electronegativity of hydrogen. If dipole-dipole attractions are important contributors to intermolecular attractions, we would expect HCl and HBr to have higher boiling points than the hydrides with the same number of electrons. Figure 1.24 shows that HCl has a slightly higher boiling point than is predicted by the trend of the boiling points for the other period three and higher hydrides, but it is not far out of line. It appears that, as molecules are tumbling about in a liquid, dipole-dipole attractions do not contribute much to the intermolecular attractions responsible for holding the molecules together. Therefore, we have to search for another explanation for the anomalous behavior of NH_3, H_2O, and HF.

The hydrogen bond The model of the water molecule we have developed emphasizes the partial separation of the centers of positive and negative charge that results from the bond dipoles and the bent geometry of the molecule. The partial charge separation produces a molecular dipole and dipolar molecules can attract one another and preferentially orient themselves toward each other with the negative ends of one near the positive ends of another, as you showed in Consider This 1.24. The data shown in Figure 1.24 demonstrate that this attraction is greatly enhanced when it involves hydrogen attached to N, O, or F at the positive end of one molecule and a highly electronegative atom (usually N, O, or F) at the negative end of the other. Models for this interaction between two water molecules are shown in Figure 1.25.

Figure 1.25.

Representations of hydrogen bonding between two water molecules.

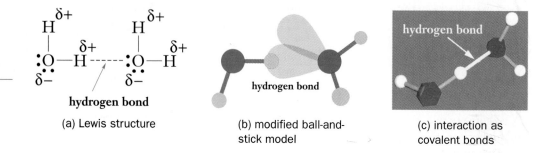

(a) Lewis structure

hydrogen bond

(b) modified ball-and-stick model

(c) interaction as covalent bonds

Experiments and calculations both show that the interaction between the hydrogen in one molecule and the lone pair on oxygen in the other has some covalent, electron-sharing, character. The model in Figure 1.25(c) may over-emphasize the amount of covalence.

The bonding between two water molecules represented in Figure 1.25 is called a **hydrogen bond.** Hydrogen bonds are formed when a hydrogen atom covalently bonded to a highly electronegative atom (N, O, and F) is attracted to a nonbonding electron pair on another highly electronegative atom (N, O, or F). The second electronegative atom may be in another molecule, or it may be in the molecule containing the hydrogen atom. The Lewis structure in Figure 1.25(a) highlights the proximity of partial positive and negative charges when the hydrogen atom on one water molecule is near a nonbonding pair of electrons on an adjacent water molecule. The modified ball-and stick model in Figure 1.25(b) shows the same interaction, with the large pink lobe representing the space where the electron pair is most likely to be found. Figure 1.25(c) represents the interaction as two covalent bonds: a shorter (and stronger) bond between the oxygen atom and one of the hydrogen atoms on the left-hand molecule, and a longer (much weaker) bond between the same hydrogen atom and the oxygen atom of the right-hand water molecule.

Figure 1.10 gave the hydrogen–oxygen covalent bond length in a single water molecule as 94 pm. As you can see in Figure 1.26, when water molecules hydrogen bond with one another, this bond stretches a bit to 100 pm, because the hydrogen atomic core is attracted to the nonbonding electron pair in the second oxygen. The length of the hydrogen bond to the second molecule is 180 pm, about twice the length of the hydrogen–oxygen covalent bond. Also note that the hydrogen atom forming the hydrogen bond is located on the line joining the two oxygen atoms. This linear geometry of the three atoms seems to form the strongest hydrogen bonds so hydrogen bonds generally form in situations where the hydrogen can lie roughly on the line between the two electronegative atoms.

Figure 1.27 shows that each water molecule can simultaneously form four hydrogen bonds. Two hydrogen bonds [(1) and (2) in the figure] link the central molecule's hydrogen atoms to nonbonding electron pairs on oxygen atoms in

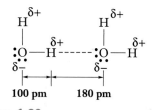

Figure 1.26.

Geometry of hydrogen bonding in water.

other water molecules. The other two hydrogen bonds [(3) and (4) in the figure] use the central molecule's nonbonding electron pairs on oxygen to link to hydrogen atoms on other molecules. A water molecule that forms four hydrogen bonds has a tetrahedral arrangement of bonds around the oxygen atom: two covalent bonds to the hydrogen atoms and two hydrogen bonds to hydrogen atoms on other water molecules.

In liquid water, hydrogen bonds are continuously broken and reformed as the moving water molecules jostle each other. Although its hydrogen-bonding partners are constantly changing, at any instant, each water molecule in the liquid has hydrogen bonds to about three other water molecules. These hydrogen bonding interactions among water molecules are strong enough to raise the boiling point of water to 100 °C, which is why water is a liquid at room temperature.

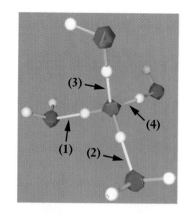

Figure 1.27.

A water molecule hydrogen bonded (arrows) to four other water molecules.

1.26 CONSIDER THIS

Does hydrogen bonding explain the boiling points of NH₃ and HF?

(a) The boiling points for NH_3 and HF (Figure 1.24) are out of line with other group V and VII compounds. Sketch models to show how hydrogen bonding can account for their relatively high boiling points.

(b) If we have a large number, N, of water molecules, there are $2N$ hydrogen atoms and $2N$ nonbonding pairs that can interact to form hydrogen bonds. Since it takes one hydrogen atom and one nonbonding pair to form each hydrogen bond, the maximum number of hydrogen bonds that can be formed among these molecules is $2N$. If we have N ammonia molecules, how many hydrogen atoms are available to form hydrogen bonds? How many nonbonding pairs? What is the maximum number of hydrogen bonds that can be formed among these N ammonia molecules? Do the same analysis for N hydrogen fluoride molecules.

(c) Do your results in part (b) help to explain the boiling points of NH_3 and HF relative to H_2O in Figure 1.24? Explain why or why not.

1.27 CHECK THIS

Predicting relative boiling points

Lewis structures for *n*-butane, ethyl methyl ether, and 1-propanol are shown. Predict the relative boiling points of these compounds. Explain your reasoning.

n-butane

ethyl methyl ether

1-propanol

1.8. Further Structural Effects of Hydrogen Bonding in Water

What kinds of hydrogen-bonded networks can you construct?

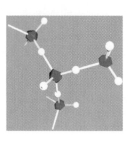

Work with a small group to construct about a dozen ball-and-stick models of water molecules. Use the shortest (pink) sticks in your set for the hydrogen–oxygen bonds. Connect one of your water molecules to four others by hydrogen bonds. Use the longer (white) sticks in your set for the hydrogen bonds, as in Figures 1.25(c) and 1.27. This picture shows how to begin such connections. Continue to add water molecules to see how a network of hydrogen-bonded water molecules might form. Can you form a ring of hydrogen-bonded water molecules? If so, how many molecules are required to form a ring without bending any of the sticks? Compare your structure to the representation shown in Figure 1.2(a). What conclusion(s) can you draw?

Ice: A hydrogen-bonded network Rings of hydrogen-bonded water molecules constructed for Investigate This 1.28 might look something like those in Figure 1.28(a). Each ring contains six oxygen atoms and six hydrogen atoms,

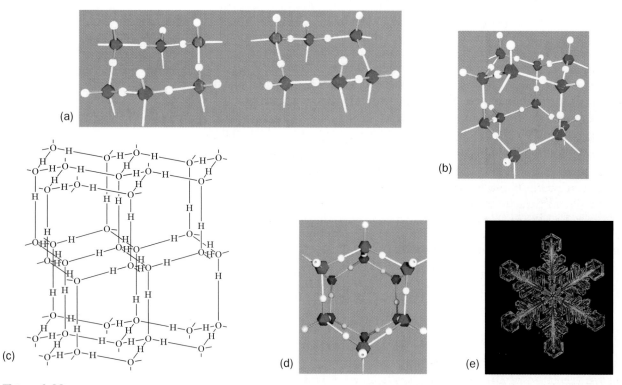

(a)

(b)

(c)

(d)

(e)

Figure 1.28.

Hydrogen-bonded networks of water molecules and the structure of ice. Panels (a) and (b) show successive stages in building a model of the ice structure. Panel (c) extends the structure to six of the cages in (b) and (d) is a top view of the structure in (b). Panel (e) shows a snowflake for comparison.

with six other hydrogen atoms sticking out from the ring. In this model, covalent H—O bonds alternate with H—O hydrogen bonds in each ring. All bond angles about oxygen are 109.5° tetrahedral angles, so the rings are puckered. Each oxygen atom has one covalent bond to a hydrogen atom and one hydrogen bond that are not part of the ring. Two ring structures can link using three hydrogen bonds, as shown in Figure 1.28(b). This cage-like structure can be extended in all three directions as shown in Figure 1.28(c).

The extended regular structure, with every water molecule forming four hydrogen bonds, is ice. The symmetry of the stacked six-member rings can be seen when viewed from the top, as in Figure 1.28(d) and Figure 1.2(a). Snowflakes [Figure 1.28(e)] reflect this six-fold symmetry. The extended hydrogen-bonded structure in ice produces more open space between water molecules than in liquid water. Consequently, as we pointed out in Section 1.1, ice is less dense than liquid water. Water is the only common substance that expands when it freezes.

Web Companion

Chapter 1, Section 1.8 — ①

Try interactive animations to investigate the structure of solid water.

②
③
④

1.29 CONSIDER THIS

How do the structures of liquid and solid water compare?

Show how breaking one of the hydrogen bonds in the ring of water molecules you made for Investigate This 1.28 allows the structure to collapse. How would such a collapse affect the ice structures shown in Figure 1.28? Is this collapse related to the relative densities of ice and water? Explain your answer.

1.30 INVESTIGATE THIS

What are the temperatures in an ice-water mixture?

Do this as a class activity and work in small groups to analyze and discuss the results. Fill a tall narrow transparent container with crushed ice and add cold water until it is full. Stir the mixture vigorously for about a minute to get the temperature of the mixture uniform at 0 °C. Place two thermometers or other temperature probes in the mixture with one measuring the temperature near the top and the other near the bottom of the mixture. Allow the system to remain undisturbed while you record the temperature of both thermometers every 5 minutes for 30–60 minutes. Plot the data.

1.31 CONSIDER THIS

How do you interpret the temperatures in an ice-water mixture?

Did you expect the results you observed in Investigate This 1.30? Why or why not? At what temperature do you think liquid water is most dense? What is the evidence for your answer?

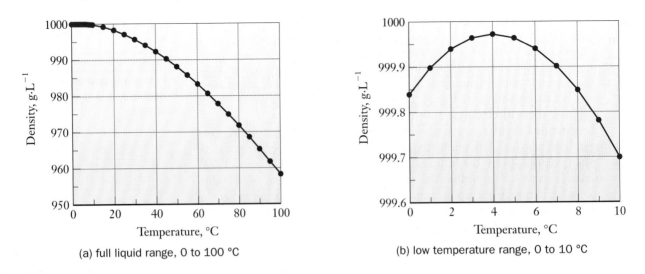

(a) full liquid range, 0 to 100 °C

(b) low temperature range, 0 to 10 °C

Figure 1.29.

Density of liquid water as a function of temperature.

Variation of liquid water density with temperature The lower density of solid ice compared to liquid water is unusual, but there is still more to the water density story. Ice melts when it has enough thermal energy (molecular motion) to begin to break some of the hydrogen bonds. As they break (at 0 °C), the open structure of the solid starts to collapse. Molecules move to take up some of the empty space, producing a liquid that is denser than the solid. As the temperature rises slightly, the increased jostling of the water molecules continues to allow them to find spaces to move into and increases the density even further [Figure 1.29(b)]. This effect reaches its maximum at 3.98 °C. As the temperature continues to increase, above 4 °C, the water molecules move faster, resulting in a larger average distance between them. The density decreases as shown in Figure 1.29. Water at 4 °C is the densest form of water (under normal pressure)—either as solid or liquid.

1.32 CONSIDER THIS

What are the connections between water density and temperature?

What is your interpretation of the results from Investigate This 1.30? Do the density properties of water and ice support your interpretation? Why or why not? Try Problems 1.62 and 1.63 to extend your analysis and to examine an important consequence of the density properties of water for aquatic life.

1.9. Hydrogen Bonds in Biomolecules

Life as we know it would not exist without the hydrogen bond. Hydrogen bonding is central to the structure and function not only of water, but also of proteins, nucleic acids, and many other molecules of biochemical importance. In this section we introduce several biomolecules that are essential constituents of living organisms. Our emphasis is on the role of hydrogen bonding, not the details of their molecular structures.

Proteins Proteins are high molecular mass compounds formed by linking hundreds or thousands of **amino acid** molecules end-to-end. Figure 1.30 shows how the linking occurs when three amino acids are strung together. The amino acid pictured in the figure is alanine, 1 of 20 that combine to form proteins. All these amino acids contain the basic structure shown in red in the figure. Each amino acid has a **side group,** a molecular fragment made up of various combinations of H, C, N, O, and S, like the methyl, $-CH_3$, group on alanine. The different side groups establish the unique identity of each amino acid. The chain of linked amino acids is called the protein *backbone.* Side groups are spaced along the backbone like charms on a bracelet.

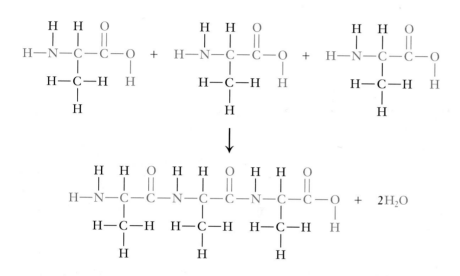

Figure 1.30.

Linking amino acids to form a chain. The atoms shown in red are part of the structure of all 20 amino acids found in proteins. A double line between atomic symbols represents two pairs of shared electrons (see Chapter 5).

Protein folding For a protein to carry out its tasks in an organism, the chain of amino acids must coil and fold upon itself into what is called its "active form." Many of the interactions that hold proteins in their active forms are hydrogen bonds that form along the protein backbone. The backbone is flexible enough to allow hydrogen bonding between a hydrogen atom covalently bonded to nitrogen and an oxygen atom that is four amino acids away along the chain. A basic structure, the **α-helix,** present in most proteins is held together by these hydrogen bonds, which are shown as heavy dots in Figure 1.31(a). The chain folds back on itself forming an overall structure such as that shown in Figure 1.31(b). Hydrogen bonds between parts of the backbone that are separated by tens or hundreds of amino acids help to hold this structure together.

When proteins are heated, the higher temperatures increase molecular motions, breaking some of the hydrogen bonds and the protein begins to unfold. Unfolding destroys the molecule's natural structure and activity; the protein becomes

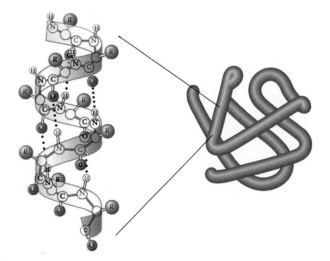

(a) short α-helical length of a much longer protein chain

(b) the entire folded protein with the location of the helix in (a) marked

Figure 1.31.

Folding a protein chain into (a) an α-helix and (b) the active form of the protein. Hydrogen bonds in (a) are shown as heavy dotted lines between the light gray hydrogen atoms and red oxygen atoms. The green balls represent side groups on the protein chain.

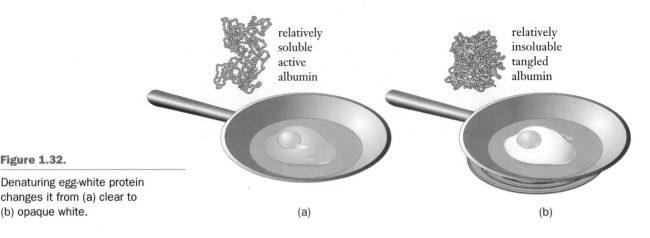

relatively
soluble
active
albumin

relatively
insoluable
tangled
albumin

Figure 1.32.

Denaturing egg-white protein
changes it from (a) clear to
(b) opaque white.

(a)   (b)

denatured. Most proteins denature above about 50 to 60 °C. If the solution hasn't been heated too hot or for too long, and is cooled slowly, the chain can sometimes refold and hydrogen bonds can reform, returning the structure to its original form. Part of the process of cooking protein-containing foods involves the heat denaturing of proteins. A familiar example of protein denaturation occurs when you cook an egg (Figure 1.32). As the molecules unfold, they become less soluble, their strands begin to tangle with one another, and the more-or-less clear protein solution [Figure 1.32(a)], changes to opaque and white [Figure 1.32(b)]. Unlike renaturing following mild heating, the cooking process cannot be reversed; you cannot unfry an egg.

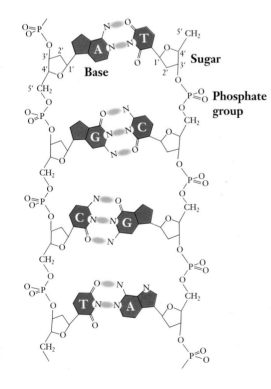

Nucleic acids The genes of all organisms are composed of **deoxyribonucleic acid (DNA).** DNA is a ladder-like molecule (Figure 1.33) with two parallel strands of alternating sugar and phosphate groups forming the "uprights," and the "rungs" of the ladder formed by hydrogen bonding between pairs of side groups on the chain. Every sugar unit on both strands has an attached side group, or **base,** which may be any one of four molecules—adenine, guanine, cytosine, or thymine (Figure 1.34). The bases pair in two ways: adenine forms two hydrogen bonds with thymine (A–T) and guanine forms three hydrogen bonds with cytosine (G–C). In living cells, the DNA ladder is twisted so that the uprights form a double helix. The order of the bases along the ladder encodes the genetic information stored in DNA.

Figure 1.33.

Hydrogen bonding in DNA. A, C, G, and **T** represent adenine, cytosine, guanine, and thymine. Pink areas represent hydrogen bonds.

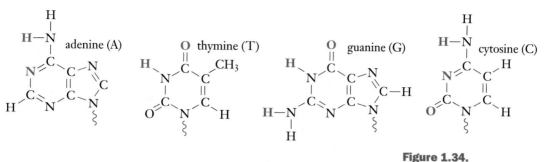

Figure 1.34.

Sites for hydrogen bonding in the bases in DNA.
Squiggly lines are the attachment to the sugar-phosphate chain. Atoms used in hydrogen bonding are in red.

1.33 INVESTIGATE THIS

How do the DNA bases fit together?

(a) Work in small groups and use your model kits to construct a model of each of the DNA bases—adenine, guanine, cytosine, and thymine (Figure 1.34). Use the light blue polyhedra for the Ns and dark gray polyhedra for the Cs in the rings. Use the standard black and red polyhedra 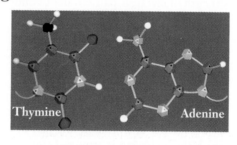 for the other Cs and Os. Use light green sticks to connect second-period elements to one another and pink sticks to connect Hs. Show how the A–T and G–C pairs can be held together by two and three hydrogen bonds, respectively.

(b) Try making the alternative pairs, A–C and G–T. Can these pairs fit together? Explain why or why not.

(c) Why don't we consider pairs like A–G, C–T, or A–A? Would there be a problem incorporating these into the DNA ladder? Explain.

At higher temperatures, increased molecular motions of the DNA molecule disrupt the hydrogen bonds, and the two linked strands separate. As with proteins, this process in which DNA loses its double-helical, hydrogen-bonded shape is called denaturation.

1.34 CONSIDER THIS

How does hydrogen bonding affect the thermal stability of DNA?

The relative numbers of G–C and A–T pairs affect the thermal stability of DNA helices. Why should changes in the G–C to A–T ratio affect the temperature at which the helices denature? How do you expect the thermal stability of DNA with a higher G–C to A–T ratio to compare to the stability of DNA with a lower G–C to A–T ratio? Explain your reasoning.

Strength of hydrogen bonds As proteins and nucleic acids carry out their functions in an organism, they must undergo reversible structural changes. Enzymes, for example, are proteins that change their shape slightly as they bind to and catalyze chemical reactions between other biological molecules. And DNA strands have to separate in order to undergo duplication (making new DNA strands) and transcription (making complementary RNA strands). These changes typically involve breaking some hydrogen bonds and forming others without breaking any covalent bonds. Covalent bonds hold the molecules of life together in relatively permanent structures under normal conditions. Though hydrogen bonds also help hold the molecules in their active forms, they allow for changes in shape that can occur with only small energy input. In other words, hydrogen bonds complement stronger covalent bonds in producing the remarkable tapestry of life. Typically, hydrogen bonds are about 5–10% the strength of electron-pair covalent bonds.

Reflection and Projection

The hydrogen-bonding model makes it possible to explain a huge range of physical and chemical properties of molecules. The essential features of the model are surprisingly simple: A hydrogen atom bound to a very electronegative atom is attracted to another very electronegative atom elsewhere. The requirements are availability of a nonbonding pair of electrons on the electronegative atom to which the hydrogen is attracted and a geometry that puts the hydrogen roughly on a line between the two electronegative atoms. The examples we have used to illustrate bulk properties of matter where hydrogen bonding is a factor (boiling points, densities of water and ice, denaturation of proteins and nucleic acids) only scratch the surface. All of chemistry is influenced by this remarkable phenomenon in more ways than can be described in any one textbook.

So far, we have not said much about the energy implications of any of the attractions between molecules. From our boiling point comparisons in Section 1.7, we found that, for many compounds, dispersion forces are the most important attractions holding the molecules together. For those polar compounds whose molecules can hydrogen bond with one another, hydrogen-bonding attractions often dominate the interactions among the molecules. And we have said that hydrogen bonds are weaker than covalent bonds, but we have not quantified any of these interactions. In the next sections, we begin a discussion of energy implications when water moves between liquid and gas phases or when it is heated or cooled. Thanks to hydrogen bonding, water is both an enormous reservoir of available energy as well as an enormous sink for waste heat. When we consider the energy associated with phase changes, we will have to deal with the problem of counting molecules, and we will look at techniques for counting by weighing in Section 1.11.

1.10. Phase Changes: Liquid to Gas

1.35 INVESTIGATE THIS

What happens when you sweat?

For this investigation, you will simulate sweating by wetting a finger. Be sure your hands are clean and dry. Wet one of your fingers and then wave your hand

continued

in the air. Do you feel any difference between the wet finger and the others? If so, what is the difference?

1.36 CONSIDER THIS

What is the purpose of sweating?

(a) What do you think caused your wet finger to feel different when you waved it in the air in Investigate This 1.35? What is required to produce this phase change? What evidence do you have to justify your answer? How could you test your explanation?

(b) How is your analysis in part (a) related to sweating?

When a liquid boils to give a gas at atmospheric pressure or evaporates to give the gas at a lower pressure, energy is required to disrupt the attractive interactions between the molecules in the liquid phase. In Section 1.7, we used this idea to correlate boiling points with the strength of these attractions. We did not, however, discuss the source of the required energy. Energy cannot be created or destroyed. If energy enters a liquid to cause boiling or evaporation, the energy has to come from somewhere. When energy is put into one substance, an equivalent amount of energy has to leave another substance.

Since energy can go into or come out of a substance, we need to have a way to indicate the direction of this change. We will use E to symbolize energy, and we will use ΔE to symbolize a *change* in energy. The Δ (uppercase Greek letter delta) symbol always means the *final value minus the initial value* of the specified quantity, for example, $\Delta E = E_{final} - E_{initial}$. When energy enters a substance, the final amount of energy in the substance, E_{final}, is larger than the initial amount of energy, $E_{initial}$, that is, $E_{final} > E_{initial}$, so ΔE is positive. Conversely, when energy leaves a substance, $E_{final} < E_{initial}$, and ΔE is negative.

1.37 CONSIDER THIS

What are the signs of the energy changes for sweating?

(a) What happened to the water on your skin when you waved your wet finger in the air in Investigate This 1.35? Does energy leave the water or enter the water to cause this process? What is the sign of the energy change in the water?

(b) Does energy leave your skin or enter your skin in this process? What is the sign of the energy change in your skin? Does your answer help explain the difference between your wet and dry fingers in the investigation?

Energy diagrams An energy diagram (Figure 1.35) is a graphical way to visualize changes in energy. Energy diagrams appear often in this book, so you need to become familiar with their interpretation. The horizontal lines represent the energy of the substances written directly above them. Energy levels are not always given absolute numeric values, because the diagrams are designed to show energy *differences*, ΔE. The head of an arrow showing the direction of change points to the final energy and the tail is at the initial energy level. An arrow pointing up, as in Figure 1.35, means that the change in energy for the process has a positive sign—$\Delta E > 0$—because the final energy value is greater than the initial energy value. Energy is required to bring about the indicated change. An arrow pointing down means that the change in energy for the process has a negative sign—$\Delta E < 0$—because the final energy value is smaller than the initial energy value. Energy is released when the indicated change occurs.

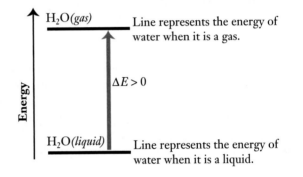

Figure 1.35.

An energy diagram for evaporation of a sample of water.

Figure 1.35 is an energy diagram for water evaporating from your skin. Since energy has to enter the water to cause the **evaporation** (vaporization), the energy of the gaseous water is higher than the energy of the liquid water before evaporation. The horizontal line representing the energy of the gas is higher on the graph than the line representing liquid. The upward-pointing red arrow in Figure 1.35 shows you that ΔE for the liquid-to-gas phase change is positive.

1.38 CHECK THIS

Interpreting an energy diagram

(a) The energy diagram shown here is for a change in the temperature of one liter of water. The arrow represents the energy change when the water changes from 50 to 25 °C. Does energy enter or leave the water when this change occurs? How is this represented on the diagram? Explain your answers.

(b) Is the $\Delta E > 0$ or is $\Delta E < 0$ for this process? How is this represented on the diagram? Explain your answers.

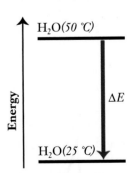

1.39 CONSIDER THIS

How are energy level diagrams correlated?

(a) Sketch an energy level diagram that represents the energy change *in your skin* when energy from your skin is used to evaporate some water. Draw an arrow that represents the direction of the change. How is the sign of ΔE for this process related to the direction of the energy-change arrow on your diagram?

(b) Assume that the amount of water that is evaporated by your skin in part (a) is the same as the amount of water represented by Figure 1.35. Should the length of the energy-change, ΔE, arrow on your diagram be longer, shorter, or the same length as the energy-change arrow in Figure 1.35. Explain your reasoning.

1.40 INVESTIGATE THIS

Do all liquids evaporate at the same rate?

Do this as a class activity and work in small groups to analyze and discuss the results. Use three identical paper towels. Simultaneously put 10 drops of water in one spot on one towel, 10 drops of methanol on a second towel, and 10 drops of hexane on the third towel. Hold the towels up so everyone can observe the wet spots. *CAUTION:* Methanol and hexane are both flammable. Time how long it takes each of the liquid spots to disappear completely.

1.41 CONSIDER THIS

What are the relative energies of vaporization of different liquids?

(a) Is ΔE for the liquids in Investigate This 1.40 positive or negative? Give your reasoning.

(b) For which liquid is the absolute value of ΔE smallest? For which is the absolute value of ΔE largest? Base your reasoning only on our model for vaporization and data from Investigate This 1.40.

(c) Sketch an energy diagram that compares the energy changes, ΔEs, for each liquid evaporating.

In Investigate This 1.40, the liquid films all gained energy from the paper towels and surrounding air. Each liquid sample gains the same amount of energy per unit time, but they do not all disappear at the same time. Therefore, different quantities of energy must be required for each sample to vaporize. What are these quantities? Our energy diagrams have not shown energy units and all our comparisons have been either directional or in terms of relative size based on experimental evidence. Before we go further, we need to define the units for energy.

The joule is named to honor James Joule (English scientist, 1818–1889), one of the first scientists to study the quantitative relationships between heat and work.

Energy units and conversions The energy unit that is probably most familiar to us is the calorie. A **calorie** was originally defined as the amount of energy required to increase the temperature of exactly 1 g of water by exactly 1 °C. This quantity of energy is called the **specific heat** of water. Scientists now use the **joule** (J) as the standard unit of energy. The joule is defined in terms of an amount of work done rather than an amount of thermal energy transferred. The relationship of joules to calories is

$$1 \text{cal} \equiv 4.184 \text{ J} \tag{1.1}$$

(The $\equiv$ symbol means *is defined as.*) Most energy values in biological systems are still given in calories. We will often give values in both units. Or, you can work them out for yourself.

1.42 WORKED EXAMPLE

Conversion of energy units

The energy required to vaporize water is $2.4 \times 10^3 \text{ J} \cdot \text{g}^{-1}$. How many calories are required to vaporize 1 g of water?

Necessary information: We need the relation between joules and calories in equation (1.1).

Strategy: A *quantity* of something tells you how much of it there is. Quantity has *units* and a *numeric value*. Expressing the quantity in different units requires a different numeric value. If you buy 1 dozen eggs or 12 eggs, for example, you get the same quantity of eggs, but the numeric values for dozen and number are different. Rewrite equation (1.1) as a unit conversion factor and use it to convert the energy of vaporization in joules to calories.

Implementation: Equation (1.1) can be rewritten as:

$$\frac{1 \text{ cal}}{4.184 \text{ J}} = 1 \tag{1.2}$$

We have been given the quantity of energy required to vaporize 1 g of water in units of joules, so we use the ratio given by equation (1.2) to convert to units of calories:

$$\text{energy of vaporization} = 2.4 \times 10^3 \text{ J} \cdot \text{g}^{-1}$$

$$= (2.4 \times 10^3 \text{ J} \cdot \text{g}^{-1})\left(\frac{1 \text{ cal}}{4.184 \text{ J}}\right)$$

$$= 5.7 \times 10^2 \text{ cal} \cdot \text{g}^{-1}$$

The unit $\text{J} \cdot \text{g}^{-1}$ is the same as J/g. The exponential form is less ambiguous and is used throughout the book.

Does the answer make sense? Equation (1.1) shows that 1 cal is a larger amount of energy than 1 J. Therefore, a quantity of energy expressed in calories will have a smaller numeric value than the same quantity expressed in joules. This is the result we got.

continued

After finishing any problem involving calculations, check to see if the numeric result is reasonable:

- Check the sign of the result. Here we are converting a positive quantity energy from one unit to another, so the answer must be positive.
- Check the units of the result. If units are always included with numeric values, the units should cancel to give the appropriate units in the result.
- Check the arithmetic. In this case, mentally doing the calculations to one significant figure yields: $2000 \div 4 = 500$, which is the right order of magnitude.
- Check the number of significant figures. Here, the datum with the greatest uncertainty is the energy in joules, which has an uncertainty of 1 part in 24. We have given the result with two significant figures, which implies an uncertainty of about 1 part in 57. Reporting only one significant figure would imply an uncertainty of 1 part in 6, which would be far too uncertain. Reporting three significant figures would imply an uncertainty of about 1 part in 570, which would not be justified by the original datum.

1.43 CHECK THIS

Conversion of energy units

(a) The amount of energy required to vaporize hexane is 92 cal·g^{-1}. How many joules are required to vaporize one gram of hexane?

(b) From the data in part (a) and in Worked Example 1.42, what conclusion can you draw about the attractions between hexane molecules relative to those between water molecules? Is your conclusion consistent with the models in Section 1.7? Explain why or why not.

Correlations with molecular models In Investigate This 1.40, you observed that 10 drops of liquid hexane evaporated faster than 10 drops of liquid water and could conclude that less energy is required to evaporate hexane than water. The data in Worked Example 1.42 and Check This 1.43 show that the energy required to vaporize a gram of hexane is only about one-sixth the amount required to vaporize a gram of water, so the conclusion you drew from the investigation is confirmed by further data. But we are not ready quite yet to use experiments and data like these to correlate our molecular models of attractions with the observable energies of vaporization.

To compare the energies of vaporization for two compounds, we need to compare the energy required to vaporize the same number of molecules of each compound. A molecule of hexane, C_6H_{14}, has 20 atoms and a molecule of water, H_2O, has 3. A water molecule has a smaller mass than a hexane molecule, so one gram of water will contain more molecules than one gram of hexane. An analogy would be the number of grapes and oranges in the same mass of each fruit. A grape has a smaller mass than an orange and Figure 1.36 shows that there are more grapes than oranges in 0.5 kg of each. When 1 g of water is vaporized, more molecules change state than when 1 g of hexane is vaporized.

Figure 1.36.

Comparison of equal masses of grapes and oranges.

More interactions have to be broken in the water sample and this might account for the greater quantity of energy required to evaporate it. In order to make comparisons that we can relate to our molecular models of matter, we need to know the amount of energy required to vaporize the same number of molecules of each compound. We need a way to count out the same number of molecules (or know the relative numbers of molecules) for our comparisons.

1.11. Counting Molecules: The Mole

Individual molecules are far too small to count using our usual laboratory apparatus, so we have to devise other ways to count them. A convenient way to measure a quantity of matter is to use a balance to determine its mass. **Mass** is a fundamental property of matter that measures the quantity of matter in a sample. The problem is that the molecules of different compounds have different masses, so the same mass of two compounds will contain different numbers of molecules. In chemistry, *the unit that expresses numbers of molecules is referred to as the* **amount** *of a substance.* To measure out the same number of molecules of two different compounds, the ratio of the masses taken must be the same as the ratio of the molecular masses.

> In everyday use, *amount* means just about any measure (mass, volume, number, and so on) that specifies how much of something we have. In chemistry, *amount* specifies number of molecules.

1.44 CONSIDER THIS

How can you count objects by measuring mass?

In many game arcades, a player gets one or more paper tickets for each game won. Players often bring large batches of tickets to the redemption center to exchange for prizes. The person at the center uses a balance to determine the mass of the tickets and awards prizes based on the mass. If a dozen tickets have a mass of 1.9 g, and a player redeems 24.1 g of tickets, how many dozens of tickets are redeemed? How many tickets are redeemed? Describe your method for solving this problem.

Mole and molar mass If you know the mass of a dozen identical objects and measure the mass of an unknown number of those same objects, you can determine how many dozens of the objects you have by dividing the measured mass by the mass of a dozen of the objects, as you did in Consider This 1.44. Instead of a dozen, the standard scientific unit for measuring amount (number) is the **mole** (abbreviated **mol**). The individual particles that make up the substance being measured can be anything: atoms, molecules, ions, grains of sand, mosquitoes, whatever. Particles the size of atoms and molecules, unlike tickets to be redeemed, are so small that they cannot be counted directly, but we can use a balance to determine the mass of an observable amount of any element or compound. Then, if we know the mass of one mole of the element or compound, we can calculate how many moles are present in the measured mass.

> The name *mole* has an "e" on the end, but the unit abbreviation *mol* does not. Mole is used in text discussions; mol is used with numeric values.

1.45 WORKED EXAMPLE

Calculating numbers of moles

A mole of water has a mass of 18.02 g. If you drink a glass of water, about 225 g, how many moles of water do you drink?

Necessary information: The problem statement contains all the needed information.

Strategy: Use the mass equivalence of a mole of water molecules to write a unit conversion factor to convert the mass of water you drank to number of moles of water.

Implementation: 1.0 mol H_2O = 18.02 g, and we rewrite this equivalence as:

$$\frac{1.000 \text{ mol } H_2O}{18.02 \text{ g}} = 1$$

Use this unit factor to convert the mass of our drink to moles:

$$225 \text{ g } H_2O = (225 \text{ g } H_2O)\left(\frac{1.000 \text{ mol } H_2O}{18.02 \text{ g}}\right) = 12.5 \text{ mol } H_2O$$

Does the answer make sense? The mass of water you drank is more than 10 times the mass of 1 mol of water, so our answer, more than 10 mol, makes sense. The three significant figures in the result are consistent with the most uncertain value in the original data, the mass of the water.

Web Companion

Chapter 1, Section 1.11

Use interactive animations to work through mass, mole, and molar mass interconversions.

1.46 CHECK THIS

Mass to moles and moles to mass conversions

(a) Ten drops of water is about 0.5 mL, or about 0.5 g of water. How many moles of water are in 10 drops of water?
(b) The mass of 1 mol of sucrose (table sugar) is 342 g. A bag of sugar from the market contains 2.26 kg of sugar. How many moles of sugar are in the bag?
(c) The mass of 1 mol of sodium chloride (table salt) is 58.5 g. A container of salt from the market contains 12.6 mol of salt. How many grams of salt are in the container? You can check your answer at the market.

Recall that the periodic table on the inside front cover of the book gives the *relative atomic masses* of the atoms of each element. The **molar mass** of any element is the mass of the element in grams that is equal to its relative atomic mass. A molar mass of an element contains a mole of atoms of the element. For example, 1.0079 g of hydrogen contains 1 mol of hydrogen atoms; the molar mass of hydrogen is 1.0079 g. Similarly, the molar mass of carbon is 12.01 g and the molar mass of oxygen is 16.00 g. A molecule contains more than one atom, so the molar mass of a molecule is the sum of the molar masses of all its constituent atoms. For example, oxygen gas consists of O_2 molecules, so a mole of oxygen gas contains 2 mol of oxygen atoms, and the molar mass of oxygen gas is 32.00 g (= 2 × 16.00 g).

This oxygen gas example reminds us to be careful to specify the atomic or molecular unit of interest when we are using molar masses. We often specify the **formula unit,** the chemical formula of the atomic or molecular unit of interest. For example, if we are interested in bromine atoms, we specify the formula unit as Br, the atomic symbol for bromine, and the molar mass of this formula unit is 79.91 g. If we are interested in elemental bromine, which is a liquid of Br_2 molecules, the formula unit is Br_2 and the molar mass in 159.82 g.

Most molecules contain more than one kind of atom and all the atomic masses in their formula unit are combined to get the molar mass. A mole of water, H_2O, for example, contains 2 mol of hydrogen atoms, (2.0158 g of hydrogen atoms) and 1 mol of oxygen atoms (16.00 g of oxygen atoms). The molar mass of water used in Worked Example 1.45 and Check This 1.46 is 18.02 g (= 2.0158 g hydrogen + 16.00 g oxygen).

1.47 WORKED EXAMPLE

Calculating molar mass

What is the molar mass of hexane, C_6H_{14}?

Necessary information: We use the given formula unit and the relative atomic masses of C and H atoms from the inside front cover.

Strategy: The number of atoms of each element in the formula unit multiplied by the molar mass of each elemental atom gives the mass of each element in a mole of formula units. The sum of these masses is the molar mass of the formula unit.

Implementation: The C_6H_{14} formula unit for hexane contains 6 carbon atoms and 14 hydrogen atoms, so in 1 mol of formula units, we have

$$6 \text{ mol C} = (6 \text{ mol C})\left(\frac{12.01 \text{ g}}{1 \text{ mol C}}\right) = 72.06 \text{ g (mass of C in one mole } C_6H_{14})$$

$$14 \text{ mol H} = (14 \text{ mol H})\left(\frac{1.0079}{1 \text{ mol H}}\right)$$

$$= 14.11 \text{ g (mass of H in one mole } C_6H_{14})$$

$$\text{molar mass } C_6H_{14} = (72.06 \text{ g}) + (14.11 \text{ g}) = 86.17 \text{ g}$$

Does the answer make sense? One way to tell whether our answer makes sense is to check the arithmetic, which you should do. You could also look up the molar mass in a reference handbook.

1.48 CHECK THIS

Calculating molar mass

(a) How many moles of carbon atoms are there in one mole of sucrose, formula unit $C_{12}H_{22}O_{11}$? How many moles of hydrogen atoms? How many moles of oxygen atoms? What is the molar mass of sucrose? Explain how you get your answers.

(b) How many moles of sodium atoms are there in 1 mol of sodium chloride, formula unit NaCl? How many moles of chlorine atoms? What is the molar mass of sodium chloride? Check your results for parts (a) and (b) with the data in Check This 1.46.

In 1859, a consistent set of relative atomic masses, based on the results of many experiments and the atomic model of matter, was proposed and soon agreed upon by scientists. Since then, many more experiments have been done that refined the original values and produced the relative atomic masses in our modern periodic table. Over these years, the amount of a substance that constitutes a mole has also been redefined and refined. At present, scientists have agreed on one of the forms of carbon, carbon-12 (also written as ^{12}C), as the standard for determining amount (moles). One mole is defined to be the number of atoms in exactly 12 g of carbon-12. This number has been determined experimentally to be 6.0221367×10^{23} and is called **Avogadro's number.** The symbol for Avogadro's number is N_A. The mass of a sample of any pure substance that contains Avogadro's number of particles is one mole of that substance. To summarize:

More history of atomic masses and the periodic table is discussed in Chapter 4, Section 4.1.

$$1 \textbf{ mole} \equiv \text{number of atoms in 12 g of } ^{12}C$$
$$= 6.0221367 \times 10^{23} \text{ particles} = N_A \qquad \textbf{(1.3)}$$

Although the mole is a number of particles, when we use moles, we rarely need to know the actual number of particles. To compare measurable amounts of substances, we almost always compare number of moles rather than number of molecules (or formula units). Figure 1.37 reminds us that comparing masses of formula units is identical to comparing molar masses of these formula units.

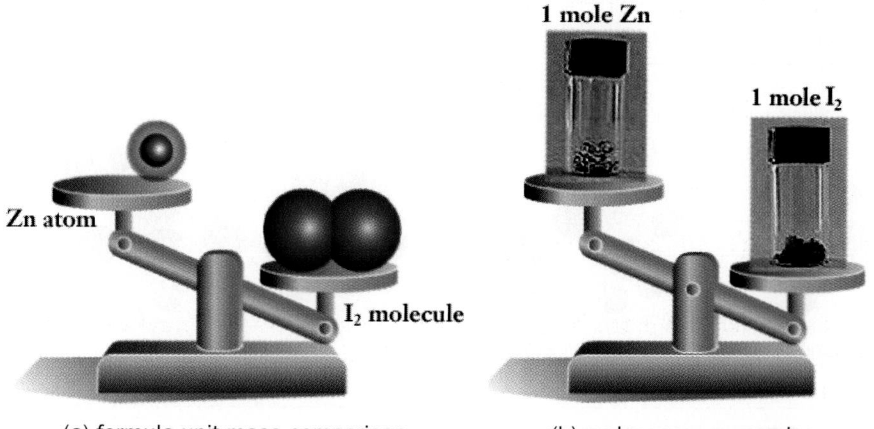

(a) formula unit mass comparison

(b) molar mass comparison

Figure 1.37.

Imaginary comparison of formula unit masses and molar masses. See the *Web Companion,* Chapter 1, Section 1.11, for interactive activities based on these comparisons.

1.49 CHECK THIS

Comparing amounts of compounds

In Check This 1.46, the number of moles of water in ten drops of water was calculated, which is the amount of water used in Investigate This 1.40. How many moles of methanol, CH_3OH, and hexane, C_6H_{14}, were used in the investigation? Assume that 10 drops of methanol has a mass of 0.4 g and 10 drops of hexane a mass of 0.3 g. For which compound did the largest number of molecules evaporate? For which did the smallest number evaporate? Explain your answers.

Molar vaporization energies   In Investigate This 1.40, you observed that it took the longest time for water to evaporate. Now, the results from Check This 1.46 and 1.49 show that there were more molecules (moles) of water to be evaporated. Was more time required just because there were more molecules to evaporate? The only way to answer this question is to measure the amount of energy that has to be supplied to cause the evaporation of a known amount (number of molecules or moles) of liquid. The results of such experiments are shown for five compounds in Table 1.2. The table gives the energies of vaporization per mole, the boiling points, and dipole moments (a measure of the relative polarities of the molecules). Figure 1.38 shows ball-and-stick molecular models for the five compounds in Table 1.2.

Table 1.2 *Vaporization energies, boiling points, and dipole moments for selected compounds.*

Compound	Line formula	Molar mass, g	Energy of vaporization, kJ·mol⁻¹	Boiling point, °C	Dipole moment, Debye
water	H_2O	18	44	100	1.87
methanol	CH_3OH	32	39	65	1.70
ethanol	CH_3CH_2OH	46	41	79	1.69
dimethyl ether	CH_3OCH_3	46	23	−25	1.30
hexane	$CH_3(CH_2)_4CH_3$	86	32	69	0

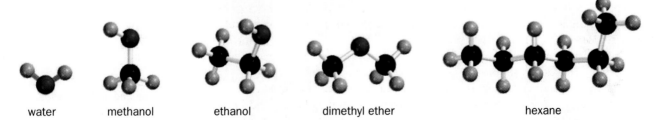

water   methanol   ethanol   dimethyl ether   hexane

Figure 1.38.

Molecular models of the compounds in Table 1.2.

1.50 CONSIDER THIS

Which compounds in Table 1.2 form hydrogen bonds?

(a) In addition to water, which compounds in Table 1.2 have molecules that can hydrogen bond to one another? How many hydrogen bonds can each molecule form?

(b) Make molecular models and drawings similar to Figure 1.25(a) to show the hydrogen bonding in the compounds identified in part (a). Give your reasoning for why the molecules of the remaining compounds can't hydrogen bond among themselves.

Vaporization energies and molecular attractions We can use molar vaporization energies to compare our molecular models for attractions among molecules, because these energies are for the same number of molecules vaporizing. Let's look first at the polar molecules in Table 1.2. Methanol and ethanol are **alcohols.** Alcohols are related to water in that both have an O—H group. For alcohols, the other oxygen bond is to an **alkyl** group, a group that contains one or more carbon atoms with every unused bond on each carbon atom occupied by hydrogen. Alcohol molecules have an O—H group and can hydrogen bond with one another, but can form a maximum of only about half as many hydrogen bonds as water, that is, about one mole of hydrogen bonds per mole of alcohol (see Consider This 1.26).

If we assume that most of the energy of vaporization of water, 44 kJ·mol^{-1}, is required to break hydrogen bonds, then about half this amount, 22 kJ·mol^{-1}, is required to break the mole of hydrogen bonds in the alcohols. This analysis suggests that breaking hydrogen bonds requires about 22 kJ·mol^{-1}. The remainder of the energy of vaporization for the alcohols, about 20 kJ·mol^{-1}, must be required to overcome the dispersion forces among the molecules.

Like alcohols, **ethers** also contain an oxygen atom, but in this case, both oxygen bonds are to alkyl groups. Ether molecules have no hydrogen atom covalently bonded to the electronegative oxygen, so the molecules of ether cannot hydrogen bond among themselves. The striking differences between the properties of ethanol and dimethyl ether (Table 1.2) are a consequence of hydrogen bonding in the ethanol. These compounds have the same molecular formula, C_2H_6O, and, therefore, the same molar mass and same number of electrons; but ethanol is a liquid and dimethyl ether is a gas at room temperature. Their boiling points differ by more than 100 °C. The energy of vaporization for the ether, 23 kJ·mol^{-1}, must be required to overcome the dispersion forces among the molecules. Since ethanol and dimethyl ether have the same molecular formula, the dispersion forces among the molecules in their liquids should be about the same, 23 kJ·mol^{-1}. This value is consistent with our reasoning that gave about 20 kJ·mol^{-1} for the dispersion forces among ethanol molecules.

Finally, consider the nonpolar molecule, hexane, which is a liquid at room temperature and whose molar vaporization energy and boiling point are comparable to those for the alcohols in Table 1.2. The attractions that hold the hexane molecules together are dispersion forces and there are enough electrons in hexane (50) to give liquid hexane some properties that are comparable to smaller hydrogen-bonding molecules. As we will see in Chapter 2, dispersion forces are important in determining how molecules that contain large nonpolar segments as well as a polar group interact with one another and with different molecules.

1.51 WORKED EXAMPLE

Vaporization energies in Investigate This 1.40

How much energy was required to vaporize the water sample in Investigate This 1.40?

Necessary information: We need the energy of vaporization of water, 44 kJ·mol^{-1}, from Table 1.2 and the number of moles of water in 10 drops, 0.03 mol, from Check This 1.46(a).

Strategy: Use the number of moles of water vaporized and the energy required for each mole to calculate the energy required for the 10-drop sample.

Implementation:

$$\text{energy to vaporize 0.03 mol H}_2\text{O} = (0.03 \text{ mol H}_2\text{O})\left(\frac{44 \text{ kJ}}{1 \text{ mol H}_2\text{O}}\right) = 1 \text{ kJ}$$

Does the answer make sense? Ten drops of water is a good deal less than a mole of water, so it will take a good deal less than the molar energy of vaporization to vaporize it, as we find.

1.52 CHECK THIS

Vaporization energies in Investigate This 1.40

(a) Use the data in Table 1.2 and your results from Check This 1.49 to calculate the energy required to vaporize the methanol and hexane samples in Investigate This 1.40.
(b) Were your observations in the investigation consistent with the energies from part (a) and the energy for water from Worked Example 1.51? Explain why or why not.

Reflection and Projection

A central concern of chemistry is to probe matter with whatever tools we have in an attempt to tease out a fundamental understanding of how nature works. In order to explain the bulk properties of matter, scientists invent and refine molecular models that are consistent with the observed properties and help to predict and interpret others as well. Section 1.10 began with consideration of why a wet finger held in the wind feels cooler than a dry finger, and moved on to consider why some liquids evaporate from a paper towel faster than others. Our interest in these phenomena was focused on the direction and magnitude of the energy changes. When a liquid is vaporized, energy has to be supplied to disrupt the attractions between the molecules and this energy has to come from somewhere.

To interpret the data from evaporating liquids in molecular terms, we found that we needed a way to assure that we were comparing the same number of molecules (or a known ratio) in each case. This necessity led to our introduction of the scientific unit for amount, the mole. The mole is Avogadro's number of anything, but our usual application is to atoms, molecules, and ions. Molar mass is the mass in grams equal to the sum of the relative atomic masses that make up the formula unit of the element or compound of interest.

When we expressed energy of vaporization as the energy required to vaporize one mole of a compound, we could compare the data and interpret similarities and differences in terms of the molecular models of dispersion forces and hydrogen bonding from our previous discussions. Dispersion forces increase as the size of molecules increase because more electrons are available to form instantaneous dipoles. All molecules have dispersion forces attracting their molecules to each other and some have other interactions that contribute to the attraction. The hydrogen bond is an especially strong polar attraction between molecules in which a hydrogen atom covalently bonded to a very electronegative atom lies approximately on a line between this electronegative atom and another. For small molecules, hydrogen bonding often dominates those physical properties that are related to attractions between molecules. All of these *intermolecular* attractions are weaker than the *intramolecular* bonding (covalent bonds) that hold atoms together in molecules.

We have discussed a few of the properties of water that depend on the extensive hydrogen bonding between water molecules, and there are several more. In the next section, we will continue our discussion of energy and water, and see another way that water is essential to life as we know it.

1.12. Specific Heat of Water: Keeping Earth's Temperature Stable

On the July 4, 1997, the Mars Pathfinder space probe landed on our planetary neighbor. The next day, a vehicle about the size of a child's wagon, the Sojourner Rover shown in Figure 1.39, left the lander and began moving about, using its various instruments to explore the Martian surface. The design team for the Mars lander had to design instruments and vehicles that could work over a large temperature range. The Martian atmospheric temperature varies about 74 °C, from −80 °C during the Martian night to about −6 °C during the Martian day. In contrast, the night-to-day difference is usually less than 25 °C on Earth, and the average temperature of Earth is about 16 °C, well above the freezing point of water.

Part of the reason that conditions on the two planets are so different is that more than 70% of Earth's surface is covered with water. This mass of water acts as a giant energy reservoir. It absorbs energy from the sun and warm air during the day, without changing its temperature very much. The water returns thermal energy to the cooler atmosphere at night. The large mass of water evens out the extremes of temperatures. Mars, without water on its surface, lacks this mechanism for modulating surface temperature changes.

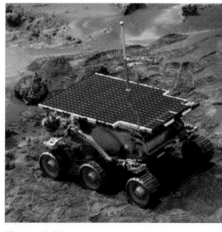

Figure 1.39.

The Sojourner Rover exploring the Martian surface.

1.53 INVESTIGATE THIS

What is observed when liquids are heated at the same rate?

Do this as a class activity and work in small groups to analyze and discuss the results. Add 120–150 g of room-temperature water to a 250-mL Styrofoam® cup. Add the same mass of room-temperature ethanol to a second identical cup. Clamp identical electrical immersion heaters and thermometers in each cup with the thermometers about 1 cm from the heaters. Read and record the temperature of the liquids every 10 seconds for about 1 minute. Simultaneously plug both heaters into electrical outlets. Continue reading and recording the temperatures every 10 seconds until one of the liquids reaches about 70 °C. Unplug the heaters.

1.54 CONSIDER THIS

How does temperature change in heated liquids?

(a) Plot the temperatures of each of the liquids in Investigate This 1.53 as a function of time. How do the plots for the water and ethanol differ? How are they the same?
(b) The same amount of energy has been added to both liquids. How can you account for any differences in the results for the two liquids?

Molar heat capacity Maintaining the earth's temperature in a range that supports life is essential for the survival of living things. One of the properties of water that is essential for temperature regulation in organisms and in the environment is its heat capacity. When energy (heat) enters a substance, the molecules begin to move a bit faster. These motions include rotation, vibration, and movement from one location to another. The increased motion is observed as an increase in the temperature of the substance. The **molar heat capacity, C,** of a substance is the amount of energy required to raise the temperature of one mole of the substance by one degree Celsius. Molar heat capacities for water and several other compounds are given in Table 1.3.

Table 1.3 *Molar heat capacity and specific heat for selected compounds.*

Compound	Formula unit	Molar mass, g	Molar heat capacity, $J \cdot mol^{-1} \cdot {}^{\circ}C^{-1}$	Specific heat, $J \cdot g^{-1} \cdot {}^{\circ}C^{-1}$
water	H_2O	18	75.4	4.18
methanol	CH_4O	32	81.0	2.53
ethanol	C_2H_6O	46	112	2.44
1-propanol	C_3H_8O	60	144	2.39
hexane	C_6H_{14}	86	195	2.27
octane	C_8H_{18}	114	254	2.23
decane	$C_{10}H_{22}$	142	315	2.22
dodecane	$C_{12}H_{26}$	170	375	2.21

You can see from Table 1.3 that the molar heat capacity is larger for higher molar mass, more complex molecules. This makes sense, since it takes more energy to make heavier molecules move about and there are more ways for complex molecules to vibrate and rotate. Although water does not seem to stand out in the listing of molar heat capacities in Table 1.3, Figure 1.40 shows that it does not fit the pattern of the other molecules in the table. The line in Figure 1.40 is the best-fit line through the hydrocarbon data and you see that the alcohols fall pretty well on the line, but that water does not. The molar heat capacity of water is higher than is predicted by the trend in the data for the other liquids.

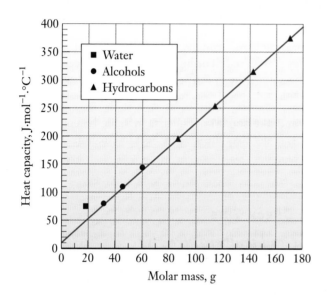

Figure 1.40.

Molar heat capacities vs. molar mass for the compounds in Table 1.3.

Specific heat Another way to analyze heat capacity data is to divide the molar heat capacity for a compound by its molar mass. This procedure gives the specific heat, which we introduced briefly in Section 1.10. As we noted there, the **specific heat, s,** of a substance is the amount of energy required to increase the temperature of exactly 1 g of the substance by exactly 1 °C. Water has a high specific heat, $s = 4.18 \text{ J} \cdot \text{g}^{-1} \cdot {}^{\circ}\text{C}^{-1}$, which is about 10 times that of metals such as copper or iron. If the same amount of heat were added to equal masses of iron and water, the increase in temperature of the iron would be about 10 times that of the water. We also see in Table 1.3 that the specific heat of water is almost double that of the other liquids, which are between 2.0 and 2.5 $\text{J} \cdot \text{g}^{-1} \cdot {}^{\circ}\text{C}^{-1}$.

1.55 WORKED EXAMPLE

Specific heat and temperature rise

In Investigate This 1.53, you added the same amount of energy to equal masses of water and ethanol. Use the results of the investigation to show which liquid has the higher specific heat.

Necessary information: We need to know that the ethanol reached a higher temperature than the water when the same amount of energy was added to each in Investigate This 1.53.

continued

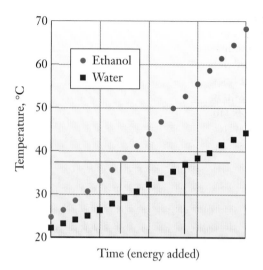

Time (energy added)

Implementation: When energy is added to a liquid that is below its boiling temperature, the temperature of the liquid increases, as we observed for both ethanol and water in Investigate This 1.53. This graph shows the results for such an experiment. The temperature of the ethanol increases more than the temperature of the water, which means that the rate of temperature increase for ethanol is greater than for water. In other words, more energy is required to increase the temperature of water a given amount than is required to increase the temperature of ethanol by the same amount. [The horizontal line is added to the figure to show that the water gets to a given temperature later (after more energy has been added) than ethanol.] Specific heat is the amount of energy required to bring about a given increase in temperature of a substance. Because more energy is required to bring about a given increase in the temperature of the water, the specific heat of the water is larger than that of ethanol.

Does the answer make sense? The data in Table 1.3 confirm that the specific heat of water is larger than the specific heat of ethanol.

1.56 CHECK THIS

Specific heat and temperature rise

If you carry out an investigation like Investigate This 1.53 with methanol and hexane in the cups, which will reach the higher temperature? Use the data in Table 1.3 and clearly explain the reasoning for your answer.

Temperature To work with heat capacities and specific heat, it is important to distinguish between temperature and thermal energy. **Temperature** is the scale of hotness or coldness of a substance. It is a measure of the average molecular motions in a substance. We commonly measure temperatures in degrees Celsius, °C, but, in scientific work, we often use the absolute temperature, T, which has units of kelvin, K. The size of the degree is the same on both scales, but their zero points are different. The relationship between absolute and Celsius temperatures is:

$$0 \text{ K} = -273.15 \text{ °C} \quad \text{or} \quad 273.15 \text{ K} = 0 \text{ °C} \tag{1.4}$$

There is a quantitative relationship between the temperature of a substance in kelvin and the motions of its molecules, but all we need here is the qualitative idea that at a higher temperature molecular motion is increased. Temperature is an **intensive property**; it does not depend upon the amount of material present. The temperature of a drop of boiling water is the same as the temperature of a liter of boiling water. The average molecular motions in the two samples are the same.

No degree symbol, °, is used with K, the symbol for temperature in kelvin. To convert a Celsius temperature, t °C, to its equivalent kelvin temperature, T K, use the relationship

$$T \text{ K} = (t \text{ °C} + 273.15 \text{ °C})\left(\frac{1 \text{ K}}{1 \text{ °C}}\right)$$

$$T \text{ K} = (t + 273.15)\text{K}$$

We usually simply use 273 for the conversion factor when Celsius temperature is given only to the nearest degree.

Thermal energy (heat) The thermal energies of the drop of boiling water and liter of boiling water are very different, however, since there are so many more molecules in a liter than in a drop. When **thermal energy,** or **heat,** measured in joules or calories, is added to a substance, it increases the motion of the atoms and molecules of the substance. Thermal energy (heat) is an **extensive property;** that is, it depends on the amount of material present. If you spill boiling water on your hand, thermal energy is transferred from the water to your hand. The damage to your hand depends on the amount of water that you spill. The thermal energy, ΔE, transferred to or from a substance can be calculated from the specific heat of the substance, the mass of the substance, and the change in temperature, $\Delta T (= T_{final} - T_{initial})$, that the substance undergoes.

1.57 WORKED EXAMPLE

Thermal energy change in water

Calculate the increase in thermal energy, ΔE, of the water in Investigate This 1.53 if 120. g of water is raised from 21.1 °C to 44.4 °C. (Note that the decimal point in the mass of the water specifies that the zero is significant, that is, the mass of the water is known to three significant figures. Another way to specify this would be to write the mass as 1.20×10^2 g.)

Necessary information: We need the specific heat of water from Table 1.3.

Strategy: The specific heat of water tells us the energy required to raise the temperature of one gram of water one degree Celsius. Multiply the specific heat by the mass of water to which energy has been added to get the energy required to raise its temperature one degree. Then multiply the result by the temperature change to account for the number of degrees the temperature was raised.

Implementation: We can summarize the strategy in a single equation:

increase in thermal energy $= \Delta E =$ (specific heat) $\times$ (mass) $\times$
(temperature change)

$$\Delta E = (4.18 \, \text{J·g}^{-1}\text{·°C}^{-1}) \times (120. \, \text{g}) \times (44.4 \, \text{°C} - 21.1 \, \text{°C})$$
$$= 11.7 \times 10^3 \, \text{J} = 11.7 \, \text{kJ}$$

Does the answer make sense? We solved the problem by reasoning from the definition and units of the specific heat. If about 4 J are required to raise the temperature of 1 g of water 1 °C, then about 500 J are required to raise the temperature of 120 g of water 1 °C. Each 1°C increase in temperature requires another 500 J of energy. The temperature change is about 20 °C, so about 10,000 J are required. The answer we got is this same order of magnitude, so it makes sense.

1.58 CHECK THIS

Thermal energy change in ethanol

Calculate the temperature change if 11.7 kJ of thermal energy is added to 120. g of ethanol.

1.59 CONSIDER THIS

Are your calculations related to Investigate This 1.53?

Compare your answer in Check This 1.58 with the temperature change for the same mass of water from Worked Example 1.57. Is your answer related to the results from Investigate This 1.53? Explain why or why not.

Explaining the high specific heat of water When thermal energy is added to water, some of the energy is used to break hydrogen bonds. The result is that the temperature rises less than if there were no hydrogen bonding and the average hydrogen bonding is now a bit weaker. Since water is a low molecular mass molecule, a small mass of water is a relatively large number of moles of water and, hence, moles of hydrogen bonds to be broken. A given mass of water can absorb a relatively large amount of thermal energy without as large an increase in temperature as the same mass of another compound.

We have discussed how the high specific heat of water helps to maintain the earth at a rather constant temperature that is suitable for life. Similarly, because your body is largely water, it is maintained at a rather constant temperature without greatly upsetting your metabolism, even when the temperature of the environment changes substantially. This is yet another way that the bent, polar structure of the water molecule is essential for life.

1.13 Outcomes Review

The next time you plunge your hands in water or watch ice cubes floating in a glass of water, you will be able to interpret the observable properties of water—and some other compounds—in terms of an atomic-molecular model and the interactions of the molecules with one another. In your mind's eye, you'll be able to see the bent, polar water molecules hydrogen bonding to their neighbors with the molecules held together as a liquid or as a solid. And you can relate the properties of different compounds to one another by using our molecular models of matter and comparisons based on counting molecules or moles. You are prepared now to go on, in Chapter 2, to learn how the structure of water molecules affects the interactions of water with other substances and to deepen your understanding of the mole concept.

Before going on, check your understanding of the ideas in this chapter by reviewing these expected outcomes of your study. You should be able to

- Describe solids, liquids, and gases in terms of their macroscopic properties and write or draw molecular-level descriptions that explain these properties [Section 1.1].
- Make drawings that show how the electrical nature of matter explains the results of electrostatic experiments [Section 1.2].
- Use the nuclear atomic model, the shell model for electrons, and the periodic table to determine the charge on the atomic core and the number of valence electrons in an atom [Section 1.2].

- Use the periodic table and the atomic shell model to predict trends in atomic size and electronegativities [Sections 1.2 and 1.6].
- Describe the relationships among different molecular models and the information that each of them provides [Sections 1.3, 1.4, 1.5, and 1.6].
- Write Lewis structures for molecules whose molecular formulas contain only first and second period elements (or analogous molecules of higher period elements) [Sections 1.4 and 1.7].
- Use drawings, physical models, and words to describe the geometry of the valence electrons and nuclei for molecules whose molecular formulas contain only first and second period elements (or analogous molecules of higher period elements) [Sections 1.5 and 1.7].
- Predict the direction and relative magnitude of bond dipoles and the direction of the resultant molecular dipole, including the effect of nonbonding electrons, for simple molecular structures [Section 1.6].
- Use drawings, physical models, and words to describe the origin of intermolecular interactions due to London dispersion forces, dipolar attractions, and hydrogen bonding [Section 1.7].
- Use intermolecular attractions to predict and/or explain trends in boiling points and energies of vaporization for a series of compounds whose molecular structures you know or can determine [Sections 1.7 and 1.11].

- Use drawings, physical models, and words to describe how the structure of the water molecule is responsible for the densities of solid and liquid water, the temperature dependence of the density of liquid water, and the consequences for life on Earth [Section 1.8].
- Describe some of the places where hydrogen bonding occurs in biomolecules and explain how hydrogen bonding is important for the functions of these molecules [Section 1.9].
- Describe and use energy diagrams to illustrate the direction of energy transfer from one substance to another when phase changes occur [Section 1.10].
- Use the relationship among energy change, temperature change, mass, and specific heat to make quantitative comparisons among substances that gain or lose thermal energy [Sections 1.10, 1.11, and 1.12].
- Use the molar mass of a compound, determined from the relative atomic masses of its constituent atoms, to calculate the number of moles in a given mass of the compound and vice versa [Section 1.11].
- Use drawings, physical models, and words to describe the molecular basis for the differences in specific heats among different compounds [Section 1.12].
- Use drawings, physical models, and words to describe the molecular basis for the differences in viscosities among different liquid compounds and for the dependence of viscosity on temperature [Section 1.14].

1.14. Extension—Liquid Viscosity

Viscosity is a measure of the resistance to flow of a liquid. Syrup and motor oil are examples of substances with high viscosity. The higher the viscosity of a liquid, the slower it flows. In this Extension, we are providing an opportunity for you to apply the models of attractions among molecules that we discussed in this chapter to interpret and predict the relative viscosities of different compounds.

1.60 INVESTIGATE THIS

How fast do different liquids flow from a pipet?

Use two identical glass Pasteur pipets for this investigation. Mark each pipet at the same place about 3 cm from the top. Draw water into one pipet until it is above the mark. Hold the pipet vertically with its tip over a container to catch

continued

the water. Time how long it takes the water to drain out of the pipet, starting from the time its top surface passes the mark. Repeat to be sure the time is reproducible to ± 1 or 2 sec. Use the second pipet to carry out the same procedure with hexane. *CAUTION:* Hexane is flammable.

1.61 CONSIDER THIS

How are viscosities related to molecular attractions?

(a) How do the outflow times for the water and hexane in Investigate This 1.60 compare? Which liquid has the higher viscosity? Explain the reasoning for your answer.

(b) Consider the effect a network of hydrogen-bonded water molecules has on the ability of individual molecules to move about in the liquid. Could restricted motion at the molecular level affect the observed flow of water from a pipet? If so, how? If not, why not?

(c) Based on our model of attractions among hexane molecules, do you expect the same kind of restricted motion at the molecular level as described for water in part (b)? Why or why not? What might be the affect on the observed flow of hexane from a pipet? Explain your reasoning.

(d) Based on your analyses in parts (b) and (c), what would you predict about the relative viscosities of hexane and water? Is your prediction consistent with your observations in Investigate This 1.60? Explain why or why not.

In Table 1.4, you can see that water has a higher viscosity than several other low-molar-mass compounds. As your analyses have shown, this is because there are many hydrogen bonds between water molecules, so the molecules resist moving past one another.

Table 1.4 *Relative viscosities of liquids at 20 °C.*

Compound	Line formula	M, g·mol^{-1}	Relative viscosity*
water	H_2O	18	1.00
methanol	CH_3OH	32	0.59
ethanol	CH_3CH_2OH	46	1.19
acetone	$CH_3C(O)CH_3$	58	0.33
ethylene glycol	$HOCH_2CH_2OH$	62	19.9
diethyl ether	$CH_3CH_2OCH_2CH_3$	74	0.23
hexane	$CH_3(CH_2)_4CH_3$	86	0.33

*Viscosities are all relative to water.

*Does hydrogen bonding explain viscosities in
other compounds?*

(a) Which of the compounds in Table 1.4 can form hydrogen bonds between
molecules in their pure liquids? Is there a correlation between viscosity and
the ability to form hydrogen bonds? Explain why or why not. The Lewis
structure for acetone is

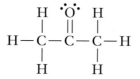

(b) Write a Lewis structure and make a molecular model of ethylene glycol (a
compound used in automobile antifreeze). What do you think is responsible
for the high viscosity of the glycol compared to the other compounds that
contain H—O bonds? Clearly present your reasoning, based on the structure of the molecule.

Figure 1.41 shows the temperature dependence of the viscosity of water. In
Section 1.12, we attributed the high molar heat capacity and specific heat of
water to hydrogen bond breaking, which uses some of the thermal energy added
to the liquid. The result is an average weakening of the hydrogen bonding as
energy is added and the temperature increases.

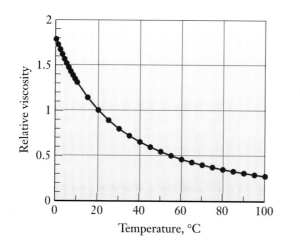

Figure 1.41.

**Relative viscosity of liquid water
as a function of temperature.**

Are the models for specific heat and viscosity of water related?

Do the data in Figure 1.41 provide support and further evidence for our explanation of the high specific heat of water just summarized? Explain why or why not.

Chapter 1 Problems

1.1. Phases of Matter

1.1. Identify each of these as being either a chemical property or a physical property. Place a *P* by all the physical properties and a *C* by all the chemical properties.
(a) _____ Water is clear and colorless.
(b) _____ Some metals react with water to produce hydrogen gas.
(c) _____ Water has a density of 1.0 g/cm³ at 4 °C.
(d) _____ Water boils at 100 °C.
(e) _____ Water is the product of a reaction between an acid and a base.
(f) _____ Water is a polar molecule.

1.2. Compare solids, liquids, and gases in each given category.

	Solids	Liquids	Gases
Definite volume?			
Definite shape?			
Fixed or changing position of molecules?			
Small or large average distance between molecules?			

1.3. Which phases of matter are referred to as "condensed phases"? What is the justification for the use of this term?

1.4. (a) Name the phase changes between each of the states of matter indicated by the arrows in this diagram.
(b) Label each of the four arrows on the diagram to indicate whether energy is released or absorbed in the process.

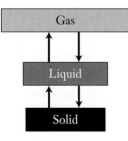

1.5. Name a substance other than water that commonly exists as a liquid at STP.

1.6. Name a substance that commonly exists as a gas at STP.

1.7. What happens if a closed glass bottle full of water is kept outside while the temperature falls below 0 °C?

1.8. One method for separating a NaCl(*aq*), sodium chloride solution, into its components is to boil the solution of salt water. In this case, water will evaporate and NaCl(*s*) will be left behind.
(a) Which property of NaCl(*s*) and water accounts for this separation?
(b) Design an apparatus that could change the water vapor back into a liquid as well as recover NaCl(*s*). Name all phase changes that occur.

1.9. Given two liquids that do not dissolve in one another (like oil and water), design an experiment that would allow you to determine which is denser.

1.10. How could you determine (experimentally) if the solid phase of a particular substance is more dense or less dense than the liquid phase of the same substance?

1.11. Consider Figure 1.3 comparing the densities of solid and liquid phases of paraffin with those of water. What additional information do you need to be able to predict what will happen if a sample of solid paraffin were dropped into liquid water? How would you find that information?

1.12. A block of ice has following dimensions: height = 20 cm, width = 20 cm, and length = 20 cm. The density of liquid water is 1.000 g·mL⁻¹ at 0 °C, and the density of ice at the same temperature is 0.917 g·mL⁻¹. Calculate the volume of the puddle of water, at 0 °C, that is left behind when the block of ice melts.

1.13. If an iron bar weighing 100.0 g is heated to a temperature above its melting point (>1535 °C), it will liquefy. What is the mass of the molten (liquid) iron? Is any additional information needed in order to answer this question?

1.2. Atomic Models

1.14. Which of these are chemical elements? How do you decide?
(a) water
(b) salt water
(c) iron
(d) iron oxide
(e) nitrogen
(f) diamond

1.15. What is the difference between core electrons and valence electrons?

1.16. How many valence electrons and core electrons do these elements possess?
(a) sodium
(d) phosphorus
(b) bromine
(e) sulfur
(c) barium

1.17. Many organisms use these ions (atoms that have lost or gained valence electrons) in their metabolism: Na^+, K^+, Mg^{2+}, Ca^{2+}, Cl^-, Br^-.
(a) Complete this table concerning these ions.

Ion	# of protons	# of electrons	# of valence electrons	Core charge	# of core electrons
Na^+					
K^+					
Mg^{2+}					
Ca^{2+}					
Cl^-					
Br^-					

(b) What patterns do you observe?

1.18. In terms of electronic structure, what is it that elements in the same period (row) of the periodic table share in common?

1.19. In terms of electronic structure, what is it that elements in the same group (column) of the periodic table share in common?

1.20. In general, do elements from the same period or elements from the same group of the periodic table have similar chemical properties? Justify your answer.

1.21. These metal ions (atoms that have lost valence electrons), Mn^{2+}, Fe^{2+}, Fe^{3+}, Cu^{2+}, and Zn^{2+}, are available for uptake by living organisms. How many protons and electrons does each ion have?

1.3. Molecular Models

1.22. Using equations modeled after Figure 1.9(c), show how to form each of these molecules from their constituent atoms. In each case, count electrons in the products to demonstrate that the octet rule is followed for all second row atoms.
(a) HF (hydrogen fluoride)
(b) NH_3 (ammonia)
(c) CH_3OH (methanol)
(d) H_2O_2 (hydrogen peroxide)

1.23. Which of these are macroscopic properties? Which are microscopic properties?
(a) the boiling point of water
(b) the HOH bond angle in water
(c) the OH bond length in water
(d) the ability of water to dissolve salt
(e) the density of water
(f) the fact that the oxygen atom of water has two pairs of nonbonding electrons

1.24. Consider these four different models for the ammonia molecule, NH_3:

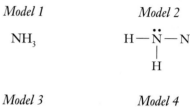

Model 1

NH_3

Model 2

Model 3

Model 4

(a) What is the name given to each type of model shown?
(b) What information can be obtained from each type of model?
(c) What other types of models can be used to represent the ammonia molecule?

1.25. Write the molecular formulas (line formulas) of the compounds represented here:

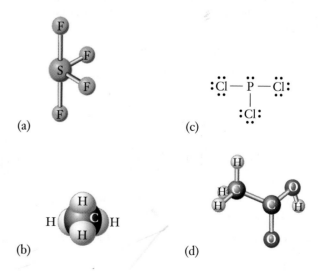

(a)

(b)

(c)

(d)

1.26. Write the molecular formulas for these compounds. (If necessary, use a reference handbook and/or other books to find out the structures.)
(a) glucose, a substance known as blood sugar
(b) nitrous oxide, a substance used as an anesthetic and as an aerosol propellant. It is commonly called laughing gas.
(c) methanol, an organic solvent and antifreeze
(d) acetylene, a gas that is used in welding torches

1.4. Valence Electrons in Molecular Models: Lewis Structures

1.27. (a) What information can be obtained from a Lewis structure?
(b) What information cannot be obtained from a Lewis structure?

1.28. Consider this electron-dot model for the ammonium ion, NH_4^+:

$$\left[\begin{array}{c} H \\ H : \overset{\cdot\cdot}{N} : H \\ H \end{array} \right]^+$$

(a) What information can be obtained from this model?
(b) What information cannot be obtained from this model?
(c) Rewrite the ammonium ion using a dash to represent each bonded pair of electrons. Does this change the information found in the model?
(d) How does this electron-dot model for the ammonium ion compare with that for methane, CH_4, given in Worked Example 1.10?

1.29. Examine Table 1.1. Silicon typically makes four covalent bonds and sulfur typically makes two covalent bonds. How many covalent bonds do you expect for each of these elements? Explain your reasoning in each case.
(a) phosphorus   (c) selenium
(b) chlorine   (d) bromine

1.30. Neon (Ne) is the element to the right of fluorine (F) in the periodic table.
(a) Examine Table 1.1. How many covalent bonds and nonbonding pairs would be expected for neon? Explain the reasoning for your answer.
(b) Ne (along with He, Ar, Kr, Xe, and Rn in the same group of the periodic table) were once known as the "inert gases." Why were they given this name?

1.31. Write the Lewis structure of ethanol, C_2H_6O. The formula may also be written C_2H_5OH or CH_3CH_2OH to make the connectivity more apparent.

1.32. Write the Lewis structures for ozone, O_3, sulfur dioxide, SO_2, and nitrite ion, NO_2^-. What do all three of these structures have in common? *Hint:* S is the central atom in SO_2 and N is the central atom in NO_2^-. The nitrite ion, NO_2^-, has an extra valence electron not furnished by the N or O atoms.

1.33. Each of these Lewis structures for NCCN has the correct number of electrons. Which is the best Lewis structure for NCCN? Explain your reasoning for rejecting the structures you did not choose. *Hint:* Multiple bonds between atoms will be discussed in Chapter 5. For the purposes of this problem, simply count each stroke (bond) as two electrons shared between the atoms it connects.

(a) $:\!\overset{\cdot}{N}\!=\!C\!=\!C\!=\!\overset{\cdot}{N}\!:$ (d) $:\!N\!\equiv\!C\!-\!C\!\equiv\!N\!:$

(b) $:\!\overset{\cdot\cdot}{N}\!-\!C\!\equiv\!C\!-\!\overset{\cdot\cdot}{N}\!:$ (e) $:\!N\!-\!\overset{\cdot\cdot}{C}\!-\!\overset{\cdot\cdot}{C}\!-\!N\!:$

(c) $:\!\overset{\cdot\cdot}{N}\!=\!C\!=\!C\!=\!N\!:$ (f) $:\!N\!=\!C\!=\!C\!-\!\overset{\cdot\cdot}{N}\!:$

1.34. Which of these Lewis structures are incorrect? In each case, explain why the structure is incorrect. Rewrite each of the incorrect structures so it is correct. *Hint:* See the hint in Problem 1.33.

(a) HOCl $H\!:\!\overset{\cdot\cdot}{\underset{\cdot\cdot}{O}}\!:\!\overset{\cdot\cdot}{\underset{\cdot\cdot}{Cl}}\!:$

(b) CS_2 $:\!S\!::\!C\!::\!S\!:$

(c) NH_3 $H\!:\!\overset{\cdot\cdot}{\underset{H}{N}}\!:\!H$

(d) $(HO)_2CO$ $H\!:\!\overset{\cdot\cdot}{\underset{\cdot\cdot}{O}}\!:\!\overset{\overset{\overset{\cdot\cdot}{O}}{C}}{\underset{\cdot\cdot}{O}}\!:\!_H$

(e) H_2Se $H\!:\!\overset{\cdot\cdot}{Se}\!:\!H$

1.5. Arranging Electron Pairs in Three Dimensions

1.35. Which of these molecules or ions has a tetrahedral (or close to tetrahedral) orientation of bonding and nonbonding electron pairs?
(a) H_2S
(b) NH_4^+
(c) NH_3
(d) NH_2^-
(e) CH_4

1.36. What is the shape of each of the molecules in Problem 1.35? Recall that the shape of a molecule describes the position of the atomic cores with respect to one another.

1.37. Figure 1.12 showed one way to stack four balls so that they are equidistant from a central point. Another way is to arrange them in a square about the point, as shown in this picture. Lines drawn between the centers of adjacent balls form a square of side $2r$, where r is the radius of a ball. The center point of the diagonal of the square is the center of the square. Use the Pythagorean theorem to find the length of the diagonal and the distance from the center of any of the balls to the center of the square.

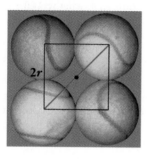

1.38. (a) A cube can be circumscribed about a regular tetrahedron, as shown in this figure. The four dots represent the centers of the four balls in Figure 1.12. The diagonal of one of the cube faces has a length $2r$, where r is the radius of a ball. The center point of any one of the cube diagonals (one is shown by the dashed line) is the center of the tetrahedron. Use the Pythagorean theorem to find the length of the cube edge and then again to find the length of the cube diagonal. Thus, show that any corner of the cube, that is, the center of any of the balls, is $\left(\dfrac{\sqrt{6}}{2}\right)r$ from the center of the tetrahedron.

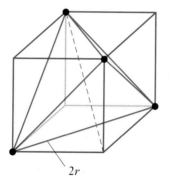

(b) How does the distance you calculated in part (a) compare to the distance of the center of each ball from the center of the square in Problem 1.37? Do your results help justify the statement in the text that "the tetrahedral arrangement puts all the balls as close as possible to the central point"? Explain why or why not.

1.39. $SiCl_4$, silicon tetrachloride, is used for the production of the very pure silicon required in many electronic devices such as transistors.
(a) Write the Lewis structure for $SiCl_4$.
(b) How does this Lewis structure compare to that of methane, CH_4?
(c) Predict the shape of the $SiCl_4$ molecule.

1.40. (a) Write the Lewis structure for borane, BH_3. Compounds like borane are sometimes called "electron deficient." How do you think "electron deficiency" is defined?
(b) 🖱 What shape do you predict for borane? You might find it useful to try a modified version of Investigate This 1.14 to help make your prediction. Also see the *Web Companion*, Chapter 1, Section 1.6.

1.6. Polarity of the Water Molecule

1.41. What is electronegativity?

1.42. Answer these questions on electronegativity.
(a) Which element has the highest electronegativity value?
(b) Where on the periodic table do you find the elements with the lowest electronegativities?
(c) Where on the periodic table do you find the elements with the highest electronegativities?
(d) What is the general trend for electronegativity as you go from the top to the bottom within a group?
(e) What is the general trend for electronegativity as you go from the left to right across a period?
(f) How do the electronegativity trends compare to the atomic size trends? Refer to Figure 1.7 for information about atomic size.

1.43. Without referring to a table of electronegativities, predict which member of each pair has the greater electronegativity. Explain the basis for your prediction in each case.
(a) F   *or*   S
(b) C   *or*   H
(c) H   *or*   O
(d) O   *or*   C

1.44. You can use the electronegativity difference between two atoms to predict the polarity of the bonds they make. Choose the pair of atoms in each case that you predict to make the more polar bond and explain how you make this prediction.
(a) H-F   *or*   H-Cl
(b) C-H   *or*   N-H
(c) K-S   *or*   Na-S
(d) O-O   *or*   N-O

1.45. What would be the consequences for life if water were a linear molecule? Consider the effect on the polarity and properties of water if it were linear. You might find it useful to draw pictures of how the linear molecules might interact with one another and compare what you get with the various interactions pictured in this chapter.

1.46. Consider the two molecules H_2O and H_2S.
(a) Compare the Lewis structures of these two molecules.
(b) Compare the molecular shape of these two molecules.
(c) Compare the bond dipoles within each molecule. *Hint:* Use the data in Figure 1.20.
(d) How do you think the overall electric dipole of these molecules compare? Explain.

1.47. Draw the Lewis structures for carbon tetrachloride, CCl_4, chloroform, $CHCl_3$, and dichloromethane, CH_2Cl_2. Clearly label the bond dipoles for each molecule. Which molecules are polar and which are nonpolar? Explain your answer.

1.48. Each of these molecules has a dipole moment = 0. In each case, explain why there is no net (molecular) dipole moment. If the molecule has bond dipoles, draw them and explain how they cancel out.
(a) N_2 (molecular nitrogen)
(b) BH_3 (borane) *Hint:* B does not satisfy the octet rule. See Problem 1.40.
(c) $SiCl_4$ (silicon tetrachloride)
(d) BeH_2 (beryllium hydride) *Hint:* Be does not satisfy the octet rule. What shape must the molecule have, in order not to have a net dipole moment? You might find it useful to try a modified version of Investigate This 1.14 to help determine this shape.
(e) CH_3CH_3 (ethane) See Check This 1.13.

1.49. Which molecule, ammonia, NH_3, or phosphine, PH_3, has the larger molecular dipole moment? Explain.

1.7. Why Is Water Liquid at Room Temperature?

1.50. Describe each of these types of intermolecular attractions:
(a) induced-dipole attractions
(b) dipole-dipole attractions
(c) hydrogen bond

1.51. How are an intramolecular covalent bond and an intermolecular hydrogen bond similar? How are they different?

1.52. What type(s) of intermolecular attractions are there between
(a) all molecules?
(b) polar molecules?
(c) a hydrogen atom in a water molecule and a nitrogen atom in ammonia (in a mixture of ammonia and water)?

1.53. Astatine, At, element 85, is radioactive and has a half-life of only 8.3 hr (see Chapter 3). Only minute traces of At have been studied. The hydride, HAt, has been detected but its physical properties are unknown. Based on the data in Figure 1.24, what would you predict for the boiling point of HAt? Explain how you make your prediction.

1.54. Methane, CH_4, and hydrogen sulfide, H_2S, do not form hydrogen bonds. Explain.

1.55. List at least three properties of water that can be attributed to the existence of the hydrogen bond. Briefly describe how each property would be affected if water did not form hydrogen bonds.

1.56. How many hydrogen bonds are possible for one water molecule in a sample of water? Illustrate your answer with a drawing of the structure of water indicating where the hydrogen bonds are possible.

1.57. (a) How many hydrogen bonds are possible for one ammonia molecule in a sample of ammonia? Illustrate your answer with a drawing of the structure of ammonia indicating where the hydrogen bonds are possible.
(b) Can all the ammonia molecules in a sample of liquid ammonia have the maximum number of hydrogen bonds you illustrated in part (a)? If not, what limits the number and how many hydrogen bonds, on average, can each ammonia molecule have?

1.58. Refer to the graph in Consider This 1.23 that relates the boiling points of a series of hydrocarbons to the number of electrons per molecule. What is the smallest hydrocarbon in the series that exists as a liquid at room temperature?

1.59. These are the boiling points of the noble gases:

Element	He	Ne	Ar	Kr	Xe	Rn
bp, °C	−269	−246	−186	−152	−107	−62

(a) Plot these data on a graph like the one in Figure 1.24. How are these data similar to those for the hydrides plotted in Figure 1.24? How are they different?
(b) Noble gas atoms are spherical and Figure 1.22 shows that tetrahedral hydrides like methane are quite symmetric and almost spherical. What is likely to be the largest factor responsible for the difference in boiling points between a noble gas and the corresponding group IV hydride?

1.60. There are often different compounds having the same formula. As you will learn in Chapter 5, these are known as *isomers*. For C_5H_{12}, there are three isomers. The common names, line formulas, structures, and boiling points of the isomers are given below. All three have the same molecular mass. Why aren't their boiling points closer together? *Hint:* The more compact a molecule, the less surface area it has for its electrons to interact with other molecules. Build models of these molecules to help visualize the surface areas of these molecules.

Pentane
bp = 36 °C

$CH_3CH_2CH_2CH_2CH_3$

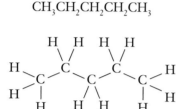

isopentane
bp = 28 °C

$(CH_3)_2CHCH_2CH_3$

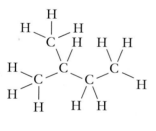

neopentane
bp = 10 °C

$C(CH_3)_4$

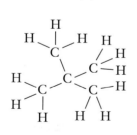

1.8. Further Structural Effects of Hydrogen Bonding in Water

1.61. If an ice cube is dropped into liquid water at 85 °C, will it float or sink (before it melts)? Justify your answer.

1.62. In Investigate This 1.30, you observed that the temperature at the bottom of the container rises to about 4 °C and remains almost constant as long as there is ice left at the top of the container. Thermal energy (heat) must be entering the container from the warmer room air (the ice does melt). You would expect the water at the bottom to continue to warm up, but the temperature at the bottom stays constant. These seem to be contradictory observations.
(a) What is the special property of water at 4 °C?
(b) If the water at the bottom warmed a bit above 4 °C, how would this property change? What would the water be likely to do? Draw pictures to illustrate what you think would happen.
(c) Would the action of the water shown in your drawings resolve the contradiction suggested above? How could you test your model experimentally?

1.63. The structure of the water molecule is the molecular basis for the survival of plant and animal life in a temperate climate lake. The seasonal "turnover" of such a lake is described in this figure.

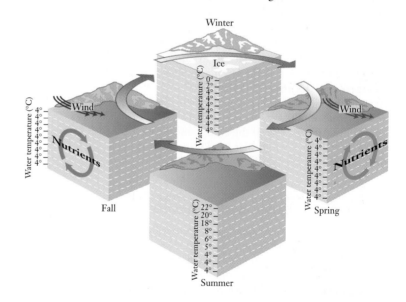

These vertical temperature changes are typical of a lake that freezes in winter. Turnovers, represented by the circling green arrows, occur in the spring and fall and mix nutrients and oxygen into the deeper waters. The turnovers are triggered by winds at the surface of the water. How would you relate your observations in Investigate This 1.30 to the changes described in this figure? Explain the connections clearly.

1.64. Consider this representation of covalent bonds and hydrogen bonds in water:

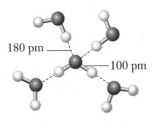

180 pm

100 pm

(a) Which bond length is associated with hydrogen bonds and which with covalent bonds?
(b) Offer a reasonable explanation for why there is a difference in bond length between these two bonds.

1.65. The melting points for methane, ammonia, water, and hydrogen fluoride are shown in this table.

Compound	mp, °C
CH_4	−182
NH_3	−77.7
H_2O	0
HF	−83.1

You can take the melting point as an indication of the relative amount of energy required to disrupt the attractions between molecules in the solid, so they are free to move about as a liquid. Develop an explanation for these data that takes into account the kinds of intermolecular attractions among molecules of each compound. Is water out of line with the rest of the compounds? Why or why not? Give a molecular level interpretation of your answer.

1.9. Hydrogen Bonds in Biomolecules

1.66. The DNA double helix, held together by the hydrogen bonds shown in Figure 1.33, is quite stiff and resistant to movement through a solution. As shown in the figure at the top of the next column, when a solution of DNA is heated, the absorption of ultraviolet light at 260 nm rises sharply over a small temperature range. At the same time the solution suddenly begins to flow more easily (more like water than like syrup). The middle of this range is usually labeled T_m (melting temperature). What is happening to the DNA to cause these changes in the solution properties? Explain your response.

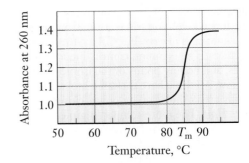

1.67. This figure below shows melting temperature, T_m (see Problem 1.66), data for a number of different double-helical DNAs plotted against the fraction of A–T pairs in the DNAs. Why are the melting temperatures a function of the fraction of A–T pairs? Does the direction of the dependence make sense? Clearly explain your reasoning. *Hint:* Review Consider This 1.34.

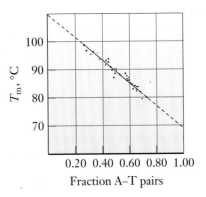

1.68. In Problem 1.66, the melting temperature of DNA in solution was defined as the temperature at which sharp changes in ultraviolet light absorption and solution flow occur. Why do you think this is called a melting temperature? Is(Are) there any analogy(ies) between what happens to DNA in these solutions and what happens when ice melts? Explain your reasoning clearly.

1.69. The proteins in most organisms are denatured at temperatures above about 60 °C. Microorganisms that live in hot springs and organisms that live near deep ocean thermal vents survive at temperatures near or above the boiling point of water, 100 °C. Their proteins are made of the same amino acids as all other organisms. What role do you think hydrogen bonds might play in helping these organisms survive?

1.70. The structure of the DNA of the thermophilic (*therme* = heat + *philos* = loving) organisms discussed in Problem 1.69 also has to be maintained in the high temperature environments where they live. What kind of A—T *vs.* G—C composition would you expect to find for the DNA in these organisms? Present your reasoning clearly. *Hint:* See Problem 1.67.

1.71. Cellulose is a long-chain molecule made up of glucose molecules bonded together as shown in this illustration. You will learn more about cellulose later. Many chains like these are hydrogen bonded to neighboring chains to form the fibers that are used to make paper and cotton and linen cloth. The hydrogen bonds make cotton cloth soft and flexible because they are easily broken and remade, which allows the fibers to change shape.

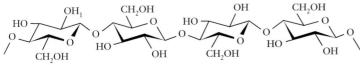

(a) How do you account for the fact that cotton clothing is easily wrinkled? Explain your reasoning clearly.
(b) How does ironing wrinkled cotton clothing restore its "press"? Use diagrams to illustrate your answer.
(c) How might you make permanent press cotton cloth? Indicate what you would try to accomplish; don't be concerned about the detailed chemical processes that might be required.
(d) Permanent press cotton clothing is not as soft as regular cotton clothing. Is this the result you might expect from your response to part (c)? Explain why or why not.

1.10. Phase Changes: Liquid to Gas

1.72. How is energy involved for a substance to change from one state to another? Explain.

1.73. Make these conversions:
(a) $4550 J = $ _____ kJ
(b) $250 J = $ _____ calories
(c) $500 Cal = $ _____ J [1 nutritional Calorie (Cal) = 1000 calories]

1.74. A phase diagram, such as this one for water, is a common way of representing phase changes. Phase diagrams are pressure-*vs*-temperature plots that show the pressures at which the phase changes of a substance occur as a function of temperature.

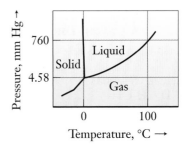

(a) What information does this phase diagram give you about the phases of water at standard atmospheric pressure of 760 mm Hg (= 1 atm)? Explain your reasoning briefly. *Hint:* Consider starting at the pressure axis, moving across the plot (increasing temperature) at a constant 760 mm Hg, and noting when phase changes occur.
(b) For a change from gaseous water to liquid water at 100 °C and 760 mm Hg, will the sign of ΔE be positive or negative? Explain your reasoning.
(c) For a change from liquid water to solid water at 0 °C and 760 mm Hg, will the sign of ΔE be positive or negative? Explain your reasoning.

1.75. Sketch the energy diagrams that describe these processes. Make sure that you draw an arrow that represents the direction of the change and that you indicate the sign of ΔE for each process.
(a) A sample of ice melting, $H_2O(s) \rightarrow H_2O(l)$, at 0 °C. $\Delta E > 0$ for the sample.
(b) The combustion of hydrogen gas in oxygen gas is one of many chemical reactions that release considerable quantities of energy. This process is described by the equation:

$$2H_2(g) + O_2(g) \rightarrow 2H_2O(l) + energy$$

(c) The decomposition of mercury oxide (HgO) occurs at high temperatures. For this process to occur, energy has to be supplied. The process is described by the equation:

$$energy + 2HgO(s) \rightarrow 2Hg(l) + O_2(g)$$

(d) The burning (or oxidation) of mercury is the reverse of the decomposition process in part (c) and is described by the equation:

$$2Hg(l) + O_2(g) \rightarrow 2HgO(s) + energy$$

(e) Burning a sample of methane in oxygen gas, for which $\Delta E < 0$. This reaction is described by the equation:

$$CH_4(g) + 2O_2(g) \rightarrow CO_2(g) + 2H_2O(l)$$

1.76. The amount of energy required to vaporize methanol is 1.22 kJ·g^{-1}. How many kcal are required to vaporize 1 g of methanol?

1.11 Counting Molecules: The Mole

1.77. Calculate the molar mass of a formula unit of these substances:
(a) dimethyl ether, CH_3OCH_3
(b) ethanol, CH_3CH_2OH

1.78. How many molecules of water are in exactly 1 mol of water? How many grams of water are in exactly 1 mol of water?

1.79. How many moles in each of these?
(a) 100.0 g of acetone
(b) 100.0 g of methanol
(c) 100.0 g of dimethyl ether
(d) 100.0 g of sucrose

1.80. How many molecules are there in 1 g of water, methanol, acetone, ethanol, and dimethyl ether?

1.81. Argon atoms have a diameter of approximately 100 pm. If a mole of argon atoms were lined up one after another, how long, in meters, would the line be? The distance from Earth to the Sun is 1.5 × 10^{10} m. What percentage of this distance would the line of Ar atoms reach?

1.82. (a) How many moles of hydrogen bonds are there in a mole of ice? Explain how you get your answer. *Hint:* If each H stopped hydrogen bonding, all the hydrogen bonds would be gone.
(b) The energy required to melt ice is 6.02 kJ·mol^{-1}. If the model for ice melting presented in the text is correct, how many moles of hydrogen bonds are broken when a mole of ice melts? What percentage of the total hydrogen bonds is this? Clearly explain how you get your answers. *Hint:* Recall that the energy required to break a hydrogen bond between two water molecules is 20–25 kJ·mol^{-1}.

1.83. In which of these compounds can the molecules hydrogen bond to themselves? Draw a diagram of the molecules of each compound hydrogen bonding among themselves.
(a) water   (d) ethanol
(b) methanol   (e) dimethyl ether
(c) acetone

1.84. The boiling points of dimethyl ether (CH_3OCH_3) and diethyl ether ($CH_3CH_2OCH_2CH_3$) are −25 °C and 35 °C, respectively. What interaction is mainly responsible for the observed difference?

1.85. Acetone, $CH_3C(O)CH_3$ (see Consider This 1.62), like dimethyl ether, has no self-hydrogen-bonding capacity, but its dipole moment is more than double that of the ether. Can this high dipole moment explain the properties

of acetone compared to the other compounds in Table 1.2? Clearly explain the reasoning for your response and include as many comparisons as possible.

Molar mass	Energy of vaporization	Boiling point	Dipole moment
58 g	32 kJ·mol^{-1}	56 °C	2.88 Debye

1.12 Specific Heat of Water: Keeping Earth's Temperature Stable

1.86. Define specific heat.

1.87. What is the difference between an intensive and an extensive property?

1.88. Is each of these properties intensive or extensive?
(a) the boiling point of water
(b) the density of water
(c) the specific heat of water
(d) the ratio of hydrogen to oxygen atoms in a sample of water
(e) the (maximum) solubility of salt in water

1.89. Convert 37.0 °C, normal human body temperature, to Kelvin.

1.90. How much thermal energy is required to raise the temperature of 1.0 g of water by
(a) 10.0 °C?   (c) 25.0 K?
(b) 25.0 °C?

1.91. How much thermal energy is required to raise the temperature of these samples by 10.0 °C?
(a) 10.0 g of water   (b) 25.0 g of water

1.92. How much thermal energy is required to raise the temperature of 20.0 g of 1-propanol by 15.0 °C?

1.93. In dry parts of the world, blowing the hot outside air through mats soaked in water before it enters a building provides "air conditioning." On what scientific principle is this system based? How does it work? What advantages and disadvantages can you see for the people and things in the building?

1.94. Would you expect evaporative cooling of your skin to be more effective on dry days or humid days? Clearly explain the reasoning for your answer.

1.95. Table 1.3 gives the heat capacity of liquid water as 75 J·mol^{-1}·°C^{-1} (18 cal·mol^{-1}·°C^{-1}). The heat capacity of solid water (ice) is 38 J·mol^{-1}·°C^{-1} (9.0 cal·mol^{-1}·°C^{-1}). Why do you think the heat capacity of the solid is less than that of the liquid? Are our models of liquid and solid water consistent with your explanation?

1.96. The heat capacity of gaseous water (steam) at one atmosphere pressure is 33 J·mol^{-1}·°C^{-1} (7.9 cal·mol^{-1}·°C^{-1}). What sort of model do you think would describe gaseous water? Why do you think the heat capacity of the gas is less than that of the liquid (see Table 1.3 or Problem 1.95)? Is your model of the gas and the model of the liquid we have discussed in this chapter consistent with your explanation?

1.97. Table 1.2 shows that 44 kJ·mol^{-1} (10.5 kcal·mol^{-1}) are required to change 1 mol of liquid water to water vapor. This value is for water near 25 °C. The energy of vaporization depends on the temperature of the water, as shown in this figure. Why does the energy required to vaporize water vary with temperature this way? Clearly explain your reasoning.

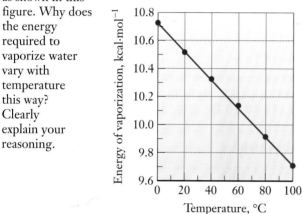

1.14. Extension—Liquid Viscosity

1.98. To which physical property of liquids does the expression "slow as molasses in January" owe its truth? Would molasses be "faster" in June? Why?

1.99. Nature exploits the properties of the hydrogen bond in many ways. Scientists also work to find ways to use this weak bond with its strength in numbers to create materials with interesting and useful properties. One group of researchers has made a compound whose molecules have "sticky" ends; each end of one molecule forms four hydrogen bonds to another to produce long chains, as represented in the figure.

(a) For solutions of this compound (in a non-hydrogen-bonding solvent) that vary in concentration from about 8 to 80 g·L^{-1}, the viscosity varies as shown in this logarithmic plot.

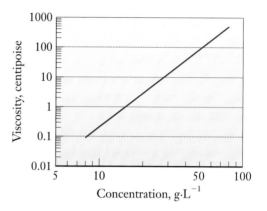

As the concentration changes by a factor of 10, by what factor does the viscosity change? Clearly explain how you might interpret this result. *Hint:* Recall that large molecules can't move rapidly in solution, so their solutions resist flow.

(b) The viscosity of these solutions is temperature dependent. Do you predict an increase or decrease in viscosity as the temperature of a solution is increased? Explain the reasoning for your prediction.

(c) The researchers also made a compound whose molecules are essentially half of one of the molecules shown above. These new molecules have only one "sticky" end. The viscosities of mixtures of 32 g·L^{-1} of the original compound with small amounts of the new compound are shown in this plot. The horizontal axis shows the decimal fraction of the mixture that is the new compound; 0.01, for example, means that 1 in 100 of the molecules in the solution are the new molecules. Clearly explain how you might interpret what is going on in the solution to produce these results.

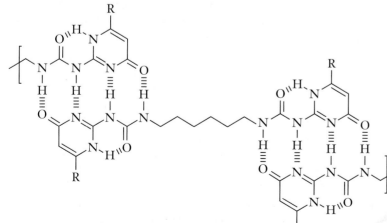

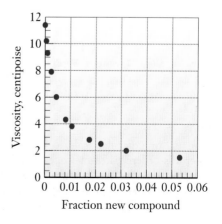

Fraction new compound

1.100. The viscosity of *n*-heptane, $CH_3(CH_2)_5CH_3$, as a function of temperature is plotted here on the same scale as Figure 1.41, which gives the corresponding data for the viscosity of water. To make the plots comparable, the values here are relative to the viscosity of heptane at 20 °C. What similarities do you observe between these data and those for water? What differences do you observe? What explanation can you provide for the similarities and differences?

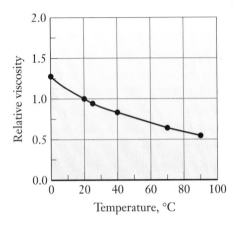

General Problems

1.101. Briefly explain why:
(a) You experience a cooling effect after walking out of the ocean onto a warm, sandy beach, especially on a breezy day.
(b) Liquids can be defined as "disordered" solids. Are there problems with this definition?
(c) Liquids can be defined as "dense" gases. Are there problems with this definition?
(d) Solid water (ice) floats on liquid water.
(e) Steam can badly burn you if it condenses to water on your skin.
(f) Lakes freeze from the top to the bottom.
(g) Water pipes break if water freezes in them.

1.102. Chemistry is everywhere. A friend has asked you if the claims here are scientifically accurate: "Keep a bottle of water at your desk and take frequent sips from it. One-third of water is oxygen, so drinking it will help keep you alert." (Source: "Working Smart," *Vitality*, October 1997 [Vitality Inc., Dallas, TX], as quoted in *C & EN* "Newscripts.") Write a response that will help your friend sort out what is true and what might not be.

1.103. One model of liquid water (an "iceberg" model) is a mixture of molecular-scale, ice-like structures among other less-ordered, less-hydrogen-bonded molecules with the molecules continually exchanging between the two forms. In a sample of liquid water, it is possible to give

extra energy of motion only to those molecules that are pointing (oriented) in the same direction. The natural rocking and jiggling of these molecules soon changes their orientation (they become more random). Scientists have measured the time required for the change and find that some of the water molecules make the change in an average of about 0.7×10^{-12} seconds (0.7 picoseconds). The rest take an average of about 13×10^{-12} seconds (13 picoseconds). The scientists concluded that liquid water acts like it is made up of two species. (Source: *Science*, 1997, 276, 658.)
(a) Is this conclusion consistent with the "iceberg" model of liquid water? What might the two species be? Clearly state the reasoning for your answers.
(b) Which of the two species changes orientation rapidly and which more slowly? Explain your reasoning. Use drawings, if they are helpful.
(c) If there are two species, why aren't they apparent to our senses in our everyday contacts with water?

1.104. Gases behave much like liquids in terms of things floating and sinking in them. What can be deduced from each of these facts:
(a) A helium-filled balloon will rise in air.
(b) A hot-air balloon will rise in (colder) air.
(c) A balloon filled with carbon dioxide will sink in air.

1.105. What will happen if an astronaut standing on the Moon lets go of a helium-filled balloon she is holding? Recall that the Moon, unlike Earth, has no atmosphere. *Hint:* See Problem 1.104(a).

1.106. Potassium acid fluoride is a salt composed of the ions, K^+ and $(FHF)^-$. The negative ion, $(FHF)^-$, can be thought of as two fluoride ions hydrogen bonded by a proton with the Lewis structure: $\left[:\!\ddot{F}\!:H\!:\!\ddot{F}\!: \right]^-$. The H—F bond length in hydrogen fluoride, HF, is 93 pm. Each of the H—F bond lengths in $(FHF)^-$ is 113 pm. Formation of $(FHF)^-$ from HF and F^- releases about 155 kJ·mol^{-1}, making it by far the strongest hydrogen bond known, although much weaker than the covalent (electron pair) bond in HF which releases about 565 kJ·mol^{-1} when it forms.
(a) What, if anything, is peculiar about the Lewis structure shown for $(FHF)^-$?
(b) Discuss the similarities and differences between this hydrogen bond and the hydrogen bond between two water molecules. Also discuss whether the $(FHF)^-$ example blurs the distinction between covalent bonds and hydrogen bonds.

1.107. (a) About 70% (by mass) of your body is water. How many moles of water does your body contain? How many molecules of water?
(b) The table in the margin on page 15 gives the elemental composition of your body. Assume that the

same number of atoms of oxygen and nitrogen are combined in molecules other than water. How many oxygen atoms (per 100,000 atoms) are combined with hydrogen atoms to make the water in your body?
(c) Use your results from parts (a) and (b) to calculate the total number of atoms in your body. How many of these atoms are nitrogen? carbon? How many moles of nitrogen does your body contain? carbon?

1.108. This table gives the names, line formulas, and energies of vaporization for most of the hydrocarbons (molecules containing only carbon and hydrogen atoms) whose boiling points are given in the figure in Consider This 1.23.

Hydrocarbon	Formula	Energy of vaporization, kJ·mol^{-1}
ethane	CH_3CH_3	15.65
propane	$CH_3CH_2CH_3$	20.13
butane	$CH_3(CH_2)_2CH_3$	24.27
pentane	$CH_3(CH_2)_3CH_3$	27.61
hexane	$CH_3(CH_2)_4CH_3$	31.92
heptane	$CH_3(CH_2)_5CH_3$	35.19
octane	$CH_3(CH_2)_6CH_3$	38.58
nonane	$CH_3(CH_2)_7CH_3$	43.76

(a) Plot these energies of vaporization as a function of the number of electrons in each molecule. Draw the best possible straight line through the points. (If you use a graphing calculator or computer-graphing program, it can construct the line for you.) Predict the energy of vaporization for decane, $CH_3(CH_2)_8CH_3$.
(b) Why is there an increase in energy required to vaporize these molecules as $-CH_2-$ groups are added? Use the slope of your line from part (a) to determine how much the energy of vaporization increases for each $-CH_2-$ group added.
(c) Assume that induced dipole attractions (dispersion forces) are directly proportional to the number of electrons in molecules with second row elements connected in a chain, like those in this table. What do you predict for the energy of vaporization of dimethyl ether? How does your prediction compare to the value given in Table 1.2? How do you explain any difference?
(d) The energies of vaporization of diethyl ether, $CH_3CH_2OCH_2CH_3$, and butanol, $CH_3CH_2CH_2CH_2OH$, are 29.1 and 45.9 kJ·mol^{-1}, respectively. Use what you have learned in the previous parts of this problem, plus the data in Table 1.2, to predict these energies and compare them with the experimental values. What attractions among the molecules must you account for in each case?

1.109. The dipole moments for NH_3, NF_3, and NCl_3 are 1.47, 0.234, and 0.39 D, respectively. The dipole moment arrow (as in Figure 1.18) for NH_3 and NCl_3 points toward the nonbonding electron pair on the nitrogen atom. The dipole moment arrow for NF_3, shown in Worked Example 1.19, points in the opposite direction.
(a) Draw pictures of the three molecules that show their shape and the relative size and direction of their bond dipole arrows.
(b) What are the relative sizes and directions of the bond dipoles in these three molecules? Do these bond dipoles help explain the relative size and direction of the molecular dipole moments (bond dipole arrows)? Explain why or why not.
(c) The nonbonding electrons on the nitrogen atom probably make about the same contribution to the molecular dipole moment in all three molecules. Does this information, together with your responses in part (b), help to explain the relative size and direction of the molecular dipole moments? Explain.

Water that falls as rain and snow drains into streams and rivers that flow to the sea. Along the way, substances from the earth's solid crust, including the calcium carbonate represented here, dissolve and are carried to the sea. Marine mollusks like these use calcium carbonate to build their solid shells.

Aqueous Solutions and Solubility

What water-soluble and water-insoluble substances do you know?

List six substances you know are soluble in water. List six substances you know are not soluble in water. Work in a small group to compare your lists and produce a combined list of all your different soluble and insoluble substances. Discuss your lists in class to find out how similar all your experiences with solubility are and to try to make some generalizations about soluble and insoluble substances.

L ife on Earth began in the seas, and the chemistry in living cells occurs in aqueous media. If we are to understand the chemistry of living things, we must understand the chemistry of substances that dissolve or do not dissolve in water. For example, a large percentage of the material in eggshells and seashells is calcium carbonate. The mollusks shown on the facing page obtain the calcium and carbonate ions needed for their shells from the seawater around them. The Latin citation that opens the chapter reminds us that water dripping on a stone or flowing in a river can slowly dissolve solid rocks composed of calcium carbonate. So why doesn't the calcium carbonate in seashells simply redissolve and return to the sea? That is one of the questions we will discuss in this chapter.

Chapter 1 focused mainly on the interactions of water molecules with one another in the pure liquid. This chapter extends that discussion to consider interactions of water with other substances. Common experience, as Consider This 2.1 points out, is that some substances, such as salt, sugar, and antifreeze, dissolve readily in water. Other substances, such as chalk, flour, and oil, are not readily soluble—even if the mixture is stirred vigorously or heated. Compounds that dissolve in water and compounds that do not dissolve in water each have characteristic properties. Among the central themes of this chapter are identification of the properties that make some compounds water-soluble and others insoluble and understanding the process by which molecules dissolve in water.

Two important ideas from Chapter 1 undergird these discussions: the polar nature of H–O bonds and the V-shaped structure of the water molecule. Together, these properties give the water molecule a permanent dipole moment. The polarity of water enables it to dissolve a variety of molecules and ions, and prevents the dissolution of others. Aqueous solutions have properties that differ from pure water. Properties such as electrical conductivity and acidity or basicity can help us characterize the interactions that occur in these solutions. After a general introduction to the solution process, we will look again at the intermolecular interactions introduced in Chapter 1—hydrogen bonding, polar attractions, and London dispersion forces—focusing this time on interactions among the molecules in solutions.

2.1. Substances in Solution

Solution, solvent, solute, dissolve, and related terms come from the Latin word *solvere* = to loosen. Solvents "loosen" solutes from their pure form and mix their components with the solvent to give solutions. *Homogeneous* is derived from Greek: *homo* = the same + *genos* = kind; *heterogeneous* is derived from *hetero* = different.

Before beginning our discussion of solutions, we need to consider the specific vocabulary that chemists use to describe solutions. Many terms relating to solutions are familiar because they are used in everyday English, but their familiarity can create a problem. In everyday usage a word often has several meanings; but in scientific use, each term has one precise meaning.

Solution nomenclature A **solution** is a homogeneous mixture of a **solvent,** the substance present in excess (water in the present discussion), and one or more **solutes,** the substances present in smaller amounts that are described as being **dissolved** in the solvent. A **homogeneous mixture** is a *uniform* mixture of two or more substances. The properties of homogeneous mixtures are the same throughout the mixture. A teaspoon of table sugar added to a glass of water and stirred until the solid is no longer visible produces a homogeneous mixture of sucrose in water, Figure 2.1(a). The liquid has the same sweetness throughout.

Figure 2.1(b), a molecular-level representation of the sugar solution, shows the essential idea: *Every* sugar molecule is surrounded by water molecules rather than sugar molecules being clumped together. By contrast, **heterogeneous** mixtures are not uniform throughout. Orange juice is a heterogeneous mixture whose composition varies from one part of the mixture to another—the liquids inside and outside of the pulp, for example.

(a) Sugar dissolved in water.

(b) The water molecules are muted so the sugar molecules are easier to see.

Figure 2.1.

(a) A homogeneous sugar solution and (b) its molecular-level representation.

2.2 INVESTIGATE THIS

What is observed when substances dissolve?

Your group has three capped vials containing (1) urea, $H_2NC(O)NH_2$, a white crystalline compound, (2) ethanol, CH_3CH_2OH, a colorless liquid with a distinctive odor, and (3) water, H_2O. Uncap the urea and water, add about half the water to the urea, recap the urea–water mixture, and swirl to dissolve the urea. Record your observations, including whether the solution becomes warm or cool. Repeat this procedure with the alcohol and the other half of the water.

2.3 CONSIDER THIS

Are there energy changes when substances dissolve?

(a) Did you observe any warming as the solutions formed in Investigate This 2.2? Cooling?

(b) Did dissolving either solute in water give off energy to the surroundings? Take up energy from the surroundings? Explain how you know.

The solution process Substances that dissolve in water may be solids, liquids, or gases. Understanding the interactions between these solutes and water is the key to understanding the formation and nature of solutions. Without yet going into the details of the interactions, we can begin to think about what has to occur for a substance to dissolve in water (or any other solvent). In any pure solid or liquid, such as sugar, urea, or ethanol, molecules attract one another strongly enough to stay in the condensed phase. For solute molecules to become mixed homogeneously among water molecules in a solution, the intermolecular attractions among the solute molecules in their pure solid or liquid state have to be broken. When the solution forms, some attractions between water molecules are broken, and new attractions between solute and water molecules are made.

Figure 2.2 shows how we analyze the dissolution process and find its energy by breaking the process into steps that take us from the solid (or liquid) that dissolves to the solution it forms with water (or other solvent). Step 1 in Figure 2.2 converts the molecules that will be dissolved to their gas phase. This step breaks the solute–solute attractions and requires an input of energy, represented by the long, upward-pointing red arrow, that we have labeled $\Delta E_{\text{sol-sol}}$. In Step 2, these gas phase solute molecules are mixed with the water molecules to form the solution. In this step, some attractions between water molecules are broken, which requires an input of energy. But new attractions form between the water and solute molecules and these release a good deal of energy. The combined effect of these two interactions is the release of energy represented by the long, downward-pointing blue arrow. The figure shows that the process of **solvation,** the formation of attractive interactions between separated solute molecules and a solvent, is a downhill process, going from higher to lower energy: $\Delta E_{\text{solvation}} < 0$.

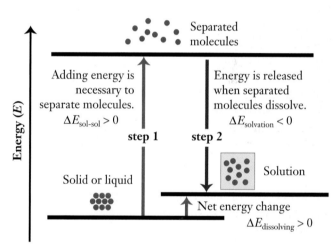

Figure 2.2 shows the uphill process, breaking solute–solute attractions, $\Delta E_{\text{sol-sol}}$, requiring more energy than is released by solvation, $\Delta E_{\text{solvation}}$. The dissolving process requires a net input of energy: $\Delta E_{\text{dissolving}} > 0$. A change that requires an input of energy is called **endothermic.** Conversely, if more energy is released by solvation than is required to break the attractions, the overall solution process releases energy. A change that releases energy is called **exothermic.**

Figure 2.2.

Energy changes in the dissolution process.

2.4 CHECK THIS

Exothermic energy diagram

(a) Sketch the energy diagram for an overall dissolution process that is exothermic.
(b) Is the dissolution of urea in water exothermic or endothermic? Explain how you know. Is the dissolution of ethanol in water exothermic or endothermic? Explain how you know.

Favorable and unfavorable factors We often categorize the factors that affect a process as favorable or unfavorable. A **favorable factor** is one that increases the likelihood that the process will proceed in the direction we are considering. An **unfavorable factor** works against the direction of the process, that is, it favors going in the reverse direction. For example, releasing the energy of solvation lowers the energy of a solute–solvent system, and lower energy favors the formation of the solution. The energy required to separate the solute molecules has to be added to the system, which raises the energy, and is therefore an unfavorable factor. In Chapter 8, we will quantify favorable and unfavorable factors, when we discuss the **entropy** of solutions.

Overall, an exothermic process releases energy and lowers the energy of the system from which the energy comes. The lower final energy favors the process in the exothermic direction. By this same reasoning, an endothermic process is unfavorable. In Investigate This 2.2, you have seen that some solutes dissolve in water exothermically and others dissolve endothermically. This observations means that the net energy change in a solution process must not be the only factor that affects solubility, and we will have to be on the lookout for others.

2.2. Solutions of Polar Molecules in Water

Table 2.1 reminds you of the characteristics of the three intermolecular interactions we discussed in Chapter 1—hydrogen bonding, polar attractions, and London dispersion forces—and illustrates each with representative molecules.

Table 2.1 *Comparisons among intermolecular attractions.*

	Hydrogen bonding	**Polar attraction**	**Dispersion forces**
requires	H atom bonded to O, N, or F	permanent dipolar molecules	all molecules
strength	20–30 kJ·mol^{-1}	2–5 kJ·mol^{-1}	can be quite large—increases with number of electrons per molecule
geometry	H atom approximately on a line between electronegative atoms	positive and negative ends aligned to attract one another	any orientation—best when molecules are side by side
example	water	dimethyl ether	octane

Source: Atkins and Jones, *Chemical Principles* (Freeman, 1999), pp. 284, 367.

2.5 INVESTIGATE THIS

Which compounds are soluble in water?

(a) Work in small groups to make models of methanol (CH_3OH), 1-butanol ($CH_3CH_2CH_2CH_2OH$), and hexane ($CH_3CH_2CH_2CH_2CH_2CH_3$). Decide which intermolecular attractions, Table 2.1, are present between the molecules in each pure compound. Use these models to predict which molecules will form strong intermolecular attractions with water.

(b) Predict which liquids (methanol, 1-butanol, or hexane) will be soluble in water. Mix 1 mL of each liquid with 1 mL of water; observe which compounds dissolve in water. If one or more of the liquids does not seem to dissolve, try adding two or three drops to 1 mL of water to find out if a small amount will dissolve.

2.6 CONSIDER THIS

Which compounds are soluble in water?

(a) Which of your solubility predictions in Investigate This 2.5(b) was correct? What were the reasons for your predictions? If any were incorrect, explain why they were incorrect.

(b) Which of the liquids is most soluble in water? Which is least soluble? Explain how you reach these conclusions.

Hydrogen bonds among unlike molecules In Chapter 1, we found that hydrogen bonding accounts for many physical properties of water. There, we considered hydrogen bonding between water molecules, but there is nothing to prevent hydrogen bonding between water and other molecules that can hydrogen bond with water, such as alcohols. The maximum number of hydrogen bonds between one methanol molecule and surrounding water molecules is three, as shown in Figure 2.3. The hydrogen covalently bonded to the methanol oxygen takes part in one of these hydrogen bonds. The two nonbonding electron pairs on the methanol oxygen form bonds with hydrogen from water to give the other two. The water molecules have other nonbonding electron pairs and covalently bonded hydrogen atoms, so they can extend the network to other water and methanol molecules.

Web Companion

Chapter 2, Section 2.2 — (1)

Try interactive visualizations of water interacting with polar and nonpolar molecules.
(2)
(3)
(4)

2.7 CHECK THIS

Hydrogen-bonded network in water–methanol solution

Two of the four water molecules in the top panel of the *Web Companion*, Chapter 2, Section 2.2.2, look like they are oriented to be able to hydrogen bond with oxygen in methanol, but only one is allowed to replace the "x."

(a) What is the problem with the orientation of the molecule that doesn't fit?

(b) How does your answer in part (a) help you understand the network of hydrogen bonds in water around the methanol molecule? Explain.

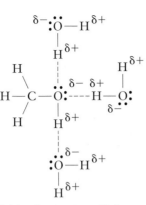

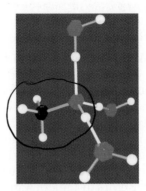

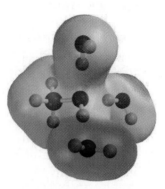

Figure 2.3.

Maximum hydrogen bonding between a methanol molecule and water molecules. Molecules are in approximately the same orientation with respect to one another in each representation.

(a) Lewis structure: Hydrogen bonds are shown as dotted lines.

(b) Molecular model: Hydrogen bonds are white and intramolecular covalent bonds are pink and blue.

(c) Charge density surface model, showing polarities

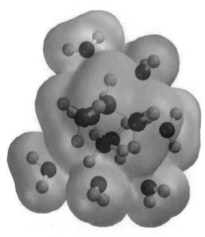

Figure 2.4.

A hexane molecule in aqueous solution.

The O–H group in methanol fits into this extensive hydrogen-bonded network and accounts for the high solubility in water that you found in Investigate This 2.5. Other alcohols dissolve in water for exactly the same reason. In fact, methanol, ethanol, CH_3CH_2OH, 1-propanol, $CH_3CH_2CH_2OH$, and 2-propanol, $CH_3CH(OH)CH_3$, are miscible with water. Two liquids are **miscible** if they form homogeneous solutions when they are mixed in any proportion.

Nonpolar solutes At the other extreme of solubility, you found that hexane is insoluble in water. Hexane is a nonpolar compound that cannot form hydrogen bonds with water. A few hexane molecules do dissolve in a volume of water, but there are no specific interactions between them and the surrounding water molecules (Figure 2.4). Rather, there are ever-changing weak attractions, London dispersion forces, between individual water and hexane molecules.

It is difficult to measure the energy change associated with the tiny amount of a hydrocarbon that dissolves in water, but it is close to zero for those that have been measured. This means that the energy of the weak interactions among water and hexane molecules is about the same as the energy required to break up the hexane–hexane attractions and some water–water attractions.

2.8 CHECK THIS

Energy diagram for hexane dissolving in water

Sketch the energy diagram for the overall solution process of hexane dissolving in water. Assume that the net energy change for the dissolution is zero.

Molecular reorganization If there is no difference in energy between a solution of hexane in water and the pure liquids, it would be logical for you to question *why* hexane is insoluble in water. The answer lies in the reorganization of the molecules that must occur in going from pure liquids to the solution. The solute molecules go from being all grouped together in the pure liquid to a solution of individual solute molecules scattered among the solvent molecules. The reorganization involved in mixing two kinds of molecules together always favors the mixed state and leads to increased solubility.

On the other hand, Figure 2.4 shows that a molecule of hexane in water takes up space in the liquid without forming strong attractions with the water molecules. The water molecules lose some of their freedom of movement when they have to be reorganized to make way for the hexane molecule. This reorganization, even though it does not require energy, works against the solubility of nonpolar molecules. The solvent reorganization is unfavorable and outweighs the favorable reorganization of mixing. The dominance of the unfavorable solvent reorganization explains why nonpolar compounds are insoluble in water. In Chapter 8, we will say more about the reasons for this behavior and develop a quantitative model, based on entropy, for these favorable and unfavorable reorganizations. For now, the observation that nonpolar compounds are relatively insoluble in water is enough to allow you to understand and make predictions about solubilities.

Intermediate cases Your results from Investigate This 2.5 show that 1-butanol is somewhat soluble in water; small amounts do dissolve. A 1-butanol molecule has an alcohol group and a four-carbon alkyl chain that is like a nonpolar hydrocarbon. The alcohol group attractions to water, shown in Figure 2.5, should help 1-butanol dissolve. (We have seen that alcohols with smaller alkyl groups are miscible with water.) On the other hand, the alkyl group is nonpolar and will not help the 1-butanol dissolve. The upshot is limited solubility; alcohols become less and less soluble as the size of the alkyl group grows.

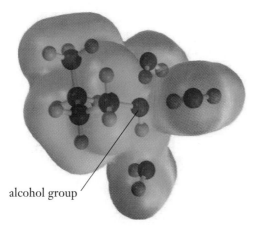

alcohol group

Figure 2.5.

A 1-butanol molecule in aqueous solution. The polar alcohol O-H group in butanol hydrogen bonds with water. The nonpolar end of the butanol, like nonpolar hexane (Figure 2.4), interacts only weakly with water molecules.

Like dissolves like From what we have seen so far, we can predict that a compound with a polar group that can hydrogen bond with water will be more soluble in water than a similar compound without the polar group: 1-pentanol, $CH_3CH_2CH_2CH_2CH_2OH$, is more soluble than pentane, $CH_3CH_2CH_2CH_2CH_3$. We can also predict that the water solubility of compounds with polar and nonpolar parts will decrease as the nonpolar part gets larger: dimethyl ether, CH_3OCH_3, is more soluble than dipropyl ether, $CH_3CH_2CH_2OCH_2CH_2CH_3$. There is a generalization that polar molecules that can hydrogen bond with water dissolve in water, a polar hydrogen-bonding solvent, but nonpolar molecules do not. This is an example of a rule you may have heard before: Like dissolves like.

2.9 CHECK THIS

Predict relative solubilities

Write Lewis structures for diethyl ether, $CH_3CH_2OCH_2CH_3$, pentane, $CH_3CH_2CH_2CH_2CH_3$, and 1-butanol, $CH_3CH_2CH_2CH_2OH$.
(a) Which of these molecules can form hydrogen bonds in their pure liquids?
(b) Which can form hydrogen bonds with water?
(c) Which compound is most soluble in water? Which is least soluble? Explain.
(d) Which compound is most soluble in mineral oil, a nonpolar liquid? Which is least soluble? Explain.

Solutes with multiple polar groups One of the soluble substances you probably listed in Consider This 2.1 is sugar. You can dissolve about 200 g of sucrose (table sugar), $C_{12}H_{22}O_{11}$, in 100 mL of water. The structure of a simpler sugar, glucose, $C_6H_{12}O_6$, which also readily dissolves in water (about 100 g of glucose per 100 mL of water), is shown in Figure 2.6. Note that the glucose molecule contains five alcohol groups as well as the oxygen in the ring. You would expect a glucose molecule to hydrogen bond strongly with water molecules.

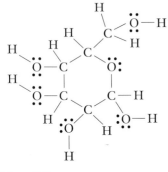

Figure 2.6.

The structural formula of glucose.

2.10 CONSIDER THIS

What are the interactions of glucose with water?

Use the structural formula in Figure 2.6 to determine the maximum number of water molecules that can be hydrogen bonded to a glucose molecule. Compare the maximum number of hydrogen bonds between water and glucose (a six-carbon compound) to the maximum number between water and methanol (a one-carbon compound). Does this comparison help to explain the high solubility of glucose? If so, how does it help?

Your comparison in Consider This 2.10 shows that, carbon-for-carbon, glucose and methanol have about the same number of attractive, hydrogen-bonding interactions with water. A complex compound whose molecules contain several polar groups that can hydrogen bond with water can be relatively soluble in water. Another comparison is between the solubility of glucose and cyclohexanol. Cyclohexanol, Figure 2.7, has only one alcohol group on the six-carbon ring and has a solubility of about 3.6 g per 100 mL of water. The much higher solubility of glucose is due to its many polar, hydrogen-bonding groups.

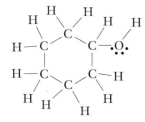

Figure 2.7.

The structural formula of cyclohexanol.

2.11 CHECK THIS

Predict relative solubilities

(a) Write out the Lewis structures for these compounds with multiple polar groups.

continued

$$H(OCH_2CH_2)_4OH \qquad HO(CH_2CH_2)_5OH$$

polyethylene glycol (PEG 200)   1,10-decanediol

(b) Which compound is more soluble in water? Explain.

Reflection and Projection

Solutions are mixtures in which solute molecules separate from each other and disperse uniformly throughout the solution. Dissolving a solute in a solvent requires energy to break the attractions among the pure solute molecules (and also some of the attractions among the solvent molecules). Some or all of this energy is supplied by the attractions between the solute and solvent molecules in the solution. Many polar molecules dissolve in water (and in one another) because of hydrogen bonding between the solute and water molecules. Polar compounds that contain no hydrogen atom attached to an oxygen or nitrogen atom cannot hydrogen bond among themselves. They can, however, hydrogen bond with water, which makes them more water soluble than comparable nonpolar compounds.

Nonpolar compounds are generally insoluble in water, mainly because the water molecules must reorganize to accommodate them. This reorganization is an unfavorable process that outweighs the favorable reorganization of mixing the solute and solvent. Compounds whose molecules contain both polar parts that can hydrogen bond with water and large nonpolar parts are less soluble than comparable compounds with smaller nonpolar parts. Compounds with several polar groups are more soluble than comparable compounds with fewer polar groups. Many complex compounds with several polar groups, including many biomolecules like sugars, dissolve readily in water.

The solutes we have discussed so far are carbon-containing polar compounds that dissolve to give solutions of individual molecules mixed among the water molecules. Among the soluble substances that you listed in Consider This 2.1, there must have been ones like table salt that are not carbon-containing compounds. We will now examine how solutions of these substances are similar to and different from the solutions we have been discussing.

2.3. Characteristics of Solutions of Ionic Compounds in Water

2.12 INVESTIGATE THIS

Which solutions conduct an electric current?

Use an electrical conductivity tester to determine whether or not solutions conduct an electric current. The tester consists of a source of electrical potential, two wires that dip into the test solution, and a meter or a signal light to signal electric current flow. Place 2–3 mL of distilled water in a small test tube or well plate. Test the conductivity of pure water. Dissolve a small amount of glucose in water in another test tube or well and test the conductivity of this sugar solution. Repeat the conductivity test with a solution of table salt in another test tube or well.

2.13 CONSIDER THIS

Why do some solutions conduct an electric current?

(a) Which of the liquids in Investigate This 2.12 conduct an electric current? How do you tell?

(b) Can you distinguish among pure water, an aqueous glucose solution, and an aqueous sodium chloride solution by their conductivities? Use sketches like these of the conductivity experiment to show what you *observed* for each solution. What molecular-level picture can you suggest to explain your answer? Illustrate your explanation on the sketches.

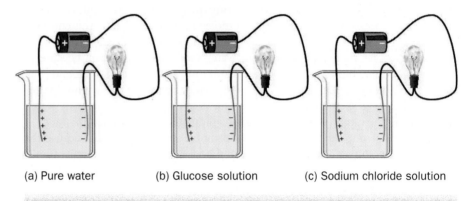

(a) Pure water   (b) Glucose solution   (c) Sodium chloride solution

Solution conductivity For a solution to give a positive electrical conductivity test (bulb lights up), electric charge has to be transported from one test wire to the other. Solutions that conduct a current must contain mobile charged particles. When an electrical potential is applied to the solution, negative particles move toward the positive wire and positive particles move toward the negative wire. Water molecules are electrically neutral and cannot transport electric charge; this case is illustrated in Figure 2.8(a). In the previous section, we pictured a solution of glucose as electrically neutral glucose molecules mixed with and hydrogen bonding with water molecules. This picture is confirmed by the electrical conductivity results: A solution of glucose, like pure water, does not conduct electricity, as illustrated in Figure 2.8(b).

Web Companion

Chapter 2, Section 2.3

Use molecular visualizations to check your understanding of solution conductivity.

① ② ③ ④

Figure 2.8.

Observed results and molecular-level interpretation of solution conductivity.

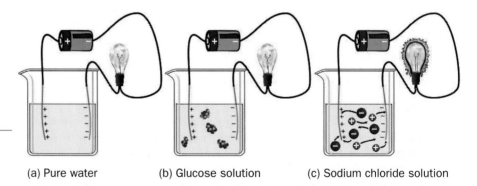

(a) Pure water   (b) Glucose solution   (c) Sodium chloride solution

Many solids, like sodium chloride, dissolve in water to form solutions that *do* conduct an electric current, as illustrated by the bulb lighting up in Figure 2.8(c). These solutions must contain charged particles that move in response to the electrical field between the wires, as illustrated in the figure. There must be equal numbers of positive and negative charges in the solution, since neither the water nor the solid had an electric charge before they were mixed to make the solution.

We have said nothing about what happens when the charged particles reach the wires. In Investigate This 2.12, you may have noticed that gas bubbles formed on the wires in the conducting solution. Formation of the gas indicates that a chemical reaction is occurring to produce the gas. For our purposes in this chapter, we are interested only in whether solutions do or do not conduct an electric current. We will return in Chapter 10 to discuss the chemical reactions that result from the flow of electric charge through conducting solutions.

Ionic solids and solutions Figure 2.9 shows a molecular-level model for the dissolving of solid compounds whose solutions conduct electricity. The solid, Figure 2.9(a), is made up of **ions,** species that have too few or too many electrons to balance the positive charge of their nuclei. A **cation (positive ion)** is an atom, or a covalently bonded group of atoms, that has lost one or more electrons. The positive charge results from there being too few electrons to balance the positive charges of nuclear protons. An **anion (negative ion)** is an atom, or a covalently bonded group of atoms, that has gained one or more extra electrons. The negative charge results from there being more electrons than are needed to balance the positive charges of the nuclear protons. The ions are held in place by strong attractive forces to their nearest neighbors, forming an orderly arrangement called an **ionic crystal.**

Cation is pronounced "CAT-ion" and anion is pronounced "AN-ion."

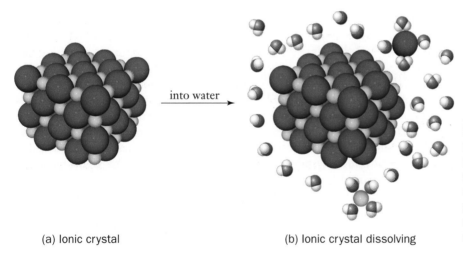

(a) Ionic crystal (b) Ionic crystal dissolving

Figure 2.9.

Model of an ionic crystal dissolving in water. The ionic crystal (a) begins to dissolve as water molecules surround the ions (b). The smaller gray spheres represent positive cations and the larger green spheres represent negative anions.

2.14 CHECK THIS

Properties of crystals

(a) If each cation has a 1+ charge and each anion a 1− charge, what is the net charge on the crystal in Figure 2.9(a)? What is the net charge on the remaining crystal in Figure 2.9(b)? Explain how you get your answers.

(b) Does the *Web Companion*, Chapter 2, Section 2.4.1, movie show properties of crystals not shown in Figure 2.9(a)? If so, what are they? Why aren't they shown in the figure?

Web Companion

Chapter 2, Section 2.5.1–2 ── ①

Use interactive visualizations to ②
examine the ionic dissolving
process.
③
④

Ion–dipole attractions in ionic solutions When a soluble ionic solid dissolves in water, the resulting solution is called an **ionic solution.** Figure 2.9(b) shows how the polar water molecules are attracted by the electrically charged ions and surround the ions that enter the solution. Once they have left the crystal and are surrounded by water molecules, the ions are free to move about. When conductivity tester wires are placed in the solution, positive ions and negative ions carry current through the solution, as shown in Figure 2.8(c).

Look closely at the model for dissolved ions in Figure 2.9(b). The figure shows that water molecules orient themselves differently around positive and negative ions. [For clarity, Figure 2.9(b) shows only a few of the water molecules that surround each ion.] Cations attract the negative (oxygen) ends of water molecules. Anions attract the positive (hydrogen) ends of water molecules. Attractions between ions and water molecules are called **ion–dipole attractions,** and they are relatively strong. Ions at the corners and along the edges of a crystal dissolve more readily than those in the crystal faces, because more water molecules can gather around the ions while they are still part of the crystal.

As an ion leaves the crystal, water molecules surround it completely, forming a **hydration layer** (*hydro* = water). Dissolved ions are often called **hydrated ions.** Although individual water molecules are constantly being exchanged in hydration layers, the hydration layer helps keep oppositely charged ions from getting close to one another. Positive and negative ions are usually separated by at least two layers of water molecules. Consequently, each dissolved ion moves almost independent of all the others.

2.15 CHECK THIS

Ionic hydration layers and crystal dissolution

(a) About how many water molecules can surround an ion to form its first hydration layer? Explain how you arrive at your answer.

(b)  Is your answer in part (a) consistent with the molecular-level representations in the *Web Companion*, Chapter 2, Section 2.5.2, animations? Explain why or why not. Crystal dissolution is represented in the movie. What interactions favor dissolving? What interactions favor the crystal remaining intact? Explain clearly how the competition among these interactions is represented.

Aqueous ionic solutions are everywhere around you and in you. Seas and oceans contain many dissolved ionic compounds; that's why they taste salty. Even freshwater streams, such as the one in the chapter opening photograph, contain some dissolved ionic compounds. Almost every liquid you drink, including tap water, is an ionic solution, and all biological fluids are ionic solutions. Yet, many solid ionic compounds, for example, seashells and your teeth, do not dissolve in water. In the next four sections, we will consider ionic compounds in more detail and find out how to predict which are soluble and which are not.

2.4. Formation of Ionic Compounds

Names and formulas Before we examine how ionic compounds form, let's consider how we name these compounds and write their formulas. Names and formulas for a few common **monatomic** (single atom) ions and **polyatomic** (many atom) ions are given in Table 2.2. Ionic compounds do not contain individual molecules, so a chemical formula such as NaCl does not represent a molecule. The formula represents a ratio of ions. For any ionic compound, the numbers needed to cancel the positive and negative charges determine this ratio. The smallest whole-number ratio of positive ions to negative ions that is electrically neutral is used as the chemical formula. Calcium chloride crystals, for example, contain Ca^{2+} and Cl^- ions. Two chloride ions are required to balance the charge on one calcium ion, and the chemical formula is written as $CaCl_2$. Ionic charges are almost never indicated in the formula. Note that the cation is always named first in the names of ionic compounds.

Table 2.2 *Names and formulas of a few common ions.*

Cations				Anions			
Monatomic		Polyatomic		Monatomic		Polyatomic	
calcium	Ca^{2+}	ammonium	NH_4^+	bromide	Br^-	carbonate	CO_3^{2-}
magnesium	Mg^{2+}	hydronium	H_3O^+	chloride	Cl^-	hydroxide	OH^-
potassium	K^+			fluoride	F^-	nitrate	NO_3^-
sodium	Na^+			oxide	O^{2-}	phosphate	PO_4^{3-}
				sulfide	S^{2-}	sulfate	SO_4^{2-}

2.16 WORKED EXAMPLE

Chemical formulas for ionic compounds

Write the chemical formulas for the ionic compounds (a) ammonium carbonate and (b) calcium phosphate.

Necessary information: We need to know, from the names, that compound (a) contains the ammonium and carbonate ions and (b) contains the calcium and phosphate ions. Formulas and charges for the ions are given in Table 2.2.

Strategy: The numbers of cations and anions must be such that the total positive charge cancels the total negative charge. The easiest ways to find the numbers are by simply inspecting the charges on the ions or by trial and error.

Implementation:
(a) The ammonium ion is 1+ and the carbonate ion is 2−. There must be two 1+ ammonium ions to balance each 2− carbonate. The formula is $(NH_4)_2CO_3$. The parentheses around the ammonium ion are necessary to distinguish the number of ammonium ions (2) from the number of hydrogen atoms in the ammonium ion (4).

continued

(b) The calcium ion is 2+ and the phosphate ion is 3−. We'll need more than one of each ion. Try two phosphate ions to give a 6− charge. We will need three calcium ions to cancel the 6−. The formula is $Ca_3(PO_4)_2$. As before, parentheses around the phosphate are required to avoid subscript confusion.

Does the answer make sense? Multiply the charge times the subscript for both the positive and negative ion to be sure that the total plus and minus charge does cancel out. The easiest mistake to make in writing formulas is to switch the subscripts; this check helps avoid that error.

2.17 CHECK THIS

Formulas for ionic compounds

Write the chemical formulas for the ionic compounds (a) ammonium chloride, (b) sodium sulfide, and (c) calcium carbonate.

Why ionic compounds form Before considering what happens to ions in solution, we will consider why ionic compounds form in the first place. Table 2.2 shows that atoms of elements in Group I (**alkali metals**) and Group II (**alkaline earths**) lose electrons to form monatomic positive ions. Atoms of elements in Groups VI (oxygen and sulfur) and Group VII (**halogens**) gain electrons to form monatomic negative ions. The keys to this behavior lie in values for the electronegativities of these elements, which are given in Figure 2.10, and in the attraction between oppositely charged ions.

	I	II	III	IV	V	VI	VII	VIII
1				hydrogen **H** 2.2				helium He —
2	lithium **Li** 1.0	beryllium **Be** 1.6	boron **B** 2.0	carbon **C** 2.6	nitrogen **N** 3.0	oxygen **O** 3.4	fluorine **F** 4.0	neon Ne —
3	sodium **Na** 0.93	magnesium **Mg** 1.3	aluminum **Al** 1.6	silicon **Si** 1.9	phosphorus **P** 2.2	sulfur **S** 2.6	chlorine **Cl** 3.2	argon Ar —
4	potassium **K** 0.82	calcium **Ca** 1.3						

Figure 2.10.

Electronegativities for the first 20 elements.

On the left-hand side of the periodic table, the elements in Groups I and II have low electronegativities. These elements have a relatively weak attraction for electrons. On the other side of the periodic table, elements in Groups VI and VII have high electronegativities. These elements have a strong attraction for electrons. Imagine a gas phase reaction between a sodium atom losing a loosely held electron and a chlorine atom attracting that electron to yield a sodium cation and chloride anion, as represented in Figure 2.11.

In Figure 2.11, we are trying to represent the overall changes that occur at the atomic level in this reaction. An electron is transferred from atomic sodium to atomic chlorine. The transfer gives an 11+ sodium nucleus surrounded by only 10 electrons; this is a sodium cation. The transfer also gives a 17+ chlorine nucleus surrounded by 18 electrons; this is a chloride anion. The sizes of the atoms and ions in Figure 2.11 are drawn to scale. The sodium cation (which has fewer electrons than the atom) is smaller than the chloride anion (which has more electrons than the atom). This is a general result: *Almost all monatomic anions are larger than any monatomic cation.*

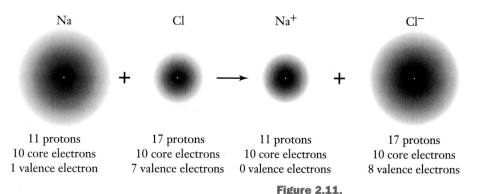

Na	Cl	Na$^+$	Cl$^-$
11 protons 10 core electrons 1 valence electron	17 protons 10 core electrons 7 valence electrons	11 protons 10 core electrons 0 valence electrons	17 protons 10 core electrons 8 valence electrons

Figure 2.11.

Molecular (atomic) level representation of an electron transfer reaction.

Chemical reaction equations Another way to write the reaction between a gaseous sodium atom and a gaseous chlorine atom to yield a gaseous sodium cation and a gaseous chloride anion is this symbolic representation:

$$Na(g) + Cl(g) \rightarrow Na^+(g) + Cl^-(g) \tag{2.1}$$

Chemical equations, like equation (2.1), are valuable shorthand representations describing chemical change. The physical state of each substance is often indicated in parentheses following the chemical formula. Gases are represented using *(g)*, liquids using *(l)*, solids using *(s)*, and aqueous solutions using *(aq)*. The right-facing arrow means "yields." Substances to the left of the arrow are called **reactants** and substances to the right are called the **products** of the reaction. When there are multiple reactants or products, they are separated by a plus sign, which is read as "and." A **balanced chemical equation** has *equal numbers of each atom* on each side of the arrow and the *net electrical charge is the same* on each side of the arrow. When we want to refer to "the chemical reaction represented by chemical equation (xx)," we will usually shorten the reference to "reaction (xx)." For example, we would refer to the "reaction of Na*(g)* and Cl*(g)* represented by chemical equation (2.1)" as "reaction (2.1)."

Web Companion

Chapter 2, Section 2.4 ───(1)

Study interactive visualizations ──(2)
of ionic crystals and crystal ──(3)
lattice energy. ──(4)

2.18 WORKED EXAMPLE

Writing a balanced chemical reaction equation

Write a balanced gas phase chemical reaction equation that represents the reaction between calcium atoms and fluorine atoms to yield calcium cations and fluoride anions.

Necessary information: Table 2.2 shows us that the calcium cation has a 2+ charge and the fluoride anion a 1− charge.

Strategy: Since a calcium atom has to lose two electrons to form a calcium cation, it must transfer these electrons to two fluorine atoms, each of which gains an electron to form a fluoride ion.

continued

Implementation: The chemical reaction representing this transfer is

$$Ca(g) + 2F(g) \rightarrow Ca^{2+}(g) + 2F^-(g)$$

Note that the numbers of atoms or ions taking part in the reaction are specified by coefficients written in front of the symbol for the species.

Does the answer make sense? The same number of calcium atoms (or ions) is represented on each side of the yields arrow and the same is true for number of fluorine atoms (and ions). The net electrical charge on the reactants is zero; none of the reactants has an electrical charge. The net electrical charge on the products is also zero; the 2+ charge on the calcium cation and the 1− charges on each of *two* fluoride anions cancel out. Equal numbers of atoms of each kind appear in the reactants and products and the net electrical charge is the same for both reactants and products. The chemical reaction equation is balanced and makes sense.

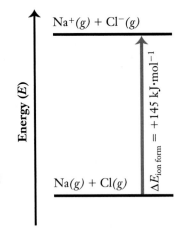

Figure 2.12.

Energy change for formation of Na⁺(g) + Cl⁻(g) from Na(g) + Cl(g).

The energy diagram labels:
$Na^+(g) + Cl^-(g)$
$Na(g) + Cl(g)$
Energy (E)
$\Delta E_{ion\ form} = +145\ kJ \cdot mol^{-1}$

2.19 CHECK THIS

Writing a balanced chemical reaction equation

Write a balanced gas phase chemical reaction equation that represents the reaction between magnesium atoms and oxygen atoms to yield magnesium cations and oxide anions.

Formation of ionic crystals Figure 2.12 shows an energy diagram for reaction (2.1), the formation of positive and negative gaseous ions from gaseous atoms. You see that the reaction is endothermic; 145 kJ·mol⁻¹ are *required* for the reaction. Energetically, the electron transfer is an uphill process. The energy change, $\Delta E_{ion\ form}$, is unfavorable for the formation of ions from atoms.

Experimentally, however, an enormous number of compounds of metal cations and nonmetal anions *do* exist. All such compounds of Group I and II metal cations with Group VI and VII anions are white ionic crystals that resemble ordinary table salt. The source of energy that makes formation of the compounds possible is the attraction of the cations and anions for one another. They combine to form extended three-dimensional **lattices,** regular repeating patterns of positive and negative ions, the ionic crystals shown in Figure 2.9. The attractions between cations and anions are represented in a two-dimensional lattice in Figure 2.13.

The energy of attraction between unlike charges (or repulsion between like charges) is

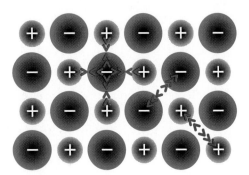

Figure 2.13.

Attractions and repulsions in an ionic lattice. Red arrowheads facing *toward* one another represent coulombic attractions and blue arrowheads facing *away* from one another represent coulombic repulsions.

$$E \propto \frac{Q_1 Q_2}{d} \tag{2.2}$$

The attraction between oppositely charged bodies is called **coulombic attraction,** to honor Charles A. de Coulomb (French physicist, 1736–1806) who first quantified this attraction. In words, equation (2.2) says that the energy *(E)* of attraction (or repulsion) between two electrically charged particles is directly proportional to the magnitude of each charge *(Q)* and inversely proportional to the distance (*d*) separating the particles. If the charges have opposite signs, the energy has a negative value, that is, it is less than zero. Attraction of opposite charges lowers the energy of the lattice (crystal) in Figure 2.13. Conversely, the energy is positive if the charges are the same; repulsion of like charges raises the energy of the lattice (crystal).

> Coulomb found that the force of interaction between charged bodies is inversely proportional to the square of the distance between them. This inverse squared relationship is called Coulomb's law. Equation (2.2), the energy associated with the interaction, is derived from Coulomb's law.

▨▨▨ 2.20 CONSIDER THIS ▨▨▨

Are attractions or repulsions stronger in a crystal?

Is the magnitude of the attractive energy, which is represented by red arrowheads in Figure 2.13, larger or smaller than the magnitude of the repulsive energy represented by blue arrowheads? Explain the reasoning for your answer.

In an ionic crystal, represented by the lattice in Figure 2.13, oppositely charged ions are closest together. Greater distances separate ions with the same charge. Positive–negative attractions between adjacent ions are strong and outweigh the repulsions between more distant like-charged ions. An ionic crystal is lower in energy than its separated ions. The **lattice energy** of an ionic crystal is the energy that would have to be supplied to convert the solid crystal to separated gaseous cations and gaseous anions:

$$Na^+Cl^-(s) \rightarrow Na^+(g) + Cl^-(g) \qquad (2.3)$$

For NaCl, the lattice energy, $\Delta E_{lattice}$, is 787 $kJ\cdot mol^{-1}$. [In equation (2.3), the charges are shown in Na^+Cl^- to emphasize that it is an ionic crystal.]

Figure 2.14 brings together the energies we have been discussing, in order to determine the energy change, $\Delta E_{xtal\ form}$, when Na^+Cl^- ionic crystals (abbreviated "xtal") are formed directly from gaseous Na and Cl atoms:

$$Na(g) + Cl(g) \rightarrow Na^+Cl^-(s) \qquad (2.4)$$

The left-hand, downward-pointing blue arrow in Figure 2.14 represents the energy change, $\Delta E_{xtal\ form}$, for reaction (2.4). Experimental difficulties make it impossible to measure the energy change for this reaction directly, but it can be calculated by combining the energy changes for reactions (2.1) and (2.3).

The upward-pointing red arrow in Figure 2.14 represents the energy change for reaction (2.1), $\Delta E_{ion\ form}$. The energy change for the formation of the NaCl crystals from the ions, the reverse of reaction (2.3), is $-\Delta E_{lattice} = -787\ kJ\cdot mol^{-1}$, and is represented by the right-hand, downward-pointing blue arrow on the energy diagram. Whenever a reaction is reversed, the energy change for the reverse reaction is equal in magnitude, but opposite in sign, to the energy change for the forward reaction. The energy change for reaction (2.4), the formation of Na^+Cl^- ionic crystals from gaseous Na and Cl atoms, is the sum of the energy changes for the steps that bring about this same change:

$$\Delta E_{xtal\ form} = \Delta E_{ion\ form} + (-\Delta E_{lattice})$$

$$= (145\ kJ\cdot mol^{-1}) + (-787\ kJ\cdot mol^{-1}) = -642\ kJ\cdot mol^{-1}$$

Reaction (2.4) forming Na^+Cl^- ionic crystals from gaseous Na and Cl atoms is exothermic.

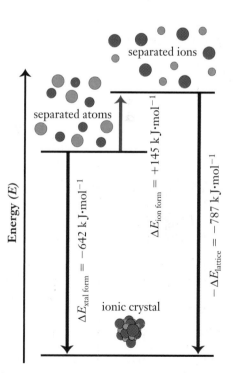

Figure 2.14.

Energy diagram for the formation of one mole of ionic crystals of NaCl.

2.21 CHECK THIS

Energy diagram for formation of one mole of CaCl₂

The lattice energy for calcium chloride, $CaCl_2$, crystals is 2260 kJ·mol⁻¹. The gaseous reaction forming the ions from the atoms,

$$Ca(g) + 2Cl(g) \rightarrow Ca^{2+}(g) + 2Cl^-(g),$$

requires 1039 kJ·mol⁻¹. Draw an energy diagram, analogous to Figure 2.14, and use it to find $\Delta E_{\text{xtal form}}$ for the formation of ionic crystals of $CaCl_2$ from the gaseous atoms.

2.5. Energy Changes When Ionic Compounds Dissolve

 ### 2.22 INVESTIGATE THIS

What temperature changes occur when solids dissolve?

Before and after photos of CaCl₂ added to water, with a digital thermometer showing the temperatures and a conductivity meter in the liquid showing before and after conductivity.

Work in small groups on this investigation. You will use three capped, numbered vials, each containing about 1 g of a white solid. The solids are (1) ammonium chloride, NH_4Cl, (2) calcium chloride, $CaCl_2$, and (3) sodium chloride, NaCl. Open one of the vials, hold the vial near the bottom, and have a partner pour about 5 mL of room temperature water into the vial. Feel the temperature of the vial and watch what happens inside. Let your partners also feel the vial. Cap the vial so you don't spill the contents. Gently shake its contents while continuing to observe any additional changes. Continue recording your observations until no further change is evident. Repeat this procedure with the other two vials containing white solids. Uncap each vial and use an electrical conductivity tester to determine whether the solutions conduct electricity.

2.23 CONSIDER THIS

What energy changes occur when solids dissolve?

(a) In Investigate This 2.22, were the three solids that were dissolved ionic? What is the evidence for your answers?

(b) When a process releases energy, the energy usually warms surrounding objects, such as the container. When a process requires energy, the energy usually comes from surrounding objects, which get cooler. In Investigate This 2.22, were the dissolving processes for the three solids exothermic or endothermic? What is the evidence for your answers?

(c) Did any of your results surprise you? Why or why not? What might be happening when these white solids are mixed with water?

Solubility: Hydration energy and lattice energy You found in Investigate This 2.22 that sodium chloride, ammonium chloride, and calcium chloride all dissolve readily in water and that their solutions are good conductors of electrical current, so the solutions must contain ions. You also found that the dissolving process was exothermic for some compounds (the solution got warm) and endothermic for others (the solution got cool). These energy changes are the subject of this section.

Attraction between ions and water in the solution, which we discussed in Section 2.3, is a major factor that favors dissolving of ionic solids. We express this attraction as the **hydration energy,** for the ions. Hydration energy is just a special name for solvation energy when the solvent is water. The hydration energy is the energy change, $\Delta E_{\text{hydration}}$, when the gaseous ions dissolve in water. For NaCl, the process is

$$\text{Na}^+(g) + \text{Cl}^-(g) \rightarrow \text{Na}^+(aq) + \text{Cl}^-(aq) \qquad \textbf{(2.5)}$$

The hydration energy for NaCl is -784 kJ·mol^{-1}. Attraction between positive and negative ions in the solid, the lattice energy, is the major factor that favors keeping the crystal intact. The balance between the hydration and lattice energies helps determine whether an ionic solid is soluble. Figure 2.15 is an energy diagram showing these two energies and the resulting net energy change, $\Delta E_{\text{dissolving}} = 3$ kJ·mol^{-1}, for dissolving solid NaCl. (Figure 2.15 is another specific example of the general energy diagram for the solution process shown in Figure 2.2.)

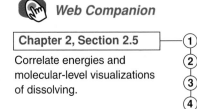

Web Companion

| Chapter 2, Section 2.5 | — ① |

Correlate energies and molecular-level visualizations of dissolving.

② ③ ④

2.24 CONSIDER THIS

How does Figure 2.15 correlate with Investigate This 2.22?

The net energy change, $\Delta E_{\text{dissolving}}$, for dissolving solid NaCl, Figure 2.15, is quite small. Is this consistent with your observations in Investigate This 2.22? Explain why or why not.

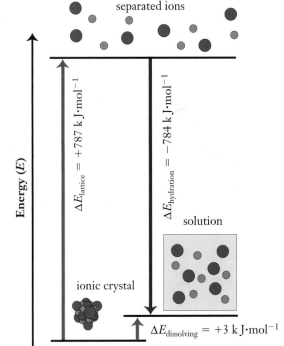

separated ions

$\Delta E_{\text{lattice}} = +787$ kJ·mol^{-1}

$\Delta E_{\text{hydration}} = -784$ kJ·mol^{-1}

Energy (E)

solution

ionic crystal

$\Delta E_{\text{dissolving}} = +3$ kJ·mol^{-1}

Figure 2.15.

Energy changes for NaCl dissolving in water.
Dissolving NaCl(s) in water is a slightly endothermic process, $\Delta E_{\text{dissolving}} = 3$ kJ·mol^{-1}.

For some ionic compounds, as you observed with CaCl$_2$ in Investigate This 2.22, the hydration energy is much greater than the lattice energy, and the energy released during dissolving, $\Delta E_{\text{dissolving}} < 0$, warms the solution quite substantially. For other ionic compounds, such as ammonium chloride, the lattice energy is greater than the hydration energies of the ions, and the energy taken in during dissolving, $\Delta E_{\text{dissolving}} > 0$, removes energy from the solution and makes it cooler. Some of the hot packs and cold packs you might have used to ease minor injuries employ dissolution reactions like these to produce their thermal effects. The lattice and hydration energies for a few ionic solids are given in Table 2.3.

Table 2.3 *Lattice and hydration energies (in kJ·mol⁻¹) for some ionic compounds.*

The entries where the cation row and anion columns intersect are $\Delta E_{\text{lattice}}$ and $\Delta E_{\text{hydration}}$ values for the compound. For example, 2440 kJ·mol⁻¹ is $\Delta E_{\text{lattice}}$ for $MgBr_2$.

	Anion					
	Cl^-		Br^-		I^-	
Cation	$\Delta E_{\text{lattice}}$	$\Delta E_{\text{hydration}}$	$\Delta E_{\text{lattice}}$	$\Delta E_{\text{hydration}}$	$\Delta E_{\text{lattice}}$	$\Delta E_{\text{hydration}}$
Li^+	861	−898	818	−867	759	−822
Na^+	787	−784	751	−753	700	−708
K^+	717	−701	689	−670	645	−625
Ag^+	916	−850	903	−819	887	−774
Mg^{2+}	2524	−2679	2440	−2626	2327	−2540
Ca^{2+}	2260	−2337	2176	−2285	2074	−2194
Sr^{2+}	2153	−2205	2075	−2142	1963	−2053

	S^{2-}		CO_3^{2-}	
Mg^{2+}	3406	−3480	3122	−3148
Ca^{2+}	3119	−3140	2804	−2817
Sr^{2+}	2974	−3030	2720	−2725

2.25 WORKED EXAMPLE

Energy change for dissolving CaCl₂ in water

Use the data in Table 2.3 to draw an energy diagram, analogous to Figure 2.15, for dissolving solid calcium chloride, $CaCl_2$, in water and find $\Delta E_{\text{dissolving}}$.

Necessary information: All the data we need are in Table 2.3.

Strategy: We use the values for $\Delta E_{\text{lattice}}$ and $\Delta E_{\text{hydration}}$ from Table 2.3 to determine the energy levels for the ionic solid crystal and the ionic solution relative to the energy of the separated ions. The energy difference between these two levels is $\Delta E_{\text{dissolving}}$.

Implementation: The energy level diagram is shown here.

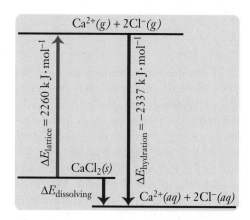

continued

$\Delta E_{dissolving}$ is the sum of the energy changes for the two steps that give the equivalent change:

$$\Delta E_{dissolving} = \Delta E_{lattice} + \Delta E_{hydration} = (2260 \text{ kJ·mol}^{-1}) + (-2337 \text{ kJ·mol}^{-1})$$

$$= -77 \text{ kJ·mol}^{-1}$$

Both the energy-level diagram and this calculation show that dissolving $CaCl_2(c)$ in water is an exothermic process.

Does the answer make sense? In Investigate This 2.22, you found that the solution got warm when solid calcium chloride was dissolved. The dissolution reaction is exothermic, as our energy level diagram predicts.

2.26 CHECK THIS

Energy change for dissolving LiI in water

Lithium iodide, LiI, is very soluble in water. Draw an energy-level diagram, analogous to Figure 2.15 and the one in Worked Example 2.25, for dissolving solid LiI in water. Does the solution get warmer or cooler as the solid dissolves? Explain your answer.

Reflection and Projection

The polar model for water developed in Chapter 1 provides the basis for understanding how water dissolves ionic compounds by hydrating (solvating) the cations and anions and keeping them apart from one another. Solutions of ions conduct an electric current because individual positive and negative ions, being almost independent of each other and surrounded by water molecules, can move and carry electrical charges through the solution. Sizes and charges of ions reflect their electron arrangements. Ionic compounds do not exist as molecules, so formulas are written as the smallest whole-number ratio of positive and negative ions that cancel charge to give an overall electrically neutral formula.

Forming an ionic compound from a metal and a nonmetal requires that the metal atoms form cations by giving up electrons. Nonmetal atoms, in turn, accept those electrons to form anions. More energy is required to remove the electrons than is returned when anions form. Ionic compounds form because their lattice energies, the energies required to break the net attractions among ions in crystal lattices, are large. The energy associated with dissolving an ionic compound in water, $\Delta E_{dissolving}$, depends on the relative magnitudes of the lattice energy, $\Delta E_{lattice}$, and hydration energy, $\Delta E_{hydration}$, for that compound.

You observed that some soluble ionic compounds dissolve exothermically and others dissolve endothermically. This result means that you cannot use the energy of the solution process to predict solubilities. Therefore, we will investigate solubilities experimentally to see if we can discover some rules or generalizations that will make it possible to predict which ionic compounds will be soluble and which will be insoluble.

2.6. Precipitation Reactions of Ions in Solution

 2.27 INVESTIGATE THIS

What reactions of ions in solution can you observe?

Use an electrical conductivity tester to test the conductivity of aqueous solutions of $CaCl_2$, Na_2SO_4, and $NaNO_3$. Then add about 1 mL of the $CaCl_2$ solution to each of two empty vials. Add about 1 mL of the Na_2SO_4 solution to the first vial. Screw on the lid and gently swirl the contents. Record any changes you observe. Add 1 mL of $NaNO_3$ solution to the second vial, screw on the lid, gently swirl the contents, and record any changes you observe. Test the conductivity of each of the two mixtures you made.

2.28 CONSIDER THIS

How do you explain the reactions of ions in solution?

How are your observations of the two mixtures in Investigate This 2.27 the same? How are they different? How do you explain any changes that occurred when you mixed the solutions to make the mixtures? Are the conductivities consistent with your explanation? Explain why or why not.

"Precipitate" and "precipitation" generally have to do with falling. Rain falls from the sky. An insoluble solid falls from solution. In chemistry, "precipitate" is used as both a verb (the action of precipitating) and as a noun (the solid formed by a precipitation reaction).

Ions in solution in living cells, including Na^+, K^+, Ca^{2+}, Fe^{2+}, Cl^-, HPO_4^{2-}, $H_2PO_4^-$, HCO_3^-, and CO_3^{2-}, are essential in transport, construction, metabolism, and energy transfer. In the previous section, we looked at the energy associated with bringing ions into solution when an ionic compound dissolves, but we did not consider the reverse of that process. Ionic compounds can come out of solution, or **precipitate,** when their solubility is exceeded. You observed the formation of a precipitate, a solid substance separating from the liquid, when you mixed the $CaCl_2$ and Na_2SO_4 solutions in Investigate This 2.27.

People who suffer from kidney stones have first-hand experience with precipitation reactions from solutions in their body. Kidney stones, Figure 2.16(a), are primarily calcium phosphate, $Ca_3(PO_4)_2$ and calcium oxalate, CaC_2O_4. Other precipitates, such as the $CaCO_3$ crystals in the balance organ of the inner ear, Figure 2.16(b), and calcium phosphate in our bones and teeth, are essential to our well-being. And shelled animals like the mollusks in the chapter opener depend on precipitation of $CaCO_3$ to form their shells. Why do some substances, such as calcium phosphate, precipitate, while others, such as sodium phosphate, remain in solution? And how do we establish the identity of a precipitate that forms when ionic solutions are mixed?

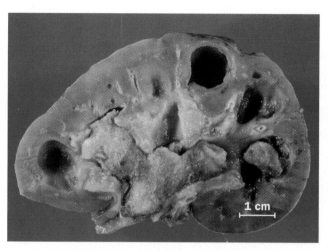

(a) Kidney stones, $Ca_3(PO_4)_3$ and CaC_2O_4

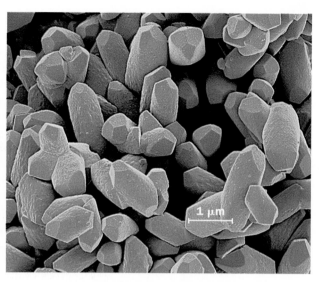

(b) Balance organ crystals, $CaCO_3$

Figure 2.16.

Two precipitates that form in humans. Note the great difference in size of these solids. The kidney stones are about 10,000 times larger than the inner ear balance organ crystals.

Formation of precipitates

Let's analyze the formation of the precipitate you observed when you mixed solutions of calcium chloride and sodium sulfate in Investigate This 2.27. When the calcium chloride solution was in a container by itself, both the calcium ions and the chloride ions remained in solution. The mixture of sodium ions and sulfate ions and the mixture of sodium ions and nitrate ions in their separate containers also remained in solution. The solid precipitate must have resulted from interactions of the ions in the mixed solution. Table 2.4 is a review of the starting conditions, the ions together in each mixture, and the result for each mixture.

Web Companion

Chapter 2, Section 2.6

Interactive animation and visualization of a precipitation reaction and ionic equations.

①
②
③
④

Table 2.4 *Results of mixing $CaCl_2$ with Na_2SO_4 and $NaNO_3$.*

	Before mixing			After mixing	
	$CaCl_2$ solution	Na_2SO_4 solution	$NaNO_3$ solution	$CaCl_2$ and Na_2SO_4	$CaCl_2$ and $NaNO_3$
positive ion(s)	$Ca^{2+}(aq)$	$Na^+(aq)$	$Na^+(aq)$	$Ca^{2+}(aq) +$ $Na^+(aq)$	$Ca^{2+}(aq) +$ $Na^+(aq)$
negative ion(s)	$Cl^-(aq)$	$SO_4^{2-}(aq)$	$NO_3^-(aq)$	$Cl^-(aq) +$ $SO_4^{2-}(aq)$	$Cl^-(aq) +$ $NO_3^-(aq)$
conductivity?	yes	yes	yes	yes	yes
precipitate?				yes (white solid)	no

Identity of the precipitate

When an ionic solid dissolves, water molecules solvate ions from the crystal lattice and take them into solution, as we have represented in Figure 2.9(b) and the *Web Companion*, Chapter 2, Section 2.5.

Precipitation is the reverse of dissolving. If an ionic compound, such as ordinary table salt, NaCl, dissolves readily, then we know that the ions—$Na^+(aq)$ and $Cl^-(aq)$ in this case—will probably not form a precipitate when they are mixed together. Initially, as shown in Table 2.4, only one kind of cation and one kind of anion is present in each container. When the contents of the calcium chloride and sodium sulfate containers are mixed, all four ions are present together. If a precipitate forms, at least one kind of cation must react with at least one kind of anion.

The four possible ionic combinations are Na_2SO_4, $CaCl_2$, $NaCl$, and $CaSO_4$. The first two possibilities can be eliminated because they were in solution together before mixing. If the third combination, NaCl, were insoluble, it would have precipitated when solutions of $CaCl_2$ and $NaNO_3$ were mixed, but no precipitate formed. $CaSO_4(s)$, calcium sulfate, is the only remaining possibility, and it is the precipitate formed when solutions of $CaCl_2$ and Na_2SO_4 are mixed (Figure 2.17). It is widely distributed as a solid in nature, and geologists call it gypsum. The construction industry uses enormous quantities of gypsum as sheet rock or dry wall in homes and offices. You may have used it as a casting material called plaster of Paris, and some chalk is $CaSO_4(s)$.

$\oplus$ = Na^+ $\bigcirc$ = SO_4^{2-}

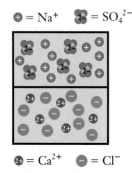

$\oplus$ = Ca^{2+} $\ominus$ = Cl^-

(a) The ions in sodium sulfate, Na_2SO_4, and calcium chloride, $CaCl_2$, solutions in separate containers

Figure 2.17.

Molecular-level representation of the precipitation of calcium sulfate.

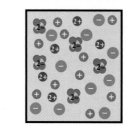

(b) An *imagined* mixture of ions in the instant before they react

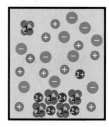

(c) Solid calcium sulfate, $CaSO_4$, represented by the aggregation of ions at the bottom of the frame

Ionic reaction equations One way to write the chemical equation that describes the precipitation reaction is

$$Ca^{2+}(aq) + 2Cl^-(aq) + 2Na^+(aq) + SO_4^{2-}(aq)$$

$$\rightarrow CaSO_4(s) + 2Na^+(aq) + 2Cl^-(aq) \tag{2.6}$$

The reaction represented symbolically by reaction equation (2.6) is represented at the molecular level in Figure 2.17.

2.29 CHECK THIS

Electrical conductivity in Investigate This 2.27

Do equation (2.6) and Figure 2.17 explain the electrical conductivities you observed in Investigate This 2.27. Explain why or why not.

An important observation from Figure 2.17 and equation (2.6) is that the sodium ions and the chloride ions play no part in the overall reaction. The ions must be present in the solutions to balance electrical charge, but they appear as independent aqueous ions on both sides of equation (2.6). The ions are shown in Figure 2.17 as independent ions before and after the reaction. Ions that are present, but do not participate in a chemical reaction are called **spectator ions.** They are often omitted when writing chemical equations:

$$Ca^{2+}(aq) + SO_4^{2-}(aq) \rightarrow CaSO_4(s) \tag{2.7}$$

Equation (2.7) is an example of a **net ionic equation,** which shows only the ions that react and the product of their reaction.

2.30 WORKED EXAMPLE

Product of a precipitation reaction and net ionic reaction

When clear, colorless solutions of cadmium nitrate, $Cd(NO_3)_2$, and sodium sulfide, Na_2S, are mixed, a beautiful orange precipitate is formed. What is the ionic compound that precipitates? Write the net ionic equation for the precipitation reaction.

Necessary information: We need to recall the solutions we used in Investigate This 2.27.

Strategy: List all the combinations of a cation and an anion (accounting for charge balance) that can be formed in the mixture and eliminate those that are soluble.

Implementation: The cations and anions present in the mixture are $Cd^{2+}(aq)$, $Na^+(aq)$, $NO_3^{2-}(aq)$, and $S^{2-}(aq)$. The four possible combinations are $Cd(NO_3)_2$, Na_2S, CdS, and $NaNO_3$. The first two are in the solutions we started with, so they must be soluble. The last, $NaNO_3$, sodium nitrate, was one of the solutions we used in Investigate This 2.27, so it must be soluble. The only possibility remaining is CdS, cadmium sulfide, which is the orange solid. The complete ionic equation is

$$Cd^{2+}(aq) + 2NO_3^{2-}(aq) + 2Na^+(aq) + S^{2-}(aq)$$
$$\rightarrow CdS(s) + 2Na^+(aq) + 2NO_3^{2-}(aq)$$

The net ionic reaction equation is

$$Cd^{2+}(aq) + S^{2-}(aq) \rightarrow CdS(s)$$

Does the answer make sense? We eliminated all the other possibilities, so cadmium sulfide must be the solid that is formed. This solid is used in oil paints to make colors that range from cadmium yellow to cadmium orange. Note that the complete ionic equation and the net ionic equation are balanced in both atoms and charge, as they must be.

2.31   CHECK THIS

Product of a precipitation reaction and net ionic reaction

When aqueous solutions of silver nitrate, $AgNO_3$, and sodium chloride, NaCl, are mixed, a white precipitate is formed. What is the ionic compound that precipitates? Write the net ionic equation for the precipitation reaction. The *Web Companion*, Chapter 2, Section 2.5, provides lab-, symbolic-, and molecular-level (interactive animations) representations of this reaction.

Equilibrium   Another important observation, shown in Figure 2.17, but not indicated by either equation (2.6) or (2.7), is that small amounts of $Ca^{2+}(aq)$ and $SO_4^{2-}(aq)$ remain in solution after the solutions are mixed. A few ions are always left behind when any ionic solid precipitates from aqueous solution. After a time, when an ionic solid is left in contact with an aqueous solution of its ions, a balance is reached between dissolving and precipitation and there is no further *net* change in the amount of solid or of ions in the solution. At the balance point, $Ca^{2+}(aq)$ and $SO_4^{2-}(aq)$ continue to react to form $CaSO_4(s)$, but elsewhere in the mixture $CaSO_4(s)$ is dissolving:

$$CaSO_4(s) \rightarrow Ca^{2+}(aq) + SO_4^{2-}(aq) \tag{2.8}$$

Dissolving, reaction (2.8), and precipitation, reaction (2.7), are still going on but at exactly the same rate, so no net change is observed. When this balance has been achieved, the solution is said to be in **equilibrium** with the solid. (We will continue to use the concept of equilibrium qualitatively, but delay a quantitative discussion until Chapter 9.)

Equilibrium can be represented by an equation like this, for the case of calcium sulfate:

$$Ca^{2+}(aq) + SO_4^{2-}(aq) \rightleftharpoons CaSO_4(s) \tag{2.9}$$

The two oppositely pointing arrows in equation (2.9) indicate that the reaction is **reversible**—it is going in the forward and reverse directions at the same time. A solution that contains its equilibrium amount of ions is called a **saturated solution.** The identity of the compound and the temperature of the solution determine the amount of an ionic compound that will dissolve to give a solution at equilibrium (saturation). The amounts of solutes in solutions are expressed as concentrations. We will return to concentrations in Section 2.8, but first we'll finish what we set out to do, figure out which ionic compounds are soluble and which are not.

2.7. Solubility Rules for Ionic Compounds

2.32   INVESTIGATE THIS

Which mixtures of ionic compounds yield precipitates?

Work together in small groups to do this investigation and to interpret its results. You will use a 24-well microtiter plate and aqueous solutions of several ionic compounds. Place the well plate on a white sheet of paper and label the

continued

rows and columns as shown in the drawing. Add about 1 mL (equivalent to a well about one-third full) of each of the specified pairs of solutions to each well. Well **A1**, for example, should have 1 mL of NaCl and 1 mL of KNO_3 added to it. Observe and record which solution pairs produce a precipitate.

KNO_3 KBr K_2SO_4 K_2CO_3 $K_2C_2O_4$ K_3PO_4

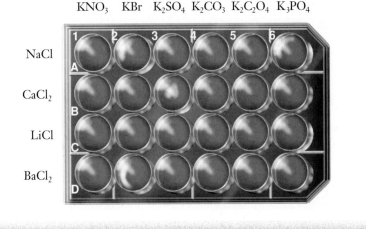

NaCl

$CaCl_2$

LiCl

$BaCl_2$

2.33 CONSIDER THIS

Which ionic compounds are soluble and which are insoluble?

(a) Several of the mixtures in Investigate This 2.32 *did not* form precipitates. What do you learn from these results about the solubility of the ionic compounds that can be formed from each mixture? What compounds *do not* form insoluble precipitates?

(b) Several of the mixtures *did* form precipitates. What are the solids formed in these mixtures? How do you know? Write a net ionic equation [like equation (2.7)] to describe the formation of each precipitate.

Although the data you have to work with are limited, you can use the results from Investigate This 2.32 to develop some generalizations about which ionic compounds are soluble and which are not. We'll begin the reasoning with a Worked Example and then you can finish it.

2.34 WORKED EXAMPLE

Solubility trends for alkali metal ions, halides, and nitrate

What generalizations can we make about the solubility of ionic compounds with alkali metal cations (Li^+, Na^+, and K^+)? What generalizations can we make about the solubility of ionic compounds with halide anions (Cl^- and Br^-) and with nitrate anion (NO_3^-)?

Necessary information: This problem uses only the observations from Investigate This 2.32.

continued

Reasoning: The anion test solutions all had K^+ as the cation, so potassium-ion compounds with these anions are soluble. No precipitates formed in any of the Na^+ cation (A-row) wells, so sodium-ion compounds with these anions are soluble. The behavior of Li^+ cation (C-row wells), another alkali metal, is more complicated. Li+ forms no precipitates with mononegative anions (Cl^-, NO_3^-, and Br^-), but Li^+ does form a precipitate with phosphate.

The cation test solutions all had Cl^- as the anion, so chloride-ion compounds with these cations are soluble, and we already know that KCl is soluble. No precipitates are formed with the Br^- anion in the 2-column wells, so bromide-ion compounds of all the cations are soluble. No precipitates are formed with the NO_3^- anion in the 1-column wells, so nitrate-ion compounds of all the cations are soluble.

Conclusions: These tentative generalizations about water solubilities are justified by the data:

- Ionic compounds of the alkali metal cations are soluble, except for some Li^+ compounds with multiply charged anions.
- Ionic compounds of the halide anions (Cl^- and Br^-) are soluble.
- Ionic compounds of the nitrate anion are soluble.

Determining the solubilities of ionic compounds requires extensive data collection under a range of conditions, but observations like those in Investigate This 2.32 are useful in establishing trends that allow us to make predictions.

2.35 CONSIDER THIS

What are solubilities for multiply charged cations and anions?

(a) What generalizations can you make about the solubility of ionic compounds with alkaline earth cations (Ca^{2+} and Ba^{2+})?
(b) What generalizations can you make about the solubility of ionic compounds with multiply charged anions (SO_4^{2-}, CO_3^{2-}, $C_2O_4^{2-}$, and PO_4^{3-})?
(c) Can you make any generalizations about the probable solubility of an ionic compound, if both the cation and anion are multiply charged? Explain your reasoning.

Solubility rules for ionic compounds We can combine much of what we have learned so far into a few simple solubility rules that work for a wide variety of ionic compounds:

- Ionic compounds with an alkali metal cation or a halide or nitrate anion are likely to be soluble.
- Ionic compounds of a monopositive cation and mononegative anion are likely to be soluble.
- Ionic compounds of a multiply charged cation and a multiply charged anion are likely to be insoluble.

There are, of course, exceptions to these broad generalizations and they do not address the intermediate case of ionic compounds in which only one of the ions is multiply charged. You can, however, use the rules to make predictions that will be correct in many cases you are likely to meet.

2.36 CHECK THIS

Predicting solubilities

(a) Sodium phosphate, Na_3PO_4, and cobalt chloride, $CoCl_2$, are both relatively soluble in water. Is this what you would predict? Explain why or why not.

(b) Consider the possible outcomes when solutions of the two ionic compounds in part (a) are mixed. Would a precipitate be likely to form from either (or both) new cation–anion pairing? Explain your reasoning. Write a net ionic equation for the reaction(s) you predict to occur. Check your predictions in Investigate This 2.38.

Although the rules above give us predictive power, they do not provide an underlying molecular basis for understanding solubility. The energetics of the solution process is one factor in determining solubility, but we have seen that the energetics for soluble compounds can be favorable in some cases and unfavorable in others. Compare the data in Table 2.3 for the alkali metal halides, all of which are quite soluble, and you will find both exothermic and endothermic overall energies for dissolving. The data in the table show that the solution process for the alkaline earth sulfides and carbonates is exothermic, that is, energetically favorable, but these compounds are all insoluble.

Reorganization factors in ionic solubility In Section 2.2, we found that nonpolar molecules and molecules with large nonpolar parts are not soluble in water, even though mixing of the solute with water is a favorable process. For these cases, the reorganization of the water molecules required to accommodate the solute is unfavorable and limits the solubility. Similar factors are at work in ionic solutions. The reorganization of the solute from ions fixed in a rigid lattice to ions mixed with water molecules and free to move about in solution is favorable. The problems arise because of the reorganization of the water molecules required to hydrate the ions. The orientation of the water molecules about the ions reduces the water molecules' freedom of movement, which makes this reorganization unfavorable.

Multiply charged ions attract water molecules more strongly and have a larger orientation effect on the water molecules than do singly charged ions. Consequently, the unfavorable reorganization that multiply charged ions impose on the water molecules is large enough to lower the solubility of these ions. The solubility of an ionic compound composed of a cation and anion that are both multiply charged is doubly unfavored and, as we observe, most of these compounds are insoluble. Ionic compounds composed of singly charged cations and anions are usually soluble. From this observation, we can infer that the favorable mixing of singly charged solute ions with water molecules outweighs their unfavorable reorganization of hydration. We will discuss these reorganization factors further in Chapter 8.

▀▀▀▀▀▀ **2.37 CONSIDER THIS** ▀▀▀▀▀▀

What might cause some exceptions to the solubility rules?

(a) The silver halides, AgCl, AgBr, and AgI, are exceptions to our solubility rules. All these ionic halides are insoluble. Ag^+ and K^+ cations are almost the same size. How would you compare their reorganization effects on water molecules? Based on your comparison, what would you predict about the solubilities of their halides? Explain.

(b) Do the data in Table 2.3 help explain the difference in solubility of the potassium and silver halides? If so, show how. If not, explain why not.

Exceptions to the rules Ag^+ and K^+ are about the same size, so we would expect that, when dissolved, both ions would have about the same effect on the reorganization of water molecules. Any difference between the solubility of silver and potassium ionic solids is likely to be a result of a difference in their dissolution energies, $\Delta E_{dissolving}$. The data in Table 2.3 show that the K^+ halides [except KI(*s*)] dissolve endothermically. This unfavorable effect for KCl(*s*) and KBr(*s*) is relatively small, less than 20 kJ·mol^{-1}, and the potassium halides are quite soluble. The silver halides also dissolve endothermically, but the unfavorable energy effect is a good deal larger, 66 to 84 kJ·mol^{-1}, and the silver halides are insoluble. This substantial energy requirement apparently greatly limits the solubility of the silver compounds. Despite exceptions like these, the simple rules stated above are useful and will lead to correct predictions in most cases.

Reflection and Projection

We accomplished what we set out to do and have a short set of simple rules you can use to predict the solubility of many ionic compounds. Coupled with the general rules for predicting the solubility of polar and nonpolar molecular compounds, these are powerful tools for understanding phenomena that involve substances dissolving in or precipitating from aqueous solutions. Along the way, we have learned about the interactions that govern the energetics of solution processes. We have also learned about another important factor that helps to determine solubilities: the effect of reorganization of solute and solvent molecules when a solution is formed. The reorganization of solute molecules mixing with solvent molecules always favors solution. The reorganization of solvent molecules to orient around solute molecules is an unfavorable factor for solution and favors the separate solute and solvent. The balance among all these factors leads to the observed solubilities and rules.

We use chemical equations as a shorthand way to represent chemical reactions. When forward and reverse reactions are balanced, which we show by the use of double arrows, $\rightleftharpoons$, we say that a reaction is at equilibrium. A solution in equilibrium with pure solute, known as a saturated solution, contains an unchanging amount of solute. The amount of solute dissolved, expressed as its concentration, is a quantitative measure of solubility. Using concentrations enables us to make meaningful comparisons among solubilities and to study and characterize many other kinds of reactions as well. Next we will turn our attention to determining concentrations.

2.8. Concentrations and Moles

2.38 INVESTIGATE THIS

Do Co^{2+}(aq) and PO_4^{3-}(aq) react to form a precipitate?

Do this as a class investigation but work in small groups to discuss and analyze the results. The reagents for this investigation are aqueous solutions of cobalt chloride hexahydrate, $CoCl_2 \cdot 6H_2O$, and sodium phosphate Na_3PO_4, whose concentrations are given in the table. *CAUTION:* Cobalt salts are somewhat toxic. Wear plastic gloves when handling these reagents. In four small centrifuge tubes, prepare the equal-volume mixtures indicated in the table. Record your observations on the appearance of the original solutions, what occurs during mixing, and the final appearance of the mixtures. Centrifuge the four mixtures to bring any precipitates to the bottom of the tubes.

Mixture 3 is shown before and after addition of phosphate and after centrifugation.

Mixture	1	2	3	4
36 g $CoCl_2 \cdot 6H_2O$ in 1 L solution	X mL	X mL		
18 g $CoCl_2 \cdot 6H_2O$ in 1 L solution			X mL	X mL
24 g Na_3PO_4 in 1 L solution	X mL		X mL	
12 g Na_3PO_4 in 1 L solution		X mL		X mL

After centrifugation, note and record the appearance of the liquid and solid in each tube and the relative amounts of solid in each tube. These results will be used and discussed throughout this section and further in Section 2.10.

2.39 CONSIDER THIS

How are concentration and amount of precipitate related?

(a) Do your results from Investigate This 2.38 confirm the prediction you made in Check This 2.36? Explain why or why not.
(b) Were the amounts of precipitate in the four mixtures the same or different? If different, which one had the most precipitate? Reasoning on the basis of your net ionic equation, is this the mixture you would predict to have the most precipitate? Explain.
(c) Is the appearance of the liquid in each mixture the same or different after centrifugation? How would you explain any similarities or differences?

Mixtures 3 and 4 are shown after centrifugation.

Concentration The **concentration** of a solution is a measure of the amount of solute present in a given quantity, often a volume, of the solution. The higher the concentration, the more solute present in the same volume of solution, as shown in Figure 2.18. We will use this concept to try to understand the observations in Investigate This 2.38.

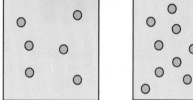

(a) Concentration = c   (b) Concentration = 2c

Figure 2.18.

Illustration of solution concentration. When twice as much sugar is dissolved, the molecular level representation shows that there are twice as many solute particles (sugar molecules) in the same volume of solution; the concentration is doubled.

Web Companion

Chapter 1, Section 1.11	①
Review moles and molecules	②
with interactive questions	③
and visualizations.	④

Some of the results from Investigate This 2.38 may seem a bit puzzling. When you compare the contents of Mixtures 3 and 4, you might expect that adding a higher concentration of Na_3PO_4 solution in Mixture 3, 24 $g·L^{-1}$ compared to 12 $g·L^{-1}$ in Mixture 4, would produce more precipitate. There is more phosphate to react and the net ionic reaction equation requires two phosphate anions for every three cobalt cations:

$$3Co^{2+}(aq) + 2PO_4^{3-}(aq) \rightarrow Co_3(PO_4)_2(s) \qquad (2.10)$$

You observe, however, that the amounts of precipitate formed in the two mixtures are the same, as shown in the photograph in Consider This 2.39. Since there is the same amount of cobalt ion in Mixtures 3 and 4, we might conclude that it all reacts and is used up to form the same amount of precipitate in these two mixtures. We will return to this idea in Section 2.10, after learning more about how to express solution concentrations in a way that is chemically meaningful.

2.40 CONSIDER THIS

Is there evidence for complete reaction of $Co^{2+}(aq)$ *in these mixtures?*

(a) In Investigate This 2.38, is there evidence for complete reaction of $Co^{2+}(aq)$ in Mixtures 3 and 4? Explain.

(b) If the $Co^{2+}(aq)$ had *not* reacted completely in these mixtures, what would you have expected to observe? Explain the experimental basis for your expectation, if any.

(c) Does a comparison of the results for Mixtures 1 and 4 reinforce your explanation in part (a)? Explain why or why not.

Moles and molarity Since chemical equations, like equation (2.10), relate numbers of ions and molecules, we need to express solution concentrations in a way that tells us about the numbers of ions and molecules in a given amount of solution. The appropriate way to express the amount of solute is in numbers of moles. Recall, from Chapter 1, Section 1.11, that the mole is the chemist's counting unit, 6.02×10^{23} particles of anything. You seldom need to use the actual number of particles, but you do need to remember that the mole is for *counting* particles. A common way of expressing the amount of solute in a volume of solution is as the number of moles present per liter of solution. The number of moles of solute per liter of solution is defined as the **molarity (M)** of the solution. A solution containing 1.00 mol of a solute in 1.00 L of solution has a concentration of 1.00 $mol·L^{-1}$, which is the same as 1.00 M. **Square brackets, [ ],** represent the molarity of a species, for example, $[Co^{2+}(aq)]$ is the molarity of $Co^{2+}(aq)$ in a solution containing the Co^{2+} cation.

One **molar mass** of a compound is the mass of the compound *in grams* that has the same numeric value as the relative molecular mass of the compound. A molar mass of a compound is a mole of the compound and contains

6.02×10^{23} molecules of the compound. To calculate the molar mass of a compound, you *multiply the relative mass of each atom* (from the inside front cover of this book) *by the number of that atom in a molecule of the compound and then sum the products to get the number of grams in a molar mass of the compound.* Ionic compounds do not contain discrete molecules, but the mass of one **formula unit,** the formula we use to specify the atom ratio in the ionic compound, is treated as if it were a molecule. For example, cobalt phosphate, formula unit $Co_3(PO_4)_2$, contains three cobalt ions for every two phosphorus atoms and every eight oxygen atoms in the ionic lattice. The molar mass of $Co_3(PO_4)_2$ is

$$\text{molar mass } Co_3(PO_4)_2 =$$

$$\left(3 \text{ mol Co} \times \frac{58.9 \text{ g Co}}{1 \text{ mol Co}}\right) + \left(2 \text{ mol P} \times \frac{31.0 \text{ g P}}{1 \text{ mol P}}\right) + \left(8 \text{ mol O} \times \frac{16.0 \text{ g O}}{1 \text{ mol O}}\right) \tag{2.11}$$

$$\text{molar mass } Co_3(PO_4)_2 =$$

$$(3 \times 58.9 \text{ g Co}) + (2 \times 31.0 \text{ g P}) + (8 \times 16.0 \text{ g O}) = 366.7 \text{ g} \tag{2.12}$$

2.41 CHECK THIS

The molar mass of cobalt chloride hexahydrate

How many atoms of each element are in a formula unit of cobalt chloride hexahydrate, $CoCl_2 \cdot 6H_2O$? What is the molar mass of cobalt chloride hexahydrate, $CoCl_2 \cdot 6H_2O$?

To understand more about the results from Investigate This 2.38, we need to know the molar concentration of the solutions used. Then we can compare the reacting numbers of moles, which are proportional to the reacting number of ions. We will calculate the reacting numbers of moles in a series of Worked Examples and Check This examples that you can use to check your understanding. A three-step calculation process is used:

1. Find the number of moles in a mass of a reactant,
2. Find the molarity of the stock solution containing this mass, and
3. Find the number of moles in each mixture.

2.42 WORKED EXAMPLE

Mass to mole conversion

How many moles of cobalt chloride hexahydrate, $CoCl_2 \cdot 6H_2O$, are in a 36-g sample of the ionic crystals?

Necessary information: We need the molar mass of $CoCl_2 \cdot 6H_2O$ from Check This 2.41.

continued

Strategy: Use the molar mass to write a unit factor for mass-to-mole conversion.

Implementation:

$$1 \text{ mol CoCl}_2 \cdot 6H_2O = 238 \text{ g CoCl}_2 \cdot 6H_2O \text{ [from molar mass]}$$

Mass-to-mole conversion:

$$36 \text{ g CoCl}_2 \cdot 6H_2O = 36 \text{ g CoCl}_2 \cdot 6H_2O \left(\frac{1 \text{ mol CoCl}_2 \cdot 6H_2O}{238 \text{ g CoCl}_2 \cdot 6H_2O} \right)$$

$$= 0.15 \text{ mol CoCl}_2 \cdot 6H_2O$$

Does the answer make sense? Thirty-six grams is much less than the molar mass, 238 g, so the sample must contain much less than 1 mol of cobalt chloride hexahydrate. Our answer is much less than 1 mol, so it is in the right direction.

2.43 CHECK THIS

Mass-to-mole conversion

How many moles of sodium phosphate, Na_3PO_4, are in a 24-g sample of the ionic crystals? Note that you first have to find the molar mass of Na_3PO_4.

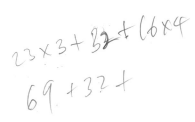

2.44 WORKED EXAMPLE

Molarity of a solution

What is the molarity of an aqueous solution of cobalt chloride hexahydrate, $CoCl_2 \cdot 6H_2O$, that contains 36 g of the solid ionic compound in 1 L of solution?

Necessary information: We need the number of moles of $CoCl_2 \cdot 6H_2O$ in 36 g of the solid, 0.15 mol, from Worked Example 2.42.

Strategy: Divide the number of moles in the solution by the volume of solution to get the $mol \cdot L^{-1} = M$.

Implementation:

$$\text{molarity of CoCl}_2 \cdot 6H_2O \text{ solution} = \left(\frac{0.15 \text{ mol CoCl}_2 \cdot 6H_2O}{1 \text{ L}} \right)$$

$$= 0.15 \text{ M}$$

Does the answer make sense? The solution we started with had 0.15 mol of solute in a liter of solution, so it must be 0.15 M.

━━━━━ **2.45** **CHECK THIS** ━━━━━━━━━━━━━━━━

Molarity of a solution

What is the molarity of an aqueous solution of sodium phosphate, Na_3PO_4, that contains 24 g of the solid ionic compound in 1 L of solution?

━━━━━ **2.46** **WORKED EXAMPLE** ━━━━━━━━━━━━

Volume-to-mole conversion

How many moles of cobalt chloride hexahydrate, $CoCl_2 \cdot 6H_2O$, are present in Mixture 1 in Investigate This 2.38? The mixtures shown all contained 6.0 mL of the cobalt solutions.

Necessary information: We need the molarity of the $CoCl_2 \cdot 6H_2O$ solution, 0.15 M, from Worked Example 2.44 and the volume of this solution, 6.0 mL, used in the investigation.

Strategy: Use the molarity to write a unit factor for volume-to-mole conversion.

Implementation:

$$1 \text{ L } CoCl_2 \cdot 6H_2O \text{ solution} = [\text{“contains” written as an equality}]$$
$$0.15 \text{ mol } CoCl_2 \cdot 6H_2O \text{ [from molarity]}$$

We used 6.0 mL (0.0060 L) of the 0.15 M $CoCl_2 \cdot 6H_2O$ solution in Mixture 1. The volume-to-mole conversion is

$0.0060 \text{ L } CoCl_2 \cdot 6H_2O$ solution

$$= 0.0060 \text{ L} \left(\frac{0.15 \text{ mol } CoCl_2 \cdot 6H_2O}{1 \text{ L solution}} \right) = 0.00090 \text{ mol } CoCl_2 \cdot 6H_2O$$

Does the answer make sense?   We used about $\frac{1}{160}$ of a liter of solution, so we expect the number of moles in the mixture to be about $\frac{1}{160}$ as many as in a liter. This is indeed what we found.

━━━━━ **2.47** **CHECK THIS** ━━━━━━━━━━━━━━━━

Volume-to-mole conversion

How many moles of sodium phosphate, Na_3PO_4, are present in Mixture 1 in Investigate This 2.38? The mixtures shown all contained 6.0 mL of the phosphate solutions.

The results from Worked Example 2.46 and Check This 2.47 show you that Mixture 1 contains the same number of moles of Na_3PO_4 as moles of $CoCl_2 \cdot 6H_2O$. Although the mixtures contain a different *mass* of each solute, their molar masses are different so the number of *moles* of each is the same. Table 2.5 summarizes the results of our previous calculations and gives the results for the other $CoCl_2 \cdot 6H_2O$ and Na_3PO_4 solutions and mixtures. We now have the information necessary to understand how the reactions going on in the mixtures produce the observed results. Before going on to this analysis in Section 2.10, let's review and generalize the sequence of calculations we have just done and practice a bit more with mass–mole–volume calculations in Section 2.9.

Table 2.5	**Molar composition of the solutions and mixtures in Investigate This 2.38.**		

(Assumes a volume of 6.0 mL of each solution is used in each mixture.)

	mol·L^{-1}	mol CoCl$_2$·6H$_2$O	mol Na$_3$PO$_4$
higher concentration CoCl$_2$·6H$_2$O	0.15		
lower concentration CoCl$_2$·6H$_2$O	0.075		
higher concentration Na$_3$PO$_4$	0.15		
lower concentration Na$_3$PO$_4$	0.075		
mixture 1		0.00090	0.00090
mixture 2		0.00090	0.00045
mixture 3		0.00045	0.00090
mixture 4		0.00045	0.00045

2.9. Mass–Mole–Volume Calculations

We can summarize the process we followed in the previous calculations with a "route map," Figure 2.19(a). The blue arrow from mass (grams) to moles represents the calculations in Worked Example 2.42 and Check This 2.43.

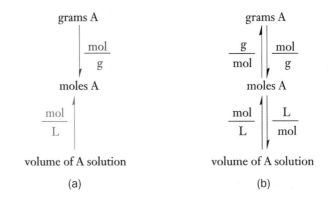

Figure 2.19.

Route map for mass–mole–volume calculations.

The unit conversion factor, mol/g, beside the arrow is a reminder that this is the way the molar mass is used to convert mass to moles. The red arrow from volume of solution to moles represents the calculations in Worked Example 2.46 and Check This 2.47. The unit conversion factor, mol/L, beside the arrow is a reminder that this is the way the molarity is used to convert volume to moles.

We can generalize the map to show that these conversions can go the other way as well, moles to mass and moles to volume, Figure 2.19(b). Again, the unit conversion factors beside the arrows remind you how the molar mass and molarity are used to make the conversions. We will often use this route map to illustrate the calculations we make. If you find it a helpful tool to direct your thinking about other problems like these, please use it there also.

2.48 WORKED EXAMPLE

Mole-to-mass conversion

We need 0.045 mol of glucose, $C_6H_{12}O_6$, for an experiment. What mass of glucose do we have to weigh out to get this amount?

Necessary information: We need relative atomic masses, so we can determine the molar mass of glucose.

Strategy: Calculate the molar mass of glucose and use it to convert moles to mass. The conversion is illustrated by the red arrow and unit factor on this route map.

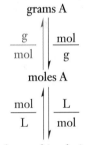

Implementation: Molar mass of glucose:

C: 6×12.0 g   =   72.0 g
H: 12×1.008 g   =   12.1 g
O: 6×16.0 g   =   96.0 g
     molar mass   =   180.1 g glucose

Mole-to-mass conversion:

$$0.045 \text{ mol glucose} = 0.045 \text{ mol glucose} \left(\frac{180.1 \text{ g glucose}}{1 \text{ mol glucose}} \right) = 8.1 \text{ g glucose}$$

Does the answer make sense?   We need about $\frac{1}{20}$ of a mole of glucose, and 8 g is about $\frac{1}{20}$ of 180 g, so our answer makes sense.

2.49 CHECK THIS

Mole-to-mass conversion

What mass of water do you have to weigh out to obtain 12.5 mol of water?

2.50 WORKED EXAMPLE

Mole-to-volume conversion

Suppose the glucose we need in Worked Example 2.48, 0.045 mol, is in an aqueous solution that is 0.10 M in glucose. What volume of the solution do we need to use?

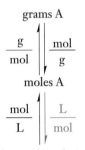

Necessary information: All the information we need is in the statement of the problem.

Strategy: Use the molarity of the solution to convert moles to volume of solution. The conversion is illustrated in red on this route map.

Implementation:

$$0.045 \text{ mol glucose} = 0.045 \text{ mol glucose} \left(\frac{1 \text{ L solution}}{0.10 \text{ mol glucose}} \right)$$

$$= 0.45 \text{ L solution}$$

Does the answer make sense? We need about $\frac{1}{20}$ of a mole of glucose, and a liter of solution contains $\frac{1}{10}$ of a mole, so about half a liter is required, as we found.

2.51 CHECK THIS

Mole-to-volume conversion

What volume of a 0.050 M solution of potassium phosphate contains 1.00×10^{-3} mol of potassium phosphate?

Preparing solutions of known molarity So far, our calculations have taken only single steps on the route map in Figure 2.19(b). You will, however, encounter problems that require multiple steps. A common example that arises many times in chemistry and biology experiments is the need to prepare a volume of aqueous solution that contains a solute at a known molarity. The procedure for making the solution is to dissolve the appropriate mass of the solute in a small volume of water and then add water to obtain the desired volume of solution. You need to calculate the appropriate mass of solute to weigh out.

2.52 WORKED EXAMPLE

Preparing a solution of known molarity

How many grams of the amino acid glycine, H_2NCH_2COOH, do you need to weigh out to prepare 250.0 mL (0.2500 L) of aqueous 0.150 M glycine solution?

continued

Necessary information: We need the relative atomic masses, so we can determine the molar mass of glycine.

Strategy: We know the desired volume of solution and its concentration, so we need to work back through moles to the mass, as shown by the red arrows on this route map.

grams A

$$\frac{g}{mol} \quad \frac{mol}{g}$$

moles A

$$\frac{mol}{L} \quad \frac{L}{mol}$$

volume of A solution

Implementation: Molar mass of glycine:

C:	2×12.01 g	$=$	24.02 g
H:	5×1.008 g	$=$	5.04 g
O:	2×16.00 g	$=$	32.00 g
N:	1×14.01 g	$=$	14.01 g
	molar mass	$=$	75.07 g glycine

Volume-to-mole conversion:

$$0.2500 \text{ L solution} = 0.2500 \text{ L solution} \left(\frac{0.150 \text{ mol glycine}}{1 \text{ L solution}} \right)$$

$$= 0.0375 \text{ mol glycine}$$

Mole-to-mass conversion:

$$0.0375 \text{ mol glycine} = 0.0375 \text{ mol glycine} \left(\frac{75.07 \text{ g glcine}}{1 \text{ mol glycine}} \right)$$

$$= 2.82 \text{ g glycine}$$

Does the answer make sense? The solution contains less than 1 mol of solute per liter and we only want one-quarter of a liter, so the required mass should be a good deal less than one molar mass, as it is.

2.53 CHECK THIS

Preparing a solution of known molarity

How many grams of glucose, $C_6H_{12}O_6$, do you need to weigh out to prepare 50.0 mL (0.0500 L) of an aqueous 2.50 M glucose solution?

2.54 CONSIDER THIS

Do the solubility rules explain the composition of seawater?

(a) Rivers on Earth, like the one in the chapter opener, have been flowing to the sea for billions of years. Everything the river waters dissolve eventually ends up in the sea. This table shows the ions that are present in highest

continued

Ionic composition of seawater

Ion	Molarity
Cl^-	0.545
Na^+	0.468
SO_4^{2-}	0.028
Mg^{2+}	0.054
Ca^{2+}	0.010
K^+	0.010
HCO_3^-	0.002

concentration in modern seawater. Do our solubility rules help explain this composition? Why or why not?

(b) Show that the number of moles of positive charge and of negative charge per liter of seawater is almost the same, that is, that seawater is an electrically neutral ionic solution.

Reflection and Projection

Chemists, biologists, and other scientists who work with solutions are often faced with the need to make mass–mole–volume interconversions like those discussed in the previous two sections. We introduced the mass–mole–volume route map as a guide that you might find helpful in directing your approach to these kinds of problems. If it is helpful, please continue to use it, but if you are more comfortable with another approach, by all means use that. Fundamental to all these conversions is the concept of the mole and, as you see on the route map, moles are central in all the conversions. Applying the mole concept to chemical reactions is our next task.

2.10. Reaction Stoichiometry in Solutions

Mixtures 1, 2, 3, and 4 after centrifugation.

2.55 CONSIDER THIS

Are concentration and amount of precipitate related?

The quantitative calculations in Section 2.8 involved the molarity of stock solutions of the reactants, $CoCl_2 \cdot 6H_2O$ and Na_3PO_4, and the number of moles of each in Mixture 1 in Investigate This 2.38. Are your answers for Consider This 2.39(b) still the same? Why or why not? Are the molar concentrations and numbers of moles in the four mixtures, from Table 2.5, page 108, consistent with your observed results? Explain.

All of the reactions and calculations in this section refer to the results from Investigate This 2.38, which are shown in Table 2.5. We have said that the net ionic reaction equation is

$$3Co^{2+}(aq) + 2PO_4^{3-}(aq) \rightarrow Co_3(PO_4)_2(s) \qquad (2.10)$$

To analyze the mixtures in which this reaction occurs, we need to know how many moles of each ion are present in the mixtures. So far, however, we have only referred to the concentrations and numbers of moles of $CoCl_2 \cdot 6H_2O$ and Na_3PO_4 in the stock solutions and mixtures. We need to account for the formation of ions when these ionic compounds dissolve, and then determine the number of moles (and concentrations) of the reactants, $Co^{2+}(aq)$ and $PO_4^{3-}(aq)$, in the mixtures.

2.56 WORKED EXAMPLE

Moles of ions in an ionic compound solution

$CoCl_2 \cdot 6H_2O(s)$ dissolves in water to form $Co^{2+}(aq)$ and $Cl^-(aq)$ ions:

$$CoCl_2 \cdot 6H_2O(s) \rightarrow Co^{2+}(aq) + 2Cl^-(aq) \qquad (2.13)$$

The six water molecules in the solid ionic compound are not shown explicitly in the products; they are part of the hydration layer for $Co^{2+}(aq)$. How many moles of $Co^{2+}(aq)$ are present in Mixture 1?

Necessary information: We need the balanced reaction equation (2.13) and the number of moles of $CoCl_2 \cdot 6H_2O(s)$ in the mixture, from Table 2.5 (repeated in this marginal table).

Strategy: The reaction equation shows us that each mole of $CoCl_2 \cdot 6H_2O(s)$ that goes into solution yields 1 mole of $Co^{2+}(aq)$ in solution. Use this relation to write a unit factor.

Mole composition of the mixtures

Mixture	$CoCl_2 \cdot 6H_2O$	Na_3PO_4
1	0.00090	0.00090
2	0.00090	0.00045
3	0.00045	0.00090
3	0.00045	0.00045

Implementation:
1 mol $CoCl_2 \cdot 6H_2O(s)$ = ["yields" written as an equality] 1 mol $Co^{2+}(aq)$

0.00090 mol $CoCl_2 \cdot 6H_2O(s)$

$$= 0.00090 \text{ mol } CoCl_2 \cdot 6H_2O(s) \left(\frac{1 \text{ mol } Co^{2+} (aq)}{1 \text{ mol } CoCl_2 \cdot H_2O(s)} \right)$$

$$= 0.00090 \text{ mol } Co^{2+}(aq) \text{ in solution}$$

Does the answer make sense? Equation (2.13) shows that the ionic compound is completely ionized in solution. The one-to-one relationship in equation (2.13) gives the same number of moles of dissolved cobalt ions as number of moles of cobalt chloride hexahydrate that dissolve.

2.57 CHECK THIS

Moles of ions in an ionic compound solution

Write the balanced chemical reaction equation for dissolving sodium phosphate, $Na_3PO_4(s)$, an ionic compound, in water. How many moles of $PO_4^{3-}(aq)$ are present in Mixture 1?

> Stoichiometry is from Greek *stoicheon* = element + *metry* = measurement and refers to quantitative relationships among atoms, molecules, or moles. We will use it to refer to reactant–product ratios in balanced reactions.

Stoichiometric reaction ratios In equation (2.13), or the equation you wrote in Check This 2.57, there is a simple one-to-one relation between the ion in its compound and the ion in solution, so you might think that writing a unit

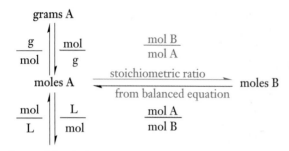

Figure 2.20.

Addition of reaction stoichiometry to the mass–mole–volume route map.

factor is a waste of effort. However, these relationships can be more complicated and we need to be sure to account correctly for them in reactions. **Stoichiometry** is the quantitative relationship of moles of reactants and products in a chemical equation. To remind you how to account for the stoichiometry of reactions, we will add a step to our route map for doing mass–mole–volume calculations, as shown in Figure 2.20. If A represents $CoCl_2 \cdot 6H_2O(s)$ and B represents $Co^{2+}(aq)$, the red arrow on the map represents the conversion we made in Worked Example 2.56.

For the mixtures in Investigate This 2.38, we can use the stoichiometric ratios from reaction equation (2.10) to calculate how much of each reactant, $Co^{2+}(aq)$ and $PO_4^{3-}(aq)$, is used up in the reaction and how much product, $Co_3(PO_4)_2(s)$, could have been formed. Reaction equation (2.10) shows that 3 mol of $Co^{2+}(aq)$ reacts with 2 mol of $PO_4^{3-}(aq)$ to yield 1 mol of $Co_3(PO_4)_2(s)$. We'll use this stoichiometric ratio of reactants to find out how many moles of $PO_4^{3-}(aq)$ are required to react with all the $Co^{2+}(aq)$ in Mixture 1.

2.58 WORKED EXAMPLE

Moles of $PO_4^{3-}(aq)$ required for complete reaction with $Co^{2+}(aq)$

How many moles of $PO_4^{3-}(aq)$ are required to react with 0.00090 mol $Co^{2+}(aq)$ (Mixture 1) by reaction equation (2.10)?

Necessary information: We need the reactant mole ratio from reaction equation (2.10).

Implementation:
3 mol $Co^{2+}(aq)$ = ["reacts with" written as an equality] 2 mol $PO_4^{3-}(aq)$

$$0.00090 \text{ mol } Co^{2+}(aq) = 0.00090 \text{ mol } Co^{2+}(aq) \left(\frac{2 \text{ mol } PO_4^{3-}(aq)}{3 \text{ mol } Co^{2+}(aq)} \right)$$

$$= 0.00060 \text{ mol } PO_4^{3-}(aq)$$

Does the answer make sense? The two-to-three mole ratio is exactly what we expect.

2.59 CHECK THIS

Moles of $Co^{2+}(aq)$ required for complete reaction with $PO_4^{3-}(aq)$

How many moles of $Co^{2+}(aq)$ are required to react with 0.00090 mol $PO_4^{3-}(aq)$ (the amount in Table 2.5 that you calculated for Mixture 1 in Check This 2.47) by reaction equation (2.10)?

Limiting reactant The result from Worked Example 2.58 shows that 0.00060 mol of $PO_4^{3-}(aq)$ is required to react with all the $Co^{2+}(aq)$ in Mixture 1. The mixture contains 0.00090 mol $PO_4^{3-}(aq)$, enough to react with all the $Co^{2+}(aq)$. [The clear, colorless solution left after centrifugation is experimental evidence that essentially all of the pink $Co^{2+}(aq)$ has precipitated.] As we would then expect, your result from Check This 2.59 shows that there is not enough $Co^{2+}(aq)$ to react with all the $PO_4^{3-}(aq)$. At least 0.00135 mol of $Co^{2+}(aq)$ is required and the mixture contains only 0.000090 mol. The amount of precipitate that can be formed by reaction (2.10) in Mixture 1 is *limited* by the amount of $Co^{2+}(aq)$ present in the mixture. The reactant that limits the amount of reaction that can occur in a particular reaction mixture is called the **limiting reactant.**

To determine the limiting reactant in a reaction mixture, you do a set of calculations like those in Worked Example 2.58 and Check This 2.59. Then you compare the results with the composition of the mixture to see which reactant is present in excess; the other reactant is the limiting reactant. Sometimes, if you know something about similar mixtures, you can reason out which is the limiting reactant in a mixture without going through all the calculations. For example, Mixture 3 has the same amount of $PO_4^{3-}(aq)$ as Mixture 1, but only half as much $Co^{2+}(aq)$. You know that there is more than enough $PO_4^{3-}(aq)$ in Mixture 1 to react with all the $Co^{2+}(aq)$ present in that mixture. Mixture 3 contains only half as much $Co^{2+}(aq)$, so $Co^{2+}(aq)$ must still be the limiting reactant in this mixture and only half as much reaction can occur. This explains why Mixture 3 produces only about half as much precipitate as Mixture 1.

2.60 CHECK THIS

Limiting reactant

(a) What is the limiting reactant in Mixture 2 from Investigate This 2.38? Try to reason out the answer and then do the least amount of calculation necessary to check your reasoning.

(b) Do your observations in Investigate This 2.38 provide any evidence to support your conclusion in part (a)? If so, explain the evidence.

Amount of product formed What mass of precipitate is produced by the reaction in each of the four mixtures in Investigate This 2.38? To show how to answer this question, we will complete our mass–mole–volume route map, which we will now call the stoichiometric route map, Figure 2.21.

Any conversion you carry out involves moles. There are two steps to the calculation of the mass of product from the known number of moles of $Co^{2+}(aq)$ that limit the amount of reaction in Mixture 1; the red arrows in Figure 2.21 show these steps. First we do the mole-mole conversion to determine the number of moles of product, $Co_3(PO_4)_2(s)$, that can be formed and then a mole-mass conversion, as we have done in the previous sections, to get the mass of $Co_3(PO_4)_2(s)$.

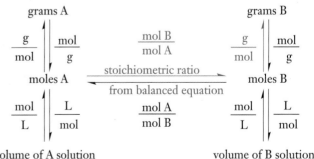

Figure 2.21.

Addition of product stoichiometry to the stoichiometric route map.

2.61 WORKED EXAMPLE

Mass of product formed in an ionic precipitation reaction

What mass of $Co_3(PO_4)_2(s)$ is produced in Mixture 1 in Investigate This 2.38?

Necessary information: We need the stoichiometry of the reaction from reaction equation (2.10), the number of moles of limiting reactant, and the molar mass of $Co_3(PO_4)_2(s)$.

Strategy: The approach is presented above and illustrated in Figure 2.21.

Implementation:

3 mol $Co^{2+}(aq)$ = ["yields" written as an equality] 1 mol $Co_3(PO_4)_2(s)$

$$0.00090 \text{ mol } Co^{2+}(aq) = 0.00090 \text{ mol } Co^{2+}(aq)\left(\frac{1 \text{ mol } Co_3(PO_4)_2(s)}{3 \text{ mol } Co^{2+}(aq)} \right)$$

$$= 0.00030 \text{ mol } Co_3(PO_4)_2(s)$$

1 mol $Co_3(PO_4)_2(s)$ = 366.7 g $Co_3(PO_4)_2(s)$ [molar mass]

0.00030 mol $Co_3(PO_4)_2(s)$

$$= 0.00030 \text{ mol } Co_3(PO_4)_2(s)\left(\frac{366.7 \text{ g } Co_3(PO_4)_2(s)}{1 \text{ mol } Co_3(PO_4)_2(s)} \right)$$

$$= 0.11 \text{ g } Co_3(PO_4)_2(s)$$

Does the answer make sense? There is slightly less than $\frac{1}{1000}$ of a mole of $Co^{2+}(aq)$ to react. We should get somewhat less than $\frac{1}{3000}$ of a mole of $Co_3(PO_4)_2(s)$, which is what we find.

2.62 CHECK THIS

Mass of product formed in an ionic precipitation reaction

(a) In reaction equation (2.10), what is the stoichiometric ratio of $Co_3(PO_4)_2(s)$ formed to $PO_4^{3-}(aq)$ that reacts?
(b) What mass of $Co_3(PO_4)_2(s)$ is produced in Mixture 2? Use your results from Check This 2.60, and the stoichiometric ratio from part (a). How does this mass compare with the amount from Mixture 1? Is this result consistent with your observations on the mixtures? Explain.

The mass of $Co_3(PO_4)_2(s)$ you calculated in Check This 2.62 is about three-quarters the mass we calculated in Worked Example 2.61. Less precipitate is formed in Mixture 2 than in Mixture 1 (but it may be hard to tell whether it is one-quarter less). $Co^{2+}(aq)$ is the limiting reactant in Mixtures 3 and 4, so the amount of precipitate in each should be the same and half as much as in Mixture 1. Thus, your observations on the precipitates in Investigate This 2.38 all make sense.

Limiting reactant

Show, without doing any further calculations, that $Co^{2+}(aq)$ is the limiting reactant in Mixture 4. Explain your reasoning.

Amount of reactant remaining There is, however, another question we can ask about the reaction mixtures: What is left in the solutions? The complete balanced ionic reaction equation for the $Co_3(PO_4)_2(s)$ precipitation is

$$3Co^{2+}(aq) + 6Cl^-(aq) + 6Na^+(aq) + 2PO_4^{3-}(aq)$$

$$\rightarrow Co_3(PO_4)_2(s) + 6Na^+(aq) + 6Cl^-(aq) \qquad \textbf{(2.14)}$$

Reaction equation (2.14) only partially represents the actual state of affairs in our mixtures. We know that the ratio of $Co^{2+}(aq)$ to $PO_4^{3-}(aq)$ in the mixtures is not three to two. In Mixtures 1, 3, and 4, $Co^{2+}(aq)$ is the limiting reactant, so some $PO_4^{3-}(aq)$ is left over in the solution, together with the spectator ions, $Na^+(aq)$ and $Cl^-(aq)$. The leftover $PO_4^{3-}(aq)$ is not represented in equation (2.14). Figure 2.22 shows a molecular-level representation of cases like these.

$\oplus = Na^+$ $\clubsuit = PO_4^{3-}$

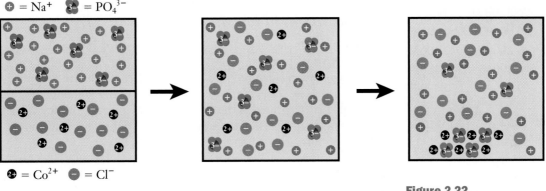

$\text{2+} = Co^{2+}$ $\ominus = Cl^-$

Figure 2.22.

Representation of the reaction between $CoCl_2$ and Na_3PO_4 with Na_3PO_4 in excess.
Molecular-level representation of $Co_3(PO_4)_2$ precipitation with equal amounts of Co^{2+} and PO_4^{3-}, a stoichiometric excess of PO_4^{3-}.

Illustrations of reactant mixtures

(a) Show (by counting) that all the solutions and the solid represented in Figure 2.22 are electrically neutral.
(b) Make a sketch, modeled after Figure 2.22, that represents the reaction in Mixture 2.

Figure 2.22 gives you a qualitative picture of what is happening at the molecular level in the mixture and shows that some $PO_4^{3-}(aq)$ is left over in the solution. To find the actual amount that is left over in Mixture 1, we subtract the number of moles that react with $Co^{2+}(aq)$ (0.00060 mol) from the original number of moles present (0.00090 mol). There is 0.00030 mol of $PO_4^{3-}(aq)$ remaining in the solution.

Concentrations of ions in a reactant mixture

The volume of Mixture 1 shown in the photos is 12 mL (0.012 L). What is the molar concentration of $PO_4^{3-}(aq)$ in this solution?

Necessary information: We need the number of moles of $PO_4^{3-}(aq)$ left in this solution, 0.00030 mol, from the preceding paragraph.

Strategy: Divide the number of moles by the volume of solution to get molarity.

Implementation:

$$\text{molarity of } PO_4^{3-}(aq) = \left[PO_4^{3-}(aq)\right] = \frac{0.00030 \text{ mol}}{0.012 \text{ L}} = 0.025 \text{ M}$$

Does the answer make sense? The $PO_4^{3-}(aq)$ is in a small volume of solution, so we expect the molarity, $mol \cdot L^{-1}$, to be larger than the number of moles, as it is.

Mole composition of the mixtures

Mixture	$CoCl_2 \cdot 6H_2O$	Na_3PO_4
1	0.00090	0.00090
2	0.00090	0.00045
3	0.00045	0.00090
3	0.00045	0.00045

Concentrations of ions in a reactant mixture

(a) How many moles of $Co^{2+}(aq)$ are left unreacted in Mixture 2? Show how you get your answer.
(b) The volume of Mixture 2 is 12 mL (0.012 L). What are the molar concentrations of $Co^{2+}(aq)$, $Na^+(aq)$, and $Cl^-(aq)$ in this solution? Use the results in Table 2.5 (repeated in the marginal table) to get the moles of $Na^+(aq)$ and $Cl^-(aq)$. What part(s) of the stoichiometric route map could you use for each ion?

All the calculations we have made on the number of moles and the molarities in the reaction mixtures have neglected the reverse reaction:

$$Co_3(PO_4)_2(s) \rightarrow 3Co^{2+}(aq) + 2PO_4^{3-}(aq) \tag{2.15}$$

In Section 2.6, we said that no ionic precipitation reaction uses up all the ions. A few ions are always left in the solution and the concentrations of ions in solution ultimately reach equilibrium with the ions in the solid. For the kind of precipitation reactions we have considered here, there is only a tiny concentration of the limiting reactant ion left in solution. The assumption that the reactions go to completion is a good one for many, but not all, cases.

Reflection and Projection

This section dealt with the application of the mole concept to ionic reactions in solution. As a specific example to guide the discussion, we worked through an analysis of the reaction mixtures and observations made in Investigate This 2.38.

In order to do this, we expanded our mass–mole–volume route map to include the stoichiometry of reactions. The centrality of moles to stoichiometry is shown clearly on the expanded map where every conversion involves moles. A balanced chemical equation provides stoichiometric mole ratios among the reactants and products of a reaction that permit you to do a variety of conversions and comparisons. We found, for example, that the number of moles of a limiting reactant controlled the amount of precipitate formed in each mixture. And, knowing the amount of each reactant used up, we could calculate the composition of the final solution (assuming complete precipitation of the ionic solid).

So far, we have discussed the aqueous solubility of polar and nonpolar liquids and solids and of ionic solids. What about the solubility of gases? A discussion of gas solubility leads to another fundamental characteristic of aqueous solutions: their acid–base properties. These are the topics of the next sections.

2.11. Solutions of Gases in Water

2.67 INVESTIGATE THIS

What are the properties of aqueous solutions of gases?

Do this as a class investigation and work in small groups to discuss and analyze the results.

(a) Use an electrical conductivity tester to test these solutions:

- Distilled or deionized water [dissolved nitrogen, $N_2(g)$, oxygen, $O_2(g)$, and a tiny bit of carbon dioxide, $CO_2(g)$, from the air]
- Carbon dioxide, $CO_2(g)$, dissolved in water
- Ammonia, $NH_3(g)$, dissolved in water
- Hydrogen chloride, $HCl(g)$, dissolved in water
- Glucose, $C_6H_{12}O_6(s)$, dissolved in water (provides a reference to previous discussions)

Which solutions conduct an electrical current? Which do not?

(b) Use a pH meter or pH paper to measure the pH of each of the solutions. Record the results and save them for analysis in Consider This 2.74, Section 2.12.

> Aqueous solutions of $CO_2(g)$, $NH_3(g)$, and $HCl(g)$ are available at the supermarket as seltzer water, household ammonia, and some liquid drain cleaners, respectively.

2.68 CONSIDER THIS

Why do some aqueous solutions of gas conduct electrical current?

(a) What conclusion(s) can you draw about the composition of the solutions of gases that conduct an electric current in Investigate This 2.67?

(b) Are all the conductivities the same?

(c) Can you use the answer in part (a) to explain the answer in part (b)? Why or why not?

Water dissolves gases just as it does solids and liquids. This is fortunate, because dissolved carbon dioxide and oxygen are essential for the plant and animal life in rivers and seas. The solubilities of the gases whose solutions you investigated are given in Table 2.6. Some gases are much more soluble than others, just as we found for the solubilities of other molecular and ionic solutes.

Table 2.6 *Aqueous solubility of gases at 25 °C.*

Solubilities are in $g \cdot kg^{-1}$ (grams of gas per kilogram of water) at a total pressure of 101 kPa (1 atmosphere).

Gas	N_2	O_2	CO_2	NH_3	HCl
Solubility	0.018	0.039	1.45	470	695

2.69 WORKED EXAMPLE

Molarity of saturated aqueous solutions of gases

Table 2.6 gives the maximum amount of each gas (at atmospheric pressure) that will dissolve in 1 kg (≈ 1.00 L) of water. Assuming that N_2 does not change the volume of the water it dissolves in, what is the molarity of a saturated solution of N_2?

Necessary information: The problem statement and Table 2.6 contain all we need.

Strategy: Convert grams of dissolved gas to moles and divide by the volume to get molarity.

Implementation:

$$0.018 \text{ g N}_2 = (0.018 \text{ g N}_2)\left(\frac{1 \text{ mol N}_2}{28.0 \text{ g N}_2}\right) = 6.4 \times 10^{-4} \text{ mol}$$

$$\text{molarity of N}_2 = \frac{6.4 \times 10^{-1} \text{ mol}}{1.00 \text{ L}} = 6.4 \times 10^{-4} \text{ M}$$

Does the answer make sense? Much less than 1 mol of nitrogen gas dissolves in a liter of water, so the low concentration makes sense.

2.70 CHECK THIS

Molarity of saturated aqueous solutions of gases

What are the molarities of saturated solutions of O_2 and CO_2 gases? Use the same approximations as in Worked Example 2.69 and show how you get your answers.

2.71 WORKED EXAMPLE

Molarity of a saturated aqueous solution of ammonia

What is the molarity of the ammonia solution in Table 2.6?

Necessary information: When this much ammonia dissolves in 1 L of water, the volume of the solution will not be 1 L, so we need the density of the solution, 0.89 kg·L^{-1}, as well as the molar mass of ammonia.

Strategy: Use the density to convert the mass of the solution to volume. Use the molar mass to convert the mass of ammonia to moles and then apply the definition of molarity.

Implementation: A quantity of 0.470 kg of ammonia is dissolved in exactly 1 kg of water, so the mass of the ammonia solution is 1.470 kg. The volume of the solution is

$$1.470 \text{ kg solution} = 1.470 \text{ kg solution} \left(\frac{1 \text{ L solution}}{0.89 \text{ kg solution}} \right)$$

$$= 1.65 \text{ L solution}$$

$$1 \text{ mol NH}_3 = 17.0 \text{ g [from molar mass]}$$

$$470 \text{ g NH}_3 = 470 \text{ g NH}_3 \left(\frac{1 \text{ mol NH}_3}{17.0 \text{ g NH}_3} \right) = 27.6 \text{ mol NH}_3$$

$$\text{NH}_3 \text{ molarity} = \left(\frac{27.6 \text{ mol NH}_3}{1.65 \text{ L}} \right) = 16.8 \text{ M NH}_3$$

Does the answer make sense? More than 27 mol of ammonia dissolves in a liter of water, so this high molarity is expected and makes sense.

2.72 CHECK THIS

Molarity of a saturated aqueous solution of hydrogen chloride

What is the molarity of the hydrogen chloride solution in Table 2.6? The density of the solution is 1.20 kg·L^{-1}.

Molecular models of the five gases are shown in Figure 2.23. The molecules of nitrogen and oxygen are nonpolar. There should be little attraction between them and water molecules and, indeed, Table 2.6 and the calculations in Worked Example 2.69 and Check This 2.70 show that these gases are not very soluble. The gas molecules in solution can readily escape into the gas phase when they are near the surface of the solution. The explanation(s) for the larger solubilities of the other three gases must lie in their interactions with water molecules.

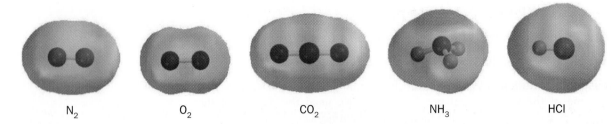

N₂   O₂   CO₂   NH₃   HCl

Figure 2.23.

Models showing the polarity of five gaseous molecules.

Web Companion

Chapter 1, Section 2.11.1 ——(1)

View and analyze molecular-level visualizations of HCl reacting with H₂O.

②
③
④

Water as a reactant Your results in Investigate This 2.67 provide a clue to the explanation. Solutions of $CO_2(g)$, $NH_3(g)$, and $HCl(g)$ in water conduct an electric current; therefore, ions must be present in these solutions. Solutions of glucose or $O_2(g)$ and $N_2(g)$ do not conduct an electric current; ions are not present in these solutions. Our model for the solubility of glucose involves multiple hydrogen bonds (Consider This 2.10) but no ions. Similarly, no ions are formed in solutions of N_2 and/or O_2. How are ions formed when molecules of CO_2, NH_3, and HCl dissolve in water?

There are differences among the three conducting solutions. The solution of dissolved HCl is a better conductor of electric current than solutions of either CO_2 or NH_3. Higher conductivity must mean that more ions are present in the HCl solution. Aqueous solutions of hydrogen chloride, HCl(*aq*), are called hydrochloric acid. Acidic aqueous solutions contain substantial concentrations of **hydronium ion**, $H_3O^+(aq)$, water molecules to which a hydrogen cation (H^+) has been added. The addition of H^+ to a molecule is referred to as **protonation** because the added H^+ is simply a proton. In HCl(*aq*), the HCl molecules are the source of protons:

$$HCl(aq) + H_2O(l) \rightarrow Cl^-(aq) + H_3O^+(aq) \tag{2.16}$$

$$:\!\overset{..}{\underset{..}{Cl}}\!:\!H_{(aq)} + :\!\overset{..}{\underset{..}{O}}\!:\!H_{(l)} \longrightarrow :\!\overset{..}{\underset{..}{Cl}}\!:\!\overset{-}{_{(aq)}} + H:\!\overset{..}{\underset{..}{O}}\!:\!\overset{+}{H}_{(aq)} \tag{2.17}$$
$$\qquad\qquad\quad H \qquad\qquad\qquad\qquad\qquad H$$

Reaction (2.16) produces equal numbers of positive hydronium ions and negative chloride ions, $Cl^-(aq)$. These ions are responsible for the conductivity of the solution.

In reaction (2.16), water is a reactant in the chemical reaction, rather than a passive solvent for the reaction. Reactions between water and CO_2, and between water and NH_3, must also produce ions that enable their solutions to conduct electric current. Explaining the behavior of substances such as carbon dioxide or ammonia in water requires the concepts and terminology of acids and bases introduced in the next two sections. We'll also use this knowledge of acids and bases to find out how high-molecular-mass biomolecules can dissolve in water.

2.73 CHECK THIS

Compare solutions of NaCl and HCl

Use the *Web Companion*, Chapter 2, Sections 2.5.2–4 and 2.11.1, movies to compare molecular-level representations of solutions of NaCl and HCl, including the way they are formed. Make a list of the similarities and differences and briefly discuss each one.

2.12. The Acid–Base Reaction of Water with Itself

2.74 CONSIDER THIS

Which aqueous solutions of gas are acidic? Which are basic?

Refer to your record of pH measurements on the solutions of gases in Investigate This 2.67. At room temperature, aqueous solutions with a pH < 7 are defined as being **acidic.** Those solutions with a pH > 7 are defined as being **basic.** Which of the solutions are acidic? Which are basic? Which have a pH of 7? Is there any correlation between the pH and the electrical conductivity of these solutions?

> Aqueous solutions in contact with air contain a small amount of $CO_2(aq)$ dissolved from the air. Does this explain the pH you observe for distilled water? Explain why or why not.

The pH scale Hydronium ion concentrations in water vary from greater than 1 M in highly acidic solutions to less than 10^{-14} M in highly basic solutions. The pH scale was invented to describe the wide range of hydronium concentrations in water without using exponential numbers. pH is defined mathematically as a logarithmic function of the hydronium ion concentration (molarity):

$$pH \equiv -\log_{10}\left[H_3O^+\right] \qquad (2.18)$$

pH is the negative of the base-10 logarithm of the hydronium ion concentration.

Another way to think about the pH scale is in terms of the exponent of 10 when hydronium ion concentration is expressed as molarity. Figure 2.24 shows graphically how the pH and hydronium ion concentration are related; as $[H_3O^+]$ gets smaller (larger negative exponent), pH increases. Keep in mind that the pH is a logarithmic function. When the pH of a solution decreases from 4 to 2, for example, the concentration of hydronium ion, $[H_3O^+]$, has increased by a factor of 100, from 0.0001 to 0.01 M.

Figure 2.24.

The pH scale and corresponding concentrations of $H_3O^+(aq)$ and $HO^-(aq)$. In addition to the numerical concentration values, the wedges and symbols in the wedges are reminders of the *relative* concentrations of the hydronium ions and hydroxide ions at low, high, and intermediate pH.

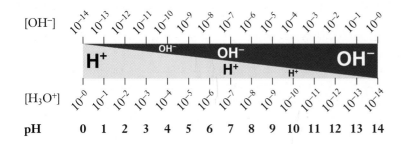

You observed that pure water does not conduct an electrical current, which suggests that there are no ions in pure water. That's almost, but not quite, true. At 25 °C, about two molecules in every billion react to form a hydronium ion, $H_3O^+(aq)$, and a **hydroxide ion**, $OH^-(aq)$:

$$H_2O(l) + H_2O(l) \rightleftharpoons H_3O^+(aq) + OH^-(aq) \qquad (2.19)$$

$$\ddot{:}\!\overset{..}{O}\!:\!H \cdots \ddot{:}\!\overset{..}{O}\!:\!H \;\rightleftharpoons\; \ddot{:}\!\overset{..}{O}\!:^{-} \cdots H\!:\!\overset{..}{O}\!:^{+}\!H \qquad (2.20)$$

A hydroxide ion is formed when a water molecule transfers an H^+ to another molecule. Figure 2.25 shows charge density models representing reaction (2.20). The proton in the hydrogen bond, the dashed line in reaction (2.20), has to move only about 80 pm to form the two ions; no electrons move between the two molecules of water.

> When an –OH group is covalently bonded to another atom, as in alcohols, it is named *hydroxy-*; the OH⁻ ion is named *hydroxide*.

Figure 2.25.

Transfer of a proton between two hydrogen-bonded water molecules.
Note the great change in charge distribution as the proton is transferred from one water molecule to the other.

 Web Companion

Chapter 2, Section 2.12.2–3 ①

Examine visualizations of proton ②
transfers among H₂O molecules. ③
④

The double arrows in Figure 2.25 and reaction equations (2.19) and (2.20) emphasize that the proton transfer is reversible. A hydronium ion can transfer the proton back to the hydroxide ion to form two water molecules. The process of transferring protons among water molecules is constant, and occurs very rapidly, so **equilibrium** is maintained between the molecules and ions. The pink and blue wedges and the H^+ and OH^- symbols in Figure 2.24 provide a visual indicator of the relative equilibrium concentrations of hydronium and hydroxide ion at different pH values.

The water molecules represented in equation (2.20) and modeled in Figure 2.25 are, in turn, hydrogen bonded to many others. This network of hydrogen-bond connections is represented in Figure 2.26. Figure 2.26(a) and (b) again show the proton transfer between two hydrogen-bonded water molecules. Figure 2.26(c) shows how a domino effect of proton transfers among the hydrogen-bonded network can lead to *separated* $H_3O^+(aq)$ and $OH^-(aq)$ ions. In pure water, there are equal concentrations of each ion. At 25 °C, $[H_3O^+] = [OH^-] = 1 \times 10^{-7}$ M, which is the middle of the pH scale in Figure 2.24.

> Experiments show that the speeds of reactions of $H_3O^+(aq)$ and $OH^-(aq)$ are 10–100 times faster than the ions could move in water. Transfer of protons from one molecule to the next, as shown in Figure 2.26, explains the observed speeds.

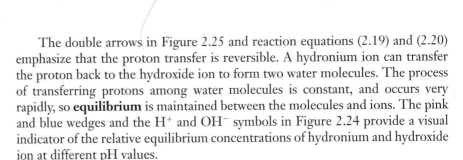

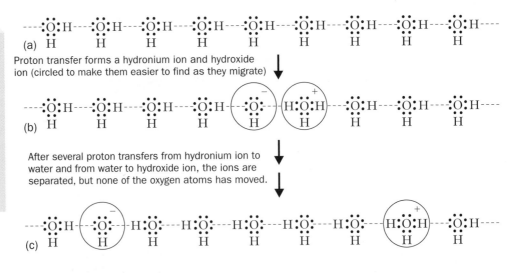

Figure 2.26.

Proton transfers in a hydrogen-bonded network of water molecules.
Each line shows the same set of molecules; only the bonding between H and O changes.

Brønsted–Lowry acids and bases Reactions (2.16) and (2.19) are similar. Hydronium ion, $H_3O^+(aq)$, is formed in each case. The formation of the hydronium ion results from the transfer of a proton (H^+) from dissolved hydrogen chloride gas to water in reaction (2.16) and from one water molecule to another in reaction (2.19). *Any molecule or ion that can transfer or donate a proton to another molecule or ion is called a* **Brønsted–Lowry acid.** *The molecule or ion that accepts the proton is called a* **Brønsted–Lowry base.** An aqueous solution is **acidic** if a Brønsted–Lowry acid in solution transfers protons to water to give a solution with more $H_3O^+(aq)$ (pH < 7) than is formed by reaction (2.19). An aqueous **basic** solution contains more $OH^-(aq)$ (pH > 7) than is formed by reaction (2.19).

In 1923, Johannes Brønsted (Danish chemist, 1879–1947) and Thomas Lowry (English chemist, 1874–1936) independently proposed the definitions for acid and base given here.

2.75 CONSIDER THIS

How many moles of hydronium ion are present in water?

If a compound is dissolved in water and the resulting solution is 10^{-4} M in $H_3O^+(aq)$, is the compound in solution a Brønsted–Lowry acid? Explain why or why not.

The hydrides of Group VI (H_2S, H_2Se, and H_2Te) and Group VII (HF, HCl, HBr, and HI) are Brønsted–Lowry acids that can donate a proton to water, as we have shown in reaction (2.16) for HCl. For example, a saturated solution of H_2S, hydrogen sulfide, in water has a pH of about 4, that is, $[H_3O^+(aq)] \approx 10^{-4}$ M. This hydronium ion concentration is higher than that in pure water, 10^{-7} M, so the solution is acidic. At least some of the hydrogen sulfide molecules have donated one of their protons to water molecules:

$$H_2S(aq) + H_2O(l) \rightleftharpoons H_3O^+(aq) + HS^-(aq) \qquad \textbf{(2.21)}$$

You probably know about several other acids besides these hydrides. Indeed, there are many more Brønsted–Lowry acids and in the next section we will introduce a large class of these acids, the oxyacids, that are especially important in biological systems.

2.76 CHECK THIS

Reactions of Brønsted–Lowry acids with water

Write the reaction equations for proton transfer from $H_2Te(aq)$ and HF(aq) to water.

2.13. Acids and Bases in Aqueous Solution

Oxyacids The great majority of Brønsted–Lowry acids you will meet are **oxyacids,** compounds in which the acidic hydrogen atom is bonded to an oxygen atom, which is, in turn, bonded to a nonmetal atom. Familiar acids like sulfuric (H_2SO_4), nitric (HNO_3), carbonic (H_2CO_3, in carbonated beverages), and acetic ($HC_2H_3O_2$, in vinegar) are oxyacids. The conventional formulas for these acids give you no hint that any of the hydrogen atoms are bonded to oxygen atoms and, indeed, little information about any of the connectivity in the molecules. Alternative structural formulas for oxyacids are shown in these reaction equations for transfer of a proton from sulfuric and carbonic acids to water:

$$(HO)_2SO_2(aq) + H_2O(l) \rightleftharpoons H_3O^+(aq) + HOSO_3^-(aq) \tag{2.22}$$

sulfuric acid   hydrogen sulfate anion

$$(HO)_2CO(aq) + H_2O(l) \rightleftharpoons H_3O^+(aq) + HOCO_2^-(aq) \tag{2.23}$$

carbonic acid   hydrogen carbonate anion

The products of the reactions of oxyacids with water are hydronium ion and an **oxyanion,** the anion formed from an oxyacid by loss of one of its acidic protons. Oxyanions are Brønsted–Lowry bases, that is, proton acceptors. Table 2.7 shows a few common oxyacids and their oxyanions. The table gives both the conventional and structural formulas for each oxyacid as well as the Lewis structure and the names of both the oxyacid and its oxyanion. The atoms of oxyacids and oxyanions are all joined by covalent bonds, including some double bonds.

Some Lewis structures involving phosphorus, sulfur, and chlorine have more than eight valence electrons around these atoms. Elements of period three and above can accommodate more than eight valence electrons. We will return to discuss these and double bonds in Chapters 5 and 6.

2.77 CHECK THIS

Structures and reactions for Brønsted–Lowry oxyacids and oxyanions

(a) Each oxyanion in Table 2.7 is formed from its corresponding oxyacid by loss of a proton. Write Lewis structures for the carbonate, sulfate, perchlorate, monohydrogen phosphate, and phosphate ions. Use the Lewis structures for the oxyacids and the structural formulas for the oxyacids and oxyanions to aid you.

(b) Write reaction equations for the transfer of a proton to water from nitric acid and from dihydrogen phosphate ion. Write each equation in three forms: with conventional formulas, with structural formulas, and with Lewis structures for all the oxyacids and oxyanions.

Ionic compounds with oxyanions Oxyanions can form ionic compounds with cations. You investigated several of these in Investigate This 2.32. When one of these compounds is dissolved in water, the acid–base properties of the solution are largely determined by the oxyanion. An important class of these ionic compounds are **hydroxides,** such as NaOH and $Mg(OH)_2$, which contain OH^-, the hydroxide ion. Solutions of these compounds always contain more $OH^-(aq)$ than pure water, so the solutions are basic:

$$Mg(OH)_2(s) \rightleftharpoons Mg^{2+}(aq) + 2OH^-(aq) \tag{2.24}$$

2.78 CHECK THIS

pH of a saturated magnesium hydroxide solution

Milk of Magnesia® is a mixture of $Mg(OH)_2(s)$ and water. About 0.01 g of $Mg(OH)_2$ will dissolve in a liter of water. What is $[OH^-(aq)]$ in this solution? What is the approximate pH?

Table 2.7 *Some common oxyacids and their oxyanions.*

Oxyacid name	Conventional oxyacid formula	Oxyacid Lewis structure	Oxyacid structural formula	Oxyanion structural formula	Oxyanion name
water	H_2O		H_2O	HO^-	hydroxide ion
hydronium ion	H_3O^+		H_3O^+	H_2O	water
nitric acid	HNO_3		$HONO_2$	NO_3^-	nitrate ion
carbonic acid	H_2CO_3		$(HO)_2CO$	$HOCO_2^-$	hydrogen carbonate ion
hydrogen carbonate ion	HCO_3^-		$HOCO_2^-$	CO_3^{2-}	carbonate ion
acetic acid (ethanoic acid)	$HC_2H_3O_2$		$CH_3C(O)OH$	$CH_3CO_2^-$	acetate ion (ethanoate ion)
perchloric acid	$HClO_4$		$HOClO_3$	ClO_4^-	perchlorate ion
hypochlorous acid	$HClO$		$HOCl$	ClO^-	hypochlorite ion
sulfuric acid	H_2SO_4		$(HO)_2SO_2$	$HOSO_3^-$	hydrogen sulfate ion
hydrogen sulfate ion	HSO_4^-		$HOSO_3^-$	SO_4^{2-}	sulfate ion
phosphoric acid	H_3PO_4		$(HO)_3PO$	$(HO)_2PO_2^-$	dihydrogen phosphate ion
dihydrogen phosphate ion	$H_2PO_4^-$		$(HO)_2PO_2^-$	$HOPO_3^{2-}$	monohydrogen phosphate ion
monohydrogen phosphate ion	HPO_4^{2-}		$HOPO_3^{2-}$	PO_4^{3-}	phosphate ion

Nucleic acids are oxyanions The **nucleic acids,** DNA and RNA, are large biomolecules that are ionic in cellular solutions. Each nitrogen base in DNA and RNA is bonded to a sugar molecule, as shown for double-stranded DNA in Figure 2.27. The sugar molecules are, in turn, bonded together by phosphate groups to form the backbone of each strand of the nucleic acid, the "uprights" of the ladder we described in Chapter 1, Section 1.9. The phosphate groups in the backbones donate their protons to water and are ionized to oxyanions. The acidic phosphate groups in the backbone give DNA and RNA their "A" (for acid). There are two negative phosphate anions for every base pair in the DNA structure; DNA and RNA are highly charged molecules. In the helical structures of the molecules, the sugar phosphate backbones are on the outside of the helix where the anionic sites are in contact with water. These information-carrying molecules can have molar masses in billions of grams, but are still somewhat water soluble because there is a strong attraction between these oxyanions and water molecules. Their water solubility is important because they and the molecules they interact with are all in the aqueous medium of living cells.

Carboxylic acids Many Brønsted–Lowry oxyacids are **carboxylic acids,** acids that contain the **carboxyl group,**

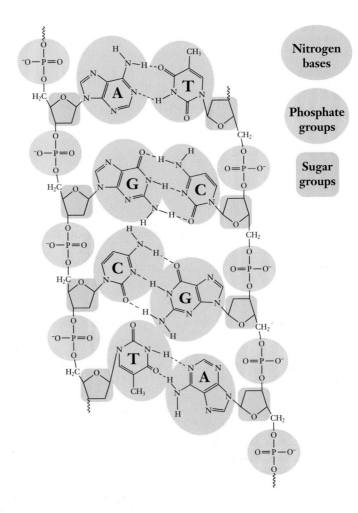

Figure 2.27.

Short segment of the ionic DNA molecule.

The phosphate groups are negatively charged and are attracted to water molecules.

When a proton leaves a carboxylic acid group, the **carboxylate ion**, is formed. The reaction for acetic (ethanoic) acid forming acetate (ethanoate) ion is

$$CH_3C(O)OH(aq) + H_2O(l) \rightleftharpoons CH_3C(O)O^-(aq) + H_3O^+(aq)$$

(2.25)

(2.26)

Most of the acids in biological systems, such as the amino acids (Figure 1.30 in Chapter 1), are carboxylic acids. These acids are sometimes called *organic* acids because the carboxyl group is often bonded to carbon-containing groups that are found in living (organic) systems. In Chapter 6, we will discuss why the hydrogen in a carboxylic acid group (and in other oxyacids as well) is acidic; for now, remember that compounds with a carboxylic acid group are Brønsted–Lowry oxyacids, proton donors.

2.14. Extent of Proton-Transfer Reactions: Le Chatelier's Principle

2.79 CONSIDER THIS

What are the reactions in an aqueous solution of carbon dioxide?

(a) When carbon dioxide gas dissolves in water, it can react with water:

$$CO_2(g) + H_2O(l) \rightleftharpoons (HO)_2CO(aq)$$ (2.27)

Does reaction (2.27) help explain the acidity of CO_2 solutions and its moderate solubility in water? Explain your responses.

(b) Your results in Check This 2.70 show that about 0.035 mol of $CO_2(g)$ dissolves in a liter of water at room temperature. The dissolved gas could undergo reaction (2.27) followed by:

$$(HO)_2CO(aq) + H_2O(l) \rightleftharpoons H_3O^+(aq) + HOCO_2^-(aq)$$ (2.23)

If reaction (2.23) goes to completion, what would be the concentration of hydronium ion in this solution? What would be the approximate pH of this solution? How does the pH measured in Investigate This 2.67 compare to this calculated value? What conclusion can you draw from this comparison about the extent of reaction (2.23)?

(c) Does your conclusion in part (b) help explain the observed electrical conductivity of the solution of dissolved carbon dioxide relative to the other solutions in the investigation? Explain why or why not.

Reaction equation (2.27) shows that carbon dioxide dissolves in and reacts with water to form an oxyacid, carbonic acid. Several other nonmetal oxides also dissolve in and react with water to yield oxyacids. Sulfur trioxide, $SO_3(g)$, for example reacts to give sulfuric acid:

$$SO_3(g) + H_2O(l) \rightleftharpoons (HO)_2SO_2(aq) \tag{2.28}$$

The stoichiometry of reaction (2.28) shows that if we dissolve 0.1 mol of $SO_3(g)$ in 1 L of water, we get a solution that is 0.1 M in sulfuric acid. If we measure the pH of this solution, we find that it is about 1. Figure 2.24 shows that this corresponds to a hydronium ion concentration, $[H_3O^+(aq)]$, of 10^{-1} M = 0.1 M. Since the hydronium ion concentration is equal to the concentration of sulfuric acid we prepared, we can conclude that the transfer of a proton from sulfuric acid to water, reaction equation (2.22), is complete. That is, in aqueous sulfuric acid solutions, all the sulfuric acid molecules transfer their first proton to water to form hydronium ion and the hydrogen sulfate anion. This is the situation represented at the molecular level in Figure 2.28(a).

$$HA \;+\; H_2O \;\rightleftharpoons\; A^- \;+\; H_3O^+$$

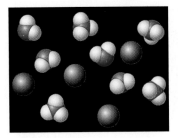

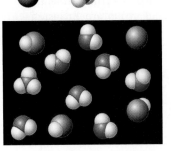

(a) Complete proton transfer   (b) Incomplete proton transfer

Figure 2.28.

Representations of the extent of proton transfer from an acid, HA, to water.

The majority of Brønsted–Lowry acids do not transfer their protons completely to water in aqueous solution. This is the situation represented at the molecular level in Figure 2.28(b). For example, you have found that about 0.035 mol of $CO_2(g)$ dissolves in a liter of water at room temperature. If reactions (2.27) and (2.23) go to completion, the hydronium ion concentration, $[H_3O^+(aq)]$, in this solution would be 0.035 M. That is, $[H_3O^+(aq)]$ would be between 0.1 M (10^{-1} M) and 0.01 M (10^{-2} M), so the pH of the solution would be in the range 1–2. Your pH measurements in Investigate This 2.67 show that a solution of carbon dioxide in water has a pH in the range 4–5. This means that $[H_3O^+(aq)]$ is much lower than it would be if reactions (2.27) and (2.23) went to completion. At equilibrium, only small amounts of $H_3O^+(aq)$ and $HOCO_2^-(aq)$ ions are formed in the solution, which accounts for the low conductivity of the carbon dioxide solution, compared to the solution of $HCl(g)$ in water, hydrochloric acid. Hydrochloric acid, like sulfuric acid, is another acid that transfers its protons completely to water. We will discuss these proton transfers more quantitatively in Chapters 6 and 9. Here we will consider the directionality of these reactions as solution conditions change.

2.80 CHECK THIS

Proton transfer from acetic (ethanoic) acid to water

(a) When the electrical conductivities of 0.1 M aqueous solutions of hydrochloric and acetic acids are tested, the acetic acid solution conducts less well. What can you conclude about the extent of reaction (2.25) compared to reaction (2.16)? Explain your reasoning.

(b) Which solution will have the lower pH, that is, be more acidic? Explain your reasoning.

2.81 INVESTIGATE THIS

Are CO$_2$ solutions affected by added acids and bases?

Do this as a class investigation and work in small groups to discuss and analyze the results.

(a) Use a 125-mL Erlenmeyer flask, a three-hole rubber stopper to fit the flask, 75 mL of a saturated solution of CO$_2$(g) in water (seltzer water), 0.6 g of solid sodium hydrogen carbonate, NaHOCO$_2$, and a few drops of universal acid–base indicator solution. Fit the three holes of the stopper with a plastic pipet about half full (3 mL) of 6 M hydrochloric acid, HCl, solution, a second plastic pipet filled completely (6 mL) with 6 M aqueous sodium hydroxide, NaOH, solution, and a glass or plastic tube to which a small balloon is tightly sealed. Place the sodium hydrogen carbonate and indicator in the flask, add the seltzer water, and insert the rubber stopper to seal the flask, as shown in this photograph. Swirl the flask gently to mix the contents. Record the color of the solution, the state of inflation of the balloon, and any other observations on the system. As a control solution, add a few drops of universal indicator to 75 mL of seltzer water in another flask and record its color.

(b) Add the hydrochloric acid to the mixture in the flask by squeezing out the contents of its pipet. Again swirl the flask gently to mix the contents and record the color of the solution, the state of inflation of the balloon, and any other observations on the system.

(c) Add the sodium hydroxide to the mixture by squeezing out the contents of its pipet. Again swirl the flask gently to mix the contents and record the color of the solution, the state of inflation of the balloon, and any other observations on the system.

2.82 CONSIDER THIS

How are CO$_2$ solutions affected by added acids and bases?

(a) Is there evidence for reaction(s) when the acid and the base are added to the mixture in Investigate This 2.81? What is the evidence in each case? *Note:* Universal indicator is red in acidic solutions, yellow to green in pH 5–8 solutions, and blue in basic solutions.

(b) Under what, if any, conditions was gas evolved by the mixture? Under what, if any, conditions was gas absorbed by the mixture? How might you explain your observations?

In Investigate This 2.81, addition of hydrogen carbonate ion, HOCO$_2{}^-$(aq), to a solution of carbon dioxide dissolved in water (carbonic acid) gives a solution with a pH of about 5–6 (a little higher than carbonic acid alone) and produces no bubbles of gas. If the (HO)$_2$CO(aq) solution equilibrium is described by equation (2.23) and very little of the (HO)$_2$CO(aq) transfers a proton to water, as we discussed above, these are the results we expect. The added HOCO$_2{}^-$(aq) reacts with some of the H$_3$O$^+$(aq) to produce a tiny bit of (HO)$_2$CO(aq). The result is a solution with a slightly higher pH (lower [H$_3$O$^+$(aq)]) than for carbonic acid alone. We might expect the additional $(HO)_2CO(aq)$ to disturb the equation (2.27) equilibrium and produce some $CO_2(g)$ by the reverse of the reaction. However, the amount of extra $(HO)_2CO(aq)$ that can be formed is so tiny that reaction (2.27) is barely disturbed and no detectable gas is produced.

Le Chatelier's principle The situation we have just described is an example of Le Chatelier's principle in action. Henry Louis Le Chatelier (French chemist, 1850–1936) worked on many industrial processes and was interested in ways to assure that reactions would proceed in the directions that produced desired products. Since many such processes involve reversible reactions, he looked for and found patterns in the behavior of equilibrium systems when they were disturbed or changed. **Le Chatelier's principle** states that *a system at equilibrium responds to a disturbance in a way that minimizes the effect of the disturbance.* The only disturbances that affect equilibria are changes in the *concentrations* of the reactants and/or products and changes in the *temperature* of the system. For example, increasing the concentration of a reactant or product causes the reacting system to respond in such a way as to use up some of the added molecules.

> Sometimes factors other than concentration and temperature are mentioned as disturbances, but you will find that they all really boil down to concentration or temperature effects.

Returning to our $(HO)_2CO(aq)$ equilibrium system to which $HOCO_2^-(aq)$ is added, we note that the disturbance to the equilibrium, reaction (2.23), is the addition of one of the reaction products, $HOCO_2^-(aq)$. Le Chatelier's principle says that the system will respond by trying to use up some of the added product. Reaction (2.23) will proceed in reverse until equilibrium is reattained with less of the products, $H_3O^+(aq)$ and $HOCO_2^-(aq)$, and more of the reactant, $(HO)_2CO(aq)$, than were present in the instant after the addition. We can represent the change this way:

$$(HO)_2CO(aq) + H_2O(l) \leftarrow H_3O^+(aq) + HOCO_2^-(aq) \qquad \textbf{(2.29)}$$

Equation (2.29) symbolizes the chemistry we discussed above when we said that $[H_3O^+(aq)]$ would be decreased a bit by addition of $HOCO_2^-(aq)$. Formation of more $(HO)_2CO(aq)$ will disturb equilibrium reaction (2.27) as well:

$$CO_2(g) + H_2O(l) \leftarrow (HO)_2CO(aq) \qquad \textbf{(2.30)}$$

As we said, the response to the minor disturbance was too small to be detectable.

This was not the case when you added a good deal of $H_3O^+(aq)$ (in the form of hydrochloric acid solution) to the reaction mixture. The disturbance to the system is again the addition of one of the reaction products. Equation (2.29) again represents the change that will occur as the system responds to minimize the effect of the disturbance by using up some of the added product. In this case, a large amount of $(HO)_2CO(aq)$ can be formed by the added acid; equilibrium (2.27) is greatly disturbed; and reaction (2.30) represents the system's response, which you observed as a copious formation of bubbles and inflation of the balloon.

2.83 WORKED EXAMPLE

Stoichiometry of addition of acid in Investigate This 2.79

Approximately what volume of $CO_2(g)$ can be produced by reaction (2.30) in Investigate This 2.81(b), if all the newly formed $(HO)_2CO(aq)$ reacts? Assume that a mole of gas at room temperature and one atmosphere pressure occupies 25 L.

continued

Necessary information: We need the mass of sodium hydrogen carbonate added, 0.6 g, the volume of 6 M hydrochloric acid added, 3 mL, the stoichiometry of the reactions, and the molar volume of a gas from the problem statement.

Strategy: Calculate the volume of gas, $CO_2(g)$, produced from the moles of gas produced. Assume all the extra $(HO)_2CO(aq)$ produced by reaction (2.29) reacts by reaction (2.30) to give $CO_2(g)$. This is a reasonable assumption because the original solution is saturated with $CO_2(g)$, which is in equilibrium with the original amount of $(HO)_2CO(aq)$. No more $CO_2(g)$ can be accommodated in the solution, so any formed by reaction (2.29) will escape and the reaction will proceed until only the original amount of $(HO)_2CO(aq)$ is left. Calculate the moles of $H_3O^+(aq)$ and $HOCO_2^-(aq)$ in the mixture to determine which is the limiting reactant and then calculate the moles of $(HO)_2CO(aq)$ formed and, hence, moles of $CO_2(g)$ produced.

Implementation: The moles of $H_3O^+(aq)$ and $HOCO_2^-(aq)$ in the mixture are

$$mol\ H_3O^+(aq) = mol\ HCl(aq) = (6\ M)(0.003\ L) = 18 \times 10^{-3}\ mol$$

$$1\ mol\ NaHOCO_2 = 84\ g$$

$$mol\ HOCO_2^-(aq) = mol\ NaHOCO_2 = \frac{0.6\ g\ NaHOCO_2}{84\ g{\cdot}mol^{-1}\ NaHOCO_2}$$

$$= 7 \times 10^{-3}\ mol$$

The hydronium and hydrogen carbonate ions react in a one-to-one ratio. Hydronium ion is in great excess, so $HOCO_2^-(aq)$ is the limiting reactant and can produce 7×10^{-3} mol $(HO)_2CO(aq)$ by reaction (2.29), which can in turn produce 7×10^{-3} mol $CO_2(g)$ by reaction (2.30).

$$volume\ CO_2(g) = [7 \times 10^{-3}\ mol\ CO_2(g)](25\ L{\cdot}mol^{-1})$$

$$= 0.175\ L \approx 180\ mL\ CO_2(g)$$

Does the answer make sense? When the hydrochloric acid was added in Investigate This 2.81(b), gas bubbles were produced in the mixture and the small balloon inflated somewhat. A 7-cm-diameter sphere has a volume of about 180 mL and your inflated balloon probably had a diameter close to this, so the answer makes sense.

2.84 CONSIDER THIS

What is the stoichiometry of addition of base in Investigate This 2.81?

(a) In Investigate This 2.81(c), how many moles of hydroxide ion (as a 6 M solution of NaOH) were added to the reaction mixture? Explain.

(b) What did you observe when the hydroxide ion was added to the mixture? What can you conclude about the direction of reaction (2.27) in this system with added hydroxide ion? Explain your reasoning.

continued

(c) Hydroxide ion reacts with acids in aqueous solution to form water by proton transfer from the acid. There are two acids for hydroxide ion to react with in the mixture from Investigate This 2.81(b); the stoichiometries of the reactions are

$$2OH^-(aq) + (HO)_2CO(aq) \rightarrow 2H_2O(l) + CO_3^{2-}(aq) \tag{2.31}$$

$$OH^-(aq) + H_3O^+(aq) \text{ [from added excess of } HCl(aq)] \rightarrow 2H_2O(l) \tag{2.32}$$

How many moles of $(HO)_2CO(aq)$ are available for the hydroxide to react with? Remember to account for both the original solution and your answer in part (b). Explain your reasoning. How many moles of $H_3O^+(aq)$ from the excess added in Investigate This 2.81(b) are present for the hydroxide to react with? Explain.

(d) Was enough hydroxide ion added to react with all the acid present? What are the major ions present in the mixture at the end of Investigate This 2.81? Explain your reasoning.

(e) Show how reaction (2.31) and Le Chatelier's principle explain your observations and support the conclusion you reached in part (b).

A Brønsted–Lowry base: Ammonia We have discussed the results of Investigate This 2.67 for the gases that dissolve to produce acidic solutions (pH < 7) and have shown how water accepts a proton from these molecules to give solutions with higher concentrations of hydronium ions than are present in pure water. To complete our discussion, we need to consider ammonia, $NH_3(g)$, which dissolves to give a basic solution (pH > 7) with a higher concentration of hydroxide ions than in pure water. Ammonia is a small polar molecule that can form hydrogen bonds with four water molecules, so its high solubility is not surprising. The solution conducts electricity, but poorly, so not many ions are present. The only source of hydroxide ion in this basic solution is water itself, so water molecules must transfer protons to ammonia, which acts as a Brønsted–Lowry base:

$$H_2O(l) + NH_3(aq) \rightleftharpoons OH^-(aq) + NH_4^+(aq) \tag{2.33}$$

$$\ddot{:}\!\!\overset{\displaystyle H}{\underset{\displaystyle H}{\ddot{O}}}\!\!:H\text{---}\overset{\displaystyle H}{\underset{\displaystyle H}{\ddot{N}}}\!\!:H \rightleftharpoons \overset{-}{\ddot{:}\!\!\overset{}{\underset{\displaystyle H}{\ddot{O}}}\!\!:}\text{---}H\overset{\displaystyle H}{\underset{\displaystyle H}{\overset{+}{\ddot{N}}}}\!\!:H \tag{2.34}$$

 Web Companion

Chapter 2, Section 2.13.4 — ①

Study and analyze molecular-level animations of NH_3 reacting with H_2O.

②
③
④

Reaction (2.33) produces a Brønsted–Lowry acid, ammonium ion, $NH_4^+(aq)$, and a Brønsted–Lowry base, hydroxide ion, as illustrated in Figure 2.29(a) with space-filling models and in Figure 2.29(b) with charge density models. The formation of ions explains why the solution conducts an electrical current and the formation of hydroxide ion makes the solution basic. The double arrows remind us that the reaction is reversible; hydroxide and ammonium ions can react to form water and ammonia. The rather poor electrical conductivity and only moderately basic solution (pH about 11) are indicators that reaction (2.33) does not go to completion, so produces low concentrations of $OH^-(aq)$ and $NH_4^+(aq)$ ions.

(a) The hydroxide ion is circled to make it easier to find as it is formed and "migrates" by further proton transfer.

(b) Charge density model of the reactions shown in (a). The water molecules are shown in different orientations.

Figure 2.29.

Formation and migration of an ammonium and hydroxide ion by proton transfers.

2.85 INVESTIGATE THIS

What happens when ammonium and hydroxide ions mix?

(a) Add a few drops of water to about 0.25 g of solid ammonium chloride, $NH_4Cl(s)$, in a small test tube. Carefully determine whether this mixture produces any odor.

(b) Repeat part (a), but use a mixture of 0.25 g $NH_4Cl(s)$ and 0.25 g $NaOH(s)$.

2.86 CONSIDER THIS

How do ammonium and hydroxide ions react?

(a) What were the similarities and differences between the observations you made in parts (a) and (b) in Investigate This 2.85? What can you conclude about any reactions occurring in either case? Explain your reasoning.

(b) Show how Le Chatelier's principle and what you know about reaction (2.33) support your conclusion in part (a).

Review the properties of solutions of gases We have now interpreted all of the results from Investigate This 2.67, based on whether molecules can react with water to produce ions in solution and, if so, how they react:

• Distilled water (with dissolved N_2 and O_2 and a tiny bit of CO_2) does not conduct an electrical current because so few ions are present.

- Aqueous solutions of HCl(*g*) conduct electricity well and have high acidity (low pH) because reaction (2.16) produces a large number of ions; the reaction goes to completion.
- Aqueous solutions of CO_2(*g*) conduct electricity poorly and have only moderate acidity [a higher pH than for HCl(*aq*)], because reaction (2.23) produces fewer ions; the reaction does not go to completion.
- Aqueous solutions of NH_3(*g*) are basic (pH > 7) and conduct electricity poorly. The solutions are basic because reaction (2.33) produces the ions NH_4^+(*aq*) and OH^-(*aq*). They are poor conductors of electricity because the reaction does not go to completion, so the concentration of ions is low. If we try to make a solution with high concentrations of NH_4^+(*aq*) and OH^-(*aq*) ions, as in Investigate This 2.85(b), the equilibrium system represented by reaction (2.33) is disturbed and responds by using up some of the excess ions to form ammonia. Some ammonia escapes as a gas and is detected by its distinctive odor.

Brønsted–Lowry acid–base pairs All of the proton-transfer reactions we have described so far involve the reaction of an acid with a base to produce a new acid and a new base:

$$HOSO_3^-(aq) + HOCO_2^-(aq) \rightleftharpoons SO_4^{2-}(aq) + (HO)_2CO(aq) \quad \text{(2.35)}$$

$$\text{acid}_1 \qquad\qquad \text{base}_2 \qquad\qquad \text{base}_1 \qquad\qquad \text{acid}_2$$

The hydrogen sulfate ion [identified as acid_1 in equation (2.35)] serves as an acid, donating a proton to a base, the hydrogen carbonate ion (identified as base_2). The acid–base reaction produces a new acid, carbonic acid (called acid_2), and a new base, sulfate ion (called base_1). In each acid–base pair (pair 1: $HOSO_3^-$ and SO_4^{2-}; and pair 2: $(HO)_2CO$ and $HOCO_2^-$), the acid differs from the base only by having a proton that the base does not have. Acid–base pairs that differ only by a proton present in the acid and not the base are called **conjugate acids and bases.**

2.87 CHECK THIS

Identifying Brønsted–Lowry conjugate acid–base pairs

Identify the Brønsted–Lowry acid and the Brønsted–Lowry base on each side of the proton transfer reactions (2.16), (2.21), (2.22), (2.23), and (2.25). For each conjugate pair, identify the proton (H^+) that differentiates the acid on one side of the reaction arrows from its conjugate base on the other side.

Another conjugate acid–base example is proton transfer from one water molecule to another:

$$H_2O(l) + H_2O(l) \rightleftharpoons OH^-(aq) + H_3O^+(aq) \quad \text{(2.19)}$$

$$\text{acid}_1 \qquad \text{base}_2 \qquad \text{base}_1 \qquad \text{acid}_2$$

One water molecule serves as a Brønsted–Lowry acid, donating a proton to the other water molecule, the Brønsted–Lowry base. The acid–base reaction produces a new Brønsted–Lowry acid, H_3O^+, and a new Brønsted–Lowry base, OH^-. Note that water can act as both an acid (proton donor) and a base (proton acceptor).

2.88 CHECK THIS

Species that can both donate and accept protons

(a) In what system besides water reacting with itself have we written water as a proton donor? Identify the conjugate acid–base pairs in the reaction.
(b) Find examples of other species in Table 2.7 that can act as both a Brønsted–Lowry acid and a Brønsted–Lowry base.

Reflection and Projection

The concepts of acids and bases in aqueous solution were introduced to help explain the water solubility of carbon dioxide, ammonia, and hydrogen chloride. Water reacts with itself to give tiny, equal concentrations of hydronium ions, $H_3O^+(aq)$, and hydroxide ions, $OH^-(aq)$. And in reverse, mixtures of $H_3O^+(aq)$ and $OH^-(aq)$ ions react stoichiometrically to give water. If an aqueous solution has a higher concentration of hydronium ions than pure water, the solution is said to be acidic. Brønsted–Lowry acids donate protons to water to form hydronium ions. Basic solutions have a higher concentration of hydroxide ions than pure water. The extra hydroxide can be added by dissolving ionic compounds like sodium hydroxide, NaOH, or by dissolving a Brønsted–Lowry base that accepts protons from water and produces hydroxide ions. All Brønsted–Lowry acid–base reactions involve two conjugate acid–base pairs. Each pair differs only by the absence of a proton in the base form.

We also introduced Le Chatelier's principle, which states that a system at equilibrium responds to a disturbance in a way that minimizes the effect of the disturbance. Le Chatelier's principle provides a way to predict the direction of the changes that occur in a system disturbed by addition or removal of a reactant or product. We applied the principle to acid–base reactions, but it can be applied to any reaction system, as we shall see in later chapters.

You can use the concepts discussed in this chapter—hydrogen bonding, polar attractions, London dispersion forces, favorable and unfavorable molecular reorientations, ionic solubilities, and acid–base reactions—to understand the water solubilities of many compounds under different solution conditions. We will continue to use these ideas and expand upon them more quantitatively as we proceed through the rest of the book.

2.15 Outcomes Review

In this chapter, we developed a few simple rules that permit you to predict solubilities or relative solubilities of molecules and ionic solids in water. To explain these rules, we discussed the factors (energy and molecular reorganization) that control the solubility of solutes in water. And we found that there is a maximum solubility (saturation) for most solutes when the pure solute and dissolved solute are in equilibrium. We examined two properties of solutions, electrical conductivity and pH, that provide evidence for molecular-level representations of aqueous solutions as solutions of hydrated molecules and/or ions. We found that water is a reactant in many solutions, especially in acid–base systems, and that acid–base reactions can affect solubilities. We defined solution concentrations in terms of molarity, $mol \cdot L^{-1}$, and applied mole concepts to solutions by analyzing the stoichiometry of precipitation and acid–base reactions.

Check your understanding of the ideas in the chapter by reviewing these expected outcomes of your study. You should be able to

- Use an energy diagram to characterize the basic steps in the dissolving process and give a molecular-level explanation for the direction of the individual and net energy changes [Sections 2.1, 2.2, and 2.5].
- Use molecular models, Lewis structures, and other representations of molecules to show how the three major attractions between like and unlike molecules—hydrogen bonding, polar attractions, and London dispersion forces—affect the solubility of a given molecular solute in water [Section 2.2].
- Give a molecular-level explanation for the favorable and unfavorable factors that determine the solubility of a given molecular solute or ionic compound [Sections 2.2 and 2.7].
- Predict the relative aqueous solubilities of a given set of molecular solutes [Section 2.2].
- Show the direction of motion of the molecules and ions in a solution being tested with an electrical conductivity tester [Section 2.3].
- Write the chemical formula of any ionic compound, given the charges on the cation and anion, and name ionic compounds of common ions, given the chemical formula [Sections 2.4 and 2.13].
- Draw an energy diagram for the formation of an ionic crystalline compound from its elemental gas phase atoms and give a molecular-level explanation for the direction of the energy changes of the individual steps [Section 2.4].
- Make a drawing showing the process of dissolving a polar solute or an ionic compound that shows how water molecules hydrate the dissolved molecules or ions [Sections 2.3 and 2.5].
- Draw an energy diagram for the dissolution of an ionic crystalline compound in water and give a molecular-level explanation for the direction of the energy changes of the individual steps [Section 2.5].

- Use lattice and hydration energies to determine whether a given ionic compound will dissolve exothermically or endothermically in water [Section 2.5].
- Predict whether a precipitate will form when two ionic solutions are mixed [Sections 2.6 and 2.7].
- Carry out these interconversions: grams ⇔ moles of a compound, moles (grams) of reactant ⇔ moles (grams) of product in a stoichiometric reaction, and volume of a solution of known concentration ⇔ moles (grams) of solute in that volume [Sections 2.8, 2.9, 2.10, and 2.14].
- Prepare (give step-by-step instructions for preparing) an aqueous solution of a specified molarity in some solute [Section 2.9].
- Determine the limiting reactant in a reaction mixture (solution) and the concentrations of all species formed or remaining in the solution when the reaction is complete [Sections 2.10 and 2.14].
- Use conductivity and/or pH data to determine whether a solute undergoes an acid–base reaction with water and, if so, write the equation for the chemical reaction [Sections 2.11, 2.12, 2.13, and 2.14].
- Use the concentration of hydronium ion, hydroxide ion, or the pH to tell whether an aqueous solution is acidic or basic [Sections 2.12, 2.13, and 2.14].
- Write the equation for the reaction between Brønsted–Lowry acids and bases and identify the Brønsted–Lowry conjugate acid–base pairs in any acid–base reaction [Sections 2.12 and 2.14].
- Draw a molecular-level diagram and/or explain in words the reactions occurring in a reacting system (dissolution, precipitation, or acid–base) at equilibrium [Sections 2.6, 2.12, 2.13, and 2.14].
- Use Le Chatelier's principle to predict and/or explain the direction of changes observed in a reaction system disturbed by addition or removal of a reactant or product [Sections 2.14 and 2.16].

2.16. Extension—CO₂ and the Carbon Cycle

 **2.89** INVESTIGATE THIS

What happens when CO₂(g) is bubbled into limewater?

Do this as a class investigation and work in small groups to discuss and analyze the results. Put 150–200 mL of limewater, a solution of $Ca(OH)_2$ in water, in a 400-mL beaker. Add a few drops bromthymol blue acid–base indicator solution

continued

to the beaker. Bubble a stream of $CO_2(g)$ into the solution in the beaker. Observe and record any changes occurring in the beaker. Stop the investigation when no further changes are observed in the solution into which $CO_2(g)$ is being bubbled.

2.90 CONSIDER THIS

What reactions occur when $CO_2(g)$ is bubbled into limewater?

(a) Bromthymol blue is a dye that changes color as the pH of the solution changes. The color changes from blue to green to yellow as the solution changes from basic to acidic. Did you observe any color changes in Investigate This 2.89? If so, did the solution change from low to high or from high to low pH as $CO_2(g)$ is bubbled in?

(b) Did you observe any precipitate formation in the investigation? If so, what do you think precipitated?

(c) If you observed both color changes and precipitation, what was the correlation, if any, between them? Can you suggest an explanation that fits all your observations?

The carbon cycle All living things require sources of carbon to build the complex molecules of life. Plants use carbon dioxide from the air or dissolved in water to photosynthesize carbon-containing food molecules. Animals eat the plants to get the carbon in these food molecules. The sources of the carbon dioxide required to build food molecules are the nonliving carbon compounds that occur in the crust of the earth. Figure 2.30 shows that the largest two reservoirs of carbon on the planet are carbon-containing minerals in rocks (such as $CaCO_3$ in limestone and marble) and fossil fuels (the remains of once-living organisms). The next largest carbon reservoir is the dissolved carbon (part of it as dissolved carbon dioxide) in the oceans. Atmospheric carbon dioxide is continuously exchanging with that in the oceans. Figure 2.30 shows how the reservoirs are related to one another and to living systems. The movement of carbon between and among living and non-living sources is called the **carbon cycle.**

Many organisms, such as the mollusks we introduced in the chapter opener, incorporate

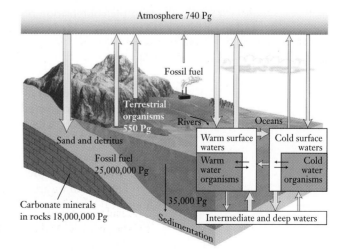

Figure 2.30.

Distribution of carbon and its movement among carbon-containing reservoirs. The thickness of the arrows shows the relative amounts of carbon moving between the reservoirs. The numeric values are in petagrams, 1 Pg = 10^{15} g ≈ 10^9 tons.

carbon as carbonates in shells. In turn, the shells of creatures that live in the ocean eventually end up on the ocean bottom. Over time, those carbonates are incorporated into carbonate rock. Large deposits of limestone, which is primarily calcium carbonate, $CaCO_3$, occur almost everywhere on Earth. Limestone is insoluble in pure water, but it has a larger solubility in water containing dissolved $CO_2(g)$. Let's see how Le Chatelier's principle helps us to understand this solubility.

Reactions in carbon dioxide–limewater solutions Analysis of the changes you observed in Investigate This 2.89 will help you understand the conditions for dissolving limestone. The solution of limewater contained $Ca^{2+}(aq)$ and $OH^-(aq)$ ions from the dissolved $Ca(OH)_2(s)$. When $CO_2(g)$ bubbled into this solution, some of it dissolved and formed carbonic acid, as we discussed in Section 2.14:

$$CO_2(g) + H_2O(l) \rightleftharpoons (HO)_2CO(aq) \tag{2.27}$$

Carbonic acid reacts with the hydroxide ion from the dissolved calcium hydroxide to form carbonate ions and water in two steps, both of which are Brønsted–Lowry acid–base reactions:

$$(HO)_2CO(aq) + OH^-(aq) \rightleftharpoons HOCO_2^-(aq) + H_2O(l) \tag{2.36}$$

$$HOCO_2^-(aq) + OH^-(aq) \rightleftharpoons CO_3^{2-}(aq) + H_2O(l) \tag{2.37}$$

[Equation (2.31) simplified this process by combining reactions (2.36) and (2.37) into a single reaction.] Carbonate ions react with calcium ions to precipitate calcium carbonate, as you observed in Investigate This 2.32 and as is predicted by our solubility rules:

$$CO_3^{2-}(aq) + Ca^{2+}(aq) \rightleftharpoons CaCO_3(s) \tag{2.38}$$

This series of acid–base and precipitation reactions explains the formation of the white precipitate in Investigate This 2.89.

2.91 CHECK THIS

Acid–base changes in carbon dioxide–limewater solutions

Are your observations on the color changes in Investigate This 2.89 consistent with the composition of the original solution and the chemistry of the series of reactions (2.27), (2.36), (2.37), and (2.38)? Explain why or why not.

As $CO_2(g)$ continued to bubble through the cloudy solution in Investigate This 2.89, you observed that the solution once again became clear (and a different color from when you began). Carbonic acid continues to be formed by reaction (2.27), but now it has no $OH^-(aq)$ with which to react. To understand what happens to the carbonic acid and the calcium carbonate, we can use Le Chatelier's principle.

The system we are considering contains a precipitate of $CaCO_3(s)$ in equilibrium with an aqueous solution containing $Ca^{2+}(aq)$, $CO_3^{2-}(aq)$, $HOCO_2^-(aq)$, and $(HO)_2CO(aq)$. One of the equilibrium reactions in this solution is

$$(HO)_2CO(aq) + CO_3^{2-}(aq) \rightleftharpoons 2HOCO_2^-(aq) \tag{2.39}$$

Our continued addition of carbon dioxide to this system forms more carbonic acid, $(HO)_2CO(aq)$, by reaction (2.27), and this is a disturbance to equilibrium reaction (2.39). We have increased the concentration of carbonic acid, $(HO)_2CO(aq)$, one of the reactants. So, according to Le Chatelier's principle, the system responds by using up the added reactant until a new equilibrium is established. Carbonic acid is used up by reaction with the carbonate ion, $CO_3^{2-}(aq)$, so some of the carbonate ion is used up as well.

Reducing the concentration of carbonate ion disturbs equilibrium reaction (2.38) by reducing the concentration of one of the reactants. Le Chatelier's principle says that the system will respond by forming more of the missing carbonate ion reactant, which it can do if some of the solid calcium carbonate dissolves. To summarize these changes, we can write the relevant chemical reactions with single arrows instead of equilibrium double arrows, in order to emphasize the directions they take due to these disturbances:

$$CO_2(g) + H_2O(l) \rightarrow (HO)_2CO(aq) \tag{2.27a}$$

$$(HO)_2CO(aq) + CO_3^{2-}(aq) \rightarrow 2HOCO_2^-(aq) \tag{2.39a}$$

$$CO_3^{2-}(aq) + Ca^{2+}(aq) \leftarrow CaCO_3(s) \tag{2.38a}$$

To emphasize that these are the same reactions we wrote previously for this system, we have used the same equation numbers with an added "a" to distinguish them from those we wrote as equilibria.

2.92 CHECK THIS

The net reaction for dissolving CaCO$_3$(s) with added CO$_2$(g)

Combine reaction equations (2.27a), (2.39a), and (2.38a) appropriately to show that the net equation for dissolving CaCO$_3$(s) with added CO$_2$(g) is

$$CO_2(g) + H_2O(l) + CaCO_3(s) \rightarrow Ca^{2+}(aq) + 2HOCO_2^-(aq) \tag{2.40}$$

Explain your reasoning for choosing this combination.

Water from the surface of the earth contains carbonic acid (dissolved carbon dioxide from the air). As this water finds its way into the rocks beneath the surface, it can dissolve small amounts of limestone by reaction sequences like those we just wrote. Although the process may take millions of years, vast limestone caverns, such as that in Figure 2.31, can be formed. Of more immediate importance, carbonates found in the soil are made soluble. Once these carbonates are converted to hydrogen carbonate ions in solution, the hydrogen carbonate ions can be absorbed by plants and converted to food molecules. As the hydrogen carbonate ion concentration builds up, equilibrium reaction (2.39) is disturbed and this sequence of reactions occurs:

$$(HO)_2CO(aq) + CO_3^{2-}(aq) \leftarrow 2HOCO_2^-(aq) \tag{2.39b}$$

$$CO_2(g) + H_2O(l) \leftarrow (HO)_2CO(aq) \tag{2.27b}$$

The CO$_2$(g) produced by reaction (2.27b) can also be used by plants and converted to food molecules. And so it goes, round and round the carbon cycle.

Figure 2.31.

A limestone cavern with stalactites hanging from the ceiling.

 **2.93 CHECK THIS**

Applying Le Chatelier's principle

Use Le Chatelier's principle to explain how an increasing concentration of hydrogen carbonate in a solution at equilibrium leads to the directions shown for reactions (2.39b) and (2.27b).

2.94 INVESTIGATE THIS

What happens when $Ca^{2+}(aq)$ and $HOCO_2^-(aq)$ react?

You have two capped tubes, each about one-quarter full of a clear, colorless liquid. One of the tubes contains an aqueous solution of calcium chloride, $CaCl_2$, and the other an aqueous solution of sodium hydrogen carbonate, $NaHCO_3$. Uncap the tubes and pour one of the solutions into the other. Observe and record all evidence that reactions have occurred.

2.95 CHECK THIS

Interpreting the reaction of $Ca^{2+}(aq)$ with $HOCO_2^-(aq)$

(a) What evidence for reactions did you observe in Investigate This 2.94? What kinds of reactions have occurred? Explain how you reach your conclusions.

continued

(b) What reaction products do you think you can identify from your observations? Is the reaction exothermic or endothermic? Explain the reasoning for your choices.

(c) Can you use Le Chatelier's principle and reactions (2.27), (2.38), and (2.39) to explain the formation of the products you identified in part (b)? Show how. Write the reactions in the directions Le Chatelier's principle indicates they will go and combine them to get the net reaction equation for the changes that occurred in Investigate This 2.94.

Recall the question we asked in the chapter opening: Why doesn't the calcium carbonate in seashells simply redissolve and return to the sea? Part of the answer is that seawater contains dissolved calcium and carbonate ions, so little *net* dissolution of the calcium carbonate can occur. But another part of the answer is that it *does* redissolve, over millions of years, by processes like reaction (2.40) in limestone caves formed from the shells and in streams flowing to the seas.

Chapter 2 Problems

2.1. Substances in Solutions

2.1. Tasting samples is unsafe and forbidden in the laboratory. If you had a solution of a sugar, how could you determine if the solute is uniformly distributed throughout the solution without tasting samples? Explain your plan.

2.2. Mixtures involving solids and liquids can be classified into two groups: homogeneous solutions and heterogeneous mixtures in which, after a time, the solid settles out of the mixture. When a solid does not truly dissolve in the liquid but forms a heterogeneous mixture, it is called a suspension. A suspension is usually shaken to redistribute the solid prior to use. Go to your local grocery and/or drug store and identify five products that are solutions and five products that are suspensions.

2.3. Draw energy diagrams for these processes. Label each energy change as endothermic or exothermic and explain why you draw it this way.
(a) A solvent changing from the liquid state to the gaseous state.
(b) Gaseous solvent molecules solvating a solute to form a liquid solution.

2.4. (a) When a certain liquid molecular substance dissolves in water, the solution feels *cool*. Sketch an energy diagram that shows the relationships among energy theoretically needed to separate the molecules in the liquid state, energy released when the molecules dissolve in water, and the net energy change in this solution process.

(b) When a certain liquid molecular substance dissolves in water, the solution feels *warm*. Sketch an energy diagram that shows the relationships among energy theoretically needed to separate the molecules, the energy released when the molecules dissolve, and the net energy change in this solution process.

2.5. The overall solution process for solution of calcium chloride, $CaCl_2$, in water is exothermic. Draw an energy diagram for this process.

2.2. Solutions of Polar Molecules in Water

2.6. Can you predict whether a substance will be soluble in water by looking at its line formula? Can you predict whether a substance will be soluble in water by looking at its structural formula? Explain your responses.

2.7. Explain what the expression "like dissolves like" means. Illustrate your explanation with appropriate examples.

2.8. For each of these compounds, write out the structural formula, using a reference handbook to find the structures you don't know. Circle all the polar bonds found in each structure. Show the direction of the bond polarity for each one. Do not consider the C–H bond to be polar.
(a) testosterone
(b) acetylsalicylic acid
(c) methyl salicylate

2.9. (a) Use Lewis structures to help explain why ethanol, CH_3CH_2OH, is miscible with water.
(b) How do you predict the solubility of 1-pentanol, $CH_3(CH_2)_4OH$, in water will compare to that of ethanol in water? Explain.

2.10. Explain in terms of intermolecular attractions each of these like dissolves like observations.
(a) Methanol (CH_3OH) is not miscible with cyclohexane (C_6H_{12}).
(b) Naphthalene ($C_{10}H_8$) is insoluble in water.
(c) Naphthalene is soluble in benzene (C_6H_6).
(d) 1-Propanol ($CH_3CH_2CH_2OH$) is miscible with water.

2.11. Predict the relative solubility of gasoline, C_8H_{18}, in water. Explain the reasoning for your prediction.

2.12. (a) Write the Lewis structures for 1-hexanol, $CH_3(CH_2)_5OH$, and 1,6-hexanediol, $HO(CH_2)_6OH$, and in each molecule, identify the region or regions where hydrogen bonding with water can occur.
(b) Which compound in (a) do you predict is more soluble in water? Explain.

2.13. Predict the relative water solubilities (from most soluble to least soluble) for each set of structures. Make models to help visualize the structures.
(a) $CH_3CH_2CH_2CH_2OH$

$CH_3CH_2CH_2CH_2CH_2OH$

$HOCH_2CH_2CH_2CH_2OH$

$HOCH_2CH_2CH_2OH$

(b) $CH_3CH_2CH_2CH_2NH_2$

$(CH_3)_2CHCH_2CH_2NH_2$

$(CH_3)_3CCH_2CH_2NH_2$

$(CH_3CH_2)_3CCH_2NH_2$

2.14. Sugars are natural products that are generally quite soluble in water. Sucrose, table sugar, is a common source of sugar in food. Fructose and lactose are two other sugars that are important in living systems. Use library resources or the Internet to find the structural formulas for these three sugars. On the basis of their structures, explain their high solubility in water. [Your response should include the reference(s) to where you found the structures.]

2.15. The fats in our bodies are composed of relatively nonpolar molecules that are almost insoluble in water. Vitamins are often classified as fat soluble or water soluble.

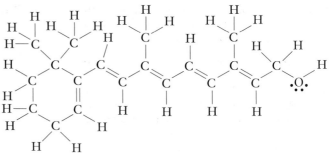

vitamin A

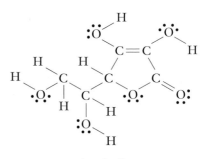

vitamin C

Consider the structures of vitamins A and C shown above. Which of these vitamins do you expect to be more soluble in aqueous systems and which in fatty tissues in the body? Use the structures to explain your reasoning.

2.16. (a) Vitamin B is water soluble and vitamins D, E, and K are fat soluble. (See Problem 2.15.) Based on this information, explain which vitamins could be stored in your body and which should be included in your daily diet.
(b) From your answers for Problem 2.15, would vitamins A and C be stored in your body or should they be included in your daily diet?
(c) With the ready availability of vitamin supplements, cases of hypervitaminosis, an illness caused by an excessive amount of vitamins, are now being diagnosed by medical doctors. Explain for which vitamins hypervitaminosis is likely to occur.

2.3. Characteristics of Solutions of Ionic Compounds in Water

2.17. In Investigate This 2.10, what had to be present in the solution in order for the light bulb to glow?

2.18. The *Web Companion*, Chapter 2, Section 2.3.1, has movies representing the $Fe^{3+}(aq)$ and $NO_3^-(aq)$ ions. How many water molecules are shown in the hydration layer for each ion? What are the similarities and differences in the arrangement and orientation of the water molecules around each of the ions? How do you explain the similarities and differences?

2.19. Solution A was prepared by mixing 0.5 g of ethanoyl chloride (acetyl chloride), $CH_3C(O)Cl$, with 100 mL of water. Solution B was prepared by mixing 0.5 g of 2-chloroethanol, $ClCH_2CH_2OH$, with 100 mL of water. Solution A conducts an electric current but solution B does not. What can you conclude about the contents of each solution? Explain the reasoning for your answer.

2.20. What ions are present in solution when these solids dissolve? Identify each type of ion as either a cation or an anion.
(a) $BaCl_2$
(b) KCl
(c) Na_3PO_4
(e) NH_4Cl
(f) Na_2S
(g) $MgSO_4$

2.21. (a) Imagine that one of your friends who is not taking this chemistry course says that salt solutions must conduct electrons, just like wires, because you can replace part of an electric circuit (as shown in the pictures in Investigate This 2.10) with a salt solution and the current will still flow. How will you answer your friend and explain how electric charge continues to flow without a flow of electrons through the solution?
(b) 🕮 Would the *Web Companion*, Chapter 2, Section 2.3.2, interactive molecular level representation of electrical conductivity in ionic solution help your explanation? Explain.

2.22. Solid sodium chloride, $NaCl$, does not conduct electricity but an aqueous solution of sodium chloride is a good conductor. Solid mercuric chloride, $HgCl_2$, does not conduct electricity and neither does its aqueous solution even though the solid is soluble. Offer a possible explanation for the difference in behavior of the aqueous solutions.

2.23. Imagine that you are a positively charged ion surrounded by large numbers of polar molecules like our simple ellipsoids with positive and negative ends (Chapter 1, Figures 1.15 and 1.16). Why would you have a problem feeling the attraction of a negatively charged ion or the repulsion of another positively charged ion? Explain your reasoning. Use drawings, if they help clarify your explanation.

2.4. Formation of Ionic Compounds

2.24. Which of these do you predict to conduct electricity when dissolved in water? Explain your reasoning in each case.
(a) $MgBr_2$
(b) CH_3OH
(c) $NaOH$
(d) CH_3OCH_3
(e) KNO_3
(f) $CH_3CH_2CH_2CH_3$

2.25. Predict the most likely charge when these elements form monatomic ions. Explain the rationale for your choice in each case.
(a) alkali metals
(b) oxygen family
(c) alkaline earth metals
(d) halogens

2.26. Write the chemical formula for the ionic compound formed by the combination of these ions.
(a) magnesium cation + bromide anion
(b) calcium cation + nitrate anion
(c) magnesium cation + sulfate anion
(d) potassium cation + oxide anion

2.27. Name these ionic compounds.
(a) Na_2SO_4
(b) $MgCl_2$
(c) $(NH_4)_2CO_3$
(d) Al_2S_3

2.28. Write the formulas for these ionic compounds.
(a) barium nitrate
(b) ammonium phosphate
(c) calcium oxide
(d) potassium sulfate

2.29. Name these ionic compounds.
(a) MgS
(b) Na_3PO_4
(c) NH_4NO_3
(d) $LiOH$

2.30. Write the formulas for these ionic compounds.
(a) calcium iodide
(b) sodium fluoride
(c) potassium carbonate
(d) barium hydroxide

2.31. Complete this grid, giving both the formula and the name of the compound formed between each pair of ions.

	CO_3^{2-}	PO_4^{3-}	F^-
Mg^{2+}			
NH_4^+			
Al^{3+}			
Na^+			

2.32. From suitable references (or labels on containers), find the chemical formulas and write chemical names for these ionic substances.
(a) Milk of Magnesia®
(b) epsom salt
(c) plaster of Paris
(d) caustic soda
(e) soda ash

2.33. Equation (2.1) can be broken down into two steps: (1) loss of an electron by a sodium atom, $Na \rightarrow Na^+ + e^-$ and (2) gain of an electron by a chlorine atom, $Cl + e^- \rightarrow Cl^-$. Write the appropriate reaction equations for formation of the common cations or anions of these elements.
(a) potassium
(b) calcium
(c) sulfur
(d) bromine

2.34. In terms of the energy of electrical attraction [equation (2.2)], explain why the ionization energy for *all* elemental atoms *always* has a positive value. For example, the energy required for the reaction, $Na \rightarrow Na^+ + e^-$, is the ionization energy, $\Delta E_{ionization} = 496 \text{ kJ·mol}^{-1}$, for sodium atoms.

2.35. Examine the lattice energies in Table 2.3. Are these data consistent with electrical attraction energies [equation (2.2)]? Explain the reasoning for your answer.

2.36. (a) Based on electrical attraction energies, in which crystal, KBr or CaBr$_2$, would the greatest forces of attraction and repulsion be observed? Explain your reasoning. Assume that the distance separating the charges is the same for both crystals.
(b) Do the data in Table 2.3 support your answer in part (a)? If so, explain how. If not, explain why not.

2.37. The lattice energy for calcium chloride, CaBr$_2$, crystals is 2176 kJ·mol^{-1}. The gaseous reaction forming the ions from the atoms,

$$Ca(g) + 2Br(g) \rightarrow Ca^{2+}(g) + 2Br^-(g),$$

requires 966 kJ·mol^{-1}. Draw an energy diagram, analogous to Figure 2.14, and use it to find $\Delta E_{xtal\ form}$ for the formation of ionic crystals of CaBr$_2$ from the gaseous atoms.

2.38. Consider these lattice energies for some ionic solids. All values are in kJ·mol^{-1}.

	F$^-$	Cl$^-$	Br$^-$
Li$^+$	1046	861	818
Na$^+$	929	787	751
K$^+$	826	717	689

(a) Use these data to discuss how the lattice energy changes with the size of the anion, keeping the size of the cation constant. *Hint:* The size of the ions in a group (column) of the periodic table increases as one goes down the group.
(b) Use these data to discuss how the lattice energy changes with the size of the cation, keeping the size of the anion constant.
(c) What generalization can be drawn about the size of ions and the lattice energies of their salts?
(d) Use your generalization from part (c) to predict how the lattice energy of CsI compares with that of NaCl.

2.39. Consider the energy diagram at the top of the next column for the formation of 1 mol of ionic crystals of MgCl$_2$.

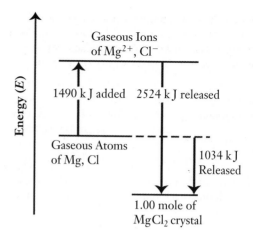

(a) What is the lattice energy for this compound? Explain how you get your answer.
(b) How much energy is required for the reaction: $Mg(g) + 2Cl(g) \rightarrow Mg^{2+}(g) + 2Cl^-(g)$? Explain how you get your answer.
(c) What is the energy associated with the formation of ionic crystals from the gaseous atoms? Explain how you get your answer.
(d) How do the three energies associated with the formation for MgCl$_2$ compare with those for NaCl, given in Figure 2.14?

2.40. (a) The energy required to remove electrons from gaseous silver atoms to form gaseous silver cations is 731 kJ·mol^{-1}. When gaseous iodine atoms gain electrons to form gaseous iodide anions, 296 kJ·mol^{-1} of energy is released. Calculate the energy for this reaction:

$$Ag(g) + I(g) \rightarrow Ag^+(g) + I^-(g)$$

(b) The lattice energy for AgI(s) crystals is 887 kJ·mol^{-1}. Draw an energy diagram analogous to Figure 2.14 and use it to find $\Delta E_{xtal\ form}$ for the formation of ionic crystals of AgI from the gaseous atoms.

2.41. The lattice energy for potassium bromide, KBr, crystals is 689 kJ·mol^{-1}. The formation of separate gaseous atoms of potassium, K(g), and bromine, Br(g), from the ionic crystal would require 594 kJ·mol^{-1}. How much energy is required for the reaction of the gaseous potassium and bromine atoms to form ions in the gas phase?

$$K(g) + Br(g) \rightarrow K^+(g) + Br^-(g)$$

Explain how you get your answer and draw an energy diagram analogous to Figure 2.14 illustrating your answer.

2.42. What is the lattice energy for magnesium fluoride, MgF$_2$? The formation of ionic crystals of MgF$_2$ from the gaseous atoms releases 1424 kJ·mol^{-1}. The reaction of the atoms to form ions in the gas phase, $Mg(g) + 2F(g) \rightarrow$ $Mg^{2+}(g) + 2F^-(g)$, requires 1533 kJ·mol^{-1}. Explain how you get your answer and draw an energy diagram analogous to Figure 2.14 illustrating your answer.

2.5. Energy Changes When Ionic Compounds Dissolve

2.43. When ammonium acetate, $NH_4C_2H_3O_2$ [= $(NH_4^+)(C_2H_3O_2^-)$], is dissolved in water, the mixture becomes quite cold. (Ammonium acetate is the salt used in some cold packs.)
(a) Is the dissolving of ammonium acetate endothermic or exothermic? Explain.
(b) What are the relative magnitudes of the crystal lattice energy and hydration energy for ammonium acetate? Use an energy diagram to explain the reasoning for your answer.
(c) Write formulas for the ions in solution using standard chemical notation.
(d) Sketch the hydrated ions in the way we have tried to show molecular-level interactions in Figure 2.9.

2.44. (a) When LiCl dissolves in water, is the process exothermic or endothermic? Use an energy diagram to explain the reasoning for your answer.
(b) When KBr dissolves in water, is the process exothermic or endothermic? Use an energy diagram to explain the reasoning for your answer.

2.45. Lithium sulfate, Li_2SO_4, is quite soluble in water (261 g·L^{-1}) while calcium sulfate, $CaSO_4$, is essentially insoluble (4.9 mg·L^{-1}). What ion–ion and/or ion–dipole interactions are responsible for this difference? Be as specific as you can.

2.46. There are many ways to describe or represent what happens when sodium sulfate, Na_2SO_4, dissolves in water. First, give an explanation in words. Next, write an ionic equation to represent the solution process. Then use a molecular-level representation to illustrate what happens when sodium sulfate dissolves in water. (See Figures 2.9, 2.17, and 2.22.)

2.6. Precipitation Reactions of Ions in Solutions

NOTE: Whenever you write a chemical reaction equation, remember to include the appropriate state notation, *(s)*, *(l)*, *(g)*, or *(aq)*, for each species in your equation.

2.47. When aqueous solutions of potassium phosphate, K_3PO_4, and calcium bromide, $CaBr_2$, are mixed, a white precipitate is formed. When tested for electrical conductivity, both starting solutions test positive. Following the mixing and precipitation, the product solution also tests positive for electrical conductivity.

(a) Prepare a table modeled after Table 2.4 to summarize what has happened.
(b) What new combinations of cations and anions are possible following mixing? One of these new combinations is the precipitate and the other is soluble in water. Which is which? Explain your reasoning.
(c) Draw a molecular-level representation similar to Figure 2.17 to illustrate what happens when the two solutions are mixed.
(d) Write a complete ionic equation describing the reaction that occurs when these solutions are mixed.
(e) Write a net ionic equation describing the reaction that occurs when these solutions are mixed.

2.48. (a) When aqueous solutions of potassium chloride, KCl, and sodium bromide, NaBr, are mixed, no precipitate is formed. What can you conclude about the water solubility of $NaCl(s)$ and $KBr(s)$? Explain your reasoning.
(b) Draw a molecular-level representation similar to Figure 2.17 to illustrate what happens when the two solutions are mixed.
(b) Write a complete ionic equation describing what occurs when these solutions are mixed. What would you write for the net ionic equation? Explain.

2.49. $Ba^{2+}(aq)$ is toxic to humans. However, when physicians need to x-ray the gastrointestinal (GI) tract—stomach and intestines—they fill the patient's GI tract with barium sulfate and water. How can it be that the patient is not harmed by this procedure?

2.50. Differences in solubility can be used to help separate cations from solutions where they are mixed together. The process is called selective precipitation. Consider this table of solubilities, and then suggest a sequence of precipitation reactions to separate Ag^+, Ba^{2+}, and Fe^{3+} from solution. Explain your approach and write a net ionic equation for each reaction that takes place.

	Cation		
Test solution	$Ag^+(aq)$	$Ba^{2+}(aq)$	$Fe^{3+}(aq)$
NaCl	ppt	no ppt	no ppt
NaOH	ppt	no ppt	ppt
Na_2SO_4	no ppt	ppt	no ppt

2.51. Aluminum nitrate, $Al(NO_3)_3$, is soluble in water. So is sodium oxalate, $Na_2C_2O_4$. When an aluminum nitrate solution is mixed with a sodium oxalate solution, a precipitate forms.
(a) What is the precipitate? State the reasoning for your prediction.
(b) Write a net ionic equation for the reaction that occurs.

2.52. A solution of lithium nitrate is mixed with a solution of sodium phosphate. A white precipitate is observed to form.
(a) What is the white precipitate? State the reasoning for your prediction.
(b) Write a net ionic equation for the reaction that occurs.

2.53. What does it mean if there is a forward arrow over a backward arrow, $\rightleftharpoons$, in an equation?

2.7. Solubility Rules for Ionic Compounds

2.54. A solution of cadmium chloride, $CdCl_2$, is mixed with a solution of ammonium sulfide, $(NH_4)_2S$. A yellow-orange precipitate is observed to form.
(a) What is the orange-yellow precipitate? State the reasoning for your prediction.
(b) Write a net ionic equation for the reaction that occurs.

2.55. Predict the products of each of these reactions between aqueous solutions. If no visible change will occur, write NO APPARENT REACTION to the right of the arrow. Give the reasoning for your prediction in each case. Write the balanced complete ionic reaction equation and the net ionic reaction equation for each case where reaction occurs.
(a) barium chloride*(aq)* + sodium sulfate*(aq)* $\rightarrow$
(b) silver nitrate*(aq)* + magnesium chloride*(aq)* $\rightarrow$
(c) strontium nitrate*(aq)* + potassium nitrate*(aq)* $\rightarrow$
(d) ammonium phosphate*(aq)* + calcium bromide*(aq)* $\rightarrow$

2.56. Write balanced net ionic equations for reactions that would be suitable for laboratory preparation of these solid ionic compounds. Suggest compounds whose aqueous solutions you could use to carry out these preparations.
(a) $BaSO_4$ (c) $Ca_3(PO_4)_2$
(b) $AgCl$ (d) CaC_2O_4

2.8. Concentrations and Moles

2.57. You have prepared 1 L of a 0.1 M solution of NaOH. Next, you accidentally spilled about 200 mL of this solution. What has happened to the concentration of the remaining solution?

2.58. You have been asked to assist with a chemical inventory of a General Chemistry stockroom and have found a 0.5-L bottle about half full of a solution labeled 0.5 M $CaCl_2$.
(a) What does this label tell you about the solution?
(b) Can you tell about how many moles of $CaCl_2$ are in the bottle? If so, show how. If not, tell what further information you need to answer the question.

(c) Can you tell about how many grams of $CaCl_2$ are in the bottle? If so, show how. If not, tell what further information you need to answer the question.

2.59. Refer to the molecular structure of vitamin C, shown in Problem 2.15.
(a) What is the molecular formula of vitamin C?
(b) What is a molar mass of vitamin C? Explain your work.
(c) How many moles of vitamin C are present in a 500-mg tablet of the vitamin? Explain your reasoning.
(d) How many molecules of vitamin C are present in each 500-mg tablet? Explain your reasoning.

2.60. Calculate the mass (in grams) of these compounds. Show your reasoning clearly.
(a) 2.5 mol of the artificial sweetener aspartame, $C_{14}H_{18}N_2O_5$
(b) 0.040 mol of aspirin, $C_9H_8O_4$
(c) 2.5×10^{23} molecules of cholesterol, $C_{27}H_{46}O$
(d) 1.2×10^{22} molecules of caffeine, $C_8H_{10}N_4O_2$

2.61. How many atoms of carbon are there in 5 mg of niacin? Show your reasoning clearly.

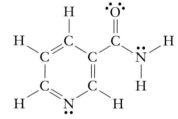

niacin

2.62. Bacteria generally contain a single molecule of DNA that encodes all their genetic information. What is the concentration, $mol \cdot L^{-1}$, of DNA in a spherical bacterium that has a diameter of 10^{-6} m = 1 μm? Clearly explain how you arrive at your answer. *Hint:* The volume of a sphere is $\frac{4}{3}\pi r^3$.

2.63. Blood serum is typically about 0.14 M in NaCl. Calculate the number of sodium ions in 50 mL of blood serum. Show your reasoning clearly.

2.9. Mass–Mole–Volume Calculations

2.64. Calculate the number of grams of solute present in each of these solutions. Show your reasoning clearly.
(a) 350 mL of 0.105 M $K_2Cr_2O_7$
(b) 50 mL of 1.0 M $FeCl_3 \cdot 6H_2O$
(c) 0.3 L of 1.70 M KCl

2.65. Calculate the molar concentration of solute present in each of these solutions. Show your reasoning clearly.
(a) 120 mL containing 4.5 g of NaCl
(b) 0.25 L containing 1.3 g of NH_4Cl
(c) 1.3 L containing 1.85 g of $AgNO_3$

2.66. (a) 5.405 g glucose, $C_6H_{12}O_6$, is dissolved in enough water to make 1.000 L of solution at 20 °C. What is the molarity of glucose in this solution? Show your reasoning clearly.

(b) How many milliliters of the solution prepared in part (a) will you need in order to obtain 0.950 mmol of glucose at 20 °C? Show your reasoning and work clearly and completely. *Hint:* 1 mol = 1000 mmol.

2.67. Two students were asked to prepare a 1.00 M solution of $CuSO_4$. One student found a bottle, labeled $CuSO_4 \cdot 5H_2O$. He weighed 159.60 g of this hydrated copper sulfate, transferred it to a 1-L volumetric flask, and dissolved it in a small quantity of water. Then, he added more water until the solution just reached the calibration mark etched on the neck of the flask and thoroughly mixed the contents of the flask. The second student followed exactly the same procedure, but she used the anhydrous salt of copper sulfate, $CuSO_4$. Which student prepared the solution with the correct concentration? Calculate the molar concentration of $CuSO_4$ in each solution.

2.68. You need about 170 mL of 0.10 M NaOH for an experiment. The concentration of this solution has to be fairly exact. Describe how to prepare the solution.

2.69. Normal saline, a solution given by intravenous injection, is a 0.90% (mass to volume %) sodium chloride solution. How many grams of sodium chloride are required to make 250. mL of normal saline solution? What is the molarity of this solution? The density of this solution is the same as water, 1.00 g·mL^{-1}. Show your reasoning clearly.

2.70. When urine is analyzed, the normal range for urea, $(NH_2)_2CO$, one of the solutes in urine, is 13–40 g·(24 hr)$^{-1}$. (For urinalysis, a patient's urine is collected over a 24-hr period to be sure that the sample is representative.) A patient's laboratory tests show a urea content of 25 g·(24 hr)$^{-1}$. Suppose the normal output of urine for this patient is 2.5 L·(24 hr)$^{-1}$. What is the molarity of the urea in the patient's urine? Explain your reasoning.

2.71. One of the ionic compounds in sports drinks is potassium dihydrogen phosphate, KH_2PO_4. The label on one of these drinks tells us that 240 mL of the solution contains 30 mg of potassium. KH_2PO_4 is the only ingredient in the solution that can provide this potassium. How many grams of KH_2PO_4 are dissolved in 240 mL of the solution? What is the molarity of the KH_2PO_4, in this solution? Clearly show and explain all the work you do to solve this problem.

2.10. Reaction Stoichiometry in Solutions

2.72. How many moles each of carbon, hydrogen, and oxygen atoms are present in 2 mol of ammonium acetate, $NH_4C_2H_3O_2$? What is the total number of moles of atoms in 2 mol of the compound? What is the total number of moles of *ions* in 2 mol of the compound?

2.73. A student is trying to prepare artificial kidney stones in the laboratory. How many grams of calcium phosphate can he make by mixing 125 mL of 0.100 M calcium chloride with 125 mL of 0.100 M sodium phosphate? Explain your reasoning and any assumptions you make in solving this problem.

2.74. What volume of 0.100 M $SO_3^{2-}(aq)$ is needed to react exactly and completely with 24.0 mL of 0.200 M $Fe^{3+}(aq)$? The equation that represents the reaction that occurs is

$$2Fe^{3+}(aq) + SO_3^{2-}(aq) + 3H_2O(l) \rightarrow$$
$$2Fe^{2+}(aq) + SO_4^{2-}(aq) + 2H_3O^+(aq)$$

2.75. Assume that you mix 50.0 mL of a solution that is 0.45 M Na_2SO_4 with 50.0 mL of a solution that is 0.36 M $BaCl_2$.
(a) How many moles of each of the four ions, Na^+, SO_4^{2-}, Ba^{2+}, and Cl^-, are present in the mixture? Explain your reasoning clearly.
(b) If the $SO_4^{2-}(aq)$ in the mixture reacts with $Ba^{2+}(aq)$ to give $BaSO_4(s)$, how many moles of $Ba^{2+}(aq)$ are required to react with all the $SO_4^{2-}(aq)$ in the mixture? Explain your reasoning clearly.
(c) If the $Ba^{2+}(aq)$ in the mixture reacts with $SO_4^{2-}(aq)$ to give $BaSO_4(s)$, how many moles of $SO_4^{2-}(aq)$ are required to react with all the $Ba^{2+}(aq)$ in the mixture? Explain your reasoning clearly.
(d) Is $Ba^{2+}(aq)$ or $SO_4^{2-}(aq)$ the limiting reactant in this mixture? Explain how you make this choice.

2.76. Predict what precipitate will form when each of the following aqueous solution mixings is carried out. Determine the limiting reagent for each reaction and the mass of the precipitate (assuming that all precipitation reactions go to completion). If there is no precipitate, then write NO APPARENT REACTION and explain your reasoning.
(a) Mix 125 mL of 0.15 M $BaBr_2$ with 125 mL of 0.15 M Na_3PO_4.
(b) Mix 85 mL of 0.40 M NH_4Cl with 65 mL of 0.50 M KNO_3.
(c) Mix 85 mL of 0.40 M $(NH_4)_2S$ with 65 mL of 0.50 M $ZnCl_2$.
(d) Mix 15.0 mL of 0.20 M $AgNO_3$ with 15.0 mL of 0.40 M NaBr.

2.77. When 50. mL of an aqueous 0.1 M $SrCl_2$ solution is mixed with 50. mL of an aqueous 0.1 M Na_3PO_4 solution, a white precipitate is formed.
(a) Write a complete ionic reaction equation for the reaction in the mixture.
(b) How many moles of chloride anion remain in solution when the precipitation is complete? How many grams of chloride is this? Explain the reasoning for your answers.
(c) How many moles of each of the other ions remain in solution when the precipitation is complete? Explain the reasoning for your answers.
(d) Use your equation from part (a) and your results from parts (b) and (c) to show that the solution is electrically neutral after the precipitation is complete. Show your reasoning clearly.

2.11. Solutions of Gases in Water

2.78. Are gases very soluble in water? Explain your reasoning.

2.79. Predict whether the noble gases (He, Ne, Ar, Kr, and Xe) have a low solubility in water (less than $1\ g \cdot L^{-1}$) or a high solubility in water (greater than $10\ g \cdot L^{-1}$). Explain clearly.

2.80. (a) Use the data in Table 2.6 for this problem. How many moles of nitrogen gas, $N_2(g)$, dissolve in 10.0 L of water when the temperature is 25 °C and the pressure of the gas is 101 kPa (one atmosphere)?
(b) How many moles of oxygen gas, $O_2(g)$, dissolve in 0.100 L of water when the temperature is 25 °C and the pressure of the gas is 101 kPa (one atmosphere)?

2.81. The solubility of $H_2(g)$ in water at 25 °C is $7.68 \times 10^{-4}\ mol \cdot L^{-1}$. When the temperature is decreased to 0 °C, the solubility of hydrogen is $9.61 \times 10^{-4}\ mol \cdot L^{-1}$. How do you account for the greater solubility at the lower temperature?

2.82. Hydrogen bromide gas, $HBr(g)$, dissolves in water to form an acidic solution. What is the name of this aqueous solution? *Hint:* What is the analogous solution of $HCl(g)$ called?

2.83. If $HBr(g)$ is bubbled into water until the solution is saturated, the resulting solution is approximately 8.9 M in $HBr(aq)$. The density of the solution is about $1.5\ kg \cdot L^{-1}$. What is the solubility expressed in $g \cdot kg^{-1}$ (as in Table 2.6)? Clearly explain your reasoning.

2.84. (a) $HCl(g)$ is quite soluble in diethyl ether $(CH_3CH_2OCH_2CH_3)$. Various reason(s) for this solubility are given: dipole–dipole interactions, hydrogen bonding, or HCl ionization. Draw a picture to illustrate each of these potential interactions of HCl and ether.
  (i) dipole–dipole
  (ii) hydrogen bonding
  (iii) HCl ionization
(b) What experiment could be done to eliminate or confirm one or more of these.

2.85. Would you expect hydrogen chloride gas, $HCl(g)$, to be more or less soluble in hexane than in water? Explain your reasoning.

2.86. Assume that the amount of a gas that dissolves in water is directly proportional to its pressure over the solution; the lower the pressure, the less gas dissolved.
(a) Use the data in Table 2.6 to figure out the masses of nitrogen and oxygen that dissolve in 1.0 L of water at 25 °C when air (80% nitrogen and 20% oxygen—mole percent) at a total pressure of 101 kPa dissolves in the water. State all your assumptions explicitly and explain clearly the method you use to arrive at your answer.
(b) What percent of the dissolved mass of gas is oxygen? Show how you get your answer.
(c) Is the mass percent of oxygen in the air greater than, less than, or the same as its mass percent in the gases dissolved in water? Explain your reasoning.

2.12. The Acid–Base Reaction of Water with Itself

2.87. What is an acid?

2.88. What is a base?

2.89. If a solution of acid A has pH of 1 and a solution of acid B has pH of 3, what can you tell about the two acid solutions?

2.90. Identify aqueous solutions with these properties as acidic or basic or neither. Explain your reasoning in each case.
(a) $pH < 7$
(b) $[H_3O^+(aq)] = 1.0 \times 10^{-7}\ M$
(c) $[OH^-(aq)] > 1.0 \times 10^{-7}\ M$
(d) $pH > 7$
(e) $[H_3O^+(aq)] > 1.0 \times 10^{-7}\ M$
(f) $[OH^-(aq)] < 1.0 \times 10^{-7}\ M$
(g) $[H_3O^+(aq)] < 1.0 \times 10^{-7}\ M$
(h) $[OH^-(aq)] = 1.0 \times 10^{-7}\ M$

2.91. Calculate the pH of each of these solutions.
(a) $[H_3O^+(aq)] = 1.0 \times 10^{-2}\ M$
(b) $[H_3O^+(aq)] = 1.0 \times 10^{-10}\ M$
(c) $[H_3O^+(aq)] = 5.0 \times 10^{-4}\ M$
(d) $[H_3O^+(aq)] = 5.0 \times 10^{-8}\ M$

2.92. Assuming that the reaction of $HCl(g)$ and water goes to completion to form $H_3O^+(aq)$ and $Cl^-(aq)$, what

is the molar concentration of HCl(*aq*) that will result in solutions having
(a) pH = 4?   (b) pH = 2?

2.93. HCl(*g*) is named hydrogen chloride, but HCl(*aq*) is named hydrochloric acid. By analogy, what are the names of HI(*g*) and HI(*aq*)? of H_2S(*g*) and H_2S(*aq*)?

2.94. 🖰 Write a brief essay describing the relationship of the two movies in the *Web Companion*, Chapter 2, Section 2.12.3, to the figure at the bottom of the page (which is similar to Figure 2.26).

2.13. Acids and Bases in Aqueous Solutions

2.95. Phosphorus pentoxide, P_2O_5(*s*), is a nonmetal oxide, which reacts with water to form a solution of phosphoric acid, $(HO)_3PO$(*aq*) [or H_3PO_4(*aq*)].
(a) Write the balanced chemical reaction equation for the reaction of phosphorus pentoxide with water.
(b) If 1.42 g of phosphorus pentoxide is mixed with 250. mL of water, what is the molarity of the resulting phosphoric acid solution? Show your reasoning and clearly state any assumptions.

2.96. Give a name for each of the following ionic compounds with oxyanions (shown with their conventional formulas). See Table 2.7 for the names of oxyanions. *Hint:* Arsenic, As, is in the same family as P and forms many analogous compounds.
(a) $Ca(HSO_4)_2$ (d) Ce_2SO_4
(b) Na_2CO_3 (e) $KHCO_3$
(c) $Al_2(HPO_4)_3$ (f) Na_3AsO_4

2.97. Draw Lewis structures (showing all nonbonding electron pairs as a pair of dots and all covalent bonds as lines) for the nitrate, ethanoate (acetate), and hydrogen sulfate oxyanions.

2.98. Identify each Brønsted–Lowry acid and base in the following reactions. If necessary, write out the complete balanced ionic equation before identifying the acids and bases. Place an A below each acid and a B below each base.
(a) $H_2S(g) + H_2O(l) \rightleftharpoons HS^-(aq) + H_3O^+(aq)$
(b) $NaOH(aq) + HCl(aq) \rightleftharpoons NaCl(aq) + H_2O(l)$
(c) $NH_3(g) + HCl(g) \rightleftharpoons NH_4^+Cl^-(s)$

2.99. Identify each Brønsted–Lowry acid and base in the following reactions. If necessary, write out the complete balanced ionic equation before identifying the acids and bases.
(a) $NO_2^-(aq) + H_3O^+(aq) \rightleftharpoons HNO_2(aq) + H_2O(l)$
(b) $2H_3O^+(aq) + 2ClO_4^-(aq) + Mg^{2+}(OH^-)_2(s) \rightleftharpoons$
$Mg^{2+}(aq) + 2ClO^-{}_4(aq) + 4H_2O(l)$
(c) $HNO_3(aq) + Al^{3+}(OH^-)_3(s) \rightleftharpoons$
(d) $HCN(aq) + NaOH(aq) \rightleftharpoons$

2.100. The Lewis structures of $HOCO_2^-$, $(HO)_2PO_2^-$, and $HOPO_3^{2-}$ are omitted from Table 2.7. Draw their Lewis structures (showing all nonbonding electron pairs as a pair of dots and all covalent bonds as lines).

2.14. Extent of Proton-Transfer Reactions: Le Chatelier's Principle

2.101. When ammonia dissolves in water, it does so as the result of an acid–base reaction. Two possible acid–base reactions of ammonia and water are

$$H_2O(l) + NH_3(g) \rightleftharpoons H_3O^+(aq) + NH_2^-(aq)$$
$$H_2O(l) + NH_3(g) \rightleftharpoons OH^-(aq) + NH_4^+(aq)$$

(a) Identify the Brønsted–Lowry acids and bases in each reaction by placing an A below each acid and a B below each base.
(b) Use reasoning based on the relative electronegativities of nitrogen and oxygen to predict which equation represents the actual acid–base reaction when ammonia gas dissolves in water. (You can check your prediction by recalling that, in Investigate This 2.63, you discovered that an aqueous ammonia solution has a pH > 7.)

2.102. When methylamine, CH_3NH_2(*g*), dissolves in water, a weak electrical conductivity is observed. Explain this observation using a balanced equation in your answer. Omit any ions/molecules that do not directly participate in the reaction.

2.103. Ethylene glycol, $HOCH_2CH_2OH$ (used in automobile antifreeze products), is miscible with water in all proportions. Will the resulting solution be basic, acidic, or neutral? Explain. Will the resulting solution display electrical conductivity? Explain.

2.104. Esterification (which is discussed in Chapter 6) is one of the most important reactions of carboxylic acids in biological systems. A simple example is the reaction of acetic (ethanoic) acid with ethanol to form ethyl acetate (a common solvent found in fingernail polish remover) and water in this equilibrium reaction:

$$CH_3C(O)OH + HOCH_2CH_3 \rightleftharpoons$$
  acetic acid   ethanol

$$CH_3C(O)OCH_2CH_3 + H_2O$$
  ethyl acetate

Use Le Chatelier's principle to predict and clearly explain the outcome of these changes:
(a) Starting with 0.1 mol of acetic acid and 0.1 mol of ethanol, would more, less, or the same amount of ethyl acetate be formed, if water is added to the reaction mixture?
(b) Would a mixture of 0.2 mol of acetic acid and 0.1 mol of ethanol form more, less, or the same amount of ethyl acetate as a mixture of 0.1 mol of acetic acid and 0.1 mol of ethanol?

2.105. Use explanations based on Le Chatelier's principle to explain or make predictions in each of the following cases.
(a) Consider This 2.79 examined the solubility of carbon dioxide in water. Why is the solubility of carbon dioxide greater in an aqueous sodium hydroxide solution than in water itself?
(b) Calcium sulfate is slightly soluble in water. If sodium sulfate (solid) is added to a saturated aqueous solution of calcium sulfate, what will probably happen to the concentration of calcium cation, $[Ca^{2+}(aq)]$? Explain.

2.16. Extension—CO_2 and Le Chatelier's Principle

2.106. Represent each of these statements as a complete balanced chemical equation.
(a) Carbonic acid is formed when carbon dioxide reacts with water.
(b) Calcium carbonate (limestone) reacts with carbonic acid to form an aqueous solution of calcium hydrogen carbonate.
(c) Calcium hydrogen carbonate reacts with calcium hydroxide to form calcium carbonate precipitate.
(d) Calcium hydrogen carbonate reacts with sodium hydroxide to form calcium carbonate precipitate and the water-soluble salt sodium carbonate.
(e) The mixing of aqueous solutions of sodium hydrogen carbonate and sodium hydroxide is exothermic. A reaction has occurred but there is no precipitate.

2.107. The names stalactite and stalagmite for the structures that grow, respectively, down from the ceiling and up from the floor of a limestone cave (Figure 2.31) are derived from the Greek word meaning "to drip." Indeed, if you examine the tip of a stalactite, you will often find a drop of liquid. The liquid is an aqueous solution containing calcium cations and hydrogen carbonate anions. See Check This 2.92.
(a) As the water evaporates from this drop, what happens to the concentration of calcium cations? of hydrogen carbonate anions? Explain your reasoning.
(b) What reaction does Le Chatelier's principle predict will occur in the evaporating drop of solution in part (a)? Clearly explain your choice.
(c) Does your answer in part (b) help explain the growth of stalactites? How about stalagmites (which grow directly under stalactites)? Give your reasoning clearly.

2.108. A gas is evolved when calcium carbonate is placed in an aqueous solution of $HCl(g)$.
(a) What is the gas? Explain how you reached this conclusion.
(b) Write the net ionic equation for the reaction that produces the gas.

(c) Show how Le Chatelier's principle and your knowledge of gas solubilities explain the observed results.

General Problems

2.109. 🔲 Sugar (sucrose) crystals are hard and they crunch and break when you put pressure on them. Grease is soft and easily smeared on a surface using little pressure. How do the *Web Companion*, Chapter 2, Section 2.2.3 and 5, molecular-level representations of sucrose and grease explain the observed macroscopic behavior of these substances? Clearly relate your explanation to the structures shown.

2.110. (a) About 2 g of calcium sulfate, $CaSO_4(s)$, dissolves in a liter of water. What are the molarities of $Ca^{2+}(aq)$ and $SO_4^{2-}(aq)$ in a saturated solution of calcium sulfate? Is seawater saturated with calcium sulfate? (See Consider This 2.54 for the composition of seawater.)
(b) An ionic compound is usually less soluble in a solution that already contains either its cation or anion. Is this effect consistent with Le Chatelier's principle? Explain why or why not.
(c) Does the effect described in part (b) influence your response to part (a)? How?
(d) Does the effect described in part (b) help explain why the calcium carbonate in seashells (see the chapter opening) does not redissolve in the sea? Explain your response.

2.111. A sample of salt water with a density of 1.02 g·mL^{-1} contains 17.8 ppm (by mass) of nitrate, $NO_3^-(aq)$. Calculate the molarity of nitrate ion in the sample of salt water (ppm = parts per million).

2.112. (a) Assume that seawater may be represented by a 3.50% by weight aqueous solution of NaCl (density 1.025 g·mL^{-1}). What is the molarity of sodium chloride in this "seawater"?
(b) Is your result in part (a) consistent with your answer to Consider This 2.54(b)? Explain why or why not.

2.113. A student prepared a solution for her biochemistry laboratory by weighing 5.15 g of a compound and dissolving it in 10.0 g of water. The concentration of this solution was 2.7 M, and its density was 1.34 g·mL^{-1}. Which of the following compounds did the student use to prepare the solution? Explain your reasoning clearly.
(a) $(NH_4)_2SO_4$ (c) CsCl
(b) KI   (d) $Na_2S_2O_3$

2.114. What percentage (approximate) of the water molecules is protonated in aqueous solutions with these pHs? Explain your reasoning. *Hint:* The molarity of water in pure water and dilute aqueous solutions is about 55.5 M.
(a) 7   (b) 6   (c) 4

2.115. Figure 2.24 shows the concentration of hydroxide ion, $[OH^-(aq)]$, as well as the concentration of hydronium ion, $[H_3O^+(aq)]$, correlated with the pH of solutions.
(a) When the pH is 3, what are the concentrations of the hydroxide and hydronium ions? What is the numeric value of the mathematical product $[H_3O^+(aq)] \cdot [OH^-(aq)]$ at this pH?
(b) For any pH you choose, what is the mathematical product $[H_3O^+(aq)] \cdot [OH^-(aq)]$? Can you think of a reason for this result?

2.116. (a) The sulfur dioxide molecule, SO_2, has a permanent dipole moment. What is the shape of the molecule? Give the reasoning for your answer.
(b) When sulfur dioxide, a nonmetal oxide, dissolves in water, the resulting solution conducts electricity. How can this be explained? Be sure to include an appropriate chemical reaction equation to justify your answer.

2.117. (a) Table 2.7 lists the name of the $HOCO_2^-$ ion (or HCO_3^-) as hydrogen carbonate ion. This ion also has the common name "bicarbonate" that is used in substances such as bicarbonate of soda, $NaHCO_3$. Explain how this name can be rationalized.
(b) TSP is the common name for a cleaning product containing sodium and phosphate ions. What is the chemical formula for the major ingredient in TSP and what do the letters TSP represent?

2.118. What volume of 0.075 M sulfuric acid, $(HO)_2SO_2(aq)$, solution will be required to react exactly and completely with the hydroxide ions in each of the following basic solutions? *Hint:* Each sulfuric acid molecule can provide two hydronium ions to the solution.
(a) 1.00 g of $KOH(s)$ dissolved in 75 mL of water
(b) 1.00 g of $KOH(s)$ dissolved in 150. mL of water

2.119. Assume that you have a 1-lb (454 g) container of drain cleaner, mostly solid sodium hydroxide, that you wish to get rid of by reaction with vinegar, about 0.9 M acetic (ethanoic) acid.
(a) Write a balanced chemical reaction equation for the reaction.
(b) What is the minimum volume of vinegar required to react completely with the drain cleaner? Explain your reasoning clearly.

2.120. Raindrops dissolve gaseous oxides of nitrogen and sulfur (formed by both natural processes and by burning fossil fuels) and form acidic solutions (acid rain), such as nitric and sulfuric acids [see equation (2.28)]. Acid rain has caused a small pond to become so acidic that most of its aquatic life has died. A community group has made a proposal to restore the pond by adding enough lime, $CaO(s)$ (quicklime), to react with the hydronium ion by this reaction stoichiometry:

$$2H_3O^+(aq) + CaO(s) \rightarrow 3H_2O(l) + Ca^{2+}(aq)$$

They have asked for your help to figure out how much lime to use.
(a) The volume of the pond is about 4.5×10^4 m^3 (1 m^3 = 1000 L) and the $H_3O^+(aq)$ concentration is 5.0×10^{-5} M (pH = 4.30). How many moles of hydronium ion does the pond contain? Explain clearly the procedure you use.
(b) How many moles of lime have to be added to react with 90% of the hydronium ion present? How many kilograms of lime is this? If lime is purchased in 50-lb bags (1 lb = 454 g), how many bags will be needed? Explain clearly, so the community group can understand.
(c) Estimate what the pH of the pond will be after the lime is added. Explain your reasoning.

2.121. To cool themselves, many animals sweat when the weather is warm (see Chapter 1, Section 1.10). Chickens, however, do not have sweat glands, so they pant to help cool themselves. In hot weather, chickens pant a lot and lose more carbon dioxide than normal (they hyperventilate). The level of dissolved carbon dioxide in their blood decreases and the hens lay eggs with thinner and more fragile shells.
(a) Why are the shells less sturdy than usual? What reaction(s) is(are) being affected?
(b) Egg farmers use a simple and inexpensive method to keep their hens' dissolved-carbon dioxide levels normal in hot weather. What do you think they do?

2.122. 🎬 The chemical equations we write to represent precipitation reactions usually look like this one:

$$Ag^+(aq) + Cl^-(aq) \rightleftharpoons AgCl(s)$$

This representation gives us little clue about what might be happening at the molecular level during the precipitation. The animated movie, *Web Companion*, Chapter 2, Section 2.6.3, showing the interaction of chloride and silver ions, provides one way to visualize the precipitation process. Describe the steps in the process and illustrate your description with chemical reaction equations. That is, try to translate the molecular-level representation to a symbolic representation.

"Twinkle, twinkle, little star,
How I wonder what you are!
Up above the world so high,
Like a diamond in the sky!"

THE STAR
JANE TAYLOR (1783–1824)

The Crab Nebula, a relatively near neighbor in the constellation Taurus, is the remnant of a supernova explosion about 6500 light years from Earth. The stellar explosion occurred about 7500 years ago. Light from the supernova reached Earth and was observed and recorded by Chinese astronomers in the year 1054. Elements were scattered into interstellar space by the explosion, which was so energetic that the new "star" could be seen during the day for more than three weeks. The graph shows the present-day abundance of elements in the universe.

Origin of Atoms

3.1 CONSIDER THIS

Where do atoms come from?

(a) Where do the atoms in your body come from?

(b) In part (a), you probably thought of several different sources for the atoms in your body. Where did the atoms in these sources come from?

(c) If you continue working back, as in part (b), toward the origin of atoms, where do you end up?

A s you have seen in the previous chapters, a unique feature that distinguishes Earth from all the other planets and moons in our solar system is the presence of abundant surface water at a temperature that keeps most of it in the liquid state. You also know that about 70% of your own body is water. Where did all this water come from? Consider This 3.1 pushes the question back further: *Where did the elemental hydrogen and oxygen atoms in water come from in the first place?* And the question applies to all the elements you have met so far: Where did the atoms of carbon, nitrogen, sodium, phosphorus, and other elements come from? For all the elements except hydrogen, and most helium and lithium, the answer is *the stars*. Indeed, as the song says, "We are stardust." In this chapter, you'll see how literally true this statement is and also how a star is "like a diamond in the sky."

A great deal of what we know about our solar system, stars, galaxies, and the universe is based on spectroscopic evidence gathered by astronomers over the past several centuries. We will introduce spectroscopy in this chapter and find in Chapter 4 that it is a thread that ties together cosmic models and the atomic models we will discuss there. The results of more than a century of spectroscopic study and modeling of the universe are summarized in the cosmic elemental abundance graph shown on the facing page. After an examination of these data, we will briefly discuss our present model for the birth and evolution of the universe that has led to the present abundance of elements in the universe and on Earth. Elemental atoms are formed in nuclear reactions, so a good deal of the chapter is devoted to these reactions and their energies.

"We are stardust," is a phrase from Joni Mitchell's *Woodstock* (popularized by Crosby, Stills, Nash, and Young).

3.1. Spectroscopy and the Composition of Stars and the Cosmos

3.2 INVESTIGATE THIS

What is the effect of a diffraction grating on white light?

Do this as a class investigation and work in small groups to analyze and discuss the results. Completely cover the stage of an overhead projector with a sheet of thin cardboard that has a "slit" about 2×25 cm cut into its center. Project the

continued

image of the slit vertically on the projection screen. Hold a sheet of transmission diffraction grating in front of the projector lens with the grating parallel to the length of the slit. (A transmission **diffraction grating** is a series of very narrow, closely spaced transparent slits in an opaque film.) Write a description of what you observe.

3.3 CONSIDER THIS

How does a diffraction grating affect different colors of light?

(a) Are different colors of light affected differently as they pass through the diffraction grating in Investigate This 3.2? If so, what color of light is most affected? Explain why you answer as you do.

(b) What do you observe about the symmetry of the image on the screen? How might you explain your observation?

Spectrum (singular; plural = **spectra,** from Latin for vision) was originally used to refer to the rainbow of colors created by dispersion of white light; see Investigate This 3.2 and Figures 3.1 and 3.2. *Spectrum* now refers to the entire range of electromagnetic wavelengths we will meet in Chapter 4.

Spectroscopy Isaac Newton (English natural philosopher and mathematician, 1642–1727) was the first scientist to observe that white light could be dispersed by a prism into the **spectrum** of colors (wavelengths of light) shown in Figure 3.1. We'll discuss the concepts of waves, light, and the origin of diffraction in Chapter 4. For our purpose here, your observation that light can be spread into a spectrum by a diffraction grating is all we need. The different colors of light are characterized by their **wavelengths,** so some of our spectra will show a wavelength scale, which is part of the discussion in Chapter 4. Newton used light from the sun for his experiments, so **spectroscopy** (spectrum + *skopeein* = to watch), the observation of spectra, began with light from a star. Modern spectroscopic experiments are usually carried out using diffraction gratings to disperse the light, as you did in Investigate This 3.2, and is also illustrated in Figure 3.2.

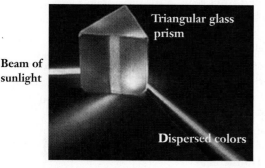

Figure 3.1.

A prism disperses white light into the visible spectrum. Violet light is bent most from the original light beam.

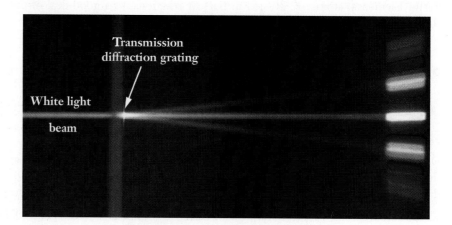

Figure 3.2.

A diffraction grating disperses white light into the visible spectrum. Violet light is bent least from the original light beam.

3.4 INVESTIGATE THIS

What do spectra of different emission sources look like?

Use a small diffraction grating or hand-held spectroscope to view the light emitted from an incandescent light bulb, a fluorescent light bulb, and two or more atomic discharge lamps, such as a neon lamp, set up by your instructor. Record your observations and work in small groups to develop descriptions of what you observe from each light source.

3.5 CONSIDER THIS

How would you characterize the spectra from different sources?

(a) What is the same in the various spectra you observed in Investigate This 3.4? What is different? How might you explain any differences?

(b) Is there a correlation between the spectra you observe and the appearance of the emission from each of the sources? Explain your response with specific examples.

Continuous and line spectra Gases energized by an electric current passing through them often emit visible light. The glow from neon lights and the discharge tubes in Investigate This 3.4 are examples of this phenomenon; scientists have investigated the spectra of many such emissions. The instruments used for these experiments are called **spectroscopes** or **spectrographs,** Figure 3.3, if they record the spectrum photographically. Usually black-and-white film is used in spectrographs, so that all emissions show up on the negative as black (exposed film) against a white background (unexposed film).

The results of these experiments were just what you found in Investigate This 3.4. The visible emissions from some glowing systems resemble the **continuous spectrum** shown in Figures 3.1, 3.2, and 3.4(a); all the wavelengths (colors) of the spectrum are present. Other glowing systems, as you have observed, emit only a few wavelengths (colors) of light; these are called **line spectra.** A few examples of visible emission line spectra are shown in Figures 3.3 and 3.4(b)–(d). Light enters a spectrograph through long, narrow rectangular slits, as you see in Figure 3.3. It is the image of the slits that is observed. If light

Figure 3.3.

Diagram of the parts of a simple spectrograph.

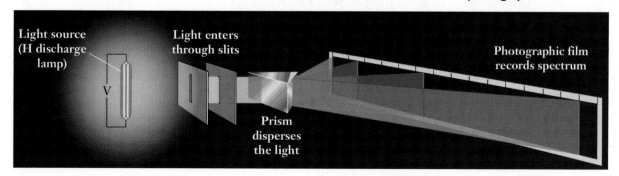

Light source (H discharge lamp)

Light enters through slits

Prism disperses the light

Photographic film records spectrum

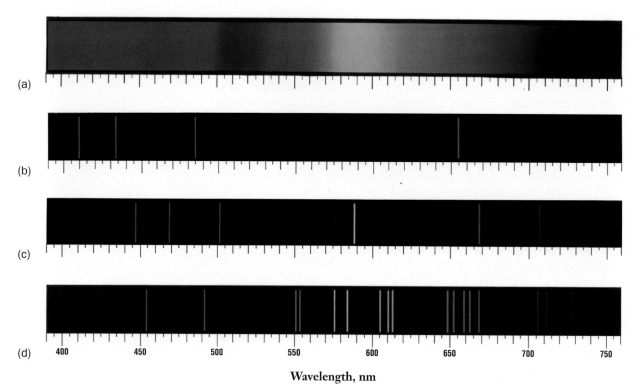

Wavelength, nm

Figure 3.4.

Visible emission spectra.
(a) is a continuous spectrum
to compare with line spectra of
(b) hydrogen, (c) helium, and
(d) barium atoms.

One prominent emission
line from the sun could not be
assigned to any element known
on Earth, so it was assumed to
be a new element and was
named "helium" (*helios* = sun).
Soon helium was also discovered on Earth; its emission
line spectrum matched the
emission line from the sun.

of a single wavelength (color) enters the slits, the image you see is a long, narrow rectangle, *a line*. A line spectrum shows that the emission source is emitting only certain wavelengths of light.

Line spectra of elements and stars The emission wavelengths in a line spectrum are characteristic of the element(s) in the sample that is(are) emitting. *Each element has a unique line spectrum;* its presence in a sample can be identified by its line spectrum. The continuous visible spectrum of the sun that Newton saw proved to be much more interesting when it was investigated with better spectrographs. There are brighter lines at several wavelengths in the sun's spectrum. When scientists compared these brighter lines with the line spectra of known elements, they found that they could match emission lines from the sun with emissions lines from elements on Earth. Thus, spectroscopy could be used to analyze the elemental composition of the sun. Further improvements in telescopes and spectrographs now permit us to do the same kind of analysis of far more distant stars and other cosmic structures, such as nebulas like the one pictured in the chapter opener.

─────────── **3.6 WORKED EXAMPLE** ───────────

Spectroscopic analysis of stellar composition

Figure 3.5 shows the spectral emission lines from the surfaces of two stars (without the continuous background emission) and line spectra for several

continued

elements. (The line spectra are positive photographic images; they are white where they are exposed and dark where not exposed.) Do hydrogen atoms occur on star X?

Necessary information: We need the emission lines from the star and from hydrogen atoms, both of which are given in Figure 3.5.

Implementation: A simple way to compare the spectra is to line them up with one another (or use a straightedge aligned at the same wavelengths on the two scales in Figure 3.5) to see if the lines in the atomic spectrum also appear in the stellar spectrum. The match up for star X and hydrogen is

The four lines in the hydrogen atom spectrum match four lines in the stellar spectrum, so we can conclude that hydrogen atoms are present on star X. The other emission lines in the spectrum of the star show that other elements are present as well (see Check This 3.7).

Does the answer make sense?   In Section 3.3, we will look at our present model for the beginning and evolution of our universe to see how scientists make sense of data like these.

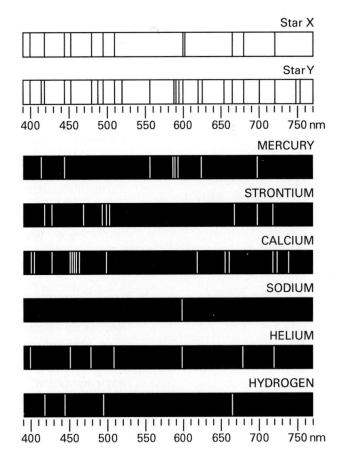

3.7 CHECK THIS

Spectroscopic analyses of stellar composition

Use the data in Figure 3.5 to answer these questions. Give the reasoning for your answers.
(a) Does hydrogen occur on star Y?
(b) What is the composition of star X?
(c) What is the composition of star Y?
(d) What elements are common to both stars?

Figure 3.5.

Stellar and elemental atomic line spectra.

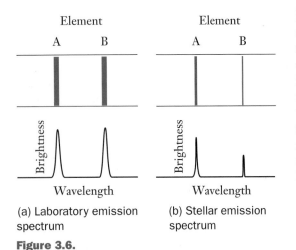

(a) Laboratory emission spectrum

(b) Stellar emission spectrum

Figure 3.6.

Laboratory and stellar emission lines from elements A and B.

(a) Represents the emissions from equal amounts of elements A and B; (b) are the same emissions from a star. The peaks on the graph beneath each spectrum show the brightness of the corresponding line: the brighter the emission, the more abundant the element.

Elemental abundance in the universe Determining the abundance of the elements in the universe (or even our little corner of it) is not an easy task. Spectroscopic studies provide most of the evidence for this determination, but they are limited in the information they can provide. It is easier and more accurate to find the *ratio* of the abundance of one element to another, rather than try to determine the absolute amount of either one. Figure 3.6 illustrates how the ratio of the amounts of two elements, A and B, might be determined from the relative brightness of their emission lines. In a laboratory sample, Figure 3.6(a), the emission lines are equally bright when there are equal amounts of A and B. By comparing these emissions to the relative brightness of the emissions of these elements from a star, Figure 3.6(b), we can determine the amount of B relative to the amount of A in the star.

3.8 CHECK THIS

Relative abundance of elements in Figure 3.6

If you arbitrarily assign the value 100 to the stellar abundance of element A in Figure 3.6(b), what value would you give the abundance of element B? Explain.

From the results of many observations similar to those in Figure 3.6 plus other kinds of data and models of the universe, scientists make estimates of the *relative* abundance of all the elements in the universe. Figure 3.7 shows these relative abundances for the first two-thirds of the periodic table and brings together a great deal of information about the universe. In order to present the information compactly, the figure presents the data on a logarithmic scale and gives all the abundances relative to silicon. The logarithmic scale is required because the lightest elements are 100 billion (10^{11}) times more abundant than heavier elements. There is no good way to express such enormous variations on a linear scale. Silicon is chosen as the reference because it is quite abundant on Earth, moderately abundant in the universe, and easily detected.

3.9 WORKED EXAMPLE

Abundance of carbon relative to iron in the universe

Use the data in Figure 3.7 to estimate the abundance of carbon relative to iron in the universe.

Necessary information: The data in the figure are given on a logarithmic scale. One unit on a log scale represents a 10-fold change (an order of magnitude) in the value.

Implementation: The values for C and Fe on the plot are close to 7 and 6, respectively. This means that the number of carbon atoms in the universe is about 10 times as large as the number of iron atoms (about 10^7 compared to 10^6).

continued

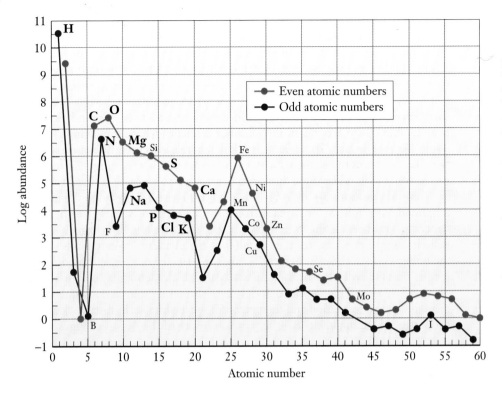

Figure 3.7.

Abundance of elements in the universe. All abundances are relative to 10^6 atoms of silicon, atomic number 14. Elements shown in larger bold type are required in substantial amounts by all organisms on Earth. Elements in smaller regular type are required in smaller amounts or by only a few organisms. The points are connected only to highlight the trends in the data. Source: R. J. P. Williams and J. J. R. Frausto da Silva, *The Natural Selection of the Elements: The Environment and Life's Chemistry* (Oxford University Press Inc., New York, 1996), p. 290.

Does the answer make sense?   The abundances increase as we go up on the abundance axis. The position of carbon is higher on the plot than the position of iron, so carbon has the higher abundance, which is what we found from the numerical values.

3.10 CHECK THIS

Relative abundances of elements in the universe

Use the data in Figure 3.7 to do these problems.
(a) Estimate the relative abundance of Ni to Co. Explain your reasoning.
(b) What is the most abundant element in the universe?
(c) Which elements are *at least* 10 times more abundant than Si?
(d) Which elements are between 10 and 100 times *less* abundant than Si?

3.11 CONSIDER THIS

What are the trends in elemental abundance?

(a) In Figure 3.7, what is the general trend in abundance as the atomic number increases?
(b) Do you see any major exceptions to the trend you described in part (a)?

Trends in elemental abundance The general trend in Figure 3.7 is decreasing abundance of the elements with increasing atomic number. The trend continues for the rest of the heavier elements that are not shown in the figure. There are two striking exceptions to the relatively smooth downward trend of the abundances as a function of atomic number. First, the abundances of Li, Be, and B (atomic numbers 3, 4, and 5) are less than we would expect, if they behaved like the elements just before and just after them. Second, there is a peak in abundance around atomic number 26 (iron) where the abundances are larger than we would expect for a smoothly decreasing curve. We also note that the abundances of the elements with even atomic numbers are higher than nearby elements with odd atomic numbers.

Several questions are raised by the data in Figure 3.7. Why do the elemental abundances follow (or deviate from) the pattern we see? Have the abundances always been as they are today? If they have changed over time, are they still changing? What is(are) the origin(s) of the elements? Many kinds of data, including some of those in Figure 3.7, are the basis for our present model of the formation and evolution of the universe and the elements that comprise it. In Section 3.3, we will give a brief description of this model and its consequences for elemental atom formation and then go on in succeeding sections to look at some properties of atomic nuclei that help explain more of the details of Figure 3.7.

Reflection and Projection

Light passing through a prism or diffraction grating is diffracted or broken up into its constituent wavelengths (colors). The fundamental discovery that each element has a unique emission line spectrum gave scientists a tool for analyzing the atomic composition of any object, including stars and other luminous objects in the heavens, that emit light. One result of stellar analysis is that hydrogen and helium are found in essentially all stars and throughout the universe. Since these are the lightest elements and the simplest atoms, it makes sense that the atomic history of the universe might begin with these atoms (or their nuclei).

It's a remarkable achievement to be able to analyze a sample that is millions or even billions of light years from Earth (or to discover a new element). The results are part of the experimental evidence that led to the development of the model for elemental synthesis and evolution we will discuss in the next sections of the chapter. Before we see how atoms of the elements are formed in stars, we will review some familiar atomic concepts in a little more detail than in Chapter 1. You may find that you do not need this review and can skip directly to Section 3.3.

3.2. The Nuclear Atom

3.12 CONSIDER THIS

What do you already know about atoms?

List at least six things you know about the composition and structure of atoms. Work in a small group to combine your individual lists into a single list that includes all the things the group knows about atomic structure. Discuss your group list with the entire class and your instructor.

The nuclear atom In our nuclear atomic model, all atoms are composed of the same three types of subatomic building blocks: **electrons, protons,** and **neutrons,** the masses and charges of which are given in Table 3.1. Protons and neutrons make up the tiny, dense **nucleus** (plural, **nuclei**) that occupies about 1 part in 10^{15} of the volume of an atom. Recall from Chapter 1 (see Figure 1.5) that, if an atom were the size of a baseball stadium, the nucleus would be about the size of a pea. The remainder of an atom's volume is "empty" space containing enough light, fast moving electrons to balance the positive nuclear charge.

Table 3.1 *Properties of subatomic particles.*

Particle	Symbol	Unit charge, e^a	Mass, g	Mass, u^b
Electron	e^- $\left(_{-1}^{0}e^-\right)$	$1-$	9.10939×10^{-28}	5.48680×10^{-4}
Proton	p $\left(_{1}^{1}p^+\right)$	$1+$	1.672623×10^{-24}	1.00728
Neutron	n $\left(_{0}^{1}n^0\right)$	0	1.674929×10^{-24}	1.00866

[a]The charges here are given *relative* to the magnitude of the charge on the electron, which is called the **elementary charge,** which is 1.60218×10^{-19} C (coulomb).

[b]An **atomic mass unit,** u, is defined as exactly $\frac{1}{12}$ the mass of a carbon-12 atom: u = 1.66054×10^{-24} g. Relative atomic masses (see the end pages of this book) may be interpreted as molar masses in grams or as atomic masses in atomic mass units.

Protons and atomic number The nuclei of *all* the atoms of a particular element have the same number of protons; for example, all oxygen atoms have eight protons. Atoms of different elements have different numbers of protons in their nuclei; whereas all oxygen atoms have eight protons, all carbon atoms have six protons. The number of protons is therefore the determining characteristic of an element. The number of protons in a nucleus is called the **atomic number, Z.** Each proton is positively charged, so the atomic number is equal to the positive charge on the nucleus.

The periodic table on the end papers of this book shows the atomic number above each elemental symbol. Since each element has a unique atomic number, the elemental symbol itself specifies the atomic number of the element. For emphasis and clarity, however, we sometimes write the atomic number (equal to the nuclear charge) as a *subscript to the left of the elemental symbol.* For example, hydrogen may be written as $_1$H, helium as $_2$He, and carbon as $_6$C, as illustrated in Figure 3.8.

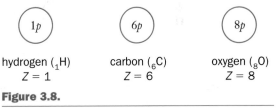

hydrogen ($_1$H)   carbon ($_6$C)   oxygen ($_8$O)
  Z = 1   Z = 6   Z = 8

Figure 3.8.

Nuclear structure and nomenclature: atomic number. The atomic number (Z) is the number of protons in the element and is unique to each element.

Neutrons and mass number Neutrons and protons each have a mass of about 1 u, as you can see in Table 3.1. The **mass number, A,** of an atom is equal to the sum of the number of protons, Z, *plus* the number of neutrons, N: $A = Z + N$. Thus, the mass number represents the total number of nuclear particles, collectively called **nucleons,** of an element. When the mass number of a nucleus is to be specified, it is written as a *superscript to the left of the atomic symbol.* For example, the helium nucleus is represented as $_2^4$He to specify that the nucleus consists of two protons (Z = 2) and a total of four nucleons, two protons and two neutrons ($A = Z + N = 4$). When the name of an element is written out, we specify its mass number by writing that number after the name, as in helium-4. Figure 3.9 illustrates our three examples.

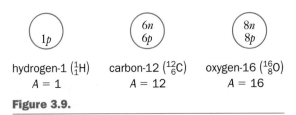

hydrogen-1 ($_1^1$H)   carbon-12 ($_6^{12}$C)   oxygen-16 ($_8^{16}$O)
  A = 1   A = 12   A = 16

Figure 3.9.

Nuclear structure and nomenclature: mass number. The atomic mass number (A) is the number of protons plus the number of neutrons (N): $A = Z + N$.

3.13 WORKED EXAMPLE

Writing atomic symbols

How many protons, electrons, and neutrons are found in an atom of potassium-41? Write its atomic symbol, including its mass number and atomic number. Note that we have specified the mass number of the potassium nucleus by naming it "potassium-41."

Necessary information: We will use the periodic table as well as the definitions of mass number, A, and atomic number, Z.

Implementation: *Determine the number of protons:* The atomic number, Z, is equal to the number of protons in the nucleus of an atom. In the periodic table, we find that potassium's atomic number is 19. Therefore, the potassium-41 nucleus contains 19 protons ($Z = 19$).

Determine the number of electrons: Since an atom of potassium-41 is neutral, the number of electrons is equal to the number of protons. A potassium-41 atom contains 19 electrons.

Determine the number of neutrons: The mass number, A, is the sum of the atomic number, Z, plus the number of neutrons, N: $A = Z + N$. We rearrange this equation to solve for the number of neutrons: $N = A - Z$. The atom name, potassium-41, gives the mass number, $A = 41$. Therefore,

$$\text{number of neutrons} = N = A - Z = 41 - 19 = 22$$

Write the atomic symbol: Use the periodic table to find the atomic symbol, K, for potassium. Write the mass number as a superscript in front of K. Write the atomic number as a subscript in front of K. The atomic symbol, including the mass number and atomic number, is $^{41}_{19}\text{K}$.

Does the answer make sense? Double check to be sure that $A = Z + N$, so you can be confident of your arithmetic.

3.14 CHECK THIS

Writing atomic symbols

How many protons, neutrons, and electrons are found in these atoms? Write their atomic symbols, including their mass numbers and atomic numbers.
(a) lead-206   (b) cobalt-59

Isotopes The number of neutrons can differ among nuclei having the same number of protons, and this difference affects the mass of the atom. Atoms with the same atomic number, Z, but different mass numbers, A, are known as **isotopes.** For example, there are three isotopes of hydrogen: $^{1}_{1}\text{H}$ has no neutrons in the nucleus, ^2_1H has one neutron, and ^3_1H has two neutrons. All three are hydrogen atoms because they have only one proton in their nuclei. A sample of hydrogen atoms naturally occurring on Earth consists of 99.985% ^1_1H and 0.015% ^2_1H; ^3_1H does not occur naturally; it is synthesized in nuclear reactors. We specify which isotope we are referring to by writing the mass number after the name: helium-4, hydrogen-3, or carbon-12 (also abbreviated as He-4, H-3, and C-12, respectively). Isotopic nuclei are illustrated in Figure 3.10.

There is no significant difference in the size of an atom when the number of neutrons is different. This is because the size of an atom is determined not by the tiny nucleus but by the much larger volume in which its electrons move, as we saw in Chapter 1 and will elaborate upon in Chapter 4. The chemical reactions of isotopes of any element are nearly indistinguishable, because **chemical properties are largely determined by the arrangement of the electrons, not the nuclei.** Substitution of one isotope for another in a reaction almost never changes the reaction; using isotopic tracers to follow the fate of particular atoms in a chemical change can provide information about the pathway for the reaction. Also, subtle variations in the way the isotopes of an element take part in a reaction under different conditions can be used to tell what the conditions were when the reaction occurred. An example of this use of isotopes is given in Section 3.9.

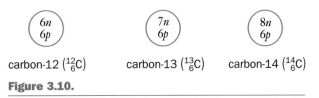

carbon-12 ($^{12}_6C$)   carbon-13 ($^{13}_6C$)   carbon-14 ($^{14}_6C$)

Figure 3.10.

Nuclear structure and nomenclature: isotopes. Isotopes of an element all have the same number of protons but different numbers of neutrons.

Geochemists make extensive use of isotopic analyses to determine the age of geological samples and to reconstruct the conditions, such as temperature, under which samples were formed.

3.15 WORKED EXAMPLE

Identifying isotopes

(a) Determine the pair of isotopes from the atoms in this list (X represents an unknown element):

$$^{54}_{24}X \qquad ^{54}_{26}X \qquad ^{56}_{25}X \qquad ^{56}_{26}X$$

(b) Identify the element that has these isotopes.

Necessary information: We will use the periodic table as well as the definitions of an isotope, mass number, A, and atomic number, Z.

Implementation:
(a) We need to identify the pair of atoms that have the same atomic number, Z, but different mass number, A. In the above list, the pair of isotopes is $^{54}_{26}X$ and $^{56}_{26}X$.

(b) To determine the element, we note that iron, Fe, has an atomic number of 26. Two isotopes of iron are $^{54}_{26}FE$ and $^{56}_{26}Fe$.

Does the answer make sense? These two isotopes of iron differ only in the number of neutrons that each contains. $^{54}_{26}Fe$ contains 28 neutrons ($N = A - Z = 54 - 26 = 28$), while $^{56}_{26}Fe$ contains 30 neutrons ($56 - 26 = 30$).

3.16 CHECK THIS

Identifying isotopes

Identify the pair of isotopes and explain why you reject the others:

(a) $^{242}_{94}\text{Pu}$ and $^{242}_{96}\text{Cm}$ (b) $^{134}_{56}\text{Ba}$ and $^{138}_{56}\text{Ba}$ (c) $^{27}_{13}\text{Al}$ and $^{27}_{14}\text{Si}$

Ions In an electrically neutral atom of an element, the number of electrons is equal to the number of protons; the net charge is zero. As you saw in Chapter 2, an **ion** is formed if the number of electrons is either fewer than or greater than the number of protons and the net charge on the ion is indicated as a *superscript to the right of the atomic symbol.* For example, the hydrogen cation is a positively charged ion because the atom has lost an electron; it can be written as $_1\text{H}^+$. The hydrogen anion is a negatively charged ion because the atom has gained an electron; it can be written as $_1\text{H}^-$. You have seen that it's possible for an atom to lose or gain more than one electron; for example, $_{12}\text{Mg}^{2+}$ has two fewer electrons than it has protons. In chemical formulas and equations, charge is nearly always shown, if it is not zero, but the mass and atomic numbers are usually omitted. For nuclear reaction equations, mass numbers, A, are shown to indicate which isotopes are meant. Often, especially when we want to balance nuclear equations, we show both mass number, A, and atomic number, Z, for all species in the reaction. You will see several examples in the remainder of the chapter.

> By convention, 1+ and 1− charges on ions are shown as + and −; the numeral one is not used.

3.17 WORKED EXAMPLE

Counting subatomic particles in ions

How many protons, neutrons, and electrons are present in the ion, $^{34}_{16}\text{S}^{2-}$?

Necessary information: We will need the definitions of ion; mass number, A; and atomic number, Z.

Implementation: *Determine the number of protons and neutrons:* Using the explanation given in Worked Example 3.13, the $^{34}_{16}\text{S}^{2-}$ nucleus contains 16 protons and 18 neutrons.

Determine the number of electrons: The superscript $2-$ shows that $^{34}_{16}\text{S}^{2-}$ is a negatively charged ion, containing two more electrons than protons. We already know that $^{34}_{16}\text{S}^{2-}$ contains 16 protons. Therefore, the number of electrons in $^{34}_{16}\text{S}^{2-}$ is 18 ($= 16 + 2$).

Does the answer make sense? Given $^{34}_{16}\text{S}^{2-}$, we can add the number of protons ($Z = 16$) and neutrons ($N = 18$) to check the mass number, A: $16 + 18 = 34$. Since this ion has a negative two charge, it contains two more electrons than protons, for a total of 18 electrons.

3.18 CHECK THIS

Explain the difference(s) between $^{206}_{82}\text{Pb}^{2+}$ and $^{208}_{82}\text{Pb}^{4+}$.

3.3. Evolution of the Universe: Stars

3.19 CONSIDER THIS

How much of the universe is hydrogen and helium?

Over 99% of the atoms in the universe are hydrogen or helium. Helium atoms make up about 7% of this total. Use the data in Figure 3.7 to prove or disprove these statements.

Most stars contain hydrogen and helium, as you discovered for the two stars in Check This 3.7. These are the simplest elements and they are ubiquitous in the present-day universe. Whatever model we develop to explain the origin and evolution of the universe must account for all this hydrogen and helium.

The Big Bang theory The most widely accepted scientific model for the beginning of the universe is called the **Big Bang theory,** according to which the universe began in a rapid expansion from an initial state of nearly infinite mass at nearly infinite density and temperature. Though evidence supports this theory, the laws of energy and matter that we are familiar with on present-day Earth fail to explain how the universe could have been so dense and so hot. Nor do they explain what happened within the first one-trillionth of a second (10^{-12} s = 1 picosecond, ps). Nevertheless, the Big Bang is the only scientific theory yet devised that explains the observable features of the universe as we know them today.

In trying to understand how these features evolved, the expansion is most significant. During the expansion, which continues to this day, the universe has cooled from an initial temperature of $>10^{27}$ K to its modern-day average value of about 3 K. As matter cooled and continued to expand, **condensations** occurred as some of the particles condensed, that is, stuck together and became more dense or concentrated. In order for particles to stick together, forces attracting them to one another must be greater than the kinetic forces (motions) that keep them apart. Lower temperatures reduced the kinetic energy (slowed the motion), so bits of matter stuck together as the temperature went down. Figure 3.11 summarizes the history of the universe as a series of condensations from the Big Bang to the formation of Earth. In this section we will focus on the evolution of stars and, in Section 3.7, return to the formation and history of Earth.

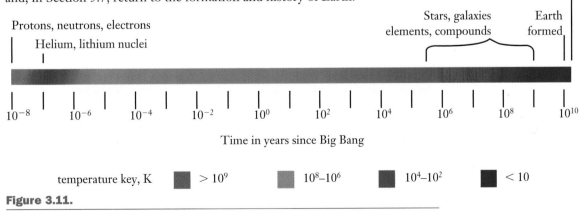

Figure 3.11.

A timeline for the condensations in the evolution of the universe. The temperature indicated by the color is the average temperature of the universe. There are a vast number of local variations, especially within and near stars and planets.

The first nuclear fusions A few seconds after the Big Bang, when the temperature had dropped to approximately 10^9 K, protons and neutrons could condense to form small amounts of $^2\text{H}^+$, $^3\text{He}^{2+}$, $^4\text{He}^{2+}$, and $^7\text{Li}^{3+}$ nuclei by reactions like these:

$$^1p^+ + {}^1n^0 \rightarrow {}^2\text{H}^+ + \gamma \text{ (gamma radiation; discussed in Section 3.4)} \quad (3.1)$$

$$^2\text{H}^+ + {}^2\text{H}^+ \rightarrow {}^3\text{He}^{2+} + {}^1n^0 \quad (3.2)$$

$$^3\text{He}^{2+} + {}^2\text{H}^+ \rightarrow {}^4\text{He}^{2+} + {}^1p^+ \quad (3.3)$$

These reactions are **nuclear fusions,** the building of heavier nuclei from two (or more) lighter particles or nuclei. Nuclear fusion often creates new elements, as in reaction (3.2) where an isotopic helium, He-3, nucleus is produced from two isotopic hydrogen, H-2, nuclei.

For nuclear fusion to occur, the positively charged nuclei have to be able to get close enough together to fuse into a single nucleus. They have to collide with enough kinetic energy to overcome their mutual **coulombic repulsion** [equation (2.2), Section 2.4)]. A minute or two after the Big Bang, the temperature had dropped further ($<10^7$ K); the nuclei didn't have enough kinetic energy to overcome this repulsion and these nuclear fusions stopped. Most nuclei were still $^1\text{H}^+$ ($^1p^+$, protons). As the expansion continued for about 10^5 years, the temperature fell below a few thousand kelvin and electrons became associated with atomic nuclei to form ions and atoms:

$$^4\text{He}^{2+} + e^- \rightarrow {}^4\text{He}^+ \quad (3.4)$$

$$^1\text{H}^+ + e^- \rightarrow {}^1\text{H} \quad (3.5)$$

The universe today is filled with vast clouds of hydrogen left over from the first seconds of its birth and the later condensation of protons with electrons.

3.20 CHECK THIS

Correlating reactions and temperatures in atom formation

In atom formation, what would be the approximate range of temperatures in which each of these reactions might occur? Explain your reasoning.
(a) $^6\text{Li}^{2+} + e^- \rightarrow {}^6\text{Li}^+$
(b) $^2\text{H}^+ + {}^4\text{He}^{2+} \rightarrow {}^6\text{Li}^{3+}$

The life and death of a star Carbon exists throughout the present-day universe, but there was no carbon present when fusion stopped in the early universe. Synthesis of carbon and other heavier elements had to await the formation of stars. For almost 14 billion years, clouds of hydrogen atoms (mixed with smaller amounts of helium and lithium) have spanned the expanding universe. Within these clouds, gravitational attraction pulled atoms together, making some regions of the clouds more dense and hotter than the rest. Collisions in these denser regions became more frequent, and occurred with ever-increasing kinetic energy, as gravity pulled the atoms closer and closer together, until nuclear fusion could again occur. This is how the first stars were born over 13 billion years ago.

Within these local concentrations of hydrogen that became stars, the temperature rose to 1.5×10^7 K. Atoms lost their electrons and the hydrogen nuclei (protons) now had enough kinetic energy to overcome the coulombic repulsion of like charges. The combination of high-energy collisions and the high density of nuclei produced nuclear fusion, beginning with this net reaction:

$$4{}^1\text{H}^+ \rightarrow {}^4\text{He}^{2+} + 2e^+ \text{ (energy released)} \tag{3.6}$$

Reaction (3.6), which we will discuss further in the next sections, is an example of nuclear burning, (hydrogen burning, in this case) because the reaction, once started, releases enough energy to keep the temperature high enough to cause more hydrogen nuclei to fuse.

The conversion of four hydrogen nuclei into one helium nucleus resulted in an overall reduction of the number of particles in the star. Just as reducing the amount of air in a balloon causes it to get smaller, fewer particles in a star produce less outward push to oppose the collapse of the star. Consequently, the core of the star became smaller due to the inward pull of gravity. Total collapse was prevented by the energy released from the fusion reactions at the core of the star, which gave the helium nuclei additional kinetic energy and more outward push (like heating the air in a balloon) to counterbalance the gravitational force.

Ultimately the temperature of the star's core rose to 10^8 K, at which temperature the energies were high enough to start helium burning to produce carbon-12 by a rapid sequence of fusion reactions involving the almost-simultaneous collision of three helium-4 nuclei:

$$^4\text{He}^{2+} + {}^4\text{He}^{2+} + {}^4\text{He}^{2+} \rightarrow {}^{12}\text{C}^{6+} \text{ (energy released)} \tag{3.7}$$

If the star was massive enough, the net conversion of helium nuclei to carbon nuclei led to a second collapse and further heating of the interior of the star to about 10^9 K. At this temperature, the kinetic energies of the nuclei were large enough for carbon burning:

$$^{12}\text{C}^{6+} + {}^{12}\text{C}^{6+} \rightarrow {}^{20}\text{Ne}^{10+} + {}^4\text{He}^{2+} \text{ (energy released)} \tag{3.8}$$

Another series of collapses, heatings, and nuclear burnings occurred in the star and progressed from neon to oxygen to silicon burning in less than a year. Silicon burning lasted about a day and involved stepwise reactions of silicon nuclei with helium nuclei, protons, and neutrons to form nuclei of the elements near iron in the periodic table. An unbalanced representation of this process is

$$^{28}\text{Si}^{14+} + ({}^4\text{He}^{2+} + {}^1p^+ + {}^1n^0) \rightarrow$$
$$^{56}\text{Fe}^{26+} \text{ (+ nearby nuclei) (energy released)} \tag{3.9}$$

The scenario outlined here is based on our present cosmological model of star evolution. The time scale of events is related to the calculated values for the speeds of the processes taking place.

This rapid series of burnings (fusions) caused a catastrophic collapse of the core of the star, as its temperature rose well above 10^9 K, and it exploded. The exploding star is called a **supernova.** The remains of these massive explosions, one of which is shown in the chapter opener, scatter atoms for great distances and continue to emit energy in the form of light and other radiation for many thousands of years. In the cataclysm of the supernova, very high concentrations of neutrons were produced and these reacted with iron and its neighboring nuclei to form all the higher atomic number nuclei by addition of neutrons and subsequent loss of electrons from unstable nuclei. The first stages in this process for some iron nuclei were

$$^{56}\text{Fe}^{26+} + 3{}^1n^0 \rightarrow {}^{59}\text{Fe}^{26+} \tag{3.10}$$

$$^{59}\text{Fe}^{26+} \rightarrow {}^{59}\text{Co}^{27+} + e^- \tag{3.11}$$

Diamonds are pure carbon. All the atoms in all diamonds, natural or synthetic, were made in stars. The nursery rhyme at the beginning of the chapter is right: twinkling stars are "like a diamond in the sky."

The final result of these processes repeated billions of times and still continuing, is the presence of all the elements (up to $Z = 92$) in the universe today. Nuclear furnaces, stars and supernovae, produced and continue to produce all the elements heavier than hydrogen. So, in addition to the twinkling light we see, stars are producing heavier atoms, including carbon.

The birth of a new star All the elements that formed in the star's interior during its billions of years, as well as the heavy elements formed in the supernova explosion, scattered throughout the galaxy in which it occurred. The remains of exploded stars drift through and between galaxies, including our own, as enormous clouds of gas and dust. The surprising fact is that galaxies contain far more of this debris than all the matter in their stars. These interstellar gas clouds are not all alike. Those of greatest interest for our story are the cooler dust clouds that furnish the raw material for the formation of new stars and planetary systems. These clouds are approximately 10^{14} times less dense than Earth's atmosphere.

Our knowledge of the composition of these dust clouds, like our knowledge of the stars, is based mostly on spectroscopic evidence.

3.21 CHECK THIS

Mass of hydrogen atoms in our Milky Way galaxy

Assume that our galaxy, the Milky Way, is a circular disk about 20,000 light years thick with a radius of about 40,000 light years.

(a) What is the volume of our galaxy in m^3? Show how you arrive at your answer. *Hint:* A light year is the distance light travels in 1 year. Light travels at a speed of 3×10^8 m·s^{-1}.

(b) A cubic meter of air at the surface of Earth contains about 2.4×10^{25} molecules which have a mass of about 1.2 kg. What is the mass of a cubic meter of hydrogen atoms that has a density 10^{14} times lower than Earth's atmosphere?

(c) If hydrogen atoms, at the density you found in part (b), fill our galaxy, what is their total mass? Explain how you get your answer.

(d) Masses in the universe are sometimes given in units of solar mass, that is, the mass of our star, the Sun, which is about 2.0×10^{30} kg. According to your result in part (c), how many solar masses of hydrogen atoms are present in our galaxy? Show how you get your answer.

(e) One estimate of the number of stars in our galaxy is 2×10^{11}. Does your result in part (d) provide any justification for the statement we made about the relative amounts of matter in stars compared to that in the gas clouds that fill the galaxy? Explain why or why not.

In the course of time, the atomic nuclei in the dust clouds cool enough to pick up electrons and form atoms and molecules. The dust clouds are rich in water and small organic molecules—the essentials for life. As new stars form, many are surrounded by enormous clouds of dust. When a large enough mass of this dust cools sufficiently, gravitational collapse leads to collections of dust grains that are drawn to one another to form larger and larger lumps of matter. Over time, planets form from these growing lumps as they hurtle through space sweeping up more dust. All the while, this matter is being held in orbit around the new star. One of these new stars is our Sun and one of the planets is Earth, whose story we will continue in Section 3.7.

Reflection and Projection

The history of the universe is a fascinating story that can lead to useful insights about the universe today. We have included a brief synopsis of the story to provide an explanation for the origins of elemental atoms and to show how spectroscopic analysis provided much of the information required to develop that explanation. The Big Bang theory provides a model for the generation of the very lightest elements within the first few moments of the universe's history.

Over the succeeding 13.7 billion years, these initially formed nuclei have interacted with one another by gravitational attraction and by collisions to form stars. In stars, the density and temperature are high enough for nuclei to fuse to form heavier nuclei, whose atoms are observed spectroscopically at the surface of the stars. We have written several of these nuclear reactions, but have not indicated what the rules are for writing and balancing them. We have also pointed out the large amounts of energy released in nuclear reactions but have said nothing about the source of this energy. In order to understand more about the evolution of the elements and their abundance in the universe and on Earth, we need to examine nuclear reactions and energetics in more detail.

3.4. Nuclear Reactions

Emissions from nuclear reactions Before we try to balance nuclear fusion reactions, we need to consider some of their unique products. Both chemical and nuclear reactions can produce heat, sound, and light. Nuclear reactions also produce various particles as well as energy. Two examples of particles are **positrons**, e^+, which have the same mass but opposite charge of electrons, e^-, and **neutrinos**, v_e ("little neutrons"), which are neutral particles with high energies but almost zero mass. Nuclear particles you are more likely to have heard of are **alpha particles, α,** which are helium atom nuclei; **beta particles, β,** which are electrons; and neutrons, $_0^1n^0$. Energy in the form of high-energy **gamma radiation, γ,** is also emitted by many nuclear reactions. Table 3.2 summarizes the emissions from nuclear processes.

Web Companion

Chapter 3, Section 3.4.1–2

Study animations of the changes that occur in nuclear emissions.

Sometimes you hear about "alpha rays" or "beta rays." Alpha and beta particles travel in straight lines from their source, like a ray of light, and they have high energies, so they were first thought to be forms of radiation (see Chapter 4). They were soon shown to be affected by electric and magnetic fields, indicating that they are charged particles, but the names stuck.

Table 3.2 *Emissions from nuclear reactions.*

Emission	Symbol	Contribution to atomic number, Z	Contribution to mass number, A
alpha	$\alpha = {}_2^4He^{2+}$	+2	+4
beta	$\beta^- = {}_{-1}^0e^-$	−1	0
gamma[a]	γ	0	0
positron	$\beta^+ = {}_{+1}^0e^+$	+1	0
neutrino	v_e	0	0
neutron	$_0^1n^0$	0	1

[a]Gamma emissions (radiation) have different amounts of energy that depend on the process that produces them; these differences are not relevant for our purposes.

━━━━━━ **3.22 CHECK THIS** ━━━━━━

Nuclear emissions

(a) 👆 Isotopic nuclei that emit beta particles do not emit positrons, and *vice versa*. Do the emissions shown in the *Web Companion*, Chapter 3, Section 3.4.2, animations exemplify this observation? Explain why or why not.

(b) Do these animations suggest that an isotopic nucleus might emit either an alpha or a beta particle? Explain why or why not.

Balancing nuclear reactions The sum of the overall sequence of fusion reactions that produce helium nuclei from hydrogen nuclei in newly forming stars, equation (3.6), is written in more complete notation as

$$4\,_1^1\text{H}^+ \rightarrow \,_2^4\text{He}^{2+} + 2\,_{+1}^{0}e^+ \text{ (energy release)} \tag{3.12}$$

To balance a nuclear reaction, you need to know that ***charge, mass number (A), and atomic number (Z) are all conserved in nuclear reactions.*** You might find it useful to denote an electron as $_{-1}^{0}e^-$ and a positron as $_{+1}^{0}e^+$, as we have done in equation (3.12), to remind yourself of their respective contributions to the atomic numbers and mass numbers (zero) in a reaction, as shown in columns 3 and 4 of Table 3.2.

In reaction (3.12), the sum of the charges on each side of the arrow is 4+. The sum of the mass numbers is four on each side; the mass of the positron is tiny and contributes zero to the mass number count. Each positron contributes +1 to the sum of the atomic numbers in the products. When this +2 contribution (from two positrons) is combined with the atomic number of helium, the sum of the atomic numbers in the products is 4, which is the same as in the reactants. A helium nucleus contains two neutrons, so reaction (3.12) must have converted protons, $_1^1 p^+$, to neutrons, $_0^1 n^0$:

$$_1^1 p^+ \rightarrow \,_0^1 n^0 + \,_{+1}^{0}e^+ \tag{3.13}$$

Web Companion

┌─────────────────────────────┐
│ Chapter 3, Section 3.4.3–4 │──①
└─────────────────────────────┘
Practice balancing nuclear   ②
reactions with these
interactive animations.   ③
                                ④

━━━━━━ **3.23 WORKED EXAMPLE** ━━━━━━

Balancing nuclear reactions: particles emitted

Balance this nuclear reaction by finding the particle emitted.

$$^{14}\text{C} \rightarrow \,^{14}\text{N} + ?$$

Necessary information: We will use the periodic table to find atomic numbers (charge on the nucleus).

Strategy: Write the reactant and product nuclei with their mass numbers, atomic numbers, and charges shown explicitly. Add these quantities on each side of the reaction arrow and determine what values the unknown particle must have, in order to make the sums the same on each side. Find the particle with these characteristics in Table 3.2 and complete the problem by writing the reaction including this particle.

continued

Implementation:

$$^{14}_{6}C^{6+} \rightarrow \, ^{14}_{7}N^{7+} + \, ?$$

	Known reactants	Known products	Unknown
mass number sum	14	14	0
atomic number sum	6	7	−1

We require a particle that contributes zero mass and an atomic number of −1 (a negatively charged particle). The particle in Table 3.2 that fits this description is the electron (beta particle), $^{0}_{-1}e^{-}$. The balanced nuclear reaction is

$$^{14}_{6}C^{6+} \rightarrow \, ^{14}_{7}N^{7+} + \, ^{0}_{-1}e^{-}$$

A carbon-14 nucleus forms a nitrogen-14 nucleus by emitting a beta particle.

Does the answer make sense? Making sense of the balanced equation means checking *again* to see that the sums of the charges, mass numbers, and atomic numbers are the same on both sides.

3.24 CHECK THIS

Balancing nuclear reactions: particles emitted

Balance these nuclear reactions by finding the particle emitted in each case.
(a) $^{150}Gd \rightarrow \, ^{146}Sm + \, ?$
(b) $^{65}Zn \rightarrow \, ^{65}Cu + \, ?$

3.25 WORKED EXAMPLE

Balancing nuclear reactions: nuclear product

Balance this nuclear reaction by finding the elemental nucleus formed.

$$^{232}Th \rightarrow \, ? + \alpha$$

Strategy: This kind of problem is solved in exactly the same way as in Worked Example 3.23. For the alpha particle, use the notation $^{4}_{2}He^{2+}$.

Implementation:

$$^{232}_{90}Th^{90+} \rightarrow \, ? + \, ^{4}_{2}He^{2+}$$

	Known reactants	Known products	Unknown
mass number sum	232	4	228
atomic number sum	90	2	88

continued

We require a product with a mass number of 228 and an atomic number of 88 (an 88+ charge). The element with atomic number 88 (from the periodic table) is radium, so the unknown product is radium-228, and the balanced nuclear reaction is

$$^{232}_{90}\text{Th}^{90+} \rightarrow {}^{228}_{88}\text{Ra}^{88+} + {}^{4}_{2}\text{He}^{2+}$$

Does the answer make sense? Check the sums of the charges, mass numbers, and atomic numbers on each side of the reaction *again*.

3.26 CHECK THIS

Balancing nuclear reactions: nuclear product

Balance these nuclear reactions by finding the elemental nucleus formed in each case.

(a) $^{18}\text{F} \rightarrow ? + \beta^+$ (b) $^{214}\text{Bi} \rightarrow ? + \beta^-$

3.27 WORKED EXAMPLE

Balancing nuclear reactions

The first human transmutation of one element into another used this reaction:

$$^{14}\text{N} + ? \rightarrow {}^{17}\text{O} + {}^{1}\text{H}$$

Balance the reaction to discover the particle used to produce the transmutation.

Strategy: Again, the basic approach is just as in the previous examples. The variation is that we don't know whether the missing reactant is another elemental nucleus or one of the particles from Table 3.2. Our analysis will give us the answer.

Implementation:

$$^{14}_{7}\text{N}^{7+} + ? \rightarrow {}^{17}_{8}\text{O}^{8+} + {}^{1}_{1}\text{H}^+$$

	Known reactants	Known products	Unknown
mass number sum	14	18	4
atomic number sum	7	9	2

We require a reactant with a mass number of 4 and an atomic number of 2 (a 2+ charge). The alpha particle, $^{4}_{2}\text{H}^{2+}$, fits this description, so the balanced nuclear reaction is

$$^{14}_{7}\text{N}^{7+} + {}^{4}_{2}\text{He}^{2+} \rightarrow {}^{17}_{8}\text{O}^{8+} + {}^{1}_{1}\text{H}^+$$

Does the answer make sense? Check the sums of the charges, mass numbers, and atomic numbers on each side of the reaction *again*.

3.28 CHECK THIS

Balancing nuclear reactions

Balance these nuclear reactions by finding the missing elemental nucleus or particle in each case.
(a) $^{13}C + n \rightarrow$?
(b) $^{1}H + {}^{1}H \rightarrow {}^{2}H +$?
(c) $^{249}Cf + {}^{18}O \rightarrow$? $+ 4n$ (This is the way we use particle accelerators to synthesize most elements beyond uranium.)
(d) $^{20}Ne +$? $\rightarrow {}^{24}Mg$

3.29 CHECK THIS

Element-forming nuclear reactions

(a) Write a balanced nuclear reaction forming nitrogen-14 from carbon-12.
(b) Write a series of balanced nuclear reactions adding alpha particles, one at a time, to form magnesium-24 from carbon-12.
(c) Write a balanced nuclear reaction forming magnesium-24 from two carbon-12 nuclei.

Positron–electron annihilation Equation (3.12) does not finish this reaction sequence in stars. We haven't accounted for the fate of the positrons. Remember that there are electrons in stars. Positrons and electrons destroy (annihilate) one another and produce energy (gamma radiation):

$$_{+1}^{0}e^{+} + {}_{-1}^{0}e^{-} \rightarrow 2\gamma \tag{3.14}$$

The energy released in reaction (3.14) is 1.0×10^8 kJ·mol^{-1}. Accounting for positron–electron annihilation in reaction (3.12) gives this net reaction:

$$4{}_{1}^{1}H^{+} + 2{}_{-1}^{0}e^{-} \rightarrow {}_{2}^{4}He^{2+} \text{ (energy released)} \tag{3.15}$$

The source of the energy released in nuclear reactions is the topic of Section 3.5.

Here on Earth, we take advantage of reaction (3.14) in **positron emission tomography** or **PET scans**. In a PET scan, positron emitting nuclei, usually ^{18}F [see Check This 3.26(a)], are used to identify active areas of the brain in the following way. Active brain cells require glucose for energy. A PET scan subject is injected with a compound that acts like glucose but contains ^{18}F. Once brain cells have taken up the compound, positrons emitted by the ^{18}F react with nearby electrons to produce γ radiation, which is detected by an array of detectors around the person's head. A computer converts this information into a picture, as in Figure 3.12, showing the location of the active brain cells, that is, those that are using the glucose-like substance.

Tomography (*tomos* = slice) refers to pictures of brain "slices." The images of the locations of the positron-emitting nuclei are constructed as though slices of the brain were being examined.

Figure 3.12.

PET scans showing active areas of the brain during seeing and hearing activities.

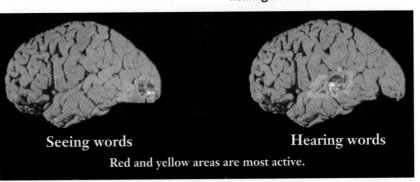

Seeing words   Hearing words
Red and yellow areas are most active.

PET scans can be used to locate regions of a patient's brain that are malfunctioning, due to injury, disease, or genetics, in order to diagnose problems and help determine effective treatments.

3.30 INVESTIGATE THIS

How do the emissions from a radioactive sample differ?

Do this as a class investigation and work in small groups to analyze and discuss the results. The most common kind of household smoke detectors contains a tiny amount of radioactive americium-241 dioxide. Use this source from a dismantled smoke detector as a sample and a radiation survey meter, Figure 3.11, that is sensitive to alpha, beta, and gamma emissions.

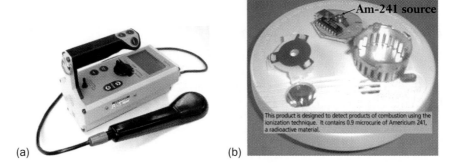

Figure 3.13.

Radiation survey meter and dismantled smoke detector with its label.

(a)　　　　　　　　　　　　　　　(b)

Measure and record the radioactivity of the sample in whatever units the meter provides. Insert a piece of thin cardboard between the sample and the probe and measure and record the amount of radioactivity detected. Repeat the measurement and recording after adding a thin sheet of aluminum to the cardboard between the sample and the probe and then after adding a sheet of lead.

3.31 CONSIDER THIS

What is(are) the emission(s) from decay of americium-241?

(a) Were there substantial changes in the radioactivity when any of the materials was inserted between the source and the detector in Investigate This 3.30? For which ones, if any?

(b) The emissions from radioactive sources have different abilities to penetrate matter. A sheet of cardboard stops alpha particles, but it requires a sheet of aluminum to stop beta particles. Only a dense metal like lead stops gamma radiation. Based on your results from Investigate This 3.30, what is (are) the emission(s) from americium-241? Give your reasoning.

Radioactive decay The only nuclear reaction we have discussed so far is fusion; but the nuclear reaction that is probably most familiar is **radioactive decay.** Radioactive decay transforms one nucleus, a **radioisotope,** into a different nucleus by emission of one or more of the particles in Table 3.2. The "radio" in these terms arises because all these emissions were once thought to be *radiation*. The nuclear emissions you have heard about before were probably from radioisotopes; an example is the americium-241 in the smoke detector in Investigate This 3.30. All the nuclear reactions in Worked Examples 3.23 and 3.25 and in Check This 3.24 and 3.26 are radioactive decays. In a radioactive decay process, the reactant nucleus is often called the **parent** and the product nucleus is called the **daughter.** For example, the nuclear equation for the radioactive decay of the uranium-235 nucleus is

$$^{235}_{92}U^{92+} \rightarrow \ ^{231}_{90}Th^{90+} + \ ^{4}_{2}He^{2+} \tag{3.16}$$

The uranium-235 parent decays to form the thorium-231 daughter by alpha emission.

3.32 CHECK THIS

Daughter nuclei in radioactive decays

Determine the daughter nucleus and write the balanced nuclear reaction for these decays.
(a) Thorium-231 decays by beta emission.
(b) $^{241}Am \rightarrow \ ?\ +$ (decay product(s) from Investigate This 3.30 and Consider This 3.31)

3.33 INVESTIGATE THIS

How can you count a moderately large sample?

For this investigation, use a sample of 150–200 (exact number unknown) coins or candies, such as M&M's®. Place your sample in a covered container, shake and invert the container to thoroughly mix the sample, and then spill it out on a large flat surface. Remove all the coins that show heads (or all the M&M's® whose "m" is showing). Place the remaining sample in the container and repeat the shaking, spilling, and selection. Repeat this procedure, keeping track of how many times you do it, until 12 ± 3 coins or candies remain. Work in your group to discuss how to use your data to determine the number of coins or candies in the original sample.

3.34 CONSIDER THIS

Can you count backward to determine the original sample size?

(a) Assume that each object you used in Investigate This 3.33 has an equal probability of ending up in either of its two states (heads or tails; "m" or no "m") when it is spilled from the container. What fraction of the sample you spilled out will be removed at each round of the procedure? Explain the reasoning for your answer.

continued

(b) Is it possible to use your answer in part (a) to work backward from the final small sample to determine the original sample size? If so, explain how and determine the original sample size. To check your procedure, count the original objects.

Web Companion

Chapter 3, Section 3.4.5–9 — ①

Use interactive animations ②
to determine the half-life of ③
a radioisotope. ④

Half-life Radioisotopes decay by well-defined reactions, like equation (3.16), and each radioisotope has a definite half-life for its radioactive decay. The **half-life** is the time required for one half of a radioactive sample to decay to its products; the half-life of uranium-235 decaying by reaction (3.16) is 7.1×10^8 years. The half-lives of different radioactive nuclei are unique and can often be used as way of dating different materials. Radioisotope half-lives vary from much less than a second to more than 10 billion years. Figure 3.14 shows graphically the fraction of a radioactive sample remaining after a given number of half-lives.

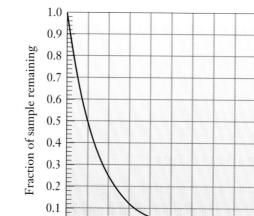

Figure 3.14.

Fraction of original radioactive sample remaining as a function of half-life. After 10 half-lives, a little less than one-thousandth of the original sample remains.

3.35 CONSIDER THIS

What fraction of a sample remains after n half-lives?

If, at time zero, a sample contains N_0 radioactive nuclei, then, after one half-life, the number of nuclei remaining unreacted, N_1, will be $(\frac{1}{2})N_0$. After another half-life has elapsed, half of this remaining sample will have reacted. The number of nuclei remaining after two half-lives, N_2, will be $(\frac{1}{2})(\frac{1}{2})N_0 = (\frac{1}{2})^2 N_0$.
(a) Show that:

$$\text{fraction of sample remaining after } n \text{ half-lives} = f_n = \frac{N_n}{N_n} = \left(\frac{1}{2}\right)^n \quad \textbf{(3.17)}$$

(b) Does equation (3.17) describe the curve in Figure 3.14? Explain why or why not.
(c) How is equation (3.17) related to your answers and results in Consider This 3.34? Can you use equation (3.17) to determine the number of objects in your original sample in Investigate This 3.33? Explain.

3.36 WORKED EXAMPLE

Sample age related to carbon-14 half-life

All living plants contain about the same percentage of carbon-14. When the plant dies, no more carbon-14 is added; the isotope decays by beta particle emission (see Worked Example 3.23) with a half-life of 5730 years. A sample of wood from an archaeological excavation had about 20% as much carbon-14 as living wood. About how old is the sample?

Necessary information: We will use equation (3.17).

Strategy: Find the number of half-lives since the sample of wood was no longer part of a growing plant or tree and multiply by the length of a half-life to get the age of the sample. Determine the number of half-lives required for the sample to have decayed to 20% of its original C-14 by applying equation (3.17).

Implementation: One way to use equation (3.17) is to take the logarithm of both sides of the equation:

$$\ln(f_n) = \ln\left(\tfrac{1}{2}\right)^n = n \cdot \ln\left(\tfrac{1}{2}\right) = -0.693n \qquad \text{(3.18)}$$

We know the decimal fraction of C-14 remaining, 0.20, so we can calculate the number of half-lives, n:

$$n = \frac{\ln(f_n)}{-0.693} = \frac{-1.61}{-0.693} = 2.3 \text{ half-lives}$$

The time since the wood was no longer part of a living plant is

$$\text{sample age} = \left(\frac{5730 \text{ yr}}{1 \text{ half-life}}\right)(2.3 \text{ half-lives}) = 13 \times 10^3 \text{ years}$$

Does the answer make sense? One-quarter (25%) of a radioactive sample remains after two half-lives have gone by. It must take a little longer for another 5% to decay, leaving only 20% (0.20) of the original sample, and this is what we found; the number of half-lives is a bit more than two. You can also check the result using Figure 3.14, which is reproduced here with a horizontal and vertical line showing how to read the graph to go from fraction remaining to number of half-lives elapsed.

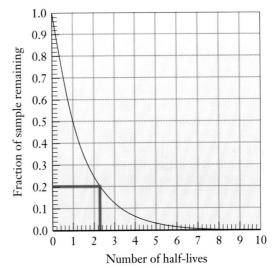

3.37 CHECK THIS

Radioisotope decay and half-life

Phosphorus-32, which is extensively used in biochemical experiments, decays by beta emission with a half-life of 14.3 days. Phosphorus-32 waste can be disposed of as trash, if it has been stored in radioactive waste containers until less than 1% of the original sample is left. About how long must a sample of phosphorus-32 be stored before it can be placed in the trash? Explain your reasoning.

═══ **3.38 CHECK THIS** ═══

Half-life determination

(a) 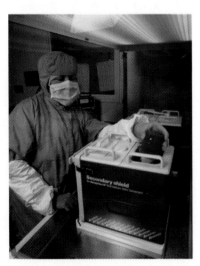 Draw a decay curve for the element undergoing radioactive decay in the *Web Companion*, Chapter 3, Section 3.4.6–9. Clearly indicate on the curve how you would measure the half-life of the element. *Hint:* Equation (3.17) might be helpful.

(b) Clearly indicate on the curve in part (a) how you would show that the half-life is a constant, no matter what the size of the sample.

Radioisotopes and human life Radioisotopes can be harmful to living things. For example, strontium-90, one of the many radioactive products from nuclear weapons explosions, decays by emission of a beta particle:

$$^{90}_{38}\text{Sr} \rightarrow ^{90}_{39}\text{Y} + ^{0}_{-1}e^{-} \tag{3.19}$$

During the 1950s, when nations were testing nuclear weapons above ground, strontium-90 was present in the particles that fell on plants, was eaten by cows, and got into their milk. Milk is a good source of calcium, which is essential for bone growth, especially in children. Because strontium and calcium are chemically quite similar (in the same family of the periodic table), some of the strontium-90 ended up in children's bones. The beta particles emitted could destroy some of the cells in the bone marrow where red blood cells are produced and could cause anemia or leukemia.

Radioisotopes are also helpful. For example, they are used extensively in medical diagnosis and treatment, as we have seen previously for PET scans. The radioisotope most commonly used in medical diagnosis is technetium-99m, $^{99m}_{43}\text{Tc}$. The "m" stands for "metastable" and means the nucleus is not in its most stable form; it decays by gamma emission with a half-life of 6 hours:

$$^{99m}_{43}\text{Tc} \rightarrow + ^{99}_{43}\text{Tc} + \gamma \tag{3.20}$$

is the decay product of molybdenum-99 (which has a half-life of about 67 hours):

$$^{99}_{42}\text{Mo} \rightarrow ^{99m}_{43}\text{Tc} + ^{0}_{-1}e^{-} \tag{3.21}$$

Molybdenum-99 is produced in nuclear reactors by neutron bombardment of Mo-98:

$$^{98}_{42}\text{Mo} + ^{1}_{0}n \rightarrow ^{99}_{42}\text{Mo} \tag{3.22}$$

The Mo-99 is packaged in small Tc-99m generators, Figure 3.15; many hospitals have these generators and routinely use $^{99m}_{43}\text{Tc}$. A few radioisotopes commonly used in medical applications are listed in Table 3.3. Radioisotopes that are medically useful can also be biologically damaging, so, whenever they are used, radioactive elements and their compounds have to be handled and disposed of with special precautions (see Check This 3.37).

> The name *technetium* comes from the Greek word for *artificial*. All Tc isotopes are unstable. The longest lived has a half-life of 2.6×10^{6} years, so any of the natural element that was present when Earth formed has long ago decayed. Tc was the first synthetic element made.

Figure 3.15.

A generator containing $^{99}_{42}\text{Mo}$ that produces $^{99m}_{43}\text{Tc}$ for medical purposes.

Table 3.3 *Some radioisotopes used in biomedical applications.*

Radioisotope	Emission	Half-life	Uses
$^{99m}_{43}$Tc, technetium-99m	γ	6 hr	imaging brain, lung, liver, spleen, bone
$^{18}_{9}$F, fluorine-18	β^+	110 min	PET scans
$^{131}_{53}$I, iodine-131	β^-	8 day	treatment of the thyroid
$^{67}_{31}$Ga, gallium-67	γ	78 hr[a]	treatment of lymphomas
$^{51}_{24}$Cr, chromium-51	γ	28 day[a]	blood tests
$^{201}_{81}$Tl, thallium-201	γ	73 hr[a]	diagnosing heart defects
$^{60}_{27}$Co, cobalt-60	β^-, γ	5.26 yr	radiation therapy for cancer

[a]Ga-67, Cr-51, and Tl-201 decay by capturing one of their core electrons to transform a proton to a neutron, e.g., ^{67}Ga $+ e^- \rightarrow {}^{67}$Zn $+ \gamma$.

3.39 CHECK THIS

Electron-capture reactions

(a) Write the electron-capture reaction for Ga-67 (see note to Table 3.3) with full symbols for all nuclei and particles to show that mass number, atomic number, and charge are conserved.

(b) Write the electron-capture reaction for Tl-201.

3.5. Nuclear Reaction Energies

Nuclear reactions convert mass to energy The unimaginably large energies produced by the stars, as well as nuclear explosions and nuclear power reactors on Earth, are evidence of the energy that can be released in nuclear reactions. The source of this energy arises is conversion of some mass to energy: *Mass is not conserved in nuclear reactions.* From his relativity theory, Albert Einstein (German-born physicist, 1879–1955) derived the **mass–energy equivalence** relationship:

$$\Delta E = \Delta m \cdot c^2 \qquad (3.23)$$

In equation (3.23), c is the **speed of light**, 3.00×10^8 m·s^{-1} and, if the change in mass, Δm (final mass minus initial mass), is in kilograms, the equivalent amount of energy, ΔE, is in joules.

3.40 CHECK THIS

Energy equivalence of mass

(a) If 0.1 mg ($= 1 \times 10^{-7}$ kg; about the mass of a tiny grain of salt) of mass is converted to energy, what is ΔE? *Hint:* The final mass is zero and the initial mass is 0.1 mg, so Δm is -0.1 mg.

(b) Is the process in part (a) exothermic or endothermic? Explain how you know.

3.41 WORKED EXAMPLE

Energy release in nuclear fusion

What is the energy change, ΔE, in the net reaction of four protons and two electrons to form a helium nucleus?

$$4{}^{1}_{1}\text{H}^+ + 2{}^{0}_{-1}e^- \rightarrow {}^{4}_{2}\text{He}^{2+} \text{ (energy released)} \tag{3.15}$$

What is the energy change when 1.00 g of the reactants undergoes reaction (3.15)?

Necessary information: We have the masses of the reactants in Table 3.1 and use a reference handbook to find that the mass of a ${}^{4}\text{He}^{2+}$ nucleus is 6.64466×10^{-24} g.

Strategy: Use the particle mass data, in kilograms, to find the total mass of reactants and products and then take the difference to find Δm, the change of mass in the reaction. Then use the mass–energy equivalence relationship, equation (3.23), to calculate ΔE, the energy change. Convert this energy for four protons and two electrons to the energy change that 1 g (0.00100 kg) of reactants would produce.

Implementation:

$$\text{reactant mass} = 4 \cdot (1.67262 \times 10^{-27} \text{ kg}) + 2 \cdot (9.1094 \times 10^{-31} \text{ kg})$$
$$= 6.69230 \times 10^{-27} \text{ kg}$$

$$\text{product mass} = 6.64466 \times 10^{-27} \text{ kg}$$

$$\Delta m = 6.64466 \times 10^{-27} \text{ kg} - 6.69230 \times 10^{-27} \text{ kg} = -0.04764 \times 10^{-27} \text{ kg}$$

$$\text{energy equivalent} = \Delta E = (-0.04764 \times 10^{-27} \text{ kg})(3.00 \times 10^8 \text{ m} \cdot \text{s}^{-1})^2$$
$$= -4.29 \times 10^{-12} \text{ J}$$

We can convert to 1 g (0.00100 kg) of reactants, using the equivalence:

$$6.69 \times 10^{-27} \text{ kg of reactants} = -4.29 \times 10^{-12} \text{ J}$$

$$\text{energy from 1.00 g of reactant} = (0.00100 \text{ kg})\left(\frac{-4.29 \times 10^{-12} \text{ J}}{6.69 \times 10^{-27} \text{ kg}}\right)$$

$$= -6.41 \times 10^{11} \text{ J}$$

Does the answer make sense? The reaction is highly exothermic. We have said several times that the energies released in nuclear reactions are large. The -640 billion J we calculated for 1 g of reactants is a large amount of energy (enough to vaporize 250 metric tons of water). Note that the mass data we used have six significant figures, but the difference between them is small and has only four significant figures.

3.42 CHECK THIS

Energy release in nuclear fusion

(a) Calculate the energy change in the nuclear fusion reaction represented in the *Web Companion*, Chapter 3, Section 3.5.1, animation. The masses of C-12 and O-16 nuclei are, respectively, 19.92101×10^{-24} g and 26.55291×10^{-24} g.

(b) How is the energy change represented in the animation?

3.43 CHECK THIS

Nuclear reaction energies

(a) Show that reactions (3.1), (3.2), and (3.3) can be combined to give the net reaction for the nucleosynthesis of helium-4 nuclei in the first few seconds after the Big Bang:

$$2\,{}^1p^+ + 2\,{}^1n^0 \rightarrow {}^4He^{2+} \tag{3.24}$$

Hint: Consider how many times reaction (3.1) must occur to provide the reactants for the succeeding reactions.

(b) What is the energy change in the nuclear reaction represented by equation (3.24)?

(c) How does your result in part (b) compare to the energy change calculated in Worked Example 3.41 for nucleosynthesis of helium-4 by a different pathway? What could be responsible for a difference in energy between the two pathways? *Hint:* Consider the conversion of a proton, ${}^1p^+ = {}^1_1H$, to a neutron, ${}^1n^0$.

Comparison of nuclear and chemical reactions Nuclear reactions are very different from chemical reactions. Chemical reactions involve only electrons. Nuclear reactions (except electron-capture) involve only nuclei and are essentially independent of what any electrons in the vicinity are doing. A large observable difference between chemical and nuclear reactions is that nuclear reactions involve far greater amounts of energy. Table 3.4 compares the energy released in chemical reactions and two kinds of nuclear reactions: The nuclear fusion reactions we have considered so far, which build heavier nuclei from lighter nuclei, and **nuclear fission** reactions, which split heavier nuclei into two lighter nuclei plus neutrons.

Fission is from the Latin *findere* = to split.

Table 3.4 *Comparison of chemical and nuclear reaction energies.*

The energy is for the reaction of 1 g of reactant in each case.

Reaction	ΔE, kJ
combustion of methane (1 g of CH_4) $CH_4(g) + 2O_2(g) \rightarrow CO_2(g) + 2H_2O(l)$	-56
spontaneous nuclear fission of uranium-235 ${}^{235}U^{92+} \rightarrow {}^{90}Sr^{38+} + {}^{143}Xe^{54+} + 2\,{}^1n^0$ (one example)	-7.0×10^7
nuclear fusion of hydrogen-1 to form helium-4 $4\,{}^1H^+ + 2e^- \rightarrow {}^4He^{2+}$	-64×10^7

3.44 WORKED EXAMPLE

Mass loss in nuclear and chemical reactions

Assuming that the energies in Table 3.4 represent mass lost in the reactions, compare the mass loss when 1 g of hydrogen is converted to helium by nuclear fusion to the mass loss when 1 g of methane is burned.

continued

Strategy: Apply the Einstein mass–energy equivalence relationship to the two relevant energies in Table 3.4.

Implementation:

mass loss in fusion = Δm

$$= \frac{\Delta E}{c^2} = \frac{-6.4 \times 10^{11}\,\text{J}}{(3.00 \times 10^9\,\text{m·s}^{-1})^2} \approx -7 \times 10^{-6}\,\text{kg}$$

$$= -7 \times 10^{-3}\,\text{g}$$

mass loss in combustion = Δm

$$= \frac{-5.6 \times 10^4\,\text{J}}{(3.00 \times 10^8\,\text{m·s}^{-1})^2} \approx -6 \times 10^{-13}\,\text{kg}$$

$$= -6 \times 10^{-10}\,\text{g}$$

Does the answer make sense? The energies are different by about seven orders of magnitude, so the mass losses should be different by seven orders of magnitude, as they are.

3.45 CHECK THIS

Mass loss in nuclear fission

Assuming that the energy of the nuclear fission reaction in Table 3.4 represents mass lost in the reaction, calculate the mass loss when 1 g of U-235 reacts as shown. How does this mass loss compare to that for the fusion of hydrogen nuclei to yield helium?

The results in Worked Example 3.44 show that the mass loss for hydrogen-to-helium nuclear fusion, reaction (3.15), is about 7 mg (7×10^{-3} g) for each gram of hydrogen reacted. This is a mass that is easily measurable with a good laboratory balance. The mass loss in the chemical reaction, combustion of methane, is about 0.6 ng (6×10^{-10} g), a mass too small to be measured on any balance. All of our usual mole and mass calculations for chemical reactions assume that *mass is conserved in chemical reactions*. Our assumption gives results that are consistent with experiments, because the reaction energies are so small that their mass equivalence is not observable.

3.46 WORKED EXAMPLE

Energy released in a hypothetical nuclear reaction

How much energy would be released in this hypothetical nuclear reaction?

$$13p + 14n \rightarrow {}^{27}\text{Al}^{13+}$$

The mass of an Al-27 nucleus is 44.7895×10^{-24} g.

continued

Necessary information: We have the Einstein mass–energy equivalence relationship, and the masses of the reactant protons and neutrons are given in Table 3.1.

Strategy: Find the difference in mass between the products and reactants in the reaction and convert the mass change to an energy change.

Implementation: Mass change in forming one Al-27 nucleus:

$$\Delta m = 44.7895 \times 10^{-27} \text{ kg} - \left[13(1.67262 \times 10^{-27} \text{ kg})\right.$$
$$\left. + 14(1.67493 \times 10^{-27} \text{ kg})\right]$$
$$= -0.4038 \times 10^{-27} \text{ kg}$$

Energy equivalent of mass change in forming one Al-27 nucleus:

$$\Delta E = (-0.4038 \times 10^{-27} \text{ kg})(3.00 \times 10^8 \text{ m·s}^{-1})^2 = -3.63 \times 10^{-11} \text{ J}$$

Does the answer make sense? Comparing this result with Check This 3.43(b), we see that there is an increased mass change (loss) as the size (mass) of the synthesized nucleus increases. Do Check This 3.47 to see if this pattern continues.

3.47 CHECK THIS

Energy released in a hypothetical nuclear reaction

How much energy would be released in this hypothetical nuclear reaction?

$$82p + 126n \rightarrow {}^{208}\text{Pb}^{82+}$$

The mass of a Pb-208 nucleus is 345.2788×10^{-24} g. Does your result fit the pattern of increasing mass loss for increasing mass of the synthesized nucleus? Explain.

Nuclear binding energies The masses of most nuclei are known to at least six decimal places. You can do the kind of calculations in Worked Example 3.46 and Check This 3.47 to determine the mass loss and energy released in the hypothetical nuclear reaction forming any nucleus from its component protons and neutrons. Since mass is lost and energy is released in these nuclear syntheses, the energy of the nucleus is lower (more negative) than the separated protons and neutrons, as illustrated in Figure 3.16. The energy zero is the separated protons and neutrons. The nuclei are more stable than their separated components. The energy released in the formation of a nucleus from its protons and neutrons is known as the **nuclear binding energy.**

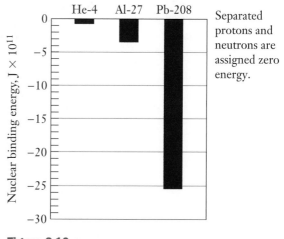

Separated protons and neutrons are assigned zero energy.

Figure 3.16.

Energy released in the synthesis of He, Al, and Pb nuclei.

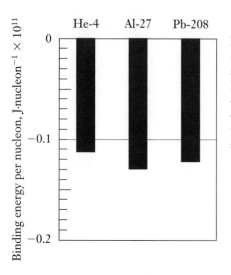

Figure 3.17.

Energy released per nucleon in the synthesis of He, Al, and Pb nuclei.

Note that the major scale divisions here are 50 times smaller than those in Figure 3.16. The binding energy per nucleon is, of course, smaller than the total binding energy for a nucleus.

The greater the mass of the nucleus, the more mass is converted to energy in its formation from protons and neutrons. In heavier nuclei, more protons and neutrons are bound together, so it makes sense that more energy will be released when they bind, as shown in Figure 3.16. But, is the same amount of energy released every time a new proton or neutron is added? To answer this question, we divide the total binding energy for each nucleus by the number of **nucleons** (protons plus neutrons = mass number, A) in the nucleus. For Pb-208, we divide -26.2×10^{-11} J by 208 nucleons and get -0.126×10^{-11} J·nucleon^{-1}. This procedure gives us the **binding energy per nucleon** shown in Figure 3.17 for our example nuclei. The binding energy per nucleon is comparable for all three nuclei and is about -0.1×10^{-11} J·nucleon^{-1} for the formation of a single nucleus. As you see, the binding energy per nucleon does vary a bit. The intermediate mass nucleus, Al-27, is somewhat more stable than the lower and higher mass nuclei, He-4 and Pb-208.

You will usually find nuclear binding energies given per mole rather than per individual nucleus. A mole of nuclei is Avogadro's number of nuclei, so we must multiply the binding energies per nucleon (for a single nucleus) by Avogadro's number, 6.022×10^{23} mol^{-1}, to get the molar binding energy per nucleon. For Pb-208 we have

molar binding energy per nucleon

$$= (-0.126 \times 10^{-11} \text{ J·nucleon}^{-1})(6.022 \times 10^{23} \text{ nucleon·mol}^{-1})$$

$$= -76.0 \times 10^{10} \text{ J·mol}^{-1} = -76.0 \times 10^{7} \text{ kJ·mol}^{-1}$$

Figure 3.18 shows the molar binding energies per nucleon, in kJ·mol^{-1}, for all nuclei. The figure provides a basis for comparison among nuclei; *the more stable nuclei are lower on the diagram.*

 Web Companion

Chapter 3, Section 3.5.4 ──①

Study animations based on the nuclear binding energy curve.
②
③
④

━━━━ **3.48 CHECK THIS** ━━━━

Binding energy per nucleon for carbon

The mass of a carbon-12 nucleus is 19.92101×10^{-24} g. What is the molar binding energy per nucleon, kJ·mol^{-1}, for carbon-12? Compare your answer with the value in Figure 3.18.

━━━━ **3.49 CONSIDER THIS** ━━━━

Trends in nuclear stability

(a) In Figure 3.18, what is the trend in nuclear stability up to iron-56?
(b) What is the trend in nuclear stability after iron-56?

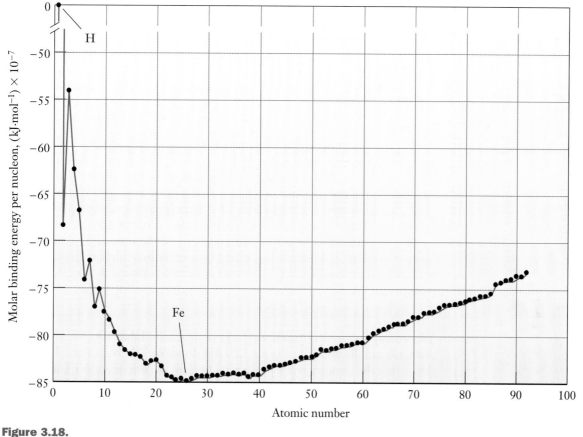

Figure 3.18.

Nuclear binding energy per nucleon. Zero energy on this plot is for hydrogen, $Z = 1$.
The lowest energy, most stable nucleus is iron, $Z = 26$.

Fusion, fission, and stable nuclei Figure 3.18 shows that iron-56 is the most stable nucleus; it lies in an energy valley that includes the metals in the middle of the fourth period of the periodic table. On the low mass (lower atomic number) side of the valley, the products of nuclear fusion reactions, such as ^{12}C (from three ^4He) or ^{16}O (from ^{12}C and ^4He), are more stable than the reactants. Figure 3.19 illustrates the energy changes in these two reactions. Highly exothermic reactions, giving products that are more stable than the reactants, are usually favored to proceed. Such reactions account for the stellar synthesis of the nuclei up to iron and neighboring metals, as outlined in Section 3.3.

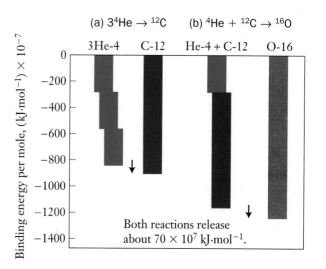

Figure 3.19.

Net energy release in two nuclear fusion reactions. Arrows indicate the energy released by the reaction.

3.50 CHECK THIS

Energy release in nuclear fusion reactions

Show how the energies represented by the bars in Figure 3.19 are obtained and check the statement that both reactions release about the same energy. *Hint:* The data and results from Check This 3.42 and 3.43 could be useful.

On the high mass (higher atomic number) side of the stability valley in Figure 3.18, nuclear fusion is not a favored process. The heavier the nucleus, the less stable it is, so fusion reactions are endothermic. This is the reason why, in the evolution of the elements (Section 3.3), the heavier nuclei are formed mainly by neutron capture and subsequent beta particle emission, rather than nuclear fusion. For heavier nuclei, fission reactions that produce lighter, more stable nuclei from heavier ones are exothermic. Two of the naturally occurring isotopes of uranium, ^{235}U and ^{238}U, undergo **spontaneous nuclear fission** to produce two lighter nuclei and two or more neutrons as products. A representative reaction for ^{235}U is

$$^{235}U^{92+} \rightarrow {}^{90}Sr^{38+} + {}^{143}Xe^{54+} + 2\,{}^1_0 n^0 \tag{3.25}$$

The energy released in reaction (3.25) is shown in Table 3.4 (for 1 g of ^{235}U reacting) and illustrated in Figure 3.20 (for 1 mol of ^{235}U reacting). Figure 3.20 shows that the strontium-90 and xenon-143 nuclei formed in this fission reaction are more stable than the reactant uranium-235 nucleus. Several isotopes of plutonium, an element synthesized in nuclear reactors, also undergo spontaneous nuclear fission.

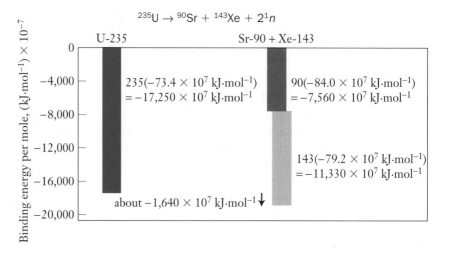

Figure 3.20.

Net energy release in a spontaneous nuclear fission reaction. The molar binding energies per nucleon used here are calculated for the isotopic nuclei in the reaction. They differ slightly from those in Figure 3.18, which are calculated for the most abundant isotope of each element.

Web Companion

| Chapter 3, Section 3.5.2–3 | ① |

Collision-induced fission reactions are represented by these animations.

② ③ ④

3.51 CHECK THIS

Fusion or fission?

Indicate whether each of these elemental nuclei would be more likely to undergo fusion or fission reactions and explain why:
(a) Pu-240   (b) Li-6   (c) Fe-56

Nuclear chain reactions In addition to spontaneous fission, these heavy nuclei also undergo **collision-induced fission** when a neutron collides with the nucleus and gives it extra energy. Reaction (3.26) is an example of a collision-induced fission initiated by collision with a neutron:

$$^{235}_{92}U^{92+} + {}^{1}_{0}n^0 \rightarrow {}^{90}_{38}Sr^{38+} + {}^{143}_{54}Xe^{54+} + 3{}^{1}_{0}n^0$$

(3.26)

The same amount of energy is released by this collision-induced fission reaction as is released by the spontaneous fission. Collision-induced fission is important because it speeds up the fission process. Note that more neutrons are produced in reaction (3.26) than are used up. These product neutrons could go on to cause more fission and even more neutrons—a **chain reaction** that is represented in Figure 3.21.

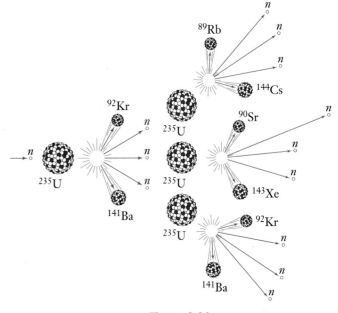

Figure 3.21.

Representation of a ^{235}U chain reaction, initiated by collision with a neutron.

3.52 CHECK THIS

Collision-induced nuclear fission reactions

Reaction (3.26) is one of the collision-induced fission reactions represented in Figure 3.21. Write balanced nuclear reaction equations for the other collision-induced fission reactions shown in the figure.

If every neutron produced by the fissions in Figure 3.21 goes on to cause another fission, the amount of energy released triples at each stage. If the mass of the ^{235}U is large enough, most of the neutrons produced collide with another ^{235}U nucleus before they can escape. In this **critical mass,** the chain continues and causes the fission of a large amount of the ^{235}U in less than 1 second. The energy released in this process is enormous and causes a massive explosion. Such uncontrolled chain reactions are the heart of nuclear bombs. To control this chain reaction, the ^{235}U is contained in such a way that other materials absorb many of the neutrons produced before they can collide with another ^{235}U nucleus. Figure 3.22 shows how this is done in a nuclear reactor. Controlled chain reactions are at the core of power and medical nuclear reactors. Uranium that has been enriched in the ^{235}U isotope is used in nuclear power reactors to produce electricity.

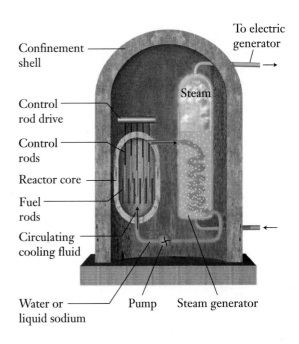

Figure 3.22.

A nuclear power reactor. The rods of neutron-absorbing material, such as Cd, are moved in and out to control the U-235 chain reaction.

3.53 CONSIDER THIS

Chain reactions

(a) Assume that you have a critical mass of U-235 that reacts as shown in this *Web Companion*, Chapter 3, Section 3.5.3, animation. If every neutron produced reacts with another U-235 nucleus within 1 microsecond, 10^{-6} s, about how long will it take for a mole or more of nuclei, 6.022×10^{23} nuclei, to be reacting at each step? Explain clearly how you get your answer. *Hint:* The sort of reasoning we used in Consider This 3.35 and Worked Example 3.36 can be applied here as well.

(b) The result in part (a) represents an uncontrolled nuclear chain reaction. Consider now a controlled chain reaction, as in a nuclear reactor, Figure 3.22. Assume that, for every 10,000 nuclear disintegrations, 10,001 neutrons go on to cause further reaction. All the other neutrons are lost or absorbed by the control rods. About how many reaction steps will it take for a mole or more of nuclei, 6.022×10^{23} nuclei, to be reacting at each step? Explain clearly how you get your answer. *Hint:* In this case, the amount of reaction increases by 1.0001 at each stage.

Reflection and Projection

We interrupted our history of the universe to look in more detail at the nuclear reactions that are the energy source for stars and the source of all the elements more massive than hydrogen. Nuclear reactions, including radioactive decay of unstable nuclei, produce enormous amounts of energy and emit particles and radiation that carry off this energy. All of the emitted particles, as well as the nuclear reactants and products, have to be accounted for in balancing nuclear reactions, which conserve charge, mass number (A = number of nucleons), and atomic number (Z). Nuclear reactions do not, however, conserve mass; the products of many nuclear reactions have less mass than the reactants. The lost mass is converted to energy, which we can calculate using the Einstein mass–energy equivalence relationship. This is the source of the energy from nuclear reactions.

Nuclei have less mass than the sum of the masses of their individual protons and neutrons. The energy equivalent of this mass loss is the nuclear binding energy, which we usually express as binding energy per nucleon, so we can compare one nucleus with another on the same basis. The binding energy per nucleon varies with atomic number and has a minimum at iron-56. Lighter elements generally undergo fusion reactions that produce heavier nuclei that bring them closer to the binding energy minimum. Heavier nuclei are more likely to undergo fission reactions that produce products closer to the binding energy minimum.

Now we will resume our story of the evolution of the universe and its atoms. Our first goal is to try to correlate the observed abundance of the elements in the universe from Section 3.1 with the relative stabilities of their nuclei from Section 3.5. Then we will go on to discuss the formation of planets, in particular, Earth and its elements.

3.6. Cosmic Elemental Abundance and Nuclear Stability

3.54 CONSIDER THIS

How are elemental abundance and nuclear stability related?

This graph is a combination of Figures 3.7 and 3.18 for the low atomic number elements. The left-hand axis is the molar binding energy per nucleon (plotted in black) and the right-hand axis is the logarithm of cosmic elemental abundance (red and blue plots for even and odd atomic numbers, respectively).

(a) Compare the binding energies of lithium, beryllium, and boron with all the others. Does your comparison help explain the abundances of these three elements? Why or why not?

(b) Do binding energies help explain the abundances of nitrogen and fluorine relative to carbon, oxygen, and neon? Why or why not?

(c) Do reactions such as (3.7) and (3.8) help explain the difference in abundance of even and odd atomic number elements? Why or why not?

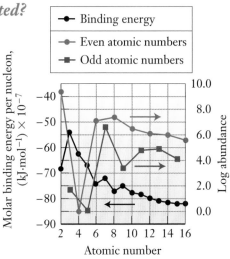

Trends in elemental abundance The evolution of the elements we discussed in Section 3.3 proceeded from lighter to heavier nuclei in sequence. The syntheses of heavier elements depend on the existence of lighter elements that can fuse (or capture neutrons) to make them. The syntheses also depend on a high enough temperature to make them possible. Many stars are not massive enough to go through the complete process outlined in Section 3.3 and these produce few heavier elements. The combination of all these factors creates the general trend of decreasing abundance with increasing atomic number that was shown in Figure 3.7 and repeated in Figure 3.23. The trend continues for the rest of the heavier elements that are not shown in the figures.

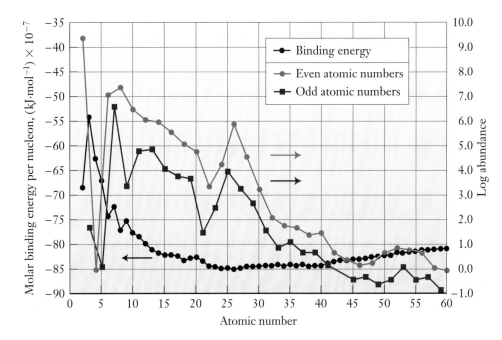

Figure 3.23.

Comparison of cosmic elemental abundance with binding energy per nucleon. Hydrogen is omitted from these plots because its only source in the universe was the Big Bang.

As we noted in Section 3.1, there are two striking exceptions to the relatively smooth downward trend of the abundances as a function of atomic number. The abundances of Li, Be, and B (atomic numbers 3, 4, and 5) are seven to eight orders of magnitude less than we would expect, if they behaved like the elements just before and just after them. Conversely, there is a peak in abundance around atomic number 26 (iron) where the abundances are about three orders of magnitude larger than we would expect for a smoothly decreasing curve. Recall that we started our discussion of nuclear reactions and their energetics by suggesting that they would help us understand more about nuclear evolution. Let's see how the binding energy data help us understand more about the cosmic abundance of the elements.

Correlations between abundance and nuclear binding energy

As you see in Figures 3.18 and 3.23, the nuclear stabilities (binding energies per nucleon) for the first several elements do not increase uniformly, but form a sort of saw-tooth pattern of alternating stabilities. In particular, you see that Li, Be, and B nuclei are less stable than either helium or carbon. Their synthesis from helium is an unfavorable process, because they are less stable than helium. Their fusion reactions to form higher atomic number nuclei are favored, since these reactions all form more stable nuclei. The combination of unfavorable conditions for synthesis, but favored reactions to form heavier, more stable nuclei, explains why the cosmic abundance of Li, Be, and B is so low.

In Section 3.3, we saw that the heaviest nuclei produced in stellar nuclear fusion processes are those in the neighborhood of iron. Now we can see why fusion stops at this point. The most stable nuclei (largest binding energies per nucleon) are those with atomic numbers near 26 (iron). Fusion to form higher atomic number nuclei is not favored, and a wholly different process has to occur in exploding stars to form the higher atomic number nuclei. If all the nucleons in the universe were fused to form the most stable nuclei, the products would be the nuclei near atomic number 26. This is why these products have built up in the cosmos and why there is a peak in cosmic elemental abundance at iron.

3.55 CHECK THIS

Cosmic elemental abundance and nuclear reactions

Figures 3.7 and 3.23 show that there is a small peak in abundance between atomic numbers 50 and 60 and the hint of another between atomic numbers 34 and 40. Do the nuclear reactions represented in Figure 3.21 suggest an origin for these peaks in abundance? Clearly explain the reasoning for your response.

3.56 CHECK THIS

The higher abundance of elements with even atomic numbers

(a) The nuclear synthesis reactions we have written [equations (3.7), (3.8), and (3.9), for example] involve combinations of alpha particles and the products

continued

of their fusion reactions to produce the lighter nuclei (up to those in the iron region). How, if at all, do these reaction sequences help explain the higher abundance of nuclei with even atomic numbers?

(b) Propose fusion reactions of hydrogen and carbon nuclei that would produce stable nitrogen nuclei and thus provide the beginning of reaction sequences that would form nuclei with odd atomic numbers. *Hint:* N-13 is unstable and decays by electron capture.

The nuclear fusion (and neutron capture) processes that produce the elements never stop. They are going on right now in billions of stars in our own galaxy, including our Sun, and in billions more stars in billions of other galaxies. The relative abundances of elements in the universe are continuously changing (although not rapidly), as the lightest nuclei are being used up to form more of the heavier ones. About 9 billion years ago, a very tiny amount of this matter collected and formed our solar system. We will end our brief history of elemental evolution by looking at the elements on earth today.

3.7. Formation of Planets: Earth

3.57 CONSIDER THIS

Where did(does) terrestrial matter come from?

(a) All the matter on Earth was made in stars (or the Big Bang), but how did it get here? Would you expect the abundances of the elements on Earth to be about the same as the cosmic abundance of the elements? Why or why not?

(b) Is matter still being added to the planet Earth? What evidence is there to justify your answer?

The planet Earth Recall, from the end of Section 3.3, that our solar system was formed from the remains of previous stars. The elements that were born in those stars became part of a growing planet, our Earth, as it circled the Sun and grew by sweeping up the dust and larger particles from which the solar system evolved. Initially, these collision processes created a hot body that was mainly molten rock and metals. When the dust was mostly used up, collisions became less frequent and Earth began to cool and differentiate by density into a core, a mantle, and a crust, as illustrated in Figure 3.24.

The radius of the core is about 3400 km.

The mantle is about 3000 km thick.

The crust is about 30 km thick.

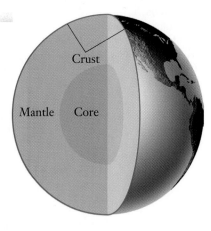

Figure 3.24.

Simplified structure of the planet Earth. The thickness of the crust is exaggerated; the crust is only about $\frac{1}{100}$th the thickness of the mantle.

The core is iron and nickel, probably solid at the center but mostly liquid. The mantle is a viscous, molten mixture of various minerals, mostly silicates such as olivine, $FeMgSiO_4$. The thin, solid crust floats on top of the mantle and is moved about (very slowly) by convection currents in the mantle. We live on the outer surface of the crust and the oceans fill the valleys where the crust is thinnest.

We can't be sure of the composition of the original dust cloud from which our solar system formed, but we can guess that it was probably similar to the composition of the universe shown in Figure 3.7. To find out more about the possible composition of the cloud, we can analyze the crust of the earth to determine its elemental composition. Comparing the cosmic and crustal compositions shown in Figure 3.25 provides clues about the early history of the planet.

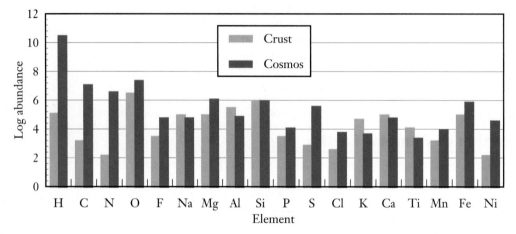

Figure 3.25.

The abundance of selected elements in Earth's crust and in the cosmos. The cosmic abundances are from Figure 3.7. Abundances are all relative to 10^6 Si atoms.
Source: Dickerson and Geis, *Chemistry, Matter, and the Universe* (W. A. Benjamin, Inc., New York, 1976), p. 165.

3.58 CONSIDER THIS

How do crustal and cosmic elemental abundances compare?

For which of the elements in Figure 3.25 are the crustal abundances two or more orders of magnitude less than the cosmic abundances? For which elements are the crustal abundance's two or more orders of magnitude greater than the cosmic abundances? What explanation can you think of for these experimental observations?

For most of the elements in Figure 3.25, the crustal and cosmic abundances are rather similar. Three notable exceptions are hydrogen, carbon, and nitrogen —all of which are much less abundant in the earth's crust than in the universe. If we assume that these elements were present at their cosmic abundance when Earth formed, their loss is one piece of evidence that the early Earth was hot. The simple compounds of these elements, such as methane (CH_4) and ammonia (NH_3), are gases or easily vaporized liquids that would have been driven off the newly forming hot planet. Sulfur, which is about a thousand times lower in abundance in the crust compared to the cosmos, also forms rather volatile compounds with hydrogen (H_2S) and other light elements and could also have been lost by vaporization.

3.59 CHECK THIS

Where is the nickel?

Nickel is also much less abundant in the crust than we would expect, but it is not likely to have vaporized from the early Earth. Look at the structure of the planet Earth shown in Figure 3.24. Where is the "missing" nickel likely to be? Explain the reasoning behind your answer.

The age of the planet Earth To estimate the age of the planet Earth, we take advantage of the properties of radioactive nuclei—their half-lives and the products formed in the decay processes. For example, many rocks contain some uranium, and the ^{238}U nucleus decays by a series of alpha and beta particle emissions to give the ^{206}Pb nucleus:

$$^{238}_{92}U^{92+} \rightarrow {}^{206}_{82}Pb^{82+} + 8{}^{4}_{2}He^{2+} + 6{}^{0}_{-1}e^{-} \tag{3.27}$$

The first step in this decay has a half-life of 4.468×10^9 years and all the subsequent steps are at least 10^4 times faster, so the half-life of the net process shown by equation (3.27) is the half-life of its first step. Pb-206 and U-238 are always found together in rocks and other isotopes of lead are usually not present.

To find the age of a rock containing U-238 and Pb-206, we assume that when the rock formed, all the metal started out as U-238. We also assume that all the Pb-206 was formed subsequently by the series of decays summarized by equation (3.27), and that no metal has been lost from the rock. If the sample originally contained a mole of U-238, 238 g, then after one half-life, there would be one-half mole, 119 g, of U-238 left and one-half mole, 103 g, of Pb-206 would have been formed. The mass ratio of Pb-206 to U-238 in the sample would be

$$\frac{\text{mass Pb-206}}{\text{mass U-238}} = \frac{103 \text{ g}}{119 \text{ g}} = 0.87$$

At the beginning, when no Pb-206 is present, this ratio is zero. The change in the mass ratio of Pb-206 to U-238 as a function of time after the beginning of the decay is shown in Figure 3.26. No rock with a mass ratio larger than about 0.85 has ever been found on Earth. The green lines in Figure 3.26 show that this is the ratio we expect for a sample that is about 4.5×10^9 years old. Since the oldest known rock on Earth is 4.5 billion years old, Earth must be at least this old, as shown in Figure 3.11.

3.60 CHECK THIS

Lead–uranium dating of a sample

Some ancient fossils were found in rock that had a Pb-206/U-238 mass ratio of 0.46. About how old are the fossils?

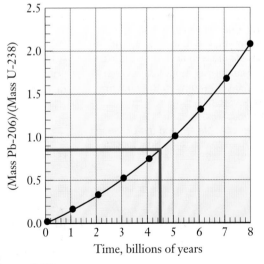

Figure 3.26.

Pb-206/U-238 mass ratio as a function of sample age. The green lines indicate the age of a rock with a Pb-206/U-238 mass ratio equal to 0.85.

The elements of life When the early Earth had cooled enough for its solid crust to form, water began to collect on the surface. The water was forced out of the hot interior of the planet Earth and/or supplied by icy comets colliding with the newly forming planet. On a planet like Earth, which had both a relatively cool crust and a relatively strong gravitational field, liquid water could be retained on the surface where it formed oceans that contained an abundance of dissolved matter. The stage was set for life to emerge. The elements on which life is based are not the most abundant on Earth. On the contrary, Figure 3.27 shows that there is little correlation between the elements that are most abundant in the earth's crust and those most abundant in organisms (humans). Note that Figure 3.27 is not based on a comparison with the abundance of silicon. Because there is so little silicon in organisms, it is not a good standard in this case.

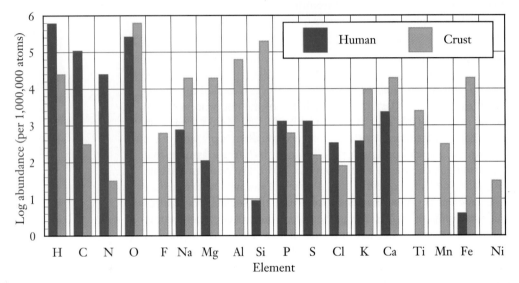

Figure 3.27.

The abundance of selected elements in a human body and in the earth's crust. The abundances are number of atoms out of every 1,000,000 atoms in the human or the crust.

Source: Dickerson and Geis, *Chemistry, Matter, and the Universe* (W. A. Benjamin, Inc., New York, 1976), p. 165.

About 99% of the atoms in living things are hydrogen, carbon, nitrogen, and oxygen. It is not surprising that hydrogen and oxygen predominate in humans (Figure 3.27), because water makes up about 70% of our mass. The fats and carbohydrates in our bodies are composed of carbon, hydrogen, and oxygen and all of our proteins contain these three elements plus nitrogen. Nucleic acids also contain these four elements. The properties of molecules constructed from these atoms seem uniquely suited to sustain life processes. We will examine some of these properties as we go on in the book. It may be that the presence of these molecules, together with water, is required for the emergence and evolution of life.

3.61 CHECK THIS

What are the roles of other elements?

(a) Figure 3.27 shows that calcium and phosphorus have about the same abundance in our bodies and are the next most abundant (along with sulfur) after hydrogen, carbon, nitrogen, and oxygen. What roles do calcium and phosphorus play that they are required in such amounts? *Hint:* Review the opening paragraphs of Section 2.6 in Chapter 2.

continued

(b) Which other elements are present to the extent of at least 50 atoms per one million atoms in our bodies? Explain how you determine which these are. Choose one of these elements and find out what its roles are in your body.

We end our history of the universe by noting that the elements of life are the most abundant elements in the cosmos (see Figures 3.7 and 3.25). They and their compounds are certain to be present in any planetary system and, under the right conditions, as on Earth, carbon-based life may have evolved around many other stars we see twinkling like diamonds in the sky.

Reflection and Projection

It's easy to get caught up in the cosmic story and lose sight of some fundamental problems. From the beginning of the book, we have emphasized interactions of positive and negative electrical charge. In this chapter, for example, we said that repulsions between positive particles have to be overcome in order for nuclear fusion to occur. But you have learned that the binding energies holding protons and neutrons together in a tiny nucleus are enormous compared to chemical energies. What can be the nature of the forces acting within the nucleus? These forces are so strong that nuclei never change during chemical reactions; chemists leave the study of these forces to physicists with high-energy particle accelerators.

The reverse problem arises when you consider the formation of atoms and molecules from nuclei and electrons. The nuclear model of the atom has the negative electrons held near the positive nucleus by their electrical attractions. Why aren't the electrons just attracted closer and closer until the atom collapses? The answer to this question governs all the chemical structures and chemical reactions around us and within us. In the next chapter we will examine the periodic properties of atoms, return to spectroscopy and a more detailed discussion of the properties of waves, and develop an atomic model that answers this question.

Section 3.8 Outcomes Review

We began the chapter by introducing spectroscopy as a method for learning about the identity and relative amounts of atoms in samples that emit light, including stars and other glowing objects in the universe. After reviewing the structure of atomic nuclei, we introduced the Big Bang theory for the formation of the universe and discussed the nuclear processes in stars that are responsible for the formation of the elements we find in the present-day universe. After practice in balancing nuclear reactions, we showed how the mass-to-energy conversion in nuclear reactions explains their enormous energy release. Nuclear binding energies determine whether a nucleus will undergo fusion or fission reactions and help explain the cosmic abundance of the elements. The elements formed in the first stars (and still being formed in present stars) are the matter from which planets, including Earth, are formed and from which life can evolve.

Check your understanding of the ideas in the chapter by reviewing these expected outcomes of your study. You should be able to

- Describe a spectrograph and how it produces the spectrum from its light source [Section 3.1].
- Distinguish between continuous and line spectra, identify an element from its line emission spectrum, and determine relative amounts of two elements by the brightness of their emission lines [Section 3.1].
- Write the complete symbol and locate the subatomic particles in a diagram of any isotope of an elemental atom or ion [Section 3.2].

- Explain cosmic condensation and nuclear fusion events as a function of the temperatures that make them possible [Section 3.3].
- Write and balance reactions along the pathway for the formation of elements *up to about* ^{56}Fe in the first stars [Sections 3.3 and 3.4].
- Write and balance reactions along the pathway for the formation of elements *beyond* ^{56}Fe in the first stars [Sections 3.3 and 3.4].
- Describe the formation of planetary systems and the source of their elementary atoms [Sections 3.3 and 3.7].
- Balance nuclear reactions (fusion, fission, and radioactive decay) by supplying the appropriate reactant or product nuclei and/or other particles [Section 3.4].
- Use half-lives to find the amount of a radioactive nucleus remaining after an elapsed time or to find the time elapsed since decay began [Sections 3.4 and 3.7].
- Determine the energy changes for mass-to-energy conversions in nuclear fusion and fission reactions and *vice versa* [Section 3.5].
- Determine nuclear binding energies from mass-to-energy conversion in formation of nuclei from protons and neutrons [Section 3.5].

- Use nuclear binding energies to predict whether a nucleus is likely to undergo fusion or fission [Section 3.5].
- Give the requirements for a nuclear chain reaction and diagram the differences between controlled and uncontrolled chain reactions [Section 3.5].
- Use abundance data to estimate abundances of elements relative to one another in the universe, on Earth, and in organisms [Sections, 3.1, 3.6, and 3.7].
- Use nuclear binding energies to explain trends in cosmic elemental abundances as well as deviations from the trends [Section 3.6].
- Use the model for planetary formation to explain the structure of the earth and the abundance of elements in the earth's crust [Section 3.7].
- Relate cosmic elemental abundances to possible evolution of life [Section 3.7].
- Describe how the age of a sample can be determined from the initial and present ratios of two isotopes, at least one of which is radioactive [Section 3.9].
- Describe how the history of a sample can be determined from the ratio of two stable isotopes of an element in the sample [Section 3.9].

3.9. Extension—Isotopes: Age of the Universe and a Taste of Honey

Age of stars and the universe One way to measure the age of the universe is to measure the age of stars. If our model, the Big Bang, is correct, the universe must be somewhat older than the oldest stars. If a star contains a long-lived radioactive isotope, its decay rate can be used to date the star, much as we discussed dating rocks in Section 3.7. In the year 2000, uranium was detected spectroscopically (Figure 3.28) in a star in a region of our Milky Way galaxy that is populated by relatively old stars. (The star is in the constellation Cetus and was observed by a telescope in the southern hemisphere.) The uranium is assumed to be U-238 (half-life = 4.468×10^9 yr).

3.62 CONSIDER THIS

Why is the uranium in stars only the U-238 isotope?

We have seen that uranium has two isotopes, U-238 and U-235 (half-life = 7.1×10^8 yr). Both of these isotopes are formed in supernova explosions (Section 3.3). Why can we assume that all the uranium in an old star is U-238? Explain your reasoning clearly.

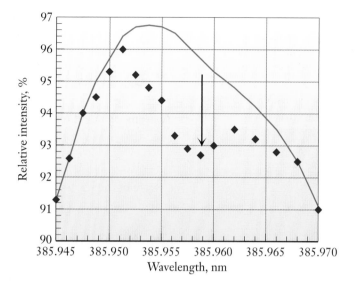

Figure 3.28.

Absorption of light by uranium ions in star CS31082–001.
The red curve is the calculated emission from the star assuming no uranium is present. The points are the experimental data and the arrow marks the center of the uranium ion absorption line. Atomic absorption spectra are discussed in Chapter 4. Compare the shape of this absorption "line" with the shapes for the emission lines shown in Figure 3.6. Source: R. Cayrel *et al.*, "Measurement of Stellar Age from Uranium Decay," *Nature*, 2001, pp. 691–2.

The amount of uranium, U-238, remaining in the star can be obtained from the data in Figure 3.28 (and the amounts of other elements from similar data at other wavelengths). In order to date the star, we also have to know how much U-238 the star contained when it formed. There is no way to know the absolute amount of uranium, or any other element, the star contained when it formed. However, astrophysicists can use models for the evolution of the universe to calculate the *relative* amounts of the heavy elements formed in supernovae. We assume that these are the ratios of these elements that get incorporated into new stars. By comparing the ratio of one element to another in a present-day star to the calculated ratio when the star was born, we can calculate the age of the star.

3.63 WORKED EXAMPLE

Star's age from its U/Ir ratio

Our model for the evolution of the universe predicts that the elements uranium and iridium are formed in a ratio of about 0.050. The U/Ir (uranium to iridium) ratio determined for the star in Figure 3.28 is 0.0079. What is the age of the star?

Necessary information: We need to know the ratios given in the problem statement, the half-life of U-238, 4.468×10^9 yr, and that iridium is not radioactive. We need Figure 3.14 or equation (3.18) to relate the fraction of U-238 left to the number of half-lives that have elapsed.

Strategy: Determine the fraction of U-238 remaining in the star and use equation (3.18) to find out how many half-lives have elapsed. The age of the star is this number of half-lives times the half-life of U-238.

continued

Implementation: When the star formed, there were 50 U atoms for every 1000 Ir atoms and now there are only 7.9 U atoms for every 1000 Ir atoms. Thus, the fraction of U remaining is

$$\text{fraction U remaining} = f_n = \frac{7.9 \text{ U}/1000 \text{ Ir}}{50 \text{ U}/1000 \text{ Ir}} = 0.158$$

Rearranging equation (3.18) to solve for the number of half-lives gives

$$n = \frac{\ln(f_n)}{-0.693} = \frac{\ln(0.158)}{-0.693} = \frac{-1.85}{-0.693} = 2.7 \text{ half-lives}$$

$$\text{age of the star} = (2.7 \text{ half-lives})[4.468 \times 10^9 \text{ yr·(half-life)}^{-1}]$$

$$= 12 \times 10^9 \text{ yr}$$

Does the answer make sense? Our stellar age suggests that the universe must be somewhat older than 12 billion years, since this star has to have been born after the first supernovae occurred. (How do we know this star is younger than the first supernovae?) Use this portion of a half-life decay curve to do another check on the lifetime calculation. The best estimate of the age of the universe is 13.7 billion years (based on measurements of the cosmic background radiation left over from the Big Bang).

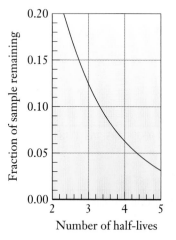

Fraction of sample remaining (y-axis): 0.00, 0.05, 0.10, 0.15, 0.20
Number of half-lives (x-axis): 2, 3, 4, 5

3.64 CHECK THIS

Star's age from its U/Os ratio

Another isotope ratio that can be used to date this star is its U/Os (uranium to osmium) ratio. The present ratio is 0.0065 and the calculated ratio for the formation of these elements is 0.054. Osmium, like iridium, is a stable element. Use these data to calculate the age of the star and compare it with the age found in Worked Example 3.63. Do they reinforce one another? Explain.

A criticism of the ratio method for stellar dating relates to the fact that uranium is a far heavier atom than either iridium or osmium. Simple losses of elements into space from the star would favor the loss of the lighter, faster moving atoms and ions compared to the heavier ones. In order to do the preceding calculations, we have assumed that the only change in the ratios is due to radioactive decay of the U-238. To answer this criticism, we could use a ratio of isotopes that are quite close in mass. Thorium-232 meets this criterion and is present in the star we are considering. However, Th-232 is also radioactive and decays with a half-life of 14.05×10^9 yr. Thus, to date the star, we have to account for the decay of both U-238 and Th-232.

We will use equation (3.18) to solve this problem but will rewrite the number of half-lives, n, as t/τ, where t is the time that has passed (which is what we are trying to calculate) and τ is the half-life of the isotope of interest. We will also write explicit expressions for the fractions of isotopes remaining. In these fractions we use atomic symbols to represent the amount of the nuclei remaining and subscripts to represent the times. Thus, for U and Th in our star, equation (3.18) gives

$$\ln\left(\frac{U_t}{U_0}\right) = -0.693\left(\frac{t}{\tau_U}\right) \tag{3.28}$$

$$\ln\left(\frac{Th_t}{Th_0}\right) = -0.693\left(\frac{t}{\tau_{Th}}\right) \tag{3.29}$$

Since t is the same time for both decays (they are in the same star), we can subtract equation (3.28) from equation (3.29) to get

$$\ln\left(\frac{Th_t}{Th_0}\right) - \ln\left(\frac{U_t}{U_0}\right) = -0.693t\left[\left(\frac{1}{\tau_{Th}}\right) - \left(\frac{1}{\tau_U}\right)\right] \tag{3.30}$$

Using the knowledge that $\ln\left(\frac{a}{b}\right) = \ln a - \ln b$, we can rewrite equation (3.30) as

$$\ln\left(\frac{U_0}{Th_0}\right) - \ln\left(\frac{U_t}{Th_t}\right) = -0.693t\left[\left(\frac{1}{\tau_{Th}}\right) - \left(\frac{1}{\tau_U}\right)\right] \tag{3.31}$$

3.65 CHECK THIS

Show that equations (3.30) and (3.31) are identical

Rewrite the logarithms of the ratios in equations (3.30) and (3.31) as differences to show that the equations are identical.

Finally, solving equation (3.31) for t and rewriting the half-life factor gives

$$t = \frac{\ln(U_0/Th_0) - \ln(U_t/Th_t)}{0.693\left(\dfrac{\tau_{Th} - \tau_U}{\tau_{Th}\tau_U}\right)} \tag{3.32}$$

3.66 WORKED EXAMPLE

Star's age from its U/Th ratio

At present, the star we have been analyzing has $\ln(U_t/Th_t) = -1.70$. One model calculation for the formation of U and Th gives $\ln(U_0/Th_0) = -0.59$. What is the age of the star?

Necessary information: We need the ratios in the problem statement and equation (3.32), as well as the half-lives: $\tau_{Th} = 14.05 \times 10^9$ yr and $\tau_U = 4.468 \times 10^9$ yr.

Strategy: Substitute these known values into equation (3.32) to get the time that has elapsed since the star formed.

continued

Implementation: Let's first calculate the denominator of equation (3.30):

$$0.693\left(\frac{\tau_{Th} - \tau_U}{\tau_{Th}\tau_U}\right) = 0.693\left(\frac{(14.05 \times 10^9 \text{ yr}) - (4.468 \times 10^9 \text{ yr})}{(14.05 \times 10^9 \text{ yr})(4.468 \times 10^9 \text{ yr})}\right)$$

$$= 0.105 \times 10^{-9} \text{ yr}^{-1}$$

Then

$$t = \text{age of the star} = \frac{\ln(U_0/Th_0) - \ln(U_t/Th_t)}{0.693\left(\dfrac{\tau_{Th} - \tau_U}{\tau_{Th}\tau_U}\right)}$$

$$= \frac{(-0.59) - (-1.70)}{0.105 \times 10^{-1}} \qquad = 11 \times 10^9 \text{ yr}$$

Does the answer make sense? Once again, the star is younger than the age of the universe. This value is consistent with the previous calculations we have made, especially considering uncertainties in both the experimental and theoretical values used.

3.67 CHECK THIS

Star's age from its U/Th ratio

(a) Another calculation (based on somewhat different assumptions about nuclear formation) for the formation of U and Th in supernovae gives $\ln(U_0/Th_0) = -0.23$. What does this value lead to for the age of the star we are studying? Is this age consistent with the others we have calculated? Explain why or why not.

(b) Considering all the values we have obtained for the age of this star, what would you report its age to be? Justify your response.

Stable isotopes and pure honey Most of the atoms in the materials around and in us are composed of two or more **stable isotopes,** that is, isotopes that do not undergo radioactive decay. Several examples of the stable isotopes found in organisms are listed in Table 3.5. Although all the isotopes of a given atom react the same way chemically, they do so at slightly different rates. In any process that involves movement of atoms, such as a chemical reaction, a heavier isotope (or molecule containing the heavier atom) moves a little more slowly than a lighter isotope. Thus some discrimination between the isotopes can occur and can provide useful markers for the kinds of processes that occurred to give the final result.

Table 3.5	*A few of the stable isotopes in biological systems.*	
Element	**Isotopes**	**Average abundance, %**
hydrogen	H-1	99.985
	H-2	0.015
carbon	C-12	98.892
	C-13	1.108
nitrogen	N-14	99.6337
	N-15	0.3663
oxygen	O-16	99.759
	O-17	0.0374
	O-18	0.2039

3.68 CHECK THIS

Isotopic combinations in molecules

Molecules of carbon dioxide, CO_2, can have several different isotopic compositions. For example, one molecule might be (C-12)(O-16)(O-17) and another (C-12)(O-18)(O-18). How many different isotopic compositions can CO_2 have? Which do you think is most common? Second most common? Least common? Justify your choices.

One example of this discrimination between isotopes is found in photosynthesis by plants. Most plants produce sugars from carbon dioxide and water by one of two photosynthetic processes called the C_3 and C_4 pathways. The subscripts refer to the number of carbons in the product of the first step that incorporates a carbon atom from carbon dioxide. For both pathways, the percentage of C-13 that ends up in the sugars is less than the percentage of C-13 in the carbon dioxide (from the air) that the plant used to synthesize the sugars. However, the amount by which the percentage of C-13 is reduced is different for the two pathways. Even though the same sugars are formed in each pathway, they have different **isotopic signatures,** that is, different ratios of the stable isotopes, $^{13}C/^{12}C$.

The $^{13}C/^{12}C$ ratio in a sugar sample is obtained by first burning the sample completely to convert all its carbon to carbon dioxide. A mass spectrometer, such as the one illustrated in Figure 3.29, is then used to measure the ratio of CO_2 with C-13 to that with C-12. The results from such analyses are usually expressed as delta, δ, values in which the sample results are compared to a standard sample of known isotopic composition. You can think of the delta value as the number of atoms per 1000 that the heavy isotope in the sample differs from the heavy isotope in the standard. The expression for $\delta^{13}C$ in terms of the sample and standard isotopic ratios is

$$\delta^{13}C = \left[\left(\frac{(^{13}C/^{12}C)_{sample}}{(^{13}C/^{12}C)_{standard}}\right) - 1\right] \cdot 1000 \qquad \textbf{(3.33)}$$

Figure 3.29.

Schematic diagram of an isotope ratio mass spectrometer. The gas sample is ionized by a beam of electrons in the ion source and then accelerated out of the source by electrically charged plates. The moving ion beam passes between the poles of a magnet, which bends its path. The path an ion follows depends on its mass. The number of ions of each mass is measured separately and the numbers compared. The entire apparatus is in a vacuum chamber, so only the ions of interest are formed and detected.

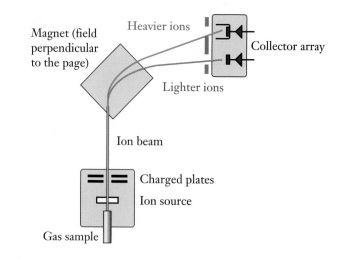

Magnet (field perpendicular to the page)

Heavier ions

Collector array

Lighter ions

Ion beam

Charged plates

Ion source

Gas sample

3.69 CHECK THIS

Determining a delta value

If the standard has a $^{13}C/^{12}C$ ratio equal to that in Table 3.5 and a sample contains 1.099% C-13, what is $\delta^{13}C$ for the sample? Does the sign of $\delta^{13}C$ reflect direction of change of the sample from the standard? Why or why not? Explain your reasoning.

The usual standard for $\delta^{13}C$ is the carbonate from a particular limestone formation in South Carolina. Relative to this standard, almost all samples from living systems have less C-13, so their $\delta^{13}C$ values are negative, as you found in Check This 3.69. In particular, the $\delta^{13}C$ values for C_3 and C_4 plants are about −25 and −10, respectively. And here is where honey fits into this story. Most flowering plants from which bees collect nectar (sugar syrup) to make honey are C_3 plants, so the honey, which is mainly a collection of these sugars, has a $\delta^{13}C$ value of −25. On the other hand, corn and sugar cane are C_4 plants, so their sugars have a $\delta^{13}C$ value of −10. Honey is an expensive product, since it is produced in relatively small quantities. Corn syrup and cane sugar syrup are rather inexpensive products. It's difficult to taste the difference between pure honey and a mixture of honey with corn syrup or cane sugar syrup. Although it's not legal to sell a mixture like this as pure honey, there is a temptation to do so. Isotope ratios help food inspectors determine whether a product labeled "honey" has been adulterated with less expensive syrups.

3.70 WORKED EXAMPLE

The amount of honey in a "honey" sample

An isotopic analysis of the sugars in a shipment of "honey" gave a $\delta^{13}C$ value of −14.3. The sample is obviously not pure honey. What percentage of the product is honey?

continued

Necessary information: We make the assumption that $\delta^{13}C$ varies linearly with the proportion of the sample that is honey (or syrup), as shown in this graph.

Strategy: Read from the graph the composition corresponding to the measured $\delta^{13}C$.

Implementation: The green lines show how to read over from -14.3 to the line and then down to the percent composition, 29% honey.

Does the answer make sense? The measured $\delta^{13}C$ value is pretty close to that for pure C_4 sugars, so the percentage of honey in the sample is low, as we found.

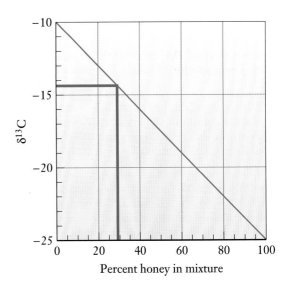

3.71 CHECK THIS

What is the legal limit for adulteration of honey?

(a) If $\delta^{13}C$ varies linearly with the proportion of honey, the percentage of honey in a sample is

$$\% \text{ honey} = \left(\frac{\delta_{\text{sample}} - \delta_{\text{syrup}}}{\delta_{\text{honey}} - \delta_{\text{syrup}}}\right) \cdot 100\% \qquad (3.34)$$

All the δ values in equation (3.34) are the $\delta^{13}C$ values for the subscripted substances. Show how this equation is derived and use it to check the answer in Worked Example 3.70. *Hint:* Write the equation of the line in slope-intercept form with the slope written in terms of the endpoints of the line and solve for the percent honey.

(b) For legal purposes, a difference of one unit in $\delta^{13}C$ from the value for pure honey is taken to represent an adulterated sample. To what percent adulteration does this difference correspond? Explain your reasoning clearly.

In this chapter, including this Extension, we have explored several uses of isotopes, including household items, medicine, geology, astrophysics, and chemical analysis. These are important uses of isotopes and there are many others, a few of which we will meet in later chapters. The overarching message in this chapter is that all of these isotopes (except the very lightest that were a product of the Big Bang) were formed and are being formed in stars and supernovae. We can understand a great deal about our universe, our galaxy, our solar system, and our planet by understanding some of the principles that guide element formation, and isotopes help us read the story of their formation.

Chapter 3 Problems

3.1. Spectroscopy and the Composition of Stars and the Cosmos

3.1. Figure 3.3 shows the parts of a simple spectrograph and the result when the light from a hydrogen discharge (hydrogen atom) lamp enters and is diffracted. Are the diagram and diffraction consistent with the information from Figure 3.1? Explain why or why not.

3.2. Draw a diagram of a simple spectrograph, modeled after Figure 3.3, that uses a transmission diffraction grating, such as shown in Figure 3.2, instead of a prism to diffract the incoming light beam. Be sure the diffraction results you show on the photographic film are consistent with the information from Figure 3.2.

3.3. A spectrometer was used to analyze a source of light and the spectrum showed a series of emission lines. What can you conclude about the light source? Explain.

3.4. Flame tests can often be used to identify the metal ions in a compound. Robert W. Bunsen (German chemist and spectroscopist, 1811–1899) invented the Bunsen burner in order to create a flame hot enough to cause light emission from metal ions in compounds that were placed in the flame. Two examples are shown here: the greenish flame from barium compounds and the scarlet flame from strontium compounds. How do the emissions from atoms discussed in Section 3.1 explain why flame tests work? Explain clearly.

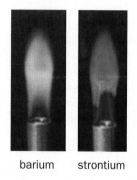

barium   strontium

3.5. More than 99% of all the atomic nuclei in the universe are hydrogen and helium. Are the data in Figure 3.7 consistent with this statement? Clearly explain why or why not.

3.6. As the previous problem states, almost all the nuclei in the universe are hydrogen and helium. Assume that almost all the mass of matter in the universe is also hydrogen and helium. About what percentage of the mass of the universe is helium? Clearly explain the reasoning you use to get your answer.

3.7. How does the mass of iron in the universe compare to the mass of carbon? Explain your reasoning.

3.8. How does the mass of iron in the universe compare to the mass of all the other fourth period transition metals, scandium through zinc, combined?

3.2. The Nuclear Atom

3.9. What do these ions, S^{2-}, Cl^-, K^+, and Ca^{2+}, have in common?

3.10. How does the mass of a proton compare to the mass of an electron?

3.11. What is an isotope?

3.12. Why are electrons not included when calculating the mass number of an isotope?

3.13. If an isotope of an element has 30 protons, 35 neutrons, and 28 electrons:
(a) What is the element?
(b) Is this an anion or a cation? Explain.

3.14. What would be the mass, in grams, of 25 protons?

3.15. Write the atomic symbols for these isotopes:
(a) $Z = 19; A = 40$
(b) $Z = 79; A = 197$
(c) $Z = 54; A = 118$
(d) $Z = 13; A = 28$
(e) $Z = 53; A = 118$
(f) $Z = 83; A = 199$

3.16. How many protons, neutrons, and electrons do these ions contain?
(a) $^{58}Ni^+$ (d) $^{37}Cl^-$
(b) $^{32}S^{2-}$ (e) $^{55}Mn^{7+}$
(c) $^{65}Zn^{2+}$ (f) $^{56}Fe^{2+}$

3.17. Complete this table. Some of the substances are ions and some are atoms.

Substance	# of protons	# of neutrons	# of electrons	Atomic number	Mass number	Nuclear symbol
rhodium		55				$^{100}_{45}Rh$
			23			$^{60}_{26}Fe^{3+}$
	88				225	$^{225}_{88}Ra$
	16	22	18			$^{?}_{16}S^{2-}$
			100			$^{248}_{100}Fm$

3.3. Evolution of the Universe: Stars

3.18. The following questions deal with the Kelvin temperature scale.
(a) Is 3 K, the present background temperature of the universe, hot or cold? Explain.
(b) The planet Earth is thought to have formed at a temperature just below 1000 K. Would Earth, at that temperature, have had liquid water on its surface? Explain.
(c) If the average daytime high temperature for a city is 85 °F, what is the temperature in kelvin? *Hint:* A temperature of 32 °F is 273 K. A degree on the Kelvin scale is 1.8 times the size of a degree on the Fahrenheit (°F) scale.

3.19. According to the Big Bang theory, stars, and eventually planets, formed when matter condensed upon cooling as the universe expanded. Give some examples of analogous phenomena showing condensation.

3.20. Put these reactions in order from the one that takes place at the lowest temperature to the one that requires the highest temperature. Explain the reasoning for each of your choices.
(a) $^{22}Ne^{10+} + {}^{22}Ne^{10+} \rightarrow {}^{43}K^{19+} + {}^{1}H^{+}$
(b) $^{13}C^{3+} + e^{-} \rightarrow {}^{13}C^{2+}$
(c) $^{12}C^{6+} + {}^{18}O^{8+} \rightarrow {}^{26}Mg^{12+} + {}^{4}He^{2+}$
(d) $^{12}C^{6+} + {}^{3}He^{2+} \rightarrow {}^{13}N^{7+} + {}^{2}H^{+}$

3.21. What kelvin temperature is necessary for nuclei to have sufficient kinetic energy to sustain nuclear fusion?

3.22. Where are the elements formed? Give examples of the processes by which elements can be formed.

3.4. Nuclear Reactions

3.23. Complete these nuclear reactions (nuclear charges have been omitted):
(a) $^{14}N + \underline{\quad} \rightarrow {}^{17}O + p$
(b) $^{13}C + neutron \rightarrow \underline{\quad} + \gamma$
(c) $^{1}H + {}^{1}H \rightarrow {}^{2}H + \underline{\quad}$
(d) $^{20}Ne + \underline{\quad} \rightarrow {}^{24}Mg + \gamma$
(e) $^{20}Ne + {}^{4}He \rightarrow \underline{\quad} + {}^{16}O$
(f) $^{27}Al + {}^{2}H \rightarrow \underline{\quad} + {}^{28}Al$

3.24. Complete these nuclear reactions (nuclear charges have been omitted):
(a) $^{97}Tc \rightarrow {}^{97}Ru + \underline{\quad}$
(b) $\underline{\quad} + {}^{4}He \rightarrow {}^{243}Bk + n$
(c) $^{249}Cf + \underline{\quad} \rightarrow {}^{263}Sg + 4n$
(d) $^{1}H + {}^{14}N \rightarrow \underline{\quad} + {}^{4}He$
(e) $n + {}^{235}U \rightarrow \underline{\quad} + {}^{94}Sr + 2n$
(f) $^{228}Ra \rightarrow \underline{\quad} + {}^{228}Ac$

3.25. Write the balanced nuclear equation for the beta decay of ^{24}Na. Include both mass and atomic numbers.

3.26. The *p–n* reaction is a common nuclear reaction. In a *p–n* reaction, a proton reacts with a nucleus to produce a new nucleus and a neutron as products.
(a) The americium-241 used in smoke detectors (Investigate This 3.30 and Consider This 3.31) is extracted from spent nuclear reactor fuel rods. It is produced in the rods by a *p–n* reaction of plutonium. What isotope of plutonium is required? Write the balanced nuclear reaction that produces americium-241.
(b) The *n–p* reaction is also common. A carbon isotope is produced by an *n–p* reaction of nitrogen-14 in the atmosphere. Cosmic rays (radiation and particles from the Sun) are the source of the protons. What isotope of carbon is produced? Write the balanced nuclear reaction for its production.

3.27. When ^{14}N captures a neutron, it decays to ^{3}H and another product. Write the balanced nuclear equation for this reaction.

3.28. Write equations to describe how the fusion of two $^{12}C^{6+}$ can lead to the formation of
(a) $^{23}Na^{11+}$ (c) $^{20}Ne^{10+}$
(b) $^{23}Mg^{12+}$ (d) $^{16}O^{8+}$

3.29. Write equations to describe how the fusion of two $^{16}O^{8+}$ can lead to the formation of
(a) $^{32}S^{16+}$ (c) $^{28}Si^{14+}$
(b) $^{31}P^{15+}$ (d) $^{24}Mg^{12+}$

3.30. Write equations to show how the three isotopes of Mg ($^{24}Mg^{12+}$, $^{26}Mg^{12+}$, and $^{27}Mg^{12+}$) are produced from the fusion of $^{12}C^{6+}$ and $^{16}O^{8+}$.

3.31. Xenon-143 decays by a series of six successive beta particle emissions to a stable isotope. Write the series of decays and identify the stable isotope.

3.32. In 1996, the skeleton of an ancient hunter was found in the mud under the water near the shore of the Columbia River in Kennewick, Washington. A tiny sample from the skeleton (collagen in the bone) was analyzed for carbon-14 and found to contain 0.362 as much of this isotope as is present in living organisms. When did the "Kennewick Man" die?

3.33. At 8:15 A.M., a PET scan patient was injected with a compound containing fluorine-18. Assuming that none of the compound is excreted, what fraction of the F-18 remains in the patient's body at noon? Explain how you solve the problem. *Hint:* See Table 3.3.

3.34. Tiny quantities of iodine are essential for the proper functioning of our thyroid gland. A common treatment for patients with enlarged thyroid glands (hyperthyroidism) is ingestion of a compound, such as NaI, containing iodine-131. The iodine concentrates in the thyroid and beta particles produced by its decay kill the thyroid cells where it has accumulated.
(a) Why are only thyroid gland cells killed by the beta emission? *Hint:* See Consider This 3.31.
(b) Assuming that none of the I-131 is excreted, how long will it take for the radioactivity to decay to 10% of its initial level? *Hint:* See Table 3.3.

3.35. Rubidium-87 decays by beta emission with a half-life of 4.9×10^{10} years.
(a) Write the balanced nuclear reaction equation for the decay of Rb-87.
(b) A rock sample from Greenland was found to have a Sr-87/Rb-87 mass ratio of 0.056. How old is the rock sample? Explain how you find the age and clearly state the assumptions you make.

3.5. Nuclear Reaction Energies

3.36. What is nuclear binding energy?

3.37. What is the difference between fusion and fission? How might you predict whether a nucleus would undergo fission or fusion?

3.38. Calculate the binding energy per nucleon for
(a) $^{20}\text{Ne}^{10+}$ (nuclear mass = 33.18913×10^{-24} g)
(b) $^{28}\text{Si}^{14+}$ (nuclear mass = 46.44450×10^{-24} g)
(c) Which nucleus is more stable? Explain your answer.

3.39. For the elemental nuclei, $^{6}\text{Li}^{3+}$ (nuclear mass = 9.98561×10^{-24} g) and $^{56}\text{Fe}^{26+}$ (nuclear mass = 92.8585×10^{-24} g), calculate, in kilojoules, the
(a) binding energy per nucleus
(b) binding energy per nucleon
(c) molar binding energy per nucleon

3.40. See the *Web Companion*, Chapter 3, Section 3.5.1–4.
(a) Draw a picture of two nuclei that would undergo nuclear fusion, for example C-12 and He-4. Be sure to indicate the components of the nucleus as protons, neutrons, or electrons. Draw the resulting nucleus, after the nuclei have undergone fusion.
(b) Why would this type of fusion occur? Explain clearly.
(c) Why would these nuclei not undergo fission? Explain clearly.

3.41. (a) The mass of a Xe-142 nucleus is $235.63075 \times 10^{-24}$ g. What is its binding energy in kJ·mol^{-1}?
(b) Use your result from part (a) and the data in Figure 3.20 to calculate the energy released in this fission reaction:

$$^{235}\text{U}^{92+} + {}^{1}n^{0} \rightarrow {}^{90}\text{Sr}^{38+} + {}^{142}\text{Xe}^{54+} + 4{}^{1}n^{0}$$

(c) How does your result in part (b) compare to the energy for the fission reaction in Figure 3.20? What might you conclude about the energies released in the fission reactions of U-235? Explain.

3.42. The energy change (in kJ·mol^{-1}) for one of the collision-induced nuclear fission reactions in Figure 3.21 is shown in Figure 3.20. The molar binding energies per nucleon for Kr-92, Rb-89, Cs-144, and Ba-141 are, respectively, -82.3, -83.7, -79.4, and -80.5×10^{7} kJ·mol^{-1}.
(a) What are the energy changes for the other nuclear fission reactions in Figure 3.21?
(b) What might you conclude about the energies released in the fission reactions of U-235? Explain.

3.43. See the *Web Companion*, Chapter 3, Section 3.4.3.
(a) Draw a nucleus of a carbon-12 atom. Clearly label the components of the nucleus as protons, neutrons, or electrons.
(b) Draw a picture of what this nucleus would look like if it separately underwent each of the main types of radioactive decay: alpha, beta, gamma, and positron emission. Identify what the resulting elemental nucleus would be in each case.
(c) Which type of radiation do you think C-12 would be like likely to emit (if any). Give an example of an elemental nucleus that would be more likely than C-12 to undergo each type of radioactive decay.

3.44. The masses of He-3 and He-4 nuclei are 5.00642×10^{-24} g and 6.644655×10^{-24} g, respectively. Is this reaction endothermic or exothermic?

$$^{3}_{2}\text{He}^{2+} + {}^{3}_{2}\text{He}^{2+} \longrightarrow {}^{4}_{2}\text{He}^{2+} + 2{}^{1}_{1}\text{H}^{+}$$

How much energy is required or released per gram of He-3 that reacts?

3.45. In March 2002, scientists from Oak Ridge National Laboratory reported that they had subjected a sample of acetone, $CH_3C(O)CH_3$, in which all the hydrogen atoms had been replaced with deuterium, ^{2}H, to intense cavitation (formation and collapse of bubbles), and had detected the formation of neutrons and tritium, ^{3}H, in the sample. A tentative explanation for this observation was that fusion of deuterium nuclei, which can occur in two equally probable ways, was taking place at the high temperatures and pressures in the collapsing bubbles:

    (i) $^2H + {}^2H \rightarrow {}^3H + p$
    (ii) $^2H + {}^2H \rightarrow {}^3He + n$

(a) What are the energy changes for reactions (i) and (ii)? The masses of H-2, H-3, and He-3 nuclei are, respectively, 3.34357×10^{-24} g, 5.00736×10^{-24} g, and 5.00642×10^{-24} g.
(b) Which set of products is more stable? Explain the reasoning for your answer.

3.6. Cosmic Elemental Abundance and Nuclear Stability

3.46. (a) The nuclear masses of helium-4 and beryllium-8 are 6.644655×10^{-24} g and 13.28949×10^{-24} g, respectively. What are the Δm and ΔE (in kJ·mol^{-1}) for this reaction?

$$^4He^{2+} + {}^4He^{2+} \rightarrow {}^8Be^{4+}$$

(b) Beryllium-8 decays by splitting into two alpha particle with a half-life of about 7×10^{-17} s. Does your result in part (a) help you understand this very short lifetime? Explain why or why not.
(c) What is ΔE (in kJ·mol^{-1}) for this reaction?

$$^4He^{2+} + {}^8Be^{4+} \rightarrow {}^{12}C^{6+}$$

(d) The sum of the two reactions in this problem is reaction equation (3.7) in the text. What is ΔE for reaction (3.7)? Explain how you get your result.
(e) The text that accompanies reaction equation (3.7) says the fusion involves "almost-simultaneous collision of three helium-4 nuclei." How is the information in this problem related to this statement? Do your results in this problem help to explain the relative abundance of beryllium in the universe? Explain your responses.

3.47. A nuclear reaction sequence that occurs in some stars is

$$^{12}C^{6+} \rightarrow {}^{13}N^{7+} \rightarrow {}^{13}C^{6+} \rightarrow {}^{14}N^{7+} \rightarrow {}^{15}O^{8+}$$
$$\rightarrow {}^{15}N^{7+} \rightarrow ({}^{12}C^{6+} + {}^4He^{2+})$$

(a) Write the balanced nuclear reactions for each step of the sequence. *Hint:* Consider fusions involving protons, ^1H, and electron-capture decays, as in Check This 3.39. Compare your reactions to those you proposed in Check This 3.56(b).

(b) Show that the *net* result of the series of reactions you wrote in part (a) is equivalent to nuclear reaction equation (3.6).
(c) The series of reactions you wrote in part (a) is usually referred to as the CNO (carbon-nitrogen-oxygen) cycle and carbon-12 is said to catalyze the formation of helium from hydrogen. Explain why carbon is assigned this role. *Hint:* A catalyst facilitates a reaction, but is not itself consumed in the net reaction.

3.7. Formation of Planets: Earth

3.48. (a) Since 70% of the mass of your body is water, we might say that you are mostly water, and the data in Figure 3.27 confirm that hydrogen and oxygen atoms are the most abundant in the human body. Is the ratio of these atoms consistent with the composition of water? Explain why or why not.
(b) Some might say that the crust of the earth is mostly sand, silicon dioxide, SiO_2. Do the data in Figure 3.27 confirm this suggestion? Explain why or why not.

3.49. During the early part of the twentieth century, chemists spent a good deal of effort determining accurate values for relative atomic masses. When lead was studied, the metal from different ores had different atomic masses. From ores that also contained uranium, the relative atomic mass was close to 206, but from other ores, the value was close to 208.

(a) What would you suggest as the reason for the difference in the lead in different ores? Is your explanation consistent with the information in this problem and this chapter? Explain your answers clearly.
(b) The relative atomic mass of lead shown in the periodic table on the end papers of the book is 207.2. Clearly explain how you account for this value.

3.9. Extension—Isotopes: Age of the Universe and a Taste of Honey

3.50. Figure 3.29 shows a schematic diagram of an instrument called a mass spectrometer, which is named by analogy with the light spectrograph (or spectrometer) represented in Figure 3.3. A light spectrometer uses a prism or diffraction grating to disperse light into its component wavelengths. What does the mass spectrometer disperse? What is the part of the mass spectrometer that is responsible for the dispersion? Explain.

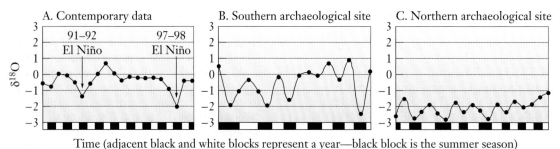

Time (adjacent black and white blocks represent a year—black block is the summer season)

Source: C. Fred T. Andrus, Douglas E. Crowe, Daniel H. Sandweiss, Elizabeth J. Reitz, and Christopher S. Romanek, "Otolith d180 Record of Mid-Holocene Sea Surface Temperatures in Peru," *Science*, 2002, pp. 1508–1511.

3.51. Since the temperature of the ocean surface has such a large impact on the earth's climate, this is one of the variables whose history scientists wish to determine as they develop climate models. Most of the approaches to discovering this history make use of differences in stable isotope ratios. Fossil catfish from the shores of Peru were the focus of one such study. These catfish deposit calcium carbonate as otoliths, "ear stones," which grow throughout the life of the fish in layers (like the rings of a tree), so annual variations can be determined. The study examined the $\delta^{18}O$ in otoliths from contemporary fish and from 6000- to 6500-year-old fossil fish from a site on the northern coast of Peru and one from farther south. The data are shown in the figure above.

(a) During an El Niño year, the sea surface temperature (SST) of the eastern Pacific Ocean off the coast of South America gets warmer than usual. How does $\delta^{18}O$ vary with temperature in these calcium carbonate otoliths? Clearly explain the reasoning for your answer.

(b) The authors conclude that "the most plausible explanation for the archaeological $\delta^{18}O_{otolith}$ values is warmer summer SSTs at [site X] and nearly tropical conditions near [site Y] in the early mid-Holocene [the last 11,000 years]." Which of the sites represented in the figure is site X and which site Y? Clearly explain the reasoning for your choices.

(c) Do the conditions at the sites you identified in part (b) make sense in terms of their geographic locations? Explain.

3.52. Potassium is relatively abundant (Figures 3.7 and 3.25), so geologists make extensive use of argon–potassium dating. Potassium-40 is radioactive and decays by two pathways: beta emission and electron capture. 10.7% of the decay occurs by the electron-capture pathway. Argon-40 formed in the electron-capture reaction is a gas and can escape from molten rock, but is trapped in the crystal lattice, if it is formed after the rock solidifies. Thus, Ar-40/K-40 dating indicates the age of the sample since it solidified.

(a) Write balanced nuclear reactions for the two modes of K-40 decay. *Hint:* See Check This 3.39.

(b) Assume that exactly 100 μg (1 g = 10^6 μg) of K-40 is present in a sample when it solidifies. After one K-40 half-life, 1.25×10^9 years, how many micrograms of Ar-40 will have been formed? What is the Ar-40/K-40 mass ratio at this time? *Hint:* 10.7% of the K-40 that reacted has become Ar-40.

(c) Calculate the Ar-40/K-40 mass ratio after two half-lives have elapsed. *Hint:* The Ar-40 formed during the second half-life adds to that formed during the first.

(d) The Ar-40/K-40 mass ratio as a function of the age of the rock (time since it solidified) is shown in this plot. Do your results from parts (b) and (c) fall on the curve? Explain why or why not?

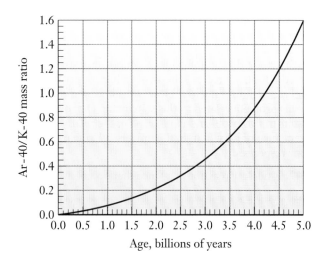

(e) A rock sample returned from the surface of our Moon had an Ar-40/K-40 mass ratio of 1.05. What is the age of the sample? What does this result suggest about the age of the Moon? Explain.

3.53. See the *Web Companion*, Chapter 3, Section 3.5–9. The plot of activity as a function of time for the radioactive decay of indium-113m on pages 7, 8, and 9 of this section of the *Web Companion* is low resolution and difficult to read. Another similar set of data for this decay is given here.

Time, min	Activity, counts per 30 s
0	68,372
30	54,852
60	45,457
90	36,901
120	29,964
150	24,570
185	18,936
229	15,020
240	12,790

(a) Plot these data, as in the *Web Companion*, and use your plot to determine the half-life for the decay. How does your value compare with the one from the *Companion*? *Hint:* One approach, which requires some mathematical manipulation, is to use a graphing calculator or computer graphing program to plot the data, find the equation of the exponential curve through the data, and use the equation of this curve to find the half-life.

(b) Apply equation (3.28) to these data (with uranium replaced by indium) and plot the data to give a linear plot. Use the equation of the line to find the half-life and compare your value with the one from the *Web Companion*.

It is possible to take advantage of the laws discovered by chemistry [the law of periodicity] without being able to explain their causes.

DMITRI I. MENDELEEV (1836–1907)

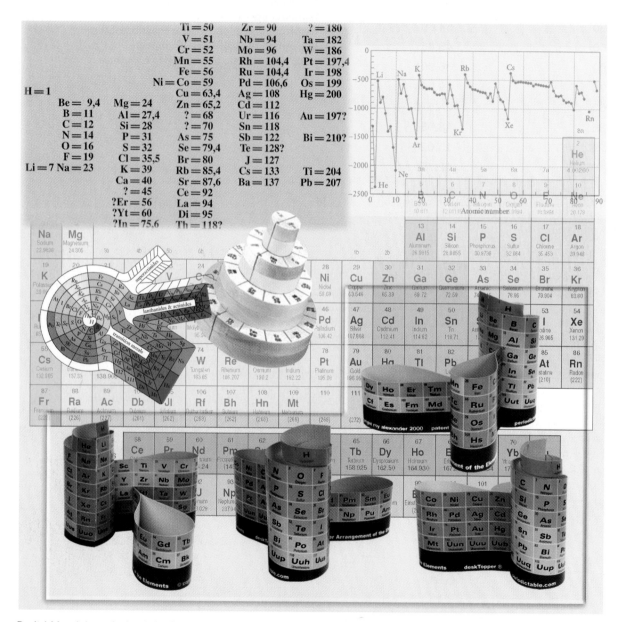

Dmitri Mendeleev devised the first periodic table (upper left) in 1869. He based his table on the repeating properties of the elements as a function of relative atomic mass. Since then, many other versions of the periodic table have been developed to help visualize the conceptual basis for periodicity. The repeating pattern of the ionization energies of elemental atoms, shown in the superimposed graph, is one of the strongest clues to the conceptual basis for periodicity, the electronic shell structure of atoms.

Structure of Atoms

I n Chapter 3, you saw that the atoms in our bodies and all the matter that surrounds us result from the Big Bang and billions of years of nucleosynthesis in stars. You also saw that the observed abundances of elemental atoms in the universe are a consequence of these processes in combination with the relative stabilities of atomic nuclei. Now we will leave nuclear transformations and focus our attention on the arrangements of electrons in atoms.

These arrangements are reflected in the structure of the periodic table of the elements, several versions of which are shown on the facing page. If life has evolved on any of the countless planets circling other stars in our universe (Chapter 3) and has reached a technological level similar to that of humans on Earth, we can predict that a periodic table of the elements similar to one of ours is part of that life. The periodic table is a succinct device that helps us remember and/or predict trends and similarities in chemical and physical properties of the elements. This understanding of periodicity is what we hope you will attain from this chapter. Our present interpretation of periodicity is based on the structure of atoms (the behavior of electrons in atoms), so, in this chapter, we develop a model of atomic structure.

The first periodic tables were devised and used before electrons and nuclei had been discovered and only about 60 elements were known. The quotation from Dmitri Mendeleev (Russian chemist, 1836–1907), who devised one of the very first periodic tables, reminds us that the empirical correlations upon which these tables were based were useful, even though the basis of the correlations was not understood. It is the search for this understanding that drove a good deal of the scientific effort described in this chapter.

We will begin by considering the kinds of correlations that led to the development of the first periodic tables and then use a more modern correlation, the ionization energies of atoms, to connect to the atomic shell model introduced in Chapter 1. The spectra of atoms that are so useful in determining the occurrence and abundance of elements in stars and throughout the universe are also useful in determining atomic structure. We will return to spectroscopy to seek evidence for atomic models that help us understand why the electron shells, attracted by the nuclei, don't collapse. As you study this chapter, we ask you to do two things: (1) Look at experimental data and try to see in them *evidence* for the shell model and (2) look at experimental data and try to interpret them as a *result* of the shell model.

4.1. Periodicity and the Periodic Table

4.1 CONSIDER THIS

Are there patterns in the molar volumes of the elements?

You can use the densities of solid and liquid elements and their relative atomic masses to calculate their volume per mole. Data for the elements known in 1869 are shown in Figure 4.1.

continued

Personal Tutor
Using tables and graphs will be important for the analyses in this chapter.

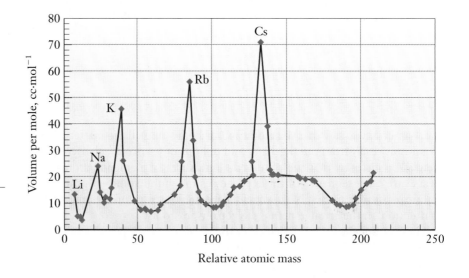

Figure 4.1.

Molar volume of the elements (1869) as a function of atomic mass. The data points are connected to make trends easier to see.

Work in small groups on these questions and discuss your findings with the whole class.

(a) How is the molar volume a measure of atomic volume?

(b) Describe any patterns in atomic size you see in Figure 4.1. Are these patterns represented in the periodic tables shown in the chapter opening illustration? Explain why or why not.

(c) Radium, an **alkaline earth metal** (Be, Mg, Ca, Sr, Ba, and Ra) with a relative atomic mass of 226, was unknown in 1869. What do you predict (from the data in Figure 4.1) for the molar volume of radium? Explain how you arrive at your answer. Use a reference handbook to find density data to check your prediction.

(d) Describe how **alkali metal** (Li, Na, K, Rb, and Cs) atoms vary in size with atomic mass. How might you use our present-day model of the nuclear atom to interpret this variation?

Origin of the periodic table The most powerful organizing concept in chemistry is **periodicity**—the pattern of repeating chemical properties—which is represented by the **periodic table.** Periodicity was conceived or discovered independently in 1869 by Mendeleev and Lothar Meyer (German chemist, 1830–1895). Their work was guided, in part, by the availability of a consistent set of relative atomic masses proposed in 1858 by Stanislao Cannizzaro (Italian chemist, 1826–1910). Scientists accepted and began using these masses because they resolved many problems among different atomic mass scales and were based on a wealth of experimental evidence that had built up over the previous century.

Mendeleev and Meyer put the known elements in order by relative atomic mass and noticed repeating patterns in their properties. Meyer, for example, calculated the ratio of elemental density to relative atomic mass and plotted the data, as in Figure 4.1. He noted the obvious peaks for the low-density alkali metals, with the denser elements in troughs between them. Another physical property he considered was boiling point (Figure 4.2). Meyer based his periodic table on patterns like the ones you see in these figures.

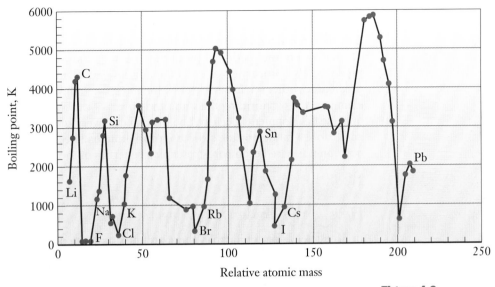

Figure 4.2.

Boiling points of the elements (1869) as a function of atomic mass. These are modern data. Meyer and Mendeleev did not know the actual values for the very low and very high boiling points. The data points are connected to make trends easier to see.

4.2 CONSIDER THIS

How do you interpret patterns in the boiling point data?

(a) Describe the pattern(s) you see in the boiling point data as a function of relative atomic mass in Figure 4.2. Note that there are gaps in the data because not all the elements had yet been discovered. Where does the pattern seem to be interrupted by missing data? Explain what you think is missing.

(b) In 1869, the **noble gases** had not yet been discovered. Their relative atomic masses and boiling points are tabulated here. If these data are added to the plot in Figure 4.2, are periodic patterns more evident or less evident? Explain your answer.

Noble gas	He	Ne	Ar	Kr	Xe	Rn
atomic mass, u	4	20	40	84	131	222
boiling point, K	4.2	27.1	87.3	119.5	165.1	211.3

Filling gaps in the periodic table In contrast to Meyer, Mendeleev focused more on chemical properties, such as reactivity and binary compounds formed by the elements. The alkali metals (elements in the first column of the periodic table), for example, all react vigorously (in some cases violently) with water to form basic solutions, as shown in Figure 4.3. They also all form one-to-one halide salts, such as NaCl, KI, and CsBr. In order to make elements with similar chemical properties align in families, Mendeleev took the bold step of leaving gaps in his periodic table for undiscovered elements. (In his first periodic table,

(a) Sodium

(b) Potassium

Figure 4.3.

Reaction of alkali metals with water. The water contains phenolphthalein, an acid–base indicator that turns red in the presence of base.

shown in the chapter opening illustration, the families are in horizontal rows. Within two years he had rotated his table so that families were in columns, as they have remained in many versions of the table.)

Based on the periodic properties of the known elements, Mendeleev predicted the properties of several undiscovered elements. Guided by these predictions, other scientists searched for and soon found several of the "missing" elements, thus confirming the value and power of the concept of periodicity. All of these developments occurred before the discovery of any subatomic particles and before a model for the structure of an atom had been developed. The organization was based on the macroscopic properties of the elements.

4.3 CONSIDER THIS

How do you predict the properties of an undiscovered element?

In Mendeleev's periodic table, there are missing elements (denoted by question marks in the chapter opening illustration) at atomic masses of 45, 68, 70, and 180. The middle two of these four elements are the ones we now know as gallium (Ga, atomic mass 69.7) and germanium (Ge, atomic mass 72.6) which, as shown by Mendeleev, are in the boron and carbon families, respectively. From the data and periodic patterns in Figures 4.1 and 4.2, predict the molar volumes and boiling points of these elements. Clearly explain how you make your predictions. Use a reference handbook to check your predictions.

Ionization energy With the discovery of electrons and as a result of experimentation on ionization of gaseous elemental ions, scientists were able to measure properties of atoms themselves and find out whether these properties also exhibited periodicity. Recall that, when we discussed the formation of ionic compounds (Chapter 2, Section 2.4), we introduced the idea that atoms can lose and gain electrons to form, respectively, positive and negative ions. For example, the ionization of a gaseous sodium atom can be represented by this reaction equation:

$$\text{Na}(g) \rightarrow \text{Na}^+(g) + e^-(g) \ (\Delta E_{\text{ionization}} = 496 \ \text{kJ·mol}^{-1}) \tag{4.1}$$

Web Companion

All Chapters ①

An interactive periodic table ②
is available by clicking on the ③
circled P in the left panel. ④

The energy required to remove an electron from a gaseous atom (or ion), $\Delta E_{\text{ionization}}$, is called its **ionization energy;** the ionization energy for sodium atoms is 496 kJ·mol^{-1}. The ionization energies for gaseous atoms of the first 20 elements are plotted in Figure 4.4 as a function of their atomic numbers. Atomic numbers now replace relative atomic mass as the independent variable, because we now know that the identity of an element depends on the number of protons in its nucleus, not on the mass of its atoms.

4.4 CONSIDER THIS

How do you interpret patterns in the ionization energy data?

(a) Describe the pattern(s) you see in the ionization energy data as a function of atomic number in Figure 4.4.

(b) Do the major features of your pattern correlate with the shell model of atoms introduced in Chapter 1, Section 1.2, especially Figure 1.7? If so, what do you think is the physical basis for the correlation? Explain your reasoning clearly.

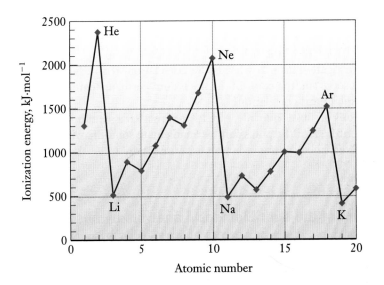

Figure 4.4.

Ionization energies for the first 20 elements. The data points are connected to make trends easier to see.

As you see, an atomic property, the ionization energy of the atoms, shows striking periodicity. The periodicity in macroscopic properties that Mendeleev and Meyer represented in their periodic tables must be a reflection of periodicity in the properties of the atoms themselves. In our nuclear model of the atom, the electrons make up the bulk of the volume of an atom and are responsible for most of its interactions with external forces. When we look at Figure 4.4, we can see that the three major peaks seem to correspond to the shells in our electron shell model from Chapter 1. However, simple shells do not account for the smaller steps between the peaks, so we will probably need to refine our model.

In the remainder of the chapter, we will explore how electrons behave in atoms and further develop our model of atomic structure. Many of the clues that guide our exploration and refinement come from the study of atomic spectra, which we introduced in the previous chapter and take up again in the next section.

4.2. Atomic Emission and Absorption Spectra

4.5 INVESTIGATE THIS

Is the visible spectrum affected by substances in the light beam?

Use the setup described in Investigate This 3.2 to produce a visible spectrum on the projection screen. Cover about half the length of the slit with a flat-bottomed Petri dish containing a solution of potassium permanganate, $KMnO_4$. Write a description comparing and contrasting the spectrum you observe from light that has passed through the solution with the spectrum from the light that has not.

Why is a spectrum affected by substances in the light beam?

(a) In Investigate This 4.5, what is the correlation, if any, between the spectra you observe and the color of the potassium permanganate solution? How do you explain this correlation?

(b) Solutions of nickel chloride, $NiCl_2$, are green. If a solution of nickel chloride is substituted for the potassium permanganate solution, what will the spectra look like? Check your prediction experimentally.

Spectrum of the sun: Elemental analysis by absorption of light

In Chapter 3, we discussed analysis of stars, including our Sun, by emission spectroscopy. By the middle of the 19th century, spectrographs had been improved to the point that another feature of the sun's spectrum was observed. Superimposed on the background of the visible spectrum are hundreds of dark lines, some of which you can see in Figure 4.5. The dark lines meant that these wavelengths were missing in the spectrum from the sun. Scientists noticed that many of the wavelengths of these dark lines corresponded to the emission wavelength lines of particular elements. Compare this observation with the experimental results from Chapter 3, shown in Figures 3.4 and 3.5 and your analysis in Check This 3.7.

Figure 4.5.

A few of the most prominent dark lines in the solar spectrum.

≈ 400 nm   ≈ 700 nm

This phenomenon of missing wavelengths can be studied in laboratories on Earth. Two spectroscopic experiments and their results are shown schematically in Figure 4.6. In the experiment represented in Figure 4.6(a), the visible emission from a sodium atom discharge lamp, like the ones you used in Investigate This 3.4, is examined in a spectrograph and the typical atomic line emission is observed. In the experiment represented in Figure 4.6(b), the continuous emission from a lamp that emits white light passes through a sample of hot gaseous sodium atoms (which are not emitting light) before it is examined in the spectrograph.

How do you interpret the results shown in Figure 4.6?

Work in small groups to compare and contrast the spectra that are obtained in the two experiments represented in Figure 4.6. What is your interpretation of the reasons for the similarities and differences between the spectra?

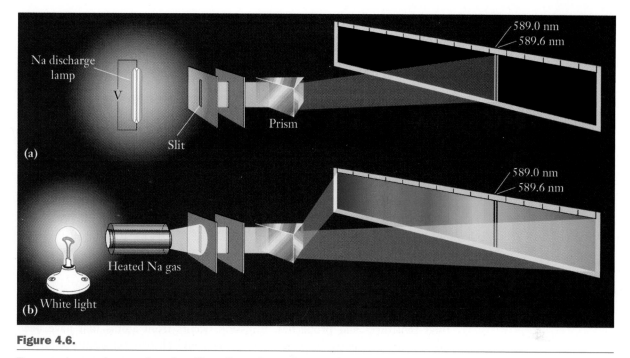

Figure 4.6.

Two spectroscopic experiments with sodium atoms. See text for explanation.

To explain the dark lines in the solar spectrum, Gustav Kirchhoff (German physicist, 1824–1887) proposed that atoms in a layer around the sun absorbed the sun's continuous emission and accounted for the missing wavelengths. Whatever process produces the emission from elemental atoms is reversible. The atoms can absorb light of the same wavelength and produce dark lines, an **absorption spectrum** [Figure 4.6(b)], where light is missing in the spectrum from a continuous source. Thus, we can use absorption spectra to analyze the composition of the cooler regions around stars and emission spectra to analyze the composition of their hotter surfaces. We also use absorption spectra to analyze the clouds of atoms and molecules between the stars and our telescopes.

4.8 CONSIDER THIS

Does light absorption explain the results in Investigate This 4.5?

(a) Why do some substances appear colored? When we look at the light passing through a colored solution, what happens to the wavelengths of light that do not reach our eyes?

(b) Instead of absorbing light at discrete wavelengths as atoms do, substances in solution usually absorb light over a range of wavelengths. Are your observations in Investigate This 4.5 an example of an absorption spectrum? Explain why or why not.

4.9 CHECK THIS

Elemental identification by absorption spectroscopy

An experiment similar to the one represented in Figure 4.6(b) gave this result:

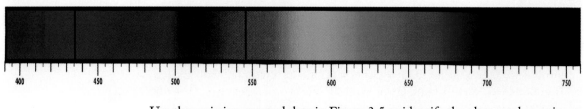

Use the emission spectral data in Figure 3.5 to identify the elemental atomic gas in the heated container. *Hint:* An atomic absorption spectrum does not necessarily absorb at *all* the wavelengths that are present in an emission spectrum of the same element.

The emission and absorption spectra of elemental atoms provide clues about their structures and properties. Emission occurs from atoms that have been given a lot of energy by a spark or a flame and absorption occurs from less energetic atoms at a lower temperature. This probably means that the emission and absorption processes in Figure 4.6 represent changes in the energy of the atoms. Perhaps an atom with extra energy can lose energy by emitting light of a certain wavelength and, conversely, perhaps an unenergetic atom can gain energy by absorbing light of the same wavelength. In order to connect wavelengths of light and atomic energies, we need to know more about the nature of light, which we will examine in the next two sections.

Reflection and Projection

The recognition by Mendeleev and Meyer that the properties of the elements repeated as a function of the atomic mass was an enormously valuable breakthrough for chemistry. This periodicity was symbolized by them and is still symbolized today by the periodic table of the elements. Keep in mind, however, that it is the periodicity underlying the table, not the table itself, that is so important for seeing correlations among and making predictions about the properties of the elements. And, as Mendeleev said, the periodicity of the elemental properties is useful, even if we do not know why the properties are periodic. But we are always curious about the "whys," because we are not content with correlations; we want more fundamental understanding of atomic structure.

We recall from Chapter 3 that a fundamental difference between the emission spectra from gaseous atoms and the emissions from other materials is that only certain wavelengths of light are emitted by energized atoms. And we have seen here that gaseous atoms also absorb only certain wavelengths of light, all of which are also emitted by the energized atoms. It seems likely that these discrete emissions and absorptions might provide a clue to the structure of atoms, especially the arrangement of electrons around the atomic nucleus. However, in order to analyze these spectra in more detail and to relate them to the energies in atoms, we need to know more about the nature of light.

4.3. Light as a Wave

How is light affected by passing through narrow openings?

(a) Hold your first and second fingers straight and together. Somewhere along the length of the fingers, most often near the palm, there will be a place where they don't quite touch. Look through the gap between your fingers at a source of light (like a ceiling light) or a bright, light-colored wall. Use the thumb and fingers of your other hand to squeeze the gap shut. Describe what you see as the gap gets narrower.

(b) Do this as class investigation and work in small groups to discuss and analyze the results. Observe the spot of light from a laser pointer on a light-colored wall or projection screen. Place a transmission diffraction grating in the laser beam. Observe and record how the spot of light from the laser is affected by the grating.

4.11 CONSIDER THIS

What happens when light passes through narrow openings?

What are the similarities between your observations in Investigate This 4.10(a) and 4.10(b)?

Diffraction of light In 1803, Thomas Young (English physicist and mathematician, 1773–1829) did experiments similar to your experiment in Investigate This 4.10(b). When he passed light of a single wavelength through slits in a barrier and observed the light on a parallel screen beyond the barrier, he saw, as in Figure 4.7, bright and dark regions that did not correspond directly to the slits. Young concluded that light moves through space as waves and, after passing through the slits, the waves form a **diffraction pattern,** the alternating bright and dark regions you observed in Investigate This 4.10 and illustrated in Figure 4.7. The bright regions result from wave crests from both slits reaching the screen at exactly the same time; the dark regions result from wave crests from one slit reaching the screen at exactly the same time as wave valleys from the other slit.

Although we can observe the diffraction patterns that are the result of light wave diffraction, we can't see light waves. Thus, it may be difficult to visualize Young's explanation for light diffraction. We can see other kinds of waves, water waves, for example, so we will use these to study wave properties and then apply the results to light waves.

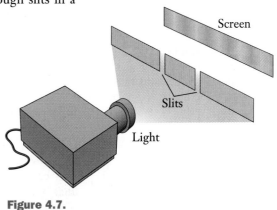

Figure 4.7.

An illustration of Thomas Young's light diffraction experiment. Light passing through the slits produces the diffraction pattern of alternating light and dark areas on the screen.

4.12 INVESTIGATE THIS

What are the characteristics of waves?

Do this as a class investigation and work in small groups to discuss and analyze the results. Use a small ripple tank to observe water wave patterns. Set up the tank so that parallel waves strike a barrier containing a single gap. Observe the wave pattern before the barrier and beyond the barrier and make a sketch of them.

4.13 CONSIDER THIS

How do you characterize water waves?

In Investigate This 4.12, how are the waves beyond the barrier the same as the ones before they strike the barrier? How are they different? Do your sketches show these similarities and differences? Why or why not?

Wave nomenclature Water waves are familiar to almost everyone because the undulations on the surface of a lake, pond, or the ocean are so easily seen. All waves, including water and light waves, are characterized by oscillations described by a sine curve, as shown in Figure 4.8. The distance from crest to crest, the **wavelength,** λ (lowercase Greek lambda), and the displacement of a crest or valley from the zero (null) level, the **amplitude,** are characteristic properties of a wave. The places where the amplitude is zero are called **nodes.**

Figure 4.8.

Components used to describe a wave. Amplitude has value and sign as the wave oscillates with the same amplitude above and below its zero (null) value.

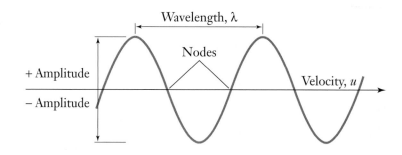

 **Web Companion**

| Chapter 4, Section 4.2.2 | ① |

Try interactive animations of waves and wave properties.

② ③ ④

The **velocity** of the wave, u, is the distance the wave moves in one second. If you were able to position yourself at some point and count the number of crests that passed that point each second, you would have measured the **frequency** of the wave, ν (lowercase Greek nu). Frequency has units of $\dfrac{1}{\text{second}}$ (or s^{-1}). The relationship between velocity, wavelength, and frequency is

$$u = \lambda \cdot \nu \tag{4.2}$$

4.14 WORKED EXAMPLE

Wavelength of water waves

Suppose you are standing on a 10-m-long dock to which a rowboat is tied. Watching the waves from the lake washing past the dock, you note that the rowboat bobs up and down three times during the 5 seconds it takes one wave crest to travel from the outer end to the shore end of the dock. What is the wavelength of the waves?

Necessary information: We need the wave velocity from the problem statement, and equation (4.2) relating the frequency and velocity to wavelength.

Strategy: Substitute the frequency and wave velocity into equation (4.2) to get wavelength.

Calculations: Each time the boat bobs up a wave crest has raised it. This occurs three times in 5 seconds, so the waves have a frequency of

$$\nu = \frac{3}{5\ \text{s}} = 0.6\ \text{s}^{-1}$$

It takes one wave crest 5 seconds to travel 10 m, so the waves travel with a velocity of

$$u = \frac{(10\ \text{m})}{(5\ \text{s})} = 2\ \text{m·s}^{-1}$$

Rearrange equation (4.2) to solve for the wavelength, λ, and substitute for ν and u:

$$\lambda = \frac{u}{\nu} = \frac{(2\ \text{m·s}^{-1})}{(0.6\ \text{s}^{-1})} = 3\ \text{m}\ \text{(one significant figure consistent with data)}$$

Does the answer make sense? Since the boat bobs three times while one wave crest travels 10 m, there must be a wave crest about every 3 m, as we have found.

4.15 CHECK THIS

Wave properties

(a) This figure is a representation of waves in a ripple tank, as in Investigate This 4.12. The figure shows the tank at one instant in time. Lighter areas are wave crests and darker areas are valleys. If the distance from crest to crest of the ripples is 8.7 mm and each point on the surface crests 1.87 times per second, what is the velocity of travel of the ripples?

(b) 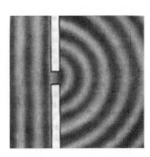 View the animations of the three different waves and answer the four questions in the *Web Companion*, Chapter 4, Section 4.2.2. Assume that the grid lines (light blue lines) are spaced 1 cm apart. What are the amplitudes, wavelengths, frequencies, and velocities of each of the three waves? Is equation (4.2) satisfied by each wave? Show why or why not.

4.16 INVESTIGATE THIS

How do water waves interact?

Do this as a class investigation and work in small groups to discuss and analyze the results. Use the ripple tank again, but replace the single-gap barrier with one that has two gaps. Observe the wave pattern beyond the barrier and make a sketch of it. In your group, agree on a way to describe the pattern(s) you observe.

4.17 CONSIDER THIS

What patterns do interacting water waves create?

(a) In Investigate This 4.16, are the water waves starting out from each of the two gaps in the barrier similar to those starting out from the single gap in Investigate This 4.12? Explain.

(b) When the water waves from each of the two gaps overlap (interact), how would you describe the observed pattern(s)? As a class, develop a common description of the pattern(s)

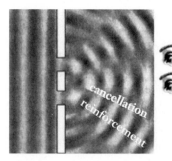

Figure 4.9.

Diffraction of overlapping water waves in a ripple tank. This shows the tank at one instant in time. Lighter areas are wave crests and darker areas are valleys. The positions of the observers refer to Check This 4.18.

observer 1

observer 2

Superimposed waves and diffraction The water wave patterns you observe when two waves overlap in Investigate This 4.16 (Figure 4.9) are examples of diffraction patterns, which you saw for light waves in Investigate This 4.10 (Figure 4.7). The processes of wave **reinforcement** and **cancellation** that created the water wave diffraction patterns are caused by **superimposition** (layering or adding together) of the two waves. When waves combine, the single wave that results is the sum of the combination.

For two identical waves, the two extreme cases of superimposition are illustrated in Figure 4.10. In Figure 4.10(a), the two waves are exactly matched to each other. They *reinforce* each other to give a wave with twice the amplitude as either of the original two. Reinforcement leads to the brighter crests and darker valleys in Figure 4.9; one ray of reinforced waves is labeled. In Figure 4.10(b), the two waves are exactly unmatched; the crest of one wave corresponds to the valley of the other. The result is that the two waves exactly *cancel* each other out. In Figure 4.9, the narrow regions without waves (one of which is labeled) are where the waves have cancelled one another. Waves of all kinds, including light waves, as we have seen, undergo this process of superimposition and diffraction.

4.18 CHECK THIS

Observed results of wave superimposition

(a) Suppose you are observer 1 in Figure 4.9 and are watching the level of the water at this position. Sketch a graph that shows the variation of the water level with time. How does your sketch correlate with the superimpositions shown in Figure 4.10? Explain.

continued

(b) Suppose you are observer 2. Make a sketch of the variation of the water level with time and answer the same question as in part (a).

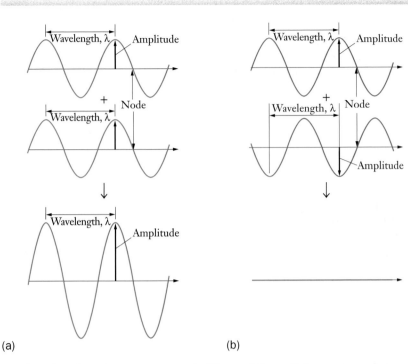

(a)   (b)

Figure 4.10.

Superimposition of waves of the same wavelength and amplitude. Nodes are matched with one another in both cases, but in (a) the crests are matched with one another and in (b) the crests are matched with valleys.

Imagine that the water waves in Figure 4.9 are light waves and the right-hand side of the illustration is a screen. In Check This 4.18, you showed that the pattern on the screen denotes places where no wave crests or valleys reach the screen and other places with strong crests and valleys. Light waves of different wavelengths are diffracted at different angles from the slits, so white light is dispersed into its spectrum of colors, as you saw in Chapter 3, Figure 3.2.

4.19 CONSIDER THIS

How are Investigate This 4.10 and 4.16 related?

(a) What are the similarities between your observations in Investigate This 4.10(b) and 4.16? Can you explain the observations in Investigate This 4.10(b) as a result of superimposition of waves? If so, what is observed when light waves reinforce one another? when they cancel?

(b) 🔖 Explain in your own words how Figures 4.7 and 4.9 are related and use the *Web Companion*, Chapter 4, Section 4.2.1, to check your response.

Electromagnetic waves Young's diffraction experiments were convincing proof of the wave nature of light. However, a basic question for 19th century scientists was, "*What is it that oscillates when light waves move through space?*" For water, the answer is easy; the surface of the water moves up and down. For light, the answer is not at all obvious. The answer developed by James Clerk Maxwell (Scottish physicist, 1831–1879) is the **electromagnetic wave theory** that we

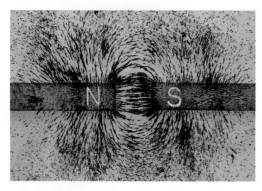

Figure 4.11.

Using iron filings to visualize a magnetic field. Iron filings scattered on a stiff piece of clear plastic align with the magnetic field of two bar magnets under the plastic with their opposite poles facing one another as indicated by the *N* and *S* designations.

still use to model many interactions of radiation (including light) with matter. Let's try to connect what we know about electric and magnetic phenomena to waves.

You know from experience and the activities in Chapter 1, Section 1.2, that an electrically charged object can attract (or repel) other charged objects from a distance. You also know from experience that a magnet can attract iron objects from a distance. To explain how electric and magnetic attractions can act at a distance, scientists developed the concepts of electric and magnetic fields. These fields exert forces on objects that are not in contact with the source of the fields. Magnetic fields are easy to visualize with iron filings, as shown in Figure 4.11. Moving electric charges, an electric current, also create magnetic fields; that is how electromagnets and electric motors work. And, conversely, magnetic fields attract and repel moving charges. Electric and magnetic phenomena are closely linked.

Electromagnetic waves (also called **electromagnetic radiation**), including visible light, are a combination of an oscillating electric field (an electric field that changes amplitude periodically like the waves represented in Figures 4.8 and 4.10) and, perpendicular to it, an oscillating magnetic field. Their wavelengths are identical, and the crests and valleys occur at exactly the same places along the direction of travel of the wave as illustrated in Figure 4.12.

Figure 4.12.

Oscillating electric and magnetic fields in electromagnetic waves. The light gray arrows indicate the direction and magnitude of the wave amplitudes.

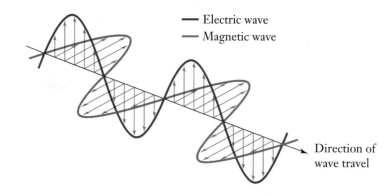

— Electric wave
— Magnetic wave

Direction of wave travel

Web Companion

Chapter 4, Section 4.2.3–4 — (1)

View animations of Figure 4.12 and the oscillating electric and magnetic fields.

(2)
(3)
(4)

Electromagnetic spectrum By the end of the 19th century, almost all the phenomena associated with light had been successfully explained by assuming an **electromagnetic spectrum** that is composed of electromagnetic radiation of different wavelengths, as shown in Figure 4.13. The wave nature of light seemed firmly established. Wave properties explained, for example, how prisms and diffraction gratings disperse white light into the spectrum of wavelengths (colors). Wave properties also provided mathematical descriptions of light waves. Light waves can't be observed the same way water waves are observed on the surface of a pond, but the mathematical model explains their observable effects, such as diffraction.

Figure 4.13 shows that the wavelengths of electromagnetic radiation vary from less than a picometer (1 pm = 10^{-12} m) to many meters. Visible light wavelengths (all the wavelengths in white light) are from about 400 nm, violet, to about 700 nm, deep red. An electromagnetic wave moves through a vacuum with a velocity of 3.00×10^8 m·s⁻¹. This is the **speed of light,** *c,* that we used in Einstein's mass–energy equivalence relationship in Chapter 3, Section 3.5.

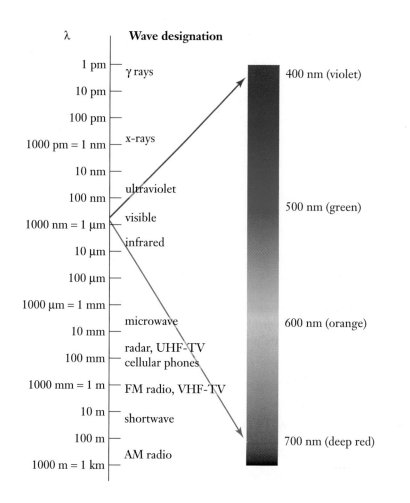

Figure 4.13.

The electromagnetic spectrum.
The spectrum is shown on a logarithmic wavelength, λ, scale. The visible region of the spectrum, a tiny part of the entire spectrum, is expanded.

4.20 CHECK THIS

Electromagnetic radiation

Explain why the *Web Companion*, Chapter 4, Section 4.2.2, animations of three waves do not represent electromagnetic radiation. *Hint:* Which property of the waves would have to be the same, if all the representations were electromagnetic radiation?

4.21 WORKED EXAMPLE

Frequency of gamma, γ, radiation

Calculate the frequency of gamma radiation, an electromagnetic wave, that has a wavelength of 2.45 pm.

Necessary information: We need equation (4.2), the wavelength, λ, of this electromagnetic wave, which is given in the problem statement, and the speed, *c*, that a light wave travels.

continued

Strategy: For electromagnetic waves, the wave velocity is c, so equation (4.2) becomes

$$c = \lambda \cdot v \tag{4.3}$$

Substitute the wavelength and speed of light into equation (4.3) and solve for v.

Implementation: Because the length unit for the speed of light is meters, the wavelength has to be converted to meters, $1 \text{ m} = 10^{12}$ pm:

wavelength of the gamma ray in meters

$$= \lambda = (2.45 \text{ pm})\left(\frac{1 \text{ m}}{10^{12} \text{ pm}}\right) = 2.45 \times 10^{-12} \text{ m}$$

frequency of the gamma ray

$$= v = \frac{c}{\lambda} = \frac{3.00 \times 10^8 \text{ m} \cdot \text{s}^{-1}}{2.45 \times 10^{-12} \text{ m}} = 1.22 \times 10^{20} \text{ s}^{-1}$$

Does the answer make sense? Equations (4.2) and (4.3) show that frequency and wavelength are inversely proportional: A wave with a short (small) wavelength has a high (large) frequency. This is the result we observe here and again in Check This 4.22.

4.22 CHECK THIS

Frequencies and wavelengths of electromagnetic radiation

(a) Calculate the frequency of green light that has a wavelength of 515 nm.
(b) Find the frequency of a wave with a wavelength of 21.11 cm. In what region of the electromagnetic spectrum do you find waves of this frequency?
(c) Based on the results from Worked Example 4.21 and parts (a) and (b) here, how would you describe the relationship of the frequency to the wavelength of electromagnetic radiation?
(d) In what region of the electromagnetic spectrum, Figure 4.13, would you find light with a frequency of $3.45 \times 10^{13} \text{ s}^{-1}$?

Source of electromagnetic radiation In Investigate This 3.4, Chapter 3, you observed light from a light bulb and looked at its spectrum. How does the hot filament in the light bulb produce electromagnetic radiation? What causes the oscillating electric and magnetic fields that make up electromagnetic radiation? As we have said, moving electrical charges create magnetic fields. Motion of charges in matter is the source of electromagnetic radiation produced by matter. As we go on, we will point out specific examples of radiation sources. One of the simplest to think about is a television or radio transmission antenna. Essentially, the antenna is a length of wire through which an electric current (moving electrons) passes back and forth, that is, oscillates. First one end of the wire is negative and then the other. The frequency of the emitted electromagnetic waves depends upon the frequency of the electronic oscillation. Different television stations use different frequencies; you tune your receiver (select a channel) to choose the one you wish to see and hear. Sometimes the frequency of electromagnetic waves, especially radio and television frequencies, is given in **hertz** (Hz), $1 \text{ Hz} = 1 \text{ s}^{-1}$.

4.23 CHECK THIS

Electromagnetic wavelength detection and emission

(a) Electromagnetic radiation detectors interact best with wavelengths that are approximately the same size as the detector. The television antennas (several lengths of metal tubing, not satellite dishes) you see on the roofs of houses and other buildings are designed to receive VHF television signals. UHF antennas, on the other hand, are short loops of wire that often are attached directly to the input on the back of the television set. Use the information in Figure 4.13 to explain why these antennas are so different.

(b) FM radio frequencies are in the range 88–108 MHz (megahertz, $1 \text{ MHz} = 10^6 \text{ Hz}$). What is the wavelength of emission (transmission) from your favorite FM station?

Emissions of electromagnetic waves from objects in the universe have been detected in all regions of the electromagnetic spectrum. At one extreme, specially designed satellite instruments have detected bursts of gamma radiation. The sources of some of these bursts appear to be near the edge of the universe; how they are created is still a mystery, but they may come from events that occurred shortly after the Big Bang that we discussed in the previous chapter. At the other end of the spectrum, the earth is continuously bathed in microwave radiation that is responsible for about one percent of the interference ("snow") on a television set that gets its signal from an antenna. This microwave background radiation is the "glow" left over from the Big Bang. Its existence and variations throughout the universe are major pieces of evidence supporting the Big Bang model and calculations of the age of the Universe.

Reflection and Projection

In this section, we have found that light acts like a wave and shares the properties of other more easily observable waves, like water waves. In particular, we can characterize light by its wavelength, frequency, and speed of travel, which are simply related as $c = \lambda \cdot v$.

We further found that light is an electromagnetic wave and that the electromagnetic spectrum of wavelengths extends over an enormous range from the longest radio waves to the shortest gamma waves, with visible light somewhere in between and accounting for only a tiny fraction of the entire range. This information about light was all known by the end of the 19th century and electromagnetic radiation seemed to be a well-understood phenomenon. It came as a surprise to scientists to find that they only knew half the story. Both parts of the story are required to help us unravel the mysteries of atomic structure, so we will continue by examining what happened at the beginning of the 20th century.

4.4. Light as a Particle: The Photoelectric Effect

 **4.24** INVESTIGATE THIS

Are the temperature and color of a hot filament related?

Do this as a class investigation and work in small groups to discuss and analyze the results. Plug a lamp with a clear glass light bulb into a variable voltage transformer. Start at zero voltage and increase the voltage slowly until the filament just begins to glow. Record the color of the filament. Hold your hand close to the light bulb and note how warm it feels. *CAUTION:* Don't touch the glass; it could be hot enough to burn you. Turn up the voltage until the filament is glowing brightly, make the same observations, and answer the same questions.

4.25 CONSIDER THIS

How are the temperature and color of a hot filament related?

In Investigate This 4.24, what, if anything, is different between the observations when the filament just begins to glow and when it is glowing brightly? What, if anything, is the same? How might you explain these observations?

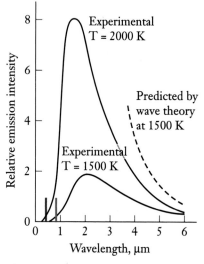

Figure 4.14.

Intensity of emission from a glowing object. The emission intensity at any wavelength is a function of the temperature of the object. The visible region of the electromagnetic spectrum is between the red (0.7 μm = 700 nm) and blue (0.4 μm = 400 nm) lines on the figure.

Emission from glowing objects The explanations of dispersion and diffraction phenomena are successes of the wave theory of light. Near the end of the 19th century, however, scientists were uncomfortably aware that the wave theory could not explain some experimental results. The observations you made in Investigate This 4.24 are an example of a phenomenon that the wave theory could not explain. The filament in a light bulb is heated by the electric current flowing through it and emits the light you see. You found that a filament at a lower temperature glows red and at a higher temperature glows white, that is, emits all visible wavelengths. The wave theory did not predict this temperature dependence of light emission. Figure 4.14 shows the prediction of the wave theory of light and the experimental results at two temperatures.

4.26 CONSIDER THIS

How are Investigate This 4.24 and Figure 4.14 related?

Which experimental curve in Figure 4.14 corresponds to the bulb with the just-glowing filament in Investigate This 4.24? Explain clearly how your observations of color and warmth of the bulb justify this response.

Planck's quantum hypothesis Max Planck (German physicist, 1858–1947) found that he could explain the emission from glowing objects if he assumed that the oscillating charges that emitted the radiation could have only certain energies:

$$E = n \cdot h \cdot \nu \tag{4.4}$$

Here n is any integer, v is the frequency of the oscillator in s^{-1}, and h is a proportionality constant with units of J·s. Thus, in this model, the oscillator energies are **quantized**; the only energies allowed are multiples of the energy **quantum**, $h\cdot v$. Further, Planck had to assume that when an oscillator emitted or absorbed energy, it did so by changing from one energy level, with n = n_i, to another energy level, with n = n_f, so that the change in energy (the energy emitted or absorbed) is

$$\Delta E = n_f \cdot h \cdot v - n_i \cdot h \cdot v = \Delta n \cdot h \cdot v \tag{4.5}$$

Planck chose the numeric value of the proportionality constant, h, to make the predictions from his model fit the experimental data in Figure 4.14. The constant has been named **Planck's constant** and its modern value is $h = 6.6256 \times 10^{-34}$ J·s.

> "Quantized," "quantum," and the more familiar "quantitative" are derived from Latin, *quantus* = how much.

4.27 CONSIDER THIS

Are emission and absorption of energy by oscillators related?

(a) If an oscillator with four quanta (plural of quantum) of energy changes to one with two quanta of energy, is energy emitted or absorbed? Use equation (4.5) to explain your response.

(b) If the oscillator in part (a) goes from having two quanta of energy to having four quanta, is energy emitted or absorbed? Explain how ΔE in this change is related to ΔE in part (a).

(c) Do you see any connection between your results in parts (a) and (b) and the experimental observations represented in Figure 4.6? Explain your response.

With $\Delta n = 1$, equation (4.5) gives the energy, E, associated with light (electromagnetic radiation) of frequency v. This relationship can also be written in terms of the wavelength of the light, by combining equation (4.5) (with $\Delta n = 1$) and equation (4.3):

$$E = \frac{h \cdot c}{\lambda} \tag{4.6}$$

4.28 CHECK THIS

Energy of electromagnetic radiation

(a) Carry out the combination of equations (4.3) and (4.5) (with $\Delta n = 1$) to get equation (4.6).

(b) Which are more energetic, γ rays or microwaves? Use equation (4.6) and the information in Figure 4.13 to answer this question. Does the energy of the electromagnetic radiation increase or decrease as you go from the bottom to the top of the scale in Figure 4.13?

(c) What is the energy (in J) of visible light with a wavelength of 414.5 nm? What color is this light? Explain your response.

Planck's assumption (hypothesis) said that the oscillators in a glowing object could have only certain energies that depend on their frequency. Hotter objects have more energy, so there could be more oscillators with higher energies. This explained why the higher temperature curve in Figure 4.14 has its maximum at a shorter wavelength (higher frequency) of emitted radiation. Because the total energy in the object is limited, the number of high-energy oscillators is limited. This explained why the emission curves finally drop toward zero intensity at shorter wavelengths. Scientists had previously thought that the energy of oscillators was associated with the amplitude of the emission, not its wavelength. Planck's hypothesis was met with a great deal of skepticism, because it seemed ridiculous to think that oscillator energies were quantized.

4.29 INVESTIGATE THIS

What is the effect of light on silver chloride?

Do this as a class investigation and work in small groups to discuss and analyze the results. Use a rectangle of absorbent paper, four colored plastic filters (red, green, blue, and colorless), 0.1 M aqueous solutions of silver nitrate, $AgNO_3$, and sodium chloride, NaCl, and a bright lamp. Near one of the long edges, label the paper with the letters *R*, *G*, *B*, and *C*, about equally spaced apart. A few centimeters beneath each label, place a drop of the silver nitrate solution. Below each drop of silver nitrate, place a drop of the sodium chloride solution so that the solutions overlap as they spread on the paper. Recall from Chapter 2 that silver ion and chloride ion react to form solid silver chloride:

$$Ag^+(aq) + Cl^-(aq) \rightarrow AgCl(s)$$

Record the appearance of the paper. Cover each pair of spots with the filter of the color corresponding to the label on the paper, as shown in the photograph. Shine a bright light on the covered paper, taking care that each pair of spots gets the same amount of illumination. After 5 minutes, remove the light and filters, examine the paper, and record its appearance.

4.30 CONSIDER THIS

Do all colors of light affect silver chloride the same way?

Silver halides are affected by light, which is why photographic film contains silver halide crystals. In damp silver chloride precipitates, we can write the reaction caused by light as

$$\left(Ag^+ \quad :\overset{\cdot\cdot}{\underset{\cdot\cdot}{Cl}}:^- \right)(s) \xrightarrow{\text{light energy}} Ag\cdot(s) + :\overset{\cdot\cdot}{\underset{\cdot\cdot}{Cl}}\cdot(aq) \qquad (4.7)$$

$$\text{silver metal} \quad \text{chlorine atom}$$

The chlorine atoms formed in this reaction can react with one another to form chlorine molecules, Cl_2, and/or with water to form chloride ions and oxygen gas, O_2. The silver forms tiny specks of silver metal that darkens the paper, as shown here.

continued

(a) Do the results from Investigate This 4.29 provide any evidence that reaction (4.7) has occurred? Explain your response.

(b) If reaction (4.7) occurs, is it the same or different for light of different colors? If it is different, what color (wavelength) light has the greatest effect? Which the least? Can you interpret these observations using the wave theory of light? Why or why not?

The photoelectric effect Five years after Planck's quantum hypothesis, his ideas were used by Albert Einstein to explain another experimental result that the wave theory had failed to explain: In the **photoelectric effect,** when light shines on a metal surface, electrons can be knocked out of the surface. The wave theory predicted that, if the light beam were bright enough (had a large enough amplitude or intensity), any frequency (wavelength) of light could knock an electron out. However, experimental results illustrated for low frequency light in Figure 4.15 show that no current flows in this circuit—that is, no electrons are knocked out—no matter how bright the light beam. Low frequencies of light cannot knock electrons out of the metal. The experimental results illustrated in Figure 4.16 show that when a higher frequency light is used, current flows. Electrons can be knocked out of the metal by high frequency light. The results for a number of different frequencies are summarized in Figure 4.17.

The photoelectric effect is the principle behind the "electric eyes" that control such things as outdoor lighting and elevator doors. The presence of light falling on a photoelectric surface causes an electric current that triggers lights to come on or go off, for example.

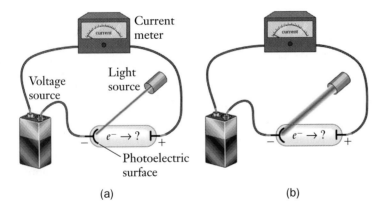

(a) (b)

Figure 4.15.

Photoelectric experiments with low frequency light of different intensities. Red represents low frequency (long wavelength— Figure 4.13) light. The thickness of the light beam shows its intensity (brightness): (a) is low intensity and (b) is high intensity.

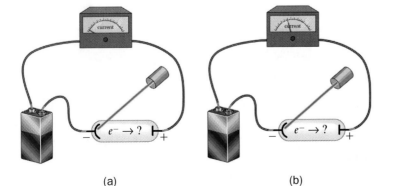

(a) (b)

Figure 4.16.

Photoelectric experiments with (a) low and (b) high frequency light. Blue, (b), represents higher frequency (shorter wavelength— Figure 4.13) light than red, (a).

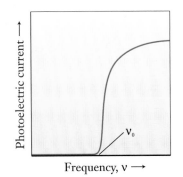

Figure 4.17.

Photoelectric current as a function of frequency of the light. The intensity (brightness) of the light is the same at all frequencies.

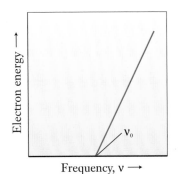

Figure 4.18.

Energy of ejected photoelectrons as a function of light frequency.

Below a minimum frequency, ν_0, no electrons are knocked out of the metal. Einstein assumed that it must take a certain minimum amount of energy to eject an electron from a particular metal surface. If Planck were right about the relationship between energy and frequency, equation (4.5), then a certain minimum frequency would be required to eject electrons. The Planck quantum hypothesis explained the data in Figure 4.17.

In addition, other photoelectric experiments showed that the maximum energy of the ejected electrons depended on the frequency of the light, as shown in Figure 4.18. Einstein reasoned that the maximum energy an ejected electron could have would be the energy provided by the light minus the energy required to eject the electron:

$$E_{electron} = E_{light} - E_{ejection} \tag{4.8}$$

He wrote the two energies on the right in terms of frequencies, ν, and Planck's proportionality constant, h:

$$E_{electron} = h \cdot \nu - h \cdot \nu_0 = h \cdot (\nu - \nu_0) \tag{4.9}$$

When Einstein analyzed the experimental data quantitatively, he found that the numerical value of the proportionality constant, h, in equation (4.9) was the same value Planck had used to explain the experimental results for emission from glowing objects. Thus, Einstein showed that light is quantized and comes in discrete energy packets, $E = h \cdot \nu$. He used the term **photon** to designate an energy packet of light.

4.31 CONSIDER THIS

How does photoelectric current depend on light intensity?

(a) Figure 4.19 shows the effect of light intensity on the photoelectric current for light with a frequency $\nu > \nu_0$. How does the model based on photons explain the results? In the photon model, what does *intensity* mean?

(b) What does the photon model predict for the maximum energy of the ejected electrons in Figure 4.19(b) compared to the electrons in Figure 4.19(a)? Explain your reasoning.

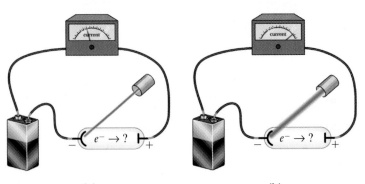

(a)   (b)

Figure 4.19.

The effect of light intensity on photoelectric current.

4.32 CONSIDER THIS

How is the darkening of silver chloride related to photons?

(a) Can you explain the results from Investigate This 4.29 using the photon (energy packet) model of light? Explain why or why not. What assumption(s) do you have to make about reaction (4.7).

(b) Reaction (4.7) requires an energy of about 4×10^{-19} J for each (Ag^+Cl^-) that reacts. Are your results consistent with this energy requirement? Explain why or why not.

4.33 CHECK THIS

Energy of photons

(a) What is the energy of a photon of green light, wavelength 515 nm? What is the energy of a mole of these photons?

(b) What is the energy of a photon of wavelength 21.11 cm, an emission from hydrogen atoms that we will discuss later?

(c) What is the energy of a photon of the frequency emitted by your favorite FM radio station [Check This 4.23(b)]? What is the energy of a mole of these photons?

The dual nature of light Since the same hypothesis and proportionality constant explained two entirely different phenomena, most scientists soon accepted Planck's original hypothesis, even though it seemed strange then and still does. We use the wave model to interpret and explain some electromagnetic phenomena. We were able to understand the diffraction of light (Section 4.3), for example, from a glowing object or a laser, in terms of a wave model. In this section, however, we found that the temperature dependence of the intensity of emission from a glowing object cannot be explained by the wave model, but requires a photon (quantum) model of light. The photon model also enables us to interpret and explain the photoelectric effect and the wavelength dependence of a chemical reaction that requires energy from light. There is no analog in the world of objects large enough to see and touch that helps us understand the dual nature of light. The justification for accepting this dual nature is that it works to give correct predictions of observable results.

4.34 CHECK THIS

Wave and photon phenomena

For each of these examples, tell whether the wave theory or photon (quantum) theory of light best explains the phenomenon and give your reason(s).

(a) The play of colors from the surface of a CD

(b) The color of a neon light

(c) Getting sunburned

(d) A rainbow

Reflection and Projection

We have found that the electromagnetic wave model of light successfully explains many phenomena, including diffraction. However, the wave model fails to explain other phenomena, such as the wavelength dependence of the intensity of emission from hot, glowing solids and the photoelectric effect. In order to explain the latter phenomena, we need to assume that electromagnetic oscillators can emit energy at only certain frequencies. The oscillators are quantized; the energy of the emitted light is directly proportional to the frequency of the oscillator. Further, in some experiments, light acts as though it has particle-like properties; the energies of the photons (light particles) are directly proportional to the frequency of the light. The proportionality constant, Planck's constant, is the same for both cases.

The dual nature of light leads to the question: Does matter also have a dual nature, both particle and wave? We will take up this question in Section 4.6, but first we will look at how the concept of energy quantization was used to develop a quantum model of atoms that explained line emissions of light from energized atoms.

4.5. The Quantum Model of Atoms

We began discussing the properties of light because we thought the emission and absorption of light by elemental atoms might provide clues about their structures and properties. Recall from Figure 4.6 that emission occurs from atoms that have been given a lot of energy by a spark or a flame and absorption occurs from less energetic atoms at a lower temperature. This probably means that the emission and absorption processes in the figure represent changes in the energy of the atoms. An atom can lose energy by emitting light with a wavelength corresponding to the amount of energy lost. The wavelength of the emitted light is given by the Planck radiation law, equation (4.6). Conversely, an atom can absorb light energy and change from a lower to a higher energy. These are the processes illustrated in Figure 4.20.

Figure 4.20.

Atomic emission and absorption of light of the same wavelength.
Light is emitted when an atom goes from a higher energy to a lower energy state. When the energy of the atom is boosted from the lower to the higher energy state, the atom absorbs light of this same wavelength.

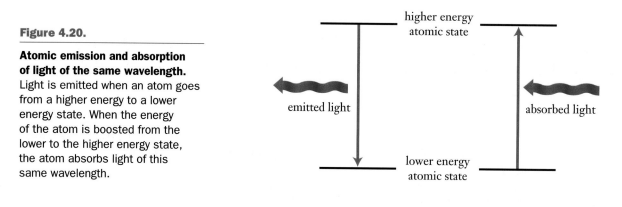

4.35 CONSIDER THIS

What property of atoms causes line emission (and absorption)?

Assume that emission and absorption spectra of atoms are the result of their electrons giving off or taking up energy.

continued

(a) If the electrons can have any energy (consistent with being part of the atom), what would their emission spectra look like? Explain.
(b) If the electrons can have only certain energies, what would their emission spectra look like? Explain.

Quantized electron energies in atoms When gaseous atoms and ions are heated to a high temperature, or when an electric spark is passed through the gas, much of the energy input is transferred to the electrons in the atoms. When the electrons give up that extra energy, it is released as electromagnetic radiation in the visible, infrared, and ultraviolet regions. Solids, such as the tungsten filament in an incandescent light bulb, also emit light when they are heated. The enormous difference, as you observed in Investigate This 3.4, is that the emission from the tungsten filament consists of *all* the wavelengths in a wide range, whereas energetic electrons in gaseous atoms and ions emit only a few wavelengths of light. Since Planck's hypothesis links energy and wavelength, this observation must mean that energetic electrons in atoms shed excess energy by emitting only certain amounts of energy.

Early in the 20th century, after the nuclear model of the atom and Planck's quantum hypothesis had been developed, Niels Bohr (Danish physicist, 1885–1962) applied these ideas to the structure of atoms. Bohr used his **quantum model** of atomic structure to calculate the energies (wavelengths) that would be emitted by an energetic hydrogen atom. The experimental values for the wavelengths of hydrogen atom emissions in the visible region of the spectrum are shown in Figure 4.21; Bohr's calculations gave exactly these same values. This agreement of a model and an experiment was convincing evidence that *electron energies in an atom are quantized*, which is why emissions from atoms are limited to a few particular energies.

Web Companion

Chapter 4, Section 4.4.1 — ①

View an animation of the H atom ② emission to give the data in ③ Figure 4.21. ④

| 410.2 nm | 434.0 nm | 486.1 nm | | 656.3 nm |

Figure 4.21.

The visible emission spectrum from atomic hydrogen.

Electron energy levels in atoms Bohr's achievement was enormous; it suggested that the quantization of electron energies in atoms is a fundamental property of atoms. Figure 4.22 is a representation of the quantized energies that Bohr calculated for the hydrogen atom. Each line, or level, in the figure indicates the relative energy of the electron in the atom, just like the other energy diagrams we have been using. The lowest energy is chosen as the starting point (E_1) and the energy levels of the electron continue to increase until the electron has so much energy that it can no longer be held by the attraction of the nucleus. When an electron receives this much energy, it separates from the original atom; the atom has been ionized. The loss of the electron leaves the ion with one unit of net positive charge. In

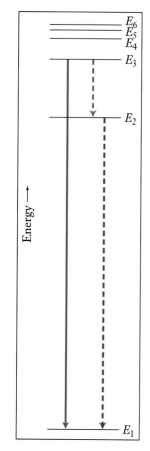

Figure 4.22.

Energy-level diagram for a one-electron atom. The energy levels are for an H atom or an ion such as He$^+$, Li^{2+}, and so on. The arrows show pathways for an electron to lose energy and return from the excited state, E_3, to the ground state, E_1. The loss of energy can occur in one step, solid arrow, or two steps, dashed arrows.

multielectron atoms, more electrons can be lost in the same way, with each loss increasing the net positive charge of the ion. Notice that the spacing between the energy levels gets closer and closer as the electron energy increases toward the ionization limit.

4.36 CHECK THIS

Experiment with a one-electron atom simulation

In the *Web Companion*, Chapter 4, Section 4.4.2, follow the directions and drag the pointer to different energy levels. Describe what occurs to the "atom." Drag the pointer to an intermediate point between energy levels. What happens? Is it possible for the electron to have an energy intermediate between two of the energy levels? Explain.

The lowest total energy of an atom is called its **ground state,** level E_1 in Figure 4.22. When an electron in the atom gains the energy to be in any one of the higher energy levels, say E_3, the atom is said to be in an **excited state** and the electron has been excited. To return to its ground-state energy, the electron has to get rid of the excess energy. There are two ways back to the ground state for this electron, as shown by the arrows in Figure 4.22. It can emit a photon with an energy equal to $E_3 - E_1$, the solid blue arrow, or it can emit one photon with energy of $E_3 - E_2$ and then a second one having an energy of $E_2 - E_1$, the dashed red and green arrows, respectively. In either case, only specific energies can be emitted and only the corresponding wavelengths of light will be observed. This is how the quantum model of electrons in atoms accounts for the observation that energized atoms or ions emit light of only a few wavelengths.

4.37 WORKED EXAMPLE

Energy level differences in the hydrogen atom

The emission wavelengths corresponding to the $E_3 - E_1$ and $E_2 - E_1$ energy differences for hydrogen, Figure 4.22, are at 102.57 and 121.57 nm, respectively. What are these energy differences in joules? in joules per mole?

Necessary information: We need the Planck relationship, equation (4.6), Planck's constant, 6.6256×10^{-34} J·s, and Avogadro's number, 6.022×10^{23} mol^{-1}.

Strategy: The energy of an emission is equal to the difference in energy between the higher and lower energy states. Use the Planck relationship to relate the wavelength of an emission to the difference in energies that produce the emission. To get the energy in J·mol^{-1}, multiply the energy for a single atom by Avogadro's number.

Implementation: Wavelengths are subscripted to show which energy levels are involved.

continued

$$E_3 - E_1 = \frac{h \cdot c}{\lambda_{3-1}} = \frac{(6.626 \times 10^{-34}\,\text{J} \cdot \text{s})(3.00 \times 10^8\,\text{m} \cdot \text{s}^{-1})}{102.57 \times 10^{-9}\,\text{m}}$$

$$= 1.938 \times 10^{-18}\,\text{J}$$

$$E_3 - E_1 \text{ (per mole)} = (1.938 \times 10^{-18}\,\text{J})(6.022 \times 10^{23}\,\text{mol}^{-1})$$

$$= 1.167 \times 10^6\,\text{J} \cdot \text{mol}^{-1}$$

$$E_2 - E_1 = \frac{h \cdot c}{\lambda_{2-1}} = \frac{(6.626 \times 10^{-34}\,\text{J} \cdot \text{s})(3.00 \times 10^8\,\text{m} \cdot \text{s}^{-1})}{121.57 \times 10^{-9}\,\text{m}}$$

$$= 1.635 \times 10^{-18}\,\text{J}$$

$$E_2 - E_1 \text{ (per mole)} = (1.938 \times 10^{-18}\,\text{J})(6.022 \times 10^{23}\,\text{mol}^{-1})$$

$$= 0.985 \times 10^6\,\text{J} \cdot \text{mol}^{-1}$$

Do the results make sense?   We see that, when the results are expressed in terms of a mole of atoms, the energies are substantial. For hydrogen, the difference in energy between the ground and first excited state is 985 kJ·mol^{-1}. This explains why so much energy is required to excite the atoms in a hydrogen discharge tube to emit light.

4.38 CHECK THIS

Energy level differences in the hydrogen atom

(a) Use the results in Worked Example 4.37 to find $E_3 - E_2$ (in joules) for the hydrogen atom.

(b) What is the emission wavelength corresponding to $E_3 - E_2$, the red arrow in Figure 4.22? Compare your answer with the emission wavelengths shown in Figure 4.21.

(c) In the *Web Companion*, Chapter 4, Section 4.4.3, drag the pointer to the blue line. Explain in your own words what the animation represents in terms of what is happening to the electron and the associated energy change(s). Explain where the photon of light comes from and why it is emitted rather than absorbed.

A basic idea to take from our discussion is that, when high-energy electrons in an atom or ion return to lower energy levels, the emitted energy takes the form of light. A light wave is produced from the difference in energy between the two energy levels. Conversely, the photons in a light wave can transfer energy to an electron in an atom *only* when the photon's energy exactly equals the *difference* in energy between two energy levels of the electron. This explains why the emission and absorption processes, Figures 4.6 and 4.20, emit and absorb light of exactly the same wavelength(s).

4.39 CHECK THIS

Predicting emission and absorption wavelengths

(a) The emissions from excited hydrogen atoms in Figure 4.21 are emissions from the E_3, E_4, E_5, and E_6 levels to return to the E_2 level. What is the $E_4 - E_3$ emission energy? At what wavelength would this emission occur? In what region of the electromagnetic spectrum (Figure 4.13) would this wavelength be found? (This emission wavelength was calculated, as you are doing, before it was found experimentally exactly where it was predicted to be.)

(b) If you were trying to detect hydrogen atoms by their absorption of energy to go from the ground state to the first excited state, what region of the electromagnetic spectrum would you need to use as your light source?

Due to limitations of the model, the only atomic energy levels the Bohr quantum model could calculate exactly were those for hydrogen and other one-electron ions like He^+ and Li^{2+}. More dismaying to chemists was the fact that the model could not be applied to molecules. And, unfortunately for our purposes, the model does not explain why atoms don't collapse. Nonetheless, the quantum model served a useful purpose in interpreting the spectral lines from atoms in gas-discharge tubes, and it led scientists to search for other models that would retain the quantization of electron energies without the shortcomings of the Bohr quantum model. That search brings us back to waves.

4.6. If a Wave Can Be a Particle, Can a Particle Be a Wave?

Wavelength of a moving particle Once the work of Planck, Einstein, and others had convinced scientists that light exhibits the properties of both waves and particles, it didn't take long to develop a wave model for the behavior of particles. In 1924, Louis de Broglie (French physicist, 1892–1987) postulated (with no experimental justification) that the behavior of moving electrons can be described as though they have wave properties. His model showed that the wavelength of a moving, wave-like particle is

$$\lambda = \frac{h}{m \cdot u} \qquad (4.10)$$

This wavelength, often called the **de Broglie wavelength,** is inversely proportional to the mass, m, and velocity, u, of the particle. The proportionality constant, h, in equation (4.10) is Planck's constant, which crops up in almost every equation dealing with atomic-level phenomena. Equation (4.10) is the bridge that relates the wave nature of electrons to the particle nature of electrons. The product, $m \cdot u$, in the denominator of equation (4.10) is the **momentum** of the particle:

$$\text{momentum} = m \cdot u \qquad (4.11)$$

Momentum measures the amount of push a moving object can exert on another object: The larger the mass, m, and faster the motion, u (velocity), the larger the momentum.

4.40 WORKED EXAMPLE

de Broglie wavelength of an electron

Calculate the de Broglie wavelength of an electron traveling at 9.4×10^5 m·s^{-1}, a velocity that is easy to attain. If the electron acts like an electromagnetic wave, to what region of the spectrum, Figure 4.13, does this wavelength correspond?

Necessary information: We need equation (4.10), Planck's constant, the velocity of the electron from the problem statement, and, from Table 3.1, its mass, 9.1×10^{-28} g.

Strategy: Substitute the mass and velocity into equation (4.10), using the correct units for mass and velocity to be compatible with Planck's constant. The energy unit in Planck's constant is the joule ($= $ kg·m^2·s^{-2}).

Implementation:

$$\lambda = \frac{h}{m \cdot u} = \frac{6.63 \times 10^{-34}\,\text{J·s}}{(9.1 \times 10^{-31}\,\text{kg})(9.4 \times 10^5\,\text{m·s}^{-1})}$$

$$= 7.8 \times 10^{-10}\,\text{m} = 0.78\,\text{nm} = 780\,\text{pm}$$

Figure 4.13 shows that electromagnetic waves in the region 100–1000 pm are in the X-ray region of the spectrum.

Does this answer make sense? The calculated wavelength of the electron is in the X-ray range, so we might expect electrons moving at this velocity to have properties like X-rays, which are used extensively by scientists to study the structure of crystals. To figure out whether this de Broglie wavelength makes sense, we have to know what happens if these electrons are substituted for X-rays in a crystal structure experiment, which is the topic of the next paragraph.

4.41 CHECK THIS

de Broglie wavelengths

(a) What is the de Broglie wavelength for an electron moving at one-tenth the speed of light? To what region of the electromagnetic spectrum does this wavelength correspond?

(b) Some baseball pitchers can throw a 145-g baseball 100 miles per hour (about 44 m·s^{-1}). What is the de Broglie wavelength for this fastball?

(c) Thermal neutrons (such as those in nuclear reactors) are neutrons moving at the speed they would have if they were atoms in a gas at room temperature, about 1 km·s^{-1}. What is the de Broglie wavelength of thermal neutrons? To what region of the electromagnetic spectrum does this wavelength correspond?

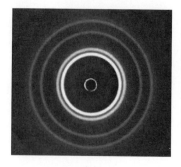

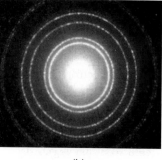

(a)

(b)

Figure 4.23.

(a) X-ray and (b) electron diffraction patterns for aluminum foil. The foil is made up of an enormous number of tiny crystals that are responsible for the diffraction. The pair of intense inner rings and pair of less intense outer rings are identical in the two patterns. A much weaker further pair of rings in the electron diffraction pattern are too weak to see in the X-ray pattern.

Experimental evidence for electron waves If moving electrons act like waves, they should show the phenomenon of diffraction, as in Figures 4.7 and 4.9. Shortly after de Broglie proposed the electron wave model, two American physicists, C. J. Davisson and L. H. Germer, interpreted the results of one of their experiments as being due to the diffraction of electrons by atoms in a crystal. The diffraction of X rays by crystals, as in Figure 4.23(a), was a well-known phenomenon. Note that the diffraction pattern from a crystal is more complicated than the pattern you observed from a grating in Investigate This 4.17(b). This is because the "gratings" in a crystal are the spacings between atoms and there are several different spacings that depend upon the geometric arrangement of the atoms in the crystal. Information from X-ray diffraction had been used to analyze the structure and spacings in crystals since the early part of the 20th century.

When electrons are accelerated to known velocities (about the same as in Worked Example 4.40) in an electric field and directed through a very thin metal foil, the electrons are diffracted, as shown in Figure 4.23(b). The patterns for X-ray diffraction and electron diffraction from the same sample are identical. The wavelength of the electron waves can be measured from the spacings of the diffraction pattern. The measured wavelength and the calculated wavelength from the de Broglie model, equation (4.10), are the same; our result in Worked Example 4.40 makes sense.

All diffraction methods, including spectroscopy, are related. Diffraction occurs when the wavelength of the waves being diffracted is comparable in size to the grating that is responsible for diffracting the waves. Since moving electrons and X rays of the same wavelength are diffracted by crystals, the spacing between atoms in the crystals must be about 100–1000 pm (see Figure 4.13). The wavelength of a fast-pitched baseball, as in Check This 4.41(b), is so short that there is no physical structure small enough to detect its wave properties. We can safely ignore the wave properties of massive objects like baseballs and automobiles. Heavy atomic- or subatomic-size particles are a different matter. Your calculation in Check This 4.41(c) shows that thermal neutrons have wavelengths comparable to X rays, that is, a size that can be diffracted by the atoms in crystals. Neutron diffraction is a powerful technique that is extensively used to explore the structures of molecules in crystals, especially biological molecules like proteins.

The results from this section and the previous one upset simple differentiations between light and matter. We see that both light and matter have wave-like and particle-like properties. The property that is observed depends on the phenomenon being investigated. Now we will use these new understandings to investigate atoms in more detail.

Reflection and Projection

The first quarter of the 20th century was both exciting and perplexing for scientists. It began with the demonstration that the well-established electromagnetic model of light was not the complete story and that light has particulate (quantum) properties as well as wave properties. It ended with the demonstration that matter, at the atomic level, also has wave properties, as shown by electron diffraction, as well as its familiar particulate properties. In the years between, Bohr and others developed a quantum model of the atom that had spectacular success interpreting atomic line spectra. The quantum model of the atom says that electrons in atoms can have only certain energies and that the loss or gain of energy by the atom can only occur by emission or absorption of light that corresponds to the difference in energy between two of these energy levels.

However, problems with the quantum model, especially in applying it to multielectron atoms and molecules, were evident. And we are still faced with the fundamental question: Why don't atoms collapse? Electrons are negatively charged and are attracted by the positively charged nucleus. What keeps electrons from being drawn closer and closer to their nuclei until all atoms collapse to the size of their nuclei? The quantum model could not explain why they were not. The advent of the de Broglie wave model for electrons provided an exciting new way to model atoms that also explained their energy quantization and got beyond the quantum model problems. In particular, the wave model leads to an explanation for why atoms don't collapse.

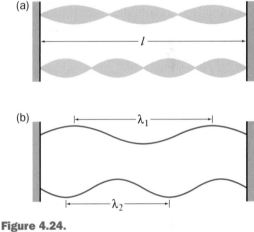

4.7. The Wave Model of Electrons in Atoms

Waves and atomic emissions Some waves, like water and light waves, are propagated through space; these are **traveling waves.** Other waves, **standing waves,** are held in place; they are constrained in some way. Vibrating guitar strings are an example of standing waves. In a guitar, the waves cannot extend beyond the posts that anchor each end of the string. The amplitude of such a wave is the maximum displacement of the string from its rest position. A node must always occur at each post. The wavelength is determined in the same way as for traveling waves—the distance between wave crests. Figure 4.24 illustrates typical standing waves in guitar strings. Figure 4.24(a) represents time exposure photographs, while Figure 4.24(b) represents instantaneous exposures.

Figure 4.24.

Standing waves in guitar strings of length _l._
(a) Represents time exposure photographs that blur the strings. (b) Represents instantaneous exposures when the waves are at maximum amplitude.

████ **4.42 CHECK THIS** ████

Standing waves on a string

Not just any wavelengths can be accommodated on the strings in Figure 4.24. The wavelengths that "fit" on a string of length _l_ are

$$\lambda = \frac{2l}{n}, \text{ where } n = 1, 2, 3, \ldots \tag{4.12}$$

(a) What is _n_ for each of the waves in the figure? If _l_ = 0.75 m, what are the wavelengths of the waves? How many nodes are there in each wave?
(b) What are _n_ and λ for the longest wavelength on a 0.75-m string? Sketch this wave, using the representations in Figure 4.24 as models.

Standing electron waves

Scientists studying atomic emission spectra in the 19th century had found mathematical expressions that described the patterns of the emission wavelengths they observed, such as those from hydrogen in Figure 4.21. There are striking similarities between these mathematical expressions for atomic emission patterns and the mathematical expressions for standing waves, like the simple series for waves on a string, equation (4.12). These similarities were part of the evidence that led scientists to apply the de Broglie model of matter waves (for the electron) to construct a wave mechanical model of atomic structure.

The electron is held in the atom by its attraction to the positive nucleus, much like a guitar string is attached to the body of the instrument. Using **wave mechanics,** the physics that provides mathematical descriptions of wave motion, we describe the electron in an atom as a standing wave. Like the standing waves of a guitar string, the waves that are associated with an electron can have only certain wavelengths. Planck's relationship, equation (4.6), specifies that an electron wave of a particular wavelength would have an energy that is inversely proportional to the wavelength. Thus, the electron (in an atom or ion) described as a wave can have only certain energies and no others. Quantization of electron energy is a natural consequence of the electron wave model.

Probability picture of electrons in atoms

The Bohr quantum model provided a picture of an atom as a nucleus with electrons in orbit about the nucleus. It would be nice to have a comparable picture of the motion of electrons that behave as waves. The problem with electrons behaving like waves, rather than like baseballs or marbles or planets, is that we can never tell exactly where the electrons are or which way they are going at a particular instant. This is because of the uncertainty principle, another property of atomic-size systems that, like waves, has no analog in the macroscopic world of baseballs and marbles. Within a year of de Broglie's electron-wave hypothesis, Werner Heisenberg (German physicist, 1901–1976) formulated the uncertainty principle. Qualitatively, the **uncertainty principle** (or **Heisenberg uncertainty principle**) states that it is impossible to measure simultaneously both the exact location and the exact momentum of an electron. Recall that the momentum of the electron is a part of the de Broglie model, equation (4.10), which closely ties the wave model to the uncertainty principle.

The best we can do for finding an electron in an atom, ion, or molecule is to determine its *probability* of being in a particular location when it is described by one of the energy states (levels) like those shown in Figure 4.22. The location of an electron near an atomic nucleus is described by a probability distribution, much as are the holes in a dartboard that represent attempts to hit the bull's-eye. The fraction of holes in each target ring, Figure 4.25(a), is the probability of a dart having landed in that ring. The likelihood of finding the electron in any particular location near the nucleus is represented in Figure 4.25(b) by the density of dots (darkness) around the nucleus (at the very center of the distribution). The probability of finding the electron decreases as distance from the nucleus increases.

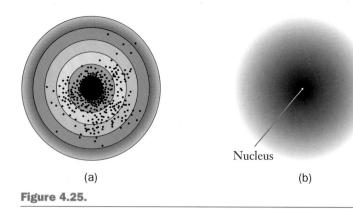

(a)   (b)

Figure 4.25.

Probability distributions. Probability of (a) hitting a dartboard bull's-eye and (b) finding an electron around a nucleus. The density of dots in a region represents the probability of (a) having hit that region with a dart or (b) finding the electron.

4.43 CHECK THIS

Probability in three dimensions

What three-dimensional geometric figure would you choose to represent the probability distribution shown two dimensionally in Figure 4.25(b)? Explain your choice.

The de Broglie wave model and Heisenberg's uncertainty principle ended scientists' attempts to describe the *path* of an electron moving from one place to another in an atom, as Bohr had done in his quantum model. Use Figure 4.24 as an analogy to visualize a standing electron wave in an atom. The instantaneous exposures of the vibrating string in Figure 4.24(b) are analogous to electrons in defined orbits, which we have had to give up, because we can't find the electron at any particular instant in time. But, we can get the electron distribution over time, as illustrated in Figure 4.25(b), and this is analogous to the blurred time exposures of the vibrating string in Figure 4.24(a).

Orbitals We need a three-dimensional model for this blurred electron wave. The simplest model is a sphere, which is probably the geometry you chose in Check This 4.43. A sphere is simple because it requires only two variables to describe it: The location of its center and its radius. We will visualize a standing electron wave in an atom as a sphere that encloses the volume where the electron is most likely to be found. The wavelength, λ, of a spherical electron wave is proportional to the radius of the sphere, R:

wavelength of a spherical electron wave $= \lambda \propto R =$ radius of the sphere

(4.13)

By the time the wave model was developed, the idea of electron *orbits* in atoms had become so widely used that we now use the word **orbital** to name electron waves (probability distributions) in atoms, like the one represented in Figure 4.25(b). Simpler pictures of orbitals, as in Figure 4.26(a), represent the volume in space within which there is roughly a 95% probability of finding an electron that has a particular energy. The spherical surface in Figure 4.27(b) shows how the simpler surfaces are related to the probability distribution. Recall from the Planck relationship, equation (4.6), that the energy of a wave is inversely proportional to its wavelength: A more constrained (smaller) electron wave has a higher energy. Before we can interpret and use orbital pictures, we need to discuss what determines electron orbital energies. This discussion will also help us understand why atoms don't collapse.

Section 4.8. Energies of Electrons in Atoms: Why Atoms Don't Collapse

We've said a lot about electron energy levels without saying what we mean by energy in this context. In a one-electron atom (or ion), which we have been using as a model, we are concerned with the kinetic energy of the moving electron and the potential energy of attraction between the positive nucleus and

Web Companion

Chapter 4, Section 4.6.1 — ①

View this animation of an ②
orbital in both representations; ③
compare with Check This 4.43. ④

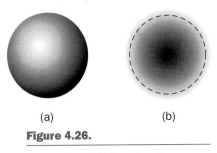

(a) (b)

Figure 4.26.

An atomic orbital; a standing spherical electron wave around a proton. (a) Orbital represented as a smooth sphere. (b) The dashed line is shown to enclose the volume within which there is 95% probability of finding the electron. The proton is at the center of the sphere.

the negative electron. The motion of the nucleus is usually not considered in atomic models, because the nuclei are so slow moving relative to the electrons that their affect on the energy is insignificant.

Kinetic energy of an electron We cannot examine energy in the same way we can handle and feel matter, but we can examine its effects. For example, you can tell the difference between a thrown baseball moving at 44 m·s^{-1} and a baseball at rest beside the pitcher's mound. The idea that an object in motion has a property that the object at rest does not have is easy enough to accept; that property is **kinetic energy.** Kinetic energy (KE) is expressed quantitatively as a function of mass (m) and velocity (u):

$$KE = \frac{m \cdot u^2}{2} \tag{4.14}$$

4.44 WORKED EXAMPLE

Kinetic energy of electrons in the diffraction experiment

What is the kinetic energy of an electron traveling at 9.4×10^5 m·s^{-1}, as in Worked Example 4.40? What would be the kinetic energy of a mole of these electrons?

Necessary information: We need equation (4.14), the mass of an electron, 9.11×10^{-31} kg, from Table 3.1, the velocity from the problem statement, and Avogadro's number, 6.022×10^{23} mol^{-1}.

Strategy: Substitute the known values for mass and velocity in equation (4.14) to get the kinetic energy for one electron and then multiply by Avogadro's number to get the kinetic energy for a mole of them.

Implementation:

$$KE \text{ (per electron)} = \frac{m \cdot u^2}{2} = \frac{(9.11 \times 10^{-31} \text{ kg})(9.4 \times 10^5 \text{ m·s}^{-1})^2}{2}$$

$$= 4.02 \times 10^{-19} \text{ J}$$

$$KE \text{ (per mole)} = (4.02 \times 10^{-19} \text{ J})(6.022 \times 10^{23} \text{ mol}^{-1})$$

$$= 2.4 \times 10^5 \text{ J·mol}^{-1} = 240 \text{ kJ·mol}^{-1}$$

Does the answer make sense? A mole of electrons has a tiny mass, 5.5×10^{-7} kg, but, when traveling at these speeds, has a substantial kinetic energy. The energy is comparable to the energies of electrons in hydrogen atoms that we found in Worked Example 4.37. This makes sense, since we imagine the electrons in atoms are moving quite rapidly about the nucleus.

4.45 CHECK THIS

Kinetic energy of more massive objects

(a) Calculate the kinetic energy of the baseball in Check This 4.41(b).

(b) Calculate the kinetic energy of a mole of the thermal neutrons in Check This 4.41(c).

(c) Compare the results in parts (a) and (b) with the kinetic energy of the electrons in Worked Example 4.44.

Potential energy of a nucleus and an electron The kinetic energy associated with motion is obvious to almost everyone. It is somewhat less obvious that two objects at rest, but physically in different locations, also differ in their energy content. A textbook placed on a high shelf has significantly more gravitational **potential energy**—*PE*, the energy resulting from position—than does a textbook on the floor. If the textbook on the shelf is knocked off, it will move under the influence of gravity to rest on the floor. The book on the floor, on the other hand, will not move up to the shelf without an input of energy.

The potential energy we are interested in is the potential energy of coulombic attraction between an electron and a proton. As you saw in equation (2.3), Chapter 2, Section 2.4, the potential energy of coulombic attraction between two charges (Q_1 and Q_2) that are a distance r apart is

$$PE \propto \frac{Q_1 Q_2}{r} \qquad \qquad (4.15)$$

For a proton and an electron, we write the unit charges as $+1$ and -1, so the potential energy is negative:

$$PE \propto \frac{(+1)(-1)}{r} = -\frac{1}{r} \qquad \qquad (4.16)$$

When the nucleus and electron are far apart (when r is large), $PE \approx 0$. As they get closer to each other, the potential energy goes farther and farther below zero. ***The closer the nucleus and electron are (the smaller the atom), the more negative their potential energy.*** This is the direction the energy must go to form a stable atom.

4.46 CONSIDER THIS

What is the effect of nuclear charge on the potential energy?

For the same distance of separation, how does the coulombic attraction of an electron and a 2+ nucleus compare to the attraction of an electron and 1+ nucleus? Which species, H or He$^+$, would you expect to be smaller? Why?

Total energy of an atom For any system, the **total energy**, E, is the sum of its kinetic and potential energies:

$$E = KE + PE \qquad (4.17)$$

A stable atom is one that has a lower total energy than the total energy associated with its nucleus and electrons when they are separated from each other. Scientists use an energy scale that defines the energy of the separated nucleus and electrons as zero. To determine the size and other properties of our wave mechanical atom, we need to find the conditions where the *total energy of the nucleus and electron wave, E, is most negative*.

The kinetic energy and the potential energy of a spherical wave are both functions of the wave radius, R. These can be calculated, and one way to do this is given in the Extension to this chapter. At this point, we need to know only the results of those calculations:

$$E = KE + PE \propto \frac{1}{R^2} - \frac{1}{R} \qquad (4.18)$$

The momentum of a matter wave, equation (4.11), is inversely related to the wavelength of the wave, equation (4.10), and the kinetic energy of a particle is directly related to the square of its momentum. These relationships give the $1/R^2$ dependence of KE. The average distance of the electron from the nucleus is related to the size of the electron wave, and the potential energy, equation (4.16), is inversely related to this distance. These relationships give the $1/R$ dependence of PE. Figure 4.27 shows the two energies, KE and PE, and their sum, the total energy, E, plotted as a function of R.

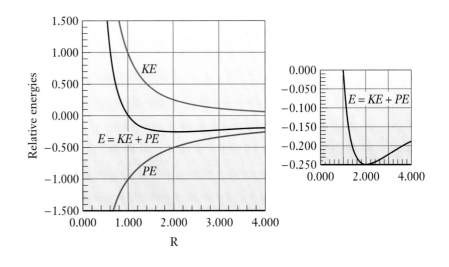

Figure 4.27.

Relative energies as functions of orbital radius, *R*, for a wave mechanical atom. Energies are for a spherical electron wave attracted to a positive nucleus. The smaller plot expands the scale for the energy range from −0.250 to 0 to show the minimum in total energy more clearly.

4.47 CONSIDER THIS

How can you interpret Figure 4.27?

Carry out the directed actions on the interactive animation in the *Web Companion*, Chapter 4, Section 4.7.1. Explain in your own words what happens

continued

as the distance between the electron and proton and the size of the electron wave decreases in terms of

(a) The potential energy. What causes the potential energy to change? What would happen if the potential energy were the only energy in the system?
(b) The kinetic energy. What causes the kinetic energy to change? What would happen if the kinetic energy were the only energy in the system?
(c) The total energy. When is the system most stable?

Why atoms don't collapse Figure 4.27 illustrates how the kinetic and potential energies of the atom vary with the size of the electron orbital (the size of the atom). The kinetic energy is always positive and increases as the orbital radius decreases. The potential energy of attraction is always negative and gets more negative as the radius of the orbital decreases: The smaller the orbital (atom), the more favorable (lower) the potential energy. This is the attraction that would, by itself, cause atoms to collapse. The kinetic energy of an electron orbital increases as the orbital gets smaller, so the kinetic energy counteracts the orbital contraction.

At some radius, the sum of the kinetic and potential energies, the total energy, goes to a minimum value. If the atom were to get any smaller, its total energy would increase and it would not be stable relative to its energy minimum. Thus, *the size of a one-electron atom is determined by a balance between the negative potential energy of attraction and the positive kinetic energy of the electron wave.* It is the kinetic energy of the electron wave that prevents the atom from collapsing.

Ionization energy for the hydrogen atom Our wave mechanical model can be made quantitative and we can calculate the energy minimum in Figure 4.27 and the size of the orbital at this point. The energy minimum is -1312 kJ·mol^{-1}, at an orbital radius near 80 pm. The experimental value for the ionization energy of the hydrogen atom, the amount of energy required to separate the electron and proton, is 1312 kJ·mol^{-1}, that is,

$$H(g) \rightarrow H^+(g) + e^- \qquad \Delta E_{ionization} = 1312 \text{ kJ·mol}^{-1} \qquad \textbf{(4.19)}$$

We can't determine the exact size of a hydrogen atom because the electron probability distribution has no sharp boundary. The electron probability envelope resembles a ball of cotton more than it does a baseball. On the other hand, we know from many different experiments that atoms have radii that are near 100 pm, so we are certainly in the right ballpark. Thus, scientists accept the wave model, because it both predicts and explains experimentally observed properties of electrons and atoms, such as electron diffraction, quantized energies, atomic size, and ionization energy.

4.48 CHECK THIS

Ionization energy and nuclear charge

Which species has the higher ionization energy, H or He$^+$? Why? Use what you learned in Consider This 4.46 to help answer these questions.

Reflection and Projection

The properties of waves have brought us to the wave model of the atom. The formation of a stable atom corresponds to reaching the lowest possible value for the total energy of the atom. The total energy of a one-electron atom is the sum of the kinetic and potential energies of the electron orbital attracted to the positively-charged nucleus. The kinetic and potential energies are related to the size of the electron orbital (the size of the atom). At some orbital size, the balance between the always-positive kinetic energy and the always-negative potential energy of attraction produces a minimum in the total energy. This is the stable atom. The fundamental concepts to remember are: *Atoms (electrons and nucleus) are held together by mutual attractions of unlike charges and their sizes are limited by the kinetic energies of the electron waves (orbitals).*

As we go on to atoms with more than one electron, the number of interactions that must be included in the energy calculations grows. More electrons mean more possibilities for changes in energy levels and much more complicated emission and absorption spectra. Analysis of these spectra requires a specialized knowledge of atomic structure that is unnecessary for our basic understanding of the properties of atoms. In the next section we will examine some other atomic properties to see what patterns we can find in the data that can provide the clues we need to develop our wave mechanical model further.

4.9. Multielectron Atoms: Electron Spin

4.49 CONSIDER THIS

What patterns do you find in atomic ionization energies?

(a) Successive ionization energies for the lightest three elements are given in Table 4.1. The ionization energy for the H atom is the energy required for reaction (4.19). For the helium atom with two electrons, the ionization energies in Table 4.1 are for these reactions:

$$He(g) \rightarrow He^+(g) + e^- \quad \Delta E_{\text{first ionization}} = 2373 \text{ kJ·mol}^{-1} \quad \textbf{(4.20)}$$

$$He^+(g) \rightarrow He^{2+}(g) + e^- \quad \Delta E_{\text{second ionization}} = 5248 \text{ kJ·mol}^{-1} \quad \textbf{(4.21)}$$

What are the reactions that correspond to the three entries for the lithium atom, Li, in the table?

(b) The first ionization energy for Li is very low compared to the first ionization energies for H and He. The first ionization energy for Li is also very low compared to its second ionization energy. What might account for this very low first ionization energy for Li? In other words, why is the electron held so loosely to its nucleus?

Table 4.1	Successive ionization energies in kJ·mol⁻¹ for H, He, and Li atoms.		
	1st electron	**2nd electron**	**3rd electron**
H	1,312		
He	2,373	5,248	
Li	520	7,300	11,808

The electron wave model for helium Helium atoms have two protons in their nuclei, so the electrons are attracted by a 2+ charge. Both electron waves can surround the nucleus and form a single, spherical electron orbital with a 2− charge. The electron waves can be thought of as reinforcing one another, just as the two waves described in Figure 4.9(a) reinforce one another. The potential energy for the helium atom is more complicated than the potential energy for the hydrogen atom. For hydrogen, we had to consider only the attraction of a proton and an electron. For helium, we must consider the attractions and repulsions among three particles. Each of the two electrons is attracted to the positive (2+) nucleus. At the same time, the two negative electrons repel each other. Without doing any calculations, you can estimate some of the effects of these differences from the hydrogen atom; in fact, you already began doing this in Check This 4.48.

The higher nuclear charge will attract each electron more strongly and draw it closer to the nucleus. Counteracting the nuclear attraction will be the repulsion between the electrons. This repulsion is not as strong, however, since each electron has only a single negative charge. Also counteracting the negative potential energy of attraction is the kinetic energy of the electron orbital, which gets more positive as the orbital gets smaller. On balance, we can predict that the larger nuclear attraction should make the helium atom smaller than the hydrogen atom and should hold the electrons more tightly. The experimental values for the ionization energies in Table 4.1 confirm that the electrons are held more tightly.

Simple calculations for the helium atom, like those for hydrogen discussed in Section 4.8, give an atomic radius about half that of hydrogen and a total minimum energy, E, of about −6700 kJ·mol⁻¹. Thus, 6700 kJ·mol⁻¹ is the energy required to remove both electrons:

$$He(g) \rightarrow He^{2+}(g) + 2e^- \tag{4.22}$$

Reaction (4.22) is the sum of reactions (4.20) and (4.21). Experimentally, the total energy required is the sum of the two ionization energies for helium, 7621 kJ·mol⁻¹, from Table 4.1. The calculated value and the experimental value do not agree exactly, but the calculation is only off by a little more than 10%. These results—a smaller electron orbital radius and a lower energy minimum for helium than for hydrogen—are what we predicted above. Recall from our previous discussions of the periodic table that atoms get smaller as you go from left to right across a period. You have just successfully crossed the first period.

The electron wave model for lithium: A puzzle Each successive element adds another unit of charge to its atomic nuclei, and another accompanying electron wave. Lithium is the next element and, based on the discussion for helium, we might expect its three electrons to be held even more tightly by the increased nuclear charge. The total ionization energy for lithium atoms, from Table 4.1, is 19,628 kJ·mol^{-1}. This is the energy required for the sum of the reactions you wrote for lithium in Consider This 4.49(a):

$$Li(g) \rightarrow Li^{3+}(g) + 3e^- \tag{4.23}$$

Our expectation is confirmed; much more energy is required to remove all the electrons from lithium than from helium. However, lithium's very low first ionization energy—even lower than that for hydrogen, which has a nuclear charge of only one—is puzzling. The second and third ionization energies are quite large, so it is clear that these electrons are tightly held. The problem is with the first electron, which seems more weakly held than expected. Either our wave model is no good or it needs some modification.

Magnetic properties of gaseous atoms We have been using the wave model because it explains the quantization of energy levels in atoms and accounts for the wave properties of electrons. Whenever a model that seems to be working fails in some way, the first approach is usually to try and fix it. Almost always, this means looking at other properties of the system that is giving the problem—the lithium atom, in this case—which brings us back to spectra. While studying atomic spectra, scientists often observed changes when the atoms were placed in a magnetic field. They also found that the paths of some atoms streaming through an inhomogeneous magnetic field (a field with curved field lines, as at the top and bottom of Figure 4.11) were unaffected by the field, while the paths of other atoms were split into two streams, as illustrated in Figure 4.29.

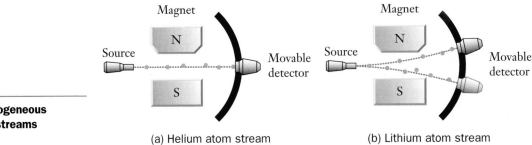

Figure 4.28.

Effect of an inhomogeneous magnetic field on streams of atoms.

(a) Helium atom stream   (b) Lithium atom stream

A stream of helium atoms, Figure 4.28(a), is an example of atoms that are not deflected in the magnetic field. They pass straight through. On the other hand, lithium atoms, Figure 4.28(b), are split into two streams by the magnetic field. Half the lithium atoms are deflected toward one pole of the magnet and half are deflected toward the other. The *lithium atoms act as though they are themselves tiny magnets*, half of which are oriented in one direction and half in the other.

Electron spin It's a magnetic property of the electrons that is responsible for the behavior of the atoms observed in these experiments. Electrons can be thought of as acting like tiny magnets; the name we use to characterize this magnetic property is **spin** (or **electron spin**). Although the name suggests images of spinning tops or figure skaters, these are probably not accurate images of electron spin. Electron spin can take two orientations in space, just like a bar magnet can be oriented with the north pole up or the north pole down, Figure 4.29. When it is important to show the spin of an electron, we use an arrow, as shown in the figure.

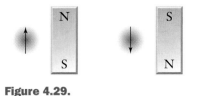

Figure 4.29.

Electron spin. The analogy of electron spin to the orientation of a magnet.

Pauli exclusion principle Electron spin is essential to our discussion of electrons in atoms because *two (and no more than two) electrons can be described by the same spatial orbital only if they have opposite spin.* The two electrons have the same energy, and are identical in every respect except for their spins. A corollary of this statement is as follows: *Two electrons of the same spin cannot be described by the same electron orbital.* This concept, developed by Wolfgang Pauli (Austrian-born physicist, 1900–1958), is called the **Pauli exclusion principle** (or **exclusion principle**). It is an essential tool that enables us to describe energy levels and regions of space occupied by electrons in multielectron atoms.

For helium, the exclusion principle tells us that the two electron waves can occupy exactly the same space if, and only if, the two electrons have opposite spins. Two electrons with opposite spin are called **spin-paired electrons.** The magnetism of spin-paired electrons cancels out, and produces no net magnetic properties in the atom or molecule where they are located. Thus, helium atoms with their spin-paired electrons pass unaffected between the poles of a magnet. We often symbolize spin-paired electrons as a pair of oppositely oriented arrows: ↑↓.

Exclusion principle applied to lithium: Puzzle solved For lithium, the two lowest-energy electron waves have paired spins, occupy exactly the same space, and surround the nucleus. The third electron is attracted to the Li^+ ion that results (the $3+$ nucleus plus the two $1-$ electrons). However, because it has the same spin as one or the other of the two electrons that are already there, it cannot share exactly the same space with them. The orbital that describes it is at a larger average distance from the nucleus than the other two. As a result, its potential energy of attraction to the nucleus is less negative, an unfavorable effect. At the same time, because the size of the orbital is larger, the kinetic energy of the third electron is less positive, a favorable effect. The net effect is a total energy that is less negative (a low ionization energy). In terms of the energy levels for the atom (see Figure 4.22), this electron will be at a higher energy than the first two electrons. Since the third electron in lithium is not spin-paired with another electron, the atom acts like a magnet, which explains its behavior in the magnetic field experiments, Figure 4.28. Taking spin and the exclusion principle into account solves the puzzle of why lithium's first ionization energy is so low, but more data are needed to tell you whether the solution is general and can be applied to elements with even more electrons.

4.10. Periodicity and Electron Shells

Patterns in the first ionization energies of the elements Since it is ionization energies that signaled a problem for our simplest atomic model, let us take up where we left off in Section 4.1 and further examine elemental ionization energies. Figure 4.4 plotted the first ionization energies for the first 20 elements. Taking a somewhat different look at the first ionization energies for

Protons also have spin and magnetic properties (different size than electrons). Proton spin is the basis of powerful techniques for probing the structure of matter, including humans via MRI, magnetic resonance imaging.

The proton magnet and electron magnet affect one another. In the hydrogen atom, the magnets can be paired or not, as shown in this energy diagram. The nuclear magnet is represented by the heavy blue arrow and the electron by the red arrow. You know that magnets resist being brought together with the same poles facing one another. This is the higher energy state.

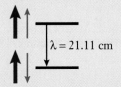

$\lambda = 21.11$ cm

The emission from H atoms that have been excited by collisions with other H atoms is detected by radio astronomy and is a basis for the estimates of the amount of hydrogen in the universe. Also see Check This 4.22 and 4.33.

Web Companion

| Chapter 4, Section 4.9.1 |
View and manipulate a three-dimensional representation
of the ionization energies.

① ② ③ ④

gaseous atoms of all the elements, Figure 4.30 plots the *negative* of the first ionization energies for gaseous atoms as a function of their atomic numbers (nuclear charge). The negative value of the ionization energy is the energy that is given off when an electron is added to an atom that is missing one electron. For example:

$$C^+(g) + e^-(g) \rightarrow C(g) \quad \Delta E = -1086 \text{ kJ·mol}^{-1} \tag{4.24}$$

The more negative the energy, the more stable the atom relative to its monocation and electron.

Figure 4.30.

The first ionization energies of atoms plotted as negative values. The lower the energy, the more stable the atom, X(g), relative to its ion and an electron, X⁺(g) + e⁻(g). The data points are connected to make trends easier to see. Values for some of the heavier elements are unknown.

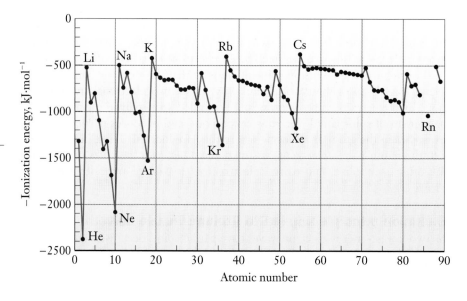

4.50 CONSIDER THIS

How do you interpret patterns in these ionization energy data?

(a) Atoms from which family (group) of the periodic table are most stable with respect to loss of an electron? Least stable? What is the relationship of these families in the periodic table that is shown on the inside front cover of the book?

(b) As you go from left to right across the plot in Figure 4.30, the nuclear charge is increasing. Higher nuclear charge should have greater attraction for the electrons. Are there places in the plot that show this effect? If so, where are they?

(c) As you go from left to right across the plot, does the attraction for the last electron generally increase, decrease, or stay about the same? Is your answer consistent with the increase in nuclear charge across the plot? If not, what factor(s) might explain why it is not?

Electron shell model for atoms In Section 4.9, we said that the third electron in lithium is not as close to the nucleus as the first two, because its spin excludes it from having the same spatial orbital (wave) as the first two. Its orbital is, on average, farther from the nucleus. As a consequence of being farther from the nucleus, this electron is not held as tightly and, as we see in Figure 4.30, not as much energy is released when it forms the Li atom by joining with the Li$^+$ ion. As the nuclear charge increases going from lithium to neon, the energy released upon formation of the atoms generally increases. This is just what we expect from the increasing attraction by the nuclei for electron orbitals that form a second shell around the nucleus.

In going from neon to sodium, there is again a large change in the energy of attraction with the last electron in sodium being even more weakly held than the last electron in lithium. If our model is correct, this means that the spin of the 11th electron excludes it from having the same spatial orbital as any of the 10 electrons already present in Na$^+$. The ionization energy data suggest that the orbital that describes the 11th electron is, on average, even farther from the nucleus and begins a third shell. This is the essence of the **shell model** for atoms—electrons behave as if they are being added in layers (shells) as atomic numbers increase. Within a shell, all the electrons are about the same distance from the nucleus. Between shells, there is a difference in the average distance from the nucleus. These concepts are summarized in Figure 4.31, an abbreviated periodic table with a shell structure shown for the first 20 elements. This model does raise a puzzling question: How can the second and third shells accommodate eight electrons and still not violate the exclusion principle? We'll return to this question in Section 4.11.

Note that hydrogen is in group IV (14) in this periodic table (and the one on the inside front cover). This position better reflects the chemistry of hydrogen compared to other elements. Other tables put hydrogen in group I (because it has one valence electron) or sometimes in group VII (because it just precedes a noble gas), but hydrogen is not chemically similar to either the alkali metals or the halogens. Its ionization energy and electronegativity are intermediate between these extremes and much closer to those of carbon. Both hydrogen and carbon have half-filled electron shells, which helps to account for their similar chemistry. We will again draw attention to their similarity in Chapter 6, Section 6.9.

Figure 4.31.

Electron shell model for the first 20 elements in the periodic table.

	I	II	III	IV	V	VI	VII	VIII
1				Hydrogen $_1$H ●				Helium $_2$He ●
core e^- / valence e^-				0 / −1				0 / −2
2	Lithium $_3$Li	Beryllium $_4$Be	Boron $_5$B	Carbon $_6$C	Nitrogen $_7$N	Oxygen $_8$O	Fluorine $_9$F	Neon $_{10}$Ne
core e^- / valence e^-	−2 / −1	−2 / −2	−2 / −3	−2 / −4	−2 / −5	−2 / −6	−2 / −7	−2 / −8
3	Sodium $_{11}$Na	Magnesium $_{12}$Mg	Aluminum $_{13}$Al	Silicon $_{14}$Si	Phosphorus $_{15}$P	Sulfur $_{16}$S	Chlorine $_{17}$Cl	Argon $_{18}$Ar
core e^- / valence e^-	−10 / −1	−10 / −2	−10 / −3	−10 / −4	−10 / −5	−10 / −6	−10 / −7	−10 / −8
4	Potassium $_{19}$K	Calcium $_{20}$Ca			**Scale** →| |← 100 pm			
core e^- / valence e^-	−18 / −1	−18 / −2						

Atomic radii Our atomic model provides an indication of the sizes, or at least the relative sizes, of atoms. We have seen that the size of an atom is a fuzzy concept, because the electron probability distribution, Figures 4.25(b) and 4.26(b), has no sharp boundary. We have tried to indicate this fuzziness in Figure 4.31 as well. Experimental measurements of atomic size are not straightforward and simple. One approach to determining the size of atoms is to measure the distance between nuclei in covalent compounds and assign part of the distance as the radius of one atom and the rest as the radius of the other atom. (We will say more about bond distances in the next chapter.) For bonds between the same two elemental atoms, **homonuclear bonds** (*homo* = same), as in F_2 or I_2, the atomic radius would be just half the bond length. For bonds between different elemental atoms, we could start with one radius from the homonuclear bonds and assign the rest of the bond distance as the radius of the other atom.

4.51 CHECK THIS

Atomic radii for carbon and chlorine

The distance between the two nuclei in the chlorine molecule, Cl_2, is 198 pm. Chlorine forms a compound with carbon, CCl_4, tetrachloromethane (carbon tetrachloride), in which the distances between the carbon and chlorine nuclei are all the same, 178 pm. What is the atomic radius of chlorine? What is the atomic radius of carbon? *Note:* Distances between nuclei are determined by several spectroscopic and diffraction techniques.

If we analyze a large number of compounds, by the method used in Check This 4.51, a reasonably consistent set of atomic radii can be derived. Someone else, using a different choice of compounds, might get a somewhat different set of atomic radii. You will find different atomic radii in different books. For our purpose here, minor variations are unimportant. We are interested in the trends and patterns for the atomic radii shown in Figures 4.31 and 4.32. Figure 4.32

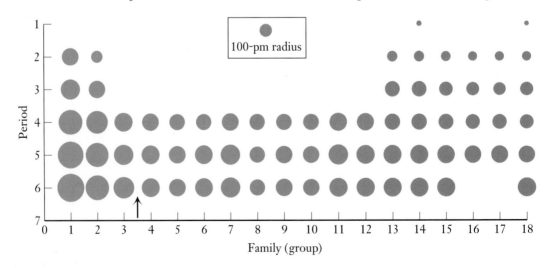

Figure 4.32.

Atomic sizes arranged by period and family. The 14 lanthanide elements (at the arrow) and the period 7 elements are not shown. For reference, a circle representing an atom with a radius of 100 pm is shown, top center.

shows the relative sizes of atoms derived from analyses of compounds containing each element (or, in a few cases, calculations based on the wave model of atoms).

Web Companion

Chapter 4, Section 4.9.2 — ①②③④

View and manipulate a three-dimensional representation of the atomic radii.

4.52 CONSIDER THIS

What are the patterns in atomic size?

(a) In Figure 4.32, how would you describe the trend in atomic radii across a period? Can you think of a reason for this trend? Explain.

(b) How would you describe the trend in atomic radii down a family (group or column)? Can you think of a reason for this trend? Explain.

(c) Does our atomic model provide a consistent explanation for these patterns? That is, is your reasoning in part (b) consistent with your reasoning in part (a)? Explain why or why not.

(d) Compare any patterns in Figure 4.32 with those in Figure 4.1. Explain why there should (or should not) be a correlation between these two plots.

Electronegativity In the previous chapters, we met **electronegativity**, a measure of the attraction that an atom (atomic core) in a molecule has for the bonding electrons it shares with another atom. We have warned that electronegativity is a concept defined by scientists and not a measurable property of an atom like its ionization energy. Linus Pauling developed his set of electronegativities based on interrelationships of the energies released in bond formation for a large number of molecules. Other scientists have developed other sets of electronegativities based on different models and data. The trends in all these values of electronegativities are the same; Figure 4.33 shows Pauling electronegativities.

Web Companion

Chapter 4, Section 4.9.3 — ①②③④

View and manipulate a three-dimensional representation of the electronegativities.

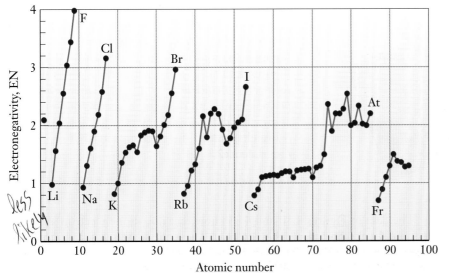

Figure 4.33.

Pauling electronegativities plotted as a function of atomic number.
The data points are connected to make trends easier to see. The gaps are the missing data points for the noble gases.

What are the patterns in electronegativity?

(a) What patterns do you observe for electronegativities in Figure 4.33? How do they vary within periods and families of the periodic table?

(b) Does Figure 4.33 provide any further evidence for an electron shell model for atoms? If so, explain clearly how you interpret the plot as evidence for the shell model. If not, explain clearly how you interpret the plot as evidence against the shell model.

(c) What relationships can you see among the plots in Figures 4.30, 4.32, and 4.33? How do you interpret these relationships?

A closer look at the electron shell model The electron shell model of atoms is based on the kinds of experimental evidence that you have been examining from the beginning of this chapter. This experimental evidence is reinforced by the wave model for atoms, which permits us to interpret the periodicity of atomic properties in terms of electron waves that are attracted as close as possible to the atomic nuclei. Restrictions on how close the waves come to the nucleus are imposed by their kinetic energies and by the exclusion principle. We interpret the shell structure as a result of the balance between the attractions and restrictions. As we said in the introduction to this chapter, we have asked you to do two things: (1) Look at experimental data and try to see in them *evidence* for the shell model and (2) look at experimental data and try to interpret them as a *result* of the shell model. We hope you have become comfortable working both ways.

What are some further properties of electron shells in atoms?

(a) How many shells are represented in Figures 4.30, 4.32, and 4.33? How many electrons are in each shell?

(b) How does the spiral periodic table shown in the center of the chapter opening illustration, and in Figure 4.34, reflect the patterns shown by the data in Figures 4.30, 4.32, and 4.33? Explain with reference to specific elements or groups of elements from the periodic table and the data. Does this periodic table and/or the data provide any evidence for "structure" (subshells) within the electron shells? If so, what is the evidence?

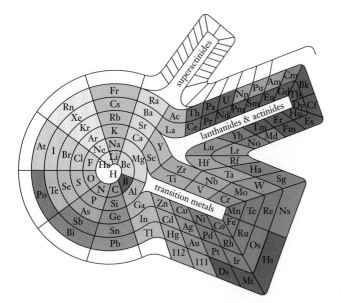

Figure 4.34.

A spiral version of the periodic table of the elements.

The data in Figures 4.30 and 4.33 show that not all the electron shells are the same. In the first place, they do not all accommodate the same number of electrons. In Section 4.9, we learned why the first shell has only two electrons.

For the succeeding shells, if you count from an alkali metal to the following noble gas—K to Kr, for example—the number of elements in periods 2 through 6 is 8, 8, 18, 18, and 32. (Period 7 is not complete; the heavier elements have yet to be synthesized.) The patterns of ionization energies and electronegativities for periods 2 and 3 are very similar, but in periods 4 and 5, the pattern is interrupted by a series of 10 elements most of which have quite similar values for these properties. Period 6 shows this same interruption with yet another longer interruption by a series of 14 elements. It appears as though we might have to amend the shell model to a subshell model, where the subshells can accommodate orbitals containing totals of 8, 10, and 14 electrons. The arms coming off the spiral form of the periodic table in Figure 4.34 help you visualize this structure.

4.55 CHECK THIS

Number of electron subshells in each shell

For each of periods 2 through 6, which subshells (the 8-, 10-, and/or 14-electron subshells) are used to build the series of elements across the period? Show how the spiral periodic table illustrates your answers.

In Figures 4.4 and 4.30, we can also see that within the 8-electron shells (or subshells) in periods 2 and 3, the first ionization energies seem to show a stair-step pattern of two, three, and three elements. We might, however, be seeing some effect of the increase in coulombic attraction as the nuclear charge increases from one elemental atom to the next. To check this possibility, we can look at successive ionization energies for an element, as we did in the previous section for He and Li to learn more about their structures. The energy required for each successive ionization has been measured for many elemental atoms, and some data for the first 20 elements are given in Table 4.2.

4.56 CONSIDER THIS

How can successive ionization energies be interpreted?

(a) What relationships are there between the numbers of core and valence electrons shown in Figure 4.31 and the data in Table 4.2? Explain your relationships with specific examples.

(b) For which atoms does Table 4.2 give the ionization energies for all the electrons in complete eight-electron shells (subshells)? Explain how you make your choices.

(c) Plot the successive ionization energies in complete eight-electron shells (subshells) for two or three of the atoms chosen in part (b). If available, use a spreadsheet program or graphing calculator to do this analysis. Do your plots show evidence for the two and three element steps that are apparent in Figures 4.4 and 4.30? If so, what is the evidence?

Table 4.2 *Successive ionization energies, in kJ·mol^{-1}, for the first 20 elements.*

The column number is the number of the electron removed. Column 1, for example, is the first ionization energies that have been plotted in Figures 4.4 and 4.31. The shading calls attention to the ionization energies before a large jump (more than a factor of 3) for the next electron lost.

Atom	1	2	3	4	5	6	7	8	9	10
$_1$H	1,312									
$_2$He	2,373	5,248								
$_3$Li	520	7,300	11,808							
$_4$Be	899	1,757	14,850	20,992						
$_5$B	801	2,430	3,660	25,000	32,800					
$_6$C	1,086	2,350	4,620	6,220	38,000	47,232				
$_7$N	1,400	2,860	4,580	7,500	9,400	53,000	64,400			
$_8$O	1,314	3,390	5,300	7,470	11,000	13,000	71,300	84,000		
$_9$F	1,680	3,370	6,050	8,400	11,000	15,200	17,900	92,000	106,000	
$_{10}$Ne	2,080	3,950	6,120	9,370	12,200	15,000	20,000	23,100	115,000	131,000
$_{11}$Na	496	4,560	6,900	9,540	13,400	16,600	20,100	25,500	28,900	141,000
$_{12}$Mg	738	1,450	7,730	10,500	13,600	18,000	21,700	25,700	31,600	35,500
$_{13}$Al	578	1,820	2,750	11,600	14,800	18,400	23,300	27,500	31,900	38,500
$_{14}$Si	786	1,580	3,230	4,360	16,000	20,000	23,800	29,200	33,900	38,700
$_{15}$P	1,012	1,904	2,910	4,960	6,240	21,000	25,400	29,900	35,900	40,960
$_{16}$S	1,000	2,250	3,360	4,660	6,990	8,500	27,100	31,700	36,600	48,100
$_{17}$Cl	1,251	2,297	3,820	5,160	6,540	9,300	11,000	33,600	38,600	43,960
$_{18}$Ar	1,521	2,666	3,900	5,770	7,240	8,800	12,000	13,800	40,800	46,200
$_{19}$K	419	3,052	4,410	5,900	8,000	9,600	11,300	14,900	17,000	48,600
$_{20}$Ca	590	1,145	4,900	6,500	8,100	11,000	12,300	14,200	18,200	20,380

Electron spin and the shell model Although the steps you find in Consider This 4.56(c) are not as pronounced as those in Figures 4.4 and 4.30, there is definite evidence that the successive ionization energies do not increase smoothly, but show two and three element steps. Another piece of evidence for more structure within the shells comes from spectroscopic studies of atoms in magnetic fields that tell us how many unpaired electron spins the atoms have. The details of these studies are not important for us here, but the results shown in Table 4.3 for the second period elements are important for our understanding of electrons in atoms.

Table 4.3 *Number of unpaired electron spins for ground-state, second period atoms.*

$_3$Li	$_4$Be	$_5$B	$_6$C	$_7$N	$_8$O	$_9$F	$_{10}$Ne
1	0	1	2	3	2	1	0

We know that any atom with an odd number of electrons, such as Li or B, must have at least one electron spin unpaired, and we might have anticipated that any atom with an even number of electrons would have them all spin paired. For Be with two electrons in the second shell, the electron spins are paired, as we anticipate. But C with two further electrons in the second shell has two unpaired electrons and N with three further electrons has three unpaired electrons. It appears as though the third, fourth, and fifth electrons added in the second shell are unpaired.

4.57 CHECK THIS

Electron spin pairing in second period elements

As three more electrons are added in O, F, and Ne, how can spin pairing explain the results shown in Table 4.3?

We could once again amend the shell–subshell model to account for these experimental results, but that is not a fruitful approach. The value of the shell model is its simplicity. As you have seen, a great deal of the data on periodic properties of the elements can be explained and/or predicted from the shell model of the atom, which is based on a simple application of the wave nature of electrons. The more we tinker with the model, the further we get from the basic concepts of electron attraction by the nucleus and the restrictions imposed by kinetic energy and electron spin on the average distance of the electron waves from the nucleus. To explain more subtle effects, such as electron-spin pairing (or lack of pairing), we need to return to the wave model of the electron in atoms and see how it is used to describe the waves in more detail.

Reflection and Projection

The wave mechanical model of the atom that accounted for the kinetic energy of an electron wave and the potential energy of attraction between electrons and the nucleus has to be modified to take into account electron spin and the exclusion principle. Only two spin-paired electrons can be described by the same electron wave (orbital). Thus, electrons in an atom, after the first two, which enclose the nucleus in a spherical electron wave, are, on average, farther from the nucleus and form successive shells of electrons. The consequences of this structure are the periodic properties of the elements that repeat as each new shell is built up from one element to the next across a period of the periodic table.

You examined several properties of atoms—ionization energies, electronegativities, and sizes—and found patterns in these properties that vary periodically. Each of these variations can be used as evidence for the atomic model or explained as a result of the atomic model. The model balances increasing electron attraction by increasing nuclear charge with restrictions on how close the electrons are to the nucleus. For example, the size of the atoms decreases across a period as the nuclear charge increases and attracts the electrons from the valence shell closer. Conversely, the size of the atoms increases as the nuclear charge increases in families (groups) of elements because a new shell, farther from the nucleus, is added in going from period to period.

Even within the patterns that are readily explained by the shell model, there are some variations that are not explained and there are other properties, such as the number of unpaired electron spins in atoms that also cannot be accounted for. To account for such observations, we need to look in more detail at electron waves in atoms.

4.11. Wave Equations and Atomic Orbitals

We introduced the idea of a wave model for electrons in Section 4.7 with the analogy to standing waves on a guitar string, Figure 4.24. The standing waves on the acoustic body of the guitar, as it is played, are more complex, Figure 4.35. As a consequence, the nodes of these standing waves and the wave equations describing the waves are more complex than the nodes on the strings of the instrument and equation (4.11) in Check This 4.37. For a three-dimensional electron wave, our only pictorial representation has been the spherical probability distribution in Figures 4.25, 4.26, and 4.27. Solutions to the wave equations for electron waves also yield more complex shapes than spheres.

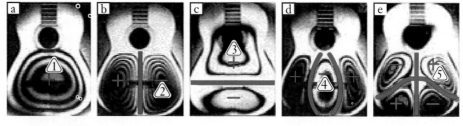

Figure 4.35.

Standing waves on the acoustic body of a guitar. Red lines and curves show the nodes. Each node traces a part of the surface that is not moving. At any instant, the surfaces on opposite sides of a node are moving in opposite directions (toward you, +, or away from you, −), just like a guitar string on either side of one of its nodes, Figure 4.24.

Schrödinger wave equation In Sections 4.8 and 4.9, you saw how the balance of kinetic and potential energies for an electron wave determines the size of the wave and its ionization energy. Soon after de Broglie introduced the wave model, scientists developed mathematical descriptions that accounted for the kinetic and potential energies of electron waves. The equations can be solved to give the total energy *and* a probability picture of electron waves. One of these equations is the **Schrödinger wave equation** (or **wave equation**) developed by Erwin Schrödinger (Austrian-born physicist, 1887–1961):

$$-\frac{h}{4\pi m}\nabla^2\psi \quad + \quad V\psi \quad = \quad E\psi \tag{4.25}$$

$$\underset{\text{energy}}{\text{kinetic}} \quad + \quad \underset{\text{energy}}{\text{potential}} \quad = \quad \underset{\text{energy}}{\text{total}}$$

The Schrödinger wave equation is a differential equation that requires calculus to solve. $\nabla^2\psi$ is a second derivative with respect to the spatial coordinates. Many mechanical systems in physics are described by wave equations like this. This is why the wave equation approach to describing atomic level systems is called wave mechanics.

The two terms on the left side of the wave equation give the kinetic and potential energies as a function of the **electron wave function,** ψ (Greek lowercase psi), a mathematical description of the wave. Their sum gives the total energy on the right. If you continue your study of chemistry, you will learn, in later courses, how to find wave functions that are solutions to the Schrödinger wave equation. For this discussion, we will present the solutions in a pictorial form and relate them to the ionization energy data in Figures 4.4 and 4.30 and Table 4.2.

Wave equation solutions: Orbitals for one-electron atoms

The wave function, ψ, depends on the spatial coordinates x, y, and z (the directions in a three-dimensional coordinate system). Thus, it is possible to calculate the value of ψ at any point in space. In wave mechanics, the square of the electron wave description, ψ^2, calculated for a given location, is proportional to the probability of finding the electron in that location. Plots of these probabilities in the region around the atomic nucleus show the shape of the electron wave (orbital) and its nodes. The lowest energy solutions to the wave equation for a one-electron atom or ion are shown in Figure 4.36. The probabilities may be represented by the density of dots on a page, as in Figure 4.25(b), but more often simpler boundary surface pictures, as in Figure 4.26(a), are used to represent orbitals. Boundary surfaces are drawn to enclose a high percentage of the electron probability (density), 95–99%, and with a shape that captures the shape of the probability distribution (orbital).

Orbitals are named, $1s$, $2s$, $2p_x$, and so on to identify their energy, shape, and orientation in space. The numerical value that begins each name is the **principal quantum number, n,** and designates the energy level, Figure 4.22, for the electron described by that orbital. In a one-electron atom or ion, all the orbitals with the same principal quantum number, n, have the same total energy. Orbitals with the same energy are called **degenerate.** The lowercase letter that follows the numeral designates the three-dimensional shape of the orbital. The letters chosen are related to spectroscopic observations on the atoms. You can remember s because these orbitals are *s*pherical. The p orbitals always come as a set of three that are mutually *p*erpendicular to each other and can be visualized as oriented along the three axes of a three-dimensional coordinate system, as shown in Figure 4.36. All three have the same energy and shape, so a subscript is used to indicate their orientation in space with respect to one another.

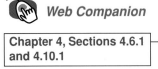

Web Companion

Chapter 4, Sections 4.6.1 and 4.10.1 — ① ② ③ ④

View lowest energy orbitals rotated in three dimensions.

Quantum refers to the quantized energies of the electrons in atoms. *Quantum number* refers to a particular energy level. The names of the orbitals were originally used for energies in the Bohr model and then were transferred to the corresponding wave mechanical solutions.

There is no negative meaning to the term *degenerate energy levels*. It simply denotes two or more solutions to the wave equation that have the same total energy.

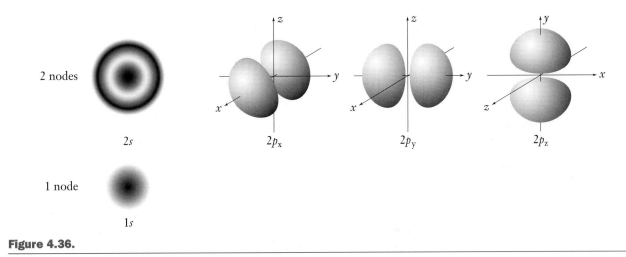

Figure 4.36.

Atomic orbitals for a one-electron atom. The s orbitals are shown as cross-sections of dot density diagrams; the p orbitals as boundary surface diagrams. All the orbitals have a node at infinity, which is not represented in the drawings. Orbitals in the second energy level all have a second node, which is represented.

4.58 CONSIDER THIS

How are orbital nodes described?

(a) What would the three-dimensional boundary surfaces for a 1*s* and 2*s* orbital look like? Could you tell them apart? Why or why not?

(b) For the orbitals in Figure 4.36, the nodes represented are surfaces. What is the shape of the nodal surface for the 2*s* orbital? For a 2*p* orbital? Give the reasoning for your answers.

It is very important to remember that an orbital is a description of an electron wave; ***orbitals do not have any existence separate from the electron.*** For *convenience*, scientists often speak about "electrons in an orbital," "adding an electron to an orbital," "half-filled orbitals," or even "empty orbitals." All of these phrases imply that orbitals are like boxes into which electrons can be placed. What is meant is that a *calculated* energy level (and corresponding spatial probability distribution) exists that could be used to describe an electron wave. If no actual electron is described by the calculation, then the orbital is "empty." If an actual electron is described by the calculation, then the electron is "in" the calculated orbital. Because these phrases are so convenient, we will use them, but always keep in mind that *orbitals are not boxes*.

4.59 CONSIDER THIS

How are principal quantum numbers, nodes, and orbitals related?

(a) What is the relationship between the number of nodes and the principal quantum number, n, in Figure 4.36? Sketch a picture of a 3*s* orbital, using the 2*s* representation in the figure as a model. Clearly explain the reasoning you use to decide how to make your sketch.

(b) All of the orbitals with n = 1 and n = 2 are shown in Figure 4.36. For n = 3, there are nine energy-degenerate spatial orbitals: one 3*s*, three 3*p*, and five 3*d* orbitals. Do you see any evidence for *d* orbitals in the structure of any of the periodic tables in the chapter opening illustration and/or in Figures 4.30 or 4.33? If so, clearly explain the evidence.

(c) What is the relationship between n and the number of energy-degenerate spatial orbitals? How many energy-degenerate spatial orbitals are there with n = 4? Explain how you reach your conclusion.

Electron configurations The 1*s* orbital is lowest in energy. A ground-state hydrogen atom has one electron described by a 1*s* orbital. The **electron configuration** of this hydrogen atom is $1s^1$. The superscript shows how many electrons in the atom are described by the orbital. A ground-state helium atom has two spin-paired electrons that can be described as superimposed 1*s* spatial orbitals; the electron configuration is $1s^2$. The energy of the $1s$ orbitals in helium is lower than the $1s$ orbital in hydrogen because of the increased nuclear charge. This description of the electron waves in hydrogen and helium is the same as the one in Sections 4.8 and 4.9.

You can imagine continuing this process of adding one electron at a time (with the appropriate change in nucleus) to build the rest of the elements. For atoms with three or more electrons, two electrons can be $1s$ electrons. Because of the exclusion principle, the rest of the electrons have to be in orbitals with n = 2 or larger. The four orbitals with n = 2 can accommodate eight electrons if there are two spin-paired electrons described by each spatial orbital. These eight electrons would account for the second shell of electrons in our shell model and a repeat of this build up for the $3s$ and $3p$ orbitals would account for the third shell of eight electrons.

But this reasoning based on the energy degeneracy of orbitals with the same principal quantum number is just the shell model again and does not help to explain the substructure of the shells that is evident in the ionization energies. Furthermore, as you found in Consider This 4.58(b), the n = 3 orbitals are ninefold degenerate in a one-electron atom, so we might expect the third shell to accommodate 18 electrons, but there are only eight elements, Na through Ne, in period three. Only when we get to period four are there 18 elements. What's wrong with the model?

Energy levels in multielectron atoms The orbitals in Figure 4.36, which are solutions to the wave equation (4.25) for a one-electron atom, are not correct for multielectron atoms. The wave equation cannot be solved exactly for an atom with two or more electrons. The calculations for multielectron atoms are all based on approximations that try to account for electron–electron repulsions and the fact that electrons cannot be told apart from one another. Simple pictorial representations of electron waves disappear in these calculations. The names of the orbitals survive to designate the energy levels derived from the calculations; these energy levels for multielectron atoms or ions are shown in Figure 4.37. The energies form a pattern like our shell structure with n = 1, 2, ... shells. No energy scale is given in the figure, because the actual energies of the levels vary from element to element (depending on the nuclear charge and total number of electrons in the atom or ion).

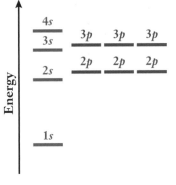

Figure 4.37.

Energy levels for multielectron atoms and ions. Relative energies are shown for the 10 lowest energy levels for multielectron atoms and ions.

4.60 WORKED EXAMPLE

The electron configuration for lithium

What is the electron configuration for the ground state of the lithium atom?

Necessary information: We need to know the relative energies in Figure 4.37 and that the lithium atom has three electrons.

Strategy: Add the electrons, one at a time, to the orbitals represented by the energy levels in Figure 4.37 in a way that makes the energy as low as possible and obeys the exclusion principle.

continued

Implementation: Add the electrons to the energy level scheme in Figure 4.37:

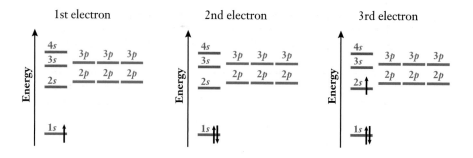

The first two electrons can spin pair at the lowest energy, $1s$, but the third electron has to go to the next higher energy, $2s$. The electron configuration is $1s^2 2s^1$.

Does the answer make sense?   This configuration has the energy and spin properties necessary to explain the behavior of the lithium atom that we discussed in Section 4.9.

4.61 CHECK THIS

The electron configuration for boron

What is the electron configuration for the ground state of the boron atom? Explain.

Electron configurations for multielectron atoms Figure 4.37 shows that the energy degeneracy among all the orbitals with the same principal quantum number no longer holds for multielectron atoms. There is still energy degeneracy among the orbitals that you can picture as having the same shape, but different orientations in space. You can use the pictures in Figure 4.36 as a reminder of the number of orbitals of the same shape and use the energy-level diagram, Figure 4.37, as the basis for writing electron configurations for atoms and ions with more than one electron.

Atoms in their ground states have the lowest total energy possible. Thus, to build atoms and figure out their electron configurations, you choose from Figure 4.27 the lowest energy level possible for each added electron. We have done this for the first 10 elements in Table 4.4. For the first 5 elements, the electron configurations are easy to predict; the last electron added goes to the lowest energy level that is not already occupied by two spin-paired electrons.

For carbon, the sixth electron could be paired with the first $2p$ electron or be in one of the other degenerate $2p$ orbitals. If two electrons are described by $2p$ orbitals, the electrons are farther apart and repel each other less when they are in different orbitals. Less electron–electron repulsion lowers the energy and

Table 4.4 *Electron configurations for the first 10 elemental atoms.*

The configurations are correlated with the first ionization energies. Steps of two-orbital, three-orbital, and three-orbital progressions in adding electrons and the corresponding ionization energies are highlighted in increasing shades of gray.

Atom	Electron configuration		First ionization energy, kJ·mol^{-1}
H	$1s^1$		1312
He	$1s^2$	$[He]^2$	2373
Li	$1s^2 2s^1$	$[He]^2 2s^1$	520
Be	$1s^2 2s^2$	$[He]^2 2s^2$	899
B	$1s^2 2s^2 2p^1$	$[He]^2 2s^2 2p^1$	801
C	$1s^2 2s^2 2p^1 2p^1$	$[He]^2 2s^2 2p^1 2p^1$	1086
N	$1s^2 2s^2 2p^1 2p^1 2p^1$	$[He]^2 2s^2 2p^1 2p^1 2p^1$	1400
O	$1s^2 2s^2 2p^2 2p^1 2p^1$	$[He]^2 2s^2 2p^2 2p^1 2p^1$	1314
F	$1s^2 2s^2 2p^2 2p^2 2p^1$	$[He]^2 2s^2 2p^2 2p^2 2p^1$	1680
Ne	$1s^2 2s^2 2p^2 2p^2 2p^2$	$[He]^2 2s^2 2p^2 2p^2 2p^2$	2080
		$[He]^2 2s^2 2p^6 = [Ne]^{10}$	

makes the atom more stable. Thus, the three electrons added going from boron to nitrogen are in separate $2p$ orbitals. An electron in one $2p$ orbital has the same energy as an electron in another of the $2p$ orbitals in the same atom. The increasing nuclear charge accounts for the steady increase in ionization energy from boron to nitrogen. The next electron added, to give oxygen, must occupy the same $2p$ orbital as an electron already present, since that is the lowest energy level available. The increased electron–electron repulsion raises the energy of the atom and makes it a bit easier to ionize. Thus, the first ionization energy of oxygen is a little lower than that of nitrogen, even though the nuclear charge has increased.

4.62 CONSIDER THIS

How are electron configurations related to ionization energies and electron spin?

(a) Write the electron configurations for the elemental atoms from Na through Ar. Use your electron configurations to make correlations between the

continued

pattern of first ionization energies in Figure 4.4 and 4.30 and the energy levels in Figure 4.37.

(b) Consider the successive ionization energies for the neon atom and ions in Table 4.2. What is the electron configuration for the atom or ion that loses an electron at each step? Is there any evidence in the pattern of ionization energies that supports the relative energy levels in Figure 4.37? Explain why or why not. *Hint:* This problem is related to what you did in Consider This 4.56(c).

(c) What is the relationship between the electron configurations in Table 4.4 and the number of unpaired electrons in these atoms, Table 4.3? Try to state the relationship as a general rule (for these cases).

(d) Use the electron configurations you wrote in part (a) and the rule you derived in part (c) to predict the number of unpaired electrons in the ground states of the third period elements. Do you find family relationships between your predictions and the data in Table 4.3? Did you expect such relationships? Explain why or why not.

Shell structure of atoms The energy levels shown in Figure 4.37 for electrons with the same principal quantum number are similar in energy, even if not degenerate. This energy-level grouping describes an electron-shell structure for atoms that also has some substructure in the shells. We can think of the orbitals discussed in this section as a refinement of the shell structure model that accounts for the variations in numbers of electrons in a shell and for the observed steps in ionization energies in periods two and three. When all the electrons are accounted for in the solutions to the wave equation for *any* isolated atom or ion, the electron probability distribution is spherical. The pictorial representation of isolated atoms we have used in Figure 4.31 and elsewhere is based on the shell structure and this spherical distribution.

Studying the properties of atoms has led us to many insights about the nature of the physical world at this very tiny scale. However, you seldom encounter atoms in any everyday application. Helium-filled balloons, mercury and sodium vapor lights (used for street lighting), and neon signs are among the few common examples. None of these applications entails chemical changes. Essentially everything you meet involves molecules, not atoms. Even in pure metals (made up of all the same kind of atom), the atoms act collectively, sort of like a gigantic molecule, as we shall see in the next chapter. Recall that chemists were disappointed with the Bohr quantum model, because it couldn't be applied to molecules. The wave mechanical model of electrons does not suffer this drawback. Wave mechanics focuses on finding electron waves that balance coulombic attraction with coulombic repulsion and kinetic energy to give the lowest possible total energy. In principle, there is no restriction on the number of nuclei that can be included in the calculations. We developed the wave mechanical model for the simpler case of atoms so that we can apply it to building molecules in the next chapter.

4.12 Outcomes Review

Many properties of the elements vary in a periodic way as they increase in atomic mass (or atomic number, in the present view) and these variations are codified in periodic tables that were first devised by Mendeleev and Meyer. The fundamental basis for this periodicity is the structure of the atom, which we developed in this chapter. In order to interpret atomic emission spectra, we need to understand the properties of light, and this means both its wave and photon (quantum) properties, which led to a quantum model of the atom. The discovery of the wave nature of electrons led to a wave description of atoms that accounts for both the kinetic and potential energy of an electron wave interacting with a nucleus. We characterized the structure of these atoms in terms of electron shells. The electron shell model is useful for understanding and predicting a great many periodic patterns in atomic and elemental properties, but it is too simple to explain more subtle periodicities and the spins of elemental atoms. For these we need the more complete description of electron waves (orbitals) based on wave equations that describe the substructure of the shells.

Check your understanding of the ideas in this chapter by reviewing these expected outcomes of your study. You should be able to

- Identify and describe periodic patterns in atomic and elemental properties and by extrapolation or interpolation predict values of these properties for elements that have not been measured [Sections 4.1 and 4.10].
- Explain the relationship between atomic emission and absorption of light [Section 4.2].
- Calculate the wavelength, frequency, or speed of propagation of a wave, given two of the three variables or the information necessary to derive them [Section 4.3].
- Describe in pictures and/or words how superimposition of waves produces diffraction patterns when waves pass through a grating [Section 4.3].
- Identify, given the characteristics of a source or detector of radiation, where in the electromagnetic spectrum the radiation will be found [Section 4.3].
- Calculate the wavelength, frequency, or energy of a photon, given two of the three variables or the information necessary to derive them [Section 4.4].
- Show how the results of photoelectric effect experiments can be explained by Planck's quantum hypothesis and how the wave model fails [Section 4.4].

- Identify and explain whether a phenomenon is a result of the wave or photon (quantum) properties of light [Section 4.4].
- Explain how the line emission and absorption of light by atoms requires that the energies of atoms be quantized [Section 4.5].
- Use an energy-level diagram for an atom and emissions from some known energy-level changes to predict the emissions from other energy-level changes [Section 4.5].
- Calculate the de Broglie wavelength of any particle of known mass and velocity and identify where in the electromagnetic spectrum it will be found [Section 4.6].
- Characterize a standing wave in terms of its amplitude, wavelength, and nodal properties [Sections 4.7 and 4.11].
- Explain why the uncertainty principle leads to a probability picture of an electron wave in an atom instead of more easily visualized orbits of an electron [Section 4.7].
- Describe in pictures and/or words how vibrating objects like strings and guitar bodies are related to the probability model of electron waves (orbitals) in atoms [Sections 4.7 and 4.11].
- Calculate or predict the direction of change for the kinetic energy, if the mass or velocity of the object changes [Section 4.8].
- Calculate or predict the direction of change for the potential energy of a system if the variables that describe the potential energy (charge and distance) are changed [Section 4.8].
- Use pictures, words, and/or equations to explain how the potential and kinetic energies of electron waves prevent atoms from collapsing and determine their size [Sections 4.8 and 4.13].
- Explain how the balance of potential and kinetic energies of electron waves together with the exclusion principle lead to a shell structure for atoms [Sections 4.9, 4.10, and 4.11].
- Show how the periodic properties of atoms and elements provide evidence for (or against) an electron shell and subshell model of atomic structure [Sections 4.10 and 4.11].
- Show how the electron shell model for atoms explains and predicts periodic properties of atoms and elements, including, but not limited to, atomic size, ionization energies, and electronegativities [Section 4.10].

- Describe the atomic orbitals derived from the Schrödinger wave equation for principal quantum numbers 1 and 2 [Section 4.11].
- Write the electron configurations for atoms of the first 20 elements, using the energy levels for multi-electron atoms and their degeneracies and accounting for the exclusion principle and electron–electron repulsion energies [Section 4.11].

- Predict, from its electron configuration, the number of unpaired electron spins a ground-state atom has [Section 4.11].
- Describe how the electron shell model and the orbitals and energies derived from solutions to the wave equation are complementary and together provide a way to explain all the periodic properties of atoms and elements discussed in the chapter [Section 4.1, 4.9, 4.10, and 4.11].

4.13. Extension—Energies of a Spherical Electron Wave

 **4.63 INVESTIGATE THIS**

How is rotation affected by size?

For this investigation, use a small, sturdy turntable and two equal masses (about 1 kg) that are easy to hold, one in each hand. Stand on the turntable and have a partner get you spinning around at a moderate speed while you are holding the masses at arm's length out to your sides. Bring the masses closer to your body and observe the effect this has on your speed of rotation. Straighten your arms again and observe your speed once more.

4.64 CONSIDER THIS

Why is rotation affected by size?

(a) To what does size refer in this title and the title of Investigate This 4.63? Explain your answer clearly.

(b) What is the relationship between size and speed of rotation? In Investigate This 4.63, is it harder to hold the masses close to or far from your body while you are spinning? Why do you think it is harder one way than another? What other experiences have you had that were similar to this investigation?

(c) Can you see a relationship between the results in Investigate This 4.63 and our discussion of the energies of an electron wave? Explain your response.

Kinetic energy of an electron wave From Section 4.8 we have

$$KE = \frac{m \cdot u^2}{2} \tag{4.14}$$

The larger the mass, m, and/or the faster the motion, u^2 (velocity squared), the larger the **kinetic energy.** The dependence of kinetic energy on mass and velocity is similar to the dependence of momentum, $m \cdot u$, on mass and velocity.

Indeed, we can write the equation for the kinetic energy of an object in terms of its momentum:

$$KE = \frac{(m \cdot u)^2}{2m} \qquad \textbf{(4.26)}$$

To get the kinetic energy of an electron wave, we substitute the momentum for an electron wave, from the de Broglie wavelength relationship, into equation (4.26):

$$m \cdot u = \frac{h}{\lambda} \text{ (de Broglie wave)} \qquad \textbf{(4.27)}$$

$$KE \text{ (electron wave)} = \frac{\left(\frac{h}{\lambda}\right)^2}{2m} = \frac{h^2}{2m \cdot \lambda^2} \qquad \textbf{(4.28)}$$

The kinetic energy of the electron wave is inversely proportional to the square of its wavelength. For a spherical electron wave, we can substitute the radius of the sphere for the wavelength in equation (4.28):

$$KE \text{ (spherical electron wave)} \propto \frac{1}{R^2} \qquad \textbf{(4.29)}$$

The smaller the radius of the electron wave (smaller volume of space occupied), the higher its kinetic energy. The observations you made in Investigate This 4.63 are analogous to this result for the kinetic energy of an electron wave. As you drew the masses closer to your body, you spun faster. As an electron is constrained to be closer to the nucleus—that is, as the size of the spherical wave decreases—the electron acts as though it moves faster; its kinetic energy increases.

4.65 CHECK THIS

Kinetic energy and the size of an electron wave

Is the relationship in equation (4.29) consistent with the curve shown for the kinetic energy in Figure 4.27? Explain your response.

4.66 WORKED EXAMPLE

Kinetic energy of an electron wave

What is the kinetic energy of an electron wave that has a wavelength, λ, of 100. pm?

Necessary information: We need the wavelength of the electron wave from the problem statement as well as the electron mass, 9.11×10^{-31} kg, from Table 3.1 and Planck's constant, 6.63×10^{-34} J·s. To be sure the units work out, we need to know that the units of joules are $kg \cdot m^2 \cdot s^{-2}$.

Strategy: Substitute the known values of mass, wavelength, and Planck's constant into equation (4.28) to find the kinetic energy.

continued

Implementation:

$$KE \text{ (electron wave)} = \frac{h^2}{2m \cdot \lambda^2} = \frac{(6.63 \times 10^{-34} \text{ J·s})^2}{2(9.11 \times 10^{-31} \text{ kg})(100 \times 10^{-12} \text{ m})^2}$$

$$= 2.41 \times 10^{-17} \text{ J}$$

Does the answer make sense? We can compare the kinetic energy for this electron wave, with the kinetic energy we got for the electron in Worked Example 4.44, 4.02×10^{-19} J, which we know from Worked Example 4.40 has a wavelength of 780 pm. The shorter wavelength electron in the present example has the higher kinetic energy. Note that, for electrons moving freely in space, the wave and particle descriptions for the kinetic energy give the same result.

4.67 CHECK THIS

Kinetic energy of matter waves

In Check This 4.41(c), you calculated the wavelength of a thermal neutron. Equation (4.28) is applicable to any matter wave. What is the kinetic energy of a thermal neutron matter wave? of a mole of neutron matter waves? How does your kinetic energy per mole compare to your result in Check This 4.45(b)? Is this the result you expect? Why or why not?

Potential energy of an electron wave From Section 4.8 we have

$$PE \propto \frac{(+1)(-1)}{r} = -\frac{1}{r} \tag{4.16}$$

Although we cannot locate the electron and measure its distance, r, from the nucleus, the radius of the spherical electron wave, R, is proportional to the average electron–nucleus distance. Thus, we can also express the **potential energy** in terms of R:

$$PE \propto -\frac{1}{R} \tag{4.30}$$

Total energy of an atom We know that the **total energy**, E, is the sum of the kinetic and potential energies:

$$E = KE + PE \tag{4.17}$$

The proportionalities in equations (4.29) and (4.30) are such that we get the result used in Section 4.8; the total energy for a nucleus–electron orbital system can be written as

$$E = KE + PE \propto \frac{1}{R^2} - \frac{1}{R} \tag{4.18}$$

In order to do quantitative calculations of energies in the atom, we would need to know the proportionality in equation (4.18). Our purpose, however, has been to understand why atoms do not collapse, and equation (4.18) provides a qualitative answer to that question. R, the radius of the electron wave, is a positive number. The total energy, E, is made up of two terms, the kinetic energy $\left(KE \propto \dfrac{1}{R^2}\right)$, which is always positive, and the potential energy $\left(PE \propto -\dfrac{1}{R}\right)$, which is always negative. The total energy depends on R and is the net balance between the positive kinetic energy and negative potential energy, as shown in Figure 4.27.

Chapter 4 Problems

4.1. Periodicity and the Periodic Table

4.1. The modern periodic table is based on the structure of atoms. This was not the evidence that Mendeleev used when he proposed the periodic arrangement of elements in 1869.
(a) Why didn't Mendeleev base his table on structure of atoms?
(b) What was the basis for Mendeleev's periodic table?
(c) Why were there missing elements in Mendeleev's periodic table?

4.2. Consider these data for the atomic radii of the elements in the first three periods.

Atomic radii, pm

Element	Radius	Element	Radius	Element	Radius
H	37	N	75	Al	118
He	32	O	73	Si	111
Li	134	F	71	P	105
Be	90	Ne	69	S	102
B	82	Na	154	Cl	99
C	77	Mg	130	Ar	97

(a) Use these data to find a periodic pattern in the atomic radii of the first three periods of main group elements.
(b) Display the data graphically.
(c) Which display, the tabulated values or your graph, best facilitates understanding of the relationships among the data? Explain your reasoning.
(d) What do you predict for the atomic radius of potassium? for calcium? What is the basis for your predictions?
(e) Use a print or Web-based reference to check the predictions made in part (d). Were your predictions accurate? Why or why not?

4.3. Figure 4.4 shows the values for the *first* ionization energies for elements with atomic numbers 1 through 20.
(a) How do you predict the value for the *second* ionization energy for helium [loss of an electron from $He^+(g)$] would compare with its first ionization energy? Write equations representing the first and second ionizations to help explain your reasoning. *Hint:* Use equation (4.1) as a model.
(b) How do you predict that the values of the *first, second, and third* ionization energies for beryllium would compare? Write equations representing the first, second, and third ionizations to help explain your reasoning.

4.4. Figure 4.4 shows the trend in first ionization energies for the first 20 elements. The diagram shown at the bottom of this column is an alternate representation of first ionization energy data.
Compare and discuss the two different representations, paying particular attention to
(i) How well the patterns across periods are represented.
(ii) How well the patterns within a family (group) are represented.
(iii) How easy it is to discern a shell structure for atoms.

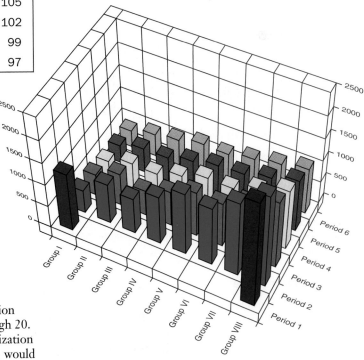

4.5. Values for the second ionization energies of the first 20 elements are repeated here from Table 4.2. These are the energies for reactions such as $Na^+(g) \rightarrow Na^{2+}(g) + e^-(g)$.

Second ionization energy (IE_2), kJ·mol^{-1}

Element	IE_2	Element	IE_2	Element	IE_2
H		O	3390	P	1904
He	5248	F	3370	S	2250
Li	7300	Ne	3950	Cl	2297
Be	1757	Na	4560	Ar	2666
B	2430	Mg	1450	K	3052
C	2350	Al	1820	Ca	1145
N	2860	Si	1580		

(a) Why is no value given for the second ionization energy of the hydrogen atom?
(b) Plot these second ionization energies as a function of atomic number, as was done for first ionization energies in Figure 4.4. How does your plot compare to Figure 4.4? How do you explain the similarities? the differences?

4.6. In the *Web Companion*, any chapter, click on the circled P in the left-hand menu to access the interactive periodic table. Click on "Plot Data."
(a) On the plot of molar mass (relative atomic mass) as a function of atomic number, can you find pairs of relative atomic masses that are "out of order," that is, decrease as the atomic number increases? Check these out on the periodic table on the inside front cover of the text to confirm that the relative atomic masses of these elemental pairs are out of order.
(b) Could the elemental pairs you found in part (a) have caused problems for Mendeleev when he constructed his periodic table? Explain the problems he might have encountered. Is there any evidence in his table (shown in the chapter opening illustration) to suggest that he had such problems?

4.2. Atomic Emission and Absorption Spectra

4.7. Why is it necessary to heat the sodium vapor sample represented in Figure 4.6(b)? *Hint:* Use a reference handbook to find the properties of sodium.

4.8. Explain why atoms at the surface of the sun emit light, whereas atoms in a layer farther away absorb light. Relate your explanation to the experiments represented in Figure 4.6.

4.3. Light as a Wave

4.9. Sketch a wave and label or define:
(a) wavelength
(b) amplitude
(c) node
(d) cycle

4.10. (a) At its closest approach to Earth, Mars is 56 million km from Earth. How long does it take a radio message from a space probe on Mars to reach Earth when the two planets are this distance apart?
(b) Does your answer in part (a) suggest problems controllers on Earth might have in maneuvering a remote-controlled vehicle, Figure 1.39, on the Martian surface? Explain.

4.11. Calculate the wavelength and identify the type of radiation that has a frequency of
(a) 101 MHz
(b) 25×10^{15} Hz
(c) 200 GHz [1 gigahertz (GHz) = 10^9 Hz]

4.12. Calculate the frequency and identify the type of radiation that has a wavelength of
(a) 400 nm
(b) 25 km
(c) 4.5×10^{-4} m

4.13. A wave has a wavelength of 0.34 m and a frequency of 0.75 s^{-1}.
(a) What is the speed of the wave?
(b) Is the wave electromagnetic radiation? Explain.

4.14. What is the wavelength of microwave radiation with a frequency of 1.145×10^{10} s^{-1}? *Note:* Microwave radiation has frequencies in the range 10^9 to 10^{12} s^{-1}.

4.15. Ultraviolet-visible (UV-Vis) spectroscopy is typically used to determine concentrations of protein samples. If a protein absorbs light at 280 nm, what frequency does this correspond to?

4.16. Radio and television antennas are designed so that the length of a crossbar is approximately equal to the wavelength of the signal received. If you know that an antenna receives a signal with frequency of 3×10^2 MHz, calculate the length of a crossbar in the antenna.

4.17. (a) In the *Web Companion*, Chapter 4, Section 4.2.4, how is the correct set of oscillating arrows related to the representation of the electromagnetic wave? Be as specific as possible in your description.
(b) How is the correct set of oscillating arrows related to the description of electromagnetic radiation given in this chapter? Explain clearly.

4.4. Light as a Particle: The Photoelectric Effect

4.18. Explain how the predictions of the wave theory of light differed from the observed distribution of wavelengths of emission from glowing bodies.

4.19. (a) Explain how the photoelectric effect was evidence for the particulate nature of light.
(b) Explain why the wave theory failed to explain the photoelectric effect.

4.20. Explain the role Einstein's explanation of the photoelectric effect played in the development of our understanding of the dual nature of light.

4.21. Classical physics assumed that atoms and molecules could emit or absorb any arbitrary amount of radiant energy. However, Planck stated that atoms and molecules could emit or absorb energy only in quanta. Give a few examples from everyday life that illustrate the concept of quantization.

4.22. What is the energy of a photon with a frequency of $1.255 \times 10^6 \ s^{-1}$?

4.23. The energy of one photon associated with a typical microwave oven is approximately $1.64 \times 10^{-24} \ J$.
(a) What is the frequency of the radiation associated with one photon of this microwave radiation?
(b) How does this frequency compare with the frequency of green light with a wavelength of 515 nm?
(c) Which has more energy—a photon of green light or a photon of microwave energy?

4.24. A 500-MHz nuclear magnetic resonance (NMR) spectrometer is routinely used to help determine the structures of biomolecules.
(a) Calculate the wavelength (in meters) of this instrument.
(b) Using Figure 4.13, determine the region of the electromagnetic spectrum in which this instrument operates.
(c) To how much energy, in $kJ \cdot mol^{-1}$, does 500 MHz correspond?

4.25. Calculate the energy of one photon and one mole of photons with a
(a) frequency of 101 MHz.
(b) wavelength of 400 nm.

4.26. A student calculated that the smallest increment of energy (quantum of energy) that can be emitted from a yellow light with wavelength of 589 nm is $4.2 \times 10^{-19} \ J$. Verify the student's answer.

4.27. In Investigate This 4.29, you explored the effect of light on silver chloride. How can this same effect be used to produce the lenses for eyeglasses that change with the light level? *Hint:* Think about reaction (4.7) in Consider This 4.30. You may want to research your prediction in print or web-based references to confirm or refute your suggestion.

4.28. Based on the results of Investigate This 4.29, why does a photographic darkroom use a red safelight?

4.29. In a *Science* article, scientists reported variations in the concentrations of thorium and potassium on the moon's surface. They used gamma ray spectrometry that recorded the gamma ray emission from potassium at approximately 1.4 MeV and from thorium at approximately 2.6 MeV. Calculate the frequencies of these two gamma ray emissions. Calculate the wavelengths of these two gamma ray emissions. *Note:* $1 \ MeV = 1 \times 10^6 \ eV$ (electron volt) and $1 \ eV = 1.602 \times 10^{-19} \ J$.

4.5. The Quantum Model of Atoms

4.30. (a) Explain what happens to electrons when an atomic emission spectrum is produced.
(b) Explain what happens to electrons when an atomic absorption spectrum is produced.
(c) How are these two types of spectra related?

4.31. Choose the best phrase to complete this sentence. Justify your choice. The intensity of a spectral line in an atomic emission spectrum can be directly related to
(i) the number of energy levels involved in the transition that gives rise to the line.
(ii) the speed with which an electron undergoes a transition from one energy level to another.
(iii) the number of electrons undergoing the transition that gives rise to the line.

4.32. Choose the best phrase to complete this sentence. Justify your choice. When electrons are excited from a ground state to an excited state,
(i) light is emitted.
(ii) heat is released.
(iii) energy is absorbed.
(iv) an emission spectrum results.

4.33. Neon lights glow because an electrical discharge is passed through a low pressure of neon in the glass tubing. In an experiment, emission from a neon light was dispersed by a prism to form a spectrum. The spectrum that was formed was not continuous but consisted of several sharp lines. Explain why a line spectrum was produced.

4.34. When energy is added to an electron in an atom it may become excited and move farther (on average) from the nucleus. That energy is released when the electron returns to its original energy. What happens when you add so much energy that the electron completely leaves the atom?

4.35. To distinguish between a solution of sodium ions and a solution of potassium ions, a simple flame emission test can be used. A drop of the solution to be tested is held in burner flame and the color of the flame is observed. (See Problem 3.4 at the end of Chapter 3.) A solution of sodium ions gives a bright yellow flame, while a solution of potassium ions gives a violet flame.
(a) To what approximate wavelengths in the electromagnetic spectrum do these emissions correspond?
(b) How, if at all, are these data related to Figure 4.6? Explain.
(c) Which ion's emission has the higher energy? Explain your answer.

4.36. The emission spectrum of mercury contains six wavelengths in the visible range: 405, 408, 436, 546, 577, and 579 nm.
(a) To what color does each of these wavelengths correspond?
(b) Which wavelength corresponds to the largest difference in energy between the atomic states responsible for the emission? Explain.
(c) Which wavelength corresponds to the smallest difference?

4.37. How do the line spectra of hydrogen, helium, mercury, and neon support the idea that the energy of electrons is quantized?

4.38. For these electron transitions (between the energy levels on this energy level diagram), say whether the transition would result in absorption or emission of a photon, and rank the relative wavelengths of the photons absorbed or emitted from shortest to longest.
(a) $E_1 \rightarrow E_2$
(b) $E_2 \rightarrow E_1$
(c) $E_4 \rightarrow E_3$
(d) $E_4 \rightarrow E_5$
(e) $E_6 \rightarrow E_2$
(f) $E_6 \rightarrow E_1$

4.6. If a Wave Can Be a Particle, Can a Particle Be a Wave?

4.39. Since de Broglie's hypothesis applies to all matter, does any object of a given mass and velocity give rise to a wave? Is it possible to measure the wavelength of any moving object?

4.40. Calculate the wavelength of a baseball of mass 0.145 kg traveling at 30 m·s^{-1}. Why does the wavelength of macroscopic objects not affect the behavior of the object?

4.41. Calculate the wavelength of an electron of mass 9.1×10^{-31} kg traveling at 1.5×10^6 m·s^{-1}. Is this wavelength significant relative to the size of an atom?

4.42. Explain why the distinction between a wave and a particle is meaningful in the macroscopic world and why the distinction becomes blurred at the atomic level. *Hint:* Consider the wavelengths of macroscopic and atomic level particles.

4.43. (a) What is the de Broglie wavelength of an alpha particle that is traveling at a velocity of 1.5×10^7 m·s^{-1}? The mass of an alpha particle is 6.6×10^{-27} kg.

(b) In what region of the electromagnetic spectrum does the wavelength of this alpha particle fall?

4.7. The Wave Model of Electrons in Atoms

4.44. The sounds you hear when musical instruments are played are created by standing waves in the instruments. The air molecules in contact with the instrument are set in motion by the standing waves and form traveling waves. The traveling waves reach us, set our eardrums vibrating, and our brains interpret the motion as music. We have seen that standing waves on a guitar string are responsible for the sound produced by the guitar. What standing waves are responsible for the sound produced by these other musical instruments? Explain your answers.
(a) drums
(b) flutes
(c) tuning forks

4.45. What contributions did these scientists make to our understanding of atomic and electronic structure?
(a) Bohr   (c) Einstein
(b) Heisenberg   (d) de Broglie

4.46. What does the Heisenberg uncertainty principle state about what we can and cannot know about an electron's behavior?

4.47. Why did the de Broglie wave model and Heisenberg's uncertainty principle give rise to a new approach in which it was not appropriate to imagine electrons moving in well-defined orbits about the nucleus?

4.48. Why is it that we can calculate exactly the position and the momentum of a rolling ball but we cannot calculate the same for an electron that exhibits wave properties?

4.8. Energies of Electrons in Atoms: Why Atoms Don't Collapse

4.49. Discuss the factors that affect the size of a hydrogen atom with regard to the kinetic and potential energy of the electron wave and the nucleus.

4.50. If particle **A** has a mass of 1.56×10^{-25} kg and particle **B** has a mass of 4.25×10^{-24} kg, which particle has the greater kinetic energy if they are both traveling at a velocity of 3.15×10^5 m s^{-1}? Explain your reasoning.

4.51. How is the potential energy of an electron wave related to its distance from the nucleus?

4.52. In Figure 4.27, why is the potential energy negative and the kinetic energy positive?

4.53. 🕸 What is the significance of the red circle in the *Web Companion*, Chapter 4, Section 4.7.1? Explain how it is related to the energies that are represented on the graph.

4.9. Multielectron Atoms: Electron Spin

4.54. Explain which property of an electron is responsible for its interaction with a magnetic field.

4.55. Examine Figure 4.28. Explain why the magnet does not seem to affect helium atoms but does affect lithium atoms.

4.56. What experiment might you perform to tell whether atoms of an element have odd numbers of electrons?

4.57. Experiments on nitrogen atoms show that the ground state atom has three electrons with unpaired spins.
(a) How many of the nitrogen atom's electrons are spin paired? Explain how you get your answer.
(b) What do the electron spin data tell you about the orbital descriptions of the three electrons with unpaired spins? Explain.

4.10. Periodicity and Electron Shells

4.58. Define the Pauli exclusion principle and how it can be used to explain the very low ionization energy of the lithium atom.

4.59. Why is the second ionization energy always greater than the first ionization energy for an atom? *closer to nucleus and more energy to remove e⁻*

4.60. (a) What is the same for each of these ions and atom: Na^+, Ne, F^-, and O^{2-}?
(b) Which of the species in part (a) has the highest ionization energy? Explain your answer. *more + attracting one e⁻*

4.61. Discuss these properties and their trends in relationship to the periodic table.
(a) ionization energy
(b) atomic radius
(c) electronegativity

4.62. What is the trend in ionization energy as you go from left to right in a row on the periodic table? How is the shell model for atoms related to this trend?

4.63. What is the trend in ionization energy as you go from top to bottom in a group on the periodic table? How is the shell model for atoms related to this trend?

4.64. How many valence electrons does each of these elements have?
(a) oxygen   (c) chlorine
(b) sodium   (d) argon

4.65. The empirical formulas of the simplest hydrides (binary compounds with hydrogen) of the second period elements are

Element	Li	Be	B	C	N	O	F	Ne
Hydride	LiH	BeH_2	BH_3	CH_4	NH_3	OH_2	FH	none

(a) Describe any pattern you see in these data.
(b) Does the hydride of hydrogen, H_2, fit the pattern(s) you described in part (a).
(c) What do you predict for the empirical formulas of the hydrides of aluminum, silicon, and phosphorus? Explain the basis of your prediction and, if possible, relate it to periodicity and the shell model of the atom.

4.66. 🖱 In the *Web Companion*, any chapter, click on the circled P in the left-hand menu to access the interactive periodic table. Click on "Plot Data."
(a) Click on the "Properties" button and then on "1st Ionization Energy" to select this property to be plotted. Compare the plot with Figures 4.4 and 4.30 to be sure you identify the same trends in all of them. What are some of these?
(b) Click on the "Properties" button and then on "2nd Ionization Energy" to select this property to be plotted. Compare the plot with the one you made in Problem 4.5. Are the conclusions you drew from the first 20 elements in Problem 4.5 valid for the rest of the elements as well? Explain why or why not.
(c) Select the "3rd Ionization Energy" to be plotted. How does this plot compare to those for the first and second ionization energies? How do you explain the similarities? the differences?

4.67. 🖱 In the *Web Companion*, any chapter, click on the circled P in the left-hand menu to access the interactive periodic table. Click on "Plot Data." Write a problem about periodicity that is based on analysis of one or more of the data plots accessible from this window. Have a classmate solve your problem.

4.68. One kind of periodic table is a three-dimensional model of eight stacked, round wooden discs with the symbols for the elements on the discs. Shown here are the symbols on the second tier of discs. How does this periodic table represent the electron shell model? Do you see any problems with this representation compared to Figure 4.31? Explain your responses.

4.69. Two views of a three-dimensional version of the periodic table are shown here (and with other views in the chapter opening illustration).

(a) What parts of this table correspond to a more conventional table, as on the inside front cover of this book? Explain the correspondences.
(b) What parts of this table correspond to the spiral table, Figure 4.34? Explain the correspondences.

4.11. Wave Equations and Atomic Orbitals

4.70. Answer each of these statements with TRUE or FALSE. If a statement is false, correct it.
(a) The Schrödinger wave equation relates the wave properties of the electron to its kinetic energy.
(b) The wave function (ψ) is a mathematical description of an allowed energy state (orbital) for an electron.
(c) The wave function (ψ) is proportional to the probability of finding the electron in a given location near the nucleus.
(d) Orbitals with the same energy are called degenerate.
(e) Orbitals can be thought of as boxes into which electrons can be placed.
(f) Each calculated energy level corresponds to a spatial probability distribution for the electron.

4.71. Verify the ground-state electron configurations for these elements. For the configurations that are incorrect, explain what mistakes have been made and write the correct electron configurations.
(a) Al $1s^2 \, 2s^2 \, 2p^4 3s^2 3p^3$
(b) P $1s^2 2s^2 2p^6 3s^2 3p^2$
(c) B $1s^2 2s^0 2p^3$

4.72. How many valence electrons are located in an *s* orbital in each of these elements?
(a) potassium   (c) magnesium
(b) fluorine   (d) boron

4.73. How many valence electrons are located in a *p* orbital in each of these elements?
(a) phosphorus   (d) bromine
(b) aluminum   (e) sulfur
(c) lithium

4.74. Atoms **A**, **B**, and **C** have the electron configurations given below. Which atom has the *largest third ionization energy*? Explain your answer.

$\mathbf{A} = 1s^2 2s^2 2p^6 3s^2$
$\mathbf{B} = 1s^2 2s^2 2p^6 3s^2 3p^4$
$\mathbf{C} = 1s^2 2s^2 2p^6 3s^2 3p^6$

4.75. Using Figure 4.37 as a template and Worked Example 4.60 as a guide, draw electron energy diagrams for the sulfur atom and the sulfide ion.

4.76. Using Figure 4.37 as a template and Worked Example 4.60 as a guide, draw electron energy diagrams for Li and Na atoms.

4.77. (a) What does the series of atoms from H through Ar have in common with the series of monocations He$^+$ through K$^+$ and the series of dications Li^{2+} through Ca^{2+}? Explain.
(b) How does your answer in part (a) relate to your results and answers to Problems 4.5 and 4.66? Explain.

4.78. In the *Web Companion*, any chapter, click on the circled P in the left-hand menu to access the interactive periodic table. Click on "Electron Configuration."
(a) How does the graphic in the electron configuration window compare to Figure 4.37? What does the vertical direction on the graphic represent? Explain.
(b) Move the cursor over the blank periodic table until the pointer is on lithium. Click to select lithium and describe what happens. How does the result compare to the graphics in Worked Example 4.60? Explain.
(c) Use the electron configuration window to compare with your results from Consider This 4.62(a). How do they compare?

4.79. In the *Web Companion*, any chapter, click on the circled P in the left-hand menu to access the interactive periodic table. Click on "Electron Configuration."
(a) Examine the electron configurations of the first series of transition metals, Sc through Cu. What conclusion(s) can you draw about the relative energies of 4*s* and 3*d* electrons? Explain how you reach your conclusion(s).
(b) How are the electron configurations and the spins of the electrons for the first series of transition metals related to the rule you found in Consider This 4.62(c)? Explain your reasoning and, if necessary, suggest how to modify your rule to make it more general, now that you have more data.
(c) Do elements in the second transition metal series, Y through Ag, follow the same electron configuration and electron spin pattern as the first series? If so, is this what you expected? Explain. If not, explain what factor(s) might cause differences.

4.13. Extension—Energies of a Spherical Electron Wave

4.80. (a) At what value of R does the total energy curve in Figure 4.27 reach a minimum?

(b) At the minimum in the total energy, what is the relationship of the kinetic energy to the potential energy of the proton–electron wave system? Show your work. *Hint:* Use the value of R you got in part (a) in equations (4.29) and (4.30). Assume that the proportionality is the same in both cases and let the proportionality constant be H.
(c) Express the value of the total energy at its minimum for the proton–electron wave system as a function of H from part (b). Use the ionization energy for the hydrogen atom to determine the value of H in $kJ \cdot mol^{-1}$.

4.81. (a) For one-electron ions such as He^+ and Li^{2+}, you can use the same analysis of the energies as we used for the hydrogen atom, with the exception that you must account for the higher nuclear charge. The potential energy, equation (4.16), depends on the nuclear charge. Write equation (4.16) for a one-electron ion which has a nuclear charge $Z+$.
(b) What is the potential energy of a spherical electron wave attracted to the nucleus in part (a)? Write this potential energy as a function of R, the radius of the wave, H, the proportionality constant from Problem 4.80(b), and Z.
(c) What is the total energy of a one-electron ion with a nuclear charge $Z+$? Write the energy as a function of R, H, and Z. *Hint:* The kinetic energy of the electron wave does not depend on the nuclear charge.
(d) Find the value of R that makes the total energy in part (c) a minimum. *Hint:* There are at least three ways to do this problem. (i) If you are familiar with calculus, you can take the derivative of E with respect to R, dE/dR, set the derivative equal to zero, and solve for R. (ii) Use a graphing calculator or computer graphing program to graph the total energy function and find the minimum by tracing the curve. (iii) Assume (correctly) that the relationship between the kinetic and potential energies at the energy minimum, which you found in Problem 4.80, is the same for all one-electron atoms and ions. Use this relationship with R as an unknown and solve for R.
(e) Using your result from part (d), write the minimum for the total energy of a one-electron ion as a function of H and Z. Use your equation to predict the ionization energies for He^+, Li^{2+}, and B^{4+}. How do your predictions compare with the experimental values in Table 4.2? Can you predict other values in the table? Explain.

General Problems

4.82. This image of a white blood cell was made by a scanning electron microscope (SEM). Consult print or Web-based resources to find out how an electron microscope differs from the traditional light-based microscope you have probably used in a biology class.

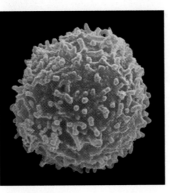

4.83. Einstein's relativity theory shows that the effective mass of a moving particle, m_{moving}, is related to its rest mass, m_{rest}, its velocity, u, and the speed of light, c, by this equation:

$$m_{moving} = \frac{m_{rest}}{\sqrt{1 - \left(\dfrac{u}{c}\right)^2}}$$

(a) What is the effective mass of a 145-g baseball thrown at $44 \; m \cdot s^{-1}$? Will a batter have to be concerned about relativistic effects in the pitched ball? Explain.
(b) What is the effective mass of the moving alpha particle in Problem 4.43(a)? How does this result affect your answer to Problem 4.43(b)? Explain. *Hint:* Use the effective mass in the de Broglie equation.
(c) What are the effective mass and the wavelength of an electron moving at 50% the speed of light?

4.84. Particle accelerators can accelerate protons, electrons, and other charged particles to extremely high energies. These high energy particles are sometimes fused with lighter elements, producing synthetic heavier elements.
(a) Calculate the wavelength (in meters) of a proton that has been accelerated to 50% the speed of light. *Hint:* See Problem 4.83.
(b) A mole of electrons is accelerated to 90% the speed of light. What kinetic energy (in $kJ \cdot mol^{-1}$) do these electrons have? What is the kinetic energy of a single accelerated electron? *Hint:* See Problem 4.83.

4.85. In the *Web Companion*, any chapter, click on the circled P in the left-hand menu to access the interactive periodic table. Click on "Electron Configuration." Examine the representation of the periodic table at the bottom of the electron configuration window.
(a) Explain the relationship of this periodic table to the one on the inside front cover of this book.
(b) Explain the relationship of this periodic table to the spiral table shown in Figure 4.34. How are corresponding parts of the atomic shell structure represented in the two tables?
(c) Explain the relationship of this periodic table to the three-dimensional table shown in Problem 4.69 and the chapter opening illustration. How are corresponding parts of the atomic shell structure represented in the two tables?

To every Form of being is
assigned . . . An active Principle.

THE EXCURSION
WILLIAM WORDSWORTH (1770–1850)

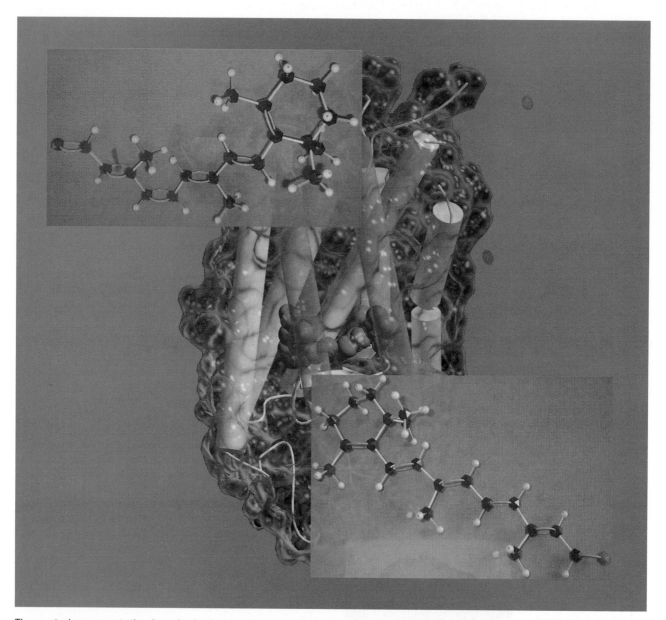

The central representation here is rhodopsin, the molecule required for vision. The gray cylinders
represent α-helices that are part of the structure of the protein opsin, a chain of about 350 amino
acids. The green structure nested among the helices is a molecule of retinal, which is bonded at its
oxygen end to one of the helices to form rhodopsin. The ball-and-stick model at the upper left also
represents the shape of the retinal molecule. When the retinal absorbs a quantum of visible light,
its shape changes to that shown on the lower right.

Structure of Molecules

When you read, you probably do not think much about the chemistry taking place in your eyes. You can read these words because the light that reaches your retina causes molecules of retinal to change shape, as shown by the ball-and-stick models on the facing page. This change in shape of a relatively small molecule forces the helices of the rhodopsin molecule to move, thus changing the shape of the whole protein molecule. A large number of these rhodopsin molecules are present in membranes in your retinas. The changes triggered by the change in shape of the retinal are responsible for the signal that is sent to your brain when light strikes your retina. The chemistry of vision is just one example of what is captured poetically in Wordsworth's words: The functions and reactivity of molecules depend on their form—their structure and bonding.

In order to understand the kinds of electron and atomic core rearrangements that have to occur to change the shape of the retinal molecule, you have to know what holds the molecule together in the first place. In Chapters 1 and 2, we introduced and used several models of molecules, including molecule names, Lewis structures, ball-and-stick models, and computer-generated electron-density models. We used these to help interpret and understand the properties of water and other molecules. We did not, however, discuss how the molecules are held together. The goal of this chapter is to help you develop an understanding of the bonding that holds atoms together to form molecules. Bond properties are the key to understanding more about the structures (shapes) of molecules and are the basis for reactions between molecules.

At the heart of molecular structure is the role of electrons in holding molecules together. ***The fundamental principle of molecular structure and bonding is that the negative electrons and positive nuclei attract one another.*** In Chapter 4, you saw that electrons in atoms behave more like vibrating guitar strings than they do like marbles or baseballs. In this chapter, we will build on those wave mechanical models for atoms to understand how valence electrons in molecular orbitals bind atomic cores together. Computer-generated electron-density representations, such as that of the water molecule, are the result of wave-mechanical calculations. The

polarity of molecules is a consequence of the unequal charge distribution described by this representation. In this chapter, we will continue to emphasize the polarity of molecules, because polarity is the basis for understanding the reactions of molecules we will examine in Chapter 6.

To set the stage for our exploration of molecular wave mechanics, we'll revisit some molecular properties we have seen previously and introduce some new molecules, many of which are isomers, and their reactions. And we will extend the writing of Lewis structures to more complicated molecules that we will come to understand using our bonding model.

5.1. Isomers

 5.1 INVESTIGATE THIS

How do different alcohols react in a Breathalyzer® test?

Do this as a class investigation and work in small groups to discuss and analyze the results. Add about 1.0 mL of acetone to each of six small test tubes and then add *1 drop* of the Breathalyzer® reagent—a solution of the orange dichromate ion, $Cr_2O_7{}^{2-}$, in sulfuric acid—to each of the test tubes. *CAUTION:* Sulfuric acid solutions are very corrosive to many materials, including skin. Dichromate is a suspected carcinogen. Be careful not to spill any reagent and wear protective plastic gloves when handling the solution. Leave one of the filled test tubes as a control and add *1 drop* or a *tiny* crystal of the following five compounds to separate test tubes containing the reagent: ethanol, *n*-butyl alcohol, *sec*-butyl alcohol, *iso*-butyl alcohol, and *tert*-butyl alcohol. Observe the solutions and record your observations.

5.2 CONSIDER THIS

Are all alcohols detected by the Breathalyzer® test?

(a) In the Breathalyzer® test for blood alcohol level, ethanol in the breath of the person being tested reacts with dichromate ion. The orange dichromate, $Cr_2O_7{}^{2-}$, is changed to the green chromic ion, Cr^{3+}, as shown in the photos. Do you find evidence for this reaction in Investigate This 5.1? Explain.

(b) Do any other alcohols react with the Breathalyzer® reagent? What is your evidence? Which alcohols?

(c) Do any alcohols *not* react with the Breathalyzer® reagent? What is your evidence? Which alcohols?

(d) Can you suggest an explanation for differences, if any, among the alcohols you tested?

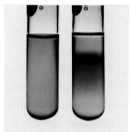

The prefixes for the butyl alcohols are short for

n = normal   *sec = secondary*
iso = isomeric   *tert = tertiary*

We have used these names to draw attention to the fact that these compounds are isomers, but will rarely use the names again.

The structural names we will usually use are based on the names of the simplest hydrocarbons:

methane	CH_4
ethane	CH_3CH_3
propane	$CH_3CH_2CH_3$
butane	$CH_3CH_2CH_2CH_3$
pentane	$CH_3CH_2CH_2CH_2CH_3$

In Chapter 1, you found that the boiling points of compounds varied with the polarity of their molecules. We attributed the higher boiling points of more polar compounds to the stronger polar attractions, especially hydrogen bonds, among the molecules in their liquid phase, which meant that more energy was required for the molecules to move into the gas phase. For nonpolar compounds, boiling points generally increase with molar mass, a rough indicator of the number of electrons in the molecules. The more electrons available, the greater the dispersion forces between molecules and the more energy required to get them into the gas phase. The boiling points and other properties of the alcohols you used in Investigate This 5.1 are given in Table 5.1. The table shows that all four of the butyl alcohols have the same molecular formula. Compounds with the same molecular formula but different properties are called **isomers**.

Table 5.1 *Selected properties of a few alcohols.*

Alcohol	Structural name	Molecular formula	Melting point, °C	Boiling point, °C	Solubility, g in 100 mL H_2O
ethyl	ethanol	C_2H_6O	−114.5	78.5	miscible
n-butyl	1-butanol	$C_4H_{10}O$	−89.8	118.0	9
sec-butyl	2-butanol	$C_4H_{10}O$	−114.7	99.5	12.5
iso-butyl	2-methyl-1-propanol	$C_4H_{10}O$	−108	108.1	10
tert-butyl	2-methyl-2-propanol	$C_4H_{10}O$	25.6	82.6	miscible

5.3 CONSIDER THIS

How do the properties of the butyl alcohol isomers differ?

(a) Are the properties of some of the isomers of butyl alcohol similar to one another? If so, which ones? Which one(s) are quite different? Explain the rationale for your choices.

(b) How do the properties of the isomeric butyl alcohols compare to the properties of ethanol (ethyl alcohol)? Are any of the differences or similarities surprising? Explain why or why not.

(c) Imagine adding a column to Table 5.1 that gives the result (positive or negative) of the Breathalyzer® test from Investigate This 5.1 for each alcohol. Would this property fit the patterns you have found in parts (a) and (b)? Explain why or why not.

You have discovered that *tert*-butyl alcohol (2-methyl-2-propanol) is strikingly different from the other $C_4H_{10}O$ isomeric alcohols. Its melting point is more than 100 °C higher than any of the others; it is a solid at room temperature, whereas the others are liquids. Its boiling point is lower than any of the others. In fact, its boiling point is almost the same as that of ethanol, even though it has a higher molar mass than ethanol. Unlike the other $C_4H_{10}O$ isomers and ethanol, *tert*-butyl alcohol does not react with the Breathalyzer® reagent.

We have seen similar differences between isomers before. Look back at Table 1.3 in Chapter 1, Section 1.11, to review the properties of ethanol, CH_3CH_2OH, and dimethyl ether, CH_3OCH_3. Ethanol is a liquid at room temperature and dimethyl ether is a gas. Ethanol is miscible with water, but only about 7 g of dimethyl ether dissolves in 100 mL of water. These compounds have the same molecular formula, C_2H_6O, and different properties, so they are isomers.

Consider one more example of isomers, three compounds with the molecular formula, C_5H_{12}. Table 5.2 gives the melting and boiling points of these isomers.

Table 5.2	Melting and boiling points of the C_5H_{12} isomers.	
Structural name	**Melting point, °C**	**Boiling point, °C**
pentane	−129.7	36.1
2-methylbutane	−159.9	27.9
2,2-dimethylpropane	−16.6	9.5

5.4 CHECK THIS

Comparison of the C_5H_{12} and $C_4H_{10}O$ isomers

(a) Do you see any similar patterns among the C_5H_{12} and $C_4H_{10}O$ isomers? If so, what are they? Explain the rationale for your response.

(b) There is some similarity in the structural names of the two sets of isomers. Do these similarities reflect the patterns you found in part (a)? Explain why or why not.

(c) The molar masses of these two sets of isomers are almost the same and the number of electrons in each set is the same. (Prove these assertions for yourself.) What factor do you think is mainly responsible for the differences in boiling points between the two sets of isomers? Explain your reasoning.

Based on what you learned in previous chapters, you have probably concluded that the large differences in properties within a set of isomeric compounds, such as the butyl alcohols, must reflect differences in their molecular structures. Some of our tasks in the rest of the chapter will be to discover what these molecular structures are, to develop the bonding model that is responsible for them, and to try to relate observed properties to the structures. We will begin by reviewing and extending one of our simplest molecular bonding models, Lewis structures, and show how you can use your molecular model kits to translate them into three dimensions.

5.2. Lewis Structures and Molecular Models of Isomers

We have seen that molecules can be thought of as collections of **atomic cores** that are held together by **valence electrons.** The valence electrons in molecules include *all* those contributed from the valence shells of the atoms that form the molecule. For example, a molecule of methane, CH_4, has eight valence electrons. The carbon atom contributes four electrons, and each of the four hydrogen atoms contributes one electron.

The Lewis model is one of the earliest bonding models of the 20th century. Lewis used the known formulas for a large number of compounds, the nuclear model of the atom, and the periodic table as the basis for his model. As you have seen, Lewis structures for molecules containing the second-period atoms, C, N, O, and F, have four pairs of valence electrons (an octet of electrons) around these atomic cores. The tetrahedral arrangement of bonds (sticks) around the atom centers in your molecular model kit is consistent with these Lewis structures. (In

Recall that the term *atomic core* is not in general use. We are using it here to remind you that every atom except hydrogen has deeply buried electrons that do not take part in chemical reactions. Later, we will use the conventional term *atom;* you will have to interpret from the context whether atomic core is what is really meant.

Section 5.9, we will discuss the logic used by van't Hoff and Le Bel, who proposed the tetrahedral arrangement around carbon more than 40 years before Lewis developed his model.) For all of the Lewis structures you have written so far, you were given a line (structural) formula or molecular model to show the connectivity of the atomic cores in the molecule.

5.5 CHECK THIS

Lewis structures of the C_2H_7N isomers

(a) Write Lewis structures for ethyl amine, $CH_3CH_2NH_2$, and dimethyl amine, CH_3NHCH_3.
(b) Use your molecular model kit to build models of these two molecules. How would you describe the shape of each molecule? Are the shapes about the same or different? Explain.

Personal Tutor
We introduced Lewis structures in Chapter 1, Section 1.4, and asked you to write several of them in Chapters 1 and 2. If necessary, you should review these and the Personal Tutor before studying this section.

Atomic connections in isomers Sometimes you will be faced with the problem of writing a Lewis structure (or structures) without knowing the connectivity of the atomic centers in the molecule(s). In these cases, you will have to choose the connectivity based on what you know about how atomic cores (centers) are connected in other molecules. We will use isomers as examples to show how to approach such problems.

5.6 WORKED EXAMPLE

Lewis structures and molecular models for C_3H_8O isomers

Write Lewis structures and build the corresponding molecular models for as many C_3H_8O isomers as you can.

Necessary information: We need to know that C, H, and O atoms have, respectively, 4, 1, and 6 valence electrons to be used in the Lewis structures. We also need to recall that a carbon atomic center usually forms two-electron bonds to *four* other atomic centers and oxygen usually forms two-electron bonds to *two* other atomic centers.

Strategy: Calculate the number of valence electrons in the molecules. Connect the four second-period atomic centers in as many ways as possible with two-electron bonds. Add all the hydrogen atoms with two-electron bonds to give four atomic centers on each carbon atom and two on the oxygen atom. Use any remaining valence electrons to complete octets on each second-period element.

Implementation: The atoms, 3 C, 8 H, and 1 O, have a total of 26 valence electrons.

One way to proceed systematically to find different connectivities is first to connect the carbon atoms to one another in as many ways as possible and then connect the oxygen atom to each carbon in turn. In the partial Lewis structures shown on the next page, we use lines to represent two-electron bonds.

continued

(a) C—C—C—O

(b) C—C—C with O double-bonded to middle C

(c) O—C—C—C

Structures (a) and (c) are identical because they can be rotated and superimposed on one another. Thus, we have only two different structures here, (a) and (b). Next, we need to consider connecting two carbon atoms to the oxygen atom. There is one such structure:

(d) C—C—O—C

Add hydrogen atoms and two-electron bonds to structures (a), (b), and (d):

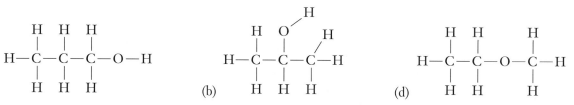

Each structure has 22 electrons in two-electron bonds. The carbon atoms have octets of electrons, but the oxygen atoms are missing 4 electrons, and we have 4 valence electrons still to distribute:

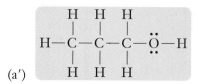 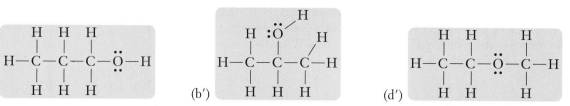

Molecular models of these three isomeric compounds are shown in Figure 5.1.

Does the answer make sense?   All the Lewis structures have 26 electrons, the usual number of atomic centers bonded to carbon and oxygen atoms, and an octet of electrons around each second-period atomic center. Since the C_2H_6O isomers are an alcohol and an ether, we would expect to find alcohol and ether molecules when a carbon and two hydrogen atoms (CH_2) are added to the C_2H_6O structure, and this is what we find. There are two isomeric alcohols because there are two different places to bond the oxygen atom to the carbon chain.

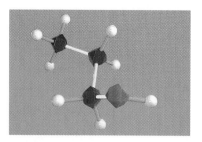 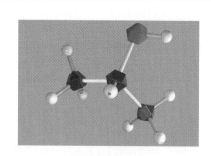 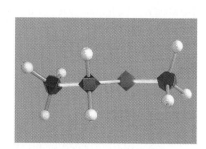

(a') 1-Propanol   (b') 2-Propanol   (d') Ethyl methyl ether

Figure 5.1.

Molecular models and structural names of the C_3H_8O isomers. Letter designations for the models are from Worked Example 5.6.

5.7 CHECK THIS

Lewis structures and molecular models for C_3H_9N isomers

Write Lewis structures and build the corresponding molecular models for as many C_3H_9N isomers as you can. Explain the procedure you use to find all the different connectivities for the Lewis structures. *Hint:* There are four isomers, three of which are analogous to the structures we found in Worked Example 5.6.

Worked Example 5.6 and Check This 5.7 demonstrate the importance of using systematic procedures for finding all the possible connectivities among the second- and higher-period elements when writing Lewis structures. In the carbon-containing molecules we have been using as examples, the number of possibilities grows rapidly as the number of carbons increases, because the carbon atoms do not have to be connected one after another, but can branch. One system for finding all the ways the carbon atoms in a set of isomers can be connected is illustrated in Worked Example 5.8.

5.8 WORKED EXAMPLE

Lewis structures and molecular models for C_5H_{12} isomers

Write Lewis structures and build the corresponding molecular models for as many C_5H_{12} isomers as you can.

Necessary information: We need to know that C and H atoms have, respectively, 4 and 1 valence electrons to be used in the Lewis structures and that a carbon atomic center forms two-electron bonds to four other atomic centers.

Strategy: The approach to this problem is the same as in Worked Example 5.6.

Implementation: The 5 C and 12 H atoms have a total of 32 valence electrons.

We begin by connecting the carbon atoms to form the longest possible chain:

(a) C—C—C—C—C

Then we remove one of the carbon atoms and add it to interior positions on the chain:

$$
\begin{array}{ccc}
& \text{C} & & & & \text{C} \\
& | & & & & | \\
\text{(b)} \ \text{C—C—C—C} & & & \text{(c)} \ \text{C—C—C—C}
\end{array}
$$

The longest carbon chain is highlighted in red. Structures (b) and (c) are identical; they can be interconverted by simply flipping one of them over. Thus, (b) and (c) represent a single structure: a four-carbon chain with another carbon atom bonded to one of the two identical interior positions.

continued

The longest continuous chain of carbon atoms can be "bent" as shown in structure (d). There may be more than one equivalent longest chain as in:

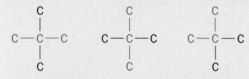

All three of these chains (and others you can find) are equivalent. Use models to prove this.

Next, we remove two carbon atoms from the longest chain and then add them first together and then separately to the interior positions on the chain:

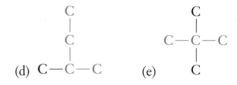

Structure (d) is identical to structures (b) and (c): a four-carbon chain (shown in red) with another carbon atom bonded to one of the two identical interior positions. Structure (e) is different from any of the others. Its longest carbon chain, three carbon atoms, is shown in red.

No further steps are possible to form different connections of the carbon atoms. Each structure uses all the carbons and eight of the valence electrons. Use the remaining 24 valence electrons to add the 12 hydrogen atomic cores with two-electron bonds:

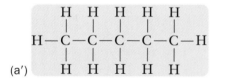

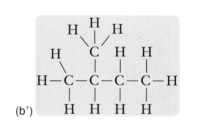

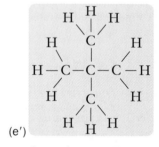

Molecular models of these three isomeric compounds are shown in Figure 5.2.

Does the answer make sense? All atom centers and valence electrons are accounted for in each Lewis structure. See Check This 5.9 for further consideration of these isomers.

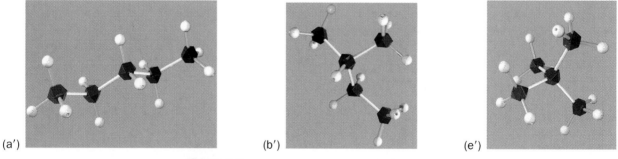

(a') (b') (e')

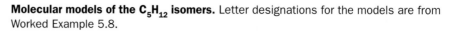

Figure 5.2.

Molecular models of the C_5H_{12} isomers. Letter designations for the models are from Worked Example 5.8.

5.9 CHECK THIS

More on the C_5H_{12} isomers

(a) Make three molecular models of structure (a), the five-carbon chain, in Worked Example 5.8. Detach a carbon atomic center and one bond stick from one end of one of the models and reattach it to make structure (b). Repeat with a second model to make structure (c). Detach two carbon atomic centers and bonds sticks from the remaining model of structure (a) and reattach them to make structure (d). Show that the three structures you have made are identical to one another, that is, that they can be superimposed on one another. This will prove that structures (b), (c), and (d) all represent the same isomer.

(b) The three Lewis structures written at the end of Worked Example 5.8 and the models in Figure 5.2 represent molecules of the compounds in Table 5.2. Which structure corresponds to which compound? Are the names a clue? Explain.

5.10 CONSIDER THIS

What are the Lewis structures and molecular models for the $C_4H_{10}O$ isomeric alcohols?

(a) Write the Lewis structures and make molecular models for the isomeric alcohols with the molecular formula, $C_4H_{10}O$.

(b) Assign the correct structural name in Table 5.1 to each of the molecular models you made in part (a). Explain how you make each assignment.

(c) Is one of the structures rather different from the others? If so, how is it different? Is the difference reflected in the properties of the compound? Explain.

(d) The molecular structure of vitamin A (retinol) is shown below. Your body converts it into the retinal molecule shown in the chapter opening illustration. Describe the similarity between the two molecules. Explain how you could use your results for the $C_4H_{10}O$ isomers to predict what would be observed if vitamin A is mixed with an acidic solution of dichromate ion, $Cr_2O_7^{2-}$, as in Investigate This 5.1.

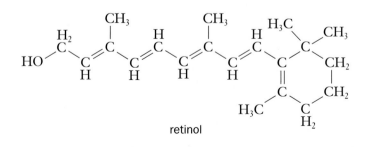

retinol

In both sets of isomers we looked at, the $C_4H_{10}O$ and C_5H_{12} compounds, one of the isomers was quite different from the others. For example, one isomer in each set melts at a much higher temperature and boils at a lower temperature than the others. And, in the case of the alcohols, this isomer did not undergo the same reaction as the others. The molecular models you made for the isomeric C_4H_8O alcohols and those shown in Figure 5.2 for the C_5H_{12} isomers show that one of the isomers in each set is quite compact, more-or-less spherical, compared to the others. On the basis of their structural names, you probably assigned these compact structures to the "odd" isomer in each set, 2-methyl-2-propanol and 2,2-dimethyl propane.

5.11 CHECK THIS

Boiling points of the C_4H_8O alcohols

Recall from Chapter 1 that alcohol molecules form hydrogen bonds to one another, which must be broken for them to vaporize. Also, recall that there are attractive dispersion forces between molecules that depend on less directed contacts with one another. The boiling point data in Table 5.1 show that 2-methyl-2-propanol vaporizes more easily than the other isomers. Do the structures of the molecules suggest an explanation for the lower boiling point of this isomer? Or, conversely, for the higher boiling points of the others? Why or why not?

Reflection and Projection

You have found that isomeric compounds have the same molecular formula but different chemical and physical properties. You have also found that there are several different ways to connect atomic cores as you write Lewis structures or make the molecular models for the molecules of isomeric compounds. The different molecular structures can give isomers very different observable properties, and you can often use the molecular structures to rationalize these differences.

So far, our arguments about structure have all rested on the molecular models we build from our model kit. How do we know these structures are valid? What is the basis for these models? In Chapter 1, Section 1.5, we suggested that four pairs of valence electrons stack in a tetrahedral array, like four balls, around a central positive atomic center, getting the negative electrons as close as possible to the positive nuclear charge, while remaining equidistant from one another. This arrangement is the basis for the tetrahedral array of holes drilled in the atomic centers that represent the second-period elements in your model kit. Why should a model based on stacking of balls around a central point be applicable to bonding in molecules? Applying the ideas from wave mechanics developed in the last chapter can help us answer this question. But keep in mind that the Lewis model, with no reference to electron waves, is a simple way to describe connectivity and you should continue to use it.

5.3. Sigma Molecular Orbitals

Molecular orbitals Descriptions of electron waves in molecules are governed by the same wave mechanical principles we discussed for atoms in Chapter 4. But molecular electron waves are necessarily more complicated. An atom, for example, has only *one* anchor point, its nucleus, for the electron waves. On the other hand, methane, a very simple molecule, has *five* nuclear centers to attract electrons (and to use as anchor points for standing waves). The standing electron waves that define regions of space where there is a high probability of finding electrons in molecules are referred to as **molecular orbitals,** paralleling the way atomic orbitals describe the location of an atom's electrons.

The increased complexity of molecules forces scientists to use approximations to calculate and describe simple pictures of molecular orbitals. For example, in the calculations, an electron wave can be limited to interaction with a pair of adjacent atomic cores, instead of with all the atomic cores in the molecule. Figure 5.3 compares the electrostatic interactions in the simplest atom, an H atom, and the simplest molecule (molecular ion), an H_2^+ ion. For an atom with one electron and one proton (atomic core), Figure 5.3(a) shows the single electrostatic interaction, *PE*, the potential energy of attraction between the proton and electron. Figure 5.3(b) shows the interactions among an electron and two protons (atomic cores). There are two attractive potential energies, PE_A and PE_B, one for each atomic core with the electron. *The negative potential energies of attraction of the two atomic cores and the electron are what hold the atomic cores together in the molecule.* This is the same electrostatic attractive potential energy that is responsible for the stability of the atom, but now there are two attractions, instead of one.

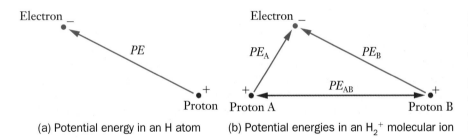

(a) Potential energy in an H atom   (b) Potential energies in an H_2^+ molecular ion

Figure 5.3.

Potential energies among atomic cores and an electron. Red single-headed arrows represent attractions between the positive cores and the electron. The double-headed blue arrow represents the repulsion between the positive cores in (b).

The total energy of the atom or the molecule is sum of these negative potential energies of attraction plus other positive contributions. There is a positive (destabilizing) contribution from the kinetic energy, *KE*, of the electron orbital; *KE* gets larger as the orbital gets smaller. The total energy for the atom is

$$E \text{ (H atom)} = PE + KE \tag{5.1}$$

The negative (stabilizing) contributions to the total energy are shaded in yellow and the positive (destabilizing) contributions are shaded in blue. In the molecule, there is also a positive contribution to the total energy from the potential energy of repulsion, PE_{AB}, between the positively charged atomic cores. The total energy for the molecule is

$$E \text{ (}H_2^+ \text{ molecular ion)} = PE_A + PE_B + PE_{AB} + KE \tag{5.2}$$

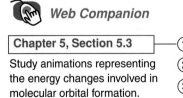

Web Companion

| Chapter 5, Section 5.3 |---①

Study animations representing
the energy changes involved in
molecular orbital formation.

②
③
④

5.12 CONSIDER THIS

How do the PEs and KE vary with the proton–proton distance?

 In the *Web Companion*, Chapter 5, Section 5.3.1, drag proton B toward proton A, as directed. Explain the effect this change in proton–proton distance has on
(a) The potential energy of the interaction between proton B and the electron.
(b) The potential energy of the interaction between proton A and the electron.
(c) The potential energy of the interaction between proton B and proton A.
(d) The kinetic energy of the electron cloud.

The potential, kinetic, and total energies for an atomic system were plotted in Chapter 4, Figure 4.27, as functions of the size of the atom. The corresponding plots for a molecule are shown in Figure 5.4 as functions of the distance between the atomic cores, the **bond length:** The shorter the bond length, the greater the attractive potential energies, PE_A and PE_B. The nearer the electron is to the atomic cores, the more favorable (more negative) are the attractive potential energies. But, also, the shorter the bond length becomes, the higher the atomic core repulsion, PE_{AB}. Figure 5.4 shows that the attractive potential predominates for this orbital. A shorter bond length increases the kinetic energy, KE, because the electron is being constrained in a smaller volume between the atomic cores. The minimum in the total energy, E, corresponds to bond formation at a bond length that gives the lowest total energy.

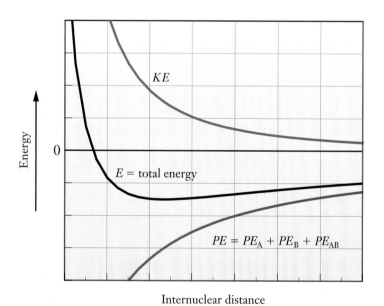

Figure 5.4.

Relative energies for a two-nuclei, one-electron molecular orbital system. As the nuclei get closer, the electron orbital is constrained to a smaller volume and its kinetic energy increases. Only relative energies and distances are shown; the plots are similar for all bonding molecular orbitals.

5.13 CHECK THIS

Potential and kinetic energy combinations

Choose three internuclear distances on the plots in Figure 5.4 and check whether equation (5.2) is obeyed in each case. What is the result?

Localized, one-electron σ (sigma) molecular orbitals

Solutions to the wave equation that account for the interactions illustrated in Figure 5.3(b) and by equation (5.2), are a set of **localized, one-electron molecular orbitals.** The lowest energy molecular orbital for the H_2^+ molecular ion is represented in Figure 5.5, together with the lowest energy orbital of the H atom for comparison. Recall from Chapter 4 that atomic s orbitals are spherically symmetric about the nucleus, as shown for the $1s$ orbital in Figure 5.5(a). The lowest energy molecular orbital shown in Figure 5.5(b) is cylindrically symmetric about the **bond axis,** the imaginary line joining the nuclei. That is, if you rotate the molecule about the bond axis, the molecular orbital shown in the figure looks the same, no matter how far the molecule is rotated. This cylindrical symmetry about the bond axis is analogous to the spherical symmetry of atomic s orbitals. Thus, σ (Greek lowercase sigma, corresponding to the Roman s) has been chosen to name these **σ (sigma) molecular orbitals.**

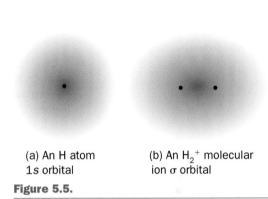

(a) An H atom
1s orbital

(b) An H_2^+ molecular
ion σ orbital

Figure 5.5.

Representation of a 1s atomic orbital and a σ molecular orbital. The size of the protons, dark dots, is greatly exaggerated.

5.14 CONSIDER THIS

What are the relative sizes of atomic and molecular orbitals?

(a) In Figure 5.5, the lowest energy σ molecular orbital for H_2^+ is shown as somewhat smaller than the H atom 1s orbital. Which orbital electron has the higher kinetic energy? Give your reasoning.

(b) How is it possible for the molecular orbital to be smaller than the atomic orbital? Clearly explain how the interactions among the protons and electron can lead to this result.

Sigma bonding orbitals **Bonding orbitals** concentrate electron density *between* atomic cores. They are called bonding orbitals because these electrons *always* contribute to lowering the energy of the molecule, thus making it more stable. Bonding molecular orbitals have lower energies than atomic orbitals in the separated atoms. The lower energy is a result of attractions between the electron and *two* atomic cores *that make the molecule more stable than its atoms.* Figure 5.6 shows this energy relationship between atoms and molecules. In **σ (sigma) bonding orbitals,** the electron density is concentrated directly between the atomic cores, as you see in Figure 5.5(b). If you imagine looking from one core (proton) toward the other, the electron orbital would block your view.

Recall the warning from Chapter 4 about the way chemists refer to electrons and orbitals. Electrons are said to "occupy" orbitals or to be "in" orbitals. We will use this shorthand at times. But keep in mind that *an orbital is a description of an electron wave and electron probability density;* it has no existence apart from the electron.

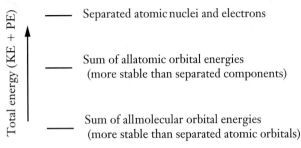

Total energy (KE + PE)

— Separated atomic nuclei and electrons

— Sum of all atomic orbital energies
(more stable than separated components)

— Sum of all molecular orbital energies
(more stable than separated atomic orbitals)

Figure 5.6.

Relative total energies for a molecule, its atoms, and their components.

This is the most favorable arrangement for coulombic attraction between the electron orbital and the nuclei. Therefore, *σ bonding orbitals form the strongest bonds and all molecules have σ bonding orbitals.*

In Figure 5.5, note that, like the proton in the H atom, the two protons in the H_2^+ molecular ion are "inside" the σ electron orbital. Though much of the electron density is between the protons, serving to bond them together, some of it surrounds the protons as well. The same is true for the second-period elements: The major portion of the sigma bonding orbital electron density is between the atomic cores. In Figure 5.7, to emphasize this point, very little of the electron density is shown "outside" the second-period atomic cores. The figure shows pictorial representations of sigma bonding molecular orbitals for second-period elements bonded to hydrogen (a proton), Figure 5.7(a), or another second-period element, Figure 5.7(b).

Figure 5.7.

Sigma bonding molecular orbitals involving second-period elements.
Second-period atomic cores are represented by a nucleus surrounded by its dense core electron orbital.

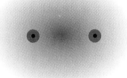

(a) σ bonding orbital between atomic cores of hydrogen and any second-period element

(b) σ bonding orbital between atomic cores of any two second-period elements

Each of the molecular orbitals pictured in Figures 5.5 and 5.7 can accommodate two electrons with opposite spins. These are the electron-pair bonds that we have been showing as a line between the elemental symbols in Lewis structures. Now we see that the line represents a roughly spherical region of electron density in the space between the atomic cores and that the attractions of the cores for the electrons in this space are what bonds the cores together. If there are two or more σ bonding orbitals attracted to a single atomic core, the exclusion principle acts, in effect, to keep the orbitals from occupying the same space. This is the basis for the simple model of connected balloons or stacked balls that we used in Chapter 1, Section 1.5, to explain molecular geometry. Using the electron wave model, we interpret the balls as sigma orbitals, Figure 5.8, and provide a justification for what was, previously, just a physical analogy.

Sigma nonbonding molecular orbitals Sigma bonding orbitals explain the lines we use for bonds in Lewis structures, but they do not explain the electrons that we show as pairs of dots associated with only one elemental symbol, for example, oxygen and nitrogen in Section 5.2. These electrons are in orbitals that are much like atomic orbitals, because they are mainly attracted by a single atomic core. Because these electrons are not between atomic cores, they do not contribute to holding atoms together, but they don't weaken the bonds either, so they are called **nonbonding electrons**. However, these electrons are in molecules and are, in principle, affected by all the atomic cores, so their orbitals

Atomic core

Figure 5.8.

Tetrahedral arrangement of four sigma orbitals around an atomic core.

are called **sigma nonbonding molecular orbitals,** and are symbolized as σ_n. In Figure 5.9, representations of a σ and a σ_n orbital are shown for comparison on a second-period atomic core. In the next section, we will begin to see how sigma orbitals determine the shape of molecules.

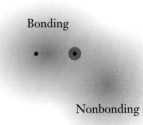

5.15 CONSIDER THIS

What is the symmetry of a σ_n orbital?

Sigma molecular orbitals have cylindrical symmetry. How would you define the axis of symmetry for a σ_n orbital? Explain your reasoning.

Figure 5.9.

Sigma bonding and nonbonding molecular orbitals on the same atomic core.

5.4. Sigma Molecular Orbitals and Molecular Geometry

5.16 CONSIDER THIS

What are the geometries of second-period hydrides?

(a) Write Lewis structures for the hydrides of carbon (CH_4), nitrogen (NH_3), and oxygen (H_2O).
(b) Make molecular models of the hydrides in part (a). Use the paddles in your model kit to represent any nonbonding electrons in the Lewis structures. How would you describe the geometry (shape) of the electron pairs around the second-period atom center in each of your models? Explain your responses.
(c) How would you describe the geometry of the atom centers with respect to one another in the model of each molecule? Explain your responses.
(d) For each hydride, are your answers to parts (b) and (c) the same? Explain why or why not.

The sigma molecular framework The distribution of electrons in molecules is governed by the same interactions we discussed for atoms. Electrons will occupy as many as possible of the lowest energy molecular orbitals; these provide the strongest attractions between the atomic cores and electrons that hold the molecule together. This means that as many σ bonding orbitals as possible are formed, each containing two spin-paired electrons. As a consequence, in the vast majority of molecules, *there is a spin-paired, two-electron, σ bonding orbital between every pair of bonded atomic cores.* This is usually referred to as a molecule's **σ framework** or **σ bonding framework.** Although other orbitals are often present as well, the σ framework is responsible for a large fraction of the stability of a molecule, which is represented by the molecular energy level in Figure 5.6. Our task now is to learn how these orbitals are related to the geometry of the molecule.

There are several possible combinations of σ and σ_n molecular orbitals around the second-period elemental atoms in molecules that interest us, but they all have one characteristic: ***The number of bonding orbitals plus the number of nonbonding orbitals always adds up to four for each second-period element (carbon through fluorine) in the molecule.*** Each of these is a two-electron orbital. Once again, you can see the connection to Lewis structures in which you write structures with four electron pairs around each second-period element. The tetrahedral geometry shown in Figure 5.8 is best for getting four electron orbitals grouped as close as possible to the positive nucleus.

Web Companion

Chapter 5, Section 5.4 — ①

Study animations of different ②
molecular shapes arising from ③
the same orbital arrangement. ④

Molecular shapes The shape or geometry of a molecule is defined by the spatial relationship of the atomic nuclei to one another. The distinction between *molecular shape* and *arrangement of orbitals* is important. The shape of a molecule is often different from the geometry of the orbitals. The examples in Figure 5.10 are a reminder of this important point. Here we have three different molecular shapes, tetrahedral, **trigonal pyramidal** (a flattened triangular pyramid), and bent, all arising from the same tetrahedral arrangement of orbitals. Study the figure carefully to learn the relationships that lead to these results.

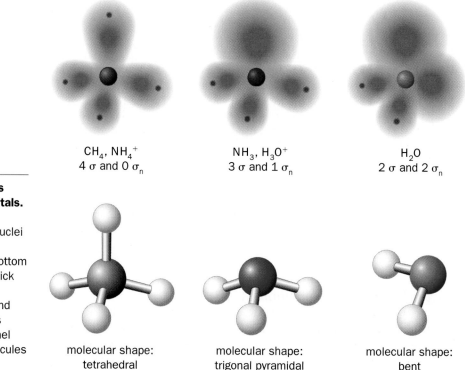

CH_4, NH_4^+
4 σ and 0 σ_n

NH_3, H_3O^+
3 σ and 1 σ_n

H_2O
2 σ and 2 σ_n

molecular shape:
tetrahedral

molecular shape:
trigonal pyramidal

molecular shape:
bent

Figure 5.10.

Molecular shapes of molecules with four sigma molecular orbitals. The top representation in each panel shows the orbitals and nuclei in perspective. The dots in the orbitals are the H nuclei. The bottom representation uses ball-and-stick models to show the molecular shape. The combination of σ and σ_n orbitals for each structure is given in the middle of each panel together with examples of molecules with each geometry.

Bond angles and bond lengths The molecular shapes in Figure 5.10 can also be characterized by their **bond angles,** the angles between imaginary lines connecting the nuclei, and **bond lengths,** the lengths of these lines. You can think of the sticks connecting the atomic centers in a ball-and-stick molecular model as representing these imaginary lines. If the orbitals retained their tetrahedral geometry in all three molecules, you would expect the H—C—H, H—N—H, and H—O—H bond angles all to be 109.5°. Experimentally, the bond angles in CH_4, NH_3, and H_2O are found to be 109.5°, 107°, and 104.5°, respectively. The agreement of the experiment with the expectation is quite good. The decreasing angle, from CH_4 to H_2O, can be rationalized as an effect of increasing number of σ_n nonbonding orbitals. Close to the second-period element, the σ_n nonbonding orbitals take up larger volumes of space than σ bonding orbitals. Bonding electrons are attracted by two positive centers, and nonbonding electrons by one. The attraction of the bonding electrons to two centers results in a more elongated orbital less concentrated at the second-period element. The regular tetrahedral arrangement of equal-sized σ bonding orbitals will be distorted when both σ bonding and σ_n nonbonding orbitals are present. The effect of the distortion is to make the angle(s) between atoms bonded to the central atom more acute, that is, smaller than 109.5°, as observed.

The bond lengths in CH_4, NH_3, and H_2O molecules are 109, 101, and 94 pm, respectively. The decreasing lengths reflect the increasing attraction, hence smaller size, for the σ orbitals on the 4+, 5+, and 6+ atomic cores of C, N, and O. A hydrogen atomic core, a proton, within a σ bonding orbital gets closer to the second row atomic core as the size of the orbital shrinks.

5.17 CHECK THIS

A fourth combination of sigma orbitals

(a) A fourth possible combination of sigma molecular orbitals is missing from Figure 5.10. An example of a molecule with this combination of orbitals is hydrogen fluoride, HF. Write the Lewis structure for HF. What is the combination of σ and σ_n orbitals for the fluorine atom? What is the shape of the molecule? What would you predict for the H—F bond length? Explain the basis for your prediction. Draw orbital structures for HF like those in Figure 5.10. Could you have predicted the shape without any reference to orbitals? Give your reasoning.

(b) 🎞 Look at the movies in the *Web Companion*, Chapter 5, Section 5.4.2. Explain clearly why HF, H_2O, and NH_3 have different molecular shapes, even though they all have four sigma molecular orbitals.

Molecules containing third-period elements Recall from Figure 3.27 in Chapter 3 that the most abundant elements in living organisms are hydrogen, oxygen, carbon, and nitrogen. The geometry of many molecules in organisms is based on the tetrahedral arrangement of sigma molecular orbitals around second-period elements that we have just discussed. Figure 3.27 also shows, however, that several other elements, including third-period elements, are present in living systems. And, if we consider the vast amount of nonliving matter on Earth, silicon, a third-period element, is the second most abundant element, after oxygen, in the crust. Silicon is an important part of the structure of many minerals and manufactured products, including all the glass we see around us. Compounds containing, phosphorus, sulfur, and chlorine are also found in minerals, in the sea, and in all organisms. Here we will examine a few simple compounds of phosphorus and sulfur to see how they are the same and/or different from the compounds of elements in the second period.

5.18 WORKED EXAMPLE

The structures of PF₃ and PF₅

Write Lewis structures and construct molecular models for a molecule of PF_3 and of PF_5.

Necessary information: We need the number of valence electrons in P and F atoms, 5 and 7, respectively. We also need to know that third- and higher-period elements are not limited to an octet of electrons around the atomic core.

Strategy: Our approach is the same as in Worked Examples 5.6 and 5.8, except that we are not limited to only four pairs of electrons around the phosphorus atomic core.

Implementation: PF_3 has 26 valence electrons and PF_5 has 40 valence electrons.

First, we connect the F atomic cores to the P atomic core with two-electron bonds:

Next, we give each of the second-period elements, the Fs, an octet of electrons:

The PF_5 structure has a total of 40 electrons in σ and σ_n orbitals, so the structure is complete as shown. The PF_3 structure has a total of only 24 electrons in σ and σ_n orbitals; the remaining two electrons can complete an octet in a σ_n orbital on the P:

Molecular models of the two structures are shown in Figure 5.11. The PF_3 structure is analogous to the NH_3 structure shown in Figure 5.10; the molecular geometry is trigonal pyramidal. The PF_5 structure is **trigonal bipyramidal.** Imaginary lines connecting the Fs form two three-sided pyramids joined at their bases.

Does the answer make sense? All valence electrons are accounted for in both structures. In the case of PF_5, this gives phosphorus five electron-pair σ-bonding orbitals. The next paragraph and Figure 5.12 examine further the geometry of these five bonds.

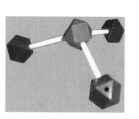

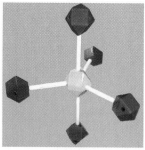

(a)   (b)

Figure 5.11.

Molecular models of (a) PF₃ and (b) PF₅.

5.19 CHECK THIS

The structure of SF₆

Sulfur hexafluoride, SF_6, is quite an unreactive gas. Write a Lewis structure for SF_6 and construct a molecular model that distributes the F atoms as symmetrically as possible around the S atom.

Stacking four balls around a central point in an arrangement that gets them all as close as possible to the point results in the tetrahedral array shown in Figure 5.8, and an explanation for the geometry around atom cores surrounded by a total of four σ and σ_n orbitals. If we examine the arrangements of five and six balls around a central point, such that all the balls are as close as possible to the point, we find the arrays shown in Figure 5.12. For five balls, three are in a triangular array with the remaining two above and below the plane of the triangle and nestling in the center of the triangle. Imaginary lines joining the centers of the five balls form a trigonal bipyramid, shown by the red lines in Figure 5.12(a). This trigonal bipyramidal array represents the five σ bonding orbitals in PF_5, Figure 5.11(b). The distances from the central point to the centers of each ball in the trigonal bipyramid are not the same; the two balls above and below the triangular plane are farther from the central point.

When we stack six balls, four are in a square arrangement with the remaining two above and below the plane of the square and nestling in the center of the square. This is an **octahedral** array. Imaginary lines joining the centers of the six balls form a regular octahedron. The distance from the central point to the center of each ball in an octahedral array is the same; all six positions (balls) around the central point are equivalent.

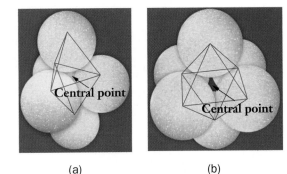

(a) (b)

Figure 5.12.

(a) Trigonal bipyramidal and (b) octahedral arrays of balls around a central point. The red structures represent lines connecting the centers of the balls.

5.20 CONSIDER THIS

How are all six octahedral positions equivalent?

(a) Use six Styrofoam® balls of the same size to construct an octahedral array of balls. Use toothpicks to hold four of the balls together in a square and then attach the other two to opposite faces of the square. Explain to other members of your group how all six balls are in equivalent positions in this arrangement. *Hint:* Close your eyes. Have a group member rotate the model around one of its axes. Open your eyes. Can you tell whether or not the model has been rotated?

(b) Is your octahedral array of balls the same as the arrangement of σ bonding orbitals in the SF_6 model you made for Check This 5.19? Should it be? Explain why or why not.

Why are there only four sigma orbitals on second-period atoms?

More than four electron-pair sigma orbitals can be accommodated around third- and higher-period elements. Why are there no molecules with more than four sigma orbitals around second-period elements? The answer involves the energies of the interactions among the atomic cores and the sigma orbital electrons. Second-period atoms have tiny cores—the nucleus and only two core electrons. When more than four two-electron sigma orbitals are stacked around this tiny core, their size prevents them all from getting as close as the four orbitals in a tetrahedral array. Thus, the potential energy of attraction between the core and each orbital is reduced. In order to get closer, the orbitals would have to become smaller, but this would increase their kinetic energy, which would further destabilize the molecule. No stable molecule is formed with more than four pairs of electrons around a second-period element.

Third-period atoms have a core that is a good deal larger—the nucleus and 10 core electrons. Five or six two-electron sigma orbitals can get essentially as close to this larger core as can four orbitals, so there is little or no loss of attraction for the larger number of electrons. The electrons are, of course, all repelling one another, so there is a point beyond which no more can be packed around a core of a given size and charge.

5.21 CHECK THIS

Sigma orbitals around third- and higher-period atoms in molecules

Fluorine forms compounds with almost every other element. The other halogens react with fluorine to form compounds containing various numbers of fluorine atoms. The most highly fluorinated compounds of each of the halogens are ClF_3, BrF_5, and IF_7. Write Lewis structures for these molecules. How many electron pairs surround each of the central atoms? Explain why Cl and Br do not react to form compounds with more F atoms.

Reflection and Projection

A summary of the main points of Sections 5.3 and 5.4 can help put the ideas in perspective.

- *A molecule is held together by the attractions between atomic cores and valence electrons.* Although repulsion between nuclei and the kinetic energy of the electron waves have to be accounted for, the attractive contributions to the potential energy make the total energy of the molecule lower than the energy of its separated atoms.
- *All molecules have sigma bonding orbitals.* There is a two-electron, σ bonding orbital between every pair of bonded atomic cores in a molecule.
- *Many molecules also have two-electron, sigma nonbonding orbitals.* Nonbonding electrons do not contribute to holding the molecule together, but σ_n orbitals, along with the σ orbitals, determine the three-dimensional geometry of molecules.

- Although the arrangement of sigma orbitals determines the geometry or shape of a molecule, *the descriptions of molecular geometry are based on the positions of the nuclei with respect to one another, not the arrangements of electron orbitals.*

These ideas, especially those that connect the stability and geometry of molecules with the simplest molecular orbitals, the sigma orbitals, are powerful. They enable you to make predictions about geometries and to understand the basis for the structures you have made with your molecular models. However, just as *s* orbitals are not the only atomic orbitals, sigma orbitals, important as they are, are not the whole story of molecular bonding. It should not be surprising that other, more complicated bonding arrangements and electron waves are also found in molecules. These are the topics of the next sections.

5.5. Multiple Bonds

5.22 INVESTIGATE THIS

Do carbon–hydrogen compounds react with permanganate?

Do this as a class investigation and work in small groups to discuss and analyze the results. Add *one drop* of each of the following three liquids to 2 mL of 95% ethanol in separate small test tubes: hexane (C_6H_{14}), hexene (C_6H_{12}), and turpentine [mostly pinene ($C_{10}H_{16}$)]. To each test tube add *one drop* of 0.1 M potassium permanganate, $KMnO_4$, solution. Observe and record the appearance of the mixtures. Discuss your results to be sure your group agrees on the observations.

5.23 CONSIDER THIS

What compounds react with permanganate?

(a) The permanganate ion, MnO_4^-, is reddish purple; manganese dioxide, MnO_2, is a dark brown/black solid. Was there any evidence that permanganate reacted with the samples in Investigate This 5.22? If so, what was the evidence?

(b) Did all the samples react? If not, which ones reacted and which did not? Are there any similarities among those that reacted? Explain your response.

5.24 WORKED EXAMPLE

The Lewis structure and molecular model for ethene

Write the Lewis structure and make a molecular model of ethene, C_2H_4 or CH_2CH_2.

continued

Necessary information: We need the number of valence electrons in C and H atoms, 4 and 1, respectively, and the rule that second-period elements have an octet, four pairs, of electrons in Lewis structures.

Strategy: The approach is the same as in Worked Examples 5.6 and 5.8.

Implementation: Ethene, C_2H_4, has 12 valence electrons.

First we will bond the carbons with a two-electron bond and then bond the four hydrogen atoms, two to each carbon (as indicated by the structural formula in the problem statement), to get

$$H-C-C-H$$
$$\;\;\;\;\;|\;\;\;|$$
$$\;\;\;\;\;H\;\;H$$

Ten of the valence electrons are used to form this σ bonding framework. The question we face is how to give each of the Cs an octet of electrons, when there are only two electrons left and there are two Cs. The answer to this dilemma is to write the remaining two electrons between the two Cs, so that each has an octet of electrons (circled in red):

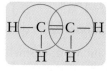

This is a correct Lewis structure. The octet rule does not forbid the sharing of more than one pair of electrons between atomic cores. Figure 5.13 shows how to use the bent bonds in your molecular model set to represent the structure of ethene.

Does the answer make sense? All the valence electrons are accounted for and all the second-period elements have an octet of electrons. We need to investigate the bonding model further to find out if this Lewis structure and the molecular model make sense.

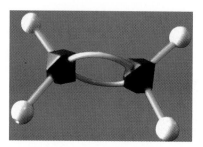

Figure 5.13.

Molecular model of the double bond in ethene using bent bonds.

5.25 CONSIDER THIS

What are the Lewis structure and molecular model for methanal (formaldehyde)?

(a) Write the Lewis structure for methanal (formaldehyde), H_2CO. How is your Lewis structure different from that for ethene in Worked Example 5.24? How is it the same?

(b) Make a molecular model of methanal. If necessary, pattern your model after the one in Figure 5.13. Explain the reasoning you use to build your model.

Double bonds and molecular properties Both the Lewis structure for ethene in Worked Example 5.24 and the one you wrote for methanal in Consider This 5.25 have two electron pairs—two lines—between the second-period atomic cores. Because there are two pairs of bonding electrons between the atoms, this is called a **double bond.** For both compounds, you used the curved connectors in your molecular model set to represent the double bond as a pair of **bent bonds,** which are required to make two connections between tetrahedral atom centers. This is the simplest way to construct a double bond with your molecular models. However, before we use this model further, we need to know if it represents the observed properties of double bonds.

5.26 CONSIDER THIS

What are the structures and reactivity of C_6H_{14} and C_6H_{12}?

(a) Write Lewis structures for C_6H_{14} and C_6H_{12} molecules that have the carbons bonded in a six-carbon chain. Is there more than one way to write a Lewis structure for either of these molecules? If so, what are the possible structures? Do your structures have any common feature(s)? What is(are) it(they)?

(b) Make molecular models corresponding to your Lewis structures in part (a). What is(are) the difference(s) among your structures? Can you correlate the molecular difference(s) with the reactivity of the C_6H_{14} and C_6H_{12} compounds in Investigate This 5.22? Explain why or why not.

(c) From your results in the investigation, what might you conclude about the structure of pinene? Explain the reasoning for your response.

(d) The structure of vitamin A (retinol) is shown in Consider This 5.10. Predict what you would observe if vitamin A were mixed with permanganate ion solution, as in Investigate This 5.22. Explain the reasoning for your prediction.

In Investigate This 5.22, you found that compounds with double bonds react with the clear, purple permanganate solution to produce a muddy brownish mixture, whereas those without double bonds (including ethanol) did not react. The representation of the double bond in Figure 5.13 (and the ones you made in Consider This 5.25 and 5.26) suggests that the electron density in a double bond is not all concentrated directly between the atomic cores, but lies outside the bond axis. This implies that the electrons in the double bond are on the "outside" of the molecule and not as tightly held as sigma electrons, so they should be better able to interact with other molecules. We predict that compounds with double bonds will be more reactive than similar compounds without double bonds, as you observed in Investigate This 5.22.

5.27 CONSIDER THIS

What is the geometry of double bonded molecules?

(a) Examine molecular models of ethene and methanal. How would you describe the geometry of the atomic centers with respect to one another? Explain your reasoning.

continued

(b) In your models, what are the H—C—H bond angles?

(c) The experimental C—C bond length in ethane, with a single *s* bonding orbital, is 154 pm. The C—C bond length in ethene, a double bond, is 133 pm. Make a molecular model of ethane to compare with your ethene model. Are the bond lengths of the models consistent with the experimental data? Propose an explanation for the shorter C—C double bond length.

(d) The experimental C—O bond length in alcohols and ethers, single bonds, is a little over 140 pm. Why is the C—O single bond shorter than the C—C single bond? The experimental C—O bond length in methanal is about 120 pm. Is this what you might expect? Explain.

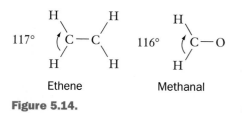

Ethene   Methanal

Figure 5.14.

All the atom centers in ethene and methanal lie in a plane. The plane of the paper is the molecular plane in these structures. Only the sigma bonds are represented.

Experiments show that the two carbon atoms and four hydrogen atoms bonded to them in ethene all lie in the same plane, that is, the doubly bonded atoms and the atoms bonded to them have **planar geometry.** You can see this planar geometry in Figure 5.13; it is the geometry you found in Consider This 5.27 for both ethene and methanal, and is represented again in Figure 5.14. The experimentally observed H—C—H bond angles in ethene and methanal are 117° and 116°, respectively. The models you examined in Consider This 5.27 were made with tetrahedral carbons, so they have H—C—H angles of 109.5°. The agreement between the observed angles and our model is not perfect, but the model is not far off.

5.28 CHECK THIS

Other bond angles in ethene and methanal

(a) What is the experimental H—C—C bond angle in ethene? The experimental H—C—O bond angle in methanal? Explain how you arrive at your answers.

(b) What is the H—C—C bond angle in your ethene model? The H—C—O bond angle in your methanal model? How do these angles compare with the experimental angles from part (a)? Explain how you arrive at your answers.

Triple bonds You have seen that second-period atoms can share two pairs of electrons to form double bonds. Can they share a third pair as well?

5.29 WORKED EXAMPLE

Lewis structure for the hydrogen cyanide molecule

Write the Lewis structure for the hydrogen cyanide molecule, HCN. Make a molecular model representing this structure and predict the geometry of the molecule.

continued

Necessary information: We need to know that H, C, and N atoms have 1, 4, and 5 valence electrons, respectively, and that second-period elements have an octet of electrons.

Strategy: The approach is the same as in Worked Examples 5.6 and 5.8.

Implementation: The HCN molecule has 10 valence electrons.

First we connect the atomic cores with two-electron bonds: H—C—N. There are six electrons left to distribute to give each second-period atom an octet and six are needed by the N alone, which would leave the C with only four electrons. If the C and N share four of the six electrons (two more bonding pairs) and the leftover pair is assigned to the N, then both C and N will have an octet: H—C≡N:.

Figure 5.15 shows how to represent the structure of HCN using the bent bonds in your molecular models. The model, with three bent bonds, forces the three atoms to lie on a line, so we predict that the molecule is linear.

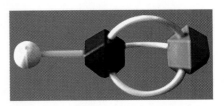

Figure 5.15.

Molecular model of the triple bond in hydrogen cyanide using bent bonds.

Does the answer make sense? The second-period atoms in our structure have octets of electrons with three pairs of bonding electrons shared between the C and N. The N also has one unshared (nonbonding) pair of electrons, which is typical for N in the molecules we have seen previously. Experimentally, the HCN molecule is found to be linear; our shape prediction is correct.

5.30 CHECK THIS

Lewis structure for the ethyne (acetylene) molecule

(a) Write the Lewis structure for the ethyne (acetylene), HCCH, molecule. Make a molecular model representing this structure and predict the geometry of the molecule.

(b) Compare your molecular model of ethyne with those for ethane and ethene, Consider This 5.27. The experimental C—C bond length in ethyne, a triple bond, is 120 pm. Is this value consistent with the single and double bond lengths and with your models? Explain. Does your explanation for bond lengths, proposed in Consider This 5.27(c), work for the triple bond as well? Explain why or why not.

The Lewis structures and molecular models of hydrogen cyanide and ethyne have three bonding electron pairs between the second-period atoms. In Figure 5.15, we have interpreted these as three bent bonds, a **triple bond.** You can see that the electron density in the triple bond is not concentrated directly between the atomic cores, so it should be available to interact with other molecules, just as in double-bonded molecules. Experimentally, we find that compounds whose molecules contain triple bonds between carbons are even more reactive than corresponding doubly bonded compounds.

Multiple bonds in higher period atoms Multiple bonds between third- and higher-period atoms are less common than between second-period atoms. This is probably because the molecular orbitals in molecules with higher-period atoms are larger and, in multiple bonds, more available to react than those in analogous second-period molecules. Reactions to form products with more σ bonding orbitals are favored and these are the compounds usually observed.

5.6. Pi Molecular Orbitals

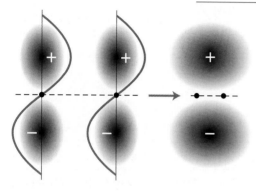

Separated atomic *p* orbitals   π Molecular orbital

Figure 5.16.

Electron wave interference (reinforcement) to form a pi bonding orbital. Reinforcement of electron wave amplitude between the atomic cores (black dots) for waves with matching + and − amplitudes. The dashed lines represent the nodal planes of the orbitals.

The bent bonds shown in Figure 5.13 and 5.15 are consistent with the reactivity and geometry of double and triple bonds and are a convenience that enables us to use tetrahedral atom centers to build models of molecules with multiple bonds. However, when we consider the wave mechanical description of the double (or triple) bond, we get a somewhat different picture. One way to visualize the wave mechanical formation of the second bond between two second-period elemental atoms is to combine electron waves from the two atoms. Recall the wave interference (diffraction) patterns you saw in Chapter 4, Section 4.3, and the explanation in Figure 4.10. When two wave crests coincide, the waves reinforce one another. Figure 5.16 represents an application of this idea to electron waves for two second-period atomic cores.

The second-period atomic cores in Figure 5.16 are bonded by a sigma orbital that is not shown in the figure. A second bonding orbital, a **π (pi) bonding molecular orbital,** is formed when the crests and valleys of electron waves in *p* orbitals on each atom reinforce one another between the atomic nuclei. (The Greek lowercase pi corresponds to the Roman *p*.) The electron density in the π orbital is not concentrated directly between the nuclei, but is parallel to the bond in two sausage-shaped lobes that are somewhat outside the bond axis. The bond axis lies in the **nodal plane** of the π orbital, the plane where the electron density goes to zero.

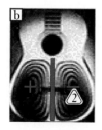

Web Companion

Chapter 5, Section 5.6 ① ② ③ ④

Try animations and exercises to help get a three-dimensional feel for sigma-pi geometries.

Pi orbital standing wave A possible confusion about a pi orbital arises because it seems to be in two separate regions of space. The two regions are parts of a whole orbital, that is, a single standing wave describing one or two electrons (depending on how many electrons are in the orbital). As an analogy, look at one of the standing waves on the acoustic body of a guitar, Figure 5.17, which repeats Figure 4.35(b). When one side of the body face is moving out toward you, the other side is moving away from you. The movements are connected. One side can't move out unless the other is moving in. This is a single wave, even though there is a node down the center where the face doesn't move at all. Similarly, the π-electron wave is a single wave. The signs shown in Figure 5.16 are analogous to the out and in movements of the guitar face.

Figure 5.17.

A standing wave on the acoustic body of a guitar.

Sigma-pi molecular geometry with one pi orbital Look again at the
first structure written in Worked Example 5.24: $H-C-C-H$.
Each carbon is surrounded by three σ bonding
electron pairs. Figure 5.18 shows how we would
represent these three electron-pair orbitals as a
trigonal planar arrangement of closely packed
balls around a central point (an atomic core).
Imaginary lines joining the centers of the three
balls form an equilateral triangle. Figure 5.19(a)
shows both of the carbons in ethene and the
associated five σ orbitals with the carbon cores
sharing one of them. This represents the sigma
framework for a molecule of ethene.

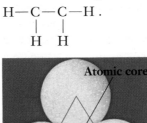

Figure 5.18.

**Trigonal planar arrangement of
three sigma orbitals around an
atomic core.**

Figure 5.19.

Molecular orbitals in ethene.
The six atomic cores define a
molecular plane (see Figure 5.14).
The σ framework is cut by this
plane. The localized π orbital lies
above and below the molecular
plane and is cut by a plane
perpendicular to the molecular plane.

(a) σ Framework   (b) Localized π orbital   (c) Orbital combination

The second bond between the carbons in ethene, the pi molecular orbital,
is a little more difficult to represent. Figure 5.19 shows how the sigma bonding
orbital framework and the pi bonding orbital combine to give the orbital
arrangement for ethene: three σ orbitals in a trigonal plane surrounding each
carbon and a π orbital in a plane perpendicular to the molecular plane and
parallel to the C—C bond axis. The planar arrangement of the atoms results
from the planar arrangement of the sigma orbitals. The reason for this planar
arrangement of σ orbitals is that this geometry gives the strongest pi bonding,
that is, the best reinforcement of the electron waves that form the pi molecular
orbital and therefore the lowest energy for the molecule. The orbital represented
in Figure 5.19(b) is called a **localized π molecular orbital,** because it is associated with (localized between) two sigma-bonded atoms.

5.31 CHECK THIS

Representations of the bonding in ethene

Load the movie and exercise "Molecular representations of ethene," in the
Web Companion, Chapter 5, Section 5.6.1. Place the cursor in the middle of
the image in the movie. Move the cursor up and down. Describe what you see.
What information can you gather from each of the different representations?
Which representation do you find most useful? Explain why.

The structure of ethene represented in Figure 5.19(c) is consistent with the properties of compounds containing double bonds that we discussed earlier. The pi bonding electrons are not concentrated directly between the carbon atomic cores and are, therefore, more accessible for reaction, thus making these compounds more reactive, as you have found. For sigma orbitals in the trigonal planar arrangement shown in Figure 5.18, the predicted H—C—H bond angle is 120°, which is close to the experimental value of 117°.

The combination of σ- and π-bonding orbitals helps to explain the shorter C—C bond length in ethene compared to the single bond in ethane. The two electrons in the σ orbital in ethene are directly between the carbon cores, just like the σ orbital in ethane. In addition, the two electrons in the π orbital, although not directly between the atomic cores, contribute a substantial extra attraction. The upshot is that the carbon atomic cores are held more closely together in ethene.

5.32 CHECK THIS

The sigma-pi representation for methanal

Explain how you would modify the illustrations in Figure 5.19 so that they represent the sigma and pi molecular orbitals in methanal. Make a sketch of your modifications.

Figure 5.20 shows another way to model the ethene structure with your molecular models. For this model, you use the dark gray, trigonal atom centers for the carbons, straight connectors for the five sigma bonding electron pairs, and four paddles, two on each carbon center, that altogether represent the pi bonding electron pair. Colored tape connects the paddles to highlight the two lobes of the pi orbital. This model is closer to the conventional representation of the sigma-pi bonding in ethene, as in Figure 5.19(c), than the model with bent bonds in Figure 5.13. However, in most cases, we will use bent bonds for models of double bonds, because they are easier to make and give much the same information about the molecular properties.

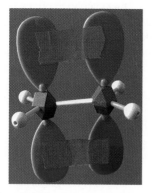

Figure 5.20.

Molecular model of ethene showing the sigma framework and pi orbital.

5.33 CHECK THIS

The sigma framework and pi molecular orbital in methanal

(a) Use your molecular models to make a model of methanal patterned after the one for ethene in Figure 5.20. How is your model similar to that for ethene? How is it different? Explain.

(b) Did you represent the sigma nonbonding electron pairs in your methanal model? If so, how? If not, can you think of a way to do so? Compare your solution to the ones others chose.

Triple bonds: Sigma-pi geometry with two pi orbitals To complete the comparison of triple bonds with double bonds, we also need to consider an interpretation in terms of sigma and pi bonding. The incomplete Lewis structure we started with in Worked Example 5.29, H—C—N, shows that the central atom, the carbon, has two sigma bonding orbitals. Figure 5.21 shows how we would represent these two electron-pair orbitals as a linear arrangement of closely packed balls on opposite sides of a central point (an atomic core).

Once again, we can consider reinforcing combinations of electron waves from the atoms creating π bonding orbitals. The central atom has two sigma orbitals and can form two pi bonding orbitals. In the sigma-pi representation, a triple bond between two atoms consists of a σ bonding orbital and two π bonding orbitals. Figure 5.22 shows the sigma-pi bonding in HCN. Sigma bonds are represented by the sticks in the ball-and-stick structure that shows how the atomic cores lie on a line. The two π orbitals are localized between the C and N atomic cores and lie in planes that are perpendicular to each other. Figure 5.22 displays one π orbital as light gray and the other π orbital as a darker gray to make it easier to see them both.

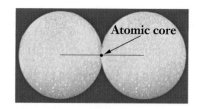

Figure 5.21.

Linear arrangement of two sigma orbitals and an atomic core.

5.34 CHECK THIS

The sigma-pi representation for ethyne

(a) Explain how you would modify the illustration in Figure 5.22 so that it represents the sigma and pi molecular orbitals in ethyne, HCCH. Make a sketch of your modifications.
(b) Explain why the C—C triple bond is shorter than the C—C double bond.

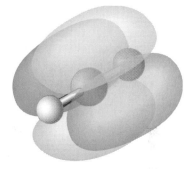

Figure 5.22.

Sigma-pi representation of the triple bond in hydrogen cyanide. The π molecular orbitals on HCN are localized between the C and N.

Reflection and Projection

In molecules, atoms of second-period elements can share one, two, or three pairs of bonding electrons to form single, double, and triple bonds. We have shown two ways to represent multiple bonds: as bent bonds between the atoms or as a σ bonding orbital plus one or two π bonding orbitals. In both representations, multiple bonds are characterized as having a good deal of electron density that is away from the bond axis and not directly between the bonded atomic cores. These electrons are readily available for interactions with other molecules, so compounds with multiple bonds are more reactive than similar compounds without multiple bonds.

Both representations of multiple bonds also explain and predict the correct geometry of molecules with multiple bonds. The generalization we made earlier in Section 5.4 is still valid for molecules with multiple bonds: *The shape of a molecule is determined by its framework of σ and σ_n bonding orbitals.* Table 5.3 summarizes the geometries we have discussed so far for molecules composed of second-period elements (and hydrogen). Often, we would like to draw these structures in a way that represents their shapes as accurately as possible. In order to do so, we need to introduce the methods chemists use to show three-dimensional structures on a two-dimensional sheet of paper. Before doing that, however, we will look at another important characteristic of molecular orbitals: the fact that they are not always localized between two atomic cores. In the next section, we will extend our discussion of pi orbitals to include molecules that cannot be described by a single Lewis structure, but require introduction of molecular orbitals that extend over three or more atoms. We will finish the section with a brief discussion of metallic bonding, in which electron orbitals extend through the entire crystal.

Table 5.3 *Combinations of localized molecular orbitals around second-period elements.*

The σ orbital geometry refers to the bold, blue elemental atom in the examples.

Number of σ and σ_n orbitals	σ Orbital geometry	Number of π orbitals	π Orbital arrangement	Examples
4	tetrahedral 109.5° angles	0	—	**CH**$_4$, **NH**$_4^+$, **NH**$_3$, **H**$_2$**O**, **H**$_3$**O**$^+$ See Figure 5.10
3	trigonal planar 120° angles	1	above and below σ framework; parallel to σ bond axis	**H**$_2$**CCH**$_2$, **H**$_2$**CO**
2	linear 180° angles	2	perpendicular to each other; parallel to σ bond axis	**HCN**, **HCCH**

5.7. Delocalized Orbitals

5.35 WORKED EXAMPLE

Lewis structure for the carbonate ion

Write the Lewis structure for the carbonate ion, CO_3^{2-}. All the oxygen atoms are equivalent.

Necessary information: Carbon and oxygen atoms have 4 and 6 valence electrons, respectively, and the ion has two more electrons than are provided by these valence electrons. The equivalent oxygen atoms must all be bonded in the same way to the carbon atom.

continued

Strategy: The approach is the same as in Worked Examples 5.6 and 5.8.

Implementation: The carbonate ion has 24 valence electrons, 22 from the atoms and 2 more that account for the 2− charge on the ion.

First we will bond the oxygen atoms to the carbon atom with two-electron bonds and then, since the oxygen atoms are equivalent, give each oxygen atom an octet of electrons:

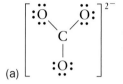

(a)

This Lewis structure uses all 24 valence electrons and has all three O atoms equivalent. However, the C atom has only six electrons (three pairs), so we need to find a way to write a structure with an octet of electrons on the C as well as on the O atoms. We can do this by using a double bond between the C atom and one of the O atoms:

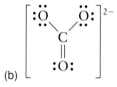

(b)

This structure gives all four second period atoms an octet of electrons, but implies that one O atom is different from the others. Since we know that the oxygen atoms are equivalent, this structure does not fit the experimental facts. Also, there is no reason to choose one O atom over another to make the double bond. Any of these three structures gives all the atoms an octet of electrons:

(b) (c) (d)

None of the four structures, (a) through (d), satisfies all the criteria for Lewis structures as well as the experimental observation that the oxygen atoms are equivalent.

Does the answer make sense? We do not yet have a satisfactory Lewis structure for the carbonate ion.

5.36 CHECK THIS

Lewis structure for the nitrate ion

Write the Lewis structure for the nitrate ion, NO_3^-. All the oxygen atoms are equivalent. Does your structure have the same problems we found for carbonate in Worked Example 5.35? Explain.

5.37 CONSIDER THIS

Do Lewis structures give correct geometries for CO_3^{2-} and NO_3^-?

Both carbonate and nitrate ions are planar. All four atoms lie in the same plane. Do the Lewis structures written in Worked Example 5.35 and Check This 5.36 predict this geometry? Show why or why not for both ions.

Worked Example 5.35 and Check This 5.36 show that, for some ions (and molecules), a single Lewis structure is not adequate to satisfy both the rules we use to write the structures and all the observed properties of the ion or molecule. For carbonate and nitrate ions, the geometry of the ions is correctly predicted by the sigma framework derived from our Lewis structures, but the equivalence of the oxygen atoms is not explained. Our Lewis structures have two single bonds and one double bond from C (or N) to O. Since exactly the same bonding is involved, the energies of the three structures are identical. Whenever we can write Lewis structures that are equivalent in energy, but have double bonds between different pairs of atoms in the molecule, none of the individual structures correctly represents the molecule.

Delocalized π molecular orbitals To account for the properties of molecules for which no single Lewis structure is correct, we need to consider pi orbitals that extend over more than two atomic cores. Figure 5.23(a) shows the **delocalized π molecular orbital,** describing two electrons, in the nitrate ion. The delocalized orbital is spread across the nitrogen core and *all three* oxygen cores in the nitrate ion. For comparison, Figure 5.23(b) shows the two-center, localized π molecular orbital in methanal.

Figure 5.23.

Delocalized and localized π molecular orbitals. Both molecules have planar σ frameworks. Because the delocalized π orbital in the nitrate ion is "shared" among three N—O pairs, each is assigned one-third of its bonding.

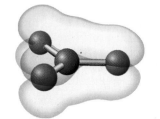
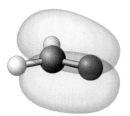

(a) Nitrate ion, NO_3^-, delocalized π   (b) Methanal, H_2CO, localized π

Bond order It's awkward always to refer to a σ or π bonding orbital between two (or more) atomic cores, so we almost always use the shorthand versions, **σ bond** and **π bond,** with the understanding that these refer to the orbitals. In methanal, the carbon and oxygen atoms are joined by a σ bond and a π bond for a total of two bonds. In the nitrate ion, there is a σ bond between the nitrogen atom and each oxygen atom, but the delocalized π orbital is spread over all four atomic cores. There are three N—O bonds among which to divide the π bonding. Each bond is assigned one-third of the π bond. Thus, each oxygen atom is joined to the nitrogen atom by a σ bond and one-third of a π bond for a total of $1\frac{1}{3}$ bonds.

The sum of the number of two-electron, bonding molecular orbitals that are shared between two atomic cores in a molecule is called the **bond order.** In molecules with only sigma orbitals, there is one, two-electron, σ orbital shared by each pair of bonded atomic cores; the *bond order is one* and this is called a **single bond.** In molecules like ethene and ethanal, a pair of atomic cores shares both a σ orbital and a π orbital; the *bond order is two* and this is called a **double bond.** This is the nomenclature we have been using to describe the bonding in Lewis structures. For molecules (or ions) with delocalized π orbitals, you have to divide up the π bonding interactions among the pairs of atomic cores. In the nitrate ion, NO_3^-, for example, each N—O bond is a $1\frac{1}{3}$ bond; the bond order is one-and-one-third. There is no simple way to express nonintegral bond orders nor is there any way to write a single Lewis structure for a molecule or ion like nitrate. A Lewis structure can't represent nonintegral numbers of electrons in bonds.

5.38 CONSIDER THIS

What are the bonding and structure of the CO_2 molecule?

(a) Write the Lewis structure for CO_2. Both oxygen atoms are bonded to the carbon.
(b) How many σ orbitals does the carbon atom have in CO_2? What geometry do you predict for the CO_2 molecule? Explain how you make your prediction.
(c) If possible, make a molecular model of the CO_2 molecule. If you can make a model, does it confirm the geometry you predicted in part (b)?

Orbital energies You can write a Lewis structure for CO_2 with two bonding pairs of electrons between the carbon atom and each of the oxygen atoms. You can also make a bent-bond model of the molecule that shows its linear structure, the three atom cores lying along a line. However, these models do not adequately describe all the properties of carbon dioxide. In particular, carbon dioxide is more stable, that is, has a lower total energy, than is predicted for a molecule with two localized π bonds between carbon and oxygen. (See Problem 5.32.) In Chapter 7, we will discuss bond formation energies quantitatively; here we will look only at why carbon dioxide has such a low total energy.

The sigma-pi bonding in carbon dioxide involves two sigma and two pi bonds on the carbon atom, which, as Table 5.3 reminds you, gives the molecule the linear structure you found in Consider This 5.38. Instead of two localized π orbitals, the wave mechanical solution for bonding in CO_2 gives two *delocalized* π molecular orbitals that extend across all three atoms, as shown in Figure 5.24. The two delocalized π orbitals lie in planes that are perpendicular to each other, just like the two localized π orbitals in hydrogen cyanide, Figure 5.22. Again, one π orbital is shown as light gray and the other as a darker gray to make it easier to see them both.

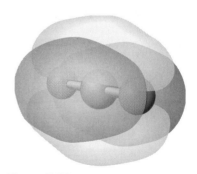

Figure 5.24.

Delocalized π molecular orbitals in carbon dioxide. One of the π orbitals is shown as a darker gray to distinguish if from the other.

5.39 CHECK THIS

Bond order in carbon dioxide

(a) When the delocalized π orbitals in Figure 5.24 are taken into account, what is the bond order for the bond between the carbon atom and one of the oxygen atoms in carbon dioxide? Explain clearly how you arrive at your answer. Does your Lewis structure from Consider This 5.38 show this same bond order? Why or why not?

(b) Carbon–oxygen single and double bonds have bond lengths of about 140 and 120 pm, respectively. The C—O bond lengths in CO_2 are 115 pm. Is this value consistent with the bonding model and properties of CO_2 just discussed? Explain why or why not.

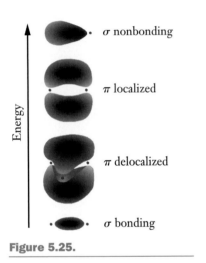

σ nonbonding

π localized

π delocalized

σ bonding

Figure 5.25.

Relative energies of molecular orbitals. Sigma nonbonding orbitals have approximately the same energy as the atomic orbitals on the atoms. Pi delocalized orbitals are spread over three or more atomic cores. The atomic cores are represented by the dots.

A molecule (or ion) with delocalized π orbitals has a lower total energy than a molecule with an equal number of localized π orbitals. The energy is lower for two reasons. First, the electrons are interacting with more than two positive atomic cores, so the potential energy of attraction is more favorable (more negative) for the delocalized electrons. Second, the delocalized orbital is larger, so the kinetic energy of the electron wave is lower (less positive). As always, when an electron is less constrained, its kinetic energy is lower and its total energy is lower (more favorable). The delocalized π orbitals in carbon dioxide, the carbonate ion, the nitrate ion, and many other molecules and ions make these molecules and ions more stable than they would be with the localized orbitals we write in their Lewis structures. Figure 5.25 is a qualitative energy-level diagram that shows the relative energies of molecular orbitals.

5.40 CHECK THIS

Structure of the ozone molecule

(a) Write a Lewis structure for the ozone molecule, O_3. What geometry does your structure predict for ozone? The three atoms are connected in a chain and the end atoms are equivalent. Is your structure consistent with these experimental facts? Explain why or why not. Would you expect ozone to have delocalized π electrons? What is the bond order between the oxygen atoms in ozone? Explain the reasoning for your responses.

(b) Oxygen–oxygen single and double bonds have bond lengths of about 148 and 120 pm, respectively. The experimental bond lengths in ozone are 128 pm. Do your results in part (a) explain this bond length? Why or why not?

Metallic properties

5.41 INVESTIGATE THIS

What happens when substances are cut?

Do this as a class investigation and work in small groups to discuss and analyze the results. Place a small piece of sodium metal in a shallow dish and try cutting it in two with a sharp knife. Record your observations, including the ease of cutting, and then return the pieces of sodium to their storage container. *CAUTION:* Sodium metal can cause severe burns by reacting with the moisture on your skin. Handle the metal with tongs.

Place a crystal of rock salt, sodium chloride, in a shallow dish and try cutting it in two with a sharp knife. Record your observations, including the ease of cutting.

5.42 CONSIDER THIS

What substances can be cut in two?

(a) In Investigate This 5.41, did sodium metal and sodium chloride behave the same or differently when you tried to cut each solid in two? Explain how they were the same or different.

(b) Try to think of a molecular-level explanation for the similarities and/or differences in the behavior of sodium metal and sodium chloride.

Recall from Chapter 2, Sections 2.3 and 2.4, especially Figures 2.9 and 2.13, that ionic solids (crystals) are three-dimensional arrays of positive and negative ions held together by coulombic attraction. These attractions are relatively strong for nearest-neighbor ions and hold the ions tightly together. When you attempt to cut an ionic crystal in two, these attractions strongly resist being broken. It takes considerable effort to cut into the crystal, which often shatters into several pieces as the crystal structure is disrupted by the applied force.

On the other hand, many metals are relatively easy to cut with strong scissors and, as you found in Investigate This 5.41, some metals, such as sodium, are soft enough to be cut easily with a knife. This difference in behavior between metals and ionic crystals must be due to a difference in the forces that hold the atoms together in a metal compared to those that hold the ions together in an ionic crystal. An obvious difference between an elemental metal, such as sodium, copper, aluminum, or iron, and an ionic crystal is that the metal contains only one kind of atom. Since most metals are relatively dense solids, the elemental spherical atoms must be packed closely together, as represented for two-dimensional packing in Figure 5.26.

Figure 5.26.

Two-dimensional close packing of spherical objects.

Three-dimensional packing of spheres

How do you have to pack the plastic balls that represent hydrogen atom centers to get them to fit in the space allotted for them in your model kit? Use drawings and words to describe the packing and to relate it to the two-dimensional case in Figure 5.26.

Metallic bonding: Delocalized molecular orbitals It is unlikely that oppositely charged ions of the same element, such as Na^+ and Na^-, are responsible for the binding in a metal. We know from our previous discussions that metals are present as positively charged ions in ionic compounds. This is because metals give up one or more of their valence electrons relatively easily; metals have low ionization energies (Chapter 4, Figures 4.4 and 4.30). Consider what can happen when metal atoms are stacked together, as represented in Figure 5.26 and your sketches for Check This 5.43. Assume that the bright spot in the center of each sphere in Figure 5.26 represents the positively charged atomic core of the metal atom and that the rest of the copper-colored sphere represents its valence electrons.

The atomic cores of the metal atoms are about as close together in this array as are the atomic cores in the molecules we have been discussing so far in this chapter. We have seen that, when the cores are this close together, the valence electrons are attracted by and interact with more than one positive center. In a uniform stack of metal atoms (a crystal of the metal), the valence electrons from one atom can interact with *all* the other atomic cores in the crystal. The electron waves, orbitals, that describe these electrons are spread out through the entire crystal. You can think of the metal crystal as a single molecule with enormously delocalized electron orbitals. This **metallic bonding model,** represented in Figure 5.27, is usually described as a *sea of electrons* surrounding the positive metal atomic cores. In a sense, all the positive centers attract all the valence electrons, which holds the structure together.

The metallic bonding model explains why it is easy to cut many metals without shattering them. Since the delocalized electron orbitals are spread out through the crystal, there is little directionality to the bonding. Adjacent atomic cores resist being pushed apart by a blade, but these pair-wise interactions are relatively weak. When enough force is applied, they are pushed apart and the electron orbitals easily readjust to the new configuration of atomic cores.

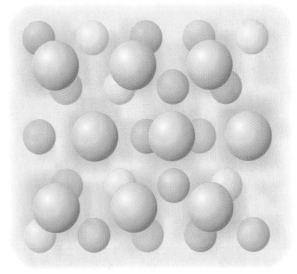

Figure 5.27.

A representation of metallic bonding. Copper-colored positive cores are immersed in a sea of electrons shown as a diffuse cloud.

More properties of metals

Metals are **ductile,** which means they can be pulled into long wires without breaking, like the copper wire shown in the marginal photograph on the next page. Metals are also **malleable,** which means they can be pounded into sheets

continued

without shattering and pressed into useful shapes such as spoons and automobile fenders. Ionic crystals like the copper sulfate pentahydrate, $CuSO_4 \cdot 5H_2O$, in the photograph, do not have these properties. How does the metallic bonding model explain these properties of metals? Why do the crystals shatter?

Other familiar properties of metals, for example, their luster or shininess, their high heat conductivity, and their high electrical conductivity, can also be interpreted with this metallic bonding model based on delocalized molecular orbitals. We will include discussions of metallic properties in problems and later chapters, as appropriate, but for now will go on to the discussion of drawing molecular structures.

5.8. Representations of Molecular Geometry

Lewis structures and ball-and-stick models simplify the molecular orbital model by using a line or a stick to represent the sausage-shaped σ bond between pairs of atoms. You can think of the pairs of dots on individual atoms in Lewis structures as two-electron σ_n orbitals. With your ball-and-stick models you can use the paddles to represent nonbonding electrons. In Figure 5.28, the orbital model, Lewis structure, and ball-and-stick models for the ammonia molecule are compared. Of the four representations shown, the Lewis structure is obviously the easiest to write, as it is written with letters, lines, and dots. However, Lewis structures are used mainly to show connectivity of atoms; the Lewis structure doesn't show the three-dimensional shape of the molecule, the trigonal pyramid formed by the nitrogen and three hydrogen atoms.

Web Companion

Chapter 5, Section 5.8

Manipulate several interactive representations of molecular geometry.

①
②
③
④

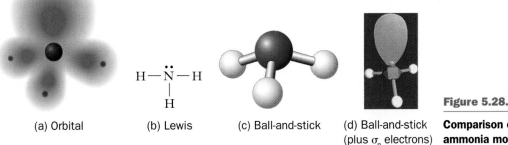

| (a) Orbital | (b) Lewis | (c) Ball-and-stick | (d) Ball-and-stick (plus σ_n electrons) |

Figure 5.28.

Comparison of models for the ammonia molecule.

Tetrahedral representation You can retain *some* of the simplicity of a Lewis structure, written much as it is in Figure 5.28(b), *and* show the three-dimensional shape of molecules. To do so, you start by analyzing the Lewis structure to see how many σ orbitals there are around the atom of interest. In ammonia, there are four σ orbitals around the nitrogen atomic core: three σ and one σ_n. You know from Section 5.4 that four σ orbitals will be arranged tetrahedrally about the nitrogen atom. To represent the three-dimensional tetrahedral structure in two dimensions requires some way to show perspective. You can do this by using different kinds of lines to represent σ bonds. Figure 5.29 shows how solid lines, dashed lines, and wedge lines are used to represent perspective and give a three-dimensional character to a drawing of the ammonia molecule. We will call this form of molecular representation a **3-D structure**.

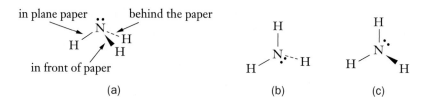

Figure 5.29.

3-D structures of the
ammonia molecule.

The kind of line you use in a 3-D structure depends upon whether the bond lies in the plane of the paper or is in front of or behind the plane, as you look at the molecule. Solid lines are bonds in the plane of the paper. Dashed lines represent bonds that extend away from you behind the paper. Wedges represent bonds that extend toward you in front of the paper. Usually, you choose to orient your structure so that as many of the bonds as possible lie in the plane of the paper in these 3-D structures. This often, but not always, makes it easier for you to interpret the structure. Note that little attempt is made to show the nonbonding electron pair in perspective. Three-dimensional structures are designed to show molecular shape, the orientation of the atoms with respect to one another, not orbital geometry. The shape is, however, determined by σ orbital geometry.

5.45 CHECK THIS

Interpreting 3-D structures

(a) Which, if any, of the 3-D structures in Figure 5.29 is easiest for you to interpret as the trigonal pyramidal structure for ammonia? Why is it easiest for you?

(b) 🖱 Load the movie, "Molecular representations of methane," in the *Web Companion*, Chapter 5, Section 5.8.1. Place the cursor in the middle of the image in the movie. Move the cursor up and down. Describe what you see. What information can you gather from each of the different representations? Which representation do you find most useful? Explain why.

(c) Draw 3-D structures for the chloromethane molecule, CH_3Cl, that are comparable to each of the 3-D structures in Figure 5.29. Make a ball-and-stick model of the chloromethane molecule. Practice your understanding of the structure of molecules by writing a description of how the drawings correlate with the actual three-dimensional model.

Trigonal planar representation The orbital model, Lewis structure, ball-and-stick model, and 3-D structure for the methanal molecule, H_2CO, are compared in Figure 5.30. To generate the 3-D structure, start with the Lewis structure and figure out how many σ orbitals there are around the carbon atomic core. In this case, there are three σ orbitals: each H—C single bond is a σ bonding orbital and the double bond is composed of a σ bonding and a π bonding orbital. You know that three σ orbitals will be in a trigonal planar arrangement around the carbon atom. It's easy to draw the planar structure, Figure 5.30(d), with all four atoms in the plane and bond angles of 120°. (Or you can draw the structure with the experimental bond angles of 116° and 122°, respectively, for the H—C—H and H—C—O angles.)

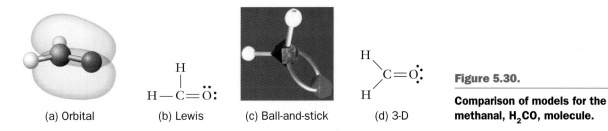

(a) Orbital   (b) Lewis   (c) Ball-and-stick   (d) 3-D

Figure 5.30.

Comparison of models for the methanal, H_2CO, molecule.

The Lewis and 3-D structures in Figure 5.30 look quite similar. Be cautious; the similarity is misleading because the Lewis structure is designed to show connectivity, not geometry. The 3-D structures show connectivity, and, in addition, molecular shape—note the planar structure and correct bond angles. In the case of planar and linear molecular structures, the Lewis and 3-D structures can be easily confused, because no wedged and dashed bonds are necessary for these 3-D structures.

5.46 CHECK THIS

3-D structures

(a) Write a Lewis structure for a molecule of methanol, CH_3OH. How many sigma molecular orbitals are there around each second-period atom in the molecule? How many of the other atoms can lie in a plane that contains the carbon and oxygen atoms? Draw a 3-D structure that has as many atoms in the plane as possible. If necessary, use wedge and dashed bonds to show where the other atoms are relative to the plane. You might find that making a ball-and-stick model will help you.

(b) Write a Lewis structure for a molecule of formic (methanoic) acid, HC(O)OH. How many sigma molecular orbitals are there around each second-period atom in the molecule? How many of the other atoms *must* lie in a plane that contains the carbon and oxygen atoms? How many of the other atoms *can* lie in this same plane? Draw a 3-D structure that has as many atoms in the plane as possible. If necessary, use wedge and dashed bonds to show where the other atoms are relative to the plane. You might find that making a ball-and-stick model will help you.

(c) Draw the 3-D structure for ethyne, HCCH.

Formic acid was first extracted from ants (*formica* = ant). It causes the sting of an ant bite.

Lewis structures and molecular shape We set out in this section to learn to translate Lewis structures into 3-D structures that show molecular shape. Molecular shape is determined by the σ orbitals and Lewis structures show you how many σ orbitals each atom has. The spatial orientation of π bonds is not shown in 3-D structures. You have to remember that a second line between atoms is a π bond that lies parallel to the bond axis and in a plane perpendicular to the molecular plane (the plane of the paper). If there are three lines between atoms, the second and third are π bonds that lie parallel to the bond axis and in two perpendicular planes that contain the bond axis; see Figures 5.22 and 5.24.

5.47 CONSIDER THIS

How do you draw 3-D structures for larger molecules?

(a) Write the Lewis structure and draw the 3-D structure for propene, CH_3CHCH_2. Propene is like ethene, Figures 5.19 and 5.20, with a H_3C- in place of one of the hydrogen atoms. Can all the carbon atoms in the molecule lie in a plane? If so, that should make them easier to draw. If they can lie in a plane, is there more than one way for them to lie in a plane? You might find it helpful to make ball-and-stick models of these molecules to help visualize in three dimensions what you're drawing on a plane.

(b) Write the Lewis structure and draw the 3-D structure for 1-butene, $CH_3CH_2CHCH_2$. Answer the same questions for 1-butene as for propene in part (a).

Consider This 5.47 probably convinced you that drawing 3-D structures for molecules with even as few as four carbons is pretty tedious. Also, most of us have a hard time actually visualizing the three-dimensional structure when there are so many dashes and wedges and elemental symbols to keep track of. Biologically important molecules are even more complex and often contain extended structures of carbon atoms, with nitrogen, oxygen, and other atoms at strategic sites. To get around these representational problems, chemists use a progressive system of shorthand to represent complex structures, with each step showing less detail. To understand structures that are represented using abbreviated symbolism, you need to be able to recognize and understand the detail that is omitted along the way. That's why you should also continue to practice writing Lewis structures, visualizing the structures they represent, making molecular models, and drawing the 3-D structures; it's the 3-D structures that are being abbreviated. We will exemplify the successive steps in this system by considering the structure of 1-butanol, one of the compounds whose properties you examined in Sections 5.1 and 5.2.

Condensed structure for 1-butanol Figure 5.31 shows a ball-and-stick model, a 3-D structure, and a condensed structure for 1-butanol. In 3-D structures, we often do not show the nonbonding electrons, as we rarely try to represent their geometry. Also, we don't need to show the lines representing bonds to hydrogen, because there is always one σ bond to hydrogen. Thus, in the **condensed structure**, we eliminate the lines to H. The H symbols are still shown associated with the atom cores to which they are bonded and placed where they will not hide essential features of the skeleton of the molecule, which is shown in red in Figure 5.31(b) and (c).

Figure 5.31.

Structural representations of the 1-butanol molecule.
The red bonds in structures (b) and (c) denote the bonds that will form the skeletal structure, Figure 5.32.

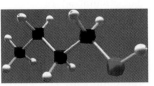

(a) Ball-and-stick model

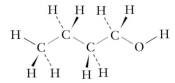

(b) 3-D structure

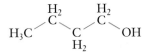

(c) Condensed structure

5.48 WORKED EXAMPLE

3-D and condensed structures for 2-propanol

The Lewis structure for 2-propanol is given in Worked Example 5.6 and a ball-and-stick model is shown in Figure 5.1(b′). Draw 3-D and condensed structures for 2-propanol.

Necessary information: All the connectivity information we need is in Worked Example 5.6.

Strategy: For the 3-D structure, choose the longest carbon chain and orient these carbon atoms so they lie in a plane. Attach other nonhydrogen atoms to the appropriate carbon(s) in the chain, using appropriate dashed or wedge bonds. Add the hydrogen atoms, using solid, dashed, or wedge bonds to give the appropriate geometry around each second- or higher-period atom.

Implementation: The 2-propanol molecule has only sigma electrons, so the geometry around each carbon atom is tetrahedral. The longest carbon chain has three carbon atoms, which we show lying in the plane of the paper. Then we add the oxygen atom bonded at the middle carbon using either a wedge or dashed bond:

We complete the 3-D structure by adding seven hydrogen atoms to give tetrahedral bonds around the carbons and bond the eighth hydrogen atom to oxygen:

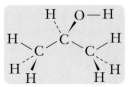

For the condensed structure, we eliminate the bond lines between hydrogen atoms and other atoms and write the Hs next to the atoms to which they are bonded:

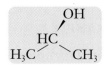

Does the answer make sense? Compare the 3-D structure to the ball-and-stick model in Figure 5.1(b′) to see whether the drawing is a good representation of the geometry of the model. Note that we could have chosen to show the oxygen atom bonded with a dashed bond (in back of the paper). This would still be the same structure shown here, because they can be converted into one another by simply rotating the molecule in space. Use molecular models to prove this.

5.49 CHECK THIS

3-D and condensed structures for the C_5H_{12} isomers

Lewis structures and ball-and-stick models for the C_5H_{12} isomers are shown in Worked Example 5.8 and Figure 5.2. Draw 3-D and condensed structures for each isomer. You might find that making ball-and-stick models will help you.

5.50 CONSIDER THIS

How many hydrogen atoms are bonded to a carbon atom?

Do any of the carbon-containing molecular structures we have shown in this or the preceding chapters have σ nonbonding orbitals on carbon? In structures such as those in Worked Examples 5.6 and 5.8, Consider This 5.10, and Figure 5.31, can you tell how many hydrogen atoms will be σ bonded to a particular carbon atom without counting the number of hydrogen atoms? Explain how.

Carbon monoxide, CO, is the single notable exception to the statement that carbon never has σ_n electrons. There are also reactive fragments of molecules (ions and free radicals), in which carbon has nonbonding electrons.

Skeletal structure for 1-butanol The final step in this process of abbreviation requires you to account for the observations and analysis you did in Consider This 5.50. *Carbon atoms in molecules do not have σ_n electrons. If a carbon atom has hydrogen atoms bonded to it, we know that the carbon atom is σ bonded to enough hydrogen atoms to give the carbon four bonding orbitals.* Therefore, we can condense the notation even more by omitting the hydrogen symbols on carbons. Further, we understand that the end of a line or the intersection of two or more lines represents a carbon atom and any attendant hydrogen atoms, so we can also omit the letter symbol for carbon. We must keep the letter symbol for oxygen (or other elements) to distinguish the atom at that position from carbon. The result is shown in Figure 5.32. To the practiced eye, this **skeletal structure** gives a reasonable idea of the three-dimensional structure of 1-butanol.

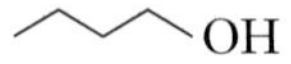

Figure 5.32.

Skeletal structure of 1-butanol.

5.51 WORKED EXAMPLE

The skeletal structure for 2-propanol

Draw the skeletal structure for 2-propanol.

Necessary information: The 3-D and condensed structures are in Worked Example 5.48.

Strategy: For the skeletal structure, start with either the 3-D or condensed structure. Eliminate all hydrogen atoms and their bonds on carbon atoms. Eliminate all the carbon Cs and extend the remaining bonds to meet where the Cs were. Retain wedge and dashed bonds, if any.

continued

Implementation: Begin with the condensed structure from Worked Example 5.48 and carry out the necessary changes:

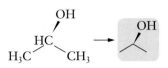

Does the answer make sense?   Make a ball-and-stick model of 2-propanol, remove all the hydrogen atoms and their bonds attached to carbon atoms, and compare the result with this skeletal structure to see that it represents the 2-propanol skeleton.

5.52 CHECK THIS

Skeletal structures for the C_5H_{12} isomers

(a) What does the line at the left end of the 1-butanol skeletal structure, Figure 5.28, represent? Explain clearly.

(b) Draw skeletal structures for the C_5H_{12} condensed structures you drew in Check This 5.48. *Note:* If you need to represent a —CH_3 group attached to a carbon chain, but not in the plane of the chain, use a dashed line or a wedge instead of just a solid line.

5.53 CONSIDER THIS

Can different skeletal structures represent the same molecule?

(a) Parts of a molecule can rotate with respect to one another without breaking any sigma bonds. This sort of motion is going on all the time. The rotation does not create a new molecule, but it does create a new shape and therefore a new skeletal structure. The skeletal structure of a different shape for 1-butanol is

OH

Make a model of 1-butanol and test whether this new shape and the one in Figure 5.32 can be interconverted without breaking any bonds.

(b) In Check This 5.52, you drew skeletal structures for the C_5H_{12} isomers. For each isomer, draw another different skeletal structure that can be obtained from the ones you already have by rotating one part of the molecule with respect to another. Is it possible to do this for all three isomers? Why or why not? You might find it helpful to use your models of the molecules.

5.54 CHECK THIS

Interpreting skeletal representations

(a) The skeletal structures of 2-methylpropane and 2-methylpropene are
$\rangle\!\!-$ and $\rangle\!\!=$, respectively. Draw condensed structures for each molecule.
What are the molecular formulas for the molecules? Explain the reasoning
for your responses.

(b) The condensed structure for retinol is shown in Consider This 5.10(d).
Draw the skeletal structure for retinol. Is your skeletal structure related to
the molecular models of retinal shown in the chapter opening illustration?
Explain your response.

Reflection and Projection

Delocalized bonding is the last bonding concept you need to interpret the bonding and behavior of the molecules you will meet in this book. When an orbital (or orbitals) spreads out and interacts with several atomic cores, the energy of the electrons is more negative (more favorable) because they are attracted by more positive charges and their kinetic energy is lower. For molecules (not metals), however, the major contributors to the favorable energy relative to their atoms are their σ bonds, which also determine their geometries. Metals lack a directional sigma-bonding framework. They are bonded by nondirectional delocalized molecular orbitals, which give them their characteristic properties, including cutting without shattering, ductility, and malleability.

One of the most important skills for any scientist working with molecular materials is the ability to visualize their structures in three dimensions. From the very beginning of this text, we have emphasized the critical role played by molecular structure in determining the properties of substances. Now that we have developed a model that explains and predicts molecular structure, it is important to be able to represent the structures in a way that aids three-dimensional visualization. Lewis structures combined with the structural rules based on the σ molecular framework are used to draw 3-D structures that are built around individual atom centers linked together to form more complex structures.

As the molecules of interest become ever larger, the focus on geometry about individual atoms becomes so complex that it can obscure, rather than clarify, structural visualization. We introduced a series of abbreviated representations for carbon-containing molecules that range from 3-D to condensed to skeletal structures. In the final step, atomic symbols for all carbon atoms and hydrogen atoms bonded to carbon disappear and you are left with intersecting lines representing the carbon skeleton of the molecule. As you use these abbreviations, always keep in mind what is missing. Continue to practice going back and forth from complete structural representations and models to the skeletal structures.

Molecular structures are, in and of themselves, interesting and often quite beautiful. In the next section, we will extend our discussion of structure to a different form of isomerism.

5.9. Stereoisomerism

How many C_4H_8 molecular structures are possible?

(a) Work in small groups on this investigation. Use your molecular models to make as many models of isomers of C_4H_8 as you can. If necessary, use bent bonds to represent multiple bonds.

(b) Draw skeletal structures of all the molecules you found in part (a). Share your results with the entire class and come to an agreement on the number of possible isomers. *Hint:* Carbon atoms can be connected in rings of three, four, or more carbons. (You can even think of the double bond as a two-membered ring.)

What are the line (condensed) formulas for the C_4H_8 isomers?

(a) Recall that line (condensed) formulas are written to try to show the connectivity among the atoms in a molecule in a compact notation on a single line. For example, one of your C_4H_8 isomers has the line formula: $CH_2C(CH_3)CH_3$. What are the line formulas of the other molecules whose models you made in Investigate This 5.55?

(b) Do all your isomers have different line formulas? If two (or more) of the models have the same line formula, why are they isomers? What is the criterion you use to call them isomers?

All of the isomers you have encountered so far in this and preceding chapters have been **structural isomers,** isomers with different connections among the atoms. These isomers all have different line formulas. Now you have found two different molecular models with the same atomic composition, C_4H_8, and the *same* line formula, $CH_3CHCHCH_3$. Compare your molecular models with the sigma-pi representations in Figure 5.33. Carbon chains are numbered beginning with the carbon atom at one end. The 2 in 2-butene indicates that the double bond connects the *second* and third carbon atoms in the four-carbon chain. The lobes of the π bonding orbital lie above and below the plane of the four carbon atoms in Figure 5.33.

Figure 5.33.

Sigma-pi bonding models of *trans-* and *cis-2-butene.* The π bonding orbital is superimposed on the σ framework to show its location between the second and third carbons with its lobes above and below the plane of the molecule.

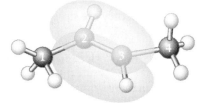

(a) *trans*-2-Butene,
b.p. = 0.9 °C

(b) *cis*-2-Butene,
b.p. = 3.7 °C

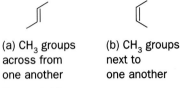

(a) CH₃ groups
across from
one another

(b) CH₃ groups
next to
one another

Figure 5.34.

Geometry of (a) *trans*-2-butene and (b) *cis*-2-butene. All four carbon atoms in each skeletal structure lie in the plane of the paper.

Cis and trans isomers The double bond in 2-butene imparts important structural properties as well as the chemical reactivity associated with multiple bonds. In Figure 5.34, we see that the two —CH_3 groups in *trans*-2-butene are on opposite sides of the double bond (*trans* = across from; one on the left and one on the right). In *cis*-2-butene, the —CH_3 groups are on the same side of the double bond (*cis* = next to; both on the right). *trans*-2-Butene cannot be changed to *cis*-2-butene (and vice versa) without breaking the π bond between the second and third carbon atoms. The same is true for your bent-bond models. The only way to convert one model into the other is to break and reform bonds. Experiments show that the conversion requires about 270 kJ·mol⁻¹, which is far more energy than the molecules have available at room temperature. If the two isomers could rotate easily around the double bond, 2-butene would have a single boiling point. The observation that *cis*-2-butene and *trans*-2-butene have different boiling points, Figure 5.33, is experimental evidence that *cis* isomers and *trans* isomers are different compounds that do not interconvert at room temperature.

5.57 CHECK THIS

Cis-trans isomers

(a) Use your molecular models to make as many models of isomers of $C_2H_2Cl_2$ as you can. Are any of your structures *cis-trans* isomers? Draw skeletal structures of all the isomers and identify the *cis-trans* isomers, if any. Explain the reasoning for your answers.

(b) The structure of retinol, vitamin A, in Consider This 5.10 is sometimes called "all-*trans*-retinol." Draw a skeletal structure and explain why this is an appropriate name for this isomer.

(c) The structural change in retinal illustrated in the chapter opening illustration is a *cis* to *trans* conversion caused by the absorption of energy from a photon of light. Draw skeletal structures of the two isomers that show where the *cis* to *trans* change occurs.

Stereo- is from Greek *stereos* = solid, three-dimensional. You are probably familiar with the word stereophonic referring to the perception of sound as coming from many directions.

Stereoisomers *Cis-trans* isomers are an example of **stereoisomers**, isomers whose atoms are connected in the same sequence, but whose shape in space is different and not interconvertible. The names of *cis-trans* stereoisomers are the same (2-butene, for example), which reflects their connectivity; the prefix *cis-* or *trans-* reflects their different shapes. If a pair of structures has the same elemental composition and the same connectivity, one way to test whether they are stereoisomers is to try to superimpose models of the molecules. If the models cannot be made to superimpose on one another without breaking and remaking bonds, they are stereoisomers.

5.58 CHECK THIS

Identifying isomers

Five skeletal structures are shown on the next page. What is the elemental composition of each? Which structures could be isomers of one another? Within a

continued

set of possible isomers, which *pairs* are structural isomers? Which are stereoisomers? Which are identical structures? Give the reasoning for your answers.

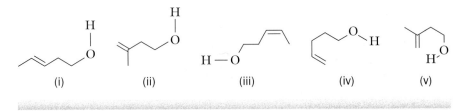

(i)   (ii)   (iii)   (iv)   (v)

Polarized light and isomerism In Chapter 4, Section 4.1, we discussed how the atomic masses proposed by Cannizzaro in 1858 led within a few years to the development of the periodic table. These atomic masses, combined with the many known mass percent compositions of compounds, enabled scientists to determine the molecular formulas for these compounds. Thus, by 1860, scientists knew the formulas for many compounds and they also knew that isomers existed; the same molecular formula could represent more than one compound. Since the isomers had different properties, they assumed that the structures had different fixed geometric shapes. They did not know what held the atoms together in these different shapes; they simply assumed that something did.

One of the properties that was different for some isomers was their interaction with polarized light. Recall from Chapter 4, Section 4.3, that a light wave consists of oscillating electric and magnetic fields perpendicular to one another. A **polarizer,** like the lenses in polarized sunglasses, allows light to pass only if the electric field is aligned in a particular direction, as in Figure 5.35. The transmitted light is said to be **polarized.** If a second polarizer is aligned in the same direction as the light, the **polarized light** will pass through. If the second polarizer is turned 90° from this alignment, none of the polarized light can get through; the polarizers are then "crossed." Figure 5.35 illustrates the principle of the **polarimeter,** an instrument invented early in the 19th century that is used to study the effect of substances on a beam of polarized light. In Figure 5.35(a) the pair of polarizers is aligned so the maximum amount of light is transmitted. Investigate This 5.59 gives you an opportunity to use a setup like Figure 5.35 to investigate the effect of water and a sugar solution on polarized light.

Web Companion

Learn more about polarized light and the polarimeter using these animations.

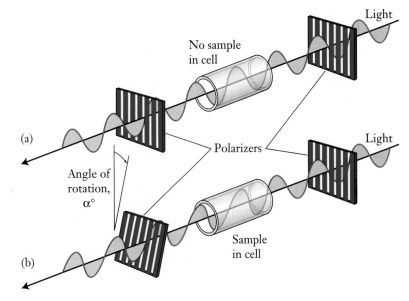

Figure 5.35.

Principle of operation of a polarimeter.
(a) The two polarizers are aligned; light passes through to the observer. (b) A sample rotates the plane of polarization by α°. The observer has to rotate the second polarizer by α° to align it with light leaving the sample.

5.59 INVESTIGATE THIS

Do solutions affect polarized light?

Do this as a class investigation and work in small groups to discuss and analyze the results. You need two, 10-cm-square sheets of polarizing filter, two glass Petri dishes about 6-cm in diameter, and a sheet of light cardboard with an 8-cm diameter hole in the center. Place the cardboard on an overhead projector and put one of the squares of polarizing filter over the hole. Make a mark on the filter near the edge of the projected image. Place the second polarizing filter on top of the first and rotate the second filter until the image on the screen is as dark as possible. Make a mark on the second filter directly over the mark on the first filter.

Fill one of the Petri dishes about half full of water and sandwich the dish between the two polarizing filters. Make sure the mark on the top filter is directly over the mark on the bottom filter and check to see whether the image on the screen is still as dark as possible in the center where the dish of water is between the filters. If the image isn't as dark as possible, rotate the top filter until the image is as dark as possible and make another mark on the top filter to indicate the new alignment of the filters. Repeat the procedure with the second Petri dish about half full of light corn syrup (a concentrated solution of sugars) from the grocery store. Record your observations.

5.60 CONSIDER THIS

What solutions affect polarized light?

(a) In Investigate This 5.59 you use the polarizers in their crossed condition because it is easier to tell when the image on the screen is darkest rather than lightest. If a sample between crossed polarizers changes the direction of polarization of the polarized light from the first filter, then the second filter has to be rotated to bring it into the crossed orientation. Does water change the direction of polarization of the light beam? What is the evidence for your answer? How would you describe the effect of water on polarized light?

(b) Does the solution of sugars change the direction of polarization of the light beam? What is the evidence for your answer? How would you describe the effect of a sugar solution on polarized light? Can you think of any further experiments you could do to find out more about the effect of water and sugar solutions on polarized light? If so, try them.

Some substances, when placed in a beam of polarized light, rotate the plane of polarization of the light. If such a sample is placed between aligned polarizers, the light passing through the sample will no longer be aligned with the second polarizer; the polarizer will have to be *rotated* to bring it back into alignment, as in Figure 5.35(b). The effect of the sample substance on polarized light is called **optical rotation.** Substances that rotate the plane of polarized light are said to be **optically active.** Different compounds rotate polarized light by different amounts. The amount of the rotation also depends on the number of molecules of the optically active compound that are in the polarized light beam. The rotation can be either clockwise (right; dextrorotatory) or counterclockwise (left; levorotatory), with respect to the light that enters the sample.

5.61 CHECK THIS

Direction of optical rotation

(a) Figure 5.35(b) shows the angle of rotation of the sample as α°. Is the rotation dextrorotatory or levorotatory? Explain your response.
(b) 🐾 The angle of rotation of the sample illustrated in the *Web Companion*, Chapter 5, Section 5.9.2, is α°. Is the rotation dextrorotatory or levorotatory? Explain your response.
(c) In Investigate This 5.59, was the rotation of the sugar solution dextrorotatory or levorotatory? Explain.

> The Latin words for *right (dexter)* and *left (laevus)* are often used to describe optical rotation: right, dextrorotatory, and left, levorotatory. Solutions of glucose and fructose are dextrorotatory and levorotatory, respectively. Their optical activity is what led to the names *dextrose* and *levulose* for these sugars.

By the middle of the 19th century, many optically active compounds had been discovered. Most of these came from living matter, like the sugars in corn syrup you used in Investigate This 5.59. While investigating the crystallization of such compounds, Louis Pasteur (French chemist, 1822–1895) observed that some solutions that were not optically active produced two different crystal forms, like those shown in Figure 5.36. Imagine that the dotted line between the two crystals is a mirror and you can see that *each is the mirror image of the other*. Pasteur separated the crystals, made solutions of both forms, and found that the solutions were optically active. The startling result was that one solution rotated polarized light clockwise and the other rotated it counterclockwise by exactly the same amount. A mixture of equal amounts of the two solutions did not rotate polarized light.

Pasteur's observations and subsequent experiments by others convinced 19th century scientists that optical isomers always came in pairs. One of the isomers rotated polarized light in one direction and the other in the opposite direction. As far as chemists could tell, all the chemical and other physical properties of optical isomers were identical; the isomers differed only in their optical rotation. The identical chemistry suggested that the isomers had basically the same arrangement of their constituent atoms, that is, they are stereoisomers. The mirror-image crystals formed by a pair of optical isomers, like those Pasteur had found, somehow seemed to reflect a fundamental mirror-image relationship between the molecular structures of the isomers.

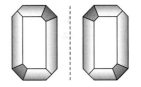

You are looking down on a face of each crystal and the facets that slope away from them.

Figure 5.36.

Mirror image crystals.

5.62 INVESTIGATE THIS

Are there stereoisomers of lactic acid?

Work in small groups and compare the structures each group member makes. The Lewis structure for lactic acid (present in sour milk and in your muscles after hard exercise) is given below. Make a molecular model of lactic acid. If you rotate parts of your model appropriately (without breaking any bonds), can you

continued

superimpose it on the models other students built? If not, how many different molecular structures are there for lactic acid? How are they related to one another? Explain using 3-D structural drawings to show the relationship(s).

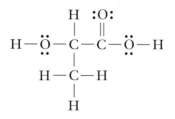

Tetrahedral arrangement around carbon In 1874, Jacobus van't Hoff (Dutch chemist, 1852–1911) and Jules Le Bel (French chemist, 1847–1930) independently proposed that a tetrahedral arrangement of four atoms or groups of atoms bonded about a central carbon atom explained the observations on optical isomers. Compounds like lactic acid represent the simplest cases, in which there is *a central carbon atom with four different atoms or groups attached*. In Investigate This 5.62, you constructed models with tetrahedral arrangements around carbon and found that there are two structures for lactic acid that are not superimposable on one another. These are **optical isomers,** stereoisomers that are mirror images of each other and differ only in their effect on polarized light.

Web Companion

Chapter 5, Section 5.9.5–7 ①
②
③
④
Study animations of optical isomers and the effects of their structure.

_____ **5.63 CONSIDER THIS** _____

Can you identify optical isomers?

These problems model the kind of logic that van't Hoff and Le Bel went through to develop the tetrahedral-carbon model.

(a) Hold your model of lactic acid by the hydrogen atom connected to the central carbon atom and look down the H—C axis toward the C. Rotate the model so that the acid group, —C(O)OH, is at the 12-o'clock position. Which group, the —CH₃ or the —OH, is at the 4-o'clock position? Make a drawing with the C in the middle, the acid at the 12-o'clock position, and the other two groups at the correct 4-o'clock and 8-o'clock positions. Next to your first drawing, draw its mirror image. Find another student with a model that is not superimposable on yours. Check to see whether your mirror-image drawing describes the other student's model.

(b) Another possible arrangement of four atoms or groups around the central carbon is a square planar arrangement. You can treat the Lewis structure in Investigate This 5.62 as a representation of a square planar lactic acid structure. Is the mirror image of this structure different from the original? Why or why not? If you have trouble deciding, construct models of the original and its mirror image. For the central atom, you will have to select one of the atom centers that has holes in a square planar arrangement. How does your answer help to decide whether a square planar structure is a possibility for lactic acid?

Although the electrical nature of matter that we discussed in Chapter 1 was well known to the scientists of the later 19th century, the substructure of the atom had yet to be discovered. In spite of this and without the many instruments available today, these scientists were able to deduce the three-dimensional structure of many carbon-containing molecules like lactic acid. Further, they knew many other compounds that had more than one carbon with four different groups attached, notably sugars, and they were able to predict the number of mirror-image isomers these compounds have. Essentially all the arrangements of groups attached to second-period atoms were known before the bonding models discussed in this chapter had been developed. Indeed, this structural knowledge helped 20th century scientists develop these bonding models.

> Late in the 19th century, there were lively controversies about the structures of metal-containing complex ions. Some of these problems were resolved by arguments based on optical rotation of the complexes. We will discuss metal-ion complexes in Chapter 6.

5.64 CHECK THIS

Properties of optical isomers

(a) View the animation of the stereoisomers of phenylalanine in the *Web Companion*, Chapter 5, Section 5.9.5. Using the Lewis structure shown on the page to model the groups on the central carbon, draw 3-D structures corresponding to the representations of the isomers at the end of the animation. Show clearly how and why these are stereoisomers and why they are optical isomers.

(b) Which 3-D structure for phenylalanine from part (a) corresponds to the isomer shown in the animation of its effect on plane-polarized light in the *Web Companion*, Chapter 5, Section 5.9.6? Describe how the animation would be different if you substituted the other isomer in the path of the polarized light.

(c) The sugars in the corn syrup in Investigate This 5.59 have more complicated structures than lactic acid and phenylalanine. The skeletal structure for one form of glucose, an ingredient in corn syrup, is

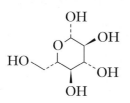

Here is a simple way to test whether this structure represents an optical isomer. Cut two identical regular hexagons out of paper. On one hexagon, put a red dot at the corner that is an oxygen atom in the glucose six-membered ring. Put a black dot at each of the other corners where an —OH group is out of the plane of the paper toward you in the glucose structure. Use the second hexagon to make a mirror image of the first. Are the two hexagons superimposable so that all dots of the same color match? What does your result tell you about the stereoisomers of glucose? Is this consistent with your results in Investigate This 5.59? Explain.

Reflection and Projection

Isomers are compounds with the same elemental composition, but different properties. The molecules from a pair of isomeric compounds can always be distinguished from one another because the isomeric molecules, or models of the molecules, cannot be superimposed on one another. Structural isomers have different connectivities among the atoms, as well as different shapes. Stereoisomers have the same connectivities, but different shapes. There are two kinds of stereoisomers, *cis-trans* isomers and optical (mirror-image) isomers. You can often distinguish *cis-trans* isomers by structural drawings, because the structures are usually planar or close to planar, and telling one side from the other is easy. These double-bonded structures are relatively rigid and it is this rigidity of the chain of double bonds in retinal that causes its change in shape (shown in the chapter opening illustration) to have such a strong effect on the shape of the much larger protein molecule to which it is bonded. Optical isomers are trickier to distinguish because the mirror-image relationship of the isomers is often not so easy to see without using models or some other physical representation of the shapes. In many carbon-containing compounds, you can tell whether optical isomerism is possible by searching the structure for carbon atoms to which four different atoms or groups are bonded.

Understanding molecular structure is important, because the reactions and interconversions of structures are what most interest chemists and biologists. In order for a molecule to react, there must be sites within it that can interact strongly with other molecules. Almost invariably, the interactions are between positive and negative centers in the reacting molecules. Such interactions are quite analogous to the reactions of cations and anions discussed in Chapter 2. In the next section, we'll examine several kinds of atomic groupings in molecules that provide sites for reaction.

5.10. Functional Groups—Making Life Interesting

Alkanes As you have seen in this and previous chapters, the structure of a molecule determines its chemical as well as physical properties. In this chapter, we have emphasized the structure and bonding in carbon-containing compounds. A large number of these compounds also contain hydrogen. In most carbon-containing molecules, hydrogen is the element that occupies all the σ orbitals that the carbon atoms are not using to bond to other elements. (This fact is what enables us to leave the hydrogen atoms off skeletal structures.)

Compounds that contain *only* hydrogen and carbon are called **hydrocarbons** to emphasize the fact that only those two elements are present. Hydrocarbons in which *every* carbon atom has tetrahedral geometry are called **alkanes.** The C_5H_{12} isomers we discussed in Sections 5.1 and 5.2 and the hexane in Investigate This 5.22 are examples of alkanes. The names of all alkanes are distinguished by the -ane ending. You should be able to recognize a structure as an alkane, but we will not present the rules for naming them or any other hydrocarbons.

In general, the chemistry of the alkanes is not very eventful because both carbon–carbon and carbon–hydrogen σ bonding orbitals are quite stable. Alkanes do burn in air, however, reacting with oxygen to produce carbon dioxide, water, heat, and light energy. Hydrocarbon fuels are of immense worldwide economic importance. These fuels, often referred to as "fossil fuels," originated from once-living organisms that decayed anaerobically (without air) over millions of years. Natural gas is about 85% methane and about 15% ethane. Propane and butane are used in portable heating devices such as patio grilles and camp stoves. Gasoline is largely a mixture of low-boiling liquid hydrocarbons in the C_7–C_9 range. Though some microorganisms use methane as a source of energy (by metabolizing it, not by burning it) or produce methane as a product of their metabolism, alkanes are not generally found in living systems. Carbon-containing molecules that *are* in living things often have nonpolar hydrocarbon parts that affect their properties and interactions with other molecules.

Functional groups The possibility for interesting reactions at the temperature of life largely depends on molecules that have electrons in less stable (higher energy) orbitals. These electrons can be attracted to positive centers in other molecules and, thus, set in motion the electron and atomic core redistributions that make up chemical reactions. The orbitals that fit this criterion are π bonding orbitals and $σ_n$ nonbonding orbitals. There will be π bonding orbitals in any molecule with multiple bonds. Hydrocarbons never have $σ_n$ nonbonding orbitals, but there will be $σ_n$ nonbonding orbitals in molecules that contain other second-period elements to the right of carbon. *Multiple bonds and the presence of atoms other than carbon and hydrogen in the molecule are responsible for the chemical reactivity of most carbon-containing molecules.* Such a center of reactivity in the molecule is called a **functional group.**

Alkenes The simplest functional group involves π bonds in hydrocarbons. The only hydrocarbons with π bonding that we will consider here are the **alkenes,** compounds that have at least one double bond between carbon atoms in the molecule. The names of alkenes end in -ene. Alkenes are more reactive than the corresponding alkanes, as you observed in Investigate This 5.22, where you found that hexene reacts with potassium permanganate while hexane, an alkane, does not. Other examples of alkenes are ethene in Figures 5.13, 5.19, and 5.20 and *trans*- and *cis*-2-butene in Figures 5.33 and 5.34. Important structural characteristics of an alkene are the planar arrangement of the atoms bonded to the carbon atoms and the strong resistance to rotation of the double bond, which results in stable *cis-trans* stereoisomers of alkenes.

Functional groups containing oxygen and/or nitrogen The greatest richness of chemistry in carbon-containing compounds lies in the variety of chemical linkages between carbon and elements such as oxygen, nitrogen, and sulfur. We will begin to consider various classes of compounds where these elements are found, but this will not be an exhaustive survey. Functional groups based on oxygen and nitrogen that you have met previously and will meet again in this text are listed in Table 5.4 with an example of each. What we want you to get here is an acquaintance that will help you recognize them and focus on their polarity, which is the key to their reactivity.

Alkynes and aromatic rings also contain π bonds. Alkynes have a triple bond between two carbon atoms. Aromatic rings are typified by benzene, C_6H_6. The six carbon atoms are in a planar ring with three π-bonding orbitals delocalized around the ring. Many biomolecules contain aromatic rings, but they are often there for rigidity rather than for reactivity.

See the *Web Companion,* Chapter 5, Section 5.7.1, for more about the bonding in benzene.

Table 5.4 *Oxygen- and nitrogen-containing functional groups.*

In each representation, the lines from the carbon atoms are bonds to either hydrogen or another carbon. The ball-and-stick representations of the examples show the σ bonding framework, not σ_n or π orbitals.

Functional group	Representation	Example
alcohol		ethanol
ether		dimethyl ether
aldehyde		propanal
ketone		propanone (acetone)
carboxylic acid		ethanoic (acetic) acid
ester		methyl ethanoate (acetate)
amine (primary)		methyl amine
amide		propanamide

5.65 INVESTIGATE THIS

How many $C_3H_8O_2$ isomers are there?

Work in small groups for this investigation and then share your results with the entire class. Use your molecular models to build as many different sigma-bonded $C_3H_8O_2$ structures as you can. Record the structures you build by making drawings showing the *connections*. One of the possible structures and its representation is

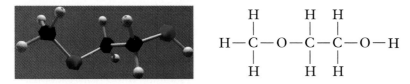

How many different structures can you build? What criteria do you use to tell that they are different from one another? Experiment with each isomer you make to see what sorts of shapes it can twist into without breaking the connections. Record your observations about the structures of the molecules.

5.66 CONSIDER THIS

What are the functional groups in $C_3H_8O_2$ isomers?

(a) Consider the drawings of the structures you made in Investigate This 5.65. Which of them contain one or more alcohol functional groups?
(b) Which structures contain a functional group (or groups) not shown in the table? How would you represent this group in the style shown in the second column of Table 5.4? Have you seen examples of such a group before? Explain.
(c) Are there any stereoisomeric pairs among your structures? Explain why or why not.

Alcohols and ethers Alcohols are compounds that contain an —OH group bonded to carbon with no other reactive atoms bonded to that carbon. An -ol ending designates the presence of the —OH group. For example, the butanol isomers in Sections 5.1 and 5.2 and the retinol (vitamin A) in Consider This 5.10 are alcohols. The endings of their names reveal that these compounds contain an alcohol functional group. Many biologically important molecules, such as the sugars you used in Investigate This 5.59, contain several alcohol groups.

Chemical reactions of alcohols are often associated with the nonbonding pairs of electrons on the alcohol oxygen. These electrons form a concentrated region of negative charge, which is attracted to the positive centers of other molecules and ions. Other reactions of alcohols occur because the electron density in the sigma bonds is pulled toward the electronegative oxygen, leaving the carbon and hydrogen as positive centers. The $^{\delta-}O-C^{\delta+}$ and $^{\delta-}O-H^{\delta+}$ bonds are both polar and the carbon and hydrogen positive centers are attracted to negative centers of other molecules and ions. In Chapter 1, you saw a result of these attractions, the hydrogen bonding among alcohol molecules. The molecule as a whole is polar, with its negative end toward the oxygen. The

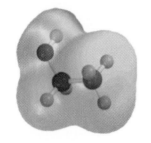

Figure 5.37.

Computer-calculated electron density distribution in ethanol.

computer-generated model in Figure 5.37 allows us to visualize the electronic distribution in the molecule and shows the charge centers we have been discussing.

Note that localized molecular orbitals disappear in the computer-calculated electron density distributions. In Figure 5.37, there is no hint of separate regions of electron density between atomic cores. The calculations that produce these pictures are based on a model of molecular bonding that accounts for the fact that electrons cannot be distinguished from one another, so that we cannot find any particular electron. Although the localized and computer models lead to different pictures, they are complementary, not contradictory. For example, both show that there is a region of high electron density associated with oxygen. Our two localized σ_n nonbonding orbitals are represented as a single region of electron density in Figure 5.37. The localized model is very successful for understanding molecular geometry and the attractions that hold molecules together. Computer modeling is very useful for understanding molecular reactions.

Ethers are compounds that contain an oxygen atom bonded to two alkyl groups. Ethers have nonbonding electron pairs on the oxygen that can form hydrogen bonds with other molecules, but ether molecules do not hydrogen bond with one another. One consequence is that ethers boil at lower temperatures than the isomeric alcohols. The oxygen atom in ethers is a negative center, as it is for alcohols, but the partial positive charge is more spread out over the alkyl groups. As a result, ethers are not as reactive as alcohols. Sometimes it's useful to think of alcohols and ethers as derivatives of water in which alkyl groups replace one hydrogen atom (alcohols) or both hydrogen atoms (ethers).

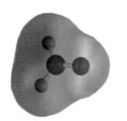

Figure 5.38.

Computer-calculated electron distribution in methanal (formaldehyde).

Carbonyl compounds Aldehydes and ketones contain a carbon–oxygen double bond called a **carbonyl group. Aldehydes** have two hydrogen atoms or a hydrogen atom and an alkyl group bonded to the carbonyl carbon, and **ketones** have two alkyl groups bonded to the carbonyl carbon. The names of aldehydes often end in -al and the names of ketones often end in -one (long oh). Pi-bonding electrons are not held as tightly as the σ-bonding electrons, so electrons in the carbonyl double bond are pulled even more toward the oxygen than those in the carbon–oxygen single bond of alcohols and ethers. Figure 5.38 shows the computer-calculated electron distribution in methanal (formaldehyde), the simplest carbonyl compound. The more extensive blue region (low electron density) at the carbon end of the molecule represents the greater shift of electron density away from the carbon atom (compared to ethanol in Figure 5.37). As with alcohols, the positive and negative ends of the carbonyl bond, $^{\delta+}C{=}O^{\delta-}$, will be attracted to complementary charges on other polar molecules and could react.

5.67 CHECK THIS

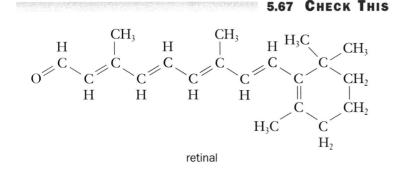

retinal

Retinol and retinal

Explain the relationship of retinol, Consider This 5.10, to retinal, shown here and in two stereoisomeric forms in the chapter opening illustration.

5.68 CONSIDER THIS

What are the interactions between an alcohol and an aldehyde?

(a) Can ethanol, Figure 5.37, and methanal, Figure 5.38, hydrogen bond with one another? If so, which charge center on each molecule will be interacting? Make a sketch, using 3-D structures to show how the molecules would interact.

(b) Make a sketch showing the other charge centers in each molecule interacting. If a partial bond were to form between the molecules in this interaction, which atom in each molecule would be bonded? Give your reasoning.

Interactions between the centers of positive and negative charge in polar molecules are similar to the interactions between positive and negative ions that we discussed in Chapter 2. Because polar molecules have only partial charges, the interactions are not as strong, but, under appropriate conditions, they can lead to bond formation of the sort you imagined in Consider This 5.68(b) and thus to new molecules. We will explore more of these interactions in Chapter 6.

Carboxyl compounds If one oxygen atom attached to a carbon atom creates a partial positive charge on carbon, two should create even more of a charge.

Carboxylic acids are characterized by a **carboxyl group,** $\overset{\delta+}{\underset{O\,\delta-}{\overset{\diagdown}{C}}}\overset{O\overline{\delta-}}{\diagup}$,

with a hydrogen atom bonded to one of the oxygen atoms, $\overset{\delta+}{\underset{O\,\delta-}{\overset{\diagdown}{C}}}\overset{O\overline{\delta-}\!-\!H}{\diagup}$ $[-C(O)OH]$.

They have a carbon atom that is positive enough to be quite reactive. The positive center at carbon is shown in the computer-calculated electron distribution for methanoic acid (formic acid), the simplest carboxylic acid, Figure 5.39. The systematic names for carboxylic acids have an -oic acid ending. You have already encountered carboxylic acids on several occasions in the text, primarily in connection with acid–base chemistry. Another kind of carboxylic acid reaction forms an **ester,** another carboxyl-group-containing compound, Table 5.4, from a carboxylic acid and an alcohol. We will discuss this reaction in Chapter 6.

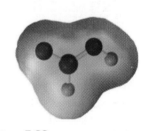

Figure 5.39.

Computer-calculated electron distribution in methanoic acid (formic acid).

5.69 CHECK THIS

Identifying functional groups

(a) What functional groups are present in retinol (Consider This 5.10)? In retinal (Check This 5.67 and chapter opener)? How many of each group are present in each molecule?

continued

(b) Find the functional group(s) in each of these molecules. Circle the atoms that define each functional group. Name each functional group. Give the molecular formula for each molecule.

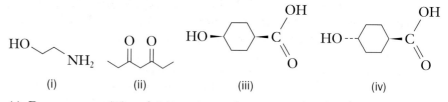

(i)   (ii)   (iii)   (iv)

(c) Do structures (iii) and (iv) represent the same molecule? Why or why not? How can you prove your answer? *Hint:* See Consider This 5.64(c).

5.70 CHECK THIS

Information from different representations of molecules

View and manipulate the representations of acetone and methylethanoate in the *Web Companion*, Chapter 5, Section 5.10.1. Which representation is most similar for the two compounds? Explain how it is similar. Explain why the similarity in electron density representations would lead to a similarity in reactivity.

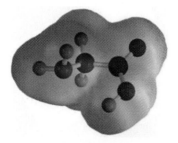

Figure 5.40.

Computer-calculated electron distribution in glycine, an amino acid.

Multiple functional groups Most biologically active compounds contain multiple functional groups. The groups are sometimes close enough together that their polarities affect one another. The amino acids found in proteins, for example, contain both a carboxylic acid group and an amine group bonded to the same carbon. Figure 5.40 shows the electron density distribution in glycine, the simplest example of these acids. Note the availability of electron pairs on the nitrogen and oxygen atoms as well as the polarization within the molecule. As we go on, you will see examples of chemical reactions of multifunctional compounds. In most cases, however, we will focus on the chemistry of an individual functional group in a compound.

5.11. Molecular Recognition

5.71 CONSIDER THIS

Can you construct a molecule-to-order?

(a) Use your molecular models to construct a pair of C_4H_8O *cis-trans* stereo-isomers. Compare your solution with those of other students. How many different *structural* isomers are represented by all these solutions?

(b) Use your molecular models to construct a pair of C_4H_8O optical stereo-isomers. Compare your solution with those of other students. How many different *structural* isomers are represented by all these solutions?

Noncovalent interactions Although we focus a great deal of attention on covalent bonds in chemistry, it is often the far weaker noncovalent interactions that control the properties of chemical and biological systems. For example, you have learned that relatively weak hydrogen bonds, typically about 5% the strength of single covalent bonds, account for the relatively high boiling points and specific heats of alcohols and water. The accuracy and reproducibility of base pairing in DNA is enhanced by hydrogen bonding. Even weaker and more subtle interactions among mixtures of polar and nonpolar molecules are responsible for the fact that "oil and water don't mix." You will see, in Chapter 8, that these interactions help form the cell membranes that separate the inside of our cells from the outside. And, at this moment, the *cis*-to-*trans* isomerization of retinal is changing the shape of opsin molecules and causing a signal to be sent to your brain (without a chemical reaction between the retinal and opsin), so that you can read this text.

Molecular recognition You may be familiar with enzymes and their role in catalyzing cellular reactions. An enzyme is a protein molecule that is quite specific in catalyzing the reaction of only one type of molecule (out of a set of very similar molecules); the reacting molecule is called its **substrate.** What is the origin of the specificity of an enzyme for its substrate? There is a region of the enzyme called the **active site (receptor),** usually in a cleft in the three-dimensional enzyme structure. Functional groups from the amino acids that make up the protein are present in the active site. In order to bind in the correct orientation for reaction, functional groups on the substrate and those in the active site must be attracted to one another by noncovalent interactions. This is the way an enzyme "recognizes" its substrate. These interactions are discussed further in Chapter 11.

5.72 CHECK THIS

One form of molecular recognition

Play the animation in the *Web Companion*, Chapter 5, Section 5.9.7, and determine which isomer fits the receptor molecular surface. Explain how this animation exemplifies the description of molecular recognition in the preceding paragraph.

Molecular recognition is not restricted to enzymes. The ways in which antibodies (also protein molecules) recognize foreign substances in our bodies, the way that one nerve cell communicates with its neighbor, and the senses of taste and smell all rely on the specific interactions of a binding site and a substrate. Recognition is becoming extremely important in nonbiological applications as well, for example, to create ordered arrays of molecules with desirable electronic properties. The more we know about the fundamental nature of molecular recognition, the better we will be able to design new materials, molecules with specific characteristics (as in the simple examples in Consider This 5.71), and devise new technologies for analysis and sensing. For example, we may soon be able to use artificial noses to detect a wide range of odors with a sensitivity well beyond the human nose. In the not-too-distant future, your refrigerator may be able to tell you when it contains spoiled food and even which food it is!

Web Companion

Chapter 5, Section 5.11 ── ①

Explore these animations ②
to see how a taste receptor ③
and its substrates interact. ④

Reflection and Projection

The sites of reactivity in carbon-containing molecules, their functional groups, are characterized by the presence of π bonding and σ_n nonbonding electrons. In hydrocarbons, the only possible functionality is the π bond. Almost all reactions of carbon-containing molecules involve functional groups containing atoms other than carbon, such as oxygen and nitrogen. These molecules all have non-bonding electrons and many also have π bonding electrons between carbon and other second-period atoms. These functional groups include alcohol, ether, carbonyl (aldehydes and ketones), carboxyl (carboxylic acids and esters), amine, and amide.

In all these functional groups, the electronegativity of oxygen or nitrogen atoms, relative to carbon atoms, polarizes the electron density toward the oxygen or nitrogen. Positive and negative centers are a characteristic of most functional groups and are responsible for their reactivity. Even this limited set of functional groups can undergo a great variety of reactions that depend on the reaction conditions and the other functional groups present in the system. In the next chapter, we will explore a variety of reactions, all of which are initiated by the attraction of positive to negative centers in molecules and ions. We finished here by examining how the shape of molecules and the interactions between their functional groups provide a means for molecular recognition.

5.12 Outcomes Review

In this chapter, we used electron wave mechanics to further develop the bonding models introduced in Chapter 1. We found that the atomic cores in all molecules are held together by attractions between their positive charges and negatively charged two-electron sigma bonding (σ) molecular orbitals located directly between them. Atomic cores in molecules may also have sigma nonbonding (σ_n) molecular orbitals. There are also molecules with multiple bonds between atoms, a σ bond and a pi (π) or two π bonds. In some cases, orbitals may be delocalized over three or more atomic centers in a molecule or over an entire crystal in metals.

The purposes for developing the bonding model are to understand and predict the shape and reactivity of molecules. The geometry (shape) of a molecule is determined by its framework of σ and σ_n electrons, and we discussed several ways, including using molecular models, to represent shapes. Isomers, both structural isomers and stereoisomers, are distinguishable from one another by different properties and different molecular shapes. The reactions of molecules generally involve attraction of a positive site in one molecule to a negative site in another. Models that help us visualize the electron distribution in molecules show the location of these sites, which are the functional groups in the molecule.

Check your understanding of the ideas in the chapter by reviewing these expected outcomes of your study. You should be able to

- Write appropriate Lewis structures for molecules whose atomic composition you know, including multiple Lewis structures representing possible isomers [Sections 5.2, 5.5, 5.8 and 5.9].
- Build molecular models based on the connectivities shown in Lewis structures of the molecules [Sections 5.2, 5.5, 5.8, and 5.9].
- Correlate patterns of chemical behavior for a series of compounds with their Lewis structures and molecular models and, based on these patterns, predict the behavior of other compounds [Sections 5.1, 5.2, 5.5, and 5.9].
- Use Lewis structures to predict multiple bonding in the molecules of a compound [Section 5.5].
- Draw and/or describe the σ, σ_n, and π orbitals for simple molecules containing elemental atoms from periods one, two, and three [Sections 5.3, 5.4, 5.6, and 5.7].
- Draw and/or describe the sigma framework of a molecule and predict the shape of the molecule [Sections 5.4 and 5.6].

- Use Lewis structures to predict whether a molecule is likely to have delocalized π orbitals [Sections 5.7 and 5.8].
- Use the delocalized molecular orbital (sea of electrons) model of metallic bonding to explain the properties of metals [Section 5.7].
- Use bonding models to determine bond order in molecules [Sections 5.7 and 5.13].
- Use bonding models to predict and/or explain the relative bond lengths in molecules [Sections 5.4, 5.5, 5.6, and 5.7].
- Write, draw, or build all the other representations of relatively simple molecules, beginning with any one of the representations: line (condensed) formula, Lewis structure, 3-D structure, condensed structure, skeletal structure, molecular model [Sections 5.2, 5.5, and 5.8].
- Use appropriate molecular representations to show whether a pair of structures are unrelated to one

another, are identical to one another, are structural isomers, or are stereoisomers (*cis/trans* and/or optical isomers) [Sections 5.1, 5.2, 5.8, and 5.9].
- Use appropriate molecular representations to show whether structural isomers and/or stereoisomers (*cis/trans* and/or optical isomers) are possible for a given molecular formula [Sections 5.2, 5.8, and 5.9].
- Identify the functional group(s) present in a molecular structure [Section 5.10].
- Use the polarities of functional groups (from electronegativities and bond polarities and/or calculated electron distributions) to predict where molecules will be likely to interact with one another [Sections 5.10 and 5.11].
- Use pictures and/or words to describe π antibonding orbitals [Section 5.13].
- Determine the bond order and magnetic properties of molecules containing electrons in π antibonding orbitals [Section 5.13].

5.13. Extension—Antibonding Orbitals: The Oxygen Story

5.73 INVESTIGATE THIS

Are N_2 and O_2 gas affected by a magnet?

Do this as a class investigation and work in small groups to discuss and analyze the results. You need a small, strong magnet, a large Petri dish about half full of a soap bubble solution, a syringe with a short needle, and sources of nitrogen gas and oxygen gas. Use an overhead projector to make your experiment easier to see. Fill the syringe with nitrogen gas, insert the needle a little below the surface of the soap solution, and blow a small, hemispherical bubble of nitrogen gas about 5–6 mm in diameter on the surface of the solution. Bring the magnet close to one side of the bubble and observe how the bubble reacts to the presence of the magnetic field. Can you move the bubble with the magnetic field from the magnet? Does the other pole of the magnet give the same result? Record your results. Repeat the experiment with oxygen gas.

5.74 CONSIDER THIS

How are N_2 and O_2 gas affected by a magnet?

Are the results for nitrogen gas and oxygen gas any different in Investigate This 5.73? If so, how do they differ? How might you explain any difference(s)?

This chapter has dealt mainly with many-atom molecules, as we have tried to prepare you to understand the structure and reactivity of the natural and synthetic substances you meet daily. In this Extension we will examine a small molecule, the diatomic oxygen molecule. Oxygen is essential to life, as it has evolved on Earth. Reactions of oxygen and fossil fuels also provide the vast amounts of energy that drive the economy of the world. Like water, oxygen is all around us and we rarely give it a second thought. Also, like water, when examined in more depth, oxygen has surprising properties. The one that you discovered in Investigate This 5.73, attraction of oxygen to a magnet, poses problems for the Lewis bonding model.

Paramagnetism of oxygen You found in your investigation that oxygen gas in a bubble can be pulled about by a magnet. The oxygen molecule is **paramagnetic.** Paramagnetic (*para* = near + magnet = nearly a magnet) substances have magnetic strengths between those of **ferromagnetic** substances (like refrigerator magnets) and **diamagnetic** (*dia* = through) substances (all electron spins paired) that are unaffected or slightly repelled by magnets. The magnetic properties of oxygen were discovered in the 19th century. Recall from Chapter 4, Section 4.9, that atoms with unpaired electron spins are attracted to the poles of a magnet. Paramagnetic substances have atoms or molecules with unpaired electron spins. Quantitative measurements on oxygen show that it has two electrons with the same spin, that is, unpaired spins. Oxygen molecules have an even number of electrons; other covalently bonded molecules with an even number of electrons do not have unpaired electrons and are not paramagnetic.

The Lewis structures for nitrogen, $:N{\equiv}N:$, and oxygen, $:\ddot{O}{=}\ddot{O}:$, both seem straightforward and predict that all electrons in both molecules will be spin-paired. Your results in Investigate This 5.73 show that nitrogen is not attracted to a magnet (it is diamagnetic), so the Lewis structure gives a correct prediction. Oxygen, however, is a strange case for which the standard Lewis rules don't fit.

Antibonding pi molecular orbitals In Section 5.6, we discussed electron wave interference and illustrated in Figure 5.16 how reinforcing waves from adjacent sigma-bonded atoms could create a π bonding molecular orbital. When we discussed wave interference in Chapter 4, Section 4.3, we said that both constructive and destructive wave interference occur; these were illustrated in Figure 4.10. In Section 5.6 we considered only the constructive interference of electron waves. Destructive interference of electron waves from two adjacent atoms is illustrated in Figure 5.41.

Figure 5.41.

Electron wave interference (destructive) to form a π* antibonding orbital. Electron wave amplitude between the atomic cores is cancelled for waves with nonmatching + and − amplitudes. Dashed lines show nodes. The cancellation has created a second nodal plane perpendicular to the bond axis.

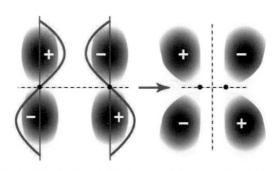

Separated atomic *p* orbitals   π* Molecular orbital

The molecular orbital that is formed by destructive interference is called a **pi antibonding orbital,** symbolized as π^* (pi star) to signal its antibonding character. You can see that the π^* orbital concentrates electron density *beyond* the two atomic cores, so its attraction to the atomic cores is in a direction that pulls them apart rather than holds them together. The π^* molecular orbitals are called antibonding because these electrons *raise* the energy of the molecule *relative to the atoms from which it is made.* Most molecules do not have any valence electrons in π^* antibonding orbitals, but two that do are essential to life. Molecular oxygen, O_2, and nitric oxide, NO, are two molecules that have electrons in π^* antibonding orbitals.

Molecular orbital model for oxygen Figure 5.42 is an approximate energy-level diagram, similar to Figure 5.25, for the bonding and antibonding valence molecular orbitals in O_2. The two nonbonding orbitals, one on each atom, are omitted from this diagram, since their four electrons don't affect the bonding energies. The energies in this diagram are given *relative to the energies of the separated atoms.* Bonding orbitals are at lower (negative) energies and antibonding orbitals are at higher (positive) energies. There are two sets of energy-degenerate orbitals, the π and π^* orbitals.

The arrows in Figure 5.42 represent the eight valence electrons in these orbitals, with the direction of the arrow representing the spins of the electrons. In accord with our model that describes electrons by the lowest energy orbitals available to them, we can accommodate six electrons in the three lowest energy orbitals: one σ bonding and two π bonding orbitals. We need to add two more electrons to complete the molecule, and the next lowest energy orbitals are two energy degenerate π^* antibonding orbitals. We place one electron in each π^* orbital, since that keeps them in different regions of space and reduces electron–electron repulsion.

Because the two antibonding electrons are in different orbitals, they do not need to be spin-paired. Recall the experimental evidence from Chapter 4, Sections 4.10 and 4.11 (especially Table 4.3 and Consider This 4.62), which showed that, when single electrons occupy energy-degenerate orbitals, they do so with unpaired spins. Thus, we predict that the ground state of oxygen will have two electrons of the same spin in the π^* orbitals, as shown in Figure 5.42. These unpaired electron spins explain the observed paramagnetism of oxygen.

In Figure 5.42, there are paired electrons in three bonding orbitals, a σ and two π, just as in the triple bonds you have seen previously. However, the two electrons in the π^* antibonding orbitals cancel the bonding energies of two of the electrons in the π bonding orbitals. Therefore, the **bond order,** that is, the effective number of two-electron bonds is two; molecular oxygen has a double bond. The molecular orbital model successfully explains (or rationalizes) both the paramagnetism and double-bond properties of molecular oxygen.

> Nitric oxide is a toxic gas, but in minute amounts it plays an essential role as a nervous system signaling molecule, a function that has been recognized only since 1990. It also has several other biological functions that are just being recognized.

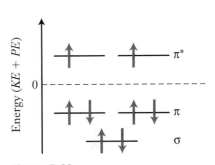

Figure 5.42.

Relative energies for oxygen molecular orbitals. Energies are relative to the separated atoms. The spins of electrons in the orbitals are shown as up and down arrows. Levels for the nonbonding orbitals, which contain four electrons, are at the zero of energy and are not shown.

5.75 CHECK THIS

Molecular orbitals in nitrogen

Assume that the energy-level diagram in Figure 5.42 is also applicable to N_2. Draw a diagram like Figure 5.42 that describes N_2. Show the spins of electrons described by each orbital as up and down arrows. What is the bond order for N_2? Are the molecular orbital and Lewis descriptions compatible? Explain.

Ground-state oxygen reacts slowly Its unpaired electrons profoundly affect the reactivity of molecular oxygen. Ground-state oxygen, with its unpaired electron spins, is quite unreactive. For ground-state oxygen, room temperature reactions go so slowly as to be almost unobservable. Combustion reactions of oxygen and fuel molecules are rapid, but they require a source of energy to get the oxygen and fuel hot, before the reaction can occur. In the laboratory, we can easily excite oxygen to states in which all its electrons are spin-paired (the two highest energy electrons in the π^* orbitals but with opposite spins). The room temperature reactions of spin-paired oxygen with carbon-containing compounds are rapid and readily destroy these compounds.

If oxygen had a spin-paired ground state, it is unlikely that life based on aerobic reactions would exist. The retinal molecule shown in the chapter opening illustration would probably be destroyed as fast as it could be formed. Recalling Wordsworth's thought from the beginning of the chapter, you might think that it is fortunate that the active principle assigned to the form (electronic state) of oxygen is to be *less* active. Oxygen is perhaps an extreme example of the connection between molecular bonding and reactivity, but it is a good example to keep in mind as you continue through the text and see many more examples of this fundamental connection.

Chapter 5 Problems

5.1. Isomers

5.1. This is the structure of 2-methyl-1-propanol, $C_4H_{10}O$.

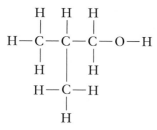

Which of these structures is also a correct representation of 2-methyl-1-propanol? Explain your decision in each case. Use models, if necessary, for your explanation.

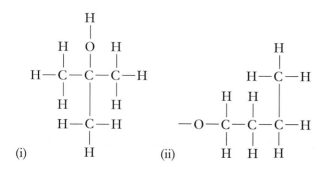

(i)   (ii)

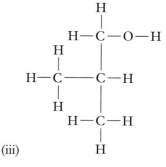

(iii)

5.2. A student followed the procedure in a laboratory manual to prepare an ionic compound, a green solid with the composition, $Co(H_2NCH_2CH_2NH_2)_2Cl_3$. She dissolved the compound in water, warmed the solution, and found that the solution turned reddish purple. She evaporated the water, recrystallized the reddish-purple crystals, and analyzed them for cobalt, chloride, and 1,2-diaminoethane, $H_2NCH_2CH_2NH_2$. She found that the reddish-purple compound had the composition, $Co(H_2NCH_2CH_2NH_2)_2Cl_3$. What conclusion(s) can you draw about the relationship(s) between the green and reddish-purple compounds? Explain your reasoning.

5.2. Lewis Structures and Molecular Models of Isomers

5.3. Both 1-pentanol and 2-pentanol have the same molecular formula, $C_5H_{12}O$.
(a) Draw the structure of each alcohol.

(b) Predict which of these two alcohols has the higher boiling point. Explain the reason for your prediction.
(c) Draw all alcohols with the formula $C_5H_{12}O$. *Hint:* There are eight total, including 1-pentanol and 2-pentanol.

5.4. Predict the relative values for the boiling points of 1-propanol and 2-propanol. Explain the reason for your prediction.

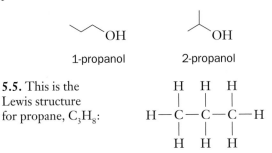

1-propanol   2-propanol

5.5. This is the Lewis structure for propane, C_3H_8:

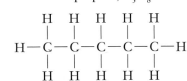

(a) Use your model kit to make this model, and then draw a representation of this structure in three dimensions.
(b) What is the shape (geometry) around the central carbon atom?
(c) Are there any other isomers for propane, C_3H_8?

5.6. This is the Lewis structure of n-pentane, C_5H_{12}.

$$H-\overset{\overset{\displaystyle H}{|}}{\underset{\underset{\displaystyle H}{|}}{C}}-\overset{\overset{\displaystyle H}{|}}{\underset{\underset{\displaystyle H}{|}}{C}}-\overset{\overset{\displaystyle H}{|}}{\underset{\underset{\displaystyle H}{|}}{C}}-\overset{\overset{\displaystyle H}{|}}{\underset{\underset{\displaystyle H}{|}}{C}}-\overset{\overset{\displaystyle H}{|}}{\underset{\underset{\displaystyle H}{|}}{C}}-H$$

(a) How many valence electrons are represented in this structure?
(b) Including electrons in atomic cores, what is the total number of electrons implicitly represented?

5.7. The boiling points of the three pentane isomers are given in Table 5.2. Make molecular models of these compounds and cover them with plastic wrap or aluminum foil to represent the surface of each molecule. Explain why n-pentane has the highest boiling point and neopentane has the lowest boiling point of this set of isomers. *Hint:* Consider the amount of surface each molecule has to interact with its neighbors in the liquids.

5.8. In Check This 5.11, you were asked to explain the boiling point data (Table 5.1) for the four $C_4H_{10}O$ alcohols. Draw the three ethers that also have the formula $C_4H_{10}O$ and predict which one will have the lowest boiling point. Explain the reasoning for your selection. *Hint:* See Problem 5.7.

5.9. In Check This 5.7 you wrote Lewis structures and made models of the four isomeric amines having the formula C_3H_9N. Predict which one will have the lowest boiling point and explain the reasoning for your selection. *Hint:* Consider all the possible intermolecular interactions in the liquids.

5.3. Sigma Molecular Orbitals

5.10. Which electrons, core or valence, are generally responsible for chemical bonding? Explain why.

5.11. Explain the difference(s) between an atomic orbital and a molecular orbital.

5.12. Explain the difference(s) between a bonding and a nonbonding molecular orbital.

5.13. Why do two hydrogen atoms combine to form a hydrogen molecule (H_2)? Explain using
(a) a Lewis structure for bonding.
(b) the MO model for bonding.

5.14. Why don't two helium atoms combine to form a helium molecule (He_2)? Explain using
(a) a Lewis structure for bonding.
(b) the MO model for bonding.

5.15. Explain why it is possible for molecules to exist, if there is always a repulsive force between positively charged nuclei.

5.16. Identify the number of valence electrons in each of these atoms or molecules:
(a) Na   (b) CO_2 (c) NH_3 (d) Se

5.17. Consider Figure 5.4, "Relative energies for a two-nuclei, one-electron molecular orbital system." It has been redrawn here with curves 1–4 labeled.

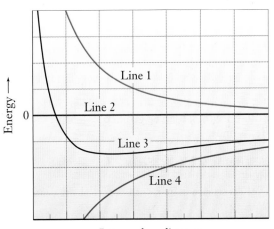

(a) What information does line 1 give you as two nuclei move toward each other?
(b) What information does line 2 give you as two nuclei move toward each other?
(c) What information does line 3 give you as two nuclei move toward each other?
(d) What information does line 4 give you as two nuclei move toward each other?

5.4. Sigma Molecular Orbitals and Molecular Geometry

5.18. In certain gas-phase reactions, methane can lose a hydrogen cation, H^+, to form the methide anion, CH_3^-.
(a) Write the Lewis structure for the methide anion.
(b) How would you describe the geometry of the sigma orbitals in the methide anion?
(c) How would you describe the geometry of the methide anion?
(d) Are your answers to parts (b) and (c) the same? Explain why or why not.

5.19. Sodium amide, $NaNH_2$, is a white crystalline ionic compound.
(a) Write the Lewis structure for the amide anion, NH_2^-.
(b) How would you describe the geometry of the sigma orbitals in the amide anion?
(c) How would you describe the geometry of the amide anion?
(d) Are your answers to parts (b) and (c) the same? Explain why or why not.

5.20. (a) How many valence electrons are there in the Lewis structure for IF_5?
(b) How many total electrons, including all of the core electrons, are there in a molecule of IF_5?
(c) Write a Lewis structure for IF_5.
(d) The geometry of the fluorine atoms in the IF_5 molecule is square pyramidal (a pyramid with a square base). Why is this different from the molecular shape of PF_5, in which the geometry of the fluorine atoms is trigonal bipyramidal, as shown in Figure 5.11(b)?

5.21. For each of these molecules or ions, write a Lewis structure, including all nonbonding pairs of electrons.
(a) I_3^- (b) BF_4^- (c) SF_4 (d) XeF_4 (e) PF_6^-

5.22. For each molecule or ion in Problem 5.21, explain why these are the observed geometries (shapes).
(a) I_3^- is linear.
(b) BF_4^- is tetrahedral.
(c) SF_4 is shaped like a see-saw, $\overline{\diagup_{\diagdown}}$, with S at the fulcrum.
(d) XeF_4 is square planar, with the Xe at the center of the square.
(e) PF_6^- is octahedral with the P in the center of the octahedron.

5.5. Multiple Bonds

5.23. Which second-period elements are capable of making double bonds under normal conditions? Why are the others not included? Explain the reasoning for your choices.

5.24. Which elements never make double bonds? Explain the reasoning for your choices.

5.25. For each molecule, write a Lewis structure and determine the molecular geometry for each carbon atom. *Hint:* Each molecule has a multiple bond. Molecular models may be helpful.
(a) C_3H_6 (Propene has one carbon–carbon double bond.)
(b) C_3H_4 (Propyne has one carbon–carbon triple bond.)
(c) C_3H_4 (Allene has two carbon–carbon double bonds.)
(d) C_2H_4O (Ethanal has one carbon–oxygen double bond.)
(e) C_2H_3N (Acetonitrile has one carbon–nitrogen triple bond.)

5.6. Pi Molecular Orbitals

5.26. What is(are) the difference(s) between a sigma (σ) and a pi (π) orbital?

5.27. Why do all molecules have sigma bonding orbitals, but not necessarily pi bonding orbitals?

5.28. If two carbon atoms are bonded by a sigma bonding orbital, why is a second bond between those two atoms a pi orbital rather than another sigma bonding orbital?

5.29. Why are molecules with multiple bonds generally more reactive than similar compounds without multiple bonds?

5.7. Delocalized Orbitals

5.30. Consider these molecules and ions. Circle each one that contains a π bond. For those molecules with a π bond, place a D next to ones where the π bond is delocalized and an L next to ones where the π bond is localized. Building models may be helpful.
(a) H_2CCH_2 (d) NH_3 (g) CH_4
(b) H_2CO (e) CO_2 (h) SO_2
(c) NO_3^- (f) $CH_3C(O)OH$ (i) H_2O

5.31. Consider the hydrogen carbonate anion, $(HO)CO_2^-$, and the ozone molecule, O_3.
(a) What is the total number of valence electrons in each species?
(b) Write the Lewis structure for each species.
(c) How many σ bonding orbitals are there around the central atom in each species?
(d) What geometry is predicted for each species?
(e) How many π bonding orbitals are there around the central atom in each species?
(f) How many localized and how many delocalized π bonding orbitals are there in each species? Explain how you arrive at your answer.
(g) What are the bond orders for carbon-to-oxygen in $(HO)CO_2^-$ and oxygen-to-oxygen in O_3? Use sketches of the delocalized orbitals, if any, to illustrate your answer. What, if any, similarities are there between the delocalized orbitals and the bond orders in these two species? Explain.

5.32. (a) We have said that molecules are likely to have delocalized pi orbitals, if there is more than one energy-equivalent way to write localized pi orbitals. Write the Lewis structure for carbon dioxide, as you did in Consider This 5.38. Would you expect the molecule to have delocalized pi orbitals? Explain why or why not.
(b) The molecular orbital picture for carbon dioxide, Figure 5.24, shows two pi orbitals delocalized over all three atomic cores in the molecule. Is this consistent with your answer in part (a)? Explain why or why not.
(c) Another way to represent the localized pi orbitals in carbon dioxide is shown in this molecular model. (The sigma nonbonding electrons are not represented in the model.) Is there more than one equivalent way to model these pi orbitals? Explain why or why not and construct a model (or models) to demonstrate your explanation.
(d) Is your answer in part (c) consistent with the delocalized pi orbital picture for carbon dioxide? Does it explain the observed bond lengths (Check This 5.39)?

5.33. Consider the nitrate anion, NO_3^-, the nitrite anion, NO_2^-, and the nitronium cation, NO_2^+.
(a) What is the total number of valence electrons in each ion?
(b) Write the Lewis structure for each ion.
(c) How many s bonding orbitals are there around the central atom in each ion?
(d) What geometry is predicted for each ion?
(e) How many π bonding orbitals are there around the central atom in each ion?
(f) How many localized and how many delocalized π bonding orbitals are there in each ion?
Explain how you arrive at your answer.
(g) What is the bond order for nitrogen-to-oxygen in each ion? Use sketches of the delocalized orbitals, if any, to illustrate your answer.
(h) What do you predict for the relative bond lengths, shortest to longest, in these ions? Explain your reasoning.

5.34. A single Lewis structure for the nitrate ion doesn't represent the distribution of the π bonding orbital over all four atom cores, an alternative that has been extensively used is to write more than one Lewis (or 3-D) structure:

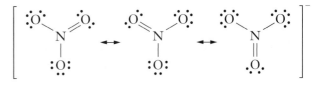

The double-headed arrows are used to indicate that the best representation is some intermediate structure (that is impossible to write).
(a) If all three of these structures existed separately, they would have identical energies. Explain why.
(b) If the structures have identical energies, then we would expect an intermediate structure based on them to have equal contributions from each structure. What fraction contribution would each make to the intermediate structure? What is your reasoning?
(c) The delocalized π bonding structure, Figure 5.23(a), led us to the conclusion that all three of the nitrogen–oxygen bonds in nitrate can be thought of as $1\frac{1}{3}$ bonds. Show how the same conclusion can be reached here, based on equal contributions from these three structures to the proposed intermediate structure.

5.35. When some material is connected between the terminals of a battery, it experiences an electrical potential difference between the two connections to the battery. If electrons can easily enter the material at the negative terminal and leave at the positive terminal, we call the material an electrical conductor. Metals are good electrical conductors. How does the delocalized molecular orbital (sea of electrons) model of metallic bonding help you understand the high conductivity of metals? Present your reasoning clearly.

5.36. You can often break a piece of wire or thin strip of metal by bending it back and forth repeatedly. This action causes dislocations of atoms in the crystal like those you see represented in Figure 5.26. Explain how this effect can eventually lead to breaking the piece of metal.

5.8. Representations of Molecular Geometry

5.37. The number of isomers possible for any particular molecular formula depends on the geometry about the atoms that make up the molecule. For each of these compounds, use your models to figure out how many isomers are possible if the four bonds about the central carbon atom (shown in **bold**) have a square planar molecular shape. How many isomers are possible if the four bonds about the central carbon atom have a tetrahedral molecular shape?
(a) **CH**$_3$**Br** (methyl bromide, a fumigant)
(b) **CH**$_2$**Cl**$_2$ (dichloromethane, a solvent used in semiconductor processing)
(c) CH$_3$CH(OH)CH$_2$OH (propylene glycol, a shampoo additive)

5.38. Skeletal structures for several molecules are given. Draw the condensed representations for each molecule. What is the molecular formula in each case?

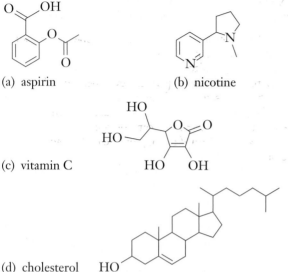

(a) aspirin

(b) nicotine

(c) vitamin C

(d) cholesterol

5.39. Conformations are forms of a molecule that differ only in that there have been rotations about single bonds. Identify which of these *pairs* of structures are conformers (that is, the same molecule) and which are actually different molecules that are not interconvertible by any combination of rotations about single bonds or motions of the molecule as a whole. Building molecular models might be helpful in some cases.

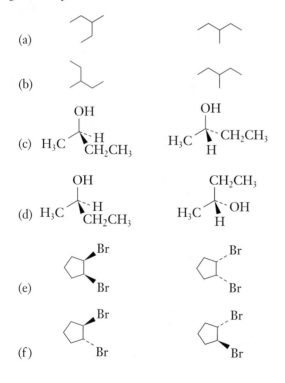

(a)

(b)

(c)

(d)

(e)

(f)

5.40. Rotations about single bonds are rapid. This allows molecules easily to assume different conformations. (See Problem 5.39.) For each of these molecules, can you find a conformation that allows all the carbon atoms to lie in a single plane? Draw 3-D structures to illustrate your answers. *Hint:* Molecular models are extremely valuable in answering this question.
(a) butane, $CH_3CH_2CH_2CH_3$
(b) 2-methylpropane, $CH(CH_3)_3$ (common name: isobutane)
(c) 1,3-butadiene, $CH_2CHCHCH_2$
(d) 1,2-propanediene, CH_2CCH_2 (common name: allene)

5.41. For each of these molecules, can you find a conformation (see Problem 5.39.) that allows all the carbon atoms AND all the hydrogen atoms to lie in a single plane? Draw 3-D structures to illustrate your answers.
(a) butane, $CH_3CH_2CH_2CH_3$
(b) 2-methylpropane, $CH(CH_3)_3$
(c) 1,3-butadiene, $CH_2CHCHCH_2$
(d) 1,2-propanediene, CH_2CCH_2

5.42. For each of these molecules, can you find a conformation (see Problem 5.39) that allows all the atomic centers to lie in a single plane? Draw the Lewis structures first to help answer this question. Draw 3-D structures to illustrate your answers.
(a) methanol, CH_3OH (wood alcohol)
(b) hydrogen peroxide, H_2O_2 (used as a disinfectant and to bleach hair)
(c) hydrogen cyanide, HCN   (poisonous gas)
(d) nitric acid, $(HO)NO_2$ (useful industrial acid)
(e) nitrous acid, $(HO)NO$ (nitrite salts are suspected carcinogens)

5.9. Stereoisomerism

5.43. Stereoisomers (isomers that share the same connectivity of their bonds but differ in their three-dimensional shapes) can be subcategorized as optical isomers (those stereoisomers that bear a mirror-image relationship to one another) and those that are not optical isomers. Identify which of these *pairs* of structures are *identical*, which are *optical isomers*, and which are *other types of stereoisomers*. Building models could be very helpful.

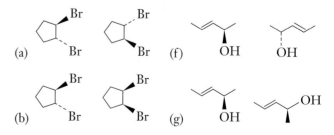

(a)

(b)

(f)

(g)

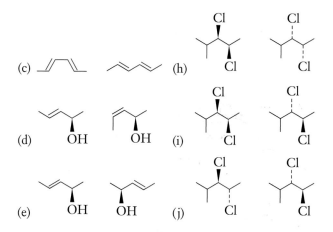

(c) ⟋⟍⟋⟍

(d) OH   OH

(e) OH   OH

(h) Cl / Cl / Cl / Cl

(i) Cl / Cl / Cl / Cl

(j) Cl / Cl / Cl / Cl

5.44. (a) Complete this table by filling in the blanks with either "the same" or "different."

Property	Structural isomers	Stereoisomers	
		Optical isomers	Other
molar mass			
boiling point			
density			
bond connectivity			

(b) For one of these data columns, all the entries are "the same." By definition, two compounds cannot be isomers if *all* of their properties are the same. What is the property that allows us to distinguish the isomers in this column?

5.45. Use your molecular models to make as many models of isomers of C_5H_{10} as you can. How many are there? (If it is easy for you to visualize molecular structures, like Lewis or condensed structures, written on a piece of paper, you can do this problem with a pencil and paper.) How does your result compare with the number of C_4H_8 isomers you obtained in Investigate This 5.55? What conclusion(s) can you draw about the relationship between number of carbons and number of isomers? Does this conclusion make sense? Explain why or why not.

5.10. Functional Groups— Making Life Interesting

5.46. (a) Use your molecular model kit to construct a model of cyclopentane. (See Investigate This 5.45.) How flexible is the ring?
(b) Remove one of the CH_2 groups and reconnect the ring to make a four-carbon ring, cyclobutane. How easy is it to do this? How flexible is the ring?

(c) Remove another CH_2 group and reconnect the ring to make cyclopropane. How easy is it to do this? How flexible is the ring? Would you expect cyclopropane to be more reactive, less reactive, or about the same as cyclopentane? Give the reasoning for your answer.

5.47. Use your model kit to make a cyclopentane model. (See Investigate This 5.45.)
(a) Replace one of the hydrogen atoms with an —OH group; the model now represents cyclopentanol. Is there more than one structure for cyclopentanol? Why or why not? If so, make separate models of each one. Are any of them optical isomers? How can you tell?
(b) Add a second —OH group on the carbon adjacent to the first —OH group. The new model is 1,2-cyclopentane-diol (*di* = two). Is there more than one structure for the diol? Why or why not? If so, make separate models of each one. Are any of them optical isomers? How can you tell?

5.48. Table 5.4 catalogs a number of important functional groups. Use the table to answer the following questions.
(a) Which functional groups contain oxygen?
(b) Which functional groups contain nitrogen?
(c) Which functional groups in the table are most likely to ionize? (You may have to refer back to Chapter 2.) Explain your reasoning.

5.49. Can you discover a mathematical relationship between the number of Cs and the number of Hs in an alkane? [Because this is a rather abstract question, we will give you an example to illustrate what we mean. If you wished to know the number of toes in a crowded room, you could simply count the number of feet and multiply by 5. Mathematically, the number of toes is a function of the number of feet. Note that this relationship is indepen-dent of the number of feet in the room (assuming no three-toed aliens or other exceptions!).]
(a) Can the number of Hs in an alkane be described as a function of the number of Cs? You will need to examine a number of alkanes to be certain that you have discovered a general relationship. Your relationship should fit branched as well as straight-chain alkanes.
(b) Can the number of Hs in an alk**ene** be described as a function of the number of Cs?
(c) Can the number of Hs in an alk**yne** be described as a function of the number of Cs?

5.50. Certain terms used in the nomenclature of carbon-containing compounds sound very much alike but do not mean the same thing at all. Clearly explain the differences between these pairs of words and draw a specific com-pound illustrating each.
(a) alkane   alkene
(b) alcohol   aldehyde
(c) ether   ester
(d) amine   amide
(e) carboxyl   carbonyl

5.51. Draw the Lewis structures and structural formulas for each of these compounds. Draw bond dipoles on the structural formulas. Identify the shape of each compound. Identify if the compound is polar or nonpolar.
(a) H_2CO (formaldehyde)
(b) $HO(O)CCH(NH_2)CH_2C(O)OH$ (aspartic acid)
(c) $H_2NCH_2CH_2NH_2$ (1,2-diaminoethane)

5.52. (a) In Problem 5.3, you drew structures for isomeric alcohols with formula $C_5H_{12}O$. There are six other structures that have the formula $C_5H_{12}O$. Draw their structures and identify the functional group(s) in these molecules.
(b) Predict which one of the six molecules in part (a) will have the lowest boiling point. Explain the reasoning for your choice.

5.11. Molecular Recognition

5.53. In a double helix of DNA, the nitrogenous base cytosine is held to the nitrogenous base guanine by hydrogen bonds. (See Investigate This 1.33.) Dashed lines in the diagram represent hydrogen bonds.

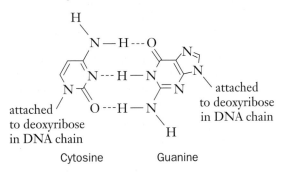

Cytosine   Guanine

(a) How many hydrogen bonds can form between thymine and adenine? Use these two structures for thymine and adenine to draw a diagram showing the hydrogen bonding that can take place between thymine and adenine.

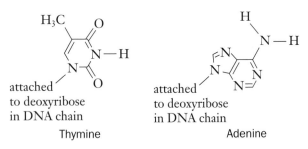

Thymine   Adenine

(b) Thymine does not pair with guanine. Offer a possible explanation for this observation.
(c) DNA has the ability to replicate (make copies of itself) accurately and reproducibly. How can this be related to base pairing? Explain your reasoning.

5.54. 🖐 Work through the *Web Companion*, Chapter 5, Section 5.11.1–3, before answering the following questions.
(a) Explain the relationship of the activity in Section 5.11.3 to the one you did in Check This 5.72.
(b) Make a model of aspartame. A ball-and-stick model is shown rotating in Section 5.11.1, and here is a 3-D representation of the molecule:

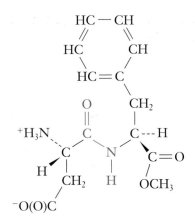

An important point to note about the structure of this dipeptide (aspartyl phenylalanine methyl ester) is that the four atoms in the center of the structure, $O=C-N-H$ (shown in red), lie in a plane. Rotation about the $C-N$ bond is very restricted. This bond acts much like a double bond because interactions between the pi bond in the carbonyl and the nonbonding electrons on the nitrogen atom delocalize the electrons over the three second period atoms and make their structure rigid. Two possible ways to construct the six-membered ring are shown in these photographs. Use whichever one you wish.

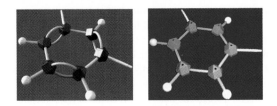

(c) As sweet receptor, cut an appropriate size hole in a piece of paper and color its edge red, blue, and green as in the sweet receptor shown in Section 5.11.3 (or simply label the edge with charges). Show that your aspartame model complements the receptor, as does the correct phenylalanine isomer in the WEB activity.
(d) Exchange the $-H$ and $-NH_3^+$, on your model (left-hand side of the structure above) and try part (c)

with the new isomer. Will this isomer still taste sweet? Explain why or why not.

(e) Undo the exchange in part (d) and then exchange the —H and one of the other groups attached to the central carbon at the right-hand side of the structure above. Try part (c) with the new isomer. Will this isomer still taste sweet? Explain why or why not. Discuss the similarities and differences between your observations in parts (d) and (e).

5.13. Extension—Antibonding Orbitals: The Oxygen Story

5.55. The double bond in O_2 and a triple bond between two carbons have almost the same bond length, 121 and 120 pm, respectively. If bond lengths get shorter as bond order increases, why is the double bond between oxygen atoms the same length as the triple bond between carbon atoms?

5.56. Fluorine (F_2) is the most reactive of the halogens. According to the Lewis model, each fluorine atom of this covalent compound has three nonbonding pairs of electrons.

(a) How many valence electrons are there in diatomic fluorine?

(b) Assume that the molecular orbital diagram shown in Figure 5.42 is also applicable to molecular fluorine. Draw a diagram like Figure 5.42 that shows how the valence electrons are assigned to the various molecular orbitals. (Remember that two nonbonding sigma orbitals are not shown in the diagram.)

(c) What is the bond order of the F-to-F covalent bond that you derive from your results in part (b)? Explain clearly.

(d) How does the molecular orbital bonding model correlate with the Lewis model for fluorine?

General Problems

5.57. Write a Lewis structure for the cyanate anion, OCN^-. Can you write more than one satisfactory structure?

(b) What geometry does(do) your Lewis structure(s) predict for the cyanate anion? How does the structure and geometry of this ion compare to the structure and geometry of the carbon dioxide molecule?

(c) Would you expect the cyanate anion to have delocalized π electrons? Explain the reasoning for your response.

(d) Answer these same questions for the isocyanate anion, ONC^-.

(e) Would you expect the cyanate or isocyanate anion to be the more stable isomer? Give the reasoning for your choice.

5.58. (a) At room temperature, beryllium fluoride, BeF_2, is a white crystalline solid and boron trifluoride, BF_3, is a gas. What conclusions can you draw about the bonding in these compounds? Explain.

(b) Write the Lewis structure for BF_3. What geometry do you predict for BF_3? What figure in the chapter helps you explain your choice of geometry? Explain. *Hint:* This compound does not obey the octet rule.

(c) Under certain conditions, BeF_2 molecules can be observed in the gas phase. Write the Lewis structure for BeF_2. What geometry do you predict for BeF_2? What figure in the chapter helps you explain your choice of geometry? Explain. *Hint:* This compound does not obey the octet rule.

5.59. Write Lewis structures for methanoic acid, $HC(O)OH$, and the methanoate anion, $HC(O)O^-$. The C—O single and double bond lengths in the acid are 134 and 120 pm, respectively. What do you predict for the C—O bond lengths in the anion? Explain the basis for your prediction.

C H A P T E R

Omnia mutantur, et nos mutamur in illis. (All things change, and we change with them.)

ATTRIBUTED TO THE EMPEROR LOTHAR I (795–855)

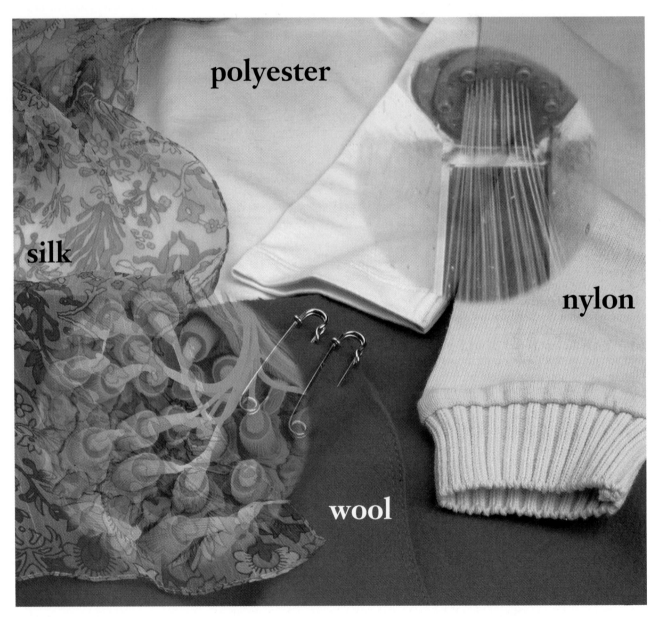

For millennia, humans have used natural polymers such as wool and silk to make cloth for clothing, shelter, and other purposes. Within the past century, our knowledge of atomic and molecular structure has led to the discovery of reactions to make synthetic polymers, such as nylon and polyester, with the same bonding as natural polymers and often with properties tailored to a particular purpose, such as clothing or containers. Spinnerets used to form nylon fibers (upper right) are human versions of the spinnerets a spider uses (lower left) to produce silk for her web.

Chemical Reactions

ore than 7000 years have passed since humans learned to purposefully create new substances that better served their needs. Early artisans discovered that the green mineral malachite was almost magically transformed in the heat of crude ovens to a reddish, lustrous metal, which we now call copper. A material that scarcely existed in the natural world had been born in the flame. It could be worked into useful shapes for tools, weapons, and ornaments. A copper–tin alloy called bronze, and then iron, became the central metals on which civilization depended. Civilizations are built on more than metal, so, while our ancestors were busily converting various stones into metals, they were just as busily converting grain into beer and bread. Taking advantage of their ability to shape their material world, hunter–gatherer societies became agrarian and mercantile societies. As the quotation on the previous page suggests, change seems to be a constant of nature and of us.

Today, we continue to convert available resources into an almost unimaginable number of items that we use every day. The illustration on the facing page shows a few examples that emphasize materials and products containing natural or synthetic (manufactured) polymers as well as the production of two polymers, silk from spider spinnerets and nylon from industrial spinnerets. Our ancestors used natural polymers like cotton, wool, and silk to make cloth and decorated the cloth with various natural dyes. In all these processes, smelting, brewing, baking, dyeing, and many others, they learned a great deal of practical chemistry by tinkering with stuff. However, they had no underlying model of matter, so they had to rely on trial and error, guided by past experience, in their search for improvements or new processes.

Only within the past few hundred years has our chemical knowledge become well enough organized to carry out studies directed at understanding the fundamental basis of the reactions upon which our civilization depends and developing entirely new materials and processes. Our predictive power has increased enormously as we have learned how molecules and ions are attracted to each other, and how atoms and electrons redistribute themselves to form the products of reactions, including the polymers shown on the facing page.

The common characteristic of all these reactions is that they begin with the attraction of positive and negative centers for one another. From the beginning of the book, we have emphasized that the fundamental basis for atom formation, molecular bonding, and molecular interactions is the attraction of positive and negative charges (or partial charges). In this chapter, we will extend this fundamental idea to interactions that lead to chemical reactions. The objective is for you to develop the ability to use what you have learned about molecular structure to make predictions about what is likely to happen when chemicals are mixed with one another.

6.1. Classifying Chemical Reactions

What chemical reactions do you know?

(a) Work in small groups for a few minutes listing all the changes you can think of that are chemical reactions. Share your list with the rest of the class and see how many different reactions the whole class has come up with. How many reactions were listed by more than one group?

(b) How did you decide whether a change is the result of a chemical reaction? List the criteria used to make this decision and to accept the reactions suggested by each group.

What changes do you observe?

Do this as a class investigation and work in small groups to discuss and analyze the results. Prepare four 250-mL beakers or clear, colorless plastic tumblers with the contents shown in this table. (A tsp is one rounded teaspoonful; dried yeast is a packet of dried baker's yeast; and $KHC_4H_4O_6$ is potassium hydrogen tartrate, also called cream of tartar.)

Beaker	1	2	3	4
Contents	dried yeast	dried yeast + tsp glucose	tsp $CaCl_2$ + tsp $NaHCO_3$	tsp $NaHCO_3$ + tsp $KHC_4H_4O_6$

To each beaker, add about 20 mL of warm (45 °C) water and swirl to mix the ingredients. Observe the beakers for about 2 minutes and record any evidence of change(s) occurring. If there is no apparent change in one or more of the beakers, wait for a few minutes and check them again. Continue this checking until you are convinced no change is going to occur.

Your lists of changes in Consider This 6.1 show that you are already quite familiar with chemical reactions. Evidence that different substances are present before and after the change was probably one of the criteria you used for deciding whether a change is a chemical reaction. For example, the evidence might be that the reaction mixture foams or bubbles, like some of the mixtures in Investigate This 6.2, indicating that a new substance, a gas, has been produced. We can define a **chemical reaction** as a change that produces molecular structures or ionic combinations that are different from those in the reactants.

Identifying reaction products Because chemical reactions produce new compounds, product identification is an important part of the process we use to find out whether a chemical reaction has occurred and, if so, what kind.

6.3 INVESTIGATE THIS

How can you analyze the gas from a reaction?

Do this as a class investigation and work in small groups to discuss and analyze the results. Repeat the reaction in the fourth beaker in Investigate This 6.2 but carry out the reaction in a stoppered, 250-mL filter flask with a length of rubber tubing attached to the filter arm, as shown in the diagram. Clamp the open end of the tubing below the surface of the liquid in a test tube about half full of limewater [a saturated solution of calcium hydroxide, $Ca(OH)_2$]. Add the solid reactants to the flask and then stopper it as soon as the water has been added. Swirl the flask to mix the reactants and observe and record what happens for a few minutes.

6.4 CONSIDER THIS

What is the gas from the $NaHCO_3 + KHC_4H_4O_6$ reaction?

(a) What did you observe as the gas from the reaction in Investigate This 6.3 bubbled into the limewater? Recall the reactions of limewater in Chapter 2, Section 2.16. What is the gas? Give the reasoning for your identification.

(b) Do you think that this gas is a reasonable product to expect from these reactants? Explain the reasoning for your answer.

The three reactions in Investigate This 6.2 that produce a gaseous product all produce the same gaseous product. The reaction in the yeast–sugar mixture, the yeast fermentation of sugar, is complex, but the fermentation of glucose (also called dextrose) can be written as

$$C_6H_{12}O_6(aq) \rightarrow 2CH_3CH_2OH(l) + 2CO_2(g) \tag{6.1}$$

Fermentation is central to the production of many alcoholic beverages and important in baking where the gaseous carbon dioxide product makes bread dough rise.

The net reaction of calcium chloride with sodium hydrogen carbonate in aqueous solution is

$$Ca^{2+}(aq) + 2HCO_3^-(aq) \rightarrow CaCO_3(s) + CO_2(g) + H_2O(l) \tag{6.2}$$

We discussed the components of this reaction in Chapter 2, Section 2.16, in relation to the carbon cycle and the formation of limestone caves.

The net reaction of sodium hydrogen carbonate with potassium hydrogen tartrate is

$$HC_4H_4O_6^-(aq) + HCO_3^-(aq) \rightarrow C_4H_4O_6^{2-}(aq) + CO_2(g) + H_2O(l) \tag{6.3}$$

Baking powders contain sodium hydrogen carbonate and a solid acid, such as potassium hydrogen tartrate. These react to give the carbon dioxide that makes cakes and cookies rise when they are baked.

6.5 CONSIDER THIS

How can you classify chemical reactions?

(a) Think about your list of reactions from Consider This 6.1 as well as the reactions from Investigate This 6.2. Can you classify them into groups of reactions that have similar properties and/or behaviors? What are the properties and/or behaviors you use?

(b) If you were given a pair of reactants, could you use your classification scheme to predict the likely outcome of a reaction between them? Why or why not? If not, what other information would you need to make your prediction?

Classifying chemical reactions The reactants in the above reactions are so different that you probably would not have predicted that each pair would produce the same product. In Consider This 6.5, you probably found that reaction classifications based on observable changes do not provide adequate information to make predictions about the likely reaction between a new set of reactants. In order to make predictions about the likely reaction, another essential piece of information you need is some idea what interactions can occur between the reactant molecules and/or ions. We have repeatedly emphasized the importance of electrical charge or partial charge (polarity) in explanations for chemical phenomena. ***Chemical reactions start when a center of positive charge in one reactant molecule or ion is attracted to a center of negative charge in another.*** Thus, understanding chemical reactions requires understanding the structure of molecules discussed in Chapter 5 and the ionic interactions introduced in Chapter 2.

Once reactant molecules have come together, a chemical reaction between them involves rearrangements of electrons and/or atoms to form products that are different from the reactants. In Chapter 2, we introduced two kinds of chemical reactions: precipitation (and its reverse, dissolution) of solid ionic salts from solution and acid–base reactions. In this chapter we will introduce other kinds of chemical reactions. Using the background on atomic and molecular structure from Chapters 2, 4, and 5, we organize chemical reactions into three broad classes: ionic precipitation, Lewis acid–base, and reduction–oxidation. Ionic precipitations occur as a result of the attraction between separated positive ions and negative ions to form ordered crystalline solids. Lewis acid–base reactions are characterized by the presence of a nonbonding pair of electrons on one of the reactants, which ends up as an electron pair bond in one of the products. (Brønsted–Lowry acid–base reactions are a special case of Lewis acid–base reactions.) Reduction–oxidation reactions are characterized by the transfer of one or more electrons from one reactant to another. You will find that reactions (6.1), (6.2), and (6.3) fit into this classification scheme.

This organization of chemical reactions is based on the electron distributions and redistributions in and between molecules and ions when reactions occur. Ours is not the only possible organization or classification of chemical reactions. Each classification has its merits and utility. In most general classification schemes, there are exceptions and specific examples that fall into more than one class. You have to accept special cases as a necessary consequence of trying to describe an enormous number of different reactions with only a few variables. Keep in mind that there are exceptions, but focusing on the overall organization will help you understand the direction of many reactions and predict their products.

6.2. Ionic Precipitation Reactions

6.6 INVESTIGATE THIS

How can you find the stoichiometry of a precipitate?

Do this as a class investigation and work in small groups to discuss and analyze the results. Label five, clean, small centrifuge tubes from 1 to 5. Mix in these tubes the volumes of 0.10 M aqueous calcium nitrate, $Ca(NO_3)_2$, and sodium oxalate, $Na_2C_2O_4$, solutions specified in this table. Centrifuge the tubes for about a minute to settle the precipitates. Record all your observations.

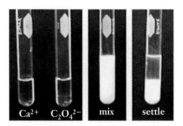

$Ca(NO_3)_2$ + $Na_2C_2O_4$ reaction

Tube number	1	2	3	4	5
Calcium nitrate, mL	0.5	1.5	2.5	3.5	4.5
Sodium oxalate, mL	4.5	3.5	2.5	1.5	0.5

6.7 CONSIDER THIS

What is the stoichiometry of the calcium oxalate precipitate?

(a) In Investigate This 6.6, what is varying in the contents of the solutions in each tube as you go from the first to the fifth?
(b) Is the amount of precipitate the same in each tube? If not, which has the most precipitate? The least? Make a rough bar graph showing the amount of precipitate in each tube as a function of tube number.
(c) What trends, if any, do you see in the graph? How would you interpret them?

Ionic precipitation reactions are a result of attractions between cations that have lost one or more electrons and anions that have gained one or more electrons. The cations of interest to us are almost always metal ions. The anions may be monatomic (such as Cl^- and S^{2-}) or polyatomic (such as PO_4^{3-} and HCO_3^-). You learned from your analysis of Investigate This 2.32, Chapter 2, that most singly charged cations and anions form relatively soluble salts. The attractions of polar water molecules to the ions—hydration energy—are strong enough to overcome the attractions of the ions to one another in the solid crystal, the lattice energy. Multiply charged anions and cations most often form relatively insoluble salts with each other. The larger charges attract one another strongly in the crystal, and the hydration energy is not enough to overcome the lattice energy. The multiple charges on aqueous calcium and oxalate ions, $Ca^{2+}(aq)$ and $C_2O_4^{2-}(aq)$, help to explain the appearance of the precipitates in Investigate This 6.6:

$$Ca^{2+}(aq) + C_2O_4^{2-}(aq) \rightleftharpoons CaC_2O_4(s) \qquad (6.4)$$

Table 6.1 gathers the solubility rules we have generated.

Table 6.1	*General solubility rules for ionic salts.*	
Cation	**Anion**	**Solubility**
singly charged	singly charged	soluble; Ag^+ is an exception
singly charged	multiply charged	alkali metal salts soluble
multiply charged	singly charged	halide and nitrate salts soluble
multiply charged	multiply charged	not soluble

6.8 CONSIDER THIS

Does ionic precipitation fit the definition of a chemical reaction?

We have said that a chemical reaction involves rearrangements of electrons and/or atoms to form products that are different from the reactants. Does reaction (6.4) fit this description? What rearrangements of electrons and/or atoms occur? *Hint:* Consider what the designation *(aq)* on the ions means.

Calcium–oxalate reaction stoichiometry There are different amounts of $Ca^{2+}(aq)$ and $C_2O_4^{2-}(aq)$ in each tube in Investigate This 6.6, so we would expect differences in the amount of precipitate that can be formed in each. The observed differences are a function of the stoichiometry of the precipitation reaction (6.4). The stoichiometry is based on charge balance. We have written the double arrows to represent the reversibility of the reaction; solid calcium oxalate does dissolve to a slight extent in water. Let's look quantitatively at the stoichiometry for one of the samples in Investigate This 6.6.

6.9 WORKED EXAMPLE

Stoichiometric calculation for calcium oxalate formation

How much precipitate can be formed in sample 5 in Investigate This 6.6?

Necessary information: We are asked "how much" without any specification of units. We'll choose to calculate the number of moles of precipitate because moles are easiest to get using the stoichiometric equation (6.4) and the volume and concentration data for the sample.

Strategy: The approach for solving problems of this kind was discussed in Chapter 2, Section 2.10, and is illustrated in the partial stoichiometry route map shown on the next page. Calculate the number of moles of each reactant present in the solution and then, using the 1:1 stoichiometry of the reaction, determine

continued

which is the limiting reactant. Finally, convert the number of moles of limiting reactant to the number of moles of precipitate it can form.

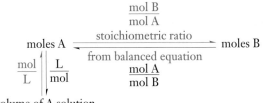

Implementation:

$$\text{mol Ca}^{2+}(aq) = (0.10 \text{ mol·L}^{-1})(4.5 \text{ mL})\left(\frac{1 \text{ L}}{1000 \text{ mL}}\right)$$

$$= 4.5 \times 10^{-4} \text{ mol Ca}^{2+}(aq)$$

$$\text{mol C}_2\text{O}_4{}^{2-}(aq) = (0.10 \text{ mol·L}^{-1})(0.5 \text{ mL})\left(\frac{1 \text{ L}}{1000 \text{ mL}}\right)$$

$$= 0.5 \times 10^{-4} \text{ mol C}_2\text{O}_4{}^{2-}(aq)$$

The number of moles of $C_2O_4{}^{2-}(aq)$ required to react with 4.5×10^{-4} mol $Ca^{2+}(aq)$ is

$$\text{mol C}_2\text{O}_4{}^{2-}(aq) \text{ required} = \left[4.5 \times 10^{-4} \text{ mol Ca}^{2+}(aq)\right]\left(\frac{1 \text{ mol C}_2\text{O}_4^{2-}}{1 \text{ mol Ca}^{2+}}\right)$$

$$= 4.5 \times 10^{-4} \text{ mol C}_2\text{O}_4{}^{2-}(aq)$$

This amount of $C_2O_4{}^{2-}(aq)$ is more than is available, so $C_2O_4{}^{2-}(aq)$ is the limiting reactant. The number of moles of precipitate, $CaC_2O_4(s)$, formed by 0.5×10^{-4} mol $C_2O_4{}^{2-}(aq)$ is

$$\text{mol CaC}_2\text{O}_4(s) = \left[0.5 \times 10^{-4} \text{ mol C}_2\text{O}_4^{2-}(aq)\right]\left(\frac{1 \text{ mol CaC}_2\text{O}_4}{1 \text{ mol C}_2\text{O}_4^{2-}}\right)$$

$$= 0.5 \times 10^{-4} \text{ mol CaC}_2\text{O}_4(s)$$

Does the answer make sense?   The concentrations of the reactants are the same, and they react in a 1:1 ratio. Since we used less of the $C_2O_4{}^{2-}(aq)$ solution, it makes sense that it is the limiting reactant and that the amount of precipitate is the same as the amount of $C_2O_4{}^{2-}(aq)$. From its molar mass, you can calculate that about 6 mg of $CaC_2O_4(s)$ is formed.

6.10 CHECK THIS

Stoichiometric calculation for calcium oxalate formation

(a) Calculate the number of moles of $Ca^{2+}(aq)$, the number of moles of $C_2O_4{}^{2-}(aq)$, and the sum of these quantities for each sample in Investigate This 6.6.

(b) Calculate the number of moles of precipitate that can be formed in the other four samples in Investigate This 6.6.

(c) How do your results in part (b) compare with the bar graph you sketched in Consider This 6.7? Which sample should give the largest amount of precipitate? Is this what you observe?

(d) What, if anything, is different about the sample with the largest amount of precipitate?

Continuous variations Notice that the *sum* of the numbers of moles of $Ca^{2+}(aq)$ and $C_2O_4{}^{2-}(aq)$ is the same for all five samples in Investigate This 6.6. What varies among the samples is the *ratio* of moles of $Ca^{2+}(aq)$ to moles of $C_2O_4{}^{2-}(aq)$. As you have found, the maximum amount of precipitate is formed in sample number 3, the sample with the 1:1 stoichiometric reaction ratio. In this sample, essentially all of the $Ca^{2+}(aq)$ and $C_2O_4{}^{2-}(aq)$ ions precipitate from the solution. In all the other samples, there is less of one or the other of these ions and less precipitate can be formed. This technique, in which the ratio of numbers of moles of two reactants is varied while their sum is held constant, is called the **continuous variation method.** Figure 6.1 is a representation of a continuous variation experiment similar to the one you did in Investigate This 6.6.

Continuous variation methods are often used to determine the stoichiometry of reactions. The maximum amount of reaction occurs when the ratio is

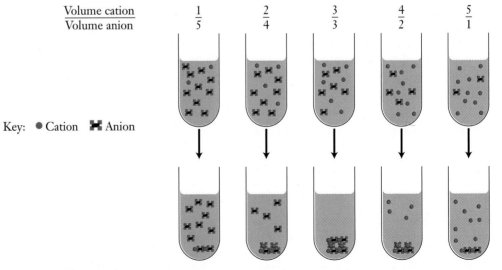

$$\frac{\text{Volume cation}}{\text{Volume anion}} \qquad \frac{1}{5} \qquad \frac{2}{4} \qquad \frac{3}{3} \qquad \frac{4}{2} \qquad \frac{5}{1}$$

Key: ● Cation ✖ Anion

Figure 6.1.

A continuous variations experiment. Solutions of the same concentration of a cation and anion that react 1:1 are mixed in the volume ratios shown. The top row represents the mixture in the instant before reaction; the bottom row after reaction to form a precipitate. The nonprecipitating counter ions are not shown.

stoichiometric. The method is applicable to any kind of reaction that produces a measurable effect that is proportional to the "number of moles of reaction." Measurable effects include formation of precipitates, production of heat, changes in color, and so on. Later in the chapter, we will introduce other techniques for determining reaction stoichiometries.

6.11 CHECK THIS

Nickel–dimethylglyoxime reaction stoichiometry

Solutions containing nickel ion, $Ni^{2+}(aq)$, are a beautiful green. When a $Ni^{2+}(aq)$ solution is mixed with a slightly basic, clear, colorless alcoholic solution of **dimethylglyoxime** (dmg), $C_4H_8N_2O_2(alc)$, a deep red solid is formed:

$$x\,Ni^{2+}(aq) + y\,C_4H_8N_2O_2(alc) \rightleftharpoons \text{red solid} \qquad (6.5)$$

A continuous variations study was carried out on this reaction with the results shown here.

Ni²⁺   dmg   mixture

Sample	1	2	3	4	5	6
$Ni^{2+}(aq)$, mol $\times 10^4$	2.55	4.89	5.98	7.50	9.89	12.6
$C_4H_8N_2O_2(alc)$, mol $\times 10^4$	18.2	15.8	14.8	13.1	10.9	8.19
mass solid, g	0.074	0.141	0.173	0.189	0.158	0.118

(a) Assuming that x and y in reaction (6.5) are small whole numbers (1, 2, or 3), use these data to figure out the likely stoichiometry of the reaction. Clearly explain your reasoning. *Note:* None of the samples contains the exact stoichiometric ratio of reactants.

(b) Based on the stoichiometry you found in part (a), how many moles of the red solid product are formed in each sample? Explain how you obtain your answers.

(c) What is the *average* total number of moles of the two reactants added together in each of the samples?

(d) Based on your stoichiometry in part (a) and result from part (c), what amount of each reactant would you have to use in this study to give the exact stoichiometric ratio? What mass of solid would you expect from this sample? Explain.

Because they are so common and usually unexciting, precipitation and dissolution reactions seem straightforward and perhaps uninteresting. Consider, however, that you see all about you solids that do not dissolve well or rapidly in water. These include the wood, masonry, metals, and fibers that make up your shelter, your means of transportation, and your clothing. Many of these are complicated ionic solids, like concrete and brick. Most materials are, in fact, rather insoluble in water. That's why the great majority of ions in biological fluids are singly-charged cations and anions, such as Na^+, K^+, and Cl^-, whose salts are water soluble. Your body also contains a good deal of Ca^{2+}, a small amount of which is a vital component of fluids, but most of which is present as the solid salts that make up your bones and teeth. It's important to recognize possible precipitation or dissolution reactions as a way of understanding materials and the way nature and humans use them.

Reflection and Projection

A **chemical reaction** is a change that involves rearrangements of electrons and/or atoms to form products that are different from the reactants. You listed several kinds of evidence you can use to tell whether a change produces new compounds; we tested one reaction; and you saw how the method of continuous variations can be used to get information about reaction stoichiometry. Chemical reactions begin when a positive center in one molecule is attracted to a negative center in another. Subsequent rearrangements of electrons and/or atoms then lead to products. We are classifying reactions into three categories: ionic precipitation reactions, Lewis acid–base reactions, and reduction–oxidation reactions. Ionic precipitation reactions and the accompanying stoichiometric calculations have been a reminder of concepts you met previously in Chapter 2. Acid–base reactions, proton transfers, were also introduced in Chapter 2. In this chapter, based on our molecular model from Chapter 5, we will first look more deeply at proton transfers and then broaden the acid–base concept to include other reactions.

6.3 Lewis Acids and Bases: Definition

The Brønsted–Lowry model of acids and bases introduced in Chapter 2, Section 2.12, is a special case of a more general model of acid–base reactions. G. N. Lewis, the same chemist who developed the Lewis bonding model that we have been using, proposed an acid–base model now called the **Lewis acid–base model.** The Lewis acid–base model, like his bonding model, is based on electron pairs. A **Lewis base** is an *electron pair donor.* A **Lewis acid** is an *electron pair acceptor.* One product of a Lewis acid–base reaction is a new molecule in which the electron pair from the Lewis base forms a covalent, electron-pair bond with the Lewis acid.

Any molecule or ion with one or more pairs of nonbonding electrons can react as a Lewis base. A molecule with a nonbonding pair of electrons is almost always polarized with the nonbonding pair as a partial negative center. Lewis bases have centers of negative charge. Therefore, cations or molecules with partial positive centers are most likely to react as Lewis acids. These generalizations help to classify a large number of reactions as Lewis acid–base reactions. But we need specific examples to make the model useful for predicting whether reactions between two molecules will occur and, if so, what the products will be.

6.12 CONSIDER THIS

How do ionic precipitation and Lewis acid–base reactions differ?

Ionic precipitation reactions such as reaction (6.4) are usually not classified as Lewis acid–base reactions. The reactants in this reaction are an anion, $C_2O_4^{2-}$, that has several pairs of nonbonding electrons, a Lewis base. The cation, Ca^{2+}, seems that it could accept a pair of electrons and act as a Lewis acid. Why don't we call the precipitation of the ionic salt, CaC_2O_4, a Lewis acid–base reaction? Explain your response clearly.

Types of Lewis acid–base reactions Many kinds of reactions can be classed as Lewis acid–base reactions; we include three types. In the example reaction for each type, the Lewis base is the first reactant. The Lewis base electron pair that is donated and the electron pair bond formed in the product are shown in red. There are three reaction types:

(1) Brønsted–Lowry acid–base reactions; proton transfer to a Lewis base:

$$:CN^-(aq) + H_2O(l) \rightleftharpoons H{-}CN(aq) + OH^-(aq) \qquad (6.6)$$

(2) Formation of metal ion complexes:

$$6:CN^-(aq) + Fe^{3+}(aq) \rightleftharpoons \left[\begin{array}{c} CN \\ | \quad CN \\ NC{-}Fe{\leftarrow}CN \\ NC^{\blacktriangledown} \quad | \\ CN \end{array} \right]^{3-}_{(aq)} \qquad (6.7)$$

(3) Nucleophile (Lewis base)–electrophile (Lewis acid) reactions:

$$CH_3\overset{..}{O}H(l) + CH_3CH_2\overset{\delta+}{C}\overset{\delta-}{H}O(l) \rightleftharpoons CH_3CH_2CHOH(l) \\ \qquad\qquad\qquad\qquad\qquad\qquad\qquad\qquad \underset{OCH_3}{\diagdown} \qquad (6.8)$$

In the next sections, we will go more deeply into each of these Lewis acid–base reaction types with further examples, investigations, and questions to help you identify and predict reaction products.

6.4. Lewis Acids and Bases: Brønsted–Lowry Acid–Base Reactions

 6.13 INVESTIGATE THIS

Do the acidities of acids differ?

Do this as a class investigation and work in small groups to discuss and analyze the results. Use a pH meter and pH electrode (or short range pH indicator papers) to determine the pH (to the nearest 0.1 pH unit) of distilled (or deionized) water and the following 0.1 M aqueous solutions: hydrochloric acid, HCl; sodium chloride, NaCl; ethanoic (acetic) acid, $CH_3C(O)OH$; and sodium ethanoate (acetate), $CH_3C(O)ONa$.

6.14 CONSIDER THIS

How do the acidities of acids differ?

List water and the solutions in Investigate This 6.13 in order of increasing pH. In which, if any, of the solutions is the concentration of hydronium ion, $H_3O^+(aq)$, about the same as in distilled water? Compared to the water, in which solutions does hydronium ion predominate? In which does hydroxide ion predominate? How might you explain these results?

Web Companion

| Chapter 6, Section 6.4 | ──①
|---|

Interactive animations illustrate
proton transfer reactions and
strong and weak acids.

②
③
④

Compare equation (6.9) to
the pH scale in Figure 2.24
(Chapter 2, Section 2.12).

Strong and weak Brønsted–Lowry acids and bases In the Brønsted–Lowry definition of acids as proton donors and bases as proton acceptors, a **strong Brønsted–Lowry acid** is one that donates all of its acidic protons to water molecules in aqueous solution. With water as the base (proton acceptor), the amount of hydronium ion, $H_3O^+(aq)$, formed is equivalent to the amount of the acid added. In Investigate This 6.13, you found that a 0.1 M HCl solution has a pH ≈ 1. We can rewrite the definition, $pH \equiv -\log_{10}[H_3O^+(aq)]$, to get the molar concentration of hydronium ion:

$$[H_3O^+(aq)] = 10^{-pH} \text{ M} \tag{6.9}$$

For our solution with pH ≈ 1.0, equation (6.9) gives $[H_3O^+(aq)] = 10^{-1}$ M = 0.1 M. The hydronium ion concentration, $[H_3O^+(aq)]$, is the same as the concentration of HCl added to the solution. HCl is a strong Brønsted–Lowry acid, which we can represent by showing "complete" donation of its proton to water:

$$HCl(aq) + H_2O(l) \rightarrow H_2OH^+(aq) + Cl^-(aq) \tag{6.10}$$

6.15 CHECK THIS

[H₃O⁺(aq)] in 0.1 M ethanoic acid solution

(a) Use your results from Investigate This 6.13 to calculate the $[H_3O^+(aq)]$ in a 0.1 M aqueous solution of ethanoic (acetic) acid. How does the $[H_3O^+(aq)]$ in this solution compare to the concentration of ethanoic acid present? What conclusion can you draw about the strength of ethanoic acid? Explain your reasoning.

(b) Explain clearly how the two molecular-level diagrams in the *Web Companion*, Chapter 6, Section 6.4.2, are related to the hydrochloric acid and ethanoic acid solutions discussed in the preceding text and part (a).

A **weak Brønsted–Lowry acid** is one that does not donate all of its acidic protons to water molecules in aqueous solution. As you have found in Check This 6.15, ethanoic acid in aqueous solution does not donate all its acidic protons to water to form hydronium ion. Ethanoic acid is a weak Brønsted–Lowry acid, which we can represent by using double arrows to indicate that its reaction with water reaches equilibrium (with the forward and reverse reactions balancing one another) before all its protons are donated:

$$CH_3C(O)OH(aq) + H_2O(l) \rightleftharpoons H_2OH^+(aq) + CH_3C(O)O^-(aq) \tag{6.11}$$

(The reaction of $HCl(aq)$ to donate its protons also reaches equilibrium, but only when there are essentially no $HCl(aq)$ molecules left; we use the one-way arrow to represent this condition.)

6.16 CHECK THIS

Fraction of proton transfer between ethanoic acid and water

Use the reaction stoichiometry from reaction expression (6.11) and your value for $[H_3O^+(aq)]$ in 0.1 M ethanoic (acetic) acid solution (from Check This 6.15)

continued

to determine the fraction of the ethanoic acid, $CH_3C(O)OH(aq)$, that transfers a proton to water. Explain your reasoning.

6.17 CONSIDER THIS

What is the molecular level interpretation of expression (6.11)?

Explain how the *Web Companion*, Chapter 6, Section 6.4.1, animation represents the double arrows in reaction expression (6.11). Reaction expressions like (6.11) may suggest that the same two product molecules formed in the forward reaction (left to right) react in the reverse reaction (right to left). Is this what the animation shows? What is the interpretation of reaction expression (6.11) shown by the animation? Explain clearly.

In aqueous solution, a **strong Brønsted–Lowry base** accepts protons from water molecules to form an amount of hydroxide ion, $OH^-(aq)$, equivalent to the amount of base added. The amide ion, $NH_2^-(aq)$ (in solutions of sodium amide, $NaNH_2(s)$, for example), is a strong base:

$$HOH(l) + NH_2^-(aq) \rightarrow HNH_2(aq) + OH^-(aq) \qquad (6.12)$$

A **weak Brønsted–Lowry base** does not accept an amount of protons equivalent to the amount of base added, so the $[OH^-(aq)]$ in a weak base solution is not equivalent to the concentration of base added. Recall from Chapter 2 that ammonia, $NH_3(g)$, dissolves readily in water but the solutions contain only modest amounts of hydroxide ion; ammonia is a weak base:

$$HOH(l) + NH_3(aq) \rightleftharpoons HNH_3^+(aq) + OH^-(aq) \qquad (6.13)$$

6.18 CHECK THIS

$[OH^-(aq)]$ in an aqueous sodium amide solution

When 0.1 mol of sodium amide is dissolved in enough water to make 1 L of solution, the pH of the solution is 13. What is the concentration of hydroxide ion, $[OH^-(aq)]$, in the solution? Is your result consistent with reaction expression (6.12)? Explain why or why not. *Note:* In aqueous solutions at 25 °C, the product of the hydronium and hydroxide ion concentrations is 10^{-14} M^2:

$$[H_3O^+(aq)][OH^-(aq)] = 10^{-14}\,M^2 \qquad (6.14)$$

Check this on the pH scale, Figure 2.24 (Chapter 2, Section 2.12). If you know the concentration of hydronium ion, you can calculate the hydroxide ion concentration, and *vice versa*.

Protons and electron pairs Reactions that can be characterized as either Brønsted–Lowry acid–base or Lewis acid–base differ only in where we focus our attention. In Brønsted–Lowry acids and bases, we focus on donating or accepting a proton. In reaction expressions (6.10) through (6.13), we have highlighted the proton that is transferred. A Brønsted–Lowry acid–base reaction always includes two conjugate acid–base pairs that differ only in the loss of a proton by the acid to form its conjugate base. In expression (6.10), for example, the conjugate acid–base pairs are {HCl(aq)/Cl$^-$(aq)} and {H_2OH^+(aq)/H$_2$O(l)}.

The Lewis model focuses on a nonbonding electron pair and the covalent bond it forms. We can rewrite the previous equations to highlight these electron pairs and covalent bonds and show the Lewis acid–base conjugate pairs. For hydrochloric acid, reaction (6.10), we write

$$\text{H—Cl}(aq) + \text{H}_2\text{O:} (l) \rightarrow \text{H}_2\text{O—H}^+(aq) + \text{:Cl}^-(aq) \qquad \text{(6.15)}$$

To emphasize the change of a nonbonding pair of electrons to a covalent bond, we show one pair of valence electrons and the covalent bond in red.

The base, H$_2$O: (l), has donated a pair of electrons to form a covalent bond with the proton from the H—Cl(aq). Expressions (6.10) and (6.15) represent the same reaction looked at from different viewpoints. The Lewis conjugate acid–base pairs are {H—Cl(aq)/:Cl$^-$(aq)} and {H$_2$O—H$^+$(aq)/H$_2$O: (l)}, which are identical to the Brønsted–Lowry conjugate pairs. But, in the Lewis case, our focus is on the electron pair, either nonbonding (in the base) or bonding (in the acid).

6.19 CHECK THIS

Writing Lewis acid–base reactions

Use reaction expression (6.15) as a model to rewrite reactions (6.11), (6.12), and (6.13), highlighting the electron pairs and covalent bonds that characterize the reactions as Lewis acid–base reactions.

Relative strengths of Lewis bases A **strong Lewis base** is one that has nonbonding electrons capable of forming a strong covalent bond with a Lewis acid. For the reactions in this section, this is a bond with a proton to form the conjugate Lewis acid. The more strongly the Lewis base holds the proton in its conjugate Lewis acid form, the less likely it is to give up this proton to another Lewis base. Thus, *the stronger the Lewis base, the weaker its conjugate acid*. We can use the results from Investigate This 6.13 to begin a ranking of the **basicity**, base strength, of several Lewis bases.

6.20 WORKED EXAMPLE

Relative basicities of chloride ion and water

In Investigate This 6.13, you found that 0.1 M hydrochloric acid has a pH ≈ 1. What is the basicity of the chloride ion, :Cl$^-$(aq), relative to water, H$_2$O: (l)?

Necessary information: We need our results from above which show that the hydrochloric acid solution contains essentially all :Cl$^-$(aq) and no H—Cl(aq).

continued

Strategy: If two Lewis bases are present in a solution and in competition for protons, the stronger base will win the competition. ***The weaker base will be less protonated and will be present mainly in its base form; there will be more of the weaker Lewis base in the solution.*** Conversely, there will be very little of its conjugate Lewis acid.

Implementation: Expression (6.15) represents the reaction in a solution containing $:Cl^-(aq)$ and $H_2O:(l)$ competing for protons. Since the solution contains $:Cl^-(aq)$ and $H_2O-H^+(aq)$, but no $H-Cl(aq)$, we know that $H_2O:(l)$ must have won the competition: $H_2O:(l)$ is a stronger base than $:Cl^-(aq)$. Since we know that the stronger the base, the weaker its conjugate acid, we also know that $H_2O-H^+(aq)$ is a weaker acid than $H-Cl(aq)$.

Does the answer make sense?   We almost never think of the chloride ion as a base, because solutions of simple chloride salts have little effect on the acid–base properties of their solutions. Therefore, to find that it is a weaker Lewis base than water makes sense.

6.21 CHECK THIS

Relative basicities of chloride and hydroxide ions

(a) In Investigate This 6.13, how does the pH of the 0.1 M sodium chloride solution compare to the pH of water? Which ion is present in higher concentration in the sodium chloride solution, $:Cl^-(aq)$ or $:OH^-(aq)$? Explain how you get your answer.

(b) The reaction of $:Cl^-(aq)$ as a Lewis base with water as a Lewis acid, is

$$H-OH(l) + :Cl^-(aq) \rightleftharpoons H-Cl(aq) + :OH^-(aq) \qquad (6.16)$$

Reaction (6.16) represents a competition for protons between two Lewis bases, $:Cl^-(aq)$ and $:OH^-(aq)$. Which one wins the competition? Which is the stronger Lewis base? Clearly explain the reasoning for your answers.

(c) Which is the stronger acid, $H-Cl(aq)$ or $H-OH(l)$? Explain.

6.22 WORKED EXAMPLE

Relative basicities of ethanoate ion and water

In Investigate This 6.13, you found that 0.1 M ethanoic acid has a pH between 2 and 3. What is the basicity of the ethanoate (acetate) ion, $CH_3C(O)O:^-(aq)$, relative to water, $H_2O:(l)$?

Necessary information: We need your results from Check This 6.16, which show that, in this ethanoic acid solution, less than 10% of the acid, $CH_3C(O)O-H(aq)$, transfers a proton to water.

continued

Strategy: The strategy for this problem is the same as for Worked Example 6.20.

Implementation: In Check This 6.19, you wrote the Lewis acid–base equation for ethanoic acid:

$$CH_3C(O)O-H(aq) + H_2O: (l) \rightleftharpoons H_2O-H^+(aq) + CH_3C(O)O: ^-(aq)$$
$$\text{(6.17)}$$

Expression (6.17) represents the reaction in a solution containing $CH_3C(O)O: ^-(aq)$ and $H_2O: (l)$ competing for protons. We know from your calculation (based on the pH measurement) that most of the $CH_3C(O)O: ^-(aq)$ is protonated as ethanoic acid $CH_3C(O)O-H(aq)$. The ethanoate ion, $CH_3C(O)O: ^-(aq)$, has won the competition: $CH_3C(O)O: ^-(aq)$ is a stronger Lewis base than $H_2O: (l)$. Because we know that the stronger the Lewis base, the weaker its conjugate acid, we also know that $CH_3C(O)O-H(aq)$ is a weaker acid than $H_2O-H^+(aq)$.

Does the answer make sense? See Check This 6.23.

6.23 CHECK THIS

Relative basicities of ethanoate and hydroxide ions

(a) The reaction of $CH_3C(O)O: ^-(aq)$ as a Lewis base with water as the Lewis acid is

$$H-OH(l) + CH_3C(O)O: ^-(aq)$$
$$\rightleftharpoons CH_3C(O)O-H(aq) + :OH^-(aq) \qquad \text{(6.18)}$$

Do you have any evidence from Investigate This 6.13 that this reaction occurs? Explain your response.

(b) The 0.1 M solution of sodium ethanoate in Investigate This 6.13 is more basic, higher pH, than water. In this solution, ethanoate and water are in competition for protons from water. Recall that the reaction of water with itself produces tiny amounts of hydronium and hydroxide ions. The reaction of ethanoate with water, expression (6.18), produces much more hydroxide. How does this observation show that the conclusion in Worked Example 6.22 makes sense?

(c) Use the pH you measured for the 0.1 M sodium ethanoate solution to determine the concentration of hydroxide ion, $[:OH^-(aq)]$, in the solution. *Hint:* Use equations (6.9) and (6.14).

(d) From your result in part (c) and the stoichiometry of reaction (6.18), calculate the concentration of unreacted ethanoate ion, $[CH_3C(O)O: ^-(aq)]$, in the solution. Which is present in higher concentration, $CH_3C(O)O: ^-(aq)$ ion or $:OH^-(aq)$ ion?

(e) Which is the stronger Lewis base, $CH_3C(O)O: ^-(aq)$ ion or $:OH^-(aq)$ ion? Clearly explain your reasoning.

The results from Worked Example 6.20 and Check This 6.21 show that chloride ion, $:Cl^-(aq)$, is a weaker Lewis base than either water, $H_2O: (l)$, or hydroxide ion, $:OH^-(aq)$. To get our relative Lewis basicities in order, we also need to know how the basicities of water and hydroxide ion compare. In Check This 6.23(b), we recalled the reaction of water with itself:

$$H-OH(l) + H_2O: (l) \rightleftharpoons H_2O-H^+(aq) + :OH^-(aq) \qquad \text{(6.19)}$$

In pure water, reaction (6.19) produces 10^{-7} M concentrations of hydronium and hydroxide ions. Almost all the water molecules remain unreacted. In the competition for protons in this solution, hydroxide has won. The hydroxide ion, $:OH^-(aq)$, is a stronger Lewis base than water, $H_2O: (l)$. Therefore, in order of *decreasing* Lewis basicity, we have

$$\text{(strongest)} :OH^-(aq) > H_2O: (l) > :Cl^-(aq) \text{ (weakest)} \qquad \text{(6.20)}$$

Their conjugate Lewis acids, in order of *increasing* acidity, are

$$\text{(weakest) } HO-H(l) < H_2O-H^+(aq) < H-Cl(aq) \text{ (strongest)} \qquad \text{(6.21)}$$

Other experiments like those in Investigate This 6.13 and more complex ones in nonaqueous solvents lead to the order of basicities and acidities for Lewis/ Brønsted–Lowry bases and acids shown in Table 6.2 on the next page.

6.24 WORKED EXAMPLE

pH of a sodium cyanide solution

Use the information in Table 6.2 to predict whether the pH of a solution of sodium cyanide, NaCN, will be quite high, quite low, or moderately high or low.

Necessary information: We need to know that NaCN dissolves to give $Na^+(aq)$ and $CN^-(aq)$ ions and that $Na^+(aq)$ is a spectator ion in this solution.

Strategy: Write the acid and/or base reactions of $CN^-(aq)$ ions and use Table 6.2 to determine what species will predominate in the solution.

Implementation: The $CN^-(aq)$ ion is a Lewis base:

$$H-OH(l) + :CN^-(aq) \rightleftharpoons H-CN (aq) + :OH^-(aq) \qquad \text{(6.22)}$$

Because $:OH^-(aq)$ is produced in this reaction, we predict that the solution will contain more hydroxide ion than hydronium ion, so the pH will be above 7. Table 6.2 shows that $:CN^-(aq)$ is a weaker base than $:OH^-(aq)$. Thus, in the solution, most of the $:CN^-(aq)$ will remain unreacted and the solution will be only moderately basic.

Does the answer make sense? In Table 6.2, we see that cyanide ion is near ammonia. We know that ammonia solutions are only moderately basic, so it makes sense that cyanide solutions should also be moderately basic.

Table 6.2 *Relative strengths of Lewis/Brønsted–Lowry bases and acids.*

The strongest bases are at the top and the weakest at the bottom, as the blue arrow shows. The strengths of the acids, shown by the red arrow, go in the reverse order. The bases and acids that can be compared in aqueous solutions are shaded. The other bases and acids are either too strong or too weak to be differentiated in aqueous solutions.

Base		Conjugate acid	
Structure	**Name**	**Structure**	**Name**
$:CH_3^-$	methide	$H-CH_3$	methane
$:NH_2^-$	amide	$H-NH_2$	ammonia
$:OCH_2CH_3^-$	ethoxide	$H-OCH_2CH_3$	ethanol
$:OH^-$	hydroxide	$H-OH$	water
$:NH_2R$	amine	$H-NH_2R^+$ (RNH_3^+)	"ammonium-like" ion
$:OCO_2^{2-}$ (CO_3^{2-})	carbonate	$H-OCO_2^-$ (HCO_3^{2-})	hydrogen carbonate
phenolate structure	phenolate	phenol structure	phenol
$:NH_3$	ammonia	$H-NH_3^+$ (NH_4^+)	ammonium
$:CN^-$	cyanide	$H-CN$	hydrogen cyanide
$:SH^-$	hydrogen sulfide	$H-SH$ (H_2S)	hydrosulfuric acid
$:OC(O)OH^-$ (HCO_3^-)	hydrogen carbonate	$H-OC(O)OH$ (H_2CO_3)	carbonic acid
$CH_3C(O)O:^-$	ethanoate (acetate)	$CH_3C(O)O-H$	ethanoic (acetic) acid
2,4-dinitrophenolate structure	2, 4-dinitrophenolate	2,4-dinitrophenol structure	2,4-dinitrophenol
$:F^-$	fluoride	$H-F$	hydrogen fluoride
$:OPO(OH)_2^-$ ($H_2PO_4^-$)	dihydrogen phosphate	$H-OPO(OH)_2$ (H_3PO_4)	phosphoric acid
$H_2O:$	water	H_2O-H^+	hydronium
$:OSO_2OH^-$ (HSO_4^-)	hydrogen sulfate	$H-OSO_2OH$ (H_2SO_4)	sulfuric acid
$:Cl^-$	chloride	$H-Cl$	hydrogen chloride
$:Br^-$	bromide	$H-Br$	hydrogen bromide
$:I^-$	iodide	$H-I$	hydrogen iodide
$:OClO_3^-$ (ClO_4^-)	perchlorate	$H-OClO_3$ ($HClO_4$)	perchloric acid

strong ↑ weak (increasing base strength)

weak ↓ strong (increasing acid strength)

Note: The conventional representation of some of the molecules and ions is shown in parentheses; compare with Table 2.7 (Chapter 2, Section 2.13).

6.25 CHECK THIS

pH of an aminium (ammonia-like) chloride solution

The active ingredients in some eye drops are rather insoluble amines, RNH_2, that are made soluble by converting them to aminium chlorides, RNH_3Cl (called amine hydrochlorides), which dissolve in water to give $RNH_3^+(aq)$ and $Cl^-(aq)$ ions. Use the information in Table 6.2 to predict approximately how high or low the pH of a solution of RNH_3Cl will be.

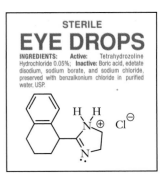

STERILE
EYE DROPS
INGREDIENTS: **Active:**   Tetrahydrozoline
Hydrochloride 0.05%; **Inactive:** Boric acid, edetate disodium, sodium borate, and sodium chloride, preserved with benzalkonium chloride in purified water, USP.

Reflection and Projection

What goes on in a chemical reaction does not depend on what we name the reaction. We can name reactions involving proton transfers as either Brønsted–Lowry or Lewis acid–base reactions and the only difference is where we place our emphasis in describing them. The reason we have introduced the Lewis acid–base model is that it is more general and can be extended to cover more reactions, as we will see in later sections. Our emphasis in the Lewis model is on a nonbonding pair of electrons on the Lewis base. We can rank Lewis bases (and hence also their conjugate acids) in order of relative basicity by observing which Lewis base predominates in a solution in which two bases compete for protons: The weaker Lewis base predominates. We can use tables of relative Lewis base (and conjugate Lewis acid) strengths, like Table 6.2, to predict which species will predominate in a solution.

The list doesn't explain *why* one Lewis base is stronger than another. If you have an explanation, you can apply it when you meet Lewis bases and acids that aren't on the list. Our task in the next section is to provide at least some of that explanation.

6.5. Predicting Strengths of Lewis/Brønsted–Lowry Bases and Acids

Electronegativity and relative Lewis base strength Electronegativity is one factor that plays a role in determining the strength of Lewis bases. The electronegativities of the elements increase going from left to right across a period of the periodic table:

(least electronegative) C < N < O < F (most electronegative)   **(6.23)**

Compare the order of electronegativities for the second-period elements with these basicities from Table 6.2:

(most basic) $:CH_3^- > :NH_2^- > :OH^- > :F^-$ (least basic)   **(6.24)**

The order of basicities is the reverse of the order of electronegativities of the atom with the nonbonding electron pair. Nonbonding electrons on atoms with high electronegativity are held more tightly and are less available for donation to form a bond with a Lewis acid, including the proton. The differences between

> The methide ion, $:CH_3^-$, may be unfamiliar, but methyl lithium, $LiCH_3$, is a reagent that is commonly used to attach a methyl group to a carbonyl carbon. You can think of methyl lithium as $(Li^+)(:CH_3^-)$.

the basicities of these anions are quite large, more than 10 orders of magnitude. We'll return in Chapter 9 to look at acid–base systems more quantitatively.

6.26 CHECK THIS

Relative Lewis base strengths

(a) Explain clearly how the charge density models shown in the *Web Companion*, Chapter 6, Section 6.5.1, help explain the order of basicities in expression (6.24).

(b) If these ions, Cl^-, SH^-, and PH_2^-, could be tested for their relative base strengths, what do you predict the order would be? What would be the relative acid strengths of the conjugate acids? Clearly explain the reasoning for your answers. Do you have any evidence that your order is correct?

6.27 CONSIDER THIS

Is electronegativity a reliable predictor of Lewis acid–base strength?

If Lewis base strength increases inversely with electronegativity, then the conjugate acids should increase in acid strength directly with electronegativity. Consider the hydrogen halides, whose charge density models are shown here (to the same scale):

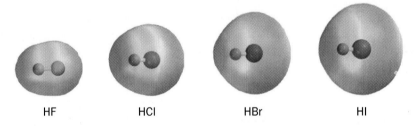

HF	HCl	HBr	HI

(a) How do the electronegativities of the halogens vary from F to I? Assuming there is a direct dependence of Lewis acid strength on the electronegativity of the halogen, what would you predict for the relative order of acid strength of the four hydrogen halides? Explain.

(b) Based on your relative order of acidity in part (a), what would you predict about the relative base strengths of the halide ions, $:F^-$, $:Cl^-$, $:Br^-$, and $:I^-$? Explain.

(c) Are your answers in parts (a) and (b) consistent with data in Table 6.2? Why or why not?

Atomic size and relative Lewis base strength Electronegativities decrease going down a column (family) of the periodic table. Therefore, based on electronegativity, your prediction for the acid strengths of the hydrogen halides in Consider This 6.27 would be decreasing acidity from HF to HI. And for the halide ions, the conjugate Lewis bases, you would predict increasing Lewis base strength from $:F^-$ to $:I^-$. However, Table 6.2 shows that the halide ions follow exactly the reverse order of Lewis basicity:

$$\text{(most basic)} :F^- > :Cl^- > :Br^- > :I^- \text{ (least basic)} \qquad (6.25)$$

Electronegativity must not be the whole story. A second factor that affects the strength of a Lewis base is its size. The second-period elements we considered in the previous paragraph are all about the same size (within about 20%), so differences in electronegativity dominate their basicity.

The halide ions increase a great deal in size down the halogen-family column of the periodic table; iodide is almost twice the size of fluoride. (You see this great size difference in the models for the hydrogen halides in Consider This 6.27.) The consequence of large size is that the valence electron charge is spread out in a much larger volume. As you have seen in Chapters 4 and 5, ions and molecules are more stable when their valence electrons occupy larger volumes. Iodide has the same number of valence electrons as fluoride, but is more stable and less reactive toward protons, because it is so much larger. The halide ions are such weak Lewis bases, except for fluoride, that their conjugate acids, the hydrogen halides, are essentially completely ionized to hydronium ions and halide ions in aqueous solutions. The order of the halide basicities (or hydrogen halide acidities) in Table 6.2 has to be determined from experiments in solvents other than pure water.

6.28 CONSIDER THIS

Does electronegativity or size dominate Lewis acid–base strength?

Charge density models for the hydrides (binary compounds of the elements with hydrogen) of the first four members of the oxygen family of elements are shown here (to the same scale):

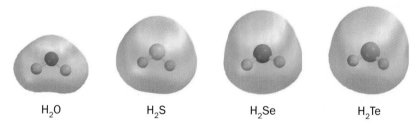

| H_2O | H_2S | H_2Se | H_2Te |

(a) Based on the electronegativities of the oxygen-family elements, what would you predict for the relative order of acid strength of these oxygen-family hydrides? Explain.

(b) Based on the sizes of the oxygen family elements, what would you predict for the relative order of acid strength of these oxygen-family hydrides? Explain.

(c) Which of these acids do you find in Table 6.2? What is the order of their acidities? Assuming this trend is maintained for all these acids, what is the relative order of their acid strengths? What is the relative order of base strengths for $:OH^-$, $:SH^-$, $:SeH^-$, and $:TeH^-$?

(d) What is the answer to the title question? Explain your reasoning.

6.29 CHECK THIS

Relative Lewis base strengths

If phosphine, $:PH_3$, reacts as a Lewis base, would you expect it to be a stronger or weaker base than ammonia? Explain the reasoning for your prediction.

Oxyacids and oxyanions: Carboxylic acids The acids we have considered so far in this section are **binary acids,** compounds of hydrogen and one other element. These are interesting and important compounds, especially as a beginning for understanding the factors that affect the strength of Lewis bases and acids, but there are only a small number of binary acids. The most common proton-donating acids are the oxyacids, which we introduced in Chapter 2, Section 2.13. Recall that the acidic protons of oxyacids are always bonded to an oxygen atom that is, in turn, bonded to another atom. The oxyanions of oxyacids are Lewis bases with one or more oxygen atoms that have nonbonding pairs of electrons that can bond to a proton. Almost all the acids in biological systems are oxyacids and most of them are carboxylic acids or derivatives of phosphoric acid.

 **6.30 INVESTIGATE THIS**

Do the acidities of alcohols and carboxylic acids differ?

Use pH paper or a pH meter to measure the pH of water and 0.1 M solutions of ethanol and ethanoic (acetic) acid in water. Is the acidity of the alcohol greater than, less than, or the same as the acidity of the carboxylic acid? Show how you use your pH data as evidence for your answer.

The water, ethanol, and ethanoic acid in Investigate This 6.30 are oxyacids:

$$H{-}OH(l) + H_2O{:}\ (l) \rightleftharpoons H_2O{-}H^+(aq) + {:}OH^-(aq) \qquad \text{(6.19)}$$

$$CH_3CH_2O{-}H(aq) + H_2O{:}\ (l)$$
$$\rightleftharpoons H_2O{-}H^+(aq) + CH_3CH_2O{:}-(aq) \qquad \text{(6.26)}$$

$$CH_3C(O)O{-}H(aq) + H_2O{:}\ (l)$$
$$\rightleftharpoons H_2O{-}H^+(aq) + CH_3C(O)O{:}^-(aq) \qquad \text{(6.17)}$$

Reactions (6.19) and (6.17) remind you of the acid–base properties of water and ethanoic acid discussed in the previous section. Reaction (6.26) is the corresponding reaction of ethanol as an oxyacid. (Electrons involved in the Lewis acid–base reaction are shown in red.)

6.31 CONSIDER THIS

How do the acidities of alcohols and carboxylic acids differ?

(a) Do your results in Investigate This 6.30 provide any evidence for or against reaction (6.26)? What is the evidence?

(b) Based on your results, how would you rank the acid strengths of water, ethanol, and ethanoic acid? How would you rank the base strengths of the hydroxide, $:OH^-(aq)$, ethoxide, $CH_3CH_2O:^-(aq)$, and ethanoate (acetate), $CH_3C(O)O:^-(aq)$, anions? Explain your reasoning.

Water and aqueous solutions of ethanol have essentially the same pH, so you can conclude that reaction (6.26) does not proceed to a measurable extent to produce hydronium ions. You know that reaction (6.17) does proceed to a small extent to give the ethanoic acid solution a moderately acidic pH. Therefore, you can conclude that ethanoic acid is a stronger acid than ethanol (or, to look at these molecules from the point of view of their conjugate bases, ethanoate anion is a weaker base than ethoxide anion). Since reaction (6.26) does not proceed as written, we can conclude that ethanol and water must have approximately the same acid strength (and ethoxide anion and hydroxide anion about the same base strength).

6.32 CHECK THIS

Relative basicities of ethanoate, ethoxide, and hydroxide anions

(a) Are the rankings in Table 6.2 consistent with the conclusions about relative basicities in the previous paragraph? Explain why or why not.

(b) Write Lewis structures for these three ions. Which Lewis structure(s) can be written in more than one equivalent way? Can you think of any reason why ethanoate should be a weaker base than ethoxide or hydroxide? Explain.

Web Companion

| Chapter 6, Section 6.4.3–4 | —① |

View animations of ethanoic acid and ethanoate ion charge densities and proton transfers.

②
③
④

Delocalized π bond in carboxylate You have found that there is only one way to write the Lewis structures of ethoxide and hydroxide, but there are two equivalent ways to write the Lewis structure of the carboxylate group of the

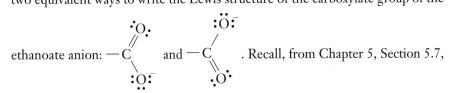

ethanoate anion: and . Recall, from Chapter 5, Section 5.7,

that when two or more equivalent Lewis structures can be written for a molecule (or ion), none of the structures is correct. The pi (π) electrons in the molecule are spread over several atoms in a delocalized orbital or orbitals. Figure 6.2 shows the overall charge density for the ethanoate (acetate) ion. Observe the equal charge density on the oxygen atoms.

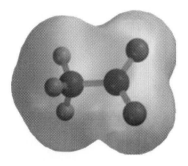

Figure 6.2.

Charge density model for the ethanoate ion.

Energetics of proton transfer Delocalization of the π electrons in the ethanoate anion lowers the energy of the ion, that is, makes it more stable relative to similar molecules that lack such electron delocalization. The stability of the ethanoate anion is the major reason why it is such a weak base compared to the ethoxide anion. Figure 6.3 shows energy-level diagrams comparing the energies for proton transfer from ethanol to water and ethanoic acid to water, reactions (6.26) and (6.17), respectively. The lower total energy of a carboxylate ion makes a large difference in the energy required for proton transfer from the carboxylic acid and the alcohol. For carboxylic acids, transfer of a proton to water requires almost no energy; the reactants and products have just about the same energy.

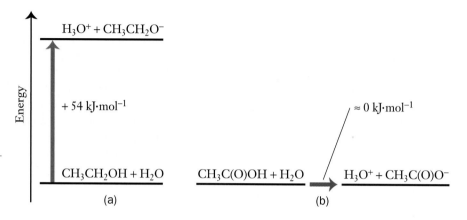

Figure 6.3.

Energy-level diagrams for proton transfer by (a) ethanol and (b) ethanoic acid.

You might wonder why, if no energy is required, only a few percent of the ethanoic acid molecules in a 0.1 M aqueous solution transfer their proton to water. Recall that you have been confronted with this same kind of puzzle before. For example, in Chapter 2, you found that both exothermic and endothermic solubility reactions occur. Energy is often a guide for *comparing similar reactions*. In Figure 6.3, the comparison is between proton transfers from electrically neutral oxyacids to water to form hydronium ion and oxyanions with a 1− charge. Energy is not, however, a reliable guide for predicting the extent of *individual* acid–base (or solubility–precipitation) reactions. The missing piece of the puzzle is the **entropy** of these reactions, which we will consider in Chapters 8 and 9.

6.33 CHECK THIS

Relative basicities of oxyanions and acidities of oxyacids

(a) How is the cyclohexanoxide anion in the *Web Companion*, Chapter 6, Section 6.5.2, similar to the ethoxide anion? How is the phenoxide anion similar to the ethanoate anion? Is your selection of the weaker base consistent with these similarities? Explain your reasoning. Is your selection of the weaker base consistent with the information in Table 6.2? Explain.

(b) Write the reaction for the transfer of a proton from carbonic acid, $(HO)_2CO(aq)$, to water. Write the Lewis structure(s) for the hydrogen carbonate anion formed in this reaction.

(c) What would you predict about the relative acidities of carbonic acid and ethanoic acid? Give the reasoning for your prediction. Is your prediction consistent with the information in Table 6.2? Why or why not?

Oxyacids and oxyanions of third-period elements Our reasoning about relative acid or base strengths based on delocalization of electrons can be extended to oxyacids and oxyanions whose molecules contain a third-period element, phosphorus, sulfur, or chlorine, to which one or more —OH groups is bonded. These compounds also often have other oxygen atoms bonded to the third-period element. Figure 6.4 shows the four oxyacids of chlorine in order from least acidic, hypochlorous acid, HOCl, to most acidic, perchloric acid, $HOClO_3$. Chloric and perchloric acids are stronger acids than the hydronium ion and transfer their protons completely to water in aqueous solution. Experiments in other solvents are required to determine the order of their acid strengths shown in the figure.

Phosphates, sulfates, and their acids are vital components of living systems and are important in many industrial processes. Perchlorates are often used in studies where the very low basicity of the anion gives desirable solution properties.

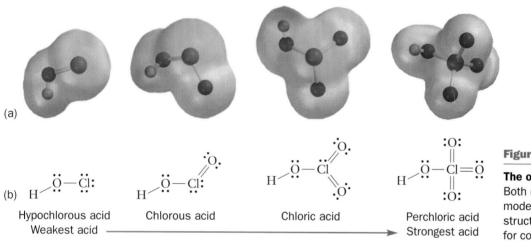

Figure 6.4.

The oxyacids of chlorine. Both (a) charge density models and (b) Lewis structures are shown for comparison.

Hypochlorous acid — Weakest acid   Chlorous acid   Chloric acid   Perchloric acid — Strongest acid

---

6.34 CONSIDER THIS

How can you explain the relative acidities of the chlorine oxyacids?

(a) Write the Lewis structures for the oxyanions formed when the chlorine oxyacids transfer a proton to a base. For which one(s) can you write two or more equivalent structures? How many more equivalent structures?
(b) Is there any correlation between your answers in part (a) and the relative strengths of the oxyacids? If so, suggest an explanation for the correlation.

You have found that there are, respectively, one, two, three, and four equivalent Lewis structures for the hypochlorite, ClO^-, chlorite, ClO_2^-, chlorate, ClO_3^-, and perchlorate, ClO_4^-, anions. The larger the number of equivalent Lewis structures, the greater the delocalization of the electrons and the more stable the oxyanion. The base strength of the oxyanions decreases as they become more stable. The perchlorate anion is the weakest base and, therefore, as we see in Figure 6.4, perchloric acid is the strongest acid. In Chapter 9, we will treat acid–base reactions more quantitatively and be able to assign numerical values to acid and base strength. The difference in acid strength going from one acid to the next in Figure 6.4 is almost a factor of a million, 10^6; perchloric acid is about 10^{17} times stronger than hypochlorous acid.

6.35 CHECK THIS

Acid–base properties of bleach solutions

Most liquid bleaches contain sodium hypochlorite, $NaOCl$, which ionizes in solution to give $Na^+(aq)$ and $ClO^-(aq)$ ions. Would sodium hypochlorite make bleach solutions acidic or basic? Give the reasoning for your answer.

6.36 CHECK THIS

Another way to predict relative strengths of oxyacids and oxyanions

(a) How many oxygen atoms are doubly bonded to chlorine in each of the oxyacids in Figure 6.4? Is there a correlation between these numbers and the relative strengths of the oxyacids? If so, formulate a rule that predicts the relative acid strengths of oxyacids (and relative base strengths of their oxyanions).

(b) What does your rule from part (a) predict about the relative basicity of the hydrogen sulfate, $HOSO_3^-$, and hydrogen sulfite, $HOSO_2^-$, oxyanions? Is this also what the delocalization of electrons in these ions predicts? Explain how you make your predictions.

Your results in Check This 6.36 show an alternative way to predict the order of acid or base strength for oxyacids and oxyanions that does not require Lewis structures for the oxyanions. Keep in mind that this rule is a correlation, not an explanation. The explanation for the basicity and acidity is the delocalization of electrons and consequent greater stability of the oxyanions.

6.37 CONSIDER THIS

How do oxyacids/oxyanions of different central atoms compare?

(a) Write the Lewis structure for phosphoric acid, $(HO)_3PO$. Write the reaction for transfer of one proton from phosphoric acid to water. What is the Lewis structure for the dihydrogen phosphate oxyanion formed in this reaction?

(b) Experimentally, we find that phosphoric and chlorous acids are about equal in acid strength. Is this what your rule from Check This 6.36(a) predicts? Is this what the delocalization of electrons in the oxyanions predicts? From this example, what conclusion might you draw about the influence of the central

continued

atom (phosphorus or chlorine here) on the strength of the oxyacids? State your reasoning clearly.

(c) Draw Lewis structures for the oxyanions of nitrous acid, HONO, and chlorous acid, HOClO. Experimentally, we find that the nitrite oxyanion is a somewhat stronger base than the chlorite oxyanion. Is this what your rule from Check This 6.36(a) predicts? Is this what the delocalization of electrons in the oxyanions predicts? From this example, what conclusion might you draw about the influence of the central atom (nitrogen or chlorine here) on the strength of the oxyacids? State your reasoning clearly.

Comparisons within and between periods In Consider This 6.37(b), you find that two third-period oxyacids whose oxyanions have the same amount of electron delocalization have about the same acid strength. As a general rule, the relative acid (or base) strengths for oxyacids (or oxyanions) of elements in the same period of the periodic table depend on the amount of electron delocalization in the oxyanions and are little affected by the identity of the central atom. Comparisons between oxyacids/oxyanions of elements from different periods of the periodic table, as in Consider This 6.37(c), show that the period does make a difference. Second-period oxyanions are stronger bases than third-period oxyanions with the same electron delocalization (as indicated by the number of equivalent Lewis structures). This observation suggests that the third-period oxyanions are more stable than their second-period counterparts. Recall that third-period atoms are larger than second-period atoms, so third-period oxyanions are somewhat larger than second-period oxyanions. Thus, delocalized electrons occupy a larger volume in the third-period oxyanions and are a bit lower in energy than the corresponding second-period oxyanions.

6.38 CHECK THIS

Relative acid and base strengths

(a) List these oxyanions in order from the weakest base to the strongest base. Explain the reasoning for your ranking.

$$CH_3CO_2^-, HOCO_2^-, NO_3^-, NO_2^-, OH^-$$

(b) These oxyanions have at least one proton available to transfer to a base. List the ions in order from the strongest to the weakest *acid*. Explain your reasoning.

$$HOSO_3^-, HOSO_2^-, (HO)_2PO_2^-, HOPHO_2^-$$

(The names of these oxyanions are hydrogen sulfate, hydrogen sulfite, dihydrogen phosphate, and hydrogen phosphite, which has one nonacidic H bonded to P.)

(c) List these acids (carbonic, phosphoric, and sulfuric) in order from the strongest acid to the weakest acid. Explain your reasoning.

$$(HO)_2CO, (HO)_3PO, (HO)_2SO_2$$

Predominant acid and base in a reaction The point of our discussion of relative strengths of acids and bases is to enable you to tell, without doing any mathematics or making any measurements, what chemical species will predominate in aqueous solutions of acids and bases. You can use Table 6.2 or the reasoning from above, for example, to determine which of two bases in a solution is the stronger. The stronger will win the competition for available protons. The concentration of the weaker base will be larger than the concentration of the stronger base. This is the reasoning we used to find relative basicities in Section 6.4. As you have seen, we can reverse the process to predict which base (and acid) will predominate in a reaction between acids and bases. There is one stumbling block to watch out for in aqueous solutions. The predominant acid–base species in aqueous solutions is water; its concentration is close to 55 M. When you are asked about the predominant bases and acids in an aqueous solution, you are almost always being asked about the acids and bases *other than water* itself.

6.39 WORKED EXAMPLE

Predicting the direction of reaction for acid–base reactions

If equal volumes of 0.1 M solutions of hydrochloric acid and sodium cyanide, NaCN, are mixed, what species will predominate in the mixture? Will the mixture be acidic or basic?

Necessary information: We have to know that NaCN(s) dissolves in water to give $Na^+(aq)$ and $CN^-(aq)$ and that $Na^+(aq)$ is a spectator ion in the mixture. From Table 6.2 we find that $CN^-(aq)$ is a stronger base than either water or $Cl^-(aq)$. We also have to recall that a 0.1 M solution of hydrochloric acid is 0.1 M in hydronium ion, $H_3O^+(aq)$.

Strategy: Write the reaction between the reactants that are mixed and use its stoichiometry to determine the species that would be present, if no further reactions occurred. Then write the acid–base reaction(s) for these species and use the relative basicities (or acidities) of the reactants and products to determine which species will predominate.

Implementation: Cyanide is a stronger base than chloride; protons from the hydronium ions in the hydrochloric acid solution will be transferred to cyanide when the solutions are mixed:

$$H_3O^+(aq) + Cl^-(aq) + Na^+(aq) + CN^-(aq)$$
$$\rightleftharpoons HCN(aq) + H_2O(l) + Na^+(aq) + Cl^-(aq) \qquad \textbf{(6.27)}$$

Note that, because it is such a weak base, chloride ion is also a spectator ion in this mixture.

Reaction (6.27) shows that $H_3O^+(aq)$ and $CN^-(aq)$ react in a 1:1 ratio. Equal volumes of 0.1 M solutions of hydrochloric acid and sodium cyanide contain the same number of moles $H_3O^+(aq)$ and $CN^-(aq)$, so the reaction uses up both reactants to produce this number of moles of HCN(aq). In a solution of HCN(aq), the acid–base reaction is

$$HCN(aq) + H_2O(l) \rightleftharpoons H_3O^+(aq) + CN^-(aq) \qquad \textbf{(6.28)}$$

continued

Since CN⁻(*aq*) is a stronger base than water, we know that it will win the competition for protons. The reactants, HCN(*aq*) [and H₂O(*l*)], will predominate in the mixture, which also contains all the unreacted Cl⁻(*aq*) + Na⁺(*aq*). Reaction (6.28) proceeds to some extent to produce small amounts of H₃O⁺(*aq*) [and CN⁻(*aq*)], so the solution will be somewhat acidic.

Does the answer make sense?   The stoichiometry of the reaction produces a solution that is identical to what we would get by dissolving HCN in salt water. The salt does not affect the acidity of the solution. HCN(*aq*) is a weak acid, so it will transfer only a few of its protons to water to produce a weakly acidic solution and a low concentration of CN⁻(*aq*). This is what we concluded.

6.40 CHECK THIS

Predicting the direction of reaction for acid–base reactions

Phenol, Table 6.2, is a solid that is moderately soluble in water at room temperature. The phenolate anion is soluble in water to a much greater extent. If you want to dissolve a lot of phenol in an aqueous solution, should you use a solution of hydrochloric acid or a solution of sodium hydroxide to try to dissolve it? Clearly explain the reasoning for your choice.

Some cases can be a little complicated to sort out. For example, both the hydrogen tartrate anion, $HOOC(CHOH)_2COO^-$(*aq*), and the hydrogen carbonate anion, $HOCO_2^-$(*aq*), can accept and donate protons. When these anions are mixed, there are two possible acid–base reactions:

$$HOOC(CHOH)_2COO^-(aq) + HOCO_2^-(aq)$$
$$\rightleftharpoons HOOC(CHOH)_2COOH(aq) + CO_3^{2-}(aq)$$

tartaric acid   carbonate
                       dianion
                                                    (6.29)

$$HOOC(CHOH)_2COO^-(aq) + HOCO_2^-(aq)$$
$$\rightleftharpoons {}^-OOC(CHOH)_2COO^-(aq) + (HO)_2CO(aq)$$

tartrate dianion   carbonic acid
                                                    (6.30)

The data in Table 6.2 show that the carbonate dianion is a stronger base than carboxylate anions like the hydrogen tartrate anion. Thus, reaction (6.29) is not likely to proceed too far toward products and can be eliminated as the predominant reaction in this mixture.

To analyze reaction (6.30), we have to compare the base strength of hydrogen carbonate anion with the base strength of a carboxylate anion, one of the ends of the tartrate dianion. In Check This 6.33, you considered this comparison from the point of view of the conjugate acids (carbonic and carboxylic acids) and found that they were of about equal strength. Thus, the bases are also about equal in strength. Without having more quantitative information about the relative base (or acid) strengths, we can reasonably conclude that reaction (6.30) is likely to proceed to yield the products, but, at equilibrium, there will be a good deal of the reactants left as well.

Recall that this is the reaction mixture from Investigate This 6.3 that produced gaseous carbon dioxide as a product. On that basis, we wrote this net equation for the reaction (with hydrogen tartrate and tartrate represented by molecular formulas):

$$HC_4H_4O_6^-(aq) + HCO_3^-(aq)$$
$$\rightarrow C_4H_4O_6^{2-}(aq) + CO_2(g) + H_2O(l) \qquad (6.3)$$

In Chapter 2, Section 2.14, we saw that when carbonic acid is formed in a solution, it can react to release carbon dioxide gas:

$$(HO)_2CO(aq) \rightleftharpoons CO_2(g) + H_2O(l) \qquad (6.31)$$

Reaction (6.31) "uses up" the carbonic acid formed by reaction (6.30) and the system adjusts by producing more, which drives the reaction toward products and explains the single direction arrow in reaction expression (6.3).

This analysis is an example of Le Chatelier's principle, which we introduced in Chapter 2, Section 2.14. **Le Chatelier's principle** says that when systems at equilibrium are disturbed they respond in a way that minimizes the effect of the disturbance. The equilibrium represented by reaction expression (6.30) is disturbed when some of the carbonic acid reacts further to produce carbon dioxide, reaction (6.31). The system responds by making more carbonic acid to compensate for the loss. This process continues until the reactants are used up. The mixture of hydrogen tartrate and hydrogen carbonate anions represents the most complicated system we will consider. We usually will focus on acid–base interactions that are easier to analyze.

Reflection and Projection

For binary Lewis/Brønsted–Lowry acids, two properties of the nonhydrogen atom in the acid explain the relative acidities and basicities shown in Table 6.2. Within a period of the periodic table, basicity decreases with increasing electronegativity and, conversely, the acidity of the conjugate acids increases. Within a family (column) of the periodic table, the basicity decreases with increasing size of the atom. In comparisons between two binary acids, the size effect, due to greater volume for the electron waves, usually dominates.

Most acids and bases are oxyacids and oxyanions in which the Lewis-base electron pairs are on an oxygen atom. Relative basicity of the oxyanions is determined by electron delocalization, which lowers the energy of the anion and makes it a weaker base. You get a qualitative measure of the amount of delocalization by counting the number of equivalent Lewis structures for the oxyanion. The identity of the central atom, for atoms in the same period, has little effect on the basicity of oxyanions. Oxyanions with central atoms from the third period are weaker bases than oxyanions of second-period atoms, because of the increased size of the third-period atoms.

Now you have the background to predict the direction of many Lewis/Brønsted–Lowry acid–base reactions. In the next sections, we will expand our view of Lewis acid–base reactions to include Lewis acids that are not proton donors.

6.6. Lewis Acids and Bases: Metal Ion Complexes

6.41 INVESTIGATE THIS

Do Lewis bases react with metal ions?

For this investigation, use a 0.1 M aqueous solution of nickel chloride, $NiCl_2$, and a 0.5 M aqueous solution of ammonia, NH_3. Record the colors of the solutions. Mix equal volumes of the two solutions. Record the color of the mixture. Does a reaction occur between the ingredients of the mixture? What is the evidence for your answer?

In Chapter 2, Section 2.5, we used ion–dipole attraction to explain how ionic solids dissolve. Water molecules were pictured as surrounding the cation with the negative end of the water dipoles oriented toward the cation. What that picture fails to convey is that, for many cations, a fixed number of water molecules arrange themselves *symmetrically* around the cation. The structure shown in Figure 6.5 is an example of a **metal ion complex.** Metal ion complexes form between cations and many different kinds of neutral molecules and ions. The molecules and ions that are attached to the central metal ion are called **ligands** because they bind to the cation.

> Ligand (and the ligaments that connect your bones) are from the Latin *ligare* = to bind.

All ligands have one or more atoms with nonbonding electron pairs that can be shared with the central metal cation. ***All ligands are Lewis bases; the central metal ion is reacting as a Lewis acid.*** The structure shown in Figure 6.5 is the chromium, Cr^{3+}, ion in aqueous solution. It is often written as $Cr(H_2O)_6^{3+}$, rather than as $Cr^{3+}(aq)$, to show the stoichiometry of the complex ion. The complex is called an **octahedral complex ion** because the geometry of the atoms bonded to the central metal ion is octahedral. The oxygen atoms are at the vertices of an imaginary octahedron around the metal ion.

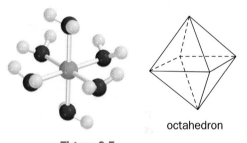

octahedron

Figure 6.5.

An octahedral complex ion, $Cr(H_2O)_6^{3+}$, and a regular octahedron.

6.42 CONSIDER THIS

How do you know a Lewis acid–base complex ion has reacted?

(a) Assume that $Ni^{2+}(aq)$ is an aqueous octahedral complex like the one shown for $Cr^{3+}(aq)$ in Figure 6.5. Make a molecular model of the nickel complex. What is one of the properties of this complex that you can infer from Investigate This 6.41? Explain.

(b) Which is the stronger Lewis base, water or ammonia? In a competition for a Lewis acid, which will win? How is your answer relevant to your observations in Investigate This 6.41? Explain.

(c) Make a molecular model of the product of the reaction in Investigate This 6.41. Explain why you think this is the product. What is one of the properties of this product?

Web Companion

Chapter 6, Section 6.6.1 — ①

Animations show the structure and reactions of metal ion complexes.

②
③
④

Colors of metal ion complexes Color is one of the most striking properties of complex ions. Most of the complexes formed by **transition metal** ions are colored. Transition metals are the metals in the middle of the periodic table shown inside the front cover. The transition is a result of valence electrons occupying a subshell (the *d* orbitals, Chapter 4, Section 4.11) with energies that are approximately the same as those of the elements at either end of the transition. The wavelengths of light these electrons absorb as they change energy within the subshell are in the visible region of the spectrum, so the ions are colored. The energies of the subshell electrons and the wavelengths absorbed are influenced by interactions with ligand electrons. You have seen that nickel(II) is green when complexed with water, $Ni(H_2O)_6^{2+}(aq)$, blue with ammonia, $Ni(NH_3)_6^{2+}(aq)$, and red with dimethylglyoxime, Check This 6.11. Much of the color you see around you in nature and in human decoration is due to metal ion complexes.

Note that we have introduced a new notation, nickel(II), for the nickel cation with a 2+ charge, that is, a nickel atom that has lost two electrons. The Roman numeral in parentheses next to the name of the element denotes the charge on the metal ion. This notation is often used for metals, especially the transition metals that form compounds with different charges on the cation. Iron, for example, commonly forms compounds in which the iron has a 2+ or 3+ charge. Two of the oxides of iron are iron(II) oxide, FeO, and iron(III) oxide, Fe_2O_3. This notation makes it easier to name compounds unambiguously, without having to remember the names formerly used for metal cations with different charges. We will use this notation when it is convenient.

> The older nomenclature, which you will still find in use, assigns different names to the cations with different charges. For example, iron(II) is called ferrous and iron(III) is called ferric, so FeO is ferrous oxide and Fe_2O_3 is ferric oxide.

6.43 INVESTIGATE THIS

Do calcium ions react with complexing ligands?

Do this as a class investigation and work in small groups to discuss and analyze the results. You will use 0.1 M aqueous solutions of calcium nitrate, $Ca(NO_3)_2$, sodium oxalate, NaC_2O_4, sodium oleate, $NaO(O)C(CH_2)_7CH=CH(CH_2)_7CH_3$ (an ingredient in some soaps), and tetrasodium **ethylenediaminetetraacetic acid (EDTA)**, $Na_4C_{10}H_{12}N_2O_8$ (used as a preservative in many foods and found in many other consumer products).

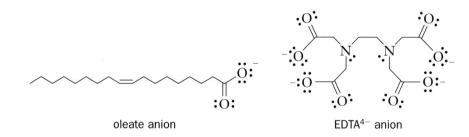

oleate anion   EDTA^{4-} anion

(a) Add about 2 mL of water and 4 drops of calcium nitrate solution to each of two test tubes. To one of the test tubes, add 5 drops of sodium oxalate solution, swirl gently to mix the solution, and record your observations. Add 5 drops of EDTA solution, gently swirl to mix the solution, and record your observations.

(b) Repeat with the other test tube, except use the oleate solution instead of the oxalate.

6.44 CONSIDER THIS

How do calcium ions react with complexing ligands?

(a) In Investigate This 6.43, what happened when oxalate or oleate was added to calcium ion solutions? Did you expect these results? Explain why or why not.

(b) What happened when EDTA solution was added to the mixtures? How can you account for these results? Does the same explanation apply to both mixtures? Explain why or why not.

Since you have seen it before, it isn't surprising to see a precipitate form when calcium and oxalate ions are mixed. Calcium oleate, $Ca[O(O)C(CH_2)_7CH=CH(CH_2)_7CH_3]_2$, is also insoluble in water, because the long nonpolar, hydrocarbon tail of oleate is not soluble in water. Anions with extended carbon frameworks often form insoluble salts. One of the ions responsible for hard water is $Ca^{2+}(aq)$. Soap scum and bathtub rings are precipitates of these metal ions with the ingredients of soap like sodium oleate.

Calcium–EDTA complex ion formation Perhaps you were surprised to observe that addition of $EDTA^{4-}(aq)$ to solutions containing precipitates of calcium oxalate and calcium oleate resulted in the disappearance of the precipitates. When the calcium oxalate precipitate goes back into solution, the calcium and oxalate ions must again be present in the solution from which they precipitated. This isn't possible (without forming a precipitate), so some reaction must have occurred to use up either the calcium ion or the oxalate ion or both. The extra ingredient in the solution, $EDTA^{4-}(aq)$, must be one of the reactants in this new reaction. It's not likely that the negatively charged $C_2O_4^{2-}(aq)$ and $EDTA^{4-}(aq)$ will react with one another; they strongly repel one another. The positive $Ca^{2+}(aq)$ cation is a likely candidate to react with the $EDTA^{4-}(aq)$.

6.45 INVESTIGATE THIS

What is the three-dimensional structure of EDTA?

Use your model kit to make a molecular model of the ethylenediaminetetra-acetate tetraanion, $EDTA^{4-}$. To make it easier to handle and visualize the structure, leave off all the hydrogen atoms, as they have been in the skeletal structure shown in Investigate This 6.43. Use the long yellow connectors in your kit to represent the nonbonding electron pairs on the nitrogen atoms and one of the nonbonding electron pairs on each charged oxygen atom. Twist your model about to show that all six of the yellow connectors (nonbonding electron pairs) can be directed toward a central point.

 Web Companion

Chapter 6, Section 6.6.2 ──①

Observe animations of $EDTA^{4-}$ ②
and the formation and structure ③
of its metal ion complex. ④

What kind of complex ion can $EDTA^{4-}(aq)$ form with $Ca^{2+}(aq)$? In Investigate This 6.45, you manipulated your $EDTA^{4-}$ model to bring nonbonding electron pairs from six of the atoms close together around a central point. If there is a positive cation at the center of this array, the electron pairs (Lewis

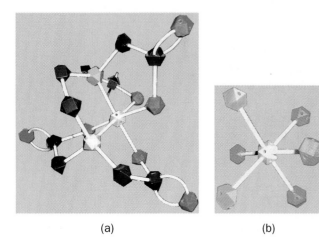

(a)   (b)

Figure 6.6.

Structure of the EDTA^{4-} complex with a central metal ion. (a) The entire complex with none of the EDTA^{4-} hydrogen atoms shown. The metal ion is gray. (b) To make the geometry clearer, only the octahedral array of atoms directly bonded to the metal ion is shown.

Chelate is from Latin *chele* = claw, like the pincer of a lobster.

bases) and the cation (Lewis acid) attract each other (bonds are made) and a stable structure is formed. Figure 6.6(a) shows this structure for a model of EDTA^{4-} wrapped around and bonded to a central metal cation.

EDTA^{4-} is only one of many ligands that contain two or more Lewis base groups that can complex with metal ions. Usually, these ligand molecules can curl about, as shown in Figure 6.6(a), so more than one of the Lewis base groups can occupy bonding positions about a metal ion. Such metal ion complexes are called **chelates,** and the ligands are called chelating ligands. Chelating ligands like EDTA are particularly effective for dissolving metal ions, such as Pb^{2+}, that have few soluble salts. EDTA is used, for example, to treat people who have lead poisoning.

Complex ion formation and solubility In a solution containing Ca^{2+}(aq), C$_2$O$_4{}^{2-}$(aq), and EDTA^{4-}(aq) [Investigate This 6.43(a)], the anions are in competition for the calcium ion. Oxalate reacts to precipitate the calcium ion as calcium oxalate:

$$Ca^{2+}(aq) + C_2O_4{}^{2-}(aq) \rightleftharpoons CaC_2O_4(s) \tag{6.4}$$

EDTA^{4-}(aq), reacts to complex (chelate) the calcium ion:

$$Ca^{2+}(aq) + EDTA^{4-}(aq) \rightleftharpoons Ca(EDTA)^{2-}(aq) \tag{6.32}$$

Evidently, under the conditions of Investigate This 6.43(a), EDTA^{4-}(aq) wins the competition.

When Ca^{2+} cations react to form Ca(EDTA)$^{2-}$(aq) complex ions by reaction (6.32), they are no longer present as Ca^{2+}(aq) ions and are not available for reaction (6.4). As EDTA^{4-}(aq) was added to the mixture containing Ca^{2+}(aq), C$_2$O$_4{}^{2-}$(aq), and CaC$_2$O$_4$(s), the Ca^{2+}(aq) in solution reacted to form Ca(EDTA)$^{2-}$(aq). Because reaction (6.4) is reversible, some of the CaC$_2$O$_4$(s) dissolved to replace the missing Ca^{2+}(aq). These reactions continued, reaction (6.32) going in the forward direction and reaction (6.4) going in reverse, until all the precipitate dissolved. Figure 6.7 is a schematic representation of precipitate formation and then dissolution of the precipitate upon addition of a complexing reagent.

Figure 6.7.

Competition between precipitate and complex formation. (a) Cation and anion mixture in the instant before reaction. (b) After the precipitation reaction. (c) The solution in the instant after a chelating ligand is added. (d) Formation of the cation–ligand complex "removes" the cation from the solution and the precipitate dissolves. The nonparticipating counter ions are not shown.

(a)   (b)   (c)   (d)

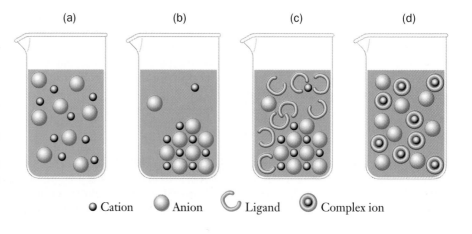

● Cation ◯ Anion ◡ Ligand ◉ Complex ion

The analysis of this competition is another application of Le Chatelier's principle. The reaction represented by expression (6.4) and Figure 6.7(b) is an ionic solid in equilibrium with its ions in solution. When the chelating ligand is added, it reacts with the cation, reaction (6.32), and reduces its concentration. This disturbs the solubility equilibrium and the solubility system responds to minimize the disturbance by adding more cation to the solution, that is, by dissolving. Le Chatelier's principle gives the direction but not the magnitude of these effects. In Chapter 9, we will examine competitive reactions like these more quantitatively.

6.46 CHECK THIS

Keeping the shine in your hair

Read the ingredients label on your shampoo (also see the label in Check This 6.25). Many contain EDTA in some form (tetrasodium EDTA, trisodium HEDTA, Edetate, and so on). What is EDTA doing there? Is your answer related to Investigate This 6.43(b)? If so, clearly explain the connection.

Metal ion complexes with four ligands The octahedral arrangement of ligands in metal ion complexes is the most common geometry you will find. Other geometries are possible and range from two to nine ligands. Complexes with four and six ligands are by far the most common. The geometry of complexes with four ligands is either square planar or tetrahedral. Examples of square planar complexes are the platinum, Pt^{2+}, complexes used in cancer treatment. One of these is $Pt(NH_3)_2Cl_2$, dichlorodiammine platinum(II); the structures of its *cis*- and *trans*-isomers are shown in Figure 6.8. Only the *cis* isomer, called cisplatin, is effective in cancer chemotherapy.

(a) *cis* isomer (cisplatin) effective for cancer treatment

(b) *trans* isomer ineffective for cancer treatment

Figure 6.8.

(a) *cis*- and (b) *trans*- isomers of $Pt(NH_3)_2Cl_2$.

6.47 CHECK THIS

Names and structures of $Pt(NH_3)_2Cl_2$

Explain why the structures of the $Pt(NH_3)_2Cl_2$ isomers are named as they are.

Tetrahedral metal ion complexes are less common, but are found in several classes of protein molecules. For example, many proteins that bind to nucleic acids are complexed to zinc ions through two sulfur atoms and two nitrogen atoms (Lewis bases with pairs of nonbonding electrons) from their amino acid side chains, as shown in Figure 6.9(a). This complex, called a ***zinc finger***, holds the protein chain in a finger-like shape that can fit into the groove of a DNA helix, Figure 6.9(b).

(a) Complexation in the zinc finger

(b) Interaction of zinc finger and DNA

Figure 6.9.

Zinc finger complex in DNA binding proteins.
(a) Two sheet-like parts of the protein (the broad arrows) and a helical region (the blue cylinder) are held together by complexing with zinc(II). (b) The finger fits in the major groove of double helical DNA.

6.48 CONSIDER THIS

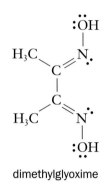

dimethylglyoxime

What is the structure of the nickel–dimethylglyoxime complex?

In Check This 6.11, you had enough data to determine the ratio of moles of $Ni^{2+}(aq)$ that react with dimethylglyoxime, $C_4H_8N_2O_2$, to give a red solid. The red solid is an electrically neutral complex, which is part of the reason it comes out of solution. The metal ion and all of the second row element atoms in the complex lie in a plane. In addition to bonding to the central metal ion, hydrogen bonding between chelate ligands is an important factor in stabilizing the complex. Construct a model of the red solid formed in the nickel(II)-dimethylglyoxime reaction.

Porphine–metal ion complexes in biological systems Zinc fingers are but one example of a vast number of metal ion complexes in biological systems. Two of the most familiar are **chlorophyll,** Figure 6.10(a), the complex that gives plants their green color, and **heme,** Figure 6.10(c) the complex that gives blood its red color. The complexing molecules in both cases are derived from the same basic **porphine** structure, Figure 6.10(b). The square array of nitrogen atoms held in place by the four joined rings is perfectly set up to complex a metal ion in its center, after two protons are lost to form the porphine dianion. The large numbers of π electrons in the double bonds of the planar porphine structure are in delocalized π orbitals that are spread over all the carbon and nitrogen atoms in the molecule. The stability of this ring system may be one of the reasons that derivatives of porphine are found as metal ion complexing ligands in many biological roles.

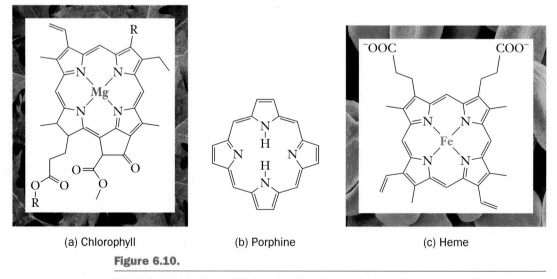

(a) Chlorophyll (b) Porphine (c) Heme

Figure 6.10.

Metal ion complexes with the porphine ring system in chlorophyll and heme.
There are no charged groups on chlorophyll. Differences in the R group at the top of the chlorophyll cause slightly different light absorptions. The R group at the bottom is a 20-carbon chain that makes chlorophyll insoluble in water but soluble in cell membranes, which also contain long hydrocarbon chains.

6.49 CHECK THIS

The porphine ring system in chlorophyll and heme

Trace the porphine ring system, Figure 6.10(b), in the structure of chlorophyll, Figure 6.10(a), and heme, Figure 6.10(c). How many π electrons are there in the porphine ring? How many π electrons are there in the porphine ring system in chlorophyll? In heme?

Chlorophyll is a Mg^{2+} *chlorin* complex and heme is a Fe^{2+} *porphyrin* complex. The ring structures are different for the two molecules, but you have found that the core of each is the porphine ring system. Chlorophyll molecules are stacked together in the membranes of chloroplasts, the organelles in green plants that absorb sunlight and use the captured energy to combine carbon dioxide and water into sugars. Heme is incorporated into a protein named globin to produce **hemoglobin,** the oxygen-carrying molecule in your blood. (See the Chapter 9 opening illustration and Figure 9.8.) Iron(II) is usually found octahedrally complexed with six ligands. In hemoglobin, one of the side groups of the protein complexes to the fifth position around the iron(II), which bonds the heme to the protein. The sixth position is where an oxygen molecule is bound when the hemoglobin is carrying oxygen from the lungs. In this oxygenated form, the complex is red. After the oxygen is left with a cell, the sixth position is occupied by a water molecule for the return to the lungs. In this deoxygenated form, the complex is bluish-purple.

Reflection and Projection

Metal ions in solution act as Lewis acids and can accept pairs of electrons donated by Lewis bases, ligands, to form metal ion complexes. Complex ion formation can compete with precipitation for metal ions and can dissolve precipitates or prevent their formation. Many ligands contain more than one functional group that act as Lewis bases; these chelating ligands form more than one bond to a metal ion to give chelate complexes. The great majority of metal ion complexes contain either four or six ligands. Four ligands form square planar or tetrahedral arrays about the central metal ion. Six ligands form an octahedral array.

All metal ions can form complexes with appropriate ligands. Complexes of transition metal ions are particularly interesting, because they are so highly colored; their variation in color with different ligands can be used to learn about the bonding in the complex and the energies of electrons in the ions. Metal ion complexes are widespread in biological systems where they perform many functions, including capture of the sun's energy for photosynthesis by plants and transport of oxygen in animals.

You're used to thinking about protons or hydronium ions as acids and it isn't too much of a leap to include metal ions as Lewis acids, electron pair acceptors. It's a somewhat larger jump to think about centers of partial positive charge in molecules as electron pair acceptors; that's the jump we'll make in the next section.

6.7. Lewis Acids and Bases: Electrophiles and Nucleophiles

 6.50 INVESTIGATE THIS

Does an acid react with an alcohol?

Do this as a class investigation and work in small groups to discuss and analyze the results. Mix about 2 g of solid salicylic acid, 5 mL of methanol, and 2 small drops of concentrated sulfuric acid in each of two 20 × 200-mm test tubes. Place one of the test tubes in a water bath at about 70 °C. After 15 minutes, remove the test tube from the water bath, allow it to cool to room temperature, and then add 2 mL of a saturated aqueous solution of sodium hydrogen carbonate to each test tube. Carefully smell the contents of both test tubes. Record your observations.

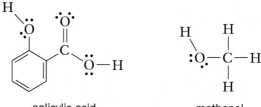

salicylic acid   methanol

6.51 CONSIDER THIS

What is the product of reaction of an acid with an alcohol?

In Investigate This 6.50, what evidence, if any, do you have that a reaction has occurred between salicylic acid and methanol? Can you identify the product of the reaction? What conditions are required for the reaction?

Nucleophiles and electrophiles Many of the reactions of the functional groups described in Chapter 5, Section 5.10, are Lewis acid–base reactions. Electronegative atoms in these groups always have nonbonding electrons and usually gather negative charge to themselves at the expense of less electronegative atoms. These centers of high electron density are Lewis bases. Some molecules have a significant positive center on carbon atoms that are bonded to oxygen or nitrogen and these act as Lewis acids. Another terminology for these reactions is in such common use that you will need to know it as well. The centers of negative charge density, Lewis bases, are called **nucleophiles** (*philos* = loving, hence "nucleus loving"). The positive centers, Lewis acids, are called **electrophiles** ("electron loving").

6.52 CHECK THIS

Centers of positive and negative charge in molecules

(a) Redraw the Lewis structures for salicylic acid and methanol, Investigate This 6.50, and label the centers of positive charge, δ+, and negative charge, δ−. Explain how you decided where to place your labels.

(b) Draw Lewis structures for each of these molecules and label the positive and negative charge centers. Explain your labeling.

$$CO_2,\ H_2O,\ (HO)_3PO,\ CH_3C(O)OH,\ HOCH_2CH_2OH,\ H_2N(CH_2)_6NH_2$$

Alcohol–carboxylic acid reaction To illustrate the nature of reactions between nucleophiles (Lewis bases) and electrophiles (Lewis acids), consider the reaction of a carboxylic acid, such as ethanoic acid, with an alcohol, such as ethanol, to form an ester and water:

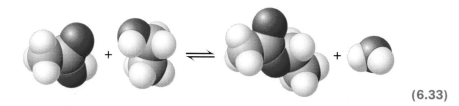

(6.33)

> Nucleophile–electrophile reactions are often slow, unless there is a catalyst available, such as H_3O^+ in Investigate This 6.50, to make the reaction more favorable. For the carboxylic acid–alcohol reaction, protonation of the
>
> acid, $H_3C-C-\overset{..}{\underset{..}{O}}H$,
>
> enhances the positive center on the carbon atom and facilitates the reaction. In order to focus on the basic nucleophile–electrophile reaction, we leave out these details in the main text.

The carbon that is attached to two oxygen atoms in the carboxylic acid is electrophilic (positive), while the oxygen in the alcohol is nucleophilic (negative). When the conditions are favorable, a nonbonding electron pair from the alcohol oxygen bonds to the carboxylic acid carbon. Following this initial Lewis acid–base reaction, the —OH group from the acid leaves (combined with a proton to form water). This second step in the overall reaction is the *reverse* of a Lewis acid–base reaction; the electrons in an electron pair bond become a nonbonding pair in the product, an ester, ethyl ethanoate. Chemists often use arrows to show where electrons are *imagined* to go in the course of such reactions (and we have used blue to show the atom centers and electrons that become part of the product water):

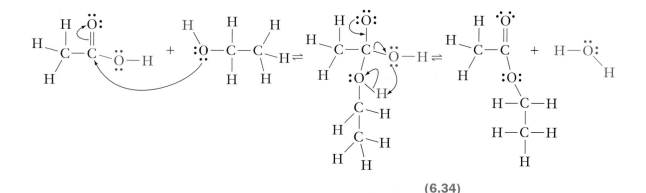

(6.34)

Our point is not to concentrate on the specifics of this particular reaction, but to recognize that electrophilic and nucleophilic centers in molecules are attracted to each other. If conditions are favorable, they can react to produce new products. As you continue through the text, you will encounter more Lewis acid–base reactions of mutually attracting nucleophilic and electrophilic centers coming together and reacting. In order to understand the interactions and the pathways of these reactions, you must recall the architecture of molecules introduced in Chapter 5.

6.53 CHECK THIS

More nucleophile-electrophile reactions

(a) In Investigate This 6.50, a carboxylic acid reacted with an alcohol. What functional group is formed as a product of this reaction? Write the Lewis structure for the reaction product, methyl salicylate. Write reaction expressions that parallel those of expression (6.34) that show how the reaction of salicylic acid (electrophile) and methanol (nucleophile) produces methyl salicylate.

(b) The 3rd and 4th reactions in Investigate This 6.2 produce carbonic acid, $(HO)_2CO(aq)$, which, as we have seen, reacts to form $CO_2(aq)$ and water:

$$(HO)_2CO(aq) \rightleftharpoons CO_2(aq) + H_2O(l) \tag{6.31}$$

CO_2 is only moderately soluble in water so it bubbles out of solution and accounts for the $CO_2(g)$ product shown in reaction expressions (6.2) and (6.3). Reaction (6.31) can be viewed as a nucleophilic–electrophilic reaction going in reverse. The *formation* of carbonic acid is the reaction of a nucleophile, water, with an electrophile, carbon dioxide. Write the first step in the carbonic acid formation reaction [analogous to the first step in reaction (6.34)] and show the rearrangement of electrons and protons that produces carbonic acid.

Condensation polymers The ester formation reaction (6.34) is also called a **condensation reaction** because the two reactants condense, come together into a single structure, in the first step of the reaction. In the overall reaction, a molecule of water is formed for every ester molecule formed. Traditionally, the overall reaction is still called a condensation because the focus is on the carbon-containing molecules. These reactions are particularly important when the reactant molecules are **polyfunctional**, containing more than one functional group that can take part in condensation reactions. An example is the reaction between terephthalic acid (1,4-benzenedicarboxylic acid) and ethylene glycol (1,2-ethanediol, a common automobile antifreeze):

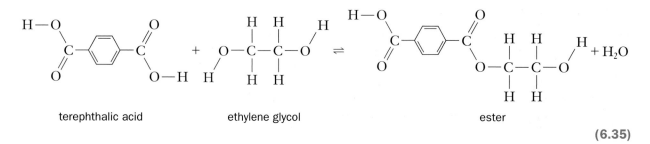

terephthalic acid   ethylene glycol   ester

(6.35)

6.54 CHECK THIS

A condensation reaction

Write the condensation step, the first step in overall reaction (6.35). Show the positive and negative (electrophilic and nucleophilic) centers that attract one another to cause the reaction.

As far as it goes, reaction (6.35) is the same as reaction (6.34) with a different acid and alcohol. But reaction (6.35) isn't as far as these reactants can go. The ester product of the reaction still has an acid and an alcohol functional group that can react, respectively, with another molecule of ethylene glycol and terephthalic acid to form a polymer molecule. **Polymers** (from Greek *polys* = many + *meros* = part) are molecules that contain many repeating units of the same kind. The individual molecules that make up the repeating units are called **monomers** (*mono* = one, single). The reaction of terephthalic acid and ethylene glycol (the monomers) is a **condensation polymerization** that produces long chains of esters, **polyesters** (*poly*meric *ester*):

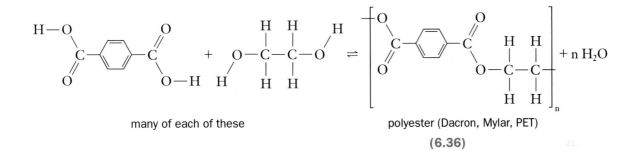

many of each of these   polyester (Dacron, Mylar, PET)

(6.36)

The structure in brackets in reaction expression (6.36) is the **polymer repeat unit** for this **condensation polymer.** The repeat unit tells you what reactant or reactants condense to make the polymer and shows what the linkage is between the reactants.

> Polyester drawn into fibers is named Dacron. Formed into thin sheets, it is Mylar (the familiar shiny material in party balloons). Molded into containers such as soft drink bottles, it is PET (polyethylene terephthalate).

6.55 CHECK THIS

Polyamide condensation polymers

(a) 🖥 Condensation polymerization is not limited to polyester formation between acid and alcohol groups. The condensation reaction of an amine with a carboxylic acid is analogous to the reaction we have shown for alcohols with carboxylic acids. Use the charge density models in the *Web Companion*, Chapter 6, Section 6.7.1, to explain which molecule is the electrophile and which the nucleophile in the condensation reaction.

continued

(b) The amine-carboxylic acid condensation product is an amide (Chapter 5, Table 5.4) and polymers with amide bonds are **polyamides.** Familiar synthetic polyamides are nylons, such as

$$\left[\begin{matrix}\text{O} \\ \| \\ \text{C} \end{matrix} - \text{CH}_2\text{CH}_2\text{CH}_2\text{CH}_2\text{C} \begin{matrix}\text{O} \\ \| \\ \end{matrix} - \begin{matrix}\text{H} \\ | \\ \text{N} \end{matrix}\text{CH}_2\text{CH}_2\text{CH}_2\text{CH}_2\text{CH}_2\text{CH}_2\begin{matrix}\text{H} \\ | \\ \text{N} \end{matrix}\right]_n$$

nylon-66

What are the reactants that condense to make this polymer? Write a reaction, analogous to reaction (6.36), for the formation of this polymer from these reactants. Explain clearly how this structure is related to the product of the amide condensation reaction shown in the *Web Companion*. Why is this polymer called nylon-66?

(c) Is(Are) the reaction(s) represented in Figure 1.30 (Chapter 1, Section 1.9), related to the condensation reactions discussed in this section? Explain why or why not.

Biological condensation polymers Proteins, starch, cellulose, and nucleic acids are all condensation polymers. Many of the linkages that build biomolecules from smaller precursors, for example, fats from glycerol and long-chain carboxylic acids, are also made by condensation reactions like those shown above. Examples of products made from natural polymers are shown in the chapter opening illustration. Wool and silk are proteins (polyamides); cotton, paper, and cardboard are cellulose, a condensation polymer of glucose illustrated in Figure 6.11.

Figure 6.11.

Structure of cellulose, a polymer of glucose.

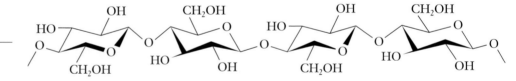

Biological condensation reactions usually occur between reactants that are *derived* from acids, alcohols, and amines. However, you can think about the reactions as though they take place between the parent acid, alcohol, or amine. In these reactions, water is formed, in addition to the condensed product. We have shown all these nucleophile–electrophile condensation reactions as reversible. Indeed, all of the condensation products, including biological molecules, can be **hydrolyzed,** or broken down by water (*hydro* = water + *lyein* = loosen, break), under appropriate conditions to reform the reactant molecules.

6.56 CHECK THIS

Water as a nucleophile

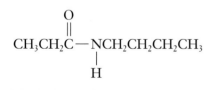

Hydrolysis is a nucleophile–electrophile reaction with water acting as the nucleophile. Write out the steps for hydrolysis of this amide. *Hint:* Think about the reverse of reaction (6.34).

Condensation reactions can also take place with non-carbon-containing Lewis bases (nucleophiles) and acids (electrophiles). An example is reaction of a substituted deoxyribose (a polyalcohol) and phosphoric acid to form a **phosphate ester:**

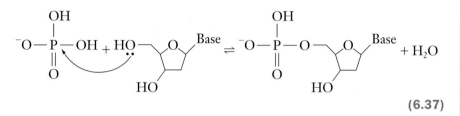

(6.37)

The Base shown on the deoxyribose represents one of the four nitrogen bases in DNA. The bond between the Base and the deoxyribose is also formed by a condensation reaction.

The electronegative oxygen atoms in phosphate draw electron density away from the central phosphorus atom and make it an electrophilic center, a Lewis acid. The nucleophilic electron pair on the alcohol is attracted to this Lewis acid, as shown by the curved arrow. Nucleic acids formed in replication (or transcription) of DNA are condensation polymers of phosphate esters:

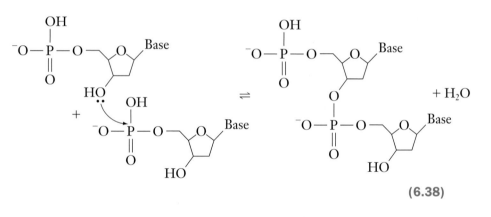

(6.38)

6.57 CHECK THIS

Polymeric structure of DNA

Compare the product of reaction (6.38) with the structure of DNA shown in Figure 2.27 (Chapter 2, Section 2.13). What is the polymer repeat unit for one strand of DNA? Why are the links between these repeat units often called phosphodiester bonds? Explain your answers.

6.8. Formal Charge

Rules for formal charge Formal charge is a useful concept to help understand the rearrangements that often occur following the condensation step in a nucleophile–electrophile reaction. **Formal charge** is the *charge an atom in a molecule or ion would have if all bonding electrons were shared equally between the bonded atoms.* Formal charges are calculated values, not experimental quantities that can be measured. The procedure for finding the formal charge on an atom in a molecule or ion is given on the next page.

- Draw the Lewis structure for the molecule.
- Assign one electron from each electron pair bond to each of the bonded atoms.
- Assign all nonbonding electrons on an atom to that atom.
- Count all the valence electrons assigned to an atom.
- Subtract this number of electrons from the positive charge on the atomic core.
- The difference, including its sign, is the formal charge on the atom.

The sum of all the formal charges in a molecule or ion has to be the overall charge on the molecule or ion. Use this sum to check your procedure.

6.58 WORKED EXAMPLE

Assigning formal charge

What are the formal charges on the atoms in the hydronium ion, H_3O^+, and in methanal, H_2CO?

Necessary information: We need to know the number of valence electrons and atomic core charges, 1, 4, and 6 for H, C, and O atoms, respectively.

Strategy: Apply the preceding rules to the two species.

Implementation: Lewis structures:

Assign the bonding electrons evenly and the nonbonding electrons to their atom:

$$3 \ H\cdot \quad \cdot\ddot{\overset{\cdot\cdot}{O}}\cdot \qquad\qquad 2 \ H\cdot \quad \cdot\ddot{\overset{\cdot\cdot}{C}}\cdot \quad \cdot\ddot{\overset{\cdot\cdot}{O}}:$$

Calculate the formal charges:

formal charge on H = 1 − 1 = 0	formal charge on H = 1 − 1 = 0
formal charge on O = 6 − 5 = +1	formal charge on C = 4 − 4 = 0
	formal charge on O = 6 − 6 = 0

To remind ourselves of the results, we can show the formal charges in the Lewis structures (omitting any zero formal charges):

Does the answer make sense? We followed the rules to get the formal charges. The sum of all the formal charges is

$$\text{for } H_3O^+: 0 + 0 + 0 + (+1) = +1 \qquad \text{for } H_2CO: 0 + 0 + 0 + 0 = 0$$

The sums give the correct overall charge on the ion and the electrically neutral molecule, so we can have some confidence that the formal charges are correct.

6.59 CONSIDER THIS

How do you assign formal charge?

(a) The isomeric ions, cyanate, NCO^-, and isocyanate, CNO^-, both exist. Write a Lewis structure for each ion. What are the formal charges on the atoms in each ion? Show that the calculated formal charges sum to the correct overall charge on the ions.

(b) The Lewis structure for the intermediate structure shown in reaction (6.34) is redrawn here. Calculate the formal charges on all three oxygen atoms and the carbon atom to which they are all bonded. All the rest of the formal charges in the structure are zero. Show that your calculated formal charges sum to the correct overall charge.

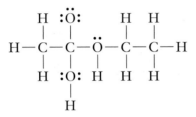

Interpretation of formal charges The +1 formal charge on oxygen in the hydronium ion, Worked Example 6.58, does not mean that the oxygen atom has a positive charge in the ion. It means that the partial charge on the oxygen atom is less negative—more positive—than it is in the neutral molecule, where the formal charge on all atoms is zero. The electronic structures we showed for molecules in Chapter 5 (Section 5.4, Figure 5.10), show identical bonding for all three hydrogen atoms in H_3O^+, with each proton in a σ bonding orbital. Only one nonbonding electron pair is left wholly for the oxygen atom, so it makes sense to consider the oxygen atom as being less negatively polarized than when it has two nonbonding pairs of electrons all to itself. The formal charge can be interpreted as an indicator of the *change* in polarization of an atom compared to its polarization in a different bonding situation.

Reactions reduce formal charges Although there is no overall charge on the intermediate structure in reaction (6.34), you found in Consider This 6.59(b) that there are nonzero formal charges on some of the oxygen atoms in the structure. But stable molecular structures usually have as many atoms as possible with zero formal charge. In ions, the charge obviously has to be somewhere; all atoms can't have zero formal charge. Electrically neutral molecules, however, rarely have formal charges. Reaction intermediates, which are postulated but usually not experimentally observable, often have formal charges on several atoms. The intermediates undergo electron and atomic core redistributions to reduce the number of formal charges; these redistributions lead to the observed products.

6.60 CHECK THIS

Using formal charge

(a) Use your answers for Consider This 6.59(a) to predict which isomeric ion, cyanate or isocyanate, is the more stable. Give the reasoning for your answer.

(b) In Consider This 6.59(b), are there any correlations you might make between formal charge and number of bonds to the oxygen atoms? Explain how you arrive at your response.

(c) In Section 6.5, we wrote oxyacid and oxyanion Lewis structures with as many double bonds to oxygen as possible [see Figure 6.4(b) and Consider This 6.34(a)]. There are alternative ways to write Lewis structures for species such as the chlorite ion:

$$\left[:\ddot{O} - \ddot{C}l - \ddot{O}: \right]^{-}$$

What are the formal charges on each atom in this structure? What are the formal charges on the atoms in the chlorite ion structure(s) you wrote for Consider This 6.34(a)? Is there a reason to favor one Lewis structure over the other? Explain why or why not.

(d) Using the model for chlorite ion in part (c), write Lewis structures for the chlorate, ClO_3^-, and perchlorate, ClO_4^-, ions. What are the formal charges on the atoms in these structures? Explain why we chose to correlate base strength with the structures you wrote for these ions in Consider This 6.34(a) rather than those you wrote here.

In reaction intermediates, oxygen is one of most common atoms with a formal charge. The three possible formal charges on an oxygen atom, -1, 0, and $+1$, are all present in the intermediate structure you analyzed in Consider This 6.59(b). You can quickly recognize these three cases by just looking at the number of bonds to the oxygen atom. Table 6.3 shows the three possibilities and, to help you remember them, their comparison to the ions water forms. Typical kinds of electron and atomic core rearrangements that change the formal charge on the oxygen atom to zero are also shown.

6.61 CONSIDER THIS

How are the correlations in Table 6.3 applied?

(a) Which of the rearrangements in Table 6.3 occur when the intermediate in reaction (6.34) goes to products? Show and explain each one clearly.

(b) Which of the rearrangements in Table 6.3 occur when water reacts with carbon dioxide to produce carbonic acid? You wrote this reaction as a nucleophile–electrophile reaction in Check This 6.53(b). Show and explain each one clearly.

(c) Write reaction (6.8) as a nucleophile–electrophile reaction and show which rearrangements in Table 6.3 occur to produce the product. Show and explain each one clearly.

Table 6.3 *Correlation between formal charge and reactions at oxygen atom centers.*

The relationship between the number of bonds to oxygen and its formal charge in molecules and reaction intermediates is also demonstrated. As a comparison, water and its ions are shown.

Formal charge on oxygen	Bonds to oxygen	Nonbonding pairs	Water species	Electron and atomic core rearrangements that bring the oxygen formal charge to zero
−1	1	3	HO^-	(a) change a nonbonding pair to bonding pair (makes a double bond) (b) gain a proton from an acid in solution to form an —OH group
0	2	2	H_2O	
+1	3	1	H_3O^+	change a bonding pair to nonbonding pair (most often with loss of H_2O or H^+ to a base[a])

[a]The base to which the proton is transferred is often not shown because it might be any of several Lewis bases, including water, in the solution mixture.

Reflection and Projection

Lewis acid–base reactions of electrophilic and nucleophilic centers in molecules are the same as other Lewis acid–base reactions: A pair of nonbonding electrons becomes an electron pair bond in the initial product. Complications arise because the initial products of these reactions undergo rearrangements that form more stable products. The rearrangements can often be understood in terms of electron and atomic center movements that minimize formal charges in the final products. Because we are usually interested in the overall reaction, the simple, underlying Lewis acid–base reaction is sometimes obscured.

For this introduction, most of our reactions were nucleophiles (alcohols and amines) reacting with the electrophilic carbon of the carboxyl group to produce condensation products (esters and amides). In these cases, the final products are a carbon-containing molecule that combines (condenses) all the carbons in both reactants into a single molecule and splits out a water molecule as a second product. When the reactants are polyfunctional, the condensations can continue adding to the ends of the condensation products to form condensation polymers. Polyesters and polyamides are used to make a vast number of products you use every day. Biological phosphate polyesters and polyamides make up your ribonucleic acids, DNA and RNA, and proteins. All these condensation polymers can be hydrolyzed—broken down by reaction with water—under appropriate conditions.

This completes our introduction to Lewis acid–base reactions that transform a nonbonding electron pair to a bonding pair. Now we will turn our attention to the third large class of reactions, those that usually involve complete transfer of one or more electrons from one reactant to another.

6.9. Reduction–Oxidation Reactions: Electron Transfer

6.62 INVESTIGATE THIS

Do ion–metal reactions between copper and silver occur?

Do this as a class investigation and work in small groups to discuss and analyze the results. Use an overhead projector to view what happens when copper wire is immersed in an aqueous silver nitrate, $AgNO_3$, solution and silver wire is immersed in an aqueous copper sulfate, $CuSO_4$, solution in adjacent wells of a 6-well microtiter plate. Watch for several minutes and record your observations.

Close-up reaction view

6.63 CONSIDER THIS

What copper and silver ion–metal reactions occur?

(a) In Investigate This 6.62, do you see any evidence for reaction in either or both of the wells? What is the evidence?

(b) Is the movie in the *Web Companion*, Chapter 6, Section 6.9.1, related to what you observe in either of the wells in Investigate This 6.62? If so, explain how they are related.

(c) Try to write a reaction expression to describe what is happening in each well.

The reactant that loses electrons is often called a **reducing agent,** since it provides electrons to the reactant that is reduced. The reactant that gains electrons is called the **oxidizing agent,** since it accepts electrons from the reactant that is oxidized. Thus, the reducing agent is oxidized and the oxidizing agent is reduced during the reaction. This can be a confusing nomenclature. We will not use it in this introduction but will return to it in Chapter 10 when reduction–oxidation reactions are examined in more detail.

In precipitation reactions, the ions retain their electrons and their identity in the solid product. In Lewis acid–base reactions, a nonbonding electron pair in the Lewis base becomes a bonding pair shared by two atoms in the product. In **reduction–oxidation reactions** (also **redox reactions,** pronounced "ree-dox"), one or more electrons is completely transferred from one reactant to the other reactant. The reactant that gains electrons is **reduced.** The reactant that loses electrons is **oxidized.** The easiest reduction–oxidation reactions to recognize are probably those in which a metal is one of the reactants, as in Investigate This 6.62. The disappearance of the metal (or the formation of another) is a sure sign that electrons have been transferred in the reaction. Metals have to lose electrons to become cations and "disappear" into solution. Metal cations have to gain electrons to get to their elemental metal form.

Direction of a reduction–oxidation reaction In Investigate This 6.62, you observed needles of silvery metal growing on the copper wire when it was placed in a solution containing silver cation, $Ag^+(aq)$. You also observed the initially clear and colorless solution become a clear pale blue color. In the other sample, there was no evidence of formation of any metal or other solid and the initially clear, blue solution containing $Cu^{2+}(aq)$ did not change. It is difficult to tell whether any of the copper metal disappears in the first sample, but the change of the solution color shows that there is some new component in the solution. The second sample shows that $Cu^{2+}(aq)$ in aqueous solution is blue.

Taken together, all this evidence suggests that aqueous silver ion and copper metal react to produce silver metal and aqueous copper ion:

$$Ag^+(aq) + Cu(s) \rightarrow Ag(s) + Cu^{2+}(aq) \qquad \textbf{(6.39)}$$

We have written reaction (6.39) as though it proceeds only in the forward direction. This is because, in Investigate This 6.62, there was no evidence of any reaction between $Ag(s)$ and $Cu^{2+}(aq)$, the reverse of reaction (6.39). As we have said previously, most reactions are to some extent reversible, but often greatly favor one direction over the other, and we express such directionality with a single arrow.

There is no simple way to predict the direction of a reduction–oxidation reaction. The factors that affect the direction of the reaction (or if a redox reaction occurs at all) include pH, the presence of complexing ligands, and the exact nature of the solvent in which the reaction occurs. We will examine these more quantitative aspects of reduction–oxidation reactions in Chapter 10. In this chapter, we will focus on recognizing reduction–oxidation reactions and writing balanced equations to describe the reactions.

Charge balance   In the reaction between $Cu(s)$ and $Ag^+(aq)$, an electron has to be gained by $Ag^+(aq)$, in order to become $Ag(s)$. The $Ag^+(aq)$ ions are reduced: *Electrons are gained.* Two electrons have to be lost by $Cu(s)$ to become $Cu^{2+}(aq)$. The $Cu(s)$ atoms are oxidized: *Electrons are lost.* Reaction expression (6.39) is not balanced in terms of the electron transfer that must occur from copper to silver ion. The imbalance in electron transfer results in an imbalance in charge on each side of the expression. As written, there is one positive charge on the left and two on the right. Charges must be balanced.

Web Companion

Chapter 6, Section 6.9.2–3 — ①
②
Study animations of these
processes that occur in the
③
copper–silver ion reaction.
④

6.64 WORKED EXAMPLE

A balanced chemical equation for the Cu/Ag⁺ reaction

Write a balanced chemical equation for the reaction of copper metal, $Cu(s)$, with silver ion in solution, $Ag^+(aq)$.

Necessary information: Reaction expression (6.39) is unbalanced but tells us the reactants and products of the reaction.

Strategy: Balanced reaction expressions have to conserve both atoms and charge. In simple reduction–oxidation reactions like this one, first balance the charge and then the atoms by inspection.

Implementation: To balance charge, we add a $Ag^+(aq)$ to the reactants:

$$2Ag^+(aq) + Cu(s) \rightarrow Ag(s) + Cu^{2+}(aq)$$

In this intermediate expression, the charges and copper atoms and ions balance, but silver atoms and ions do not. To balance the silvers, we add an atom of silver metal to the product side:

$$2Ag^+(aq) + Cu(s) \rightarrow 2Ag(s) + Cu^{2+}(aq) \qquad \textbf{(6.40)}$$

Does the answer make sense?   Equation (6.40) still describes the reaction we observed and it is balanced in atoms and charge. The two electrons lost by each $Cu(s)$ atom are transferred to two $Ag^+(aq)$ ions. Electron transfer is also balanced.

6.65 CHECK THIS

Balancing simple reduction–oxidation reaction expressions

Balance these reduction–oxidation reactions. They are shown as reversible, since we have no evidence for the favored direction.

(a) $Mg^{2+}(aq) + Cu(s) \rightleftharpoons Mg(s) + Cu^{2+}(aq)$ (6.41)

(b) $Fe^{3+}(aq) + Sn^{2+}(aq) \rightleftharpoons Fe^{2+}(aq) + Sn^{4+}(aq)$ (6.42)

(c) $Zn(s) + H^+(aq) \rightleftharpoons Zn^{2+}(aq) + H_2(g)$ (6.43)

$[H^+(aq)$ is a way to represent $H_3O^+(aq).]$

(d) $Cu^{2+}(aq) + I^-(aq) \rightleftharpoons CuI(s) + I_2(s)$ (6.44)

$[CuI$ contains the copper(I) ion.$]$

6.66 INVESTIGATE THIS

Does Cu(s) react with nitric and/or hydrochloric acid?

Do this as a class investigation and work in small groups to discuss and analyze the results. *CAUTION:* Concentrated nitric and hydrochloric acids are powerfully corrosive; handle them with respect and with rubber or plastic gloves. Some products of these reactions are toxic and must be kept in the plastic bag and disposed of properly in a fume hood.

Cu + HNO₃

(a) Place a 400-mL beaker of water, a 100-mL beaker containing 5 mL of concentrated nitric acid, HNO_3, and 0.25 g of copper wire into a 1-gal zip-seal plastic bag. Seal the bag and then drop the copper wire into the beaker of nitric acid. Observe and record any changes that occur. When it appears that no further change is occurring, pour the entire contents of the 100-mL beaker into the 400-mL beaker of water. Observe and record any changes that occur.

(b) Repeat this procedure with concentrated hydrochloric acid, HCl, instead of nitric acid.

6.67 CONSIDER THIS

What are the products of Cu(s) reactions with nitric and hydrochloric acid?

(a) In Investigate This 6.66, do you see any evidence for reaction of copper with either nitric or hydrochloric acid? What is the evidence? If there is a reaction, what reaction products can you identify? Explain how you make the identification(s).

(b) What conclusions can you draw about the reaction between copper and nitric acid? Between copper and hydrochloric acid?

The predominant ions in nitric acid are hydronium, $H_3O^+(aq)$, and nitrate, $NO_3^-(aq)$, formed by the acid–base reaction of HNO_3 with water. Nitrogen and oxygen form several gaseous compounds, one of which is reddish-brown nitrogen dioxide, $NO_2(g)$. In Investigate This 6.62, you observed that aqueous solutions containing $Cu^{2+}(aq)$ cation are blue. Your observations in Investigate This 6.66, together with the information in this paragraph suggest that we can write the unbalanced reaction between copper and nitric acid as

$$Cu(s) + H^+(aq) + NO_3^-(aq) \rightarrow Cu^{2+}(aq) + NO_2(g) \qquad \textbf{(6.45)}$$

In reaction (6.45) [and reaction (6.43) in Check This 6.65], we have written the hydronium ion as $H^+(aq)$. You will usually find it easier to balance reduction–oxidation reactions if you have fewer atoms to worry about. If you wish, you can add the "missing" H_2O [from $H_3O^+(aq)$] to the reactions after they are balanced.

6.68 CONSIDER THIS

What is the reaction of Cu(s) with $H_3O^+(aq)$?

Reaction (6.43) in Check This 6.65(c) proceeds as written: Zinc metal reacts with hydronium ion to give hydrogen gas and zinc cation, $Zn^{2+}(aq)$, in solution. We can write an analogous (unbalanced) reaction for copper:

$$Cu(s) + H^+(aq) \rightarrow Cu^{2+}(aq) + H_2(g) \qquad \textbf{(6.46)}$$

Should we include this possibility as part of reaction expression (6.45)? What evidence do you have that copper does or does not react with hydronium ion? Explain your reasoning.

Reaction expression (6.45) is not balanced in terms of either atoms or charge. The only change in charge that is evident from the expression is that copper metal has been oxidized (lost two electrons) to become $Cu^{2+}(aq)$ cation. But there is no obvious atom that has been reduced (gained electrons) in the reaction. Many reduction–oxidation reactions involve nonmetals and look something like reaction (6.45). We need a way to keep track of electrons, so we can balance electron transfers in these reactions. And we need a way to tell whether a reaction *is* a reduction–oxidation. Defining and using oxidation numbers can solve both problems.

Oxidation numbers The **oxidation number** for an atom is *defined* as the difference between the number of electrons in the neutral atom and the number of electrons in the atom as it exists in a compound or ion. If the atom has lost electrons, its oxidation number is positive. If the atom has gained electrons, its oxidation number is negative. The definition of oxidation number is easy to apply to elemental atoms and monatomic ions: The oxidation number is the charge on the atom or ion. In reaction (6.45), for example, copper metal, Cu, loses two electrons (is oxidized) in becoming Cu^{2+}. The number of electrons in Cu is two more than in Cu^{2+}, so Cu^{2+} is assigned an oxidation number of $+2$, which is equal to the charge on the ion. Sometimes, elemental atoms are written with a superscript zero, Cu^0, for example, to denote that their oxidation number is, by definition, zero.

6.69 CHECK THIS

Oxidation numbers for elements and monatomic ions

(a) The oxidation numbers for $Fe(s)$, $Fe^{3+}(aq)$, $Cl_2(g)$, and $Cl^-(aq)$ are 0, +3, 0, and −1, respectively. Show how the definition of oxidation number gives the preceding values.

(b) What are the oxidation numbers for $V^{4+}(aq)$, $H^-(g)$, $P_4(s)$, $N_2(g)$, and $Na^+(aq)$? Explain.

Assigning oxidation numbers to atoms in molecules and polyatomic ions requires a procedure for dividing the valence electrons among the atoms so we can apply the oxidation number definition. Here is the procedure we will use:

> In Chapters 1 and 4, we pointed out the similarity of hydrogen and carbon properties and chemistry, which we emphasize by placing hydrogen in the carbon family of the periodic table. We extend the similarity here by treating the atoms as equivalent when assigning oxidation numbers.

- Write a Lewis structure for the molecule or ion.
- For each electron pair bond between identical atoms, assign one electron to each atom. For the purposes of these assignments, carbon and hydrogen are assumed to be identical.
- For each electron pair bond between unlike atoms, assign all the electrons to the more electronegative atom.
- Assign all nonbonding valence electrons on an atom to that atom.
- Count all the valence electrons assigned to an atom, n_e.
- Subtract the number of valence electrons, n_e, from the positive charge on the atomic core, Z_{core}. The difference, including its sign, is the oxidation number, *ON*, for the atom.

$$ON = Z_{core} - n_e \qquad (6.47)$$

As a check on the application of this procedure, the sum of the oxidation numbers on all the atoms has to be the overall charge on the molecule or ion. Note that, like formal charge, these oxidation numbers are calculated values, not experimentally measurable quantities. There is no universally accepted procedure for assigning oxidation numbers in molecules and polyatomic ions. As long as you use one procedure consistently, your results will be internally consistent and you will be able to keep track of electron transfers.

6.70 CHECK THIS

The procedure and the definition of oxidation number

Note that our procedure for assigning oxidation numbers in multiatomic species ignores inner shell electrons, which never take part in electron transfer reactions and always are associated with their atomic core. Explain how our procedure, which accounts only for valence electrons and the atomic core charges, is consistent with the definition of oxidation number.

6.71 WORKED EXAMPLE

Assigning oxidation numbers

What are the oxidation numbers for the atoms in NO_3^- and NO_2?

Necessary information: We need to know that the core charges (= number of valence electrons) for N and O atoms are 5 and 6, respectively. It is useful to know that two molecules of NO_2 can react to form O_2NNO_2, which is bonded through the nitrogen atoms. For molecules with more than one equivalent Lewis structure, any one of them can be used to assign oxidation numbers.

Strategy: Apply the procedure for assigning oxidation numbers.

Implementation: Lewis structures for NO_3^- and NO_2:

NO_2 has an odd number of valence electrons, 17, so there is no way to write a structure that satisfies the octet rule on all the atoms. We have shown the odd electron on the N, because this explains how two NO_2 molecules can combine to form O_2NNO_2 using the odd electron from each for a bond between the N atoms.

Assign the bonding electrons to the more electronegative atom and nonbonding electrons to their atom:

Calculate the oxidation numbers:

ON for N = 5 − 0 = +5 ON for N = 5 − 1 = +4

ON for O = 6 − 8 = −2 ON for O = 6 − 8 = −2

The structures with oxidation numbers shown on each atom are

Does the answer make sense? Check the procedure by summing oxidation numbers:

for NO_3^-: (−2) + (−2) + (−2) + 5 = −1

for NO_2: (−2) + (−2) + 4 = 0

The sums give the correct overall charges so we have some assurance that we have assigned the oxidation numbers correctly.

6.72 CHECK THIS

Assigning oxidation numbers

Assign oxidation numbers to each of the atoms in these molecules and ions:

$$H_2O, \ NH_3, \ (HO)_3PO, \ PO_4^{3-}, \ HOCl, \ Cl_2, \ C_2H_6, \ CH_3OH, \ HC(O)OH$$

Interpreting oxidation numbers The oxidation number for an atom in a molecule or ion is a measure of its share of the valence electrons compared to what it has as a neutral atom. If the oxidation number of an atom increases, it has been **oxidized** (has lost electrons). If the oxidation number of an atom decreases, it has been **reduced** (has gained electrons). Changes in oxidation numbers signal electron transfers in a reaction. *In all reduction–oxidation reactions, one atom is oxidized (oxidation number increases going from reactants to products) and another atom is reduced (oxidation number decreases).* Table 6.4 compares the descriptions of oxidation and reduction in terms of electron loss or gain and oxidation number increase or decrease.

Table 6.4 *Comparison of electron and oxidation number change in oxidation and reduction.*

Process	Electrons	Oxidation number
Oxidation of an atom	*loss*	*increase*
Reduction of an atom	*gain*	*decrease*

6.73 CHECK THIS

Reduction–oxidation reactions

(a) Show that reaction (6.45) is a reduction–oxidation reaction by showing that one of the atoms in the reactants is oxidized and another is reduced. *Hint:* Use the results from Worked Example 6.71.

(b) Assign oxidation numbers to all the atoms in this reaction expression.

$$HCl(aq) + H_2O(l) \rightleftharpoons H_3O^+(aq) + Cl^-(aq) \qquad (6.48)$$

Is reaction (6.48) a reduction–oxidation reaction? Which atom(ion) is reduced? Which is oxidized? Explain the reasoning for your answer.

(c) Assign oxidation numbers to the iodine atoms(ions) in this reaction expression:

$$H^+(aq) + H_2O_2(aq) + I^-(aq) \rightleftharpoons H_2O(aq) + I_2(aq) \qquad (6.49)$$

Is reaction (6.49) a reduction–oxidation reaction? Which atom(ion) is reduced? Which is oxidized? Explain the reasoning for your answer.

===== **6.74** CONSIDER THIS =====

Are there patterns in oxidation numbers?

(a) List the oxidation number for the oxygen atoms in Worked Example 6.71 and Check This (6.72) and (6.73). What conclusion can you draw about the oxidation number of oxygen atoms in compounds? Explain how you reach your conclusion. Can you think of a reason why this pattern is observed?

(b) List the oxidation number for the hydrogen atoms in Check This (6.72) and (6.73). What conclusion can you draw about the oxidation number of hydrogen atoms in compounds? Explain how you reach your conclusion. Can you think of a reason why this pattern is observed?

An alternative way to assign oxidation numbers The procedure we have been using to assign oxidation numbers is based on electronegativity differences (except for hydrogen bonded to carbon) and always works to give consistent assignments. Your analyses in Consider This 6.74 show that oxygen in its compounds has an oxidation number of -2, except in peroxides, when its oxidation number is -1. Hydrogen, in the compounds we have considered so far, always has an oxidation number of $+1$, except when it is bonded to carbon, when its oxidation number is zero. These patterns lead to the set of rules, shown in Table 6.5, for assigning oxidation numbers without writing a Lewis structure and dividing valence electrons between the atoms. The rules do not work in every case you meet, but they are often useful for a quick appraisal of whether a reaction is a reduction–oxidation.

Table 6.5 *An alternative set of rules for assigning oxidation numbers, ON.*

Rule
1. The *ON* for atoms in elements is 0.
2. The *ON* for monatomic ions is the ionic charge.
3. The *ON* for oxygen atoms in most compounds is -2.   The *ON* for oxygen atoms in peroxides is -1.
4. The *ON* for hydrogen atoms in many compounds is $+1$.   The *ON* for a hydrogen atom bound to carbon is zero.
5. For neutral compounds, *ON*s must sum to zero.
6. For polyatomic ions, *ON*s must sum to charge.

===== **6.75** CHECK THIS =====

Oxidation number for oxygen in peroxides

Use electronegativies and the definition of oxidation number to explain how the bonding in peroxides, such as hydrogen peroxide, H_2O_2, gives these oxygen atoms a -1 oxidation number.

6.76 WORKED EXAMPLE

Assigning oxidation numbers

Use the rules in Table 6.5 to assign oxidation numbers for the atoms in NO_3^- and NO_2.

Necessary information: All the information we need is in Table 6.5.

Strategy: Apply the applicable rules in Table 6.5. For these cases, this means Rule 3 and either Rule 5 or Rule 6.

Implementation: The sum of the oxidation numbers in NO_3^- has to be -1:

$$(ON \text{ for } N) + 3(ON \text{ for } O) = -1$$

Substitute -2 for $(ON$ for O$)$ and solve for $(ON$ for N$)$:

$$(ON \text{ for } N) = -1 - 3(-2) = +5$$

The sum of the oxidation numbers in NO_2 has to be 0:

$$(ON \text{ for } N) + 2(ON \text{ for } O) = 0$$

$$(ON \text{ for } N) = 0 - 2(-2) = +4$$

Does the answer make sense? The oxidation numbers based on the rules are the same as those assigned from the definition in Worked Example 6.71, so they are consistent.

6.77 CHECK THIS

Assigning oxidation numbers

Use the rules in Table 6.5 to assign oxidation numbers for each of the atoms in these molecules and ions: H_2O, NH_3, $(HO)_3PO$, PO_4^{3-}, $HOCl$, Cl_2, C_2H_6, CH_3OH, $HC(O)OH$. How do your answers compare to what you got in Check This 6.72? Can you explain any differences?

Reflection and Projection

Our third large class of reactions, reduction–oxidation reactions, involves complete transfer of one or more electrons from one reactant to another. The reactant that loses electrons is oxidized and the reactant that gains electrons is reduced. In simple cases, usually with monatomic ions and elements, you can easily tell that reactants have gained and lost electrons and thus identify the reaction as reduction–oxidation. In more complex cases, usually those with molecules and/or polyatomic ions, you usually will need to assign oxidation numbers to all the atoms in the reactants and products, in order to identify a reaction as a reduction–oxidation. The atom that decreases in oxidation number in the reaction is reduced and the atom that increases in oxidation number is oxidized. You can often use the rules in Table 6.5 to make the decision whether a reaction is reduction–oxidation. If the rules lead to problems or ambiguities, go back to the definition of oxidation numbers to assign them appropriately.

You can usually balance simple reduction–oxidation reaction equations by inspection, as we demonstrated early in this section. You have probably noticed, however, that we have not yet balanced the more complex reaction of copper with nitric acid. Balancing reduction–oxidation reaction equations is the topic of the next section and we will finish the chapter with a brief look at reduction–oxidation reactions of carbon-containing molecules, including sugar fermentation.

6.10. Balancing Reduction–Oxidation Reaction Equations

In a balanced reduction–oxidation reaction equation, the total number of electrons lost by one reactant must equal the total number of electrons gained by another. Many such reactions are not easy to balance by inspection, but require a systematic procedure. Two procedures are in wide use: the oxidation-number method and the half-reactions method. We will introduce both and you can choose to use whichever is easier for you. Initially, we will restrict our consideration to reactions carried out in acidic solutions. Reaction equations in acidic solutions can always incorporate water and/or hydronium ion, two species that are present in high concentration in these solutions. Balancing equations for reactions in basic solutions requires a slightly modified procedure, which we will introduce at the end of the section.

The oxidation-number method Table 6.6 presents the systematic procedure for balancing reduction–oxidation reaction equations by the **oxidation-number method.** The method uses the change in oxidation numbers for the atom that is oxidized and the atom that is reduced to determine the number of electrons lost and gained, respectively.

Table 6.6 *Oxidation-number method for balancing reduction–oxidation equations.*

This procedure applies to acidic and neutral solutions.

Step	Procedure
1.	Write all reactants and products, except H^+ and H_2O, in the form of a chemical equation.
2.	Assign oxidation numbers to those atoms that undergo changes in oxidation number.
3.	Adjust coefficients for the reactant and product containing the atom being oxidized to balance that atom. Adjust coefficients for the reactant and product containing the atom being reduced to balance that atom.
4.	Compare the number of electrons being released by the atom undergoing oxidation to the number of electrons being gained by the atom undergoing reduction. While maintaining the coefficient ratios from Step 3, adjust coefficients to equalize the number of electrons associated with oxidation and reduction.
5.	If the number of oxygen atoms differs between the left and right sides of the equation, add water molecules to the side needing more oxygen atoms.
6.	If the number of hydrogen atoms differs between the left and right sides of the equation, add hydrogen ions (H^+) to the side needing more hydrogen atoms.
7.	Check the equation to be certain that all atoms are balanced and that the net charge is the same on both sides of the equation.

6.78 WORKED EXAMPLE

Balancing redox equations by the oxidation-number method

Use the oxidation-number method to balance the equation for the reaction of copper with concentrated nitric acid.

$$Cu(s) + H^+(aq) + NO_3^-(aq) \rightarrow Cu^{2+}(aq) + NO_2(g) \tag{6.45}$$

Necessary information: We need to know, from Section 6.9, that the oxidation numbers for Cu, Cu^{2+}, N in NO_3^-, and N in NO_2 are 0, +2, +5, and +4, respectively.

Strategy: Follow the steps outlined in Table 6.6.

Implementation:

Step 1: The reactants containing atoms that are oxidized and reduced are Cu and NO_3^-; these atoms appear in the products as Cu^{2+} and NO_2.

$$Cu(s) + NO_3^-(aq) \rightarrow Cu^{2+}(aq) + NO_2(g) \tag{6.50}$$

Step 2: Oxidation numbers are shown in red above the atoms.

$$\overset{0}{Cu}(s) + \overset{+5}{N}O_3^-(aq) \rightarrow \overset{+2}{Cu^{2+}}(aq) + \overset{+4}{N}O_2(g) \tag{6.51}$$

Step 3: The number of copper atoms in the reactants and products is the same. The number of nitrogen atoms in the reactants and products is the same. No adjustment of the coefficients in expression (6.51) is necessary at this step.

Step 4: We can show the gain and loss of electrons as

gain 1 electron

$$\overset{0}{Cu}(s) + \overset{+5}{N}O_3^-(aq) \rightarrow \overset{+2}{Cu^{2+}}(aq) + \overset{+4}{N}O_2(g)$$

lose 2 electrons

$$\tag{6.52}$$

To balance the gain and loss of electrons, two nitrate ions need to be reduced to two nitrogen dioxide molecules:

gain 2 electrons

$$\overset{0}{Cu}(s) + 2\overset{+5}{N}O_3^-(aq) \rightarrow \overset{+2}{Cu^{2+}}(aq) + 2\overset{+4}{N}O_2(g)$$

lose 2 electrons

$$\tag{6.53}$$

Step 5: In expression (6.53), there are six oxygen atoms in the reactants and four in the products. We add two water molecules to the products to balance the oxygen atoms:

$$Cu(s) + 2NO_3^-(aq) \rightarrow Cu^{2+}(aq) + 2NO_2(g) + 2H_2O(l) \tag{6.54}$$

Step 6: In expression (6.54), there are four hydrogen atoms in the products, but none in the reactants. We add four hydrogen ions (hydroniums) to the reactants to balance hydrogen atoms:

$$Cu(s) + 4H^+(aq) + 2NO_3^-(aq) \rightarrow Cu^{2+}(aq) + 2NO_2(g) + 2H_2O(l)$$

$$\tag{6.55}$$

continued

Step 7: In reaction expression (6.55), atoms are balanced and charge is balanced. There is a net charge of $+2$ [$= 4(+1) + (-2)$] on the reactant side and $+2$ on the product side.

Does the answer make sense? Reaction expression (6.55) describes the observed changes in the reaction of copper with concentrated nitric acid and is balanced in atoms, charge, and electrons transferred. The answer makes sense.

6.79 CHECK THIS

Balancing redox equations by the oxidation-number method

(a) In less concentrated nitric acid solutions, Cu^0 is oxidized to Cu^{2+}, but the gaseous reaction product is nitric oxide, NO, rather than NO_2. Assign the oxidation number for N in NO and use the oxidation-number method to write the balanced equation for the reaction.

(b) Use the oxidation-number method to balance the reaction of hydrogen peroxide with iodide:

$$H_2O_2(aq) + I^-(aq) \rightarrow H_2O(l) + I_2(aq) \qquad \text{(6.56)}$$

(c) Use the oxidation-number method to balance the reaction of thiosulfate anion, $S_2O_3^{2-}$, with iodine [a reaction often used in analyses of reactants that produce iodine by oxidizing iodide, as in reaction (6.56)]:

$$2\begin{bmatrix} :\ddot{O}: \\ \| \\ :\ddot{O}-S-\ddot{S}: \\ \| \\ :\ddot{O}: \end{bmatrix}^{2-}_{(aq)} + I_2(aq) \rightarrow \begin{bmatrix} :O: \qquad :O: \\ \| \qquad \| \\ :\ddot{O}-S-\ddot{S}-\ddot{S}-S-\ddot{O}: \\ \| \qquad \| \\ :O: \qquad :O: \end{bmatrix}^{2-}_{(aq)} + 2I^-(aq) \qquad \text{(6.57)}$$

As you first use the oxidation-number method for reduction–oxidation equation balancing, you will probably find it helpful to show the loss and gain of electrons explicitly, as we did in expressions (6.52) and (6.53) in Worked Example 6.78. With more experience, you can probably do the gain and loss in your head and adjust the coefficients without showing the intermediate expressions.

6.80 INVESTIGATE THIS

What happens when bleach and iodide are mixed?

Do this as a class investigation and work in small groups to discuss and analyze the results. Mix 1 mL of 6 M aqueous sulfuric acid, $(HO)_2SO_2$, solution with 20 mL of 20% aqueous potassium iodide, KI, solution in a small flask. Record your observations. Add a few drops of household bleach solution to the flask and swirl to mix. Record your observations.

6.81 CONSIDER THIS

What is the reaction between bleach and iodide?

(a) Did you observe any evidence for reaction between acid and potassium iodide solution in Investigate This 6.80? Between bleach and acidic potassium iodide solution? What is the evidence in each case?

(b) Can you identify any product or products of the reaction? Explain your response. Can you tell whether the reaction is reduction–oxidation? Why or why not?

(c) The active ingredient in most common household bleaches is hypochlorite ion, OCl^-. If the reaction in Investigate This (6.80) is a reduction–oxidation, is hypochlorite probably reduced or oxidized in the reaction? What is the reasoning for your answer? What might be the product(s) formed from the hypochlorite? Could you see any evidence for this(these) product(s) in the investigation? Why or why not?

The half-reactions method The **half-reactions method** for balancing reduction–oxidation reaction equations is based on splitting the reaction in two, considering the reduction reaction and oxidation reaction separately, and then recombining them so that the number of electrons lost in the oxidation is the same as the number of electrons gained in the reduction. You do not have to consider oxidation numbers explicitly in the balancing procedure outlined in Table 6.7, but you may have to use them to determine which species is reduced and which oxidized, in order to write the appropriate half-reactions.

Table 6.7 *Half-reactions method for balancing reduction–oxidation reactions.*

This procedure applies to acidic and neutral solutions.

Step	Procedure
1.	Write all reactants and products except H^+ and H_2O in the form of separate chemical equations for the oxidation and reduction half-reactions.
2.	For each half-reaction, adjust coefficients for atoms of all elements other than hydrogen and oxygen.
3.	For each half-reaction, if the number of oxygen atoms differs between the left and right sides of the equation, add water molecules to the side needing more oxygen atoms.
4.	For each half-reaction, if the number of hydrogen atoms differs between the left and right sides of the equation, add hydrogen ions (H^+) to the side needing more hydrogen atoms.
5.	For each half-reaction, add the number of electrons, e^-, needed to balance the charge. One half-reaction (the reduction) will need electrons added to the left side, the other half-reaction (the oxidation) will need electrons added to the right side.
6.	Multiply each half-reaction by the minimum factor required to equalize the number of electrons in each half-reaction.
7.	Add the equations for the half-reactions together, canceling electrons and excess water molecules or hydrogen ions.

6.82 WORKED EXAMPLE

Balancing redox equations by the half-reactions method

In Investigate This 6.80, the red-brown color formed in the reaction is due to elemental iodine, I_2, formed by oxidation of I^-. (The I_2 combines with excess I^- to form a complex ion, I_3^-, which is what you observe; but we will focus on the I_2.) Hypochlorite ion, OCl^-, is the species that is reduced and the product of the reduction is chloride ion, Cl^-. (There is no visible evidence for this product; other experiments are required to confirm that it is formed.) Thus, we can write the reactants and products as

$$I^-(aq) + OCl^-(aq) \rightarrow I_2(aq) + Cl^-(aq) \tag{6.58}$$

Use the half-reactions method to balance this reaction.

Necessary information: We need to know that $I^-(aq)$ reacts to give $I_2(aq)$ and $OCl^-(aq)$ reacts to give $Cl^-(aq)$.

Strategy: Follow the steps outlined in Table 6.7.

Implementation:

Step 1: The half-reactions are

$$I^-(aq) \rightarrow I_2(aq) \tag{6.59}$$

$$OCl^-(aq) \rightarrow Cl^-(aq) \tag{6.60}$$

Step 2: There are two iodine atoms on the product side of expression (6.59) and only one on the reactant side, so the coefficient on $I^-(aq)$ must be 2. The number of chlorine atoms in the reactants and products is the same in expression (6.60).

$$2I^-(aq) \rightarrow I_2(aq) \tag{6.61}$$

$$OCl^-(aq) \rightarrow Cl^-(aq) \tag{6.60}$$

Step 3: In expression (6.60), there is one oxygen atom in the reactants and none in the products. Add a water molecule to the products to balance the oxygen atoms.

$$2I^-(aq) \rightarrow I_2(aq) \tag{6.61}$$

$$OCl^-(aq) \rightarrow Cl^-(aq) + H_2O(l) \tag{6.62}$$

Step 4: In expression (6.62), there are two hydrogen atoms in the products and none in the reactants. Add two hydrogen ions (hydronium) to the reactants to balance hydrogen atoms.

$$2I^-(aq) \rightarrow I_2(aq) \tag{6.61}$$

$$2H^+(aq) + OCl^-(aq) \rightarrow Cl^-(aq) + H_2O(l) \tag{6.63}$$

Step 5: In expression (6.61), there are two negative charges on the reactant side and no charges on the product side. Add two electrons, e^-, to the product side.

$$2I^-(aq) \rightarrow I_2(aq) + 2e^- \tag{6.64}$$

continued

In expression (6.63), the net charge on the reactant side is $+1$ $[=2(+1) + (-1)]$ and on the product side is -1. Add two electrons to the reactant side to make the net charge on each side -1.

$$2H^+(aq) + OCl^-(aq) + 2e^- \rightarrow Cl^-(aq) + H_2O(l) \tag{6.65}$$

Step 6: The number of electrons in each half-reaction is the same, so no multiplicative factor is required to make them the same.

Step 7: Add the two half-reaction equations (6.64) and (6.65) to cancel the electrons. There are no excess water or hydrogen ions to be cancelled.

$$2I^-(aq) \rightarrow I_2(aq) + 2e^- \tag{6.64}$$

$$\underline{2H^+(aq) + OCl^-(aq) + 2e^- \rightarrow Cl^-(aq) + H_2O(l)} \tag{6.65}$$

$$2H^+(aq) + OCl^-(aq) + 2I^-(aq) \rightarrow Cl^-(aq) + H_2O(l) + I_2(aq) \tag{6.66}$$

Does the answer make sense? Reaction equation (6.66) describes the observed changes in the reaction of bleach (OCl^-) with an acidic solution of iodide ion and is balanced in atoms, charge, and electrons transferred. The answer makes sense.

6.83 CHECK THIS

Balancing redox equations by the half-reactions method

(a) One way to analyze iron(II) ion in solution is by reaction with dichromate ion, $Cr_2O_7^{2-}$:

$$Fe^{2+}(aq) + Cr_2O_7^{2-}(aq) \rightarrow Fe^{3+}(aq) + Cr^{3+}(aq) \tag{6.67}$$

Explain how you know this is a reduction–oxidation reaction and balance it by the half-reactions method.

(b) Household bleach—hypochlorite solution—always has a warning label, something like: "Hazard. Do not mix with other household chemicals such as toilet bowl cleaners, rust removers, acids . . .." These chemicals are often quite acidic. When hypochlorite ion solution is mixed with hydronium ion solution the products are chlorine gas, Cl_2, and oxygen gas, O_2. Explain how you know this is a reduction–oxidation reaction, write the reactants and products, and balance the reaction equation by the half-reactions method. *Hint:* Water has an oxygen atom with a -2 oxidation number and can decompose to O_2 with a zero oxidation number.

(c) In some reduction–oxidation reactions, the same element, in different molecules, is both reduced and oxidized. One example that is commonly used in chemical analysis is the reaction of iodate ion, IO_3^-, with iodide ion, I^-, in acidic solution:

$$IO_3^-(aq) + I^-(aq) \rightarrow I_2(aq) \tag{6.68}$$

Balance this reaction equation by the half-reactions method.

Reduction–oxidation reactions in basic solutions In basic solutions, the concentration of hydronium ion is so small that we cannot consider it as a reactant. Hydroxide ions, $OH^-(aq)$, and water are the predominant species that we can use as sources of oxygen and hydrogen, respectively, to balance reduction–oxidation reaction equations. There are several ways to account for this change. By far the easiest is first to balance the reaction equation as if it occurs in acidic solution, using either of the two methods we have just introduced. Then add the same number of $OH^-(aq)$ to both sides of the equation so that the added $OH^-(aq)$ is equivalent to the $H^+(aq)$ and "reacts" with it to produce water. Finish up by eliminating any excess water molecules and then rechecking for atom and charge balance.

6.84 WORKED EXAMPLE

Balancing redox equations in basic solution

To make household bleach chlorine, $Cl_2(g)$, is reacted with an aqueous solution of sodium hydroxide [$Na^+(aq)$ and $OH^-(aq)$]. The reaction products are $OCl^-(aq)$ and $Cl^-(aq)$:

$$Cl_2(g) \rightarrow OCl^-(aq) + Cl^-(aq) \tag{6.69}$$

Write the balanced equation for this reaction in basic solution.

Necessary information: The reactants and products are given in expression (6.69).

Strategy: Use the half-reaction method to balance the equation as though it occurs in acidic solution. Add enough $OH^-(aq)$ to both sides of the equation to convert any $H^+(aq)$ to water.

Implementation: We will go through the steps of the half-reaction procedure without comment. Be sure you can explain each step for yourself.

Step 1: $Cl_2(g) \rightarrow OCl^-(aq)$ $Cl_2(g) \rightarrow Cl^-(aq)$

Step 2: $Cl_2(g) \rightarrow 2OCl^-(aq)$ $Cl_2(g) \rightarrow 2Cl^-(aq)$

Step 3: $Cl_2(g) + 2H_2O(l) \rightarrow 2OCl^-(aq)$ $Cl_2(g) \rightarrow 2Cl^-(aq)$

Step 4: $Cl_2(g) + 2H_2O(l)$
$\rightarrow 2OCl^-(aq) + 4H^+(aq)$ $Cl_2(g) \rightarrow 2Cl^-(aq)$

Step 5: $Cl_2(g) + 2H_2O(l)$
$\rightarrow 2OCl^-(aq) + 4H^+(aq) + 2e^-$ $Cl_2(g) + 2e^- \rightarrow 2Cl^-(aq)$

Step 6: unnecessary

Step 7: $Cl_2(g) + 2e^- \rightarrow 2Cl^-(aq)$

$$\underline{Cl_2(g) + 2H_2O(l) \rightarrow 2OCl^-(aq) + 4H^+(aq) + 2e^-}$$

$$2Cl_2(g) + 2H_2O(l) \rightarrow 2OCl^-(aq) + 2Cl^-(aq) + 4H^+(aq) \tag{6.70}$$

continued

There are four $H^+(aq)$ among the products of reaction expression (6.70), so we will add four $OH^-(aq)$ to both sides of the equation to convert these $H^+(aq)$ in the product to an equivalent amount of water:

$$2Cl_2(g) + 2H_2O(l) + 4OH^-(aq) \rightarrow 2OCl^-(aq) + 2Cl^-(aq) + \underline{4H^+(aq) + 4OH^-(aq)}$$
$$= 4H_2O(l)$$

Eliminating excess water gives the balanced reaction expression in basic solution:

$$2Cl_2(g) + 4OH^-(aq) \rightarrow 2OCl^-(aq) + 2Cl^-(aq) + 2H_2O(l)$$
$$Cl_2(g) + 2OH^-(aq) \rightarrow OCl^-(aq) + Cl^-(aq) + H_2O(l) \qquad \text{(6.71)}$$

Does the answer make sense? Reaction expression (6.71) describes the reaction discussed in the problem presentation, is reduced to simplest form, and is balanced in atoms, charge, and electrons transferred. The answer makes sense.

6.85 CHECK THIS

Balancing redox equations in basic solution

(a) Balance reaction expression (6.69) in basic solution by the oxidation-number method. *Hint:* Write Cl_2 twice on the reactant side and let one of them be reduced and the other oxidized. Add all the Cl_2 molecules together after the equation is balanced.

(b) Hydrogen peroxide, H_2O_2, can oxidize manganous ion, Mn^{2+}, to manganese(IV) dioxide, MnO_2 in basic solution:

$$H_2O_2(aq) + Mn^{2+}(aq) \rightarrow MnO_2(s) \qquad \text{(6.72)}$$

Balance this equation for reaction in basic solution. *Hint:* Assume that the product of reduction of the peroxide is water.

6.11. Reduction–Oxidation Reactions of Carbon-Containing Molecules

 ### 6.86 INVESTIGATE THIS

Does methanal react with silver ion?

Do this as a class investigation and work in small groups to discuss and analyze the results. Place 4 mL of a basic solution of silver diammine complex, $Ag(NH_3)_2^+(aq)$, in a clean 200-mm test tube. *CAUTION:* This basic solution is caustic and can harm clothes and flesh. Silver ion solutions can stain clothing and flesh. Handle with care and rubber or plastic gloves. Add to the test tube a few drops of an aqueous solution of methanal (formaldehyde), H_2CO. Swirl the mixture in the test tube for two or three minutes. Record your observations.

6.87 CONSIDER THIS

What is the reaction of methanal with silver ion?

In Investigate This 6.86, what is the evidence, if any, that methanal reacts with silver ion? What products of the reaction can you identify? How would you classify the reaction? Explain.

Oxidation of methanal by silver ion The most obvious observation in Investigate This 6.86 is the formation of silver metal. It was also a product of the reaction with copper in Investigate This 6.62. Once again, the silver metal signals the presence of a reactant that can provide the electrons required to produce the Ag^0 from the Ag^+. The silver is formed when methanal is added to the mixture, so methanal is probably the source of electrons and must itself be oxidized. What are the possible oxidation products of methanal?

6.88 CONSIDER THIS

What oxidation numbers are available for carbon?

(a) What is the oxidation number of each of the atoms in each of these compounds?
methane, CH_4
methanol, CH_3OH
methanal, CH_2O
methanoic (formic) acid, $HC(O)OH$
carbon dioxide, CO_2

(b) Is there a correlation between the number of bonding pairs of electrons between carbon and oxygen atoms and the oxidation number of carbon? If so, state a rule you can use to assign oxidation numbers to a carbon atom, based on the number of bonds to oxygen atoms.

(c) When methanal is oxidized in the reaction with silver ion in Investigate This 6.86, what are the possible oxidation products? Do you have any evidence for or against any of the possible products? Explain your reasoning.

We can use the results of your analysis from Consider This 6.88 to write the reactants and products for methanal oxidation by silver ion. Although silver ion is present in the solution as its diammine complex, $Ag(NH_3)_2^+(aq)$, we will write it as $Ag^+(aq)$. The ammonia is not oxidized or reduced, and it makes the equations simpler if it is left out. You saw no gas produced by the reaction, so the most likely oxidized product from methanal is methanoic acid:

$$Ag^+(aq) + CH_2O(aq) \rightarrow Ag(s) + HC(O)OH(aq) \qquad (6.73)$$

In basic solution, a possible balanced reaction is

$$2Ag^+(aq) + CH_2O(aq) + 2OH^-(aq)$$
$$\rightarrow 2Ag(s) + HC(O)OH(aq) + H_2O(l) \qquad (6.74)$$

Balancing the methanal-silver ion reaction equation

(a) Show how to balance expression (6.73) to get (6.74) in basic solution.
(b) What will happen to methanoic acid in basic solution? Write the reaction equation for this reaction. *Hint:* See Section 6.4.
(c) Combine your equation from part (b) with the balanced reaction (6.74) to get the overall reaction of methanal with silver ion that accounts for the fate of methanoic acid in basic solution.

Yeast fermentation of glucose In Investigate This 6.2 you observed the formation of foam as glucose was fermented by yeast to produce carbon dioxide and ethanol:

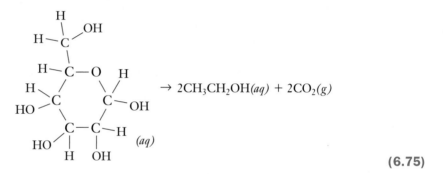

$$\rightarrow 2CH_3CH_2OH(aq) + 2CO_2(g)$$

(6.75)

Reaction (6.75) is reaction (6.1) rewritten to show the connectivity of the atoms in glucose.

Oxidation numbers for carbons in glucose fermentation

Assign oxidation numbers for all the carbon atoms in the reactant and products in the glucose fermentation reaction. Use your rule from Consider This 6.88(b), if this makes the task easier.

The carbon atoms in glucose all have oxidation numbers of +1 or +2, but two carbon atoms in the fermentation products have an oxidation number of +4 and two others an oxidation number of zero. The fermentation of glucose represents an *internal reduction–oxidation reaction*. Electrons are transferred from one carbon atom to another to yield an oxidized product, carbon dioxide, and a reduced product, ethanol. Expression (6.75) is the net result of two series of biochemical reactions: glycolysis and fermentation. **Glycolysis** (*glyco* = sugar + *lyein* = loosen, break; sugar breaking) is a metabolic pathway that oxidizes glucose to two molecules of pyruvic acid. The balanced half-reaction for glycolysis is

$$\tag{6.76}$$

The molecule that accepts the electrons lost by glucose in reaction (6.76) is **nicotinamide dinucleotide (NAD⁺),** an oxidant used by almost all organisms. The reduction of NAD^+ to **reduced nicotinamide dinucleotide (NADH)** can be represented by this half-reaction (where the reactive part of the molecule is enclosed in the rectangle and R is the rest of the molecule, which is not oxidized or reduced):

> In 1897, Hans and Eduard Buchner were making cell-free extracts of yeast for medicinal purposes and discovered that these extracts rapidly fermented sugar. This was the first demonstration that living cells are not necessary to carry out a metabolic reaction sequence.

$$\tag{6.77}$$

6.91 CHECK THIS

Net glycolysis reaction

(a) Check to see that half reactions (6.76) and (6.77) are balanced with respect to atoms and electrons.
(b) Write the balanced equation for oxidation of glucose to pyruvic acid by NAD^+. You may substitute the abbreviations NAD^+ and NADH for the structures in equation (6.77).

Other changes occur during glycolysis, including the production of adenosine triphosphate, ATP, the compound that organisms use to provide energy for most life processes. We will say more about energy and ATP in the next chapter. As glycolysis proceeds, organisms face a problem; NAD^+ gets used up because there is only a tiny amount of it in a cell. If oxygen is available, other metabolic processes oxidize the NADH back to NAD^+ and glycolysis can keep going to oxidize more glucose and produce more ATP. Yeast can live in the absence of oxygen; in order to do so, they need a way to regenerate their NAD^+ from

NADH. **Fermentation,** the reduction of pyruvic acid to produce ethanol and carbon dioxide, is the pathway yeast use to regenerate their NAD^+:

$$
\begin{array}{l}
\text{H} \quad :\!\ddot{O}\!: \quad \cdot\ddot{O}\cdot \\
| \qquad \| \qquad \nearrow \\
\text{H}-\text{C}-\text{C}-\text{C} \qquad\qquad + 2\text{H}^+(aq) + 2e^- \rightarrow \\
| \qquad\qquad\quad \backslash \\
\text{H} \qquad\qquad \cdot\ddot{O}\!:\!-\text{H}\,(aq)
\end{array}
$$

$$
\text{H}-\overset{\displaystyle \text{H} \quad \text{H}}{\underset{\displaystyle \text{H} \quad \text{H}}{\text{C}-\text{C}}}-\ddot{\text{O}}\!:\,(aq) + \ddot{\text{O}}\!=\!\text{C}\!=\!\ddot{\text{O}}\,(g)
$$

(6.78)

6.92 CHECK THIS

Reduction–oxidation of glucose

(a) Write the balanced equation for the reduction of two molecules of pyruvic acid by NADH to produce ethanol and carbon dioxide.

(b) Show that the sum of the balanced equation from Check This 6.91(b) and the one you wrote in part (a) gives reaction (6.75), the overall fermentation (reduction–oxidation) reaction for glucose.

(c) What happens to the NAD^+ and NADH in the summation you made in part (b)? Does this explain why the cell needs only a small amount of these molecules for its reduction–oxidation reactions?

All life depends upon reduction–oxidation reactions to provide the energy required to survive and to carry out processes such as silk production by spiders shown in the chapter opening illustration. The fermentation pathway is an inefficient use of glucose for energy. Total oxidation of glucose can produce about 20 times more energy than is available from fermentation. In order to understand these energy considerations, you need to know more about energy, the subject of the next chapter. Then, in Chapter 10, we will look again at the energy released by reduction–oxidation reactions and the way that organisms capture energy.

6.12 Outcomes Review

We have divided chemical reactions into three large classes: precipitation reactions, Lewis acid–base reactions, and reduction–oxidation reactions. In precipitation reactions, ions lose their bonding to waters of hydration and form crystals held together by attractions between cations and anions. The electrons on the cations and anions remain with the same atomic core throughout the reaction.

In Lewis acid–base reactions, a nonbonding electron pair on a Lewis base is attracted to a positive center on a Lewis acid and the electrons form a new covalent bond between the two reactants. Brønsted–Lowry acid–base reactions are a special class of Lewis acid–base reactions in which the new covalent bond is to a proton. We showed how to account for and predict the relative strengths of these acids and bases. Lewis acid–base reactions also occur between ligands, Lewis bases, and metal ions,

Lewis acids, to which the bases bond to form metal ion complexes, many of which are important in biological systems. Finally Lewis acid–base reactions occur between nucleophiles, Lewis bases, and electrophiles, Lewis acids, to form a variety of final products from the rearrangements that occur after the initial bond formation. Our examples were condensation reactions, many of which are used by both man and nature to create polymers.

Reduction–oxidation reactions involve complete transfer of one or more electrons from one reactant to another. The reactant that loses electrons is oxidized and the reactant that gains electrons is reduced. Oxidation numbers identify reduction–oxidation reactions and can be used to balance their reaction equations. Another method for balancing is to break the reaction into half-reactions, a reduction and an oxidation, balance these separately, and then recombine them.

Check your understanding of the ideas in the chapter by reviewing these expected outcomes of your study. You should be able to

- Classify a chemical change as a precipitation, Lewis acid–base, or reduction–oxidation reaction based on known reactants and products, or from experimental observations on the change [Sections 6.2, 6.4, 6.6, 6.7, 6.9, and 6.11].
- Predict probable precipitation reactions based on the cation and anion charges [Section 6.2].
- Use the stoichiometry of continuous variations studies to determine the formula of a reaction product or the ratio in which reactants react [Section 6.2].
- Define and give examples of three classes of Lewis acid–base reactions: Brønsted–Lowry proton transfers, metal ion complexation, and nucleophile–electrophile reactions [Sections 6.3, 6.4, 6.6, and 6.7].
- Use the observed pH and stoichiometry of a solution to determine the relative basicity (acidity) of the species in the solution [Section 6.4].
- Predict the relative basicities (acidities) of a series of Lewis/Brønsted–Lowry bases (acids) of known structure [Section 6.5].
- Use relative basicities (acidities) to predict the predominant species in solutions of Lewis/Brønsted–Lowry bases (acids) [Sections 6.4 and 6.5].
- Recognize the formation of a metal ion complex between a Lewis base and a metal ion in solution by observations on mixtures of the reactants (or in competitions between the Lewis base and another reactant for the metal ion) and suggest a structure for the complex [Section 6.6].

- Identify the nucleophilic and electrophilic sites in a pair of reactants that react to form a condensation product, including condensation polymers [Section 6.7].
- Predict the product of a nucleophile–electrophile reaction, including condensation polymerization, between reactants with the functional groups we have introduced [Sections 6.7 and 6.8].
- Use formal charge to explain the rearrangements that some reaction intermediates undergo or to explain the relative stability of different isomeric Lewis structures [Section 6.8].
- Identify the molecules or ions that are reduced and oxidized and their respective products in a reduction–oxidation reaction [Sections 6.9, 6.10, and 6.11].
- Assign oxidation numbers to all the atoms in a given molecule or ion [Sections 6.9 and 6.11].
- Balance a given reduction–oxidation reaction in acidic or basic solution by inspection, the oxidation-number method, or the half-reactions method [Sections 6.9, 6.10, and 6.11].
- Use your knowledge of the oxidation numbers of atoms in various molecules or ions to predict possible reduced or oxidized products from a reaction [Sections 6.10 and 6.11].
- Use oxidation numbers to show that a given reaction is an internal reduction–oxidation [Section 6.11].
- Explain how a titration is carried out and used to determine the amount of a reactant present in a solution of unknown concentration [Section 6.13].
- Use titration data to determine the amount of a reactant present in a solution of unknown concentration [Section 6.13].

6.13. Extension—Titration

6.93 INVESTIGATE THIS

How can you analyze an acid–base reaction?

Do this as a class investigation and work in small groups to discuss and analyze the results. Add two drops of bromocresol green acid–base indicator solution to about 20 mL of a 0.10 M aqueous solution of ethanoic (acetic) acid, $CH_3C(O)OH$, and swirl to mix well. Use a calibrated pipet to add 1.0 mL of this solution to each of eight wells in a 24-well microtiter plate. Leave the solution in the first well as it is. Add 0.20 mL of 0.10 M aqueous sodium hydroxide, NaOH, solution to the second well, 0.40 mL to the third, 0.60 mL to the fourth, and so on to 1.40 mL to the eighth well. Use an overhead projector to view the plate. Record the color in each of the wells.

6.94 CONSIDER THIS

How can you follow an acid–base reaction?

What is the pattern of colors that you observe in the wells in Investigate This 6.93? How can you explain the pattern?

Ethanoic acid reaction with hydroxide In Investigate This 6.93, hydroxide ion (from the NaOH solution) is added to ethanoic acid. The acid–base reaction that can occur is

$$CH_3C(O)OH(aq) + OH^-(aq) \rightleftharpoons H_2O(l) + CH_3C(O)O^-(aq) \quad \textbf{(6.79)}$$

Table 6.2 shows that hydroxide ion is a stronger base than ethanoate. If equal numbers of moles of ethanoic acid and hydroxide are mixed, the predominant base in the mixture, after reaction occurs, is the ethanoate ion, $CH_3C(O)O^-(aq)$. What is the predominant base if *unequal* numbers of moles of ethanoic acid and hydroxide are mixed?

If fewer moles of hydroxide than ethanoic acid are mixed, reaction (6.79) proceeds until the hydroxide is used up, but then no more ethanoate can be formed. Ethanoate will be the predominant base in the mixture and there will also be ethanoic acid left over. If more moles of hydroxide than ethanoic acid are mixed, reaction (6.79) proceeds until the ethanoic acid is used up. This leaves a solution containing substantial concentrations of both hydroxide and ethanoate ions.

6.95 CHECK THIS

Acid–base equivalence in Investigate This 6.93

In which well in Investigate This 6.93 is the number moles of hydroxide added equal to the number of moles of ethanoic acid at the start? Show how you get your answer. Where is this well in the pattern of colors you observed in the investigation?

Titration When base is added to an acid (or vice versa), as in Investigate This 6.93, the **equivalence point** of the addition is when the number of moles of base added is equal to the number of moles of acid originally present. If you have some way to recognize the equivalence point, you can determine the number of moles of acid in an unknown sample by keeping track of the number of moles of base you need to add to get to the equivalence point. This kind of analysis is called a **titration.** Titrations are another way to determine reaction stoichiometry, if the concentrations of both reactants are known.

In Investigate This 6.93, you observed that the **acid–base indicator** (bromocresol green) is yellow in acetic acid solution, the first well. You also observed that the indicator is blue in solutions that contain more hydroxide than acid, the wells after the equivalence point. In wells to which some base had been added, but before the equivalence point, the indicator is green (a mixture of blue and yellow). These wells contain ethanoate ion formed by reaction (6.79) as well as unreacted ethanoic acid; the wells are at some intermediate acidity. At the

Acid–base indicators are highly colored dyes whose color is sensitive to the pH of the solution. There are many different acid–base indicators, each of which changes color at a different pH. The pH test paper you have used contains these dyes that give the color you use to get the pH.

equivalence point the indicator color in the well is blue. The equivalence point with this indicator is signaled by the appearance of the blue color (the disappearance of the green color). More accurate experiments are required to prove this assertion, but the proper choice of indicator makes it possible to determine the equivalence point in many acid–base titrations.

A common use of titrations is to find the concentration of a reactant in a solution of unknown concentration. The traditional laboratory apparatus for carrying out titrations is illustrated in Figure 6.12. The long, graduated glass tube with a valve at one end to control the outflow of liquid is a **buret**. The buret usually contains a reactant solution of accurately known concentration, called the **titrant**. The idea is to add the titrant in small portions to a vessel containing a *measured volume* of the other reactant, as you modeled in Investigate This 6.93, whose concentration is not known. When the equivalence point is reached, the volume of titrant required, its concentration, and the volume of the unknown solution are used to calculate the unknown concentration. For acid–base titrations, the equivalence point is often signaled by an indicator that changes color.

(a) Before the equivalence point   (b) At the equivalence point

Figure 6.12.

An acid–base titration with phenolphthalein indicator. The flask contains ethanoic acid of unknown concentration and a few drops of phenolphthalein indicator, which is colorless in acid and red in base. The titrant is sodium hydroxide solution of known concentration. The titrant is added slowly while the flask is swirled. The addition is stopped just as the solution turns light pink.

6.96 WORKED EXAMPLE

An acid–base titration

A 25.00 mL volume of an ethanoic acid solution of unknown concentration is titrated with 0.105 M sodium hydroxide with phenolphthalein as the indicator. The buret reading at the beginning of the titration is 3.45 mL and the reading at the end (just as the titrated solution turns pink) is 22.67 mL. What is the concentration of the ethanoic acid?

Necessary information: All the volume and molarity information we need are available from the problem statement. Burets are graduated beginning with zero at the top, so the volume delivered (in milliliters) is the final reading minus the initial reading. We know from equation (6.79) that hydroxide and ethanoic acid react in a 1:1 ratio.

Strategy: Determine the volume of hydroxide added and use its molarity to calculate the number of moles of hydroxide that react. Since the acid and base react in a 1:1 ratio, the number of moles hydroxide used is the number of moles of ethanoic acid in the original sample. Use the number of moles ethanoic acid, the volume of the sample, and the definition of molarity to get the concentration.

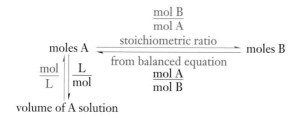

continued

Implementation: The volume of base used in the titration is

$$\text{vol base} = (22.67 \text{ mL}) - (3.45 \text{ mL}) = 19.22 \text{ mL}$$

$$\text{moles base} = (19.22 \text{ mL})\left(\frac{1 \text{ L}}{1000 \text{ mL}}\right)(0.105 \text{ M}) = 2.02 \times 10^{-3} \text{ mol}$$

$$\text{moles acid} = \text{moles base} = 2.02 \times 10^{-3} \text{ mol}$$

$$\boxed{\text{concentration of acid}} = \frac{\text{mol acid}}{\text{vol acid}} = \frac{2.02 \times 10^{-3} \text{ mol}}{0.02500 \text{ L}} = \boxed{0.0807 \text{ M}}$$

Does the answer make sense? The volumes of acid and base that react are similar, so the concentrations of the solutions should be similar, as we found. The volume of base used is smaller than the volume of acid, so the acid is less concentrated, as we found.

6.97 CHECK THIS

An acid–base titration

(a) If two drops of phenolphthalein are added to 50.0 mL of an aqueous solution of ethylenediamine (1,2-diaminoethane), $H_2NCH_2CH_2NH_2$, will the solution be colored or colorless? Explain the reasoning for your answer.

(b) When the solution in part (a) is titrated with a 0.500 M solution of hydrochloric acid, a color change of the solution occurs when 25.0 mL of the acid has been added. What is the color change observed? What is the concentration of the ethylenediamine solution? The reaction of ethylenediamine with hydronium ion is

$$2H_3O^+(aq) + H_2NCH_2CH_2NH_2(aq)$$
$$\rightleftharpoons {}^+H_3NCH_2CH_2NH_3^+(aq) + 2H_2O(l) \qquad \textbf{(6.80)}$$

6.98 INVESTIGATE THIS

What other titrations are possible?

Work in small groups to do this investigation and to discuss and analyze the results. Each group will use two small test tubes and three labeled thin-stem plastic pipets containing, respectively, water, 0.0050 M aqueous **ascorbic acid** (vitamin C) solution, and 0.0050 M aqueous iodine, I_2, solution.

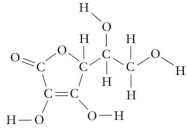

ascorbic acid

(a) In one test tube, mix 20 drops of water and 3 drops of iodine solution. Record your observations. In the second test tube, mix 20 drops of ascorbic acid solution and 3 drops of iodine solution. Record your observations.

(b) Continue to mix iodine solution, 2–3 drops at a time, into the ascorbic acid solution while keeping a count of the *total* number of drops added. Stop adding at the first sign that the ascorbic acid has all reacted. Record the number of drops required.

6.99 CONSIDER THIS

What is the stoichiometry of the iodine–ascorbic acid reaction?

(a) In Investigate This 6.98, what evidence do you have that a reaction occurs between ascorbic acid and iodine? Explain.

(b) How many drops of the iodine solution were required to react with all the ascorbic acid in 20 drops of the ascorbic acid solution? How did you know when all the ascorbic acid had reacted? Assuming that the size of drops is the same for both solutions, what can you conclude about the stoichiometric reaction ratio between iodine and ascorbic acid? Clearly explain the reasoning for your answer.

Iodine–ascorbic acid reaction Iodide ion in water, $I^-(aq)$, is colorless. If the iodine in the iodine solution is reduced to iodide, the color disappears:

$$I_2(aq) + 2e^- \rightarrow 2I^-(aq) \tag{6.81}$$

The reaction with ascorbic acid in Investigate This 6.98 must furnish the two electrons. Your results from Consider This 6.99 show that iodine reacts 1:1 with ascorbic acid. Each ascorbic acid molecule must transfer two electrons to reduce a molecule of I_2 to $2I^-$. The product of this oxidation of ascorbic acid is **dehydroascorbic acid** (*dehydro* means "without hydrogen"):

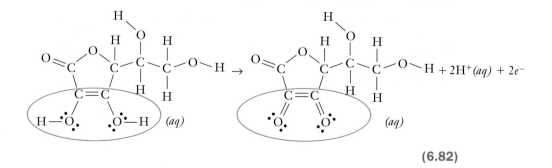

$$\tag{6.82}$$

Titration with iodine, as in Investigate This 6.98, is one of the methods for determining the amount of ascorbic acid (vitamin C) in foods.

Ascorbic acid oxidation

(a) Show how two electrons are lost in going from reactant to product(s) in reaction (6.82).

(b) Write the balanced chemical equation for the reaction of ascorbic acid with iodine.

In principle, any reaction that goes essentially to completion with well-defined stoichiometry can be used in a titration to analyze one of the reactants. For efficiency, we also want reactions that are rapid, so the titration can be done quickly. The usual limitation on possible titration reactions is finding a way to detect the equivalence point. Color changes are most convenient, but there are also instrumental methods that can often be used when appropriate indicators or other color changes are not available.

Chapter 6 Problems

6.1. Classifying Chemical Reactions

6.1. In each of these cases, indicate whether a chemical reaction (or reactions) occurs and explain how you know.
(a) Water is boiled in a teakettle.
(b) Boiling water is poured into a bowl containing a package of instant oatmeal and stirred to make a hot cereal breakfast.
(c) A glass of a carbonated soft drink is left overnight and tastes flat the next morning.
(d) A glass of ice cubes and water is left overnight and there are no ice cubes in the water the next morning.

6.2. In each of these cases, indicate whether a chemical reaction (or reactions) occurs and explain how you know.
(a) A match is dropped on the floor.
(b) A match is struck and used to start a barbeque.
(c) A piece of paper is folded to make a paper airplane.
(d) A piece of paper is torn into many small pieces to make confetti.

6.3. In each of these cases, indicate whether a chemical reaction (or reactions) occurs and explain how you know.
(a) An acorn buried and forgotten by a squirrel grows into an oak tree.
(b) A bottle of milk left too long in the refrigerator turns sour.
(c) Equal volumes of solutions of blue food coloring and yellow food coloring are mixed and the resulting solution is green.

(d) A few drops of bromocresol green solution are added to 20 mL of a colorless 0.1 M solution of sodium acetate and the resulting solution is blue. When 10 mL of a colorless solution of 0.1 M hydrochloric acid is added, the mixture is green. When a further 10 mL of the hydrochloric acid is added, the mixture is yellow.

6.4. A few small pieces of dry ice (solid carbon dioxide, $-78\ °C$) are dropped into a tall glass cylinder containing a red solution of dilute ammonia to which has been added a few drops of phenol red acid–base indicator solution. Bubbles of gas are rapidly evolved, a white fog forms above the solution, and, after a short time, the solution in the cylinder turns yellow. What evidence, if any, do you have for physical and chemical reactions occurring in this system? Explain your reasoning carefully.

6.2. Ionic Precipitation Reactions

6.5. Which of these would you predict to be insoluble ionic solids when placed in water? Explain the basis for your predictions.
(a) $NaCl$
(b) $CaCO_3$
(c) Na_2CO_3
(d) $BaCl_2$
(e) $BaSO_4$

6.6. A 25.0 mL volume of 0.100 M sodium sulfate, Na_2SO_4, is added to 25.0 mL of 0.200 M barium chloride, $BaCl_2$. The mixing of these two solutions results in the formation of a white precipitate. How many grams of the precipitate can be formed? What is the limiting reactant?

6.7. Precipitation reactions often play a role in quantitative analysis. Consider these procedural steps and measurements taken to determine the mass percent of iron in a dietary iron tablet.

Procedural steps	Measurements taken
1. The mass of 20 tablets is determined. Then the tablets are ground into a fine powder.	Total mass of 20 tablets = 22.131 g
2. The mass of a sample of the powder is measured. Then the powder is dissolved in nitric acid, converting all iron present to soluble Fe(III) ions.	Mass of sample of powder = 2.998 g
3. The iron is precipitated from the solution through the addition of aqueous ammonia. Then the solid is separated and strongly heated to drive off all water. The iron is now present as the dry solid $Fe_2O_3(s)$.	Mass of dry, solid Fe_2O_3 = 0.264 g

(a) Use these data to determine the average mass of iron in a dietary iron tablet.
(b) What is the mass percent of iron in a dietary iron tablet?
(c) What are some of the assumptions made in using this procedure to determine the average mass of iron in a dietary iron tablet?

6.8. A 0.649-g sample containing only potassium sulfate, K_2SO_4, and ammonium sulfate, $(NH_4)_2SO_4$, is dissolved in water and then treated with excess barium nitrate, $Ba(NO_3)_2$, solution to precipitate all of the sulfate ion as barium sulfate, $BaSO_4$.
(a) Write the net ionic equation for the precipitation reaction.
(b) If 0.977 g of precipitate is formed, how many moles of sulfate ion were in the original sample?
(c) What is the mass percent of K_2SO_4 in the original sample? *Hint:* Let w = mass of K_2SO_4 in the sample and write expressions for the number of moles of sulfate present as K_2SO_4 and $(NH_4)_2SO_4$.

6.9. Use Figure 6.1, an idealized representation of a continuous variations experiment, to see how continuous variations experiments are analyzed by graphical procedures.

(a) Each sample contains the same total number of moles of the cation and anion. Calculate the mole fraction of cation in each sample, that is, the decimal fraction of the total moles that is cation in each sample.
(b) Assume that the "precipitate" shown in each sample is proportional to the amount of precipitate you would weigh from that sample. Make a graph of the amount of precipitate formed in each sample as a function of the mole fraction of cation.
(c) What does your graph in part (b) look like? Does it have straight-line segments? If you draw the straight lines, where do they intersect? How is the intersection related to the 1:1 stoichiometry of the reaction?
(d) If you did not know the stoichiometry, could you use the graphical analysis to find it? Explain the procedure you would use to do so.

6.10. Analyze the data in Check This 6.11 by the graphical method outlined in Problem 6.9. What is the stoichiometry of the reaction between nickel ion and dimethylglyoxime? Is this the same result you got in Check This 6.11? Why or why not?

6.3. Lewis Acids and Bases: Definition

6.11. Identify the ions or molecules in this list that you predict to be Lewis bases. Explain your reasoning for each selection.
(a) NH_3
(b) H^+
(c) NH_4^+
(d) OH^-
(e) CN^-
(f) CH_3NH_2
(g) H_3O^+
(h) CH_3^-

6.12. (a) Do all Lewis bases share a common characteristic? If so, explain what it is.
(b) Do Lewis acids also share a common characteristic? If so, explain what it is.

6.4. Lewis Acids and Bases: Brønsted–Lowry Acid–Base Reactions

6.13. Identify which reactant can be classified as a Brønsted–Lowry acid or base, and as a Lewis acid or base in this reaction. In each case, give the reason for your choice.

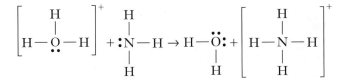

6.14. Lactic acid, $CH_3CH(OH)C(O)OH$, accumulates in our muscle tissue when we exercise strenuously. This acid transfers only a fraction of its acidic protons to water in a Brønsted–Lowry acid–base reaction. Write a chemical equation for this reaction and identify the Brønsted–Lowry acids and bases.

6.15. A 0.05 M solution of perchloric acid, $HOClO_3$, in water has a pH of 1.3. If sodium perchlorate, $NaClO_4$, is dissolved in water, will the solution pH be higher, lower, or the same as the water before the salt was added? Explain your response.

6.16. A 0.1 M solution of sodium cyanide, NaCN, in water has a pH of 11. If hydrogen cyanide gas, HCN, is dissolved in water to give a 0.1 M solution, will the solution pH be higher, lower, or about the same as the water before the gas was dissolved? Approximately what pH would you expect the solution to have? Explain your responses.

6.17. Hydronium ions are able to pass only very slowly from one side to the other of biological membranes. A difference in pH from one side to the other of membranes is produced by metabolism and is required for synthesis of the ATP cells need to keep functioning. Any substance that can carry protons from one side to the other of the membrane can interrupt the synthesis of ATP. The requirement for such a substance is that it be soluble in water in a form that can react with hydronium but less soluble in water (more soluble in nonpolar solvents like the inside of a membrane) after it has reacted.
(a) 2,4-Dinitrophenol (see Table 6.2) is one of the substances that has been used for this purpose in many studies of metabolism. Why is 2,4-dinitrophenol a good candidate for this purpose?
(b) In the 1930s, some people used 2,4-dinitrophenol as a weight-reducing drug. (This use was discontinued after several deaths occurred.) What is the basis for this use of the compound?

6.18. Consider the reaction between boric acid and water:

$$B(OH)_3(s) + 2H_2O(l) \rightleftharpoons [B(OH)_4]^-(aq) + H_3O^+(aq)$$

(a) Rewrite this equation using Lewis structures. Then analyze the reactants in terms of Lewis acid–base definitions.
(b) Explain why the formula for boric acid is more correctly written as $B(OH)_3$ instead of H_3BO_3.
(c) Some acids, such as sulfuric, $(HO)_2SO_2$, and phosphoric, $(HO)_3PO$, can donate more than one proton to a Lewis (or Brønsted–Lowry) base. These are called polyprotic acids. Explain why boric acid acts as a monoprotic (one proton) rather than polyprotic acid.

6.19. Examine the information given and rank these bases from weakest to strongest: $CH_3C(O)O^-$ (ethanoate or acetate), NO_3^- (nitrate), OH^- (hydroxide). Clearly explain the reasoning, based on this experimental evidence, for your ranking.

0.1 M $CH_3C(O)OH$, pH = 2.9

0.1 M HNO_3, pH = 1.0

0.1 M $NaNO_3$, pH = pH of water in which it was dissolved

0.1 M $CH_3COO^-Na^+$ (sodium ethanoate), pH = 8.9

0.1 M NaOH, pH = 13.0

6.5. Predicting Strengths of Lewis/Brønsted–Lowry Bases and Acids

6.20. Select the member of each pair that you predict to be the stronger base. Explain the basis for your selections.
(a) NH_3 or PH_3
(b) HP^{2-} or S^{2-}
(c) HP^{2-} or O^{2-}
(d) CH_3O^- or CH_3S^-

6.21. If equal volumes of 0.1 M aqueous solutions of sodium hydroxide, NaOH, and ammonium chloride, NH_4Cl, are mixed, what Lewis acid and what Lewis base will predominate in the mixture? Explain your reasoning. *Note:* Ammonium chloride dissolves to give a solution of ammonium cations and chloride anions.

6.22. What ions predominate in the solution formed when equal volumes of 0.10 M NH_3 and 0.10 M HCl are mixed? Explain your reasoning.

6.23. How do you think the basicity of the dihydrogen arsenate anion, $(HO)_2AsO_2^-$, compares to the basicity of the dihydrogen phosphate anion, $(HO)_2PO_2^-$? Explain your response.

6.24. Compare the Lewis structures of $CH_3C(O)O^-$ and $CH_3C(O)S^-$.
(a) Are there delocalized electrons in these ions? If so, where? Explain.
(b) Which ion do you predict to be the stronger base? Explain the basis for your prediction.

6.25. (a) A 0.1 M solution of potassium amide, KNH_2, in water has a pH of 13. Write a chemical equation for the acid–base reaction that occurs in this solution. What are the predominant Brønsted–Lowry acids and bases in the solution? How would you characterize the base strengths of the bases in this solution? Explain your responses.
(b) A 0.1 M solution of ethylamine, $CH_3CH_2NH_2$, in water has a pH of about 11. Write a chemical equation for the acid–base reaction that occurs in this solution. What

are the predominant Brønsted–Lowry acids and bases in the solution? How would you characterize the base strengths of the bases in this solution? Explain your responses.

(c) How are the reactions you write and your interpretations of the experimental evidence in parts (a) and (b) the same and different? Explain clearly.

6.26. Consider these structural reaction expressions showing the stepwise formation of carbonic acid, H_2CO_3, from water and carbon dioxide:

(a) For each step analyze the Lewis acid–base reaction that is represented. Identify which reactant (or part of the intermediate structure) is acting as the Lewis acid and which as the Lewis base in each step.

(b) We have previously written the overall chemical equation for this reaction as

$$H_2O(l) + CO_2(g) \rightleftharpoons H_2CO_3(aq)$$

What advantages or disadvantages does this equation have relative to the structural reactions for understanding the Lewis acid–base reaction? Explain your reasoning.

6.27. Consider this reaction:

$$H_2O(l) + HCO_3^-(aq) \rightleftharpoons H_3O^+(aq) + CO_3^{2-}(aq)$$

(a) Use the structural representations shown in Table 6.2 to rewrite the equation for this reaction. Identify all Lewis acids and bases in the reaction and the conjugate pairs of Lewis acids and bases.

(b) Does the position of the equilibrium for this reaction lie mostly toward the product side or mostly toward the reactant side? Explain your reasoning.

6.6 Lewis Acids and Bases: Metal Ion Complexes

6.28. At pH 7, aluminum hydroxide, $Al(OH)_3$, is quite insoluble. As the pH increases, $Al(OH)_3$ slowly dissolves into solution. Write a chemical reaction to explain this observation. *Hint:* Note that aluminum is in the boron family (column) of the periodic table and see Problem 6.18.

6.29. When a solution of ethylenediamine (1,2-diaminoethane), $H_2NCH_2CH_2NH_2$, en, is added slowly to a solution of copper(II) ion, the original light blue color becomes a much deeper blue and then changes to a beautiful magenta. A group of students prepared 12 samples for a continuous variations study of these color changes. They mixed the volumes of 0.020 M ethylenediamine and 0.020 M copper sulfate solutions given in this table. They measured the amount of light absorbed (the absorbance, A) by each solution at two wavelengths of light, 640 nm (red) and 560 nm (green). Their experimental results are also given in the table.

en, mL	0	1	2	3	4	5	6	7	8	9	10	11
Cu, mL	12	11	10	9	8	7	6	5	4	3	2	1
A_{640}	0.044	0.095	0.150	0.200	0.256	0.301	a	0.261	0.187	b	b	b
A_{560}	0.001	0.026	0.050	0.075	0.107	0.146	a	0.307	0.411	0.312	0.208	0.105

aThe group spilled some of this sample and didn't record its absorbance readings.
bNo readings taken.

(a) Use graphical analyses to determine all you can about the complexes formed by copper(II) ion with ethylenediamine. In particular, try to figure out how many ethylenediamine molecules chelate to each copper ion. Clearly explain the reasoning you use in your analyses and to reach your conclusions. *Hint:* See Problems 6.9 and 6.10.

(b) If the 6:6 solution had been mixed before some was spilled, would it have been appropriate for the students to use what was left to get the data for this sample? Why or why not? If it had not been spilled, what would the absorbance readings have been for this sample? Explain how you get your answers.

(c) Use your model kit to make a model of the copper–ethylenediamine complex(es).

6.30. As a 4.0 M aqueous solution of ethylenediamine (en, $H_2NCH_2CH_2NH_2$) is added slowly to 200 mL of a stirred 0.10 M aqueous solution of nickel(II) ion, the original green color of the nickel(II) ion changes through light blue to purple. The color stops changing after 15 mL of the en solution has been added. What conclusion(s) can you draw about the complex(es) formed between the nickel(II) and en? Clearly state the reasoning for your conclusions.

6.31. One method of treating lead or mercury poisoning is to administer chelating agents that effectively sequester (bind and "hide") metal ions, preventing their incorporation into bone or blood. One such agent is EDTA, discussed in Section 6.6. Another is the ligand called "BAL," which stands for "British AntiLewisite." Its chemical name is 2,3-dimercaptopropanol and this is its Lewis structure:

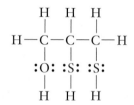

(a) Use your model kit to make a molecular model of this molecule and then sketch the model showing the bond angles revealed by the model. *Hint:* As you did in Investigate This 6.45, leave off the hydrogen atoms to make the model easier to handle, visualize, and sketch.
(b) Use the model to explain how this molecule can engage in a Lewis acid–base reaction to sequester metal ions such as Pb^{2+}.

6.7. Lewis Acids and Bases: Electrophiles and Nucleophiles

6.32. (a) Explain how both Brønsted–Lowry acid–base reactions and Lewis acid–base reactions illustrate the importance of charge in understanding chemical reactions.
(b) Why are *both* definitions necessary and useful for understanding chemical reactions?

6.33. Consider this possible reaction:

$$CH_3C(O)OH + NH_3 \rightleftharpoons CH_3C(O)NH_2 + H_2O$$

(a) Which reactant is the electrophile? Explain.
(b) Which reactant is the nucleophile? Explain.
(c) What other reaction might these reactants undergo? Explain which reaction is more likely to be observed.

6.34. Proteins are important biological condensation polymers that result from the linking of amino acids.

(a) Draw the structural formula for the tripeptide (short polymer of amino acids) formed when three glycine molecules, $H_2NCH_2C(O)OH$, are linked together by amide bonds.
(b) Write the reaction expression for formation of the glycine tripeptide.

6.35. ✏ Explain in detail how the movie animation in the *Web Companion*, Chapter 6, Section 6.7.2, describing the formation of an amide bond is related to the formation of the ester bond shown in reaction (6.34). Does the animation help to clarify the steps in the reaction sequence? Explain why or why not.

6.36. Identify the nucleophile, the electrophile, and any spectator ions in each of these balanced equations. If necessary, rewrite the reactants showing all the valence electrons.
(a) $NaOH + CH_3-I \rightarrow NaI + CH_3-OH$
(b) $(CH_3)_3N + CH_3-I \rightarrow (CH_3)_4N^+ + I^-$
(c) $(CH_3)_3C-Br + CH_3CH_2OH$
$\rightarrow (CH_3)_3C-O-CH_2CH_3 + HBr$
(d) $NH_3 + CH_3C(O)OCH_2CH_3$
$\rightarrow CH_3C(O)NH_2 + CH_3CH_2OH$
(e) $CH_3NH_2 + CH_3C(O)Cl$
$\rightarrow CH_3C(O)NHCH_3 + HCl$

6.37. Many esters have pleasant odors. Examine the formulas of these esters (and their odors) and write the formula of the alcohol and the acid from which each ester is made. *Note:* C_6H_5 represents the phenyl group, a benzene ring with one of the hydrogen atoms replaced by another atom or group, as in phenol (*phen*yl alcoh*ol*), Table 6.2.
(a) $CH_3C(O)OCH_2CH_2CH(CH_3)_2$ (oil of banana)
(b) $CH_3CH_2CH_2C(O)OCH_2CH_3$ (pineapple)
(c) $CH_3C(O)O(CH_2)_7CH_3$ (orange)
(d) $C_6H_5C(O)OCH_3$ (ripe kiwi)
(e) $CH_3C(O)OCH_2C_6H_5$ (jasmine)

6.38. Many esters have pleasant odors. Write the formula of the ester that is formed by reaction of each of these acid and alcohol combinations.
(a) $CH_3(CH_2)_2C(O)OH + CH_3(CH_2)_4OH$ (apricot)
(b) $CH_3C(O)OH + CH_3CH_2OH$ (some nail polish removers)
(c) $CH_3C(O)OH + (CH_3)_2CH(CH_2)_2OH$ (pear)
(d) $CH_3CH_2CH_2C(O)OH + CH_3OH$ (apple)
(e) $HC(O)OH + CH_3CH_2OH$ (rum)

6.39. Alcohols can react with carbonyl compounds, aldehydes and ketones, as well as with carboxylic acids. The products of these reactions are called hemiacetals or hemiketals. Consider the reaction of methanol with propanal:

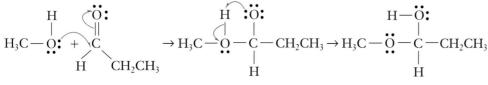

a hemiacetal

Describe the similarities and differences between these reaction steps and those shown in reaction (6.34) and in Problem 6.26. What do you think it is about the structures of a carboxylic acid and an aldehyde that makes their reactions with an alcohol somewhat different?

6.40. If a molecule contains both an alcohol and an aldehyde functional group, then an *intra*molecular reaction can occur, that is, a reaction between electrophilic and nucleophilic sites on the same molecule.

Intramolecular reactions like this are often much faster than intermolecular reactions. In part, this is because the nucleophile and electrophile are held close together; they are parts of the same molecule.
(a) Draw the product you would get from the intramolecular reaction of the alcohol and aldehyde in the molecule above. If this skeletal structure is confusing, draw out the structure, showing all the carbon and hydrogen atoms. *Hint:* See Problem 6.39. Use your molecular models to model the reaction, if that helps you visualize what will happen.
(b) Compare the structure of the product you get in part (a) with the structure of glucose shown in reaction (6.75) and in Problem 6.41. What structure(s) in the two molecules is(are) the same? Explain.

6.41. Cellulose and starch are condensation polymers of glucose. The condensation reaction between two glucose molecules is represented in this equation:

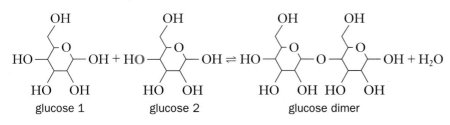

glucose 1   glucose 2   glucose dimer

(a) Many of the condensation reactions we wrote in Sections 6.7 and 6.8 involved an alcohol or an amine, Lewis bases, reacting with a positive carbon atom, a Lewis acid, double bonded to an oxygen atom. Is there a carbon atom in the glucose molecule that might react as such a Lewis acid? Which one? Explain the reasoning for your choice.
(b) Based on your choice in part (a) show which pair of nonbonding electrons in the reactants is attracted to this positive carbon atom to start the reaction toward the product shown. On the basis of this interaction, show where the oxygen atom that is highlighted in the reactants will end up in the products. Explain how you make your choice.

6.42. Occasionally, there is more than one possible nucleophilic (or electrophilic) reaction site in a molecule or ion. This means that more than one possible product is possible. An example is the reaction of sodium nitrite (NaNO$_2$) with isoamyl bromide (3-methyl-1-bromo-butane) to form isoamyl nitrite, the active ingredient in smelling salts.

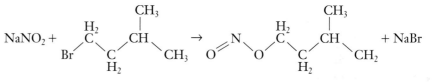

(a) Write the Lewis structures for the nitrite anion and the product, isoamyl nitrite.
(b) Identify the nucleophile, the electrophile, and any spectator ions in this reaction. *Another product is also formed.*
(c) Examine the Lewis structure you wrote for the nitrite anion in part (a). Can you identify more than one nucleophilic site?
(d) Draw the structure for the other product formed in the reaction of sodium nitrite with isoamyl bromide.
(e) Is there any reason to expect the oxygen atoms in the nitrite ion to be less nucleophilic than the oxygen atom in hydroxide, for example? Explain why or why not. *Hint:* Consider electron delocalization.

6.43. Phosphatidylcholine is a common component of your cell membranes.

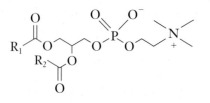

The units that make up this molecule are glycerol, $OHCH_2CH(OH)CH_2OH$, two carboxylic acids, $R_1C(O)OH$ and $R_2C(O)OH$, choline, $HOCH_2CH_2N(CH_3)_3^+$, and phosphoric acid. R_1 and R_2 are hydrocarbon chains, usually containing 15, 17, or 19 carbons.
(a) Write the overall reaction that forms phosphatidylcholine from its five units.
(b) Four bonds are required to hold the five units together. What kind of bond is each one of these four? Explain your answers.

6.8. Formal Charge

6.44. Calculate the formal charge for each atom in these substances.
(a) NH_4^+
(b) NO_3^-
(c) NH_3
(d) H_2O

6.45. Write Lewis structures with formal charges for each atom in these anions:
(a) CO_3^{2-}
(b) NO_2^-
(c) PO_4^{3-}
(d) HPO_4^{2-}
(e) $H_2PO_4^-$

6.46. Is there more than one possible way to write Lewis structures for any of the ions in Problem 6.45? If so, calculate the formal charges for each atom in each structure. Which structure(s) are most favored? Explain why.

6.47. The formal charges are shown for ozone and nitric acid. Fill in the missing electron pairs and show that the formal charges are correct.

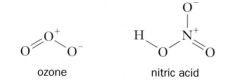

ozone   nitric acid

6.9. Reduction–Oxidation Reactions: Electron Transfer

6.48. Consider the reaction:
$$3Sn^{2+}(aq) + 2Bi^{3+}(aq) \rightarrow 3Sn^{4+}(aq) + 2Bi(s)$$
(a) How could you tell that a reaction had occurred upon mixing solutions of the reactants? How would you know that it is a reduction–oxidation reaction? Explain.
(b) Which element is oxidized?
(c) Which element is reduced?
(d) How many electrons are lost by each ion of the element that is oxidized?
(e) How many electrons are gained by each ion of the element that is reduced?

6.49. A piece of sodium metal, $Na(s)$, is placed in water. A vigorous reaction occurs in which hydrogen gas is evolved and the metal disappears. Upon testing the resulting solution with phenolphthalein, the solution turns pink. [See Figure 4.3(a).]
(a) How could you tell that a reaction had occurred upon placing the sodium in the water? How would you know that it is a reduction–oxidation reaction? Explain.
(b) Write the net ionic equation for the reaction described above.
(c) Which reactant is oxidized?
(d) Which reactant is reduced?

6.50. Identify these changes in charge as oxidation or reduction.
(a) $Zn \rightarrow Zn^{2+}$
(b) $S^{2-} \rightarrow S$
(c) $Fe^{3+} \rightarrow Fe^{2+}$
(d) $Ag^+ \rightarrow Ag$

6.51. Given these observations:
  (i)   Mg metal reacts with $Zn^{2+}(aq)$ to yield $Mg^{2+}(aq)$ and Zn metal.
  (ii) Zn metal reacts with $Cu^{2+}(aq)$ to yield $Zn^{2+}(aq)$ and Cu metal.
  (iii) Mg metal reacts with $Cu^{2+}(aq)$ to yield $Mg^{2+}(aq)$ and Cu metal.
(a) Write the balanced oxidation–reduction reaction for each observation.
(b) Which metal is most likely to undergo an oxidation? Explain.
(c) Which metal cation is most likely to undergo a reduction? Explain.

6.52. For these reactions: (i) assign oxidation numbers to each atom; (ii) identify the element that is oxidized; and (iii) identify the element that is reduced.
(a) $V_2O_5 + 2H_2 \rightarrow V_2O_3 + 2H_2O$
(b) $2K + Br_2 \rightarrow 2K^+ + 2Br^-$
(c) $N_2 + 3H_2 \rightarrow 2NH_3$

6.53. (a) What is the oxidation number for the sulfur atom in sulfur trioxide, SO_3?
(b) What is the oxidation number for the sulfur atom in the sulfite ion, SO_3^{2-}? in the sulfate ion, SO_4^{2-}?
(c) Sulfur trioxide dissolves in water to give an acidic solution. This is not a redox reaction. What is the acid that is produced? Explain the reasoning for your answer.

6.54. Determine the oxidation numbers of the nitrogen atoms in
(a) NH_3
(b) NO
(c) NO_2^-
(d) NO_3^-
(e) Does the location of nitrogen near the middle of the second period of the periodic table help to explain your results in parts (a) through (d)? Explain why or why not.

6.55. What is the oxidation number of the metal in these metal-ion complexes?
(a) $PtCl_4^{2-}$
(b) $Cu(NH_3)_4^{2+}$
(c) $Fe(CN)_6^{3-}$
(d) MoS_4^{2-}
(e) $Zn(OH)_4^{2-}$

6.10. Balancing Reduction–Oxidation Reaction Equations

6.56. For each of these oxidation–reduction reactions, balance the reaction and identify the reactant oxidized and the reactant reduced.
(a) $MnO_4^-(aq) + SO_3^{2-}(aq) + H^+(aq)$
$\rightarrow Mn^{2+}(aq) + SO_4^{2-}(aq) + H_2O$
(b) $NO_3^-(aq) + Zn(s) + H^+(aq)$
$\rightarrow NH_4^+(aq) + Zn^{2+}(aq) + H_2O$
(c) $Cl_2 + OH^-(aq) \rightarrow ClO_3^-(aq) + Cl^-(aq) + H_2O$

6.57. Zinc–air cells use oxygen from the air to create electrochemical energy. Ambient oxygen gas, O_2, is converted in aqueous solution to hydroxide ion, OH^-, and zinc metal is converted to zinc oxide, ZnO, during discharge of a zinc–air cell.
(a) What is being oxidized in this reaction and what is being reduced as this cell discharges? What are the changes in oxidation number taking place?
(b) Write an overall chemical equation for the reaction that takes place in this cell.
(c) How many moles of electrons are being transferred per mole of zinc converted to zinc oxide?

6.58. These redox reactions occur in aqueous acidic solutions. Identify which element is oxidized and which

reduced in each case. Explain how you make these identifications. Balance each reaction by both the oxidation-number and half-reactions methods and compare the results. Which method do you prefer? Explain the reason(s) for your preference.
(a) $S^{2-}(aq) + NO_3^-(aq) \rightarrow NO_2(g) + S_8(s)$
(S_8 rings are the common form of elemental sulfur.)
(b) $Hg^{2+}(aq) + NO_2^-(aq) \rightarrow Hg(l) + NO_3^-(aq)$
(c) $Fe^{2+}(aq) + NO_3^-(aq) \rightarrow Fe^{3+}(aq) + NO(g)$

6.59. These redox reactions occur in aqueous basic solutions. Identify which element is oxidized and which reduced in each case. Explain how you make these identifications. Balance each reaction by either the oxidation-number or half-reactions method, whichever you prefer.
(a) $S^{2-}(aq) + I_2(s) \rightarrow SO_4^{2-}(aq) + I^-(aq)$
(b) $MnO_4^-(aq) + C_2O_4^{2-}(aq) \rightarrow MnO_2(s) + CO_2(g)$
(c) $Bi(OH)_3(s) + Sn(OH)_3^-(aq)$
$\rightarrow Bi(s) + Sn(OH)_6^{2-}(aq)$

6.60. The reaction that occurs when solid cobalt(II) sulfide dissolves in nitric acid can be represented as

$$CoS(s) + NO_3^-(aq) \rightarrow Co^{2+}(aq) + NO(g) + S_8(s)$$
(S_8 rings are the common form of elemental sulfur.)

(a) Identify which element is oxidized and which reduced in this reaction. Explain how you make these identifications.
(b) Balance the reaction. *Hint:* You might find it easier to treat elemental sulfur as S to balance the equation and then multiply by eight to get S_8 as the product.

6.11. Reduction–Oxidation Reactions of Carbon-containing Molecules

6.61. Lactic acid, $CH_3CH(OH)COOH$, is produced in your muscles during strenuous exercise and converted by your cells into pyruvic acid, $CH_3C(O)COOH$, as the muscles rest. Is this conversion an oxidation or a reduction reaction? Explain your reasoning.

6.62. Cider (apple juice) can ferment to form some ethanol and carbon dioxide in solution. This hard (ethanol-containing) cider can react with oxygen from the air to form cider vinegar, a dilute solution of ethanoic (acetic) acid. Write a balanced equation for the vinegar-forming reaction. *Hint:* It might be helpful to consider that an oxygen molecule can gain four electrons and produce two oxide ions.

6.63. When wood is burned in a wood stove, water is always one product. However, carbon dioxide, carbon monoxide, and/or free carbon particles are among the carbon-containing products that form when wood cellulose burns. Wood cellulose is a polymer of glucose, $C_6H_{12}O_6$.
(a) Write three separate equations showing the different carbon-containing products (plus water) that can form if one mole of glucose in wood burns with oxygen gas from the air.
(b) Use these three equations to help explain why it is dangerous to use a wood stove in an area that is not properly ventilated.

6.64. *With respect to the carbon atom on the right in each structure*, identify each of these reactions as an oxidation, a reduction, or neither.

(a) $H_2C{=}CH_2 \rightarrow H_3C{-}C\overset{\displaystyle O}{\underset{\displaystyle OH}{\diagup\!\!\diagdown}}$

(b) $\overset{\displaystyle HO}{\underset{\displaystyle OH}{\diagdown}}H_2C{-}CH_2 \rightarrow H_2C{=}CH_2$

(c) $H_3C{-}C\overset{\displaystyle O}{\underset{\displaystyle H}{\diagup\!\!\diagdown}} \rightarrow H_2C{=}CH_2$

(d) $H_2C{=}CH_2 \rightarrow H_3C{-}\overset{\displaystyle OH}{\overset{\displaystyle |}{CH_2}}$

6.13. Extension—Titration

6.65. Solid potassium permanganate, $KMnO_4$, dissolves to form an intensely colored pink solution in water. $KMnO_4(aq)$ solutions are often used to oxidize other substances. Depending on the pH of the solution, the permanganate ion, $MnO_4^-(aq)$, may be converted to the colorless $Mn^{2+}(aq)$ ion, to brown $MnO_2(s)$ in neutral or slightly alkaline solution, or to pale green $MnO_4^{2-}(aq)$ in highly alkaline solutions.
(a) Identify the oxidation number of manganese in each of these species and explain why the permanganate ion gains electrons in each of these possible reactions.

(b) Write the reduction half-reaction for each of these changes.
(c) In each of these three cases, what visual clues might you have to determine if the reduction reaction was completed? These visual changes can make potassium permanganate useful in titration reactions to determine the concentration of species that reduce the permanganate ion. What limitations are there to these visual methods for finding the endpoint of a titration?

6.66. The reaction of dichromate ion with iron(II) ion is a standard method for analyzing iron in samples such as ores:

$$Cr_2O_7^{2-}(aq) + Fe^{2+}(aq) \rightarrow Cr^{3+}(aq) + Fe^{3+}(aq)$$

A chromium atom has six valence electrons; dichromate is

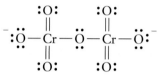

(a) What is the oxidation number of the chromium atoms in the dichromate anion? Explain how you get your answer.
(b) Balance the reaction equation (in acidic solution).
(c) A 0.178-g sample of iron ore is dissolved in acid and treated to convert all the iron present to the iron(II) oxidation state. When titrated with a 0.0100 M solution of dichromate, 34.57 mL of the dichromate solution is required to reach the equivalence point. What is the mass percent of iron in the ore?

General Problems

6.67. When potassium dichromate, $K_2Cr_2O_7$, an orange solid, is dissolved in an aqueous basic solution, a yellow solution containing chromate ion, $CrO_4^{2-}(aq)$, is formed. Write a balanced chemical equation that describes the reaction forming the chromate from dichromate. Is this a precipitation, Lewis acid–base, or oxidation–reduction reaction? Clearly state the reasoning for your response.
Hint: See Problem 6.66 for the structure of the dichromate anion.

6.68. Consider this table of data.

Metal ion	Ratio of ionic charge/ionic radius
Li^+	1.5
Ca^{2+}	2.1
Mg^{2+}	3.1
Al^{3+}	6.7

(a) Use these data to predict which ion is expected to act as the strongest Lewis acid toward water in aqueous solution. Explain your reasoning.
(b) How can you rationalize the relationship between the ratio of ionic charge to ionic radius for Ca^{2+} and Mg^{2+}?
(c) Given that sodium ion acts as a spectator ion in aqueous solution and has no significant acid–base properties, what do you predict will be the ratio of its ionic charge to its ionic radius?

6.69. 🖑 (a) To make nylon fibers, the solid polymer is melted and then extruded through very fine holes in spinnerets, as shown in the chapter opening illustration. The fine strands from the spinnerets are wound into threads that are used to make nylon fabric. The strength of nylon fibers is largely a result of the many hydrogen bonds between long individual polymer molecules aligned parallel to one another as illustrated in the *Web Companion*, Chapter 6, Section 6.7.2, movie. How do you think the extrusion process works to produce these aligned, hydrogen-bonded polymers?
(b) Most silk is obtained from silkworms, which produce essentially one kind of protein polymer to make their cocoons. Spiders, on the other hand, need silk for a variety of tasks and produce several different kinds of silk, some of which are stronger than an equivalent steel wire. They do this using aqueous suspensions of different protein polymers and extruding different mixtures of these polymers, depending on the task for which it is used. The fibers are formed as the mixtures are extruded through their spinnerets, as shown in the chapter opening illustration. Why do you think spiders have evolved to use mixtures like this, rather than producing each type of silk as a separate unique polymer?

Watt and Stephenson whispered in the ear of mankind their secret, that *a half-ounce of coal will draw two tons a mile . . .*

THE CONDUCT OF LIFE, "WEALTH"
RALPH WALDO EMERSON (1803–1882)

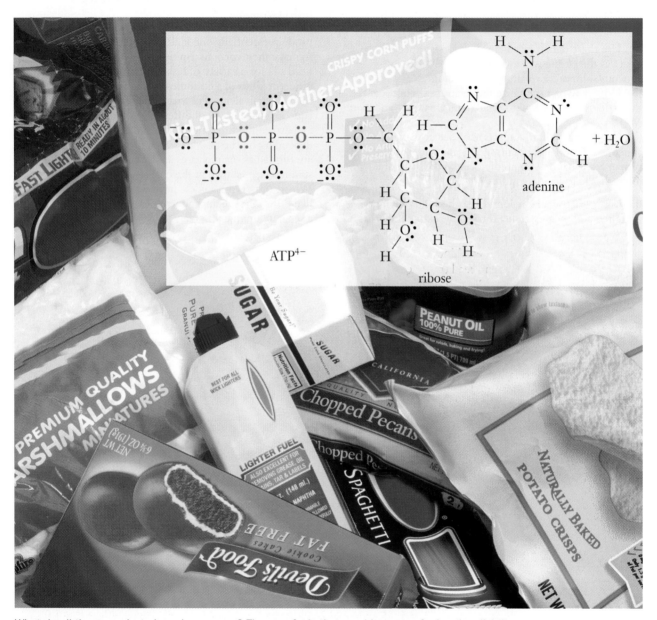

What do all these products have in common? They are fuels that provide energy for heating, lighting, and life. The overall reaction of food molecules with oxygen that is carried out in your body can be carried out in the laboratory by burning the food. In your body, a good deal of the energy released in this reaction is captured to produce adenosine triphosphate, ATP, molecules that you use to provide the energy for many of your life processes.

Chemical Energetics: Enthalpy

I n the mid-19th century, Emerson marveled that "a half-ounce of coal [would] draw two tons a mile. . . ." He was referring to the steam locomotive that burned a small amount of coal to boil water to drive pistons to produce the power to move large loads (Figure 7.1). From the perspective of the early 21st century, it is difficult to imagine the excitement that accompanied the development of engines that could replace waterpower and draft animals to do humanity's heavy lifting and hauling. Engines that transformed chemical energy into mechanical energy fueled the Industrial Revolution, which transformed everyday life for most people on Earth.

Now we take for granted vehicles that use the energy released when hydro-carbons burn and nuclear power plants that use the energy released from the fission of uranium-235 (Chapter 3, Section 3.5) to produce electricity. It is easy to lose sight of the fact that *everything we do requires a source of energy*. A variety of these energy sources, including some that supply the metabolic energy you need to stay alive, are shown in the illustration on the facing page. Every change that occurs involves the exchange of energy among substances. The same principles of energy exchange govern everything from the movement of locomotives to the movement of your eyes as you read this page.

These principles were discovered by scientists and engineers trying to understand what governed the amount of movement that heat (steam) engines could produce. The branch of science they developed to quantify the rela-tionships between thermal and mechanical energy is called **thermodynamics.**

The name *thermodynamics* (*therme* = heat + *dynamis* = power) reflects its practical origin.

Figure 7.1.

George Stephenson's steam locomotive, *The Rocket*. Invented in 1829, this locomotive introduced innovations shared by all steam locomotives since that time.

Our emphasis in this chapter will be on the law of conservation of energy, also known as the first law of thermodynamics, which provides us with powerful tools for investigating heat and work and relating them to energy changes in chemical reactions.

7.1. Energy and Change

7.1 INVESTIGATE THIS

What happens when foods are put in a flame?

Do this as a class investigation and work in small groups to discuss and analyze the results. Investigate several foods, like those shown in the chapter-opening illustration, that can provide you with energy. Use several pieces of uncooked spaghetti, one mini-marshmallow, a large potato chip, a pan partially filled with water, and a flame (matches, lighter, candle, or burner).

Take a full-length piece of spaghetti and hold it by one end over the pan of water. Bring the flame under the free end of the spaghetti and hold it there until the appearance of the spaghetti begins to change—then remove the flame. If necessary, extinguish the spaghetti by immersing it in water. Record your observations.

Repeat this procedure with a marshmallow held on the end of another full-length piece of spaghetti and then again with a large potato chip held in the flame.

You might investigate other foods as well.

7.2 CONSIDER THIS

Is energy involved when foods are burned?

(a) What changes did you observe in Investigate This 7.1? Was there any evidence that energy was involved in these changes? What evidence? What kinds of energy? Do your observations help you answer the question posed in the chapter opener?

(b) One function of food is to provide metabolic energy. Based on your observations for burning different foods, do you think the results are in any way indicative of their different food energy values? Compare your observations and interpretations with those of your classmates.

Your body does not burn food for energy in the way you burned foods in Investigate This 7.1, but the result is the same. When carbon–hydrogen–oxygen compounds, such as those in the carbohydrates and fats in the foods used in the investigation, are burned completely, the final products are carbon dioxide and water. When carbon–hydrogen–oxygen compounds are metabolized completely for energy, the final products are carbon dioxide and water. The *change is the same* in each case.

Measuring energy Although the caloric content of a food, as displayed on its Nutrition Facts label, is a measure of the amount of energy available from oxidation of the food, it is *not* determined by feeding the food to someone and measuring her or his increase in energy. The energy value is determined by burning the food under carefully controlled conditions. The thermal energy released (in Calories per gram) during complete combustion is the value that typically appears on the food label.

In chemistry, the international energy unit, the joule, J, is most often used to report energy measurements. We will express all energies in joules or kilojoules, kJ (1 kJ = 1000 J). An older metric unit, the calorie (lowercase c), is also still widely used, especially in the life sciences. The **calorie** is now defined as 4.184 J. The nutritional Calorie (uppercase C) on food labels, which is what you count when dieting, is 1000 cal = 4184 J.

Energy conservation and chemical changes Overall chemical reactions, especially those in living systems, typically proceed through several steps in the complete journey from reactants to products. Intermediate reactions are often difficult to study directly, because the transitional compounds are so rapidly converted to the next products. However, it is possible to measure accurately the overall energy change for a reaction without knowing anything about the individual steps, and therein lies much of the power of thermodynamics.

During the Industrial Revolution, scientists, engineers, and inventors were seeking the most efficient means of converting thermal energy to mechanical energy. Scientist–inventors, including James Watt and George Stephenson (heroes of the Emerson quotation that opens the chapter) had begun to build locomotives and engines by improving the atmosphere-powered steam engine invented in 1705 by Thomas Newcomen. These practical thermodynamicists wanted to convert as much thermal energy to mechanical energy as possible.

Along the way, these scientists discovered a great deal about heat, work, and other forms of energy. James Joule (British physicist, 1818–1889) discovered the fundamental principle that *when any change occurs, energy is always conserved.* Energy does not appear from nowhere nor does any of it disappear. Only the *form* of the energy changes (as between mechanical energy and thermal energy). This principle, the **first law of thermodynamics,** applies to chemical reactions as well as other kinds of change.

The results of many measurements over many years have been gathered, tabulated, and published, making it possible for you to compare quantitatively the energy changes associated with different processes. Using only tabulated energy data, for example, you can determine the amount of energy associated with the bonds in molecules and you can predict energy changes for chemical reactions that have never been carried out in the laboratory. With the conceptual and mathematical tools presented in this chapter, you can learn how to do these analyses.

7.3 CONSIDER THIS

How is the release of thermal energy related to chemical bond energy?

In Investigate This 7.1, you observed energy released in the form of heat and light from burning foods. What conclusion(s) can you draw about the strengths of the chemical bonds of the reactants and products in these reactions? Explain your reasoning. Remember that breaking bonds requires energy and forming bonds releases energy.

You have already begun your analyses by considering the energy released when foods are burned. The products are the same, water and carbon dioxide, when fats and carbohydrates are burned. However, fatty foods generally burn more steadily and release more energy than carbohydrates—starches and sugars. The greater the energy released, the greater the difference between the energy released in product bond formation and the energy required for reactant bond breaking. Since more net energy is released when fats burn, the bond breaking in fats must take less energy relative to product bond formation than bond breaking in carbohydrates. You might conclude that the bonds in fats are weaker than the bonds in carbohydrates. We will extend this discussion of bond energies in Section 7.7. First, let us consider two forms of energy, heat and work, and their interconnections.

7.2. Thermal Energy (Heat) and Mechanical Energy (Work)

7.4 INVESTIGATE THIS

What happens when a pinwheel is held over a flame?

Do this as a class investigation and work in small groups to discuss and analyze the results. Use a candle, an open-ended metal cylinder clamped vertically, and a pinwheel.

(a) Hold your hand about 15 cm above the flame and note how warm the combustion gases coming from the flame feel. Hold the pinwheel about 15 cm above the flame and observe its motion, if any.

(b) Move the metal cylinder over the candle so that it serves as a chimney. Wait a few moments for the metal to become warm. Hold your hand above the top of the chimney and note how warm the combustion gases coming from the flame feel. Hold the pinwheel above the top of the chimney and observe its motion, if any.

7.5 CONSIDER THIS

What causes a pinwheel over a flame to turn?

(a) In Investigate This 7.4, did you feel a difference in the temperature of the combustion gases without and with the chimney in place over the flame? If so, which felt warmer? Try to explain why there was a temperature difference or why there was not.

(b) Did the pinwheel turn faster without or with the chimney in place over the flame? Try to explain why there was a difference or why there was not.

(c) Is there any correlation between your answers in parts (a) and (b)? If so, are your explanations consistent with the correlation? Explain how.

Two familiar forms of energy are **heat** (**thermal energy** transferred from a warmer to a cooler object) and **work** (**mechanical energy** that moves an object from one place to another). The usual symbols we use for quantities of heat and work are *q* (heat) and *w* (work). In this chapter, we will consider the thermal and mechanical energy transfers associated with chemical reactions. Transfers of thermal and mechanical energy from one substance or place to another can be measured in the laboratory and related to the chemical reactions being studied.

Directed and undirected kinetic energy Thermal energy and mechanical energy are two types of *kinetic* energy. Thermal energy results from the *undirected motions* of individual atoms, ions, and molecules. The molecules are moving chaotically in every direction. Mechanical energy results from *directed motion*, the net movement of the atoms, ions, and molecules in an object in the same direction. The chimney and pinwheel that you used in Investigate This 7.4 illustrate the difference between undirected kinetic energy and directed kinetic energy. We can use Figure 7.2 to help understand this example.

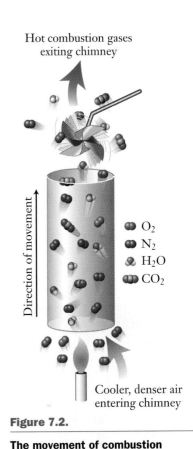

Figure 7.2.

The movement of combustion gases. The velocity (speed and direction) of each molecule is shown by its motion tail, whose length indicates the speed. The molecules rising in the chimney are a combination of combustion products and molecules from the air. The molecules are moving in all directions, but there is net upward motion as they are buoyed up by the denser gases that enter the chimney at the bottom.

(Figure labels: Hot combustion gases exiting chimney; Direction of movement; O_2; N_2; H_2O; CO_2; Cooler, denser air entering chimney)

7.6 CHECK THIS

The difference between warm and cool gases

If we had not labeled the cooler gas in Figure 7.2, would you still be able to tell that it is cooler? Explain why or why not.

The energy released by the flame is mostly thermal energy. (Energy is also released as light, but we are not considering light here; it represents only a small part of the total energy.) The energy in the molecular motions of the hot combustion gases is their thermal energy. Thermal energy is transferred through collisions to the other gases in the air. The expanding hot gases are buoyed up the chimney by cooler, denser air that enters at the bottom of the chimney. The net upward movement of the gases depicted in Figure 7.2 is directed kinetic energy and can do work. The rising gases do work on the pinwheel, causing it to turn.

7.7 CONSIDER THIS

How do thermal and mechanical energies interact?

Clearly, the rising combustion gases from the flame produce a change in the pinwheel when it is placed above the chimney in Investigate This 7.4. If the gases produce a change in the pinwheel, would you expect the spinning pinwheel, in turn, to produce any change in the gases? If so, what kind of change? How might the properties of the emerging gases differ between when the pinwheel is there and when it is not? Discuss the possibilities in small groups and then share your ideas with the class.

The pinwheel turns because the moving gas molecules strike the vanes and transfer some of their kinetic energy to them. These gas molecules now have lower kinetic energy, so they are cooler than before they struck the pinwheel and did work on it. Such interplays of heat and work interest engineers who concern themselves with how to build machines that generate maximum work (*w*, directed kinetic energy) and minimum heat (*q*, undirected kinetic energy) from a change in the energy source. Chemists, on the other hand, concern themselves with understanding why chemical reactions occur as they do. That understanding, gained through the study of heat and work in chemical systems, provides the keys to making better materials, understanding life processes, producing energy more efficiently, minimizing environmental damage, and much more.

Chemical reaction energies In Chapter 5, you learned that the total energy of a molecule is a complex interplay of potential and kinetic energies that involves attractions and repulsions of electrical charges and the sizes of electron waves (orbitals). When molecules react to form different molecules, the total energy of the products is almost always different than the total energy of the reactants. Our problem is that the total energy of a molecular system cannot be measured directly. What we have to do is measure the energy *changes* that occur and try to infer from these changes something about the total energies of the molecules involved. We measure the energy changes in chemical reactions by measuring their effects on the properties of substances that are easy to observe, such as the temperature of a mass of water. We will concentrate on thermal energy and start by considering how it is exchanged between one substance and another.

7.3. Thermal Energy Transfer (Heat)

7.8 INVESTIGATE THIS

Can radiation change the temperature of water?

Do this as a class investigation and work in small groups to discuss and analyze the results. Use a bright incandescent light, two flat-sided containers filled with water, and two thermometers set up as shown in the photograph. Record the temperature of the water in the containers at the beginning and after 10 minutes in the light beam. Remove the container nearer the light source and read the temperature in the remaining container after another 10 minutes.

7.9 CONSIDER THIS

How does radiation change the thermal energy of water?

(a) How did the temperature of the water in the two containers in Investigate This 7.8 compare after 10 minutes in the light beam? Is there evidence that the thermal energy of the water in one or both of the containers changed? Explain.

(b) After 10 further minutes in the light beam (without the nearer container), what happened to the temperature of the water in the remaining container? Is there evidence for further change in the thermal energy of the water in this container? Explain why you think you obtained the results you did.

The early thermodynamicists conceived of heat as a fluid that flowed from one substance to another. The fluid was called *caloric* from the Latin word for heat. This imaginary fluid is the obvious source of the name calorie and of such terminology as *heat flow*. Our present understanding is that thermal energy, the chaotic motion of atoms and molecules, can be transferred from one substance to another in two ways: by electromagnetic radiation emitted by the warmer body and absorbed by the cooler, or by contact of the two substances. To be consistent with our present model, we will use terminology such as *thermal energy transfer*, or just *heat*, and not use the term *flow* when we discuss energy changes.

Thermal energy transfer by radiation You may have had the experience of sitting in front of bonfire or a fire in a fireplace and having your face feel hot while your back felt cool. What you were feeling on your face was the **radiant energy** from the glowing fire. Thermal energy was being transferred to you by the long wavelength, **infrared electromagnetic radiation,** from the flame. The radiant energy was absorbed by the molecules in your skin and caused them to move faster (increased their temperature). The thermal energy of your face was greater than it would have been if it were sensing just the temperature of the air. Radiation travels in straight lines and, since your body was in the way, couldn't reach your back, which simply sensed the temperature of the air.

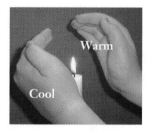

Radiant heating

Review Chapter 4, Sections 4.3 and 4.4, and Figures 4.13 and 4.14, for electromagnetic radiation wavelengths and energy emission by glowing bodies.

7.10 CONSIDER THIS

How does radiation transfer energy to water?

(a) Explain your results from Investigate This 7.8 in terms of absorption of infrared radiation from the light beam by the water. Can you see any difference in the light beam before and after it enters the water? Why or why not?

(b) If you were to place your hand in the light beam before it passed through the containers of water and then in the beam after it passed through the water, would you feel any difference? Explain your response. Try the experiment.

Thermal energy transfer by contact The molecules in a warmer substance are moving more rapidly, on the average, than those in a cooler substance. In collisions between molecules of the two substances the more energetic, warmer molecules will cause the molecules of the cooler substance to start moving faster. For example, we said that the hot combustion product gases in Investigate This 7.4 transfer some of their thermal energy to the surrounding air by collisions with the air molecules. In this energy transfer process, the cooler substance will get warmer. And, since energy is conserved, the warmer substance will get cooler. Two substances at the same temperature exchange thermal energy, but there is no *net* energy transfer and thus no change in their temperatures.

These energy transfers take place by two pathways: conduction and convection. You have experienced thermal energy transfer by conduction and convection, Figure 7.3, even if these are not familiar terms. When you touch a pan of water heated on a hot stove, thermal energy from the burner is *conducted* to your skin by the metal or glass walls of the pan. In **conduction,** Figure 7.3(a), thermal energy from a hot source causes particles in the material in contact with the source to vibrate faster. These particles, in turn, collide with adjacent particles and start them vibrating faster and so on through the conductor. The particles stay fixed in place as the energy is transferred.

Conduction is from Latin that means "to *lead* together." Convection is from Latin that means "to *carry* together."

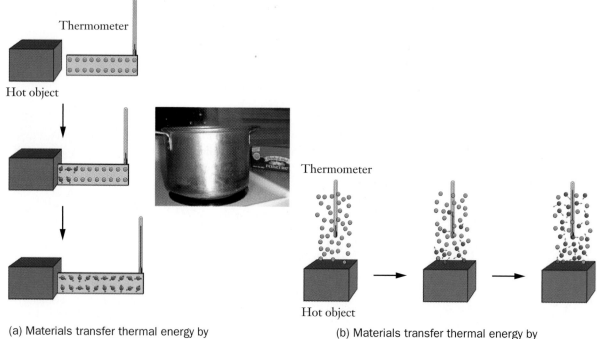

(a) Materials transfer thermal energy by conduction as vibrational energy (red) is transferred from one particle to the next.

(b) Materials transfer thermal energy by convection as the more energetic particles (red) move from one place to another.

Figure 7.3.

Conduction and convection of thermal energy.

In a fluid medium (gas or liquid), the particles can move about and **convection,** Figure 7.3(b), can transfer energy from one place to another. The particles initially in contact with the hot source move away from it and set up bulk movement in the fluid. As the pan of water is heated, the cooler denser water from the top sinks to the bottom displacing the warmed less dense water; *convection* currents are set up in the water which help transfer thermal energy from the burner throughout the liquid.

7.11 CONSIDER THIS

How are conduction and convection the same? different?

(a) What is(are) the difference(s) between conduction and convection? Explain how the difference(s) is(are) illustrated at the molecular level in Figure 7.3.

(b) Could the results in Investigate This 7.8 be caused by convection and/or conduction of thermal energy? What tests might you do to find out?

A similarity between conduction and convection is that atoms or molecules have to contact one another to transfer kinetic energy (motion) from one to another. In conduction, the motions are vibrations of the particles in the lattice and in convection the motions are mainly translations through space. In a metal pan on a stove, Figure 7.3, thermal energy is transferred from the burner to the entire pan without any atoms moving from one lattice position to another in the electric heating element or the pan. In the water in the pan and in the air and combustion gases in Figure 7.2, faster moving molecules move from one place to another and, if they collide with other molecules, can get them moving faster as well. Transfer of thermal energy does not require a mysterious substance (caloric); the transfer occurs by way of electromagnetic radiation or particles actually contacting one another.

Reflection and Projection

The concepts of thermodynamics were discovered or formulated mainly by scientists and inventors seeking ways to maximize the amount of work and minimize the amount of heat involved in an energy change. Heat and work are manifestations of the kinetic energy of molecular motion. An important thermodynamic concept, the first law of thermodynamics, is that energy is conserved in every process/change that occurs, including chemical reactions.

Thermal energy can be transferred from one substance to another by radiation, conduction, and/or convection. The temperature of the substance that loses the thermal energy decreases. The temperature of the substance that gains the thermal energy increases. Energy is conserved in these transfers. Often, however, we can bring about a change in the thermal energy of a substance without transferring thermal energy from another substance. One way to do this is by doing work on the substance of interest. Since there is more than one way to bring about changes in the thermal energy of a substance, we need to examine how overall changes depend upon the pathway for change. We also have to be clear about what parts of the environment we need to take into account when analyzing a change. These are the topics of the next sections.

7.4. State Functions and Path Functions

7.12 INVESTIGATE THIS

What happens when you rub your hands together?

Put the palms of your hands together and rub them together briskly for 3–4 seconds. Besides the rubbing, what sensation do you feel on your palms?

7.13 CONSIDER THIS

What change occurs when you rub your hands together?

(a) Rubbing your hands together, as you did in Investigate This 7.12 and as you have done when your hands are cold, warms your skin. What change must have occurred to the molecules in your skin to cause this sensation? Explain your reasoning.

(b) How do you think rubbing your hands together brings about the change in part (a)?

(c) Is there another way you could get the same sensation without rubbing your hands on something? What would bring about the change in the molecules in this second case? Explain.

One of the characteristics of thermodynamics is careful definition of the properties of thermodynamic variables such as temperature, thermal energy, heat, work, and so forth. Usually we will take an approach that depends on what seems to make sense and can be described in terms of everyday experiences or easily observable effects. But there are some distinctions that can help you understand why we use different variables that seem to stand for the same quantity. One fundamental distinction is between variables that are state functions and variables that are path functions.

> The term *state function* can be confusing, since we also commonly refer to different physical *states* (solid, liquid, or gas) of matter. The description of its present condition, the thermodynamic state of a system, has to take into account the physical states of its components, but more fundamentally, it must account for the energy relationships among the components.

State functions   Variables whose value depends only on the state (present condition) of the system under study and not at all on how it got into that state are **state functions.** Some important state functions are temperature, pressure, and volume; we will meet others as we go along. A physical analogy for state functions is altitude above sea level. If you are on top of Pike's Peak, Figure 7.4, your altitude above sea level, 3.8 km, is the same whether you drove to the top, walked to the top, or took the cog railway. State functions are like altitude. It makes no difference how the collection of molecules arrived at their particular state. What is important is that they are in that state.

Figure 7.4.

Two ways to get to the top of Pike's Peak. An auto road winds up the mountain in a series of switchbacks. The cog railway climbs more steeply and more directly.

If the state of a collection of molecules changes, the changes in its state functions depend only on the initial and final state of the molecules. Again using the altitude analogy, if you descend Pike's Peak from 3.8 km above sea level (initial state: height = h_i) and end your descent 2.1 km above sea level (final state: height = h_f), your change in state (altitude), Δh, is

$$\Delta h = h_f - h_i = 2.1 \text{ km} - 3.8 \text{ km} = -1.7 \text{ km} \tag{7.1}$$

When the final state (h_f in this analogy) has a less positive value than the initial state (h_i), the numeric value for the change of state (Δh) has a negative sign, just as you have often seen for changes in energy in previous chapters.

7.14 CONSIDER THIS

How do different pathways for descent of Pike's Peak compare?

Imagine that you descend Pike's Peak 1.7 km in altitude by car and a friend takes the cog railway down. Will you both travel the same linear distance on the ground, as measured, for example, by the odometer on your car? Explain why or why not.

Path functions Numeric values for some variables are dependent on the path taken to get from the initial to the final state; these are **path functions.** In Consider This 7.14, for example, to descend 1.7 km, you will travel a longer distance in a car down the winding road than your friend will coming down the straighter cog railway. The actual distance traveled is a path function, whereas the difference in altitude is a state function. In thermodynamic systems, heat and work are path functions. When you rubbed your hands together in Investigate This 7.12, they felt warmer. The molecules in your skin were made to move more rapidly by the mechanical work you did rubbing your hands together. You could bring about this same change by holding your hands near a source of thermal energy, such as a flame. These two possible pathways from the initial state of skin at about 37 °C to warmer skin at about 40 °C are shown in Figure 7.5.

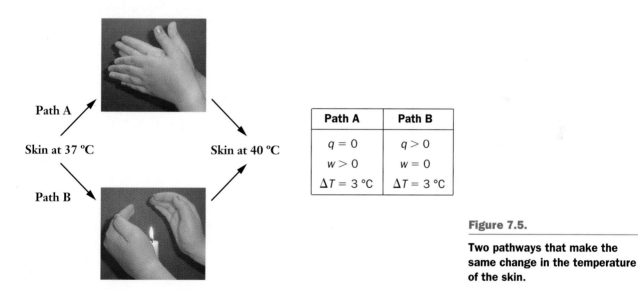

Path A	Path B
$q = 0$	$q > 0$
$w > 0$	$w = 0$
$\Delta T = 3\ ^\circ C$	$\Delta T = 3\ ^\circ C$

Figure 7.5.

Two pathways that make the same change in the temperature of the skin.

The change in the temperature of the skin, $\Delta T = 3\ ^\circ C$, is the same for both pathways, but the work and heat are not. Along path A, work (moving your hands) was done on the skin to raise its temperature. There was no thermal energy transferred to the skin from a warmer source; q is zero for path A. Along path B, the temperature of the skin changes as thermal energy is transferred by radiation, convection, and conduction from the flame to the skin. No mechanical energy moved the skin molecules; w is zero for path B. The chart in Figure 7.5 compares the two pathways.

How does the energy of the skin molecules change in Figure 7.5?

We have said that temperature is a measure of the average energy of the molecules in a system. When the energy of the molecules increases, the temperature increases. How do the energy changes of the skin molecules compare for the two pathways in Figure 7.5? Explain how you reach your conclusion.

In Section 7.1, we said that burning foods produces the same change as complete metabolism of the foods. The initial and final states are the same in both cases. However, the pathways are quite different and the amounts of heat and work are different, as we will find in Section 7.10.

7.5. System and Surroundings

You may have noticed that we often refer to "the system of interest," "the chemical reaction system," or some other reference to the "system." In Consider This 7.15, we referred to the "average energy of the molecules in a system." A thermodynamic analysis of a change always requires that we distinguish the collection of molecules that we are studying, the **system,** from its **surroundings,** everything else. Let's look at the different types of systems we will encounter.

The system in Figure 7.5

What is the system in Figure 7.5 that is referred to in Consider This 7.15?

Open, closed, and isolated systems If we were studying the contents of the aquaria in Figure 7.6, we could consider the aqueous solution, fish, plants, and gravel as our system. Everything else (the glass of the aquarium, the surrounding air, the table upon which the aquarium rests, the room, the building, the planet, the entire universe) is the surroundings. In the aquarium in Figure 7.6(a), energy and matter can move between the system and the surroundings. Atmospheric gases dissolve in the water. Thermal energy is exchanged by conduction and convection with the table and the air. Light (radiant) energy enters the system. Work could be done on the water by stirring it. These must be accounted for as part of the surroundings. This aquarium is an **open system** because matter and energy can be exchanged with the surroundings. All living things are open systems. They must take in matter from their surroundings to survive.

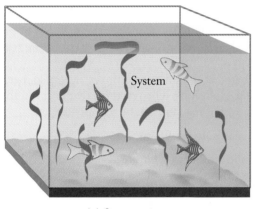

(a) Open system

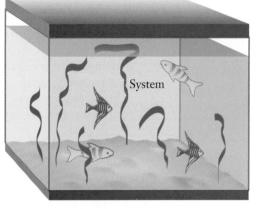

(b) Closed system

Figure 7.6.

System (the contents of the aquaria) and surroundings (everything else).

If we were to seal the top of the aquarium, as shown in Figure 7.6(b), no exchange of matter could occur with the surroundings. (We would now include the small amount of air trapped above the water as part of the system.) Thermal energy could still enter or leave the system, and light energy could still be absorbed and, if we had sealed a stirrer inside, we could still do work on the system by stirring the water. The sealed aquarium is a **closed system.** Closed systems do not exchange matter with their surroundings. Work can be done on them or they can do work on the surroundings, if their walls are not rigid (a moving stirrer paddle counts as a moving "wall"). Heat can enter or leave a closed system, depending on the temperature of the surroundings, if their walls are not thermally insulated.

If we were to wrap the entire sealed aquarium in Figure 7.6(b) in an opaque thermally insulating blanket, no heat could enter or leave the system. No exchange of matter could occur between the sealed system and the surroundings. If there is no stirrer in the thermally insulated and sealed aquarium, then no work could be done on the system. A system that can exchange neither work nor heat with its surroundings is called an **isolated system.** Table 7.1 summarizes the matter, heat, and work transfers that can occur between the surroundings and the different kinds of systems.

Table 7.1 *Transfers that can occur between a system and its surroundings.*

Type of system	Matter	Heat	Work
Open	yes	yes	yes
Closed	no	yes	yes
Isolated	no	no	no

7.17 CHECK THIS

Closed and open systems

(a) Think about the flame in Investigate This 7.4. Consider the fuel that burns as the system. Is the burning fuel a closed or open system? Explain the reasoning for your response.

(b) Ammonium nitrate, $NH_4NO_3(s)$, an important fertilizer, has also been responsible for some devastating explosions. The reactions that occur in the explosion are complex, but the overall reaction can be represented as

$$2NH_4NO_3(s) \rightarrow 2N_2(g) + 4H_2O(g) + O_2(g)$$

Consider an exploding sample of solid ammonium nitrate as the system. Is this a closed or open system? Explain the reasoning for your response.

A fuel burning or a sugar undergoing aerobic metabolism in an organism are examples of processes occurring in open systems. At least one of the reactants, oxygen, has to be supplied from the surroundings in both cases. By contrast, explosions generally can be classified as occurring in closed systems. All the atoms necessary for the reaction are together in a single compound (ammonium nitrate or trinitrotoluene, TNT, for example) or a mixture of compounds (a hydrogen–oxygen mixture, for example). We will meet further examples of both kinds of systems in this and the succeeding chapters.

7.18 CONSIDER THIS

How do systems and surroundings interact?

(a) If you hold a beaker containing a sample of ammonium chloride, $NH_4Cl(s)$, dissolving in water, your hand feels cold. (Recall Investigate This 2.22 in Chapter 2.) If the ammonium chloride and water are the system, what are the relevant surroundings? How do the surroundings interact with the system? Is the system open, closed, or isolated? Explain your responses.

(b) If the same process as in part (a) is carried out in a well-insulated container, what are the relevant surroundings? How do the surroundings interact with the system? Is the system open, closed, or isolated? Explain your responses.

(c) Imagine that you drop a hot piece of metal into a well-insulated container of room temperature water. If the metal is the system, what are the surroundings? How do the surroundings interact with the system? What do you expect will happen to the system and the surroundings? Is the system open, closed, or isolated? Explain your responses.

(d) Imagine carrying out the same process as in part (c), but now take the metal plus the water as the system. How do the surroundings interact with the system? What changes do you expect will occur in the system and the surroundings? Is the system open, closed, or isolated? Explain your responses.

(e) 🖱 Compare the figure in the *Web Companion*, Chapter 7, Section 7.6.1, with Figure 7.7 on page 457. Click on the cups and classify the parts of the set up as system or surroundings.

Reflection and Projection

In the last two sections we have made some important distinctions between different kinds of thermodynamic functions, between a system and its surroundings, and among the interactions of the system of interest with its surroundings. We found that the values for some functions, heat and work, in particular, can be different when a change occurs by different pathways. Functions that depend upon how a change is carried out are called *path functions*. The values of other functions depend only on the initial and final states of the system undergoing change and not on the pathway for the change. Functions that depend only on the initial and final state of the system are called *state functions*.

We will define what the systems are, as we use and analyze them in the remainder of this chapter and the rest of the text. We do this so that we can also be clear about what the surroundings are that we have to account for in our analysis of changes in the system. The matter, heat, and work transfers that can occur between the surroundings and open, closed, and isolated systems (Table 7.1) become clearer as they are applied to real chemical systems. We will begin by examining thermal energy transfers in more detail. The change in temperature of one substance can be used determine the amount of thermal energy it has gained (or lost) from another source, a chemical reaction, for example. This application of the first law is called calorimetry and is the topic of the next section.

7.6. Calorimetry and Introduction to Enthalpy

7.19 INVESTIGATE THIS

What happens when an acid and base are mixed?

For this investigation, use a small test tube and separate plastic pipets containing 1 M sodium hydroxide (NaOH) solution and 1 M hydrochloric acid (HCl) solution. Add about 1 mL of the NaOH solution to the test tube and then about 1 mL of the HCl solution. Gently touch the test tube near the bottom where the solution is. Record all your observations.

7.20 CONSIDER THIS

What causes the changes when an acid and base are mixed?

What was (were) your observation(s) when you touched the test tube after the acid and base were mixed in Investigate This 7.19? What process or reaction could be responsible for your observations? Explain how you reach your conclusion.

In Investigate This 7.19, we can designate the acid and base molecules (ions) that react with one another as the system. The surroundings are the water in which the acid and base are dissolved, the test tube, the air around it, and, when you touch the test tube, your finger. Everything else in the universe is also a part of the surroundings, but you use your common sense to make a judgment about what parts of it can have an effect on the system and with which parts the system can exchange thermal energy.

Since the amount of thermal energy transferred depends on the pathway for the process, we often use subscripts on q to designate the conditions under which the thermal energy transfer is measured. We are mostly interested in reactions that take place in open containers on the laboratory bench, as in Investigate This 7.19, or in biological systems. Since these reactions occur under conditions of constant atmospheric pressure, we designate the associated constant-pressure thermal energy transfer as q_P **(heat at constant pressure).**

Enthalpy The 19th-century thermodynamicists introduced a useful new energy function called **enthalpy, H.** The **enthalpy change** for a process or reaction, ΔH, *has the same numerical value as q_P for the process.* ΔH, which can be determined by measuring the thermal energy transfer at constant pressure, is the difference between the final value for the enthalpy, H_{final}, and the initial value of the enthalpy, $H_{initial}$:

$$\Delta H = H_{final} - H_{initial} = q_P \tag{7.2}$$

Note that the quantity q_P is the *difference* between enthalpies, not enthalpy itself. As you will see in Section 7.10, ΔH is not just another name for constant-pressure thermal energy transfer. The enthalpy of a chemical system corresponds to a reservoir of thermal energy.

When thermal energy is transferred *to* a system, q_P is positive—thermal energy is added to the system. The enthalpy of the system has increased, $H_{final} > H_{initial}$; therefore, from equation (7.2), ΔH is positive. Positive values for ΔH and q_P are characteristic of **endothermic** changes. On the other hand, if thermal energy *leaves* the system we are studying, q_P is negative—thermal energy is subtracted from the system. The enthalpy of the system decreases, $H_{final} < H_{initial}$, and ΔH is negative. Negative values for ΔH and q_P are characteristic of **exothermic** changes. Although total enthalpies cannot be measured directly, the amount of heat can be experimentally determined, as you will see in the rest of this section. A measured value of q_P enables us to determine the enthalpy change, ΔH, for the system under investigation.

The thermal energy effect of a reaction can often be determined by touching the reaction vessel to determine whether the reaction is exothermic (it will feel warm to the touch, because thermal energy is leaving the reacting system and is being added to the surroundings, including your skin) or endothermic (it will feel cool to the touch, because thermal energy is being added to the reacting system and is being taken from the surroundings). Several kinds of devices, called **calorimeters,** have been developed to measure thermal energy changes quantitatively.

7.21 CHECK THIS

Is the reaction of an acid with a base exothermic or endothermic?

(a) Write the net ionic reaction that occurs in Investigate This 7.19.

(b) Is this reaction exothermic or endothermic? Give your reasoning.

Constant-pressure calorimetry

7.22 INVESTIGATE THIS

What happens when urea dissolves in water?

Do this as a class investigation and work in small groups to discuss and analyze the results. Set up a simple constant-pressure calorimeter, like the one illustrated in Figure 7.7, containing 100. mL of room temperature water. While gently stirring the water, record its temperature every 15 seconds for about 2 minutes. Add 6.0 g of urea, $H_2NC(O)NH_2(s)$, to the water and continue stirring and recording the temperature of the solution for another three minutes.

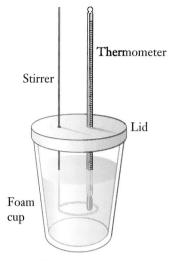

Thermometer

Stirrer

Lid

Foam cup

Figure 7.7.

A simple constant-pressure calorimeter.

7.23 CONSIDER THIS

What is ΔT when urea dissolves in water?

What were the initial and final temperatures of the water and solution in the calorimeter in Investigate This 7.22? What was $\Delta T = T_f - T_i$ for the urea dissolution reaction?

Figure 7.7 illustrates a simple **constant-pressure calorimeter,** a device that can be used to measure thermal energy changes in aqueous reactions. The setup requires only an insulated container, such as a Styrofoam® cup, an insulating lid, a thermometer, and perhaps a stirrer. Thermal energy produced (exothermic) or consumed (endothermic) by the reaction is transferred to or from the aqueous solution. You measure the resulting temperature change of the solution with the thermometer. If the calorimeter is perfect, no thermal energy will leave the calorimeter and none will enter from the outside. Under these conditions, the sum of all the thermal energy transfers occurring in the calorimeter must be zero:

$$q_{P(reaction)} + q_{P(solution)} = 0 \qquad (7.3)$$

Energy is conserved; the transfer of thermal energy *from* one component of the calorimeter contents must result in the transfer of thermal energy *to* another component. If the reaction is exothermic, then $q_{P(reaction)}$ has a negative value. Consequently, in order to satisfy equation (7.3), $q_{P(solution)}$ must have a positive value. A positive value for $q_{P(solution)}$ means that thermal energy is added to the solution, that is, it gets warmer.

7.24 CONSIDER THIS

What is observed for an endothermic reaction in a calorimeter?

(a) If an endothermic reaction is carried out in a calorimeter, is $q_{P(reaction)}$ positive or negative? Explain.

continued

(b) If the reaction is endothermic, what is the sign of $q_{P(\text{solution})}$? Will you observe the water to become warmer or cooler? Explain your reasoning.

(c) Is the dissolution of urea in water exothermic or endothermic? Explain how you can use your observations from Investigate This 7.22 to answer this question.

In practice, calorimeters are never perfect. The solution absorbs most of the heat from an exothermic reaction and provides most of the heat for an endothermic reaction, but some of the heat remains unaccounted for. This heat goes to heating (or cooling) the calorimeter itself (the container, thermometer, and stirrer). For our purposes, however, we are going to neglect this deviation from ideality. We will assume that all the thermal energy released or absorbed by the reaction in our calorimeter is absorbed or furnished by the solution, as in equation (7.3). The error we introduce by assuming that the calorimeter is ideal is usually only a few percent.

7.25 CHECK THIS

Energy transfers in a calorimeter

Carry out the animated reaction and the analysis of changes in the *Web Companion*, Chapter 7, Section 7.6.2. State in your own words what is going on in this calorimeter, the reasons for the observed changes, and the roles, if any, that thermal energy transfer by radiation, conduction, and convection play in these changes.

Temperature and thermal energy change In a calorimetric experiment, you measure the temperature change of the liquid. You then relate the temperature change to the thermal energy changes in the calorimeter. The equation linking the thermal energy transferred to a substance to the temperature change of the substance was derived in Chapter 1, Worked Example 1.57, although we didn't use these symbols there:

$$q_{P(\text{substance})} = (m)(c)(\Delta T) \tag{7.4}$$

Here, m is the mass of the substance (in grams), c is the specific heat of the substance (in joules per gram per degree Celsius temperature change, $J \cdot g^{-1} \cdot {}^{\circ}C^{-1}$), and ΔT is the temperature change in degrees Celsius, $T_f - T_i$. Recall that specific heat is defined as the energy required to raise the temperature of 1 g of a substance by 1 °C. The specific heat of a substance is a characteristic of that substance, just as its boiling point and melting point are. The specific heat of water is $4.18 \ J \cdot g^{-1} \cdot {}^{\circ}C^{-1}$. In most of our calculations, we will assume that the specific heat of dilute aqueous solutions is the same as that for water.

7.26 WORKED EXAMPLE

Determination of $q_{P(reaction)}$ for a reaction in a calorimeter

A 100.0 mL volume of 0.105 M acetic acid solution (0.0105 mol) and 100.0 mL of 0.12 M ammonia solution (0.012 mol) were mixed in a calorimeter like the one in Figure 7.7. The reaction that occurs is

$$CH_3C(O)OH(aq) + NH_3(aq) \rightarrow NH_4^+(aq) + CH_3C(O)O^-(aq) \quad (7.5)$$

The solutions were both at 22.50 °C before mixing. After mixing, the temperature rose to 23.15 °C. What is $q_{P(reaction)}$ for the reaction in the calorimeter? Assume that the specific heat of the solution is the same as the specific heat of water, $4.18\ J \cdot g^{-1} \cdot {}^{\circ}C^{-1}$, and that the density of the solutions is $1.00\ g \cdot mL^{-1}$.

Necessary information: We need equations (7.3) and (7.4).

Strategy: Substitute the experimental quantities into equation (7.4) to find $q_{P(solution)}$ and then substitute $q_{P(solution)}$ in equation (7.3) to get $q_{P(reaction)}$.

Implementation: The change in temperature is

$$\Delta T = (23.15\ {}^{\circ}C) - (22.50\ {}^{\circ}C) = 0.65\ {}^{\circ}C$$

The mass of the mixed solution in which reaction occurs is the sum of the masses of the original solutions, each of which is 100. g [= (100.0 mL) (1.00 g·mL^{-1})]; the mass of the mixed solution is 200. g. Thus, $q_{P(solution)}$ is

$$q_{P(solution)} = (m)(c)(\Delta T) = (200.\ g)(4.18\ J \cdot g^{-1} \cdot {}^{\circ}C^{-1})(0.65\ {}^{\circ}C) = 5.4 \times 10^2\ J$$

Rearrange equation (7.3) to get $q_{P(reaction)}$:

$$q_{P(reaction)} = -q_{P(solution)} = -5.4 \times 10^2\ J$$

Does the answer make sense?   To increase the temperature of 200. g of water one degree requires 836 J [= (200. g)(4.18 J·g^{-1}·°C^{-1})(1 °C)]. Here we found that a smaller amount of thermal energy, about 540 J, warms 200. g of solution somewhat less than one degree, so the answer is in the right direction and of the right magnitude.

7.27 CHECK THIS

Determination of $q_{P(reaction)}$ for dissolution of urea in a calorimeter

Use your data from Investigate This 7.22 to determine $q_{P(reaction)}$ for the dissolution of 6.0 g of urea in 100. mL of water. *Hint:* Assume that the density of water is 1.00 g·mL^{-1} and that the specific heat of the solution is the same as the specific heat of water. Remember that the mass of the solution is the mass of the water plus the solute.

Calculating molar enthalpy change In Worked Example 7.26 and Check This 7.27, values for $q_{P(\text{reaction})}$ were determined from experimental calorimetric data. We have said that $\Delta H_{\text{reaction}} = q_{P(\text{reaction})}$, so we also have $\Delta H_{\text{reaction}}$ values for these reactions under the conditions studied. Usually, however, we express enthalpy changes per mole of reaction, $kJ \cdot mol^{-1}$, where the coefficients in the balanced reaction equation are taken to be molar quantities. We can convert our measured values for $q_{P(\text{reaction})}$ (in kJ) to $\Delta H_{\text{reaction}}$ (in $kJ \cdot mol^{-1}$) by accounting for the number of moles of reactants that reacted in our experimental systems.

7.28 WORKED EXAMPLE

Determination of $\Delta H_{\text{reaction}}$ for a reaction in a calorimeter

Use the result from Worked Example 7.26 to calculate $\Delta H_{\text{reaction}}$ when 1 mol of acetic acid reacts with 1 mol of ammonia by reaction (7.5).

Necessary information: We need the volume and concentration data from Worked Example 7.26 to determine the number of moles of acetic acid and ammonia that reacted to give $q_{P(\text{reaction})} = -5.4 \times 10^2\,\text{J}$.

Strategy: Use the data in Worked Example 7.26 to find the number of moles of reactants that react and then figure out what $q_{P(\text{reaction})} = \Delta H_{\text{reaction}}$ would be if one mole reacted.

Implementation: Acetic acid and ammonia react in a one-to-one mole ratio and the data in Worked Example 7.26 show that there was an excess of ammonia (0.012 mol) compared to acetic acid (0.0105 mol) in the mixture. The acetic acid is the limiting reactant, so 0.0105 mol of each reactant reacts to yield $q_{P(\text{reaction})} = -5.4 \times 10^2\,\text{J}$.

$$\Delta H_{\text{reaction}} = q_{P(\text{reaction})} \text{ (per mole)} = \left(\frac{-5.4 \times 10^2\,\text{J}}{0.0105\,\text{mol}} \right) = -51\,\text{kJ} \cdot \text{mol}^{-1}$$

Does the answer make sense? About one hundredth of a mole of reaction occurs in the calorimeter and transfers about 540 J to the solution and calorimeter. One mole of reaction (one hundred times the amount of reaction in the calorimeter) would produce one hundred times as much thermal energy, about 54 kJ. This is approximately the result we got.

7.29 CHECK THIS

Determination of $\Delta H_{\text{reaction}}$ for dissolution of urea in a calorimeter

Use your result from Check This 7.27 and the data in Investigate This 7.22 to calculate $\Delta H_{\text{reaction}}$ when one mole of urea is dissolved to yield a solution of the same molarity as the final solution in the investigation. *Hint:* Assume that the volume of the final solution in the investigation is 100. mL.

Handling significant figures properly in experimental measurements is essential, if the results are to be trusted. Worked Examples 7.26 and 7.28 are good exercises in one of the more troublesome aspects of determining the number of significant figures to include in the results of calculations. In both cases, the numerical values for the experimental measurements were quite accurate, but the final result depended on a small difference between two large values (the measured temperatures). As you review the calculations to check your understanding of the concepts, also check your understanding of why the significant figures are reported as they are.

7.30 CONSIDER THIS

How can you get more accurate values for $\Delta H_{reaction}$?

Suppose you decide to try to get a more accurate value of $\Delta H_{reaction}$ for the reaction between aqueous acetic acid and ammonia solutions by doubling the amount of reactants used in the experiment described in Worked Example 7.26. When you carry out the reaction with 200. mL of each reactant solution you find that the temperature change is 0.66 °C.

(a) Is this the result you would expect? Why or why not?

(b) Use these data to determine $\Delta H_{reaction}$ (watch your significant figures) and compare it to the value from Worked Example 7.28. Is this the result you would expect? Why or why not?

(c) 🖱 Is the *Web Companion*, Chapter 7, Section 7.6.3, related to parts (a) and (b)?

Improving calorimetric measurements Trying to improve the precision of a calorimetric measurement by increasing the volumes of reactant solutions used does not work. As you discovered in Consider This 7.30, the amount of thermal energy produced increases, but, at the same time, the amount of solution increases by the same proportion. The increased thermal energy produces essentially the same temperature change in this greater mass of solution. You could work with more concentrated solutions, but they can also produce problems. There might be thermal effects from diluting the solutions when they are mixed or the specific heats of the solutions may be significantly different from water. If you do not account for such factors, your answer may be more precise (more significant figures) but less accurate (further from the true value).

The factor that can make the biggest improvement in calorimetric measurements is more precise temperature measurement, that is, use of a more sensitive temperature-measuring device. For the simple calorimeters we have been discussing, more sensitive thermometers are not justified. If the temperature measurements are improved, losses of thermal energy to the surroundings are large enough to show up as important factors. Better temperature measurements are justified and required in more sophisticated calorimeters.

7.31 CHECK THIS

Effect of increasing the amount of another calorimetric reaction

If the calorimetric experiment in Investigate This 7.22 is carried out by dissolving 12 g of urea in 100. mL of water, the final temperature of the solution is about 7 °C lower than the initial temperature of the water.

(a) Is this the result you would expect? Why or why not?

(b) How does this experiment differ from the one suggested in Consider This 7.30? Does the difference help explain the result? Why or why not?

Reflection and Projection

Calorimetry is an important application of the first law of thermodynamics. One common use is the determination of the thermal energy released or gained by chemical reactions in aqueous solution. The thermal energy transfers are between the system (the reacting molecules) and their surroundings (the liquid solution). The thermal energy transfer to the surroundings can be calculated from the specific heat of the liquid and the measured values for the mass and temperature change of the liquid. If the reaction is carried out in an insulated container (calorimeter) the sum of the thermal energy transfers in the calorimeter is zero.

Calorimetric measurements on reactions in solution are almost always carried out in open containers at a constant pressure of one atmosphere. The thermal energy (heat) transferred in the reaction is usually called q_P to indicate that the process is carried out at constant pressure. A new thermodynamic variable, enthalpy *(H)*, was introduced. A change in the system usually results in a change in the enthalpy, ΔH, which can be equated to the measured q_P for the change at constant pressure. Enthalpy changes in chemical reactions are a result of the breaking and making of bonds in the reactions. The enthalpies required to break bonds can be determined in various ways, including calorimetric experiments. The next sections show how these enthalpies can be used to calculate the enthalpy changes for reactions.

7.7. Bond Enthalpies

 7.32 INVESTIGATE THIS

What happens when yeast is added to hydrogen peroxide?

Do this as a class investigation and work in small groups to discuss and analyze the results. Put 2–3 g of dried baker's yeast into a 100-mL graduated cylinder. Add about 10 mL of 3% hydrogen peroxide, H_2O_2, solution to the cylinder. *CAUTION:* 3% hydrogen peroxide can damage both skin and clothes. Wear protective gloves and avoid splashing the liquid. Add about 1 mL of liquid

continued

dishwashing detergent and swirl gently to mix the contents of the cylinder. Gently touch the outside of the cylinder near the bottom, so you can feel any thermal effects. Observe any changes in either the yeast or the solution. Record your observations. Set the cylinder and its contents aside to use in Investigate This 7.35.

7.33 CONSIDER THIS

What reaction occurs when yeast is added to hydrogen peroxide?

What evidence do you have from Investigate This 7.32 that a reaction occurs when yeast is added to hydrogen peroxide? What do you think the product(s) of the reaction could be? How could you test for the presence of this(these) product(s)?

Chemical reactions: Bond breaking and bond making One of the goals in studying chemistry is to gain the ability to predict likely outcomes for chemical reactions and to calculate accompanying energy changes. In Chapter 5, we introduced covalent bonds, focusing mainly on the role of valence electrons in holding the atoms together. In chemical reactions of covalent compounds, at least some of the covalent bonds in the reactants break, the atoms rearrange themselves, and new covalent bonds form in the products. *The bond-breaking process requires an input of energy, and the bond-forming process releases energy.*

- *If stronger bonds replace weaker ones, the overall reaction is exothermic.*
- *If weaker bonds replace stronger ones, the overall reaction is endothermic.*

Experimental determinations of enthalpy changes for thousands of reactions provide the basis for enormous predictive power. Using only enthalpy data tables (introduced below), you can figure out whether any reaction you might imagine would release energy or consume energy.

We will use the reaction of hydrogen peroxide, Investigate This 7.32, as an example for analysis. You probably observed the formation of gas as a product of the reaction when the yeast was added to the peroxide. Some of the gas was trapped in the foam that filled the rest of the cylinder. You could also feel that the solution warmed up as the reaction occurred. The reaction is exothermic. If we can find out what the gaseous product(s) is(are), we will know more about the bond breakings and bond formations that occur in the reaction.

Hydrogen and oxygen are the only elements in the aqueous hydrogen peroxide, $H_2O_2(aq)$, solution; the most likely gases that can be formed are molecular hydrogen, $H_2(g)$, and/or molecular oxygen, $O_2(g)$. Possible reactions that could form these gases are

$$2H_2O_2(aq) \rightarrow 2H_2O(l) + O_2(g) \tag{7.6}$$

$$H_2O_2(aq) \rightarrow H_2(g) + O_2(g) \tag{7.7}$$

The bond rearrangements in reactions (7.6) and (7.7) are illustrated in Figure 7.8.

Figure 7.8.

Bonding rearrangements for possible reactions of hydrogen peroxide. The reactions produce (a) water and molecular oxygen gas or (b) molecular oxygen and hydrogen gases.

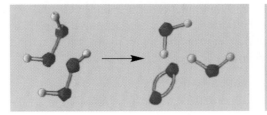

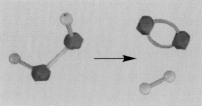

(a) $2H_2O_2(aq) \rightarrow 2H_2O(l) + O_2(g)$

(b) $H_2O_2(aq) \rightarrow H_2(g) + O_2(g)$

7.34 CHECK THIS

Properties of the possible H_2O_2 reaction products

(a) What is the composition of the gas produced by reaction (7.6)? What are its properties?
(b) What is the composition of the gas produced by reaction (7.7)? What are its properties?
(c) How could you test the product gas to distinguish between these possibilities?

7.35 INVESTIGATE THIS

What is the gas formed in the hydrogen peroxide reaction?

(a) Do this as a class investigation and work in small groups to discuss and analyze the results. Test the flammability of the gas trapped in the foam formed in Activity 7.32. *WARNING:* Place the cylinder behind a safety shield that protects others as well as yourself when the gas is tested. Light a long fireplace match, blow it out, and quickly insert the glowing match tip a few centimeters into the mouth of the cylinder while carefully observing the results.
(b) What gas is formed in the reaction? Explain the reasoning for your answer.

7.36 CONSIDER THIS

Which bonds break and form in the hydrogen peroxide reaction?

(a) What bonds are broken and what bonds are formed as the reactant is converted to products in reaction (7.6)? Make models of the reactants. Take them apart and reassemble them to make the products. How does the number of bonds broken compare with the number of bonds formed? How many two-electron bonds are present in the reactants? In the products?
(b) Do the same kind of model exercise and answer the same questions for reaction (7.7).

Bond enthalpy Before we can continue the analysis of the energetics of the hydrogen peroxide reaction, you will need to know more about the energies in chemical bonds. Then you can find out if your conclusion in Investigate This 7.35(b) is consistent with the exothermicity of the reaction observed in Investigate This 7.32. The model for covalent bonding developed in Chapter 5 is based on the idea that most chemical bonds consist of shared pairs of electrons being closely held between pairs of atoms (atom cores). The carbon atom in methane, CH_4, for example, has four electron-pair bonds to the four hydrogen atoms. The question for us here is: How energetically independent are the four bonds? We characterize the strength of a chemical bond in terms of its **bond enthalpy,** *the enthalpy required to break the bond.* If one carbon–hydrogen bond is broken, how are the bond enthalpies of the remaining three carbon–hydrogen bonds affected?

This is an important question. If the bond enthalpy for each bond is independent of all the other bonds in the molecule, we should be able to find enthalpy changes for chemical reactions using nothing but bond enthalpies. We would add together the enthalpies required to break the bonds in the reactants and subtract the enthalpies released when the bonds in the products form. If bond enthalpies are independent of one another, this calculation should give the same enthalpy change as that obtained by measuring the enthalpy change for the reaction with a calorimeter.

Homolytic is from homos = same + lyein = loosen.

Homolytic bond cleavage The independence of bond enthalpies can be tested by measuring the energy (enthalpy) it takes to break covalent bonds one after another in a gaseous molecule. (We study molecules in the gas phase because they are relatively far apart and independent of one another.) These experiments are carried out in such a way that each atom keeps one electron from each pair of valence electrons in the bond that is broken. This symmetric bond breaking is called **homolytic bond cleavage.** Table 7.2 shows the homolytic bond cleavage reactions and enthalpies for successive removal of each hydrogen atom from methane.

Table 7.2	*Enthalpy required for successive homolytic bond cleavages in methane.*

All reactants and products are gases.

Homolytic bond cleavage reactions	ΔH, kJ·mol^{-1}

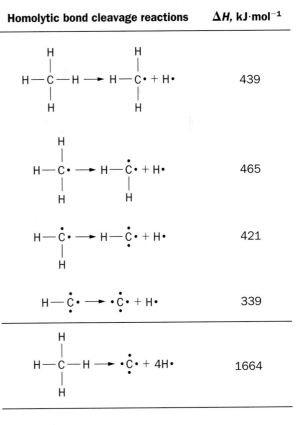

	439
	465
	421
	339
	1664

7.37 CHECK THIS

Overall homolytic bond cleavage reaction of methane

Show how to combine the four individual homolytic bond cleavages for methane, Table 7.2, to give the overall reaction and overall ΔH of reaction. Are all the bonding electrons in methane accounted for in the products? Explain your answer.

Free radicals are molecules with one or more one-electron, σ-nonbonding orbitals. Free radicals are quite reactive species. All of the reaction products in Table 7.2 are free radicals. One important free radical in biological systems is nitric oxide, NO, $\cdot \ddot{N}=\ddot{O}\!:$. NO plays a great many roles as a signaling molecule in our bodies. The reaction of oxygen with C=C and C≡C bonds often produces free radicals like $R_2COO\cdot$, which are responsible for spoiling food (rancid butter), for some of the tissue damage in arthritis, and for the aging of cells and organisms, among other things.

Average bond enthalpies If bond enthalpies were independent of one another, all the values in Table 7.2 would be the same. As you see, the enthalpy required for breaking the C—H bonds in methane is not the same for each bond, which means that the bond enthalpies are not totally independent of one another. But the **average bond enthalpy**, $416 \text{ kJ·mol}^{-1}\left(=\dfrac{1664 \text{ kJ·mol}^{-1}}{4}\right)$, is within 20% of all the individual bond enthalpies. When similar experiments are done to remove hydrogen atoms from many other carbon-containing compounds, the experimental bond enthalpies average 414 kJ·mol^{-1}. Average bond enthalpies for several pairs of atoms in a variety of compounds are tabulated in Table 7.3.

Table 7.3 *Average bond enthalpies for homolytic bond cleavage.*

Values are in kJ·mol⁻¹ of bonds for gaseous compounds at 25 °C.

	H	C	N	O	P	S
H—	**436.4**	414	393	460	326	368
C—	414	347	276	351	263	255
C=		620	615	745ᵃ		477
C≡		812	891	**1071**		
N—	393	276	193	176	209	
N=		615	418			
N≡		891	**941.4**			
O—	460	351	176	142	502	
O=		745ᵃ		**498.7**		469
P—	326	263	209	502	197	
S—	368	255				268
S=		477		469		352
Cl—	**431.9**	338				
Br—	**366.1**	276				
I—	**298.3**	238				

Note: Values for diatomic molecules (in bold italic) have four significant figures because they are not averaged over different compounds.
ᵃThe C=O bond enthalpy in CO_2 is 799 kJ·mol⁻¹ of bonds. Recall from Chapter 5 that CO_2 has delocalized pi (π) bonds that lead to stronger bonding; the bond enthalpy is larger than for a simple C=O bond.

The bond enthalpy for a particular pair of atoms is in the cell where the row of the atom on the left (showing the correct number of bonds) intersects the column of the atom to which it is bound. The C=N bond enthalpy, for example, is read where the C= row intersects the N column, a value of 615 kJ·mol⁻¹. The same value is read beginning with the N= row and moving across to the C column.

7.38 CONSIDER THIS

How do bond enthalpies for single and multiple bonds compare?

(a) Examine the average bond enthalpies in Table 7.3. How do the bond enthalpies of multiple bonds (such as C=C, and C≡C) compare to the

continued

bond enthalpies of the corresponding single bonds (C—C)? Are all double bonds less than twice as strong as single bonds between the same two atoms or are they all more than twice as strong? Give examples to justify your answer.

(b) Do you see any consistent patterns in the strengths of multiple bonds relative to the strengths of single bonds? Give examples to justify your answer.

Patterns among average bond enthalpies Before we go on to use average bond enthalpies to examine reactions, let's analyze the data in Table 7.3 to see what kinds of patterns there are among the bond enthalpies. These might help us gain further insight into the structure and bonding in molecules. We will focus on *bond formation* instead of bond breaking. For example, Table 7.3 shows that breaking an oxygen-oxygen single bond requires an input of 142 kJ·mol^{-1}. The reverse process, forming an oxygen-oxygen single bond releases 142 kJ·mol^{-1}; ΔH for the bond formation is -142 kJ·mol^{-1}.

The covalent compounds that often interest us are those found in living systems and all of them involve bonds among carbon, nitrogen, and/or oxygen, so we will limit our analysis to these atoms. Table 7.4 shows the enthalpies of bond formation for carbon–carbon, nitrogen–nitrogen, and oxygen–oxygen bonds. These are values from Table 7.3 with negative signs because bond formations are exothermic.

Table 7.4 *Enthalpies of formation for C–C, N–N, and O–O bonds.*

All energies are in kJ·mol^{-1} of the specified bond.

Bond order	Carbon–carbon	Nitrogen–nitrogen	Oxygen–oxygen
Single	−347	−193	−142
Double	−620	−418	−498.7
Triple	−812	−941.4	

As we expect from our bonding model, the enthalpy released in bond formation increases with bond order for all three atoms: the higher the bond order, the stronger the bond between the same atoms. There is a significant difference, however, between carbon and the other two atoms when we compare the *relative* enthalpy release for single bond formation with that for multiple bond formation. For example, forming an oxygen–oxygen double bond releases almost four times as much enthalpy as the formation of four oxygen–oxygen single bonds. Thus, replacing two oxygen–oxygen single bonds by a double bond is quite an exothermic reaction. For carbon, however, forming the double bond releases only about 90% as much enthalpy as the sum of two carbon–carbon single bonds. Replacing two single carbon–carbon bonds with a double bond is an endothermic reaction. These and other comparisons are shown graphically in Figure 7.9 for carbon, oxygen, and nitrogen.

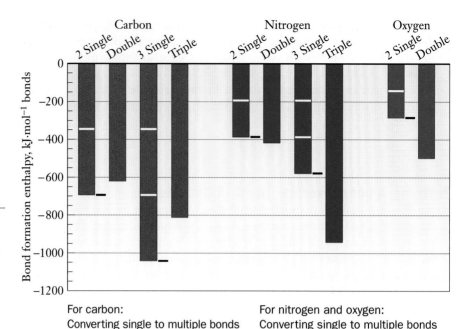

Figure 7.9.

Comparing bond formation enthalpies for equivalent numbers of bonding electrons. Comparisons are for single *vs.* multiple bonds for C, N, and O. Single bond enthalpy bars are stacked to make the comparisons clear and the horizontal black bars call attention to the sum of the single bond enthalpies.

For carbon:
Converting single to multiple bonds is endothermic. Converting multiple to single bonds is exothermic.

For nitrogen and oxygen:
Converting single to multiple bonds is exothermic. Converting multiple to single bonds is endothermic.

In Figure 7.9, note that, for carbon, the enthalpy released in forming a multiple bond is always less than the sum of the enthalpies released in forming an equivalent number of single bonds. Just the reverse is true for nitrogen and oxygen. The comparisons in Table 7.4 and Figure 7.9 can help us understand the direction of chemical change without doing any calculations. For example, under the proper conditions, ethene (ethylene) polymerizes to give **polyethylene** chains:

$$2nH_2C{=}CH_2 \rightarrow {+}(CH_2{-}CH_2{-}CH_2{-}CH_2)_n^{-} \qquad (7.8)$$

The product chain continues in both directions. Count the number of bonding electrons between carbons pictured in Figure 7.10; the electrons from the double bond in ethene form two single bonds in polyethylene. This is an exothermic process; the two single bonds formed are more stable than the double bond they replace.

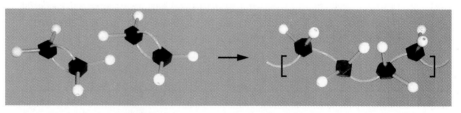

Figure 7.10.

Ethene polymerization.
The brackets enclose the same part of the chain as the parentheses in equation (7.8).

The reactants represent a large number of ethene molecules; only two are shown. The bent bonds are retained in the product polymer to show how each double bond in the ethenes has become two single bonds in the polymer.

7.39 CHECK THIS

Formation of polyethylene

Make several models of ethene molecules and "react" them to form poly-ethylene. How does your polyethylene "product" compare to the product shown in reaction (7.8) and Figure 7.10?

Another example comes from the airbag used in automobiles. Solid sodium azide, NaN_3, is the reactant that produces the gas that fills the airbag. The reaction of the azide gives nitrogen gas as a product:

$$2Na^+ + 2\,\overset{-\,\bullet\bullet}{\text{:}N}=\overset{+}{N}=\overset{\bullet\bullet\,-}{N\text{:}} \rightarrow 2Na\bullet + 3\,\text{:}N\equiv N\text{:} \qquad (7.9)$$

In this case, three triple bonds replace four double bonds between nitrogen atoms. The reaction is rapid and exothermic; more stable nitrogen–nitrogen multiple bonds are formed.

7.40 CHECK THIS

Sulfur–sulfur bonds

The most stable form of elemental oxygen is the diatomic molecule with a multiple bond between the oxygen atoms. The most stable form of sulfur, the third period element beneath oxygen in the periodic table, is the S_8 molecule, a ring of eight atoms bonded by single bonds.
(a) Write Lewis structures for S_2 and S_8.
(b) Make four molecular models of S_2 using oxygen or carbon atom centers to represent sulfur atom centers. Convert these four S_2 molecules to an S_8 mole-cule. How many two-electron bonds are there in the reactants? In the products?
(c) Sketch a bond formation enthalpy diagram (using the data in Table 7.3 and modeled on Figure 7.9) to show the relative enthalpies of bond formation for four moles of S_2 compared to one mole of S_8. Is the reaction you mod-eled in part (b) exothermic or endothermic?
(d) Does your result in part (c) help explain why oxygen and sulfur molecules are so different? Explain why or why not.

Bond enthalpy calculations Now we will go on to use bond enthalpies to analyze reactions that involve bonds between unlike atoms and to estimate numeric values for enthalpies of reaction. Although average bond enthalpies cannot provide precise values for reaction enthalpies, they are a good starting point, if other reaction enthalpy data are not available. When the enthalpy change for a reaction that has never been run can be estimated to within 10–20%, for example, the information can help researchers decide whether to pursue an investigation or to abandon it.

Let's go back to estimate the energetics of the hydrogen peroxide reaction of Investigate This 7.32. In order to do this, we have to recognize a problem. All the average bond enthalpies in Table 7.3 refer to gas-phase reactions, but the reactions we want to analyze, reactions (7.6) and (7.7), involve reactants and products in solution as well as gases. For our estimates of the energetics of the hydrogen peroxide reaction we will analyze two analogous gas phase reactions:

$$2H_2O_2(g) \rightarrow 2H_2O(g) + O_2(g) \tag{7.10}$$

$$H_2O_2(g) \rightarrow H_2(g) + O_2(g) \tag{7.11}$$

We cannot expect the results of our calculations to give an accurate value for the actual reaction. However, they might provide corroboration for your conclusion that reaction (7.6) [reaction (7.10)] is the reaction that occurred in Investigate This 7.32.

7.41 WORKED EXAMPLE

Reaction enthalpy from average bond enthalpies

Use the average bond enthalpies from Table 7.3 to calculate the enthalpy change for the decomposition of hydrogen peroxide to give water and oxygen by reaction (7.10).

Necessary information: We need to know which bonds are broken in the reactants and which bonds are formed in the products; this is what you found in Check This 7.36(a).

Strategy: Use average bond enthalpies to calculate the enthalpy required to break the bonds in the reactants and the enthalpy released when the bonds in the products are formed. To keep track of which bond enthalpy is under scrutiny, bond enthalpies will be symbolized as BH, with a subscript denoting the bond being analyzed. BH_{H-O}, for example, symbolizes the bond enthalpy for *breaking* a hydrogen–oxygen bond and $-BH_{H-O}$ symbolizes the bond enthalpy for *forming* a hydrogen–oxygen bond. One way to proceed is to imagine all the bonds in the products being broken homolytically to produce atoms, for which the enthalpy change is $\Sigma BH_{reactants}$:

$$2H-\overset{..}{\underset{..}{O}}-\overset{..}{\underset{..}{O}}-H \rightarrow 4H\cdot + 4\cdot\overset{..}{\underset{..}{O}}\cdot$$

$$2H_2O_2(g) \rightarrow 4H(g) + 4O(g) \quad \Sigma BH_{reactants} \tag{7.12}$$

The summation symbol, Σ, indicates that the bond enthalpies for breaking all the bonds in the reactants are added together. Then the atoms are recombined to form the products, for which the enthalpy change is $\Sigma(-BH_{products})$:

$$4H\cdot + 4\cdot\overset{..}{\underset{..}{O}}\cdot \rightarrow 2H-\overset{..}{\underset{..}{O}}-H + \overset{..}{\underset{..}{O}}=\overset{..}{\underset{..}{O}}$$

$$4H(g) + 4O(g) \rightarrow 2H_2O(g) + O_2(g) \quad \Sigma(-BH_{products}) \tag{7.13}$$

The sum of reactions (7.12) and (7.13) is the reaction of interest, reaction (7.10); $\Delta H_{reaction}$ is the sum of the enthalpy changes for reactions (7.12) and (7.13):

$$\Delta H_{reaction} = \Sigma BH_{reactants} + \Sigma(-BH_{products}) \tag{7.14}$$

continued

Figure 7.11 summarizes these equations and the strategy in an enthalpy-level diagram (which is exactly the same as an energy diagram, but with enthalpy as the energy function).

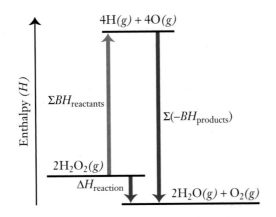

Figure 7.11.

Enthalpy-level diagram for
$2H_2O_2(g) \rightarrow 2H_2O(g) + O_2(g)$
The reaction is assumed to be exothermic, as you found experimentally in Investigate This 7.32.

Implementation: Table 7.5 shows the combinations of average bond-breaking and bond-forming enthalpy changes for reactions (7.12) and (7.13).

Table 7.5   *Bond-breaking and bond-forming enthalpy changes for* $2H_2O_2(g) \rightarrow 2H_2O(g) + O_2(g)$.

	Reactant bond breaking				Product bond formation		
Bond	**Number of bonds**	**BH, kJ·mol^{-1}**	**ΣBH, kJ**	**Bond**	**Number of bonds**	**BH, kJ·mol^{-1}**	**$\Sigma(-BH)$, kJ**
O—O	2	142	284	O=O	1	499	−499
H—O	4	460	1840	H—O	4	460	−1840
	$\Sigma BH_{reactants} =$ $2BH_{O-O} + 4BH_{H-O} = 2124$				$\Sigma(-BH_{products}) =$ $-BH_{O=O} - 4BH_{H-O} = -2339$		

Table 7.5 gives us the numerical values for the sums in equation (7.14) and substituting these values into the equation gives $\Delta H_{reaction}$:

$$\Delta H_{reaction} = (2124\,\text{kJ}) + (-2339\,\text{kJ}) = -215\,\text{kJ}$$

Does the answer make sense?   We set up an *imaginary* pathway for reaction (7.10) that involved enthalpy changes we could calculate from average bond enthalpies. The sum of the enthalpy changes along this pathway from products to reactants is the same as the enthalpy change for the direct reaction. The calculated enthalpy change is negative; we predict that the reaction is exothermic. This is consistent with what you found experimentally. Note that this enthalpy change is for the reaction as written, with two moles of hydrogen peroxide reacting. The enthalpy change for one mole reacting is half this value:

$$H_2O_2(g) \rightarrow H_2O(g) + \tfrac{1}{2}O_2(g)$$

$$\Delta H_{reaction} = -108\,\text{kJ (for one mole of } H_2O_2) \qquad\qquad (7.15)$$

7.42 CHECK THIS

Reaction enthalpy from average bond enthalpies

(a) Use average bond enthalpy data from Table 7.3 to determine the enthalpy change, $\Delta H_{reaction}$ for reaction (7.11):

$$H_2O_2(g) \rightarrow H_2(g) + O_2(g) \tag{7.11}$$

(b) Draw an enthalpy-level diagram, modeled after Figure 7.11, for this reaction.

(c) 🐾 Describe how the interactive exercise and resulting enthalpy-level diagram in the *Web Companion*, Chapter 7, Section 7.7.3, are related to the calculations and enthalpy-level diagrams in parts (a) and (b) and in Worked Example 7.41 and Figure 7.11.

7.43 CHECK THIS

The hydrogen peroxide decomposition reaction

(a) How does the enthalpy change for reaction (7.11) in Check This 7.42, compare to the enthalpy change for reaction (7.10) in Worked Example 7.41?

(b) Is reaction (7.6) or reaction (7.7) responsible for your observations in Investigate This 7.32 and 7.35? Summarize all the experimental and calculated results that lead you to your conclusion.

Accuracy of bond enthalpy calculations The results of the experiments and the calculated enthalpies of reaction in Worked Example 7.41 and Check This 7.42 are all consistent with reaction (7.6) being the exothermic decomposition reaction for hydrogen peroxide in the presence of yeast. The question remains whether the quantitative results of calculations based on average bond enthalpies for gas phase reactions are accurate enough to justify their use as more than a qualitative indicator of exothermicity or endothermicity, especially for reactions that involve solutions, liquids, and solids, as well as gases.

Hydrogen peroxide, hydrogen, oxygen, and water are compounds that have been extensively studied, and we can obtain accurate values for the enthalpy changes for reaction (7.6) (written here for one mole of hydrogen peroxide), and reaction (7.7):

$$H_2O_2(aq) \rightarrow H_2O(l) + \tfrac{1}{2}O_2(g) \quad \Delta H_{reaction} = -95 \text{ kJ} \cdot (\text{mol } H_2O_2)^{-1} \tag{7.6}$$

$$H_2O_2(aq) \rightarrow H_2(g) + O_2(g) \quad \Delta H_{reaction} = +191 \text{ kJ} \cdot (\text{mol } H_2O_2)^{-1} \tag{7.7}$$

The bond enthalpy calculations for these reactions (in the gas phase) produced reaction enthalpies that were in the right direction (exothermic and endothermic) and within about 50% of these experimental values.

Our calculated results could have been much better, if we had accounted for the enthalpy required to get hydrogen peroxide from solution into the gas phase and the enthalpy released when gaseous water condenses to a liquid. The calculations would not have been a great deal more difficult. However, we set out to find out how useful reaction enthalpy calculations based solely on average bond enthalpies can be. The answer appears to be that bond enthalpy calculations are quite useful in giving us the direction of a reaction enthalpy change, but that we cannot depend on them to give accurate numerical values, especially if the reactants and products are not all gases. The great advantage of bond enthalpies is that the modest set of values in Table 7.3 provides all the information you need to calculate reaction enthalpies for an enormous number of reactions, even ones that you only imagine—like reaction (7.7). In the next section, we will discuss another method for obtaining reaction enthalpies that gives accurate numerical values, but requires a much more extensive data table.

Reflection and Projection

The two-electron, covalent bonds in molecules are not completely independent of one another. However, the energies (enthalpies) required to break similar bonds (H–C, for example) in the same or different gaseous molecules are not too different. We can, therefore, use the average of these enthalpies as a reasonable approximation of the bond enthalpy, tabulate the results for different kinds of bonds, and use these values to calculate enthalpy changes for reactions. Enthalpies of reaction are calculated by summing up the bond enthalpies for all the reactant bonds that are broken, summing up all the bond enthalpies for product bonds that are formed (negative values), and combining these positive and negative sums to get the enthalpy change for the reaction:

$$\Delta H_{reaction} = \Sigma BH_{reactants} + \Sigma(-BH_{products}) \qquad (7.14)$$

You can't expect enthalpy changes calculated from average bond enthalpies to be as accurate as direct calorimetric measurements (when these are possible). However, these calculated values can usually be used to determine whether a reaction will be exothermic or endothermic. Calculated values of reaction enthalpies are all you can get if the reactants and/or products are only imagined molecules. You also saw how bond enthalpies help to understand why oxygen and nitrogen tend to form simple diatomic molecules with double and triple bonds, respectively, whereas carbon preferentially forms extended chains of single bonds.

Useful as bond enthalpies are, calculations based on them are not accurate enough for all purposes, especially for analyzing complex reactions in solution. A different approach that is not dependent on assumptions about the bonding properties of molecules is required.

7.8. Standard Enthalpies of Formation

Bond enthalpies are extraordinarily useful for predicting enthalpy changes for reactions that are experimentally inaccessible, but they are limited by the requirement that reasonable enthalpy estimates require reactants and products to be in the gas phase. The alternative approach is to use standard enthalpies of formation. The **standard enthalpy of formation** of a compound is the enthalpy change when one mole of the compound is formed at a pressure of 1 bar ($= 10^5$ kg·m^{-1}·s^{-2} = 100 kPa) from its elements in their standard states. The **standard state** of an element is its most stable form at 1 bar and a specified temperature (usually 25 °C). We have to specify the element in its most stable form, because many elements exist in more than one structure at 1 bar and 25 °C. Different structures of an element are called **allotropes**. Carbon, for example, is an element that exists in several allotropic solid forms, Figure 7.12, that differ in enthalpy. **Graphite** is the most stable allotrope, and it is chosen as the standard state for carbon.

The standard state pressure used to be defined as one atmosphere pressure, about 101 kPa. For most purposes, tiny differences between enthalpies of formation tabulated for one atm and one bar are negligible; values found in older tables and handbooks are still usable.

Figure 7.12.

Enthalpy-level diagram for carbon allotropes. ΔH values are for one mole of carbon atoms in each form. Graphite is made up of many layers; a tiny area of one layer is shown here. The diamond structure extends in three dimensions and can form large crystals.
Buckminsterfullerene, C_{60}, is a discrete molecular allotrope of carbon. A family of these carbon-cage, fullerene molecules ("buckyballs"), with larger and smaller numbers of carbon atoms, has been made and analyzed.

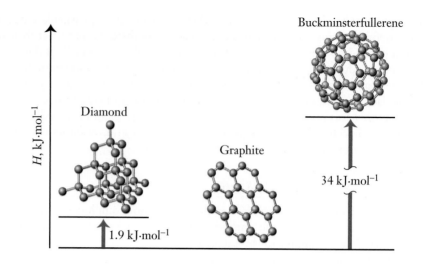

7.44 WORKED EXAMPLE

Formation of a compound from its standard-state elements

Write the reaction equation for the formation of one mole of methanol, CH_3OH, in its standard state from its elements in their standard states.

Necessary information: We need to know the standard states of all the reacting elements and the product. The standard states for the elements are C(*graphite*), $H_2(g)$, and $O_2(g)$. Methanol is a liquid at 25 °C and 1 bar, $CH_3OH(l)$.

Strategy: Write an expression for the formation reaction, using the elements and product in their standard states, and then balance the equation for one mole of product.

Implementation:

unbalanced: $C(graphite) + O_2(g) + H_2(g) \rightarrow CH_3OH(l)$

balanced: $C(graphite) + \frac{1}{2}O_2(g) + 2H_2(g) \rightarrow CH_3OH(l)$

Does the answer make sense? The reactants are elements in their standard states and the product is methanol in its standard state. The balanced reaction equation corresponds to the definition of the reaction whose enthalpy change is the enthalpy of formation of methanol. Note that the coefficients are chosen so that one mole of product is formed, even when fractional coefficients are necessary for reactant elements.

7.45 CHECK THIS

Formation of compounds from their standard-state elements

Write the reaction equations for the formation of one mole of dimethylamine, $(CH_3)_2NH(g)$, and 1 mol of ethylamine, $CH_3CH_2NH_2(g)$, in their standard

continued

states (shown with their formulas) from their elements in their standard states. How do the reactants for the two cases compare? What can you conclude about the reaction equations for the formation of isomers?

In Worked Example 7.44, we wrote the reaction equation for the formation of 1 mol of methanol in its standard state from its elements in their standard states. The standard enthalpy of formation for methanol is the enthalpy change for the reaction we wrote:

$$C(graphite) + \tfrac{1}{2}O_2(g) + 2H_2(g) \rightarrow CH_3OH(l)$$

$$\Delta H_f^\circ = -238.7 \text{ kJ·mol}^{-1} \qquad (7.16)$$

Thermodynamic tables use ΔH_f° **(standard enthalpy of formation)** to designate ΔH values for reactions under standard conditions. The superscript "°" in ΔH_f° designates the standard state pressure of one bar, and the subscript "f" reminds us that the value is for an enthalpy of formation.

Appendix B lists standard enthalpies of formation for many compounds, including those used in problems in this text. Although temperature is not specified as part of the standard state, tables of enthalpies of formation are almost always compiled for compounds at 298 K (25 °C), as in Appendix B. The bond enthalpies included in Table 7.3 are also all given for 25 °C. In order to set up enthalpy tables that everyone can use, scientists have agreed on the reference point for enthalpy measurements. The choice of reference point is the elements in their standard states, which are all arbitrarily *assigned* an enthalpy of formation of zero. Since only *differences* in enthalpy are measurable, the choice of zero for the enthalpy of formation of the elements is made as a matter of convenience.

In addition to the accuracy of the data, enthalpies of formation greatly simplify calculations involving complex molecules because the particulars of chemical bonding are not factors in the calculations. The following Worked Examples 7.46 and 7.49 illustrate the use of standard enthalpies of formation to obtain the standard enthalpy changes for chemical reactions, and the Check This problems provide an opportunity for you to practice the procedures.

7.46 WORKED EXAMPLE

Standard enthalpy change for an isomerization reaction

Use Appendix B data to calculate the standard enthalpy of reaction for the **isomerization,** change of one isomer to another, of 1 mol of cyclobutane, *cyclo*-$C_4H_8(g)$ (a four-carbon ring) to 1 mol of 1-butene, $CH_2{=}CHCH_2CH_3(g)$ (an alkene isomer of cyclobutane) at 25 °C:

$$cyclo\text{-}C_4H_8(g) \rightarrow CH_2{=}CHCH_2CH_3(g) \qquad \Delta H^\circ_{reaction} = ? \qquad (7.17)$$

Necessary information: We need standard enthalpies of formation for the reactant and product. The values from Appendix B are 26.65 and 1.17 kJ·mol^{-1}, respectively, for cyclobutane and 1-butene. Note that formation of these isomers from the elements is endothermic.

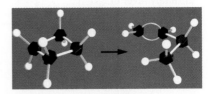

Molecular representation of the cyclobutane isomerization

continued

Strategy: Write the enthalpy of formation reaction equations for the reactant and product and combine them appropriately to obtain the desired reaction equation (7.17). The enthalpies of formation will be combined in the same way to get the standard reaction enthalpy.

Implementation:

$$4C(graphite) + 4H_2(g) \rightarrow cyclo\text{-}C_4H_8(g)$$
$$\Delta H_f^\circ(\text{cyclo}) = 26.65 \text{ kJ·mol}^{-1} \tag{7.18}$$

$$4C(graphite) + 4H_2(g) \rightarrow CH_2{=}CHCH_2CH_3(g)$$
$$\Delta H_f^\circ(\text{butene}) = 1.17 \text{ kJ·mol}^{-1} \tag{7.19}$$

Reverse reaction equation (7.18), including the sign of the enthalpy change, and add it to reaction equation (7.19) to get reaction equation (7.17):

$cyclo\text{-}C_4H_8(g) \rightarrow \cancel{4C(graphite)} + \cancel{4H_2(g)}$	$\Delta H^\circ = (1 \text{ mol})[-\Delta H_f^\circ(\text{cyclo})]$	$-(7.18)$
$\cancel{4C(graphite)} + \cancel{4H_2(g)} \rightarrow CH_2{=}CHCH_2CH_3(g)$	$\Delta H^\circ = (1 \text{ mol})[\Delta H_f^\circ(\text{butene})]$	(7.19)

$$cyclo\text{-}C_4H_8(g) \rightarrow CH_2{=}CHCH_2CH_3(g) \tag{7.17}$$

$$\Delta H^\circ_{\text{reaction}} = (1 \text{ mol})[-\Delta H_f^\circ(\text{cyclo})] + (1 \text{ mol})[\Delta H_f^\circ(\text{butene})]$$

Which can be rewritten as

$$\Delta H^\circ_{\text{reaction}} = (1 \text{ mol})[\Delta H_f^\circ(\text{butene})] - (1 \text{ mol})[\Delta H_f^\circ(\text{cyclo})] \tag{7.20}$$

$$\Delta H^\circ_{\text{reaction}} = (1 \text{ mol})[1.17 \text{ kJ·mol}^{-1}] - (1 \text{ mol})[26.65 \text{ kJ·mol}^{-1}] = -25.48 \text{ kJ}$$

Figure 7.13 is an enthalpy-level diagram showing another way to combine standard enthalpies of formation to get the standard enthalpy of the reaction, as in equation (7.20).

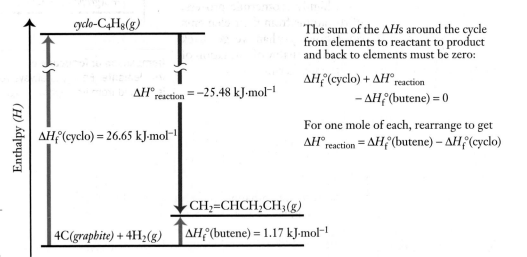

The sum of the ΔHs around the cycle from elements to reactant to product and back to elements must be zero:

$$\Delta H_f^\circ(\text{cyclo}) + \Delta H^\circ_{\text{reaction}}$$
$$- \Delta H_f^\circ(\text{butene}) = 0$$

For one mole of each, rearrange to get
$$\Delta H^\circ_{\text{reaction}} = \Delta H_f^\circ(\text{butene}) - \Delta H_f^\circ(\text{cyclo})$$

Figure 7.13.

Enthalpy-level diagram for isomerization of cyclobutane to 1-butene.

Does the answer make sense? The enthalpy level diagram, Figure 7.13, clearly illustrates that 1-butene is a lower energy, more stable compound than cyclobutane. The reaction to form the 1-butene isomer should be exothermic, as we have calculated it to be.

7.47 CHECK THIS

Standard enthalpy change for an isomerization reaction

(a) Make molecular models to illustrate the isomerization of cyclopentane, $cyclo$-$C_5H_{10}(g)$ (a five-carbon ring) to 1-pentene, CH_2=$CHCH_2CH_2CH_3(g)$ (an alkene isomer of cyclopentane):

$$cyclo\text{-}C_5H_{10}(g) \rightarrow CH_2\text{=}CHCH_2CH_2CH_3(g) \quad \Delta H°_{reaction} = ? \quad (7.21)$$

(b) Use Appendix B data to calculate $\Delta H°_{reaction}$ for reaction (7.21) at 25 °C.
(c) Sketch an enthalpy level diagram, modeled after Figure 7.13, for this reaction to show the combination of standard enthalpies of formation that gives the standard enthalpy of reaction.

We have pointed out many times that bond formation is always exothermic: The atoms bonded together are lower in energy than the separated atoms. You might, therefore, have wondered about the positive standard enthalpy of formation of the isomers, cyclobutane and 1-butene, in Worked Example 7.46. How can the formation of these compounds from their elements be endothermic, if bond formation is always exothermic? The answer lies in the difference between formation of a molecule from its separated atoms (always exothermic) and from the atoms in their elemental standard states, as illustrated for cyclobutane in Figure 7.14.

Figure 7.14 shows that the formation of cyclobutane from its separated atoms is a highly exothermic process. However, the formation of the atoms from their elements requires a bit more energy (enthalpy) than we get back from compound formation, so the enthalpy of formation of cyclobutane from its *elements* is endothermic.

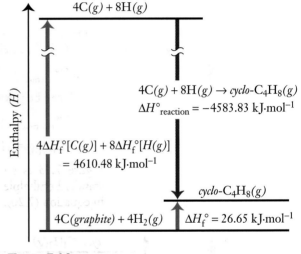

Figure 7.14.

Comparison of formation of *cyclo*-C_4H_8 from its atoms and its elements. Energy is always released when a molecule is formed from its separated gas-phase atoms.

7.48 CHECK THIS

Comparison of formation of 1-butene from its atoms and its elements

(a) Use Appendix B data to construct an enthalpy-level diagram, modeled after Figure 7.14, comparing the enthalpy changes in formation of 1-butene from its atoms and its elements.
(b) Explain why there are similarities and differences between your enthalpy-level diagram in part (a) and the one shown in Figure 7.14.

Equation (7.20), Figure 7.13, and your equivalent formulations in Check This 7.47 show that the standard enthalpy change for isomerization reactions is the difference between the standard enthalpy of formation of the product and the standard enthalpy of formation of the reactant (with the number of moles of reactant and product accounted for):

$$\Delta H^\circ_{reaction} =$$
$$(\text{mol product})\left[\Delta H_f^\circ(\text{product})\right] - (\text{mol reactant})\left[\Delta H_f^\circ(\text{reactant})\right] \quad \textbf{(7.22)}$$

Let's see how this approach works for more complicated reactions.

7.49 WORKED EXAMPLE

Standard enthalpy change for a reaction

Use Appendix B data to calculate the standard enthalpy of reaction, $\Delta H^\circ_{reaction}$, for the complete oxidation of methanol, $CH_3OH(l)$ at 25 °C:

$$CH_3OH(l) + \tfrac{3}{2}O_2(g) \rightarrow CO_2(g) + 2H_2O(l) \quad \Delta H^\circ_{reaction} = ? \quad \textbf{(7.23)}$$

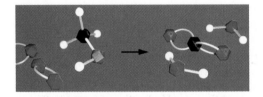

Molecular representation
of the methanol oxidation

Necessary information: The standard enthalpies of formation for methanol, oxygen, water, and carbon dioxide are, respectively, −238.9, 0.0 (element in its standard state), −285.8, and −393.5 kJ·mol^{-1}.

Strategy: Use the standard enthalpies of formation and the stoichiometry of the reaction to get the overall standard enthalpy changes for formation of the reactants, $\Delta H_f^\circ(\text{reactants})$, and the products, $\Delta H_f^\circ(\text{products})$. Combine these to get the standard enthalpy change for the reaction, $\Delta H^\circ_{reaction}$.

Implementation: For the reactants [$O_2(g)$ is included to be complete]:

$$C(graphite) + 2H_2(g) + \tfrac{1}{2}O_2(g) \rightarrow CH_3OH(l) \qquad \Delta H^\circ = (1\ \text{mol})\left[\Delta H_f^\circ(CH_3OH)\right]$$
$$\tfrac{3}{2}O_2(g) \rightarrow \tfrac{3}{2}O_2(g) \qquad \Delta H^\circ = \left(\tfrac{3}{2}\ \text{mol}\right)\left[\Delta H_f^\circ(O_2)\right]$$

$$C(graphite) + 2H_2(g) + 2O_2(g) \rightarrow CH_3OH(l) + \tfrac{3}{2}O_2(g) \tag{7.24}$$
$$\Delta H_f^\circ(\text{reactants}) = (1\ \text{mol})\left[\Delta H_f^\circ(CH_3OH)\right] + \left(\tfrac{3}{2}\ \text{mol}\right)\left[\Delta H_f^\circ(O_2)\right] \tag{7.25}$$
$$\Delta H_f^\circ(\text{reactants}) = (1\ \text{mol})\left[-238.9\ \text{kJ·mol}^{-1}\right] + \left(\tfrac{3}{2}\ \text{mol}\right)\left[0.0\ \text{kJ·mol}^{-1}\right] = -238.9\ \text{kJ}$$

For the products:

$$C(graphite) + O_2(g) \rightarrow CO_2(g) \qquad \Delta H^\circ = (1\ \text{mol})\left[\Delta H_f^\circ(CO_2)\right]$$
$$2H_2(g) + O_2(g) \rightarrow 2H_2O(l) \qquad \Delta H^\circ = (2\ \text{mol})\left[\Delta H_f^\circ(H_2O)\right]$$

$$C(graphite) + 2H_2(g) + 2O_2(g) \rightarrow CO_2(g) + 2H_2O(l) \tag{7.26}$$
$$\Delta H_f^\circ(\text{products}) = (1\ \text{mol})\left[\Delta H_f^\circ(CO_2)\right] + (2\ \text{mol})\left[\Delta H_f^\circ(H_2O)\right] \tag{7.27}$$
$$\Delta H_f^\circ(\text{products}) = (1\ \text{mol})\left[-393.5\ \text{kJ·mol}^{-1}\right] + (2\ \text{mol})\left[-285.8\ \text{kJ·mol}^{-1}\right] = -965.1\ \text{kJ}$$

continued

Figure 7.15 is an enthalpy-level diagram that shows how the standard enthalpies of formation of the reactants and products combine to give the standard enthalpy of reaction:

$$\Delta H°_{reaction} = \Delta H_f°(\text{products}) - \Delta H_f°(\text{reactants}) \qquad (7.28)$$

$$= -965.1 \text{ kJ} - (-238.9 \text{ kJ}) = -726.2 \text{ kJ}$$

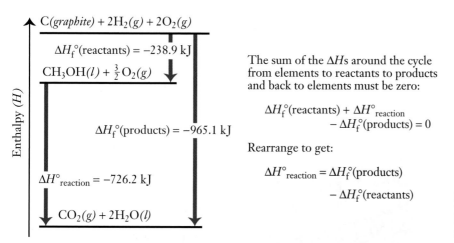

The sum of the ΔHs around the cycle from elements to reactants to products and back to elements must be zero:

$$\Delta H_f°(\text{reactants}) + \Delta H°_{reaction} - \Delta H_f°(\text{products}) = 0$$

Rearrange to get:

$$\Delta H°_{reaction} = \Delta H_f°(\text{products}) - \Delta H_f°(\text{reactants})$$

Figure 7.15.

Enthalpy-level diagram for oxidation of methanol.

Does the answer make sense?   Without further calorimetric measurements, we do not know whether the calculated *numeric* value is correct. However, we know that methanol burns in air and gives off thermal and light energy, so the reaction must be exothermic, as our result shows.

7.50 CHECK THIS

Standard enthalpy change for a reaction

(a) Use Appendix B data to calculate the standard enthalpy of reaction, $\Delta H°_{reaction}$, for the complete oxidation of glucose, $C_6H_{12}O_6(s)$ at 25 °C:

$$C_6H_{12}O_6(s) + 6O_2(g) \rightarrow 6CO_2(g) + 6H_2O(l) \qquad \Delta H°_{reaction} = ? \quad (7.29)$$

(b) Draw an enthalpy-level diagram, modeled after Figure 7.15, for this reaction.

Look back at Worked Examples 7.46 and 7.49 at your work in Check This 7.47 and 7.50, and at equation 7.22 and examine the final form of the equations used to calculate the standard enthalpies of reaction. In all cases, the standard enthalpy of reaction was obtained by subtracting the standard enthalpy of formation of the reactants from the standard enthalpy of formation of the products. This result can be expressed as

$$\Delta H°_{reaction} = \Sigma[n_j(\Delta H_f°)_j]_{products} - \Sigma[n_j(\Delta H_f°)_j]_{reactants} \qquad (7.30)$$

In equation (7.30), the subscripts "*j*" refer to the individual reactant or product compounds; n_j is the coefficient in the chemical equation for the *j*th compound and $(\Delta H_f°)_j$ is the standard enthalpy of formation for that compound. The reactant and product sums include all the reactant and product compounds, respectively.

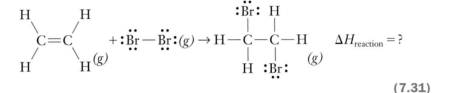

7.51 CHECK THIS

Bond enthalpy and enthalpy of formation comparison

Compare the enthalpy change for a reaction calculated from bond enthalpies and standard enthalpies of formation.

(a) The bond dissociation enthalpy of bromine is 193 kJ·mol^{-1}. Use this knowledge and bond enthalpies to estimate the standard enthalpy change, $\Delta H°_{reaction}$, for the gas-phase reaction of bromine with ethene to yield 1,2-dibromoethane:

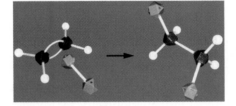

$$(7.31)$$

Molecular representation of ethene bromination

(b) The standard enthalpy of formation of 1,2-dibromoethane is −43.1 kJ·mol^{-1}. Use this value and others from Appendix B to calculate $\Delta H°_{reaction}$ for reaction (7.31). *Hint:* Remember that bromine is not in its standard state in this reaction.

(c) Compare your results for parts (a) and (b). What might account for any difference?

7.9. Harnessing Energy in Living Systems

When you set a marshmallow on fire in Investigate This 7.1, you were rearranging the chemical bonds in the sugar(s) of the marshmallow and the oxygen from the air. The carbon in the sugar was oxidized to carbon dioxide and the hydrogen combined with more oxygen to form water. Equation (7.29) in Check This 7.50 showed this overall reaction for one sugar, glucose. In this combustion the enthalpy difference between the reactants and products was released as thermal energy to the environment around the burning marshmallow. The pathway for conversion of reactants to products, although complex, is relatively direct and involves no other reactants. This is not the case when that same glucose, perhaps eaten as a toasted marshmallow, is metabolized in your body. Glucose metabolism is a long, complicated, but highly organized, process that requires a large number of enzymes and several biochemical pathways. Energy from the oxidation of glucose is released piecemeal as the oxidation process proceeds.

Coupled reactions During the oxidation process some of the energy released is "lost" as thermal energy that helps maintain your body temperature at 37 °C. Some of the energy is captured by coupling a reaction that *releases* energy to one that *requires* energy. The basis for these **coupled reactions** is that the reaction providing the energy does not proceed unless the reaction needing the energy input also occurs. In our bodies, the exothermic oxidation of glucose is coupled at several points along the biochemical oxidation pathway with the endothermic formation of adenosine triphosphate, ATP^{4-}, from adenosine diphosphate, ADP^{3-} (Figure 7.16):

$$ADP^{3-}(aq) + HOPO_3{}^{2-}(aq) + H^+(aq) \rightarrow ATP^{4-}(aq) + H_2O(l)$$

$$\Delta H°_{\text{ATP form}} \approx +21 \text{ kJ·mol}^{-1}$$

(7.32)

For simplicity, we are writing the hydronium ion as $H^+(aq)$, as we did in reduction–oxidation reactions in Chapter 6, Sections 6.9, 6.10, and 6.11.

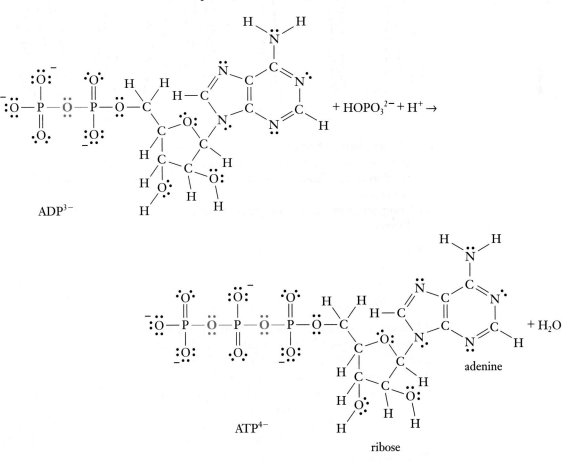

Figure 7.16.
Reaction to form ATP^{4-} from ADP^{3-} and phosphate.

The links between phosphate groups in ATP^{4-} and ADP^{3-} (labeled in red in Figure 7.16) are acid anhydride functional groups. An **acid anhydride** (from *anhydrous* = without water) is a functional group formed when two acid molecules condense to form a single unit with the loss of water, as shown in Figure 7.16. Formation of acid anhydrides from the acids is always an endothermic process. Furthermore, the grouping of four negative charges that repel one another in ATP^{4-} helps to explain why an input of energy is required to form it from ADP^{3-} and $HOPO_3{}^{2-}$.

The most direct way that coupling occurs to form ATP^{4-} is by the transfer of a phosphate group from one of the oxidized products of glucose oxidation to an ADP^{3-}. Two consecutive reactions in the glucose oxidation pathway are

$$^{-2}O_3POCH_2CHOHCHO + NAD^+ + HOPO_3^{2-} \rightarrow$$
$$^{-2}O_3POCH_2CHOHC(O)OPO_3^{2-} + NADH + H^+ \qquad \text{(7.33)}$$

$$^{-2}O_3POCH_2CHOHC(O)OPO_3^{2-} + ADP^{3-} \rightarrow$$
$$^{-2}O_3POCH_2CHOHC(O)O^- + ATP^{4-} \qquad \text{(7.34)}$$

In reaction (7.33), an enzyme catalyzes the oxidation of an aldehyde group to a carboxylic acid and its conversion to a mixed acid anhydride in which two different acids (phosphoric and carboxylic) have condensed with the loss of water and formation of an anhydride bond between the acids. The compound that is reduced is NAD^+, nicotinamide dinucleotide (see Chapter 6, Section 6.11). The energy released by the reduction-oxidation reaction provides the energy required for endothermic formation of the anhydride.

In reaction (7.34), another enzyme catalyzes the transfer of a phosphate group from the mixed acid anhydride to form a phosphoric acid anhydride. Since one acid anhydride bond is broken and another is formed, the reaction overall involves very little enthalpy change. The *combination* of reactions (7.33) and (7.34) represents a coupling of the energy released in one of the glucose oxidation steps to the formation of ATP^{4-} from ADP^{3-} using the mixed anhydride formed during the oxidation to couple the reactions.

7.52 CHECK THIS

The phosphate bond to ribose in ATP^{4-} and ADP^{3-}

The phosphate groups in ATP^{4-} and ADP^{3-} are bonded by acid anhydride links. What kind of functional group links the first phosphate group to ribose? *Hint:* See Chapter 6, Section 6.7.

Energy captured as ATP^{4-} In our bodies, coupling of the glucose oxidation pathway to the formation of ATP^{4-} can produce about 36 moles of ATP^{4-} for each mole of glucose oxidized. We can estimate what this means in terms of enthalpy changes by combining reaction (7.29) and reaction (7.32) taken 36 times, and their standard enthalpy changes:

$$C_6H_{12}O_6(s) + 6O_2(g) \rightarrow 6CO_2(g) + 6H_2O(l) \qquad \Delta H°_{\text{glucose oxidation}} \quad \text{(7.29)}$$

$$36ADP^{3-}(aq) + 36HOPO_3^{2-}(aq) + 36H^+(aq) \rightarrow 36ATP^{4-}(aq) + 36H_2O(l) \qquad 36\Delta H°_{\text{ATP form}} \quad \text{(7.32)}$$

$$C_6H_{12}O_6(s) + 6O_2(g) + 36ADP^{3-}(aq) + 36HOPO_3^{2-}(aq) + 36H^+(aq) \rightarrow$$
$$6CO_2(g) + 42H_2O(l) + 36ATP^{4-}(aq) \qquad \text{(7.35)}$$

$$\Delta H°_{\text{coupled reaction}} = \Delta H°_{\text{glucose oxidation}} + 36\Delta H°_{\text{ATP form}} \qquad \text{(7.36)}$$

$$\Delta H°_{\text{coupled reaction}} = -2801 \text{ kJ} + (36 \text{ mol})(21 \text{ kJ}\cdot\text{mol}^{-1}) = -2045 \text{ kJ}$$

The $\Delta H°_{\text{glucose oxidation}}$ comes from your calculation in Check This 7.50. Overall, the enthalpy change for the coupled reaction is -2045 kJ for every mole of glucose oxidized to produce 36 moles of ATP^{4-}. This result means that, under standard conditions, 756 kJ out of the 2801 kJ from glucose oxidation would go into making ATP^{4-}. The rest of the energy is released as heat to the surroundings. Under these conditions, the percentage of enthalpy converted to the ATP^{4-} required for other biological processes is about 27% $\left[\left(\dfrac{756\text{ kJ}}{2801\text{ kJ}}\right) \times 100\% = 27\%\right]$.

It is interesting to compare this biological oxidation with combustion. Steam engines that evolved from *The Rocket*, Figure 7.1, were rarely more than about 10% efficient in converting the thermal energy of combustion to work and present-day standard internal combustion engines are about 30% efficient.

7.53 CHECK THIS

Other biological fuels

What other biological fuel compounds did you investigate in Investigate This 7.1? What others are represented in the chapter opening illustration? Which seem to be the best fuels? Explain your answer.

Using the energy captured as ATP The reverse of reaction (7.35), the **hydrolysis** (breaking down by water; see Chapter 6, Section 6.7) of ATP^{4-} to ADP^{3-}, is exothermic:

$$ATP^{4-}(aq) + H_2O(l) \rightarrow ADP^{3-}(aq) + HOPO_3{}^{2-}(aq) + H^+(aq)$$

$$\Delta H°_{\text{ATP hydrolysis}} \approx -21 \text{ kJ·mol} \tag{7.37}$$

Figure 7.17 is a mechanical analogy showing how the enthalpy released by glucose oxidation is coupled through the synthesis and hydrolysis of ATP^{4-} to other energy-requiring biological reactions that provide such things as locomotion, information processing, and synthesis of new biological molecules.

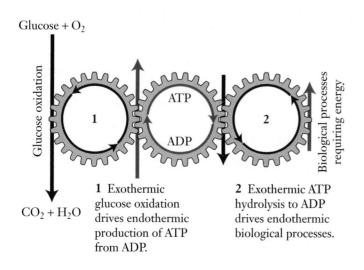

1 Exothermic glucose oxidation drives endothermic production of ATP from ADP.

2 Exothermic ATP hydrolysis to ADP drives endothermic biological processes.

Figure 7.17.

Glucose oxidation coupled via ATP–ADP to energy-requiring reactions. Enthalpy change arrows are not to scale, but show the direction and relative magnitudes of the changes. Thermal energy is "lost" to the surroundings at each transformation, glucose-to-ATP and ATP-to-biological processes.

For example, joining amino acids to form a protein is an endothermic process. Consider the simple case of bonding two glycines, $H_2NCH_2C(O)OH$, to form diglycine, $H_2NCH_2C(O)NHCH_2C(O)OH$, and water:

$$2Gly(aq) \rightarrow Gly{-}Gly(aq) + H_2O(l)$$

$$\Delta H°_{Gly-Gly} = +8 \text{ kJ·mol}^{-1}$$

(7.38)

If this reaction is coupled to the hydrolysis of ATP^{4-}, we can write the coupled reaction as the sum of reactions (7.37) and (7.38):

$$2Gly(aq) + ATP^{4-}(aq) \rightarrow$$

$$Gly{-}Gly(aq) + ADP^{3-}(aq) + HOPO_3^{2-}(aq) + H^+(aq)$$

$$\Delta H°_{reaction} = -13 \text{ kJ·mol}^{-1}$$

(7.39)

Figure 7.18 shows how this exothermic coupling is represented on an enthalpy-level diagram.

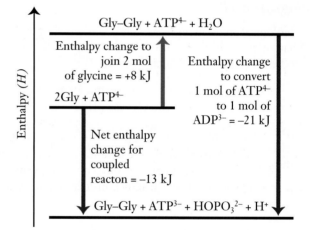

Figure 7.18.

Enthalpy coupling of diglycine synthesis to ATP^{4-} hydrolysis.

Protein synthesis in organisms is much more complicated than reactions (7.38) and (7.39) suggest, but the overall energetics are reasonably represented here. The enzymes and ribonucleic acids that catalyze protein synthesis do not work unless ATP available to be hydrolyzed as part of the overall process. This is, as we said above, the essence of reaction coupling. The energy from glucose oxidation (and the oxidation of other fuel molecules, like fats and proteins) sustains all life processes, but the energy is furnished indirectly through ATP^{4-}. Knowledge of energy relationships in living systems is crucial to understanding those systems, and that knowledge is built, in part, on measuring and interpreting reaction enthalpies. We will return to this discussion in the next chapters as we add to our understanding of thermodynamics and coupled reactions.

7.54 CHECK THIS

Other pathways for ATP hydrolysis

(a) In some coupling reactions ATP^{4-} hydrolysis takes another pathway:

$$ATP^{4-}(aq) + H_2O(l) \rightarrow AMP^{2-}(aq) + (O_3POPO_3)^{4-}(aq) + 2H^+(aq)$$

$$(O_3POPO_3)^{4-}(aq) + H_2O(aq) \rightarrow 2HOPO_3^{2-}(aq)$$

continued

The enthalpy change for each of these hydrolysis reactions is about the same as the ATP^{4-} hydrolysis to give ADP^{3-} and phosphate, reaction (7.37). The sum of these two reactions is

$$ATP^{4-}(aq) + 2H_2O(l) \rightarrow AMP^{2-}(aq) + 2HOPO_3^{2-}(aq) + 2H^+(aq)$$

What is $\Delta H°$ for this reaction combination? What advantage (if any) is there for an organism to use this combination pathway for ATP^{4-} hydrolysis compared to reaction (7.37)? Explain.

(b) How might you explain why the enthalpy changes for reaction (7.37) and for each of the two individual hydrolysis reactions in part (a) are about the same?

Reflection and Projection

The standard enthalpy of formation of a compound is defined as the enthalpy of reaction for formation of the compound in its standard state from the most stable form of its elements in their standard states. The standard state is one bar pressure and the tabulated values are usually given for 298 K (25 °C). Enthalpies of formation are experimental values independent of any bonding model and can be combined to give accurate values for enthalpies of reaction. An example from the complex metabolism of glucose coupled to the formation of ATP^{4-} shows how much information you can get from enthalpies of formation and reaction, even in the absence of detailed knowledge of the actual reactions.

Sometimes, however, the way a reaction is carried out determines the changes we observe. This is particularly the case for reacting systems involving both heat and work. Up to this point, we have quantitatively analyzed only thermal energy transfers; next we will briefly consider systems that also involve mechanical energy transfers.

7.10. Pressure–Volume Work, Internal Energy, and Enthalpy

The *sum of the potential and kinetic energy* in a collection of molecules is a state function called its **internal energy.** We have been using the symbol E to represent internal energy. When a change takes place in the collection of molecules, the resulting collection has a new value for its internal energy. Using the symbols E_i to represent the initial (beginning) state and E_f to represent the final (ending) state, we write the change in internal energy, ΔE **(internal energy change),** as

$$\Delta E = E_f - E_i \qquad (7.40)$$

When the internal energy of a collection of molecules changes, in a chemical reaction, for example, the energy change ($E_f - E_i$ or ΔE) can always be expressed as some combination of heat and work. To satisfy the law of conservation of energy, all the energy must be accounted for. This leads directly to the mathematical statement for the **first law of thermodynamics:**

$$\Delta E = q + w \qquad (7.41)$$

Heat and work that enter a system are positive; they increase the internal energy. Heat and work that leave a system are negative; they decrease the internal energy.

A practical example of the way the first law is applied is the internal combustion engine in an automobile, which is a reminder of why thermodynamics was developed in the first place. When the mixture of gasoline and air explodes inside an engine cylinder, as in Figure 7.19, the rearrangement of atoms to form new molecules produces a large decrease in internal energy. The value for ΔE is negative when energy is released in the reaction, as is the case here. Expanding gases move the piston; work (w) leaves the system of reacting gases to move the automobile. The rest of the energy released by the exploding gasoline–air mixture is released as heat (q), which does not contribute to the motion of the automobile. One task for automotive engineers is to design engines that convert the highest possible percentage of the energy released to work.

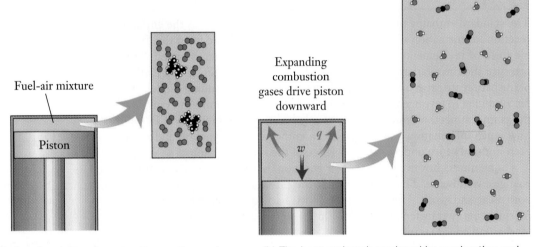

(a) A cylinder of an internal combustion engine and a representation of the reactants in the cylinder

(b) The heat and work produced by combustion and a representation of the reaction products

Figure 7.19.

Heat and work in the cylinder of an internal combustion engine.
Overall combustion reaction:

$$2C_8H_{18}(g) + 25O_2(g) \rightarrow 16CO_2(g) + 18H_2O(g)$$

7.55 CONSIDER THIS

How is the first law of thermodynamics used?

Combustion in the cylinder of a small engine is found to produce 5 kJ of heat and 2 kJ of work during each power stroke. After redesigning the engine, engineers find that they can produce 3 kJ of work during each power stroke between the same initial and final states as in the original design. How much heat is produced during each power stroke in the redesigned engine? Explain the signs you assign to all the variables in solving this problem.

Definition of work From physics, we have the definition of **work** (w) as the product of the force (F) acting on an object multiplied by the distance (d) the object is moved in the direction of the force:

$$w = F \times d \tag{7.42}$$

As we noted above, work, like heat, has directionality. *Work entering a system of interest is given a positive sign*, since it increases the energy of the system. The kind of work that most interested the 18th- and 19th-century thermodynamicists was the work of gases, steam in particular, pushing on pistons. In this section and in Section 7.13, we will discuss the work done by or on gases produced or consumed in chemical reactions and return to another important form of work, electrical work, in Chapter 10.

Constant volume and constant pressure reactions When a reaction that produces gases is carried out in a closed container with rigid walls, the added gas increases the pressure inside the container, but nothing moves, because the gas is trapped in the container. This reaction is carried out at *constant volume*. In a constant volume system, no work is done by the gas, $w = 0$. By contrast, when the same reaction is carried out in a container with a movable piston as one wall, Figure 7.20, work is done pushing back against the constant external atmospheric pressure on the piston. The reaction is carried out at *constant pressure*; for the cases of most interest to us, this is the pressure of the atmosphere.

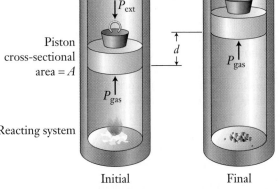

Piston cross-sectional area = A

Reacting system

Initial   Final

Figure 7.20.

Constant pressure reaction apparatus. A reaction that produces gas is carried out at constant pressure in a cylinder with a movable piston. A constant force on the piston, the external pressure, is represented by a weight sitting on the piston.

 *Web Companion*

Chapter 7, Section 7.11.1

Work with an animation of a reaction carried out at constant *P* and at constant *V*.

①
②
③
④

In order to determine the amount of work done in a constant pressure reaction, we'll analyze the changes that occur when the reaction occurs in the experimental setup in Figure 7.20. The pressure of the atmosphere, P_{ext}, acts like a weight resting on top of the piston. Before the reaction occurs, the absolute value of the pressure of gas inside the piston, $|P_{gas}|$, is equal to the absolute value of the external pressure, $|P_{gas}| = |P_{ext}|$. When the reaction occurs, gas is produced in the cylinder, it pushes against the piston and raises the weight until, once again, $|P_{gas}| = |P_{ext}|$.

Pressure–volume work To relate the pressure of the atmosphere, P_{ext}, to the force, F_{ext}, acting on the piston we need to know that pressure is defined as force per unit area:

$$P_{ext} = \frac{F_{ext}}{A}$$

Rearranging the definition gives the force:

$$F_{ext} = P_{ext} \times A \tag{7.43}$$

Before we can use this force to calculate the work, we need to define the directionality of the work we will calculate. We are interested in how much work is done *on* our reacting system, since work done on the system will increase its energy. The force acting on the system is the atmospheric pressure, P_{ext}. You can see in Figure 7.20 that P_{ext} is acting in a direction *opposite* to the movement of the piston when gas is produced. Therefore, the work done on the reacting system is negative:

$$w \text{ (on the reacting system)} = -F_{ext} \times d \tag{7.44}$$

Substituting the force from equation (7.43) into equation (7.44) gives

$$w = -P_{ext} \times A \times d \tag{7.45}$$

The cross sectional area of the piston, A, times the distance the piston moves, d, is the change in volume, ΔV, of the gas in the cylinder:

$$\Delta V = A \times d \tag{7.46}$$

Finally, we substitute ΔV from equation (7.46) into equation (7.45) to get the **pressure–volume work:**

$$w = -P_{ext} \times \Delta V \tag{7.47}$$

The sign of the pressure-volume work done on or by a system can be confusing. It takes work to raise (lift) a weight; work is being done *on* the weight to raise it. The only place for that work to come from is the reacting system. Therefore, *work has to leave the reacting system in order to raise the weight*. Indeed, equation (7.47) shows that when gas is produced, $\Delta V > 0$ (positive change in volume), work leaves the reacting system, $w < 0$. If a reaction uses up gas, $\Delta V < 0$ (negative change in volume), work enters the reacting system, $w > 0$. Next, we will look at how heat, pressure–volume work, and internal energy are related to enthalpy.

7.56 WORKED EXAMPLE

Pressure–volume work at constant pressure

A reaction was carried out at constant atmospheric pressure in a cylinder like the one in Figure 7.20. The volume of gas in the cylinder before reaction was 0.756 L and the volume of gas after the reaction was 0.983 L. The atmospheric pressure was 1 bar, 100. kPa. How much work was done on the reacting system during this change?

Necessary information: To keep energy units straight, we need to know that $1 \text{ kPa·L} = 1 \text{ J}$.

Strategy: Calculate the change in volume, ΔV, and use equation (7.47) to get the work.

Implementation:

$$\Delta V = (0.983 \text{ L}) - (0.756 \text{ L}) = 0.227 \text{ L}$$

$$w = -P_{ext} \times \Delta V = -(100. \text{ kPa})(0.227 \text{ L}) = -22.7 \text{ J}$$

Does the answer make sense?   Since the volume is increased by the reacting system, work *leaves* the system, that is, the system does work in order to push back the piston. Work leaving the system is negative, so the sign of our result is correct and shows that work is done *by* not *on* the reacting system.

7.57 CHECK THIS

Pressure–volume work at constant pressure

For a reaction carried out at constant atmospheric pressure in a cylinder like the one in Figure 7.20, the final volume was found to be 0.521 L *less* than the initial volume. If the atmospheric pressure on the piston was 103 kPa, how much work was done on the reacting system?

Enthalpy revisited and defined Whether a reaction is carried out at constant volume or at constant pressure, the change in internal energy, ΔE, is the same. Using the subscripts V and P to designate constant volume and constant pressure, respectively, we have

At constant volume:

$$\Delta E = q_V + w_V = q_V + (-P\Delta V) = q_V - 0 = q_V \qquad \textbf{(7.48)}$$

At constant pressure:

$$\Delta E = q_P + w_P = q_P + (-P\Delta V) = q_P - P\Delta V \qquad \textbf{(7.49)}$$

For a constant volume system, $\Delta V = 0$, so $(-P\Delta V) = 0$. Equation (7.48) shows that the value of q_V (**heat at constant volume**) is a direct measure of the change in internal energy. However, we have said that we are usually interested in reactions at constant pressure. When the thermal energy transfer, q_P, is measured for a reaction at constant pressure, a pressure–volume work term must be taken into account, if gases are involved as reactants and/or products. If gases are not involved, then ΔV will be nearly zero and the $-P\Delta V$ term in equation (7.49) will also be close to zero. In these cases $q_V \approx q_P$.

Recall that when we introduced enthalpy in Section 7.6, we said that we would explain why scientists devised this special function. The reason is that, even when gases are involved, chemists are almost never interested in pressure–volume work. The focus of chemistry is on how the energy stored in chemical bonds changes in the course of a chemical reaction. The work term in the first law only makes things more complicated. The way to avoid dealing with the work term is to define a new thermodynamic quantity, and that is what was done with enthalpy *(H)*:

$$H \equiv E + PV \qquad \textbf{(7.50)}$$

The change in enthalpy, ΔH, for a reaction at constant pressure is

$$\Delta H = \Delta E + P\Delta V \qquad \textbf{(7.51)}$$

Equations (7.49) and (7.51) can be combined to give

$$\Delta H = q_P \qquad \textbf{(7.52)}$$

Equation (7.52) shows that q_P is a measure of the *change* in enthalpy. We can get ΔH directly from an experimental measurement of the thermal energy transfer at constant pressure, as we have been doing throughout the chapter.

We have been treating enthalpy as a state function (as when we sum the enthalpy changes around a cycle to zero in Figures 7.13 and 7.15). Enthalpy is defined in terms of variables that are state functions (internal energy, pressure, and volume), so enthalpy is a state function. We can measure the enthalpy change in a system that undergoes a change by measuring q_P, the thermal energy transferred in a constant pressure pathway that brings about this change. The enthalpy change is the same no matter how the specified change in the system is carried out; ΔH is a function only of the initial and final states of the system. The thermal energy transferred in the change, q, is a path function and depends on the way the change is carried out. Only for a constant pressure pathway is the heat equal to the enthalpy change.

7.58 CONSIDER THIS

How can ΔE (or ΔH) be the same for a reaction at constant volume and constant pressure?

Consider an exothermic reaction that produces a gaseous product from reactants in solution. When the reaction is run in a constant volume calorimeter, you find the increase in temperature of the solution is greater than when the same amount of reaction occurs at constant pressure.

(a) Is q, the thermal energy transfer *to the reaction*, positive or negative? Explain the reasoning for your answer.

(b) Is q larger (more positive) for the reaction in the constant volume or constant pressure system? Explain the reasoning for your answer.

(c) What is the sign of the work, w, done *on* the system in the constant pressure reaction system? Explain the reasoning for your answer.

(d) Show how your answers in parts (b) and (c) can combine to make ΔE the same, whether the reaction is carried out at constant volume or at constant pressure. Show the same for ΔH.

This analysis explains why enthalpy has also been called *heat content*. However, this term is misleading. Heat is not "contained" in a body, but is the thermal energy transferred between two bodies in a change.

Recall that the internal energy change is the change in the sum of potential and kinetic energies in going from reactants to products. The enthalpy change represents the thermal energy component of this change. To see this, consider that, if you subtract the pressure-volume work done on the system, $-P\Delta V$, from ΔE in equation (7.49), you get the enthalpy change, ΔH:

$$\Delta E - (-P\Delta V) = [q_p + (-P\Delta V)] - (-P\Delta V)$$

$$\Delta E + P\Delta V = q_p = \Delta H$$

To make these ideas more concrete, let's consider some chemical reaction systems.

7.59 WORKED EXAMPLE

Comparing ΔE and ΔH

Are ΔE and ΔH the same for the oxidation of methanol? If not, is work done *on* the reaction system or *by* the reaction system?

$$CH_3OH(l) + \tfrac{3}{2}O_2(g) \rightarrow CO_2(g) + 2H_2O(l) \qquad (7.23)$$

Necessary information: In addition to the balanced chemical equation (7.23) with the phases of the reactants and products given, we need the definition of work done on the system, $-P\Delta V$.

continued

Strategy: Does a volume change occur in the reaction? If it does, then ΔE and ΔH are not the same. Use the sign of the volume change to determine whether work was done on or by the system.

Implementation: In reaction equation (7.23) there are more moles of gaseous reactants, $\frac{3}{2}$ mole, than moles of gaseous products, 1 mole. Thus, there is a change in volume: ΔE and ΔH are not the same.

Since the number of moles of gas decreases, ΔV is negative and $-P\Delta V$ is positive. Work is done on the system in this change.

Does the answer make sense? In a reaction with a change in the number of moles of gas, the volume changes and ΔE and ΔH are not the same. A decrease in volume of the reacting system means work is done on it, as you saw earlier in this section.

7.60 CHECK THIS

Comparing ΔE and ΔH

For which of these chemical reactions are ΔE and ΔH the same. Where pressure–volume work is done, indicate whether work is done on the reaction system or by the reaction system:
(a) $2H_2(g) + O_2(g) \rightarrow 2H_2O(l)$
(b) $C_6H_{12}O_6(s) + 6O_2(g) \rightarrow 6CO_2(g) + 6H_2O(l)$
(c) $CaCO_3(s) \rightarrow CaO(s) + CO_2(g)$
(d) $H_3O^+(aq) + OH^-(aq) \rightarrow 2H_2O(l)$

7.61 WORKED EXAMPLE

Comparing ΔE and ΔH quantitatively

When one mole of methanol is oxidized, reaction equation (7.23), under standard conditions at a constant pressure of 1 bar, 1.2 kJ of work is done on the system. How do $\Delta E°_{reaction}$ and $\Delta H°_{reaction}$ compare for this reaction?

Necessary information: We need equation (7.49) and $\Delta H°_{reaction} = -726.2$ kJ from Worked Example (7.49).

Strategy: Substitute the heat, $q_p(= \Delta H°_{reaction})$ and the work, w_p, into equation (7.49) to get $\Delta E°_{reaction}$.

Implementation:

$$\Delta E°_{reaction} = q_p + w_p = \Delta H°_{reaction} + w_p = -726.2 \text{ kJ} + 1.2 \text{ kJ} = -725.0 \text{ kJ}$$

We see that $\Delta E°_{reaction} \approx \Delta H°_{reaction}$. The values differ by less than 0.2%.

Does the answer make sense? The thermal energy transfer from the reacting system is quite large compared to the work done on the system. The work term has little affect on the relative values of the energy and enthalpy changes. This is a general result for most highly exothermic or endothermic reactions, even when gases are involved: $\Delta E°_{reaction} \approx \Delta H°_{reaction}$.

7.62 CHECK THIS

Comparing ΔE and ΔH quantitatively

(a) When 1 mol of 1,2-dibromoethane decomposes to give ethene and gaseous bromine, under standard conditions at a constant pressure of 1 bar, 2.5 kJ of work is done by the system. How do $\Delta E°_{reaction}$ and $\Delta H°_{reaction}$ compare for this reaction?

$$CH_2BrCH_2Br(g) \rightarrow CH_2CH_2(g) + Br_2(g)$$

Hint: This is the reverse of reaction (7.31) in Check This 7.51.

(b) How good is the approximation, $\Delta E°_{reaction} \approx \Delta H°_{reaction}$, for this reaction system? Under what conditions will the approximation not be valid? Explain your reasoning.

When there is no change in gas volume, $\Delta V = 0$, little pressure–volume work is done either on or by the system. In this case, $\Delta E \approx \Delta H$; the internal energy change and the enthalpy change are essentially the same. Also, when the thermal energy transfer is large, we can usually equate the energy and enthalpy changes. In Section 7.13, you can investigate the difference between a chemical change that is carried out under constant volume and constant pressure conditions.

Reflection and Projection

Changes in the internal energy of a system are a result of heat and work transfers to the system (E increases) or from the system (E decreases). The sum of the heat and work transferred to or from a system is its change in internal energy, ΔE, which can be determined by measuring the transfer of thermal energy, q_V, to or from the system in a constant volume change. For constant pressure systems, we must also account for the pressure–volume work to determine the changes in E. Pressure–volume work transfers are positive if the system is compressed or the number of moles of gas decreases (the volume decreases) and negative if the system expands or the number of moles of gas increases (the volume increases).

Enthalpy, H, was introduced in order to avoid dealing with pressure–volume work. ΔH can be determined by measuring the transfer of thermal energy, q_P, to or from the system in a constant pressure change. Enthalpy, like energy, pressure, volume, and temperature, is a state function; the change in enthalpy for a change in a system depends only on the initial and final states of the system. For most reactions of practical importance, like combustions or reactions in solution, the difference between ΔH and ΔE is so small that it usually can be neglected; that is, $\Delta H \approx \Delta E$.

The energetics of reactions are critically important for understanding and using them, but the energy (or enthalpy) doesn't tell us everything, as we will remind you in the next section.

7.11. What Enthalpy Doesn't Tell Us

Is it really possible to obtain enough energy from half an ounce of coal (14 g) to move two tons a mile? The citation from Emerson at the beginning of the chapter says that it is. We don't have enough information about *The Rocket*, Stephenson's locomotive and the loads it pulled to answer this question directly, but we can use what we know about modern engines to get some idea whether it was really possible. This is what we often have to do, substitute a problem (pathway) we know how to solve for another that we don't have enough information to solve.

There are small automobiles that can travel about 50 miles on the energy of combustion from one gallon (3.7 L) of gasoline or one mile on 74 mL. Assume that the car is a two-ton load (an overestimate) and its engine is our "locomotive." *The engine is our system.* Engines run in cycles and return to the same state at the end of each cycle. The internal energy of an engine is the same at the end of a mile of travel as it was at the beginning; $\Delta E = 0$ for the engine. Therefore, from the first law of thermodynamics, the maximum amount of work we can get *out* of an engine to move a load is equal to the amount of heat that we put into the engine from the fuel it burns, $q = w$. Table 7.6 provides a comparison between the heat available to *The Rocket* and the automobile.

Table 7.6 *Comparison of thermal energy available to* **The Rocket** *and an automobile.*

Fuel	Coal (carbon)	Octane (C_8H_{18})
Mass of fuel for one mile	14 g	$(74 \text{ mL})\left(\dfrac{0.8 \text{ g}}{1 \text{ mL}}\right) = 60 \text{ g}$
Moles of fuel	1.2 mol	0.5 mol
ΔH of combustion	400 kJ·mol^{-1}	5400 kJ·mol^{-1}
Thermal energy produced	480 kJ	2700 kJ

If *The Rocket* had been as fuel efficient as a modern car, it would have required $\left(\dfrac{2700 \text{ kJ}}{480 \text{ kJ}}\right) \times 0.5 \text{ oz} = 3 \text{ oz}$ of coal to move the load a mile. Steam engines, especially the earliest ones were not nearly this efficient, but even if the difference was a factor of 10, only 30 oz, about 2 pounds, of coal would have been required to move 2 tons, 4000 pounds, a mile. The steam engine was a marvelous invention. Forgive Emerson for stretching the marvel a bit.

Our comparison of thermal energies in Table 7.6 is valid, but the statement that the amount of work available is equivalent to the thermal energy of fuel combustion is not. Not all the heat (undirected molecular motion) that goes into an engine can be converted to work (directed motion). The first law of thermodynamics, conservation of energy, does not provide any explanation for this limitation on conversion of one form of energy to another. What other puzzles are posed by the first law?

In Section 7.9, we emphasized coupling endothermic to exothermic reactions, in order to drive the energy-requiring reactions in living systems. This emphasis might lead you to believe that the criterion for a reaction to occur is that it be exothermic. When stronger chemical bonds replace weaker ones, the products *are* more stable (lower potential energy), and it seems reasonable that a chemical reaction should be likely. Many observed reactions are exothermic, but many endothermic reactions also occur. You have seen several examples in this and in previous chapters.

In everyday use, "spontaneous" usually means impulsive or arising without forethought and often carries the implication of happening quickly. In science, spontaneous simply indicates that a change is possible or feasible and can occur; it implies nothing about speed.

We observe changes taking place around us all the time. We call many of these changes **spontaneous,** which in chemical terms means that the change is possible and can occur without apparent external cause. The process may be slow, but, given enough time, it does occur. *Speed of change is not a criterion for spontaneity*. Two examples of spontaneous change are dissolving ammonium chloride, $NH_4Cl(s)$, or calcium chloride, $CaCl_2(s)$, in water (Chapter 2, Investigate This 2.22). In these cases, if you just add the solute to some water, the dissolution occurs without any further intervention on your part. The enthalpy change for dissolving ammonium chloride in water is positive:

$$NH_4Cl(s) \rightarrow NH_4^+(aq) + Cl^-(aq) \quad \Delta H^\circ = +14.8 \text{ kJ} \quad \text{(7.53)}$$

The enthalpy change for dissolving calcium chloride in water is negative:

$$CaCl_2(s) \rightarrow Ca^{2+}(aq) + 2Cl^-(aq) \quad \Delta H^\circ = -82.9 \text{ kJ} \quad \text{(7.54)}$$

The sign of the enthalpy change is reversed when a reaction is written in reverse. An exothermic reaction going in reverse is endothermic; the initial and final states are reversed. Since both endothermic and exothermic reactions can be spontaneous, the sign of ΔH can't be used to predict the way reactions actually go; enthalpy alone does not enable us to understand or predict the direction of spontaneous chemical change. However, understanding the directionality of change is essential to explaining, for example, the driving forces in the organization of living cells. To address the problem of directionality of change, we have to account for changes in both the system *and* its surroundings. And we need another thermodynamic state function—**entropy**—the subject of the next chapter.

7.12 Outcomes Review

Everything you do, including thinking, requires energy. In this chapter we examined various forms of energy and transfers of energy between chemical and physical systems. We learned that the first law of thermodynamics is a statement of the principle of conservation of energy. Changes in the internal energy of a system, ΔE, are combinations of thermal and mechanical energy (heat and work) transfers to or from the system. We found that the same change in internal energy could be brought about by different heat and work transfers. A change in internal energy is a function only of the initial and final states of the system, but heat and work transfers depend on the path taken from the initial to the final state of the system.

We introduced a new state function called enthalpy, H, that accounts for the thermal energy component of an internal energy change when pressure-volume work is involved in the change. Calorimetric measurements at constant pressure provide a measure of ΔH for reactions. We can use these enthalpy data (among others) to get bond enthalpies and enthalpies of formation for compounds. Using these values, we can calculate the enthalpy change for essentially any reaction. We cannot, however, use energy to predict the direction of changes; we will pursue that goal in the next chapter.

Check your understanding of the ideas in the chapter by reviewing these expected outcomes of your study. You should be able to

- Identify the forms of energy transferred in physical and chemical changes and show how energy is conserved in the changes [Sections 7.1, 7.2, 7.3, 7.4, and 7.10].
- Draw molecular level representations of thermal energy (undirected kinetic energy) and mechanical energy (directed kinetic energy) transfers [Sections 7.2, 7.3, and 7.10].

- Identify whether a thermal energy transfer occurs by radiation, conduction, and/or convection [Section 7.3].
- Define and identify the variables that are functions of state and those that are functions of the path for a given change [Sections 7.4 and 7.10].
- Identify the system and the relevant surroundings for a given change [Section 7.5].
- Define and identify open, closed, and isolated systems [Section 7.5].
- Use the data from calorimetric measurements to calculate the thermal energy transferred to or from a reacting system [Section 7.6].
- Use the data from constant pressure calorimetric measurements to calculate the enthalpy change for the reacting system [Sections 7.6 and 7.13].
- Define and give examples of homolytic bond cleavage reactions [Section 7.7].
- Write equations for homolytic bond cleavage and homolytic bond formation for compounds, use bond enthalpies to calculate the enthalpy changes associated with these processes, and obtain the enthalpy change for gas phase reactions [Section 7.7].
- Draw enthalpy-level diagrams that show how bond enthalpies combine to give the enthalpy change for a reaction [Section 7.7].
- Use bond enthalpies to predict whether, for given atoms, reactions will favor singly bonded or multiply bonded products [Section 7.7].
- Define standard states for elements and compounds and write the equations whose enthalpy changes are the standard enthalpies of formation of the compounds [Section 7.8].

- Use standard enthalpies of formation to calculate the standard enthalpy change for a reaction [Section 7.8].
- Draw enthalpy-level diagrams that show how standard enthalpies of formation combine to give the standard enthalpy change for a reaction [Section 7.8].
- Define coupled reactions and identify examples based on the definition [Section 7.9].
- Show whether coupling between two reactions would be an energetically favorable combination [Section 7.9].
- State the first law of thermodynamics in terms of internal energy, heat, and work and use it to analyze a change that occurs by different pathways [Section 7.10].
- Explain the difference between ΔE and q_V and between ΔH and q_P [Section 7.10].
- Calculate ΔE, given ΔH and appropriate pressure and volume or P-V work data for a reaction, and *vice versa* [Sections 7.10 and 7.13].
- Determine whether a process is consistent with (allowed) by the first law of thermodynamics [Section 7.11].
- Use the kinetic–molecular model of gases to explain the observed effects of changes in P, V, n, or T [Section 7.13].
- Calculate the final value for P, V, n, or T, given their initial values and the changes in three of the variables [Section 7.13].
- Use the ideal gas equation to calculate the P-V work done on or by a chemical reaction system [Section 7.13].

7.13. Extension—Ideal Gases and Thermodynamics

 7.63 INVESTIGATE THIS

What happens when a gas is compressed or heated?

Do this as a class investigation and work in small groups to discuss and analyze the results. Draw about 40 mL of air into a 50-mL plastic syringe and connect the outlet of the syringe to a U-tube made of plastic tubing that is about half full of colored water (for visibility). Adjust the syringe plunger so the liquid levels are the same in both sides of the U-tube, as shown in the illustration.

(a) Push the syringe plunger in to decrease the volume of air in the syringe by several milliliters. Record your observations on the liquid levels in the U-tube.

continued

(b) Readjust the syringe plunger until the U-tube liquid levels are the same again. While holding the volume constant, warm the syringe by holding it in your hand or directing a warm stream of air onto it. Record your observations on the liquid levels in the U-tube.

7.64 CONSIDER THIS

What causes changes when a gas is compressed or heated?

(a) How, if at all, did the liquid levels in the U-tube change when you decreased the volume in the syringe in Investigate This 7.63? What change(s) in the properties of the gas sample was (were) responsible for any change in the liquid levels? Explain the connection clearly.

(b) How, if at all, did the liquid levels in the U-tube change when you warmed the air in the syringe? What change(s) in the properties of the gas sample was (were) responsible for any change in the liquid levels? Explain the connection clearly.

In Investigate This 7.63, the liquid-containing U-tube is a **manometer,** a device for measuring gas pressures. When the liquid levels are the same in both arms of the manometer, the gas pressures pushing down on the liquid surfaces are the same, as shown in Figure 7.21(a). When the levels are unequal, a higher gas pressure is pushing down on the liquid in the arm with the lower level. This higher pressure is equal to the pressure of gas pushing down on the other arm plus the pressure of the column of liquid that is above the lower level, Figure 7.21(b).

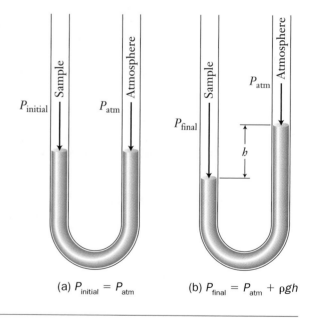

(a) $P_{initial} = P_{atm}$ (b) $P_{final} = P_{atm} + \rho gh$

Figure 7.21.

Manometric measurement of a gas sample. The manometer in Investigate This 7.63 is represented here. (a) Shows the equal pressures of the sample and the atmosphere when the liquid levels are equal. (b) Shows how the pressure of the sample is determined when the liquid levels are not equal. ρgh is the pressure in kPa due to the column of liquid with a density of ρ kg·L^{-1} and height of h meters acted on by gravitational acceleration, $g = 9.80$ m·s^{-2}.

7.65 CONSIDER THIS

How are manometer readings interpreted and quantified?

Suppose you have set up a syringe and manometer as shown in the illustration in Investigate This 7.63. After moving the syringe plunger you observe that the liquid level in the manometer arm attached to the syringe is 15.7 cm higher than the level in the open arm.

(a) What conclusion can you draw about the gas pressure in the syringe? What motion of the syringe plunger produced this pressure? Explain your responses.

(b) If the atmospheric pressure in the room is 100.4 kPa, what is the pressure of the gas in the syringe? *Hint:* The density of liquid (water) in the manometer is $1.00 \text{ kg} \cdot \text{L}^{-1}$.

Kinetic–molecular model of gases In Investigate This 7.63, you found that the level of liquid in the arm of the manometer connected to the syringe went down when the gas was compressed or warmed. Both compression and warming of a gas sample increase its pressure. Why? To answer this question we use a molecular-level model of gases. We have been using the model implicitly since Chapter 1, where we presented illustrations and animations (in the *Web Companion*) of molecules in the gas phase. In this section, we are going to use the model to explain the observed behavior of gases, which we will describe by the ideal gas equation. Then we will apply the equation to a chemical reaction occurring at either constant volume or constant pressure and see how it can account quantitatively for the results.

In our model for gases, we assume that molecules in the gas phase are far apart, so they do not often interact with one another, and that they are moving about randomly. We assume that the molecules act like hard spheres (marbles, for example) in their collisions with the walls of their container. In these the collisions a molecule pushes on the wall and the wall pushes back with an equal and opposite push. Finally, we assume that the average kinetic energy of the molecules is proportional to the temperature of the gas (in kelvin). This is called the **kinetic–molecular model** of gases because it is based on molecular motion ("kinetic" refers to motion).

When we measure the pressure of a gas sample, we are measuring the effect of the collisions of the moving gas molecules with the walls of their container (including the surface of the liquid in our manometer). Two factors can cause the pressure to change. First, the number of molecules colliding with the walls in a given time might change. The number of molecules hitting the walls per unit time is proportional to the number of molecules per unit volume of the gas. Second the average speed with which the molecules strike the walls might change. If the average kinetic energy of the molecules changes, their average speed changes.

> Kinetic and related terms (from Greek *kineein* = to move) also appears in the many other contexts: kinetic energy is the energy associated with motion; kinematics is the branch of physics that deals with motion; kinesthesiology is the study of muscular movement in animals; and telekinesis is the yet-to-be-proved movement of objects by use of thought alone.

7.66 CONSIDER THIS

How does kinetic–molecular theory apply to Investigate This 7.63?

(a) How does the number of molecules per unit volume of gas change in part (a) of Investigate This 7.63? Are your manometric observations consistent with this change? Use the kinetic–molecular model of gases to explain your response.

continued

(b) How does the average kinetic energy of the gas molecules change in part (b) of Investigate This 7.63? Are your manometric observations consistent with this change? Use the kinetic-molecular model of gases to explain your response.

(c) Imagine that you remove some of the molecules from a flask containing a gaseous sample (keeping the temperature constant). If the flask is connected to a manometer, what would you expect to observe? Use the kinetic–molecular model of gases to explain your response.

The ideal-gas equation Your results in Investigate This 7.63 and your analyses in Consider This 7.66 provide you the direction and the kinetic–molecular explanation for pressure changes that occur when you change the volume, temperature, and number of molecules in a gaseous sample. You found that the pressure increases when the volume decreases (an inverse relationship), that the pressure increases when the temperature increases (a direct relationship), and that the pressure will decrease if the number of molecules in the sample decreases (a direct relationship). We can show these relationships in a tentative equation that combines them all:

$$P \propto \frac{nT}{V} \tag{7.55}$$

Here, n is the number of moles of gas, a measure of the number of molecules in the sample.

7.67 CHECK THIS

Relationships among the properties of a gas sample

Explain how equation (7.55) embodies the relationships in the preceding paragraph.

Equation (7.55) was derived from quantitative studies of gases begun in the 17th century and was a part of the evidence used to develop the kinetic–molecular model during the 19th century. Equation (7.55) implies that the effects of changes in the properties of a gas sample are independent of one another and independent of the identity of the gas. The kinetic–molecular model for gases is consistent with this independence. The model treats all gases as hard spheres, so the identity of the gas makes no difference. The model assumes that the average kinetic energy is the same for all samples of gas molecules at the same temperature. The average kinetic energy does not depend on the size of the sample or the number of molecules per unit volume.

The relationships among the variables in equation (7.55) are usually written in this form:

$$PV = nRT \tag{7.56}$$

Equation (7.56) is called the **ideal-gas equation** and the proportionality constant, **R**, is called the **gas constant.** The value of the gas constant is

8.314 $J \cdot mol^{-1} \cdot K^{-1}$, when pressure is given in kPa and volume in liters. The ideal-gas equation is an approximation to the behavior of real gases. Molecules are not hard spheres, but are somewhat squishy, so collisions between them and with surfaces are not ideal. Molecules in the gas phase are not entirely independent of one another: They take up space and attract one another by polar and dispersion forces. The attractions depend on the distance between molecules and the time they spend near one another, so they are less important when there are few molecules per unit volume (low pressure) or moving rapidly (high temperature). The ideal-gas equation provides a good approximation for the behavior of most gases near or below atmospheric pressure and at room temperature or above.

Pressure measurements are still often given in atmospheres (atm), so another useful value for R is 0.08206 $L \cdot atm \cdot mol^{-1} \cdot K^{-1}$.

7.68 WORKED EXAMPLE

Using the ideal-gas equation

The pressure in an automobile tire reads 38 psi (pounds per square inch) on a tire gauge at the start of a trip when the tire is at 21 °C. At the end of the trip, the tire has warmed to 42 °C. Assuming that the volume of the tire does not change and that no air leaves the tire, what is the air pressure in the tire at the end of the trip? What would the tire gauge read?

Necessary information: We need to know that a tire gauge reads the pressure of the tire *above* atmospheric pressure, which is about 15 psi. We need the ideal-gas equation and the temperature in kelvin, which we get by adding 273 to the numeric value of the Celsius temperature.

Strategy: Convert the temperatures to kelvin, calculate the initial pressure of air in the tire, and apply the ideal-gas equation to the initial and final conditions to find the final tire pressure. Since the number of moles of gas in the tire and the volume of gas in the tire do not change during the trip, one way to do this problem is to rearrange the ideal-gas equation to put all the constant terms on one side. Since these terms do not change, equate the initial and final values of the variable side of the equation and solve for the unknown value.

Implementation:

initial temperature: (21 + 273) K = 294 K

final temperature: (42 + 273) K = 315 K

initial pressure: 38 psi + 15 psi = 53 psi

Rearranging the ideal-gas equation to group the constant terms on the right gives

$$\frac{P}{T} = \frac{nR}{V} = \text{a constant}$$

Therefore,

$$\frac{P_i}{T_i} = \frac{P_f}{T_f} = \frac{53 \text{ psi}}{294 \text{ K}} = \frac{P_f}{315 \text{ K}}$$

$P_f = 57$ psi   The tire gauge will read 42 psi (= 57 psi − 15 psi)

continued

Does the answer make sense? You know from Investigate This 7.63 that the pressure of a sample of gas increases when it is warmed up (at constant volume). This is also the result of this calculation, so it makes sense. The kelvin temperature increased by about 7% ($\approx 21/304$) and the pressure also increased by about 7% ($\approx 4/55$), as we expect from their direct relationship. Note that we can express the pressure in any units that are convenient for the problem, as long as only initial and final *ratios* are involved.

7.69 CHECK THIS

Using the ideal-gas equation

Imagine that you do an experiment like the one in Investigate This 7.63, but instead of a U-tube manometer, you use an electronic pressure sensor interfaced to a computer to measure the pressure. If you start with the syringe plunger at a volume of 44.7 mL at 1.00×10^2 kPa, to what volume will you have to move the plunger to get a gas pressure of 1.50×10^2 kPa, if the temperature of the gas stays the same.

7.70 WORKED EXAMPLE

Using the ideal-gas equation

A 525-mL sample of gas initially at 167.4 kPa and 212 °C undergoes a change to 101.3 kPa and 37 °C. What is the final volume of the gas?

Necessary information: We need the ideal-gas equation and the conversion from °C to K.

Strategy: Convert the temperatures to kelvin and use an approach similar to Worked Example 7.68. Rearrange the ideal-gas equation to put all the constant terms on one side, equate the initial and final values of the variable side of the equation, and solve for the unknown value.

Implementation: In this problem, the number of moles of gas does not change, so write the ideal gas equation as

$$\frac{PV}{T} = nR = \text{a constant}$$

Therefore, $\dfrac{P_i V_i}{T_i} = \dfrac{P_f V_f}{T_f}$

and $V_f = \left(\dfrac{P_i}{P_f}\right)\left(\dfrac{T_f}{T_i}\right)V_i = \left(\dfrac{167.4 \text{ kPa}}{101.3 \text{ kPa}}\right)\left(\dfrac{310 \text{ K}}{485 \text{ K}}\right)(525 \text{ mL}) = 555 \text{ mL}$

Does the answer make sense? Consider the changes in pressure and temperature separately and then combine the effects. If the pressure on a gas is *decreased*, its volume will *increase* in inverse proportion. In this case, the pressure is reduced to about $\frac{3}{5}$ of its initial value, which would result in an increase in

continued

volume by a factor of about $\frac{5}{3}$. If the temperature of a volume of gas is *decreased*, its volume will *decrease* in direct proportion. In this case, the temperature is reduced to about $\frac{3}{5}$ of its initial value, which would result in a decrease of the volume by a factor of $\frac{3}{5}$. The pressure and temperature effects on the volume are in opposite directions and just about cancel one another. The volume should not change very much and this is the result we got, so it makes sense.

7.71 CHECK THIS

Using the ideal-gas equation

(a) To a sample of gas at 57.8 kPa and 289.2 K in a rigid (constant volume) container is added a second gas. The final pressure and temperature of the gas mixture are 95.8 kPa and 302.7 K, respectively. What is the *ratio* of the number of moles of added gas to the number of moles of gas originally present? *Hint:* All ideal gases act the same. A mixture of ideal gases can be treated as if all the molecules are the same.

(b) If the container in part (a) has a volume of 547 mL, how many moles of each gas are present in the final mixture? Explain how you get your answer.

Analysis of constant volume and constant pressure processes

A reason for introducing gas behavior and the ideal-gas equation is so that we can use them to compare ΔE and ΔH quantitatively for reactions carried out at constant volume and constant pressure. These analyses are applicable only to reactions that involve gases as reactants and/or products. For many other reactions, as we showed in Section 7.10, $\Delta E \approx \Delta H$.

7.72 INVESTIGATE THIS

Do reactions in open and capped containers differ?

Do this as a class investigation and work in small groups to discuss and analyze the results. Use a plastic soft drink bottle, a balance, a watch or timer, a plastic syringe with a short needle, two $\pm0.1\ °C$ thermometers that read from 0 to 100 °C, and a 24/40 ribbed rubber septum. Solid sodium hydrogen carbonate (baking soda, $NaHCO_3$) and 6 M hydrochloric acid (HCl) solution are the reactants in this system. The set up for the investigation is shown in the photograph.

(a) Add 8.4 g of $NaHCO_3(s)$ to the plastic bottle. Place one thermometer in the bottle, with the thermometer bulb at the bottom. Use the other thermometer to measure the temperature of a solution of 6 M HCl(*aq*) to the nearest 0.1 °C. Use the syringe to dispense 20.0 mL of the HCl solution into the plastic bottle. Record temperature and time as you swirl the mixture to be sure all the solid contacts the acid solution. While continuing to mix, record the temperature of the mixture approximately every 10 seconds for about 2 minutes.

continued

(b) Repeat the experiment with these changes: After the $NaHCO_3(s)$ and thermometer are in the bottle, seal the bottle with a septum cap. Put the syringe needle through the septum cap to inject the acid into the bottle and then remove the needle. *IMPORTANT:* When the reaction is complete and you have made all your observations, remove the needle from the syringe and insert the needle through the septum cap to release gas pressure from inside the bottle.

7.73 CONSIDER THIS

Why do reactions in open and capped containers differ?

Hydrogen carbonate ion reacts with hydronium ion to produce carbon dioxide gas and water. The reaction with solid sodium hydrogen carbonate (baking soda) is

$$NaHCO_3(s) + H_3O^+(aq) \rightarrow CO_2(g) + Na^+(aq) + 2H_2O(l) \qquad (7.57)$$

(a) In Investigate This 7.72, what was the maximum change in temperature (from the starting temperature of the acid) in the open bottle? In the capped bottle? Were these results surprising? Why or why not?

(b) Which one of the reactions in Investigate This 7.72, was carried out at constant volume? At constant pressure?

(c) Does your answer to part (b) help you interpret the results you reported in part (a)? Why or why not?

Web Companion

Chapter 7, Section 7.11.1 — ①

Use an interactive animation ②
of Investigate This 7.72 to
help analyze the changes. ③
④

You found that the temperature change when reaction (7.57) is carried out in an uncapped container is different from the temperature change for the reaction carried out in a capped container; the thermal energy transfers are different in the uncapped and capped containers. The reaction that occurs is the same in both containers, so the enthalpy change, ΔH, must be the same in both cases. This is also true for ΔE. Since energy is conserved, some other form of energy must also be different in the two cases. What is different, as you saw in Section 7.10, is the work.

Reaction (7.57) produces carbon dioxide gas. In the capped container, the added gas increases the pressure inside the container, but nothing in the surroundings moves, because the gas is trapped. This reaction is carried out at constant volume; no work is done on or by the reacting system. When the reaction is carried out in the open container, the carbon dioxide pushes back the atmosphere as it fills the bottle and pushes the air out. The reacting system does work pushing back the atmospheric gases. The reaction is carried out at constant atmospheric pressure.

You have found that reaction (7.57), the hydrogen carbonate–acid reaction, is endothermic; q_V and q_P are both positive quantities. The thermal energy transfer is larger when the reaction is run at constant pressure: $q_P > q_V$. Since the reaction produces gas, ΔV is positive and $w = -P\Delta V$ is negative. The heat and work terms in equation (7.49) must combine (work canceling some heat) to give the same value for ΔE as is obtained in equation (7.48) for the constant volume reaction:

$$\Delta E = q_V = q_P + w = q_P - P\Delta V \qquad (7.58)$$

The analysis here is similar to what you did in Consider This 7.58. The directions (signs) of the quantities in equation (7.58) are shown in Figure 7.22. To find out whether the relationships among q_V, q_P, and w shown in the figure (and asserted in Consider This 7.58) are true, we need to determine their numeric values.

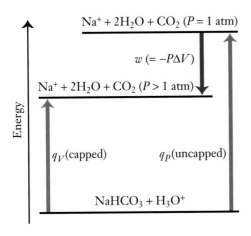

$Na^+ + 2H_2O + CO_2$ ($P = 1$ atm)

w ($= -P\Delta V$)

$Na^+ + 2H_2O + CO_2$ ($P > 1$ atm)

q_V(capped) q_P(uncapped)

$NaHCO_3 + H_3O^+$

Energy

Figure 7.22.

Relationships among q_V, q_P, and w for reaction (7.57).

7.74 CONSIDER THIS

What are q_V and q_P for the hydrogen carbonate–acid reaction?

Use your liquid volume and temperature data from Investigate This 7.72 (or the data from the *Web Companion*, Chapter 7, Section 7.11.1) to estimate values of q_V and q_P for reaction (7.57). Assume that no thermal energy is transferred to or from the contents of the bottle during the reactions. Treat this setup just like the calorimeters discussed in Section 7.6 and use your data to do calculations like the ones there. In the absence of any other information, assume that the specific heat of the solution is the same as for water.

Pressure–volume work calculation Your calculations in Consider This 7.74 provide estimates of the numeric values of q_V and q_P for reaction (7.57), but to test equation (7.58) we still need a value for the work term, $-P\Delta V$. The easiest approach is to use the ideal gas equation, $PV = nRT$, to describe the gas, because we can then express the pressure–volume work in terms of variables we already know, the temperature and the number of moles of reactants and products. Write the ideal gas equation for the initial conditions (number of moles of reactant gases, n_i) and the final conditions (number of moles of product gases, n_f):

Initial: $PV_i = n_i RT$

Final: $PV_f = n_f RT$

Subtracting the initial conditions from the final conditions yields

$$P(V_f - V_i) = (n_f - n_i)RT$$

$$P\Delta V = (\Delta n)RT \tag{7.59}$$

Finally, substituting for $P\Delta V$ in equation (7.58) gives

$$q_V = q_P - (\Delta n)RT \tag{7.60}$$

7.75 WORKED EXAMPLE

Converting q_P (or ΔH) to q_V (or ΔE)

In Check This 7.51, you calculated $\Delta H°_{\text{reaction}} = -126.3$ kJ for bromination of 1 mol of ethene at 298 K:

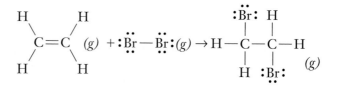

What is $\Delta E°_{\text{reaction}}$ at 298 K for this reaction?

Necessary information: We need to keep in mind that q_V and q_P are, respectively, measures of ΔE and ΔH.

Strategy: Determine Δn for the reaction and substitute in equation (7.60).

Implementation: If 1 mol of gaseous ethene reacts, 1 mol of gaseous bromine also reacts and one mole of gaseous product is formed. Thus 2 mol of gaseous reactants go to 1 mol of gaseous product and $\Delta n = (1 \text{ mol}) - (2 \text{ mol}) = -1$ mol.

$$q_V = q_P - (-1 \text{ mol})(8.314 \text{ J·mol}^{-1}\text{·K}^{-1})(298 \text{ K}) = q_P + 2.5 \text{ kJ}$$

Substitute $\Delta E°_{\text{reaction}} = q_V$ and $\Delta H°_{\text{reaction}} = q_P$ to get

$$\Delta E°_{\text{reaction}} = -126.3 \text{ kJ} + 2.5 \text{ kJ} = -123.8 \text{ kJ}$$

Does the answer make sense?   The reaction is exothermic, so thermal energy leaves the reacting system. Since the number of moles of gas decreases in the reaction, the volume change is negative, $\Delta V < 0$. If the change occurs at constant pressure, work is done on the reacting system and mechanical energy is added to the system. For a change at constant volume, no work is done on or by the system, but the change in internal energy is the same as for the constant pressure process. Since no work is added to the constant volume system, less thermal energy has to leave the system for the same change in internal energy. This is the result we got: $|q_V| < |q_P|$. Compare the numerical values here with those in Check This 7.62; be sure you can explain the correlations. Draw a diagram, modeled after Figure 7.22, to show the relationships among q_V, q_P, and w for this system.

7.76 CONSIDER THIS

What is w for the hydrogen carbonate–acid reaction?

(a) How many moles of $NaHCO_3$ did you use in the reactions in Investigate This 7.72?
(b) How many moles of H_3O^+ (HCl) did you use in the reactions?
(c) How many moles of CO_2 were formed in the reactions?
(d) What is $P\Delta V$ for the reaction carried out at constant pressure?
(e) Within the uncertainties of the experimental measurements, are your results from part (d) and from Consider This 7.74 consistent with equation (7.58)? Explain why or why not.

7.77 CHECK THIS

$\Delta E_{reaction}$ and $\Delta H_{reaction}$ for the hydrogen carbonate–acid reaction

Values of $\Delta E_{reaction}$ and $\Delta H_{reaction}$ are usually reported for a mole of reactant undergoing the reaction. Use your results from Consider This 7.74 and the number of moles reacting from Consider This 7.76 to get $\Delta E_{reaction}$ and $\Delta H_{reaction}$ for one mole of $NaHCO_3$ reacting according to reaction equation (7.57). Explain your procedure.

Chapter 7 Problems

7.1. Energy and Change

7.1. A slice of cheese pizza typically contains 180 Calories of food energy. The human heart requires about 1 J of energy for each beat.
(a) Assuming that when you metabolize the pizza, all of the food energy will be used to keep your heart beating, how many heartbeats can this slice of pizza sustain? About how many minutes would this keep your heart going? *Hint:* 1 Calorie = 1 kcal.
(b) Assume the efficiency of use of this food energy is about the same as that calculated for capturing the energy from glucose oxidation, Section 7.9. What are your answers for part (a) in this case? Explain your reasoning.

7.2. Pasta is largely starch, and a deep-fried potato chip is mostly starch and fat.
(a) What do your observations from Investigate This 7.1 tell you about the energy value of these foods? Do they combine exothermically with oxygen? What is the evidence and reasoning for your response?
(b) Starch is a polymer of glucose. Is this composition consistent with your answers in part (a) and other data from Investigate This 7.1? Explain your reasoning.

7.3. Predict how the results of Investigate This 7.1 would differ if a baked, rather than a fried, potato chip were used. What will happen if a raw potato slice is used? What do you predict will be observed if a chip using Olestra® (a fat substitute) is used? In each case, explain the reasons for your prediction.

7.2. Thermal Energy (Heat) and Mechanical Energy (Work)

7.4. Several different types of energy have been discussed in this chapter. These include kinetic energy, work, potential energy, and heat. How are these terms related?

7.5. Figure 7.2 illustrates a model for the movement of combustion gases.
(a) Draw a similar diagram for a tethered helium-filled balloon. *Hint:* Unlike Figure 7.2, there will only be one type of particle, representing helium atoms.
(b) Will the helium atoms in the tethered balloon exhibit any directed motion? Why or why not? What will happen to the motion of the helium atoms in the balloon when the tether is released? Explain briefly.

7.6. Consider a drop of water in a waterfall. At the very top of the fall, the molecules in the drop are moving randomly. While it is falling (pulled down by gravity), its molecules have a net downward motion. When it hits the pool at the bottom, the molecular motion again becomes random.
(a) The water at the bottom of the waterfall is a little warmer than the water at the top. What can you say about the change in internal energy of your drop of water as it goes from the top of the waterfall to the bottom?
(b) Has work been done or has thermal energy been transferred to cause the change in the internal energy of the water drop? Explain where the work and/or thermal energy come from.

7.7. When heat is added to H_2O to do work, why does the vapor form of water generate more work than either ice or liquid water?

7.8. Why is it important to you, as a consumer, for automotive engineers to design engines that convert the highest possible percentage of energy to work?

7.3. Thermal Energy Transfer (Heat)

7.9. For each case, is thermal energy being transferred by radiation or by a contact process of conduction and/or convection? Briefly justify your choice.
(a) A silver spoon at room temperature, when placed in a cup of hot water, becomes too warm to touch.
(b) An apple pie is baked in an electric oven.
(c) The water in an outdoor swimming pool cools from 25 °C to 20 °C as the summer season changes into autumn.
(d) You sunbathe with your back exposed to the sun, until your back feels very warm.

7.10. At the end of a skating season, an indoor ice rink was closed. Without any outside cooling, how will the roof and the ice rink floor reach a common temperature? Will it be a slow or fast process? Explain your answer.

7.11. After being immersed in cold water, the core body temperature of an individual may be abnormally low. Design three methods, one using convection, the second using conduction, and the third using radiation to transfer heat to warm up an individual as quickly and safely as possible. Which method will be the most efficient in this case?

7.4. State Functions and Path Functions

7.12. Your monthly bank statement gives the opening balance for your account, details the transactions that have occurred, and reports the ending balance. Which aspects of this statement correspond to state functions and which to path-dependent functions?

7.13. The quarterback starts a play on his own 35-yard line, but is pushed back by the defense to his own 25-yard line. He then successfully completes a pass to the opponent's 40-yard line. What is the number of yards gained in the play? Is the number of yards gained a state function or a path-dependent function?

7.14. Today's most advanced fossil-fuel burning power plants producing electricity operate at an efficiency of about 42%. Suggest some reasons why a higher percentage of the energy from burning the fossil fuels is not converted to work. Do you think it is theoretically possible to convert 100% of the energy into work? Explain your reasoning.

7.15. Consider two different cases in which a fully charged battery becomes totally discharged.

> Case 1: The battery is used to provide power to a flashlight.

> Case 2: The battery is used to provide power for a child's toy car.

(a) Will the change from the battery being fully charged to being totally discharged be a state function or a path-dependent function in each case? Explain your reasoning.
(b) Will there be any thermal energy transfer by radiation or by contact in either use of the battery? Explain your reasoning?

7.16. Consider a C-172 airplane (small private plane), a helicopter, a parachute jumper, and a hawk, all presently at 1892 feet above the Clemson, South Carolina, airport, which is at an altitude of 892 feet. Each plans a landing at the airport. Comment on the state functions and path functions for each of these airborne objects.

7.5. System and Surroundings

7.17. Which ending(s) for this sentence is(are) correct? The enthalpy change for a system open to the atmosphere is _____. Explain your reasoning in each case.
(a) dependent on the identity of the reactants and products
(b) zero
(c) negative
(d) q_P
(e) positive

7.18. Give real-life examples of a(n)
(a) open system.
(b) closed system.
(c) isolated system.
(d) exothermic process.
(e) endothermic process.

7.19. A properly stoppered Thermos® bottle containing coffee is nearly an isolated system. The stopper prevents water vapor from escaping, while the vacuum construction keeps heat from being transferred to the surroundings. How would you turn this isolated system into an open system? A closed system?

7.20. Earth is sometimes described as a closed system, particularly when considering regional or global environmental problems. For example, acid rain may fall at great distance from the original site of the emissions responsible for the observed effect. Ozone depletion near the South Pole is linked to the emission of chlorofluorocarbons from uses in industrial nations. While this view does convey a sense of the many connections that affect environmental issues, is it correct to think of the earth as a closed system from the standpoint of thermodynamics? Why or why not?

7.21. This diagram shows an experiment in which a hot silver bar is added to a beaker of water at room temperature.

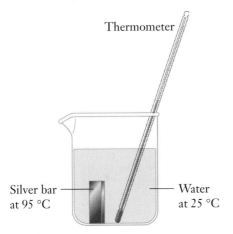

Thermometer

Silver bar at 95 °C — Water at 25 °C

(a) If the object of the experiment is to find the specific heat of the silver, define the system and the surroundings.
(b) If you already know the specific heat of the silver, the same experimental setup might be used to determine if there is heat loss through the walls of the glass beaker or from the thermometer itself. Define the system and surroundings in that experiment.
(c) Does this experimental diagram illustrate an open, closed, or isolated system? Give the reasoning behind your choice.

7.6. Calorimetry and Introduction to Enthalpy

7.22. In a calorimetric measurement, why might it be important to know the heat capacity of the calorimeter? The calorimeter heat capacity is the amount of thermal energy the calorimeter (container, thermometer, etc.) transfers per degree temperature change.

7.23. Choose the best response to complete the final sentence. Explain why you make this choice and what is wrong with each of the others. The equation for decomposition of gaseous ammonia, $NH_3(g)$, is

$$NH_3(g) \rightarrow \tfrac{1}{2}N_2(g) + \tfrac{3}{2}H_2(g) \quad \Delta H = -45.9 \text{ kJ}$$

This reaction equation and enthalpy change indicate that the *formation* of gaseous ammonia _____.
(a) evolves 45.9 kJ for each mole of ammonia formed
(b) evolves 23 kJ for each mole of nitrogen used
(c) absorbs 45.9 kJ for each mole of ammonia formed
(d) absorbs 23 kJ for each mole of nitrogen used
(e) is an exothermic process

7.24. A 10.0-g sample of an unknown metal was heated to 100.0 °C and then added to 20.0 g of water at 23.0 °C in an insulated calorimeter. At thermal equilibrium, the temperature of the water in the calorimeter was 25.0 °C. Which of the metals in this list is most likely to be the unknown? Explain your reasoning.

Specific heat data	
$H_2O(l)$	$4.184 \text{ J·g}^{-1}\text{·°C}^{-1}$
$Au(s)$	$0.13 \text{ J·g}^{-1}\text{·°C}^{-1}$
$Ag(s)$	$0.22 \text{ J·g}^{-1}\text{·°C}^{-1}$
$Al(s)$	$0.90 \text{ J·g}^{-1}\text{·°C}^{-1}$

7.25. A student received 55.0 g of unknown metal from his laboratory instructor. To identify this metal, he decided to determine its specific heat. He heated the metal sample to a temperature of 98.6 °C in a water bath. Then he transferred it slowly to a calorimeter containing 100. mL of water at 24.6 °C. The maximum temperature reached by the calorimeter system was 25.8 °C. His lab instructor, who was watching the experiment, explained to the student that he had introduced a large amount of error and asked him to repeat the experiment with the correct procedure. In the second trial, the temperature of the water bath was 98.4 °C and the initial temperature of water in the calorimeter was 24.3 °C. The maximum temperature after the sample was immersed in the calorimeter was 26.5 °C. What error had the student made? Explain how you know.

7.26. A student mixed 100. mL of 0.5M HCl and 100. mL of 0.5M of NaOH in a Styrofoam® cup calorimeter. The temperature of the solution increased from 19.0 °C to 22.2 °C. Is this process exothermic or endothermic? Assuming that the calorimeter absorbs only a negligible quantity of heat, that the density of the solution is 1.0 g·mL^{-1}, and that its specific heat is $4.18 \text{ J·g}^{-1}\text{·°C}^{-1}$, calculate the molar enthalpy change for the reaction:

$$HCl(aq) + NaOH(aq) \rightarrow H_2O(l) + NaCl(aq)$$

7.27. Instant cold packs, such as those used to treat athletic injuries, contain ammonium nitrate and a separate pouch of water. When the pack is activated by squeezing to break the water pouch, the ammonium nitrate dissolves in water. This is an endothermic reaction and the change in enthalpy is 25.7 kJ·mol^{-1} of ammonium nitrate dissolved.
(a) Why does this endothermic reaction produce a cold sensation on your skin?
(b) The cold pack contains 125 g of water to dissolve 50.0 g of ammonium nitrate. What will be the final temperature of the activated cold pack, if the initial room temperature is 25 °C? Assume that the specific heat of the solution is the same as that for water, 4.184 J·g^{-1}·°C^{-1}.
(c) What other assumptions do you make in carrying out the calculation in part (b)?
(d) The final temperature is about 7 °C. Does this result justify the assumptions in parts (b) and (c)? If not, explain which assumption(s) might not be valid.

7.7. Bond Enthalpies

7.28. Ethanol, by itself or in blends, is an important alternative fuel for gasoline-powered engines. The molecular equation for the complete combustion of ethanol is

$$C_2H_5OH(l) + 3O_2(g) \rightarrow 2CO_2(g) + 3H_2O(l)$$

(a) Rewrite the balanced equation using full Lewis structures for each reactant and product molecule.
(b) Use your model set to build a model of each reactant and product molecule in this reaction.
(c) Refer to the Lewis structures, the models, and Table 7.3 to determine the total enthalpy input that will be required to break all the bonds in the reactants.
(d) Refer to the Lewis structures, the models, and Table 7.3 to determine the total enthalpy that will be released in forming all the bonds in the products.
(e) What is the net enthalpy change in this combustion reaction? Is your result likely to be an overestimate or underestimate of the experimental $\Delta H°_{\text{reaction}}$? Explain your reasoning.
(f) Ethanol is also a fuel when consumed by humans. Will your result in part (e) also give an estimate for ethanol's fuel value in our bodies? Why or why not?

7.29. Ethanol, CH_3CH_2OH, and dimethyl ether, CH_3OCH_3, are isomers. Use average bond enthalpies to determine which of the isomers is the more stable. By how much? Clearly explain your reasoning and the calculations you do to get your result.

7.30. Consider this reaction in which two amino acids join to form a peptide bond, releasing a water molecule:

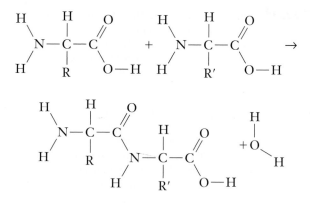

Use average bond enthalpies to estimate the enthalpy change for this reaction.

7.31. The glucose molecule can exist in a number of different forms, including the open-chain and cyclic forms represented below. By convention, the carbon atoms in the chain are numbered 1 through 6 beginning at the top, as shown. Carbon-1 in the cyclic form is the one at the far right of the structure. To make the cyclic form, the oxygen atom bonded to carbon-5 in the linear form becomes the ring oxygen in the cyclic form where it is bonded to both carbon-1 and carbon-5. Which is the more stable, the open-chain or the cyclic structure? How much more stable? Show your reasoning and calculations clearly.

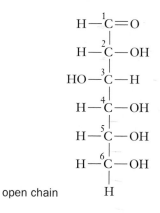

open chain

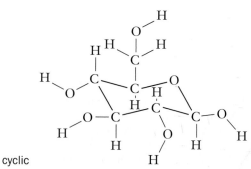

cyclic

7.32. Two isomers with the molecular formula CH_2N_2 are

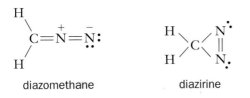

diazomethane   diazirine

Diazomethane and diazirine are both gases. Under appropriate conditions, each can react to give ethene and nitrogen:

$$2CH_2N_2(g) \rightarrow H_2C=CH_2(g) + 2N_2(g)$$

(a) Use average bond enthalpies to calculate $\Delta H°_{reaction}$ for diazomethane and for diazirine undergoing this reaction. Are the reactions exothermic or endothermic?
(b) Construct enthalpy-level diagrams (like Figure 7.11) for the two reactions. Which isomer is the more stable?

Are there any factors that might complicate use of bond enthalpies for these molecules?
(c) Does our molecular bonding model help explain the relative stability? Explain why or why not.

7.33. This table gives the molecular formulas and enthalpies of combustion for three common sugars found in living organisms. What conclusions can you draw from these data about the bonding in these molecules? Explain your response.

Sugar	$\dfrac{\Delta H_{combustion}}{kJ \cdot mol^{-1}}$	Formula
Fructose	$C_6H_{12}O_6$	-2812
Galactose	$C_6H_{12}O_6$	-2803
Glucose	$C_6H_{12}O_6$	-2803

7.34. Disaccharides are sugars formed by the combination of two simpler sugars. A maltose molecule, for example, is a combination of two glucose molecules:

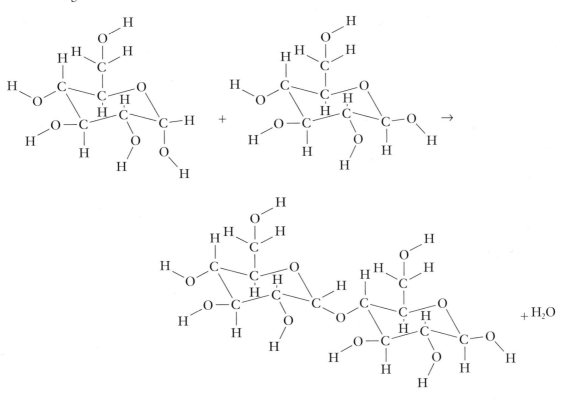

(a) Estimate $\Delta H°_{reaction}$, using bond enthalpies. Is it necessary to do any numerical calculations to make this estimate? Explain why or why not.
(b) Use your result from part (a) and the data from the Problem 7.33 to estimate the enthalpy of combustion of maltose. The experimental value is $-5644\ kJ \cdot mol^{-1}$. How well does your estimate compare to the experimental value? How do you explain the agreement or lack of agreement?

7.35. The reaction between hydrazine and hydrogen peroxide in rocket engines is

$$H_2NNH_2(g) + 2HOOH(g) \rightarrow N_2(g) + 4H_2O(g)$$

(a) Use average bond enthalpies to estimate the standard enthalpy change (in kJ per mole of hydrazine) for the reaction.
(b) Compare the energy released in the hydrazine reaction to the energy that would be obtained if two moles of ammonia were oxidized by hydrogen peroxide.

$$2NH_3(g) + 3HOOH(g) \rightarrow N_2(g) + 6H_2O(g)$$

(c) Which provides the most energy per gram of fuel, hydrazine or ammonia?

7.36. In the *Web Companion*, Chapter 7, Section 7.7.2, you select the bonds that are broken in the reactants and made in the products.
(a) Use the data in the *Web Companion*, Chapter 7, Section 7.7.3, to calculate the enthalpy change for the process you selected in Section 7.7.2.
(b) How does your result in part (a) compare with the analysis in Section 7.7.3 where all reactant bonds are broken and all product bonds formed?
(c) Explain why the comparison in part (b) comes out the way it does.

7.37. Carbon–hydrogen bond-dissociation enthalpies are shown for two different hydrogen atoms in the reactions of propene and 2-butene shown here. The enthalpies for the two different bond dissociations are quite different. The dissociation of a carbon–hydrogen bond on a carbon that is next to a doubly bonded carbon, reactions (1) and (3), requires less energy.

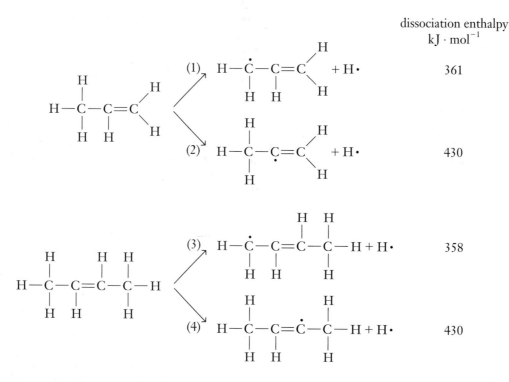

dissociation enthalpy
kJ · mol^{-1}

(1) 361

(2) 430

(3) 358

(4) 430

(a) Are the three-carbon and four-carbon free radicals formed in reactions (1) and (3), respectively, more stable (lower enthalpy) or less stable (higher enthalpy) than those formed in reactions (2) and (4)? Clearly explain the reasoning for your response, using enthalpy-level diagrams, if these aid your explanation.
(b) The unpaired electron in the three-carbon and four-carbon free radicals formed in reactions (1) and (3) is often shown as participating in delocalized π bonding (see Chapter 5, Section 5.7), with the two π electrons from the adjacent double bond. Write Lewis structures supporting this model. Do the bond-dissociation enthalpies and your interpretation in part (a) support this model? How would delocalization of the unpaired electron orbital affect the energy of the radical? Explain your reasoning clearly.

7.8. Standard Enthalpies of Formation

7.38. You usually see sulfur in the form of a light yellow powder, the rhombic crystalline allotrope of sulfur. If you melt this solid (m. p. 112.8 °C), pour the liquid into a filter paper cone, and then unfold the cone as the sample cools, you find that the liquid has solidified in dark yellow needles. This is the triclinic crystalline allotrope of sulfur. After several days, the needles become covered by a light yellow layer. Which allotrope of sulfur is probably its most stable form under standard conditions? Explain your answer.

7.39. Urea (Latin *urina* = urine), $H_2NC(O)NH_2$, is a water-soluble compound made by many organisms, including humans, to eliminate nitrogen. Use standard enthalpies of formation to find the enthalpy for the reaction producing urea from ammonia and carbon dioxide.

$$2NH_3(g) + CO_2(g) \rightarrow H_2NC(O)NH_2(s) + H_2O(l)$$

Show all your work so your procedure is clear.

7.40. In Problem 7.39, you used enthalpies of formation to calculate the enthalpy for making urea from ammonia and carbon dioxide. How could you estimate the enthalpy of vaporization of urea? Explain the procedure you would use and then carry it out. *Hint:* What other way could you estimate the enthalpy of the reaction forming urea? How do the two ways differ?

7.41. During fermentation of fruit and grains, glucose is converted to ethanol and carbon dioxide according to this reaction:

$$C_6H_{12}O_6(s) \rightarrow 2C_2H_5OH(l) + 2CO_2(g)$$

(a) Use standard enthalpies of formation to calculate $\Delta H°_{reaction}$.
(b) Is this reaction exothermic or endothermic?
(c) Which has higher enthalpy, the reactant or the products?
(d) Calculate ΔH for the formation of 5.0 g of $C_2H_5OH(l)$.
(e) What quantity of heat is released when 95.0 g of $C_2H_5OH(l)$ is formed at constant pressure?

7.42. Consider this gas phase reaction in which methanoic (formic) acid reacts with ammonia. The product formamide contains a peptide bond and there is the release of a water molecule:

(a) Use average bond enthalpies to estimate the enthalpy change for this reaction. How does this $\Delta H°$ compare to the $\Delta H°$ you calculated in Problem 7.30? Would you have expected this result? Why or why not?
(b) Use standard enthalpies of formation to calculate the change in enthalpy for this reaction, assuming reactants and products are gases in their standard states. The enthalpies of formation of gaseous methanoic acid and formamide are -379 and -186 $kJ·mol^{-1}$, respectively.
(c) Compare the results of your calculations in parts (a) and (b). Offer some reasons to help explain your comparison.

7.43. The combustion of ammonia is represented by this equation:

$$4NH_3(g) + 5O_2(g) \rightarrow 4NO(g) + 6H_2O(g)$$

Experimentally, we find $\Delta H°_{reaction} = -905$ kJ for the reaction as written.
(a) Use the standard enthalpies of formation for $NO(g)$ and $H_2O(g)$ and the standard enthalpy change for the reaction to calculate the standard enthalpy of formation for $NH_3(g)$.
(b) How does your result in part (a) compare with the standard enthalpy of formation for $NH_3(g)$ in Appendix B? Explain why they are the same (or different).

7.44. In the process known as coal gasification, coal can be reacted with steam and oxygen to produce a mixture of hydrogen, carbon monoxide, and methane gases. These gases are desirable as fuels and they can also serve as the starting material for the synthesis of other organic substances such as methanol, used for the production of synthetic fibers and plastics. Consider these reactions that can take place in the process of coal gasification.

Reaction 1: $C(s) + H_2O(g) \rightarrow H_2(g) + CO(g)$

Reaction 2: $C(s) + \frac{1}{2}O_2(g) \rightarrow CO(g)$

Reaction 3: $C(s) + 2H_2O(g) \rightarrow CH_4(g) + O_2(g)$

Use standard enthalpies of formation to find the enthalpies of reaction for each of these reactions. If it is possible to control the reaction conditions to favor one or more of these reactions, which is the most energetically favorable? The least energetically favorable? Explain your reasoning.

7.9. Harnessing Energy in Living Systems

7.45. What is a coupled reaction? What is its importance in biological reactions?

7.46. What is the role of ATP in biological reactions?

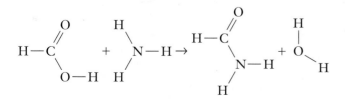

7.47. Coupled reactions take place in living systems, but coupled processes have many applications in industry as well. For example, waste heat from a power plant can be captured and used to do work, such as desalting seawater. (a) Other than the availability of seawater, what do you think might limit the usefulness of this example of a coupled process? (b) Can you think of any other examples of energetically coupled processes from your personal experience or other courses you have taken? Briefly describe such a process. You might find it useful to research this topic using Web-based resources.

7.48. Many ATP^{4-}-coupled reactions require that Mg^{2+} be present as well. What do you think is the function of Mg^{2+} in these reactions? *Hint:* See Chapter 6, Section 6.6.

7.49. The aerobic reaction sequence following glycolysis [see equation (6.77) in Chapter 6, Section 6.11] involves the complete oxidation of pyruvic acid, $CH_3C(O)C(O)OH$ ($=C_3H_4O_3$). This reaction is coupled to the formation of ATP^{4-}, according to this overall reaction:

$$2C_3H_4O_3(l) + 5O_2(g) + 30ADP^{3-}(aq)$$
$$+ 30HOPO_3^{2-}(aq) + 30H^+(aq) \rightarrow$$
$$6CO_2(g) + 34H_2O(l) + 30ATP^{4-}(aq)$$

(a) Calculate the $\Delta H°$ for the uncoupled oxidation of pyruvic acid to CO_2 and H_2O. (b) Calculate the $\Delta H°$ for the coupled reaction and determine the percentage of energy converted to ATP^{4-}.

7.10. Pressure–Volume Work, Internal Energy, and Enthalpy

7.50. An inventor claims to have built a device that is able to do about 0.8 kJ of work for every 1 kJ of thermal energy put into it. He is looking for investors to buy shares in his company, so he can commercialize his invention. Would you invest in his company? Explain why or why not.

7.51. The volume of a gas is decreased from 10. to 1.0 L at a constant pressure of 5.0 atmospheres. Calculate the work associated with the process. Is work done *on the gas* or *by the gas*? (1 L·atm = 101.3 J)

7.52. A hot air balloon is inflated by using a propane burner to heat the air in the balloon. If during this process 1.5×10^8 J of heat energy cause the volume of the balloon to change from 5.0×10^6 L to 5.5×10^6 L, what are the values of q, w, and ΔE for this process? Assume that the balloon expands against a constant pressure of 1.0 atmosphere (1 L·atm = 101.3 J).

7.53. Photographic flash bulbs, once more common than they are now, produced light from the reaction of either zirconium or magnesium wire with oxygen in the bulb, for example:

$$2Mg(s) + O_2(g) \rightarrow 2MgO(s) + \text{energy}$$
$$\text{(light and heat)}$$

They became quite warm when used, but only rarely did they actually explode! Discuss the changes in ΔE, q, and w for the chemical reaction that takes place in a flash bulb.

7.54. A gaseous chemical reaction occurs in which the system loses heat and contracts during the process. What are the signs, "+" or "−", for $P\Delta V$, ΔE, and ΔH? Explain your reasoning.

7.55. Consider two experiments in which a gas is compressed in a closed syringe by pushing the plunger against the trapped gas. Both experiments use the same size syringe.

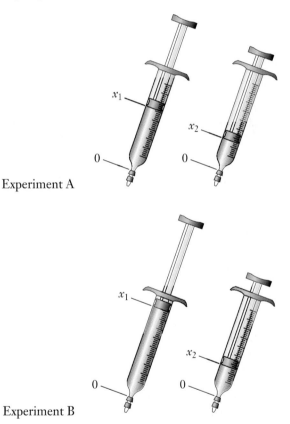

Experiment A

Experiment B

(a) How does the pressure–volume work performed in Experiment A compare to that performed in Experiment B? Assume the external pressure is the same in both experiments. Explain your reasoning. (b) Assume that a third experiment is performed using a syringe with a wider diameter than the one used in the first two experiments. How would the pressure–volume work compare with that in Experiments A if Δx is the same? Assume the pressure is the same in both experiments.

7.56. Consider the changes that occur in this reaction:

$$3H_2(g) + N_2(g) \rightarrow 2NH_3(g)$$

Is work done by the system (gaseous reactants and product) or on the system? Explain your answer. Assume that this reaction occurs at constant temperature and pressure.

7.57. Most of the nitrogen used to inflate an automobile airbag is produced by the reaction of sodium azide:

$$2[Na^+][N_3^-](s) \rightarrow 2Na(g) + 3N_2(g)$$

A driver's side airbag might typically contain about 95 g of sodium azide.
(a) A mole of nitrogen gas occupies about 25 L at 25 °C. How large is the airbag this nitrogen will inflate?
(b) How much work is done by the nitrogen as it inflates the airbag?
(c) If this exothermic reaction is investigated in a constant volume calorimeter, will the measured thermal energy release be greater or less than the thermal energy released at constant pressure? How large will the difference be? Clearly explain your reasoning.

7.58. For this reaction, $\Delta H° = -484$ kJ for the formation of 2 mol of water:

$$2H_2(g) + O_2(g) \rightarrow 2H_2O(g)$$

When 0.5 mol of $H_2(g)$ was reacted with 0.3 mol of $O_2(g)$, at a constant pressure of 1.0 atm, the change in volume was -6.1 liters. Calculate how much work was done and the value of $\Delta E°$ for this reaction. Is work done on the system or by the system?

7.11. What Enthalpy Doesn't Tell Us

7.59. Indicate whether each of these statements is true or false. If a statement is false, write the correct statement.
(a) Consider a 10-mL sample of pure water at 25 °C and 1 atm pressure (state A). The sample is cooled to 1 °C and then the pressure is reduced to 0.5 atm (state B). It takes 5 hours to carry out the change from state A to state B. The sample is then heated and the pressure is raised to 1 atm. In one minute the water is back at 25 °C (the sample is back at state A). The internal energy change in going from state A to B is equal to, but opposite in sign to the internal energy change going from state B to A.
(b) The work done in changing from state A to B and the work done in changing from state B to A in part (a) are numerically the same but opposite in sign.
(c) The enthalpy change for a change of state of a system is independent of the exact state of the reactants or products.
(d) At constant pressure the amount of heat absorbed or evolved by a system is called the enthalpy change, ΔH.
(e) If volume does not change, the amount of heat released during a change of state of a system is equal to the decrease in internal energy of that system.
(f) If a reaction is spontaneous, it is always exothermic.
(g) If the enthalpy change for this reaction, $N_2(g) + O_2(g) \rightarrow 2NO(g)$, is 180.5 kJ, then the enthalpy change for this reaction, $\frac{1}{2}N_2(g) + \frac{1}{2}O_2(g) \rightarrow NO(g)$, is 90.2 kJ.

7.60. When 1 mol of $NH_4Cl(s)$ is dissolved in water, forming $NH_4^+(aq)$ and $Cl^-(aq)$, the change in enthalpy is 14.8 kJ·mol^{-1}. When 1 mol of $CaCl_2(s)$ is dissolved in water, forming $Ca^{2+}(aq)$ and $Cl^-(aq)$, the change in enthalpy is -82.9 kJ·mol^{-1}.
(a) Which beaker felt cool to touch after the salt dissolved in water? Which one felt warm?
(b) What would you have to do to maintain the temperature of each beaker at 25 °C?
(c) To determine $\Delta H°$ for dissolution of salts in water, what factors must you consider?

7.61. Given that $\Delta H_f°$ for $Cl^-(aq) = -167.4$ kJ·mol^{-1}, use the enthalpy of reaction data from Problem 7.60 to calculate $\Delta H_f°$ for $NH_4^+(aq)$ and for $Ca^{2+}(aq)$.

7.62. A 19th-century chemist, Marcellin Berthelot, suggested that all chemical processes that proceed spontaneously are exothermic. Is this correct? Give some examples that justify your answer.

7.13. Extension— Ideal Gases and Thermodynamics

7.63. Consider two samples of gas in identical size containers. Which of these statements is true? Explain your reasoning for each choice. Rewrite the false statements to make them true.
(a) If the temperature of the two samples is the same, then the pressure of gas in each container is the same.
(b) If the temperature and pressure of the two samples are the same, then the number of moles of gas in each container is the same.
(c) If the gas in each container is the same and the temperature and number of moles of gas in each container are the same, then the number of collisions with the wall per unit time is the same in both containers.

7.64. Molecules of all gases at the same absolute (Kelvin) temperature have the same average kinetic energy of translation, $KE = \frac{1}{2}mu^2$, where m is the mass of a molecule and u is the average speed. If a gas is in a container that has one or more tiny holes, the molecules will be able to escape through the holes. Faster moving molecules escape more readily.
(a) A rubber balloon was inflated with helium gas and floated in the air at the end of its string. The next day, the balloon was somewhat smaller and no longer floated. Explain why. *Hint:* When a thin sheet of rubber is stretched, tiny holes form in the sheet.
(b) Two identical rubber balloons were blown up to the same size, one with helium gas and the other with air. The next day, both balloons were somewhat smaller, but not the same size. Which gas was in the larger balloon? Explain your answer.

7.65. A change involving gases is carried out at constant pressure in a cylinder with a piston, like the one illustrated in Figure 7.20. The final position of the piston is higher in the cylinder than it was initially. What possible change(s) might occur in the system enclosed in the cylinder to cause the observed change in the position of the piston? Clearly explain how the change(s) you suggest would account for the movement of the piston.

7.66. Problem 7.57(a) says a mole of nitrogen gas occupies about 25 L at 25 °C (and 1 atm pressure). Show that this is true. What is the volume for a mole of oxygen? Explain.

7.67. A sample of gas is put into a rigid (fixed volume) container at −3 °C and a pressure of 38.3 kPa. The container is then placed in an oven at 267 °C.
(a) What pressure would you expect to measure for the gas in the container at this higher temperature? Explain.
(b) The measured pressure of the gas in the container at 267 °C is 133.2 kPa. How does this experimental value compare with the pressure you calculated in part (a)? If they are different, what factor(s) could account for the difference? Clearly explain your reasoning.

7.68. Analysis of a liquid hydrocarbon sample shows that the ratio of carbon to hydrogen atoms in the compound is one-to-one. A 0.237-g sample of the liquid is placed in a 327-mL flask attached to a pressure sensor. The air is pumped out of the flask and then the flask is sealed and heated to 150 °C, at which temperature the liquid has all vaporized and the pressure in the flask is 33.2 kPa.
(a) How many moles of the hydrocarbon sample are in the flask? Explain how you get your answer.
(b) What is the molar mass of the hydrocarbon? What is the molecular formula of the hydrocarbon? Explain how you get your answers.

7.69. A mineral sample contains several different ionic compounds, including calcium carbonate, $CaCO_3(s)$. A 1.587-g sample of the mineral was heated to decompose the calcium carbonate to $CaO(s)$ and $CO_2(g)$. The gas was collected and measured and had a pressure of 83.35 kPa at 295.3 K in a 127.5 mL collection container.
(a) How many moles of gas were evolved by the sample? Explain how you get your answer.
(b) How many moles of $CaCO_3(s)$ were present in the mineral sample. Explain.
(c) What percentage of the mineral sample is $CaCO_3(s)$. Explain how you get your answer.

7.70. The fermentation of glucose produces ethanol and releases carbon dioxide. This equation represents the reaction:

$$C_6H_{12}O_6(s) \rightarrow 2C_2H_5OH(l) + 2CO_2(g)$$

(a) Do $\Delta E°$ and $\Delta H°$ have the same value for this reaction? Explain your reasoning.
(b) If $\Delta E°$ and $\Delta H°$ differ, which is larger? By how much? Explain clearly.

General Problems

7.71. A constant-pressure calorimeter similar to the one shown in Figure 7.7 can be used to study the reaction of magnesium with hydronium ion:

$$Mg(s) + H_3O^+(aq) \rightarrow Mg^{2+}(aq) + H_2(g)$$

A 100.0 mL volume of 0.5 M hydrochloric acid, $HCl(aq)$, was put in the calorimeter; its temperature was 19.32 °C. Then, 0.1372 g of Mg metal was added to the acid and the temperature observed until it reached a maximum of 25.69 °C.
(a) What is $\Delta H_{reaction}$, in kJ per mole of Mg reacted?
(b) Calculate $\Delta E_{reaction}$, in kJ per mole of Mg reacted.
(c) How could you measure $\Delta E_{reaction}$ directly? Are data like those in this problem precise enough to measure any difference between $\Delta E_{reaction}$ and $\Delta H_{reaction}$? Explain why you answer as you do.

7.72. Several elegant experiments have been done to determine the amount of work that *molecular motors* from living cells can do. The motor in the illustration consists of seven proteins, three α, three β, and a γ, that come together as shown. The motor is too small (about 10 × 10 × 8 nm) to be observed, so a long actin filament is attached to the motor shaft, the γ protein, and movement of the filament is observed under a microscope. The rotation of the shaft is in 120° steps. Each step appears to require the coupled hydrolysis of one ATP molecule, reaction (7.37). Calculations based on the observed motion indicate that the motor is able to produce a force of at least 100×10^{-12} N (newton). In one complete revolution, the top of the motor shaft moves about 1 nm.

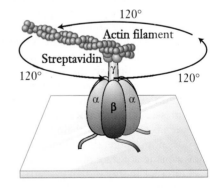

(a) How much work does the motor do in one complete revolution? (1 J = 1 N·m).

(b) How much energy (enthalpy) is provided by ATP during one complete revolution?

(c) What is the approximate efficiency of this motor? Explain how you arrive at your conclusion.

7.73. Calculate the $\Delta H°_{reaction}$ for the hydrolysis of a molecule of glycylglycine to two molecules of glycine using these $\Delta H°_{combustion}$ data for glycylglycine and glycine. Assume that the nitrogen atoms in both molecules are oxidized to NO_2. Explain your procedure.

$\Delta H°_{combustion}$ for glycine,
$H_2NCH_2C(O)OH$, $= -981$ kJ·mol⁻¹

$\Delta H°_{combustion}$ for glycylglycine,
$H_2NCH_2C(O)NHCH_2C(O)OH$,
$= -1,996$ kJ·mol⁻¹

7.74. The diagram shows a glass demonstration version of a fire syringe (also called a fire piston). A tiny amount of a flammable material, cloth or paper, is placed at the closed end of the syringe. The plunger is inserted and then thrust quickly into the piston, compressing the air inside to about 1/20th of its initial volume. The temperature of the air rises almost instantly to about 1000 K and ignites the flammable material, which burns with a bright flash.

(a) The internal energy of a gas depends on its temperature. When the temperature of the gas changes, its change in internal energy depends on its constant-volume molar heat capacity (C_V), the number of moles of gas, and the temperature change: $\Delta E = n \times C_V \times \Delta T$. C_V for air is 21 J·mol⁻¹. What is ΔE for the gas compressed in a fire syringe that is 15 cm long and has a cross-sectional area of 0.20 cm²? One mole of gas at 300 K occupies about 25 L.

(b) The compression in a fire syringe is so rapid that the gas has no time to transfer thermal energy to its surroundings during the compression. How much work

Wisp of cotton

is done on the gas during the compression? What pressure (in atm) on the plunger is required to obtain this much work? Explain clearly how you obtain your answer. (1 L·atm = 101.3 J)

(c) To get a better "feel" for the effort required to operate a fire syringe, calculate the force in pounds required to produce the pressure you calculated in part (b). Use the fact that a pressure of 1 atm = 15 pound·in.⁻². Show clearly how you do the necessary unit conversions. Is the force required a reasonable amount for a human to produce?

(d) Make a molecular level drawing showing the motion of the molecules of air in the syringe before, during, and immediately after the compression. Clearly show how they obtain the increase in energy you calculated in part (a).

(e) Why does the amount of flammable material in the syringe have to be small? If the temperature of the gas reaches 1000 K, couldn't it set fire to a larger quantity? *Hint:* Consider the heat capacity (or specific heat) of a piece of solid cloth or paper.

> Fire syringes made of hollow wood like bamboo have apparently been used to start fires for several centuries. A fire piston was patented in England in 1807. The diesel engine works on this same principle: The air in the engine cylinder is compressed quickly and the fuel is injected just before the compression stroke is complete.

7.75. Given these atomization reactions and their corresponding enthalpy changes, calculate $\Delta H°$ for the combustion of 1 mol of $PH_3(g)$ to yield $H_2O(g)$ and $P_2O_5(g)$. Show your reasoning clearly.

$PH_3(g) \rightarrow P(g) + 3H(g)$	$\Delta H° = 965$ kJ·mol⁻¹
$O_2(g) \rightarrow 2O(g)$	$\Delta H° = 490$ kJ·mol⁻¹
$H_2O(g) \rightarrow 2H(g) + O(g)$	$\Delta H° = 930$ kJ·mol⁻¹
$P_2O_5(g) \rightarrow 2P(g) + 5O(g)$	$\Delta H° = 3382$ kJ·mol⁻¹

Things fall apart; the centre cannot hold;
Mere anarchy is loosed upon the world.

W. B. YEATS (1865–1939),
THE SECOND COMING

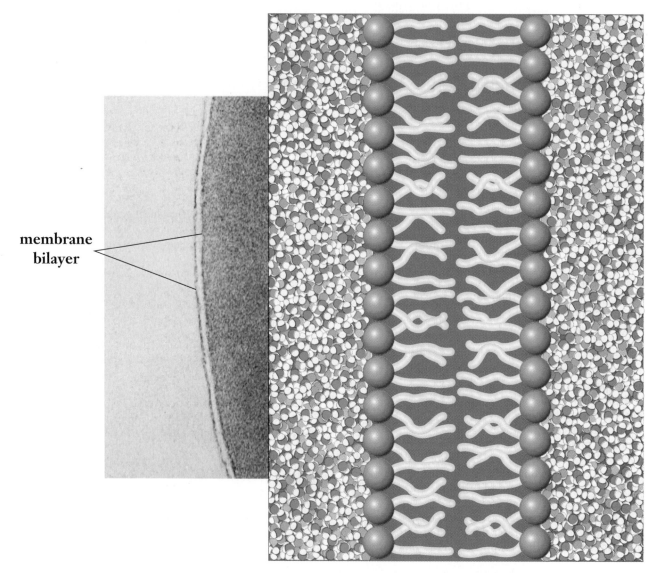

membrane
bilayer

The electron micrograph is of a thin layer of a red blood cell that has been specially prepared to show the membrane that encloses the cell. The two parallel layers of the membrane are labeled. The drawing shows a simple model of a tiny portion of a phospholipid bilayer cell membrane with water molecules on both sides of the membrane, as they would be in an intact cell. The yellow "fingers" in the central part of the bilayer represent the hydrocarbon chains of long-chain carboxylic acids connected by ester bonds to the polar "heads" of the phospholipid molecules.

Entropy and Molecular Organization

The lines from Yeats do seem to describe many aspects of the world as we experience it. Is this really the way things are? The notion of anarchy (disorder) is certainly supported by substantial evidence: apples rot, iron rusts, glass breaks, paper tears, batteries run down. The natural order of things *does* seem to favor *dis*order. On the other hand, consider something as familiar as snowflakes on a winter day. The complex crystals shown in Figure 8.1 originate from ordinary water freezing in the atmosphere and falling to earth. How is it that some things become so disorganized while others become so highly organized?

Living systems represent a huge degree of organization. Complex molecules are arranged into complex arrays to form cells. Cells are organized into complex arrays to form organs. Organs are organized into complex arrays to form plants and animals. The illustration on the facing page represents a tiny portion of the molecules in a cell membrane. Long hydrocarbon chains of the membrane are stacked next to one another in the center of the figure. Water molecules are less organized than the membrane molecules and can move about randomly on both sides of the membrane Water molecules can also pass through the membrane from one side to the other. The net direction of movement of water molecules through a cellular membrane depends upon the concentrations of the solutions on either side of the membrane. Water molecules can move through the membrane from the less concentrated solution to the more concentrated solution; this movement of the water molecules is called **osmosis.**

There is little or no enthalpy (or energy) change in the formation of cell membranes or when osmosis occurs. At the end of the previous chapter, we pointed out that the first law of thermodynamics (conservation of energy) cannot explain the direction of changes. Spontaneous processes can occur endothermically, exothermically, or without appreciable input or output of enthalpy. Enthalpy changes do not provide a way to predict the direction in which a process will occur. We have also seen that molecular organization seems to increase in some changes—precipitation, for example—and decrease in others, such as dissolution. Therefore, molecular organization alone cannot be directing the changes. The central objective for this chapter is to identify and characterize the property of systems that is responsible for change. We will find that *entropy* is the thermodynamic quantity that tells us the direction processes will go spontaneously. Let's begin our discussion with the processes of mixing and osmosis.

Figure 8.1.

Snowflake.

8.1. Mixing and Osmosis

 8.1 INVESTIGATE THIS

Does water move through a hollow carrot filled with syrup?

Do this as a class investigation, but work in small groups to discuss the results. Bore a cylindrical hole about $\frac{3}{4}$ the length of a large carrot. Insert a 10-cm length of a clear, rigid plastic straw or glass tubing a short distance into the hole and seal it to the carrot. Add enough dark sugar syrup—such as dark corn syrup or pancake syrup—to the cavity in the carrot to bring the liquid level a centimeter or two above the carrot. Mark the level of the syrup on the tubing. Support the carrot and tubing upright in a container of water. The water level should be just about to the top of the carrot. Record your observations on the set up every five minutes or so.

8.2 CONSIDER THIS

What direction does water move through a hollow carrot?

(a) Did the column of liquid in the tubing rise or fall in Investigate This 8.1? Does the volume of liquid in the cavity of the carrot increase or decrease? Explain your reasoning.

(b) Play the animation in the *Web Companion*, Chapter 8, Section 8.4.1. How does the animation compare to what you observed in Investigate This 8.1? Explain.

(c) What is your explanation for your observations from the investigation and the animation from the *Web Companion*? Compare your interpretation with that of other students.

Web Companion

Chapter 8, Section 8.3.1–2 — (1)

View a movie and molecular-level animation of the mixing process.
(2)
(3)
(4)

Mixing Mixing is familiar to you in many changes you have observed, for example, milk mixing into coffee or sugar dissolving and mixing into tea. Figure 8.2 illustrates such a change. When a drop of concentrated dye solution is added to a container of water, the dye molecules begin to mix into the water and become diluted. If we wait long enough, the dye molecules become dispersed uniformly throughout the entire container. Mixing to form a uniform solution is the spontaneous direction of this process. As Figure 8.2 shows, no matter how long you wait, you never observe a homogeneous solution of dye molecules unmixing, that is, all coming together in a single drop somewhere in the container.

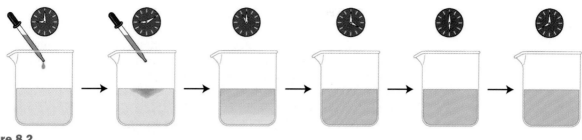

Figure 8.2.

A drop of dye mixing into water. By four o'clock, the soluble dye has formed a uniform mixture with the water. The resulting uniform solution will not spontaneously unmix, no matter how long you wait.

Osmosis Osmosis is fundamentally a mixing process: A concentrated solution becomes less concentrated when additional solvent mixes into it. This happens when a barrier (usually a thin sheet of material called a **membrane**) allows solvent molecules to pass through, but prevents passage of solute ions or molecules. For example, Figure 8.3(a) represents a concentrated solution on one side of a membrane and an equal level of a less concentrated solution (or pure solvent, as shown in the figure) on the other side. After some time, the level of the more concentrated solution will rise, and the level of the less concentrated solution will fall [Figure 8.3(b)]. The solvent has passed through the membrane, mixing with and diluting the concentrated solution.

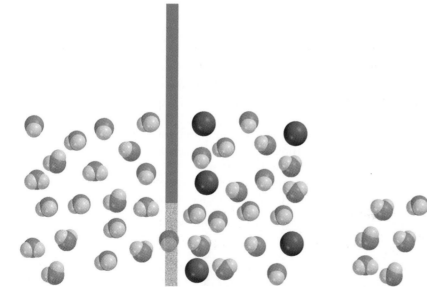
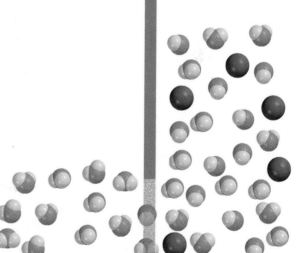

(a) The initial liquid levels are the same on both sides of the semipermeable membrane.

(b) After a time, as water passes through the membrane, the level of the solution rises and the solute is diluted.

Figure 8.3.

Changes that occur in osmosis.

A permeable (*per* = through + *meare* = to pass) barrier allows everything to pass through and an impermeable (*im* = not) barrier allows nothing to pass through. A semipermeable barrier is in between.

Membranes that allow some species to pass through readily while allowing others to go through very slowly or not at all are called **semipermeable.** Semipermeable membranes surround plant and animal cells—the model of a cell membrane in the chapter-opening illustration is one example. (We will say more about biological membranes in Section 8.11.) If osmosis occurs in Investigate This 8.1, water must pass through the cell membranes on its way to diluting the sugar solution in the center of the carrot. The rising column of solution is evidence that the solution volume is increasing due to the osmosis of water through the cell membranes.

Why is it that we observe mixing, but not unmixing, of homogeneous solutions? Why is it that osmosis always takes place in a direction that moves solvent so that it dilutes a more concentrated solution? The answer to these questions is that a mixed state or a more dilute state is more probable (more likely to occur) than an unmixed or concentrated state. In order to see why this is so, we need to know more about probability and how molecular arrangements are related to probability. Then we will return to see how these ideas are applied to understanding mixing and osmosis.

8.2. Probability and Change

Lotteries and other games of chance are based on probabilities. The games are always designed so that the probability that the players will lose is greater than the probability that they will win.

We often use the terms *chance* and **probability** in a casual way, but probability has a precise mathematical meaning that we can use to understand change. Some changes will always occur, and the probability of their occurrence is assigned a value of 1. Some changes will never occur, and the probability of their occurrence is assigned a value of 0. Everything else is in between, and more care is therefore required to define the probability.

8.3 CONSIDER THIS

What is the probability of change?

What numerical value (0, 1, or somewhere in between) would you assign to the probability of each of these changes being observed? Explain each of your answers.
(a) A book pushed off a table falls to the floor.
(b) A solution of sugar in water separates into pure solid sugar and liquid water.
(c) A glass of liquid water sitting on your desk freezes.
(d) An ice cube sitting on your desk melts.
(e) A discarded piece of paper moves from your wastebasket to your desktop.
(f) A teaspoon of sugar dissolves in a teaspoon of water.

With a little help from mathematical probability, we can use our model of the atomic and molecular world to understand the direction of change. More importantly, the model enables us to predict changes that will occur in new situations where we have no previous observations to go on. The molecular-level explanatory and predictive model for physical and chemical changes is straightforward: *If a system can exist in more than one observable state (mixed or unmixed, for example), spontaneous changes will be in the direction toward the state that is most probable.* The hard part is to figure out the probability of each possible observable state of the system, so that we can apply the probability model.

Probability The number of ways that the molecules and the energy in a system can be arranged to give a particular state of the system is a measure of the probability that this state will be observed. For the moment, we will focus on molecular arrangements and then take up energy arrangements in Section 8.5. Think about the drop of dye mixing with water in Figure 8.2. Before mixing, all the dye molecules are together in a drop, and any particular dye molecule must be in this tiny volume. After mixing, that same dye molecule can be anywhere in the total volume of the solution. In the mixture, the dye molecules are spread among the water molecules. The number of possible arrangements of the dye molecules increases, making the mixed state more probable than the state before mixing. The system always changes toward the mixed state, and the soluble dye molecules are never observed to separate from the solution.

The molecular arrangements we will consider are fundamentally arrangements of molecules among quantized energy levels (see Problem 8.96). Here, however, we take a simpler approach and think of them as arrangements in space, which lead to the same conclusions about directionality of change.

If we start in Figure 8.2 with 300 mL of water, we have about 17 mol $\left(= \dfrac{300 \text{ g}}{18 \text{ g·mol}^{-1}}\right)$ of water. Seventeen moles of water is about $(17 \text{ mol}) \times (6 \times 10^{23} \text{ molecules·mol}^{-1}) = 10^{25}$ molecules of water. Imagine that the 300 mL of water is divided into 10^{25} tiny boxes, each containing a water molecule. When the drop of dye is added, the number of boxes increases slightly, and dye molecules now occupy a small fraction of them. We assume that either a water molecule or a dye molecule can occupy each of the tiny boxes. Even if only one in a million of the molecules is a dye molecule, there are about 10^{19} dye molecules in the mixture.

To count the number of arrangements of this many molecules requires using statistical methods that obscure the simple, basic idea of counting. To get at the basic idea, we will consider simple model systems that contain only a few particles (molecules) and then generalize to more realistic systems. To use our counting results we make a *fundamental assumption: Each distinguishably different molecular arrangement of a system is equally probable.* This is a postulate that cannot be proved and is only accepted because it predicts results that we observe. A new arrangement is **distinguishable (distinguishable arrangement)** from another if you turn your back, someone rearranges the molecules, and you can tell the new arrangement from the original when you look again at the system. Exchanging identical objects does not produce a new arrangement.

8.3. Counting Molecular Arrangements in Mixtures

8.4 INVESTIGATE THIS

How are distinguishable arrangements counted?

Let's try an example of the kind of counting we'll need to do for this simple model. Use five labeled boxes and three identical, unlabeled objects (such as the candies shown here) to place in them.

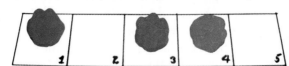

(a) How many different, *distinguishable* arrangements can you make of three candies among the five boxes, if each box can hold only one candy? As you try new arrangements, make a table to list and count the different ways

continued

of placing candies in the boxes. The arrangement above might be shown as (1, 0, 1, 1, 0). How many distinguishable arrangements are there? Can you tell that you have found all possibilities? Compare your method for finding the number of distinguishable arrangements with methods used by your classmates. Did you miss any arrangements or duplicate any?

(b) For a more challenging investigation, find the number of distinguishable arrangements for three objects in six boxes. Again, compare your results with those of other students.

Mixing model

Mixing model One strategy you might have used for finding the distinguishable arrangements in Investigate This 8.4 is to start with the three objects in boxes 1, 2, and 3; this is one arrangement (1, 1, 1, 0, 0). Moving the object in box 2 to box 4 creates a second arrangement (1, 0, 1, 1, 0), which is the example shown in the investigation. Moving the object into box 5 creates a third arrangement (1, 0, 1, 0, 1). If you start with the (1, 1, 1, 0, 0) arrangement and interchange the objects in boxes 1 and 2, you will get a (1, 1, 1, 0, 0) arrangement; this is not a new arrangement. Your result for the investigation should have been 10 distinguishable arrangements.

In order to relate what you did in Investigate This 8.4 to a realistic system such as the mixing of a dye into water, we need to create an analogous model for the dye solution. Figure 8.4 shows an exploded two-dimensional view of 15 of the 10^{25} boxes available for a water molecule or a dye molecule. We will use these 15 boxes, as you used the 5 boxes in the investigation, to model a two-dimensional mixing process as the dye molecules spread throughout the solution.

Web Companion

Chapter 8, Section 8.3.3–5 — ①

Practice counting arrangements ②
in an interactive animation ③
of mixing. ④

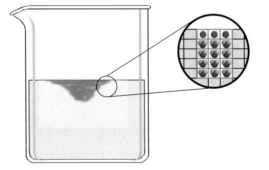

Figure 8.4.

Model of a drop of dye molecules added to water. Green dots represent dye molecules; water molecules are space-filling models.

Arrangements for the unmixed system

Arrangements for the unmixed system The dye-mixing model contains three dye molecules (the solute, indicated by green dots) and 12 water molecules (the solvent, indicated by the space-filling models) in the 15 boxes. When the dye is first added, the molecules occupy only the top layer of boxes. Only one arrangement, the one shown in Figure 8.4, is possible for the three dye molecules. The number of distinguishable molecular arrangements associated with a particular observable state is usually given the symbol W. For the dye molecules (solute) in this model, there is only one way to arrange three molecules in the top three boxes, so $W_{solute} = 1$. Similarly, there is only one way of arranging the 12 identical water (solvent) molecules in the remaining 12 boxes (each contains one molecule, as shown in Figure 8.4), so $W_{solvent} = 1$.

Think of W as the number of **W**ays of arranging identical particles (molecules) or identical energy quanta, as we will show in Section 8.5.

Any arrangement of the solute particles can be paired with any arrangement of the solvent particles. The total number of molecular arrangements associated with a given state of the system, W_{system}, is the *mathematical product* of the number of distinguishable arrangements of solute and solvent particles for that state:

$$W_{system} = (W_{solute}) \times (W_{solvent}) \qquad (8.1)$$

For our 15-box system, in the state that has 3 solute particles in the top 3 boxes and 12 solvent molecules in the remaining 12 boxes, W_{system} is

$$W_{system} = (W_{solute}) \times (W_{solvent}) = 1 \times 1 = 1 \qquad (8.2)$$

This result makes sense; if every dye molecule looks like every other dye molecule, and every water molecule looks like every other water molecule, then the arrangement in Figure 8.4 is the only one possible.

Arrangements for the mixed system As the dye molecules and water molecules begin to mix, some dye molecules move to the next layer of boxes, and those that leave are replaced in the first layer by water molecules. With the three dye molecules confined to the top 2 layers, there are 20 distinguishable arrangements, as you probably discovered if you completed the challenge activity, Investigate This 8.4(b). All 20 possible arrangements are shown in Figure 8.5.

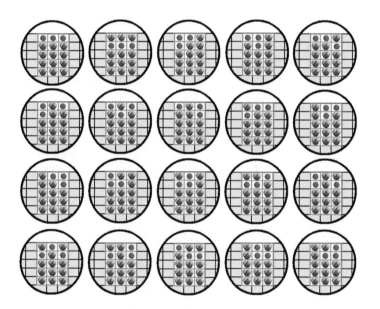

Figure 8.5.

Model for arrangements of dye molecules in water. Dye molecules have mixed from Figure 8.4 into the top two layers.

When the three dye molecules have been arranged among the top 6 boxes, there are only 12 boxes left for the 12 water molecules, and there is only one distinguishable way to distribute these molecules among the boxes. Thus, as before, $W_{solvent} = 1$. For our 15-box system, in the state that has 3 solute particles distributed among the top 6 boxes and 12 solvent molecules in the remaining 12 boxes, W_{system} is

$$W_{system} = (W_{solute}) \times (W_{solvent}) = 20 \times 1 = 20 \qquad (8.3)$$

Our assumption is that the more probable state will be the one having the largest number of distinguishable arrangements. The state with the dye molecules occupying any of 6 boxes instead of only 3 is the more probable. Solute molecules spread out (giving more arrangements) rather than clump together (with fewer possible arrangements). The increased number of distinguishable arrangements makes the mixed state more probable and, therefore, more favored.

As mixing continues, the number of distinguishable arrangements (*W*) increases rapidly. The 3 dye molecules have 84 distinguishable arrangements in the top 3 layers, 220 in the top four, and 455 in all 15 boxes. The plot in Figure 8.6 shows *W* as a function of the number of boxes available per molecule in our sample. For example, a value of 5 on the *x*-axis represents the case of 3 molecules in 15 boxes. The *x*-axis is proportional to the *volume* (expressed as the number of boxes) that is available to the solute particles in the system. As the molecules spread through the solution, the number of arrangements rapidly increases.

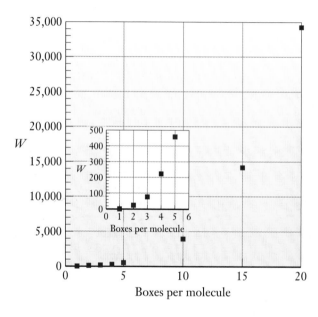

Figure 8.6.

W as a function of the number of boxes per molecule for three dye molecules. No continuous curve is shown because only integer numbers of boxes are possible for this system.

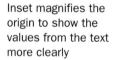
Inset magnifies the origin to show the values from the text more clearly

---

8.5 CHECK THIS

Unmixing

Play the animation in the *Web Companion*, Chapter 8, Section 8.3.6. Explain what the animation represents and how you can conclude that unmixing is an unfavorable process.

8.4. Implications for Mixing and Osmosis in Macroscopic Systems

One important diffusion process is the movement of oxygen from your arterial bloodstream (where its concentration is high) into your cells (where the oxygen concentration is low).

Diffusion The process modeled in the previous section corresponds to diffusion in a liquid or gas. **Diffusion** is the movement of molecules or ions from a region of higher concentration (many particles in a small volume) into regions of lower concentration (few particles in a large volume). In the model system, Figures 8.4 and 8.5, boxes that initially contain only water molecules represent the lower concentration region.

Diffusion always occurs in the same direction, with solute particles moving from regions of higher concentration toward regions of lower concentration.

Figure 8.6 shows that the number of molecular arrangements increases continuously as the volume (number of boxes) available to solute molecules increases. For a given solution, the largest number of molecular arrangements occurs when the particles are spread throughout its entire volume. Diffusion ultimately spreads particles uniformly throughout the available volume. This is the most favored arrangement, as you have observed for actual mixtures, such as milk in coffee, sugar in tea, or a dye in water.

The result represented graphically in Figure 8.6 applies qualitatively to many analogous systems. Expansion of a gas from a small volume to a larger volume, for example, increases the volume (number of boxes in the model) available to each gas molecule. The expanded state of the system is favored, since the value for W increases as the volume increases. Consequently, gases always expand to fill any container, occupying the maximum volume available.

Molecular model for osmosis In osmosis, the solute molecules are prevented from diffusing, because they cannot pass through the semipermeable membrane into the pure solvent. However, if the solvent molecules move into the solution, the volume of the solution increases and there is a larger volume for the solute molecules to move about in. Our model from Section 8.3 for countable numbers of molecules can be used to explain osmosis as a result of mixing of solute and solvent molecules. The change shown in Figure 8.3 is represented schematically in Figure 8.7 for a countable number of solvent and solute molecules. We start with nine molecules on each side of the semipermeable membrane (shown in yellow between the two liquids). On the left are nine solvent molecules in nine boxes and on the right are three solute molecules mixed with six solvent molecules in nine boxes. The numbers of possible molecular arrangements given in the figure for the solvent (W_{solv}) and solution (W_{soln}) are from Figure 8.6.

Web Companion

Chapter 8, Section 8.4.2–4	①
Reinforce this discussion with interactive, molecular-level animations of osmosis. ② ③ ④

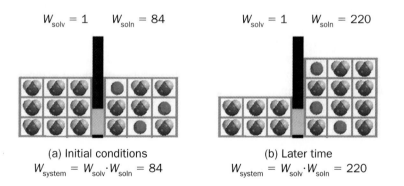

$W_{solv} = 1$ $W_{soln} = 84$ $W_{solv} = 1$ $W_{soln} = 220$

(a) Initial conditions
$W_{system} = W_{solv} \cdot W_{soln} = 84$

(b) Later time
$W_{system} = W_{solv} \cdot W_{soln} = 220$

Figure 8.7.

Increase of molecular arrangements in osmosis. Compare with Figure 8.3. The semipermeable membrane is represented in yellow here.

If three solvent molecules pass through the membrane from the solvent side to the solution side, there are now 12 boxes on the right occupied by three solute molecules and nine solvent molecules. The total number of molecular arrangements increases by almost a factor of three for this process, so the change is favorable and the process can occur as shown. Your results from Investigate This 8.1 show that the process does, in fact, occur in macroscopic systems. There, the water passed through the membranes of the carrot cells to dilute the solution, and raise its level by increasing its volume, in the center of the carrot. At some point, the net movement of water stops. We shall return in Section 8.13 to discuss osmosis in more quantitative terms to understand why the net movement of water stops. For now, however, we will go on to complete our more qualitative discussion of molecular and energy arrangements.

8.6 CHECK THIS

Transfer of solvent from solution to pure solvent

In the *Web Companion*, Chapter 8, Section 8.4.5–7, if we start from the same initial condition as in Figure 8.7(a), how does the number of molecular arrangements for the system change when three solvent molecules pass from the solution to the pure solvent side of the membrane? Explain how the numeric values are obtained. Would you expect to observe this "reverse" osmosis? Why or why not?

Reflection and Projection

Everyone has experienced mixing processes and observed that homogeneous solutions do not "unmix." You may also have had some experience with osmosis, but might not have made a connection between mixing and osmosis. We have now made the connection through a molecular-level model that provides an explanation for the direction of these processes. Our approach is to consider an imaginary system of just a few molecules (or objects) and calculate how many distinguishable ways there are to arrange these molecules in a limited number of possible positions. Our assumption is that change is in the direction that produces an increase in distinguishable arrangements among the molecules of a system. We find that the number of arrangements increases as the volume available to the molecules increases.

At this point, our probability analysis is incomplete because it is based only on arrangements of particles and has not accounted for the role of enthalpy—the energy content of a substance. In Chapter 7, we saw that enthalpy changes, thermal energy transfers, were an outcome of chemical and physical changes. We will consider the roles of the two change factors—molecular arrangements and thermal energy transfers—together in Section 8.7 and following sections. To prepare for that discussion, we need to consider energy arrangements among the particles in a system.

8.5. Energy Arrangements Among Molecules

8.7 INVESTIGATE THIS

How is a different kind of distinguishable arrangement counted?

Try this example of another kind of counting, using four soft candies of different colors and two toothpicks. How many distinguishable ways can you arrange zero,

continued

one, or two toothpicks stuck into the four candies if *any number of toothpicks can be placed in the same candy*? The pictures in this list show a few possible arrangements:

number of toothpicks	toothpicks in candies	arrangement

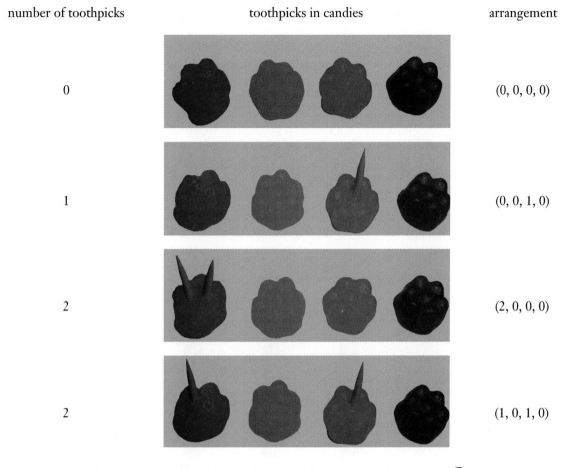

0		(0, 0, 0, 0)
1		(0, 0, 1, 0)
2		(2, 0, 0, 0)
2		(1, 0, 1, 0)

(a) How many arrangements are there for zero, one, and two toothpicks? How will you know that you have found all the possibilities? Compare your results with what other students get.

(b) For a more challenging activity, find the number of distinguishable arrangements for three toothpicks among four candies.

Web Companion

Chapter 8, Section 8.5.1–2 ── ①

Use the interactive animations of a metallic crystal and check your counting results.

② ③ ④

Model for energy arrangements A solid made up of identical atoms in a regular array, such as a metallic crystal, is the simplest system we can use to count different arrangements of energy associated with individual atoms. The atoms are located at known positions in the crystal and are not free to move about. Consequently, they are distinguishable from one another by their position, just as the candies are distinguishable in Investigate This 8.7. Recall from the discussion in Chapter 4, Sections 4.4 and 4.5, that atomic systems can only absorb (or release) energy in certain amounts called **quanta.** For this discussion, we will assume the quanta are all the same size (that is, that they all are the same amount of energy). This is a good model for an atomic solid. In an atomic solid, any atom in the array can absorb as many quanta as are available and, just like the toothpicks in Investigate This 8.7, we can't tell one identical quantum from another.

At the molecular level, all energy transfers are quantized. Macroscopic energy transfers *appear* to be continuous, because energy quanta are so small. Many quanta must be exchanged before the transfer can be detected.

Counting energy arrangements Investigate This 8.7 is similar to Investigate This 8.4, but there is a fundamental difference. In Investigate This 8.4, the pattern of boxes was fixed, and the objects (molecules) could move among them to produce different arrangements. Only one object could reside in any particular box. For Investigate This 8.7, the candies are identifiable by color (and/or position) and variable numbers of toothpicks can be distributed to each one. Similarly, an atom in a crystal, identified by its location, can have different numbers of quanta. Some atoms have no quanta. Others have one. Still others have two, and so on. Just as moving molecules around produced distinguishable arrangements, quanta moving among atoms produce distinguishable arrangements. Figure 8.8 shows all possible arrangements for zero or one quantum distributed among four atoms in a solid, and two of the possible arrangements for two quanta. Compare these arrangements with those you found in Investigate This 8.7.

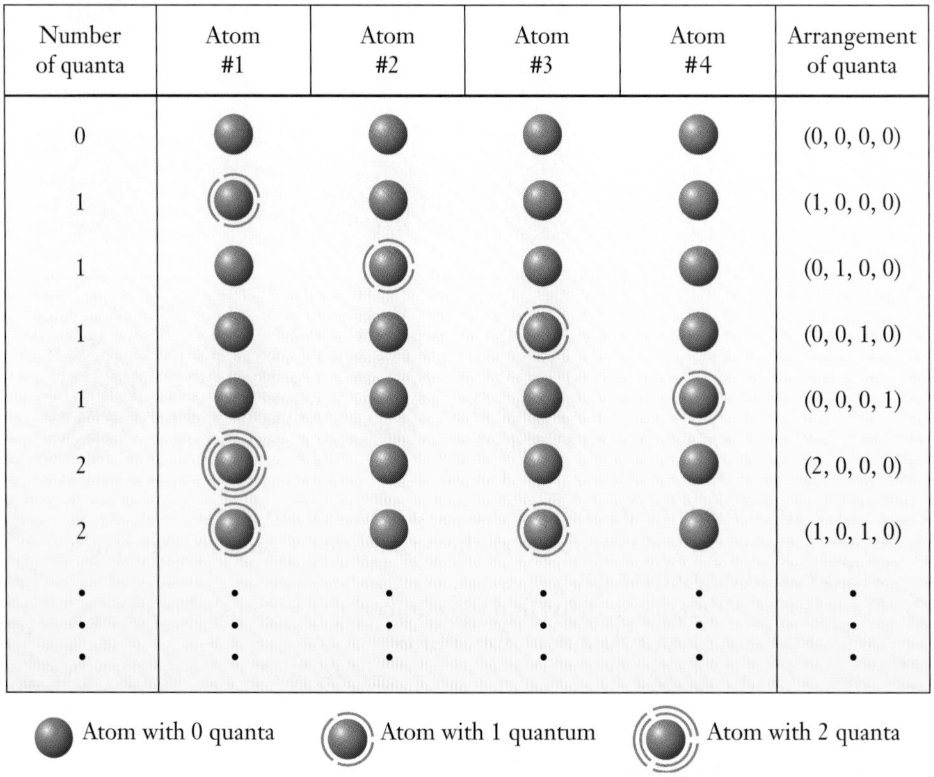

Number of quanta	Atom #1	Atom #2	Atom #3	Atom #4	Arrangement of quanta
0					(0, 0, 0, 0)
1	◎				(1, 0, 0, 0)
1		◎			(0, 1, 0, 0)
1			◎		(0, 0, 1, 0)
1				◎	(0, 0, 0, 1)
2	◉				(2, 0, 0, 0)
2	◎		◎		(1, 0, 1, 0)
⋮	⋮	⋮	⋮	⋮	⋮

● Atom with 0 quanta ◎ Atom with 1 quantum ◉ Atom with 2 quanta

Figure 8.8.

Arrangements of zero, one, and two quanta in a four-atom solid.

Investigate This 8.7 demonstrates that the number of possible energy arrangements increases as the number of quanta increases. This is like the systems in Section 8.3, where we showed that the number of possible molecular arrangements increases as the volume of space available per particle increases. Using the approach of Investigate This 8.7, we can find the numbers of **distinguishable energy arrangements,** W, for increasing numbers of quanta. The results are shown in Figure 8.9, a plot of W as a function of energy (number of quanta) in a four-atom solid.

A four-atom system is too small actually to see, but we can use the results in Figure 8.9 to understand the direction of changes in observable systems. Let's begin by considering an energy transfer in four-atom systems that can be related to an observable result you already know: When a warm object and a cool object are placed in contact, thermal energy is transferred from the warm object to the cool object until they reach the same temperature.

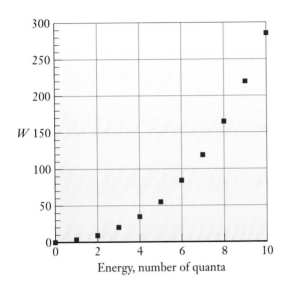

Figure 8.9.

Plot of *W* as a function of energy content of a four-atom solid.
No continuous curve is shown because only integer numbers of quanta are physically possible.

8.8 WORKED EXAMPLE

Energy transfer between solids with different energies

Imagine two identical four-atom solids, one having four quanta of energy, and the other having eight. (a) What observable property differs between these two solids? (b) Allow two quanta to be transferred from the solid with eight quanta to the solid with four quanta. Use the data from Figure 8.9 to determine whether this energy transfer is spontaneous.

Necessary information: We need Figure 8.9 and the method for obtaining total numbers of distinguishable arrangements that we used in Section 8.3. We need to recall that temperature is a measure of the energy content of a substance.

Strategy: (a) Relate the number of energy quanta in each solid to a measurable property of the solid. (b) An energy transfer is spontaneous if the change increases the number of possible arrangements of energy quanta. Compare the total number of distinguishable arrangements of the energy quanta before and after the assumed energy transfer.

Implementation: (a) Adding energy to an object increases its energy content, producing an increase in temperature. Extracting energy from an object decreases its energy content, producing a decrease in temperature. Temperature is the measurable property that distinguishes the two solids. The four-atom solid with eight energy quanta has a higher temperature than the four-atom solid with four energy quanta. When two energy quanta are transferred, giving each four-atom solid six quanta of energy, the temperatures of the two solids are the same. (b) The total number of arrangements in a system, W_{tot}, is the product of the

Web Companion

Chapter 8, Section 8.5.3–7 ①
Use interactive animations that ②
model a process like the one ③
in Worked Example 8.8. ④

continued

number of arrangements possible in each component of the system. In this case, there are two objects:

$$W_{tot} = W_{obj\ 1} \times W_{obj\ 2}$$

The total number of arrangements is the *product* of arrangements for each object because any energy arrangement in object 1 can occur in conjunction with any energy arrangement in object 2.

For the initial state,

$$W_i = W_{(8\ quanta)} \times W_{(4\ quanta)}$$

For the final state,

$$W_f = W_{(6\ quanta)} \times W_{(6\ quanta)}$$

From Figure 8.9, read values of W for four-atom solids with four, six, and eight quanta.

$$W_{(4\ quanta)} = 35$$

$$W_{(6\ quanta)} = 84$$

$$W_{(8\ quanta)} = 165$$

Calculate the total number of initial and final arrangements:

$$W_i = W_{(4\ quanta)} \times W_{(8\ quanta)} = 35 \times 165 = 5775$$

$$W_f = W_{(6\ quanta)} \times W_{(6\ quanta)} = 84 \times 84 = 7056$$

There are more ways to arrange the energy quanta when both solids have six quanta compared to when one solid has four and the other has eight. The transfer of quanta from the warmer to the cooler solid increases the total number of arrangements of quanta, so the transfer is spontaneous.

Does the answer make sense?   This problem represents energy transfer from a warm object (more energy quanta) to an identical cooler object (fewer energy quanta); both objects reach the same intermediate temperature (with each having the same number of energy quanta). This is the direction of observed changes in macroscopic systems.

8.9 CHECK THIS

Energy transfer from a cooler to a warmer solid

Imagine two identical four-atom solids, one having two quanta of energy, and the other having six. Allow one quantum to be transferred from the solid with two quanta to the one with six. Use the data from Figure 8.9 to determine whether this energy transfer is spontaneous. Explain how your result applies to observations on macroscopic systems.

The transfer of energy from a warm to a cool object, Worked Example 8.8, increases the number of ways that energy quanta can be distributed between the two objects. Since the transfer increases the number of arrangements of the quanta, our probability model predicts that the energy transfer is spontaneous. The prediction agrees with our observations that energy is always transferred from a warm object to a cool object when they are placed in contact. In Check This 8.9, you found that transfer of energy from a cool to a warm object decreases the number of arrangements of the quanta. Our model predicts that this energy transfer is nonspontaneous. You never observe a warm object getting warmer when placed in contact with a cooler object.

8.6. **Entropy**

The ideas derived from the arrangement activities using atomic-scale systems in the previous sections also apply to systems that are large enough to study directly. For this purpose, we introduce a new thermodynamic quantity, **entropy, S,** which is a measure of the number of different ways that particles or quanta of energy can be arranged. (You can think of entropy as a measure of "Spreadedness.") Entropy is directly related to W, the number of possible arrangements of particles and energy: *The larger the number of possible arrangements, the larger the entropy of the system.* The entropy of a system is proportional to the logarithm of W:

$$S \propto \ln W \qquad (8.4)$$

We have seen in the previous sections that the numbers of arrangements, W, for the components of a system combine multiplicatively to give the total number of arrangements for the system. But we want thermodynamic functions to combine additively, as you have seen for enthalpy in Chapter 7. The logarithms of W do combine additively; take equation (8.1) as an example:

$$\ln W_{\text{system}} = \ln \left[(W_{\text{solute}}) \times (W_{\text{solvent}}) \right] = \ln W_{\text{solute}} + \ln W_{\text{solvent}} \qquad (8.5)$$

Entropies, defined in terms of $\ln W$, are additive. The proportionality constant between S and $\ln W$ is the **Boltzmann constant** and is given the symbol k. The full definition of S is

$$S \equiv k \ln W \qquad (8.6)$$

The Boltzmann constant has a value 1.68×10^{-23} J·K^{-1}. (The Boltzmann constant appears everywhere in models of molecular behavior as a proportionality constant relating the energy of molecules to their temperature in kelvin.) The $\ln W$ term has no units, so entropy has the same units as the Boltzmann constant, energy per degree. Equation (8.6) applies to individual molecules. For macroscopic systems, we use moles to specify the amount of substance and replace k by $N_A k$ (Avogadro's number times the Boltzmann constant). The new proportionality constant is the **gas constant, R:**

$$R = N_A k = (6.023 \times 10^{23} \text{ mol}^{-1})(1.68 \times 10^{-23} \text{ J·K}^{-1})$$

$$= 8.314 \text{ J·K}^{-1}\text{·mol}^{-1} \qquad (8.7)$$

R has units of J·K^{-1}·mol^{-1}, so molar entropy has units of J·K^{-1}·mol^{-1}.

Recall that we have met and used the gas constant previously, in Chapter 7, Section 7.13.

Net entropy: **The second law of thermodynamics** In the preceding sections, we have described two different origins for W. In the first, the different distinguishable arrangements in a system were a result of the distribution of molecules. We will call entropy calculated on that basis **positional entropy.** In the second, the distinguishable arrangements are a result of the distribution of

energy quanta among molecules. We will call entropy calculated on that basis **thermal entropy.**

Since k is a constant, S depends only on the number of distinguishable molecular and/or energy arrangements of a system—the more arrangements, the higher the entropy. Higher entropy is associated with higher probability. **Net entropy change** is the sum of all the positional and thermal entropy changes *in the surroundings as well as in the system* being studied. *Observed changes always take place in a direction that increases net entropy.* This is the **second law of thermodynamics.** There is no way to prove the second law of thermodynamics. It is based on experience and observation. No spontaneous change has ever been observed that fails to follow the second law, just as no change has ever been observed in which energy is not conserved (the first law).

Absolute entropy Entropy, like enthalpy, is a state function. The entropy of a system in a particular state is determined entirely by the positional and thermal contributions to the number of distinguishable arrangements of the system. The entropy is independent of the pathway by which the system got into this state. Recall that enthalpy is a function of the internal potential energy of a system; we have no way of knowing an absolute value for enthalpy. Thermodynamic tables, Appendix B, for example, give enthalpies of formation, ΔH_f°, the enthalpy *change* for the formation of the compound from its elements in their standard states. The elements are a *reference system*, which is *defined* to have an enthalpy of formation of zero. We *can* determine *absolute* values for entropies. Tabulated values for entropies of compounds and elements in their standard states are given as S° (no "Δ"). The reference point for entropies is **absolute zero** ($T = 0$ K), the temperature at which $W = 1$, that is, where there is a single arrangement for particles (perfect order) and all particles are in their lowest energy state. When $W = 1$, for perfectly ordered systems, then $S_0^\circ = 0$.

> Absolute zero has never been experimentally achieved, that is, $W = 1$ has not been attained for any real system. Temperatures of about 0.0000000005 K have been achieved.

Reflection and Projection

The discussions of distinguishable arrangements in the preceding sections leads to the definition of entropy as a measure of the number of different ways that particles or quanta of energy can be arranged. Change can occur if the change results in an increase in the possible arrangements of particles and/or quanta (both of which increase entropy). When the results are applied to the observable properties of substances, several important generalizations can be made:

- For soluble substances, diffusion of a component from a region of high concentration into one of lower concentration increases the entropy, ultimately producing a homogeneous mixture.
- There are more energy arrangements and greater entropy for systems in which two objects have the same temperature than when one object is warmer than the other. Transfer of energy from a warm body (more energy quanta per molecule) to a cool body (fewer energy quanta per molecule) increases the entropy of the overall system.
- The reference point for measuring entropy is absolute zero, at which temperature the entropy of a perfectly ordered substance is zero. Entropy values in thermodynamic tables are values for *absolute* entropy.

Changes such as those referred to in this list are often described as being driven by the entropy change, because the most probable condition is always the state of highest entropy. The notions that are embedded in the concept of entropy and the second law of thermodynamics become clearer when we study them in relation to real systems. We will do this in the next several sections, as we consider phase changes, other chemical changes, and colligative properties, including osmotic pressure.

8.7. Phase Changes and Net Entropy

8.10 CONSIDER THIS

What are the relative probabilities of different phases?

Consider the sequence of changes of state as a substance goes from solid to liquid to gas, as represented by this diagram. In the solid, a molecule occupies a fixed lattice position. It cannot

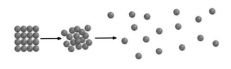

exchange places with any of its neighbors. As a liquid or a gas, the molecules are not in fixed positions; they move. In the liquid, the collection of molecules occupies a definite volume. In the gas, the molecules move to fill the entire volume of their container.

(a) Draw three squares representing the "volume" available to a molecule in the solid, liquid, and gas shown here. If the volume per molecule determines the number of molecular arrangements in a system, which phase has the largest number of arrangements available? What is the implication of your answer for the relative probabilities of the three phases?

(b) 🖱 Explain how the movies and interactive questions in the *Web Companion*, Chapter 8, Section 8.7.1–3, are related to the diagram here and your responses in part (a).

As you have learned, the molecules (or atoms) in a solid are not free to move about, except to vibrate in place. The volume available to each molecule is only slightly larger than the molecule itself. In a liquid, the molecules can move about anywhere within the liquid volume, so the volume available to each molecule is the entire volume of the liquid. Similarly, molecules in the gas phase can move about in the entire volume occupied by the gas. Because gases occupy larger volumes than liquids, gases always have more distinguishable arrangements than liquids, which, in turn, always have more distinguishable arrangements than solids. Therefore, for a given substance, the positional entropy of its gas is greater than the entropy of its liquid, which is greater than the entropy of its solid. When a gas expands, the molecules are farther apart, and the expanded gas has more distinguishable arrangements, because all the molecules have a larger volume to move in. Expansion of gases, as well as phase changes from solid to liquid, liquid to gas, and solid to gas, increase the positional entropy of the system.

███████ **8.11 CHECK THIS** ████████████████████████

Direction of positional entropy changes

For each of these changes, tell whether the positional entropy of the system increases, decreases, or stays the same and explain why.

(a) Ice melts to form liquid water.
(b) Water vapor in the air condenses to form a cloud.
(c) Equal volumes of ethanol and water are mixed to form a homogeneous solution.
(d) Water is poured from a graduated cylinder into a beaker.
(e) A piece of dry ice (solid carbon dioxide) sublimes to form carbon dioxide gas.
(f) Ethene (ethylene) molecules are polymerized to form polyethylene.

A dilemma The preceding discussion leads to a dilemma. If the liquid phase is more probable (has more molecular arrangements) than the solid phase, and if the gaseous phase is more probable still, why are substances ever found in condensed phases? To answer this question, we have to consider the *net* entropy change for these changes (and all other changes as well). Recall that the net entropy change for a process is the sum of all the positional and thermal entropy changes for the process.

Melting and freezing water To see how positional and thermal entropy combine to determine the direction of change, let's examine two familiar processes: the **melting** of ice and the **freezing** of water at a constant pressure of 1 bar, standard conditions. Our approach will be to consider these changes at temperatures where we know they occur spontaneously and determine the relationship between the positional and thermal entropy changes that *must* be true, in order to give a positive net entropy change in each case. Then we will look at our models for positional and thermal entropy changes to see how they can be reconciled with the required relationships.

Melting changes a solid to a liquid, a change of the system from a small number of molecular arrangements (solid water) to a larger number (liquid water). The positional entropy of the water molecules increases in this change, $S°_{liquid} > S°_{solid}$, and favors the formation of the liquid:

$$\Delta S_{system\ (s \rightarrow l)} = \Delta S°_{system\ (s \rightarrow l)} = S°_{liquid} - S°_{solid} > 0 \qquad (8.8)$$

For ice melting, the standard entropy change, $\Delta S°_{s \rightarrow l}$, is 22.0 J·K^{-1}·mol^{-1}. But melting does not occur in an isolated piece of ice. Melting occurs only if energy is supplied from the surroundings to break some of the hydrogen bonds that hold the water molecules in position in the crystal.

███████ **8.12 CONSIDER THIS** ████████████████████████

What happens in an ice–water mixture at different temperatures?

The melting point of ice is 273 K. The temperature of a mixture of ice and liquid water is 273 K. Figure 8.10 shows a container of this mixture sitting on a large block of metal that is at a temperature T.

continued

(a) If $T > 273$ K, what will you observe happening to the ice–water mixture? (This is equivalent to placing the container on a room-temperature countertop.) Will energy leave the block and enter the ice–water system or will energy transfer be from the ice water to the block? Explain your response.

(b) If $T < 273$ K, what will you observe happening to the ice–water mixture? (This is equivalent to placing the container in the freezing compartment of a refrigerator.) Will energy leave the block and enter the ice–water system or will energy transfer be from the ice water to the block? Explain your response.

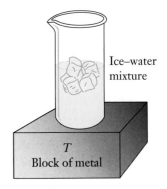

Ice–water mixture

T
Block of metal

Figure 8.10.

Ice–water mixture in contact with a large block of metal.

You know that the ice in a glass of ice water will melt, if the glass is left on a counter. In order to melt the ice, energy must enter the mixture to break hydrogen bonds in the solid ice. The standard enthalpy change for ice melting is $\Delta H°_{s\rightarrow l} = 6.00$ kJ·mol^{-1}. For the process in Consider This 8.12(a), this energy (enthalpy) has to come from the surroundings, which we take to be the block of metal at $T > 273$ K. For every mole of ice that melts, the enthalpy change in the surroundings is

$$\Delta H_{surr\,(s\rightarrow l)} = -\Delta H°_{s\rightarrow l} = -6.00 \text{ kJ·mol}^{-1} \qquad (8.9)$$

The surroundings lose energy, the thermal entropy of the surroundings decreases; $\Delta S_{surr\,(s\rightarrow l)} < 0$.

We assume the ice–water mixture and metal block together are an isolated system: The only thermal transfers are between these components.

Net entropy change for ice melting

The net entropy change for the ice melting is the sum of the positional entropy change for the system (ice–water mixture) and the thermal entropy change for the surroundings (block of metal):

$$\Delta S_{net\,(s\rightarrow l)} = \Delta S_{system\,(s\rightarrow l)} + \Delta S_{surr\,(s\rightarrow l)} \qquad (8.10)$$

Under the given conditions, $T > 273$ K, melting is the observed process, so this net entropy change must be positive. Figure 8.11 shows the relationship between the positional and thermal entropies which *must* be true for $\Delta S_{net\,(s\rightarrow l)} > 0$, that is, for the final entropy to be larger than the initial entropy. The arrows in the figure show that $|\Delta S_{system\,(s\rightarrow l)}| > |\Delta S_{surr\,(s\rightarrow l)}|$. The magnitude of $\Delta S_{system\,(s\rightarrow l)}$ must be greater than the magnitude of $\Delta S_{surr\,(s\rightarrow l)}$.

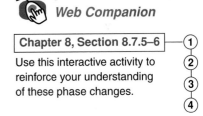

Web Companion

Chapter 8, Section 8.7.5–6 — ①

Use this interactive activity to reinforce your understanding of these phase changes.

②
③
④

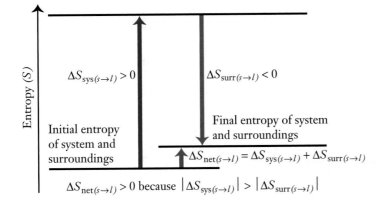

Entropy (S)

$\Delta S_{sys(s\rightarrow l)} > 0$

$\Delta S_{surr(s\rightarrow l)} < 0$

Initial entropy of system and surroundings

Final entropy of system and surroundings

$\Delta S_{net(s\rightarrow l)} = \Delta S_{sys(s\rightarrow l)} + \Delta S_{surr(s\rightarrow l)}$

$\Delta S_{net(s\rightarrow l)} > 0$ because $|\Delta S_{sys(s\rightarrow l)}| > |\Delta S_{surr(s\rightarrow l)}|$

Figure 8.11.

System, surroundings, and net entropy changes for melting ice at $T > 273$ K. On these entropy diagrams, blue arrows represent positional entropy change in the system, red arrows represent thermal entropy changes in the thermal surroundings, and green arrows represent the net entropy change (the sum of the positional and thermal entropy changes).

8.13 CHECK THIS

Net entropy change for water freezing

Figure 8.12 shows the relationship between the positional and thermal entropies which *must* be true, if $\Delta S_{net\ (l \to s)} > 0$, for water freezing at $T < 273$ K. This is the process in Consider This 8.12(b). Freezing water is the reverse of melting ice, so $\Delta S_{system\ (l \to s)} = -\Delta S_{system\ (l \to s)}$. Thus, the arrows representing ΔS_{system} in Figures 8.11 and 8.12 are the same length, but opposite in direction. Also, because freezing is the reverse of melting, $\Delta H_{surr\ (l \to s)} = -\Delta H_{surr\ (l \to s)}$.

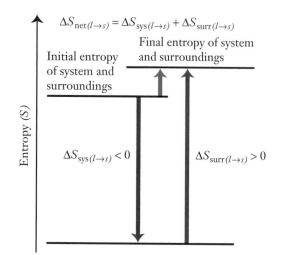

$$\Delta S_{net(l \to s)} = \Delta S_{sys(l \to s)} + \Delta S_{surr(l \to s)}$$

Final entropy of system and surroundings

Initial entropy of system and surroundings

Entropy (S)

$\Delta S_{sys(l \to s)} < 0$ $\Delta S_{surr(l \to s)} > 0$

Figure 8.12.

System, surroundings, and net entropy changes for freezing water at $T < 273$ K. The colors of the arrows have the same meaning as in Figure 8.11.

(a) Explain why the arrows representing $\Delta S_{system\ (l \to s)}$ and $\Delta S_{surr\ (l \to s)}$ go in the directions shown in Figure 8.12.

(b) Figure 8.11 showed the relationship between $\left| \Delta S_{system\ (s \to l)} \right|$ and $\left| \Delta S_{surr\ (s \to l)} \right|$, in order for $\Delta S_{net\ (s \to l)} > 0$. Write the appropriate relationship between $\left| \Delta S_{system\ (l \to s)} \right|$ and $\left| \Delta S_{surr\ (l \to s)} \right|$, in order for $\Delta S_{net\ (l \to s)} > 0$, that is, for the final entropy to be larger than the initial entropy, as shown in Figure 8.12.

Magnitude of thermal entropy changes Figures 8.11 and 8.12 show that the positional (system) entropy changes for melting and freezing have opposite signs. This is what we expect, since melting increases positional entropy and freezing decreases it. Similarly, the thermal (surroundings) entropy changes for melting and freezing have opposite signs. The thermal entropy of the surroundings decreases as energy is lost from the surroundings to melt ice. The thermal entropy of the surroundings increases as energy is gained by the surroundings when freezing water releases it. What may seem peculiar about these changes is that the magnitude of the thermal entropy change in the surroundings is different for the different temperatures of the surroundings, above and below 273 K. Since it is this difference that is responsible for ice melting when $T > 273$ K and water freezing when $T < 273$ K, we need to examine these thermal entropy changes in more detail.

8.14 CONSIDER THIS

How does thermal entropy change depend on temperature?

(a) Use the data from Figure 8.9 and the definition of entropy, equation (8.6), to calculate the change in entropy, ΔS, for the four-atom solid when it loses one quantum of energy to go from 10 to 9 quanta. Do the same for the loss of one quantum to go from 5 to 4 quanta. For which case is the *change* in entropy larger?

(b) Which change in part (a) takes place at the higher temperature? Explain how you know.

(c) The energy change in each case in part (a) is the same, one quantum, but the entropy changes and temperatures are different for the two cases. Propose a relationship that describes the relative sizes of entropy changes as equal energy changes are made at different temperatures. *Hint:* Recall the units of entropy.

Quantitative expression for thermal entropy change The transfer of one quantum of energy from an object having many quanta has less effect on the number of distinguishable arrangements than does the transfer of one quantum from an object having fewer quanta. The entropy decrease for loss of energy from an object at a higher temperature is smaller than the entropy decrease for the loss of the same amount of energy from the same object at a lower temperature. This result is general. *For a given energy change in a substance, the change in thermal entropy, ΔS, is always greater at a lower temperature.*

You might find an analogy helpful in remembering this important result. Think of energy in terms of money. Suppose you have two friends, one of whom has $1000 and the other only $10. If you give each friend $5, which one will be affected more by getting the extra money? If, instead of giving each friend money, you ask each to contribute $5 to buy food for a small party, which one would be affected more by giving up the money? In both cases, the friend who has less money will be more affected by getting or giving the $5. The friend with less money (energy) corresponds to a system at lower temperature (less money per person). The greater monetary effect for your less wealthy friend is analogous to the greater change in entropy at a lower temperature.

For constant pressure systems, we express thermal entropy change quantitatively as

$$\Delta S = \frac{\Delta H}{T} \tag{8.11}$$

This is the *form* of the relationship you should have written for Consider This 8.14(c): For a given change in energy (enthalpy), the thermal entropy change is inversely proportional to the temperature. Equation (8.11) reminds you that the units of entropy are $J \cdot K^{-1} \cdot mol^{-1}$. We could start from $S = k \ln W$, the definition of entropy, apply it to more realistic systems than you did in Consider This 8.14, and derive equation (8.11). In practice, we use equation (8.11) to express ΔS for energy transfers, because ΔH and T are easily measured quantities and finding all the ways to arrange thermal energy quanta in macroscopic systems is hard. We will not try to prove or derive equation (8.11), but will show examples of its application. We begin by continuing our present examples, melting ice and freezing water.

8.15 WORKED EXAMPLE

Net entropy change for melting ice with surroundings at 283 K

If the temperature of the block of metal in Figure 8.10 is 283 K, what is the net entropy change for melting one mole of ice in the ice–water mixture under standard conditions?

Necessary information: We need to know that, for melting ice, $\Delta S^\circ_{s \to l} = 22.0 \, \text{J} \cdot \text{K}^{-1} \cdot \text{mol}^{-1}$ and $\Delta H^\circ_{s \to l} = 6.00 \, \text{kJ} \cdot \text{mol}^{-1}$.

Strategy: Calculate the thermal entropy change for the surroundings from equation (8.11) and add it to the positional entropy change for the system to get the net entropy change.

Implementation: Since the energy required to melt the ice leaves the surroundings, $\Delta H_{surr} = \Delta H^\circ_{s \to l} = -6.00 \, \text{kJ} \cdot \text{mol}^{-1}$. The thermal entropy change for the surroundings is

$$\Delta S_{surr} = \frac{\Delta H_{surr}}{T_{surr}} = \frac{-6.00 \times 10^3 \, \text{J} \cdot \text{mol}^{-1}}{283 \, \text{K}} = -21.2 \, \text{J} \cdot \text{K}^{-1} \cdot \text{mol}^{-1}$$

The positional entropy change for the system is

$$\Delta S_{system} = \Delta S^\circ_{s \to l} = 22.0 \, \text{J} \cdot \text{K}^{-1} \cdot \text{mol}^{-1}$$

The net entropy change for melting ice with the surroundings at 283 K is

$$\Delta S_{net} = \Delta S_{system} + \Delta S_{surr} = 22.0 \, \text{J} \cdot \text{K}^{-1} \cdot \text{mol}^{-1} + (-21.2 \, \text{J} \cdot \text{K}^{-1} \cdot \text{mol}^{-1})$$
$$= 0.8 \, \text{J} \cdot \text{K}^{-1} \cdot \text{mol}^{-1}$$

Does the answer make sense? Our result shows that the net entropy change is positive for ice melting at 283 K, so the process is spontaneous. Since the temperature of the surroundings is above 273 K, we know from experience that ice does melt. This problem is an example of the general case illustrated in Figure 8.11.

8.16 CHECK THIS

Net entropy change for freezing water with surroundings at 263 K

If the temperature of the block of metal in Figure 8.10 is 263 K, what is the net entropy change for freezing one mole of water in the ice–water mixture? Does your result predict that this process is spontaneous? Is there evidence that it is? Explain why or why not.

The dilemma resolved Now we have the answer that resolves the dilemma raised by the results of Consider This 8.10 and posed at the beginning of the section: Why are substances ever found in condensed phases? An input of enthalpy *from* the surroundings is required to change a system from a low entropy condensed phase to a higher entropy phase; $\Delta H_{surr} < 0$. The familiar changes are solid to liquid (melting), liquid to gas (vaporization), and solid to gas (sublimation). Both the gain of positional entropy by the system ($\Delta S_{system} > 0$) and the loss of thermal entropy by the surroundings must be accounted for, as they are in this expression for the net entropy change:

$$\Delta S_{net} = \Delta S_{system} + \Delta S_{surr} = \Delta S_{system} + \frac{\Delta H_{surr}}{T} \qquad (8.12)$$

The thermal entropy change for the surroundings is negative, but its numerical value decreases as the temperature increases. Thus, *at higher temperatures,* ΔS_{system} dominates equation (8.12), the net entropy change is positive, and *changes to phases of higher positional entropy are favored.*

From the preceding discussion, you can conclude that three factors influence the direction of a phase change.

- If the **positional entropy** of the system increases, the phase change is favored. Gases have greater positional entropy than liquids and liquids have greater positional entropy than solids.
- If *thermal energy* is transferred from the system to the surroundings, the phase change is favored. Exothermic changes (gases condensing to liquids or liquids freezing to solids) always increase the **thermal entropy** of the surroundings.
- Entropy change in the thermal surroundings is inversely proportional to the *temperature* of the surroundings.

8.17 CONSIDER THIS

What are net entropy changes for ice and water changes at 273 K?

(a) Use equation (8.12) to determine the net entropy change for melting a mole of ice at 273 K and standard pressure, if it is in contact with surroundings that are also at 273 K. Would you expect any of the ice to melt? Explain why or why not.

(b) Do the same calculation for a mole of water freezing at 273 K when the surroundings are at 273 K. Would you expect any of the water to freeze? Why or why not?

Phase equilibrium In Consider This 8.17, you were asked to predict what change, if any, would occur when ice or water at 273 K (the system) is placed in contact with surroundings at 273 K. You found in part (a) that, under these conditions, ΔS_{net}(ice melting) = 0.0 J·K^{-1}. And in part (b), you found that ΔS_{net}(water freezing) = 0.0 J·K^{-1} for water at 273 K in contact with surroundings at 273 K. Neither the liquid nor the solid phase is favored at 273 K. There will be no *net* change in the amount of ice or water in the system and no *net* transfer of energy between the ice and/or water system and the surroundings. The system is said to be in **equilibrium.** There is no change in entropy for the change in either direction; ΔS_{net} = **0.** Under equilibrium conditions, the entropy changes in the system and the surroundings are exactly balanced for a change in either direction.

8.8. Gibbs Free Energy

Focusing on the system Under standard conditions, the direction of a phase change can be determined using standard enthalpies and entropies in equation (8.12) to calculate the net entropy change for the system and surroundings. However, we are not usually interested in what is happening in the surroundings. In many cases, the surroundings are simply a convenient place to which to transfer thermal energy when the change is exothermic or a convenient source of thermal energy when the change is endothermic. The thermodynamic relationships that enable us to focus on changes in the system without explicitly considering the surroundings grow out of the work of J. Willard Gibbs (American mathematical physicist, 1839–1903). The concepts Gibbs developed enable us to use only system variables to determine which changes in the system can occur and which changes are impossible. A new state function that resulted from Gibbs' work is called the **Gibbs free energy, G.** Our task in this section is to discover what the Gibbs free energy is and how it is related to the ideas we have already developed.

The entropy change in the thermal surroundings depends only on the thermal energy transferred to or from the system at constant pressure and temperature. The actual process that occurs in the system to produce this energy transfer is irrelevant. Since thermal energy lost by the system is gained by the surroundings and vice versa, we can write

$$\Delta H_{surr} = -\Delta H^{\circ}{}_{system} \tag{8.13}$$

We have already used this relationship (for $\Delta H_{system} = \Delta H^{\circ}{}_{phase\ change}$) in Worked Example 8.15 and, following that example, you used it in Check This 8.16 and Consider This 8.17. We do not have to restrict ourselves to standard conditions but can also consider changes that occur at any constant pressure, that is, $\Delta H_{surr} = -\Delta H_{system}$. For these more general conditions, equation (8.12) for the net entropy change can be written as

$$\Delta S_{net} = \Delta S_{system} - \frac{\Delta H_{system}}{T_{surr}} \tag{8.14}$$

ΔS_{net} as a function of system variables If changes occur in systems at the same temperature as the thermal surroundings, we have $T_{system} = T_{surr} = T$ and equation (8.14) becomes

$$\Delta S_{net} = \Delta S_{system} - \frac{\Delta H_{system}}{T} \tag{8.15}$$

Equation (8.15) provides us a powerful tool for understanding and predicting the direction of change in chemical systems because it gives the net entropy change (whose sign tells us the direction of change) in terms of system variables only. As you use equation (8.15), keep in mind its restrictions: The system changes have to be measured at constant pressure and temperature and the temperatures of the system and thermal surroundings have to be the same.

8.18 WORKED EXAMPLE

Formation of $N_2O_4(g)$ under standard conditions at 298 K

The formation of $N_2O_4(g)$, dinitrogen tetroxide, from $NO_2(g)$, nitrogen dioxide, is represented by this equation:

$$2NO_2(g) \rightleftharpoons N_2O_4(g)$$

For the formation of 1 mol of $N_2O_4(g)$ at 298 K, the standard entropy and enthalpy changes are -175.8 J·K^{-1} and -57.2 kJ, respectively (from Appendix B). Do these values seem to make sense in terms of positional entropies and bond formations? Is the reaction to form $N_2O_4(g)$ spontaneous under standard conditions?

Necessary information: We need to recall that energy is released when bonds are formed.

Strategy: Use our models for entropy change to decide whether the entropy and enthalpy changes make sense. Then, substitute the numeric values for entropy, enthalpy, and temperature into equation (8.15) to get the net entropy change for $N_2O_4(g)$ formation and look at its sign to determine if the reaction is possible.

Reasoning: The reaction changes two moles of gaseous molecules to one mole. The formation of fewer product molecules means that the atoms in the products have less freedom to move about separately. For example, any nitrogen in the NO_2 reactant can be found anywhere in the system, but after the reaction, pairs of nitrogens must be found together in half as many N_2O_4 molecules. Thus, a decrease in number of moles of gas in the reaction gives a negative value for the positional entropy change for the reaction, which is consistent with the experimental value.

The formation of the bond between two NO_2 molecules to give N_2O_4 releases energy, so the reaction must be exothermic, as the experimental value shows.

Implementation:

$$\Delta S_{net} = \Delta S_{system} - \frac{\Delta H_{system}}{T} = -175.8 \text{ J·K}^{-1} - \left(\frac{-57.2 \times 10^3 \text{ J}}{298 \text{ K}} \right)$$

$$= -175.8 \text{ J·K}^{-1} + 191.9 \text{ J·K}^{-1}$$

$$\Delta S_{net} = 16.1 \text{ J·K}^{-1}$$

This positive value for ΔS_{net} means that the formation of N_2O_4 from NO_2 under standard conditions at 298 K is spontaneous.

Does the answer make sense?   The large exothermicity of the reaction leads to a large positive change in the thermal entropy of the surroundings [accounted for as $(-\Delta H_{system}/T)$], which compensates for the negative change in the positional entropy of the reacting system. Since the thermal entropy change in the surroundings will decrease as the temperature increases, the reaction might not be favored at a higher temperature.

<hr>

8.19 CHECK THIS

Formation of $N_2O_4(g)$ under standard conditions at 350 K

Is the reaction in Worked Example 8.18 spontaneous under standard conditions at 350 K? Assume that the standard entropy and enthalpy changes for the reaction are the same at 350 K and 298 K. Is your conclusion the same as in Worked Example 8.18? Why or why not?

<hr>

Shift to an energy perspective Equation (8.15) can be rearranged by multiplying through by T to get

$$T\Delta S_{net} = T\Delta S_{system} - \Delta H_{system} \tag{8.16}$$

the terms in equation (8.16) have units of energy. Absolute temperatures are always positive, so the sign of $T\Delta S_{net}$ is determined by the sign of ΔS_{net} and is positive (> 0) for reactions that are spontaneous and negative (< 0) for those that are nonspontaneous. However, the *conventional* way we deal with energies is to *express changes that lead to a lower total energy* (the favorable direction) *as negative*, since the final energy is lower than the initial. Multiplying through equation (8.16) by -1 yields

$$-T\Delta S_{net} = -T\Delta S_{system} + \Delta H_{system} = \Delta H_{system} - T\Delta S_{system} \tag{8.17}$$

<hr>

8.20 CHECK THIS

Criterion for direction of change

In terms of $-T\Delta S_{net}$, what are the criteria for spontaneous and nonspontaneous reactions?

<hr>

Gibbs free energy and the direction of change For our purposes in this chapter and the remainder of the book, we could use the numeric values of $-T\Delta S_{net}$ as the criterion for spontaneity and equilibrium. However, for several reasons, some historical and some practical, a new thermodynamic function, the **Gibbs free energy** (**G,** or simply the **free energy**) is usually used. The Gibbs free energy for a system is defined as

$$G \equiv H - TS \tag{8.18}$$

In equation (8.18), all the variables refer to the system. *For a change in a system at constant pressure and temperature*, the change in Gibbs free energy, ΔG, is

$$\Delta G = \Delta H_{system} - T\Delta S_{system} \tag{8.19}$$

If we compare equation (8.19) with equation (8.17), we see that

$$\Delta G = -T\Delta S_{net} \tag{8.20}$$

Thus, in terms of ΔG, equation (8.19) gives the same criteria for changes that are spontaneous and nonspontaneous as you found for $-T\Delta S_{net}$ in Check This 8.20:

If ΔG ($= -T\Delta S_{net} = \Delta H_{system} - T\Delta S_{system}$) < 0,

the change is spontaneous. (8.21)

If ΔG ($= -T\Delta S_{net} = \Delta H_{system} - T\Delta S_{system}$) > 0,

the change is nonspontaneous. (8.22)

If ΔG ($= -T\Delta S_{net} = \Delta H_{system} - T\Delta S_{system}$) $= 0$,

the system is at equilibrium. (8.23)

8.21 WORKED EXAMPLE

Free energy change for urea formation at 298 K

The formation of urea from ammonia and carbon dioxide is represented by this equation:

$$2NH_3(g) + CO_2(g) \rightleftharpoons H_2NC(O)NH_2(s) + H_2O(l)$$

For the formation of 1 mol of urea at 298 K, the standard entropy and enthalpy changes are -424 J·K^{-1} and -133.3 kJ, respectively. Calculate the standard Gibbs free energy change, $\Delta G°$ (as usual, "°" denotes the change under standard conditions), for formation of 1 mol of urea at 298 K. Does the free energy show that this change is spontaneous?

Necessary information: We need the criteria represented by equations (8.21) through (8.23).

Strategy: Substitute the values for T, $\Delta H°_{system}$, and $\Delta S°_{system}$ into equation (8.19) and look at the sign of $\Delta G°$ to see if urea formation is spontaneous under standard conditions at 298 K.

Implementation:

$$\Delta G° = \Delta H°_{system} - T\Delta S°_{system} = -133.3 \times 10^3 \text{ J} - (298 \text{ K})(-424 \text{ J·K}^{-1})$$

$$\Delta G° = -133.3 \times 10^3 \text{ J} + 126 \times 10^3 \text{ J} = -7 \times 10^3 \text{ J} = -7 \text{ kJ}$$

A negative value for the free energy change, $\Delta G = \Delta G°$ in this example, means that the formation of urea under standard conditions at 298 K is spontaneous.

Does the answer make sense? We have no independent way to know whether this answer is correct, without knowing something about the reaction. Many organisms excrete nitrogenous waste by converting ammonia to urea by this reaction, so it is likely to be a spontaneous reaction under these conditions, although perhaps not under others.

Free energy change for urea formation at 325 K

Calculate the standard Gibbs free energy change, $\Delta G°$, for formation of 1 mol of urea at 325 K. Assume that the standard entropy and enthalpy changes for the reaction are the same at 325 K and 298 K. Is your conclusion the same as in Worked Example 8.21? Why or why not?

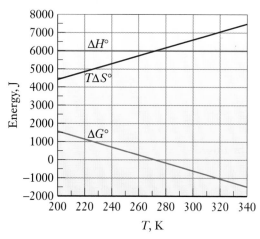

Figure 8.13.

$\Delta H°$, $T\Delta S°$, and $\Delta G°$ vs. T for the change $H_2O(s) \rightarrow H_2O(l)$ at standard pressure.

The relationships in equations (8.21) through (8.23) are illustrated graphically in Figure 8.13 for the solid-to-liquid phase change for water under standard conditions. ΔH and ΔS are assumed to be constant over the temperature range shown. You have made this same assumption in Check This 8.19 and 8.22. The assumption is not strictly true. Entropies and enthalpies do change with temperature, but the changes (for liquids and solids) are not large and can usually be neglected, if the range of temperatures is only a few tens of degrees.

8.23 WORKED EXAMPLE

Graphical relationship of $\Delta H°$, $T\Delta S°$, and $\Delta G°$ vs. T

Show how the values for $\Delta G°$, at 240 K and 300 K, in Figure 8.13 are derived from the plotted values for $\Delta H°$ and $T\Delta S°$.

Necessary information: We need equation (8.19).

Strategy: Read the $\Delta H°$ and $T\Delta S°$ values at these temperatures from the graph and then calculate $\Delta H° - T\Delta S°$ to get $\Delta G°$.

Implementation:

At 240 K: $\Delta H° = 6000$ J, $T\Delta S° = 5300$ J, and $\Delta G° = 6000 - 5300 = 700$ J

At 300 K: $\Delta H° = 6000$ J, $T\Delta S° = 6600$ J, and $\Delta G° = 6000 - 6600 = -600$ J

Does the answer make sense? The values we calculated are the values for $\Delta G°$ on the plot.

8.24 CHECK THIS

Significance of the graphical relationship of $\Delta H°$, $T\Delta S°$, and $\Delta G°$ vs. T

What is the significance of the point where the $\Delta H°$ and $T\Delta S°$ lines cross in Figure 8.13? Under what condition(s) is the solid-to-liquid phase change spontaneous? Under what condition(s) is the change nonspontaneous? Explain the reasoning for your answers.

Criterion for equilibrium The criterion for equilibrium in equation (8.23) is a generalization of what we discussed at the end of Section 8.7 for phase equilibria. Under standard conditions, $\Delta S = \Delta S^{\circ}_{net} = 0$ for a phase change at the temperature where the two phases are in equilibrium. More generally, *systems are at equilibrium when $\Delta G = -T\Delta S_{net} = 0$ for a change from the initial to final state under any constant pressure and temperature conditions.* For the solid-to-liquid phase change in water under standard conditions, Figure 8.13 shows that $\Delta G = \Delta G^{\circ} = 0$ at 273 K. We know solid and liquid water are in equilibrium at this temperature.

8.25 CHECK THIS

The melting point of benzene

The standard enthalpy of formation and standard entropy of solid benzene, C_6H_6, are 38.4 kJ·mol^{-1} and 135 J·K^{-1}·mol^{-1}, respectively. The corresponding values for liquid benzene are 49.0 kJ·mol^{-1} and 173 J·K^{-1}·mol^{-1}. Use these data to calculate ΔH° and ΔS° for the reaction:

$$C_6H_6(l) \rightarrow C_6H_6(s)$$

Use equation (8.23) to estimate the freezing temperature of benzene under standard conditions, assuming that ΔH° and ΔS° do not change with temperature. See Table 8.2, page 561, to check your result.

Why a new thermodynamic variable? In a more detailed treatment of thermodynamics, the Gibbs free energy plays a broader role than we are presenting here, and it is convenient to have a single variable (that accounts for enthalpy and entropy) to make the mathematics easier. From a practical point of view, tables of thermodynamic properties (Appendix B) always list standard free energies of formation (the difference in free energy between the compound in its standard state and its elements in their standard states) together with standard enthalpies of formation and standard entropies. Thus, it is easy to get ΔG° ($= -T\Delta S^{\circ}_{net}$) for a reaction. For many reactions in biological systems, enthalpies and entropies of reaction can't be measured, but the free energy of reaction can be determined from equilibrium measurements, as we will see in Chapter 9.

Why "free" energy? Equation (8.19), $\Delta G = \Delta H_{system} - T\Delta S_{system}$, is a succinct combination of the first and second laws of thermodynamics. For any reaction, the internal energy change (exclusive of the pressure–volume work) is included in the enthalpy term. The entropy term takes account of the change in molecular arrangements that occurs during the reaction. What is "left over" is the **free energy,** that is, the energy that is *available* to do other forms of work in the surroundings. In Chapter 7, Section 7.11, we did a calculation to figure out the amount of work available from the thermal energy released by burning fuel. For that calculation, we equated the amount of work to the thermal energy change of the reaction, but noted that it is not possible to convert thermal energy completely into work. Equation (8.19) now shows that it is entropy, the second law, that limits this conversion and also gives the maximum amount of work that *is* available, the free energy. No real device can actually produce this much work in the surroundings, because there are always energy losses in its internal workings.

Reflection and Projection

The last three sections have taken the qualitative idea of changes proceeding in the direction that produces larger numbers of distinguishable molecular arrangements and expressed it in quantitative terms using entropy and enthalpy values for the ice–water phase change as concrete examples. The switch we made from a perspective that includes both the system and surroundings to a focus on the system alone requires close attention to the restrictions placed on the application of equation (8.15). The changes for which the equation can be used must take place at constant temperature and pressure and the temperature of the system and surroundings must be the same.

A new thermodynamic function, the Gibbs free energy, G, was introduced. Under constant temperature and pressure conditions, the Gibbs free energy change, ΔG, for a change in a system is $\Delta G = \Delta H_{system} - T\Delta S_{system}$. We can predict whether a change is spontaneous by determining the sign of its free energy change. If $\Delta G < 0$, the change is spontaneous. If $\Delta G > 0$, the change is nonspontaneous. If $\Delta G = 0$, the initial and final states are in equilibrium.

We will use both free energy and entropy, whichever is more convenient in a particular case, to help understand physical and chemical changes. The next sections deal with chemical reactions and then further examples of phenomena, some of which may be surprising, but many of which are familiar, that will provide practice using our criteria for the direction of change.

8.9. Thermodynamic Calculations for Chemical Reactions

Standard molar entropies of many compounds are given in Appendix B. These values are derived from statistical calculations based on equation (8.6), from calorimetric experiments, and from equilibrium measurements. Appendix B also provides values for the standard molar Gibbs free energies of formation of these compounds. The values for entropies and free energies can be combined, just as those for energies and enthalpies are combined, to calculate entropy and free energy changes for processes of interest. In Chapter 7, we combined tabulated values for standard enthalpies of formation to obtain the standard enthalpy changes for reactions. Here, we'll do the same for standard entropy and standard free energy changes.

Note: The standard enthalpies and free energies in Appendix B are expressed in kilojoules, kJ, and the standard entropies are in joules, J, because the numeric values are most convenient in these units.

General equation for calculating entropy changes The overall standard entropy change for a reaction is the sum of all the standard entropies of the products minus the sum of all the standard entropies of the reactants. The mathematical expression for the difference of sums is

$$\Delta S^\circ_{reaction} = \Sigma\left[n_j(S^\circ)_j\right]_{products} - \Sigma\left[n_j(S^\circ)_j\right]_{reactants} \tag{8.24}$$

In words, equation (8.24) tells us: "Multiply the number of moles of each product by the standard entropy for that product. Add all the product entropy terms together. Multiply the number of moles of each reactant by the standard entropy for that reactant. Add all the reactant energy terms together. Finally, subtract the reactant sum from the product sum." Equation (8.24) for the standard entropy change of a reaction is identical in form to equation (7.30), Chapter 7, Section 7.8, for the standard enthalpy change of a reaction.

Entropy change for glucose oxidation An example we have used before is the oxidation of glucose to give carbon dioxide and water:

$$C_6H_{12}O_6(s) + 6O_2(g) \rightarrow 6CO_2(g) + 6H_2O(l) \qquad (8.25)$$

In terms of molecules, equation (8.25) tells us that seven reactant molecules combine to produce twelve product molecules, as represented in Figure 8.14. Atoms that were bonded to one another in the reactants (the carbon atoms in glucose, for example) are less constrained in the products. For example, the six carbon atoms, originally all in one glucose molecule, Figure 8.14(a), are freed to spread throughout the volume available in six separate carbon dioxide molecules, Figure 8.14(b). The number of molecular arrangements available to the products of the reaction is much greater than the number of molecular arrangements available to the reactants. Without doing any calculations, we can conclude that the collective entropy of the products is larger than the entropy of the reactants. The entropy change for the reaction described in equation (8.25) should be positive, and it may be quite large.

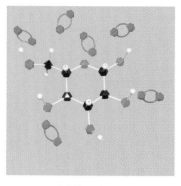

(a) Reactants

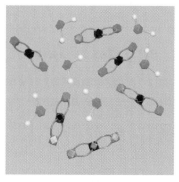

(b) Products

Figure 8.14.

Molecular-level representation of the oxidation of a glucose molecule.

8.26 WORKED EXAMPLE

Entropy change for oxidation of glucose

Calculate the standard entropy change when one mole of glucose is oxidized by oxygen, reaction (8.25), at 298 K. Compare the result with the prediction that the entropy change should be positive and quite large.

Necessary information: We need equation (8.24) and reaction (8.25) and, from Appendix B, the standard molar entropies, $S°$, at 298 K for each reactant and product.

Compound	$S°$, J·K^{-1}·mol^{-1}
$C_6H_{12}O_6(s)$	212.1
$O_2(g)$	205.1
$CO_2(g)$	213.7
$H_2O(l)$	69.91

continued

Strategy: Use equation (8.24) and the stoichiometry of reaction (8.25) to combine the standard molar entropies to get the standard entropy change for the reaction. Compare the result with our prediction that the entropy change should be large and positive.

Implementation:

$$\Delta S°_{\text{reaction}} = \Sigma\left[n_j(S°)_j\right]_{\text{products}} - \Sigma\left[n_j(S°)_j\right]_{\text{reactants}}$$

$$= \left[(6 \text{ mol } CO_2)(213.7 \text{ J·K}^{-1}\text{·mol}^{-1}) + (6 \text{ mol } H_2O)(69.91 \text{ J·K}^{-1}\text{·mol}^{-1})\right]$$

$$- \left[(1 \text{ mol } C_6H_{12}O_6)(212.1 \text{ J·K}^{-1}\text{·mol}^{-1}) + (6 \text{ mol } O_2)(205.1 \text{ J·K}^{-1}\text{·mol}^{-1})\right]$$

$$= +259.0 \text{ J·K}^{-1}$$

Does the answer make sense? The large, positive standard entropy change is consistent with our prediction that there are more ways to arrange the product molecules than there are to arrange the reactant molecules.

8.27 CHECK THIS

Predict the direction of the entropy change for glucose fermentation

Humans have known how to produce ethanol by fermentation (Chapter 6, Section 6.11) for almost seven thousand years. The reaction for glucose fermentation is

$$C_6H_{12}O_6(s) \rightarrow 2C_2H_5OH(l) + 2CO_2(g)$$

Do you expect the entropy change for this reaction to be positive or negative? Relatively large or relatively small? Explain your responses.

8.28 CHECK THIS

Entropy change for glucose fermentation

Use standard entropies, $S°$, from Appendix B to calculate $\Delta S°_{\text{reaction}}$ for glucose fermentation. Do your calculated results agree with your prediction in Check This 8.27?

Free energy change for glucose oxidation Both the entropy change in the system and the thermal entropy change for energy transfer between system and surroundings must be considered to determine whether a reaction can proceed in the direction written. We cannot be sure that glucose oxidation, reaction (8.25), can occur until we know the free energy change, $\Delta G_{\text{reaction}}$, for the reaction. We can use standard enthalpies of formation from Appendix B and equation (7.30) to calculate $\Delta H°_{\text{reaction}}$:

$$\Delta H°_{\text{reaction}} = \Sigma\big[n_j(\Delta H_f°)_j\big]_{\text{products}} - \Sigma\big[n_j(\Delta H_f°)_j\big]_{\text{reactants}} \qquad (7.30)$$

We then substitute $\Delta H°_{\text{reaction}}$ and the $\Delta S°_{\text{reaction}}$, calculated in Worked Example 8.26, into equation (8.19) to calculate $\Delta G°_{\text{reaction}}$ ($=\Delta H°_{\text{reaction}} - T\Delta S°_{\text{reaction}}$).

8.29 WORKED EXAMPLE

Free energy change for oxidation of glucose

Use the standard enthalpy and entropy of reaction to calculate the standard free energy change when one mole of glucose is oxidized by oxygen at 298 K.

Necessary information: You calculated $\Delta H°_{\text{reaction}} = -2801$ kJ in Chapter 7, Check This 7.50, and we calculated $\Delta S°_{\text{reaction}} = 259.0$ J·K^{-1} in Worked Example 8.26.

Strategy: Substitute $\Delta H°_{\text{reaction}}$ and $\Delta S°_{\text{reaction}}$ into equation (8.19) to get $\Delta G°_{\text{reaction}}$.

Implementation:

$$\Delta G°_{\text{reaction}} = (-2801\text{ kJ}) - (298\text{ K})(+259.5\text{ J·K}^{-1})$$
$$= (-2801\text{ kJ}) - (+77.3\text{ kJ}) = -2878\text{ kJ}$$

Does the answer make sense? The information we get from the enthalpy and entropy is that both $\Delta H°_{\text{reaction}}$ and $\Delta S°_{\text{reaction}}$ favor the oxidation of glucose. Thermal energy is transferred to the surroundings as the reaction progresses and the products of the reaction can be arranged in more ways than the reactants. The large, negative value for the standard free energy change confirms that there is a large driving force for the oxidation of glucose, which is an essential reaction in the metabolism of both plants and animals.

8.30 CHECK THIS

Free energy change for fermentation of glucose

Use the standard enthalpies of formation from Appendix B to calculate $\Delta H°_{\text{reaction}}$ for the fermentation of glucose (Check This 8.27). Combine this $\Delta H°_{\text{reaction}}$ with your $\Delta S°_{\text{reaction}}$ from Check This 8.28 to get $\Delta G°_{\text{reaction}}$ for glucose fermentation. Do you expect this reaction to proceed under standard conditions? Why or why not?

Appendix B lists standard free energies of formation, $\Delta G_f°$, and we can use them to calculate $\Delta G°_{\text{reaction}}$ directly:

$$\Delta G°_{\text{reaction}} = \Sigma\big[n_j(\Delta G_f°)_j\big]_{\text{products}} - \Sigma\big[n_j(\Delta G_f°)_j\big]_{\text{reactants}} \qquad (8.26)$$

Equation (8.26) has the same form as equation (7.30) and equation (8.24).

8.31 WORKED EXAMPLE

Free energy change for oxidation of glucose

Use standard free energies of formation to calculate the standard free energy change when one mole of glucose is oxidized by oxygen at 298. K.

Necessary information: We need equations (8.25) and (8.26) and, from Appendix B, the standard free energies of formation, ΔG_f°, at 298 K for each reactant and product.

Compound	ΔG_f°, kJ·mol^{-1}
$C_6H_{12}O_6(s)$	−910.52
$O_2(g)$	0
$CO_2(g)$	−394.36
$H_2O(l)$	−237.13

Strategy: Use equation (8.26) and the stoichiometry of reaction equation (8.25) to combine the standard molar free energies of formation, ΔG_f°, to get the standard free energy change for the reaction, $\Delta G^\circ_{reaction}$.

Implementation:

$$\Delta G^\circ_{reaction} = \Sigma\left[n_j(\Delta G_f^\circ)_j\right]_{products} - \Sigma\left[n_j(\Delta G_f^\circ)_j\right]_{reactants}$$

$$= \left[(6\text{ mol CO}_2)(-394.36\text{ kJ·mol}^{-1}) + (6\text{ mol H}_2O)(-237.13\text{ kJ·mol}^{-1})\right]$$

$$- \left[(1\text{ mol C}_6H_{12}O_6)(-910.52\text{ kJ·mol}^{-1}) + (6\text{ mol O}_2)(0\text{ kJ·mol}^{-1})\right]$$

$$= -2878.4\text{ kJ}$$

Does the answer make sense? This direct calculation of the standard free energy of reaction and the combination of the standard enthalpy and entropy of reaction in Worked Example 8.29 give the same numeric result. The oxidation of glucose under standard conditions has a large driving force.

8.32 CHECK THIS

Free energy change for fermentation of glucose

Use the standard free energies of formation from Appendix B to calculate $\Delta G^\circ_{reaction}$ for glucose fermentation (Check This 8.27). How does this result compare with your result in Check This 8.30?

Entropy change for dissolving ionic solids We found, in Chapter 2, that dissolving ionic solids in water is an endothermic process for some compounds and exothermic for others. Entropy can help us understand the direction of these processes. For ions in solution, a concentration has to be specified for standard entropies and standard enthalpies and free energies of formation; the usual choice is 1 molal. **Molality (*m*)** is defined as the number of moles of solute dissolved in one kilogram of solvent. The entropy, enthalpy of formation, and free energy of formation of 1 m hydronium ion are the reference point and defined as zero. (For aqueous solutions, molality and molarity are almost the same: 1 $m \approx$ 1 M.)

Your bones and teeth are partly made of solids containing calcium ions, Ca^{2+}, and phosphate ions, PO_4^{3-}. The simplest compound of these two ions is calcium phosphate, $Ca_3(PO_4)_2$. Although this compound has a low solubility in water, we can write the reaction for calcium phosphate dissolving and consider standard enthalpy and standard entropy changes for dissolving:

$$Ca_3(PO_4)_2(s) \rightleftharpoons 3Ca^{2+}(aq) + 2PO_4^{3-}(aq) \qquad (8.27)$$

Equation (8.27) describes the reaction of a solid ionic compound, in which the ions are constrained to particular locations in the crystal, dissolving to yield ions that can move about in the entire volume of the solution; dissolution of an ionic solid is represented in Figure 8.15. More arrangements are possible for ions in aqueous solution than in crystalline solids, so we *expect* the standard entropy change for reaction (8.27), $\Delta S^\circ_{reaction}$, to be positive. Since calcium phosphate is quite insoluble at room temperature, reaction (8.27) does not proceed to an appreciable extent. Thus, we can conclude that $\Delta G^\circ_{reaction}$ ($= \Delta H_{reaction} - T\Delta S^\circ_{reaction}$) is probably positive. In order for $\Delta G^\circ_{reaction}$ to be positive, $\Delta H^\circ_{reaction}$ must be positive (an endothermic reaction) *and* numerically larger than $T\Delta S^\circ_{reaction}$. Otherwise, $\Delta G^\circ_{reaction}$ would be negative (and the solid would be soluble). Let's see what the thermodynamic data tell us.

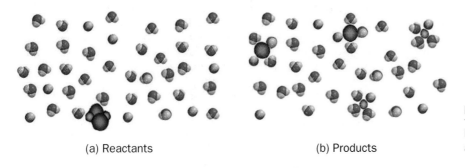

(a) Reactants   (b) Products

Figure 8.15.

Molecular-level representation of an ionic solid dissolving in water.

8.33 WORKED EXAMPLE

Free energy change for dissolving calcium phosphate

Use standard enthalpies of formation and entropies to calculate the free energy change for dissolving one mole of calcium phosphate in water to form $Ca^{2+}(aq)$ and $PO_4^{3-}(aq)$ at 298 K.

Necessary information: We need equations (7.30), (8.19), and (8.24), reaction (8.27), and, from Appendix B, the standard molar enthalpies of formation, ΔH_f°, and standard molar entropies, S°, at 298 K for each reactant and product.

Species	ΔH_f°, kJ·mol^{-1}	S°, J·mol^{-1}·K^{-1}
$Ca_3(PO_4)_2(s)$	−4121	+236
$Ca^{2+}(aq)$ (1 m)	−543	−56
$PO_4^{3-}(aq)$ (1 m)	−1277	−221

continued

Strategy: Use equations (7.30) and (8.24) and the stoichiometry of reaction (8.27) to combine the standard molar enthalpies of formation and standard molar entropies to get the standard enthalpy and entropy changes for the reaction. Substitute these into equation (8.19) to get the standard free energy change for the reaction, $\Delta G°_{reaction}$.

Implementation: The standard enthalpy of reaction is

$$\Delta H°_{reaction} = \left[(3 \text{ mol } Ca^{2+})(-543 \text{ kJ·mol}^{-1}) + (2 \text{ mol } PO_4^{3-})(-1277 \text{ kJ·mol}^{-1})\right]$$
$$- \left[(1 \text{ mol } Ca_3(PO_4)_2)(-4121 \text{ kJ·mol}^{-1})\right] = -62 \text{ kJ}$$

The standard entropy of reaction is

$$\Delta S°_{reaction} = \left[(3 \text{ mol } Ca^{2+})(-56 \text{ J·mol}^{-1}·K^{-1}) + (2 \text{ mol } PO_4^{3-})(-221 \text{ J·mol}^{-1}·K^{-1})\right]$$
$$- \left[(1 \text{ mol } Ca_3(PO_4)_2)(+236 \text{ J·mol}^{-1}·K^{-1})\right] = -846 \text{ J·K}^{-1}$$

The standard free energy change for the reaction is

$$\Delta G°_{reaction} = \Delta H°_{reaction} - T\Delta S°_{reaction} = -62 \text{ kJ} - (298 \text{ K})(-846 \text{ J·K}^{-1}) = +190 \text{ kJ}$$

Does the answer make sense? $\Delta G°_{reaction}$ is positive, as we expected for the standard free energy change for dissolving a compound that has a low solubility in water. The surprise here is that the signs of $\Delta S°_{reaction}$ and $\Delta H°_{reaction}$ are *opposite* from what we had predicted. The entropy for the solid is actually *greater* than the entropy for the ions in solution. In Chapter 2, Section 2.7, we found that dissolving ionic solids with multiply-charged ions is usually an exothermic process. If we had recalled this, we should have been skeptical about our *prediction* for the sign of $\Delta H°_{reaction}$ in this case, even before doing the calculation.

8.34 CHECK THIS

Free energy change for dissolving barium carbonate

Barium carbonate is an insoluble solid:

$$BaCO_3(s) \rightleftharpoons Ba^{2+}(aq) + CO_3^{2-}(aq)$$

(a) Use the values for standard molar enthalpies of formation and standard molar entropies in Appendix B to calculate the standard enthalpy and entropy changes for this reaction. Is the sign of the entropy change surprising? Why or why not?

(b) Calculate the standard free energy change for the reaction. Is the sign of the free energy change surprising? Why or why not?

Entropies of ions in solution The discussion so far has treated ions dissolving into solution as if they were set free to move about independent of everything else in the solution. On the other hand, recall from Chapter 2 that the dissolving process involves ion–dipole attractions between the ions and the water molecules of the solution. The ions exist in aqueous solution with a sheath of water molecules surrounding them in a more-or-less orderly array, as shown in Figure 8.15(b). The equation for dissolving calcium phosphate is more accurately written as

$$Ca_3(PO_4)_2(s) + (m + n)H_2O(l) \rightleftharpoons 3[Ca(H_2O)_m]^{2+} + 2[PO_4(H_2O)_n]^{3-}$$

(8.28)

Figure 8.15 is a schematic representation of the decrease in arrangements of water molecules that occurs when an ionic solid dissolves in water. Before dissolution, Figure 8.15(a), the water molecules are free to move about in the liquid. In the solution, Figure 8.15(b), water molecules around the hydrated ions are less free to move about because they are held next to one another by their attraction to the ions. The decrease in arrangements (decrease in entropy) is greater for multiply charged ions, as you can see by comparing the entropies for various aqueous ions in Appendix B.

Often, values for m and n in reaction (8.28) are only approximately known, so the "*(aq)*" we usually write usefully represents the fact that the ions are solvated with some number of water molecules. However, writing the reaction as is done in reaction (8.28) makes it clearer that there is organization on the product side as well as on the reactant side. In Chapter 2, we referred to this ordering as an unfavorable rearrangement of the water molecules. Now we see that the result of the ordering is a decrease in the entropy of the solution, compared to what we would predict based on mixing alone. Rather than assuming that dissolved ions in a mixture have greater net entropy than the pure ionic solid and water, always check tabulated entropy values.

Significance of ΔG and $\Delta G°$ Note that all of the above calculations give us the *standard* **free energy change, $\Delta G°_{reaction}$,** for a reaction of the reactants in their standard states to give products in their standard states. Few reactions are actually carried out this way; $\Delta G°_{reaction}$ is a *guide* to help us understand the direction of reaction. We really need $\Delta G_{reaction}$ (ΔG), the **free energy change** under the actual reaction conditions to tell for sure whether the reaction is possible or not. In Chapter 9, we will discuss how to get $\Delta G_{reaction}$ and the significance of $\Delta G°_{reaction}$.

Reflection and Projection

By examining the number and physical states of the reactants and products of reactions, we made predictions about the sign of the entropy and enthalpy changes for the reactions. We tested these predictions by calculating standard entropy, enthalpy, and free energy changes for the reactions. In many cases, these predictions are valid. However, the qualitative idea that dissolved ions should always have greater entropy than the separated pure solid and the solvent proves to be untrue. In order to interpret entropies of solution, we need to use the solubility model developed in Chapter 2. The organization of water molecules about dissolved ions has to be taken into account and is especially important when the ions are multiply charged.

Positional entropy plays a fundamental role in controlling changes in these systems. It is not only ionic solutes that have an effect on the organization of water molecules. In the next sections, we will examine the consequences of the interactions of water with nonpolar molecules and molecules with both polar and nonpolar regions.

8.10. Why Oil and Water Don't Mix

 8.35 INVESTIGATE THIS

What happens when oil and water are mixed?

For this investigation, use a small test tube and three plastic pipets containing these liquids: cooking oil (a clear, pale yellowish liquid), water, and a liquid dish-washing detergent.

(a) Place a few milliliters of water in the test tube and add about $\frac{1}{2}$ mL of oil. Cap the tube with your finger, mix by inverting the tube several times, and allow the tube to sit quietly for several seconds. Record your observations when no further change seems to be occurring.

(b) Add about $\frac{1}{2}$ mL of the dishwashing detergent to the test tube. Cap the tube with your finger and mix by inverting the tube several times. Try not to create a lot of suds. Allow the tube to sit quietly and observe. Record your observations when no further change seems to be occurring. We will return to this result in Consider This 8.40.

8.36 CONSIDER THIS

Why don't oil and water mix?

In Investigate This 8.35(a), you investigated the solubility of oil in water. In Chapter 2, after you investigated the solubility of hexane in water (Investigate This 2.3), we explained the insolubility of nonpolar compounds in terms of unfavorable rearrangements of water molecules around any dissolved nonpolar molecule. How would you construct this explanation in terms of entropy?

As you know from Investigate This 8.35 and common experience, oil and water don't mix. Nonpolar compounds, such as cooking oil, are insoluble in water. Now that we have net entropy change (or free energy change) as a criterion for the directionality of processes, we can try to understand the factor(s) that make oil and water immiscible. We need to account for both thermal and positional entropy changes for the dissolution of nonpolar solutes in water. Let us look at thermal entropy first.

Dispersion forces are the attractive interactions that hold nonpolar molecules together in their solid and liquid states. Dispersion forces exist for all atoms and molecules and, in hydrocarbons, they are the largest attractive forces. In polar molecules, attractions among the permanent dipoles are usually the strongest forces. Therefore, mixing hydrocarbon molecules and water molecules requires breaking some hydrogen bonds between the water molecules and disrupting dispersion interactions between hydrocarbon molecules. Disrupting attractions requires an input of energy. New attractions that occur among the mixed molecules release energy. When we measure the enthalpy change for dissolving hydrocarbons in water, $\Delta H_{solution}$ turns out to be very nearly zero in most cases. The disruptive and attractive interactions cancel one another. Since the enthalpy change for dissolving nonpolar molecules in water is about zero, the thermal entropy change (equation 8.11) is also about zero.

Oriented water molecules If there is no change in thermal entropy for dissolving a hydrocarbon in water, it must be the positional entropy that is unfavorable and causes the insolubility. We know that mixtures of two substances always have higher positional entropy than the unmixed substances. Therefore, for hydrocarbons and water, there must be some other unfavorable entropy effect that is larger than the entropy of mixing. Figure 8.16 illustrates the origin of this unfavorable entropy effect.

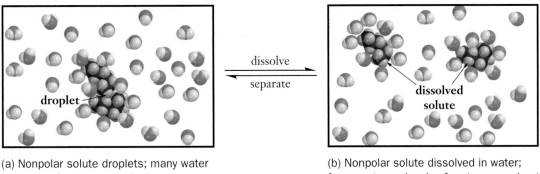

(a) Nonpolar solute droplets; many water molecules free to move about

(b) Nonpolar solute dissolved in water; fewer water molecules free to move about

Figure 8.16.

Decrease in arrangements of water molecules when a nonpolar solute dissolves. Compare the effects of ionic and nonpolar solutes on water molecules in Figure 8.15 and here.

There is experimental spectroscopic evidence that water molecules next to nonpolar molecules are oriented and lose some of their freedom of motion. When the nonpolar molecules are grouped together, not as many water molecules are restricted in their motion, Figure 8.16(a). If the nonpolar molecules mix into the water, more water molecules are needed to group around the individual nonpolar molecules, so more water molecules are restricted in their motion, Figure 8.16(b). Thus, fewer molecular arrangements of water molecules are possible. The loss of molecular arrangements reduces the entropy of the water molecules and makes the overall positional entropy change for mixing oil and water negative; oil and water don't mix.

Clathrate is derived from a Greek word meaning "lattice"; one substance is mixed into the crystal lattice of another.

Clathrate formation Although hydrocarbons do not dissolve well in liquid water, the relatively ordered structure that water forms around them is much like ice. Thus, at lower temperatures, water can form ice-like structures around low-molecular-mass hydrocarbons. The cage structure shown in Figure 8.17 is typical for nonpolar molecules in water. Such mixtures, called **clathrates,** are characterized by molecules of the solute distributed throughout crystals of the solvent. Water, as a semisolid host crystal, forms clathrates with many nonpolar molecules. One well-studied clathrate is formed between water and methane, CH_4. Methane produced by microorganisms is trapped as water clathrates at the bottom of seas and oceans and is a source of energy for other organisms as well as a potential source of fuel for humans. Figure 8.18 shows a methane clathrate mound with marine worms feeding on the methane.

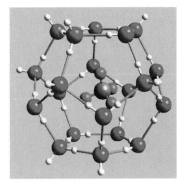

Figure 8.17.

Ice clathrates are cages of water molecules around hydrocarbon molecules.

Figure 8.18.

Organisms feeding on methane clathrate at the bottom of the sea.
(a) Methane clathrate is the yellowish blobs on the side of the rock. (b) Close-up view of the clathrate shows pink worms, about 3–5 cm in length, that live on this energy source.

(a)

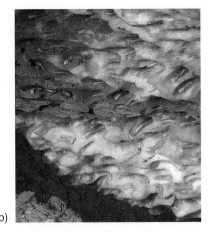

(b)

8.11. Ambiphilic Molecules: Micelles and Bilayer Membranes

8.37 INVESTIGATE THIS

How does light interact with solutions?

Do this as a class investigation and work in small groups to discuss and analyze the results. Fill two transparent, colorless plastic or glass containers with water. To one of the containers add about one teaspoon of table sugar. To the other add 3–4 drops of liquid soap or detergent. Use separate stirrers to stir the contents of the containers until the solid and/or liquid disappear and are thoroughly mixed with the water. Allow the containers to sit undisturbed for several minutes so the liquids are no longer swirling. Darken the room and direct a narrow beam of light through one of the solutions from the side while observing from the top. Repeat the observation with the other solution. Record your observations, especially noting any differences between the two solutions.

8.38 CONSIDER THIS

What causes light to interact with a solution?

Individual solute molecules in a solution are too small to affect the path of a beam of light. If there are clumps of solute molecules, the clumps may be large enough to interfere with a light beam and scatter it in all directions. How do you interpret any difference(s) between the observations you made on the two solutions in Investigate This 8.37?

In order to understand how cell membranes, like those represented in the chapter opening illustration, form, we will look at the properties of the molecules that form them. In particular, we'll examine how they interact with water and the part water plays in creating the membrane structure. We'll start with a similar but simpler model system, detergent molecules, which are less complicated than membrane molecules. In addition to helping us understand membranes, detergents have interesting and useful properties themselves, as you have observed in Investigate This 8.35(b) and 8.37.

Ambiphilic molecules The structure of one kind of **detergent molecule**, $R-OSO_3^-$ (dodecyl sulfate), is shown in Figure 8.19. One end of the structure is highly polar, the ionized sulfate group, $-OSO_3^-$, and the rest of the molecule, $R-$, is a long, nonpolar hydro-

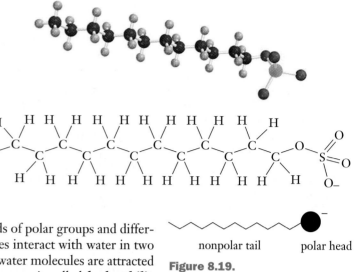

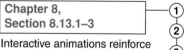

nonpolar tail   polar head

Figure 8.19.

Different representations of a detergent molecule, dodecyl sulfate.

carbon tail. Different detergents have different kinds of polar groups and different hydrocarbon chain lengths. Detergent molecules interact with water in two ways. The polar end of the detergent and the polar water molecules are attracted to one another. The polar end (head) of the detergent is called **hydrophilic** (*hydro* = water + *philos* = loving) and is soluble in water. The nonpolar hydrocarbon tail acts like an oil. Hydrocarbons are called **hydrophobic** (*hydro* = water + *phobos* = fearing or hating) because they don't mix with water to form solutions. Molecules like detergents that have both hydrophilic and hydrophobic regions are called **ambiphilic** (*ambi* = both + *philos* = loving).

Micelle formation Although we didn't use the term *ambiphilic* in Chapter 2, in Investigate This 2.3, we found that such molecules have limited solubility in water. Their polar (hydrophilic) regions are soluble; their nonpolar (hydrophobic) regions are not. One way an ambiphile can satisfy both the hydrophilic and hydrophobic interactions is to form **micelles,** as shown in Figure 8.20.

The diameter of the polar head of a detergent (or soap) molecule is relatively large compared to the diameter of its hydrocarbon tail, making the molecule shaped sort of like a thin cone, as shown in Figure 8.20(a). If the molecules cluster with their hydrophobic tails together, Figure 8.20(b), water molecules are freed from their interactions with the nonpolar tails. Freeing these water molecules increases the entropy of the solution, which favors clustering. Notice

Web Companion

Chapter 8,
Section 8.13.1–3

(1)
(2)
(3)
(4)

Interactive animations reinforce this explanation and illustration of micelle formation.

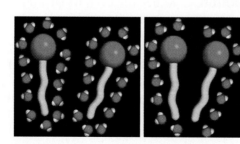

(a) Representations of an ambiphilic soap molecule (stearate)

(b) Ambiphilic molecular clustering frees water molecules and increases the entropy of the system.

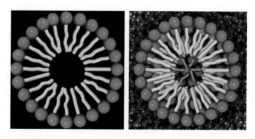

(c) Micelles are formed as more of the cone-shaped ambiphilic molecules cluster.

Figure 8.20.

Formation of a micelle from an ambiphilic soap molecule.

that packing the cone-shaped detergent molecules makes the clusters form curved structures. Imagine this packing continuing in three dimensions to form a roughly spherical closed structure, Figure 8.20(c), which is the shape micelles take.

The polar heads in the micelle form an outer layer in contact with water, and the hydrocarbon tails are tucked inside interacting with one another and freeing many water molecules. Micelle formation increases the net entropy of the solution by causing an increase in entropy of the water that is no longer organized around the hydrocarbon tails. A detergent micelle contains about 50–100 molecules and is about 3–5 nm in diameter. There can be a large number of micelles in the solution, so substantial amounts of detergent can mix with water this way. Although you can't see individual micelles, they are large enough to scatter light that shines through the mixture.

8.39 CHECK THIS

Evidence for micelle formation

Are your observations for the solutions in Investigate This 8.37 consistent with this last statement about light scattering? Explain why or why not.

8.40 CONSIDER THIS

How do ambiphilic molecules interact with nonpolar solutes?

In Investigate This 8.35(b), you found that a mixture of water, oil, and detergent appears cloudy, but does not separate into a layer of oil on water. What interactions among the water, ambiphilic detergent, and nonpolar oil molecules could be responsible for this behavior? Draw a sketch showing your model of these interactions.

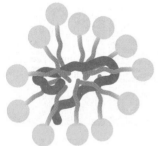

Figure 8.21.

A nonpolar molecule in the hydrophobic interior of a detergent micelle.

Detergent action Consider what can happen if nonpolar molecules such as oil are present in an aqueous mixture that also contains ambiphilic detergent molecules. The hydrophobic tail of the detergent molecules and the nonpolar solute can interact to free some of the solvating water molecules, which is the kind of interaction we have illustrated in Figures 8.16 and 8.20(b). As micelles form, the nonpolar molecules associated with their hydrophobic tails become incorporated into the interior of the micelle, as illustrated in Figure 8.21. In Investigate This 8.35(b), the detergent added to the oil and water mixture formed micelles containing the oil molecules. Since the oil molecules could not get together to form droplets of oil, they could not form a separate layer. When you use detergent (or soap) and water to wash your hands or soiled laundry, oils on your hands or on the soiled cloth become incorporated into micelles and are rinsed away with the water. The entropy increase in the solution, as micelles are formed, is responsible for your clean hands and clothes.

Phospholipid bilayers The ambiphilic molecules that make up cell membranes are **phospholipids. Lipids** (*lipos* = fat) are esters of long-chain carboxylic acids (fatty acids) with glycerol. Plants and animals use lipids for a variety of purposes including membrane structures and energy storage as fats and oils. Phospholipids, Figure 8.22, have two long hydrophobic tails and a polar head that contains charged phosphate and ammonium-like groups. The polar head is comparable in diameter to the two tails, so the overall shape of a phospholipid is approximately a cylinder, as shown in Figure 8.23(a). If the phospholipid molecules cluster together with their hydrophobic tails together, Figure 8.23(b), water molecules are freed from their interactions with the nonpolar tails. Freeing of the water molecules increases the entropy of the solution and favors clustering.

When cylindrical phospholipids pack together like cans on a supermarket shelf, a layer of molecules is formed, but this still leaves the ends of the hydrophobic tails available to interact with water. More water can be freed and entropy increased, if two layers come together with their hydrophobic sides facing to form a **phospholipid bilayer,** Figure 8.23(c) and the chapter opening illustration. Phospholipid bilayers are not rigid and can bend around and close on themselves, Figure 8.24. This closed structure is called a **liposome** or **lipid vesicle.** Liposomes can vary in size, but small ones are about 25 nm in diameter and contain about 2700 phospholipid molecules. Note that the liquid on the interior of a liposome is an aqueous solution.

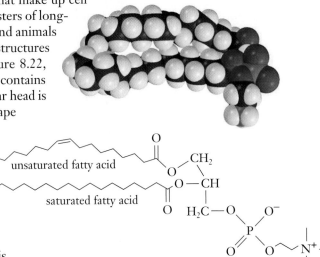

unsaturated fatty acid

saturated fatty acid

Figure 8.22.

A phospholipid molecule, phosphatidyl choline.

Web Companion

Chapter 8, Section 8.13.1 — ①

View an animation of bilayer formation illustrated in Figure 8.23.

②
③
④

 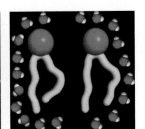

(a) Representations of a phospholipid molecule

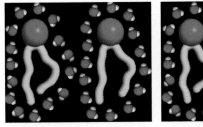

(b) Ambiphilic molecular clustering frees water molecules and increases the entropy of the system.

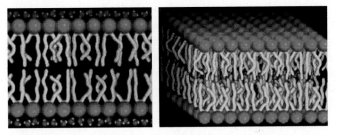

(c) Further clustering of the cylindrical molecules gives a layer, and two layers combine to form a bilayer.

Figure 8.23.

Formation of a phospholipid bilayer.

Inner aqueous compartment

Bilayer membrane

Figure 8.24.

Formation of a liposome from a phospholipid bilayer.

Getting through a phospholipid bilayer membrane Like those making up liposomes, **phospholipid bilayer membranes** enclose all living cells, and their membranes have the ambiphilic properties we have been describing. Liposomes are used to investigate some of the properties of cell membranes, such as how well various solutes can pass through them To do this, investigators make liposomes in an aqueous solution that contains the solute of interest. The solution on the inside of the liposome will contain the solute. After separating the liposomes from the solution, they are placed in a new solution that does not contain the solute of interest. Then the time it takes for the solute molecules to move out of the liposome into the surrounding solution is measured. From these experiments, we can determine the relative permeability of membranes for various molecules. A few results of such studies are given in Table 8.1.

Table 8.1 *Relative membrane permeabilities for representative molecules and ions.*

Molecule	Relative permeability
water, H_2O	10,000
glycerol, $C_3H_8O_3(aq)$	100
glucose, $C_6H_{12}O_6(aq)$	1 (assigned)
chloride ion, $Cl^-(aq)$	0.001
potassium ion, $K^+(aq)$	0.0001
sodium ion, $Na^+(aq)$	0.00002

A bilayer membrane is not a static structure; phospholipid molecules can shift about within their side of the bilayer. As the shifting and jostling of the phospholipid molecules occurs, small holes or pores form and disappear in the membrane. Each of these transient pores provides an opportunity for a molecule near the surface of the membrane to enter the bilayer or even cross it entirely. The molecules present in the largest amount on both sides of a membrane are water, so a water molecule is the most likely to enter a membrane pore. As we've noted, interaction of a water molecule with the hydrocarbon tails is not very favorable, so any water molecules entering a membrane are likely to leave. They can either return through the side where they entered or pass all the way through the membrane and leave on the other side.

Both polarity and size are significant in whether a molecule can cross the membrane. Small polar molecules like glycerol cross less easily, but still quite readily. Larger polar molecules like sugars and amino acids cross even more slowly. The transient pores in the membrane are small, so these larger molecules simply have a hard time entering the membrane in the first place. Ions like $Na^+(aq)$, $K^+(aq)$, $Ca^{2+}(aq)$, and $HOCO_2^-(aq)$ don't easily cross phospholipid bilayer membranes at all because the water molecules associated with the ions have to cross the membrane with them and this cluster is larger than most of the pores. Thus size as well as their unfavorable interaction with the nonpolar interior of the membrane keeps hydrated ions out.

8.41 CONSIDER THIS

How do the properties of bilayer membranes help explain osmosis?

In Investigate This 8.1, we used a carrot to set up an investigation of osmosis. Osmosis involves transfer of solvent molecules through a semipermeable membrane and we found that an increase in entropy as solvent molecules dilute a solution explains the observed direction of the process. Do the relative permeabilities of membranes in Table 8.1 help to explain why the carrot could be used for this investigation? Why or why not?

Reflection and Projection

It seems that positional entropy should increase when mixtures form that provide larger volumes for one or both of the components to move in. We found, however, that ion hydration reduces the positional entropy of aqueous solutions and is the reason that ionic solids with multicharged ions are rather insoluble. Similarly, we find that the organization of water molecules around nonpolar solutes also reduces the positional entropy of these aqueous solutions and is the reason that oil and water do not dissolve in one another but form separate layers when mixed.

The outcomes are more complex when ambiphilic molecules are mixed with water, because their polar (hydrophilic) parts interact energetically favorably with water via polar interactions, while their nonpolar (hydrophobic) parts do not. The entropy of the mixture would be increased if the nonpolar parts could be separated from the water. Two ways to maintain the energetically favorable polar interactions and increase the entropy of the mixture are for the ambiphilic molecules to cluster into micelles or bilayers. Detergent molecules with a single nonpolar tail generally form micelles with the nonpolar tails on the inside and the polar heads on the outside of a roughly spherical cluster. Phospholipid molecules with double nonpolar tails generally form lipid bilayers: two flexible sheets of the molecules with their nonpolar tails facing one another on the inside of the bilayer and the polar heads in contact with water on the outsides of the bilayer.

Phospholipid bilayers can close on themselves and form the membranes that surround all living cells as well as the organelles (structures inside the cells) in eukaryotic cells. Molecules can cross from one side to the other of a phospholipid bilayer membrane, but they do so at different speeds, depending on their size and polarity. Thus, bilayer membranes are semipermeable and osmosis takes place through them. We began the chapter discussing osmosis and now return to it and to a group of other properties, colligative properties, including osmotic pressure, all of which you can understand by analyzing their entropies.

8.12. Colligative Properties of Solutions

 8.42 INVESTIGATE THIS

What is the freezing point of salt water?

Do this as a class investigation and work in small groups to discuss and analyze the results. Fill a 250-mL beaker with crushed ice. Add about 50 mL of water and place a wooden stirrer and a thermometer in the beaker. Stir the mixture vigorously and record the temperature of the ice–water mixture. Add about 90 g of salt, NaCl. Again stir vigorously for 2–3 minutes and record the temperature.

8.43 CONSIDER THIS

How does salt affect the freezing point of water?

(a) In Investigate This 8.42, how do the temperatures compare before and after addition of the salt?

(b) How does the entropy of a salt solution compare to the entropy of pure water? Could the difference be responsible for the result you noted in part (a)? Explain.

Freezing point of a solution In Section 8.7, we considered phase changes in pure substances. To explain the results from Investigate This 8.42, we need to consider the behavior of solutions of nonvolatile solutes. **Nonvolatile solutes** are solutes that cannot evaporate from the solution. Ionic solids (table salt) and polar covalent molecular solids (sugar) are examples. Our discussion of osmosis earlier in the chapter has already begun this consideration, and now we will extend that discussion. The fundamental concept we build on is that mixing a soluble solute with a solvent makes the positional entropy of the solution greater than the positional entropy of the pure solvent. In Section 8.4, especially in Figure 8.7, you saw how this concept (expressed in terms of numbers of arrangements instead of entropy) explains the direction of osmosis. Since freezing and melting involve energy, the change in freezing temperature in Investigate This 8.42 must also involve energy. Thermal entropy also has to be accounted for in our explanation.

Entropy change for freezing from solution Figure 8.25 shows the change that occurs when a solution of a nonvolatile solute begins to freeze.

Figure 8.25.

Changes that occur when a solution begins to freeze.

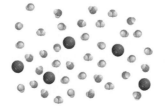

(a) Solution of a nonvolatile solute

(b) Freezing produces pure solid solvent

Note that the solid that freezes out of the solution is pure solvent. For example, when an aqueous solution of salt is frozen, the solid that forms is pure ice. We know that the entropy of a solution is greater than the entropy of pure liquid solvent, $S_{\text{solution}} > S_{\text{solvent}}$, and also that the entropy of a pure liquid is greater than that of a pure solid, $S_{\text{solvent}} > S_{\text{solid}}$. Thus, we can write (at the same temperature):

$$S_{\text{solution}} > S_{\text{solvent}} > S_{\text{solid}} \tag{8.29}$$

The solution has the greatest number of distinguishable arrangements of particles. The liquid has fewer arrangements, and the solid has the fewest of all. The entropy decrease to form solid solvent from a solution is always greater than the entropy decrease to form solid solvent from the pure liquid solvent. These relationships are shown graphically in Figure 8.26.

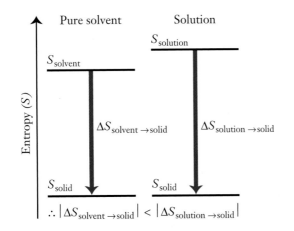

Figure 8.26.

Positional entropy changes for freezing a pure solvent and a solution.

When the solution is in equilibrium with the solid solvent, $\Delta G_{\text{solution}\rightarrow\text{solid}}$ is zero:

$$\Delta G_{\text{solution}\rightarrow\text{solid}} = 0 = \Delta H_{\text{solution}\rightarrow\text{solid}} - (T_{\text{solution}\rightarrow\text{solid}})\Delta S_{\text{solution}\rightarrow\text{solid}} \tag{8.30}$$

Equation (8.30) can be rearranged to give

$$\Delta H_{\text{solution}\rightarrow\text{solid}} = (T_{\text{solution}\rightarrow\text{solid}})\Delta S_{\text{solution}\rightarrow\text{solid}} \tag{8.31}$$

When pure liquid solvent is in equilibrium with the solid solvent, there is a similar equation:

$$\Delta H_{\text{solvent}\rightarrow\text{solid}} = (T_{\text{solvent}\rightarrow\text{solid}})\Delta S_{\text{solvent}\rightarrow\text{solid}} \tag{8.32}$$

The enthalpy change (ΔH) accompanying freezing is independent of whether the pure solid comes from the pure liquid solvent or from a solution, that is:

$$\Delta H_{\text{solvent}\rightarrow\text{solid}} = \Delta H_{\text{solution}\rightarrow\text{solid}} \tag{8.33}$$

Consequently, the right-hand sides of equations (8.31) and (8.32) are also equal. Setting them equal and rearranging gives

$$\frac{\Delta S_{\text{solvent}\rightarrow\text{solid}}}{\Delta S_{\text{solution}\rightarrow\text{solid}}} = \frac{T_{\text{solution}\rightarrow\text{solid}}}{T_{\text{solvent}\rightarrow\text{solid}}} \tag{8.34}$$

Figure 8.26 shows us that the ratio of entropy changes, $\dfrac{\Delta S_{\text{solvent}\to\text{solid}}}{\Delta S_{\text{solution}\to\text{solid}}}$, is less

than one. Therefore, $\dfrac{T_{\text{solution}\to\text{solid}}}{T_{\text{solvent}\to\text{solid}}}$ is less than one, so $T_{\text{solution}\to\text{solid}}$ must be *less*

than $T_{\text{solvent}\to\text{solid}}$. The freezing point *decreases* when a solute is added to a pure liquid, as you found in Investigate This 8.42. This phenomenon is known as **freezing point lowering** (also called freezing point depression).

> *Colligative* is from Latin meaning to bind together or act together. Each solute molecule or ion in the solution acts together with all the others to produce the observed effect, freezing point lowering, for example.

Colligative properties Note that none of the properties of the solute entered into this discussion of freezing point lowering of solutions. Likewise, the only property of the solute that is relevant to osmosis is that the solute not be able to pass through the semipermeable membrane. *Properties of solutions that do not depend on the identity of the nonvolatile solute (or solutes), but only on how many of them are present in the solution, are called* **colligative properties.** Colligative properties include freezing point lowering, **boiling point elevation** (solutions boil at a higher temperature than the pure solvent), and **osmotic pressure** (osmosis raises the level of the solution, so the pressure on the solution side of the membrane is higher than on the solvent side). Figure 8.27, which also repeats Figures 8.3 and 8.25, shows the change associated with each of these properties. The important thing to observe in each case is that the change involves a solution and its pure solvent.

(a) Freezing produces pure solid solvent.   (b) Boiling (vaporization) produces pure gaseous solvent.

Figure 8.27.

Changes that occur for solutions of nonvolatile solutes.

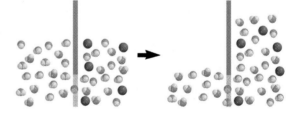

(c) Pure solvent diffuses across the membrane and dilutes solution

Every molecule or ion dissolved in a solution is included in the determination of the number of ways to arrange solute and solvent particles *(W)*. Consequently, the entropy of a solution depends *only* on the relative numbers of solute and solvent particles, regardless of the identity of the solute particles. Equation (8.34) shows that the freezing point is inversely proportional to the entropy change for freezing solid solvent from the solution. Since the entropy change depends only on the numbers of particles, the freezing point also depends only on the number of particles.

Quantifying freezing point lowering Relative numbers of solute and solvent particles (equivalent to relative numbers of moles) are easy to determine in the laboratory, as is the difference in freezing point between a solution and the pure solvent. For water, 1.0 mol of a molecular solute dissolved in 1 kg of water lowers the freezing point of the solution to -1.86 °C. If the number of solute particles is increased to 2.0 mol (again dissolved in 1 kg of water), the freezing point is lowered to -3.72 °C. The relationship between freezing point lowering and solution concentration is

$$\Delta T_{fp} = k_{fp} \cdot m \qquad (8.35)$$

Equation (8.35) shows that ΔT_{fp} ($= T_{fp\,(solution)} - T_{fp\,(solvent)}$), the change in freezing point of the solvent, is directly proportional to the molality of the solution. The freezing point of the solvent is always higher than that of the solution, so ΔT_{fp} is always negative. The proportionality factor, k_{fp} **(freezing point lowering constant)**, is a characteristic of the solvent. Values for k_{fp} have been determined for many solvents; Table 8.2 gives you an idea of the range of values for a few solvents.

The "experimental" results given here are for an ideal solute that simply mixes among the water molecules without interacting with them. Values for k_{fp} are determined by extrapolating data for real solutions to the limit of high dilution. The simple equations we use here are valid only for dilute solutions.

The value for k_{fp} can be derived from equations (8.31) and (8.32) combined with the relationship for the difference between the entropies of a solution and the pure solvent. The result is that k_{fp} depends on the molar mass, enthalpy of fusion, and temperature of melting of the solvent.

Table 8.2

Freezing point lowering constants.[a]

Solvent	T_{fp}, °C	k_{fp}, °C·m^{-1}
water	0.00	-1.86
benzene	5.50	-5.12
cyclohexane	6.5	-20.2
naphthalene	80.2	-6.9
camphor	179.8	-40.0

[a]k_{fp} is related to a temperature *difference*. The size of the difference is the same for both °C and K.

As an example, a 0.10 *m* aqueous solution of glucose in water will have a freezing point lowering of

$$\Delta T_{fp} = (-1.86 \text{ °C·}m^{-1})(0.10 \text{ }m) = -0.19 \text{ °C}$$

The freezing point of the solution will be

$$T_{fp\,(solution)} = T_{fp\,(solvent)} + \Delta T_{fp} = 0.00 \text{ °C} + (-0.19 \text{ °C}) = -0.19 \text{ °C}$$

As you can see, sensitive thermometers are required to measure freezing point lowering for aqueous solutions.

8.44 WORKED EXAMPLE

Determining a freezing point lowering constant

The freezing point of benzophenone, $C_{13}H_{10}O$, is 321.7 K (48.6 °C). In an experiment, 5.231 g of solid benzophenone was mixed with 0.127 g of

continued

biphenyl, $C_{12}H_{10}$, and the mixture was heated to melt the benzophenone and dissolve the biphenyl. The freezing point of the mixture was 320.2 K. Use these data to calculate the freezing point lowering constant for benzophenone.

Necessary information: We need equation (8.35), the definition of molality, and the molar mass of biphenyl.

Strategy: Use the masses of biphenyl and benzophenone and the molar mass of biphenyl to calculate the molality of the solution of biphenyl in benzophenone. Then substitute the molality and the freezing point lowering into equation (8.35) and solve for k_{fp}.

Implementation: The molar mass of biphenyl is 154.2 g·mol^{-1}, so the number of moles of biphenyl is

$$0.127 \text{ g of biphenyl} = (0.127 \text{ g}) \left(\frac{1 \text{ mol biphenyl}}{154.2 \text{ g}} \right) = 8.24 \times 10^{-4} \text{ mol}$$

The molality, m, of the solution of biphenyl in benzophenone is

$$\frac{8.24 \times 10^{-4} \text{ mol biphenyl}}{5.231 \times 10^{-3} \text{ kg benzophenone}} = 1.574 \times 10^{-1} \, m$$

The freezing point lowering is

$$\Delta T_{fp} = T_{fp \text{ (solution)}} - T_{fp \text{ (solvent)}} = 320.2 \text{ K} - 321.7 \text{ K} = -1.5 \text{ K} = -1.5 \,°\text{C}$$

The freezing point lowering constant is

$$k_{fp} = \frac{\Delta T_{fp}}{m} = \frac{-1.5 \,°\text{C}}{1.574 \times 10^{-1} \, m} = -9.5 \,°\text{C}\cdot m^{-1}$$

Does the answer make sense? What we have to go on are the values in Table 8.2 for the freezing point lowering constants of other compounds. The melting point and molar mass of benzophenone are similar to those of naphthalene, and the freezing point lowering constant we calculated is also similar, so we are probably in the right ballpark. Use a handbook or Web-based reference to check this freezing point lowering (or freezing point depression) constant for benzophenone.

8.45 CHECK THIS

Determining molar mass of a solute from freezing point lowering

Suppose you dissolve 0.243 g of a newly synthesized compound in 7.567 g of benzene and find that the freezing point of the solution is 5.12 °C.
(a) What is the molality of the solution? Give the reasoning for your answer.
(b) What is the molar mass of the new compound? Give the reasoning for your answer.

Effect of solute ionization on colligative properties Since freezing point lowering is a colligative property and depends on the number of particles of solute, we need to know how many particles are produced per mole of the formula unit when a substance dissolves. For example, 1.0 mol of calcium chloride formula units, $CaCl_2$, dissolved in 1 kg of water produces 1.0 mol of $Ca^{2+}(aq)$ ions and 2.0 mol of $Cl^-(aq)$ ions. Each cation and anion has the same effect on the freezing point, so *one* mole of $CaCl_2$ depresses the freezing point as much as *three* moles of a dissolved molecular compound that does not ionize.

8.46 WORKED EXAMPLE

Freezing point lowering by an ionic solute

What is the freezing point of a solution prepared by dissolving 0.10 mol of magnesium sulfate, $MgSO_4$, in 1.00 kg of water?

Necessary information: We need the k_{fp} for water from Table 8.2 and we need to know that magnesium sulfate dissolves to yield magnesium cations and sulfate anions.

Strategy: Calculate the total molality of solute particles. Then use equation (8.35) to get the freezing point lowering and the freezing point of the solution.

Implementation: A quantity of 0.10 mol of magnesium sulfate dissolves to give 0.10 mol of magnesium cations and 0.10 mol of sulfate anions. The total number of moles of solute particles is, therefore, 0.20 mol and the solution is 0.20 *m*. The freezing point lowering is

$$\Delta T_{fp} = (-1.86 \,°C \cdot m^{-1})(0.20 \, m) = -0.37 \,°C$$

The freezing point of the solution is

$$T_{fp \, (solution)} = T_{fp \, (water)} + \Delta T_{fp} = 0.00 \,°C + (-0.37 \,°C) = -0.37 \,°C$$

Does the answer make sense? The freezing point lowering for the ionic solution is two times larger than the freezing point lowering we calculated for a 0.10 *m* glucose solution. Since this ionic solution has twice as many solute particles, this is the result we would expect.

8.47 CHECK THIS

Freezing point lowering by a polar solute in a nonpolar solvent

The freezing point of a 0.50 *m* solution of ethanoic acid, $CH_3C(O)OH$, in benzene is 3.96 °C.

(a) From the freezing point value, what is the molality of particles in the solution? Explain.

(b) How does your result in part (a) compare to the molality of the solution whose freezing point was measured? How can you explain any difference between these molalities? *Hint:* Consider possible hydrogen bonding interactions. (See Problem 8.54.)

Boiling point elevation When we discussed freezing point depression in solutions, we saw that the entropy of solutions was greater than the entropy of pure liquids, which was, in turn, greater than the entropy of solids. The explanation of higher boiling points for solutions of nonvolatile solutes follows the same line of reasoning applied to vaporization of pure solvents and solutions. In this case, the entropy of the solution is greater than the entropy of the pure liquid solvent and the entropy of the pure gaseous solvent is larger than either of the condensed phases:

$$S_{gas} > S_{solution} > S_{solvent} \tag{8.36}$$

8.48 CONSIDER THIS

Is the entropy change larger for boiling a solution or pure solvent?

(a) Use the data in equation (8.36) to draw a diagram (modeled after Figure 8.26) showing the entropy changes for boiling a solution and its pure solvent.

(b) What is the size of $|\Delta S_{solvent \rightarrow gas}|$ relative to $|\Delta S_{solution \rightarrow gas}|$? Does this relationship help you understand why a solution has a higher boiling point than its pure solvent? Explain. You might find it helpful to write a relationship analogous to equation (8.34).

Changes in boiling point are based on molality, just as were changes in freezing point:

$$\Delta T_{bp} = k_{bp} \cdot m \tag{8.37}$$

Table 8.3 lists **boiling point elevation constants, k_{bp}**, for the three liquid solvents included in Table 8.2. The boiling point elevation constant is positive, since ΔT_{bp} (= $T_{bp\,(solution)} - T_{bp\,(solvent)}$) and $T_{bp\,(solution)} > T_{bp\,(solvent)}$.

Table 8.3 *Boiling point elevation constants.*

Solvent	T_{fp}, °C	k_{fp}, °C·m^{-1}
water	100.00	0.512
benzene	80.0	2.53
cyclohexane	80.7	2.79

8.49 CHECK THIS

Boiling point of a solution

Calculate the boiling point of a 0.25 m biphenyl solution in benzene.

8.13. Osmotic Pressure Calculations

Osmosis was discussed in Sections 8.1 and 8.4. The explanation for osmosis in terms of molecular arrangements (Section 8.4) is parallel to the explanations for freezing point lowering and boiling point elevation, given in terms of entropies in the previous section. Now we will consider osmosis, in particular, osmotic pressure, more quantitatively, just as we did for those other colligative properties.

Our molecular model in Section 8.4 shows that the positional entropy of the solution continues to increase as the solution becomes more dilute, because the number of possible arrangements of the solute molecules increases. In Investigate This 8.1, we might expect water to continue to pass through the membrane (carrot) to the solution until the water is used up. That does not happen. Water finally does stop its *net* movement through the semipermeable membranes. The column of solution exerts pressure on the solution that is greater than the atmospheric pressure on the pure water. The column of solution stops rising when the pressure at the bottom of the column is large enough to stop the net movement of water into the solution. The pressure on the solution that is required to *just* stop net movement of water is called the **osmotic pressure,** symbolized by the uppercase Greek letter pi, Π.

When the solution level stops rising, the movement of water molecules through the membrane is the same in both directions. This means that the net entropy change for a solvent molecule passing from the solvent to the solution must be the same as that for the reverse process, a solvent molecule passing from the solution to the solvent. We know that entropy of mixing is gained when a solvent molecule enters the solution. There must be the same gain in entropy when a solvent molecule leaves the high pressure solution to enter the low pressure solvent.

The high pressure on the solution compresses the liquid. You can think of the compression as making the boxes occupied by the *solution* molecules in Figure 8.7(b) smaller. This means that the volume available to molecules in the solution is smaller. A smaller volume per molecule decreases the entropy of the solution. Net flow of solvent stops when the entropy decrease due to the compression by the osmotic pressure exactly compensates the entropy increase due to mixing. Liquids are not very compressible, so we would expect the pressure required to cause this decrease in entropy to be quite high. That is, we predict high osmotic pressures.

Quantifying osmotic pressure Since it is a colligative property of solutions of nonvolatile solutes, the osmotic pressure of a solution depends only on the number of molecules or ions present in the solution and not on their identity. When dealing with osmosis, *molarity* is used to express solute concentration rather than *molality*. The relationship between osmotic pressure and the molar concentration, c, of molecules or ions in a solution is

$$\Pi = cRT \qquad (8.38)$$

T is the temperature in kelvin and R is the gas constant. If the total concentration of solute particles in a solution is 0.10 M, the osmotic pressure at 25 °C (= 298 K) is

$$\Pi = cRT = (0.10 \text{ mol·L}^{-1})(0.08205 \text{ atm·L·mol}^{-1}\text{·K}^{-1})(298 \text{ K}) = 2.4 \text{ atm} \qquad (8.39)$$

Osmose is derived from a Greek verb meaning to impel or force.

Recall that molarity and molality are almost identical for dilute aqueous solutions.

R has units of energy·mol^{-1}·K^{-1}. The energy units used for R determine the osmotic pressure units. Osmotic pressure is often given in atmospheres, for which $R = 0.08205$ atm·L·mol^{-1}·K^{-1}.

If you were to use water in a barometer to measure 1 atm pressure, you would have to use a barometer about 10.4 m high. An osmotic pressure of 2.4 atm is equivalent to a column of water about 25 m high! This is the height of a very tall tree or a seven-story building. Our prediction about the magnitude of osmotic pressures is correct.

Osmotic pressures are large and easy to measure, even for dilute solutions, which makes it possible to use osmotic pressure measurements to determine molar masses of compounds that are not very soluble in water. The technique is commonly used for large molecules encountered in polymer chemistry, biology, and biochemistry. An apparatus for measuring osmotic pressure of a solution is diagrammed in Figure 8.28. An aqueous solution to be measured is placed in the left-hand compartment and water is placed in the right compartment. A membrane that is permeable only to water separates the two compartments. Instead of letting the solution rise in the tube, an external pressure, P, is applied. The external pressure is increased until the net flow of water between the compartments stops. At this point $\Pi = P - P_{atm}$.

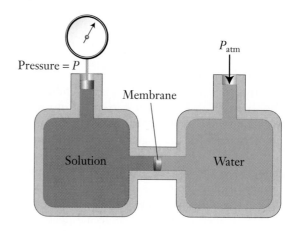

Figure 8.28.

Device for measuring osmotic pressure.

8.50 WORKED EXAMPLE

Molar mass determination by osmotic pressure measurement

Suppose you have discovered a new protein and need to know its molar mass. Determine the molar mass of the protein using these data from an osmotic pressure measurement: the osmotic pressure is 121 Pa at 310 K for a solution of 5.6 mg of the protein in 11.0 mL of solution.

Necessary information: We need to know equation (8.39) and the definition of molarity. The osmotic pressure is given in Pa (pascal), so we need to know R in units of pascals, liters, moles, and kelvin: $R = 8.314 \times 10^3 \ Pa \cdot L \cdot mol^{-1} \cdot K^{-1}$.

Strategy: The molar mass can be calculated from the number of moles of protein in 5.6 mg of protein. Equation (8.39) shows that the osmotic pressure gives the molarity of the solution used for its measurement. Molarity is $mol \cdot L^{-1}$, so the molarity and volume of solution give the number of moles of protein in the solution.

continued

Implementation: The molarity of the solution is

$$c_{soln} = \frac{\Pi}{RT} = \frac{121\ \text{Pa}}{(8.314 \times 10^3\ \text{Pa·L·mol}^{-1}\text{·K}^{-1})(310\ \text{K})} = 4.69 \times 10^{-5}\ \text{M}$$

The moles of solute in V_{soln} liters of solution is

$$\text{mol solute} = c_{soln}V_{soln} = (4.69 \times 10^{-5}\ \text{M})(11.0 \times 10^{-3}\ \text{L}) = 5.16 \times 10^{-7}\ \text{mol}$$

The definition of moles of solute is rearranged to get the molar mass:

$$\text{molar mass} = \frac{\text{mass solute}}{\text{mole solute}} = \frac{5.6 \times 10^{-3}\ \text{g}}{5.16 \times 10^{-7}\ \text{mol}} = 1.08 \times 10^4\ \text{g·mol}^{-1}$$

Does the answer make sense?   A molar mass of about 10,800 g·mol^{-1} is reasonable for a relatively small protein.

8.51 CHECK THIS

Osmotic pressure of an ionic solution

Seawater contains many ions. For simplicity, assume that it can be represented as a 0.50 M solution of sodium chloride, NaCl. What is the osmotic pressure (in atmospheres) of seawater at 20 °C? Remember that each mole of sodium chloride produces two moles of ions in the solution.

Reflection and Projection

Colligative properties of solutions have many applications in everyday life. Next time you use salt to melt ice from sidewalks or add antifreeze to an automobile radiator to prevent the coolant from freezing or boiling over, stop to consider that you are harnessing the power of entropy. The idea that mixtures are more probable than pure substances reflects our common experience, and the explanation in terms of ways of arranging particles and thermal quanta is consistent with that experience.

   Sometimes, however, changes seem to deviate from our usual experiences or phenomena do not seem to be consistent with the sort of entropic ideas we have developed. To understand such cases, we have to examine the systems in greater detail and often discover that the examination deepens our concept of entropy and increases our ability to use the concept in new situations. We hope that the examples presented in this chapter will help as you begin to think through such new situations.

8.14. The Cost of Molecular Organization

Recall that we started the chapter with a question: "How is it that some things become so disorganized while others become so highly organized?" It is *positional* organization and disorganization (water forming an intricate snowflake or sugar dissolving in iced tea) that usually catches our attention. However, we are not always aware of the subtle positional organization of molecules, as in water around ionic and molecular solutes, and we cannot see the organization and disorganization of energy quanta that must also be accounted for in answering the question.

In all spontaneous changes, the net entropy increases. This means that the number of arrangements of molecules and energy quanta at the end of the change is always greater than the number of arrangements before the change. There are fewer molecular arrangements for organized molecular systems than for disorganized systems. Therefore, when a change leads to greater positional organization of some molecules (water in a snowflake or phospholipids in a membrane), you have to look deeply into all the positional and energetic changes that occur elsewhere during the change to find the positional and/or energetic changes that increase the overall number of arrangements.

In many cases, energy released by a change is responsible for a large increase in the overall number of arrangements. That is, a large increase in thermal entropy of the surroundings produces a large increase in net entropy (or decrease in free energy of the system). One striking example of this situation is the oxidation of glucose, which we examined In Section 8.9 Worked Examples 8.29 and 8.31. In this case, $\Delta G°_{reaction} = -2878$ kJ per mole of glucose oxidized. The lion's share of this free energy change is the enthalpy released to the surroundings, 2801 kJ, which increases their thermal entropy. Recall that we found, in Chapter 7, Section 7.9, that the oxidation of glucose in biological systems is coupled to the production of ATP:

$$C_6H_{12}O_6(s) + 6O_2(g) \rightarrow 6CO_2(g) + 6H_2O(l) \qquad \Delta G°_{reaction} \quad \textbf{(8.40)}$$

$$36ADP^{3-}(aq) + 36HOPO_3^{2-}(aq) + 36H_3O^+(aq) \rightarrow 36ATP^{4-}(aq) + 72H_2O(l) \quad \Delta G_{ATP\ form} \quad \textbf{(8.41)}$$

$$C_6H_{12}O_6(s) + 6O_2(g) + 36ADP^{3-}(aq) + 36HOPO_3^{2-}(aq) + 36H_3O^+(aq) \rightarrow$$

$$6CO_2(g) + 78H_2O(l) + 36ATP^{4-}(aq) \qquad \Delta G°_{coupled} \quad \textbf{(8.42)}$$

The standard free energy change for the production of ATP under standard cellular conditions is positive, approximately 31 kJ per mole of ATP formed. Thus, $\Delta G°_{ATP\ form} = 1100$ kJ ($=36 \times 31$ kJ) for reaction (8.41). For the coupled reaction:

$$\Delta G°_{coupled} = \Delta G°_{reaction} + \Delta G°_{ATP\ form} = -2878\ kJ + 1100\ kJ = -1778\ kJ \qquad \textbf{(8.43)}$$

This is an enormous driving force that is equivalent to a net entropy increase of almost 6000 $J \cdot K^{-1}$ for the coupled system. Some of the energy that would have gone into the surroundings to increase their thermal entropy is captured in the coupled process and converted to chemical potential energy in the bonding of ATP.

In subsequent coupled processes, the chemical potential energy stored in ATP is used to drive other reactions as ATP is hydrolyzed to ADP and phosphate, the reverse of reaction (8.41). These coupled reactions also have negative free energy (positive net entropy) changes. Among these coupled reactions are those that synthesize the phospholipid molecules used to build the cellular membranes. As we have seen, the formation of the membranes is accompanied by an entropy increase when water is freed from its interactions with the phospholipids; it requires essentially no energy. Energy was, however, provided by the oxidation of glucose far back up the chain of reactions that formed the phospholipids.

Thus, the cost of all the organization and complex processes of life, whether driven by coupling with reactions like ATP hydrolysis or by more subtle positional entropy effects, is the large increase in thermal entropy when fuel (food) molecules are oxidized. As we go on in the book, we will further discuss ways the free energy released in these oxidations is captured.

But this analysis still leaves a question: Where does all the glucose come from in the first place? Green plants synthesize vast quantities of glucose every day from carbon dioxide and water. This is the reverse of reaction (8.40), so the standard free energy change for the synthesis reactions is 2878 kJ, a large *positive* free energy change. Plants must couple the synthesis with some other process(es) that has(have) an even larger *negative* free energy change.

Synthesis of glucose in plants begins with the absorption of several quanta of visible light energy from the sun. This energy is ultimately released as free energy to drive the synthesis. The light energy itself is a result of the energy released in the sun's nuclear fusion reactions that we discussed in Chapter 3. Even though positional entropy decreases in a fusion reaction, the enormous release of energy increases the thermal entropy of the surroundings, so the net entropy change is very large and positive. Ultimately, therefore, the biological organization we see on earth is "paid for" by the net entropy increase in the sun's nuclear fusion reactions. Electromagnetic radiation from the sun couples this increase to the synthesis of glucose on Earth.

8.15 Outcomes Review

In this chapter, we introduced the thermodynamic function, entropy, that is responsible for the direction of all spontaneous changes. We found that net entropy is a measure of the number of distinguishable arrangements of atoms, molecules, and energy quanta in a system and its surroundings and that net entropy always increases in spontaneous changes. The net entropy is the sum of positional and thermal entropies of the system and surroundings. The examples we used to understand and analyze the role of net entropy were phase changes in pure compounds and in solutions, osmosis, oxidation of glucose, solubility of ionic and molecular solutes, and the formation of micelles, phospholipid bilayers, and cell membranes. We also introduced another thermodynamic function, the Gibbs free energy, that is directly related to net entropy changes, but is more convenient for some purposes. Net entropy change and free energy can be used interchangeably to understand and analyze actual or hypothetical changes.

Check your understanding of the ideas in the chapter by reviewing these expected outcomes of your study. You should be able to

- Relate the relative probability of two outcomes to the number of ways each outcome can be achieved [Sections 8.2, 8.3, and 8.4].

- Calculate the number of distinguishable arrangements of a small number of identical particles (molecules) among a limited number of distinguishable locations [Section 8.3].

- Relate mixing to osmosis and be able to explain both in terms of increasing number of distinguishable molecular arrangements [Sections 8.1 and 8.4].

- Describe the conditions required for osmosis, predict the direction of osmosis, and calculate the osmotic pressure for aqueous systems [Sections 8.1, 8.4, and 8.13].

- Calculate the number of distinguishable arrangements of a small number of identical energy quanta among a limited number of distinguishable atoms [Section 8.5].

- Predict the direction and final outcome of energy transfer between two systems containing a countable number of energy quanta distributed among a countable number of distinguishable atoms [Section 8.5].

- Relate the results for countable systems, both matter and energy distributions, to real systems [Sections 8.4 and 8.5].

- Use the definition of entropy in terms of distinguishable arrangements to predict the relative entropies of different systems, for example, phases of matter or reactants and products of a reaction [Sections 8.7, 8.8, 8.9, 8.10, 8.11, 8.12, and 8.16].

- Distinguish between positional and thermal entropy changes for a process and combine them to determine net entropy change for the process [Sections 8.6, 8.7, 8.8, 8.9, 8.10, 8.11, 8.12, and 8.16].
- Use positional entropy changes and the relative thermal entropy changes for the same energy change at different temperatures to analyze and predict the direction of phase changes in pure compounds and solutions [Sections 8.7 and 8.12].
- State the criterion for equilibrium in chemical systems and relate the state of equilibrium to the positional and thermal entropy changes occurring [Sections 8.7 and 8.8].
- Explain the relationship of Gibbs free energy change to net entropy change for a process and use the sign and/or magnitude of either one to predict whether the process is possible, not possible, or in equilibrium [Sections 8.8 and 8.9].
- Calculate the free energy change for a process in a system using values for the enthalpy and entropy changes for the process [Sections 8.8 and 8.9].
- Use standard enthalpies and standard free energies of formation, and standard entropies (from tabulated values) to calculate the standard enthalpy, free energy, and entropy changes for a reaction [Section 8.9].
- Explain in words, equations, diagrams, and/or molecular-level sketches the origin and direction of positional and thermal entropy changes for dissolving ionic and molecular solutes in water and predict the observable outcomes [Sections 8.9 and 8.10].
- Explain in words, equations, diagrams, and/or molecular-level sketches the origin and direction of positional and thermal entropy changes for formation of micelles and phospholipid bilayer membranes by ambiphilic molecules [Section 8.11].
- Use words and/or molecular level sketches to describe the structure and properties of micelles, bilayer membranes, and liposomes [Section 8.11].
- Calculate freezing point lowering, boiling point elevation, and osmotic pressure for solutions [Sections 8.12 and 8.13].
- Use experimental values for colligative properties to determine the concentration of solutes in the solution and/or the molar mass of the solute [Sections 8.12 and 8.13].
- Connect the formation of organized collections of molecules to increases in positional and/or thermal entropy in the system and surroundings that drive the organization [Sections 8.7, 8.10, 8.11, and 8.14].
- Predict the direction of the positional entropy change(s) that must occur to produce the observed effects of heating or cooling a system, such as a rubber band [Section 8.16].

8.16. Extension—The Thermodynamics of Rubber

8.52 INVESTIGATE THIS

What happens when a stretched rubber band is heated?

Do this as a class investigation and work in small groups to discuss and analyze the results. Use large paper clips or some other means to attach a 0.5-L bottle of water to one end of a rubber band that is about 0.8 cm wide and 10 cm long. Attach the other end of the rubber band to a sturdy support (such as the back of a chair or a coathook) so that the rubber band and bottle hang freely without rubbing on anything. The full bottle of water should stretch the rubber band to about three times its relaxed length. Measure and record the length (in mm) of the rubber band from one end to the other. Use a hair dryer to warm the rubber band. Note whether its length changes, and how. Record the length when the rubber band is warm. Let the rubber band cool and record its length again.

8.53 CONSIDER THIS

What is the direction of change when a rubber band is heated?

Did the length of the rubber band in Investigate This 8.52 change in the direction you expected? Explain why or why not. Discuss the results, and possible entropy changes, with other members of your class.

Rubber bands are familiar objects, but some of their properties are not so familiar and are surprising. An entropy analysis of these properties leads to interesting insights about the behavior of rubber molecules. In Investigate This 8.52 you found that warming a stretched rubber band causes it to contract. The net entropy change for the contraction must be positive, $\Delta S_{net} > 0$, or equivalently, the free energy change must be negative, $\Delta G < 0$:

$$\Delta G = \Delta H_{rubber} - T\Delta S_{contraction} < 0 \qquad (8.44)$$

The enthalpy change of the rubber is a result of a transfer of thermal energy *to* the rubber from the surroundings, so $\Delta H_{rubber} > 0$. Since ΔH_{rubber} is *positive*, ΔG can only be *negative* if $T\Delta S_{contraction}$ is *positive* and numerically *larger* than ΔH_{rubber}. In other words, the entropy of the rubber must *increase* as the rubber band contracts.

Molecular structure of rubber To explain the increase in positional entropy of the rubber, we must take into account the structure of rubber molecules. Figure 8.29 illustrates rubber molecules in their relaxed state and in their stretched state. Rubber molecules contain between 4,000 and 20,000 carbon atoms linked together to form long chains of covalently bonded hydrocarbons. In their relaxed state, these rubber molecules are tangled within themselves and with other rubber molecules, somewhat as diagrammed in Figure 8.29(a). The number of possible arrangements for random tangling is enormous.

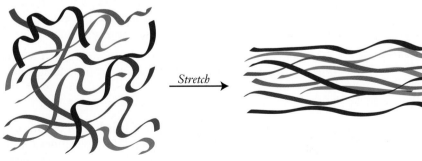

(a) Relaxed
(more arrangements,
higher entropy)

$S_{relaxed} > S_{stretched}$

(b) Stretched
(fewer arrangements,
lower entropy)

Figure 8.29.

Schematic illustrations of relaxed and stretched carbon chains in rubber.

When rubber is stretched a moderate amount, no chemical bonds are broken. Instead, the molecules are forced to begin untangling. An arrangement of long chains of molecules lying more-or-less parallel to each other, as in Figure 8.14(b), is lower in entropy than the tangled mass of molecules. Fewer arrangements are possible for the more orderly array:

$$\Delta S_{contraction} = S_{relaxed} - S_{stretched} > 0 \qquad (8.45)$$

The change in the organization of the molecules explains the positive value for the entropy change when a rubber band contracts. The entropy change is large enough that $T\Delta S_{contraction}$ is numerically larger than ΔH_{rubber}, producing a negative value for the free energy change, ΔG in equation (8.44). The surprising contraction of rubber when it is warmed is a result of the positional entropy of long carbon chains.

8.54 INVESTIGATE THIS

What happens when a rubber band is stretched?

Hold a rubber band, such as the one used in Investigate This 8.52, by its ends in its relaxed position. Place the rubber band against the sensitive part of your lip and allow it to come to the same temperature as your lip. Quickly stretch the rubber band to three or four times its relaxed length. Note any changes in the temperature of the rubber band. Now hold the stretched rubber band against your lip and allow the rubber band to come to the same temperature as your lip. Without letting go of the ends, quickly let the rubber band contract to its relaxed length. Note any changes in the temperature of the rubber band. Repeat the experiment to be sure of your observations.

Some people find that their foreheads are more sensitive to temperature change than their lips.

8.55 CONSIDER THIS

Does the model of rubber explain the rubber band results?

Does the rubber warm, cool, or stay the same temperature when stretched? When it contracts? Do the relative magnitudes and signs of the enthalpy and entropy changes in equation (8.44) explain the observations? Explain why or why not.

The contraction of a stretched rubber band when it is heated is a bit surprising, since most solids expand when they are heated. The implications for the positional and thermal entropy changes that must occur lead to a better understanding of the molecular structure and properties of rubber. You will find that entropy analyses are often useful in cases like this.

Chapter 8 Problems

8.1. Mixing and Osmosis

8.1. Which processes involve an *increase* in organization for the system specified? Which involve a *decrease* in organization? Explain your reasoning briefly.
(a) ice cream melting
(b) students taking their places in a classroom with fixed seating
(c) a shrub forming spring flowers
(d) obtaining pure water from seawater
(e) opening and shuffling a new deck of cards
(f) raking leaves into a single pile

8.2. Explain why a water soluble dye spreads out when it is mixed with water resulting in a uniformly colored solution. Will the reverse process ever happen?

8.3. A few drops of green food coloring are added to some cookie dough.
(a) Are these drops of food coloring likely to disperse on their own throughout the cookie dough? Explain your reasoning.
(b) Compared to your answer in part (a), are drops of green food coloring more or less likely to disperse throughout a glass filled with water? Explain your reasoning.

8.4. The membranes of red blood cells are semi-permeable. Moving water out of the blood cell causes the cell to shrivel, a process known as *crenation*. Moving water into the blood cell causes the cell to swell and rupture, a process known as *hemolysis*. People needing body fluids or nutrients and who cannot be fed orally are given solutions by intravenous (or IV) infusion. An IV feeds nutrients directly into veins. Explain what would happen if an IV having a higher concentration than the solution within the red blood cell was given.

8.5. Explain the role of osmosis in each of these processes.
(a) Swimming in a freshwater lake makes your body waterlogged and increases your need to urinate.
(b) Cucumbers wrinkle, lose water, and shrink when placed in brine (concentrated saltwater solution).
(c) Prunes and raisins swell in water.
(d) You get thirsty and your skin wrinkles when you swim in the ocean.
(e) Salt and sugar are used to preserve many foods.
(f) Many saltwater fish die when placed in fresh water and vice versa.

8.2. Probability and Change

8.6. Choose the best answer(s) and explain your choice(s). When a dye solution mixes with water in an unstirred container, the mixing occurs because
(i)   The mixed state has a higher number of distinguishable arrangements than the unmixed state.
(ii)   The mixed state has a lower heat content than the unmixed state.
(iii)   The mixed state is more probable than the unmixed state.

8.7. What numerical value (0, 1, or somewhere in between) would you assign to the chance of each of these changes being observed? Briefly discuss your reasoning, giving any conditions under which the probability could change.
(a) A broken eggshell reforming into an unbroken eggshell.
(b) A drop of food coloring dispersing throughout a cup of water at room temperature.
(c) Drawing the ace of spades from a shuffled deck of cards.
(d) Finding all of the pepperoni slices on one half of a pizza surface.
(e) A large oak tree falling to the ground and then returning to the upright position.
(f) Recovering six pairs of clean socks from the dryer after washing six pairs of dirty socks.

8.8. Think about the examples from your daily experiences that would illustrate the statement: "If a system can exist in more than one observable state, any changes will be in a direction toward the state that is most probable." Which state will be the most probable for
(a) water at -5 °C?
(b) water at 130 °C?
(c) sugar crystals in a glass of hot water?
(d) a drop of perfume in a room?
(e) iron filings in a magnetic field? (See Chapter 4, Section 4.3, Figure 4.11.)

8.3. Counting Molecular Arrangements in Mixtures

8.9. (a) Find the number of possible distinguishable arrangements for two identical objects in the five labeled boxes (one object per box) in Investigate This 8.4.
(b) How does the number of arrangements you found in part (a) compare with the number of possibilities found in Investigate This 8.4? Is this the result you expected? Why or why not? Explain why the results come out as they do.

8.10. (a) Refer to Figure 8.5. How many distinguishable arrangements are there if two, rather than three, soluble dye molecules are allowed to mix in the top two layers (six boxes) of the system? Show your set of arrangements pictorially.
(b) Compare your result for two dye molecules mixing into six boxes with that shown in the figure for three dye molecules mixing into six boxes. Explain the relationship.
(c) Given your comparison in part (b), what conclusion can you draw about the effect of number of particles on the number of distinguishable arrangements for mixing into a fixed volume (number of boxes)? Does this result make sense? Explain.

8.11. In Figure 8.6, as the number of boxes per molecule increases 10-fold from 2 to 20, does the number of possible arrangements for three solute molecules in the cells also increase 10-fold? If it does not, by what factor does the number of arrangements increase? Explain your reasoning, using information from Figure 8.6.

8.12. For systems like those in Section 8.3, the number of distinguishable arrangements, W, of n identical objects among N labeled boxes is

$$W = \frac{N!}{(N-n)!n!}$$

The exclamation point, "!", denotes a factorial. $N!$, pronounced "en factorial," is the product of all the integers from N to 1. For example, 5! (5 factorial) = $5 \cdot 4 \cdot 3 \cdot 2 \cdot 1 = 120$. (Many scientific calculators have a factorial function key that gives you factorials of numbers up to 69—limited by the maximum value the calculator can display.)
(a) Use the formula to calculate the number of distinguishable arrangements of 3 objects in 6 boxes. Is your result the same as the number of arrangements shown in Figure 8.5? Explain why or why not.
(b) Calculate the number of distinguishable arrangements for the other mixtures represented in Figure 8.6 and compare your results with the figure.
(c) Calculate the number of distinguishable arrangements for a mixture of 5 objects in 20 boxes and for a mixture of 10 objects in 20 boxes. Do your results reinforce the conclusion you drew in Problem 8.10(c)? Explain why or why not.

8.4. Implications for Mixing and Osmosis in Macroscopic Systems

8.13. Imagine a living plant cell surrounded by a semipermeable membrane through which water can pass but sucrose cannot. The concentration of sucrose

in this cell is 0.5%. By which of these processes can you increase the concentration of sucrose in the cell? Explain the reasoning for your choice and the reason(s) you reject the others.
  (i)   Place the cell in pure water.
  (ii) Place the cell in a sucrose solution with a concentration greater than 0.5%.
  (iii) Place the cell in a sucrose solution with a concentration less than 0.5%.
  (iv) Any of the above will work.

8.14. (a) Starting with the initial arrangement of solute and solvent molecules shown in Figure 8.7(a), will the number of molecular arrangements increase or decrease after osmosis takes place? Explain.
(b) Draw a diagram modeled after Figure 8.7(b) to show what will happen if six of the nine solvent molecules pass from the solvent into the solution. How many distinguishable arrangements of the solution, the solvent, and the total system are possible after this change has occurred? Is the change likely to occur? Explain why or why not.
(c) Is the process in part (b) osmosis or reverse osmosis? Explain.

8.15. The concentration of solute particles in a red blood cell is about 2%. Problem 8.4 states that red blood cells are susceptible to crenation or hemolysis. Red blood cells would probably shrivel the most when immersed in which of these solutions? Explain the reasoning for your choice and the reason(s) you reject the others. (Sucrose cannot pass through the membrane of blood cells.)
  (i)   1% sucrose solution
  (ii) 2% sucrose solution
  (iii) 3% sucrose solution
  (iv) distilled water

8.5. Energy Arrangements Among Molecules

8.16. (a) If energy is transferred from a warmer object in contact with an identical, but cooler, object, how does the number of energy arrangements after the transfer compare to the number before the transfer? Explain.
(b) When net energy transfer stops, how will the final temperatures of the objects compare with the original two temperatures? Explain.

8.17. Consider two identical four-atom solids, one having 4 quanta of energy and the other having 10.
(a) What observable property differs between these two solids?
(b) Allow 3 quanta of energy to be transferred from the solid with 10 quanta to the solid with four quanta of energy. Use the data from Figure 8.9 to determine

whether this energy transfer is likely, explaining your reasoning.

(c) Compare the result from part (b) to the result obtained in Worked Example 8.8. In both cases, quanta were transferred from a warmer four-atom solid to a cooler four-atom solid. Are both transfers likely to take place? Explain the reasoning for your response.

8.18. Choose the best answer. Explain the reasoning for your choice and the reason(s) you reject the others. Which statement describes why thermal energy moves from a hot body to a cold body?

(i)   There is more thermal energy in the final state.

(ii) There is more thermal energy in the initial state.

(iii) The number of distinguishable arrangements for the energy quanta is higher in the final state.

(iv) The number of distinguishable arrangements for the energy quanta is higher in the initial state.

8.19. For systems like those in Section 8.5, the number of distinguishable arrangements, W, of n identical quanta among N atoms in a solid is

$$W = \frac{(N + n - 1)!}{(N - 1)!n!}$$

See Problem 8.12 for information about factorials.

(a) Use the formula to calculate the number of distinguishable arrangements of 2 quanta among 4 atoms. Is your result the same as the number of arrangements you found for two toothpicks in four candies in Investigate This 8.7? Explain why or why not.

(b) Calculate the number of distinguishable arrangements for the cases represented in Figure 8.9 and compare your results with the figure.

(c) Consider the transfer of four quanta of energy from an 8-atom solid with an initial 16 quanta to an identical 8-atom solid that initially has 10 quanta. Is this transfer likely to occur? Why or why not? Is there another transfer that is more likely to occur? If so what is it and why is it more likely?

8.6. Entropy

8.20. State the second law of thermodynamics in words and as an equation.

8.21. What is the difference between *positional* entropy and *thermal* entropy?

8.22. The second law of thermodynamics states that observed changes always take place in a direction that *increases net entropy*. Is this statement true if only thermal entropy is considered? If only positional entropy is considered? Give the reasons for your answers.

8.23. Why is it possible to tabulate absolute entropies, but not absolute enthalpies? That is, why does Appendix B list $S°$ (not $\Delta S°$) and $\Delta H_f°$ (not $H°$)?

8.24. Answer each of these statements either true or false. If you decide that the statement is false, write the correct statement.

(a) The entropy of a mole of solid mercury is higher than a mole of mercury vapor.

(b) If the net entropy increases during a process, the process is spontaneous.

(c) The entropy of a system depends on the pathway it took to its present state.

(d) The more positive its net entropy change, the faster the process.

(e) The entropy of 1 L of a 0.1 M aqueous solution of NaCl is lower than the entropy of the undissolved 0.1 mol of NaCl and 1 L of water.

(f) The entropy is larger for a system of two identical blocks of copper in which the temperature of each block is 70 °C than for the same blocks of copper if one has a temperature of 20 °C and the other is at 120 °C.

(g) There is no reference point for measuring entropy.

8.25. Predict whether the positional entropy change in each of these processes will be positive or negative. Explain your reasoning in each case.

(a) precipitation of $BaSO_4$ upon mixing $Ba(NO_3)_2(aq)$ and $H_2SO_4(aq)$

(b) cooling of ammonia gas from room temperature to -50 °C, which results in $NH_3(g) \rightarrow NH_3(l)$

(c) formation of ammonia in the reaction:
$N_2(g) + 3H_2(g) \rightarrow 2NH_3(g)$

(d) decomposition of $CaCO_3$ in the reaction:
$CaCO_3(s) \rightarrow CaO(s) + CO_2(g)$

(e) formation of NO in the reaction:
$N_2(g) + O_2(g) \rightarrow 2NO(g)$

8.26. Occasionally one hears that evolution violates the second law of thermodynamics. The claim is that the incredible precision with which the human body is arranged could not have evolved from less complex organisms without violating the second law of thermodynamics. What is the problem with this argument?

8.27. Do you think it is scientifically right to blame the second law of thermodynamics for clutter in a friend's workspace?

8.28. For each of these isomeric pairs of molecules, predict which one of the pair will have the higher entropy under the same pressure and temperature conditions. Explain the basis of each prediction.

(a) butanoic acid   *or*   dioxane

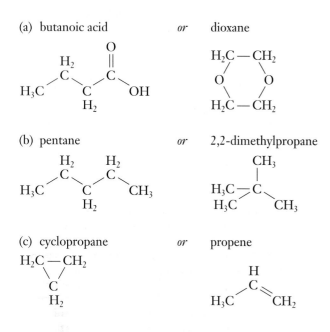

(b) pentane   *or*   2,2-dimethylpropane

(c) cyclopropane   *or*   propene

8.7. Phase Changes and Net Entropy

8.29. Consider this statement: "Observed phase changes always take place in a direction that increases entropy." Is this statement true? Critically analyze this statement for scientific accuracy, explaining your reasoning.

8.30. Which member of each pair has the higher entropy? In each case, explain your reasoning.
(a) $H_2O(s)$ *or* $H_2O(l)$
(b) $CaCl_2 (aq)$ *or* $CaCl_2 (s)$
(c) 5.0 g $H_2O(l)$ at 1 °C   *or* 5.0 g $H_2O(l)$ at 60 °C
(d) $H_2O(g)$ *or* $H_2O(l)$
(e) $CO_2 (g)$ *or* $CO_2 (s)$, dry ice

8.31. The positional entropy change for water freezing is negative. Explain how it is possible for water to freeze.

8.32. For each of these changes, explain if the positional entropy of the system increases, decreases, or stays the same.
(a) Solid CO_2 (dry ice) changes to carbon dioxide gas.
(b) Liquid ethanol (C_2H_5OH) freezes.
(c) Mothballs made of naphthalene ($C_{10}H_8$) sublime.
(d) Plant materials burn to form carbon dioxide and water.

8.33. Consider an isolated system where a balloon containing steam, $H_2O(g)$, is immersed into a thermally isolated container of liquid water at 10.0 °C.
(a) What will happen to the steam in the balloon?
(b) Does the entropy of the water in the balloon increase or decrease? Explain.
(c) Does the entropy of the water outside the balloon increase or decrease? Explain.
(d) Is the net entropy change for this isolated system positive or negative? Explain.

8.34. (a) The melting point for H_2S is 187 K (−86 °C). Use a diagram similar to Figure 8.11 to explain the entropy changes for the system, the surroundings, and the net entropy change for melting H_2S at temperatures above 187 K.
(b) How will the entropy diagram differ if liquid H_2S changes to solid H_2S at 187 K?
(c) The phase change for water changing from solid to liquid takes place at 0 °C (273 K). Explain why the two substances have such different melting points. *Hint:* H_2S forms only weak hydrogen bonds.

8.35. Arrange the compounds ammonia, methane, and water in order of increasing entropy of vaporization, $\Delta S_{l \to g}$. Explain the reasoning for your ordering. *Hint:* Review Chapter 1, Section 1.7.

8.36. The complete combustion of ethanol, $C_2H_5OH(l)$, is represented by this reaction expression:

$$C_2H_5OH(l) + 3O_2(g) \to 2CO_2(g) + 3H_2O(l)$$

For the combustion of 1 mol of ethanol under standard conditions, the standard entropy and standard enthalpy changes are −139 J·K⁻¹ and −1367 kJ·mol⁻¹, respectively.
(a) Comment on these values in terms of positional entropies.
(b) Comment on these values in terms of bond formation.
(c) Considering positional entropy, is this reaction favored to produce carbon dioxide and water under standard conditions of temperature and pressure? Why or why not?

8.37. Consider the process: $Br_2(l) \to Br_2(g)$
(a) Calculate $\Delta H°$ and $\Delta S°$ for this phase transition. Use the data in Appendix B.
(b) Calculate $\Delta S°_{net}$ for this phase transition at

    (i) 0.0 °C   (ii) 40 °C   (iii) 70 °C

(c) At what temperature does the bromine begin to boil? Explain the reasoning you use to determine the boiling point.

8.38. Use the $\Delta H°$ and $\Delta S°$ for $H_2O(l) \to H_2O(g)$ to show how temperature influences whether the increase in the entropy of the system will favor the phase change.

8.39. (a) Use the standard entropies and standard enthalpies of formation for liquid and gaseous ethanal, CH_3CHO, from Appendix B to find its standard entropy of vaporization, $\Delta S°_{l \to g}$, and standard enthalpy of vaporization, $\Delta H°_{l \to g}$.
(b) What is the boiling point of ethanal? Explain the reasoning you use to get your result, including any assumptions you make.

8.40. (a) Use the standard enthalpies of formation for liquid and gaseous methanol, CH_3OH, from Appendix B to find its standard enthalpy of vaporization, $\Delta H°_{l \to g}$.
(b) The boiling point of methanol is 65.0 °C. Use your result from part (a) to find the standard entropy of vaporization, $\Delta S°_{l \to g}$, for methanol. Explain the reasoning you use to get your result, including any assumptions you make.
(c) Use the data in Appendix B to calculate $\Delta S°_{l \to g}$ for methanol. How does your value from these data compare to your result in part (a)? How would you explain any difference?

8.8. Gibbs Free Energy

8.41. Consider the complete combustion of ethanol, $C_2H_5OH(l)$, given in Problem 8.36. For the combustion of one mole of ethanol at standard conditions, the standard change in Gibbs free energy is $-1325 \text{ kJ·mol}^{-1}$ at 298 K.
(a) Does the value of the Gibbs free energy change predict this reaction will take place at 298 K? Explain.
(b) At what temperature will this reaction be at equilibrium under standard conditions? Justify your response.
(c) Discuss why the numerical value you calculated in part (b) is likely to be in error.

8.42. Consider the reaction:

$$2H_2O_2(g) \to 2H_2O(g) + O_2(g)$$

For this reaction at 298 K, $\Delta H = -106 \text{ kJ}$ and $\Delta S = 58 \text{ J·K}^{-1}$. Would you expect $H_2O_2(g)$ to be stable or likely to decompose at 298 K? Explain your reasoning.

8.43. A certain reaction is nonspontaneous at 300 K. The entropy change for the reaction is 130 J·K^{-1}.
(a) Is the reaction endothermic or exothermic? Explain your response.
(b) What is the minimum value of ΔH for this reaction? Explain the reasoning for your answer.

8.44. The standard enthalpy of formation ($\Delta H_f°$) and the standard free energy of formation ($\Delta G_f°$) of elements are zero. The standard absolute entropy ($S°$) is not zero. Explain why.

8.45. Changes can be characterized by the signs of ΔH and ΔS, for the system, as illustrated in this diagram.

For example, processes with positive values for both ΔH and ΔS fall in the quadrant of the diagram labeled "B."

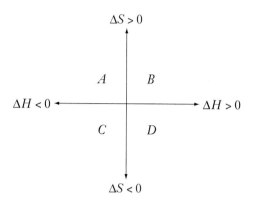

(a) Which quadrant(s), A, B, C, or D, represents changes or processes that are always spontaneous? Explain your choice(s).
(b) Which quadrant(s) represents changes or processes that are never spontaneous? Explain your choice(s).
(c) Two of the quadrants represent processes that could *possibly* be spontaneous, depending on the temperature at which the processes are carried out. Which quadrants are these? Which quadrant represents processes that could possibly be spontaneous at high temperatures? Which quadrant represents processes that could possibly be spontaneous at low temperatures? Clearly explain the reasoning for your choices.
(d) Which quadrant represents the freezing of a liquid? Explain.
(e) Which quadrant represents the melting of a solid? Explain.

8.46. At a certain temperature, ΔG for the reaction $CO_2(g) \rightleftharpoons C(s) + O_2(g)$, is found to be 42 kJ. Which of these statements is correct? Explain the reasoning for your choice and the reason(s) you reject the others.
   (i)   The system is at equilibrium.
   (ii) The process is not possible.
   (iii) CO_2 will be formed spontaneously.
   (iv) CO_2 will decompose spontaneously.

8.47. Consider the oxidation of methane:
$CH_4(g) + 2O_2(g) \to CO_2(g) + 2H_2O(g)$.
(a) By inspection, predict the change of entropy. Explain the basis of your prediction.
(b) Use the values of $\Delta H_f°$, $\Delta G_f°$, and $S°$ from Appendix B to calculate the change in standard enthalpy, free energy, and entropy for this reaction at 25 °C and 1 bar. Explain the procedures you use.
(c) Was your prediction in part (a) close to the calculated value in part (b)? Explain why or why not.

8.48. This reaction, $2H_2(g) + O_2(g) \rightarrow 2H_2O(l)$, is known to have a large negative free energy change at room temperature. Does this mean that if you mix hydrogen and oxygen at room temperature you should be able to see the reaction occur? Explain.

8.49. For this reaction, $C(\textit{graphite}) + 2H_2(g) \rightarrow CH_4(g)$, $\Delta H° = -74.8$ kJ and $\Delta S° = -80.8$ J·K⁻¹.
(a) Plot, as was done in Figure 8.13, $\Delta H°$, $T\Delta S°$, and $\Delta G°$ as a function of temperature for the reaction under standard conditions. What assumptions must you make to construct your plot?
(b) Use your plot to estimate the temperature at which this reaction is in equilibrium under standard conditions. Explain how you find the temperature.
(c) Is the temperature you found in part (b) likely to be an accurate prediction? Explain why or why not.

8.50. What is the free energy change, ΔG, when one mole of water at 100 °C and 100 kPa pressure (standard pressure) is converted to gaseous water at 100 °C and 100 kPa pressure? The standard enthalpy change for this process is 40.6 kJ. Explain your answer.

8.51. We have shown that observed changes occur in the direction that increases the net entropy of the universe (or at least our tiny corner of it), ΔS_{net}, the sum of positional and thermal entropy changes for the process. You will find other writers who say that observed changes occur in the direction that minimizes the energy of the system undergoing change and maximizes its entropy. Explain clearly how these two viewpoints are actually the same. (We have used the first because it makes the connection between higher probability and direction of change more explicit.)

8.9. Thermodynamic Calculations for Chemical Reactions

8.52. For this combustion reaction, $CH_3COCH_3(g) + 4O_2(g) \rightarrow 3CO_2(g) + 3H_2O(l)$, at 298 K, $\Delta G = -1784$ kJ and $\Delta H = -1854$ kJ.
(a) Will the value of $T\Delta S$ for this reaction be about the same magnitude as for ΔG and ΔH? Make a qualitative estimate of this value.
(b) Comment on what your estimated relative value for $T\Delta S$ tells you about the factor that contributes most to the net entropy change, ΔS_{net}, for this reaction.
(c) Calculate the value of ΔS for this reaction, reporting your answer in J·K⁻¹.

8.53. For this reaction, $4Ag(s) + O_2(g) \rightarrow 2Ag_2O(s)$, at 298 K, $\Delta H° = -61.1$ kJ and $\Delta S° = -0.132$ kJ·K⁻¹.
(a) Assuming that the reaction takes place at constant pressure, is it spontaneous at 25 °C? Explain.
(b) Is it going to be spontaneous at 273 °C? Explain the reasoning for your response and clearly state any assumptions you make.

8.54. The thermodynamic data in this table have been obtained for the gas-phase dimerization reaction of methanoic acid (dotted lines represent hydrogen bonds):

Species	$\Delta H_f°$ kJ·mol⁻¹	$S°$ J·K⁻¹·mol⁻¹
HC(O)OH(g)	−362.63	251.0
[HC(O)OH]₂(g)	−785.34	347.7

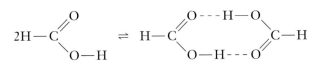

(a) Calculate $\Delta H°$, $\Delta S°$, and $\Delta G°$ for the gas-phase dimerization at 298 K. Is the formation of the dimer from the monomers spontaneous under these conditions? Explain.
(b) Calculate the enthalpy change per hydrogen bond formed in the gas phase. Show your reasoning clearly.
(c) Why is it not useful to estimate the free energy or entropy of hydrogen bond formation in the same way you did the enthalpy in part (b)?

8.55. Use data from Appendix B to calculate $\Delta H°$, $\Delta S°$, and $\Delta G°$ for each of these dissolution reactions:

$$KCl(s) \rightleftharpoons K^+(aq) + Cl^-(aq)$$

$$AgCl(s) \rightleftharpoons Ag^+(aq) + Cl^-(aq)$$

(a) Do your results support the solubilities we discussed in Chapter 2, Section 2.7? Explain why or why not.
(b) Do you predict that the solubility of KCl(s) should increase, decrease, or stay the same as the temperature is increased? Justify your answer thermodynamically.
(c) Do you predict that the solubility of AgCl(s) should increase, decrease, or stay the same as the temperature is increased? Justify your answer thermodynamically.

8.56. (a) For $3C_2H_2(g) \rightleftharpoons C_6H_6(l)$, use data from Appendix B to find $\Delta G°$ at 298 K for the conversion of ethyne, $C_2H_2(g)$, to benzene, $C_6H_6(l)$. In which direction is the reaction spontaneous under standard conditions? Explain.
(b) Should it be possible to find a temperature where this direction is reversed? Give a thermodynamic answer based on the reactants and check it with data from Appendix B.

8.57. (a) Hydrazine, $N_2H_4(l)$, and its derivatives are used as rocket fuels. Several different oxidizers can be used with hydrazine. Use data from Appendix B to calculate $\Delta H°$, $\Delta S°$, and $\Delta G°$ at 298 K for the reaction with $H_2O_2(l)$: $N_2H_4(l) + 2H_2O_2(l) \rightarrow N_2(g) + 4H_2O(l)$.
(b) Is the reaction spontaneous? Explain.

(c) In a rocket engine, the reactants will get hot. Are higher temperatures more favorable or less favorable for the reaction? Justify your answer thermodynamically.

8.58. Consider the reaction that converts pyruvic acid to ethanal and carbon dioxide:

$$CH_3COC(O)OH(l) \rightleftharpoons CH_3CHO(l) + CO_2(g)$$

(a) Use the data from Appendix B to calculate $\Delta H°$, $\Delta S°$, and $\Delta G°$ for this reaction at 298 K. Are your results consistent with equation (8.19)? Explain why or why not.
(b) Is this reaction spontaneous under standard conditions? Explain.
(c) If the pressure of carbon dioxide is raised to 100 atm (about 10,000 kPa), how will the entropy of carbon dioxide be affected? Give the reasoning for your response. *Hint:* Recall what the ideal gas equation, Chapter 7, Section 7.13, equation (7.56), tells you about the relationship of the pressure to the volume of a gas and how the entropy of a system varies with the volume available to the molecules.
(d) How will the change in pressure of carbon dioxide affect the free energy change for the reaction? Will the reaction be more or less likely to proceed as written? Explain.
(e) Assuming that the reaction is in equilibrium before the carbon dioxide pressure is increased, what does Le Chatelier's principle predict about the effect on the equilibrium of increasing the pressure? Is this prediction consistent with your response in part (d)? Explain why or why not.

8.59. When water vapor is passed over hot carbon, this reaction occurs:

$$C(s) + H_2O(g) \rightarrow CO(g) + H_2(g)$$

The product mixture is called "water gas" or "synthesis gas," since it can be used as a starting material to synthesize many carbon-containing molecules.
(a) Use data from Appendix B to calculate $\Delta H°$, $\Delta S°$, and $\Delta G°$ for this reaction at 298 K. Is the reaction spontaneous under standard conditions? Explain.
(b) Will increasing the temperature make the reaction more favorable or less favorable in the direction written? Justify your answer thermodynamically.
(c) Will the reaction be favored in the direction written by increasing or decreasing the pressure of gases in the system? Justify your answer thermodynamically.
(d) Given your answers in parts (b) and (c), explain what conditions you would recommend to optimize the formation of products:

(i)   low temperature and low pressure
(ii) low temperature and high pressure
(iii) high temperature and low pressure
(iv) high temperature and high pressure

8.10. Why Oil and Water Don't Mix

8.60. The cellulose molecules (polymers of glucose) that make up paper are polar and hydrophilic. How does the structure of fat explain how water beads up and does not get absorbed by waxed paper (paper impregnated with fatty substances like this)?

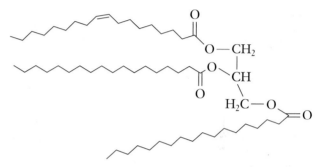

8.61. (a) Do you think the positional entropy change for the formation of the methane–water clathrate from the pure compounds is positive or negative? Clearly explain the reasoning for your answer.
(b) What reason(s) can you think of for the clathrate formation being more favorable at lower temperatures? Clearly explain the reasoning for your answer.
(c) What reason(s) can you think of for the clathrate formation being more favorable at higher pressures of methane? Clearly explain the reasoning for your answer. *Hint:* The molar entropy of a gas decreases as its pressure increases at constant temperature. [Does this make sense to you? See Problems 8.58(c) and 8.93(e).]

8.62. Benjamin Franklin was always interested in the world around him. He noted, for example, that a teaspoon of oil (about 5 mL) poured on the surface of a calm pond spread out to form an oil patch about $\frac{1}{2}$-acre (about 4000 m²) in area. What do these data tell you about the size, in nanometers, of an oil particle (molecule)? Carefully note and explain any assumptions you make in answering this question.

8.11. Ambiphilic Molecules: Micelles and Bilayer Membranes

8.63. Suppose you have several ambiphilic compounds with the same polar head but nonpolar tails of different lengths. How will the lengths of the tails affect the ability of the compounds to form micelles in water? Use drawings to help make your explanation clear.

8.64. The diagram here shows liquid water with a less dense nonpolar liquid floating on it (oil on water).

nonpolar liquid

water

(a) Suppose you have an ambiphilic compound that can't form micelles in water because its nonpolar tail is too long. If you add just a little bit of this ambiphilic compound to the system above, where do you think the molecules of the compound will be in the system? As part of your answer draw a diagram to show where the ambiphilic molecules are and how they are interacting with the water and nonpolar liquid.

(b) If you add more of the ambiphilic compound to the system, where will the molecules of the compound be in the system? As part of your answer draw a diagram to show where the ambiphilic molecules are and how they are interacting with the water and nonpolar liquid. *Hint:* Consider a reversed micelle. What is reversed? How does this help solve the problem?

8.65. Suppose you have an ambiphilic compound that will form reversed micelles in the nonpolar liquid in Problem 8.64. Do you think it would be possible to remove solutes from the water by getting them to dissolve in the reversed micelle in the nonpolar liquid? (This would be sort of like the action of a detergent in water.) What kind of properties would the polar head of the ambiphilic ompound need, in order to get a solute out of the water?

> Experiments are being done with reversed micelles and liquid CO_2 as the nonpolar liquid to see if such extractions from water can be done without generating a lot of waste solvents that could pollute the environment.

8.66. Hepatocytes, the predominant cells in your liver, are roughly cubical with edge lengths of about 15×10^{-6} m (15 μm).

(a) About how many phospholipid molecules, Figure 8.22, are required to form the cell membrane? State all the assumptions you make to solve this problem and explain your solution method clearly.

(b) How many moles of phospholipid are required to form the membrane and what is the mass of the phospholipid in the membrane? State any further assumptions you make to solve this problem and explain your solution method clearly.

8.67. Mayonnaise has four main ingredients: oil, vinegar (mostly water), lemon juice (mostly water), and egg yolk. An egg yolk contains a lot of phospholipids, proteins, and fats to nourish the chick embryo in a fertilized egg. What is the purpose of the egg yolk in the mayonnaise? Clearly explain your reasoning.

8.68. To carry out their functions, membranes need to stay fluid, so that they retain their permeability and the proteins embedded in them can move about as necessary. One way that cold-blooded organisms keep their membranes fluid at different temperatures is to change the proportion of saturated and unsaturated fatty acids, Figure 8.22, in the phospholipid membrane molecules. Compared to a fish that lives in the tropics, would you expect a fish that lives in cold Arctic waters to have a higher or lower proportion of unsaturated fatty acids in its membranes? Clearly explain the reasoning for your response, including molecular level drawings, if that helps your explanation.

8.12. Colligative Properties of Solutions

8.69. Oil-rich countries in the Middle East once considered towing icebergs from the Arctic Ocean to help solve their water shortages. Wouldn't the icebergs be made of salt water like the ocean water? Explain.

8.70. Explain why $S_{\text{solution}} > S_{\text{solvent}} > S_{\text{solid}}$.

8.71. Why does a colligative property depend only on the number of solute particles and not their chemical properties?

8.72. Which solution will have the lowest freezing point? Explain your answer.
  (i)   1 M NaCl
  (ii) 1 M $CaBr_2$
  (iii) 1 M $Al(NO_3)_3$

8.73. Explain why a mixture of antifreeze and water, instead of pure water, is typically placed in an automobile's radiator.

8.74. Which aqueous solution has the largest ΔT_{fp}? Explain your reasoning.
  (i)   1.5 *m* solution of glucose, $C_6H_{12}O_6$
  (ii) 1.5 *m* solution of sodium chloride
  (iii) 1.5 *m* solution of magnesium chloride

8.75. What is the freezing point of a solution made by dissolving 5.00 g of sucrose, $C_{12}H_{22}O_{11}$, in 100.0 g of water?

8.76. A solution is made by mixing 10.00 g of n-decane, $CH_3(CH_2)_8CH_3$, with 100.00 g of benzene.
(a) What is the molality of this solution?
(b) Calculate the freezing point depression, ΔT_{fp}, for this solution.
(c) What is the freezing point of this mixture?

8.77. These solutions of a nonvolatile, nonionizing solute are prepared in benzene and in cyclohexane. Determine which solution will freeze at a lower temperature.

Explain your procedure.
  (i)   0.30 *m* solution in benzene
  (ii) 0.10 *m* solution in cyclohexane

8.78. (a) An aqueous solution of 1.0 *m* sodium chloride is predicted to have a freezing point of −3.72 °C. Explain why. (b) However, the actual freezing point is −3.53 °C. Why is the experimental freezing point lowering less than the predicted value?

8.79. What is the freezing point of a solution made by dissolving 5.00 g of sodium sulfate, Na_2SO_4, in 100.0 g of water?

8.80. Sorbitol is a sweet-tasting substance that is sometimes used as a substitute for sucrose. Its formula is $C_6H_{14}O_6$. What will be the freezing point of an aqueous solution containing 1.00 g sorbitol in 100.0 g of water?

8.81. Calculate the change in the freezing point of water when 3.5 mg of hemoglobin (molar mass = 64,000 g) is dissolved in 5.0 mL of water. Would this method be accurate in determining molar masses of biomolecules of this size?

8.13. Osmotic Pressure Calculations

8.82. Both glucose solutions and physiological saline solutions [NaCl(*aq*)] are given to patients through intravenous injections. The average osmotic pressure of blood serum is 7.7 atm at 25 °C. To prevent osmosis through the semi-permeable membrane of a red blood cell, the glucose solution or physiological saline solution given must have the same osmotic pressure as blood serum. (We say that these solutions are isotonic with blood serum.) Calculate the concentration of glucose solution and saline solution that are isotonic with blood serum.

8.83. Suppose you dissolve 0.150 g of a newly isolated protein in 25.0 mL of water. The solution has an osmotic pressure of 0.00342 atm at 277 K. What is the molar mass of this protein?

8.84. Suppose you have a solution separated from pure water by a membrane that is only permeable to water, as shown in this diagram.

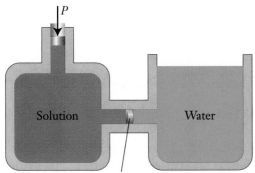

Membrane permeable to water

The solution will have a certain osmotic pressure, Π, determined by the solutes it contains. What would happen if a pressure, *P*, greater than this osmotic pressure is applied to the solution by a piston, as shown in the diagram? Use drawings to illustrate what is happening at the molecular level.

8.85. The process suggested in Problem 8.84 is called reverse osmosis. What is it that is reversed? What practical uses can you think of for reverse osmosis?

8.86. Fresh water can be prepared from seawater by the process of reverse osmosis. (See Problem 8.84.) This method involves applying a pressure greater than the osmotic pressure to the seawater to force the flow of pure water from the seawater through the semipermeable membrane to the pure water. Calculate the minimum pressure that must be applied to a seawater sample to produce reverse osmosis. Assume that the concentrations of seawater's solutes are as follows: $[Cl^-]$ = 18,000 ppm; $[Na^+]$ = 10,500 ppm; $[Mg^{2+}]$ = 1200 ppm; $[SO_4^{2-}]$ = 870 ppm; $[K^+]$ = 379 ppm. The abbreviation ppm stands for parts per million. One ppm is 1 mg of the solute in 1000 g of solvent.

8.87. There is a brand of time-release plant fertilizer called Osmocote®. The package says, "The unique resin coated granules are . . . easy to use. Within one week after application the soil moisture causes the granules to swell into plump capsules of liquefied plant food, which continuously release nutrients for approximately 9 months." What sort of properties does the resin coating have to have? Why do you think the company that makes this product decided on the name "Osmocote"?

8.14. The Cost of Molecular Organization

8.88. We can consider a growing plant (the system) to be an example of decreasing entropy. Small molecules, like CO_2 and H_2O, are built into complex, but orderly, arrangements of macromolecules. Which of these statements applies to this example? Explain the reasoning for your choice and the reason(s) you reject the others.
  (i)   The second law of thermodynamics—that net entropy always increases in observed processes—is being violated.
  (ii) The second law of thermodynamics is not being violated because the entropy of the surroundings is increasing.
  (iii) The second law of thermodynamics is not being violated because the entropy of the surroundings is decreasing.
  (iv) Plant growth is so complex that the laws of thermodynamics cannot be applied.
  (v)   None of these statements applies to this example.

8.89. Consider a fertilized chicken egg in an incubator, a constant temperature and pressure environment. The eggshell and its membrane are permeable to atmospheric gases. In about 3 weeks, the egg will hatch into a chick.
(a) Take the egg as the system. Is this an open, closed, or isolated system? Explain. *Hint:* Review Chapter 7, Section 7.5, if necessary.
(b) In the fertilized egg, proteins from the hen are formed into a highly ordered chick. Does the entropy of the system increase or decrease? Briefly explain why the development of the chick is or is not consistent with the second law of thermodynamics.
(c) What happens to the energy of the system as the chick develops? What forms of energy contribute to the change in energy (if any) of the system?

(d) Does the free energy of the system increase, decrease, or remain the same during development? How do you know?

8.16. Extension—Thermodynamics of Rubber

8.90. Suppose you are investigating a biological polymer that undergoes a reaction to give a product, polymer → product, and find that, at a given temperature, the reaction is spontaneous and endothermic.
(a) What can you tell about the signs of ΔH, ΔS, and ΔG for this reaction? Explain your reasoning.
(b) What do your results tell you about the structure of the product relative to the polymer? Explain your reasoning.

General Problems

8.91. The questions in this problem are based on the data in this table of values for the enthalpies of formation and entropies of several hydrocarbons in the gas phase. (Many of these compounds are liquids at 298 °C, so the gas-phase values here are calculated in standard ways from other experimental data.)

Enthalpies of formation and entropies for selected gas-phase hydrocarbons.

Name	Molecular formula	Condensed structural formula	ΔH_f°, kJ·mol^{-1}	S°, J·K^{-1}·mol^{-1}
ethene	C_2H_4	$H_2C=CH_2$	52.28	219.8
propene	C_3H_6	$H_2C=CHCH_3$	20.41	266.9
cyclopropane	C_3H_6	Cs bonded in an equilateral triangle	53.30	237.4
1-butene	C_4H_8	$H_2C=CHCH_2CH_3$	1.17	307.4
cyclobutane	C_4H_8	Cs bonded in a nonplanar square	26.65	265.4
1-pentene	C_5H_{10}	$H_2C=CHCH_2CH_2CH_3$	−20.92	347.6
2-methyl-1-butene	C_5H_{10}	$H_2C=C(CH_3)CH_2CH_3$	−36.62	342.0
3-methyl-1-butene	C_5H_{10}	$H_2C=CHCH(CH_3)_2$	−28.95	333.5
2-methyl-2-butene	C_5H_{10}	$H_3CCH=C(CH_3)_2$	−42.55	338.5
trans-2-pentene	C_5H_{10}	$H_3CCH=CHCH_2CH_3$	−31.76	342.3
cyclopentane	C_5H_{10}	Cs bonded in a nonplanar pentagon	−77.24	292.9
1-hexene	C_6H_{12}	$H_2C=CH(CH_2)_3CH_3$	−41.7	386.0
cyclohexane	C_6H_{12}	Cs bonded in a nonplanar hexagon	−123.1	298.2
1-heptene	C_7H_{14}	$H_2C=CH(CH_2)_4CH_3$	−62.16	424.4
1-octene	C_8H_{16}	$H_2C=CH(CH_2)_5CH_3$	−82.93	462.8
1-nonene	C_9H_{18}	$H_2C=CH(CH_2)_6CH_3$	−103.5	501.2
1-decene	$C_{10}H_{20}$	$H_2C=CH(CH_2)_7CH_3$	−124.1	539.6
1-undecene	$C_{11}H_{22}$	$H_2C=CH(CH_2)_8CH_3$	−144.8	578.1
1-dodecene	$C_{12}H_{24}$	$H_2C=CH(CH_2)_9CH_3$	−165.4	616.5

(a) Use average bond enthalpies from Chapter 7, Section 7.7, Table 7.3 to calculate the difference you would expect between the enthalpies of formation of the C_3, C_4, C_5, and C_6 cyclic compounds and their linear alkene isomers. Do you predict the cyclic or linear isomers to be more stable? How does your calculated difference compare with the experimental values from the table here? Is your prediction correct? Explain your reasoning. Use your molecular models to construct models of the cyclic and linear isomers. How, if at all, do the models affect your explanation and reasoning?

(b) Compare the entropies of the C_3, C_4, C_5, and C_6 cyclic compounds and their linear alkene isomers. Are the differences in the direction you would have predicted? Explain the basis for your prediction. Use the molecular models you constructed in part (a) to develop a rationale for the relative magnitudes as well as direction of the entropy differences. Clearly explain the basis for your reasoning.

(c) Are there trends in the data for the linear alkene isomers? Can you illustrate these graphically and draw more quantitative conclusions? Predict ΔH_f° and S° for 1-tetradecene, $C_{14}H_{28}$. (What does a model of this compound look like?) Explain how you arrive at your predictions. How confident are you of these predictions?

(d) Can you use an analysis similar to that in part (c) to predict ΔH_f° and S° for cyclooctane, C_8H_{16}? (What does a model of this compound look like?) Explain how you arrive at your predictions. How confident are you of these predictions?

(e) Write a discussion of the trends and correlations in the data for all the C_5H_{10} alkene isomers in the table. Use models as a basis for rationalizing and explaining what you find.

8.92. An inventor claims to have discovered a catalyst that will break water down to hydrogen and oxygen gases at room temperature without an input of energy. Use data from Appendix B and thermodynamic arguments to counsel possible investors whether or not to invest their money to commercialize this catalyst.

8.93. Liquids evaporate; molecules leave the liquid and enter the gas phase. The pressure of the gas molecules in equilibrium with the liquid is the vapor pressure of the liquid at that temperature. Solutions of nonvolatile solutes have lower vapor pressures than the pure solvent; vapor pressure lowering is another colligative property.

(a) Which entropy is higher, that for pure solvent, $S_{solvent}$, or that for a solution, $S_{solution}$? Explain your response.

(b) The same input of enthalpy, ΔH_{vap}, is required to vaporize a given amount of liquid solvent from the pure solvent and from a solution. At a given temperature, how do the thermal entropy changes for vaporization of the same amount of liquid, $\Delta H_{vap}/T$, compare for pure solvent and a solution? Explain.

(c) At equilibrium, the net entropy change, ΔS_{net}, must be zero for both vaporization of a pure solvent and vaporization of a solution. Given your answer in part (b), how must the positional entropy changes for (solvent → gas) and (solution → gas) compare? Explain.

(d) Given the relative values in part (a) and your answer in part (c), which gas must have the higher entropy, the one in equilibrium with pure solvent or the one in equilibrium with the solution? Explain your reasoning. *Hint:* A diagram similar to Figure 8.26 might be useful.

(e) The entropy of a gas is inversely related to its pressure; higher pressure, more compressed gas has a lower entropy than the same gas at a lower pressure. Which gas in part (d) has the lower pressure? Is this result consistent with vapor pressure lowering by a nonvolatile solute. Explain the reasoning for your answers.

(f) A beaker of water and a beaker of a 1 M aqueous glucose solution are placed side by side inside a sealed, transparent container held at constant temperature. If you observe this system for many days, what do you predict you will observe about the liquid levels in the two beakers? Explain the reasoning for your answer.

8.94. The top diagram is a schematic representation of the binding of a substrate molecule (the squiggle) in the active site on an enzyme (the fat tilted "C"). In order to bind, the substrate has to fold into a shape that fits the site, as shown on the right of the diagram. The bottom diagram shows the interaction of the substrate with a metal ion (the little lozenge shape) in solution. Binding to the metal ion causes the substrate to fold as shown.

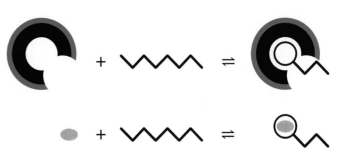

(a) If you consider only the changes represented by the top diagram, what do you predict about the entropy change for this process? Clearly explain your reasoning.

(b) How will the presence of the metal ion in the solution affect the entropy of binding of the substrate with the enzyme? What assumptions do you make? Explain your reasoning.

8.95. For our analysis of mixing in Section 8.3, we focused on the arrangements of solute molecules. Let's see what the results are if we focus on the arrangements of the solvent molecules. Consider the 15-box model system represented in Figures 8.4 and 8.5.

(a) How many distinguishable arrangements of 12 solvent molecules are there in the 12 lower boxes in Figure 8.4? For each possible arrangement, there is only one arrangement of the solute molecules in the remaining 3 boxes at the top. What is the total number of possible arrangements for this system?

(b) If the 12 solvent molecules must occupy all the lower 9 boxes and can also occupy any of the 6 upper boxes, how many distinguishable arrangements are there? Explain your reasoning. For each arrangement of solvent molecules, there is only one arrangement of the solute molecules in the remaining three empty boxes. What is the total number of possible arrangements for this system? How does your answer compare with Figure 8.5?

(c) If the 12 solvent molecules must occupy all the 6 lower boxes and can also occupy any of the 9 upper boxes, how many distinguishable arrangements are there? (The formula presented in Problem 8.12 might be useful.) What is the total number of possible arrangements for this system? How does your answer compare with the results in Figure 8.6 and its accompanying text? Explain.

(d) If the 12 solvent molecules may occupy any of the boxes, how many distinguishable arrangements are there? What is the total number of possible arrangements for this system? How does your answer compare with the results in Figure 8.6 and its accompanying text?

(e) Does it make any difference whether you take the solvent or solute point of view to calculate the total number of arrangements of molecules in the system? Why or why not?

8.96. At the molecular level, all energies are quantized, including the kinetic energy of motion of molecules in a gas. The energy levels for this motion are so close together that the number of energy levels available is many times more than the number of molecules. Thus, the probability that more than one molecule will occupy any energy level is essentially zero: There will be either one or zero molecules in each energy level. As the volume of the gas increases, the energy levels get even closer

together. We can use our countable-systems approach to examine the consequences for the entropy (the number of arrangements of molecules among the energy levels) and the average energy of the molecules. Consider these energy levels in a 12-unit range of energy for three different total volumes of gas, where $V_1 < V_2 < V_3$:

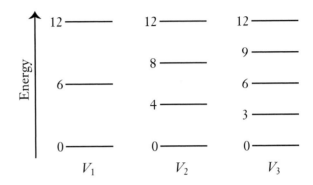

(a) Our gas consists of three identical molecules. If there are either one or zero molecules in each energy level, there is only one distinguishable way for the molecules to be arranged among the energy levels in V_1. There must be one molecule in each level. (Rearranging the identical molecules among the three levels does not produce a new distinguishable arrangement.) In terms of the energy units on the diagram, what is the total energy of the molecules in this system? What is the average energy per molecule?

(b) If there are either one or zero molecules in each energy level, show that there are four distinguishable ways for the molecules to be arranged among the energy levels in V_2. What is the total energy of the molecules in each of these arrangements? Since each of these arrangements is equally probable (our fundamental assumption in Section 8.2), each will contribute one quarter to the total energy of the system. What is the total energy of the system? Explain your reasoning. What is the average energy per molecule?

(c) If there are either one or zero molecules in each energy level, how many distinguishable ways are there for the molecules to be arranged among the energy levels in V_3? What is the total energy of the molecules in each of these arrangements? What is the total energy of the system? What is the average energy per molecule? Clearly explain your reasoning for each response.

(d) Do the results for the average energy per molecule in the three volumes surprise you? Why or why not? At constant temperature, the energy, E, of an ideal gas (Section 7.13) does not depend on its volume. Are your results consistent with this property? Explain why or why not.

(e) How are your results in part (c) and those in Investigate This 8.4 related, if at all? The marginal note on page 517 implies that what we have called positional entropy is fundamentally based on arrangements of molecules among quantized energy levels. Explain how your results in this problem relate to positional entropy.

. . . the phenomena of chemical equilibrium play a capital role in all operations of industrial chemistry.

HENRI LOUIS LE CHATELIER, 1850–1936

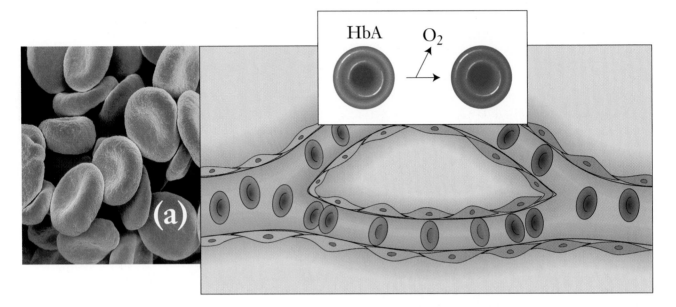

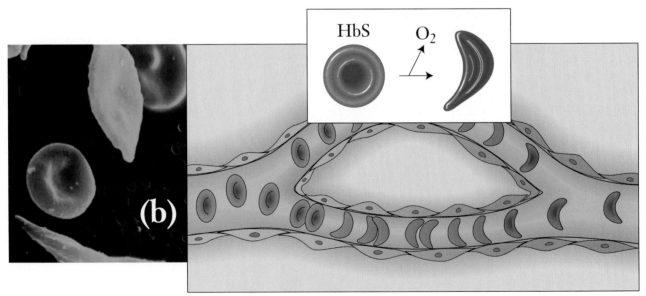

The electron micrographs on the left show red blood cells (a) from a normal adult with normal hemoglobin, HbA, and (b) from a person with sickle-cell disease, whose red blood cells contain sickle-cell hemoglobin, HbS. The illustrations on the right depict the fate of the cells as they pass through capillaries and release their oxygen to the surrounding cells. The uptake and release of oxygen in the lungs and capillaries, respectively, as well as the sickling of the cells are results of equilibrium processes. One of our tasks in this chapter will be to discover the factors that control equilibria and understand the conditions that favor one direction or the other for the systems illustrated here.

Chemical Equilibria

H Le Chatelier was a French chemist who worked on problems in metallurgy, cements, explosives, and other important industrial processes and products. His research led him to the conclusion in the opening quotation, but he could have expanded his comment on the central role of equilibrium to include all of chemistry: industrial reactions, laboratory studies, and living systems. Although few of the systems we see around us are at equilibrium, their changes toward equilibrium are what we observe. Thus, an understanding of equilibrium can help us understand and predict the direction of changes.

Examples of equilibria in a biological system are illustrated on the facing page. Our red blood cells are packed with the oxygen-carrying protein **hemoglobin.** Hemoglobin has a molar mass of about 66,000 g and each molecule can bind up to four oxygen molecules. In the lungs, where the oxygen pressure is high, oxygen binds to the heme groups on hemoglobin. When the blood reaches the capillaries, where the oxygen pressure is low (since it is used up by our cells), the process is reversed and hemoglobin releases oxygen. Binding of oxygen is an equilibrium process that favors bound oxygen when the oxygen concentration is high and release of oxygen when the oxygen concentration is low.

Normal blood cells are shaped like tiny candy mints, both before and after oxygen is released, as shown in part (a) of the opening illustration. Normal adult hemoglobin molecules, HbA, have little attraction for one another and shift about in the blood cell without distorting its shape as it passes through tiny capillaries. By contrast, the red blood cells of persons who suffer from **sickle-cell disease** have a mutant form of hemoglobin. The mutant, or HbS, molecules attract one another when the hemoglobin is deoxygenated and form long, stiff chains of HbS that distort the red blood cells into the sickle-like shape shown in part (b) of the illustration. The sickled cells get stuck in the capillaries and restrict the flow of blood and hence of oxygen to the cells, resulting in a great deal of pain and sometimes death. The mutation responsible for sickle-cell disease was discovered in experiments that depend on acid–base equilibria in proteins. Attractions between HbS molecules are a result of these equilibria.

In Chapter 8, we found that a system and its surroundings are in equilibrium, if the net entropy change for any change in the system and its surroundings is zero. There we developed this idea for phase changes. In this chapter, our task is to extend it to equilibria in chemical reactions. Equilibria in all types of chemical reactions are fundamentally the same and can all be treated in exactly the same way. The change in Gibbs free energy, ΔG, is sometimes more convenient to use than net entropy change. Almost all arguments about directionality in biochemical reactions are based on free energies and we will look at some examples, including equilibria like those involved in oxygen transport by hemoglobin.

We will begin with a qualitative examination of the nature of chemical reactions at equilibrium to see what variables influence the equilibrium state. Then we will introduce an empirical approach to equilibrium constants, equilibrium constant expressions, and equilibrium system calculations. Finally, we will relate thermodynamics and equilibria and discover the power of this combination in a few chemical and biological systems.

9.1. The Nature of Equilibrium

9.1 INVESTIGATE THIS

How can we model a system at equilibrium?

(a) Do this as a class investigation with two student investigators and work in small groups to discuss and analyze the results. Each investigator has a plastic basin about 10 cm deep. One investigator has a 100-mL beaker and the other a 250-mL beaker. Add water to fill one of the basins about two-thirds full; the other basin is empty at the start. In unison, the investigators dip their beakers into their basins and transfer any water in their beakers to the other investigator's basin. While continuing the water transfers for a few minutes, observe and record the water levels in the two basins.

(b) Repeat the procedure in part (a), except start with all the water in the other basin.

9.2 CONSIDER THIS

What is dynamic equilibrium?

(a) After several transfers of water in Investigate This 9.1, what do you observe about the water levels in the two basins? Does it make a difference which basin initially contains water? What would you predict the result to be if the water was initially distributed so some was in each basin? Explain your responses.

(b) Explain how the system of water transfer between basins can be considered a model for a chemical reaction system. Why is this called a *dynamic* equilibrium system?

Identifying systems at equilibrium It is easy to be fooled about whether a chemical reaction system is at equilibrium. The *observable* properties of a system at equilibrium are unchanging, just as the liquid levels in Investigate This 9.1 are constant after the reactions have gone on for a time. The reactions in each direction are balanced and the net amounts of products and reactants are constant. The analogy in Investigate This 9.1 shows that the reactions continue, but the amount of product and reactant (water in the two basins) remains constant. This is a state of **dynamic equilibrium:** The observable state of the system is unchanging (amounts of reactants and products remain the same), but the reactions continue to change reactants to products and products back to reactants.

But a chemical system may only *appear* to be unchanging because the reactions are so slow that no change occurs during the time you make your observations. For example, consider wood, which is composed largely of cellulose, a polymer of glucose. You know that both the enthalpy and entropy of combustion of glucose greatly favor its oxidation (Chapter 8, Section 8.9).

There is a strong driving force (large net entropy increase) to produce carbon dioxide and water from wood and oxygen. Yet wooden structures surrounded by a sea of atmospheric oxygen have existed for centuries. You can see the driving force in action if you get the reaction (combustion) started by heating the wood with a match. Wood and oxygen mixtures are not in equilibrium, but you can't tell this by just looking at the unchanging mixtures.

As in the wood–oxygen case, you can try disturbing an unchanging reaction system to see if it responds in a way that is consistent with being at equilibrium. Recall that **Le Chatelier's principle** states that a system at equilibrium responds to a disturbance in a way that minimizes the effect of the disturbance. We have been using this principle to predict how systems at equilibrium will adjust to changes in concentration or temperature and we will continue to use it in this chapter. For wood and oxygen, the disturbance (energy from the match flame) causes such an enormous change (transformation of the wood to new molecules) that you can be certain the system was *not* in equilibrium before the disturbance.

 9.3 INVESTIGATE THIS

Do solutions of Fe(NO₃)₃ and KSCN react when mixed?

(a) Carry out this activity in small groups; discuss your observations and interpretations as you proceed. To each of three adjacent wells in a 24-well plate add about 1 mL (one-third of a well) of 0.002 M potassium thiocyanate, KSCN, solution. To each of the three wells, add *1 drop* of 0.2 M iron(III) nitrate, Fe(NO₃)₃, solution. Record your observations on the appearance of the two reagent solutions and of their mixtures.

(b) To one of the mixtures, add one more drop of the iron(III) nitrate solution. Record your observations on the appearance of this mixture compared to the others.

(c) To one of the remaining original mixtures, add a *tiny* crystal of solid KSCN and observe what happens as the solid dissolves. Record your observations on the appearance of this mixture compared to the others.

9.4 CONSIDER THIS

How do solutions of Fe(NO₃)₃ and KSCN react when mixed?

(a) In Investigate This 9.3(a), do you have any evidence that aqueous solutions of Fe(NO₃)₃ and KSCN react? Explain.

(b) What did you observe when more Fe(NO₃)₃ solution was added to one of the wells in Investigate This 9.3(b)? How do you interpret any change that occurred?

(c) What did you observe when more KSCN was added to one of the wells in Investigate This 9.3(c)? How do you interpret any change that occurred?

9.5 CHECK THIS

Concentrations in the Fe(NO₃)₃–KSCN mixture

A drop of solution from most droppers is about 0.05 mL of liquid. Thus, adding one drop of 0.2 M $Fe(NO_3)_3$ solution to 1 mL of water in Investigate This 9.3(a) dilutes the $Fe^{3+}(aq)$ by a factor of 20. What is the concentration of $Fe^{3+}(aq)$ in the $Fe(NO_3)_3$–KSCN mixtures you made in Investigate This 9.3(a)—assuming that no reaction occurs to use it up. What is the concentration of $SCN^-(aq)$ in this mixture? Which ion is present in higher concentration?

 Web Companion

Chapter 9, Section 9.1.1–2 — ①

View a movie of the reaction and animations of the separate ionic solutions.

② ③ ④

$Fe^{3+}(aq)$ reaction with $SCN^-(aq)$ Solutions of KSCN and $Fe(NO_3)_3$ are ionic. Mixtures of the two solutions contain $K^+(aq)$, $SCN^-(aq)$, $Fe^{3+}(aq)$, and $NO_3^-(aq)$ ions. All of these ions must be colorless, since the individual reagent solutions are colorless. However, as you found in Investigate This 9.3, their mixture is red or red-brown. Potassium nitrate, $KNO_3(s)$, is a white, crystalline solid that dissolves to give clear, colorless solutions; thus reaction of the potassium, $K^+(aq)$, and nitrate, $NO_3^-(aq)$, ions is not responsible for the color. The other combination of ions that might be responsible for the red color is $Fe^{3+}(aq)$ with $SCN^-(aq)$. In fact, these ions react to form a red, one-to-one **metal ion complex** (a Lewis acid–base complex, Chapter 6, Section 6.6):

$$Fe^{3+}(aq) + SCN^-(aq) \rightarrow Fe(SCN)^{2+}(aq) \tag{9.1}$$

The thiocyanate ion displaces one of the water molecules complexed with the iron(III) ion. This reaction (excluding the waters of hydration) is illustrated in Figure 9.1 and would explain your observations when the KSCN and $Fe(NO_3)_3$ solutions are mixed in Investigate This 9.3(a). You found in Check This 9.5 that the $Fe^{3+}(aq)$ in this mixture is present in excess over the $SCN^-(aq)$, as shown in Figure 9.1.

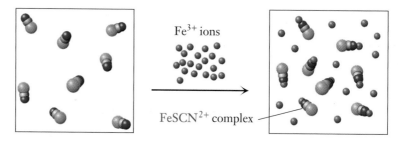

Figure 9.1.

Stoichiometric reaction between $SCN^-(aq)$ and excess $Fe^{3+}(aq)$ to form $Fe(SCN)^{2+}(aq)$.

Equilibrium in the $Fe^{3+}(aq)$–$SCN^-(aq)$ system In Investigate This 9.3(b), more $Fe^{3+}(aq)$ is added to an $Fe(NO_3)_3$–KSCN mixture, and the light orange solution becomes a deeper orange-red. More of the red species, $Fe(SCN)^{2+}(aq)$, must have been formed. Since the addition of more $Fe^{3+}(aq)$ produces more $Fe(SCN)^{2+}(aq)$, some $SCN^-(aq)$ must have been left unreacted in the original mixture. The results suggest that reaction (9.1) reaches equilibrium with free $SCN^-(aq)$ ions remaining in the solution:

$$Fe^{3+}(aq) + SCN^-(aq) \rightleftharpoons Fe(SCN)^{2+}(aq) \tag{9.2}$$

This alternative view of the reaction and the two additions of $Fe^{3+}(aq)$ are illustrated at the molecular level in Figure 9.2(a). Figure 9.2(b) shows the observed colors in these solutions.

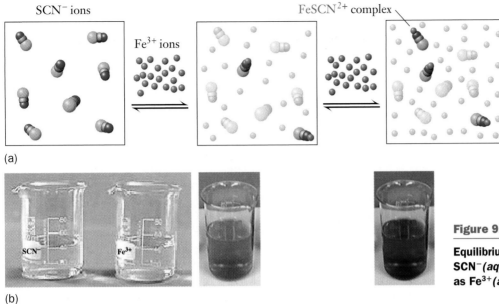

SCN⁻ ions

Fe³⁺ ions

FeSCN²⁺ complex

(a)

(b)

SCN⁻   Fe³⁺

Figure 9.2.

Equilibrium among $Fe^{3+}(aq)$, $SCN^-(aq)$, and $Fe(SCN)^{2+}(aq)$ as $Fe^{3+}(aq)$ is added.

The double **equilibrium arrows** in equation (9.2) and Figure 9.2 indicate that the reaction proceeds in both directions. We can explain our observations in Investigate This 9.3, if we assume that the reaction reaches an equilibrium state (balance of driving forces) in which there is still $SCN^-(aq)$ left in the mixture after the first addition of $Fe^{3+}(aq)$. The $Fe^{3+}(aq)$ in the second addition can then react with some of this remaining $SCN^-(aq)$ to form more $Fe(SCN)^{2+}(aq)$. The second addition of $Fe^{3+}(aq)$ is a disturbance to the iron(III)-thiocyanate ion system that was in equilibrium. Le Chatelier's principle states that the system will respond in a way that minimizes the effect of the disturbance, that is, to use up some of the extra $Fe^{3+}(aq)$ to form more $Fe(SCN)^{2+}(aq)$.

Web Companion

Chapter 9, Section 9.1.3

View animations of the formation and dissociation of the $Fe(SCN)^{2+}(aq)$ complex.

① ② ③ ④

9.6 CHECK THIS

Dynamic equilibrium animation

(a) Explain how you chose the phrase to complete the sentence in the *Web Companion*, Chapter 9, Section 9.1.4.

(b) On average, how many metal ion complexes are present in the tiny volume of solution represented in this animation? Explain how you get your answer and how it relates to part (a).

9.7 CHECK THIS

Adding $SCN^-(aq)$ to a $Fe(NO_3)_3 - KSCN$ solution

(a) In Investigate This 9.3(c), you added a tiny bit of KSCN(s) to your $Fe(NO_3)_3 - KSCN$ solution. Is the response of the system to this disturbance consistent with the assumption that reaction (9.2) is in equilibrium? Use Le Chatelier's principle to explain why or why not.

(b) Draw a diagram, modeled after the right-hand part of Figure 9.2(a), to illustrate your response in part (a).

Where is thiocyanate found in nature? Collect about 1 mL of your saliva, dilute with an equal amount of water, mix, and filter the liquid. To the filtrate, add a few drops of the 0.2 M iron(III) solution you used in Investigate This 9.3. What do you observe? What can you conclude about thiocyanate in biological fluids?

9.8 INVESTIGATE THIS

What conditions affect CoCl₂ in solution?

Do this as a class investigation and work in small groups to discuss and analyze the results. Place about 2 g of solid $CoCl_2 \cdot 6H_2O$ in a *dry* 125-mL Erlenmeyer flask. Add about 50 mL of rubbing alcohol—91% isopropyl alcohol (2-propanol)—to the flask and swirl to dissolve the solid.] *WARNING:* Cobalt compounds are somewhat toxic; wear disposable gloves when handling the solid and the solutions. Alcohols are flammable; extinguish all flames before doing this investigation. Divide the solution equally among four *dry*, 200-mm test tubes. Observe and record the appearance, especially the colors, of all the solids, liquids, and solutions.

(a) Add water a few drops at a time to the Co^{2+} ion solution in one of the test tubes. Mix thoroughly by swirling after each addition. Stop adding water when the solution becomes a lavender color, as in this photograph. Repeat this addition of water (to get a lavender color) with another of the solutions. Finally add this same amount of water to a third one of the solutions and then add about 2 mL more water. Record the colors of the solutions in all four test tubes.

(b) Place the test tube with one of the lavender solutions from part (a) in a beaker of ice water. Place the test tube with the other lavender solution in beaker of 80–90 °C water. After two or three minutes, observe and record the colors of the solutions. Remove the test tubes from the water baths and allow them to return to room temperature. After two or three minutes, observe and record the colors of the solutions.

9.9 CONSIDER THIS

How do solvents and temperature affect CoCl₂ in solution?

(a) What color is the solution when solid $CoCl_2 \cdot 6H_2O$ is dissolved in alcohol in Investigate This 9.8? Is this the color you expected? Explain why or why not.

(b) What color is the solution to which you added the most water in Investigate This 9.8(a)? Is this the color you expected to see for a solution of $CoCl_2 \cdot 6H_2O$? Explain why or why not.

(c) In Investigate This 9.8(b), what changes do you observe when the lavender solutions are cooled and heated? Is there a correlation between these changes and the observations you made in part (a) of the investigation? Can you suggest a way to explain the correlation?

The Co^{2+} ion solutions in Investigate This 9.8 are pink or blue or lavender (a mixture of the pink and blue), depending on the proportions of water and alcohol in the solvent. The pink and blue colors represent different species containing the Co^{2+} cation, so we conclude that different solvents cause changes in the equilibria involving this cation. The chemistry of this system is a little more complicated than the $Fe^{3+}(aq)$–$SCN^-(aq)$ system because we have to be more careful to consider the interaction of water with Co^{2+} cations. To simplify our discussion, we'll ignore many intermediate reaction steps because they do not affect the overall conclusions.

Co^{2+} complexes with water and chloride Solid cobalt chloride is usually written as $CoCl_2 \cdot 6H_2O$, but $(Co(H_2O)_6)Cl_2$ is more accurate. Six water molecules are bonded to each Co^{2+} cation (another Lewis acid–base metal ion complex) and it is this complex ion, $Co(H_2O)_6^{2+}$, that is responsible for the deep red-purple color of the solid. Water molecules in coordination complexes are often rather loosely bonded to the metal ion and easily come off when the complex is in solution. In an aqueous solution, there are so many water molecules nearby that another quickly takes the place of the one that was lost, so the Co^{2+} cation in aqueous solution is present mainly as the pink (red) $Co(H_2O)_6^{2+}$ complex.

When the solid $(Co(H_2O)_6)Cl_2$ is dissolved in 91% alcohol, only 9% of the solvent is water. A water molecule lost from the $Co(H_2O)_6^{2+}$ complex is not so rapidly replaced. Other reactions can occur, including reactions of the cobalt cations with the chloride anions:

$$Co(H_2O)_6^{2+}(aq) + 4Cl^-(aq) \rightleftharpoons CoCl_4^{2-}(aq) + 6H_2O(l) \qquad \text{(9.3)}$$

When chloride ions replace water molecules bound to the Co^{2+} cation, there is a change in the geometry that gives the new complex, $CoCl_4^{2-}$, a blue color. The blue complex is formed as the solid $(Co(H_2O)_6)Cl_2$ dissolves in alcohol because the lack of water in the solvent favors the loss of water from $Co(H_2O)_6^{2+}$.

9.10 CONSIDER THIS

What occurs when water is added to alcoholic Co^{2+} solutions?

Addition of water to your alcoholic solutions of cobalt chloride in Investigate This 9.8(a) changed their color. Does reaction (9.3) explain the changes you observed? Why or why not? Is the response of the system to this disturbance consistent with the assumption that reaction (9.3) is in equilibrium? Use Le Chatelier's principle to explain why or why not.

Temperature and the Co^{2+}–water–chloride system In Investigate This 9.8(b), you changed the temperatures of the Co^{2+} solutions and observed that this disturbance changes the colors of the solutions. When you removed energy from the lavender solution, by cooling it, the solution turned pink. When you added energy to the lavender solution, by warming it, the solution turned blue. When the solutions returned to their original energy states, back to room temperature, they once again became lavender. These observed color changes can be correlated with changes in the relative amounts of the $Co(H_2O)_6^{2+}$ and $CoCl_4^{2-}$ complexes in the solutions, as in Figure 9.3.

When the energy of the system decreases, more of the pink $Co(H_2O)_6^{2+}$ ion is formed. When the energy of the system increases, more of the blue $CoCl_4^{2-}$ ion is formed. We can conclude that it requires an input of energy to form the products in reaction (9.3). The reaction, as written, is endothermic, $\Delta H_{reaction} > 0$. These qualitative observations don't provide enough information to get a numerical value for $\Delta H_{reaction}$, but this simple experiment immediately gives you its sign.

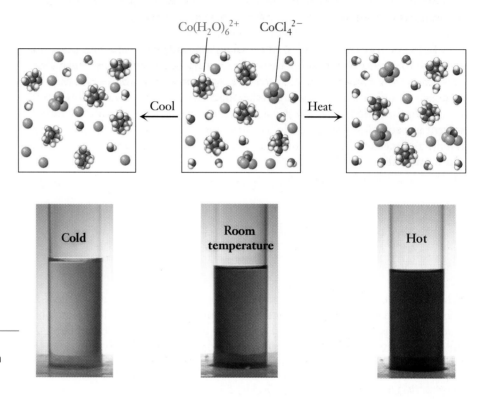

Figure 9.3.

Response of the Co(H$_2$O)$_6$$^{2+}$–CoCl$_4$$^{2-}$ equilibrium to heating and cooling.

9.11 CHECK THIS

Temperature and Le Chatelier's principle

Adding energy to the $Co(H_2O)_6^{2+}$–$CoCl_4^{2-}$ equilibrium system (raising the temperature) is a disturbance to the system. Use Le Chatelier's principle to explain the observed effect on the endothermic reaction (9.3).

Reflection and Projection

In order to get some insight into chemical equilibrium, we began this section with an investigation of an analogous physical system that illustrates the dynamic nature of chemical equilibrium. At equilibrium, the chemical reactions that convert reactants to products and vice versa continue to occur, but both proceed at the same rate, so there is no net change in the concentrations in the system.

What we have learned in this section about the factors that affect equilibrium systems has been qualitative. From these specific examples, let us suggest three generalizations:

- Increasing the concentration of a reactant or product causes the reacting system to adjust in such a way as to try to use up the added molecules.
- Solution equilibria are affected by the composition and properties of the solvent.
- The way an equilibrium system adjusts to an increase in the energy of the system tells you whether the reaction, as written, is **endothermic** (more products formed) or **exothermic** (more reactants formed).

These important and useful generalizations can give you a great deal of insight into systems you haven't previously studied. You can consider these generalizations as applications of Le Chatelier's principle: A system at equilibrium responds to a disturbance in a way that minimizes the effect of the disturbance. *Le Chatelier's principle gives the direction but not the magnitude of the effects.* In order to get further insight into equilibrium systems and to make comparisons among them, we need to make our analyses more quantitative. As our treatments get more quantitative, keep in mind that the results have to be consistent with the qualitative, directional arguments you can make without the calculations.

9.2. Mathematical Expression for the Equilibrium Condition

9.12 INVESTIGATE THIS

What is the pH of an acetic acid solution?

Do this as a class investigation and work in small groups to discuss and analyze the results. Use a pH meter and pH electrode to measure the pH (± 0.02 units) of a 0.10 M aqueous solution of acetic (ethanoic) acid. Record the pH of the sample. Did you expect this result? Explain.

The investigations you did in Section 9.1 give you a feeling for how products and reactants are distributed in an equilibrium system and how the distribution changes when the system is disturbed. However, you did not determine the actual concentration of any species present in the solutions. In Investigate This 9.12, you have now determined the pH of an acetic (ethanoic) acid solution and, from the pH, you can calculate the hydronium ion concentration, $[H_3O^+(aq)]$, in the solution. If all of the acetic acid had transferred its protons to water, the $[H_3O^+(aq)]$ would have been about 10^{-1} M and the pH would have been about 1. However, the measured pH is nearer 3, so $[H_3O^+(aq)] \approx 10^{-3}$ M. In Chapter 2, we used this same result to argue that not all the acetic acid molecules transfer their protons to water; in other words, acetic acid is a weak acid. We can write a reversible, equilibrium reaction for the reaction of acetic acid with water:

$$CH_3C(O)OH(aq) + H_2O(aq) \rightleftharpoons H_3O^+(aq) + CH_3C(O)O^-(aq) \quad \textbf{(9.4)}$$

We have written $H_2O(aq)$ as a reminder that we are interested in a solution, not pure liquid water.

━━━━━ **9.13 CONSIDER THIS** ━━━━━

How is an equilibrium system characterized quantitatively?

Many dyes change color reversibly as the pH of the solution they are in changes, and some of these dyes are used as acid–base indicators. (See Chapter 6, Section 6.13.) For example, we can write the equilibrium reaction of a dye, D, with hydronium ion as

$$D(aq) + H_3O^+(aq) \rightleftharpoons HD^+(aq) + H_2O(aq)$$

The concentrations of D(aq) and HD$^+(aq)$ can be analyzed spectroscopically. This table shows the results of a series of experiments with the same total dye concentration and varying [H$_3$O$^+(aq)$].

pH	[H$_3$O$^+(aq)$], M	[D(aq)], M	[HD$^+(aq)$], M
7.00	1.0×10^{-7}	5.2×10^{-5}	undetectable
5.00	1.0×10^{-5}	4.1×10^{-5}	1.1×10^{-5}
4.40	4.0×10^{-5}	2.5×10^{-5}	2.7×10^{-5}
4.00	1.0×10^{-4}	1.4×10^{-5}	3.8×10^{-5}
2.00	1.0×10^{-2}	undetectable	5.2×10^{-5}

(a) What is the total dye concentration in these solutions? Explain your answer.
(b) What relationships do you see among the varying [H$_3$O$^+(aq)$] and [D(aq)] and [HD$^+(aq)$]? Show how Le Chatelier's principle explains these relationships.
(c) How does the ratio of the acid and base form of the dye vary with [H$_3$O$^+(aq)$]?
(d) Try to find a mathematical relationship among [H$_3$O$^+(aq)$], [D(aq)], and [HD$^+(aq)$] that has the same numeric value for all three solutions in which these concentrations are known.

The equilibrium constant expression Before we can further analyze equilibrium systems, we need to know more about how the concentrations in equilibrium systems are related to one another. To make our discussion more general, let's write a general equation for a reversible chemical reaction:

$$a\mathbf{A} + b\mathbf{B} \rightleftharpoons c\mathbf{C} + d\mathbf{D} \tag{9.5}$$

Equation (9.5) says that a fixed number (represented by a) of molecules or ions of **A** react with b molecules or ions of **B** to form c molecules or ions of **C** and d molecules or ions of **D**. The italicized lowercase letters represent the coefficients in the chemical equation, and the bold uppercase letters represent the formulas for reactants and products. At first, we will consider only reactions in which all species are in the same phase (in a solution or as a mixture of gases).

Experiments to determine equilibrium concentrations in solution were first carried out in the middle of the 19th century. The mathematical relationship of these concentrations to one another was determined empirically in 1863 by the Norwegian chemists C. M. Guldberg and P. Waage. Guldberg and Waage found that, as long as all species are in the same phase, the relationship of the equilibrium molar concentrations is

$$K = \left(\frac{[\mathbf{C}]^c [\mathbf{D}]^d}{[\mathbf{A}]^a [\mathbf{B}]^b} \right)_{eq} \tag{9.6}$$

The ratio of equilibrium molar concentrations shown in equation (9.6) is the **equilibrium constant expression.** The subscript *eq* on the ratio reminds us that these are equilibrium concentrations. The numeric value of the equilibrium constant expression is denoted by the **equilibrium constant, *K*.**

In the equilibrium constant expression, concentrations of products appear in the numerator, with each raised to a power equal to its coefficient in the chemical equation. Concentrations of reactants, written in the same way, appear in the denominator. If the system is at equilibrium and a concentration is changed, by adding more of one reactant, for example, all the concentrations will change as the reaction proceeds to use up some of the added reactant (Le Chatelier's principle) until equilibrium is reestablished and the ratio in equation (9.6) is again equal to *K*.

> Guldberg and Waage called the mathematical relationship in equation (9.6) the *law of mass action,* and you will still occasionally see that term used.

9.14 WORKED EXAMPLE

Equilibrium constant expression and equilibrium constant

Write the equilibrium constant expression for the reaction of $Fe^{3+}(aq)$ with $SCN^-(aq)$ to form the $Fe(SCN)^{2+}(aq)$ complex, reaction equation (9.2). Assume that the number of symbols for each of the species in Figure 9.2 is directly proportional to the molarity of that species in the solution. If the solution shown after the first addition of $Fe^{3+}(aq)$ is at equilibrium, what is the numeric value of *K*, the equilibrium constant for the reaction?

Necessary information: We need reaction equation (9.2), equation (9.6), and the molarities of ions (proportional to numbers of symbols) in the solution: $[Fe^{3+}(aq)] = 18c$; $[SCN^-(aq)] = 6c$; and $[Fe(SCN)^{2+}(aq)] = 2c$. Here, c is the proportionality constant between molarity and numbers of symbols in the molecular-level drawing.

Strategy: Write the equilibrium constant expression corresponding to reaction equation (9.2) and substitute the molarities to get a numeric value for *K*.

Implementation: For the reaction equation $Fe^{3+}(aq) + SCN^-(aq) \leftrightharpoons Fe(SCN)^{2+}(aq)$, all the stoichiometric coefficients are unity, so the equilibrium constant expression and equilibrium constant are

$$K = \frac{[Fe(SCN)^{2+}(aq)]}{[Fe^{3+}(aq)][SCN^-(aq)]} = \frac{2c}{(18c)(6c)} = \frac{1}{54c}$$

Does the answer make sense?   Without other information about the reaction system, you have no way to check this result. You can obtain further information in Check This 9.15.

9.15 CHECK THIS

Equilibrium constant expression and equilibrium constant

(a) If the solution shown after the second addition of $Fe^{3+}(aq)$ in Figure 9.2 is at equilibrium, what is the numeric value of K, the equilibrium constant for the reaction of $Fe^{3+}(aq)$ with $SCN^-(aq)$ to form the $Fe(SCN)^{2+}(aq)$ complex? Make the same assumptions as in Worked Example 9.14.

(b) How does your result in part (a) compare to the result in Worked Example 9.14? Does the comparison suggest that the results make sense? Explain why or why not.

(c) The numeric value of the equilibrium constant for this reaction is about $1.3 \times 10^2 \ M^{-1}$ at room temperature. What is the value of the proportionality constant, c, in our calculations? Explain how your get your value.

9.16 CONSIDER THIS

What can equilibrium constant expressions tell us?

(a) Write the equilibrium constant expression for the reaction of acetic acid transferring protons to water, reaction (9.4).

(b) Assume that you add more acetic acid to the solution in Investigate This 9.12. Use the equilibrium constant expression you wrote in part (a) to explain whether the pH of the solution would increase, decrease, or stay the same.

(c) Explain whether your answer in part (b) is consistent with Le Chatelier's principle.

(d) In Consider This 9.13(c), you found a mathematical relationship among $[H_3O^+(aq)]$, $[D(aq)]$, and $[HD^+(aq)]$ that had a constant value. Write the equilibrium constant expression for the reaction in Consider This 9.13 and compare it to the relationship you found. What are the similarities and differences? How can you explain any differences?

Equation (9.6) is particularly useful for substances in solution, but *the ratio of molarities is a constant only in dilute solutions*. For accurate work, when concentrations are much above 0.1 M, activities rather than concentrations must be used in the equilibrium constant expression. You can think of the **activity** of a species as an "effective" concentration. In ionic solutions, for example, attractions between the negative and positive ions usually make the activity of the ions lower than their molar concentrations. Oppositely charged, neighboring ions attract one another and are not as free to interact with the other species in solution. In terms of reactivity, the ions appear to be present at lower concentration than they really are. Several solution models are available to calculate corrected ionic activities, but the calculations add a layer of complexity that is unnecessary for an understanding of equilibrium principles. In this text, we will stick mostly with dilute solutions and use equilibrium constant expressions based on concentration, but with the warning that experimental and calculated results may be different.

Standard states in solution Many calculations using equilibrium constants require using logarithms, but logarithms have meaning only when applied to dimensionless quantities, quantities without units. Immediately we have a problem with pH. We defined pH as the negative logarithm of the hydronium ion concentration:

$$\text{pH} \equiv -\log[\text{H}_3\text{O}^+(aq)] \tag{9.7}$$

But the hydronium ion concentration has units of molarity, so this equation is mathematically flawed.

Chemists avoid the dimensioned-number trap by defining a **standard state** for substances in whatever physical state they are found. This is the same thermodynamic standard state we introduced in Chapter 7, Section 7.8, and also used in Chapter 8. For substances in solution, the standard state is defined to be a 1 molal (m) concentration. In dilute aqueous solutions, however, molality and molarity are almost the same, so we will take the standard state concentration in aqueous solutions to be 1 M. Now, **pH** can be defined as the *negative logarithm of the ratio of the molar concentration of hydronium ion to the standard state:*

$$\text{pH} \equiv -\log\left(\frac{[\text{H}_3\text{O}^+(aq)]}{1\ \text{M}}\right) \tag{9.8}$$

In equation (9.8), both the numerator and denominator of the argument for the logarithm have units of molarity, so we are dealing with a dimensionless ratio that is *numerically equal* to the value of concentration expressed in moles per liter. From now on, we will use square brackets, [ ], *only* when dealing with concentrations including units. We will use *parentheses*, ( ), in equilibrium constant expressions to denote that each term is the ratio of the molar solution concentration to the standard state concentration. Using this convention, equation (9.8) is

$$\text{pH} \equiv -\log(\text{H}_3\text{O}^+(aq)) \tag{9.9}$$

In Consider This 9.16, you wrote the equilibrium constant expression for acetic acid transferring protons to water in terms of molar concentrations. Let's abbreviate acetic acid as HOAc and acetate ion as OAc$^-$ and write the equilibrium constant expression in terms of ratios of molar concentrations of solutes to their 1 M standard state concentrations:

$$K = \left(\frac{(\text{H}_3\text{O}^+(aq))\ (\text{OAc}^-(aq))}{(\text{HOAc}(aq))\ [\text{H}_2\text{O}(aq)]}\right)_{eq} \tag{9.10}$$

We have a problem in equation (9.10) because the equilibrium constant ratio still contains the concentration of water, $[\text{H}_2\text{O}(aq)]$, in the solution. [You probably met this same problem in Consider This 9.16(d).] The standard state for water is pure water, $\text{H}_2\text{O}(l)$. Pure water is 55.5 M ($\approx$1000 g·L^{-1}/ 18.0 g·mol^{-1}). To be consistent with our molarity ratios for the solutes, we can write for the solvent, $(\text{H}_2\text{O}(aq)) = [\text{H}_2\text{O}(aq)]/55.5$ M. In dilute aqueous solutions, the water concentration is still very close to 55.5 M, so $(\text{H}_2\text{O}(aq)) \approx 1$. The equilibrium constant expression then becomes

$$K_a = \left(\frac{(\text{H}_3\text{O}^+(aq))\ (\text{OAc}^-(aq))}{(\text{HOAc}(aq))\ (\text{H}_2\text{O}(aq))}\right)_{eq} = \left(\frac{(\text{H}_3\text{O}^+(aq))\ (\text{OAc}^-(aq))}{(\text{HOAc}(aq))}\right)_{eq} \tag{9.11}$$

This equilibrium constant is given the symbol K_a to remind us that it refers to an acid reacting to transfer a proton to water.

Not everyone is careful to make the distinction between concentration terms with units and terms that are ratios to the standard state. You will often see equilibrium constant expressions written using square brackets for the terms. When you see such expressions, check to see that concentration is expressed as molarity. If it is, simply strike the concentration units and deal with the pure number as the ratio of concentration to the standard state.

Other standard states The standard state for gases is defined to be a pressure of 1 bar, P°. One bar is defined as 10^5 kg·m^{-1}·s^{-2}. Thus, for a gas at a pressure P in bar, the ratio P/P° = (gas) is the numeric value to be used in equilibrium constant expressions. Pure solids or liquids are treated in a way that is analogous to water in dilute aqueous solutions. Because there is no way to vary the composition of a pure solid or liquid, the ratio to be used in equilibrium constant expressions is unity. All these standard state definitions and the resulting dimensionless concentration ratios to be used in equilibrium constant expressions are presented in Table 9.1.

Table 9.1 *Standard states and dimensionless concentration ratios.*

Physical state	Standard state	Dimensionless ratio
gas, P bar	gas at P° = 1 bar	P/P°
pure liquid	pure liquid	1
pure solid	pure solid	1
solute in aqueous solution, c M	c° = 1 M[a]	c/c°
solvent in dilute solution	pure liquid	1

Note. The external pressure is 1 bar for every standard state.
[a]Remember that the thermodynamic standard state unit for solutes in solution is *molal* concentration. Because molality and molarity are almost the same in dilute aqueous solutions and molarity is more familiar, we use molarity as the standard state unit.

9.3. Acid–Base Reactions and Equilibria

All equilibrium systems can be analyzed in essentially the same way. Equilibria in solutions of weak acids, weak bases, or both are particularly important, so we have begun our quantitative discussion of equilibria and equilibrium constant calculations with these systems. (You have already encountered several acid–base reactions in Chapters 2 and 6.) In later sections, we will consider implications and extensions of acid–base equilibria, particularly in systems involving biologically important molecules, and extend the ideas to other equilibria.

The acetic acid–water reaction To see how the concepts in Section 9.2 are applied, we'll continue our discussion of the reaction of acetic acid with water and the result from Investigate This 9.12. For all our analyses, we will start with a balanced chemical reaction equation (or equations) and the equilibrium constant expression derived from it. For acetic acid in aqueous solution, we have already written reaction equation (9.4) and equation (9.11):

$$CH_3C(O)OH(aq) + H_2O(aq) \rightleftharpoons H_3O^+(aq) + CH_3C(O)O^-(aq) \quad \textbf{(9.4)}$$

$$K_a = \left(\frac{(H_3O^+(aq))\,(OAc^-(aq))}{(HOAc(aq))\,(H_2O(aq))}\right)_{eq} = \left(\frac{(H_3O^+(aq))\,(OAc^-(aq))}{(HOAc(aq))}\right)_{eq} \quad \textbf{(9.11)}$$

If we express all solute concentrations as molarities, all their standard states are 1 M. Thus, as we said in Section 9.2, we can substitute the numerical values of the equilibrium molarities into equation (9.11) to determine the equilibrium constant for the transfer of a proton from acetic acid to water.

9.17 WORKED EXAMPLE

Acid equilibrium constant for acetic acid

The pH of a 0.050 M solution of acetic acid was measured to be 3.02. What is the equilibrium constant, K_a, for reaction (9.4)?

Necessary information: We need equation (9.11) and the conversion, $(H_3O^+(aq)) = 10^{-pH}$.

Strategy: Substitute concentrations into the equilibrium constant expression (9.11), to get K_a. The pH of the solution gives us $(H_3O^+(aq))$. The starting concentration of acetic acid is known and reaction (9.4) shows that 1 mol of acetate ions forms for each mole of hydronium ions formed. To simplify the calculations, assume 1 L of solution, so that the numeric value of the molarity, M, is equal to the number of moles of each species in the solution. Also assume starting with pure water, pH 7, to which enough acetic acid, HOAc, is added to make a 0.050 M solution.

Implementation:
In pure water at 298 K, pH 7.00: $(H_3O^+(aq)) = 10^{-7.00} = 1.0 \times 10^{-7}$.
In the final solution, pH 3.02: $(H_3O^+(aq)) = 10^{-3.02} = 9.6 \times 10^{-4}$.
　　A useful way to keep track of what is going on in chemical systems like this is to use a **change table,** a table you construct to show the initial number of moles of each species before reaction, the change in moles when the reaction occurs, and the final number of moles. Here is the change table for this system:

Species	HOAc(aq)	OAc⁻(aq)	H₃O⁺(aq)
Initial mol	0.050	0	1.0×10^{-7}
Change in mol	9.6×10^{-4} reacts	9.6×10^{-4} formed	9.6×10^{-4} formed
Final mol	$0.050 - 9.6 \times 10^{-4}$	9.6×10^{-4}	9.6×10^{-4}

　　The amount of acetic acid that reacts is only about 2% (1 part in 50). Usually, we neglect changes that are less than about 5%, unless we have data that are good enough to justify taking them into account. In this case, we neglect the amount of HOAc that reacts, so the concentration of HOAc(aq) is ≈ 0.050 M and $(HOAc(aq)) \approx 0.050 = 5.0 \times 10^{-2}$. Now we have all the data we need to substitute in equation (9.11) to get K_a:

$$K_a = \left(\frac{(H_3O^+(aq))(OAc^-(aq))}{(HOAc(aq))}\right)_{eq} = \frac{(9.6 \times 10^{-4})(9.6 \times 10^{-4})}{(5.0 \times 10^{-2})} = 1.8 \times 10^{-5}$$

continued

Does the answer make sense?   We have known since Chapter 2, Section 2.13, that acetic acid is a weak acid that does not transfer all its protons to water in aqueous solution. Here we find that only about 2% of the acetic acid molecules transfer their protons. This means that reaction (9.4) does not proceed very far toward products before the system comes to equilibrium. Since not much product is formed, we expect the equilibrium constant for the reaction to be small, much less than 1. This is the answer we got; it makes sense.

9.18 CHECK THIS

Acid equilibrium constant for acetic acid

Use your experimental value for the pH of the 0.10 acetic acid in Investigate This 9.12 to calculate a value of K_a for acetic acid. If you find it helpful, use a change table to keep track of the chemical changes that take place. How does your value compare to the one we calculated in Worked Example 9.17 from data for a more dilute solution? If the two values differ significantly (more than a factor of two), suggest a possible explanation.

Solution pH for an acid of known K_a   In Worked Example 9.17 and Check This 9.18, we used measured values of pH and knowledge of the composition of the solutions to calculate an acid equilibrium constant by substitution into an equilibrium constant expression. The problem you will meet more often is calculating the pH of a solution of a weak acid of known concentration and known acid equilibrium constant. This sort of problem might arise if you needed an acidic solution of a particular pH and wished to know what acid would give the appropriate pH. The approach is the same as above: write the balanced equilibrium reaction and corresponding equilibrium constant expression and then substitute known values and solve for the unknown pH.

9.19 WORKED EXAMPLE

pH of an aqueous solution of lactic acid

Calculate the pH of a 0.050 M solution of lactic acid, $CH_3CH(OH)C(O)OH$, which we will abbreviate as HOLac. The K_a for lactic acid is 1.4×10^{-4}.

Necessary information: We need the definition $pH = -\log(H_3O^+(aq))$, the balanced equation for the acid reaction (transfer of a proton from HOLac to H_2O), and the corresponding equilibrium constant expression:

$$HOLac(aq) + H_2O(aq) \rightleftharpoons H_3O^+(aq) + OLac^-(aq) \tag{9.12}$$

$$K_a = \left(\frac{(H_3O^+(aq))\,(OLac^-(aq))}{(HOLac(aq))} \right)_{eq} = 1.4 \times 10^{-4} \tag{9.13}$$

continued

Strategy: Let x stand for the number of moles per liter of $H_3O^+(aq)$ produced by reaction (9.12) when equilibrium is reached. Equation (9.12) shows that one $OLac^-(aq)$ ion is produced for every $H_3O^+(aq)$ formed in the solution; there are x moles per liter of $OLac^-(aq)$ in the equilibrium solution. Construct a change table, substitute into equation (9.13), and solve for x.

Implementation: The change table is

Species	HOLac(aq)	OLac⁻(aq)	H₃O⁺(aq)
Initial mol	0.050	0	1.0×10^{-7}
Change in mol	x reacts	x formed	x formed
Final mol	$0.050 - x$	x	x

$$K_a = 1.4 \times 10^{-4} = \left(\frac{(H_3O^+(aq))\,(OLac^-(aq))}{(HOLac(aq))} \right)_{eq} = \left(\frac{(x)\,(x)}{0.050 - x} \right)$$

Before trying to solve this equation for x, let's use a little chemical reasoning to simplify the arithmetic. The small equilibrium constant tells us that lactic acid is a weak acid, so the amount that reacts to form $H_3O^+(aq)$ and $OLac^-(aq)$ is small; that is, x is a small quantity. Let us *assume* that it is small enough to neglect relative to 0.050, so that we have $(HOLac(aq)) \approx 0.050$. We must *check this assumption* after we have found a value for x.

$$K_a = 1.4 \times 10^{-4} \approx \left(\frac{(x)\,(x)}{0.050} \right) = \frac{(x)^2}{0.050}$$

$$(x)^2 = 0.050(1.4 \times 10^{-4}) = 7.0 \times 10^{-6}$$

$$(x) = 2.6 \times 10^{-3} = (H_3O^+(aq)) = (OLac^-(aq))$$

$$pH = -\log(H_3O^+(aq)) = -\log(2.6 \times 10^{-3}) = 2.59$$

Check the assumption: Our value for x, 2.6×10^{-3}, is about 6% of 0.050 (3 parts in 50). This result is on the borderline of acceptability. In Worked Example 9.17, we said that about 5% is the "discrepancy" we are willing to accept between our simplifications and our calculated results (or measured values). In most cases, we settle for a result like we got here because we are only interested in a ballpark estimate of the pH.

Does the answer make sense? Comparing the acid equilibrium constants, we see that lactic acid, $K_a = 1.4 \times 10^{-4}$, is a stronger acid (larger K_a) than a cetic, $K_a = 1.8 \times 10^{-5}$ (Worked Example 9.17). We would, therefore, expect that a lactic acid solution would be somewhat more acidic (lower pH) than an acetic acid solution of the same concentration. The pH we calculated for the 0.050 M lactic acid solution, 2.59, is lower (more acidic) than that for a 0.050 M acetic acid solution, 3.02 (Worked Example 9.17). The answer makes sense.

benzoic acid

pH of an aqueous solution of benzoic acid

Benzoic acid, $K_a = 6.4 \times 10^{-5}$, is sparingly soluble in water. A saturated solution of benzoic acid is 0.028 M. What is the pH of this solution? What assumption(s) do you make to get your answer? How good is(are) your assumption(s)?

In Worked Example 9.19, we made the simplifying assumption that a negligible amount of the original weak acid reacted to transfer its protons to water. You probably made the same assumption in Check This 9.20. In both cases, the assumption leads to a result that is good to about 5 or 6%. If we wanted to be more mathematically correct, we could go back to the equilibrium constant expressions, rearrange them as quadratic equations and use the quadratic formula to solve them. For two reasons, we will not do more elaborate arithmetic for problems like these. First, the arithmetic tends to take us too far from the underlying chemistry of the systems. Second, and more fundamentally, equilibrium constants and the activities of ions in solution (which we only approximate by using molarities) are rarely known accurately enough to justify using more elaborate methods. If we really need to know the pH of a lactic acid or benzoic acid solution, we use a pH electrode and pH meter to measure it. The calculations give us a good idea what to expect, but we are not surprised when the experimental measurement is a bit different.

The autoionization of water You might have noticed that, when we constructed the change tables in Worked Examples 9.17 and 9.19, we showed all of the hydronium ion, $H_3O^+(aq)$, in the final equilibrium solution coming from the added weak acid. This means that, in the equilibrium solution, we neglected the $H_3O^+(aq)$ from the reaction of water molecules with themselves:

$$H_2O(l) + H_2O(l) \rightleftharpoons H_3O^+(aq) + OH^-(aq) \tag{9.14}$$

Reaction (9.14) is often referred to as the **autoionization** of water because it represents the reaction of one water molecule with another to produce a pair of ions. Water autoionization occurs in all aqueous solutions as well as in pure water. The equilibrium constant expression for this equilibrium reaction is

$$K_w = \left(\frac{(H_3O^+(aq))\,(OH^-(aq))}{(H_2O(l))\,(OH_2O(l))} \right)_{eq} \tag{9.15}$$

> The curly brackets in equation (9.16) have no special significance except to enclose the concentration product. They are used to avoid possible confusion with parentheses and square brackets, which do have special meaning.

$$K_w = \{(H_3O^+(aq))\,(OH^-(aq))\}_{eq} \tag{9.16}$$

K_w is the **water autoionization constant.** We know that the pH of pure water is 7.00 at 298 K, so $(H_3O^+(aq))$ in the equilibrium reaction (9.14) is $10^{-7.00}$. Since one $OH^-(aq)$ is produced for each $H_3O^+(aq)$, we also have $(OH^-(aq)) = 10^{-7.00}$. Substitution into equation (9.16) gives

$$K_w = \{10^{-7.00} \times 10^{-7.00}\} = 10^{-14.00} = 1.00 \times 10^{-14} \tag{9.17}$$

The equilibrium constant expression (9.16), with $K_w = 1.00 \times 10^{-14}$, *must* be true in all aqueous solutions at 298 K. If the solution contains other sources of $H_3O^+(aq)$ or $OH^-(aq)$ (such as added acids or bases), the concentrations of $H_3O^+(aq)$ or $OH^-(aq)$ from these sources must be accounted for in equilibrium constant expression (9.16).

9.21 WORKED EXAMPLE

Extent of water autoionization in an acetic acid solution

What is the contribution of reaction (9.14) to the concentration of $H_3O^+(aq)$ in a 0.050 M acetic acid solution with a pH of 3.02 (as in Worked Example 9.17)?

Necessary information: We need to know the conversion: $(H_3O^+(aq)) = 10^{-pH}$ and the equilibrium constant expression (9.16), with $K_w = 1.00 \times 10^{-14}$.

Strategy: The only source of $OH^-(aq)$ in this solution is water autoionization, reaction (9.14). Thus, the concentration of $H_3O^+(aq)$ *produced by water autoionization* must be equal to the concentration of $OH^-(aq)$. Substitute the measured $(H_3O^+(aq))$ in this solution into equation (9.16) to get the concentration of $OH^-(aq)$ in the solution and, hence, the concentration of hydronium ion contributed by water autoionization.

Implementation: $(H_3O^+(aq)) = 10^{-pH} = 10^{-3.02} = 9.6 \times 10^{-4}$

Substitute this result into equation (9.16) to get $(OH^-(aq))$:

$$K_w = 1.00 \times 10^{-14} = \{(H_3O^+(aq))\,(OH^-(aq))\}_{eq} = \{9.6 \times 10^{-4}\,(OH^-(aq))\}_{eq}$$

$$(OH^-(aq)) = \frac{1.00 \times 10^{-14}}{9.6 \times 10^{-4}} = 1.04 \times 10^{-11}$$

$$\therefore [OH^-(aq)] = 1.04 \times 10^{-11}\ M$$

The concentration of $H_3O^+(aq)$ produced by autoionization of water is $[H_3O^+(aq)]_{autoionization} = [OH^-(aq)] = 1.04 \times 10^{-11}\ M.$

Does the answer make sense? The concentration of $H_3O^+(aq)$ produced by autoionization of water is about eight orders of magnitude (10^8 times) smaller than the total concentration of $H_3O^+(aq)$ in the solution. Transfer of protons from acetic acid to water produces essentially all the $H_3O^+(aq)$ in the solution, as we assumed in Worked Example 9.17. Note that the amount of $H_3O^+(aq)$ produced by water autoionization in this solution is four orders of magnitude (10^4 times) smaller than in pure water. Le Chatelier's principle shows us that this makes sense. Addition of extra $H_3O^+(aq)$ from the acetic acid disturbs the equilibrium represented by equation (9.14). The system responds by "using up" a very tiny amount of the added $H_3O^+(aq)$, which also uses up most of the $OH^-(aq)$ and accounts for its low concentration in the solution.

9.22 CHECK THIS

Extent of water autoionization in an lactic acid solution

(a) What is the contribution of reaction (9.14) to the concentration of $H_3O^+(aq)$ in a 0.050 M lactic acid solution with a pH of 2.59 (as in Worked Example 9.19)?

(b) Compare your answer in part (a) with the result for the acetic acid solution in Worked Example 9.21. Does the difference between the two results make sense? Explain your response.

9.23 INVESTIGATE THIS

What is the pH of an acetate ion solution?

Do this as a class investigation and work in small groups to discuss and analyze the results. Use a pH meter and pH electrode to measure the pH (±0.02 units) of a 0.10 M aqueous solution of sodium acetate, $CH_3C(O)ONa$ (or NaOAc). Record the pH of the sample.

9.24 CONSIDER THIS

Does the pH of the acetate ion solution make sense?

Is the pH of the acetate ion solution, $OAc^-(aq)$, in Investigate This 9.23 about what you expected? Explain why or why not.

The acetate ion–water reaction As you have seen, most acetic acid molecules in aqueous solution do not transfer their protons to water. This must mean that acetate ions have an attraction for protons. When acetate ions are added to water, the water molecules will donate protons to the acetate:

$$H_2O(aq) + CH_3C(O)O^-(aq) \rightleftharpoons CH_3C(O)OH(aq) + OH^-(aq) \quad \textbf{(9.18)}$$

If reaction (9.18) occurs, aqueous solutions of acetate ion should be basic, that is, have a pH greater than 7. In Investigate This 9.23, your result for the 0.10 M sodium acetate solution (containing sodium cations and acetate anions), was probably a pH in the range 8.5 to 9.0. Reaction (9.18) proceeds to some extent, but it must not go far, because the solution is only weakly basic. Let's quantify the reaction by determining its equilibrium constant.

The equilibrium constant expression for reaction (9.18) is

$$K_b = \left(\frac{(HOAc(aq))\ (OH^-(aq))}{(OAc^-(aq))} \right)_{eq} \quad \textbf{(9.19)}$$

In equation (9.19), we have used our abbreviations for acetic acid and acetate ion and have given the equilibrium constant the symbol K_b to remind us that it refers to a base reacting to accept a proton from water.

9.25 WORKED EXAMPLE

Base equilibrium constant for acetate ion

The pH of a 0.050 M solution of sodium acetate was measured to be 8.73. What is the equilibrium constant, K_b, for reaction (9.18)?

Necessary information: We need to know the conversion: $(H_3O^+(aq)) = 10^{-pH}$, the equilibrium constant expression (9.16), with $K_w = 1.00 \times 10^{-14}$, and equilibrium constant expression (9.19).

continued

Strategy: Substitute concentrations into equilibrium constant expression (9.19) to get K_b. The pH of the solution gives $(H_3O^+(aq))$ and equation (9.16) gives $(OH^-(aq))$. We know the concentration of acetate ion we started with and reaction (9.18) shows that 1 mol of acetic acid forms for each mole of hydroxide ions formed. Assume we start with pure water, pH 7, to which enough sodium acetate, NaOAc, is added to make a 0.050 M solution.

Implementation:
In pure water at 298 K, pH 7.00: $(H_3O^+(aq)) = (OH^-(aq)) = 10^{-7.00} = 1.0 \times 10^{-7}$.
In the final solution, pH 8.73: $(H_3O^+(aq)) = 10^{-8.73} = 1.9 \times 10^{-9}$.

$$(OH^-(aq)) = \frac{K_w}{(H_3O^+(aq))} = \frac{1.00 \times 10^{-14}}{1.9 \times 10^{-9}} = 5.3 \times 10^{-6}$$

The change table for this system is

Species	OAc⁻(aq)	HOAc(aq)	OH⁻(aq)
Initial mol	0.050	0	1.0×10^{-7}
Change in mol	5.3×10^{-6} reacts	5.3×10^{-6} formed	5.3×10^{-6} formed
Final mol	$0.050 - 5.3 \times 10^{-6}$	5.3×10^{-6}	5.3×10^{-6}

The amount of acetate ion that reacts is only about 0.01%, which is negligible. The concentration of $OAc^-(aq)$ is 0.050 M, so $(OAc^-(aq)) = 0.050$.

$$K_b = \left(\frac{(HOAc(aq))\,(OH^-(aq))}{(OAc^-(aq))}\right)_{eq} = \left(\frac{(5.3 \times 10^{-6})\,(5.3 \times 10^{-6})}{0.050}\right)$$

$$= 5.6 \times 10^{-10}$$

Does the answer make sense? Since the solution is only weakly basic, reaction (9.18) must not proceed far toward products before equilibrium is established. The small value of the equilibrium constant, K_b, is consistent with this experimental result.

9.26 CHECK THIS

Base equilibrium constant for acetate ion

Use your experimental value for the pH of the 0.10 sodium acetate solution in Investigate This 9.23 to calculate a value of K_b for the acetate ion. How does your value compare to the one we calculated in Worked Example 9.25 for data from a more dilute solution? If the two values differ significantly (more than a factor of two), suggest a possible explanation.

Relationship between K_a and K_b for a conjugate acid–base pair
In any solution that contains acetic acid and acetate ion, the concentrations in the solution must satisfy *both* equations (9.11) and (9.19) simultaneously. To see the consequences of this requirement, calculate the product, $K_a \cdot K_b$:

$$K_a \cdot K_b = \left(\frac{(H_3O^+(aq))\ (OAc^-(aq))}{(HOA(aq))} \right)_{eq} \left(\frac{(HOAc(aq))\ (OH^-(aq))}{(OAc^-(aq))} \right)_{eq} \quad \textbf{(9.20)}$$

Since we are considering a single solution, the concentration ratio of acetic acid in the first quotient cancels that in the second. The same is true for the acetate ion concentration ratio:

$$K_a \cdot K_b = \left\{ (H_3O^+(aq))\ (OH^-(aq)) \right\}_{eq} = K_w \quad \textbf{(9.21)}$$

K_a and K_b for any **conjugate acid–base pair** are related to one another by equation (9.21); their product is always K_w. If you know either the conjugate acid or the conjugate base equilibrium constant, you can calculate the other.

9.27 CHECK THIS

$K_a \cdot K_b$ for the acetic acid–acetate ion pair

Use your experimental values for K_a and K_b from Check This 9.18 and 9.26 to calculate $K_a \cdot K_b$. Within the uncertainties of your values, is your result consistent with equation (9.21)?

---

pK You will often find values for equilibrium constants given as **pK** instead of K. The p in this nomenclature is the same as the p in pH. The p means to take the negative logarithm (base 10) of the value represented by the other part of the symbol. For example, pH is the negative logarithm of the hydronium ion concentration ratio: $pH = -\log(H_3O^+(aq))$. pK is the negative logarithm of the equilibrium constant: $pK = -\log K$. Using p values makes some calculations easier. Equation (9.21), for example, can be transformed as follows:

$$-\log(K_a \cdot K_b) = -\log K_w \quad \textbf{(9.22)}$$

The right-hand and left-hand sides of equation (9.22) can be expressed as pK values:

$$-\log(K_a \cdot K_b) = -(\log K_a + \log K_b) = (-\log K_a) + (-\log K_b) = pK_a + pK_b$$
$$-\log K_w = pK_w = -\log(1.0 \times 10^{-14}) = -(-14.00) = 14.00$$
$$\therefore\ pK_a + pK_b = pK_w = 14.00 \quad \textbf{(9.23)}$$

The pK_a values for several common acids are given in Table 9.2. To get the pK_b value for the conjugate base of one of the acids, subtract pK_a from 14.00.

9.28 CHECK THIS

Using the relationship of pK_a to pK_b

When ammonia is dissolved in water, the water transfers protons to the ammonia to form a basic solution:

$$NH_3(aq) + H_2O(aq) \rightleftharpoons NH_4^+(aq) + OH^-(aq) \quad \textbf{(9.24)}$$

Write the K_b expression for reaction (9.24). Use the data in Table 9.2 to get pK_b and K_b for the reaction.

Table 9.2 pK_a values for several common acids at 298 K.

Many acids transfer more than one proton to water. Their successive pK values are labeled pK_{a1}, pK_{a2}, and so on.

Compound	Reaction	pK_a
sulfuric acid, pK_{a1}	$(HO)_2SO_2 + H_2O \rightarrow HOSO_3^- + H_3O^+$ transfer is "complete"	< 0
oxalic acid, pK_{a1}	$HO(O)CC(O)OH + H_2O \rightleftharpoons HO(O)CC(O)O^- + H_3O^+$	1.25
sulfuric acid, pK_{a2}	$HOSO_3^- + H_2O \rightleftharpoons SO_4^{2-} + H_3O^+$	1.99
phosphoric acid, pK_{a1}	$(HO)_3PO + H_2O \rightleftharpoons (HO)_2PO_2^- + H_3O^+$	2.15
glycine,[a] pK_{a1}	$^+H_3NCH_2C(O)OH + H_2O \rightleftharpoons {}^+H_3NCH_2C(O)O^- + H_3O^+$	2.35
aspartic acid[b]	$-CH_2C(O)OH + H_2O \rightleftharpoons -CH_2C(O)O^- + H_3O^+$	3.90
oxalic acid, pK_{a2}	$HO(O)CC(O)O^- + H_2O \rightleftharpoons {}^-O(O)CC(O)O^- + H_3O^+$	4.27
glutamic acid[b]	$-CH_2CH_2C(O)OH + H_2O \rightleftharpoons -CH_2CH_2C(O)O^- + H_3O^+$	4.42
acetic acid	$CH_3C(O)OH + H_2O \rightleftharpoons CH_3C(O)O^- + H_3O^+$	4.76
histidine[b]		6.00
carbonic acid, pK_{a1}	$(HO)_2CO + H_2O \rightleftharpoons HOCO_2^- + H_3O^+$	6.36
phosphoric acid, pK_{a2}	$(HO)_2PO_2^- + H_2O \rightleftharpoons (HO)PO_3^{2-} + H_3O^+$	7.20
"tris"	$(HOCH_2)_3CNH_3^+ + H_2O \rightleftharpoons (HOCH_2)_3CNH_2 + H_3O^+$	8.08
cysteine[b]	$-CH_2SH + H_2O \rightleftharpoons -CH_2S^- + H_3O^+$	8.36
ammonium ion	$NH_4^+ + H_2O \rightleftharpoons NH_3 + H_3O^+$	9.24
boric acid	$B(OH)_3 + 2H_2O \rightleftharpoons B(OH)_4^- + H_3O^+$	9.24
glycine,[a] pK_{a2}	$^+H_3NCH_2C(O)O^- + H_2O \rightleftharpoons H_2NCH_2C(O)O^- + H_3O^+$	9.78
phenol		9.98
carbonic acid, pK_{a2}	$HOCO_2^- + H_2O \rightleftharpoons CO_3^{2-} + H_3O^+$	10.33
tyrosine[b]		10.47
lysine[b]	$-(CH_2)_4NH_3^+ + H_2O \rightleftharpoons -(CH_2)_4NH_2 + H_3O^+$	10.69
phosphoric acid, pK_{a3}	$(HO)PO_3^{2-} + H_2O \rightleftharpoons PO_4^{3-} + H_3O^+$	12.15
arginine[b]		12.48
water	$H_2O + H_2O \rightleftharpoons H_3O^+ + OH^-$	14.00

[a]Typical pK_a values for the carboxylic acid and ammonium groups on the same carbon in amino acids are 1.5 to 2.5 and 9.5 to 10.5, respectively. These groups are used to make the amide bonds between amino acids in proteins. Only the amino acid acidic and basic side groups are left to give proteins their acid–base properties.

[b]Only the side groups are represented for these amino acids. The rest of the molecule is like glycine.

9.29 CONSIDER THIS

What is the relationship of pH and pOH?

(a) From equation (9.16), derive an equation that gives the sum of pH and **pOH** for any aqueous solution.

(b) What is the pOH of a solution with pH = 8.73? Use your pOH to determine $(OH^-(aq))$. Is your $(OH^-(aq))$ the same as that in Worked Example 9.25? Explain why or why not.

Reflection and Projection

About 140 years ago, Guldberg and Waage showed that the results for several well-studied solution equilibria could be described in terms of equilibrium constants and equilibrium constant expressions that are ratios of the equilibrium concentrations of the products to the reactants. The form of the equilibrium constant expression for a reaction is related to its balanced chemical equation and the stoichiometric coefficients in the equation. Many further experimental studies have confirmed Guldberg and Waage's empirical observations and have extended equilibrium constant expressions to reactions involving gases and pure solids and liquids, as well as solutions. In order to make all the equilibrium constants consistent with one another, concentrations in equilibrium constant expressions are replaced by ratios of concentrations to a set of standard states, Table 9.1, for gases, pure solids and liquids, and solutes in solution.

Equilibrium constants can be calculated from measurements of concentrations in reactions at equilibrium in solution. Our first applications have been to aqueous acid–base reactions. We derived equilibrium constant expressions for K_a, K_b, and K_w and found that $K_a \cdot K_b = K_w$ (or $pK_a + pK_b = pK_w$) for conjugate acid–base pairs. Now we need to consider how to figure out what will happen in solutions that contain substantial concentrations of more than one acid and/or base. We will mainly focus on solutions of conjugate acid–base pairs (buffer solutions) and their effect on other species in solution, including biological molecules.

9.4. Solutions of Conjugate Acid–Base Pairs: Buffer Solutions

9.30 INVESTIGATE THIS

What are the pHs of conjugate acid-base pair solutions?

Do this as a class investigation and work in small groups to discuss and analyze the results. Make up the three samples in this table in three clean, dry, labeled sample vials or test tubes large enough to accommodate a pH electrode to measure the pH.

Sample	#1	#2	#3
0.10 M HOAc	9.0 mL	5.0 mL	1.0 mL
0.10 M NaOAc	1.0 mL	5.0 mL	9.0 mL

Use a pH meter and pH electrode to measure the pH of each solution to ±0.02 pH unit and record the results.

9.31 CONSIDER THIS

Do the pHs of conjugate acid–base pair solutions make sense?

As you do these analyses, consider your results from Investigate This 9.12 and 9.23, as well as Investigate This 9.30. Describe any correlations you find between the pH of the solutions and their composition. Do your correlations make sense in terms of the acid–base properties of acetic acid and acetate ion? Explain why or why not.

To understand the pH of solutions that contain several different acids and bases—biological solutions, for example—they must all be accounted for. In many cases, such solutions contain just a few predominant acid–base conjugate pairs that maintain the system at the appropriate pH. When we study the reactions of molecules, especially biomolecules, in the laboratory, we almost always use such conjugate acid–base pairs to control the pH of the solutions studied.

We can use the acid–base equilibrium systems introduced in Section 9.3 to prepare solutions of known and constant pH. These **buffer solutions** are used to control pH at a relatively constant value. Buffer solutions contain substantial concentrations of *both* the conjugate acid and conjugate base of a conjugate acid–base pair. If hydronium ion, $H_3O^+(aq)$, is added to a buffer solution, it will react with the conjugate base and be used up. The pH of the solution will not change as much as it would have without the buffer. A similar argument holds for the addition of hydroxide ion, $OH^-(aq)$, and its reaction with the conjugate acid of the buffer. The solutions whose pHs you measured in Investigate This 9.30 are buffer solutions, they contain substantial concentrations of both acetic acid and acetate ion, a conjugate acid–base pair. Before we examine quantitatively how a buffer solution works, let's analyze your results from the investigation.

We know the composition of the solutions and the pH of each in Investigate This 9.30. We can use these data to get K_a for acetic acid. You used the data from Investigate This 9.12 to determine this K_a, but analysis of the solutions in Investigate This 9.30 can give a more accurate value, because tiny amounts of impurities do not affect the pH as much.

9.32 WORKED EXAMPLE

K_a for acetic acid

The pH of a solution prepared by mixing 10.0 mL each of 0.050 M aqueous solutions of acetic acid and sodium acetate (similar to Sample #2 in Investigate This 9.30) is 4.77. Find K_a and pK_a for reaction (9.4), transfer of a proton from acetic acid to water.

Necessary information: We need the $(H_3O^+(aq)) = 10^{-pH}$ conversion. We also need to know that the number of moles of acetic acid (or acetate ion) in the stock solution that was added, $M_{stock} \cdot V_{stock}$, equals the number of moles of acetic acid in final mixture, $M_{mixt} \cdot V_{mixt}$:

$$M_{stock} \cdot V_{stock} = M_{mixt} \cdot V_{mixt} \qquad (9.25)$$

continued

Strategy: As in previous examples, substitute concentrations into the equilibrium constant expression (9.11) to get K_a. The pH of the solution gives $(H_3O^+(aq))$. Calculate the molarities of acetic acid and acetate ion in the mixture and construct a mole change table based on one liter of solution to get the values needed to substitute in equation (9.11). Since the solution is somewhat acidic, protons must have been transferred from HOAc to water.

Implementation: $(H_3O^+(aq)) = 10^{-4.77} = 1.7 \times 10^{-5}$

$$\text{acetic acid: } M_{mixt} = \frac{M_{stock} \cdot V_{stock}}{V_{mixt}} = \frac{0.050 \text{ M} \cdot 0.010 \text{ L}}{0.020 \text{ L}} = 0.025 \text{ M}$$

$$\text{acetate ion: } M_{mixt} = \frac{M_{stock} \cdot V_{stock}}{V_{mixt}} = \frac{0.050 \text{ M} \cdot 0.010 \text{ L}}{0.020 \text{ L}} = 0.025 \text{ M}$$

Species	HOAc(aq)	OAc⁻(aq)	H₃O⁺(aq)
Initial mol	0.025	0.025	1.0×10^{-7}
Change in mol	1.7×10^{-5} reacts	1.7×10^{-5} formed	1.7×10^{-5} formed
Final mol	$0.025 - 1.7 \times 10^{-5}$	$0.025 + 1.7 \times 10^{-5}$	1.7×10^{-5}

The amounts of acetic acid that react and acetate ion that form are negligible compared to the amounts present initially, so we can neglect them when we calculate K_a:

$$K_a = \left(\frac{(H_3O^+(aq)) \, (OAc^-(aq))}{(HOAc(aq))} \right)_{eq} = \frac{(1.7 \times 10^{-5}) \, (0.025)}{0.025} = 1.7 \times 10^{-5}$$

$$pK_a = -\log(1.7 \times 10^{-5}) = 4.77$$

Does the answer make sense? This value for K_a agrees well with the value we calculated in Worked Example 9.17, and the pK_a agrees well with the value in Table 9.2.

9.33 CHECK THIS

K_a for acetic acid

(a) Use your experimental data for each of the three samples in Investigate This 9.30 to find three K_a and pK_a values for reaction (9.4), transfer of a proton from acetic acid to water. How do the three values compare? Is this the result you expected? Why or why not?

(b) How do these values compare with the one you got in Check This 9.18? Which do you think is the more reliable? Present your argument clearly.

We can analyze the data from Worked Examples 9.17 and 9.32 in terms of Le Chatelier's principle. With only acetic acid added to make the solution in Worked Example 9.17, the concentration of $H_3O^+(aq)$ formed by transfer of protons from acetic acid to water, reaction (9.4), is about 10^{-3} M. In Worked Example 9.32, acetate ion is added to the solution. In response to this increase in concentration of one of the products of reaction (9.4), some of the added acetate ion reacts with $H_3O^+(aq)$ and lowers its concentration. We find that the concentration of $H_3O^+(aq)$ in this example is about 10^{-5} M, a 100-fold reduction from the solution without added acetate ion. Le Chatelier's principle explains the direction of the change in pH when acetate ion is added to an acetic acid solution.

Conjugate base-to-acid ratio The calculations we made above in Worked Example 9.32 and Check This 9.33 are not difficult, but there is another way to treat these systems that can make the calculations even easier. We will do this for a generic acid, HA, and its conjugate base, A^-:

$$HA(aq) + H_2O(aq) \rightleftharpoons H_3O^+(aq) + A^-(aq) \tag{9.26}$$

The equilibrium constant expression for reaction (9.26) is

$$K_a = \left(\frac{(H_3O^+(aq))\,(A^-(aq))}{(HA(aq))}\right)_{eq} \tag{9.27}$$

Take the negative logarithm of both sides of equation (9.27) to get

$$-\log K_a = pK_a = -\log(H_3O^+(aq))_{eq} - \log\left(\frac{(A^-(aq))}{(HA(aq))}\right)_{eq} \tag{9.28}$$

$$pK_a = pH - \log\left(\frac{(A^-(aq))}{(HA(aq))}\right)_{eq} \tag{9.29}$$

Equation (9.29) expresses the relationship among pK_a, pH, and the *ratio* of the conjugate base to conjugate acid concentrations in the solution. For some purposes, you need only the ratio of concentrations, not the actual concentrations, to characterize a solution. In Worked Example 9.32, we made the (good) approximation that changes in the concentrations of the acid and base were negligible compared to the concentrations initially present. This approximation is valid, if K_a is relatively small (not much reaction occurs) and the conjugate acid and base concentrations are relatively large. As a rule of thumb, if $100K_a$ is smaller than *both* the acid and the base concentration, you can use the approximation. In solutions that meet this condition, you can use the base-to-acid ratio rather than actual concentrations, in equation (9.29).

9.34 CHECK THIS

Conjugate base-to-acid ratios and pK_a for acetic acid

(a) For each mixture in Investigate This 9.30, the base-to-acid ratio is the ratio of the volume of base to volume of acid used to make the mixture. Show why this is so. Use these volume ratios in equation (9.29) to get a value of pK_a for acetic acid from each mixture.

(b) How do your results in part (a) compare with those you got in Check This 9.33? Is this what you expect? Explain why or why not.

9.35 INVESTIGATE THIS

How does a buffer respond to added $H_3O^+(aq)$ or $OH^-(aq)$?

Do this as a class investigation and work in small groups to discuss and analyze the results. Use four clean, dry, labeled sample vials or test tubes large enough to accommodate a pH electrode. Put 10 mL of water in two of the vials. In each of the other two make a mixture of 5.0 mL 0.10 M aqueous acetic acid solution and 5.0 mL of 0.10 M aqueous sodium acetate solution.

(a) Measure and record the pH of one of the water samples. Add *1 drop* of 1.0 M aqueous hydrochloric acid solution, mix the solution, and again measure and record the pH. Repeat this procedure with one of the acetic acid–acetate mixtures in place of the water.

(b) Repeat the procedure with the other water sample and the other acetic acid–acetate solution, but this time use *1 drop* of 1.0 M aqueous sodium hydroxide solution instead of hydrochloric acid.

9.36 CONSIDER THIS

How does buffer pH respond to added $H_3O^+(aq)$ or $OH^-(aq)$?

(a) In Investigate This 9.35(a), how large is the *change* in pH when a drop of hydrochloric acid is added to 10 mL of water? To 10 mL of an acetic acid–acetate ion mixture? Do these results make sense? Explain why or why not.

(b) In Investigate This 9.35(b), how large is the *change* in pH when a drop of sodium hydroxide is added to 10 mL of water? To 10 mL of an acetic acid–acetate ion mixture? Do these results make sense? Explain why or why not.

How acid–base buffers work A buffer reduces the effect of some disturbance to a system. Your results from Investigate This 9.35 show why solutions of a weak acid and its conjugate base are called buffers. Addition of a drop of 1.0 M hydrochloric acid changes the pH of water from about 6 (the usual pH of water containing a little dissolved carbon dioxide from the air) to near 2, a pH change of about 4 units. The same amount of hydrochloric acid added to the acetic acid–acetate solution causes a pH drop of about 0.1 unit. Why is the effect on the pH of the buffer solution so much smaller? The molecular-level drawings in Figure 9.4 can help you interpret these effects.

If one drop (about 0.05 mL) of 1.0 M aqueous hydrochloric acid is added to 10 mL of water, the acid is diluted by a factor of about 200. In this solution, $[H_3O^+(aq)] \approx 0.005$ M and the pH will be about 2.3, as you observed in Investigate This 9.35(a) and as is represented in Figure 9.4(a). When the same amount of hydrochloric acid is added to the acetic acid–acetate solution, there is a base, acetate ion, to react with the hydronium ions from the hydrochloric acid. The pH doesn't change much, which means that not much hydronium ion is left unreacted.

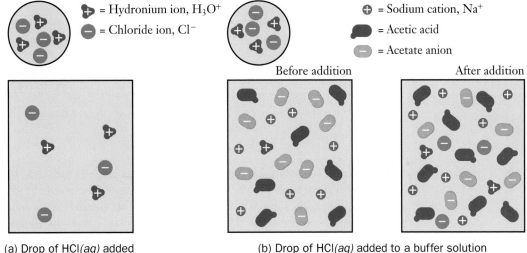

(a) Drop of HCl(*aq*) added
to pure water

(b) Drop of HCl(*aq*) added to a buffer solution
of acetic acid and sodium acetate

Figure 9.4.

Molecular-level representation of acid–base buffer action.

Acetate must be the predominant proton acceptor; acetate is a stronger base than water, as you can recall by reviewing Table 6.2 in Chapter 6, Section 6.4. The presence of a base stronger than water in the buffer solution makes the effect of adding $H_3O^+(aq)$ much smaller; there is only a small decrease in pH. Figure 9.4(b) illustrates that most but not all of the added hydronium ion reacts with acetate anion to form acetic acid. At equilibrium, there is a little more hydronium ion in the solution than there was before the hydrochloric acid was added.

9.37 CHECK THIS

Stoichiometry in buffer solutions

(a) Before the hydrochloric acid is added in Figure 9.4(b), what are the positive and negative species in the solution? Is the solution electrically neutral? Explain.

(b) After the hydrochloric acid is added in Figure 9.4(b), what are the positive and negative species in the solution? Is the solution electrically neutral? Explain.

To analyze buffer action further, it is useful to rearrange equation (9.29) to a form that gives the pH of a solution as a function of known values for pK_a and the conjugate base-to-acid concentration ratio:

$$pH = pK_a + \log\left(\frac{(A^-(aq))}{(HA(aq))}\right)_{eq} = pK_a + \log\left(\frac{(\text{conjugate base})}{(\text{conjugate acid})}\right)_{eq} \quad (9.30)$$

We can use equation (9.30), often called the **Henderson–Hasselbalch equation,** to explain the buffering action when hydronium or hydroxide ions are added to a buffer solution. It can also help us figure out how to make buffer solutions with a desired pH.

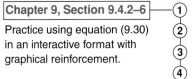

Web Companion

Chapter 9, Section 9.4.2–6

Practice using equation (9.30) in an interactive format with graphical reinforcement.

① ② ③ ④

Buffer pH Before doing quantitative calculations, let's see what equation (9.30) tells us qualitatively. Consider a buffer solution in which the conjugate acid and conjugate base concentrations are the same, $[HA(aq)] = [A^-(aq)]$. In this solution,

$$\left(\frac{(A^-(aq))}{(HA(aq))}\right)_{eq} = 1, \text{ so}$$

$$pH = pK_a + \log\left(\frac{(A^-(aq))}{(HA(aq))}\right)_{eq} = pK_a + \log(1) = pK_a \qquad (9.31)$$

If a solution contains equal concentrations of a conjugate acid–base pair, the pH of the solution is equal to the pK_a of the acid. If the conjugate base-to-acid ratio in a solution is not unity, Table 9.3 shows the *direction* that the pH of the solution varies from pK_a. Note from the table that when the relative amount of conjugate base increases, the solution becomes more basic ($[H_3O^+(aq)]$ decreases); when the relative amount of conjugate acid increases, the solution becomes more acidic ($[H_3O^+(aq)]$ increases). This correlation should help you remember the direction of change of the hydronium ion concentration, $[H_3O^+(aq)]$, and the pH when solution conditions are changed.

Table 9.3 *Correlation of buffer pH with pK_a and conjugate base-to-acid ratio.*

$\left(\dfrac{\text{(conjugate base)}}{\text{(conjugate acid)}}\right)_{eq}$	pH relative to pK_a	$[H_3O^+(aq)]$ relative to reference solution
= 1; equal concentrations	pH = pK_a	reference solution
> 1; base predominates	pH > pK_a	$[H_3O^+(aq)]$ decreases; more basic
< 1; acid predominates	pH < pK_a	$[H_3O^+(aq)]$ increases; more acidic

9.38 CHECK THIS

pH and the conjugate acid–base ratio

(a) Show how to get the results in Table 9.3 for $[HA(aq)] > [A^-(aq)]$ and $[HA(aq)] < [A^-(aq)]$. Give your reasoning clearly. *Note:* The logarithm of a number less than one is negative.

(b) 🔵 Practice the concepts in Table 9.3 by working through the *Web Companion*, Chapter 9, Section 9.4.2–6. After tracing the curve on page 6, write an explanation of how the curve relates to the graphic representation of the (conjugate base)/(conjugate acid) ratio shown below the graph. Explain how the curve and graphic representation relate to your results in Investigate This 9.30 and 9.35.

9.39 WORKED EXAMPLE

pH change when hydronium ion is added to a buffer solution

An aqueous buffer solution that is 0.050 M in both acetic acid and acetate ion has a pH of 4.76 (= pK_a for acetic acid). If 0.05 mL of 1.0 M aqueous HCl solution is added to 10 mL of this buffer solution, what is the pH of the resulting solution?

Necessary information: We will need to use equation (9.30).

Strategy: Assume that all the added hydronium ion reacts with acetate ion to form acetic acid. Each mole of added hydronium reacts with 1 mol of acetate ion to form 1 mol of acetic acid. Using this stoichiometry, construct a change table to get the final concentrations of acetic acid and acetate ion and substitute these in equation (9.30) to get the final pH of the solution.

Implementation:
Initial mol HOAc(aq) = (0.050 M)(0.010 L) = 5.0×10^{-4} mol.
Initial mol OAc$^-(aq)$ = (0.050 M)(0.010 L) = 5.0×10^{-4} mol.
Mol $H_3O^+(aq)$ added = (1.0 M)(0.00005 L) = 5.0×10^{-5} mol.

Species	HOAc(aq)	OAc$^-(aq)$	$H_3O^+(aq)$
Initial mol	5.0×10^{-4}	5.0×10^{-4}	(pH = 4.76)
Change in mol	5.0×10^{-5} formed	5.0×10^{-5} reacts	?
Final mol	5.5×10^{-4}	4.5×10^{-4}	?

$$pH = pK_a + \log\left(\frac{(\text{conjugate base})}{(\text{conjugate acid})}\right)_{eq}$$

$$= 4.76 + \log\left(\frac{4.5 \times 10^{-4} \text{ mol}/0.010\text{L}}{5.5 \times 10^{-4} \text{ mol}/0.010\text{L}}\right)$$

$$pH = 4.76 - 0.09 = 4.67$$

Does the answer make sense? The reaction of hydronium ion with acetate ion formed some acetic acid and used up some acetate ion, so the conjugate acid and base concentrations are no longer equal; there is more acetic acid than acetate ion. Table 9.3 shows us that the pH should decrease and the calculation bears this out. However, even though 10% of the acetate has reacted, the pH changes by less than 0.1 unit. The buffer system resists change in the pH of the solution. How does this result compare with your observation in Investigate This 9.35?

9.40 CHECK THIS

pH change when hydroxide ion is added to a buffer solution

An aqueous buffer solution that is 0.05 M in both acetic acid and acetate ion has a pH of 4.76. If 0.05 mL of 1.0 M aqueous sodium hydroxide solution is added to 10 mL of this buffer solution, what is the pH of the resulting solution? Show

continued

your work clearly. How does your result compare with your observation in Investigate This 9.35?

Preparing a buffer solution To make a buffer with a desired pH, we choose a conjugate acid–base pair that has about the same pK_a as the desired pH. For this conjugate acid–base pair a base-to-acid ratio of unity will produce a solution close to the desired pH. Table 9.3 shows the *direction* you need to change the conjugate base-to-acid ratio to get the desired pH and equation (9.30) gives you a way to calculate the ratio that is required. For example, if you need a pH 7.85 buffer solution, Table 9.2 shows that "tris," with a $pK_a = 8.08$, would be a good choice for the buffer. Since the desired pH is more acidic than the pK_a, the conjugate base-to-acid ratio has to be less than unity; that is, you need more conjugate acid than base. We find the required conjugate base-to-acid ratio by rearranging equation (9.30) and solving for the ratio:

"Tris" is an abbreviation for the compound tris(hydroxy-methyl)aminomethane, a commonly used buffer in biochemical studies.

$$\log\left(\frac{(\text{conjugate base})}{(\text{conjugate acid})}\right)_{eq} = \text{pH} - pK_a \tag{9.32}$$

$$\log\left(\frac{(\text{conjugate base})}{(\text{conjugate acid})}\right)_{eq} = 7.85 - 8.08 = -0.23$$

$$\left(\frac{(\text{conjugate base})}{(\text{conjugate acid})}\right)_{eq} = 10^{-0.23} = 0.59$$

9.41 CONSIDER THIS

How do you prepare a buffer solution of known concentration?

(a) Suppose you want to prepare a pH 7.85 "tris" buffer solution that contains 0.050 M *total* concentration of conjugate acid and conjugate base. What are the individual concentrations of the acid and base that will give this pH and a total concentration of 0.050 M? Clearly show how you get your answer.

(b) The acid form of "tris" is an ammonium-like ion that is available as its chloride salt; the molar mass of the salt is 157.6 g. The base, molar mass = 121.1 g, is also available as a pure solid. What masses of the acid salt and the base must be dissolved in a liter of solution to prepare the buffer in part (a)? Show all your work clearly.

Buffer solutions are probably the most widely used application of acid–base chemistry. All studies of biological systems are carried out in buffered solutions (or in living organisms whose internal solutions are buffered by several conjugate acid–base pairs). Many chemical reactions are studied in buffered solutions so that variations in the concentration of hydronium ion (and/or hydroxide ion) don't confuse interpretation. We use calculations based on equation (9.30) to find the pH of buffer solutions of a given composition and to determine the composition required to prepare a buffer solution of a desired pH. We can also use this equation to interpret the acid–base behavior of biomolecules, as we will see in the next section.

9.5. Acid–Base Properties of Proteins

9.42 INVESTIGATE THIS

How is a protein solution affected by addition of $H_3O^+(aq)$?

Do this as a class investigation and work in small groups to discuss and analyze the results. Place 250 mL of a basic 0.25% aqueous solution of casein (a protein found in milk) in a 400-mL beaker. Stir the solution at medium speed with a magnetic stirrer and a magnetic stir bar. Use a pH electrode and pH meter to monitor and read the pH of the solution. Add 3 M hydrochloric acid solution several drops at a time to the stirred casein solution. After each addition, record your observations on the appearance of the solution and its pH. After the pH of the solution has reached about 2, reverse the pH change by adding 3 M sodium (or potassium) hydroxide solution a few drops at a time to the stirred casein solution. After each addition, record your observations on the appearance of the solution and its pH. Repeat to see if your observations are reproducible.

9.43 CONSIDER THIS

How are protein solubility and pH related?

(a) Did any of your observations in Investigate This 9.42 surprise you? Why or why not?

(b) How was the solubility of the protein (casein) related to the pH of the solution? At what pH was the most precipitate formed? Look at the data for amino acids in Table 9.2. Can you think of a way to relate your observations in Investigate This 9.42 to the properties of the amino acids that are combined to make the protein?

Recall from Chapter 2 that almost all biological molecules can act as proton donors and/or proton acceptors. We used this property to explain the solubility of these high molecular mass *ionic* molecules in aqueous solutions. The results from Investigate This 9.42 demonstrate that the solubility of casein is pH dependent. Proteins all react somewhat differently to pH changes, but most show behavior similar to that of casein; their solubilities vary with pH. We have enough information from our discussion of acid–base equilibria to be able to understand this variation.

Acid–base side groups on proteins We have seen previously (Chapter 1, Section 1.9 and Chapter 6, Section 6.7) that **proteins** are polymers of amino acids. The amine group and carboxylic acid group bonded to the same carbon on one amino acid form amide bonds with two neighboring amino acids to build the polymer chain. When this happens, these groups are no longer available to accept or donate protons. Acidic or basic **amino acid side groups** remain free to accept or donate protons. These groups give proteins their acid–base properties. The seven amino acid side groups with acid–base properties are noted with a superscript b in Table 9.2.

> Some of the acid–base side groups on proteins are buried inside the folded structure and do not contribute to interactions with the surrounding solution.

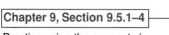

Most real proteins have many acidic and basic side groups. Accounting simultaneously for the acid–base properties of all these side groups is complicated. To illustrate how the pH of a protein solution affects the charge on the protein, we use the greatly simplified protein model in Figure 9.5. The two side groups shown are a carboxylic acid and an amine, which have quite different pK_a values. We will use equation (9.32) to analyze the base-to-acid ratio for each side group as a function of pH and the applicable pK_a:

$$\log\left(\frac{[-COO^-]}{[-COOH]}\right)_{eq} = pH - pK_{a1} \tag{9.33}$$

$$\log\left(\frac{[-NH_2]}{[-NH_3^+]}\right)_{eq} = pH - pK_{a2} \tag{9.34}$$

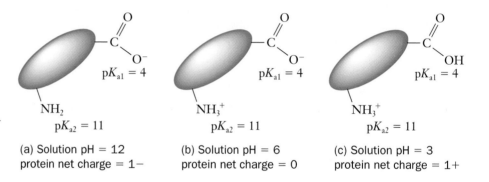

Figure 9.5.

Simplified model protein (the ellipsoid) with only two acid–base side groups.

(a) Solution pH = 12
protein net charge = 1−

(b) Solution pH = 6
protein net charge = 0

(c) Solution pH = 3
protein net charge = 1+

For the protein in a solution with a pH of 12, Figure 9.5(a), the right-hand side of both equations, (9.33) and (9.34), is positive because pH > pK_a. When the logarithm of the base-to-acid ratio is positive, the ratio is greater than unity, and the base form predominates. Thus, the protein is shown in Figure 9.5(a) with both groups in their base form. The amine is uncharged and the carboxylate ion has a negative charge, so the **net charge** (sum of the charges of the side groups) on the protein is 1−.

9.44 CONSIDER THIS

How do side group charges on our model protein vary with pH?

(a) Why are the side groups on the model protein shown as they are in Figures 9.5(b) and 9.5(c) at solution pH 6 and 3, respectively? Clearly explain your answers.

(b) Explain how we obtain the net charges shown for the protein in Figures 9.5(b) and 9.5(c).

Isoelectric pH for proteins In solutions with a high pH (basic solutions), proteins usually carry a net negative charge. Conversely, in solutions with a low pH (acidic solutions), proteins usually carry a net positive charge. At some intermediate pH, the net charge on a protein must be zero, as depicted for the model protein in Figure 9.5(b). The pH of the solution in which a protein has zero net charge is called the **isoelectric pH** (*iso* = same). At the isoelectric pH (sometimes called the **isoelectric point**), the net positive and net negative electric charges on the protein are the same, so the overall net charge is zero.

To see how these charges affect the properties of proteins, consider the solution of casein in Investigate This 9.42. At high pH, all the casein molecules are negatively charged. At low pH, they are all positively charged. In either case, the molecules repel one another and have no tendency to clump together and form a precipitate. At an intermediate pH, the isoelectric pH for casein, the molecules have no net charge and do not repel one another. The molecules do, however, have regions of positive and negative charge and the negative region on one molecule can attract the positive region on another. These attractions can bring together clumps of protein molecules and precipitate the protein from solution.

9.45 CONSIDER THIS

What is the isoelectric pH for casein?

(a) Are your observations from Investigate This 9.42 consistent with the picture of protein behavior just presented? Why or why not?
(b) What would you estimate the isoelectric pH of casein to be? Explain how you make your estimate.

9.46 CHECK THIS

Relating a Web Companion *animation to the casein investigation*

(a) Follow the directions in the *Web Companion*, Chapter 9, Section 9.4.3, and observe the curve on the graph and the animation below the graph. Explain the correlation between the graphical plot and the animation.
(b) Explain how the graphical plot and the animation are related to your results in Investigate This 9.42. *Hint:* Work through the previous pages of this section of the *Web Companion*.
(c) What, if any, correlations do you see between the graphics on this page of the *Web Companion* and buffer solutions? Could these graphics be used to illustrate buffering action? Explain the reasoning for your responses.

Electrophoresis A widely used technique for separating and analyzing biological molecules, **electrophoresis** utilizes the acid–base principles we have been discussing. One type of electrophoresis apparatus is illustrated in Figure 9.6. Separation of molecules occurs within a thin slab of gel that is in contact with buffer solutions at both ends. Charged electrodes in the buffer solutions set up an electrical potential from one end of the gel to the other. Electrically charged sample molecules migrate through the gel under the influence of the electric field.

Electrophoresis (pronounced ee-lek′-troe-fah-ree′-sis) ıs a combination of electric + *phorein* = to carry; carrying electricity or electrical charge.

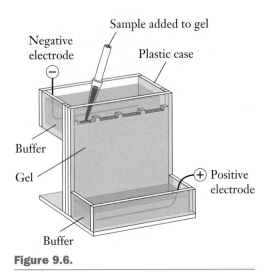

Negative electrode

Sample added to gel

Plastic case

Buffer

Gel

(+) Positive electrode

Buffer

Figure 9.6.

Diagram of a gel electrophoresis apparatus.

The apparatus shown in Figure 9.6 is designed to separate negatively charged molecules as they migrate from the negative (top) electrode toward the positive (bottom) electrode. Molecules with a higher negative charge migrate faster than those with a lower charge and move farther down the gel in a given time. Thus, molecules with different charges are separated from one another. We have seen above that the charge on a protein depends on the pH of the solution in which it is dissolved. The buffer pH for electrophoresis separations of proteins is usually chosen to be high enough so that the proteins will have a negative charge.

Acids, bases, and sickle-cell hemoglobin The chapter opening illustration shows what happens to the red blood cells (erythrocytes) of persons with sickle-cell disease when the hemoglobin in the cells is deoxygenated. Figure 9.7 shows the results of an electrophoretic analysis of sickle-cell hemoglobin, HbS, and normal adult hemoglobin, HbA. At the pH 8.6 chosen for the analysis, both proteins have a net negative charge. However, HbA moves farther toward the positive end of the gel; it migrates faster. HbA must, therefore, have a higher net negative charge than HbS.

9.47 CONSIDER THIS

What is the difference between HbA and HbS?

The negatively charged side groups on hemoglobin at pH 8.6 are carboxylate anions. Does the electrophoresis result in Figure 9.7 suggest that HbS has more or fewer carboxylate side groups than HbA? Explain how you arrive at your conclusion.

Figure 9.7.

Results of electrophoresis of normal (HbA) and sickle-cell (HbS) hemoglobins at pH 8.6.

The hemoglobin molecule, Figure 9.8, consists of four separate protein chains (two α chains and two β chains) that are stacked together in a tetrahedral arrangement. After the difference in charge between HbA and HbS was discovered, further analysis of the separate chains showed that they were identical except for one amino acid (out of 146) on each of the β chains. The sixth amino acid in the β chains of HbA is **glutamic acid,** but HbS has **valine** at this position. Glutamic acid, Table 9.2, has a carboxylic acid side group, but valine has a nonpolar, $(CH_3)_2CH-$, side group. The loss of the acid side group explains the electrophoresis results, which you analyzed in Consider This 9.47.

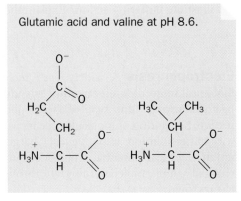

Glutamic acid and valine at pH 8.6.

In the folded protein chains, these side groups are on the outside of the molecule, as shown in Figure 9.8. HbS has a nonpolar, hydrophobic site where HbA has a polar, hydrophilic site.

We saw in Chapter 8, Section 8.11, that the entropy of aqueous solutions increases when nonpolar parts of molecules come together and release the water molecules that are "frozen" around them. The extra nonpolar site on one HbS molecule comes together with a nonpolar site on another and the process continues to form long insoluble chains of HbS molecules that are responsible for the shape and the rigidity of sickled red blood cells.

Reflection and Projection

Solutions that contain substantial concentrations of both members of a conjugate acid–base pair are buffer solutions that resist changes in pH when disturbed by the addition of hydronium ion or hydroxide ion. Buffers work because systems at equilibrium respond to a disturbance in a way that minimizes the effect on the equilibrium, Le Chatelier's principle. Addition of hydronium ion to a buffer solution, for example, disturbs the equilibrium by increasing the concentration of hydronium, so the conjugate acid–base-hydronium ion system is no longer in equilibrium. Some of the base reacts with some of the added hydronium to reduce the hydronium concentration, increase the conjugate acid concentration, and decrease the conjugate base concentration. The net effect of these changes is to restore the conjugate acid–base–hydronium ion equilibrium. The new equilibrium amount of hydronium ion is lower than it would have been for the same initial amount of hydronium ion added to pure water.

We applied these conjugate acid–base principles to the acid–base properties of proteins. The ratio of a protein's conjugate base-to-acid forms is controlled by the pH of the solution in which it is dissolved. We study a protein's acid–base properties by observing the effects of changing the solution pH. Among other properties, we can determine the isoelectric pH (net zero charge and minimum solubility) and can learn about the differences among proteins by electrophoresis.

Everything we have done with acids and bases has been directly based on the equilibrium concepts developed in the first part of the chapter. We will now see how these concepts are applied to another familiar chemical reaction, dissolution (and precipitation) of ionic solids in water and in solutions containing additional species.

Figure 9.8.

Molecular model of hemoglobin.
The α and β chains are at the top and bottom, respectively. The arrow pointing to the green dot locates where valine is substituted for a glutamic acid on one HbS β chain. This location on the other β chain is hidden from view.

9.6. Solubility Equilibria for Ionic Salts

 **9.48 INVESTIGATE THIS**

Is silver chromate insoluble?

Do this as a class investigation and work in small groups to discuss and analyze the results. Use 10.0 mL of an aqueous 0.0015 M solution of silver nitrate,

continued

$AgNO_3$, in a 25-mL beaker on an overhead projector stage. Add one drop of an aqueous 0.0015 M solution of potassium chromate, K_2CrO_4, to the silver nitrate solution and swirl the mixture. *WARNING:* Cr(VI) compounds are suspect carcinogens. Wear disposable gloves when handling the solutions. Record your observations. Keep track of the amount you add as you continue to add potassium chromate solution, *1 drop at a time*, until a precipitate forms and makes the solution cloudy.

9.49 CONSIDER THIS

How insoluble is silver chromate?

In Investigate This 9.48, what volume of potassium chromate solution did you have to add in order for precipitation of silver chromate to begin? How do you explain this observation? *Note:* One drop of aqueous solution is about 0.05 mL.

Solubility product Mixing the two solutions in Investigate This 9.48 gives a solution containing $K^+(aq)$, $Ag^+(aq)$, $CrO_4^{2-}(aq)$, and $NO_3^-(aq)$ ions. The precipitate that forms must be $Ag_2CrO_4(s)$. The dissolution and precipitation of ionic salts are usually represented by dissolution reactions:

$$Ag_2CrO_4(s) \rightleftharpoons 2Ag^+(aq) + CrO_4^{2-}(aq) \tag{9.35}$$

The equilibrium reaction expression for this dissolution reaction is

$$K = \left(\frac{(Ag^+(aq))^2\,(CrO_4^{2-}(aq))}{(Ag_2CrO_4(s))} \right)_{eq} \tag{9.36}$$

Recall, Table 9.1, that the concentration ratio for a pure solid is unity, so $(Ag_2CrO_4(s)) = 1$. Thus, equation (9.36) becomes

$$K_{sp} = \{(Ag^+(aq))^2\,(CrO_4^{2-}(aq))\}_{eq} \tag{9.37}$$

The equilibrium constant expression for dissolution of an ionic compound is just the product of the molar concentrations of the ions in the solution raised to their appropriate stoichiometric powers. The subscript sp reminds us that this equilibrium constant, K_{sp}, is the **solubility product** for the solid ionic compound. Keep in mind that *the solubility product is only valid when the ions in solution are in equilibrium with the solid.*

9.50 WORKED EXAMPLE

Solubility product, K_{sp}, for silver chromate, $Ag_2CrO_4(s)$

Suppose that, in Investigate This 9.48, a cloudy solution formed on addition of the fifth drop of 0.0015 M potassium chromate solution. Estimate K_{sp}, the

continued

solubility product constant, for silver chromate, $Ag_2CrO_4(s)$. (Your results may be different; you can use the strategy here to get an estimate from your observation.)

Necessary information: We need to know the concentrations of the solutions used in Investigate This 9.48; the $Ag^+(aq)$ solution is 0.0015 M and the $CrO_4^{2-}(aq)$ solution is 0.0015 M. The volumes of solution used are 10.0 mL and 0.25 mL (the assumed five drops), for the $Ag^+(aq)$ and $CrO_4^{2-}(aq)$, respectively.

Strategy: Substitute $[Ag^+(aq)]$ and $[CrO_4^{2-}(aq)]$ (molar concentrations in the solution when solid is also present) into equation (9.37) to get K_{sp}. Assume that the solid *just* begins to form when the 0.25 mL of $CrO_4^{2-}(aq)$ has been added, so that a negligible number of moles of the $Ag^+(aq)$ and $CrO_4^{2-}(aq)$ will have reacted to form the observed precipitate. With this assumption, use the number of moles of $CrO_4^{2-}(aq)$ added and moles of $Ag^+(aq)$ originally present to calculate their concentrations in 10.25 mL of the mixture containing a trace of solid.

Implementation:
Mol $Ag^+(aq)$ originally present = $(10.0 \times 10^{-3}$ L$)(0.0015$ M$) = 1.5 \times 10^{-5}$ mol.
Mol $CrO_4^{2-}(aq)$ added = $(2.5 \times 10^{-4}$ L$)(0.0015$ M$) = 3.8 \times 10^{-7}$ mol.
    Only a trace amount of these ions has reacted to give precipitate, so these are the moles of these ions in equilibrium with the trace amount of solid. The concentrations of the ions are

$$\text{final molar concentration of } Ag^+(aq) = \frac{1.5 \times 10^{-5} \text{ mol}}{0.01025 \text{ L}} = 1.5 \times 10^{-3} \text{ M}$$

$$\text{final molar concentration of } CrO_4^{2-}(aq) = \frac{3.8 \times 10^{-7} \text{ mol}}{0.01025 \text{ L}}$$

$$= 3.7 \times 10^{-5} \text{ M}$$

$$K_{sp} = \{(Ag^+(aq))^2 \ (CrO_4^{2-}(aq))\}_{eq} = (1.5 \times 10^{-3})^2(3.7 \times 10^{-5}) = 8 \times 10^{-11}$$

Does the answer make sense? The small value for the solubility product shows that silver chromate is not very soluble, as we have observed. But is our numeric value correct? The validity of the result depends on whether our strategic assumption is any good. What if the concentrations were on the borderline of solubility when four drops of potassium chromate solution had been added, so that just a tiny amount more would have caused precipitation? For this case, we would have calculated $[Ag^+(aq)] = 1.5 \times 10^{-3}$ M, $[CrO_4^{2-}(aq)] = 3 \times 10^{-5}$ M, and $K_{sp} = 7 \times 10^{-11}$, which is almost the same as the value we calculated above. (What do you get from your results for Investigate This 9.48?) Although solubility products are often given with two (or even more) significant figures, these are usually not justified under actual solution conditions where the activities of ions are only approximated by their molarities and the existence of ion complexes that keep the ions in solution is not accounted for. Also, in experiments such as Investigate This 9.48, slow formation of solids is not uncommon.

9.51 CHECK THIS

Solubility product, K_{sp}, for silver chromate, $Ag_2CrO_4(s)$

Assume that you carry out an experiment like that in Investigate This 9.48 but use a buret to add an aqueous 5.0×10^{-3} M solution of silver nitrate, $AgNO_3$, to 100 mL of a stirred, aqueous 2.0×10^{-4} M potassium chromate, K_2CrO_4, solution. You detect the first permanent cloudiness (deep red precipitate of Ag_2CrO_4) when you have added 9.8 mL of the silver solution. Estimate the numeric value for the solubility product constant, K_{sp}. Explain how you get your answer.

Solubility and solubility product The **solubility** of an ionic salt in water is the number of moles of the solid that dissolve in one liter of solution to form a saturated solution. A **saturated solution** is one in which as much solid as possible has dissolved; the dissolved ions are in equilibrium with undissolved ionic solid. You can calculate the solubility product for an ionic salt from its solubility and vice versa. Such calculations do not agree exactly with values determined in other ways, because the simple solubility product relationships exemplified by equation (9.37) do not take into account ionic interactions in solution. The calculations usually agree within one or two orders of magnitude, so the results are generally in the right ballpark. The solubility of insoluble ionic compounds varies a great deal. Table 9.4 gives the solubility products for several cation–anion combinations to give you an idea of the range of solubilities.

Table 9.4 *Solubility products, K_{sp} and pK_{sp}, for several insoluble ionic compounds.*

Compound	K_{sp}	pK_{sp}	Compound	K_{sp}	pK_{sp}
AgCl	1.8×10^{-10}	9.74	$Cu(IO_3)_2$	1.4×10^{-7}	6.85
AgBr	5.0×10^{-13}	12.30	CuC_2O_4	2.9×10^{-8}	7.5
AgI	8.3×10^{-17}	16.08	CuS	8×10^{-37}	36.1
Ag_2CrO_4	1.2×10^{-12}	11.92	$Fe(OH)_2$	7.9×10^{-16}	15.1
Ag_2S	8×10^{-51}	50.1	FeS	8×10^{-19}	18.1
$Al(OH)_3$	3×10^{-34}	33.5	$Fe(OH)_3$	1.6×10^{-39}	38.8
$BaCO_3$	5.0×10^{-9}	8.30	$MgCO_3$	3.5×10^{-8}	7.46
$BaSO_4$	1.1×10^{-10}	9.96	$Mg(OH)_2$	7.1×10^{-12}	11.15
BaC_2O_4	1×10^{-6}	6.0	$PbCl_2$	1.7×10^{-5}	4.78
$CaCO_3$	6.0×10^{-9}	8.22	$PbBr_2$	2.1×10^{-6}	5.68
$CaSO_4$	2.4×10^{-5}	4.62	PbI_2	7.9×10^{-9}	8.10
$Ca(OH)_2$	6.5×10^{-6}	5.19	$PbSO_4$	6.3×10^{-7}	6.20
CaC_2O_4	1.3×10^{-8}	7.9	PbS	3×10^{-28}	27.5

9.52 WORKED EXAMPLE

Solubility product, K_{sp}, for magnesium hydroxide, $Mg(OH)_2(s)$

A saturated solution of magnesium hydroxide (Milk of Magnesia®) contains 0.009 g of $Mg(OH)_2$ per liter of solution. What are the solubility and solubility product for $Mg(OH)_2$?

Necessary information: We need the molar mass of $Mg(OH)_2$, 58 g·mol^{-1}. We'll use the stoichiometry of the dissolution reaction and solubility product constant:

$$Mg(OH)_2(s) \rightleftharpoons Mg^{2+}(aq) + 2OH^-(aq) \tag{9.38}$$

$$K_{sp} = \{(Mg^{2+}(aq))(OH^-(aq))^2\}_{eq} \tag{9.39}$$

Strategy: Use the mass of $Mg(OH)_2$ that dissolves in a liter of solution and the molar mass to calculate the solubility, s, in mol·L^{-1}. For every mole of $Mg(OH)_2(s)$ that dissolves the solution contains 1 mol of $Mg^{2+}(aq)$ and 2 mol of $OH^-(aq)$. Since s mol·L^{-1} dissolve, the concentration of $Mg^{2+}(aq)$ is s M and the concentration of $OH^-(aq)$ is $2s$ M. These values are substituted into the solubility product expression (9.39) to get K_{sp}.

Implementation:

$$\text{molar solubility of } Mg(OH)_2 = s = \frac{0.009 \text{ g·L}^{-1}}{58 \text{ g·mol}^{-1}} = 1.6 \times 10^{-4} \text{ M}$$

$$K_{sp} = (s)\cdot(2s)^2 = 4s^3 = 4(1.6 \times 10^{-4})^3 = 1.6 \times 10^{-11} \approx 2 \times 10^{-11}$$

Does the answer make sense? Using the stoichiometry of the solution reaction, we converted solubility to the solubility product. Our result is in the right range, about three times larger than the value in Table 9.4. This gives you an idea of the kind of agreement you can expect for ionic equilibria calculated without corrections for ionic interactions. When you do these kinds of calculations, be careful to use the stoichiometric coefficients in the reaction equation correctly to get the concentrations of each ion from the stoichiometry of the reaction and then again as exponents in the solubility product expression.

9.53 CHECK THIS

Solubility of silver phosphate, Ag_3PO_4

The solubility product constant, K_{sp}, for silver phosphate, $Ag_3PO_4(s)$, is 2.8×10^{-18}.
(a) Write the dissolution reaction and the solubility product expression for $Ag_3PO_4(s)$.
(b) What is the molar solubility, s, of $Ag_3PO_4(s)$ in water? *Hint:* The strategy and calculations are the reverse of those in Worked Example 9.52. Use

continued

the stoichiometry of the reaction to write the concentrations of the ions in terms of s. Substitute these concentrations in the solubility product expression and solve for s.

(c) One handbook gives the solubility for $Ag_3PO_4(s)$ as 0.0065 g·L^{-1} and another gives 0.065 g·L^{-1}. Which handbook value is more consistent with your result? Explain your reasoning.

Common ion effect **Gravimetric analysis** is a method to analyze an ionic sample by precipitating the ion of interest as an insoluble salt with an appropriate counterion and determining the mass of the product. Suppose we wish to analyze a solution containing barium ion by precipitating it as its sulfate, $BaSO_4(s)$. The dissolution reaction and solubility product expression for $BaSO_4(s)$ are

$$BaSO_4(s) \rightleftharpoons Ba^{2+}(aq) + SO_4^{2-}(aq) \tag{9.40}$$

$$K_{sp} = \{(Ba^{2+}(aq))(SO_4^{2-}(aq))\}_{eq} = 1.1 \times 10^{-10} \tag{9.41}$$

How are we going to make sure almost all the $Ba^{2+}(aq)$ gets precipitated? Let's think backwards for a moment. If we *dissolve* $BaSO_4(s)$ in water, its solubility is s mol·L^{-1} and the solubility product tells us that $s^2 = 1.1 \times 10^{-10}$, so $s = 1.0 \times 10^{-5}$ mol·L^{-1} (about 0.002 g·L^{-1}). This is a low solubility, but for analytical purposes, we would like to find a way to reduce the solubility of the solid even further. If we add extra $SO_4^{2-}(aq)$ [say by adding $Na_2SO_4(s)$ to the solution], the system should respond to this disturbance by using up some of the added anion. Reaction (9.40) will go in reverse to use up some of the added $SO_4^{2-}(aq)$, which also precipitates more of the $Ba^{2+}(aq)$. Figure 9.9 is a molecular-level representation of this procedure for gravimetric analysis. Let's see how the analysis works out quantitatively.

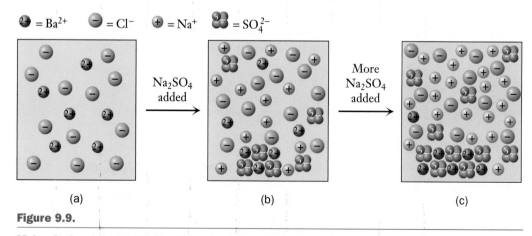

(a)   (b)   (c)

Figure 9.9.

Molecular-level representation of gravimetric analysis for barium ion. (a) The solution to be analyzed. In (b), a stoichiometric amount of sulfate ion is added and some $BaSO_4(s)$ precipitates. When more sulfate ion is added, in (c), the precipitation of $BaSO_4(s)$ is almost complete.

9.54 WORKED EXAMPLE

Solubility of $BaSO_4(s)$ in 0.10 M $SO_4^{2-}(aq)$ solution

Find the molar solubility of $BaSO_4(s)$ in an aqueous 0.10 M solution of sodium sulfate, Na_2SO_4, that is, a solution that is 0.10 M in $SO_4^{2-}(aq)$ anion.

Necessary information: We need the solubility product for $BaSO_4(s)$, equation (9.41).

Strategy: Let s be the molar solubility of $BaSO_4(s)$ in the solution, construct a change table for the dissolution reaction, substitute the final molar concentrations into equation (9.41), and solve for s. We will assume, as we often have, that we have 1 L of solution, so that numbers of moles and molarity have the same numeric value.

Implementation: The change table is

Species	$BaSO_4(s)$	$SO_4^{2-}(aq)$	$Ba^{2+}(aq)$
Initial mol	0	0.10	0
Change in mol	solid added; s dissolves	s goes into solution	s goes into solution
Final mol	solid still present	$0.10 + s$	s

If the $BaSO_4(s)$ were dissolving in pure water, we know from the preceding discussion that s would be 1.0×10^{-5} M. This is the maximum s can be and it is only 0.01% of 0.10 M, so $0.10 + s \approx 0.10$:

$$K_{sp} = \{(Ba^{2+}(aq))(SO_4^{2-}(aq))\}_{eq} = (s)(0.10)$$

$$s = \frac{K_{sp}}{0.10} = \frac{1.10 \times 10^{-10}}{0.10} = 1.1 \times 10^{-9} \text{ mol·L}^{-1}$$

Does the answer make sense? The solubility of $BaSO_4(s)$ in a solution that already contains $SO_4^{2-}(aq)$ is lower (10,000-fold lower) than in pure water. This is exactly what we reasoned above on the basis of Le Chatelier's principle and the calculations bear out our prediction.

9.55 CHECK THIS

Solubility of $Cu(IO_3)_2(s)$ in water and 0.25 M $Cu^{2+}(aq)$ solution

(a) What is the molar solubility of $Cu(IO_3)_2(s)$ in water? Use the K_{sp} from Table 9.4.
(b) What is the molar solubility of $Cu(IO_3)_2(s)$ in a 0.25 M aqueous solution of copper nitrate, $Cu(NO_3)_2$? How does your answer compare to your result from part (a)? Is this what you expected? Why or why not?
(c) Without doing any calculations, predict whether the solubility of $Cu(IO_3)_2(s)$ in a 0.25 M aqueous solution of sodium iodate, $NaIO_3$, will be larger, smaller, or the same as the solubility you calculated in part (b). Give your reasoning clearly.

In the systems we have just discussed, the ionic solid and the solution had an ion in common. The lowered solubility of the ionic solid in such a solution is an example of the **common ion effect,** the response of an ionic equilibrium system to an increase in concentration of one of the ions involved in the equilibrium. In Worked Example 9.54, since there is already $SO_4^{2-}(aq)$ in the solution, less $BaSO_4(s)$ has to dissolve to satisfy the solubility product. The common ion effect is another example of Le Chatelier's principle and the application of equilibrium principles, not a new concept. Figure 9.10 illustrates the common ion effect on the molecular level.

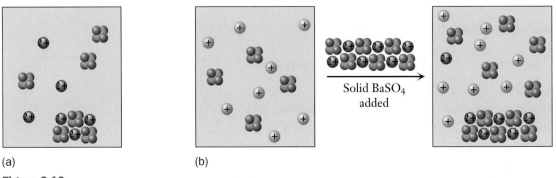

(a)   (b)

Figure 9.10.

Common ion effect on solubility of $BaSO_4(s)$ in a solution containing $SO_4^{2-}(aq)$. (a) A solution of $BaSO_4(s)$ dissolved in pure water. (b) The same amount of $BaSO_4(s)$ added to a solution already containing $SO_4^{2-}(aq)$ ions; less solid dissolves.

9.56 CHECK THIS

Compare Figures 9.9 and 9.10

(a) What is the same and what is different about Figures 9.9(b) and 9.10(a)? Focus on the similarities and discuss why they are the same.

(b) How could you make the solution in Figure 9.10(a) identical to the one in Figure 9.9(b)? What does your answer imply about the solubility of an ionic compound in a solution that contains ions but not a common ion? Explain clearly.

(c) What is the same and what is different about Figures 9.9(c) and the solution that results when the solid is added in Figure 9.10(b)? Focus on the similarities and discuss why they are the same.

(d) How could you make the resulting solution in Figure 9.10(b) identical to the one in Figure 9.9(c)? Does your answer reinforce the conclusion(s) you suggested in part (b)? Explain why or why not.

We do not often add a solid to a solution containing a common ion. However, in gravimetric analysis, we are interested in the solubility of a precipitate in a solution containing an excess of the counterion used to precipitate the ion of interest. This is exactly the same system we have been discussing, except that, in this case, the solid is formed when two solutions are mixed. We have to account for this reaction to determine how much of the common ion is present in the final solution. Figures 9.9 and 9.10 and your analysis in Check This 9.56 exemplify these cases.

You usually see the common ion effect associated with ionic solubilities. However, if you go back to our discussion of acid–base equilibria in Section 9.3, you will find several examples of the effect of changing the concentration of an ion that is common to both the reactants and products. One example would be the effect on the equilibrium concentrations when acetate ion is added to a solution that already contains acetate ion and acetic acid (a compound of acetate ion and H^+). Le Chatelier's principle makes it clear that the common ion effect is found everywhere in ionic equilibria.

9.57 CONSIDER THIS

Does a common ion always decrease solubility?

Mercuric iodide, $HgI_2(s)$, is a sparingly soluble, $s = 1.3 \times 10^{-4}$ M, red-orange solid. Here are some solubility data for $HgI_2(s)$ dissolving in solutions containing $I^-(aq)$ (from dissolved KI):

$[I^-(aq)]$, M	0.05	0.20	0.50	1.0	2.0
s, M	0.025	0.10	0.25	0.50	1.0

(a) How does the solubility of $HgI_2(s)$ vary as the $[I^-(aq)]$ in the solution varies? Is this what you expect from the common ion effect? Explain why or why not.
(b) What explanation can you suggest to explain these data?

Although solubility reactions look simple and we have treated them simply, they are not so simple. Many solute–solvent and solute–solute interactions occur in ionic solutions that affect these reactions. An example we did not discuss is the formation of cation–anion complexes, such as $BaSO_4(aq)$ or $Cu(IO_3)^+(aq)$, during dissolution and precipitation reactions. A related example is formation of higher complexes, such as $HgI_3^-(aq)$ and $HgI_4^-(aq)$ in the mercury(II)-iodide system described in Consider This 9.57. Accounting for all these reactions and their equilibria is required for an accurate characterization of a solubility system. We will continue to neglect these complexities but warn you that they exist and that they make the simple calculations only approximations (sometimes quite poor approximations).

Reflection and Projection

We introduced the solubility of ionic compounds in water in Chapter 2 and used specific examples in subsequent chapters to illustrate various aspects of chemical reactivity. In Investigate This activities, you often produced precipitates by mixing solutions of ions that reacted to form an insoluble solid. You could have attained the same end result (more slowly) by dissolving the solid in a solution of the counterions from the solutions you mixed to get a precipitate. It doesn't matter how the mixture of solid and solution are produced; at equilibrium, the product of the ionic concentrations in the solution (with appropriate stoichiometric exponents) is a constant, the solubility product, K_{sp}. You already knew that insoluble ionic compounds vary in their solubility. The solubility product quantifies the variability and Table 9.4 shows the great range of solubilities. The solubility product and solubility product expression also provide you the means to calculate the effects of added common ions on a system at solubility equilibrium.

In all the equilibrium calculations we have done, we have relied on the empirical relationship among the reactant and product concentrations, equation (9.6), that was determined by Guldberg and Waage in the 19th century. Together with the balanced chemical reaction, this equilibrium constant expression has been the starting point for all of our discussions. In Chapter 8, however, we stated the criterion for a system at equilibrium in terms of net entropy change, ΔS_{net}, or Gibbs free energy change, ΔG. For a system at equilibrium with respect to a change, we found that ΔS_{net}, or ΔG, is zero. How is the empirical equilibrium constant expression related to the thermodynamic criterion for equilibrium? That is the subject of the next section.

9.7. Thermodynamics and the Equilibrium Constant

In Chapters 7 and 8, we learned how to use the data in Appendix B, standard enthalpies of formation, $\Delta H_f°$, standard entropies, $S°$, and standard Gibbs free energies of formation, $\Delta G_f°$, to calculate standard enthalpy, entropy, and free energy changes for reactions. Also, in Chapter 8, we learned that for a change in a reaction system, ΔS_{net} or ΔG is equal to zero for the change under equilibrium conditions. However, we have not learned how to calculate changes in these thermodynamic quantities under nonstandard state conditions. What we need to learn is how to convert values we know, $\Delta G°_{reaction}$, for example, to the corresponding values under nonstandard state conditions, $\Delta G_{reaction}$.

Free energy of reaction and the reaction quotient In Section 9.11, we show how to begin from the definition of entropy, Chapter 8, Section 8.6, and arrive at the relationship between $\Delta G_{reaction}$ and $\Delta G°_{reaction}$ for a gas phase reaction. At the end of Section 9.11, we suggest that the relationship for the gas phase reaction can be applied to any reaction, including our general equation for a reversible reaction:

$$a\mathbf{A} + b\mathbf{B} \rightleftharpoons c\mathbf{C} + d\mathbf{D} \tag{9.5}$$

The relationship applied to reaction equation (9.5) is

$$\Delta G_{reaction} = \Delta G°_{reaction} + RT\ln\left(\frac{(C)^c(D)^d}{(A)^a(B)^b}\right) \tag{9.42}$$

In equation (9.42), R and T are, respectively, the molar gas constant and the temperature in kelvin.

9.58 CHECK THIS

Agreement of units in equation (9.42)

Show that the units of the term on the far right of equation (9.42) are the same as those for the free energy change for a mole of reaction.

Equation (9.42) is the relationship we need to find the free energy change for a reaction under nonstandard state conditions. The ratio in parentheses in equation (9.42) is called the **reaction quotient** and is usually symbolized as *Q:*

$$Q = \left(\frac{(C)^c(D)^d}{(A)^a(B)^b}\right) \qquad (9.43)$$

$$\Delta G_{reaction} = \Delta G°_{reaction} + RT\ln Q \qquad (9.44)$$

The reaction quotient should look familiar to you, since it has exactly the same *form* as the equilibrium constant expression for reaction (9.5). You write the reaction quotient for any reaction just as you would write its equilibrium constant expression.

9.59 WORKED EXAMPLE

Reaction quotient and free energy for a gas phase reaction

Bromine, Br_2, and chlorine, Cl_2, gases can react to form the interhalogen compound (a diatomic species of two different halogen atoms), BrCl:

$$Br_2(g) + Cl_2(g) \rightleftharpoons 2BrCl(g) \qquad (9.45)$$

(a) Write the reaction quotient for reaction (9.45).
(b) The standard molar free energies of formation (at 298 K) of $Br_2(g)$, $Cl_2(g)$, and BrCl((g) are 3.14, 0.00, and -0.88 kJ·mol^{-1}, respectively. What is the free energy change for reaction (9.45) at 298 K, if the pressures of $Br_2(g)$ and $Cl_2(g)$ are each 0.120 bar and the pressure of BrCl(g) is 0.240 bar? Is the reaction spontaneous under these conditions? Why or why not?

Necessary information: In addition to the information in the problem statement, we need equation (9.42) [or, equivalently, equations (9.43) and (9.44)], the definition of the concentration ratio for gases from Table 9.1, and the value of R, 8.314 J·mol^{-1}·K^{-1}.

Strategy: Use the reaction equation (9.45) to write the reaction quotient (the same form as the equilibrium constant expression). Calculate $\Delta G°_{reaction}$ and Q from the data in the problem, combine them in equation (9.42) to determine $\Delta G_{reaction}$, and, from the sign of $\Delta G_{reaction}$, decide whether the reaction is spontaneous under the specified conditions.

Implementation: (a) The reaction quotient is

$$Q = \frac{(BrCl(g))^2}{(Br_2(g))(Cl_2(g))}$$

(b) $\Delta G°_{reaction} = (2\ mol)(-0.88\ kJ·mol^{-1}) - (1\ mol)(3.14\ kJ·mol^{-1})$
$$- (1\ mol)(0.00\ kJ·mol^{-1})$$

$$\Delta G°_{reaction} = -4.90\ kJ$$

$$Q = \frac{(0.240\ bar/1\ bar)^2}{(0.120\ bar/1bar)(0.120\ bar/1\ bar)} = 4.00$$

$$\Delta G_{reaction} = \Delta G°_{reaction} + RT\ln Q$$
$$= -4.90\ kJ + (8.314\ J·mol^{-1}·K^{-1})(298\ K)\ln(4.00)$$

$$\Delta G_{reaction} = -4.90\ kJ + 3.43\ kJ = -1.47\ kJ$$

Since $\Delta G_{reaction}$ is negative, the reaction is spontaneous (can occur) under the specified conditions.

continued

Does the answer make sense? Without further information about this reaction (or other reactions for comparison), we cannot determine if the reaction is actually spontaneous under these conditions, although the existence of inter-halogen compounds shows us that it *can* be.

9.60 CHECK THIS

Reaction quotient and free energy for a gas phase reaction

Consider reaction (9.45) with the same $Br_2(g)$ and $Cl_2(g)$ pressures as in Worked Example 9.59 and a $BrCl(g)$ pressure of 0.500 bar. Is the reaction spontaneous under these conditions? Why or why not?

Standard free energy of reaction and the equilibrium constant

Recall that a reaction is in equilibrium, if $\Delta G_{reaction} = 0$ at the temperature and pressure in the system. Under these equilibrium conditions, we can rewrite equations (9.42) and (9.44) as

$$\Delta G_{reaction} = 0 = \Delta G°_{reaction} + RT\ln\left(\frac{(C)^c(D)^d}{(A)^a(B)^b}\right)_{eq}$$

$$= \Delta G°_{reaction} + RT\ln Q_{eq} \qquad (9.46)$$

As you can see, the equilibrium value of the reaction quotient, Q_{eq}, is just the equilibrium constant expression:

$$Q_{eq} = \left(\frac{(C)^c(D)^d}{(A)^a(B)^b}\right)_{eq} = K \qquad (9.47)$$

$$\Delta G°_{reaction} = -RT\ln K \qquad (9.48)$$

Equation (9.48) provides an enormously useful relationship between the thermodynamics of a reaction and the observable composition of the chemical system at equilibrium. It is valid for any reaction, as long as we express the concentrations of the reactants and products as concentration ratios that are consistent with the standard states listed in Table 9.1, Section 9.2. We can calculate the equilibrium constant for a reaction without doing any experimental measurements on the reaction system, if we know $\Delta G°_{reaction}$. One way to get $\Delta G°_{reaction}$ is to calculate it from the standard free energies of formation in Appendix B. Conversely, if we measure the equilibrium constant for a reaction, we can calculate $\Delta G°_{reaction}$ without doing any calorimetric measurements. If we do calorimetric measurements and/or measure the temperature dependence of the equilibrium constant, we can also obtain $\Delta H°_{reaction}$ and $\Delta S°_{reaction}$. We will explore these possibilities in the rest of the chapter.

9.61 INVESTIGATE THIS

How much urea will dissolve in water?

Use a thin-stem plastic pipet and two capped, graduated conical plastic tubes containing, respectively, 4.0 g of urea, $H_2NCONH_2(s)$, and water. Read and record the volume of water to ± 0.1 mL. Use the pipet to transfer 4.0 mL of water from the water-containing tube to the tube containing urea, cap the tube, and invert several times to dissolve as much urea as possible. Add water one drop at a time to the urea solution, with *vigorous* and *thorough* mixing between additions, until the urea is *just* dissolved. Return all unused water from the pipet to the tube containing pure water. Measure and record the volume of liquid in each tube to ± 0.1 mL.

9.62 CONSIDER THIS

What is the solubility of urea in water?

(a) What is the volume of your just-saturated urea solution in Investigate This 9.61? What volume of water did you use to make this solution?
(b) How many moles of urea are dissolved in this just-saturated urea solution? What is the molarity of the urea, $[H_2NCONH_2(aq)]$, in the solution?
(c) How many moles of water are in this just-saturated urea solution? (Assume that the density of water is 1.0 $g \cdot mL^{-1}$.) What is the molarity of water, $[H_2O(aq)]$, in the solution?

The dissolution reaction for urea, Investigate This 9.61, is

$$H_2NCONH_2(s) + H_2O(aq) \rightleftharpoons H_2NCONH_2(aq) \qquad (9.49)$$

The equilibrium constant expression for the dissolution is

$$K_{diss} = \left(\frac{(H_2NCONH_2(aq))}{(H_2NCONH_2(s))(H_2O(aq))} \right)_{eq} \qquad (9.50)$$

Since the urea is a pure solid, we know that $(H_2NCONH_2(s)) = 1$. In Consider This 9.62(b), you found the molar concentration of urea, $[H_2NCONH_2(aq)]$, and you know the standard state of a solute is 1 M, so you can calculate $(H_2NCONH_2(aq))$. A saturated solution of urea in water is far from dilute, so we have to take into account the molarity of water, $[H_2O(aq)]$, from Consider This 9.62(c), to determine $(H_2O(aq))$:

$$(H_2O(aq)) = \frac{[H_2O(aq)]}{55.5 \text{ M}} \qquad (9.51)$$

9.63 CONSIDER THIS

What are $\Delta G°$, $\Delta H°$, and $\Delta S°$ for dissolution of urea in water?

(a) Use your results from Consider This 9.62(b) and 9.62(c) to calculate K_{diss} for the dissolution of urea in water. Explain your approach clearly.

(b) What is $\Delta G°_{reaction}$ for the dissolution of urea in water? Explain your reasoning.

(c) In Chapter 7, Investigate This 7.22, you dissolved 6.0 g of urea (0.10 mol) in 100. mL of water in a Styrofoam® cup and analyzed the results in Consider This 7.23 and Check This 7.27 and 7.29. Is the dissolution reaction exothermic or endothermic? What is $\Delta H°_{reaction}$ for this dissolution reaction? Explain clearly how you get your answer.

(d) What is $\Delta S°_{reaction}$ for the dissolution of urea in water? Explain how you get your answer.

The reasoning and calculations you did in Consider This 9.63 are applications of two important thermodynamic equations:

$$\Delta G°_{reaction} = -RT\ln K \qquad\qquad (9.48)$$

$$\Delta G°_{reaction} = \Delta H°_{reaction} - T\Delta S°_{reaction} \text{ (from Chapter 8)} \qquad (9.52)$$

The example shows how an equilibrium constant measurement and a calorimetric measurement for a reaction provide enough data to obtain the standard state changes for the free energy, enthalpy, and entropy of the reaction. In the next section we will see how measurements of the equilibrium constant at two or more different temperatures can give the same information.

9.8. Temperature Dependence of the Equilibrium Constant

 9.64 INVESTIGATE THIS

How does temperature affect the solubility of $PbI_2(s)$?

Do this as a class investigation and work in small groups to discuss and analyze the results. Use a magnetic stir bar and a magnetic stirrer–hot plate combination to stir 100 mL of ice-cold water in a 250-mL beaker. Add 5 mL of a 1 M aqueous lead nitrate, $Pb(NO_3)_2$, solution and then 2 mL of ice-cold 0.05 M aqueous potassium iodide, KI, solution to the stirred water. *WARNING:* All lead compounds are toxic. Wear disposable gloves when handling the solutions. Watch for the appearance of a precipitate of lead iodide, $PbI_2(s)$. When a precipitate has formed, turn on the hot plate, heat the solution to about 70 °C, and record your observations. Remove the beaker from the heat, allow the solution to cool, and record your observations.

9.65 CONSIDER THIS

Is the dissolution of $PbI_2(s)$ exothermic or endothermic?

What did you observe in Investigate This 9.64 when you heated the mixture containing $PbI_2(s)$? When you allowed the heated solution to cool? Use your observations and Le Chatelier's principle to explain clearly whether $PbI_2(s)$ dissolution is exothermic or endothermic.

$$PbI_2(s) \rightleftharpoons Pb^{2+}(aq) + 2I^-(aq) \tag{9.53}$$

In Consider This 9.65, you used Le Chatelier's principle to analyze your experimental observations on the temperature dependence of the solubility of $PbI_2(s)$. We can make the analysis more quantitative by using the temperature dependence of the equilibrium constant. To get this temperature dependence, we combine equations (9.48) and (9.52), two equations for $\Delta G°_{reaction}$, to give

$$\Delta G°_{reaction} = -RT\ln K = \Delta H°_{reaction} - T\Delta S°_{reaction} \tag{9.54}$$

We rearrange the right-hand equality of equation (9.54) to solve for $\ln K$:

$$\ln K = -\left(\frac{\Delta H°_{reaction}}{R}\right)\frac{1}{T} + \frac{\Delta S°_{reaction}}{R} \tag{9.55}$$

Although $\Delta H°_{reaction}$ and $\Delta S°_{reaction}$ vary with temperature, their variation is not large, so we will assume that they are constant. At two different temperatures T_1 and T_2, with equilibrium constants K_1 and K_2, respectively, we have

$$\ln K_1 = -\left(\frac{\Delta H°_{reaction}}{R}\right)\frac{1}{T_1} + \frac{\Delta S°_{reaction}}{R} \tag{9.56}$$

$$\ln K_2 = -\left(\frac{\Delta H°_{reaction}}{R}\right)\frac{1}{T_2} + \frac{\Delta S°_{reaction}}{R} \tag{9.57}$$

Subtracting equation (9.56) from (9.57) gives

$$\ln K_2 - \ln K_1 = \ln\left(\frac{K_2}{K_1}\right) = -\left(\frac{\Delta H°_{reaction}}{R}\right)\frac{1}{T_2} + -\left(\frac{\Delta H°_{reaction}}{R}\right)\frac{1}{T_1}$$

$$= -\left(\frac{\Delta H°_{reaction}}{R}\right)\left(\frac{1}{T_2} - \frac{1}{T_1}\right) \tag{9.58}$$

$$\ln\left(\frac{K_2}{K_1}\right) = \left(\frac{\Delta H°_{reaction}}{R}\right)\left(\frac{T_2 - T_1}{T_2 \cdot T_1}\right) \tag{9.59}$$

9.66 CHECK THIS

Rearranging equation (9.58)

Show how equation (9.59) is obtained from equation (9.58).

9.67 WORKED EXAMPLE

$\Delta H°_{reaction}$ from solubility temperature dependence

One handbook gives the aqueous solubility of $PbI_2(s)$ as 0.44 g·L^{-1} at 0 °C and 4.1 g·L^{-1} at 100. °C. Use these data to find K_{sp} at both temperatures and then to determine $\Delta H°_{reaction}$.

Necessary information: We need the molar mass of PbI_2, 461 g·mol^{-1}, and the relationship of the solubility product to the molar solubility, s, of $PbI_2(s)$: $K_{sp} = 4s^3$.

Strategy: Use the two values of s to get K_{sp} at 0 °C (273 K) and 100 °C (373 K) and then substitute into equation (9.59) and solve for $\Delta H°_{reaction}$.

Implementation:

at 273 K: $\quad K_{sp}(273\ \text{K}) = 4\left(\dfrac{0.44\ \text{g·L}^{-1}/461\ \text{g·mol}^{-1}}{1\ \text{M}}\right)^3 = 3.5 \times 10^{-9}$

at 373 K: $\quad K_{sp}(373\ \text{K}) = 4\left(\dfrac{4.1\ \text{g·L}^{-1}/461\ \text{g·mol}^{-1}}{1\ \text{M}}\right)^3 = 2.8 \times 10^{-6}$

$$\ln\left(\frac{K_2}{K_1}\right) = \ln\left(\frac{2.8 \times 10^{-6}}{3.5 \times 10^{-9}}\right) = \left(\frac{\Delta H°_{reaction}}{R}\right)\left(\frac{T_2 - T_1}{T_2 \cdot T_1}\right)$$

$$= \left(\frac{\Delta H°_{reaction}}{8.314\ \text{J·mol}^{-1}\cdot\text{K}^{-1}}\right)\left(\frac{(373\ \text{K}) - (273\ \text{K})}{(373\ \text{K})(273\ \text{K})}\right)$$

$$\Delta H°_{reaction} = (8.314\ \text{J·mol}^{-1}\cdot\text{K}^{-1})\left(\frac{(373\ \text{K})(273\ \text{K})}{(373\ \text{K}) - (273\ \text{K})}\right)\cdot\ln\left(\frac{2.8 \times 10^{-6}}{3.5 \times 10^{-9}}\right)$$

$$= 56\ \text{kJ·mol}^{-1}$$

This is the standard enthalpy change for dissolving one mole of $PbI_2(s)$.

Does the answer make sense? The dissolution is endothermic, as you reasoned in Consider This 9.65 from the observed increasing solubility with increasing temperature. Note that the solubilities given in the problem statement confirm that the solubility of $PbI_2(s)$ increases with temperature. We have another check on our answer: the data for enthalpies of formation in Appendix B. We calculate $\Delta H°_{reaction}$ for 1 mol of $PbI_2(s)$ dissolving:

$$\Delta H°_{reaction} =$$

$$(1\ \text{mol})\cdot\Delta H_f°(Pb^{2+}(aq)) + (2\ \text{mol})\cdot\Delta H_f°(I^-(aq)) - (1\ \text{mol})\cdot\Delta H_f°(PbI_2(s))$$

$$\Delta H°_{reaction} = (1.7\ \text{kJ}) + (-110.3\ \text{kJ}) - (-175.1\ \text{kJ}) = 67\ \text{kJ}$$

The $\Delta H°_{reaction}$ we got from solubilities and the $\Delta H°_{reaction}$ calculated from table values agree within about 20%. Considering the problems with solubilities that we have pointed out, this is reasonable agreement. Both values show that the reaction is endothermic. Note that all the reactants and products must be included in the calculation, even though they do not all appear in the solubility product expression. This is another reminder that the solubility product expression is only valid when the solution is in equilibrium with the solid.

9.68 CHECK THIS

$\Delta H°_{reaction}$ *from solubility temperature dependence*

(a) One handbook gives the solubility of $Ca(OH)_2(s)$ as 1.85 g·L^{-1} at 0 °C and 0.77 g·L^{-1} at 100 °C. Use these data and Le Chatelier's principle to predict the sign of $\Delta H°_{reaction}$ for the dissolution of $Ca(OH)_2(s)$. Explain your reasoning clearly.

(b) Use the data in part (a) to find K_{sp} at both temperatures and then to determine $\Delta H°_{reaction}$. Is the sign of $\Delta H°_{reaction}$ consistent with your answer in part (a)? Explain why or why not.

(c) Check your result in part (b) using the values for standard enthalpies of formation in Appendix B. Do your two values for $\Delta H°_{reaction}$ agree reasonably well? Explain.

As you see, measurements of the equilibrium constant for a reaction at two different temperatures provide enough information to obtain $\Delta H°_{reaction}$ for the reaction. This value can then be combined with one (or both) of the equilibrium constant values to get $\Delta S°_{reaction}$, as well as $\Delta G°_{reaction}$ and the equilibrium constant at other temperatures.

9.69 WORKED EXAMPLE

$\Delta S°_{reaction}$ *and* $\Delta G°_{reaction}$ *from solubility temperature dependence*

Use the results from Worked Example 9.67 to get $\Delta S°_{reaction}$ for the dissolution of $PbI_2(s)$ in water. What are $\Delta G°_{reaction}$ and K_{sp} for this reaction at 298 K (25 °C)?

Necessary information: We'll need all the equalities in expression (9.54) and, from Worked Example 9.67, $K_{sp} = 3.5 \times 10^{-9}$ at 273 K and $\Delta H°_{reaction} = 56 \times 10^3$ J·mol^{-1}. We also need to assume that $\Delta H°_{reaction}$ and $\Delta S°_{reaction}$ do not vary with temperature.

Strategy: Substitute the values for K_{sp} and $\Delta H°_{reaction}$ into expression (9.54) and solve for $\Delta S°_{reaction}$. Then substitute $\Delta H°_{reaction}$ and $\Delta S°_{reaction}$ into expression (9.54) to get $\Delta G°_{reaction}$ at 298 K. Finally, substitute $\Delta G°_{reaction}$ at 298 K into expression (9.54) to get K_{sp} at 298 K.

Implementation:

$$-RT\ln K_{sp} = \Delta H°_{reaction} - T\Delta S°_{reaction}$$

$$\Delta S°_{reaction} = R\ln K_{sp} + \frac{\Delta H°_{reaction}}{T}$$

$$= (8.314 \, \text{J·mol}^{-1}\text{·K}^{-1})\cdot\ln(3.5 \times 10^{-9}) + \frac{56 \times 10^3 \, \text{J·mol}^{-1}}{273 \, \text{K}}$$

$$\Delta S°_{reaction} = -162 \, \text{J·mol}^{-1}\text{·K}^{-1} + 205 \, \text{J·mol}^{-1}\text{·K}^{-1} = 43 \, \text{J·mol}^{-1}\text{·K}^{-1}$$

continued

For $\Delta G°_{reaction}$ and K_{sp} at 298 K, we have:

$$\Delta G°_{reaction} = \Delta H°_{reaction} - T\Delta S°_{reaction}$$

$$= (56 \times 10^3 \text{ J·mol}^{-1}) - (298 \text{ K})(43 \text{ J·mol}^{-1}\text{·K}^{-1})$$

$$\Delta G°_{reaction} = 43 \times 10^3 \text{ J·mol}^{-1} = 43 \text{ kJ·mol}^{-1}$$

$$\Delta G°_{reaction} = -RT\ln K_{sp}$$

$$\ln K_{sp} = -\frac{\Delta G°_{reaction}}{RT} = -\left(\frac{43 \times 10^3 \text{ J·mol}^{-1}}{(8.314 \text{ J·mol}^{-1}\text{·K})(298 \text{ K})}\right) = -17.4$$

$$K_{sp} = 3 \times 10^{-8}$$

Does the answer make sense? The K_{sp} at 298 K is intermediate between the values at 273 K and 373 K (from Worked Example 9.67) and closer to the value at 273 K, as we would expect. Our calculated solubility product is about three times larger than the value in Table 9.4, which is good agreement. We can check the thermodynamic values by calculating them from the data in Appendix B. We get 67 J·mol⁻¹·K⁻¹ and 44 kJ·mol⁻¹, respectively, for $\Delta S°_{reaction}$ and $\Delta G°_{reaction}$. The standard entropy changes differ somewhat, but both show that the change is positive. The values for the standard free energy change are the same.

9.70 CHECK THIS

$\Delta S°_{reaction}$ and $\Delta G°_{reaction}$ from solubility temperature dependence

(a) Use your results from Check This 9.68 to get $\Delta S°_{reaction}$ for the dissolution of $Ca(OH)_2(s)$ in water. What are $\Delta G°_{reaction}$ and K_{sp} for this reaction at 298 K (25 °C)?

(b) Check your result in part (a), using the values for standard entropies and free energies of formation in Appendix B. Do the two values for $\Delta S°_{reaction}$ and $\Delta G°_{reaction}$ agree reasonably well? Explain. How does your value for K_{sp} compare with the value in Table 9.2?

Graphical determination of thermodynamic values If you measure the equilibrium constant for a reaction at several temperatures, you can plot the data to obtain $\Delta H°_{reaction}$ and $\Delta S°_{reaction}$, and, hence, $\Delta G°_{reaction}$. Equation (9.55) is the equation of a straight line if $\ln K$ is plotted as a function of $\frac{1}{T}$. The slope of the line is $-\left(\frac{\Delta H°_{reaction}}{R}\right)$ and the intercept $\left(\text{at } \frac{1}{T} = 0\right)$ is $\frac{\Delta S°_{reaction}}{R}$. As an example, consider the temperature dependence of the equilibrium pressure of a gas, A, in equilibrium with liquid A. The vaporization reaction equation and equilibrium constant expression are

$$A(l) \rightleftharpoons A(g) \qquad\qquad\qquad (9.60)$$

$$K = \frac{(P_A(g))}{(A(l))} = \qquad\qquad\qquad (9.61)$$

The equilibrium constant for vaporization is the numeric value of the equilibrium vapor pressure (in bar). Table 9.5 gives the vapor pressures as a function of temperature for two isomeric liquids.

Table 9.5 *Vapor pressure variation with temperature for two $C_2H_2Cl_2$ isomers.*

Vapor pressure, torr	1.0	10.0	40.0	100.	400.	760.
		Temperature, °C				
cis-1,2-dichloroethene	−58.4	−29.9	−7.9	9.5	41.0	59.0
trans-1,2-dichloroethene		−38.0	−17.0	−0.2	30.8	47.8

9.71 WORKED EXAMPLE

Graphical determination of $\Delta H°_{reaction}$ and $\Delta S°_{reaction}$

Plot the data in Table 9.5 and use the plot to determine $\Delta H°_{reaction}$, $\Delta S°_{reaction}$, and the vapor pressure at 298 K for the vaporization of *cis*-1,2-dichloroethene.

Necessary information: We need the equilibrium constant expression (9.61) and equation (9.55). We also need to know that 1 bar = 750 torr and temperature in K = °C + 273.1.

Strategy: Convert vapor pressures (v.p.) in torr to bar and temperatures in °C to K. Then calculate and plot $\ln K$ [= $\ln\{(v.p.)/1\text{ bar}\}$] as a function of $1/T$, determine the slope and intercept of the line through the points, and use these to calculate $\Delta H°_{reaction}$ and $\Delta S°_{reaction}$.

Implementation: This table shows the converted data that are plotted in Figure 9.11.

v.p., torr	1.0	10.0	40.0	100.	400.	760.
v.p., bar	0.0013	0.0133	0.0533	0.133	0.533	1.013
ln$K$	−6.65	−4.320	−2.932	−2.017	−0.629	0.0129
T, K	214.7	243.2	265.2	282.6	314.1	332.1
$\frac{1}{T}$, K^{-1}	0.00466	0.00411	0.00377	0.00354	0.00318	0.00301

continued

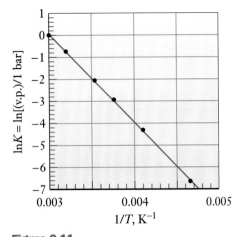

Figure 9.11.

Plot of ln(v.p.) vs. $\dfrac{1}{T}$ for

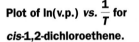

cis-1,2-dichloroethene.

The equation of the line (from the computer graphing software) in Figure 9.11 is

$$\ln K = (-4.03 \times 10^3 \text{ K})\left(\frac{1}{T}\right) + 12.22$$

Therefore, we get

$$\text{slope} = -\left(\frac{\Delta H°_{\text{reaction}}}{R}\right) = (-4.03 \times 10^3 \text{ K})$$

$$\Delta H°_{\text{reaction}} = (8.314 \text{ J·mol}^{-1}\text{·K}^{-1})(4.03 \times 10^3 \text{ K})$$

$$= 33.5 \times 10^3 \text{ J·mol}^{-1} = 33.5 \text{ kJ·mol}^{-1}$$

$$\text{intercept} = \frac{\Delta S°_{\text{reaction}}}{R} = 12.22$$

$$\Delta S°_{\text{reaction}} = (8.314 \text{ J·mol}^{-1}\text{·K}^{-1})(12.22) = 102 \text{ J·mol}^{-1}\text{·K}^{-1}$$

To get the vapor pressure at 298 K, we substitute in equation (9.55):

$$\ln K = \ln(\text{v.p.}) = -\left(\frac{\Delta H°_{\text{reaction}}}{R}\right)\left(\frac{1}{T}\right) + \frac{\Delta S°_{\text{reaction}}}{R}$$

$$= (-4.03 \times 10^3 \text{ K})\left(\frac{1}{298 \text{ K}}\right) + 12.22$$

$$\ln(\text{v.p.}) = -1.303; \therefore \text{ vapor pressure at 298 K} = 0.272 \text{ bar} (= 204 \text{ torr})$$

Does the answer make sense?   We have known since Chapter 1 that energy is required to vaporize a liquid, and in Chapter 8 we learned that the entropy of a substance increases when it goes from the liquid to the gas phase. Therefore, the positive signs for $\Delta H°_{\text{reaction}}$ and $\Delta S°_{\text{reaction}}$ make sense. To judge whether the sizes of the changes make sense, you can compare them with other simple carbon compounds (such as methanol or ethanol) in Appendix B and find that the values are fairly similar. Since 298 K (25 °C) lies between two of the temperatures in Table 9.5, we expect the vapor pressure to lie between the corresponding vapor pressures, as it does.

9.72 CHECK THIS

Graphical determination of $\Delta H°_{\text{reaction}}$ and $\Delta S°_{\text{reaction}}$

(a) Plot the data in Table 9.5 and use the plot to determine $\Delta H°_{\text{reaction}}$, $\Delta S°_{\text{reaction}}$, and the vapor pressure at 298 K for the vaporization of *trans*-1,2-dichloroethene.

(b) How do your results in part (a) compare with the values for *cis*-1,2-dichloroethene from Worked Example 9.71? Are the *directions* of any differences consistent with the structures of the two isomers? Explain why or why not. *Hint:* Consider the polarities of the isomers.

Reflection and Projection

The free energy change for a reaction can be expressed in terms of the standard free energy change and a reaction quotient that combines the concentration ratios of all the reactants and products. At equilibrium, where the free energy change is zero, the numeric value of the equilibrium reaction quotient is the equilibrium constant for the reaction. The equilibrium constant is directly related to the standard free energy change, which can be calculated from tabulated standard free energies of formation. We can apply these concepts (developed for gas phase equilibria in Section 9.11) to all reaction systems, if we are careful about how we define and use standard conditions.

Measuring the equilibrium constant for a reaction at two or more temperatures provides enough information to determine the standard enthalpy and entropy changes for the reaction. These values can be combined to find the standard free energy change and equilibrium constant for the reaction at other temperatures. The interrelationships of the equilibrium constant and thermodynamic variables provides us with numerous ways to analyze and understand equilibria, including those in biological systems, as we will discuss in the next section.

9.9. Thermodynamics in Living Systems

In a living cell, Figure 9.12, almost no reaction is at equilibrium. This is as it must be. A cell at equilibrium is unchanging, that is, dead. Only systems that are not in equilibrium can change. One reason why reactions in living cells are not in equilibrium is that many of the reactions are relatively slow, even though there are enzymes present to speed them up. Another reason is that so many competing reactions are going on that a molecule needed for one reaction might be used up by another before it can react by the first pathway.

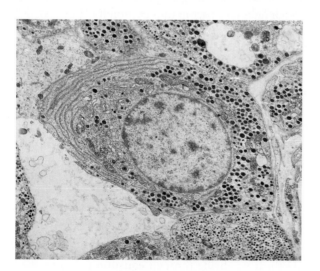

Figure 9.12.

Electron micrograph of a thin section of a hormone-secreting cell. A large number of chemical reactions go on continuously in all of the organelles in an intact cell.

Free energy for cellular reactions You might wonder why, if equilibria are not established for most reactions in living systems, we are at all interested in the equilibria for these reactions. We wish to find the free energies of reactions in living cells, because these help us understand the directionality of these life processes.

One approach is to carry out the reaction of interest in a test tube in the laboratory, allow it to come to equilibrium, determine its equilibrium constant, and thus get the standard free energy change for the reaction. We can then use this standard free energy, together with the reaction quotient under cellular conditions, to get the free energy change for the cellular reaction that is not at equilibrium.

Hydrolysis of ATP As an example of this approach, consider a reaction we have discussed before, the hydrolysis of adenosine triphosphate, ATP:

$$ATP^{4-}(aq) + 2H_2O(aq) \rightleftharpoons ADP^{3-}(aq) + HOPO_3^{2-}(aq) + H_3O^+(aq)$$
(9.62)

The structures of ATP^{4-} and ADP^{3-} are shown in Chapter 7, Figure 7.16. For reaction (9.62), the reaction quotient expressed in terms of molar concentrations is

$$Q = \frac{[ADP^{3-}(aq)]\,[HOPO_3^{2-}(aq)]\,[H_3O^+(aq)]}{[ATP^{4-}(aq)]}$$
(9.63)

We know that, at equilibrium, the numeric value of Q calculated from the concentrations in the solution is equal to the numeric value of K, the equilibrium constant. Unfortunately, these concentrations are difficult to obtain.

We encounter several problems trying to get the concentrations to substitute in equation (9.63). The pH in living cells is close to 7. Look at Table 9.2 and you will find that the conjugate acid–base pair, $(HO)_2PO_2^-(aq)$ and $HOPO_3^{2-}(aq)$, has a $pK_a = 7.20$. The conjugate base-to-acid ratio depends on the pH of the solution, so the concentration of $HOPO_3^{2-}(aq)$ depends on the pH of the solution. Exactly this same problem also occurs for both $ATP^{4-}(aq)$ and $ADP^{3-}(aq)$. The proton transfer to water from the end phosphate in each molecule also has a pK_a around 7. Therefore, the phosphate-containing compounds in reaction (9.62) are each present in at least two ionic forms and we don't know which ones take part in the reaction (or if they all do).

Equilibrium constant for hydrolysis of ATP Since we lack information about the specific concentrations in these systems, one approach is simply to use the sum of all the forms present in the system. For example, suppose we add 2×10^{-6} mol of ATP to 1.00 milliliter of a reaction mixture. We say that the ATP concentration is 0.002 M and symbolize this total as [ATP], without regard to how much of each of its specific ionic forms is present. P_i is commonly used to symbolize the sum of the concentrations of the various forms of inorganic phosphate $[(HO)_2PO_2^-(aq), HOPO_3^{2-}(aq)$, and so on]. Experiments (at 298 K) have shown that

$$K = \left(\frac{[ADP(aq)]\,[P_i(aq)]\,[H_3O^+(aq)]}{[ATP(aq)]}\right)_{eq} \approx 10^{-2}$$
(9.64)

Biochemists usually take solutions at pH 7 as their reference standard (instead of pH 0). If we divide through equation (9.72) by the hydronium ion concentration (= 10^{-7} M), we get

$$\frac{K}{[H_3O^+(aq)]} = \frac{K}{10^{-7}M} = K' = \left(\frac{[ADP(aq)]\,[P_i(aq)]}{[ATP(aq)]}\right)_{eq} \approx \frac{10^{-2}}{10^{-7}} = 10^5$$
(9.65)

Another complication in studying ATP and ADP is the dependence of their reactions on the presence of Mg^{2+}. ATP and ADP form complexes with Mg^{2+}, and it is these complexes that actually take part in enzyme-catalyzed reactions. Laboratory studies are usually carried out with added Mg^{2+} to mimic the conditions for ATP and ADP in a living cell.

Molarities instead of concentration ratios are used in equation (9.64) and subsequent equations because the standard states for collective species like ATP(aq) and P_i(aq) are not well defined. We will continue to express their equilibrium constants as dimensionless quantities, although this is not strictly correct.

K' is the equilibrium constant for ATP hydrolysis *at pH* 7. The standard free energy change for hydrolysis of a mole of ATP *at pH* 7 is

$$\Delta G^{\circ\prime}_{\text{hydrolysis}} = -RT \ln K' = -(8.314\,\text{J·mol}^{-1}\text{·K}^{-1})(298\text{ K})\ln(10^5)$$

$$= -29\,\text{kJ·mol}^{-1} \tag{9.66}$$

This nomenclature, K' and $\Delta G^{\circ\prime}$, for systems with pH 7 as the standard state for hydronium ion will appear again in Chapter 10.

9.73 CONSIDER THIS

What is the equilibrium cellular concentration of ATP?

If the cellular concentrations of ADP and P_i are about 0.0002 M and 0.001 M, respectively, what ATP concentration is in equilibrium with these concentrations of hydrolysis products? Is your result consistent with the $\Delta G^{\circ\prime}_{\text{hydrolysis}}$? Explain why or why not.

Free energy changes for cellular ATP reactions The concentrations of ATP, ADP, and P_i vary from one part of a cell to another, but the concentration of ATP is usually about ten times higher than ADP and the P_i concentration is about 0.001 M. The free energy change for ATP hydrolysis under these conditions is

$$\Delta G'_{\text{hydrolysis}} = \Delta G^{\circ\prime}_{\text{hydrolysis}} + RT \ln Q'$$

$$= \Delta G^{\circ\prime}_{\text{hydrolysis}} + RT \ln \left\{ \left(\frac{[\text{ADP}(aq)]}{[\text{ATP}(aq)]} \right) [P_i(aq)] \right\}$$

$\Delta G'_{\text{hydrolysis}}$

$$= -29\,\text{kJ·mol}^{-1} + (8.314\,\text{J·mol}^{-1}\text{·K}^{-1})(298\text{ K})\ln\left\{ \left(\frac{1}{10} \right)(0.001) \right\}$$

$$\Delta G'_{\text{hydrolysis}} = -29\,\text{kJ·mol}^{-1} + (-23\,\text{kJ·mol}^{-1}) = -52\,\text{kJ·mol}^{-1} \tag{9.67}$$

Under cellular conditions, the free energy change for ATP hydrolysis is almost twice as large as the standard free energy change. There is a strong driving force for ATP hydrolysis that can be *coupled* with other reactions (which would otherwise not be favored) to make life possible.

In Chapter 7, Section 7.9, we calculated the percentage of the *enthalpy* released in the complete oxidation of glucose that is captured by organisms to synthesize ATP. And, in Chapter 8, Section 8.14, we compared the *standard* free energy change, ΔG°, for glucose oxidation coupled to ATP synthesis. The percentage of the free energy release, $\Delta G_{\text{reaction}}$, that is captured is a more appropriate measure of the efficiency of an organism. $\Delta G^{\circ}_{\text{reaction}}$ is -2878.42 kJ for the oxidation of 1 mol of glucose:

$$C_6H_{12}O_6(s) + 6O_2(g) \rightarrow 6CO_2(g) + 6H_2O(l) \tag{9.68}$$

For the purposes of this approximate calculation, we will assume that $\Delta G_{\text{reaction}} \approx \Delta G^{\circ}_{\text{reaction}}$ for the oxidation of glucose.

The synthesis of ATP from ADP and P_i is the reverse of hydrolysis. In the cell, the free energy change for ATP synthesis has the opposite sign but the same numeric value as the free energy change for hydrolysis: $\Delta G'_{\text{synthesis}} = 52 \text{ kJ·mol}^{-1}$. Recall from Section 7.9 that about 36 mol of ATP can be formed for every mole of glucose oxidized. Synthesis of 36 mol of ATP requires about 1870 kJ. Under conditions in the cell, about 65% $\left(= \left[\frac{1870}{2880}\right] \times 100\%\right)$ of the free energy released in the oxidation of glucose can be captured in the synthesis of ATP. The free energy available from glucose oxidation is used efficiently to maintain the ATP hydrolysis reaction far from equilibrium and, thus, provide the strong directional drive for life processes.

9.74 CONSIDER THIS

How do you analyze a coupled reaction?

We can rewrite equation (9.62), *at pH* 7, as

$$\text{ATP}(aq) + \text{H}_2\text{O}(l) \rightleftharpoons \text{ADP}(aq) + \text{P}_i(aq)$$

$$\Delta G^{\circ\prime}_{(9.69)} = -RT \ln K' = -29 \text{ kJ·mol}^{-1} \tag{9.69}$$

Many of the compounds used and formed in metabolic pathways are phosphate esters. An example is glycerol phosphate, $\text{HOCH}_2\text{CHOHCH}_2\text{OPO}_3^{2-}$ (abbreviated as R-OPO_3^{2-}). These esters hydrolyze to the alcohol and phosphate; R-OPO_3^{2-} hydrolyzes to glycerol, R-OH:

$$\text{R-OPO}_3^{2-}(aq) + \text{H}_2\text{O}(l) \rightleftharpoons \text{R-OH}(aq) + \text{P}_i(aq)$$

$$\Delta G^{\circ\prime}_{(9.70)} = -9 \text{ kJ·mol}^{-1} \tag{9.70}$$

Our bodies need the phosphate ester, but the equilibrium favors the alcohol. ATP can transfer a phosphate group to the alcohol to make the phosphate ester:

$$\text{R-OH}(aq) + \text{ATP}(aq) \rightleftharpoons \text{R-OPO}_3^{2-}(aq) + \text{ADP}(aq) \tag{9.71}$$

Reaction (9.71) represents a direct coupling of two reactions to become one.

(a) What are $\Delta G^{\circ\prime}_{\text{reaction}}$ and K' for reaction (9.71)? Show clearly how you get your answers. *Hint:* If necessary, review Section 7.9 or 8.14 to recall how thermodynamic quantities are combined when reactions are coupled.

(b) If reaction (9.71) reaches equilibrium when $\left(\dfrac{[\text{ATP}(aq)]}{[\text{ADP}(aq)]}\right) = 10$ (conditions in a cell), what is the ratio of R-OPO_3^{2-} to R-OH? Is this ratio advantageous to the cell? Explain your reasoning.

Your result in Consider This 9.74(b) demonstrates how the free energy released by ATP hydrolysis can be coupled to a reaction that requires an input of free energy to displace its equilibrium in a direction that is advantageous to the organism. This free energy is available because the ATP hydrolysis reaction is so far out of equilibrium. The observed concentration of ATP, about 2×10^{-3} M, in your cells is 10^9 times larger than the equilibrium concentration you calculated in Consider This 9.73.

How is the concentration of ATP maintained so far from equilibrium when it is continually hydrolyzed to drive such a large number of other reactions? The answer lies in the continuing supply of fuel molecules entering the cell and the speed of the reactions within the cell. The cell is an open system that exchanges matter, including the molecules that serve as fuels, with its surroundings. The synthesis reactions that use the fuel go fast enough to regenerate ATP as it is depleted, so its concentration is maintained approximately constant and far from equilibrium. The speed (rate) of reactions is the topic of Chapter 11.

Equilibria in oxygen transport Oxygen is required by your cells to metabolize fuels like glucose, as in equation (9.68). The reaction shown in the chapter opening illustration is the release of the required oxygen from the oxygenated form of hemoglobin as red blood cells pass through your body's capillaries. We pointed out in the introductory discussion that the binding of oxygen to hemoglobin, Hb, is an equilibrium process with an equilibrium constant K_1:

$$Hb + O_2(g) \rightleftharpoons Hb(O_2) \quad K_1 \tag{9.72}$$

However, oxygen binding to hemoglobin is more complicated than this because four molecules of oxygen can bind to hemoglobin, one to each of the α and β subunits shown in Figure 9.8. Thus, there are three further successive oxygen-binding reactions, each adding an oxygen to the previous hemoglobin–oxygen complex and each with its own equilibrium constant:

$$Hb(O_2) + O_2(g) \rightleftharpoons Hb(O_2)_2 \quad K_2 \tag{9.73}$$

$$Hb(O_2)_2 + O_2(g) \rightleftharpoons Hb(O_2)_3 \quad K_3 \tag{9.74}$$

$$Hb(O_2)_3 + O_2(g) \rightleftharpoons Hb(O_2)_4 \quad K_4 \tag{9.75}$$

The subscripts on the equilibrium constants denote the number of bound oxygen molecules in the product.

One way to characterize the binding of oxygen to hemoglobin is shown in Figure 9.13. This is a plot of the percent oxygen saturation as a function of the pressure of oxygen gas in contact with the protein. For example, an oxygen saturation of 25% means that one-fourth the maximum amount of oxygen is bound, or that, on the average, each hemoglobin molecule has one bound oxygen molecule. This is an average, so some hemoglobins will have more than one bound oxygen and some will have none.

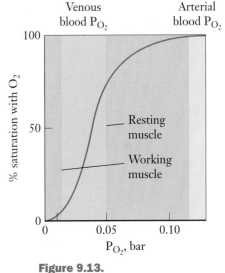

Figure 9.13.

Oxygen-saturation curve for hemoglobin.

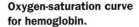

9.75 CHECK THIS

Transport of oxygen by hemoglobin

(a) What is the percent oxygen saturation of hemoglobin when it is in the arterial blood (blood that has recently been through the lungs)? In the venous blood (blood that has passed through oxygen-requiring tissues)?

(b) How do your answers in part (a) explain how hemoglobin is able to transport oxygen from the lungs to the tissues where it is required to oxidize the fuel molecules that provide energy for the cells in these tissues?

Note that the oxygen-saturation plot for hemoglobin, Figure 9.13, is shaped sort of like an s. The oxygen-saturation curve starts up slowly and then the upward slope increases as more oxygen is bound and finally bends over as the system nears saturation. The shape is usually called *sigmoidal* because the Greek letter sigma, σ, is equivalent to the Roman s. Such curves appear often in studies of biological systems. Sigmoidal responses like the one for oxygen binding to hemoglobin indicate that there is some kind of cooperation among the reactions that cause the response. This usually means a reaction that changes one part of the protein makes it easier for reactions to occur in other parts of the protein.

In the case of hemoglobin, the four protein chains in the deoxygenated form, Hb, are held together in a shape that is unfavorable for oxygen to bind to the protein. The chains are held together by both hydrophobic effects and by attractions between oppositely charged amino acid side groups on the different chains. In order for the first oxygen molecule to bind, several of these interactions have to be disrupted, which requires an input of energy. Energy is released by the binding of the oxygen to the heme group in the hemoglobin. The *net* free energy change for disrupting the interactions between chains and binding successive oxygens is shown on a free energy diagram in Figure 9.14.

Figure 9.14 shows that the net free energy change for binding the first oxygen to hemoglobin is more positive (less favorable) than for binding the others. The release of energy by the binding is about the same for each oxygen, but the interactions that must be disrupted are fewer for the second and third oxygen, which bind with about the same affinity. Binding of the fourth oxygen can be favorable, if the solution conditions are appropriate.

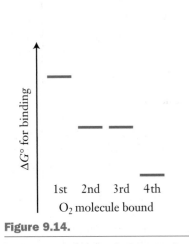
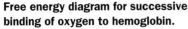

Figure 9.14.

Free energy diagram for successive binding of oxygen to hemoglobin.

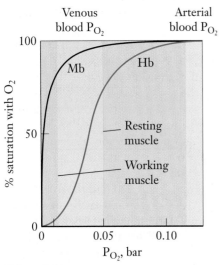

Figure 9.15.

Comparison of the oxygen-saturation curves for hemoglobin and myoglobin.

9.76 CHECK THIS

Relative sizes of oxygen-binding equilibrium constants

(a) List the oxygen-binding equilibrium constants for reactions (9.72) through (9.75) in order of increasing size. Clearly explain how you decide on your order.

(b) Hemoglobin transports oxygen inside blood vessels, but does not actually contact the cells that need the oxygen. The oxygen is transferred (through capillary and cell membranes) to myoglobin. Myoglobin, Mb, is a protein that is similar to the individual chains in hemoglobin. Mb stores oxygen in your cells, especially muscles, and carries it to where the cell needs it. Use the information in Figure 9.15 to explain why oxygen can be transferred from hemoglobin to myoglobin at the low effective oxygen pressures, less than 0.05 bar, in the capillaries.

In this chapter, we have limited our discussion to only a few equilibrium systems and the relationships between equilibria and thermodynamics. Even without further elaboration, you can see how important the concept of equilibrium is and can appreciate Le Chatelier's realization that it is central to all of chemistry. We will use equilibrium concepts from this chapter in the final two chapters of the book and you will meet equilibria often in all sciences.

9.10 Outcomes Review

Many chemical systems that appear unchanging have reached a state of equilibrium where the forward and reverse reactions that characterize their chemistry are going at the same speed. We cannot detect any net change by observing these systems, but the reactions continue at the molecular level; equilibrium is dynamic. Le Chatelier's principle states that, if we disturb an equilibrium system by changing one or more concentrations or adding or subtracting energy, the system responds in a way that minimizes the disturbance. The equilibrium constant and equilibrium constant expression provide ways to quantify this qualitative (directional) principle and we applied them to acid–base, solubility, gas phase, and biochemical reactions.

We connected the equilibrium constant to the thermodynamic variables introduced in the previous two chapters and found that this combination allows us to calculate equilibrium constants from thermodynamic data alone and conversely to use equilibrium measurements to obtain thermodynamic data. A particularly useful result of these connections is the ability to analyze the temperature dependence of equilibria quantitatively.

Check your understanding of the ideas in this chapter by reviewing these expected outcomes of your study. You should be able to

- Use Le Chatelier's principle to predict the direction of the response of an equilibrium system to changes in the concentration(s) of reactants or products [Sections 9.1, 9.3, 9.4, 9.5, 9.6, and 9.9].
- Use Le Chatelier's principle and the response of an equilibrium system to heating or cooling to tell whether the reaction is endothermic or exothermic [Sections 9.1 and 9.8].
- Explain how Le Chatelier's principle is a consequence of the dynamic nature of chemical equilibrium [Section 9.1].
- Write the equilibrium constant expression using appropriate concentration ratios, hence defining the equilibrium constant, K, for a balanced chemical reaction [Sections 9.2, 9.3, 9.6, 9.7, and 9.9].
- Find K_a and pK_a for a weak acid (or K_b and pK_b for a weak base) when you have the initial concentrations in the solution and the pH or pOH at equilibrium [Sections 9.3 and 9.4].
- Find the pH and/or pOH of a solution of a weak acid or its conjugate base when you have the initial concentrations in the solution and the K_a or pK_a for the acid [Sections 9.3 and 9.4].

- Find the concentrations of a weak acid and its conjugate base in a solution when you have the initial concentrations in the solution and the K_a or pK_a for the acid [Sections 9.3, 9.4, and 9.5].
- Find the pH of a buffer solution when you have the initial concentrations of the weak acid and its conjugate base in the solution and the K_a or pK_a for the acid [Section 9.4].
- Tell how to prepare a buffer solution of a specified pH and specified total concentration of an appropriately selected weak acid–base conjugate pair [Section 9.4].
- Predict the direction of change of the net charge on a protein and its consequent behavior in electrophoresis if one amino acid is substituted for another in its structure [Section 9.5].
- Find the solubility product, K_{sp} and pK_{sp}, for an ionic compound when you have data for the solubility of the compound and vice versa [Sections 9.6 and 9.8].
- Find the solubility of an ionic compound of known K_{sp} or pK_{sp} in a solution containing a stoichiometric excess of one of the ions [Section 9.6].
- Find $\Delta G^\circ_{reaction}$ for a reaction when you have the equilibrium constant, K, for the reaction [Section 9.7, 9.8, and 9.9].
- Find the equilibrium constant, K, for a reaction when you have ΔG_f° for the reactants and products [Sections 9.7].
- Find $\Delta H^\circ_{reaction}$, $\Delta G^\circ_{reaction}$, and $\Delta S^\circ_{reaction}$ for a reaction when you have calorimetric data for the reaction and its equilibrium constant, K [Section 9.7].
- Find $\Delta H^\circ_{reaction}$, $\Delta S^\circ_{reaction}$, and $\Delta G^\circ_{reaction}$ for a reaction when you have values of K for the reaction at two or more temperatures [Section 9.8].
- Find $\Delta G^\circ_{reaction}$ and K for a reaction at a temperature T when you have $\Delta H^\circ_{reaction}$ and $\Delta S^\circ_{reaction}$ for the reaction [Section 9.8].
- Use the reaction quotient, Q, to find the free energy change, $\Delta G_{reaction}$, for a reaction that is not at equilibrium when you have the standard free energy change, $\Delta G^\circ_{reaction}$, or K for the reaction and the concentrations in the nonequilibrium system [Sections 9.9 and 9.11].
- Find $\Delta G^\circ_{reaction}$ or $\Delta G_{reaction}$ for a coupled reaction when you have $\Delta G^\circ_{reaction}$ or $\Delta G_{reaction}$ for the individual reactions that are coupled [Section 9.9].
- Calculate the entropy of gases at nonstandard pressures [Section 9.11].
- Show how the entropy of gases at nonstandard pressures is related to the free energy and reaction quotient for a gas-phase reaction [Section 9.11].

9.11. Extension—Thermodynamic Basis for the Equilibrium Constant

9.77 CONSIDER THIS

What are some properties of nitrogen oxides?

Nitrogen reacts with oxygen to form several oxides. One of the oxides of nitrogen is **nitrogen dioxide,** NO_2.

(a) Write a Lewis structure for NO_2; nitrogen is the central atom in the molecule. Can all three atoms have an octet of electrons in their valence orbitals? Why or why not?

(b) Do you expect NO_2 to be reactive or unreactive? Why or why not?

(c) Two molecules of NO_2 can react to form N_2O_4, **dinitrogen tetroxide:**

$$2NO_2(g) \rightleftharpoons N_2O_4(g) \tag{9.76}$$

Write a Lewis structure for N_2O_4.

(d) Do you expect the entropy change for reaction (9.76) to be positive or negative? Give your reasoning.

(e) Do you expect reaction (9.76), formation of a bond between two $NO_2(g)$ molecules, to be exothermic or endothermic? Give your reasoning.

Most of the reactions we have discussed in this chapter occur in aqueous solutions. However, we begin our treatment of the relationship between equilibrium and thermodynamic variables with gases because we can use our model for positional entropy from Chapter 8 to express the entropy of a gas as a function of its pressure. Except when they collide, molecules in the gas phase have little influence on one another. Thus, we do not have to be concerned about solvent effects, which are so important in solution reactions. Our example will be the reaction of $NO_2(g)$ molecules to form $N_2O_4(g)$, reaction (9.76) from Consider This 9.77. This simple reaction has the enormous advantage that one of the gases is a deep reddish-brown color, Figure 9.16, so we can tell a good deal about the reaction system, just by looking at it, as we did with the colored solution reactions in Section 9.1.

Figure 9.16.

Effect of temperature on equilibrium mixtures of $NO_2(g)$ and $N_2O_4(g)$. In (a), both sample tubes are at room temperature. In (b), one sample tube is in ice water and the other in hot water.

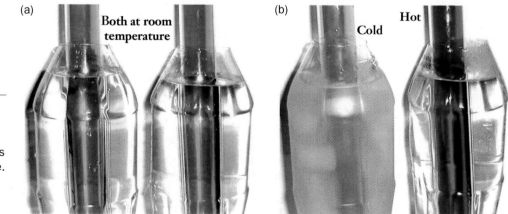

(a) Both at room temperature

(b) Cold / Hot

9.78 CONSIDER THIS

Which of the gases is colored: $NO_2(g)$ or $N_2O_4(g)$?

(a) The two sealed tubes shown in Figure 9.16(a) contain identical, equilibrium mixtures of $NO_2(g)$ and $N_2O_4(g)$. When energy is added to this equilibrium system, Figure 9.16(b), is there more or less of the colored gas in the new equilibrium mixture? Give your reasoning clearly.

(b) *Assuming* that the colored gas is $NO_2(g)$, what do Le Chatelier's principle and your answer in part (a) tell you about the energetics of reaction (9.76)? Is the reaction exothermic or endothermic? Clearly explain your answer.

(c) *Assuming* that the colored gas is $N_2O_4(g)$, what do Le Chatelier's principle and your answer in part (a) tell you about energetics of reaction (9.76)? Clearly explain your answer.

(d) Which of your answers, part (b) or part (c), agrees with your answer in Consider This 9.77(e)? Is the colored gas $NO_2(g)$ or $N_2O_4(g)$? Explain your choice.

Changes under nonstandard conditions In Chapter 8, we found that the net entropy change for a reaction under standard conditions (one bar pressure) can be written in terms of the standard entropy and enthalpy changes for the reaction:

$$\Delta S°_{net} = \Delta S°_{reaction} - \frac{\Delta H°_{reaction}}{T} \tag{9.77}$$

Most reactions do not involve reactants and products in their standard states. We must use $\Delta H_{reaction}$ and $\Delta S_{reaction}$, enthalpy and entropy changes under non-standard conditions, to get ΔS_{net} for the reaction under the actual conditions:

$$\Delta S_{net} = \Delta S_{reaction} - \frac{\Delta H°_{reaction}}{T} \tag{9.78}$$

To analyze a gas phase reaction like $NO_2(g)$ forming $N_2O_4(g)$, we need to know how the enthalpy and entropy of each gas varies with its pressure at constant temperature, that is, $\Delta H_{P\,change}$ and $\Delta S_{P\,change}$ for

$$gas\ (P°, T) \rightarrow gas\ (P, T) \tag{9.79}$$

We will assume that the gases are ideal, Chapter 7, Section 7.13. The enthalpy of an ideal gas, like its total energy, is a function only of temperature, so $\Delta H_{P\,change} = 0$ for the constant temperature process in equation (9.79). Thus, $\Delta H_{reaction} = \Delta H°_{reaction}$. Since the reaction is formation of a chemical bond, you know that the reaction must be exothermic; the product is at a lower energy than the reactants.

> The enthalpy of *ideal* gases is a function only of temperature. (See Chapter 8, Problem 8.96.) Real gases can usually be treated as ideal, if they are not at high pressure.

9.79 CHECK THIS

$\Delta H°_{reaction}$ for $2NO_2(g) \rightleftharpoons N_2O_4(g)$

Use the standard enthalpies of formation, $\Delta H_f°$, in Appendix B to calculate $\Delta H°_{reaction}$ for reaction (9.76). Does your answer confirm your reasoning in Consider This 9.77(e) and 9.78(d)? Explain why or why not.

The entropy of a gas *does* depend on its pressure. In Chapter 8, Section 8.3, we found that the number of arrangements of molecules in space, W, increases as the volume per molecule (or total volume, V) increases. The relationships among W, V, and P $(= nRT/V)$ for a gas are

$$W \propto V \propto \frac{1}{P} \tag{9.80}$$

The positional entropy change, $\Delta S_{P\,change}$, for one mole of gas going from standard pressure, $P° = 1$ bar with a volume $V°$, to a pressure P with a volume V, reaction (9.79), is

$$\Delta S_{P\,change} = R\ln W_V - R\ln W_{V°} = R\ln\left(\frac{W_V}{W_{V°}}\right) = R\ln\left(\frac{V}{V°}\right) = R\ln\left(\frac{P°}{P}\right) \tag{9.81}$$

At a pressure P, the entropy of a gas, S_{gas}, is its standard entropy, $S°_{gas}$, plus the entropy change for reaction (9.79), $\Delta S_{P\,change}$:

$$S_{gas} = S°_{gas} + \Delta S_{P\,change} = S°_{gas} + R\ln\left(\frac{P°}{P}\right) = S°_{gas} - R\ln\left(\frac{P}{P°}\right) \tag{9.82}$$

Recall from Section 9.2, Table 9.1, that $(P/P°)$ is the dimensionless ratio of pressures (in bar) we use in thermodynamics. The logarithmic terms we meet here are mathematically correct. Since $P° = 1$ bar, the numeric value of $(P/P°)$ is (P). We write the pressure in parentheses to remind ourselves that it represents the ratio. Using this nomenclature, equation (9.82) becomes

$$S_{gas} = S°_{gas} - R\ln(P) \tag{9.83}$$

9.80 CONSIDER THIS

What is the entropy of a compressed gas?

(a) The standard entropy of carbon dioxide gas (at 298 K) is 213.7 J·K^{-1}·mol^{-1}. If carbon dioxide is compressed to a pressure of 2.8 bar, what is the entropy of one mole of the gas? Does the direction of the change from the standard entropy make sense? Explain why or why not.

(b) If the gas is expanded to a pressure less than one bar, does its entropy increase or decrease relative to its standard entropy? Explain the reasoning for your answer.

Entropy change for a gas phase reaction Let's determine the entropy change for the reaction of two moles of $NO_2(g)$ at a pressure P_{NO_2} to give one mole of $N_2O_4(g)$ at a pressure $P_{N_2O_4}$. We will use equation (9.83) to find the entropy for each gas at these pressures and then combine the entropies to get, $\Delta S_{reaction}$ for reaction (9.76):

$$\Delta S_{reaction} = S_{N_2O_4} - 2\,S_{NO_2}$$

$$= \{S°_{N_2O_4} - R\ln(P_{N_2O_4})\} - 2\{S°_{NO_2} - R\ln(P_{NO_2})\} \tag{9.84}$$

$$\Delta S_{reaction} = \Delta S°_{reaction} - R\{\ln(P_{N_2O_4}) - \ln(P_{NO_2})^2\}$$

$$\Delta S_{reaction} = \Delta S°_{reaction} - R\ln\left(\frac{(P_{N_2O_4})}{(P_{NO_2})^2}\right) \tag{9.85}$$

9.81 CONSIDER THIS

What is $\Delta S_{reaction}$ for $2NO_2(g) \rightleftharpoons N_2O_4(g)$?

(a) Use the standard entropies in Appendix B to find $\Delta S°_{reaction}$ for reaction (9.76) at 298 K. Is this the result you predicted in Consider This 9.77(d)? Explain why or why not.

(b) What is $\Delta S_{reaction}$ for a system in which the $NO_2(g)$ pressure is 0.243 bar and the $N_2O_4(g)$ pressure is 0.437 bar? Is the difference between $\Delta S°_{reaction}$ and $\Delta S_{reaction}$ in the direction you expect? Explain why or why not. (Remember that standard pressure is 1 bar.)

Free energy change for a gas reaction The change in Gibbs free energy, $\Delta G_{reaction}$, for a reaction at constant temperature and pressure is

$$\Delta G_{reaction} = -T\Delta S_{net} = \Delta H_{reaction} - T\Delta S_{reaction} \tag{9.86}$$

We can account for pressure dependence of the free energy in gas reactions by substituting for $\Delta S_{reaction}$ from equation (9.85). Also, recall that $\Delta H_{reaction} = \Delta H°_{reaction}$. These substitutions give

$$\Delta G_{reaction} = \Delta H°_{reaction} - T\Delta S°_{reaction} + RT\ln\left(\frac{(P_{N_2O_4})}{(P_{NO_2})^2}\right) \tag{9.87}$$

The first two terms on the right-hand side of equation (9.87) are equal to the standard free energy change for the reaction, $\Delta G°_{reaction}$, so we have

$$\Delta G_{reaction} = \Delta G°_{reaction} + RT\ln\left(\frac{(P_{N_2O_4})}{(P_{NO_2})^2}\right) \tag{9.88}$$

9.82 CONSIDER THIS

What is $\Delta G_{reaction}$ for $2NO_2(g) \rightleftharpoons N_2O_4(g)$?

(a) Use the standard enthalpy of reaction from Check This 9.79 and standard entropy of reaction from Consider This 9.81 to calculate $\Delta G°_{reaction}$ for reaction (9.76).

(b) Use your answer in part (a) to calculate $\Delta G_{reaction}$ under the conditions stated in Consider This 9.81(b). Is reaction (9.76) spontaneous under these conditions? Explain why or why not.

The reaction quotient The ratio in parentheses in equation (9.88) is the **reaction quotient**, Q:

$$\Delta G_{\text{reaction}} = \Delta G°_{\text{reaction}} + RT\ln\left(\frac{(P_{N_2O_4})}{(P_{NO_2})^2}\right) = \Delta G°_{\text{reaction}} + RT\ln Q \tag{9.89}$$

The reaction quotient has the same *form* as the equilibrium constant expression for reaction (9.76). At this point we make a great logical leap and suggest that we might generalize the reasoning we have gone through for gases and apply equation (9.89) to all reactions, including our generalized reaction:

$$a\mathbf{A} + b\mathbf{B} \rightleftharpoons c\mathbf{C} + d\mathbf{D} \tag{9.5}$$

The result is

$$\Delta G_{\text{reaction}} = \Delta G°_{\text{reaction}} + RT\ln\left(\frac{(\mathbf{C})^c(\mathbf{D})^d}{(\mathbf{A})^a(\mathbf{B})^b}\right) \tag{9.42}$$

Equation (9.42) is the equation from which we began our discussion of the relationship between thermodynamics and the equilibrium constant in Section 9.7. Our justification for accepting the application of equation (9.89) to all reactions is its success in analyzing and explaining equilibrium systems, a few of which we discussed in Sections 9.7, 9.8, and 9.9.

Chapter 9 Problems

9.1. The Nature of Equilibrium

9.1. If $NH_3(aq)$ is slowly added to a light blue solution of $Cu^{2+}(aq)$, a pale blue precipitate, $Cu(OH)_2(s)$, forms. When additional $NH_3(aq)$ is added, the precipitate dissolves and a clear, deep blue-violet solution results. The color results from the formation of $[Cu(NH_3)_4]^{2+}(aq)$.
(a) Write balanced chemical equations for these two equilibrium reactions, labeling the colors of the species containing copper ion.
(b) Will acidic or basic conditions favor the formation of $Cu(OH)_2(s)$ from $Cu^{2+}(aq)$? Use the first equation you wrote in part (a) to help explain your reasoning.
(c) Will mildly acidic conditions favor the formation of $[Cu(NH_3)_4]^{2+}(aq)$ from $Cu(OH)_2(s)$? Use the second equation you wrote in part (a) to help explain your reasoning.
(d) Are these equilibrium reactions also redox reactions? Explain your reasoning.

9.2. Explain how Figure 9.1 would be different if $Fe^{3+}(aq)$ and $SCN^-(aq)$ formed a one-to-two metal-to-ligand complex. Make a sketch of the new figure.

9.3. Consider this reaction at equilibrium:

$$Cd^{2+}(aq) + 6CN^-(aq) \rightleftharpoons Cd(CN)_6{}^{2-}(aq)$$
colorless   colorless

(a) If the concentration of $CN^-(aq)$ is increased, how do you expect the concentrations of the other species to change? Explain your reasoning.

(b) In Investigate This 9.8(b), you were able to determine how temperature influenced the equilibrium between two cobalt(II) complexes. The same approach would not be effective with the equilibrium considered here. Explain why not.

9.4. Consider this equilibrium reaction: $H_2(g) + Cl_2(g) \rightleftharpoons 2HCl(g)$. Discuss the validity of each of these statements.
(a) When this reaction has reached a state of equilibrium, no further reaction occurs.
(b) When equilibrium is established, the number of moles of reactants equals the number of moles of products for this reaction.
(c) The concentration of each substance in the system will be constant at equilibrium.

9.5. Consider this equilibrium reaction: $CO(g) + 2O_2(g) \rightleftharpoons 2CO_2(g)$. For this reaction at equilibrium, predict the effect of raising its temperature. Explain the reasoning for your prediction. *Hint:* Consider the bond enthalpies in Table 7.3, Chapter 7, Section 7.7.

9.6. Consider this equilibrium reaction: $CO(g) + 2H_2(g) \rightleftharpoons CH_3OH(g)$. For this reaction at equilibrium, predict the effect of each of these changes on the equilibrium. In each case explain the reasoning for your prediction.
(a) Some of the $CH_3OH(g)$ is removed while the volume is held constant.
(b) Some $CO(g)$ is added to the system while the volume is held constant.

(c) The system is compressed to a smaller volume.

(d) Some $N_2(g)$ is added to the system while the volume is held constant.

9.7. Consider this Lewis acid–base reaction between boric acid and water (Lewis acid–base reactions were discussed in Chapter 6.):

$$B(OH)_3(s) + 2H_2O(aq) \rightleftharpoons [B(OH)_4]^-(aq) + H_3O^+(aq)$$

Assume that the reaction is in equilibrium with all species present at a given temperature.

(a) How do you predict this equilibrium system will change, if more $B(OH)_3(s)$ is added to the mixture? Explain your reasoning.

(b) How do you predict this equilibrium system will change, if more $H_3O^+(aq)$ is added to the mixture? Explain your reasoning.

(c) How do you predict this equilibrium system will change, if $OH^-(aq)$ is added to the mixture? Explain your reasoning.

9.8. Consider the reaction equation for a weak acid, HA*(aq)*, transferring its proton to a water molecule:
$HA(aq) + H_2O(aq) \rightleftharpoons H_3O^+(aq) + A^-(aq)$

(a) Describe how you interpret this reaction equation at the molecular level. Be as complete as possible in describing the fate of each of the species represented.

(b) 🐾 Play the animation in the *Web Companion*, Chapter 9, Section 9.4.1, that represents the reaction of a weak acid in water. Is this representation consistent with the description that you wrote for part (a)? If not, what is(are) the difference(s)? Does the animation provide any insight into a reaction that is missing in the equation? Explain.

9.2. Mathematical Expression for the Equilibrium Condition

9.9. About one percent of the world's energy resources are used to produce hydrogen and then ammonia from hydrogen and nitrogen. The reaction equation for the synthesis of ammonia is $N_2(g) + 3H_2(g) \rightleftharpoons 2NH_3(g)$. Write the equilibrium constant expression for the ammonia synthesis. If you want to calculate a numeric value for the equilibrium constant that is consistent with thermodynamic standard states, what units should be used for the dimensionless concentration ratio for each species?

9.10. An instructor used the following demonstration to show the difference between the dissociation of a strong acid and weak acid in water. She put the probes of a conductivity detector in 10.0 mL of distilled water. The detector was a simple type in which a small bulb lights up if the solution conducts electricity. In distilled water the bulb did not light up. She added 1 drop of 1 M HCl to the water and the bulb lit brightly. She rinsed the probe and put it into another 10.0-mL sample of distilled water. To this sample she added a 1 M solution of weak acid one drop at a time. When 50 drops of this acid had been

added, the bulb lit as brightly as with the HCl. Estimate the percent dissociation and the dissociation constant of the weak acid.

9.11. (a) Write the equilibrium constant expression for $CaCO_3(s) \rightleftharpoons CaO(s) + CO_2(g)$. What units must be used for each of the species, in order to make the equilibrium constant, K, dimensionless and consistent with the standard state definitions in Table 9.1? Explain.

(b) If the equilibrium pressure of $CO_2(g)$ is 0.37 bar in a mixture of these species, what is the numeric value of K? Explain the reasoning for your answer.

9.12. This problem refers to the equilibrium systems represented by the molecular-level drawings in Figure 9.2.

(a) The result of adding four more thiocyanate anions, $SCN(aq)^-$, to the equilibrium solution in the center of the figure is the formation of a total of three iron–thiocyanate complexes at equilibrium in the solution. Using the same assumptions as in Worked Example 9.14, calculate the equilibrium constant for this system. Explain your procedure.

(b) Is your equilibrium constant from part (a) consistent with those from Worked Example 9.14 and Check This 9.15? Explain why or why not.

(c) Is the result of adding the thiocyanate anion that is described in part (a) consistent with your observations in Investigate This 9.3(c)? Explain why or why not.

9.13. A gaseous mixture of HI, I_2, and H_2 in a 1.00-L container was held at a high temperature, 500 K, long enough for the reaction, $H_2(g) + I_2(g) \rightleftharpoons 2HI(g)$, to come to equilibrium and then rapidly cooled to "freeze out" the equilibrium mixture. Analysis of the cooled mixture gave these results: 2.21×10^{-3} mol HI, 1.46×10^{-3} mol I_2, and 2.09×10^{-5} mol H_2.

(a) What was the pressure, in bar, of each gas in the equilibrium mixture at 500 K? *Hint:* Recall the ideal gas equation, $PV = nRT$, with $R = 8.314 \times 10^{-2}$ L·bar·K^{-1}·mol^{-1}.

(b) Write the equilibrium constant expression for the reaction: $H_2(g) + I_2(g) \rightleftharpoons 2HI(g)$.

(c) What is the numeric value of the equilibrium constant for the reaction in part (b)? Explain your approach.

9.14. Acid–base indicators are weak acids: $HIn(aq) + H_2O(aq) \rightleftharpoons H_3O^+(aq) + In^-(aq)$. They are also dye molecules for which the colors of $HIn(aq)$ and $In^-(aq)$ are different. You can often analyze an acid–base indicator solution spectrophotometrically to determine the molar concentrations of the conjugate acid and base. For one such indicator, the concentrations were $[HIn(aq)] = 6.3 \times 10^{-5}$ M and $[HIn(aq)] = 8.7 \times 10^{-5}$ M in a solution with pH = 7.41.

(a) Write the equilibrium constant expression for the above proton transfer reaction.

(b) What is the numeric value of the equilibrium constant for the proton transfer reaction? Explain clearly how you get your answer.

9.3. Acid–Base Reactions and Equilibria

9.15. As you work through this problem and the next, use the information in this table.

Acid name	Formula	K_a
acetic acid	$CH_3C(O)OH$	1.8×10^{-5}
chloroacetic acid	$ClCH_2C(O)OH$	1.4×10^{-3}

(a) Write an equation for the acid equilibrium (proton transfer to water) for each acid.
(b) Write the equilibrium constant expression for K_a for each reaction in part (a).
(c) How will the pH of a 0.10 M solution of chloroacetic acid compare with that for a 0.10 M solution of acetic acid? Explain your prediction.

9.16. Use the K_a values given in the previous problem.
(a) Calculate the pH of a 0.050 M solution of acetic acid.
(b) Calculate the pH of a 0.050 M solution of chloroacetic acid.
(c) Discuss whether or not your results in parts (a) and (b) confirm the predictions made in part (c) of the previous problem.

9.17. The pK_w for water at 25 °C is 14.00 and is used as the basis for the pH scale. At 60 °C, the pK_w for water is 12.97.
(a) The pK_w for water refers to this reaction: $H_2O(l) + H_2O(l) \rightleftharpoons H_3O^+(aq) + OH^-(aq)$. Based on the two values for pK_w, is the autoionization reaction exothermic or endothermic? Explain your reasoning.
(b) If an aqueous solution at 25 °C has $[H_3O^+] = 1.0 \times 10^{-7}$ M, is the solution acidic, neutral, or basic? What is the pH of the solution?
(c) If an aqueous solution at 60 °C has $[H_3O^+] = 1.0 \times 10^{-7}$ M, is the solution acidic, neutral, or basic? What is the pH of the solution? How do your answers compare to those in part (b)? How do you explain the differences and similarities?

9.18. The position of equilibrium lies to the right (products are favored over reactants) in each of these reactions:
  (i) $N_2H_5^+(aq) + NH_3(aq) \rightleftharpoons NH_4^+(aq) + N_2H_4(aq)$
  (ii) $NH_3(aq) + HBr(aq) \rightleftharpoons NH_4^+(aq) + Br^-(aq)$
  (iii) $N_2H_4(aq) + HBr(aq) \rightleftharpoons N_2H_5^+(aq) + Br^-(aq)$

Based on this information, what is the order of acid strength for species acting as acids in these reactions? Explain the reasoning for your order.

9.19. Calculate the pH of 0.1 M solutions of each of these acids and bases:
(a) lactic acid, $K_a = 1.4 \times 10^{-4}$
(b) benzoic acid, $K_a = 6.5 \times 10^{-5}$
(c) aniline, $K_b = 4.3 \times 10^{-10}$
(d) hydrazine, $K_b = 8.9 \times 10^{-7}$

9.20. Calculate the pH of a 0.250 M solution of aqueous ammonia. *Hint:* See Table 9.2 for the pK_a of the ammonium ion, the conjugate acid.

9.21. Estimate the percent ionization (the percent of the acid molecules that transfer their proton to water) of an aqueous 0.350 M hydrogen fluoride, HF, solution. Does your result justify using the approximation that a negligible amount, less than 5%, of the acid reacts? Explain. The K_a of HF is 6.46×10^{-4}.

9.22. The pH of lemon juice is about 3.4. Assume that all the acid in the juice is citric acid, H_3Cit. For the reaction, $H_3Cit(aq) + H_2O(aq) \rightleftharpoons H_3O^+(aq) + H_2Cit^-(aq)$, $K_a = 7.4 \times 10^{-4}$. Assume that none of the other protons in $H_2Cit^-(aq)$ is transferred to water in the lemon juice. What is the concentration of citric acid in lemon juice? Show clearly how you get your answer and state any further assumptions you make.

9.23. When simple salts of small, highly charged cations are dissolved in water, the resulting solutions are acidic. For example, a 0.1 M solution of aluminum chloride, $AlCl_3(s)$, in water has a pH of 2.9. To understand what is happening in these solutions, consider this reaction:
$Al(H_2O)_6^{3+}(aq) + H_2O(aq) \rightleftharpoons$
$\qquad\qquad H_3O^+(aq) + Al(OH)(H_2O)_5^{2+}(aq)$
(a) Explain why the hydrated aluminum cation, $Al^{3+}(aq) = Al(H_2O)_6^{3+}(aq)$, is a weak acid. That is, explain what attractions and repulsions make the transfer of the proton more favorable in this case compared to $Na^+(aq)$, which shows no detectable acidity.
(b) What are the equilibrium constant expression and the numeric value of the equilibrium constant, K_a, for the proton transfer reaction.
(c) The pK_a for $Fe^{3+}(aq)$ is approximately 2.5. Which will be more acidic, a 0.1 M aqueous solution of Fe(III) or Al(III)? Explain your reasoning.

9.24. (a) Calculate K_a for an acid, HA, whose 0.215 M solution has a pH of 4.66. Does your result justify using the approximation that a negligible amount, less than 5%, of the acid reacts? Explain.
(b) What would be the pH of a 0.175 M aqueous solution of the salt NaA, the sodium salt of the conjugate base of HA, assuming the salt dissociates completely to $Na^+(aq)$ and $A^-(aq)$ ions in solution. Explain the reasoning for your answer.

9.25. (a) What is pK_b for the oxalate dianion, $^-O(O)CC(O)O^-(aq)$. Explain how you get your answer. *Hint:* See Table 9.2.
(b) About 35 g of sodium oxalate, the disodium salt of the oxalate dianion, dissolve in one liter of water. What is the pH of this solution? Explain.

9.26. (a) What are the pH and the percent ionization (the percent of the acid molecules that transfer their proton to water) of an aqueous 0.100 M methanoic acid,

HC(O)OH*(aq)*, solution? Explain your reasoning. *Note:* $K_a = 1.8 \times 10^{-4}$ for methanoic acid.
(b) Predict, using Le Chatelier's principle, whether the percent ionization of 0.050 M methanoic acid will be larger, smaller, or the same as for the 0.100 M solution in part (a). Explain your reasoning.
(c) Calculate the percent ionization for the 0.050 M methanoic acid solution. Was your prediction in part (b) correct? Explain why or why not.

9.27. Explain why we almost always can disregard the hydronium ion contributed by the autoionization of water when we calculate the pH of solutions that contain an acid. Under what conditions would we have to take into account the autoionization of water?

9.4. Solutions of Conjugate Acid–Base Pairs: Buffer Solutions

9.28. Which of these solutions contain buffer systems? Explain your reasoning.
   (i)   a solution of $HClO_4$ and $NaClO_4$
   (ii) a solution of Na_2CO_3 and $NaHCO_3$

9.29. How do the concentrations of acetate ion, $[OAc^-(aq)]$, and hydronium ion, $[H_3O^+(aq)]$, change when sodium acetate, NaOAc*(s)*, is dissolved in an aqueous solution of acetic acid? Explain your reasoning.

9.30. How do the concentrations of ammonium ion, $[NH_4^+(aq)]$ and hydroxide ion $[OH^-(aq)]$ change when ammonium bromide, $NH_4Br(s)$, is dissolved in a solution of aqueous ammonia? Explain your reasoning.

9.31. Calculate the pH of a 0.250 M HOAc solution in which sufficient solid NaOAc is dissolved to make the final sodium ion concentration, $[Na^+(aq)]$, 1.50 M. Assume that the volume of the solution does not change. Explain your reasoning.

9.32. Calculate the concentrations of HOAc and OAc^- in a 0.25 M acetate buffer solution at pH = 5.36. Explain your reasoning. *Hint:* Recall that the *sum* of the concentrations of acetate ion and acetic acid in this solution is 0.25 M.

9.33. (a) What is the pH of a buffer mixture composed of 0.15 M lactic acid ($HC_3H_5O_3$, $K_a = 1.4 \times 10^{-4}$) and 0.10 M sodium lactate?
(b) What is the pH halfway to the equivalence point in a titration of 0.15 M lactic acid with 0.15 M sodium hydroxide? How does this pH compare with the pH you calculated in part (a)? Which pH is higher? Does this result make sense? Explain your reasoning.

9.34. (a) What is the pH of an aqueous solution that is 0.551 M phenol, HOC_6H_5, and 0.377 M sodium phenolate, $NaOC_6H_5$? Explain. *Hint:* See Table 9.2.
(b) What will be the pH of the mixture in part (a) after 0.100 mol of HCl gas is dissolved in 1.00 L of

the solution? Assume that the volume of the solution does not change. Explain your reasoning.
(c) What will be the pH of the mixture in part (a) after 0.125 mol of solid KOH is dissolved in 1.00 L of the solution? Assume that the volume of the solution does not change. Explain your reasoning.

9.35. Some liquid household bleaches are about 5% by weight aqueous solutions of sodium hypochlorite, NaOCl, which completely dissociates to its ions in the solution. Hypochlorous acid, HOCl*(aq)*, has a pK_a of 7.5.
(a) What is the molarity of sodium hypochlorite in bleach? Show how you get your answer. *Hint:* Assume that liquid bleach has the same density as water.
(b) Write the reaction that hypochlorite anion, $OCl^-(aq)$, undergoes with water. Write the equilibrium constant expression for this reaction. What is the usual symbol we use to designate the equilibrium constant for a reaction like this? What is the numeric value of this equilibrium constant? Explain your reasoning.
(c) What is the pH of a liquid bleach solution? Explain.
(d) If you wish to change the pH of bleach solution to 6.5, should you add sodium hydroxide or hydrochloric acid? Explain your reasoning.
(e) What is the ratio of the conjugate base to the conjugate acid in a bleach solution whose pH has been adjusted to 6.5? Explain.

9.36. Calculate the $H_2CO_3(aq)$ concentration in human arterial blood, given that the pH of blood is 7.41 and the $HCO_3^-(aq)$ concentration is 26×10^{-3} M. Explain the method you use to get your answer. *Note:* K_{a1} for $H_2CO_3(aq)$ is 8.0×10^{-7} at 37 °C, normal human body temperature.

9.37. (a) What volume of 6.0 M hydrochloric acid do you have to add to 250. mL of 0.10 M aqueous sodium acetate solution to give a solution with a pH of 4.30? Explain your reasoning.
(b) What are the concentrations of acetic acid and acetate ion in the pH 4.30 solution you prepared in part (a)? Is this a buffer solution? Explain your reasoning.

9.38. Combinations of phosphoric acid, $(HO)_3PO$, and hydrogen phosphate and phosphate salts can be used to prepare a variety of buffer solutions. Assume that you have available the acid, sodium dihydrogen phosphate, $Na(HO)_2PO_2(s)$, potassium hydrogen phosphate, $K_2(HO)PO_3(s)$, and sodium phosphate, $Na_3PO_4(s)$. Which of these compounds would you choose to make buffers with pH values of 2.00, 7.00, and 12.00? In each case, explain your choice.

9.5. Acid–Base Properties of Proteins

9.39. What is the isoelectric point of a protein? What are the properties of the protein at the isoelectric point? How are the solution conditions related to the isoelectric point?

9.40. (a) Write the equilibrium constant expressions for K_{a1} and K_{a2} for the amino acid glycine. *Hint:* See Table 9.2 for the reaction equations for these equilibria. (b) Using the data in Table 9.2, calculate the isoelectric pH of glycine. Explain the procedure you use. *Hint:* Use an appropriate combination of the equilibrium constant expressions from part (a) and the condition that, at the isoelectric pH, the concentrations of the net positive and net negative forms of the amino acid are the same. (c) 🐾 Show clearly how the calculation of the isoelectric pH for the dipeptide in the *Web Companion*, Chapter 9, Section 9.5.4, is related to the procedure you used to answer part (b).

9.41. As we have seen, sickle-shaped blood cells result from just two minor changes in the total of 574 amino acids in hemoglobin. In both cases, glutamate is replaced by valine. Make sketches of a protein (see Figure 9.5) representing hemoglobin and show how the change of the two amino acids brings about the clumping together of hemoglobin that causes the change in shape of the blood cells. *Hint:* Consider formation of micelles and membranes represented in Figures 8.20 and 8.23 and the *Web Companion*, Chapter 8, Section 8.13.

9.42. You have isolated two small proteins, A and B, that you believe differ in only one amino acid. You think that one protein has a cysteine at position where the other has a histidine. When you analyze the proteins by electrophoresis (see Figures 9.6 and 9.7) at pH 8.6, you find that protein B moves farther toward the positive end of the gel. (a) Is this result consistent with your hypothesis about the difference between the two proteins? If so, which protein has the cysteine and which the histidine? Explain the reasoning for your choice. (b) If you cannot decide which protein is which from electrophoresis at this pH, what further experiment could you do to differentiate the two? Even if you can differentiate the two proteins, is there a further experiment you could do that would help to confirm your identification?

9.43. Explain how the observations you made in Investigate This 9.42 are related to sickle-cell disease.

9.44. Electrophoretic analysis is not limited to biomolecules such as proteins and nucleic acids. Other species that differ in charge, metal ion complexes, for example, can also be analyzed and separated by various forms of electrophoresis. In one experiment, a mixture of Fe(III) and Co(II) ions in an 8 M solution of hydrochloric acid was analyzed by paper electrophoresis (a strip of filter paper substitutes for the gel in Figure 9.6). Fe(III) forms a light yellow chloride complex and Co(II) forms a blue chloride complex (as you observed in Investigate This 9.8). After several minutes of electrophoresis, a yellow spot on the paper was closer to the positive end of the paper than a blue spot.

(a) Based on this observation, what conclusions can you draw about the stoichiometry of the Fe(III) and Co(II) chloride ion complexes? Explain your reasoning. (b) What would you hypothesize about the structures of the Fe(III) and Co(II) chloride ion complexes? Explain your reasoning. *Hint:* Consider what you learned in Chapter 6, Section 6.6, as well as in this chapter.

9.6. Solubility Equilibria for Ionic Salts

9.45. Use K_{sp} values for these ionic compounds to calculate the concentration of each ionic species in a saturated solution of the compound. Explain your reasoning for each.
(a) $BaSO_4$ (c) $Mg(OH)_2$
(b) Ag_2S (d) $PbBr_2$

9.46. How much solid calcium carbonate, $CaCO_3(s)$, will be obtained if 100. mL of a solution saturated with calcium carbonate is evaporated to dryness? Explain.

9.47. Calculate the mass of calcium fluoride, $CaF_2(s)$, that will dissolve in 100. mL of water. Explain your method. Assume that there is no volume change. The pK_{sp} of CaF_2 is 10.57.

9.48. Find the molar solubility and K_{sp} for $PbCrO_4(s)$, if 5.8×10^{-5} g dissolve in exactly 1 L of water. Explain your reasoning.

9.49. Calculate the pH and the concentrations of $Fe^{2+}(aq)$ and $OH^-(aq)$ when as much $Fe(OH)_2(s)$ as possible is dissolved in these solutions. Explain your reasoning and any assumptions you make in each case.
(a) pure water
(b) 0.25 M NaOH
(c) 0.25 M $Fe(NO_3)_2$

9.50. Determine if a precipitate will form when 0.0025 mol of $Ba^{2+}(aq)$ (from $Ba(NO_3)_2$) and 0.00030 mol of CO_3^{2-} (from Na_2CO_3) are mixed in 250. mL of distilled water. Explain the reasoning for your response.

9.51. A solution of $Na_2SO_4(aq)$ is added dropwise to another solution containing 2.5×10^{-3} M $Pb^{2+}(aq)$ and 0.05 M $Ca^{2+}(aq)$. What is the composition of the precipitate that will form first? Explain. *Hint:* What is the minimum concentration of sulfate anion, $[SO_4^{2-}(aq)]$, required to begin the precipitation of each of the cations?

9.52. (a) Because of its low solubility and opaqueness to x-rays, a suspension of $BaSO_4(s)$ is used routinely in x-ray analyses of the intestinal tract. How much $Ba^{2+}(aq)$ is in 200 mL of a saturated barium sulfate solution given to a patient to drink? (b) The barium cation, $Ba^{2+}(aq)$, is rather toxic. How do your results in part (a) relate to the safety of the barium sulfate used in x-ray analysis?

9.53. If a patient is slightly allergic to $Ba^{2+}(aq)$, which one of these methods would you choose to decrease the $Ba^{2+}(aq)$ concentration in order to minimize allergic discomfort when ingesting the $BaSO_4$ solution in the previous problem? Explain your choice and why you reject the others. *Note:* The dissolution reaction, $BaSO_4(s) \rightleftharpoons Ba^{2+}(aq) + SO_4^{2-}(aq)$, is endothermic.
  (i)   Heat the saturated solution.
  (ii) Add sodium sulfate, Na_2SO_4.
  (iii) Add additional $BaSO_4$.

9.54. Determine the K_{sp} for a metal salt with formula $M(OH)_2$, if its saturated solution has a pH of 8.11. Explain your reasoning.

9.55. Determine the pK_{sp} for Ag_3AsO_4, if 0.000562 g dissolves in 1 L of water.

9.56. Silver chloride, $AgCl(s)$, is white and silver chromate, $Ag_2CrO_4(s)$, is red. When chemists wish to analyze a solution for chloride ion they often titrate it with a standard solution of silver ion. A few drops of potassium chromate solution are added to the unknown chloride solution to act as an indicator. Explain how the chromate ion can act as an indicator. *Hint:* Examine Table 9.4 and consider the stoichiometry of the two solids.

9.57. The pK_{sp} for $Ni(OH)_2(s)$ is 17.2 and the pK_{sp} for $Fe(OH)_2(s)$ is 13.8. If you slowly add an aqueous solution of sodium hydroxide to a stirred aqueous solution containing 0.001 M $Ni^{2+}(aq)$ and 0.06 M $Fe^{2+}(aq)$, at what pH will a precipitate first begin to form? What will the precipitate be? Explain your reasoning. *Hint:* What is the minimum concentration of hydroxide, $[OH^-(aq)]$, required to begin the precipitation of each of the cations?

9.7. Thermodynamics and the Equilibrium Constant

9.58. Consider this reaction, $H_2(g) + I_2(g) \rightleftharpoons 2HI(g)$, at equilibrium in a rigid (constant volume) reaction vessel.
(a) If the pressure of $I_2(g)$ is increased (by adding more I_2 to the vessel), how will the pressures of $H_2(g)$ and $HI(g)$ be affected as a new equilibrium is established? Explain your reasoning.
(b) If the pressure of $HI(g)$ is reduced (by removing some HI from the vessel), how will the rate of the reverse reaction compare to the forward reaction before a new equilibrium is established? Which direction will the equilibrium shift? Explain your responses.

9.59. Use free energies of formation, ΔG_f°, from Appendix B to determine which of these two reactions has the larger equilibrium constant? Explain your choice.
  (i) $H_2(g) + \frac{1}{2}O_2(g) \rightleftharpoons H_2O(l)$
  (ii) $H_2(g) + \frac{1}{2}O_2(g) \rightleftharpoons H_2O(g)$

9.60. Urea, $(NH_2)_2CO(s)$, is an ingredient in many fertilizers and is occasionally used instead of salt to hasten the melting of ice on highways. The reaction for the industrial synthesis of urea is: $CO_2(g) + 2NH_3(g) \rightleftharpoons (NH_2)_2CO(s) + H_2O(l)$. Calculate the standard free energy change and equilibrium constant for this reaction at 298 K. Is the reaction spontaneous under standard conditions? Explain.

9.61. Consider the reaction $H_2(g) + I_2(g) \rightleftharpoons 2HI(g)$ at 500 K, for which Problem 9.13 provides the data you require to calculate the equilibrium constant, K. What is the standard free energy change for this reaction at 500 K? Explain.

9.62. Much of the ammonia, $NH_3(g)$, we synthesize (see Problem 9.9) is used to make ammonium nitrate for fertilizer. The first step in making the nitrate (as nitric acid) is the oxidation of ammonia:
$4NH_3(g) + 5O_2(g) \rightleftharpoons 4NO(g) + 6H_2O(l)$
(a) Use the data in Appendix B to determine the standard enthalpy change, $\Delta H^\circ_{reaction}$, for the oxidation of ammonia at 298 K. Is your result consistent with what you would get from average bond enthalpies (Chapter 7, Table 7.3)? Explain. *Note:* The NO bond enthalpy is 630. $kJ \cdot mol^{-1}$.
(b) Use the data in Appendix B to determine the standard entropy change, $\Delta S^\circ_{reaction}$, for the oxidation of ammonia at 298 K. Is this the result you would predict from looking at the changes that occur in this reaction? Explain why or why not.
(c) Use the $\Delta H^\circ_{reaction}$ and $\Delta S^\circ_{reaction}$ values you calculated in parts (a) and (b) to determine the standard free energy change, $\Delta G^\circ_{reaction}$, for this reaction at 298 K. Use the data in Appendix B to determine the $\Delta G^\circ_{reaction}$ from standard free energies of formation, ΔG_f°. Are the two values for $\Delta G^\circ_{reaction}$ the same? Explain why or why not.
(d) What is the equilibrium constant for the ammonia oxidation reaction at 298 K? Are products or reactants favored? Explain.

9.63. When calcium carbide, $CaC_2(s)$, a grayish solid, is dropped into water, the solution gets quite hot, gas bubbles out of the solution, and a white precipitate forms. The reaction that occurs, $CaC_2(s) + 2H_2O(l) \rightleftharpoons Ca(OH)_2(s) + C_2H_2(g)$, forms ethyne (acetylene), a flammable gas that coal miners once used in carbide lamps to light their way in the mines.
(a) What is the sign of $\Delta H_{reaction}$ for the ethyne-forming reaction? Explain how you know.
(b) What is the sign of $\Delta S_{reaction}$ for the ethyne-forming reaction? Explain how you know.
(c) What is the sign of $\Delta G_{reaction}$ for the ethyne-forming reaction? Explain how you know.
(d) Use the data in Appendix B to calculate $\Delta H^\circ_{reaction}$, $\Delta S^\circ_{reaction}$, and $\Delta G^\circ_{reaction}$ at 298 K. Are your results consistent with your previous answers? Explain why or why not.
(e) Review the calculations in part (d). What is the major factor responsible for the size of the free energy change for this reaction?

9.64. Sketch a graph of $\Delta G_{reaction}$ as a function of the composition of a reaction mixture from a reactant-rich composition through equilibrium to a product-rich composition. Explain the reasoning you use to draw your graph.

9.65. At 298 K, the equilibrium constant, K, for the decomposition of calcite (chalk), $CaCO_3(s)$, to give calcium oxide, $CaO(s)$, and carbon dioxide, $CO_2(g)$, is 1.6×10^{-23}.
(a) Based on the equilibrium constant for this reaction, is there likely to be a higher concentration of $CO_2(g)$ near the chalkboard in a classroom compared to that in the rest of the room? Explain why or why not.
(b) Data for $CaO(s)$ are not given in Appendix B. Use the data that are available there and, K, the equilibrium constant above, to estimate ΔG_f° for $CaO(s)$. Explain your procedure.

9.66. Compounds **A**, **B**, and **C** were dissolved in water in a reaction vessel and allowed to come to equilibrium. The graph shows the variation of concentration with time for this reaction: $3\mathbf{A}(aq) \rightleftharpoons \mathbf{B}(aq) + 2\mathbf{C}(aq)$

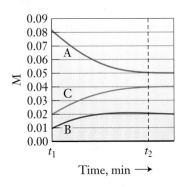

(a) Estimate the value of the reaction quotient, Q, at time t_1, the time the compounds were mixed.
(b) Is the reaction shifting in the forward or reverse direction to attain equilibrium? How can you tell?
(c) Is it reasonable to assume that equilibrium is reached at t_2? Explain why or why not.
(d) Estimate the value of the equilibrium constant, K, at t_2. Explain your procedure.
(e) Estimate the value of the standard free energy change for the reaction. Explain your reasoning.

9.67. The equilibrium constant, K, for the reaction $H_2(g) + I_2(g) \rightleftharpoons 2HI(g)$ at 700 K is 54.
(a) What is the standard free energy change for this reaction at 700 K? Explain.
(b) What is the reaction quotient, Q, in a mixture in which the pressures of $H_2(g)$, $I_2(g)$, and $HI(g)$ are, respectively, 0.040 bar, 0.35 bar, and 1.05 bar? Explain.
(c) What is the free energy of reaction under the conditions in part (b)? Is the sign of the free energy consistent with the relative values of Q and K? Explain why or why not.

(d) Is the mixture in part (b) likely to react to form more products, more reactants, or is it at equilibrium? Explain how you know.

9.68. Is methanol decomposition, $CH_3OH(l) \rightleftharpoons CO(g) + 2H_2(g)$, likely to occur spontaneously at 298 K and one bar pressure? Use thermodynamic reasoning to show why or why not.

9.69. Is the reaction of ethene with water, $CH_2CH_2(g) + H_2O(l) \rightleftharpoons CH_3CH_2OH(l)$, likely to occur spontaneously at 298 K and one bar pressure? Use thermodynamic reasoning to show why or why not.

9.8. Temperature Dependence of the Equilibrium Constant

9.70. Explain how temperature, pressure, and concentration affect a system at equilibrium.

9.71. Explain how the position of equilibrium is affected when these changes occur for this exothermic reaction, $2SO_2(g) + O_2(g) \rightleftharpoons SO_3(g)$, initially at equilibrium.
(a) Some $SO_3(g)$ is removed.
(b) The pressure is decreased.
(c) The temperature is decreased.
(d) Some $O_2(g)$ is added.

9.72. Consider the equilibrium reaction $N_2(g) + O_2(g) \rightleftharpoons 2NO(g)$. Experimental measurements show that higher temperatures favor the formation of $NO(g)$.
(a) Do you expect this reaction to be exothermic or endothermic in the direction written? Explain your choice.
(b) Use average bond enthalpies from Table 7.2, Chapter 7, Section 7.7, to calculate the reaction enthalpy for this reaction. Does your result agree with the prediction you made in part (a)? Explain why or why not. *Note:* The NO bond enthalpy is 630. $kJ \cdot mol^{-1}$.
(c) Does the equilibrium constant for this reaction increase or decrease with increasing temperature? Justify your choice.
(d) How does the standard free energy of this reaction change as the temperature increases? How do you know? If the standard enthalpy of reaction is constant, how do you account for the direction of change in the standard free energy? Do the data in Appendix B support your explanation? Show why or why not.

9.73. (a) Use the results from Worked Example 9.67 to calculate the solubility product (K_{sp}), molar solubility, and solubility in $g \cdot L^{-1}$, for $PbI_2(s)$ at 20 °C.
(b) One handbook gives a $PbI_2(s)$ solubility of 0.63 $g \cdot L^{-1}$ at 20 °C. How does your value from part (a) agree with this datum?

9.74. The solubility of $CaSO_4(s)$ at 303 K and 373 K is 2.09 $g \cdot L^{-1}$ and 1.62 $g \cdot L^{-1}$, respectively.
(a) What are the numeric values of K_{sp} for $CaSO_4(s)$ at 303 K and 373 K? Explain.

(b) What are $\Delta H°_{\text{reaction}}$ and $\Delta S°_{\text{reaction}}$ for the dissolution of $CaSO_4(s)$? Explain your procedure and any assumptions you make to get your answers.

(c) What are $\Delta G°_{\text{reaction}}$ and K_{sp} for the dissolution of $CaSO_4(s)$ at 298 K? Discuss the agreement between your results and the data in Appendix B and Table 9.4.

9.75. The coal gas (or water gas) reaction, $C(s) + H_2O(g) \rightleftharpoons CO(g) + H_2(g)$, produces a gaseous fuel mixture from a solid fuel. The product mixture is also a useful starting material for synthesis of carbon-containing compounds, such as methanol.

(a) Calculate $\Delta H°_{\text{reaction}}$, $\Delta S°_{\text{reaction}}$, and $\Delta G°_{\text{reaction}}$ for the coal gas reaction at 25 °C.

(b) At what temperature would $\Delta G°_{\text{reaction}}$ be zero? Explain your reasoning and state any assumptions you make.

(c) What is the equilibrium constant for the coal gas reaction at the temperature you calculated in part (b)? Explain.

9.76. (a) How is $\Delta G_{\text{reaction}}$ for a phase change, such as liquid $\rightleftharpoons$ gas, at equilibrium at one bar pressure related to $\Delta G°_{\text{reaction}}$ at the equilibrium temperature? Explain your reasoning.

(b) Use the data in Appendix B to estimate the temperature at which rhombic and monoclinic sulfur are in equilibrium, $S_{(s,\ rhombic)} \rightleftharpoons S_{(s,\ monoclinic)}$, at 1 bar. Explain your reasoning and state any assumptions you make.

9.77. Consider the reaction $H_2(g) + I_2(g) \rightleftharpoons 2HI(g)$ at 500 K, for which Problem 9.13 provides the data required to calculate the equilibrium constant, K, and Problem 9.61 asks for the standard free energy change, $\Delta G°_{\text{reaction}}$, for the reaction at 500 K.

(a) If the equilibrium constant for this reaction is 794 at 298 K, what are $\Delta H°_{\text{reaction}}$ and $\Delta S°_{\text{reaction}}$? Explain your reasoning and state any assumptions you make.

(b) Based on your results from part (a), what is $\Delta G°_{\text{reaction}}$ at 298 K? How does this value for $\Delta G°_{\text{reaction}}$ compare with the one you calculate from the data in Appendix B? Discuss your comparison.

9.78. For the vaporization of ethanol, $C_2H_5OH(l) \rightleftharpoons C_2H_5OH(g)$, the standard entropy change, $\Delta S°_{\text{reaction}}$, is 122 $J·K^{-1}·mol^{-1}$. The normal boiling point of ethanol at one bar pressure is 78.5 °C (351.6 K).

(a) Estimate $\Delta H°_{\text{reaction}}$ for ethanol vaporization. Explain your reasoning and state any assumptions you make.

(b) Write the equilibrium constant expression for ethanol vaporization. Explain why the numeric value of the equilibrium constant, K, is one at 78.5 °C. What is $\Delta G°_{\text{reaction}}$ for ethanol vaporization at 78.5 °C? Explain.

(c) What are the numeric values of K and $\Delta G°_{\text{reaction}}$ for ethanol vaporization at 25 °C? Explain your reasoning and state any assumptions you make.

(d) What is the vapor pressure (the pressure of the gas in equilibrium with the liquid) of ethanol at 25 °C? Explain your reasoning.

9.79. The variation of equilibrium vapor pressure for the isomeric alcohols, 1-propanol and 2-propanol, is given in this table.

Vapor pressure, torr	1.0	10.0	40.0	100.	400.	760.
	Temperature, °C					
1-propanol	−15.0	14.7	36.4	52.8	82.0	97.8
2-propanol	−26.1	2.4	23.8	39.5	67.8	82.5

(a) Plot these data and use the plots to determine $\Delta H°_{\text{reaction}}$ and $\Delta S°_{\text{reaction}}$ for the vaporization of each of these alcohols. *Note:* 1 bar = 750 torr and K = °C + 273.1.

(b) How, if at all, are the relative values of $\Delta H°_{\text{reaction}}$ related to the structures of the two alcohols? Explain.

(c) How, if at all, are the relative values of $\Delta S°_{\text{reaction}}$ related to the structures of the two alcohols? Explain.

(d) Which alcohol is the more volatile? Why is this isomer more volatile?

9.80. The pK_w for water is 14.0 at 25 °C and 12.97 at 60 °C. Use these data to determine $\Delta H°_{\text{reaction}}$, $\Delta S°_{\text{reaction}}$, and $\Delta G°_{\text{reaction}}$ for water ionization. Explain your reasoning and state any assumptions you make. Discuss the agreement (or lack thereof) between these values and those you calculate from the data in Appendix B.

9.81. Phosphorus pentachloride, $PCl_5(s)$, sublimes readily, $PCl_5(s) \rightleftharpoons PCl_5(g)$. The equilibrium pressure of the gas is 40 torr at 102.5 °C and 400 torr at 147.2 °C.

(a) Use these data to determine $\Delta H°_{\text{reaction}}$ and $\Delta S°_{\text{reaction}}$ for the sublimation of $PCl_5(s)$. Explain your reasoning and state any assumptions you make. *Note:* 1 bar = 750 torr.

(b) Assuming that the solid does not melt, at what temperature is the sublimation pressure 760 torr (1 atm)?

9.82. (a) Consider a spontaneous endothermic reaction during which the entropy of the system increases. What is the sign of the free energy change for this reaction? Explain.

(b) Consider a spontaneous exothermic reaction during which the entropy of the system decreases. What is the sign of the free energy change for this reaction? Explain.

(c) If the temperature is changed, could the reaction in part (a) become nonspontaneous? The reaction in part (b)? Explain why or why not for both cases.

9.83. Might this reaction, $CH_3OH(g) \rightleftharpoons CO(g) + 2H_2(g)$, occur spontaneously at about 1000 K and one bar pressure? Use thermodynamic reasoning to explain why or why not.

9.84. (a) Assume that rubbing alcohol is pure 2-propanol and use your results from Problem 9.79 to find the vapor pressure of 2-propanol at 35 °C, skin temperature. How is this result related to the use of rubbing alcohol to cool your skin? Explain.
(b) If 12 g of this rubbing alcohol evaporates from your skin, what is the change in the thermal energy of your body? Explain.

9.9. Thermodynamics in Living Systems

The data in this table may be needed in Problems 9.85 through 9.88. The free energies in the table are for the hydrolysis at pH 7 of the phosphorylated compounds to give the nonphosphorylated compound plus phosphate:

$$\text{compound-P} + H_2O \rightleftharpoons \text{compound} + P_i$$

Free energy of hydrolysis at pH 7 and 25 °C

Compound	$\Delta G^{\circ\prime}$, kJ·mol^{-1}
phosphoenolpyruvate, PEP	−62
creatine phosphate	−43
ATP (gives ADP)	−30
glucose-1-phosphate, Glu-1-P	−21
glucose-6-phosphate, Glu-6-P	−14
glycerol-3-phosphate, G3P	−9

9.85. Consider this isomerization reaction, glu-6-P $\rightleftharpoons$ glu-1-P, at pH 7 and 25 °C.
(a) Write the hydrolysis reactions for glu-6-P and glu-1-P. Show how to combine these reactions to give the isomerization reaction.
(b) Calculate $\Delta G^{\circ\prime}_{\text{reaction}}$ for the isomerization reaction. Explain your procedure.
(c) If this isomerization reaches equilibrium, what is the molar *ratio* of glu-6-P to glu-1-P in the reaction mixture? Explain your reasoning. *Hint:* Consider the equilibrium constant and equilibrium constant expression.

9.86. Consider this reaction: ATP + Pyr $\rightleftharpoons$ ADP + PEP (where Pyr = pyruvate).
(a) Combine hydrolysis reactions from the table to give the reaction of interest and calculate $\Delta G^{\circ\prime}_{\text{reaction}}$ and the equilibrium constant, K, for the reaction at 25 °C.
(b) In many living cells, the molar *ratio* of ATP to ADP is about 10. If this reaction is at equilibrium in such a cell, what is the equilibrium molar ratio of pyruvate to phosphoenolpyruvate? Explain.

9.87. For each of these reactions, imagine that you start with a mixture of all the reactants and products at the same concentration in a pH 7 solution at 25 °C. For each case, tell which direction the reaction will go

in order to approach equilibrium and give the reasoning for your choice.
  (i)   ATP + creatine $\rightleftharpoons$ ADP + creatine phosphate
  (ii) ATP + Glu $\rightleftharpoons$ ADP + Glu-6-P

9.88. The stoichiometry of the glycolysis (sugar breaking) pathway in organisms is

$$\text{Glu} + 2\text{ADP} + 2P_i + 2\text{NAD}^+ \rightleftharpoons$$
$$2\text{Pyr} + 2\text{ATP} + 2\text{NADH} + 2H^+$$

The overall standard free energy change, $\Delta G^{\circ\prime}_{\text{reaction}}$, for glycolysis is about −84 kJ at 25 °C. Organisms also have a gluconeogenesis (making new sugar) pathway to produce glucose from pyruvate with a stoichiometry we can represent as

$$2\text{Pyr} + 6\text{ATP} + 2\text{NADH} + 2H_2O + 2H^+ \rightleftharpoons$$
$$\text{Glu} + 6\text{ADP} + 6P_i + 2\text{NAD}^+$$

The overall free energy change for gluconeogenesis, $\Delta G^{\circ\prime}_{\text{reaction}}$, is about −38 kJ at 25 °C.
(a) Why isn't the gluconeogenesis pathway simply the reverse of the glycolysis pathway? Would there be a thermodynamic problem reversing glycolysis? Explain.
(b) Are the free energy changes for glycolysis and gluconeogenesis consistent with the differences between the overall reactions? Explain why or why not.
(c) Look again at Figure 7.17, Chapter 7, Section 7.9. The coupling shown in the figure was related to enthalpies of reaction, but the coupling is more accurately a representation of free energies of reaction. Discuss where glycolysis and gluconeogenesis would fit into the diagram if the arrows represent free energy changes.

9.89. In cells that need free energy faster than oxygen can get to them for complete glucose oxidation, such as our muscles during rapid strenuous exercise, pyruvate, $CH_3COC(O)O^-$, is reduced to lactate, $CH_3CHOHC(O)O^-$ (Lac):
$$\text{Pyr} + \text{NADH} + H^+ \rightleftharpoons \text{Lac} + \text{NAD}^+$$
(a) The reduction of pyruvate to lactate is a fermentation reaction and is required to enable glycolysis to proceed in the absence of oxygen to complete the glucose oxidation pathway. What is required for glycolysis (see preceding problem) that is provided by this fermentation reaction? Explain. *Hint:* See Chapter 6, Section 6.11.
(b) Combine the overall glycolysis reaction with this fermentation reaction to give the net reaction equation for the conversion of a glucose molecule to two lactates.
(c) If the standard free energy change, $\Delta G^{\circ\prime}_{\text{reaction}}$, for the fermentation reaction written above is about −25 kJ at 25 °C, what is the overall $\Delta G^{\circ\prime}_{\text{reaction}}$ for the glucose to lactate reaction you wrote in part (b)? Explain.
(d) What is the equilibrium constant expression for the conversion of glucose to lactate that you wrote in part (b)? Explain.

(e) What is the free energy of reaction, $\Delta G'_{reaction}$ (not the standard free energy), for the glucose to lactate reaction if the concentrations in a cell are [Glu] = 5×10^{-3} M, [Lac] = 5×10^{-5} M, [ATP] = 2×10^{-3} M, [ADP] = 2×10^{-4} M, and [P_i] = 1×10^{-3} M? Is the reaction spontaneous under these conditions? Explain.

9.11. Extension—Thermodynamic Basis for the Equilibrium Constant

9.90. (a) If the pressure of a gas is changed from pressure P_1 to pressure P_2 at constant temperature, show that the change in entropy of the gas is: $\Delta S = R\ln(P_1/P_2)$. *Hint:* Consider equation 9.83.
(b) What is the sign of ΔS, if a gas is compressed? Show that the result you get by applying the equation from part (a) is the same as you get by reasoning from the entropy discussions in Chapter 8.

9.91. Consider this reaction, $2NO_2(g) \rightleftharpoons N_2O_4(g)$, at equilibrium at 298 K in a reaction vessel whose size can be changed. Initially, the pressures of $NO_2(g)$ and $N_2O_4(g)$ are 0.225 bar and 0.438 bar, respectively.
(a) Write the equilibrium constant expression for this reaction and calculate the equilibrium constant, K, at 298 K.
(b) Imagine that the volume of the reaction vessel is doubled while holding the temperature of the contents constant. If no reaction occurs, what is the pressure of each gas in the mixture? What is the numeric value of the reaction quotient, Q, in this mixture? Explain.
(c) How must the reaction mixture in part (b) change to make Q move toward K and reattain equilibrium? Is this the same direction of pressure adjustments you would predict from Le Chatelier's principle? Explain why or why not.
(d) Calculate the pressures of the two gases at equilibrium in the new volume. *Hint:* You will need the quadratic equation or some estimation method, perhaps a spreadsheet.

General Problems

9.92. Consider the reaction $H_2(g) + I_2(g) \rightleftharpoons 2HI(g)$, for which you determined K at 500 K in Problem 9.13. If 4.00×10^{-3} mol of HI are placed in a 450-mL reaction vessel heated to 500 K, how many moles of I_2 will be present at equilibrium? Explain clearly the procedure you use to obtain your answer. *Hint:* Recall the ideal gas equation, $PV = nRT$, with $R = 8.314 \times 10^{-2}$ L·bar·K^{-1}·mol^{-1}. Take into account the amount of HI(g) that reacts and use the quadratic formula or an estimation procedure to solve the resulting equation.

9.93. How do the concentrations of acetate ion, OAc$^-$(aq), and H_3O^+(aq) change when hydrochloric acid, HCl(aq), is added to an aqueous solution of acetic acid? How would you explain the change in terms of the common ion effect?

9.94. Fluorobenzene, C_6H_5F, and related fluorinated compounds, are important starting materials for the synthesis of many fungicides, drugs, and agricultural chemicals. Recently, this efficient process for making fluorobenzene has been reported:

[The HF(g) product is later recycled with oxygen to regenerate the CuF$_2$(s) with only water as a by-product.] At 350 °C, the conversion of benzene to fluorobenzene is 5% and at 450 °C, the conversion is 30%.
(a) What is the equilibrium constant expression for this reaction? Assuming that the reaction is stoichiometric and comes to equilibrium in the reactor at about 0.1 bar total pressure of reactants plus products, what are the equilibrium constants for the reaction at 350 °C and 450 °C? Explain clearly the procedure you use to obtain your answers.
(b) What is the standard enthalpy of reaction, $\Delta H°_{reaction}$, for this process? Explain and state the assumptions you make.
(c) What is the standard entropy of reaction, $\Delta S°_{reaction}$, for this process? Explain and state the assumptions you make.
(d) If this process could be carried out at even higher temperatures, how high do you predict the temperature would have to be to convert 50% of the benzene to fluorobenzene? Explain your reasoning. *Hint:* What is the equilibrium constant for 50% conversion? How is the equilibrium constant related to the quantities in parts (b) and (c)?

9.95. A system at equilibrium responds to a disturbance in a way that minimizes the effect(s) of the disturbance (Le Chatelier's Principle). The disturbances that affect an equilibrium are changing the concentrations of the species in the reaction or changing the temperature of the system. Explain how qualitative, directional predictions based on Le Chatelier's Principle are related to the equilibrium constant expression, equation (9.47), and to the thermodynamics of equilibria embodied in equation (9.55).

... one must separate things in order to unite them. One must put them into their places as carefully as one handles fire and water . . .

I CHING
(WILHELM BAYNES TRANSLATION)

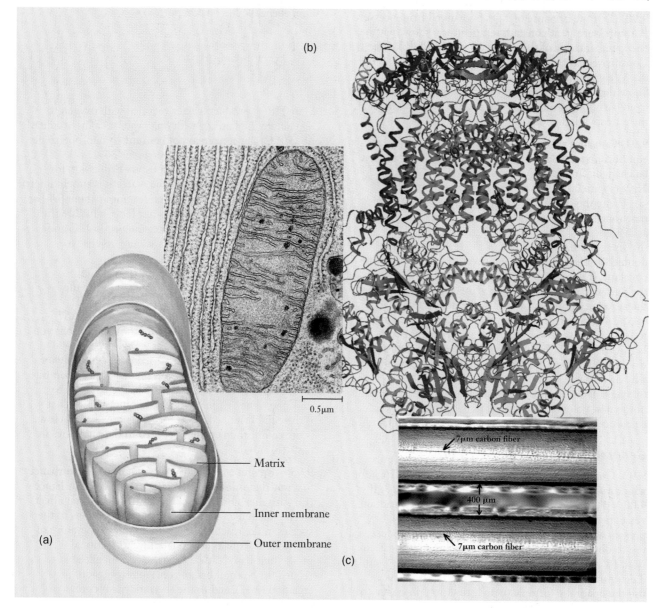

(b)

0.5 μm

Matrix

Inner membrane

Outer membrane

(a)

(c)

7 μm carbon fiber

400 μm

7 μm carbon fiber

(a) A cutaway drawing and a transmission electron micrograph of a mitochondrion, the organelle where energy for eukaryotic cells is produced by coupling the oxidation of fuel molecules to the production of ATP. (b) Computer-generated model of the cytochrome bc_1 protein structure, which consists of 11 separate proteins and iron-ion complexes. A large number of these protein structures are located in the inner membranes of mitochondria and are responsible for part of the coupling pathway for energy conversion. (c) A tiny experimental electrochemical fuel cell that uses the partial oxidation of glucose by oxygen to produce an electric current. The goal is to make these energy sources small enough to power therapeutic devices implanted in our bodies.

Reduction–Oxidation: Electrochemistry

Combustion, the *oxidation* of fuels by molecular oxygen, supplies the majority of the energy that supports the economy of the world. All multicelled and many single-celled organisms also depend on the energy from *oxidation* of biological fuel molecules by molecular oxygen. The redox (reduction–oxidation) processes in living cells take place quietly and in many steps. They do not create the large amounts of light and heat we get when we burn a fuel, as in Chapter 7, Investigate This 7.1. Nevertheless, the *overall* thermodynamics of combustion and cellular oxidation are the same; we discussed them in Chapters 7 and 8. One goal of this chapter is to find out how organisms harness redox reactions to provide their energy.

The illustration on the facing page shows the structure of **mitochondria,** the powerhouses of living cells. Within the inner mitochondrial membrane, the protein complex cytochrome bc$_1$ (also illustrated) plays a central role in **respiration,** the process of biological oxidation of fuel molecules. The chemistry that occurs in this protein complex can be represented by this reaction involving the oxidation of reduced ubiquinone (ubiquinol), UQH$_2$, to ubiquinone, UQ, by Fe(III):

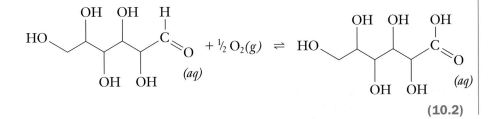

$$\text{UQH}_2 + 2H_2O + 2Fe^{3+} \rightleftharpoons \text{UQ} + 2H_3O^+ + 2Fe^{2+} \qquad (10.1)$$

Reaction (10.1) in living cells is easier to analyze and understand if we study it first in a different kind of cell. An **electrochemical cell** is a device that uses the free energy change of a redox reaction to move electrons through an external electric circuit to do work. The batteries you use to power flashlights and electronic devices like cell phones and portable CD players are electrochemical cells. The chapter-opening illustration includes a photomicrograph of a miniature electrochemical fuel cell that is based on the partial oxidation of glucose:

$$\text{(glucose)}_{(aq)} + \tfrac{1}{2}O_2(g) \rightleftharpoons \text{(gluconic acid)}_{(aq)} \qquad (10.2)$$

10.1 CONSIDER THIS

What atoms are oxidized and reduced
in reactions (10.1) and (10.2)?

(a) In reaction (10.1), which reactant atom (or ion) is oxidized? Which reactant atom (or ion) is reduced? Explain how you know. If necessary, review Sections 6.9–6.11.

(b) Answer the same questions for reaction (10.2).

Before we analyze reactions (10.1) and (10.2) in more detail, we will examine several simpler systems that will provide the concepts and practice we need to make sense of these two. In order to understand how an electrochemical or living cell works, we often need to consider the reduction and oxidation reactions separately. In many cases, these reactions can actually be separated in space; so we separate the cell into its half reactions to learn more about it, as the opening quotation from the *I Ching* suggests. We will begin our discussion by reexamining the kind of investigations we carried out in Chapter 2, where we tested solutions for the presence of ions by finding out whether the solutions conducted an electric current. This discussion will help us relate electrical measurements to the chemistry occurring in reduction–oxidation reactions and will introduce important uses of electrochemistry.

10.1. Electrolysis

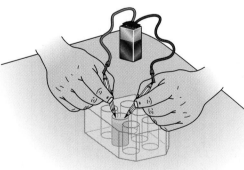

10.2 INVESTIGATE THIS

What happens when electric current passes through water?

(a) Do this as a class investigation and work in small groups to discuss and analyze the results. Half fill one well of a six-well plate with distilled water and add one or two crystals of magnesium sulfate, $MgSO_4 \cdot 7H_2O$ (Epsom salt), and several drops of universal acid–base indicator. The color of the indicator ranges from red in acidic solutions through green in neutral solutions to blue in basic solutions. Use an overhead projector to show the well on a screen and note and record the color of the solution. Use thin carbon rods (pencil lead) as **electrodes,** electrical conductors used to make electrical contact between the solution and the external electrical components. Using wires with alligator clips, connect one end of each electrode to a 9-V battery. Record which electrode is connected to the positive and which to the negative terminal of the battery. Immerse the other ends of the electrodes in the solution and hold them about 2 cm apart in the solution for 30–45 seconds and then remove them. Note and record any evidence for chemical reactions that you see occurring.

(b) After you remove the electrodes from the solution, stir the solution gently and record any evidence for chemical reactions that you see occurring.

10.3 CONSIDER THIS

What chemical reactions does an electric current cause?

(a) What evidence of chemical reaction(s) did you observe in Investigate This 10.2(a) when the carbon electrodes were in the solution? Was a reaction associated with only one or with both of the electrodes? What do you think the product(s) of the reaction(s) is(are)? Explain the reasoning for your answers.

(b) What evidence of chemical reaction(s) did you observe in part (b) of the investigation? What do you think the product(s) of the reaction(s) is(are)? Explain your reasoning.

In Chapter 2, Section 2.3, when you tested solutions for electrical conductivity, you probably noticed that bubbles of gas were produced at the wires dipped in conducting solutions. In Investigate This 10.2, you also observed bubbles of gas produced at the carbon electrodes in the solution. In addition, the solution near each electrode changed color, indicating that the pH in these parts of the solution was changing. These pieces of evidence suggest that the electric current from the battery is causing chemical reactions at or near the surface of the electrodes as the ions in the solution conduct the current through the solution.

The acid–base indicator color changes show that the solution became more acidic near the electrode connected to the positive terminal of the battery and more basic near the electrode connected to the negative terminal of the battery. If you carry out this same investigation using sodium chloride (or many other ionic compounds) instead of magnesium sulfate in the solution, the results are the same: Gas is produced at each electrode and the same pH changes occur. This information tells us that it is not the ionic compounds in the solution that are reacting at the electrodes. Water is the only other species in these solutions, so it must be the water whose atoms are being reduced and oxidized as a result of the electric current flow through the solution.

Electrolysis of water Water molecules react with electrons from the battery at the negative electrode in the solution. The hydrogen atoms are reduced from an oxidation number of +1 in $H_2O(l)$ to zero in $H_2(g)$, hydrogen gas:

$$2H_2O(l) + 2e^- \text{ (from electrode)} \rightarrow H_2(g) + 2OH^-(aq) \qquad \textbf{(10.3)}$$

Reaction (10.3) explains both the production of gas and formation of a basic solution at the negative electrode in Investigate This 10.2. Recall that, in Chapter 6, Section 6.10, we introduced half reactions as a method for balancing reduction–oxidation reaction equations. A **half reaction** represents either the reduction or the oxidation taking place in a reduction–oxidation reaction. Reaction (10.3) is a reduction half reaction.

At the positive electrode, electrons leave the reacting solution. Oxygen atoms are oxidized from a −2 oxidation number in water to zero in $O_2(g)$, oxygen gas:

$$2H_2O(l) \rightarrow O_2(g) + 4H^+(aq) + 4e^- \text{ (to electrode)} \qquad \textbf{(10.4)}$$

This oxidation half reaction, reaction (10.4), explains both the production of gas and formation of an acidic solution at the positive electrode in Investigate This 10.2.

In reaction (10.4), as in Chapter 6, Sections 6.9–6.11, we have used $H^+(aq)$ instead of $H_3O^+(aq)$. We will continue to use $H^+(aq)$ in this chapter in order to simplify reduction–oxidation equations.

Figure 10.1.

Representation of electrical conduction by an ionic solution.

════════ **10.4 CONSIDER THIS** ════════

What happens to the $Mg^{2+}(aq)$ and $SO_4^{2-}(aq)$?

(a) One possible result of reactions (10.3) and (10.4) would be to create a negatively charged solution of $OH^-(aq)$ ions around the negative electrode and a positively charged solution of $H^+(aq)$ around the positive electrode in Investigate This 10.2. Charge separation is an unfavorable process requiring a great deal of energy. Another possibility is movement of the $Mg^{2+}(aq)$ and $SO_4^{2-}(aq)$ ions to compensate for the production of the $OH^-(aq)$ and $H^+(aq)$ ions. Does the movement of the positive and negative ions shown in Figure 10.1 [a copy of Figure 2.8(c) in Chapter 2, Section 2.3] explain how charge separation is avoided? Why or why not?

(b) Sketch a new version of Figure 10.1 that explains, at the molecular level, the conductivity of ionic solutions, the flow of electrons between the electrodes outside the solution, and the observed production of bubbles at the electrodes.

In Investigate This 10.2, the overall reaction caused by the flow of electric charges in the circuit is the sum of half reactions (10.3) and (10.4). Before we add the half reactions, however, we have to account for the fact that an **electric circuit** is a closed pathway for the movement of *electric charge*. The same number of electrons must leave one electrode and enter the other electrode in an electric circuit. The flow of electric charges (ions and electrons) represented in Figure 10.1 and which you have shown in your sketch for Consider This 10.4(b) represents an electric circuit. Note carefully that *the charge carriers are not the same in all parts of an electric circuit.* Ions are the charge carriers in the solution and electrons are the charge carriers in the electrodes and wires. Inside the battery, ions are again the charge carriers, as we will find in Sections 10.2 and 10.3. To equalize the number of electrons leaving and entering the electrodes, we multiply reaction (10.3) by two before adding to reaction (10.4). The electrons will thus cancel out:

$$2[2H_2O(l) + 2e^- \text{ (from electrode)} \rightarrow H_2(g) + 2OH^-(aq)]$$

$$2H_2O(l) \rightarrow O_2(g) + 4H^+(aq) + 4e^- \text{ (to electrode)}$$

$$6H_2O(l) \rightarrow 2H_2(g) + O_2(g) + 4H^+(aq) + 4OH^-(aq) \qquad (10.5)$$

When you removed the electrodes and stirred the solution in Investigate This 10.2(b), you mixed the hydronium and hydroxide ions together, they reacted to form water, and the entire solution returned to its original color:

$$4H^+(aq) + 4OH^-(aq) \rightarrow 4H_2O(l) \qquad (10.6)$$

Combining reactions (10.5) and (10.6) gives us the net reaction caused by the electric current:

$$2H_2O(l) \rightarrow 2H_2(g) + O_2(g) \qquad (10.7)$$

We often use the term **redox reaction** to refer to a net **red**uction–**ox**idation reaction. Redox reaction (10.7) represents the **electrolysis,** the breaking up (*lysis*) of a molecule (water in this case) by the action of the electric current flowing between the electrodes. An **electrolytic cell** is a solution (or ionic liquid) containing electrodes at which electrolysis reactions occur when an electric current from an external source flows between the electrodes.

10.5 CONSIDER THIS

How do you analyze the gases from water electrolysis?

This diagram shows a simple electrolytic cell you could use to trap the gases produced by electrolysis of water. Before electrolysis began, the entire apparatus was filled with electrolyte solution. At which electrode is oxidation occurring? At which electrode is reduction occurring? Clearly explain the reasoning for your answers. Which electrode (right or left) is attached to the negative terminal of the current source? Explain your choice.

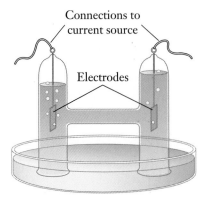

Connections to current source

Electrodes

10.6 CHECK THIS

Reactions in an electrophoresis apparatus

Look back at the diagram of an electrophoresis apparatus in Figure 9.6 (Chapter 9, Section 9.5). The gel, which is made using an ionic buffer solution, is in contact at each end with an ionic buffer solution. When the electrodes are connected to a source of electric current, gas bubbles are observed at the electrodes in the buffer wells. What gas(es) is(are) formed at each of the electrodes? Explain clearly and completely.

Anode and cathode We call the electrodes in electrochemical cells the anode and cathode. The **anode** is the electrode at which oxidation occurs. The *anode acts as an electron acceptor* from the chemical reaction at the electrode. The **cathode** is the electrode at which reduction occurs. The *cathode acts as an electron donor* to the chemical reaction at the electrode. In the external circuit, electrons flow from the anode to the cathode. This nomenclature applies both to electrolytic cells in which an electric current causes chemical reactions and to galvanic or voltaic cells in which chemical reactions are used to produce an electric current. Our discussion of galvanic cells will begin in Section 10.2.

 ## 10.7 INVESTIGATE THIS

What other reactions can occur in an electrolytic cell?

Do this as a class investigation and work in small groups to discuss and analyze the results. Fill one well of a six-well plate about one-third full of 0.1 M aqueous copper sulfate, $CuSO_4$, solution. Use thin carbon rods as electrodes and wires with alligator clips to connect them to a 6-V battery. Record which electrode is attached to the positive and which to the negative terminal of the battery. Immerse the electrodes in the solution and hold them about 2 cm apart for about 45 seconds and then remove them. Note and record any evidence for chemical reactions that you observe.

10.8 CONSIDER THIS

What are the products of electrolysis of a CuSO$_4$ solution?

(a) In Investigate This 10.7, what reaction do you think occurs at the cathode in the electrolysis cell? What is your evidence?

(b) What reaction do you think occurs at the anode? What is your evidence?

(c) What movement of species do you think occurs in the electrolytic solution? Explain the reasoning for your answer.

Stoichiometry of electrolysis When we discussed the stoichiometry of the chemical reactions that occurred in the electrolysis of water in Investigate This 10.2, we assumed that the number of electrons leaving the cathode (to reduce water) was equal to the number that entered the anode (from oxidation of water). This assumption is consistent with the chemical evidence that the moles of H$^+$(aq) and OH$^-$(aq) produced are the same. To better understand an electrochemical cell we need to be able to relate the number of moles of reaction products formed to the number of moles of electrons that enter (and leave) the cell.

We can get the number of moles of reaction products from water electrolysis by titrating the acid and/or base formed or by measuring the volume, pressure, and temperature of the hydrogen and/or oxygen formed. The reaction at the cathode in Investigate This 10.7 suggests a simpler technique. The deposit of solid metal on the cathode from a solution containing the Cu^{2+}(aq) ion indicates that the reduction reaction is

$$Cu^{2+}(aq) + 2e^- \rightarrow Cu(s) \tag{10.8}$$

The reduction process in reaction (10.8) is called **electrodeposition** because the metal is deposited by the action of the electric current through the cell. We can determine the number of moles of Cu(s) formed [and Cu^{2+}(aq) reduced] by simply weighing the cathode before and after reaction to get the mass of the electrodeposited metal. Under carefully controlled **electroplating** conditions, many metals, including copper, nickel, silver, and gold, deposit in a thin layer (plate) on the cathode.

10.9 CHECK THIS

The oxidation reaction in Investigate This 10.7

Bubbles of gas are produced at the anode in Investigate This 10.7. What is the gas? What is the oxidation reaction that produces the gas? What, if any, are the other products of this reaction? What experimental tests could you do to find out whether your responses are correct? Explain.

To determine the number of moles of electrons that enter a cell, we need to make electrical measurements that count electrons. Recall from Table 3.1, Chapter 3, that the unit of electric charge is the **coulomb,** symbolized by **C.**

The charge on an electron is 1.60218×10^{-19} C, so the charge on a mole (Avogadro's number) of electrons is

$$\text{charge per mole of } e^- = (6.0221 \times 10^{23} \text{ electron·mol}^{-1}) \times$$
$$(1.60218 \times 10^{-19} \text{ C·electron}^{-1})$$

$$\text{charge per mole of } e^- = 96{,}485 \text{ C·mol}^{-1} = 9.6485 \times 10^4 \text{ C·mol}^{-1} \quad \textbf{(10.9)}$$

The charge per mole of electrons, 9.6485×10^4 C·mol^{-1}, is called the **Faraday,** symbolized by F, to honor Michael Faraday (British scientist, 1791–1867), who developed the laws describing electrochemical stoichiometry. If we know the amount of charge that flows in the electric circuit connected to our electrolysis cell, we can calculate the number of moles of electrons that enter the cell. One way to measure the amount of charge that flows in an electric circuit is to measure the current, in **amperes,** symbolized by **A,** caused by the flow. The relationship between coulombs and amperes is $1 \text{ A} = 1 \text{ C·s}^{-1}$, so the amount of charge that flows in a circuit is the amperage of the current times the length of time the current flows:

$$(\text{amount of charge, C}) = (\text{current, A})(\text{time, s}) \quad \textbf{(10.10)}$$

10.10 WORKED EXAMPLE

Stoichiometry of nickel electrodeposition

Suppose we do a nickel electroplating experiment and find that 0.272 g of nickel is deposited on the cathode when a current of 0.246 A is passed through the electrodeposition cell for 3640. s (about an hour). Are these data consistent with the stoichiometry of this reaction?

$$\text{Ni}^{2+}(aq) + 2e^- \rightarrow \text{Ni}(s) \quad \textbf{(10.11)}$$

Necessary information: We need the molar mass of nickel, 58.69 g·mol^{-1}, and the value of the Faraday, 9.6485×10^4 C·mol^{-1}.

Strategy: Compare the number of moles of electrons that entered the cell to the number of moles of nickel deposited to see if the ratio of electrons to nickel is two, as reaction (10.11) predicts.

Implementation: The number of coulombs of charge that enter the cell is

$$\text{number of coulombs} = (0.246 \text{ A})(3640. \text{ s}) = 895 \text{ C}$$

The number of moles of electrons represented by this charge is

$$\text{moles of electrons} = (895 \text{ C})\left(\frac{1 \text{ mol}}{96{,}485 \text{ C}}\right) = 0.00928 \text{ mol}$$

The number of moles of nickel deposited is

$$\text{moles of nickel} = (0.272 \text{ g})\left(\frac{1 \text{ mol}}{58.71 \text{ C}}\right) = 0.00463 \text{ mol}$$

$$\text{The stoichiometric ratio} = \frac{\text{mol electrons}}{\text{mol nickel}} = \frac{0.00928 \text{ mol}}{0.00463 \text{ mol}} = 2.00$$

continued

Does the answer make sense? The ratio of moles of electrons that enter the cell to the moles of nickel deposited is two, as predicted by the reaction equation. The answer makes sense and helps to confirm that half reactions like reactions (10.8) and (10.11) do represent the electrochemical stoichiometry of reactions that occur in an electrochemical cell.

10.11 CHECK THIS

Electrochemical stoichiometry

Electrolytic cells can be connected in series so that the same current passes through more than one cell and the reactions in them can be directly compared. This is the way Faraday carried out many of his quantitative experiments. He often used electrodeposition of silver as his comparison; the half reaction for the reduction of $Ag^+(aq)$ is

$$Ag^+(aq) + e^- \rightarrow Ag(s) \tag{10.12}$$

(a) What do you predict will be the *ratio* of the mass of silver to mass of nickel deposited in an experimental setup with nickel and silver electrodeposition cells in series? Clearly explain your reasoning.

(b) If in the setup in part (a) a current of 0.106 A is passed for 7200. s (2 hours), what mass of silver and what mass of nickel will be deposited? Is the *ratio* of these masses what you predicted in part (a)? Explain why or why not.

Applications of electrolysis Electroplating of metals is an important application of electrolysis. Often, for example, the metal trim of motor vehicles is plated with chromium both to make it shiny and to make it resistant to corrosion. Jewelry made of less expensive metals can be gold plated to make it look like gold but be more affordable than items made with a great deal more gold. Flatware (eating utensils—knives, forks, and spoons) is often called "silverware" because expensive flatware used to be made mostly of silver (alloyed with other metals for strength). Nowadays, most silverware is silver plate, that is, a thin layer of silver electroplated onto a utensil made mostly of steel. The utensil looks like silver (and tarnishes like silver) but is less expensive than an old-fashioned silver utensil.

Far more important, however, are electrolytic processes for obtaining pure metals, such as aluminum and copper. Pure aluminum is a reactive metal and reacts rapidly with oxygen in the air to form aluminum oxide:

$$4Al(s) + 3O_2(g) \rightarrow 2Al_2O_3(s) \quad \Delta G° = -3164.6 \text{ kJ} \tag{10.13}$$

This is an enormously favorable reaction that forms a very thin, unreactive protective coating on aluminum metal. The coating prevents oxygen and many other corrosive substances, including strong acids, from reacting with the metal beneath it. Objects made of aluminum resist corrosion and retain their metallic luster for a long time. On the negative side, the great stability of $Al_2O_3(s)$ makes it very difficult to obtain pure $Al(s)$ from its ores where it is usually present as its oxide.

Therefore, although aluminum is the most abundant metal in the earth's crust (Chapter 3, Figure 3.26), it was rare and expensive until 1886 when two young inventors, Charles M. Hall (American, 1863–1914) and Paul L. T. Héroult (French, 1863–1914), simultaneously and independently invented the same electrolytic process for obtaining aluminum metal from its oxide. A diagram of the kind of apparatus that is now used to obtain aluminum from its oxide is shown in Figure 10.2.

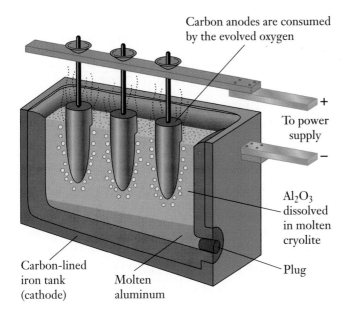

Figure 10.2.

Diagram of an electrolytic cell used to produce aluminum metal.

In the electrolytic furnace shown in Figure 10.2, $Al_2O_3(s)$ is dissolved in a molten salt, Na_3AlF_6 (cryolite), at a temperature near 1000 °C and electrolyzed:

cathode: $Al^{3+}(solvated) + 3e^- \rightarrow Al(l)$ (10.14)

anode: $2O^{2-}(solvated) \rightarrow O_2(g) + 4e^-$ (10.15)

overall: $4Al^{3+}(solvated) + 6O^{2-}(solvated) \rightarrow 4Al(l) + 3O_2(g)$ (10.16)

The species in the cryolite solution are more complex than half reactions (10.14) and (10.15) show, but the expressions represent the electrochemical stoichiometry of the electrode reactions. At the high temperatures in the electrolytic furnace, the oxygen gas reacts with the carbon anodes to form $CO_2(g)$, so the anodes are consumed and have to be replaced periodically. Aluminum metal, with a melting point 660 °C, is more dense than the molten cryolite solution, so it sinks to the sloping bottom of the vessel and is drawn off daily.

10.12 WORKED EXAMPLE

Electrolytic production of aluminum metal

Suppose an electrolytic furnace like one of those shown at the right produces one metric ton, 1000 kg, of aluminum metal per day. What current, in amperes, must be used to produce this much metal?

continued

Necessary information: The process inside each of these furnaces is what is shown in Figure 10.2. We will need the molar mass of Al, 26.98 g·mol^{-1}, and $F = 9.6485 \times 10^4$ C·mol^{-1}. We also need to remember that there are 60 s·min^{-1}, 60 min·hr^{-1}, and 24 hr·day^{-1}.

Strategy: Find the number of moles of Al produced per day and hence the number of moles of electrons required. Use the Faraday to convert moles of electrons to coulombs and then find the current that has to flow for one day to provide this many coulombs.

Implementation: The number of moles of Al produced per day is

$$\text{(mol Al)·day}^{-1} = \left(\frac{1000 \text{ kg Al}}{1 \text{ day}}\right)\left(\frac{1000 \text{ g}}{1 \text{ kg}}\right)\left(\frac{1 \text{ mol Al}}{26.98 \text{ g}}\right) = \frac{3.71 \times 10^4 \text{ mol Al}}{1 \text{ day}}$$

Each mole of aluminum produced requires 3 mol of electrons, equation (10.14); the number of coulombs of charge on these electrons is

$$\text{coulombs·day}^{-1} = \left(\frac{3.71 \times 10^4 \text{ mol Al}}{1 \text{ day}}\right)\left(\frac{3 \text{ mol } e^-}{1 \text{ mol Al}}\right)\left(\frac{9.6485 \times 10^4 \text{ C}}{1 \text{ mol } e^-}\right)$$

$$= \frac{1.07 \times 10^{10} \text{ C}}{1 \text{ day}}$$

The current required to pass this many coulombs through the electrolytic cell in one day is

$$\text{current} = \left(\frac{1.07 \times 10^{10} \text{ C}}{1 \text{ day}}\right)\left(\frac{1 \text{ day}}{24 \text{ hr}}\right)\left(\frac{1 \text{ hr}}{60 \text{ min}}\right)\left(\frac{1 \text{ min}}{60 \text{ s}}\right)$$

$$= 1.24 \times 10^5 \text{ C·s}^{-1} = 124{,}000 \text{ A}$$

Does the answer make sense? This is a very high current and may not seem reasonable. However, the furnaces shown in the figure do draw currents between 50,000 and 150,000 A. Contrast this with contemporary single-family houses whose electric circuits carry a total of only 200 to 400 A. Aluminum electrolytic refining plants are often located near an electricity generating plant in order to avoid current losses through long distance transmission.

Copper metal is occasionally found in the earth's crust in its metallic form, but the vast majority of copper is bound in its ores as oxides or sulfides. These are treated in various reduction–oxidation chemical processes to produce metal that is about 99% pure. One of the most important uses for copper metal is as an electrical conductor; almost all of the electrical wiring you see and use is copper. Copper that is only 99% pure conducts electricity much less well than 99.99% pure copper, so it must be further purified by electrolysis, as shown schematically in Figure 10.3.

The reactions of copper in the electrolytic cell are

cathode: $Cu^{2+}(aq) + 2e^- \rightarrow Cu(s)$ **(10.17)**

anode: $Cu(s) \rightarrow Cu^{2+}(aq) + 2e^-$ **(10.18)**

overall: $Cu(s)$ (from anode) $\rightarrow Cu(s)$ (to cathode)   **(10.19)**

The top arrow in the inset of Figure 10.3 denotes this overall reaction. The net effect is to transfer copper atoms from the impure anode (99% copper) to the

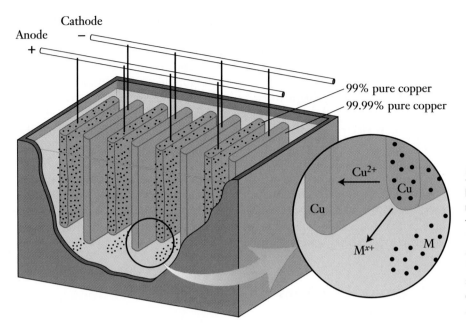

Figure 10.3.

Diagram of an electrolytic cell used to purify copper. The dots in the diagram represent impurities. During electrolysis, impurities such as Fe and Zn become oxidized to ions in solution. Other impurities such as Ag, Au, Pt, and Pd are not oxidized and fall to the bottom as a sludge.

cathode, which starts out as a thin sheet of pure copper and builds up to a thick sheet of 99.99% pure copper. The sludge that collects on the bottom of the cell is periodically collected and further refined to yield the valuable and expensive metals listed in the figure legend. In some cases, these recovered metals can be sold for enough to pay the entire cost of the electricity required to purify the copper.

10.13 CHECK THIS

Electrolytic purification of copper metal

The many cells shown here are being used to purify copper. In these cells, the cathodes start out as sheets of pure copper about 1 m square and less than 1 mm thick. When the cathodes are removed from the cell after 14 days of electrolysis, they are about 2 cm thick. If the original mass of a cathode is about 7 kg and the final mass is 170 kg, what current, in amperes, is required to purify the copper? Show your work clearly.

Reflection and Projection

From the very beginning of the book, we have been using the electrical nature of matter to help explain many observable chemical properties. It is not at all surprising, therefore, to find that electricity interacts with matter to bring about chemical changes. The half reactions introduced in Chapter 6 help to make sense of the chemistry that occurs at the electrodes in an electrolytic cell when it is connected to a source of electric current. At the cathode, electrons are supplied to the cell and a reducing half reaction occurs. At the anode, electrons leave the cell and an oxidizing half reaction occurs. The electric current is carried inside the cell by the movement of ions that maintain charge balance at both electrodes.

Half reactions also provide a way to keep track of the electrochemical stoichiometry of both the chemical species and the electrons that take part in the reactions. The key to this stoichiometry for electrons is the Faraday, 9.6485×10^4 C·mol^{-1}, the charge on a mole of electrons (or any other singly charged particle). We can use current flow through a cell, 1 A = 1 C·s^{-1}, as a measure of the number of moles of electrons and relate moles of electrons to moles of chemical reaction products, as we did for electrodepositions.

The results of electrolytic reactions are all around you in the aluminum and copper and many electroplated objects you see and use daily. However, you rarely witness electrolytic reactions themselves. On the other hand, you use the reduction–oxidation reactions in galvanic cells all the time. These cells are the power sources for almost all portable electronic devices: calculators, laptop computers, cell phones, radios, CD and tape players, and handheld computer games. In galvanic cells, we are interested in the electric current created by chemical reactions. In the next section, we will begin our discussion of galvanic cells with some simple reduction–oxidation reactions that we can use to uncover further properties of electrochemical cells.

10.2. Electric Current from Chemical Reactions

10.14 INVESTIGATE THIS

How do you get electricity from a chemical reaction?

(a) Do Investigate This 6.62 (page 400) or review your notes and discussion from that investigation.

(b) Do this as a class investigation and work in small groups to discuss and analyze the results. Fill one well of a 24-well plate about two-thirds full of 0.10 M aqueous silver nitrate, AgNO$_3$, solution. Place a silver wire in the solution with one end sticking out of the well. Fill an adjacent well about two-thirds full of 0.10 M aqueous copper sulfate, CuSO$_4$, solution and place a copper wire in the solution with one end sticking out of the well. Connect one lead (pronounced "leed") from a digital multimeter to one of the wires and the other lead to the other wire. Set the multimeter to read current (one of the ampere scales) and record the current reading from the meter.

(c) Roll a 1 × 5-cm rectangle of filter paper into a tight, 5-cm-long cylinder and fold it into a square U shape. Thoroughly wet the paper with a 10% aqueous solution of potassium nitrate, KNO$_3$. Place one leg of the U in the silver nitrate well and the other leg in the copper sulfate well and again record the current reading from the meter.

10.15 CONSIDER THIS

How do you get electricity from a chemical reaction?

(a) What were your current readings for parts (b) and (c) of Investigate This 10.14? Propose a molecular-level explanation for what you observed.

(b) What were your results in Investigate This 6.62? What reaction did you see occurring in that investigation? How do you think that reaction is related to your observations in parts (b) and (c) of Investigate This 10.14?

In Chapter 6 and in the previous section, we wrote half reactions as either reductions or oxidations, depending on the reaction we were trying to balance or the reaction that was going on at an electrode. From this point on, however, we will use the standard electrochemical *convention: **Half reactions are written as reductions.*** For the silver–copper reaction from Investigate This 6.62, the half reactions are

$$Ag^+(aq) + e^- \rightleftharpoons Ag(s) \tag{10.20}$$

$$Cu^{2+}(aq) + 2e^- \rightleftharpoons Cu(s) \tag{10.21}$$

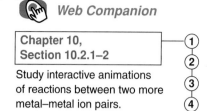

Web Companion

Chapter 10,
Section 10.2.1–2

① ② ③ ④

Study interactive animations of reactions between two more metal–metal ion pairs.

10.16 CONSIDER THIS

How are reduction half reactions combined?

(a) Explain how reactions (10.20) and (10.21) can be combined, with appropriate stoichiometric factors, to give

$$Cu(s) + 2Ag^+(aq) \rightleftharpoons Cu^{2+}(aq) + 2Ag(s) \tag{10.22}$$

Reaction (10.22) is the overall reaction observed in Investigate This 10.14(a) (Investigate This 6.62).

(b) What happens to the electrons when you combine the half reactions in part (a)? How is this analogous to what you had to do to get reaction (10.5) in Section 10.1? Explain.

Separated half reactions The net result of reaction (10.22) is to transfer two electrons from copper metal to two silver ions to produce metallic silver and one copper ion in solution. The overall reaction equation contains no electrons— free electrons do not exist in aqueous solutions. Is the idea of electron transfer just a convenient way to explain the products of the reaction? Here is where the *I Ching* quotation that opens this chapter becomes relevant. In Investigate This 10.14(b) and 10.14(c), we separated the components of the two half reactions, (10.20) and (10.21), in order to see if there is experimental evidence for electron transfer (a current flow) that unites the half reactions. The setup for Investigate This 10.14(b) is shown schematically in Figure 10.4.

Web Companion

Chapter 10,
Section 10.2.3–6

① ② ③ ④

Study interactive animations of an experiment similar to that represented in Figure 10.4.

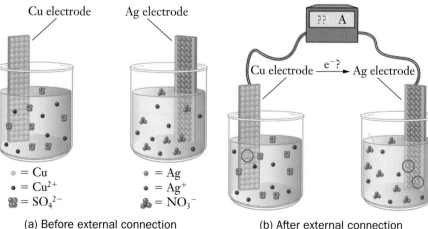

(a) Before external connection

(b) After external connection

● = Cu
● = Cu²⁺
▓ = SO₄²⁻

● = Ag
● = Ag⁺
▓ = NO₃⁻

Figure 10.4.

Does electron transfer occur between separate silver and copper wells? In (b), red rings highlight the loss of one copper atom and gain of two silver atoms at the respective electrodes.

Figure 10.4 reminds us that one of the wells contains $Ag^+(aq)$ and $Ag(s)$, the reactant and product (excluding electrons) of reaction (10.20). The other well contains $Cu^{2+}(aq)$ and $Cu(s)$, the reactant and product of reaction (10.21). The external electrical connection in Figure 10.4(b) would allow electrons to move from one electrode to the other. A possible result is represented at the molecular level in Figure 10.4(b). A copper atom from the copper electrode loses two electrons to the rest of the metal and enters the solution as a $Cu^{2+}(aq)$ ion. The electrons travel through the external part of the circuit to the silver electrode, which gives the electrode an extra two electrons. These electrons are transferred to two silver ions in the solution at the surface of the metal, thus converting them to silver atoms on the surface of the electrode.

10.17 CONSIDER THIS

What are the net charges on the solutions in Figure 10.4?

(a) Explain how Figure 10.4 represents reaction equation (10.22).
(b) What is the net charge (number of positive charges minus number of negative charges) in the silver well in Figure 10.4(a)? In Figure 10.4(b)? What is the net charge in the copper well in Figure 10.4(a)? In Figure 10.4(b)? Explain how these charges are the result of the process just described.

Your result from Investigate This 10.14(b) shows that no current flows between the electrodes in the setup shown in Figure 10.4(b). You know from your results in Investigate This 10.14(a) that reaction (10.22) occurs when copper metal is placed in contact with a solution containing silver ion. It seems as though the process we show in Figure 10.4 should be able to occur to produce a current (electron flow) through the meter. Why does no current flow?

You found in Consider This 10.17(b) that the process we imagined in Figure 10.4 leads to a net positive charge in the solution of $Cu^{2+}(aq)$ and $SO_4^{2-}(aq)$ ions and a net negative charge in the solution of $Ag^+(aq)$ and $NO_3^-(aq)$ ions. In Consider This 10.4, you analyzed a similar situation for the electrolysis of water. To avoid charge separation, ions in the solution migrate to compensate for the possible build up of charge near the electrodes. In the setup shown in Figure 10.4, there is no pathway for ions to migrate between the solutions. Electrons cannot flow between the electrodes outside the solutions unless the electric *circuit* is completed within the solutions.

Salt bridge We need a way to complete the circuit between the solutions, so we can test whether electron transfer occurs. When the reactants are in contact with one another, the ions are all in the same solution and can migrate freely; reaction (10.22) occurs, and we can observe the products. What we need in our separated system is a pathway for the charges in the separate wells to interact with one another without mixing. A **salt bridge,** a connection (bridge) between two half-reaction systems, provides this pathway for "salt" (equivalent net positive and negative ions) to move between the systems. In Activity 10.14(c), your salt bridge was a filter paper soaked in potassium nitrate solution, as represented in Figure 10.5.

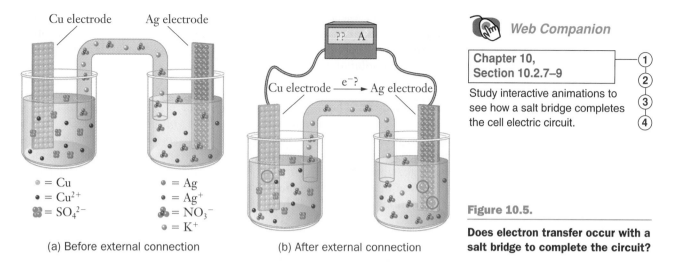

Web Companion

Chapter 10,
Section 10.2.7–9

Study interactive animations to
see how a salt bridge completes
the cell electric circuit.

Figure 10.5.

**Does electron transfer occur with a
salt bridge to complete the circuit?**

Figure 10.5(a) is the same as Figure 10.4(a), except that there is a salt bridge between the wells in Figure 10.5. Once again, we can imagine a transfer of two electrons from the copper side to the silver side of the setup. Now, when a $Cu^{2+}(aq)$ ion is formed, two nitrate ions, $NO_3^-(aq)$, can move out of the salt bridge to keep the $Cu^{2+}(aq)$ solution electrically neutral. And in the other well, when two $Ag^+(aq)$ ions are reduced to silver metal, they can be replaced by $K^+(aq)$ ions from the salt bridge to keep the $Ag^+(aq)$ solution electrically neutral. These changes are represented in Figure 10.5(b). The net charges are zero in all three solutions; there is no charge separation.

10.18 CHECK THIS

Ion migrations in a salt bridge

(a) Ions do not always move out of a salt bridge. For example, in Figure 10.5(b) a $Cu^{2+}(aq)$ ion might enter the salt bridge, instead of two $NO_3^-(aq)$ leaving it. Make a drawing showing how this alternative still makes the net charges in all three solutions zero. How would you detect the migration of $Cu^{2+}(aq)$ ions into the salt bridge?

(b) Are there other ion migrations that could occur, together with electron transfers between the electrodes through the external circuit, that also maintain the electrical neutrality of all the solutions? If so, give examples. How could you detect such migrations?

(c) 🖐 Think again about the electrophoresis apparatus in Figure 9.6 (Chapter 9, Section 9.5). The samples on the gel are ionic; the proteins we detect are negatively charged ions. Explain how the gel in an electrophoresis apparatus acts much like a salt bridge. Does the salt bridge animation in the *Web Companion*, Chapter 10, Section 10.2.9, reinforce your explanation? Explain.

In Investigate This 10.14(c), you found that, when a salt bridge connects the solutions, an electric current flows in the external connection between the electrodes. There is a net transfer of electrons between the electrodes through their external connection. The movement of ions through the salt bridge and the solutions carries an equivalent amount of electric charge. The combination of electron flow between the electrodes outside the solution and the movement of ions in the solution completes the electric circuit.

Galvanic cells The electrochemical cell you set up in Investigate This 10.14(c) is a galvanic cell. In a **galvanic cell** (also called a **voltaic cell**), a chemical redox reaction produces electrons that move through an electrical conductor connected between the two electrodes. As you have observed, the half reactions that make up an overall electrochemical cell reaction can be spatially separated, *if* ions can migrate between them through some form of salt bridge. The separate solutions and electrodes for each half reaction in a galvanic cell are called **half cells.**

Electrochemical cell notation You can imagine how tedious it would be, if we had to draw something like Figure 10.5(a) or (b) every time we wanted to describe a new cell. There is a shorthand notation for electrochemical cells that we will use from now on. In this **line notation** for electrochemical cells, the cell in Figure 10.5 is written on a single line like this:

$$\text{Cu}(s) \mid \text{Cu}^{2+}(aq,\ 0.1\ M) \parallel \text{Ag}^{+}(aq,\ 0.1\ M) \mid \text{Ag}(s) \tag{10.23}$$

A vertical line, $\mid$, denotes a boundary between distinct chemical phases. For example, in cell (10.23), the boundary between the solid copper electrode and the $\text{Cu}^{2+}(aq)$ solution is shown. Since the half-cell solutions and the salt bridge solution are quite different, there is a boundary between the half-cell solution and the salt bridge solution at each end of the salt bridge. The bridge is shown as double vertical lines, $\parallel$. If, in a particular case, it matters what is in the salt bridge, then its contents are specified between the two lines. In our example, the contents are not significant, so the bridge appears simply as a pair of lines. The concentration and/or state of each species is indicated in parentheses after the species.

You could also imagine writing the copper–silver cell like this:

$$\text{Ag}(s) \mid \text{Ag}^{+}(aq,\ 0.1\ M) \parallel \text{Cu}^{2+}(aq,\ 0.1\ M) \mid \text{Cu}(s) \tag{10.24}$$

Although this cell representation doesn't look wrong, chemists have chosen, *by convention, to write the anode on the left in electrochemical line notation,* as in cell (10.23). If we assume that everyone uses this convention, we would assume that whoever wrote cell (10.24) meant to indicate that oxidation, the anodic reaction, is occurring in the half cell that is written on the left and reduction on the right. The overall cell reaction would then be

$$2\text{Ag}(s) + \text{Cu}^{2+}(aq) \rightleftharpoons 2\text{Ag}^{+}(aq) + \text{Cu}(s) \tag{10.25}$$

Reaction (10.25) is the reverse of reaction (10.22). But we have seen above that it is reaction (10.22) that occurs when $\text{Cu}(s)$ is placed in a solution containing $\text{Ag}^{+}(aq)$. Reaction (10.25) is misleading because it does not show the favored direction of the reaction. Therefore, we say that cell (10.24) is incorrect because it implies incorrect chemistry.

10.19 WORKED EXAMPLE

Line notation for the electrolytic aluminum refining cell

Use line notation to represent the electrolytic aluminum refining cell depicted in Figure 10.2 and accompanying text, including half reactions (10.14) and (10.15).

Necessary information: We learn from Figure 10.2 that both electrodes are carbon, that the cathode is in contact with a pool of molten aluminum, and that oxygen gas is produced at the anode. The solution in the cell is Al_2O_3 dissolved in molten cryolite. For simplicity, we assume that the ions of interest are Al^{3+} and O^{2-}.

Strategy: Write the anode on the left, solution in the middle, and cathode on the right.

Implementation: The line notation for the cell is

$$\text{(anode) C} \mid O_2(g) \mid O^{2-}(solvated), \text{cryolite}(l), Al^{3+}(solvated) \mid Al(l) \mid \text{C (cathode)} \tag{10.26}$$

Does the answer make sense? The species shown in half reactions (10.14) and (10.15) are all represented in our notation, as well as the physical setup shown in Figure 10.2. Note that all species are present in the same solution, so there is no salt bridge in this cell.

10.20 CHECK THIS

Interpreting the line notation for a copper–zinc cell

The correct line notation for a copper–zinc cell, with the metal ions each at 0.10 M concentration, is

$$Zn(s) \mid Zn^{2+}(aq, 0.10\,M) \parallel Cu^{2+}(aq, 0.10\,M) \mid Cu(s) \tag{10.27}$$

(a) Write the anodic and cathodic half reactions for this cell. Write the overall cell reaction. Explain the reasoning for your answers.
(b) If a piece of copper metal is placed in a solution of zinc ion, what do you predict will be observed? If a piece of zinc metal is placed in a solution of copper ion, what do you predict will be observed? Explain your reasoning.

We have learned how to set up a galvanic cell, an electrochemical cell in which a redox reaction produces a flow of electrons in an external circuit, and we have a succinct notation for describing these cells. The batteries you use to power flashlights, portable CD players, and cell phones operate on these same principles. Each consists of two electrodes—one at which an oxidation occurs and another at which a reduction occurs. In order for electrons to flow between the electrodes in the external part of the circuit (making the filament in a flashlight bulb glow and give off light, for example) there has to be a corresponding migration of ions inside the battery to complete the circuit.

One meaning of *battery* is a group of the same things. In electricity, it used to mean a series of galvanic cells connected together to provide more power, but it has now also come to mean an individual electrochemical cell as well as a set of connected cells.

To understand more about how these cells work, including how different combinations of half cells behave and what variables affect their performance, we need a quantitative measure that we can use to characterize electrochemical cells. The measure we use is the cell potential, which we often call voltage. You are probably familiar with a variety of cells (batteries) that have different voltages. As we go on, we will look at what differentiates one cell from another and learn more about their chemistry.

10.3. Cell Potentials

10.21 INVESTIGATE THIS

What are cell voltages for different half-cell combinations?

Do this as a class investigation and work in small groups to discuss and analyze the results. As shown in this diagram, fill one of the central wells of a 24-well plate, about two-thirds full of 10% KNO_3 solution. Fill a well adjacent to the KNO_3 well about two-thirds full of 0.10 M aqueous silver nitrate, $AgNO_3$, solution and insert a silver wire as an electrode. Fill two other adjacent wells about two-thirds full of 0.10 M aqueous zinc sulfate, $ZnSO_4$, and 0.10 M aqueous copper sulfate, $CuSO_4$. As electrodes, place a strip of zinc metal in the zinc ion solution and a copper wire in the copper ion solution. Use filter paper soaked in 10% KNO_3 solution to make salt bridge connections between the KNO_3 well and each of the other wells. This setup interconnects all three half cells through the central KNO_3 well.

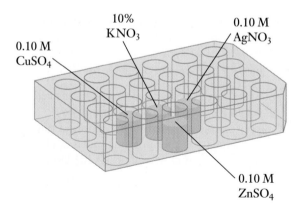

(a) Connect the input leads from a digital voltmeter to the Cu and Zn electrodes. Record the meter reading, including its units and sign, and which lead is connected to the Cu electrode and which to the Zn. Record the cell voltage to the nearest millivolt. Reverse the connections and record the same information as you did for the previous setup.
(b) Repeat the entire procedure in part (a) for the other two pairs of half cells, Cu with Ag, and Zn with Ag.

10.22 CONSIDER THIS

Are cell voltages for different half-cell combinations related?

(a) For each of the three half-cell combinations you investigated in Investigate This 10.21, you recorded two voltage readings from the digital voltmeter. How are the two readings for a combination related to one another? What was different about the setup for the two readings? Are the answers to the preceding two questions related? Explain why or why not.

(b) Consider the positive voltages you measured for the three half-cell combinations. Is there any pattern in your results? If so, how might you explain it? Does the same pattern hold for the negative readings? Explain why or why not.

Usually, a positive reading on a meter will not have a sign shown (just as we usually write positive numbers without a plus sign), but a negative reading will have a negative sign shown.

Cell potential, *E* An **electrical potential difference** between two parts of an electrical system (between two electrodes, for example) is a measure of the potential energy available to move electrons between the two parts. Electrical potential difference is measured in **volts,** symbolized by **V,** and is often called **voltage.** An electrochemical cell uses the free energy of a chemical redox reaction to produce an electrical potential difference between two electrodes. We will say more about the relationship between electrical potential difference and free energy in Section 10.5. The electrical potential difference in volts is symbolized by *E*, which stands for **electromotive force (emf),** the force that drives electrons from one electrode toward the other. *E* is also called the **cell potential.** An italic uppercase *E* is the accepted symbol for both cell potential (emf) and internal energy. You can usually tell which is meant by the context where the symbol appears. In this chapter, *E* is used only for emf.

The sign of *E* Measuring the cell potential (emf) of an electrochemical cell seems easy and straightforward. In Investigate This 10.21, you connected a digital voltmeter to the electrodes and read the meter for each half-cell combination. You also reversed the connections and found that the numeric reading for the cell was the same but that its sign was reversed. The meter reading (voltage) has a *sign* as well as a numeric value. The sign you read depends on how you connect the leads from the voltmeter to the electrodes. But reversing the connections of the measuring device, the voltmeter, can't affect the direction of the reaction occurring in the cell. This ambiguity about the sign of the cell potential could create communications problems. We need a way assure that the signs we report for cell potentials will be interpreted the same way by everyone.

We choose *conventions* for connecting a voltmeter to the electrodes, so that the reading, *E*, will be positive. Voltmeters have inputs that are coded in some way to indicate which input lead should be connected to the anode and which to the cathode of the cell to be measured. Usually the input leads are color coded, black to be connected to the anode and red to be connected to the cathode. On the meter, the inputs are often labeled "−" (or common) for the input from the anode (from which electrons enter the external circuit) and "+" for the input from the cathode. When the input leads are connected correctly, the reading on the voltmeter will be positive, as illustrated for the zinc–copper cell in Figure 10.6. If you get a negative reading, you have connected the leads to the wrong electrodes; switching the connections produces the same reading but with a positive sign (or no sign).

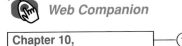

Web Companion

Chapter 10,
Section 10.3.1–2

①
②
③
④

Use this animation to check your understanding of the conventions for cell potentials.

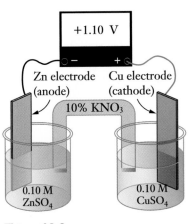

Figure 10.6.

A zinc–copper cell correctly connected to measure its cell potential, *E*.

10.23 CONSIDER THIS

What are the cell potentials and reactions in Investigate This 10.21?

(a) Are your results (meter connections and sign and magnitude of E) for the zinc–copper cell in Investigate This 10.21 consistent with the illustration in Figure 10.6? Explain why or why not.

(b) Explain the relationship between the cell represented by the line notation in Check This 10.20 and the cell illustrated in Figure 10.6. What is the cell reaction for the cell in Figure 10.6? For the zinc–copper cell in Investigate This 10.21? Explain.

(c) Use your measurements from Investigate This 10.21 to find the cell potentials and write the cell reactions for the copper–silver and zinc–silver cells. Explain your reasoning.

Combining redox reactions and cell potentials Any two half cells can be connected through a salt bridge to create an electrochemical cell. We can measure its cell potential and, by determining which electrode is the anode and which the cathode, can write the cell reaction that is producing this electron driving force, as you have done for three combinations of half cells in Consider This 10.23. In previous chapters, you have seen several examples where we combined two chemical reaction equations to get a third reaction equation. The usual reason for doing this was to combine the known values of thermodynamic variables for the two reactions to get the values for the combination reaction. Let's see if we can do similar combinations of redox reactions to get cell potentials for unmeasured (or unmeasurable) reactions. We will develop and apply the procedure to the cells in Investigate This 10.21.

You constructed three cells:

$$\text{Zn}(s) \mid \text{Zn}^{2+}(aq, 0.10\ M) \parallel \text{Cu}^{2+}(aq, 0.10\ M) \mid \text{Cu}(s) \qquad (10.27)$$

$$\text{Cu}(s) \mid \text{Cu}^{2+}(aq, 0.10\ M) \parallel \text{Ag}^{+}(aq, 0.10\ M) \mid \text{Ag}(s) \qquad (10.28)$$

$$\text{Zn}(s) \mid \text{Zn}^{2+}(aq, 0.10\ M) \parallel \text{Ag}^{+}(aq, 0.10\ M) \mid \text{Ag}(s) \qquad (10.29)$$

These cells are written correctly (anode—oxidation—on the left, as you found in Investigate This 10.21), so the cell reactions are, respectively,

$$\text{Zn}(s) + \text{Cu}^{2+}(aq) \rightleftharpoons \text{Zn}^{2+}(aq) + \text{Cu}(s) \qquad (10.30)$$

$$\text{Cu}(s) + 2\text{Ag}^{+}(aq) \rightleftharpoons \text{Cu}^{2+}(aq) + 2\text{Ag}(s) \qquad (10.22)$$

$$\text{Zn}(s) + 2\text{Ag}^{+}(aq) \rightleftharpoons \text{Zn}^{2+}(aq) + 2\text{Ag}(s) \qquad (10.31)$$

We will write the cell potentials for these three cell reactions as $E_{\text{Zn}|\text{Cu}}$, $E_{\text{Cu}|\text{Ag}}$, and $E_{\text{Zn}|\text{Ag}}$, respectively. The subscripts remind you which half cells have been combined in each case and which is the anode (written on the left).

We can combine any two of these reaction equations (with appropriate signs) to give the third. Let's combine reactions (10.30) and (10.22) to get (10.31). We add reaction (10.22) to reaction (10.30):

$$Zn(s) + Cu^{2+}(aq) \rightleftharpoons Zn^{2+}(aq) + Cu(s) \qquad (10.30)$$

$$\underline{Cu(s) + 2Ag^{+}(aq) \rightleftharpoons Cu^{2+}(aq) + 2Ag(s) \qquad (10.22)}$$

$$Zn(s) + 2Ag^{+}(aq) \rightleftharpoons Zn^{2+}(aq) + 2Ag(s) \qquad (10.31)$$

If cell reactions can be combined to yield a new cell reaction, we might suppose that their cell potentials can be combined in the same way to get the cell potential for the new reaction. In the present case, the combination would be

$$E_{Zn|Cu} + E_{Cu|Ag} = E_{Zn|Ag} \qquad (10.32)$$

10.24 CHECK THIS

Combining cell potentials and interpreting cell reactions

(a) Are your cell potentials from Consider This 10.23 consistent with equation (10.32)? Explain why or why not.

(b) Use cell reaction (10.30) to explain what will be observed if a strip of zinc metal is placed in a solution of copper(II) ion. What will be observed if a strip of copper metal is placed in a solution of zinc(II) ion?

(c) Use cell reaction (10.31) to explain what will be observed if a strip of zinc metal is placed in a solution of silver(I) ion. What will be observed if a strip of silver metal is placed in a solution of zinc(II) ion?

(d) How are your predicted observations in part (c) related to your actual observations in Investigate This 6.62? Do reactions (10.22) and (10.31) help explain this relationship? Why or why not?

Equation (10.32) is a specific case of the general result: *The cell potential for a redox reaction that is the sum (or difference) of two other redox reactions is the sum (or difference) of the cell potentials for the two reactions.* Cell potentials are intensive quantities, like temperature and pressure. Each half cell can transfer electrons with a certain force or electric potential, and the cell potential is the resultant of these two forces. The cell potential does not depend on the size of the half cells or on the amount of charge transferred (as long as the concentrations in the half cells don't change significantly). Combining one of these half cells with a different half-cell partner does not change its electron transfer potential. Electron transfer potentials for individual half cells—half-cell potentials—would be useful, if we could obtain them. We would then be able to combine half-cell potentials to calculate a cell potential without constructing an actual cell.

Reflection and Projection

Electrochemical cells (galvanic cells) move the electrons that are transferred in a redox reaction through an external, electrically conducting pathway. The interfaces between the external pathway and the oxidation and reduction that occur in the redox reaction are the cell electrodes, the anode and cathode. Oxidation occurs at the anode and reduction at the cathode. In order for electrons to flow from the anode to the cathode outside the cell, there must be migrations of ions within the cell itself to complete the electric circuit and prevent charge separation.

We characterize an electrochemical cell by its cell potential, E, measured in volts. We measure E with a voltmeter, being careful to connect the meter to the cell electrodes correctly. When two electrochemical cell reactions are combined (added or subtracted) to yield a third cell reaction, the cell potentials for the two reactions can be combined in the same way to yield the cell potential for the third reaction. Cell potentials are intensive quantities, so no stoichiometric factors enter into their combination to give the cell potential for the combination reaction. We can calculate the cell potential for any reaction that can be written as a combination of other reactions whose cell potentials are known. The calculated cell potential can then be used to make predictions about the combination reaction without having to carry it out or make a cell. The only problem with this approach is the need to find cell reactions with measured cell potentials that can be combined to yield the reaction of interest. It would be useful to have some standard way of cataloging redox reactions to make this search easier. The catalog we use is a list of standard half-cell potentials.

10.4. Half-Cell Potentials: Reduction Potentials

No electrical potential can be measured for any single half reaction; only the cell potential for a combination of two half cells can be measured. For half-cell potentials, the best we can get is a set of *relative* values, but these will allow us to make the combinations we discussed at the end of the previous section. We get these relative half-cell potentials by first choosing one half cell as a reference. Then we combine all other half cells with the reference and measure all their cell potentials in turn. Each cell potential is a combination of the half-cell potential for the reference and the half-cell potential for the one we are measuring. To generate a catalog of relative half-cell potentials, we *arbitrarily* assign a value for the half-cell potential of the reference half reaction, and then relate all others to this choice. Because we have chosen, by convention, to write all half reactions as reductions, half-cell potentials are called **reduction potentials.**

The standard hydrogen electrode Chemists have chosen to use the **standard hydrogen electrode (SHE)** as the reference half cell and half reaction and to *assign* the SHE a standard reduction potential, $E°(H^+, H_2)$, of exactly 0 V at 298 K. The half cell and half cell reaction are

$$\text{Pt}(s) \mid H_2(g, \textit{1 bar}) \mid H^+(aq, \textit{1 M}) \qquad E°(H^+, H_2) \equiv 0 \text{ V} \qquad \textbf{(10.33)}$$

$$2H^+(aq) + 2e^- \rightleftharpoons H_2(g) \qquad \textbf{(10.34)}$$

This potential is labeled a **standard reduction potential,** denoted by the superscript °, because all species in the reaction are at their standard state concentrations (see Table 9.1, Chapter 9, Section 9.2).

The SHE, Figure 10.7, is made by bubbling hydrogen gas at 1 bar pressure ($= 100$ kPa ≈ 1 atm), over a specially prepared platinum electrode immersed in a solution with a hydronium ion concentration of unity, $[H^+(aq)] = 1$ M. The SHE is not easy to set up and maintain. Other reference electrodes, whose standard reduction potentials have been determined, are often used to measure

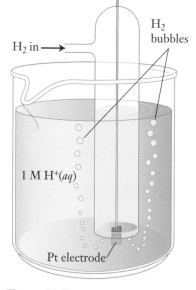

Figure 10.7.

Representation of a standard hydrogen electrode.

Labels in figure: H$_2$ in →; H$_2$ bubbles; 1 M H$^+$(aq); Pt electrode new half-cell potentials. No matter how a half cell is measured in practice, its standard reduction potential is still given relative to the SHE.

Measuring a reduction potential To measure the standard reduction potential for the silver–silver ion half cell, we would set up this cell (which is illustrated in Figure 10.8):

$$Pt(s) \mid H_2(g,\ 1\ bar) \mid H^+(aq,\ 1\ M) \parallel Ag^+(aq,\ 1\ M) \mid Ag(s) \tag{10.35}$$

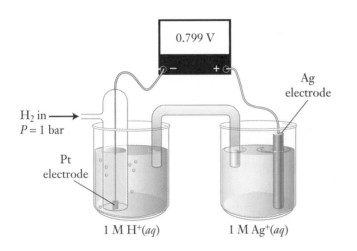

Figure 10.8.

Measuring the standard reduction potential for the silver–silver ion half cell.

By convention, ***the SHE is taken as the anode for these determinations.*** Since the reaction at the anode is assumed to be an oxidation, we have to subtract the anode half reaction (10.34), which is written as a reduction, from the cathode half reaction (10.20) to get the reaction in the cell we have written and constructed. With appropriate stoichiometric factors to cancel the numbers of electrons we get

$$2\left[Ag^+(aq) + e^- \rightleftharpoons Ag(s)\right] \qquad\qquad \mathbf{2 \times (10.20)}$$
$$-\left[2H^+(aq) + 2e^- \rightleftharpoons H_2(g)\right] \qquad\qquad -\ (\mathbf{10.34})$$
$$\overline{2Ag^+(aq) + H_2(g) \rightleftharpoons 2Ag(s) + 2H^+(aq)} \qquad\qquad (\mathbf{10.36})$$

The cell potential for the SHE-silver cell, $E_{H|Ag}$, is the difference between the reduction potentials for reactions (10.20) and (10.34):

$$E^\circ_{H|Ag} = E^\circ(Ag^+, Ag) - E^\circ(H^+, H_2) \tag{10.37}$$

The reduction potentials for both half cells and the overall cell potential are shown as standard potentials because all species in cell (10.35) are at their standard state concentration. Since $E^\circ(H^+, H_2)$ is defined as zero, we get

$$E^\circ_{H|Ag} = E^\circ(Ag^+, Ag) \tag{10.38}$$

Equation (10.38) exemplifies this rule: ***The standard reduction potential for any half cell is equal to the measured potential of a cell made by coupling the half cell of interest (with all species at standard state concentration) to the SHE to create a cell in which the SHE is taken to be the anode.*** The measured cell potential, $E^\circ_{H|Ag}$, of cell (10.35) is 0.799 V, as shown in Figure 10.8; so we have

$$Ag^+(aq) + e^- \rightleftharpoons Ag(s) \qquad E^\circ(Ag^+, Ag) = 0.799\ V \tag{10.39}$$

10.25 CHECK THIS

Standard reduction potentials

Assume that you have constructed this cell and find that the cell potential is 0.337 V:

$$Pt(s) \mid H_2(g, \textit{1 bar}) \mid H^+(aq, \textit{1 M}) \parallel Cu^{2+}(aq, \textit{1 M}) \mid Cu(s) \qquad \textbf{(10.40)}$$

(a) Write the half reaction that occurs at the cathode. What is the standard reduction potential for this half reaction? Explain your reasoning.

(b) Use the information in this problem to explain what will be observed if a piece of copper metal is placed in a solution containing hydronium ions, say, hydrochloric acid.

10.26 INVESTIGATE THIS

Do metals react with acids?

Do this as a class investigation and work in small groups to discuss and analyze the results. Fill each of five wells in one row of a 24-well plastic plate about two-thirds full of 1 M hydrochloric acid, HCl. Repeat with 1 M sulfuric acid, H_2SO_4, in five wells of another row. You have 10 metal samples (as 2-cm lengths of wire or thin strips), two each of Ag, Cu, Fe, Mg, and Zn. Place a different metal in each HCl well. Observe what happens and note any indication of reaction between a metal and the acid. Repeat with the five metals in H_2SO_4.

10.27 CONSIDER THIS

What are the reactions of metals with acids?

(a) What evidence for reactions between the metals and the acids did you observe in Investigate This 10.26? Did all five metals react the same way? Did the metals react the same way in both acids? How do you interpret the similarities and differences among the results?

(b) Write reaction equations for those cases where you observe reaction. Explain your choice of reactants and products in each case.

Signs of reduction potentials We made an *arbitrary* choice of the SHE as our standard reference electrode. Are we guaranteed that it will really be the anode in every cell we make? This would be the same as saying that there is no reducing reaction we can couple with the SHE that will reduce $H^+(aq)$. That is, no substance would react with hydronium ion, $H^+(aq)$, to produce hydrogen gas, $H_2(g)$. But you know this is wrong. In Investigate This 10.26, you found that Fe, Mg, and Zn react with solutions containing hydronium ion, $H^+(aq)$ (acidic solutions) to produce a gas (hydrogen).

Consider the cell we can construct with a SHE and a standard $Zn^{2+}-Zn$ half cell:

$$Pt(s) \mid H_2(g, \, 1 \, bar) \mid H^+(aq, \, 1 \, M) \parallel Zn^{2+}(aq, \, 1 \, M) \mid Zn(s) \qquad \textbf{(10.41)}$$

The cell reaction implied by writing the cell like this is

$$Zn^{2+}(aq) + H_2(g) \rightleftharpoons Zn(s) + 2H^+(aq) \qquad \textbf{(10.42)}$$

The reaction you *observed* in Investigate This 10.26 is the reverse: Zinc metal reacts with hydronium ion (acid solution) to produce hydrogen gas. If we measure the cell potential for cell (10.41) by connecting the anode input of our voltmeter to the SHE, the reading we get is -0.763 V. The negative reading tells us that we have connected the meter backwards (that the SHE is not truly acting as the anode). However, if we *apply the rule* above, we get

$$Zn^{2+}(aq) + 2e^- \rightleftharpoons Zn(s) \qquad E°(Zn^{2+}, Zn) = -0.763 \text{ V} \qquad \textbf{(10.43)}$$

We interpret this negative standard reduction potential as telling us that the driving force (the potential) for Zn^{2+} reduction, *relative to the SHE*, is unfavorable or weak. Conversely, the driving force for the reverse reaction, Zn oxidation, *relative to the SHE*, is quite strong, because it has a positive potential. This is the reaction you observed when you added zinc metal to acid.

10.28 CHECK THIS

What are the signs of reduction potentials for other metals?

Are the standard reduction potentials for Cu, Fe, and Mg positive or negative? What is the experimental evidence for your answers? State your reasoning clearly.

Table of standard reduction potentials With the preceding interpretation of a negative cell potential, we can apply the rule for standard reduction potentials of half reactions in all cases and get both positive and negative values for standard reduction potentials. Appendix C is a table of standard reduction potentials, and Table 10.1 gives a few of these that are relevant to this section. The sign depends on the driving force for reduction relative to the SHE. Positive signs mean strong driving forces for reduction. The reactant in the half reaction is easily reduced. Negative signs mean weak driving forces for reduction but strong driving forces for oxidation. The product in the half reaction is easily oxidized. The larger the absolute value of the standard reduction potential, the larger is the driving force. The values in the tables are arranged in order from the most positive to the most negative standard reduction potential.

The direction of redox reactions Figure 10.9 shows how to use relative values of standard reduction potentials to predict redox reactions. Imagine a cell made by coupling two half cells (half reactions) from Appendix C or Table 10.1.

Table 10.1 *Standard reduction potentials for a few half reactions.*

Half-cell reaction	$E°$, V
$PbO_2 + SO_4^{2-} + 4H^+ + 2e^- \rightleftharpoons PbSO_4 + 2H_2O$	1.658
$Ag^+ + e^- \rightleftharpoons Ag$	0.799
$O_2 + 2H_2O + 4e^- \rightleftharpoons 4OH^-$	0.403
$Cu^{2+} + 2e^- \rightleftharpoons Cu$	0.337
$2H^+ + 2e^- \rightleftharpoons H_2 + 2H_2O$	0.000
$PbSO_4 + 2e^- \rightleftharpoons Pb + SO_4^{2-}$	−0.356
$Fe^{2+} + 2e^- \rightleftharpoons Fe$	−0.41
$Zn^{2+} + 2e^- \rightleftharpoons Zn$	−0.763
$Al(OH)_3 + 3e^- \rightleftharpoons Al + 3OH^-$	−2.33
$Mg^{2+} + 2e^- \rightleftharpoons Mg$	−2.38

The half cell with the more positive standard reduction potential (higher in the list) will be the cathode of the cell. The half reaction in this half cell will proceed as a reduction. The half cell with the less positive standard reduction potential (lower in the list) will be the anode of the cell. The half reaction in this half cell will proceed in reverse, as an oxidation. The standard cell potential is the difference between the more positive and less positive reduction potential. (In Section 10.6, we will take up the case that the concentrations in the cell are not standard concentrations. It is possible that a prediction based on standard reduction potentials could be wrong, but it is a good starting point.) The greater the difference in the standard reduction potentials for the half reactions, the greater the driving force for the overall reaction.

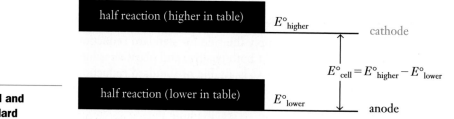

Figure 10.9.

Predicting cell potential and cell reaction from standard reduction potentials.

$$\text{cell reaction} = \text{half reaction (higher)} - \text{half reaction (lower)}$$

The species on the left of a reduction half reaction is an **oxidizing agent**. The strongest oxidizing agents are the *reactants* in the reactions at the top of the listing in Appendix C. The product of each half reaction is a **reducing agent**. The strongest reducing agents are the *products* in the half reactions at the bottom of the listing. Thus, the prediction in the previous paragraph can also be stated this way: *An oxidizing agent (reactant) in a half reaction higher on the list has the potential to oxidize any reducing agent (product) lower on the list.* These ideas are noted in Appendix C and applied to a practical electrochemical cell in Worked Example 10.29.

Web Companion

Chapter 10,
Section 10.7.1

Check your understanding of combining reduction potential with this interactive exercise.

① ② ③ ④

10.29 WORKED EXAMPLE

The aluminum–air cell

An aluminum–air cell consists of an aluminum metal electrode and a second electrically conducting electrode that is permeable to gases. Oxygen from the air passes through the permeable electrode and reacts at the surface exposed to the electrolyte solution. The solution in the cell is an aqueous solution of NaCl or NaOH. The half-cell reactions for an aluminum–air cell are

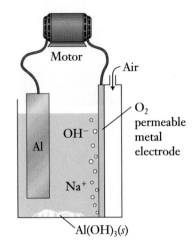

$$O_2(g) + 2H_2O(l) + 4e^- \rightleftharpoons 4OH^-(aq) \qquad \textbf{(10.44)}$$

$$Al(OH)_3(s) + 3e^- \rightleftharpoons Al(s) + 3OH^-(aq) \qquad \textbf{(10.45)}$$

Write the cell reaction and the line notation for the aluminum–air cell. What is the cell potential, if all components are at unit activity? As the cell is discharged, what changes occur?

Necessary information: From Table 10.1 (or Appendix C), we need the standard reduction potentials for half reactions (10.44) and (10.45): 0.403 V and −2.33 V, respectively.

Strategy: Combine the half reactions and their standard reduction potentials by subtracting the less positive half reaction from the more positive (with appropriate stoichiometry to cancel the electrons). The combination gives the cell reaction and its standard cell potential. Once the anode reaction is known, write the line notation for the cell. The cell reaction gives the information needed to figure out the changes that occur as the reaction proceeds.

Implementation: The appropriate half-reaction combination is

$$3\left[O_2(g) + 2H_2O(l) + 4e^- \rightleftharpoons 4OH^-(aq)\right] \qquad 3 \times \textbf{(10.44)}$$

$$\underline{-4\left[Al(OH)_3(s) + 3e^- \rightleftharpoons Al(s) + 3OH^-(aq)\right]} \qquad -4 \times \textbf{(10.45)}$$

$$3O_2(g) + 4Al(s) + 6H_2O(l) \rightleftharpoons 4Al(OH)_3(s) \qquad \textbf{(10.46)}$$

Combining the standard reduction potentials for the half cells gives the standard cell potential:

$$E°_{O_2|Al} = E°(O_2, OH^-) - E°(Al(OH)_3, Al) = 0.403 \text{ V} - (-2.33 \text{ V})$$

$$= 2.73 \text{ V} \qquad \textbf{(10.47)}$$

Oxidation of the aluminum metal occurs, so the aluminum is the anode and the cell is

$$Al(s) \mid Al(OH)_3(s) \mid \text{aqueous ionic solution} \mid O_2(g) \mid \text{inert metal} \qquad \textbf{(10.48)}$$

The ions in the solution are not part of the cell reaction equation. An ionic solution is necessary to facilitate ionic movement. Hydroxide ions migrate from the cathode, where they are produced, toward the anode, where they react to form $Al(OH)_3(s)$.

The cell reaction uses up the aluminum metal electrode and water from the solution to form $Al(OH)_3(s)$. Oxygen from the air is also used, but the air provides an almost inexhaustible source of oxygen. Recharging an aluminum–air

continued

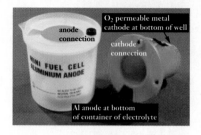

cell involves replacing the spent aluminum electrode, adding more water, and removing the solid aluminum hydroxide that has formed.

Does the answer make sense? This picture shows an aluminum–air cell. Reduction potentials predict the cell reaction and standard cell potential for the combination of the two half reactions in this electrochemical cell. The cell reaction helps us understand the chemistry that occurs when the cell is used. These cells are used in some emergency lighting systems and are being considered as power sources for portable electronic devices like cell phones.

10.30 CHECK THIS

The lead-acid battery

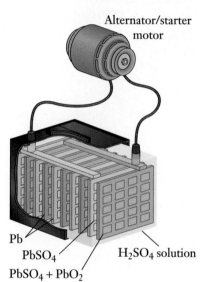

(a) The half-cell reactions for the lead-acid battery used in automobiles are

$$PbO_2(s) + SO_4^{2-}(aq) + 4H^+(aq) + 2e^- \rightleftharpoons PbSO_4(s) + 2H_2O(l) \quad (10.49)$$

$$PbSO_4(s) + 2e^- \rightleftharpoons Pb(s) + SO_4^{2-}(aq) \quad (10.50)$$

Write the cell reaction. What is the cell potential, if all components are at unit activity? Use the standard reduction potentials in Table 10.1 (or Appendix C). Modern automobiles almost all use 12-V batteries. How do you get 12 V from an automobile battery?

(b) The electrodes in the battery are grids of lead suspended in a sulfuric acid solution. When the battery is fully charged, one set of electrode grids is filled with $PbSO_4(s)$ and the other with a mixture of $PbSO_4(s)$ and $PbO_2(s)$. Write the line notation for the cell. As the cell is discharged, what changes occur at the electrodes?

(c) If the cell reaction is run in reverse, what changes occur at the electrodes? The reaction is run in reverse by applying an electrical potential that is larger than the cell potential and in the opposite direction. That is, the cell is treated like an electrolytic cell, as in Section 10.1. Remember that the cathode of an electrolytic cell is the electrode at which reduction occurs. Which electrode is this in this case? The car's alternator is the source of this potential and recharges the battery while the engine is running.

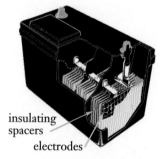

Reflection and Projection

We have learned how to catalog redox reactions by defining standard reduction potentials, $E°$, for all half cells relative to the standard hydrogen electrode. A listing of these reduction potentials in order from most positive to most negative, Appendix C, provides a tool for predicting the direction of redox reactions. In a redox reaction, the half reaction that is higher in the table will proceed as written, as a reduction. The half reaction lower in the table will go in reverse as an oxidation.

The standard reduction potentials are for all components of the half reaction at their standard state concentrations. So far, we have no way to figure out what will happen to cell or half-cell potentials when the concentrations are not standard concentrations. Also, up to now, our rules for combining cell reactions and cell potentials and for combining half reactions and standard reduction potentials to get standard cell potentials have been based on experimental results for observed chemical reactions and corresponding cell potentials. To go further in our analysis of electrochemical cells and redox reactions, we need to consider the thermodynamics of electrochemical cells. We will begin by considering the work available from electrochemical cells and the associated change in free energy for the cell reaction.

10.5. Work from Electrochemical Cells: Free Energy

10.31 INVESTIGATE THIS

How can we get work from a galvanic cell?

(a) Do this as a class investigation and work in small groups to discuss and analyze the results. Fill one well of a six-well plate about half full of aqueous 1.0 M zinc sulfate, $ZnSO_4$, solution. Bend a 1×8-cm piece of zinc sheet into an L so that the longer leg can rest flat on the bottom of the well with the shorter leg up and out of the well. Place the sheet in the well with the $ZnSO_4$ solution. Repeat this procedure in a second adjacent well, using a solution of aqueous 1.0 M copper sulfate, $CuSO_4$, solution and a 1×8-cm piece of copper sheet. Finish making a galvanic cell by connecting the half cells with a filter paper salt bridge that has been soaked in 10% aqueous potassium nitrate, KNO_3, solution. Turn a digital multimeter to its voltage scale and properly connect the leads from a digital multimeter to the copper and zinc electrodes. Note the electrode to which each lead is attached. Record the reading, including its sign and units, to the nearest millivolt.

(b) Disconnect one lead from the multimeter and then connect a small electric motor with an attached fan blade between this lead and the electrode, as shown schematically in Figure 10.5. Turn the multimeter to read current (one of the ampere scales) and record the current reading in the circuit when the fan is going.

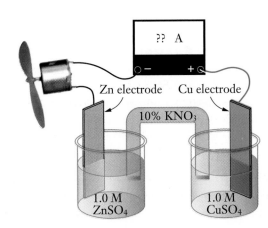

?? A

Zn electrode   Cu electrode

10% KNO_3

1.0 M $ZnSO_4$

1.0 M $CuSO_4$

Figure 10.10.

A zinc–copper galvanic cell connected to a motor and current meter. The meter and motor are connected so that all current from the cell passes through both of them.

===== **10.32 CONSIDER THIS** =====

What reaction in the zinc–copper galvanic cell produces work?

(a) Can the zinc–copper galvanic cell you made in Investigate This 10.31(a)—illustrated in Figure 10.10—do useful work in the surroundings? What observations lead to your conclusion?

(b) In your zinc–copper galvanic cell, which electrode, copper or zinc, is the cathode? Which is the anode? Base your reasoning on your observations in Investigate This 10.31(a).

(c) When written as reductions, the half reactions in the copper and zinc half cells of your galvanic cell are

$$Cu^{2+}(aq) + 2e^- \rightleftharpoons Cu(s) \qquad \text{(10.21)}$$

$$Zn^{2+}(aq) + 2e^- \rightleftharpoons Zn(s) \qquad \text{(10.43)}$$

What combination of equations (10.21) and (10.43) gives the overall reaction in your zinc–copper galvanic cell? What is the reasoning for your answer?

Electrical work As you demonstrated in Investigate This 10.31(b), galvanic cells can do work on their surroundings when electrons flow between the electrodes. You harness this work when you use batteries to power portable electrical devices. You know that different devices require different numbers and different kinds of batteries. This is because different amounts of work are required to run each one. How much work is available from an electrochemical cell and what factors affect the available work?

An electrochemical cell uses the free energy of a chemical redox reaction to produce an electrical potential difference between two electrodes. We are interested in the work that the cell can produce in the external part of the circuit. This **electrical work**, symbolized by w_{elec} and measured in joules (J), is defined as the product of the amount of charge (in coulombs, C) that passes through the circuit times the electrical potential difference, E, (in volts, V) between the electrodes:

> Electrical measurements are usually more accurate than mechanical ones like acceleration, so joules are often measured in electrical units:
>
> 1 joule = 1 volt·coulomb

$$w_{elec} = (\text{amount of charge, C})(\text{potential difference, V}) \qquad \text{(10.51)}$$

===== **10.33 WORKED EXAMPLE** =====

Electrical work

A timer used to time a chemistry examination runs on a 1.35 V battery. The current through the timer is 0.385 mA (milliamp). How much electrical work does the cell produce to run the timer for a 50-minute examination? How many moles of electrons flow through the timer circuit?

Necessary information: We need equation (10.51) and need to know that there are 60 seconds in a minute, that 1000 mA = 1 A (= 1 C·s^{-1}), and that $F = 9.6485 \times 10^4$ C·mol^{-1}.

continued

Strategy: Use the current and the time it flowed to determine the amount of charge that flowed and then substitute into equation (10.51) to get the work. Use F to convert the amount of charge to moles of electrons.

Implementation:

$$\text{current} = (0.385 \text{ mA})\left(\frac{1 \text{ A}}{1000 \text{ mA}}\right) = 3.85 \times 10^{-4} \text{ A} = 3.85 \times 10^{-4} \text{ C·s}^{-1}$$

$$\text{amount of charge} = (3.85 \times 10^{-4} \text{ C·s}^{-1})(50 \text{ min})(60 \text{ s·min}^{-1}) = 1.16 \text{ C}$$

$$w_{\text{elec}} = (1.16 \text{ C})(1.35 \text{ V}) = 1.57 \text{ J}$$

$$\text{mol } e^- = (1.16 \text{ C})\left(\frac{1 \text{ mol}}{96,485 \text{ C}}\right) = 1.20 \times 10^{-5} \text{ mol}$$

Does the answer make sense? Without some frame of reference, it's hard to tell whether one isolated result like this makes sense. As you calculate the electrical work under other conditions, keep this result in mind for comparison. In this case, the electricity is being used to run a tiny solid-state device that requires little power, so a small amount of work (energy) makes sense.

10.34 CHECK THIS

Electrical work

Use your data for the zinc–copper cell voltage and the current drawn by the motor in Investigate This 10.31 to find out how much work the cell would have to produce to run the motor for one minute. How many moles of electrons would flow through the motor?

We can calculate the work *available* from an electrochemical cell without doing current and time measurements. Let n be the number of moles of electrons that flow in the external circuit per mole of the stoichiometric reaction in the cell. An equivalent definition of n is the number of electrons in each of the half reactions that are combined to give the overall cell reaction. For the zinc–copper cell we have been considering, there are two electrons in each half reaction, equations (10.21) and (10.43), so $n = 2$. Note that n does not have units. The amount of charge that flows per mole of reaction is nF coulombs. If E is the cell potential, the electrical work available per mole of cell reaction is

$$w_{\text{elec}}\left[\text{J·(mol of reaction)}^{-1}\right] = nFE \tag{10.52}$$

10.35 CONSIDER THIS

Why are there so many sizes of batteries?

The display of batteries in this picture is bewildering in its variety. These cylindrical 1.5-V batteries come in sizes from AAAA to D. Is there a difference in the amount of work available from these different batteries? How does equation (10.52) help answer this question?

For an electrochemical cell with a certain E, we can obtain almost any amount of work by building a big enough cell and producing enough moles of reaction to get as many multiples of nFE as we want. In the real world, we would prefer not to use huge cells; consider the problems you would have with a portable CD player. The solution is to use half cells that can be connected to produce a substantial potential difference so that we can obtain a substantial amount of work without many moles of reaction. This is a practical reason why values of E and factors that affect them are so important. They will also prove to be important in understanding more about electron transfer reactions, including those in the electron transport pathway in living organisms.

10.36 WORKED EXAMPLE

Amount of work available from an electrochemical cell

The battery (cell) referred to in Worked Example 10.33 is a mercury cell, for which the overall cell reaction can be represented as

$$Zn(s) + HgO(s) \rightarrow ZnO(s) + Hg(l) \tag{10.53}$$

If $HgO(s)$ is the limiting reagent and the battery contains 1.0 g of $HgO(s)$, how long can it run the timer in Worked Example 10.33?

Necessary information: We need equation (10.52) and the result from Worked Example 10.33; work required to run the timer for 50 minutes is 1.67 J. We also need the molar mass of HgO (216.6 g·mol^{-1}) and $F = 9.6485 \times 10^4$ C·mol^{-1}. We need to see that $n = 2$; two moles of electrons are transferred for each mole of reaction (10.53). Two moles of electrons are produced when 1 mol of $Zn(s)$ (oxidation number 0) is oxidized to 1 mol of zinc(II) (oxidation number $+2$) in $ZnO(s)$. The same reasoning applies to the number of moles of electrons required to reduce mercury(II) in $HgO(s)$ to $Hg(l)$.

Strategy: Use equation (10.52) to find the work available from 1 mol of reaction (10.53). The total work available from the cell is obtained from the total moles of reaction (= number of moles of HgO in the cell) and compared to the work required to run the timer for 50 minutes to find the total time the timer could be run.

Implementation: Since $n = 2$, the work available per mole of reaction is

$$w_{elec} \text{ (per mol reaction)} = nFE = 2(9.6485 \times 10^4 \text{ C·mol}^{-1})(1.35 \text{ V})$$

$$= 2.61 \times 10^5 \text{ J·mol}^{-1}$$

The number of moles of reaction available from the cell is

$$\text{mol of reaction} = \text{mol of HgO}$$

$$= (1.0 \text{ g HgO})\left(\frac{1 \text{ mol HgO}}{216.6 \text{ g}}\right) = 4.6 \times 10^{-3} \text{ mol}$$

The work available from the cell is

$$w_{elec} = (2.61 \times 10^5 \text{ J·mol}^{-1})(4.6 \times 10^{-3} \text{ mol}) = 1.2 \times 10^3 \text{ J}$$

continued

The length of time that this amount of work will run the timer is

$$\text{time} = (1.2 \times 10^3 \, J)\left(\frac{50 \, \text{min}}{1.67 \, J}\right) = 3.6 \times 10^4 \, \text{min} = 25 \, \text{days}$$

Does the answer make sense? The cell lasts long enough to time 720 fifty-minute exams. We expect the batteries (cells) in electronic devices to provide a considerable amount of use before replacement. It makes sense that the cell in this timer should last through this many uses.

10.37 CHECK THIS

Amount of work available from an electrochemical cell

In Consider This 10.32(c), you wrote the zinc–copper cell reaction:

$$Zn(s) + Cu^{2+}(aq) \rightarrow Zn^{2+}(aq) + Cu(s) \tag{10.30}$$

Assume that each well of your zinc–copper cell in Investigate This 10.31 contains 5.0 mL of solution. If you run the fan until one-half of the $Cu^{2+}(aq)$ ion is used up, for approximately how long will the fan run? Show clearly how you get your answer.

Free energy and cell potential The sign of E, the cell potential, tells us the direction of an electron transfer reaction. The magnitude of E is a measure of the electrical work, equation (10.52), available to us *from* the reaction. In Chapter 8, Section 8.8, we found that the **free energy** change for a reaction also tells us the direction and work available (the "free" energy) from the reaction. This parallelism suggests that there is a relationship between E and $\Delta G_{reaction}$ for redox reactions. The maximum work available *from* a system is $-\Delta G_{reaction}$ for the change. For electron transfer reactions we can write

$$w_{elec}\left[J \cdot (\text{mol of reaction})^{-1}\right] = nFE = -\Delta G_{reaction} \tag{10.54}$$

$\Delta G_{reaction}$ in equation (10.54) is the free energy change per mole of stoichiometric reaction. If $\Delta G_{reaction}$ is negative, the reaction is favored in the direction written. Equation (10.54) shows that a negative $\Delta G_{reaction}$ means a favorable (positive) E for the reaction. Table 10.2 summarizes the relationships between $\Delta G_{reaction}$ and E. Under standard conditions, equation (10.54) shows that

$$nFE^\circ = -\Delta G^\circ_{reaction} \tag{10.55}$$

Table 10.2

Relationships among $\Delta G_{reaction}$, E, and favored direction of reaction.

$\Delta G_{reaction}$	E	Reaction as written is
< 0	> 0	favored
> 0	< 0	not favored
$= 0$	$= 0$	at equilibrium

Combining redox reactions and cell potentials The relationship between the cell potential and the free energy change in equation (10.54) can help us understand the basis for combining cell potentials that we introduced in Section 10.3. Let's take the following combination of cell reactions as an example:

$$Zn(s) + 2Ag^+(aq) \rightleftharpoons Zn^{2+}(aq) + 2Ag(s) \tag{10.31}$$

$$-2\left[Fe(CN)_6^{4-}(aq) + Ag^+(aq) \rightleftharpoons Fe(CN)_6^{3-}(aq) + Ag(s)\right] \quad -2 \times \text{(10.56)}$$

$$\overline{Zn(s) + 2Fe(CN)_6^{3-}(aq) \rightleftharpoons Zn^{2+}(aq) + 2Fe(CN)_6^{4-}(aq)} \tag{10.57}$$

The free energy changes *per mole of reaction* combine in the same way as the reactions:

$$\Delta G_{Zn|Fe} = \Delta G_{Zn|Ag} - 2\Delta G_{Fe|Ag} \tag{10.58}$$

We can use the relationship between free energy and cell potential, equation (10.54), to rewrite equation (10.58) in terms of cell potentials:

$$-n_{Zn|Fe}FE_{Zn|Fe} = -n_{Zn|Ag}FE_{Zn|Ag} - 2(-n_{Fe|Ag}FE_{Fe|Ag}) \tag{10.59}$$

From the reaction equations, $n_{Zn|Fe} = n_{Zn|Ag} = 2$; that is, two moles of electrons are transferred per mole of cell reactions (10.31) and (10.57). However, $n_{Fe|Ag} = 1$, since only one mole of electrons is transferred in reaction (10.56). Substituting these n values in equation (10.59) gives

$$-2FE_{Zn|Fe} = -2FE_{Zn|Ag} + 2FE_{Fe|Ag}$$

And dividing through by $-2F$, we get

$$E_{Zn|Fe} = E_{Zn|Ag} - E_{Fe|Ag} \tag{10.60}$$

Thus, to calculate the free energy changes when we combine reactions, we need to take account of the number of times each reaction enters into the combination. However, as we found experimentally and then generalized in Section 10.3, the number of times each reaction enters into the combination is not relevant for the combination of cell potentials.

Reflection and Projection

The flow of electrons from an electrochemical cell through an external circuit can be harnessed to do useful work. The amount of work is the product of the cell potential, E, and the amount of charge that flows. For one mole of cell reaction, the charge transferred is nF coulombs, where n is the number of moles of electrons transferred per mole of cell reaction. Since the free energy change for a system is also related to the work it can do on the surroundings, we can relate free energy change to cell potential, $nFE = -\Delta G_{reaction}$, which provides a fundamental thermodynamic basis for understanding electrochemical cells.

None of the cell potentials and free energy changes for cell reactions that we have written in this section have been standard cell potentials or standard free energies. The relationships in Table 10.2, for example, are for the reactions under whatever conditions exist in the cell or in a mixture of the reactants and products. Now we need to find out how to go from standard cell potentials and the related standard free energy changes to cell potentials and free energy changes for reactions under nonstandard conditions.

10.6. Concentration Dependence of Cell Potentials: The Nernst Equation

10.38 INVESTIGATE THIS

Does the silver–copper cell potential depend on [Ag⁺]?

Do this as a class investigation and work in small groups to discuss and analyze the results. As shown in the diagram, fill one of the central wells of a 24-well plate about two-thirds full of 10% KNO₃ solution. Fill three other wells adjacent to the KNO₃ well about two-thirds full of 1.00 M, 0.100 M, and 0.0100 M aqueous silver nitrate, AgNO₃, solution. Place a clean silver wire in the 0.0100 M silver solution as an electrode. Fill a fourth adjacent well about two-thirds full of 0.10 M aqueous copper sulfate, CuSO₄, solution, and place a clean copper wire in the solution as an electrode. Use filter paper soaked in 10% KNO₃ solution to make salt bridge connections between the KNO₃ well and each of the other wells.

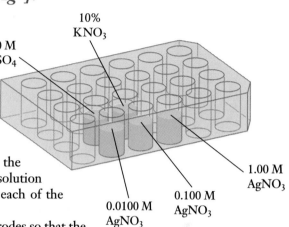

(a) Connect the input leads from a digital voltmeter to the electrodes so that the meter reading is positive. Note which lead is connected to the silver electrode and which to the copper. Record the cell potential to the nearest millivolt.

(b) Remove the silver electrode from the 0.0100 M silver solution, wipe it dry with a paper towel, place it in the 0.100 M silver solution, and measure the new cell potential.

(c) Remove the silver electrode from the 0.100 M silver solution, wipe it dry with a paper towel, place it in the 1.00 M silver solution, and measure the new cell potential.

10.39 CONSIDER THIS

How does the silver–copper cell potential depend on [Ag⁺]?

(a) What were the cell potentials for the silver–copper cells in Investigate This 10.38(a), (b), and (c)? Which electrode was the anode in each cell?

(b) Reasoning from Le Chatelier's principle, would you predict changes in the silver–copper cell potential if the concentration of the silver ion is changed? If so, what would be the direction of the change with increasing silver ion concentration? Clearly state your reasoning.

(c) Are your results from Investigate This 10.38 consistent with your prediction in part (b)? Explain why or why not.

Recall equation (9.44) from Chapter 9, Section 9.7:

$$\Delta G_{reaction} = \Delta G°_{reaction} + RT \ln Q \qquad (10.61)$$

The **reaction quotient,** Q, accounts for the actual concentrations in the system of interest and enables us to calculate $\Delta G_{reaction}$ from $\Delta G°_{reaction}$ and the known

If necessary, review the reaction quotient and its relationship to free energy in Chapter 9.

concentrations. We can use the relationship in equations (10.54) and (10.55) to substitute cell potentials for free energy changes in equation (10.61):

$$-nFE = -nFE° + RT\ln Q \tag{10.62}$$

Equation (10.62) is usually rearranged to

$$E = E° - \frac{RT}{nF}\ln Q \tag{10.63}$$

Equation (10.63) is called the Nernst equation, in honor of Walter Nernst (German chemical physicist, 1864–1941) who did extensive work on electrochemical systems and thermodynamics. The **Nernst equation** provides us what we seek for a cell reaction, a relationship among the standard cell potential, the concentrations in the cell, and the cell potential under nonstandard concentration conditions.

Although equation (10.63) is useful as it stands, you will almost always see it (and use it) in an equivalent mathematical form:

$$E = E° - \frac{2.303\, RT}{nF}\log Q \tag{10.64}$$

We get equation (10.64) from equation (10.63) by substituting the mathematical equality: $\ln Q = 2.303 \log Q$. The advantage of equation (10.64) is that concentrations are almost always expressed in scientific notation as powers of 10, and these are easier to manipulate if we are using base-10 logarithms (log). Equation (10.64) is particularly useful when dealing with hydronium ion concentrations, as we will see.

10.40 CHECK THIS

Nernst equation at 25 °C (298 K)

Show that, at 298 K (room temperature), the Nernst equation (10.64) becomes

$$E = E° - \frac{(0.05916\text{ V})}{n}\log Q \approx E° - \frac{(0.059\text{ V})}{n}\log Q \tag{10.65}$$

Hint: Recall that $R = 8.314\,\text{J·K}^{-1}\text{·mol}^{-1}$.

10.41 WORKED EXAMPLE

Nernst equation applied to a copper–cadmium cell

Cell potential measurements at 25 °C on a series of copper–cadmium cells with varying cadmium ion concentration, c, are given in this tabulation:

c, M	0.100	0.0500	0.0100	0.0050
E, V	0.737	0.745	0.767	0.775

Use a graphical method to determine whether these data are consistent with the Nernst equation and, if so, calculate $E°$ for the cell:

$$Cd(s) \mid Cd^{2+}(aq, c\,M) \parallel Cu^{2+}(aq, 0.10\,M) \mid Cu(s) \tag{10.66}$$

continued

Necessary information: We will need the Nernst equation (10.65) at 25 °C (298 K) and the appropriate reaction quotient, Q.

Strategy: Write the cell reaction for cell (10.66) and use it to write Q. Use this Q to write the Nernst equation in a form that relates E linearly to a function of $(Cd^{2+}(aq))$. Plot the data to see if they fall on a straight line and, if so, use the equation of the line to find $E°$ for the cell reaction.

Implementation: The cell reaction for cell (10.66) is

$$Cd(s) + Cu^{2+}(aq) \rightleftharpoons Cd^{2+}(aq) + Cu(s) \qquad (10.67)$$

The reaction quotient is

$$Q = \frac{(Cd^{2+}(aq))(Cu(s))}{(Cd(s))(Cu^{2+}(aq))} = \frac{(Cd^{2+}(aq))}{(Cu^{2+}(aq))}$$

Recall that the concentration ratio for a pure solid is unity, so $(Cd(s)) = 1$ and $(Cu(s)) = 1$. The Nernst equation (for $n = 2$) is

$$E = E° - \frac{(0.059\ V)}{n}\log Q = E° - \frac{(0.059\ V)}{2}\log\left[\frac{(Cd^{2+}(aq))}{(Cu^{2+}(aq))}\right] \qquad (10.68)$$

A $\log(x/y)$ term can be rewritten as a sum, $\log(x) + \log(1/y)$. Also, $\log(1/y) = -\log(y)$, so we can rewrite this Nernst equation in the form:

$$E = E° + (0.030\ V)\log(Cu^{2+}(aq)) - (0.030\ V)\log(Cd^{2+}(aq))$$

This is the equation of straight line of E plotted as a function of $\log(Cd^{2+}(aq))$, with a slope $= -(0.030\ V)$ and an intercept $= E° + (0.030\ V)\log(Cu^{2+}(aq))$. Figure 10.11 is a plot of the data.

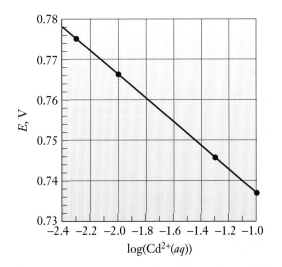

Figure 10.11.

Plot of E vs. log(Cd^{2+}(aq)).

The equation of the line in Figure 10.11 (generated by the computer plotting program) is

$$E = -(0.0297\ V)\log(Cd^{2+}(aq)) + 0.707\ V$$

continued

The slope of the line, $-(0.0297 \text{ V}) = -(0.030 \text{ V})$, is the value predicted; the results are consistent with the Nernst equation. We can use the intercept and the known $(Cu^{2+}(aq))$ to calculate $E°$:

$$0.707 \text{ V} = E° + (0.0297 \text{ V})\log(Cu^{2+}(aq)) = E°-0.0297 \text{ V}$$

$$E° = 0.737 \text{ V}$$

Does the result make sense?   Combine the standard reduction potentials for the copper and cadmium half reactions in Appendix C to see that our value of $E°$ is correct. Another qualitative check on the data is to apply Le Chatelier's principle to the variation of the measured cell potentials. In the cell reaction, $Cd^{2+}(aq)$ is a reaction product, so we would predict that decreasing its concentration would favor the cell reaction as written, that is, would increase the cell potential. This is what we observe.

10.42 CHECK THIS

Nernst equation applied to silver–copper cells

(a) Use a graphical method to determine whether your data from Investigate This 10.38 are consistent with the Nernst equation and, if so, calculate $E°$ for the silver–copper cell. Compare your experimental value for $E°$ with the one calculated from the data in Table 10.1. *Hint:* Recall that $\log(x)^a = a \cdot \log(x)$.

(b) Is the *direction* of the variation in the cell potentials with silver ion concentration in Investigate This 10.38 consistent with Le Chatelier's principle? Explain why or why not.

10.43 CHECK THIS

Three problems from the Web Companion

 Work through the first three pages of the *Web Companion*, Chapter 10, Section 10.8.1–4. Write out and explain the calculations required to solve the problems on page 4.

10.44 INVESTIGATE THIS

Does the silver-quinhydrone cell potential depend on pH?

(a) Do this as a class investigation and work in small groups to discuss and analyze the results. As shown in the diagram, fill one of the central wells of a 24-well plate, about two-thirds full of 10% KNO_3 solution. Fill a well adjacent to the KNO_3 well about two-thirds full of 0.100 M aqueous silver nitrate, $AgNO_3$, solution and insert a silver wire as an electrode. Fill two other adjacent wells about two-thirds full of pH 7.00 and pH 3.00 acid–base

continued

buffer solutions. To each buffer well add a tiny amount of solid quinhydrone. Place a platinum wire in the pH 7.00 well as an electrode and to stir the solution. Use filter paper soaked in 10% KNO_3 solution to make salt bridge connections between the KNO_3 well and each of the other wells. Connect the input leads from a digital voltmeter to the electrodes so that the meter reading is positive. Note which lead is connected to the silver electrode and which to the platinum. Record the cell potential to the nearest millivolt.

(b) Remove the platinum wire from the pH 7.00 quinhydrone well, wipe it off, and place it in the pH 3.00 quinhydrone well. Stir the solution in the well and again connect the digital voltmeter. Note which lead is connected to which electrode and record the cell potential.

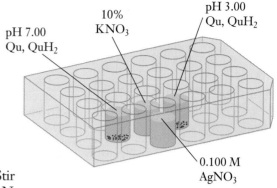

pH 7.00
Qu, QuH$_2$

10%
KNO$_3$

pH 3.00
Qu, QuH$_2$

0.100 M
AgNO$_3$

10.45 CONSIDER THIS

How does the silver-quinhydrone cell potential depend on pH?

(a) What were the cell potentials for the silver-quinhydrone cells in Investigate This 10.44? What would you expect the cell potential to be, if the quinhydrone half cell was prepared with a pH 5.00 solution? Explain the reasoning for your answer.

(b) Which electrode is the anode in Investigate This 10.44(a)? In 10.44(b)? Which electrode would be the anode if the quinhydrone half cell were prepared with a pH 5.00 solution? Explain the reasoning for your answer.

(c) **Quinhydrone** is a 1:1 combination of **quinone, Qu,** and **hydroquinone, QuH$_2$,** molecules. It is sparingly soluble in water and yields solutions that contain equal concentrations of quinone and hydroquinone:

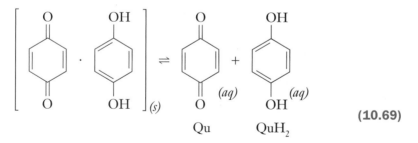

$$(10.69)$$

Quinone can be reduced to hydroquinone in aqueous solution:

$$Qu(aq) + 2H^+(aq) + 2e^- \rightleftharpoons QuH_2(aq) \qquad (10.70)$$

[Note that this half reaction and the structures of quinone and hydroquinone are closely related to reaction (10.1) and the structures of ubiquinone and ubiquinol in living cells.] Write the line notation and the cell reaction for the cell you made in Investigate This 10.44(a). Be sure your answers are consistent with the identity of the anode and cathode for the cell.

(d) Does Le Chatelier's principle explain the variation of the cell potential with pH that you found in Investigate This 10.44? Explain why or why not.

The quinhydrone half cell is different than the others we have been constructing. The platinum electrode does not appear in half-cell reaction (10.70). In the quinhydrone half cell, there are electron acceptors, the Qu molecules, and electron donors, the QuH_2 molecules. For reaction (10.70), we can imagine that a QuH_2 molecule next to the surface of the platinum electrode donates two electrons to the metal and two protons to water to form hydronium ions, H_3O^+, and a Qu molecule; QuH_2 has been oxidized. The result is that the platinum electrode has two extra electrons, which flow to the silver electrode (where silver ions are reduced). The platinum electrode is the anode, as you found in Consider This 10.45(b). The role of the platinum electrode is to sense the potential created by the quinhydrone solution and provide a pathway for electrons to leave (or enter) the half cell.

The Nernst equation and pH The chemistry of the cell reaction in Investigate This 10.44 is useful in photographic film developing. Photographic film contains tiny crystals of silver salts. Developing an exposed film involves reducing some of the silver ion, Ag^+, in these crystals to silver metal, Ag, which is half reaction (10.20). Some photographic developers contain hydroquinones as reducing agents, so we know that QuH_2 can reduce Ag^+. To get the developer reaction, we double half reaction (10.20) and subtract half reaction (10.70):

$$2\left[Ag^+(aq) + e^- \rightleftharpoons Ag(s)\right] \qquad \textbf{2} \times \textbf{(10.20)}$$

$$-\left[Qu(aq) + 2H^+(aq) + 2e^- \rightleftharpoons QuH_2(aq)\right] \qquad \textbf{– (10.70)}$$

$$2Ag^+(aq) + QuH_2(aq) \rightarrow 2Ag(s) + Qu(aq) + 2H^+(aq) \qquad \textbf{(10.71)}$$

Reaction (10.71) is the silver-quinhydrone cell reaction you wrote in Consider This 10.45(c). Let's see if the Nernst equation applied to this reaction will help us understand our data from Investigate This 10.44. The reaction quotient for reaction equation (10.71) is

$$Q = \left[\frac{(Ag(s))^2(Qu(aq))(H^+(aq))^2}{(QuH_2(aq))(Ag^+(aq))^2}\right] \qquad \textbf{(10.72)}$$

We know that $(Ag(s)) = 1$. The stoichiometry of equation (10.64) shows that equal amounts of quinone and hydroquinone dissolve, so $(Qu(aq)) = (QuH_2(aq))$. Therefore, in equation (10.72), the *ratio* $(Qu(aq))/(QuH_2(aq)) = 1$. Substituting these values into equation (10.72) gives

$$Q = \left[\frac{(H^+(aq))^2}{(Ag^+(aq))^2}\right] = \left[\frac{(H^+(aq))}{(Ag^+(aq))}\right]^2 \qquad \textbf{(10.73)}$$

Here, we have rewritten the quotient of squared terms as the square of the quotient of the terms. Substitution of this expression for Q into equation (10.65), with $n = 2$, gives

$$E = E° - \frac{(0.059\ V)}{2}\log\left[\frac{(H^+(aq))}{(Ag^+(aq))}\right]^2$$

$$= E° - \frac{2(0.059\ V)}{2}\log\left[\frac{(H^+(aq))}{(Ag^+(aq))}\right]$$

$$E = E° - (0.059 \text{ V})\log\left[\frac{(H^+(aq))}{(Ag^+(aq))}\right]$$

$$E = E° + (0.059 \text{ V})\log(Ag^+(aq)) - (0.059 \text{ V})\log(H^+(aq)) \qquad \text{(10.74)}$$

10.46 CHECK THIS

Deriving equation (10.74)

Show how equation (10.74) is obtained from the preceding equation.

We can substitute $pH = -\log(H^+(aq))$ in the last term in equation (10.74) to get

$$E = E° + (0.059 \text{ V})\log(Ag^+(aq)) + (0.059 \text{ V})pH$$

$$E = \text{constant} + (0.059 \text{ V})pH \qquad \text{(10.75)}$$

Equation (10.75) shows that the cell potential for a silver-quinhydrone cell is a linear function of pH and should increase by 0.059 V for every increase of one unit in the pH of the quinhydrone solution. Before the invention of the glass pH electrode that is used today, quinhydrone half cells were often used to obtain the pH of solutions.

10.47 CHECK THIS

pH and cell potentials for silver-quinhydrone cells

(a) Are your values for the cell potentials of the silver-quinhydrone cells in Investigate This 10.44 consistent with equation (10.75)? Explain your response.
(b) Is equation (10.75) consistent with your explanation based on Le Chatelier's principle in Consider This 10.45(d)? Explain why or why not.
(c) Use your data from Investigate This 10.44 to determine $E°$ for the silver-quinhydrone cell. Explain your method.

$E°$ and the equilibrium constant, K Since $-nFE° = \Delta G°_{rxn} = -RT\ln K$, a numeric value for $E°$ enables us to calculate the equilibrium constant for a cell reaction:

$$\ln K = \frac{nFE°}{RT} \quad \text{or} \quad \log K = \frac{nFE°}{2.303RT} \qquad \text{(10.76)}$$

You can also derive equation (10.76) from equation (10.64) by recalling that, at equilibrium, $\Delta G_{reaction} = 0$, so $E = 0$ ($= \Delta G_{reaction}/nF$). The reaction quotient for a reaction at equilibrium is $Q_{eq} = K$, so equation (10.64) becomes equation (10.76). Dead batteries have cell potentials equal to zero. Dead cells have reached equilibrium; they can furnish no more work.

10.48 WORKED EXAMPLE

Equilibrium constant for reaction of silver ion with hydroquinone

The standard cell potential for the silver-quinhydrone cell is 0.100 V. What is the equilibrium constant for the reaction of silver ion with hydroquinone at 25 °C?

$$2Ag^+(aq) + QuH_2(aq) \rightleftharpoons 2Ag(s) + Qu(aq) + 2H^+(aq)$$

Necessary information: We need equation (10.76), and we need to recognize that this reaction is the cell reaction for a silver-quinhydrone cell, equation (10.71).

Strategy: Since the reaction of interest is the cell reaction in a silver-quinhydrone cell, substitute the standard cell potential into equation (10.76) and solve for K.

Implementation:

$$\ln K = \frac{nFE^\circ}{RT} = \frac{2(96,485 \text{ C·mol}^{-1})(0.100 \text{ V})}{(8.314 \text{ J·mol}^{-1}\text{·K}^{-1})(298 \text{ K})} = 7.79 \therefore K = 2.4 \times 10^3$$

Does the answer make sense? We said that hydroquinone is used as a film developer to reduce silver ion. The relatively large value of the equilibrium constant for reaction (10.71) means that the products predominate at equilibrium, which is just what we expect for a reaction that is used to reduce silver ion in a practical application.

10.49 CHECK THIS

Conditions for developing photographic film

If you want to be sure that a good deal of the silver ion in a film is reduced by a hydroquinone developer, would you want the developer to be at a relatively low or relatively high pH? Explain. *Hint:* Write the equilibrium constant expression (or reaction quotient) for the developer reaction and see how pH affects the ratios of the other species in the system.

10.7. Reduction Potentials and the Nernst Equation

Can we apply the Nernst equation to half-cell reactions and reduction potentials as well as cell reactions and cell potentials? Remember that standard reduction potentials are really the potentials of complete cells in which the half cell of interest is coupled with the SHE. To see what we can learn about how half-cell potentials depend on concentrations, let's apply the Nernst equation to a cell that couples a nonstandard half cell with the SHE. As an example, consider this cell:

$$Pt(s) \mid H_2(g, \textit{1 bar}) \mid H^+(aq, \textit{1 M}) \parallel Ag^+(aq, \textit{c M}) \mid Ag(s) \tag{10.77}$$

The cell reaction and Nernst equation (at 298 K) for reaction (10.77) are

$$2Ag^+(aq) + H_2(g) \rightleftharpoons 2Ag(s) + 2H^+(aq) \tag{10.36}$$

$$E_{H|Ag} = E°_{H|Ag} - \left(\frac{0.059 \text{ V}}{2}\right)\log Q_{H|Ag} \tag{10.78}$$

Rewrite the cell potentials in terms of the reduction potentials for this cell and write out $Q_{H|Ag}$:

$$E(Ag^+, Ag) - E°(H^+, H_2)$$

$$= E°(Ag^+, Ag) - E°(H^+, H_2) - \left(\frac{0.059 \text{ V}}{2}\right)\log\left[\frac{(H^+(aq))^2}{(H_2(g))(Ag^+(aq))^2}\right]$$

$E°(H^+, H_2)$ is written on the left side of this equation because the cell we are considering has the SHE as the anode and its reduction potential is $E°(H^+, H_2)$. Substituting $E°(H^+, H_2) = 0$, $(H_2(g)) = 1$, and $(H^+(aq)) = 1$ into this equation gives

$$E(Ag^+, Ag) = E°(Ag^+, Ag) - \left(\frac{0.059 \text{ V}}{2}\right)\log\left[\frac{1}{(Ag^+(aq))^2}\right] \tag{10.79}$$

10.50 CHECK THIS

Simplifying equation (10.79)

Show that equation (10.79) can be written as

$$E(Ag^+, Ag) = E°(Ag^+, Ag) - \left(\frac{0.059 \text{ V}}{1}\right)\log\left[\frac{1}{(Ag^+(aq))}\right] \tag{10.80}$$

Nernst equation for a half reaction Equation (10.80) is the way you would write the Nernst equation for the silver ion reduction half reaction, equation (10.20), if you left out the electrons. We can rationalize leaving out the electrons; they don't actually exist free in the solution, but only in the solid electrodes, which do not affect the numeric value of Q. Remembering that electrons are left out, you can use the Nernst equation to write the concentration dependence of a reduction potential. The value of n to use is the number of electrons in the half-reaction equation.

10.51 WORKED EXAMPLE

Oxygen gas reduction potential as a function of pH

The half reaction and standard reduction potential for the reduction of oxygen gas are

$$O_2(g) + 4H^+(aq) + 4e^- \rightleftharpoons 2H_2O(l) \quad E°(O_2, H_2O) = 1.229 \text{ V} \tag{10.81}$$

Use Le Chatelier's principle to predict whether the reduction potential for oxygen increases or decreases as the hydronium ion concentration decreases.

continued

Use the Nernst equation to get the reduction potential for oxygen at 298 K at pH 7 and at pH 14.

Necessary information: We need equation (10.65), the Nernst equation at 298 K.

Strategy: Apply Le Chatelier's principle to half reaction (10.81), recalling that the reduction potential is a measure of the driving force for the reaction. Write Q for the half reaction and use it in the Nernst equation with appropriate values for (H$^+$(aq)) to get $E(O_2, H_2O)$ at pH 7 and 14.

Implementation: H$^+$(aq) is a reactant in the reduction half reaction, so reducing [H$^+$(aq)] from 1 M (standard conditions) to 10^{-7} M (pH 7) will cause the reverse reaction to be favored and the reduction potential should decrease. A further reduction to 10^{-14} M (pH 14) will further reduce the potential.

The Nernst equation for half reaction (10.81), with (H$_2$O(l)) = 1, is

$$E(O_2, H_2O) = E°(O_2, H_2O) - \left(\frac{0.059 \text{ V}}{4}\right)\log\left[\frac{1}{(O_2(g))(H^+(aq))^4}\right]$$

Under standard conditions, the oxygen gas pressure is 1 bar, so (O$_2$(g)) = 1 and we can rewrite the equation as

$$E(O_2, H_2O) = E°(O_2, H_2O) - \left(\frac{0.059 \text{ V}}{4}\right)\log\left[\frac{1}{(H^+(aq))^4}\right]$$

$$E(O_2, H_2O) = E°(O_2, H_2O) - (0.059 \text{ V})\log\left[\frac{1}{(H^+(aq))}\right]$$

$$E(O_2, H_2O) = E°(O_2, H_2O) - (0.059 \text{ V})[-\log(H^+(aq))]$$

$$= E°(O_2, H_2O) - (0.059 \text{ V})\text{pH}$$

Therefore, $E(O_2, H_2O)$(at pH 7) = 1.229 V − (0.059 V)(7) = 0.816 V

$E(O_2, H_2O)$(at pH 14) = 1.229 V − (0.059 V)(14) = 0.403 V

Does the answer make sense? Le Chatelier's principle predicts a decrease in reduction potential with increasing pH (lower hydronium ion concentration) and this is what we calculate using the Nernst equation. In Appendix C, the oxygen reduction potential is given for standard conditions, pH 0, and also for pH 7 where E = 0.816 V. The result at pH 14 represents reaction (10.81) with (OH$^-$(aq)) = 1 M. Recall from Section 6.10, Chapter 6, that, to convert a redox reaction equation in acidic solution to one in basic solution, we add enough hydroxide ions to convert all the hydronium ions to water and then cancel redundant waters. Applying this conversion to half-reaction equation (10.81), we add four OH$^-$(aq) to each side:

$$O_2(g) + 4H^+(aq) + 4OH^-(aq) + 4e^- \rightleftharpoons 2H_2O(l) + 4OH^-(aq)$$

$$O_2(g) + 4H_2O(l) + 4e^- \rightleftharpoons 2H_2O(l) + 4OH^-(aq)$$

$$O_2(g) + 2H_2O(l) + 4e^- \rightleftharpoons 4OH^-(aq) \qquad \text{(10.82)}$$

The standard reduction potential for half reaction equation (10.82) is denoted $E°(O_2, OH^-)$ and we know from the solution conditions that $E°(O_2, OH^-) = E(O_2, H_2O)$(at pH 14) = 0.403 V. Check this value in Appendix C.

10.52 CHECK THIS

Hydronium ion reduction potential as a function of pH

The half reaction and standard reduction potential for the reduction of hydronium ion are

$$2H^+(aq) + 2e^- \rightleftharpoons H_2(g) \quad E°(H^+, H_2) = 0.000 \text{ V} \tag{10.83}$$

Use Le Chatelier's principle to predict whether the reduction potential for the hydronium ion increases or decreases as the hydronium ion concentration decreases. Use the Nernst equation to get the reduction potential for the hydronium ion at 298 K at pH 7 and at pH 14. How do your results compare with the values in Appendix C? Explain the comparisons you make.

10.53 INVESTIGATE THIS

Can pH change the direction of a cell reaction?

Do this as a class investigation and work in small groups to discuss and analyze the results. Use a setup similar to the one in Investigate This 10.44. Use three wells grouped around one containing 10% aqueous KNO_3 solution to prepare three half cells: (1) a solution that is 0.10 M in aqueous potassium ferricyanide, $K_3Fe(CN)_6$, and 0.10 M in aqueous potassium ferrocyanide, $K_4Fe(CN)_6$, with a platinum wire electrode; (2) a quinhydrone solution in 1 M HCl (pH ≈ 0) with a platinum wire electrode; and (3) a quinhydrone solution in a pH 7 buffer solution with a platinum wire electrode. Use filter paper salt bridges soaked in 10% aqueous KNO_3 solution to connect each half cell to the KNO_3 well.

Connect one lead of a digital voltmeter to the electrode in the $Fe(CN)_6^{3-}$–$Fe(CN)_6^{4-}$ solution and the other lead to the electrode in the pH ≈ 0 quinhydrone half cell. Record which electrode is the anode and the cell potential to the nearest millivolt. Switch the lead from the pH ≈ 0 quinhydrone half cell to the pH 7 quinhydrone half cell and repeat the recording.

10.54 CONSIDER THIS

Does pH change the direction of a cell reaction?

(a) Write the correct cell notation and cell reaction for each of the cells you tested in Investigate This 10.53.
(b) Does Le Chatelier's principle explain any differences between the reactions in part (a)? Explain why or why not.

Reduction potentials under biological conditions In Consider This 10.45, you saw that the structures and redox reactions are similar for the quinone–hydroquinone pair and the ubiquinone–ubiquinol pair (UQ and UQH_2) from reaction (10.1). In Investigate This 10.53, you coupled the quinone-hydroquinone pair (the quinhydrone half cells) with a half cell containing iron complexes in their +3 and +2 oxidation states. In living cells, the ubiquinone–ubiquinol pair in mitochondria is also coupled to half reactions of iron complexes in their +3 and +2 states.

The cell potential for a cell that involves the quinone–hydroquinone pair depends on the pH of the quinhydrone solution. The same is true for the ubiquinone–ubiquinol pair in biological systems. In Investigate This 10.53, the *direction* of the cell reaction was reversed by the change in pH from about 0 to 7. Standard reduction potentials are given for all species at unit concentration, which is pH 0 for hydronium ion, but most biological fluids have a pH near 7. Therefore, as you saw in Investigate This 10.53, using standard reduction potentials as a guide to the direction of reaction at pH 7 may be misleading. If we want to make predictions about redox reactions in biological systems, we have to correct the cell potentials to account for pH effects.

Worked Example 10.51 and Check This 10.52 show that there is no problem doing this correction, but it means we lose some of the easy predictive power illustrated in Figure 10.9 for a table of standard reduction potentials. When we compare the half reactions that make up a redox reaction of biological interest, we would like to be able to look at a table and know that the half reaction higher in the table is *likely* to proceed as a reduction, in the reaction. A solution to this problem is a table of standard reduction potentials, $E^{\circ\prime}$ **[standard reduction potential (pH 7)]**, for half cells in which all components are at unit concentration, *except* $[H^+(aq)] = 10^{-7}$ M (pH 7). For biological systems, comparisons of $E^{\circ\prime}$ values give better predictions than E° values. The table of standard reduction potentials in Appendix C gives values of both E° and $E^{\circ\prime}$ for many half reactions that involve the hydronium ion as a reactant or product. For some biological reactants, only $E^{\circ\prime}$ values are given. The reactions at low pH are irrelevant, if the biological molecules, such as proteins, do not function in high concentrations of hydronium ion.

10.55 CHECK THIS

Predictions you can make about biological redox reactions

(a) In aerobic organisms, a key metabolic reaction is the redox reaction between oxidized and reduced nicotinamide adenine dinucleotide (NAD$^+$, NADH, respectively) and oxidized and reduced ubiquinone (UQ, UQH$_2$, respectively). Use the values in Appendix C to determine the standard cell potential at pH 7, $E^{\circ\prime}_{UQ|NAD}$, and $\Delta G^{\circ\prime}_{reaction}$ for this reaction:

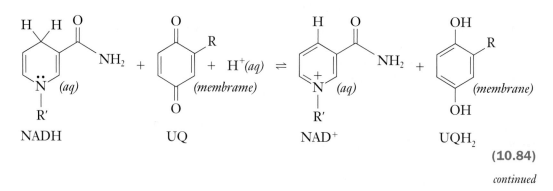

NADH   UQ   NAD$^+$ UQH$_2$

(10.84)

continued

(b) Is reaction (10.84) favored in the direction written at pH 7? Explain your response. What is the equilibrium constant for reaction (10.84) at pH 7?

(c) Show that all electron pairs in the reactants are accounted for in the products.

Reflection and Projection

The potential of an electrochemical cell depends on the concentrations of the ions and molecules in the cell solution(s). Applying Le Chatelier's principle to the cell reaction and relating more positive cell potentials to a greater driving force toward products predicts the direction of the dependence. Quantitatively, we can use the relationship of the free energy change for a cell reaction and the cell potential to obtain the Nernst equation, which relates the cell potential to the standard cell potential and the logarithm of the reaction quotient, Q, for the cell reaction. Standard cell potentials, like the standard free energy changes for cell reactions, can be used to calculate equilibrium constants for redox reactions. We can also apply the Nernst equation to half reactions by disregarding the electrons, which are never present free in the solution.

A special case that is particularly relevant to biological systems is the variation of reduction potentials with pH. Since biological redox reactions generally occur in solutions near pH 7, standard reduction potentials at pH 7, $E^{\circ\prime}$, are used to make predictions for these reactions. Other changes in solution conditions, besides direct changes in the concentrations of species that appear in the redox reaction, can also affect cell potentials and reduction potentials. Metal–ion complexation, for example, is significant in biological systems. In the next section, we will discuss systems that contain several species that can be oxidized and reduced and see how the redox reactions can be coupled to one another.

10.8. Coupled Redox Reactions

10.56 INVESTIGATE THIS

Does copper ion react with sugars?

Do this as a class investigation and work in small groups to discuss and analyze the results. Benedict's reagent is an aqueous solution of copper sulfate, $CuSO_4$, sodium carbonate, Na_2CO_3, and sodium citrate, $Na_3(C_6H_5O_7)$. Add 5 mL of Benedict's reagent to each of two small, clean, labeled test tubes. To one of the test tubes, add 2 drops of ethanol. Add one or two *small* crystals of glucose to the other test tube. Swirl to mix and dissolve the alcohol and sugars. Note the color and appearance of each mixture. Place the test tubes in a boiling water bath for about 5 minutes. Note the color and appearance of the mixtures after heating.

━━━━━━━ **10.57** CONSIDER THIS ━━━━━━━

What is the reaction of copper ion with glucose?

(a) What changes did you observe taking place in the solutions of Benedict's reagent with ethanol and glucose in Investigate This 10.56? Did both compounds give the same results? If not, what were the differences?

(b) The colors of Cu^{2+} ions complexed in solution are at the blue end of the spectrum. Citrate ion is a good complexing agent for Cu^{2+} ions. The oxide of the Cu^+ ion, Cu_2O, is an insoluble red solid. How might you account for the changes that occurred in Investigate This 10.56? Explain as fully as possible.

$$CH_2COO^-$$
$$|$$
$$HO-C-COO^-$$
$$|$$
$$CH_2COO^-$$
citrate ion

The food we eat is the source of the electrons for our metabolic redox pathways. All of our food traces its origin to the glucose photosynthesized by green plants from carbon dioxide and water. To study the metabolic pathway that gathers electrons from our food is difficult without specialized equipment and instrumentation. As an alternative, we will focus on a simpler reaction of glucose that is not part of the metabolic pathway but exemplifies its role as a reducing agent.

Glucose is oxidized to gluconic acid In Investigate This 10.56, glucose reduced copper(II) to copper(I). The sugar is oxidized in the reaction (but not all the way to carbon dioxide and water). A carbon-containing molecule is usually most easily oxidized at the carbon that has the highest oxidation number. For glucose, this is carbon-1, the aldehyde carbon in the open-chain form of the molecule:

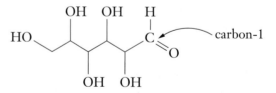

This carbon has an oxidation number of $+2$. The other carbons with $-OH$ groups attached, the alcoholic carbons, have oxidation numbers of $+1$. Alcohols do not reduce the copper(II) in Benedict's reagent.

━━━━━━━ **10.58** CHECK THIS ━━━━━━━

Alcohols in Benedict's reagent

What evidence do you have to support or refute the statement that alcohols do not reduce copper(II) in Benedict's reagent?

The oxidation of an aldehyde produces a carboxylic acid. The product of glucose oxidation at the aldehyde group is gluconic acid, which, at pH 7 and above, is present in its conjugate base form, gluconate ion. The half reaction for reduction of gluconate ion to glucose is

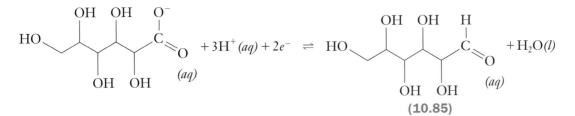

(10.85)

In abbreviated form, with R = $C_5H_{11}O_5$, reaction (10.85) is

$$RCOO^-(aq) + 3H^+(aq) + 2e^- \rightleftharpoons RCHO(aq) + H_2O(l)$$

$$E^{\circ\prime}(RCOO^-, RCHO) = -0.44 \text{ V} \qquad \textbf{(10.86)}$$

Since the standard reduction potential at pH 7, is negative, -0.44 V, the oxidation of glucose, the reverse of reaction (10.86), is favored.

Benedict's reagent is basic (because it contains sodium carbonate), so we need to write the half reaction equation (10.86) in terms of hydroxide ion instead of hydronium ion. Using the method outlined in Chapter 6, Section 6.10, and recalled in Worked Example 10.51, we get for the reaction in basic solution:

> We have been using $-C(O)O^-$ as the notation for the carboxylate ion, but in this section and in Appendix C, for ease of writing, we use $-COO^-$, another common notation.

$$RCOO^-(aq) + 2H_2O(l) + 2e^- \rightleftharpoons RCHO(aq) + 3OH^-(aq)$$

$$E^{\circ\prime}(RCOO^-, RCHO) = -0.44 \text{ V} \qquad \textbf{(10.87)}$$

$E^{\circ\prime}(RCOO^-, RCHO)$ in basic solution is the reduction potential when $[OH^-(aq)] = 10^{-7}$ M. This is pH 7, so the reduction potential at pH 7 is the same for half reactions (10.86) and (10.87). At higher hydroxide ion concentrations, half reaction (10.87) is driven to the left, which decreases the reduction potential and makes the oxidation of glucose even more favorable. Now we need to show that copper(II) can oxidize glucose under the conditions in the Benedict's reagent.

The Cu²⁺–Cu⁺ redox system In Investigate This 10.56, you observed a change from a clear, blue solution to a cloudy reddish-brown liquid when Benedict's reagent reacted with glucose. As you probably concluded in Consider This 10.57, the reaction is reduction of Cu^{2+} to Cu^+ (as its oxide precipitate). In the Benedict's reagent, the copper(II) is present as a citrate complex, $Cu(cit)^-(aq)$. The half reaction for reduction of the complex is

$$Cu(cit)^-(aq) + e^- \rightleftharpoons Cu^+(aq) + cit^{3-}(aq)$$

$$E^{\circ}(Cu(cit)^-, Cu^+) = -0.04 \text{ V} \qquad \textbf{(10.88)}$$

The standard free energy change for reaction (10.88) is $\Delta G^{\circ}(Cu(cit)^-, Cu^+) = nFE^{\circ}(Cu(cit)^-, Cu^+) = +4 \text{ kJ·mol}^{-1}$. The standard cell potential and free energy suggest that the reduction of copper(II) in this solution is not favored. However, we have not yet accounted for formation of copper(I) oxide, $Cu_2O(s)$, the reddish solid. Precipitation of $Cu_2O(s)$ lowers the concentration of $Cu^+(aq)$, thus driving reaction (10.88) to the right and making the reduction potential more positive.

10.59 WORKED EXAMPLE

Standard reduction potential for copper(II) in Benedict's reagent

The reaction of $Cu^+(aq)$ ion with hydroxide to precipitate $Cu_2O(s)$ can be represented as

$$2Cu^+(aq) + 2OH^-(aq) \rightleftharpoons Cu_2O(s) + H_2O(l)$$

$$K \approx 4 \times 10^{12} \quad \Delta G°_{reaction} = -72 \text{ kJ·mol}^{-1} \tag{10.89}$$

Determine the overall half reaction for copper(II) reduction in Benedict's reagent and the standard free energy change and standard reduction potential for the reduction.

Necessary information: We need equation (10.55) and equation (10.88) and its standard free energy change, $+4 \text{ kJ·mol}^{-1}$.

Strategy: Combine reaction (10.89) with equation (10.88), using appropriate stoichiometric factors, to cancel $Cu^+(aq)$, and get the overall reaction for reduction of copper(II) to form solid copper(I) oxide. Combine the free energies of reactions (10.88) and (10.89) to find the free energy change for the overall reaction and thence its standard reduction potential.

Implementation: The appropriate combination of reactions (10.88) and (10.89) is

$$2[Cu(cit)^-(aq) + e^- \rightleftharpoons Cu^+(aq) + cit^{3-}(aq)] \quad 2\Delta G°(Cu(cit)^-, Cu^+) = +8 \text{ kJ·mol}^{-1}$$

$$\underline{2Cu^+(aq) + 2OH^-(aq) \rightleftharpoons Cu_2O(s) + H_2O(l) \quad \Delta G°_{reaction} = -72 \text{ kJ·mol}^{-1}}$$

$$2Cu(cit)^-(aq) + 2OH^-(aq) + 2e^- \rightleftharpoons Cu_2O(s) + 2cit^{3-}(aq) + H_2O(l) \tag{10.90}$$

$$\Delta G°(Cu(cit)^-, Cu_2O) = -64 \text{ kJ·mol}^{-1}$$

$$E°(Cu(cit)^-, Cu_2O) = -\frac{\Delta G°_{(Cu(cit)^-,Cu_2O)}}{nF} = -\frac{(-64 \times 10^3 \text{J·mol}^{-1})}{2(9.6485 \times 10^4 \text{C·mol}^{-1})} = 0.33 \text{ V}$$

Does the answer make sense? Although we have no information to use to test whether the numerical results are correct, the reduction potential shows that the standard reduction potential for copper(II) citrate complex to copper(I) oxide is favorable and this is the reaction we observe.

10.60 CHECK THIS

Reduction potential for copper(II) in Benedict's reagent

(a) Use Le Chatelier's principle to predict the direction of change of the reduction potential, if the hydroxide ion concentration is reduced in the solution in which reaction (10.90) occurs.

(b) Write the reaction quotient for half reaction (10.90).

(c) The pH of Benedict's reagent is about 12. What is $E(Cu(cit)^-, Cu_2O)$ in this solution, if all other species are present at their standard state concentrations? Explain your approach. Is your result consistent with your prediction in part (a)? Explain why or why not.

Redox reaction in the Benedict's test We can combine half reactions (10.87) and (10.90) to write the reduction of copper(II) by glucose in Benedict's reagent:

$$2Cu(cit)^-(aq) + RCHO(aq) + 5OH^-(aq) \rightleftharpoons$$
$$Cu_2O(s) + RCOO^-(aq) + 2cit^{3-}(aq) + 3H_2O(l) \qquad (10.91)$$

The cell potential for reaction (10.91) is

$$E = E(Cu(cit)^-, Cu_2O) - E(RCOO^-, RCHO) \qquad (10.92)$$

In Check This 10.60(c), you found that $E(Cu(cit)^-, Cu_2O)$ at pH 12 is about 0.21 V. We do not have $E(RCOO^-, RCHO)$ at pH 12, but we know $E^{\circ\prime}(RCOO^-, RCHO) = -0.44$ V at pH 7 and we reasoned above that $E(RCOO^-, RCHO)$ at pH 12 is even more negative. The *minimum* value for the cell potential is

$$E = 0.21 \text{ V} - (-0.44 \text{ V}) = 0.65 \text{ V}$$

The reduction of copper(II) by glucose in Benedict's reagent has a large driving force and explains the positive test for glucose as a reducing agent in Investigate This 10.56.

10.61 INVESTIGATE THIS

Can you characterize the Blue-Bottle reaction?

Work in small groups on this investigation and discuss your results and hypotheses with the entire class. Your large, capped vial contains an aqueous solution that is 0.14 M glucose, 0.50 M potassium hydroxide, and 10^{-5} M methylene blue. *CAUTION:* Solutions of hydroxide are caustic and can harm skin and clothing. Take care not to spill any of the solution; work over a paper towel to catch any drops that may escape. Shake the vial two or three times. What, if any, changes do you observe? Watch to see whether anything further happens. When the solution has returned to its original appearance, shake the vial again and see if your observations are reproducible. Try variations. Uncap the vial and recap it before shaking. Shake the vial more than two or three times. Keep a record of the conditions and observations for all your trials.

10.62 CONSIDER THIS

Can you interpret the Blue-Bottle reaction?

(a) As a class, make a table of the various conditions used and observations recorded on the Blue-Bottle reaction in Investigate This 10.61. Try to reach a consensus about the factors that affect the reaction and the direction and magnitude of the effects.
(b) Methylene blue is a dye used as a bacteriological stain and as an indicator in some chemical analyses. The dye is blue in its oxidized form (MB$^+$) and colorless in its reduced form (MBH):

continued

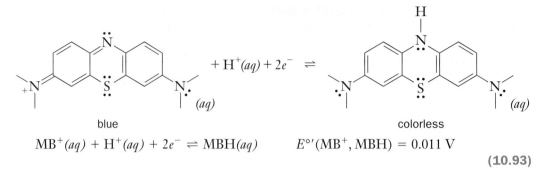

blue   colorless

$$MB^+(aq) + H^+(aq) + 2e^- \rightleftharpoons MBH(aq) \qquad E^{\circ\prime}(MB^+, MBH) = 0.011 \text{ V}$$

(10.93)

Using these properties of methylene blue and the redox properties of glucose in basic solution, can you develop a consistent explanation that accounts for the effects listed in part (a)?

The Blue-Bottle redox reaction(s) Half reaction (10.93) accounts for the disappearance of the blue color that you observe in the Blue-Bottle reaction, Investigate This (10.61). There must be a reducing agent in the solution that reduces the blue, oxidized form of methylene blue, MB^+. You can make the blue color return by shaking the solution. There must be an oxidizing agent in the shaken solution that can oxidize the colorless, reduced form of methylene blue, MBH, back to the blue form. You probably have decided that the reducing agent is glucose (as in the Benedict's reagent reaction) and the oxidizing agent is molecular oxygen. If these choices are correct, they should be able to help us answer several questions about the observations.

10.63 CONSIDER THIS

Can you explain the timing of the Blue-Bottle reaction?

(a) Why does the blue color persist for several seconds in the Blue-Bottle reaction and then suddenly fade away?
(b) Why does shaking bring the blue color back?
(c) How is the system able to repeat the cycle over and over?
(d) As you experimented with the Blue-Bottle reaction, you probably discovered that the more shakes you give the bottle, the longer the blue color persists, but it still seems to fade just as suddenly at the end. What is the explanation for these observations?

10.64 CHECK THIS

Methylene blue, MB⁺, reduction in basic solution

The Blue-Bottle reaction solution is basic. Rewrite half reaction (10.93) to show the half reaction in terms of hydroxide ion instead of hydronium ion. What is $E^{\circ\prime}(MB^+, MBH)$ for this half reaction? Explain how you get your answer.

The oxidized form of methylene blue, MB^+, is so highly colored that only a tiny amount is present in the Blue-Bottle reaction system to give the blue color. The disappearance of the blue color signals the reduction of this small amount of methylene blue by glucose, RCHO, which is present in substantial concentration:

$$MB^+(aq) + RCHO(aq) + 2OH^-(aq) \rightleftharpoons$$

$$MBH(aq) + RCOO^-(aq) + H_2O(l) \qquad \textbf{(10.94)}$$

Using standard reduction potentials from equation (10.87) and Check This 10.64, we get the standard cell potential for reaction (10.94) at pH 7:

$$E^{\circ\prime}{}_{MB|RCHO} = E^{\circ\prime}(MB^+, MBH) - E^{\circ\prime}(RCOO^-, RCHO)$$

$$= (0.011\ V) - (-0.44\ V) = 0.45\ V \qquad \textbf{(10.95)}$$

10.65 CHECK THIS

MB^+–glucose redox reaction in basic solution

How does $E^{\circ\prime}{}_{MB|RCHO}$ vary as a function of hydroxide ion concentration? Is reaction (10.94) more or less favored in more basic solutions? Give the reasoning for your answers.

Reaction (10.94) explains the disappearance of the blue color in the Blue-Bottle reaction, but the *way* it disappears is odd. In the blue solution, we start with the reactants in expression (10.94), so the reaction should continuously reduce methylene blue, and you might expect to see the blue color fade continuously. In fact, the blue color persists for a time and then fades suddenly. Let us assume that reaction (10.94) is going on whenever MB^+ and RCHO are present, that is, when the solution is blue. If this assumption is true, then the only way that the solution can remain blue is for the reduced form of methylene blue, MBH, to be oxidized back to the blue form as fast as it is formed. When the oxidizing agent has been used up, the solution quickly fades to colorless. Both the reduction of MB^+ and oxidation of MBH seem to be fast. To explain the observations, the oxidation must be faster. (Speeds of reactions are the subject of Chapter 11.)

The oxidizing agent that explains the observations is oxygen gas dissolved in the solution:

$$O_2(g) + 2H_2O(l) + 4e^- \rightleftharpoons 4OH^-(aq)$$

$$E^{\circ\prime}(O_2, OH^-) = 0.816\ V \qquad \textbf{(10.96)}$$

Shaking mixes the solution with the air in the container and some nitrogen and oxygen dissolve. Only a little bit of oxygen dissolves, but since there is only a tiny bit of methylene blue present, a little oxygen is all we need to return all the MBH to the oxidized form. As the MB^+ is reduced by glucose, the oxygen reoxidizes the MBH until the little bit of oxygen is used up. Then the dye color suddenly fades away. Another shake again adds oxygen and the process is repeated. The evidence for oxygen as the oxidizing agent seems to fit. We just need to be sure that oxygen can oxidize reduced methylene blue in basic solution.

10.66 WORKED EXAMPLE

Oxidation of methylene blue by oxygen

Write the reaction for oxidation of reduced methylene blue by oxygen in basic solution and determine $E^{\circ\prime}{}_{O_2|MB}$. Does the driving force for the reaction increase or decrease as the solution becomes more basic? Estimate $E^{\circ}{}_{O_2|MB}$ at 298 K when all species, including hydroxide ion, are at standard concentration.

Necessary information: We need the results from Check This 10.64 [reaction (10.93) in basic solution], the information from equation (10.96), and the Nernst equation at 298 K.

Strategy: Combine half reactions (10.93) and (10.96), both in basic solution, by subtracting the less positive half reaction (with appropriate stoichiometry to cancel the electrons). The combination gives the cell reaction and its standard cell potential. Applying Le Chatelier's principle to the cell reaction gives the direction the cell potential changes as hydroxide ion concentration increases. Use the Nernst equation to find $E^{\circ}{}_{O_2|MB}$, the potential under standard conditions with $[OH^-(aq)] = 1$ M (pH 14).

Implementation: The appropriate combination of half reactions and potentials is

$O_2(g) + 2H_2O(l) + 4e^- \rightleftharpoons 4OH^-(aq)$	$E^{\circ\prime}(O_2, OH^-) = 0.816$ V	(10.96)
$-2[MB^+(aq) + H_2O(l) + 2e^- \rightleftharpoons MBH(aq) + OH^-(aq)]$	$-E^{\circ\prime}(MB^+, MBH) = -0.011$ V	$-2 \times$ (10.93)
$O_2(g) + 2MBH(aq) \rightleftharpoons 2OH^-(aq) + 2MB^+(aq)$	$E^{\circ\prime}{}_{O_2|MB} = 0.805$ V	(10.97)

Hydroxide ion is a product of reaction (10.97), so by Le Chatelier's principle, increasing $[OH^-(aq)]$ should decrease the cell potential. There will be a lower driving force for the reaction as written.

The Nernst equation for reaction (10.97) is

$$E^{\circ\prime}{}_{O_2|MB} = E^{\circ}{}_{O_2|MB} - \left(\frac{0.059 \text{ V}}{4}\right)\log\left[\frac{(OH^-)^2(MB^+)^2}{(O_2(g))(MBH)^2}\right]$$

We have found that $E^{\circ\prime}{}_{O_2|MB} = 0.805$ V, when $(OH^-(aq)) = 10^{-7}$ and all other species are at unit concentration. Substituting these values into this equation and solving for $E^{\circ}{}_{O_2|MB}$ gives

$$E^{\circ}{}_{O_2|MB} = (0.805 \text{ V}) + \left(\frac{0.059 \text{ V}}{2}\right)\log(10^{-7})$$

$$= (0.805 \text{ V}) + (-0.21 \text{ V}) = 0.60 \text{ V}$$

Does the answer make sense? You know from your investigations that it is probably oxygen that oxidizes methylene blue in the Blue-Bottle reaction, so the substantial positive cell potential for this reaction makes sense. Le Chatelier's principle predicts that increasing the hydroxide ion concentration should decrease this potential and the Nernst equation confirms the prediction. The lower cell potential is still highly favorable for this oxidation.

10.67 CHECK THIS

Oxidation of methylene blue by oxygen in the Blue-Bottle reaction

Use the Nernst equation to estimate $E_{O_2|MB}$ under the conditions of the Blue-Bottle reaction in Investigate This 10.61. Assume that the ratio of oxidized to reduced methylene blue is about 10 during the blue phase of the reaction and that the pH of the solution is 14. Assume the pressure of oxygen in the air is 20 kPa. (The standard pressure for a gas is 1 bar = 100 kPa.) Is the oxidation of MBH favored? Is your conclusion changed, if the ratio of oxidized to reduced methylene blue is about 1000? Explain.

Methylene blue couples glucose oxidation to oxygen reduction

The combination of reactions (10.94) and (10.97) alternately reduce and oxidize methylene blue. The net reaction is the sum of these reactions (with appropriate adjustment for stoichiometry) and the cell potential is the sum of the individual cell potentials:

$$O_2(g) + 2RCHO(aq) + 2OH^-(aq) \rightleftharpoons 2H_2O(l) + 2RCOO^-(aq)$$

$$E^{\circ\prime}{}_{O_2|RCHO} = 1.27 \text{ V} \qquad\qquad \textbf{(10.98)}$$

Methylene blue does not appear in the net reaction (10.98). Without it, however, the reaction does not occur. Methylene blue **couples** the oxidation of glucose to the reduction of oxygen. **Coupled reactions,** which we introduced in Section 7.9, Chapter 7, are so common, especially in biochemical systems, that it is useful to have a better way to represent their pathways than with net reactions. Net reactions can be misleading, because they do not show the coupling that makes the net reaction possible. Figure 7.17 in Section 7.9, Chapter 7, is a mechanical analogy representing coupled reactions, but we need a more chemical representation.

Figure 10.12 represents the methylene blue coupling we have just analyzed. The species that is oxidized and the species that is reduced in each individual reaction are shown at the tails of two curved arrows. The arrows touch one another on their way to the respective oxidized and reduced products. The two tangent curves represent the reaction between these two reactants. Other species required for the reaction or produced by it are shown, respectively, at the tail and head of a straight arrow that is also tangent to the two curved arrows. The representation makes explicit the *cycling of the coupling species between its reduced and oxidized forms.* Each form is both a product (at the head of a curved arrow) and a reactant (at the tail of a curved arrow), so the arrows form a closed loop. Any species that cycles will not appear in the net reaction expression, as we found in reaction (10.98).

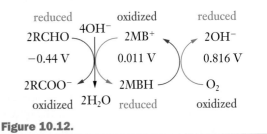

Figure 10.12.

Methylene blue coupling of glucose oxidation to oxygen reduction. The standard reduction potentials at pH 7 are given with each half reaction in the series. Oxidations are shown as blue curved arrows and reductions as red curved arrows.

10.68 CONSIDER THIS

What sort of redox coupling occurs in biological systems?

In most organisms, the first stage in the oxidation of glucose is a series of reactions called **glycolysis** (*glyco* = sugar + *lysis* = breaking). The oxidizing agent is NAD^+ (see Check This 10.55). In the overall series of reactions, each molecule of glucose produces two molecules of pyruvate ion, $C_3H_3O_3^-$ (the carboxylate ion of pyruvic acid):

$$C_6H_{12}O_6 + 2NAD^+ \rightleftharpoons 2C_3H_3O_3^- + 2NADH + 4H^+ \qquad (10.99)$$

Some of the energy of the reaction is used to produce two molecules of ATP from ADP and phosphate ion, P_i: $2ADP + 2P_i \rightleftharpoons 2ATP$.

Reaction (10.99) uses up NAD^+, which has to be regenerated in order to keep the reaction going to produce more ATP. Some organisms, such as yeast, regenerate the NAD^+ by alcoholic fermentation:

$$C_3H_3O_3^- + NADH + 2H^+ \rightleftharpoons CO_2 + CH_3CH_2OH + NAD^+ \quad (10.100)$$

(a) Write the net reaction for conversion of glucose to carbon dioxide and ethanol by glycolysis followed by alcoholic fermentation. What happens to NAD^+ and NADH in the net reaction? How do you explain this result?

(b) Draw a coupling diagram, modeled on Figure 10.12, that represents the combination of reactions (10.99) and (10.100). What is the coupling species? Explain your response. Include the ATP production as a straight arrow, like the one in Figure 10.12.

In order to couple two redox half reactions, the reduction potential for the coupling half reaction has to be intermediate between the reduction potentials of the reactions that are coupled. We have illustrated this in Figure 10.12 for the Blue-Bottle reactions. Note that as you proceed left-to-right from one half reaction to the next, the potentials become successively more positive. Glucose, RCHO(*aq*), can be oxidized by MB^+(*aq*) and in turn, MBH(*aq*) can be oxidized by O_2(*g*).

Coupled reactions in a biofuel cell In an electrochemical **fuel cell**, a fuel is oxidized at the anode and oxygen is reduced at the cathode. The reactions in a fuel cell are the same as combustions, except that a great deal of the free energy change of the reaction is converted to electrical energy instead of heat and light. Fuel cells require catalysts as a part of the electrodes to catalyze the desired oxidations and reductions. Hydrogen is the fuel in the fuel cells that were used to produce electrical power for the Apollo missions to the moon and are now used in the space shuttle. An advantage of hydrogen as the fuel is that the only product of the cell reaction is water (which is collected and used in the spacecraft). A disadvantage of present-day hydrogen fuel cells is that they have to be operated at quite high temperatures, in order to make the electrode catalysts work efficiently.

Since fuel cells are an efficient means to convert chemical energy to electrical energy, scientists are doing research to develop catalysts to oxidize more easily handled fuels such as alcohols or gasoline and to work at lower temperature. Biological catalysts, enzymes, offer one possible solution to these problems, since they work at the temperature of living organisms and usually act on

An exploding fuel cell ruined the Apollo 13 mission and almost cost the astronauts' lives.

readily available fuels, including glucose. The miniature glucose fuel cell shown in the chapter opening illustration and in Figure 10.13 is one such device.

In this glucose fuel cell, the electrodes are very fine carbon fibers that appear as thin lines dwarfed by the 1-mm-wide wells in which they are suspended. Each electrode is coated with a polymeric material containing complexed osmium metal ions and the enzyme, glucose oxidase or laccase, required for the oxidation and reduction reactions, respectively. Many factors affect the electrical output of this cell, but you can use what you have learned to analyze the oxidation that produces the electrical energy.

Figure 10.13.

Photomicrograph of a miniature glucose–air fuel cell. The connections of the carbon-fiber electrodes to the external circuit are not shown in this photograph.

10.69 CHECK THIS

Reduction potentials in a glucose–air fuel cell

The half reactions that power the glucose-air fuel cell shown in Figure 10.13 are

$$O_2(g) + 4H^+(aq) + 4e^- \rightleftharpoons 2H_2O(l) \quad E°(O_2, H_2O) = 1.229 \text{ V} \quad (10.81)$$

$$RCOO^-(aq) + 3H^+(aq) + 2e^- \rightleftharpoons RCHO(aq) + H_2O(l)$$

$$E°'(RCOO^-, RCHO) = -0.44 \text{ V} \quad (10.86)$$

(a) The solution in the cell is buffered at pH 5. Calculate the reduction potentials at 298 K for these reactions at this pH, assuming that all other species are at unit activity.

(b) What is the overall cell reaction and the cell potential under the conditions in part (a)? Use Le Chatelier's principle to predict how the cell potential would change if the activity of oxygen were decreased.

Figure 10.14 shows a series of coupled redox reactions that illustrates how the biofuel cell in Figure 10.13 operates. The enzymes glucose oxidase (GOx) and laccase (LAC) contain complexed metal ions that change oxidation number as they gain or lose electrons in the redox reactions they catalyze. In turn, they lose or regain these electrons in their interactions with the osmium ions that are part of the polymer material in which the enzymes are embedded. The osmium ions on the cathode and anode have different reduction potentials because they are differently complexed with ligands from the polymer.

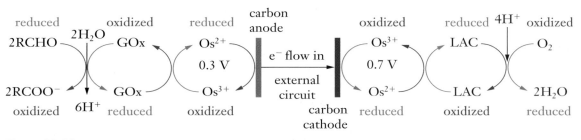

Figure 10.14.

The coupled series of reactions in the biofuel cell in Figure 10.13. GOx is the enzyme glucose oxidase and LAC is the enzyme laccase. The cell potential is about 0.4 V. Only the stoichiometries of the glucose oxidation and oxygen reduction are shown.

10.70 CHECK THIS

Intermediate potentials in the biofuel cell

(a) Use your results from Check This 10.69 to add to Figure 10.14 the reduction potentials for the gluconate–glucose and oxygen–water reactions.

(b) What would be the ideal reduction potentials for the glucose oxidase and laccase? Ideal potentials would provide about the same driving force for both the reduction and oxidation these enzymes undergo. Explain your choices.

(c) Approximately how efficient is this 0.4 V cell? That is, how much of the free energy released by the glucose–oxygen reaction can do work in the external circuit? Explain your reasoning.

The TCA cycle is also called the citric acid cycle because citric acid (see Consider This 10.57) is the tricarboxylic acid in the cycle.

Glucose oxidation in aerobic organisms In aerobic organisms, including you, the pathway for glucose oxidation is glycolysis, net reaction (10.99), followed by a series of reactions called the tricarboxylic acid (TCA) cycle. The fate of the glucose carbons in this overall pathway is

$$C_6H_{12}O_6 \xrightarrow{\text{glycolysis}} 2C_3H_3O_3^- \xrightarrow{\text{TCA cycle}} 6CO_2 \qquad \textbf{(10.101)}$$

The notations over the reaction arrows remind us that these are not individual reactions, but a series of reactions that has been given a collective name. Carbons are conserved in the pathway represented by expression (10.101), but oxygen and hydrogen are not. There is more oxygen in the product than in the reactant, so oxygen must come into the reaction from another source. Almost universally, when a source of oxygen or hydrogen atoms is required in living systems, the source is water. The stoichiometry of glucose oxidation could be written

$$C_6H_{12}O_6 + 6H_2O \xrightarrow{\text{glycolysis + TCA cycle}} 6CO_2 + 24H^+ + 24e^- \quad \textbf{(10.102)}$$

No molecular oxygen is involved in the glycolysis and TCA pathways, but the electrons ultimately end up on oxygen. **Electron transport** is the pathway by which electrons from glucose are transferred to oxygen. Not all the electrons take the same route to oxygen. However, since our aim is only to show how mitochondria use coupled redox reactions to extract energy from glucose oxidation, we will neglect the complications and reduce the electron transport pathway to a few steps. If you study metabolism further, you will amend and add to these steps, but the outcome is the same.

The electrons from glucose reduce NAD^+:

$$C_6H_{12}O_6 + 12NAD^+ + 6H_2O \xrightarrow{\text{glycolysis + TCA cycle}}$$
$$6CO_2 + 12NADH + 12H^+ \quad \textbf{(10.103)}$$

The net reaction represented by expression (10.103) requires more than a dozen steps with a different enzyme to catalyze each step. The electrons given up by glucose end up on NADH molecules in the matrix of mitochondria. The NADH reduces ubiquinone that is associated with the cytochrome bc_1 protein complex in the inner mitochondrial membrane (chapter opening illustration):

$$UQ + NADH + H^+ \rightleftharpoons UQH_2 + NAD^+ \qquad \textbf{(10.104)}$$

Reaction (10.104) is reaction (10.84) rewritten with abbreviations for the biomolecules.

10.71 CHECK THIS

Role of NAD$^+$-NADH in the glucose oxidation pathway

Assume that all the NADH produced by net reaction (10.103) undergoes reaction (10.104). What is the net outcome of reactions (10.103) and (10.104)? Write the net reaction that describes this outcome. What happens to NAD$^+$ and NADH in the net reaction? Explain the role that NAD$^+$ and NADH play in this part of the glucose oxidation pathway.

Within the cytochrome bc_1 complex, there is a redox reaction between UQH$_2$ and Fe^{3+} [recall reaction (10.1)] in a heme complex associated with cytochrome c_1:

$$2Fe^{3+}(cyt\ c_1) + UQH_2 \rightleftharpoons$$
$$2Fe^{2+}(cyt\ c_1) + UQ + 2H^+(\text{intermembrane space}) \quad \textbf{(10.105)}$$

The protein complex is oriented in the membrane in such a way that the hydronium ions produced by reaction (10.105) end up in the intermembrane space of the mitochondrion. The net result of reactions (10.103) through (10.105) is to transfer hydronium ions across the membrane, as represented in Figure 10.15. The hydronium ion concentration in the matrix is lower than in the intermembrane space. This difference in hydronium ion concentrations sets up an **electrochemical gradient,** a difference in both electrical charge and pH across the inner mitochondrial membrane. The free energy stored in the electrochemical gradient is what other proteins in the mitochondrial membrane use to drive the synthesis of ATP from ADP and P$_i$. Recall once again the words from the *I Ching* at the beginning of this chapter, "... one must separate things in order to unite them."

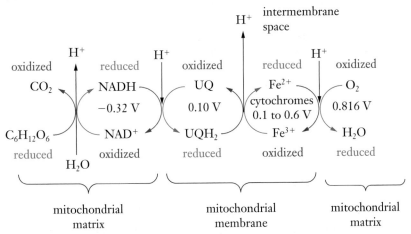

Figure 10.15.

Coupled reactions in the aerobic glucose oxidation pathway.
The stoichiometry of the reactions is not represented on this diagram. All the reactions occur either in the mitochondrial matrix or in the inner mitochondrial membrane. The net effect of the oxidation of glucose is to move hydronium ions from the matrix to the intermembrane space of the mitochondrion.

Reaction (10.105) shows that transfer of hydronium ions across the membrane will stop if all the Fe^{3+} in cytochrome c_1 is reduced. The rest of the **electron transport pathway** is a sequence of redox reactions involving other cytochromes containing both complexed iron and copper ions. At each step, the reduced product of the preceding step is reoxidized. The end of the pathway is a redox reaction between molecular oxygen and Fe^{2+} in cytochrome a_3:

$$O_2 + 4Fe^{2+}(cyt\ a_3) + 4H^+ \rightarrow 2H_2O + 4Fe^{3+}(cyt\ a_3) \quad \textbf{(10.106)}$$

All the steps involving the cytochromes have been lumped together in Figure 10.15. Reaction (10.106), on the matrix side of the membrane, further reduces the hydronium ion concentration in the matrix and helps to maintain the electrochemical gradient across the membrane.

10.72 CHECK THIS

Net reaction for oxygen reduction

Beginning with one NADH molecule from reaction (10.103), the net reaction represented by Figure 10.15 is

$$NADH + 3H^+(\text{matrix}) + \tfrac{1}{2}O_2 \rightarrow$$
$$NAD^+ + 2H^+(\text{intermembrane space}) + H_2O \quad \textbf{(10.107)}$$

(a) Write the sequence of reactions that gives net reaction (10.107).
(b) What is the standard cell potential for reaction (10.107) at pH 7? Explain how you get your answer.

Coupled reactions, including coupling that does not involve redox reactions, are common in living systems. The electron transport pathway is present in all aerobic organisms. A similar pathway is part of the photosynthetic process in green plants. The net result of the electron transport pathway in mitochondria is to create an electrochemical gradient that stores energy from the glucose oxidation. If you take a biochemistry course, you will learn how the electrochemical gradient is harnessed to produce the ATP you need to function and to continue to learn even more.

Reflection and Projection

In this last section, we introduced the redox chemistry of glucose in living systems and simpler test tube reactions. This discussion provided further examples of how the Nernst equation helps us understand the dependence of cell potentials and reactions on the composition of the solution and emphasized the role of coupled redox reactions. When we looked at the role of coupled reactions in living organisms, we left out or simplified many of the details of the pathway for storing the free energy of glucose oxidation as an electrochemical gradient. Our intent has been to show the importance of redox and electrochemical processes for living organisms and to apply the concepts developed for electrochemical cells to an analysis of redox reactions in living cells.

Analyses of equilibrium and electrochemical systems are probably the two most important applications of thermodynamics in chemistry. Questions about the direction and extent of reactions that we raised early in the book are qualitatively and quantitatively answered by these analyses. However, at least one large question remains: Why are some reactions so slow that they seem not to occur, even though their calculated free energy changes are favorable? The oxidation of glucose at room temperature is an example. Green plants synthesize enormous amounts of glucose every day; it becomes a part of their structure as cellulose or is stored as starch. The free energy for oxidation of glucose is large and negative and there is plenty of oxygen available in the atmosphere to react with it. Yet it can last for hundreds (even thousands) of years, unless it is heated very hot or is eaten by an organism and broken down by the pathways we have just discussed. Thermodynamics cannot tell us how fast a favored process will occur. To understand the speeds of chemical reactions, we need to know more about what affects the speeds, which is the topic of the next chapter.

10.9 Outcomes Review

In this chapter, we introduced two kinds of electrochemical cells: electrolytic and galvanic. In electrolytic cells we use an electric current to produce chemical reactions by reducing and oxidizing species in solution. In galvanic cells oxidation and reduction half reactions are spatially separated and the free energy of a chemical reaction drives electrons through an external circuit, where they can be used to do work. The quantitative redox relationships that follow from electrochemical cell measurements also characterize redox reactions that occur between reactants mixed in the same solution. Redox reactions in living cells obey these same relationships, which can be used to help understand and explain observed stoichiometric relationships in organisms.

Check your understanding of the ideas in the chapter by reviewing these expected outcomes of your study. You should be able to

- Use evidence from experimental observations to write probable half reactions for the reductions and oxidations taking place in an electrolytic cell [Section 10.1].
- Describe and draw molecular-level diagrams of the processes going on and the flow of charge in an electrochemical cell [Sections 10.1 and 10.2].
- Use the Faraday and relationships among time, electric current, cell potential, and cell reaction stoichiometry to calculate the amounts of products from electrolysis or the amount of work available from the reactants in a galvanic cell [Sections 10.1 and 10.5].
- Show how to connect two metal–metal ion half cells to make a galvanic cell and explain the role of each component of the cell [Section 10.2].
- Use the known cell reaction to identify the anode and cathode of an electrochemical cell [Sections 10.1, 10.2, 10.3, 10.4, and 10.5].
- Translate a physical cell setup to the conventional line notation for cells and vice versa [Section 10.2].
- Use the known sign for a galvanic cell potential to identify the direction of the cell reaction and the anode and cathode of the cell [Sections 10.3 and 10.4].
- Determine an unknown cell potential for a cell reaction by combining cell reactions with known cell potentials to give the desired cell reaction and its cell potential [Sections 10.4, 10.5, 10.6, 10.7, and 10.8].

- Use a table of standard reduction potentials to predict the direction of any redox reaction (for which data are given) and determine its standard cell potential [Sections 10.4, 10.5, 10.6, 10.7, and 10.8].
- Describe how an electrode senses the redox half reaction in a half cell in which the reduced and oxidized species are both present as dissolved ions and/or molecules [Section 10.6].
- Determine free energy change for a cell reaction from cell potential and vice versa [Sections 10.5, 10.6, 10.7, and 10.8].
- Apply Le Chatelier's principle to predict the direction of change of cell potentials as concentrations in the half cells are changed [Sections 10.6, 10.7, 10.8, and 10.10].
- Use the Nernst equation, which relates the cell potential, the standard cell potential, and the reaction quotient for the cell reaction, to determine any one of these quantities, if the other two are known [Sections 10.6, 10.7, and 10.8].
- Use the standard cell potential to determine the equilibrium constant for a cell reaction and vice versa [Sections 10.6 and 10.8].
- Use the Nernst equation to convert standard reduction potentials to reduction potentials under nonstandard conditions and vice versa [Sections 10.6, 10.7, and 10.8].
- Use a table of standard reduction potentials at pH 7 to predict the direction and standard cell potential at pH 7 for redox reactions in biological systems [Sections 10.7 and 10.8].
- Combine free energy changes for redox reactions, complexation equilibria, and solubility equilibria to get the free energy changes and reduction potentials for the net redox reactions that occur in systems with such interrelated reactions [Sections 10.8 and 10.10].
- Convert a stepwise series of reactions to a diagram that shows the coupling of reactions and vice versa [Section 10.8].
- Use standard reduction potentials to determine the probable sequence of a series of coupled redox reactions and vice versa [Section 10.8].

10.10. Extension—Cell Potentials and Non-Redox Equilibria

10.73 INVESTIGATE THIS

Does addition of ethylenediamine affect E(Cu²⁺, Cu)?

(a) Do this as a class investigation and work in small groups to discuss and analyze the results. Use a setup similar to the one in Investigate This 10.44. Use three wells grouped around one containing 10% aqueous KNO_3 solution to prepare three half cells. Fill two of the wells about two-thirds full of aqueous 0.10 M copper sulfate, $CuSO_4$, solution. Fill the third well about two-thirds full of aqueous 0.010 M $CuSO_4$ solution. Place a clean copper wire in each well as an electrode and use filter paper salt bridges soaked in 10% KNO_3 to connect each half cell to the KNO_3 well. Use a digital voltmeter to read the cell potential to the nearest millivolt and determine which electrode is the anode for each of the three cells you can make by combining pairs of half cells.

(b) To one of the wells containing 0.10 M $CuSO_4$ solution, add *one drop* of ethylenediamine (1,2-diaminoethane, $H_2NCH_2CH_2NH_2$) and stir the mixture with the electrode to make sure it is uniform. Record your observations. Use the digital voltmeter to read the cell potentials and determine which electrode is the anode for the two cells you can make by connecting the half cell with added ethylenediamine to each of the other half cells.

10.74 CONSIDER THIS

Why does addition of ethylenediamine affect E(Cu²⁺, Cu)?

(a) Are the cell potentials you measured in Investigate This 10.73(a) consistent with one another? How do you interpret any differences you observe?

(b) What were the cell potentials you measured in Investigate This 10.73(b)? What was the direction of the change in cell potential from the values you got in part (a)? What do you think could be happening in the solution to cause a change in this direction?

Concentration cells In Section 10.8, we combined redox equilibria (or standard cell potentials) and equilibria for non-redox reactions to get information about new reactions. In this section, we will use cell potentials to determine equilibrium constants for non-redox reactions going on in the cell. The cells you prepared in Investigate This 10.73 are concentration cells. In a **concentration cell,** the half reactions in the half cells are identical and the cell potential depends on the *difference* in concentrations in the half cells. The half-cell reaction and standard reduction potential for all the half cells you made are

$$Cu^{2+}(aq) + 2e^- \rightleftharpoons Cu(s) \quad E° = 0.337 \text{ V} \tag{10.108}$$

A cell made with two identical copper half cells will have no driving force to transfer electrons in either direction. The reduction potential for both half cells is the same and the measured cell potential, $E_{Cu|Cu}$, is zero.

10.75 CONSIDER THIS

Do the copper concentration cells behave as expected?

(a) Does your measured cell potential for identical copper half cells in Investigate This 10.73(a) support the prediction that the cell potential is zero? (Tiny differences between half cells can produce small cell potentials, but these are usually less than 1 or 2 mV.) What is $E°_{Cu|Cu}$ for identical copper half cells? Explain your answer.

(b) Do you get the same result for cells with nonidentical copper half cells? Why or why not?

$Cu^{2+}(aq)$ is on the reactant side of half reaction (10.108), so Le Chatelier's principle tells us that the reduction potential will decrease, if $[Cu^{2+}(aq)]$ is decreased:

$$E(\text{higher } [Cu^{2+}]) > E(\text{lower } [Cu^{2+}])$$

The lower the concentration, the greater will be this difference. Since the half cell with the lower concentration has the lower reduction potential, its electrode will be the anode of the concentration cell, as you found in Investigate This 10.73(a). One of the cells you made was

$$Cu(s) \mid Cu^{2+}(0.010\ M) \parallel Cu^{2+}(0.10\ M) \mid Cu(s) \qquad \textbf{(10.109)}$$

The cell reaction is

$$Cu^{2+}(0.10\ M) + Cu(s,\ anode) \rightleftharpoons Cu(s,\ cathode) + Cu^{2+}(0.010\ M)$$
$$\textbf{(10.110)}$$

Reaction (10.110) can be interpreted as a decrease in concentration of a more concentrated solution and an increase in concentration of a less concentrated solution. This change is analogous to diffusion of ions from higher toward lower concentration. We know that entropy and free energy favor this change, and now we see that the cell potential for a concentration cell is another measure of this directionality. The Nernst equation, at 298 K, for reaction (10.110) is

$$E_{Cu|Cu} = E°_{Cu|Cu} - \left(\frac{0.059\ V}{2}\right)\log\left(\frac{\text{lower } [Cu^{2+}]}{\text{higher } [Cu^{2+}]}\right)$$

$$= 0 - \left(\frac{0.059\ V}{2}\right)\log\left(\frac{0.010\ M}{0.10\ M}\right) \qquad \textbf{(10.111)}$$

$$E_{Cu|Cu} = -\left(\frac{0.059\ V}{2}\right)(-1) = 0.030\ V$$

10.76 CHECK THIS

Cell potentials in Investigate This 10.73

Do your results from Investigate This 10.73(a) support the prediction that the cell potential for the cells with unequal copper ion concentrations is about 0.030 V? Why or why not?

Cell potentials and complexed ions An important point to recognize about half reaction (10.108) is that $Cu^{2+}(aq)$ is the reactant that is reduced. If other forms of copper in the +2 oxidation state are present in the solution, they do not take part *directly* in the reaction. For example, $Cu^{2+}(aq)$ can form complexes with ethylenediamine, $H_2CH_2CH_2NH_2$ (abbreviated en):

$$Cu^{2+}(aq) + en(aq) \rightleftharpoons Cu(en)^{2+}(aq) \tag{10.112}$$

$$Cu(en)^{2+}(aq) + en(aq) \rightleftharpoons Cu(en)_2^{2+}(aq) \tag{10.113}$$

$$\overline{Cu^{2+}(aq) + 2en(aq) \rightleftharpoons Cu(en)_2^{2+}(aq)} \tag{10.114}$$

You might have analyzed data from continuous variation experiments on this system in Chapter 6, Problem 6.29. The stepwise formation, reactions (10.112) and (10.113), is evident in such experiments. In Investigate This 10.73(b), addition of ethylenediamine to one of the copper half cells produced the magenta color of the $Cu(en)_2^{2+}(aq)$ complex. Complex formation "uses up" $Cu^{2+}(aq)$, so its concentration is lowered in the half cell to which ethylenediamine is added, as shown schematically in Figure 10.16.

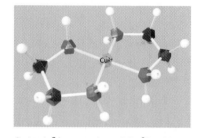

$Cu(en)_2^{2+}$, complex of Cu^{2+} with ethylenediamine (en).

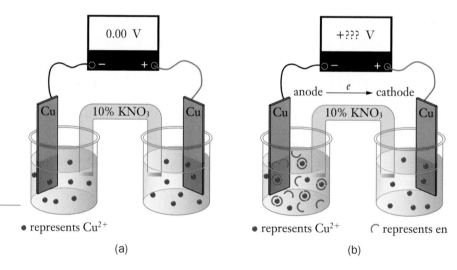

(a)

• represents Cu^{2+}

• represents Cu^{2+} ⌒ represents en

(b)

Figure 10.16.

Addition of ethylenediamine to one half cell of a copper concentration cell.

10.77 CONSIDER THIS

How does addition of ethylenediamine affect [$Cu^{2+}(aq)$]?

(a) Use your results from Investigate This 10.73(a) to predict the *direction* of the effect on the cell potential, if adding ethylenediamine to one half of a copper concentration cell decreases the [$Cu^{2+}(aq)$] in that half cell. State your reasoning clearly.

(b) Are your results from Investigate This 10.73(b) consistent with your prediction in part (a)? Explain why or why not.

continued

(c) Write the line notation for the cell obtained by combining a half cell to which ethylenediamine is added with the identical half cell without ethylenediamine. Explain the reasoning for your choice of anode. Does Figure 10.16(b) correctly represent the cell you wrote? Explain why or why not.

We can use cell potentials to learn more about step-wise and overall equilibria, like those represented in reactions (10.112) through (10.114). Worked Example 10.78 shows how to use a cell potential measurement in a silver|silver concentration cell to calculate the overall equilibrium constant for the formation of the silver diammine complex, $Ag(NH_3)_2^+(aq)$:

$$Ag^+(aq) + 2NH_3(aq) \rightleftharpoons Ag(NH_3)_2^+(aq) \qquad \textbf{(10.115)}$$

10.78 WORKED EXAMPLE

K for formation of $Ag(NH_3)_2^+(aq)$

Consider a cell setup like the one in Investigate This 10.73(b), except with silver half cells each containing 2.0 mL of aqueous 0.10 M solutions of silver nitrate, $AgNO_3$, and silver metal electrodes. When one drop (0.05 mL) of 15 M aqueous ammonia, NH_3, solution is added to one of the half cells, the measured cell potential is 0.325 V. The half cell containing ammonia is the anode. What is the equilibrium constant for reaction (10.115)?

Necessary information: We will need the Nernst equation and the stoichiometry and equilibrium constant expression for reaction (10.115).

Strategy: To get K for reaction (10.115), we need the concentrations of $[Ag^+(aq)]$, $[NH_3(aq)]$, and $[Ag(NH_3)_2^+(aq)]$ in the half cell to which ammonia is added. Use the cell potential and the Nernst equation to find $[Ag^+]$ (omitting the state designation). Then use the reaction stoichiometry and the amounts of reactants that are mixed in the half cell to find $[NH_3]$ and $[Ag(NH_3)_2^+]$, assuming that addition of ammonia makes a negligible change in the volume of the solution. Substitute all the concentrations into the equilibrium constant expression to get K.

Implementation: For this concentration cell, n is one, because silver ion reduction requires only one electron. Let (Ag^+) be the silver ion concentration ratio in the half cell with added ammonia. The Nernst equation for the concentration cell is

$$E_{Ag|Ag} = 0.325 \text{ V} = -(0.059 \text{ V})\log\left[\frac{\text{lower } (Ag^+)}{\text{higher } (Ag^+)}\right]$$

$$= -(0.059 \text{ V})\log\left[\frac{(Ag^+)}{0.10}\right]$$

Solve for the ratio of concentrations and then for (Ag^+) {= numerical value of $[Ag^+]$}:

$$\log\left[\frac{(Ag^+)}{0.10}\right] = \frac{0.325 \text{ V}}{-0.059 \text{ V}} = -5.51; \quad \left[\frac{(Ag^+)}{0.10}\right] = 3.1 \times 10^{-6}$$

$$(Ag^+) = 3.1 \times 10^{-7}$$

continued

Only a tiny amount of silver ion is left uncomplexed in solution. Essentially all the silver ion is present as $Ag(NH_3)_2{}^+$. We had 0.10 M silver ion to start with, and all the silver must still be present, so we have

$$(Ag(NH_3)_2{}^+) = 0.10$$

For the ammonia, we calculate the number of moles added and subtract the number of moles complexed with silver ion to find the number of moles left as $NH_3(aq)$ in solution.

$$\text{mol } NH_3 \text{ added} = (0.05 \times 10^{-3} \text{ L})(15 \text{ M}) = 7.5 \times 10^{-4} \text{ mol } NH_3 \text{ added}$$

$$\text{mol } NH_3 \text{ complexed} = (2.0 \times 10^{-3} \text{ L})(0.10 \text{ M})\left(\frac{2 \text{ mol } NH_3}{1 \text{ mol complex}}\right)$$

$$= 4.0 \times 10^{-4} \text{ mol } NH_3 \text{ complexed}$$

$$\text{mol } NH_3 \text{ unreacted} = (\text{mol } NH_3 \text{ added}) - (\text{mol } NH_3 \text{ in the complex})$$

$$= 3.5 \times 10^{-4} \text{ mol } NH_3 \text{ unreacted}$$

$$[NH_3] = \left(\frac{3.5 \times 10^{-4} \text{ mol}}{2.0 \times 10^{-3} \text{ L}}\right) = 0.18 \text{ M}; \quad (NH_3) = 0.18$$

The equilibrium constant expression and equilibrium constant for reaction (10.115) are

$$K = \left[\frac{(Ag(NH_3)_2^+)}{(Ag^+)(NH_3)^2}\right]_{eq} = \frac{(0.10)}{(3.1 \times 10^{-7})(0.18)^2} = 1.0 \times 10^7$$

Does the answer make sense? We do not know whether the numerical answer is correct until we look in a table of stability constants for metal ion complexes to see that it is. However, we know that the equilibrium constant will be large (much greater than unity), because essentially all the silver ion gets complexed, that is, reaction (10.115) greatly favors the products at equilibrium. Cell potentials can be used to determine equilibrium constants for reactions that affect a cell concentration, but are not, themselves, redox reactions. This is a powerful method for determining equilibrium constants, especially those involving metal ions.

10.79 CHECK THIS

K for formation of Cu(en)₂²⁺(aq)

Use your data from Investigate This 10.73 to determine the equilibrium constant for reaction (10.114). The density of liquid ethylenediamine, $C_2H_8N_2$, is 0.9 g·mL⁻¹, and 1 drop is about 0.05 mL. How does the equilibrium constant for this complexation reaction compare to that for the formation of the silver diammine complex? Which is the stronger complex? Give the reasoning for your answer.

Chapter 10 Problems

10.1. Electrolysis

10.1. Passing an electric current through an aqueous solution of sodium chloride, NaCl, containing universal pH indicator produces gases and changes in the color of the indicator at both electrodes. The observations are different for low and high concentrations of salt.

    (i)   In a 0.1 M NaCl solution, electrolysis produces a basic solution at the cathode and an acidic solution at the anode. The anode gas does not burn; both gases are odorless.

    (ii) In a 6 M NaCl solution, electrolysis produces a basic solution at the cathode and the solution at the anode loses color (is bleached). The gas produced at the anode causes a choking sensation when inhaled.

(a) Write the half reactions occurring at the cathode and anode in the dilute solution. Show how your reactions explain all the observations.

(b) Repeat part (a) for the concentrated solution.

10.2. (a) When iodine, I_2, dissolves in a clear, colorless aqueous solution of potassium iodide, KI, a clear, orange solution of triiodide, I_3^- (which can be thought of as $I_2 \cdot I^-$), is formed. When an aqueous solution of KI is electrolyzed, bubbles of gas are formed at the cathode and the solution around the anode becomes orange. What are the cathodic and anodic half reactions and net cell reaction for this electrolysis? Explain your reasoning.

(b) Sodium thiosulfate, $Na_2S_2O_3$, added to a triiodide solution reacts immediately with I_3^- to produce iodide ion, I^-, and decolorizes the solution. If a little $Na_2S_2O_3$ is added to a KI solution before electrolysis, bubbles begin to form immediately at the cathode when electrolysis is begun, but the solution around the anode remains clear and colorless for a period of time and then begins to turn orange. Are these observations consistent with the reactions you wrote in part (a)? Explain why or why not.

10.3. Elemental sodium is too reactive to be found naturally in an uncombined form. The elemental metal can be produced by the electrolysis of molten sodium chloride.

(a) How long would it take to produce 23 g of metallic sodium if a 12-A current is passed through a molten sodium chloride electrolysis cell? Explain.

(b) Chlorine gas is also produced in this electrolysis. Identify the half reactions occurring at the cathode and anode. Write the net cell reaction. How much chlorine gas is produced during the electrolysis in part (a)? Explain your reasoning.

10.4. The Dow process is used to produce pure magnesium metal by electrolysis of molten magnesium chloride. Chlorine gas is also produced in the process.

(a) Write the half reactions that occur at the cathode and anode.

(b) Which of the half reactions in part (a) is the oxidation and which the reduction? Is your identification consistent with the definition of the cathode and anode of an electrochemical cell? Explain.

(c) What is the net cell reaction in the Dow process?

(d) How many kilograms of magnesium metal could be produced if a 95,000-A current is passed through a molten magnesium chloride cell for 8.0 hours?

(e) A temperature of almost 1000 K is required to melt magnesium chloride. Wouldn't it be easier to electrolyze the solid salt at a lower temperature? Why isn't this done?

10.5. Gold is often electroplated onto other, less expensive, metals to produce gold-plated jewelry. The electrolyte solution usually contains gold(III) as its cyanide complex, $Au(CN)_4^-(aq)$. The reaction at the surface to be plated is

$$Au(CN)_4^-(aq) + 3e^- \rightarrow Au(s) + 4CN^-(aq)$$

(a) In the electroplating cell, the object to be gold plated is one electrode and a sheet of gold is the other. Which is the cathode and which the anode in this cell? What are the cathodic and anodic half reactions? Explain your responses.

(b) How many grams of gold can be electroplated on a bracelet by a 2.5-A current passing through the electroplating cell for 7.5 minutes?

(c) Commercial electroplating cells are not 100% efficient; some of the electrical energy is dissipated as heat and in side reactions. If the bracelet in part (b) gained 0.65 g during the plating, how efficient was the process? Explain.

10.6. Two electrolysis cells were set up in series so that all the current that passed through the first also passed through the second. The first cell contained an aqueous solution of silver nitrate and silver electrodes. The second cell contained an aqueous solution of a cobalt complex ion and platinum electrodes. After current had passed for some time, the cathodes in each cell were removed, washed, dried, and weighed. The silver cathode gained 0.2789 g and the platinum cathode gained 0.0502 g.

(a) Based on the results from the silver cell, how many moles of electrons had passed through the cells? Explain.

(b) What is the oxidation number of the cobalt in the cobalt complex? Explain the reasoning for your answer.

10.7. How much time will it take to electroplate 0.0353 g of chromium from an aqueous potassium dichromate, $K_2Cr_2O_7$, solution, with a current of 0.125 A passing through the cell? Explain the reasoning for your answer.

10.8. When Thomas Edison first sold electricity commercially near the end of the 19th century, he needed a method to measure the amount of electricity each customer used, so he invented a way, based on electrolysis, to measure electrical use. A fraction of the current that entered the customer's house or business passed through an electrolysis cell containing a solution of zinc sulfate and zinc electrodes. Each month, a "meter" reader would visit, remove the cathode, wash and dry it, weigh it, and then calculate the coulombs used.
(a) If 8% of the current passed through such a meter and the cathode increased in mass by 57 g in a 30-day period, how many coulombs had the customer used? On the average, what was the current flow (amperes) during this month? Clearly explain how you get your answers.
(b) The electricity that Edison sold was direct current (DC), that is, one of the prongs of a plug was always positive and the other negative. Today, our electricity is supplied as alternating current (AC) with the polarity (positive and negative signs) changing 60 times per second. Would Edison's meter work today? Why or why not?

10.2. Electric Current from Chemical Reactions

10.9. Write the cathodic and anodic half reactions, the net cell reaction, and the correct line notation for a cell based on each of these unbalanced reaction equations. Explain your reasoning.
(a) $Mn(s) + Ti^{2+}(aq) \rightleftharpoons Mn^{2+}(aq) + Ti(s)$
(b) $U(s) + V^{2+}(aq) \rightleftharpoons U^{3+}(aq) + V(s)$
(c) $Zn(s) + Ni^{2+}(aq) \rightleftharpoons Zn^{2+}(aq) + Ni(s)$
(d) $Mg(s) + Cr^{3+}(aq) \rightleftharpoons Mg^{2+}(aq) + Cr(s)$

10.10. Write the cathodic and anodic half reactions and the net cell reaction for the galvanic cells represented by these correct line notations. Explain your reasoning.
(a) $Cu(s) \mid Cu^{2+}(aq) \parallel Cu^{+}(aq) \mid Cu(s)$
(b) $Co(s) \mid Co^{2+}(aq) \parallel Ag^{+}(aq) \mid Ag(s)$
(c) $Al(s) \mid Al^{3+}(aq) \parallel Fe^{2+}(aq) \mid Fe(s)$
(d) $Pb(s) \mid Pb^{2+}(aq) \parallel Cu^{2+}(aq) \mid Cu(s)$

10.11. Consider a cell made by placing a manganese rod, $Mn(s)$, in a solution of manganese(II) sulfate, $MnSO_4(aq)$, that is connected by a salt bridge to a solution of chromium(II) sulfate, $CrSO_4(aq)$, containing a chromium rod, $Cr(s)$. The manganese electrode is negative. Write the cathodic and anodic half reactions, the net cell reaction, and the correct line notation for the cell. Which direction do anions migrate in the cell? Which direction do electrons travel through a wire connecting the electrodes? Explain.

10.12. When a clean, shiny strip of magnesium metal, Mg, is immersed in a clear, blue aqueous solution of copper sulfate, $CuSO_4$, the metal in contact with the solution quickly turns dark and the color of the solution begins to fade.
(a) Write a net ionic reaction for this process and show how it explains the observations.

(b) Sketch the setup of an electrochemical cell that would take advantage of this reaction to provide a flow of electrons in an external circuit. Write the cathodic and anodic half reactions, the net cell reaction, and the correct line notation for this cell.

10.13. When a strip of iron metal, Fe, is immersed in an aqueous solution of chromium(III) chloride, $CrCl_3$, no apparent changes occur. When a strip of chromium metal, Cr, is immersed in an aqueous solution of iron(II) nitrate, $Fe(NO_3)_2$, a dark deposit forms on the surface of the metal in contact with the solution.
(a) Sketch the setup of an electrochemical cell that would take advantage of the information above to provide a flow of electrons in an external circuit. Explain the rationale for your setup.
(b) Write the cathodic and anodic half reactions, the net cell reaction, and the correct line notation for your cell. Explain why it is correct.

10.14. If you draw power too rapidly from a flashlight battery by connecting it to a device that draws a lot of current, the battery quickly runs down. If you remove the battery and let it rest for a while, it often seems to recover and be usable once again. The rapid drain on the battery causes it to become polarized and resting allows it to depolarize.
(a) What do you think happens in the battery that makes it polarized? What happens when it depolarizes? Relate your explanation to Figures 10.4 and 10.5. (The dictionary defines *polarization* as the separation of positive and negative charge.)
(b) The "solution" inside a flashlight battery is a thick paste of ionic and molecular solids in water. Ions do not move as rapidly in this medium as they do in an aqueous solution. Is this condition consistent with your explanation in part (a)? Explain.

10.15. You can make a simple galvanic cell by sandwiching a thick piece of filter paper soaked with salt solution between a sheet of copper and a sheet of zinc. An ammeter connected to the metal electrodes shows that this cell produces an electric current. (You may have seen "lemon batteries" or "potato batteries" that use the same design, two different metals stuck into the same ionic solution— the liquid in and around the cells in the lemon or potato.) As current flows, the zinc electrode loses mass and hydrogen gas is produced at the copper electrode.
(a) What are the half reactions going on at each electrode in this cell? Explain your reasoning.
(b) What is the net cell reaction in this cell?
(c) Write the correct line notation for this cell. Explain your choice of cathode and anode.

10.16. Consider an electrochemical cell based on this spontaneous reaction:

$$Zn(s) + Cl_2(g) \rightarrow Zn^{2+}(aq) + 2Cl^{-}(aq)$$

The chlorine gas bubbles into the cell solution and reacts at the surface of a graphite (carbon) electrode. Both electrodes dip into the same solution.
(a) What are the half reactions going on in this cell?
(b) Which electrode is the anode in this cell? Explain the reasoning for your choice.
(c) Make a sketch of this cell and write the line notation that describes it.
(d) Which direction do electrons move in a wire connecting the two electrodes?
(e) How many moles of $Cl_2(g)$ must react to produce an electric current of 0.12 amp for exactly 7 days? Explain the reasoning for your answer.

10.3. Cell Potentials

10.17. The measured voltage of an electrochemical cell made by connecting an $Fe^{2+}|Fe$ half cell with a $Pb^{2+}|Pb$ half cell is positive when the anode input from the voltmeter is connected to the lead metal electrode. What is the cell reaction? Explain how you get your answer. What experiment(s) could you do to test whether your cell reaction is correct?

10.18. Consider the cell potentials under standard conditions, $E°$, for these reactions:

$$Ni(s) + 2Ag^+(aq) \rightleftharpoons Ni^{2+}(aq) + 2Ag(s)$$

$$E° = 1.03 \text{ V}$$

$$Cu(s) + 2Ag^+(aq) \rightleftharpoons Cu^{2+}(aq) + 2Ag(s)$$

$$E° = 0.46 \text{ V}$$

Show how to obtain the following cell reaction and its standard cell potential.

$$Ni(s) + Cu^{2+}(aq) \rightleftharpoons Ni^{2+}(aq) + Cu(s)$$

10.19. When a clean strip of lead is placed in a 0.2 M aqueous solution of silver nitrate, the surface of the metal in contact with the solution soon turns dark and then small silvery needles begin to grow on the surface. This behavior is similar to what you observe when a strip of copper metal is immersed in an aqueous solution of silver nitrate.
(a) Using these data, can you predict what will happen if a strip of copper is placed in a solution of lead nitrate? What will happen if a strip of lead is placed in a solution of copper nitrate? Explain your responses.
(b) These cells were made and the cell potentials measured:

$$Pb(s) \mid Pb^{2+}(aq, 0.2 \text{ M}) \parallel Ag^+(aq, 0.2 \text{ M}) \mid Ag(s)$$

$$E = 0.90 \text{ V}$$

$$Cu(s) \mid Cu^{2+}(aq, 0.2 \text{ M}) \parallel Ag^+(aq, 0.2 \text{ M}) \mid Ag(s)$$

$$E = 0.44 \text{ V}$$

What is the cell potential for a cell made by combining the $Pb^{2+}|Pb$ and $Cu^{2+}|Cu$ half cells? Write the cell in

correct line notation. What is the net cell reaction in this cell? Explain how you get your answers.
(c) Given the information in part (b), what are your answers to part (a)? Are your answers different than before? Explain why or why not.

10.20. Consider a cell made by placing a nickel wire, $Ni(s)$, in a 0.12 M aqueous solution of nickel(II) sulfate, $NiSO_4(aq)$, that is connected by a salt bridge to a 0.12 M aqueous solution of lead(II) nitrate, $Pb(NO_3)_2(aq)$, containing a strip of lead, $Pb(s)$. When the black (anode) lead from a digital voltmeter is connected to the lead electrode, the meter reading is -0.10 V.
(a) Write the cathodic and anodic half reactions, the net cell reaction, and the correct line notation for this cell. Clearly explain your reasoning.
(b) If a strip of lead were placed in a 0.12 M aqueous solution of nickel(II) nitrate, what do you predict would be observed? Explain.

10.4. Half-Cell Potentials: Reduction Potentials

10.21. What is the standard cell potential for the cell you sketched in Problem 10.12? Explain.

10.22. Consider an electrochemical cell made by connecting an $Ag^+|Ag$ half cell to a $Fe^{2+}|Fe$ half cell via a salt bridge.
(a) Which electrode will be the cathode and which the anode? Write the correct line notation for this cell. Explain how you arrive at your answers, including any assumptions.
(b) Write the net cell reaction for this cell and show how to calculate its standard cell potential.

10.23. List these metals, Al, Ca, Mg, Na, and Zn, in order of increasing strength as reducing agents. Explain how you decide the order.

10.24. List these molecular species, Br_2, Cl_2, F_2, I_2, and O_2, in order of increasing strength as oxidizing agents. Explain how you decide the order.

10.25. Explain whether each of these statements describes a spontaneous or nonspontaneous process. For those that are not spontaneous, rewrite the statement to describe the spontaneous process. Write a net cell reaction and calculate the standard cell potential for the spontaneous process in each case.
(a) Chromium metal, $Cr(s)$, reduces lead(II) ion, $Pb^{2+}(aq)$, to lead metal, $Pb(s)$.
(b) In basic solution mercury metal, $Hg(l)$, reduces cadmium hydroxide, $Cd(OH)_2(s)$, to cadmium metal, $Cd(s)$, and forms mercury(II) oxide, $HgO(s)$.
(c) Nickel(II) ion, $Ni^{2+}(aq)$, is reduced to nickel metal, $Ni(s)$, by hydrogen gas, $H_2(g)$, in acidic solution.
(d) Sodium metal, $Na(s)$, reduces water, $H_2O(l)$, at pH 7 to form hydrogen gas $H_2(g)$.

10.26. Halogens are somewhat soluble in solutions of their corresponding halide, bromine, Br_2, in aqueous Br^-, for example. These halogen-halide solutions of chlorine, bromine, and iodine are, respectively, very pale yellow-green, orange, and red. Solutions of the halide ions alone are clear and colorless. Mixing halogen-halide solutions with different halide solutions gave these results:

    (i) $Cl_2/Cl^- + Br^- \rightarrow$ orange solution
    (ii) $Cl_2/Cl^- + I^- \rightarrow$ red solution
    (iii) $Br_2/Br^- + Cl^- \rightarrow$ orange solution
    (iv) $Br_2/Br^- + I^- \rightarrow$ red solution

(a) Write the redox reaction, if any, that occurs in each case. Use the table of reduction potentials to explain why the reactions you write should occur.
(b) Predict the result you would observe, if you mixed an I_2/I^- solution with a chloride ion solution. Explain the basis of your prediction.
(c) Predict the result you would observe, if you bubbled chlorine gas, $Cl_2(g)$, into an aqueous solution of sodium bromide, $NaBr(aq)$. Explain the basis of your prediction.

10.27. One commercial source of bromine, $Br_2(aq)$, is brine that contains sodium bromide, $NaBr(aq)$. Might it be possible to oxidize the bromide to bromine by bubbling oxygen gas, $O_2(g)$, through acidified brine? Explain why or why not.

10.28. Before 1886, when Charles Hall and Paul Heroult discovered how to produce aluminum by electrolysis, the metal was extremely expensive. It was produced chemically by a redox reaction between molten aluminum chloride, $AlCl_3$, and sodium metal (which was also quite expensive but could be obtained by electrolysis—see Problem 10.3):

$$AlCl_3(l) + 3Na(s) \rightarrow Al(s) + 3NaCl(s)$$

What is the cell potential that corresponds to this redox reaction? What assumption(s) do you have to make to get your answer?

10.29. Consider a cell made by placing a sheet of zinc metal, $Zn(s)$, in an aqueous solution of zinc sulfate, $ZnSO_4(aq)$, that is connected by a salt bridge to an aqueous solution of sodium chloride, $NaCl(aq)$, containing a coil of platinum wire, $Pt(s)$, over which chlorine gas, $Cl_2(g)$, is bubbled.
(a) Which electrode is the cathode and which the anode in this cell? Explain the reasoning for your choice.
(b) Write the cathodic and anodic half reactions and the net cell reaction for this cell. Explain.
(c) Write the correct line notation for this cell and show how to calculate its standard cell potential.

10.30. Samples containing iron can be analyzed by converting all the iron to iron(II) ion, $Fe^{2+}(aq)$, in acidic aqueous solution and then titrating the iron(II) with a solution of potassium permanganate, $KMnO_4(aq)$. The permanganate ion, $MnO_4^-(aq)$, oxidizes the iron(II) to

iron(III) ion, $Fe^{3+}(aq)$; the manganese(VII) in permanganate is reduced to the manganese(II) ion, $Mn^{2+}(aq)$. Permanganate ion is an intense violet color and the products of the reaction are essentially colorless, so the equivalence point of the titration is signaled when the sample solution first turns very light pink.
(a) Write the balanced redox reaction for the titration reaction.
(b) What is the standard cell potential for this reaction? Explain.
(c) How many moles of permanganate ion are required to react with one mole of iron(II) ion? If 23.46 mL of 0.0200 M $KMnO_4(aq)$ is required to titrate a sample containing iron(II) ion, how many moles of iron are in the sample? Show your work.

10.31. One way to synthesize small quantities of chlorine gas for use in the laboratory is to drop hydrochloric acid solution, $HCl(aq)$, onto crystals of potassium permanganate, $KMnO_4(s)$.
(a) Write the balanced redox reaction for this synthesis reaction.
(b) What is the standard cell potential for this reaction? Explain.

10.32. Use the data in Appendix C to predict whether a reaction will or will not occur between each of these pairs of reactants. Explain the basis of each prediction. Write balanced redox reaction equations and give standard cell potentials for those that will occur.
(a) $Fe^{3+}(aq)$ and $Br^-(aq)$
(b) $Fe^{3+}(aq)$ and $I^-(aq)$
(c) $Fe^{2+}(aq)$ and $Br_2(aq)$
(d) $Fe^{2+}(aq)$ and $Ag^+(aq)$

10.33. Use the data in Appendix C to predict whether these reactions will or will not occur. Explain the basis of each prediction. Write balanced redox reaction equations and give standard cell potentials for those that will occur.
(a) $Cu^{2+}(aq) + H_2O_2(aq) \rightarrow Cu(s) + O_2(g)$
(b) $Cu(s) + NO_3^-(aq) + H^+(aq) \rightarrow Cu^{2+}(aq) + NO(g)$
(c) $MnO_2(s) + I^-(aq) \rightarrow Mn^{2+}(aq) + I_2(aq)$
(d) $Cr_2O_7^{2-}(aq) + CH_3CH_2OH(aq) + H^+(aq) \rightarrow$
$$Cr^{3+}(aq) + CH_3CHO(aq)$$

10.5. Work from Electrochemical Cells: Free Energy

10.34. Consider a cell in which the cell reaction is

$$Zn(s) + Pb^{2+}(aq) \rightarrow Zn^{2+}(aq) + Pb(s)$$

Under the conditions in the cell, the measured cell potential is 0.660 V.
(a) What is the free energy change, $\Delta G_{reaction}$, for the cell reaction under these conditions? Is this reaction spontaneous under these conditions? Explain.

(b) If 0.125 mol of Zn(s) react in the cell, what is the maximum value of the work produced by the cell? Explain the reasoning for your answer.

(c) What mass of Pb(s) is formed when this amount of work is produced? Explain.

10.35. In a copper–zinc cell (see Consider This 10.32), zinc metal is oxidized to zinc ion, so the zinc electrode loses mass when electrons are drawn from the cell. In a large copper–zinc cell, a new zinc electrode had a mass of 486.5 g and after long use was found to have a mass of 215.8 g.

(a) How many moles of electrons had the cell produced during this time? Explain.

(b) About how much electrical work had the cell produced during this time? Explain the reasoning and the assumptions you make to get your answer.

10.36. Consider a cell made by placing a silver wire, Ag(s), coated with solid silver chloride, AgCl(s), into a 1 M aqueous hydrochloric acid solution, $H^+(aq)$ and $Cl^-(aq)$, that also contains a platinum electrode, Pt(s), over which hydrogen gas, $H_2(g)$, at 1 bar pressure is bubbled. The cell potential, measured with the anode lead from the digital voltmeter connected to the platinum electrode, is 0.22 V.

(a) Write the line notation for this cell. Explain your reasoning.

(b) Write the cathodic and anodic half reactions (as reductions) and the net cell reaction for this cell. Explain your choices of reaction.

(c) Use the data above to determine the standard potential for this cell and the standard free energy change for the reaction, $\Delta G°_{reaction}$. Explain your reasoning.

(d) What are the standard reduction potentials for the half reactions? Explain.

10.37. The standard free energy change, $\Delta G°_{reaction}$, is −418.8 kJ for this redox reaction:

$$O_2(g) + 2H_2S(aq) \rightleftharpoons 2H_2O(l) + 2S(s)$$

(a) What are the half reactions (as reductions) that make up this overall reaction?

(b) What is the standard cell potential for the reaction? Explain how you get your result.

10.38. Oxalic acid, HOOCCOOH(s), is often used to determine the concentration of permanganate ion, $MnO_4^-(aq)$, solutions, using this redox reaction:

$$5HOOCCOOH(s) + 2MnO_4^-(aq) + 6H^+(aq) \rightarrow$$
$$10CO_2(g) + 2Mn^{2+}(aq) + 8H_2O(l)$$

The standard free energies of formation, $\Delta G°_f$, for $MnO_4^-(aq)$ and $Mn^{2+}(aq)$, are −425 kJ·mol^{-1}, and −223 kJ·mol^{-1}, respectively, and values for the other species are in Appendix B.

(a) Show how to calculate the standard free energy change for permanganate–oxalic acid reaction. Is the reaction spontaneous under standard conditions? Explain.

(b) Calculate the standard cell potential for this reaction. Explain your method.

(c) Show how to determine the standard reduction potential for this half reaction:

$$2CO_2(g) + 2H^+(aq) + 2e^- \rightleftharpoons HOOCCOOH(s)$$

10.39. Write the cathodic and anodic half cell reactions (as reductions) and the net cell reaction for each of these cells. Also determine the standard cell potential and standard free energy change for each cell reaction. Explain the reasoning for your answers.

(a) $Cr(s) \mid Cr^{2+}(aq) \parallel Cr^{3+}(aq) \mid Cr(s)$

(b) $Pt(s) \mid Fe^{3+}(aq), Fe^{2+}(aq) \parallel Cr_2O_7^{2-}(aq), Cr^{3+}(aq) \mid Pt(s)$

(c) $Zn(s) \mid Zn^{2+}(aq) \parallel Fe^{2+}(aq) \mid Fe(s)$

(d) $Hg(l) \mid Hg_2Cl_2(s) \mid Cl^-(aq) \parallel Hg_2^{2+}(aq) \mid Hg(l)$

10.6. Concentration Dependence of Cell Potentials: The Nernst Equation

10.40. Consider this experimental setup: A thin sheet of copper metal is immersed in a beaker containing a 1 M aqueous solution of copper sulfate. A thin sheet of iron is immersed in another beaker containing a 1 M solution of iron(II) nitrate. A salt bridge connects the two beakers and then an electrically conducting wire connects the metal sheets.

(a) Which metal sheet will lose mass and which will gain mass? Explain the reasoning for your answer.

(b) Which solution will become more concentrated in its metal ion? Explain.

(c) At 298 K, what is the initial electrical potential between the two pieces of metal? Will this potential increase, decrease, or remain constant as the reactions proceed? Explain your reasoning.

10.41. The measured cell potential is positive for this cell:

$$Ni(s) \mid Ni^{2+}(aq, 0.1\ M) \parallel$$
$$Fe^{3+}(aq, 0.1\ M), Fe^{2+}(aq, 0.1\ M) \mid Pt(s)$$

(a) What is the net cell reaction? Explain your reasoning.

(b) Will the cell potential increase, decrease, or stay the same, if the concentration of Fe^{2+} is increased? Explain your reasoning.

(c) Will the cell potential increase, decrease, or stay the same, if the concentration of Ni^{2+} is increased? Explain your reasoning.

(d) If a strip of nickel metal is immersed in an equimolar aqueous solution of Fe^{3+} and Fe^{2+} ions, what, if any, reaction would you expect to occur? If a reaction occurs, what ions would you expect to find in the solution? What would be the relative amounts of the ions? Explain.

10.42. The measured cell potential, E, for this cell is 0.225 V at 298 K:

$$Cd(s) \mid Cd^{2+}(aq, \text{??} M) \parallel Ni^{2+}(aq, 1.0 M) \mid Ni(s)$$

(a) Write the cathodic and anodic half reactions (as reductions) and the net cell reaction for this cell. Explain your choices.
(b) What is $E°$ for this cell? Explain.
(c) What is the concentration of cadmium ion, $[Cd^{2+}(aq)]$, in this cell? Explain the method you use to get your answer.
(d) What is the equilibrium constant for the net cell reaction? Explain your reasoning.

10.43. Check the appropriate column in the table to indicate the effect of each change on the potential of this cell:

$$Pt(s) \mid H_2(g) \mid H^+(aq), SO_4^{2-}(aq) \mid PbSO_4(s) \mid Pb(s)$$

In each case, explain the reasoning for your choice using the net cell reaction, Le Chatelier's principle, and/or the Nernst equation.

Change in the cell	Increase	Decrease	No effect
increase in pH of the solution			
dissolving Na_2SO_4 in the solution			
increase in size of the Pb electrode			
decrease in H_2 gas pressure			
addition of water to the solution			
increase in the amount of $PbSO_4$			
dissolving a bit of NaOH in the solution			

10.44. Check the appropriate column in the table to indicate the effect of each change on the potential of this cell:

$$Cu(s) \mid Cu^{2+}(aq) \parallel Cr_2O_7^{2-}(aq), Cr^{3+}(aq), H^+(aq) \mid Pt(s)$$

In each case, explain the reasoning for your choice using the net cell reaction, Le Chatelier's principle, and/or the Nernst equation.

Change in the cell	Increase	Decrease	No effect
decrease in pH in the Cr solution			
decrease in size of the Cu electrode			
addition of water to the Cu solution			
addition of water to the Cr solution			
dissolving $Cr(NO_3)_3$ in the Cr solution			
dissolving $K_2Cr_2O_7$ in the Cr solution			

10.45. The pH of a solution was measured by using a sample of the solution as the electrolyte in a quinhydrone half cell coupled to a silver half cell:

$$Pt(s) \mid H_2Qu(aq, c), Qu(aq, c), H_3O^+(pH) \parallel$$
$$Ag^+(aq, 0.10 M) \mid Ag(s)$$

The measured cell potential, E, is 0.256 V at 298 K.
(a) What is the standard cell potential, $E°$, for this cell? Explain.
(b) Show how to determine the pH of the solution.

10.46. Consider a cell in which the net reaction is $Zn(s) + Hg_2^{2+}(aq) \rightleftharpoons Zn^{2+}(aq) + Hg(l)$.
(a) Write the correct line notation for the cell.
(b) When the ionic concentrations are $[Hg_2^{2+}(aq)] = 0.010$ M and $[Zn^{2+}(aq)] = 0.50$ M, the cell potential is 1.51 V, measured at 298 K. Show how to use these data to calculate the standard cell potential, $E°$, and the equilibrium constant, K, for the cell reaction.
(c) If 1 g of mercury is poured into 100. mL of an aqueous 1.0 M solution of zinc sulfate, what will be the concentrations of ions in the solution at equilibrium? Explain your reasoning.

10.47. A cell potential, E, of 0.25 V was measured for this cell at 298 K:

$$Sn(s) \mid Sn^{2+}(aq, 0.10 M) \parallel$$
$$Sn^{4+}(aq, 0.010 M), Sn^{2+}(aq, 1.0 M) \mid Pt(s)$$

(a) Write the net cell reaction and corresponding reaction quotient, Q, for this cell. Explain your reasoning.
(b) Using only the data in this problem, show how to calculate the standard cell potential, $E°$, and standard free energy change, $\Delta G°_{reaction}$, for the cell reaction.
(c) What are the equilibrium constant expression and equilibrium constant, K, for the reaction? Explain.
(d) If 1 g of tin metal is added to 50. mL of an aqueous 0.005 M solution of tin(IV) ion, what will be the concentrations of tin species in the solution at equilibrium? Explain your reasoning.

10.48. The saturated calomel electrode is a half cell that is often used as a reference for measuring the reduction potentials of other half cells. [Calomel is an old name for mercury(I) chloride, Hg_2Cl_2.] The half-cell reaction and reduction potential for the electrode are

$$Hg_2Cl_2(s) + 2e^- \rightleftharpoons 2Hg(l) + 2Cl^-$$

(saturated aqueous KCl solution)

$$E = 0.241 V$$

When a saturated calomel electrode is connected to a $Co^{2+} \mid Co$ half cell in which the concentration of Co^{2+} is 0.050 M, the measured cell potential at 298 K is 0.561 V and the cobalt half cell is the anode.
(a) What is E for this $Co^{2+} \mid Co$ half cell? Show how you get your answer.

(b) Write the net cell reaction and determine the free energy change, $\Delta G_{reaction}$, for the reaction in this cell. Explain your reasoning.
(c) What are $E°$ and $\Delta G°_{reaction}$ for a $Co^{2+}|Co$ half cell? Explain how you get your result.

10.49. Mercury batteries (cells) are often used in electronics, watches, and medical devices like pacemakers. The half reactions (as reductions) in these cells can be represented as

$$ZnO(s) + H_2O(l) + 2e^- \rightleftharpoons Zn(s) + 2OH^-(aq)$$

$$HgO(s) + H_2O(l) + 2e^- \rightleftharpoons Hg(l) + 2HO^-(aq)$$

A digital voltmeter connected to a mercury cell measures a voltage of 1.35 V, when the anode input is connected to the electrode corresponding to the zinc metal.
(a) What is the cell reaction? Explain how you get your answer.
(b) Le Chatelier's principle suggests that a cell's potential should decrease as the concentrations of products builds up and reactants get depleted. Why? However, the voltage of a mercury cell is quite constant until almost the end of its useful life. Does the cell reaction you wrote in part (a) help explain this observation? Why or why not?
(c) Write the reaction quotient, Q, for the cell reaction in part (a). Does the reaction quotient help explain the cell's constant potential? Why or why not?

10.50. Consider a cell in which the net reaction is $Al(s) + Fe^{3+}(aq) \rightleftharpoons Al^{3+}(aq) + Fe(s)$. A cell potential, E, of 1.59 V was measured at 298 K for this cell when $[Fe^{3+}(aq)] = 0.0050$ M and $[Al^{3+}(aq)] = 0.250$ M.
(a) Write the net cell reaction and corresponding reaction quotient, Q, for this cell. Explain your reasoning.
(b) Using only the data in this problem, show how to calculate the standard cell potential, $E°$, and standard free energy change, $\Delta G°_{reaction}$, for the cell reaction.
(c) What are the equilibrium constant expression and equilibrium constant, K, for the reaction? Explain.
(d) The thermite reaction, $2Al(s) + Fe_2O_3(s) \rightarrow Al_2O_3(s) + 2Fe(l)$, is a spectacular reaction that produces molten iron as a product. Use your result in part (b) to estimate the free energy change for the thermite reaction. Explain your reasoning. The thermal energy required to melt iron (starting from room temperature) is about 81 kJ·mol^{-1}. Does it make sense that molten iron is produced? Explain.

10.51. Consider this electrochemical cell:

$$Mg(s) \,|\, Mg^{2+}(aq,\ 0.60\ M) \,\|\, Cu^{2+}(aq,\ 0.60\ M) \,|\, Cu(s)$$

(a) What is the cell potential for this cell at 298 K? Explain your reasoning.
(b) Suppose you connect the electrodes with a conducting wire and allow the cell reaction to reach equilibrium. What is the cell potential at equilibrium? What assumption(s) must you make about the cell contents? Explain.
(c) What is the $[Mg^{2+}(aq)]/[Cu^{2+}(aq)]$ ratio at equilibrium? Give your reasoning.

10.52. Consider this design for a zinc-chlorine cell:

$$Zn(s) \,|\, Zn^{2+}(aq,\ sat'd\ ZnCl)_2,\ Cl^-(aq,\ sat'd\ ZnCl_2) \,|\ \\ Cl_2(g,\ P) \,|\, Pt(s)$$

(a) What is the standard cell potential at 298 K for the net reaction in this cell? Explain how you get your answer.
(b) Write the reaction quotient expression, Q, for this cell reaction.
(c) Assume that solid zinc chloride, $ZnCl_2(s)$, has been dissolved to make the electrolyte solution in this cell and that some solid remains undissolved in the cell. Approximately 3 mol of $ZnCl_2(s)$ dissolve per liter of saturated solution. If chlorine gas is bubbled over the platinum electrode at 1.0 bar pressure, what is the cell potential for this cell? Explain your reasoning. (At these high concentrations, it is unlikely that calculations based on ideal solution behavior will give the correct cell potential, but it will probably be in the right direction.)
(d) How many moles of $Zn(s)$ will have to react to produce 75 kJ of work from this cell? How many moles of $Cl_2(g)$ will react? Explain.
(e) What are the concentrations of ions in the electrolyte solution after 75 kJ of work are produced by this cell? Explain. *Hint:* Remember that the electrolyte solution is saturated with $ZnCl_2(s)$.

10.53. A silver–zinc cell can be represented as

$$Zn(s) \,|\, ZnO(s) \,|\, KOH(aq,\ 40\%) \,|\, Ag_2O(s) \,|\, Ag(s)$$

(a) What is the cathodic half reaction? Explain why you write your reaction as you do.
(b) What is the anodic half reaction? Explain.
(c) What is the net cell reaction?
(d) What is the purpose of the electrolyte, the KOH solution, in this cell? Is your response consistent with the net cell reaction in part (c)? Explain.

10.7. Reduction Potentials and the Nernst Equation

10.54. Use Le Chatelier's principle to predict how the reduction potential for methylene blue reduction, $MB^+(aq) + H^+(aq) + 2e^- \rightleftharpoons MBH(aq)$, reaction (10.93), should vary as the pH of the half reaction is varied. This table gives reduction potential data for methylene blue at different pHs. Do these data confirm your prediction? Explain why you answer as you do.

pH	0	1	2	3	4	5	6	7	8	9	10
E, V	0.53	0.47	0.38	0.29	0.21	013	0.07	0.01	−0.03	−0.07	−0.11

10.55. Household bleach is an approximately equimolar solution of hypochlorite ion, OCl^-, and chloride ion, Cl^-, at about pH 8. (Sodium ion is the cation.) Addition of acid to bleach will produce hypochlorous acid, $HClO$, which is a strong oxidizing agent:

$$2HClO(aq) + 2H^+(aq) + 2e^- \rightleftharpoons Cl_2(g) + 2H_2O(l)$$

$$E° = 1.630 \text{ V}$$

(a) Labels on bleach warn you never to add acid, because "hazardous gas can be released." What is present for hypochlorous acid to oxidize? What is the hazardous gas produced? Write the net redox reaction that occurs and determine the standard free energy change for the reaction. Explain your reasoning.
(b) What is the equilibrium constant expression and equilibrium constant for the net redox reaction you wrote in part (a)?
(c) What is the standard reduction potential for the above reaction at pH 14 and 298 K? Show how you get your answer.
(d) Why do you not have to worry about the net redox reaction in part (a) when the bleach solution is basic? Show clearly why the reaction is not favored under the conditions in a bottle of bleach. The concentration of hypochlorite ion in bleach is a little less than one molar.

10.56. (a) What is the cell potential for this cell at 298 K?

$$Pb(s) \mid Pb^{2+}(aq, 1.0 \text{ M}) \parallel H^+(aq, 1.0 \text{ M}) \mid H_2(g, 1 \text{ bar}) \mid Pt(s)$$

(b) Suppose that enough sodium sulfate, $Na_2SO_4(s)$, is dissolved in the lead ion solution to make the final sulfate ion concentration 0.50 M. Lead sulfate, $PbSO_4(s)$, is not a very soluble salt, $K_{sp} = 6.3 \times 10^{-7}$. Much of the lead ion will be precipitated and the concentration left in the solution will be determined by the final concentration of sulfate ion and the solubility product. What is the concentration of lead ion in this solution? Explain how you get your answer.
(c) What is the reduction potential for the lead half cell under the conditions in part (b)? Explain.
(d) What is the cell potential for this cell under the conditions in part (b)? Explain.

10.57. The measured cell potential, E, for this cell (the Clark cell) is 1.435 V at 298 K:

$$Zn(s) \mid Zn^{2+}(aq, 1.00 \text{ M}), SO_4^{2-}(aq, 1.00 \text{ M}) \mid$$
$$Hg_2SO_4(s) \mid Hg(l)$$

Use this information, together with data from Appendix C, to calculate the standard reduction potential for this half reaction: $Hg_2SO_4(s) + 2e^- \rightleftharpoons 2Hg(l) + SO_4^{2-}(aq)$. Clearly explain your method.

10.58. Reduced nicotinamide adenine dinucleotide, NADH, absorbs light in the near ultraviolet region of the spectrum, around 360 nm. The oxidized form, NAD^+, does not absorb light at these wavelengths. A standard procedure for determining ethanol concentrations in the blood is to react the sample with NAD^+ and use a spectrophotometer to measure the amount of NADH that is formed:

$$NAD^+(aq) + CH_3CH_2OH(aq) \rightleftharpoons$$
$$NADH(aq) + CH_3CHO(aq) + H^+(aq)$$

$$E°' = -0.12 \text{ V}$$

(a) An enzyme, alcohol dehydrogenase, in the reaction mixture can catalyze this reaction in both directions. What is the standard free energy change for this reaction at pH 7? Which direction of reaction is favored under standard conditions at pH 7? Explain.
(b) The reaction mixture also contains a reagent that reacts with aldehydes but not alcohols. What is the purpose of this reagent? How does its presence affect the free energy for the reaction? Use the reaction quotient to explain your answer.
(c) How would changes in the pH of the reaction mixture affect the free energy? Is there any reason that pH change could not be used to make the reaction favorable for the analysis? (Recall the denaturing effect of pH changes on many proteins.)

10.59. A possible method for producing the work needed to power electronic devices implanted in our bodies is to implant zinc and platinum electrodes with the device. In oxygenated and ionic body fluids, the zinc metal would be oxidized at the anode and oxygen would be reduced at the cathode.
(a) Write the cathodic and anodic half reactions and the net cell equation for such a cell.
(b) What is the standard cell potential for the cell reaction in part (a) under biological conditions?
(c) If a current of 35 μA (1 μA = 10^{-6} A) is drawn continuously from such a cell, how often would a 4.5-g Zn electrode have to be replaced? Does this seem like a reasonable length of time for an implant to last before replacement or maintenance? Explain.

10.60. (a) Show how to use standard reduction potentials and standard free energies of formation to find the standard free energy changes for each of these three reactions and their sum.

(i) $NO_3^-(aq) + 4H^+(aq) + 3e^- \rightleftharpoons NO(g) + 2H_2O(l)$
(ii) $NO(g) + \frac{1}{2}O_2(g) \rightleftharpoons NO_2(g)$
(iii) $H_2O(l) \rightleftharpoons \frac{1}{2}O_2(g) + 2H^+(aq) + 2e^-$
(iv) $NO_3^-(aq) + 2H^+(aq) + e^- \rightleftharpoons NO_2(g) + H_2O(l)$

(b) What is the standard reduction potential for half reaction (iv)? Explain.

10.61. Another way to synthesize small quantities of chlorine gas for use in the laboratory (see Problem 10.31) is to drop hydrochloric acid solution, $HCl(aq)$, onto manganese dioxide, $MnO_2(s)$.

(a) Write the balanced redox reaction for this synthesis reaction.
(b) What is the standard cell potential for this reaction? Is the reaction spontaneous under standard conditions? Explain.
(c) If 6 M hydrochloric acid is used, what are the reduction potentials for the two half reactions that make up the redox reaction you wrote in part (a)? What is the cell potential for the overall reaction? Is the reaction spontaneous under these conditions? Explain.

10.8. Coupled Redox Reactions

10.62. The vanilla used as a flavoring in foods contains vanillin, an aldehyde:
(a) Most aldehydes give a positive Benedict's test. Write the equation for the reaction of Benedict's reagent with vanillin.
(b) A drop of "pure vanilla extract" from the supermarket shelf gives a positive Benedict's test. The ingredients label reads "vanilla bean extractives in water, alcohol, and corn syrup." Can you assume that the bottle contains vanillin? Why or why not?

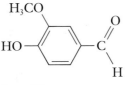

10.63. Many techniques for chemical analysis can be considered to be coupled reactions. For example, bleach (hypochlorite solution) can be analyzed using two reactions in sequence:
(i) $HClO(aq) + 3I^-(aq) + H^+(aq) \rightleftharpoons$
$$Cl^-(aq) + I_3^-(aq) + H_2O(l)$$
(ii) $I_3^-(aq) + 2S_2O_3^{2-}(aq) \rightleftharpoons 3I^-(aq) + S_4O_6^{2-}(aq)$
(Bleach cannot be analyzed directly by reaction with thiosulfate, $S_2O_3^{2-}$, because the reaction leads to a variety of products that depend on the reaction conditions.)
(a) Are both of these reactions favored in the direction written? Show how you obtain your answers.
(b) What is the coupling half reaction in this sequence? What property(ies) does this half reaction have that makes it a good coupling reaction for these redox reactions? Explain.
(c) What is the net reaction for this sequence?
(d) A 5.00-mL sample of bleach was reacted with iodide, and 32.56 mL of 0.200 M thiosulfate solution was required to react stoichiometrically with the product. What is the concentration of hypochlorite in the bleach? Show how you get your answer.

10.64. In Investigate This 6.86, Chapter 6, Section 6.11, you reacted an ammoniacal, basic solution of silver ion with an aldehyde and observed the formation of silver metal, as a mirror, on the inside surface of the test tube. The ammoniacal,

basic silver ion solution is called Tollens' reagent and the appearance of the silver mirror is a positive Tollens' test for aldehydes. If glucose, RCHO, had been used instead of the aldehyde, you would have observed the same reaction:
(i) $2Ag^+(aq) + RCHO(aq) + 3OH^-(aq) \rightleftharpoons$
$$2Ag(s) + RCOO^-(aq) + 2H_2O(l)$$
Tollens' reagent has a pH of about 12. In basic solution, silver ion precipitates as silver oxide. The reaction can be represented as
(ii) $2Ag^+(aq) + 2OH^-(aq) \rightleftharpoons$
$$Ag_2O(s) + H_2O(l) \quad K \approx 10^{16}$$
To make Tollens' reagent, ammonia is added to complex $Ag^+(aq)$ [as $Ag(NH_3)_2^+(aq)$] and keep its concentration *just low enough* that it doesn't precipitate as the oxide.
(a) What is highest concentration of silver ion, $[Ag^+(aq)]$, that can be present in the Tollens' reagent solution without precipitating by reaction (ii)? Show your reasoning.
(b) What is the reduction potential for silver ion, $E(Ag^+, Ag)$, if the silver ion is at the concentration you calculated in part (a)?
(c) What is the minimum cell potential for reaction (i), if RCHO is glucose? What is the free energy change per mole of reaction? Show your reasoning clearly.

10.65. The standard reduction potentials, $E^{\circ\prime}$, for the cytochromes in the electron transport pathway in mitochondria are given alphabetically in this table. All the half reactions involve iron(III), complexed by heme and the protein, being reduced to iron(II). What is the order of the cytochromes, from the one that oxidizes ubiquinol to the one that is oxidized by oxygen? Give the reasoning for your order.

Cytochrome	$E^{\circ\prime}$, V
a	0.29
a_3	0.55
b	0.077
c	0.254
c_1	0.22

10.66. The reduction of 3-phosphoglycerate, 3PG (a carboxylate, $RCOO^-$), to glyceraldehyde-3-phosphate, G3P (an aldehyde, RCHO), by NADPH, is required in photosynthesis:

(i)

$$O=C-O^- \\ | \\ HC-OH \\ | \\ H_2C-OPO_3^{2-}$$

3PG = $RCOO^-$

$+ NADPH + 2H_3O^+ \rightleftharpoons$

$$O=C-H \\ | \\ HC-OH \\ | \\ H_2C-OPO_3^{2-}$$

G3P = RCHO

$+ NADP^+ + 3H_2O$

NADPH is almost identical to NADH, except for an extra phosphate group (noted by the P) in NADPH. (Organisms generally use NADH as a reducing agent in energy-producing pathways and NADPH as a reducing agent in synthetic pathways.)

(a) What is the standard free energy change for reaction (i) at pH 7 and 298 K? Is the reaction favored in the direction written? Explain why or why not.

(b) An alternative pathway to reaction (i) is a sequence of two reactions that first produce 1,3-diphosphoglycerate, 1,3-DPG ($RCOO-PO_3^{2-}$) and then reduce it to G3P:

(ii) $RCOO^- + ATP^{4-} \rightleftharpoons RCOO-PO_3^{2-} + ADP^{3-}$

(iii) $RCOO-PO_3^{2-} + NADPH + H^+ \rightleftharpoons$
$$RCHO + NADP^+ + HOPO_3^{2-}$$

What is the standard free energy change for reaction (iii) at pH 7 and 298 K?

(c) What is the net reaction for the pathway in part (b)? What is the standard free energy change for the net reaction? The standard free energy change for reaction (ii) is about 19 kJ·mol^{-1}. Is the net reaction favored in the direction written? Explain why or why not.

(d) The conditions in a photosynthesizing plant cell are far from standard. Assume that they are approximately these: [ATP]/[ADP] = 10; [NADPH]/[NADP$^+$] = 10; and [HOPO$_3^{2-}$] = 10^{-2} M. Calculate the [G3P]/[3PG] ratio at equilibrium under these conditions for reaction (i) and the net reaction you wrote in part (c). Which pathway produces the greater amount of G3P at equilibrium? Explain why.

(e) Why are the reactions in part (b) called "coupled"?

10.10. Extension—Cell Potentials and Non-Redox Equilibria

10.67. Consider a half cell consisting of a silver wire placed in a solution saturated with silver thiocyanate, AgSCN(s), to which KSCN has been added to make [SCN$^-$(aq)] = 0.10 M. When this half cell is connected via a salt bridge to a standard hydrogen electrode, the cell potential at 298 K is 0.45 V with the SHE as the anode.

(a) What is the concentration of silver ion in the silver half cell? Explain your reasoning.

(b) What is the solubility product, K_{sp}, for AgSCN(s). Explain.

10.68. Consider a half cell that contains a metal electrode, an insoluble salt of the metal cation, and a solution of the anion of the salt, for example, Ag(s), AgCl(s), and Cl$^-$(aq). The half reaction and half cell can be written as

$$AgCl(s) + e^- \rightleftharpoons Ag(s) + Cl^-(aq)$$

and

$$Ag(s) \mid AgCl(s) \mid Cl^-(aq)$$

(a) Show that this half reaction is the sum of the silver ion-silver half reaction and the solubility reaction for silver chloride.

(b) Calculate the standard free energy changes for the two reactions you added in part (a). From Table 9.4, the solubility product, K_{sp}, for silver chloride is 1.8×10^{-10}. What is the standard free energy change for the half reaction above? Explain.

(c) What is the standard reduction potential for the half reaction above? How does your result compare with the value given in Appendix C?

10.69. The measured cell potential, E, for this cell is 0.608 V at 298 K:

$$Ag(s) \mid AgBr(s) \mid Br^-(aq, 0.050\ M) \parallel$$
$$Ag^+(aq, 0.200\ M) \mid Ag(s)$$

(a) What is the solubility product for AgBr(s)? Explain your reasoning. *Hint:* See Problem 10.68.

(b) Show how to calculate the standard reduction potential for this half reaction:

$$AgBr(s) + e^- \rightleftharpoons Ag(s) + Br^-(aq)$$

10.70. A silver–silver concentration cell was set up with 10.0 mL of 0.050 M silver ion, Ag$^+$, in each half cell. The potential of this cell was 0.000 V at 298 K. After 0.500 g of sodium thiosulfate pentahydrate, Na$_2$S$_2$O$_3$·5H$_2$O, was dissolved in one of the half cells and thoroughly mixed, the cell potential was 0.618 V and the half cell to which the thiosulfate was added was the anode. Silver ion forms a complex with thiosulfate:

$$Ag^+(aq) + 2S_2O_3^{2-}(aq) \rightleftharpoons Ag(S_2O_3)_2^{3-}(aq)$$

(a) Write the line notation for this concentration cell. Explain.

(b) What is the equilibrium constant for complex formation? Show how you get your answer.

(c) Show how to calculate the standard reduction potential for this half reaction:

$$Ag(S_2O_3)_2^{3-}(aq) + e^- \rightleftharpoons Ag(s) + 2S_2O_3^{2-}(aq)$$

10.71. Plot the cell potential for this cell at 298 K as a function of pH in the range 0 to 14.

$$Pt(s) \mid H_2(g,\ 1\ bar) \mid H^+(aq,\ pH) \parallel$$
$$Cl^-(aq,\ 1.00\ M) \mid AgCl(s) \mid Ag(s)$$

(a) Does your plot look the way you expected? Explain why or why not.

(b) If the cell potential is 0.435 V, what is the pH in the hydrogen half cell? Explain.

10.72. (a) Write the Nernst equation for the reduction of methylene blue, reaction (10.93).

(b) Plot the data in Problem 10.54 to find out whether they obey the Nernst equation you wrote for methylene blue reduction. Under what conditions, acidic or basic, are the data most consistent with the Nernst equation? Explain why you answer as you do.

(c) The structure of the reduced form of methylene blue shows that the molecule has two amine groups that are proton acceptors, that is, the reduced form is a base. The oxidized form is a very much weaker base. Why? Do these properties help explain your plot? Show why or why not.

10.73. You can think of solutions in the mitochondrial matrix and intermembrane space as the solutions in the two half cells of a concentration cell that depends on pH (such as two $H^+|H_2$ half cells) connected by a salt bridge. The electrical potential difference between the two solutions is the same as it would be between two actual half cells.
(a) If the pH in the matrix is one unit higher than the pH in the intermembrane space, what is the potential difference (cell potential) between them? Explain your method.
(b) What is the free energy change equivalent to this potential difference?
(c) What should be the spontaneous direction of transfer of hydronium ions in this system? Are the cell potential in part (a) and the free energy change in part (b) consistent with this direction? Explain why or why not.

General Problems

10.74. The Daniell or gravity electrochemical cell shown in this diagram was widely used to provide electricity for doorbells and railroad telegraph offices in the 19th century. A less dense aqueous solution of zinc sulfate, $ZnSO_4$, floats on a more dense aqueous solution of copper sulfate, $CuSO_4$, and serves to keep the half-cell solutions separated. Ions can still diffuse across the interface to prevent charge separation. The zinc electrode is the anode in this cell; the negative sign shows that electrons leave the cell at this electrode.

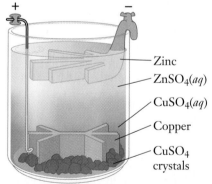

- Zinc
- $ZnSO_4(aq)$
- $CuSO_4(aq)$
- Copper
- $CuSO_4$ crystals

(a) What is the cell reaction? If the measured potential difference between the electrodes is 1.15 V, how much work can the cell do for one mole of cell reaction?
(b) What changes in cation concentrations occur as the cell is used? What ion(s) probably migrate across the interface to maintain charge neutrality in the solutions? What is the purpose of the crystals of $CuSO_4$ at the bottom of the container?
(c) The voltage of the cell slowly decreases as it is used. What do you think might cause the decrease in electrical potential? Consider Le Chatelier's principle and the changes that occur as the cell is used.

10.75. The standard cell potential, $E°$, for the Daniell cell, Problem 10.74, is 1.10 V.
(a) As it is usually set up, a fresh cell has a cell potential of about 1.15 V. What is the concentration ratio, $\dfrac{[Zn^{2+}(\text{upper layer})]}{[Cu^{2+}(\text{lower layer})]}$, for this cell? Show how you get your answer.

(b) The solid copper sulfate pentahydrate, $CuSO_4 \cdot 5H_2O$, at the bottom of the container (usually called a battery jar) keeps the lower solution saturated with copper sulfate. The density of this solution is about 1.27 $g \cdot mL^{-1}$. What is the concentration (molarity) of Cu^{2+} in this solution? What is the concentration of Zn^{2+} ion in the upper solution of a fresh cell? Explain how you get your answers.
(c) Assume that the volumes of the solutions in the upper and lower layers of the cell are both about 2.5 L. If the data from Problem 10.35 apply to this cell, what is the concentration of Zn^{2+} ion in the upper solution of the used cell? Show your reasoning.
(d) What is the cell potential for the used cell? Explain. Can you think of any reason(s) why the cell might no longer be useful?

10.76. It seems like a lot of trouble to construct a salt bridge to connect two half reactions. Why not simplify the system by setting up one of the half reactions in a tube with a porous glass disc at the end (readily available from scientific suppliers), the other in a beaker, and immersing the porous end of the tube in the solution in the beaker, as shown in this diagram? This will allow migration of ions through the porous disc with no need for a salt bridge.

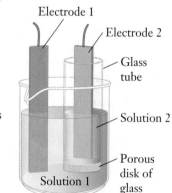

Electrode 1
Electrode 2
Glass tube
Solution 2
Porous disk of glass
Solution 1

(a) What advantages can you think of for this arrangement? What disadvantages? What happens when species migrate between the solutions?
(b) Can you think of reaction(s) for which this setup would work well?
(c) Can you think of reaction(s) for which this setup would not work well?

10.77. Explain how the Al–air cell, Worked Example 10.29, can be thought of as a fuel cell. An effort is being made to develop this cell into the power source for cell phones and other electronic devices. The cell would never need charging, but you would have to carry a supply of fuel, in case the cell ran out. Would it be safe to carry around a supply of fuel for the cell? Explain why or why not.

CHAPTER

. . . the ultimate objective of a kinetic study is to arrive at a reaction mechanism.

REACTION KINETICS
KEITH J. LAIDLER (1916–)

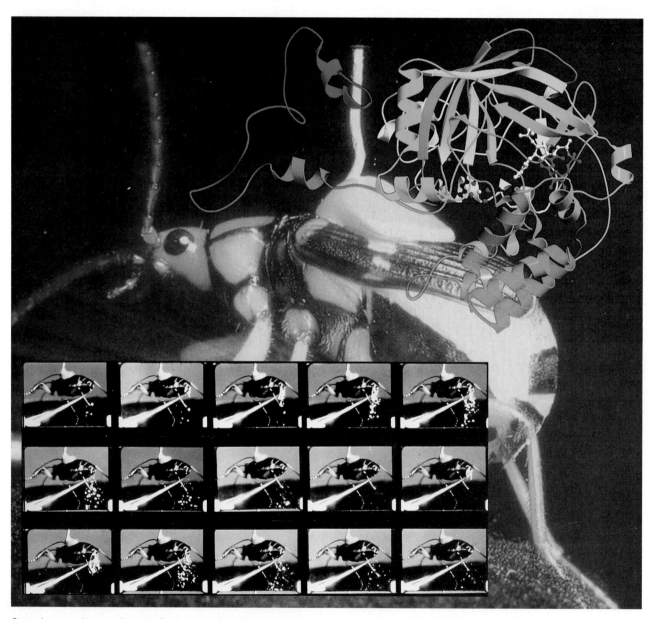

Superimposed on an image of a bombardier beetle is part of a stroboscopic movie showing the beetle emitting its defensive spray (the tiny white droplets in some frames). The time interval between these pictures is 0.25 ms (1/4000th of a second). The chemistry of the beetle's defense involves hydrogen peroxide reactions and decomposition. The molecular structure is a protein, the enzyme catalase, which catalyzes decomposition of hydrogen peroxide to water and oxygen. The active enzyme has four of these molecules stacked together.

Reaction Pathways

The series of stroboscopic pictures on the facing page represent a time study of the way **bombardier beetles** defend themselves from predators. Within a millisecond of being provoked, the beetle sprays a hot (100 °C), foul-smelling mist at its attacker. The discharge of spray is accompanied by an audible "pop," which also startles the attacker. The challenge for chemists (and biologists) was to discover what chemical reaction (or reactions) produces the energy required to heat the beetle's spray to the boiling point of water and how this reaction can occur so quickly. The energy question is one that can be answered by the thermodynamics discussed in Chapter 7. The question of speed requires an understanding of reaction pathways, or mechanisms, as they are called in the opening quotation. These pathways are the topic of this chapter.

The reactants the beetle uses are common in aerobic organisms, hydrogen peroxide, H_2O_2, and hydroquinones. In most parts of a cell, hydrogen peroxide can do damage to other molecules, so an enzyme called **catalase** (the structure shown on the facing page) catalyzes its decomposition to water and oxygen before it can do any harm. Catalase is found in almost all organisms. The decomposition of hydrogen peroxide is quite exothermic (as you found in Chapter 7, Section 7.7) and, if it takes place fast enough, can heat an aqueous reaction solution to its boiling point. We shall investigate the pathway required to produce this speed.

Studying the change of systems as a function of time is a part of all sciences. Physicists study acceleration; biologists study the growth and decline of populations; geologists study the motion of continents; and chemists study chemical reactions—the changes of one set of substances into another. Each discipline measures *rates,* the quantitative time dependence of the changes they are interested in studying. These studies usually lead to *rate laws* that describe the changes and their rates as functions of the variables that affect the rates, such as the concentrations of the reactants in the bombardier beetle. Chemists who study rates use rate laws and other information from the study of reactions to understand the pathways by which reactions occur, so that they can control reactions and design useful new reactions.

In chemistry, the study of reaction rates and reaction pathways is often called **kinetics,** as you find in Laidler's quotation that opens the chapter. When chemists speak about the kinetics of a reaction, they mean the combination of rate(s) and pathway(s) that lead to the observed chemical changes. Kinetics, like rate, reminds us of change as a function time.

> Recall that when we introduced the kinetic–molecular model of gases in Chapter 7, we noted that *kinetic* and related terms (from Greek *kineein* = to move) also appear in many other contexts: Kinetic energy is the energy associated with motion. Here kinetics is a term for the study of the motion of molecules from reactants to products.

11.1. Pathways of Change

11.1 INVESTIGATE THIS

What happens when phenolphthalein and hydroxide mix?

(a) Work in small groups to investigate this question and analyze the results. You have a capped vial containing about 5 mL of either a 1.5 or 0.75 M aqueous solution of sodium hydroxide, NaOH, and a thin-stem plastic pipet containing a 0.1% (3×10^{-3} M) solution of phenolphthalein acid–base indicator. *CAUTION:* Solutions of hydroxide are caustic and can harm skin and clothing. Take care not to spill any of the solution; work over a paper towel to catch any drops that may escape. Observe and record any changes that occur when you uncap the vial, add *one drop* of the phenolphthalein solution, recap the vial, and swirl to mix the base and indicator. Observe the mixture for a minute or two until you think no more change will occur.

(b) Repeat the addition of *one drop* of phenolphthalein and your observations to see if they are reproducible.

11.2 CONSIDER THIS

What reactions occur when phenolphthalein and hydroxide mix?

(a) What did you observe when phenolphthalein and hydroxide solutions were mixed in Investigate This 11.1? Were the changes reproducible? Explain how they were or were not.

(b) Were any of the observed changes what you expected? Why did you expect them? Were any of the observed changes surprising? Why were you surprised?

(c) How did the observations in the 1.5 and 0.75 M NaOH solutions compare? How do you interpret the similarities and differences?

You have observed reactions that occur rapidly and others that take more time. For example, in our analysis of the Blue-Bottle reaction, Chapter 10, Section 10.8, the observations could be explained by competition between two reactions: oxidation of reduced methylene blue by oxygen and reduction of oxidized methylene blue by glucose. The oxidation is very rapid; reduced methylene blue is reoxidized so fast that we can detect no change in its concentration (blue color) until the oxygen in the solution is essentially all gone. Then the slower reduction reaction can compete favorably and we see the blue solution fade to colorless. Evidently the concentrations of various reactants can make a difference in the reactions we observe, and this is one of the factors we study to understand reaction pathways.

Effect of reactant concentration on rate In Investigate This 11.1, you observed that phenolphthalein turns red the instant it is mixed with a solution containing hydroxide ion:

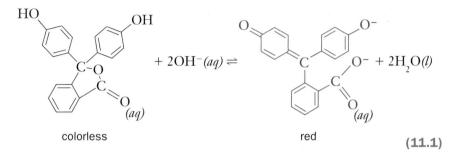

colorless   red

(11.1)

Reaction (11.1) is an acid–base proton transfer reaction. The instantaneous color change from colorless to red is your experimental evidence that the reaction is very fast. Experiments requiring specialized instruments have shown that proton transfer reactions are about as fast as reactions in solution can be. These reactions occur almost every time the proton donor (acid) and proton acceptor (base) meet one another in the solution. The **rate** of a chemical reaction is a measure of how rapidly the products are formed or the reactants disappear, so knowledge of the rate of a reaction can help us understand the molecular-level pathway of the reaction, as this proton transfer example shows. In Section 11.2, we will examine how rates are measured and quantified.

The proton transfer reaction is so rapid that you cannot detect any difference between the reactions in different base concentrations, but you observed a slower change in the mixture of phenolphthalein and hydroxide ion as the initial red color of the solution faded to colorless in a minute or two. Addition of more phenolphthalein restored the red color, which again faded to colorless. Apparently the red form of phenolphthalein undergoes some reaction that produces a colorless product or products. You also found that the color disappeared more quickly when the hydroxide concentration was higher. This result suggests that hydroxide ion is involved in a reaction with phenolphthalein that produces a colorless product. Since the hydroxide ion and the red form of phenolphthalein are both negatively charged, they repel one another and only infrequently come close enough to react. Thus, it makes sense that a reaction between them would be slow, but faster when there are more hydroxide ions, so that the number of possible interactions between the reactants is increased. A reaction that is consistent with these observations and interpretations is

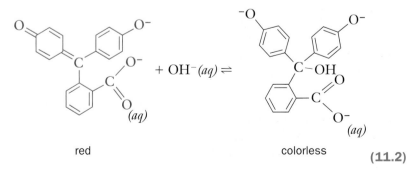

red   colorless

(11.2)

═══════ **11.3 CONSIDER THIS** ═══════

Are our pathways consistent with the phenolphthalein structure?

(a) One Lewis structure (without nonbonding electrons) for the red form of phenolphthalein in basic solution is shown in reactions (11.1) and (11.2). Draw another Lewis structure that has the same energy but with one of the negative charges on a different oxygen atom. Can you draw more than one such structure?

(b) Recall that when electrons in a molecule have a larger volume to move in, the energy of the molecule is lower. The electron energy levels will also be closer together and the molecule will be able to absorb electromagnetic radiation (light) at lower energies (longer wavelengths). How do these ideas and your structures from part (a) help to explain the color change for phenolphthalein when it is dissolved in basic solution?

(c) Compare the structures of the colorless form of phenolphthalein in reaction (11.1) and the colorless product in reaction (11.2). Does the comparison help you explain why the product of reaction (11.2) is colorless? Why or why not?

(d) Another Lewis structure for the red form of phenolphthalein is shown here. Does this structure help you visualize the reaction between the red form of phenolphthalein and hydroxide ion as the reaction of an electrophile with a nucleophile? Explain. If necessary, review Chapter 6, Sections 6.7 and 6.8.

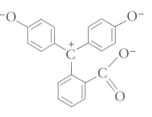

(e) Is the Lewis structure in part (d) of lower, higher, or the same energy as those you drew in part (a)? Explain your answer.

(f) All the possible Lewis structures you can draw for a molecule contribute to the overall properties of the molecule. Higher energy structures contribute less to the properties than lower energy structures. Do these ideas and your answer to part (e) add to our explanation for why the reaction of the red form of phenolphthalein with hydroxide might be slow? Explain.

🧤 **11.4 INVESTIGATE THIS**

Does temperature affect chemical reactions?

Do this as a class investigation and work in small groups to discuss and analyze the results. In two clear plastic cups or beakers, prepare two water baths: a room temperature bath and a bath at about 35 °C. Place a capped plastic vial

continued

containing about 5 mL of a 1.0 M aqueous solution of sodium hydroxide, NaOH, in each of the baths and let them sit for a few minutes to reach the temperature of the bath. *CAUTION:* Solutions of hydroxide are caustic and can harm skin and clothing. Take care not to spill any of the solution; work over a paper towel to catch any drops that may escape. Uncap each vial, add *one drop* of a 0.1% solution of phenolphthalein, recap, invert to mix, and replace in the temperature bath. Record how long it takes for the red phenolphthalein color to disappear in each sample. Repeat the addition of phenolphthalein solution if you wish to test the reproducibility of your results.

11.5 CONSIDER THIS

How does temperature affect chemical reactions?

(a) Did temperature affect the rate of the decolorizing reaction between phenolphthalein and hydroxide ion in Investigate This 11.4? If so, did the rate of reaction increase or decrease as temperature increased? Explain the reasoning for your answer.

(b) Did you expect to see any effect of temperature on the rate of the acid–base reaction when phenolphthalein was first added to the basic solution? Explain why or why not.

Effect of temperature on rate Almost all chemical reactions go faster at higher temperatures. This is what you found for the decolorizing reaction of phenolphthalein with hydroxide in Investigate This 11.4. Because increasing the temperature of a system increases its energy, we can probably conclude that the energy of a reacting system plays a role in the rate of reaction. Knowing what this role is will help us relate the temperature dependence of reactions to the possible pathways they take from reactants to products. What other factors might affect reaction rates?

11.6 INVESTIGATE THIS

How do added species affect the decomposition of H_2O_2?

Do this as a class investigation and work in small groups to discuss and analyze the results. Add about 10 mL of 3% ($\approx$ 1 M) aqueous hydrogen peroxide solution to each of nine small test tubes containing, respectively, about 0.5 g of (1) nothing, (2) raw potato, (3) raw apple, (4) uncooked beef or liver, (5) dried baker's yeast, (6) solid potassium iodide (KI),(7) solid potassium chloride (KCl), (8) solid ferric chloride or nitrate [$FeCl_3 \cdot 6H_2O$ or $Fe(NO_3)_6 \cdot 9H_2O$], and (9) solid cupric sulfate ($CuSO_4 \cdot 5H_2O$). Record what you observe happening in each test tube. Gently touch each test tube to see if there is any evidence for energy transfer to or from the mixture.

11.7 CONSIDER THIS

What is the effect of added species on the decomposition of H_2O_2?

(a) Did you observe evidence for chemical reactions occurring in any of the mixtures in Investigate This 11.6? Which mixtures? Were there any patterns in your observations?

(b) What reaction(s) do you think occurs in these mixtures? Is(Are) the reaction(s) the same in all cases? Do you have any information from investigations in previous chapters to help answer these questions? Explain the reasoning for your answers.

Effect of catalysts on rate In Investigate This 11.6, you observed that 3% hydrogen peroxide alone, test tube (1), shows no evidence of reaction to form new substances. In several of the other test tubes, you observed the evolution of gas (bubbles) and perhaps changes in temperature or changes in color not attributable to the colored solids. Let's restrict our attention to the formation of gas, which occurred in several of the systems.

In Chapter 7, Section 7.7, we investigated the decomposition of hydrogen peroxide in the presence of yeast. There, we found experimentally that the gaseous product is oxygen:

$$2H_2O_2(aq) \rightarrow O_2(g) + 2H_2O(l) \tag{11.3}$$

The gas evolved by the samples in Investigate This 11.6 is oxygen. In all cases, the reaction is the same, the decomposition of $H_2O_2(aq)$. However, the reaction rate depends on the presence of other substances in the reaction mixture. With nothing added to the solution, the decomposition is too slow to observe on a time scale of a few minutes. When certain other substances are added to the solution, you observed rapid hydrogen peroxide decomposition.

What causes the difference in rate with the added ingredients? Evidently, something in the other ingredient in these samples increased the reaction rate and made it observable. If the only *net* change that occurs in these samples is the more rapid hydrogen peroxide decomposition, we say that the other ingredients **catalyze** reaction (11.3). Evidently, a variety of salts and biological substances can act as **catalysts** for this reaction. We will consider catalysis in more detail in Sections 11.4 and 11.10. Knowing how catalysts affect a reaction rate helps us understand the possible pathway(s) for the reaction.

11.8 CONSIDER THIS

What catalyzes hydrogen peroxide decomposition?

(a) In Investigate This 11.6, potassium iodide, KI, was one of the added substances that catalyzed $H_2O_2(aq)$ decomposition. What experimental evidence do you have that shows it was the added iodide ion, $I^-(aq)$, not the added potassium ion, $K^+(aq)$, that affected the reaction? Explain your reasoning.

continued

(b) One use of drugstore hydrogen peroxide solution is as an antiseptic (*anti* = against + *sepsis* = rotting, putrefaction). When peroxide is applied to cuts and other open wounds, it fizzes vigorously. Is this what you might expect (predict) to happen? Why or why not?

By carrying out a few simple investigations using only visual observations and comparisons, you have been able to learn a good deal about the rates of chemical reactions. Some reactions, such as acid–base proton transfers, are so fast that they seem to go instantaneously under all conditions. Other reactions, such as the electrophilic center of the red form of phenolphthalein reacting with a nucleophile, hydroxide ion, are slow enough to permit easy observation of the effects of changing reaction conditions. Still other reactions, such as the decomposition of aqueous hydrogen peroxide, go so slowly that no reaction is detectable, unless other species, catalysts, are added to the reaction mixture.

You have found that the concentration of species in a solution, the temperature of the reaction mixture, and the addition of catalysts can affect reaction rates. In order to understand how these effects are used to learn more about reaction pathways, we need to get numerical values for rates, so they can be compared quantitatively. However, the ultimate purpose of these experiments is not to generate numbers but, as the quotation at the beginning of the chapter reminds us, to learn more about how and why reactions go the way they do.

11.2. Measuring and Expressing Rates of Chemical Change

To quantify rate, we define it as the change in some variable related to the reaction as a function of time, that is, the ***extent of change per time***. We'll use the $H_2O_2(aq)$ decomposition, reaction (11.3), as our first example for quantifying the change that occurs in a chemical reaction as a function of time. Since a gas is produced in the reaction, we might choose to measure the volume of gas produced (at constant pressure) or the pressure of the gas produced (at constant volume) at various times after the reaction starts. It's easier to automate the collection of pressure data, as shown in Figure 11.1, so we will use that as our first example.

Web Companion

Chapter 11,
Section 11.2.1–6

Work through this interactive analysis of another example reaction.

① ② ③ ④

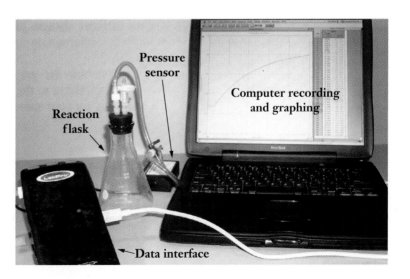

Pressure sensor

Computer recording and graphing

Reaction flask

Data interface

Figure 11.1.

Apparatus for measuring the pressure change in the $H_2O_2(aq)$ decomposition.

The reaction is carried out in a flask connected to a pressure sensor. An electrical signal from the sensor goes to a data interface where the electronics interpret the signal as a pressure in kPa. The data interface output goes to a computer where the data are stored and graphed as they are being produced. In one such experiment, we added 0.025 g of dried yeast to 15 mL of 0.175 M $H_2O_2(aq)$ in the reaction flask just before connecting it to the pressure sensor. Atmospheric pressure in the system before reaction was 102.7 kPa. Data points were taken every second, so they form a continuous curve on a graph at the scale shown in Figure 11.2.

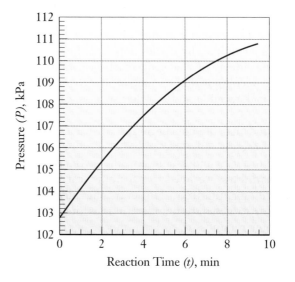

Figure 11.2.

P vs. *t* curve for $O_2(g)$ production from $H_2O_2(aq)$ decomposition.

━━━━━ **11.9 CONSIDER THIS** ━━━━━

What are the pressure changes in $H_2O_2(aq)$ decomposition?

(a) What is the pressure change (in kPa) in the 2-minute interval from 0 to 2 minutes in the reaction system described by Figure 11.2? What is the change in the 2-minute interval from 4 to 6 minutes? Explain how you get your answers.

(b) Are the pressure changes in part (a) the same or different? Do you see any problem for determining the decomposition reaction rate? Explain why or why not.

(c) 👆 Explain how the graphic in the *Web Companion*, Chapter 11, Section 11.2.2, is related to the questions in parts (a) and (b) here. Also explain the reasoning for your answer to the fill-in-the-blank box.

Rate of reaction The curve plotted in Figure 11.2 shows the increase in pressure in the reaction flask as a function of time as $O_2(g)$ is produced by $H_2O_2(aq)$ decomposition. The extent of the change in pressure, ΔP, in a given time interval, Δt, is the rate of oxygen production during that interval:

$$\text{rate of } O_2(g) \text{ production} = \frac{\Delta P}{\Delta t} = \text{slope of the } P \text{ vs. } t \text{ curve} \qquad (11.4)$$

Equation (11.4) reminds us that the rate of oxygen production is equal to the **slope** of the *P vs. t* curve. However, your results in Consider This 11.9 show that ΔP is different for the same time interval at different times during the reaction. It is obvious that the curve in Figure 11.2 is not linear but bends downward as the reaction time increases; in other words, the slope of the curve decreases as the reaction proceeds. This is shown quantitatively in Figure 11.3.

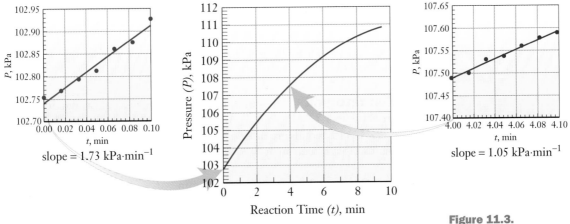

slope = 1.73 kPa·min⁻¹

slope = 1.05 kPa·min⁻¹

Figure 11.3.

Slopes of the *P vs. t* curve for $O_2(g)$ production during $H_2O_2(aq)$ decomposition. The smaller plots show the points for 0.10-minute (6-second) intervals starting at the beginning of the reaction and at 4 minutes. The slopes are for the best straight line through the experimental points.

11.10 CHECK THIS

Slopes of reaction rate curves

If we used calculus to analyze the data in Figure 11.2, we would express the rate as dP/dt, the slope of the *tangent* to the curve at any point. Explain how the animation in the *Web Companion*, Chapter 11, Section 11.2.3, illustrates this sort of analysis. How are the results related to Figure 11.3? For our purposes, small finite differences, equation (11.4), that approximate the tangent slope are completely adequate.

Initial reaction rates If the rate changes as a reaction proceeds, what rate should we use to characterize the reaction? There is no universal answer to this question. Often we choose to use the rate determined as near to the beginning of the reaction as possible. This is called the **initial reaction rate.** An advantage of this choice is that almost none of the reactants will have been used up, so it is easy to characterize the concentrations or amounts of reactant present. A more subtle advantage is that almost no products will have been formed. As products build up in a reaction, they often affect the reaction. The most obvious problem is for a reversible reaction in which the products can react to re-form the reactants. The major disadvantage of initial reaction rates is linked to the advantages. If only a tiny amount of the reactants have reacted or products been formed, measuring the small changes can be a challenge. If they can be measured, initial reaction rates are the easiest to interpret.

11.11 CONSIDER THIS

What effect does the reverse reaction have on reaction rate?

(a) If you are following a reversible reaction by measuring the formation of a product that can react to re-form the reactants, how will your measurements be affected? Will the rate of product formation appear slower or faster than if the reaction were not reversible? Explain.

(b) Do you think that reaction (11.3) might go in reverse? Why or why not? If it did, would that be a possible explanation for its observed decrease in rate with time?

Reaction rate units The initial reaction rate for $O_2(g)$ production shown in Figure 11.3 is 1.73 kPa·min^{-1}. Any convenient time units can be used to express a rate, but seconds are most often used in chemistry and biology. For the experiment we have been analyzing, we have

$$\text{rate of } O_2(g) \text{ production} = (1.73 \text{ kPa·min}^{-1})\left(\frac{1 \text{ min}}{60 \text{ s}}\right)$$

$$= 2.88 \times 10^{-2} \text{ kPa·s}^{-1} \tag{11.5}$$

Any convenient units can be used to express what is changing as a function of time in a reaction, but, when possible, we usually try to use moles, the fundamental unit of amount in chemistry.

11.12 WORKED EXAMPLE

Changing pressure units to moles

The volume of gas in the reaction flask and tubing in Figure 11.1 is 2.7×10^2 mL and the results in Figure 11.2 are for a reaction at 295 K. What is the rate of $O_2(g)$ production in mol·s^{-1}?

Necessary information: We describe gases with the ideal gas equation, $PV = nRT$, where the gas constant, R, is 8.314 kPa·L·mol^{-1}·K^{-1}. The rate of $O_2(g)$ production is 2.88×10^{-2} kPa·s^{-1}.

Strategy: To convert kPa to moles, find n/P, that is, the ratio of number of moles of gas corresponding to the pressure of the gas in the given volume. Rearrange the ideal gas equation to get this ratio, substitute known values to find a numeric value for the ratio, and then use that value to convert rate in kPa·s^{-1} to mol·s^{-1}.

Implementation: For our apparatus we have

$$\frac{n}{P} = \frac{V}{RT} = \frac{2.7 \times 10^{-1} \text{ L}}{(8.314 \text{ kPa·L·mol}^{-1}\text{·K}^{-1})(295 \text{ K})} = 1.10 \times 10^{-4} \text{ mol·kPa}^{-1}$$

continued

Converting rate in $kPa \cdot s^{-1}$ to $mol \cdot s^{-1}$ gives

$$\text{rate of } O_2(g) \text{ production} = (2.88 \times 10^{-2} \, kPa \cdot s^{-1})\left(\frac{1.10 \times 10^{-4} \, mol}{1 \, kPa}\right)$$

$$= 3.2 \times 10^{-6} \, mol \cdot s^{-1}$$

Does the answer make sense? A mole of gas at room temperature and atmospheric pressure occupies about 25 L. The volume of gas we are measuring is one-hundredth this much, and the pressure of the $O_2(g)$ produced is only a few kPa, compared to atmospheric pressure of 101 kPa. The combination of small volume and low pressure means that a total of much less than one mole of gas is produced by the reaction, so the low rate of production makes sense.

11.13 CHECK THIS

Changing pressure units to moles and assessing stoichiometry

(a) After about 35 minutes, the pressure in the reaction represented in Figure 11.2 stops increasing [no more $O_2(g)$ is produced]; the pressure in the flask remains constant at 114.5 kPa. What is the pressure of the $O_2(g)$ produced by the reaction? Explain the reasoning for your answer.

(b) Use your result from part (a) to find the number of moles of $O_2(g)$ produced by the reaction represented in Figure 11.2. Show clearly how you obtain your answer.

(c) If the decomposition reaction (11.3) goes to completion, how many moles of $O_2(g)$ would have been produced by the $H_2O_2(aq)$ in the reaction flask? Within the uncertainty of the data, what can you conclude about the reaction going to completion? Explain your reasoning.

When measuring a reaction rate, we almost always choose to follow the change that is most convenient to measure—the pressure of gas in our present example. But how do we relate what we measure to other changes in the reacting system? For example, how is the rate of production of $O_2(g)$ related to the rate of decomposition (loss) of $H_2O_2(aq)$ in the experiment we have been analyzing? To answer this question, the fundamental relationship we need is the stoichiometry of reaction (11.3), which tells us that 1 mol of $O_2(g)$ is produced for every 2 mol of $H_2O_2(aq)$ that decompose. Before we can relate the rate expressed in $(mol\ O_2\ produced) \cdot s^{-1}$ to the rate expressed in $(mol\ H_2O_2\ decomposed) \cdot s^{-1}$, we need to know that, by convention: **All rates are expressed as positive quantities.** Since $H_2O_2(aq)$ is being used up in this reaction, $\Delta(mol\ H_2O_2)$ is a negative quantity. Therefore, in the rate expression, $\Delta(mol\ H_2O_2)$ is given a negative sign so that the *rate* will be positive:

$$\frac{-\Delta(mol\ H_2O_2)}{\Delta t} = -\frac{\Delta(mol\ H_2O_2)}{\Delta t} = 2\left(\frac{\Delta(mol\ O_2)}{\Delta t}\right) \qquad (11.6)$$

━━━━━ **11.14 CONSIDER THIS** ━━━━━

How are the stoichiometries of rate expressions related?

(a) Why is the factor of two required in equation (11.6) to equate the rate expressions for $H_2O_2(aq)$ decomposition? Explain your reasoning.

(b) 👆 Explain how the concentration *vs.* time plots in the *Web Companion*, Chapter 11, Section 11.2.4, are related to the correct rate expressions. If the change in concentration of the iodide ion, I^-, could be measured in this reaction, how would the rate be expressed in terms of $\Delta[I^-]$? Explain.

Equation (11.6) makes it look like the rate for the reaction is different, depending on what product or reactant species is being considered. We need to define a single rate that characterizes a reaction, no matter what reactant or product change is measured. Consider a general reaction:

$$aA + bB \rightarrow cC \tag{11.7}$$

The rate of reaction (11.7), expressed in moles, is defined in this way:

$$\text{rate of reaction (11.7)} \equiv -\left(\frac{1}{a}\right)\frac{\Delta(\text{mol A})}{\Delta t} = -\left(\frac{1}{b}\right)\frac{\Delta(\text{mol B})}{\Delta t} = \left(\frac{1}{c}\right)\frac{\Delta(\text{mol C})}{\Delta t}$$

$$\tag{11.8}$$

For the rest of the chapter, when we refer to the **reaction rate,** we will mean rate defined so as to account for the stoichiometry of the reaction. We will still sometimes refer to the rate of formation of a product or rate of disappearance of a reactant, but to get the reaction rate, we have to account for the stoichiometry of that reactant or product in the reaction equation.

━━━━━ **11.15 CHECK THIS** ━━━━━

Rate of $H_2O_2(aq)$ decomposition

(a) What is the rate for reaction (11.3) expressed in terms of $\Delta(\text{mol } H_2O_2)$? What is the rate expressed in terms of $\Delta(\text{mol } O_2)$? Explain why you write the rate expressions as you do.

(b) For the reaction represented by Figure 11.2, what is the numeric value of the initial rate, expressed in moles? Explain your choice.

Rates expressed in molarity The rates of reactions in solution are almost always expressed as changes in molarity, rather than changes in moles. This makes sense because we want a value for the rate that can be applied to any size system: a living cell, a test tube, or an industrial-scale bioreactor. Molarity is an intensive variable (independent of the size of the system); rates expressed as $M \cdot s^{-1}$ will be independent of the amount of reaction mixture we are considering. If the volume of reacting solution is v L, the rate of hydrogen peroxide decomposition, reaction (11.3), expressed in terms of hydrogen peroxide molarity, $[H_2O_2(aq)]$, is

$$\text{rate} = -\left(\frac{1}{2}\right)\frac{\Delta[H_2O_2(aq)]}{\Delta t} = -\frac{1}{2}\left[\frac{\Delta(\text{mol } H_2O_2)}{\Delta t}\right]\left(\frac{1}{v\,L}\right) \tag{11.9}$$

11.16 WORKED EXAMPLE

Rate of $H_2O_2(aq)$ decomposition in $M \cdot s^{-1}$

What is the numeric value in $M \cdot s^{-1}$ for the initial rate of $H_2O_2(aq)$ decomposition for the reaction represented by Figure 11.2? If the rate of the reaction were the same as this initial rate for the entire reaction, how long would it take to decompose all the $H_2O_2(aq)$ in the sample?

Necessary information: The volume and concentration of the reacting solution are 0.015 L and 0.175 M, respectively. The initial reaction rate, defined in equation (11.8) and calculated in Check This 11.15, is 3.2×10^{-6} mol·s^{-1}.

Strategy: To get the rate in $M \cdot s^{-1}$, divide the rate in mol·s^{-1} by the volume of reacting solution, as shown in equation (11.9). Dividing the molarity of the solution by the rate in $M \cdot s^{-1}$ gives the time in seconds required for all the $H_2O_2(aq)$ in the sample to react, if the reaction were to go continuously at the initial rate.

Implementation:

$$\text{rate (in M·s}^{-1}) = \frac{3.2 \times 10^{-6} \text{ mol·s}^{-1}}{0.015 \text{ L}} = 2.1 \times 10^{-4} \text{ M·s}^{-1}$$

Time required for all the $H_2O_2(aq)$ in the sample to react at this rate:

$$\text{time for complete reaction} = \frac{0.175 \text{ M}}{2.1 \times 10^{-4} \text{ M·s}^{-1}} = 8.3 \times 10^2 \text{ s} \approx 14 \text{ min}$$

Does the answer make sense? We know that the reaction slows down as it proceeds, so 14 minutes will be an *underestimate* of the time the reaction actually takes to go to completion. Recall from Check This 11.12 that the complete reaction takes about 35 minutes, so our result here is consistent with the experimental observation.

11.17 CHECK THIS

Rate of $H_2O_2(aq)$ decomposition in a bombardier beetle

In one species of bombardier beetle, a 7.4 M hydrogen peroxide solution reacts in about a millisecond, 10^{-3} s. (See the chapter-opening illustration for the experimental evidence.) What is the approximate rate, in $M \cdot s^{-1}$, of the decomposition reaction in the beetle? How does this rate compare to the initial rate we measured (by analyzing Figures 11.2 and 11.3) in the presence of yeast? How might you explain any difference?

Reflection and Projection

We used just visual observations and comparisons on reacting systems to discover three factors that affect reaction rates: the concentrations of reacting species, the temperature of the reaction mixture, and the presence of catalysts— substances that increase the rate of reaction. The data necessary to determine the rate of a reaction are measurements of the change in some variable related to the reaction as a function of time. The rate of reaction is the slope of the curve on a variable *vs.* time plot of the data. We approximate the slope as the change in the variable over a small interval of time. The rate at the beginning of a reaction, the initial rate, is especially useful for analysis of the reaction because only a tiny amount of the reactants have been used up, and not enough of the products have been formed to interfere with or change the reaction of interest.

Rates are always expressed as positive quantities, so the change with time for a reactant that is used up is given a negative sign. Numeric values for the rate of a reaction can be given in any convenient units that express the amount of a product formed (or a reactant used up) per unit time of reaction. You can use the stoichiometry of the reaction to convert rates from one unit to another.

We know that changing the concentrations of reactants can change the rate of a reaction. In the next sections, we will see how to correlate these changes to discover more about the details of reaction pathways.

11.3. Reaction Rate Laws

Dependence of $H_2O_2(aq)$ decomposition rate on $[H_2O_2(aq)]$ The experiment we analyzed in Section 11.2 used yeast to catalyze $H_2O_2(aq)$ decomposition. Because yeast is a living organism, it is somewhat difficult to control exactly how much of what we add is active in catalyzing the decomposition. In order to make comparisons easier, let's analyze results from a series of $H_2O_2(aq)$ decomposition experiments catalyzed by iodide ion, $I^-(aq)$, added as potassium iodide, KI. Initial rates of reaction from plots like the one in Figure 11.3 give the data in Table 11.1. As you see, the initial reaction conditions, the concentrations of $H_2O_2(aq)$ and $I^-(aq)$, are different from one experiment to another. There are many ways to analyze data like these, but you should start out by looking for trends and patterns and then consider how you might quantify any patterns you find.

One trend that is easy to see is that the initial rates increase as you read down the table. Also, the initial concentration of $H_2O_2(aq)$, $[H_2O_2(aq)]_0$, increases

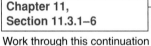

Web Companion

Chapter 11,
Section 11.3.1–6

① ② ③ ④

Work through this continuation of the interactive analysis of another example reaction.

Table 11.1 *Initial concentrations and rates of $H_2O_2(aq)$ decomposition catalyzed by $I^-(aq)$.*

Experiment	$[H_2O_2(aq)]_0$, M	$[I^-(aq)]_0$, M	$-\left(\dfrac{1}{2}\right)\dfrac{\Delta[H_2O_2(aq)]}{\Delta t}$, $M \cdot s^{-1}$
1	0.0850	0.100	0.7×10^{-4}
2	0.170	0.100	1.4×10^{-4}
3	0.250	0.100	2.1×10^{-4}
4	0.350	0.100	2.9×10^{-4}
5	0.350	0.200	6.0×10^{-4}
6	0.350	0.300	9.0×10^{-4}
7	0.350	0.400	11.7×10^{-4}

down the table for experiments 1 through 4. For these four experiments, $[I^-(aq)]_0$ is the same, so it appears that another pattern is increasing rate with increasing $[H_2O_2(aq)]$ when $[I^-(aq)]$ is constant. Looking more quantitatively at the data, note that the rate about doubles when the $[H_2O_2(aq)]$ doubles; compare experiments 1 and 2 or 2 and 4, for example. If the rate and $[H_2O_2(aq)]$ are directly proportional, then a plot of the rate *vs.* $[H_2O_2(aq)]_0$ should be a straight line. Figure 11.4 is a plot of the data from experiments 1 through 4 in Table 11.1.

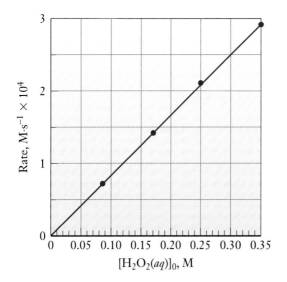

Figure 11.4.

Initial rate of $H_2O_2(aq)$ decomposition as a function of $[H_2O_2(aq)]_0$.

The plot in Figure 11.4 looks linear, and the computer-generated equation for the best straight line through the data points is

$$-\left(\frac{1}{2}\right)\frac{\Delta[H_2O_2(aq)]}{\Delta t} = \{(8.3 \times 10^{-4}\,s^{-1})[H_2O_2(aq)]\} - 0.007 \times 10^{-4}\,M\cdot s^{-1}$$

(11.10)

Within the uncertainty of the data (the rates are only reproducible to about $\pm\,0.1 \times 10^{-4}\,M\cdot s^{-1}$), the intercept, $-0.007 \times 10^{-4}\,M\cdot s^{-1}$, is 0. We can write the rate as a function of $[H_2O_2(aq)]$ as

$$-\left(\frac{1}{2}\right)\frac{\Delta[H_2O_2(aq)]}{\Delta t} = (8.3 \times 10^{-4}\,s^{-1})\,[H_2O_2(aq)]$$

(11.11)

This result makes sense: When there's no H_2O_2, its rate of decomposition is 0.

The dependence of the rate of $H_2O_2(aq)$ decomposition on $[H_2O_2(aq)]$, equation (11.11), can explain the curvature of the plot in Figure 11.2, if the decomposition catalyzed by yeast also depends on $[H_2O_2(aq)]$. As the decomposition proceeds, $[H_2O_2(aq)]$ decreases and equation (11.11) predicts that the reaction should slow down, as we found in Section 11.2. The concentration dependence predicted by equation (11.11) can be observed in a single experiment that is run far enough to use up a significant amount of peroxide.

11.18 CONSIDER THIS

How does the rate of $H_2O_2(aq)$ decomposition depend on $[I^-(aq)]$?

(a) Consider the data in Table 11.1 that show the rate of $H_2O_2(aq)$ decomposition at different initial concentrations of iodide ion, $[I^-(aq)]_0$. What patterns and relationships do you find?

(b) Does a relationship analogous to equation (11.11) relate the rate of reaction to $[I^-(aq)]$? If so, find the quantitative dependence of rate on $[I^-(aq)]$.

(c) What do your results in parts (a) and/or (b) tell you about the reaction rate when $[I^-(aq)]$ goes to zero? Does your answer make sense? Explain.

11.19 CHECK THIS

Rates of the Web Companion *reaction*

Work through the *Web Companion,* Chapter 11, Section 11.3.1–2, and then answer these questions. If necessary, review the *Companion,* Chapter 11, Section 11.2.6, for the technique used to study this reaction.

(a) Clearly explain the reasoning for your choice of the relationship between rate and Δt on the first page.

(b) What is the ratio of rates of reaction when $[(NH_4)_2S_2O_8]$ is doubled? Explain and cite specific examples.

(c) What is the ratio of rates of reaction when $[KI]$ is doubled? Explain and cite specific examples.

Rate laws Our analysis leading to equation (11.11) and your results in Consider This 11.18 show that the initial rate of decomposition of $H_2O_2(aq)$ is directly proportional to both $[H_2O_2(aq)]$ and $[I^-(aq)]$:

$$-\left(\frac{1}{2}\right)\frac{\Delta[H_2O_2(aq)]}{\Delta t} \propto [H_2O_2(aq)] \quad \text{(when } [I^-(aq)] \text{ is constant)} \qquad \textbf{(11.12)}$$

$$-\left(\frac{1}{2}\right)\frac{\Delta[H_2O_2(aq)]}{\Delta t} \propto [I^-(aq)] \quad \text{(when } [H_2O_2(aq)] \text{ is constant)} \qquad \textbf{(11.13)}$$

These results suggest that we might combine the proportionalities into a single rate equation that includes both:

$$-\left(\frac{1}{2}\right)\frac{\Delta[H_2O_2(aq)]}{\Delta t} = k[H_2O_2(aq)][I^-(aq)] \qquad \textbf{(11.14)}$$

An equation like (11.14) that expresses *the dependence of the rate of a reaction on the concentrations that affect the rate* is called a **rate law.** The proportionality constant, k, in the rate law is called the **rate constant** for the reaction. The relationships expressed in equations (11.11), (11.12), and (11.13), as well as your analysis in Consider This 11.18, are all contained in the rate law, equation (11.14).

11.20 WORKED EXAMPLE

Rate constant for $H_2O_2(aq)$ decomposition catalyzed by $I^-(aq)$

Use the numeric value in equation (11.11), which is the result from Figure 11.4, to find the numeric value of the rate constant, k, in equation (11.14).

Necessary information: We need the value 8.3×10^{-4} s^{-1} from equation (11.11) and we need $[I^-(aq)]_0 = 0.100$ M for all the reactions represented by the data in Figure 11.4.

Strategy: Comparison of equation (11.14) with equation (11.11) shows that the numeric factor in equation (11.11) can be equated to $k[I^-(aq)]_0$ for reactions with $[I^-(aq)]_0$ held constant at 0.100 M. Solve the equation for k.

Implementation: Write and solve the equation:

$$8.3 \times 10^{-4} \, s^{-1} = k[I^-(aq)]_0 = k \, (0.100 \, M);$$

$$\therefore k = \frac{8.3 \times 10^{-4} \, s^{-1}}{0.100 \, M} = 8.3 \times 10^{-3} \, M^{-1} \cdot s^{-1}$$

Does the answer make sense? Substitute this value of k and data for the experiments in Table 11.1 into equation (11.14) to find out if you successfully predict the observed rates. Also consider the units of k, $M^{-1} \cdot s^{-1}$. When multiplied by two concentration terms with units of molarity, M, the result is a value with units of $M \cdot s^{-1}$, which are the units of rate, as they must be.

11.21 CHECK THIS

Rate constant for $H_2O_2(aq)$ decomposition catalyzed by $I^-(aq)$

(a) Use the numeric value you found in Consider This 11.18(b) for the dependence of the rate of $H_2O_2(aq)$ decomposition on $[I^-(aq)]$ to find the numeric value of the rate constant, k, in equation (11.14).

(b) How does your value in part (a) compare with the value for k we got in Worked Example 11.20? Is the comparison what you expected? Explain why or why not.

All rate laws are experimental. Note that the rate law, equation (11.14), could not have been predicted or derived from the stoichiometry of reaction (10.3). The stoichiometric equation involves two molecules of hydrogen peroxide as reactants, and iodide ion does not appear in the equation at all. The rate law, however, includes the concentrations of both hydrogen peroxide and iodide ion, and there is no hint that two hydrogen peroxides are involved. This is a specific example of a general rule: *There is no necessary relationship between the stoichiometry of a reaction and the rate law for the reaction.* The stoichiometric equation for a reaction is simply a bookkeeping device to keep track of the overall change that occurs. The pathway the reactants take as they change to products determines the rate law, as we will see in Section 11.4.

Rate law for the Web Companion *reaction*

In the *Web Companion*, Chapter 11, Section 11.3.3–6, work through the determination of the rate law for the reaction. Is this rate law another example of the general rule about the relationship between rate laws and stoichiometry? Explain why or why not; be specific.

Reaction order   A further bit of nomenclature for rate laws has to do with the exponent or power to which each of the concentration terms is raised. For example, although we usually do not show exponents of unity, we could write equation (11.14) as

$$-\left(\frac{1}{2}\right)\frac{\Delta[H_2O_2(aq)]}{\Delta t} = k\left[H_2O_2(aq)\right]^1\left[I^-(aq)\right]^1 \tag{11.15}$$

The power to which the concentration of some reactant concentration in a rate law is raised is called the **order of reaction** (or **reaction order**) with respect to that reactant. In expression (11.15) [as well as in (11.14)], the reaction is first order with respect to $H_2O_2(aq)$ and also first order with respect to $I^-(aq)$. The **overall reaction order** is the sum of the orders with respect to the individual species. The decomposition of $H_2O_2(aq)$ in the presence of $I^-(aq)$ is second order overall. *The order of a reaction with respect to any of its components must be determined experimentally.*

In our example, we take the linear dependence of the rate on $[H_2O_2(aq)]$, Figure 11.4 and equations (11.10) and (11.11), as evidence that the rate depends on the first power of $[H_2O_2(aq)]$. Orders with respect to individual species may be zeroth order (no dependence on the concentration), first order (by far the most common and the one we will focus on in this chapter), or second order (dependence on the square of the concentration, which is relatively rare). Nonintegral orders are also found, but we will not introduce any such examples.

Rate law for reaction of OCl⁻(aq) with I⁻(aq) in basic solution

Hypochlorite ion, $OCl^-(aq)$, and iodide ion, $I^-(aq)$, react in basic solution:

$$OCl^-(aq) + I^-(aq) \rightarrow OI^-(aq) + Cl^-(aq) \tag{11.16}$$

The graph on the next page represents the formation of $OI^-(aq)$ as a function of time during the first second of reactions carried out in aqueous 1.0 M NaOH solution with the specified initial concentrations of $OCl^-(aq)$ and $I^-(aq)$. What rate law for reaction (11.16) can be deduced from these data?

Necessary information: The graphical results and the concentrations are all we need.

Strategy: Take the relative slopes of the concentration *vs.* time curves (lines) as a measure of the relative initial rates of reaction. Compare the ratios of

continued

relative rates to the ratios of concentrations to determine the dependence of the rate on the concentrations.

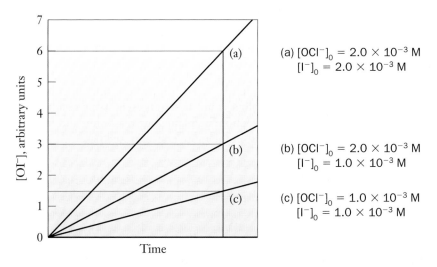

(a) $[OCl^-]_0 = 2.0 \times 10^{-3}$ M
$[I^-]_0 = 2.0 \times 10^{-3}$ M

(b) $[OCl^-]_0 = 2.0 \times 10^{-3}$ M
$[I^-]_0 = 1.0 \times 10^{-3}$ M

(c) $[OCl^-]_0 = 1.0 \times 10^{-3}$ M
$[I^-]_0 = 1.0 \times 10^{-3}$ M

Implementation: On the graph, we have drawn horizontal lines from the experimental curves to the $[OI^-]$ (vertical) axis at the same time of reaction for each curve. These horizontal lines intersect the axis at 1.5, 3.0, and 6.0, respectively, for reactions (c), (b), and (a). We can write the rate of reaction (b), for example, as

$$\text{rate (b)} = \frac{\Delta[OI^-]}{\Delta t} = \frac{[OI^-]_t - [OI^-]_0}{\Delta t} = \frac{3.0 - 0}{\Delta t} = \frac{3.0}{\Delta t}$$

The rates of the other two reactions are obtained in the same way, and since the time is the same in each case, the rates of reaction are in the ratios:

$$\frac{\text{rate (b)}}{\text{rate (c)}} = \frac{3.0}{1.5} = 2.0; \quad \frac{\text{rate (a)}}{\text{rate (b)}} = \frac{6.0}{3.0} = 2.0; \quad \frac{\text{rate (a)}}{\text{rate (c)}} = \frac{6.0}{1.5} = 4.0$$

The difference between reactions (b) and (c) is that the initial $[OCl^-]$ is doubled in (b). Since the rate is also doubled by this change, the reaction rate is directly proportional to $[OCl^-]$.

The difference between reactions (a) and (b) is that the initial $[I^-]$ is doubled in (a). Since the rate is also doubled by this change, the reaction rate is directly proportional to $[I^-]$.

With this information we can write the rate law as

$$\text{rate} = \frac{\Delta[OI^-]}{\Delta t} = k[OCl^-][I^-] \tag{11.17}$$

Does the answer make sense? If you compare reaction (a) to reaction (c), you see that both concentrations are doubled in (a) compared to (c) and the rate of (a) is four times the rate of (c). This is what the rate law predicts: If both concentrations are doubled, the rate is quadrupled. For this reaction system, we need to be careful to specify that our result applies only to basic solutions, because, if the two reactants are mixed in neutral or acidic solutions, elemental iodine, I_2, is one of the products. Evidently, $OH^-(aq)$ and/or $H^+(aq)$ ions affect the pathway of this redox reaction, as we saw for some other reactions in Chapter 10, Sections 10.7 and 10.8.

━━━━ **11.24 CHECK THIS** ━━━━

Order of reaction for [OH⁻(aq)] with phenolphthalein

(a) In Investigate This 11.1, you observed the red color of phenolphthalein in solutions of two different concentrations of base. The intensity of the color in these solutions is directly related to the concentration of the dianion whose structure is shown in equation (11.1). If you had measured the concentration of the dianion, $[phth^{2-}]$, as a function of time in your experiments, which one of these plots could best represent your results during the first few seconds of the reaction? Explain your reasoning and tell which line corresponds to your result in the more concentrated base.

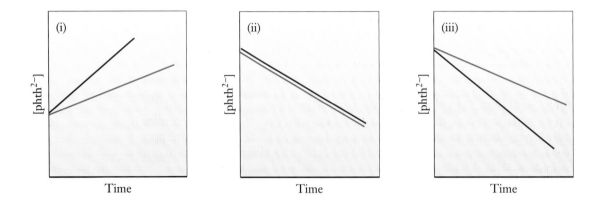

(b) How would you write the rate law for the dependence of the phenolphthalein reaction on $[OH^-(aq)]$? You will have to make an assumption about the order of the reaction with respect to $[OH^-(aq)]$. Clearly explain the reasoning for your choice.

11.4. Reaction Pathways or Mechanisms

Once we have the rate law for a reaction, what does it tell us about the pathway or mechanism of the reaction? The **reaction pathway** or **reaction mechanism** is our attempt to describe in as much detail as possible the changes in arrangements of electrons and atomic cores that lead from reactants to products. Reaction mechanisms help us understand why reactions of similar compounds can lead to different kinds of products and how to make modifications to obtain desirable products. To design more efficient and environmentally sound processes for making the materials we depend on in our daily lives, we need to know the mechanisms of the reactions involved. Knowledge of biological reaction mechanisms is critical to the search for new drugs and therapies to treat disease.

The first step in developing a reaction mechanism is determining which molecules in the system have to come together to bring about the reaction. Your first thought might be that the stoichiometry of the reaction would provide this information. But the reaction analyzed in Section 11.3, $I^-(aq)$-catalyzed decomposition of $H_2O_2(aq)$, already shows you that this isn't true. You know from that analysis that the reaction depends on the amount of iodide ion in the solution, $[I^-(aq)]$. It seems reasonable to suppose that $I^-(aq)$ must come together with the other reactant, $H_2O_2(aq)$, in order to have an effect on the reaction. But $I^-(aq)$ does not appear in the stoichiometric equation (11.3) for $H_2O_2(aq)$ decomposition.

The rate law for the $I^-(aq)$-catalyzed decomposition of $H_2O_2(aq)$, equation (11.14), *does* involve $[I^-(aq)]$ and $[H_2O_2(aq)]$. But we know that two molecules of $H_2O_2(aq)$ react for every molecule of $O_2(g)$ produced and the rate law suggests only a one-to-one meeting of $H_2O_2(aq)$ and $I^-(aq)$. Developing reaction pathways or mechanisms that can untangle apparent contradictions like these is often a challenge. We have to find a step-by-step description of the reaction that is consistent with *both* the stoichiometry and the rate law.

11.25 INVESTIGATE THIS

Do reaction changes occur in steps?

Do this as a class investigation and work in small groups to discuss and analyze the results. To each of six wells of a 24-well plate add 1.0 mL of an aqueous 0.10 M nickel sulfate, $NiSO_4$, solution. Then add the volumes of 0.50 M aqueous ethylenediamine, en, $H_2NCH_2CH_2NH_2$, and water shown in this table.

Well #	1	2	3	4	5	6
en, mL	0.0	0.2	0.4	0.6	0.8	1.0
water, mL	1.4	1.2	1.0	0.8	0.6	0.4

Use an overhead projector to project the plate and record the colors of the solutions in each well.

11.26 CONSIDER THIS

Does reaction of $Ni^{2+}(aq)$ with ethylenediamine occur in steps?

(a) How many moles of $Ni^{2+}(aq)$ ion are present in each well in Investigate This 11.25? How many moles of ethylenediamine (en) are added to each well? What is the ratio of number of moles of en to moles of $Ni^{2+}(aq)$ in each well?

(b) Based on your analysis in part (a) and observations in Investigate This 11.25, what is the stoichiometry of the reaction of $Ni^{2+}(aq)$ with en? Explain the reasoning for your answer.

(c) What, if any, evidence do you have that the reaction of $Ni^{2+}(aq)$ with en occurs in steps? Explain your reasoning.

Reactions occur in steps Let's consider why most reactions must take place in a series of steps. Formation of the metal–ligand complex (see Chapter 6, Section 6.6) between $Ni^{2+}(aq)$ (or $Ni(H_2O)_6^{2+}$) and ethylenediamine (en) is an apparently simple reaction. When solutions of $Ni^{2+}(aq)$ and excess en(aq) are mixed, reaction occurs immediately (as fast as the reactants are mixed):

$$Ni(H_2O)_6^{2+} + 3en(aq) \rightarrow Ni(en)_3^{2+}(aq) + 6H_2O(l) \qquad \textbf{(11.18)}$$

(This is a dynamic equilibrium system, but for the purposes of this discussion we are interested only in the formation of the complex.) The likelihood that the four reacting species, one $Ni(H_2O)_6^{2+}$ and three en(aq) will come together simultaneously is very small. If the reaction depended upon such meetings, it would be much slower than is observed. A series of meetings and reactions between two species at a time is much more likely. These individual steps in the overall reaction are called **elementary reactions.** Elementary reactions take place just as they are written; their rate laws can be written from their stoichiometry. For example, a likely first step (elementary reaction) in overall reaction (11.18) is

$$Ni(H_2O)_6^{2+} + en(aq) \rightarrow Ni(H_2O)_4(en)^{2+}(aq) + 2H_2O(l) \qquad \textbf{(11.19)}$$

The elementary reaction (11.19) represents an encounter (meeting) between two species to form a combined species. The probability of an encounter between two reactants in a mixture is directly proportional to the concentration of each species, so the rate of encounters depends on the product of their concentrations, that is, $[Ni(H_2O)_6^{2+}][en(aq)]$ in our example. However, as we will discuss in Section 11.6, not all encounters lead to reaction. Therefore, we have to include a proportionality constant, the rate constant, to account for the fraction of encounters that result in formation of product. In this case, the rate law for formation of the combined species is

$$\frac{\Delta\left[Ni(H_2O)_4(en)^{2+}\right]}{\Delta t} = k\left[Ni(H_2O)_6^{2+}\right]\left[en(aq)\right] \qquad \textbf{(11.20)}$$

11.27 CONSIDER THIS

What are the elementary reactions
for Ni(en)$_3^{2+}$(aq) formation?

(a) Write a series of elementary reactions of two species at a time that will add up to yield net reaction (11.18). You can think of your series of reactions as the pathway or mechanism for the reaction.

(b) Is your experimental evidence from Investigate This 11.25 consistent with the mechanism you wrote in part (a)? Explain why or why not.

Bottlenecks: Rate-limiting steps If a reaction takes place in a stepwise fashion, its overall rate cannot be any faster than the slowest elementary reaction in the sequence. A mechanical analogy may make this idea easier to understand. Consider connecting four garden hoses together to make one long hose to reach to the back of a garden: Three possible ways to connect the hoses are shown in

Figure 11.5. The sections of hose have diameters of 3/8 in., 1/2 in. (two of these), and 3/4 in. The rate of water flow in the individual hoses is two gallons per minute in the 3/8-in. hose, four gallons per minute through the 1/2-in. hoses, and eight gallons per minute in the 3/4-in. hose.

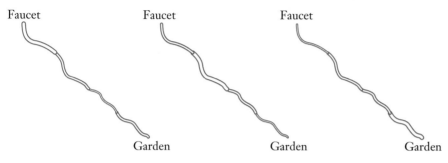

Figure 11.5.

Possible connections of four hoses to make one long hose.

11.28 CONSIDER THIS

What is the water flow through each connection in Figure 11.5?

Which connection in Figure 11.5 would give the largest flow of water to the garden? Is there another connection (not shown) that would provide an even larger flow of water? Give the reasoning for your responses.

The flow of water through the four connected hoses in Figure 11.5 is two gallons per minute, no matter where you put the smallest-diameter, 3/8-in., hose. No more water can get through the entire hose than the amount that can get through this smallest diameter section. The 3/8-in. section of hose is a **bottleneck** in the flow; it is the *rate-limiting* section of the hose. In a sequential series of elementary reactions, the overall reaction cannot be faster than the slowest step in the sequence. This slowest step is the **rate-limiting step** or **rate-determining step** in the elementary reaction series. *The rate law describes the rate-determining step.* The ions and molecules that appear in the rate law are those that must have come together before and/or during the rate-determining step. The rate law is one of the most important pieces of experimental evidence we have to work with in developing a reaction pathway or mechanism.

Other ions and molecules that take part in the reaction *after* the rate-determining elementary reaction will not appear in the rate law because these subsequent steps are faster than the rate-determining step. However, to be an acceptable mechanism, the sum of all the elementary reactions in the sequence must give the overall reaction with the correct experimental stoichiometry. This is how a reaction mechanism can be consistent with both the rate law and the reaction stoichiometry.

Applying the concept of rate-limiting steps

Let's see how these ideas about a rate-limiting step can be applied to the $I^-(aq)$-catalyzed decomposition of $H_2O_2(aq)$. Our rate law suggests that the rate-limiting step in the reaction is

Web Companion

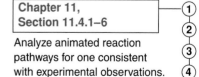

Chapter 11, Section 11.4.1–6

Analyze animated reaction pathways for one consistent with experimental observations.

an elementary reaction between $H_2O_2(aq)$ and $I^-(aq)$, since these are the species that appear in the rate law. A possible rate-determining reaction is

$$H_2O_2(aq) + I^-(aq) \rightarrow \left[H-\overset{\overset{\displaystyle \ddot{\ddot{I}}}{|}}{\underset{\displaystyle \ddot{}}{O}} \overset{\frown}{\ddot{O}} \underset{\displaystyle H}{} \right]^- \rightarrow HOI(aq) + OH^-(aq) \qquad \text{slow}$$

(11.21)

11.29 CONSIDER THIS

What are the properties of the intermediate in reaction (11.21)?

(a) What is the formal charge on each atom in the intermediate shown in reaction (11.21)? Show how you get your answers.
(b) In Chapter 6, Section 6.8, we said that "intermediates undergo electron and atomic core redistributions to reduce the number of formal charges; these redistributions lead to the observed products." Is this statement applicable to this intermediate? Show why or why not.
(c) Is reaction (11.21) a redox reaction? If so, which reactant is the reducing agent and which the oxidizing agent? Which atoms change oxidation number. Give your reasoning clearly.

The intermediate shown in reaction (11.21) probably has only transient existence and represents one stage in a continuous transition as the two reactants come together and the iodide ion displaces or substitutes for the hydroxide ion. This is a reversible reaction that favors the reactants (the reverse reaction is rapid), so only a small amount of the products are formed. However, the $HOI(aq)$ formed can react rapidly with $H_2O_2(aq)$:

$$HOI(aq) + H_2O_2(aq) \rightarrow O_2(g) + I^-(aq) + H^+(aq) + H_2O(l) \quad \text{rapid}$$

(11.22)

In neutral solution, pH 7, reaction (11.22) has an $E^{\circ\prime}$ of about 0.5 V, so the driving force, free energy change, for reaction is large. The hydronium ion, $H^+(aq)$, formed in reaction (11.22) reacts rapidly with the hydroxide ion formed in reaction (11.21):

$$H^+(aq) + OH^-(aq) \rightarrow H_2O(l) \quad \text{rapid}$$

(11.23)

11.30 CHECK THIS

Summation of elementary reactions

Show that the sum of elementary reactions (11.21), (11.22), and (11.23) is the overall reaction (11.3) for the decomposition of hydrogen peroxide to produce molecular oxygen and water. Why does $I^-(aq)$ not appear in the stoichiometric equation?

11.31 WORKED EXAMPLE

Pathway for reaction of OCl⁻(aq) with I⁻(aq) in basic solution

In Worked Example 11.23, we analyzed some rate data for this reaction:

$$OCl^-(aq) + I^-(aq) \rightarrow OI^-(aq) + Cl^-(aq) \tag{11.16}$$

Further experiments at different concentrations of hydroxide ion show that the reaction is slower at higher concentrations of base. The revised rate law is

$$\frac{\Delta[OI^-]}{\Delta t} = k_{expt} \frac{[OCl^-][I^-]}{[OH^-]} \tag{11.24}$$

A suggested mechanism for stoichiometric reaction (11.16) is

$$OCl^-(aq) + H_2O(l) \rightleftharpoons HOCl(aq) + OH^-(aq) \quad \text{rapid equilibrium, } K_{eq} \tag{11.25}$$

$$HOCl(aq) + I^-(aq) \rightarrow HOI(aq) + Cl^-(aq) \quad \text{slow, } k \tag{11.26}$$

$$HOI(aq) + OH^-(aq) \rightleftharpoons OI^-(aq) + H_2O(l) \quad \text{rapid equilibrium} \tag{11.27}$$

The sum of the series of reactions (11.25), (11.26), and (11.27) is the stoichiometric reaction (11.16). Rapid equilibrium (11.27) greatly favors the products; $OI^-(aq)$ is a weak base. Show that this pathway is consistent with the rate law and find out how k_{expt}, the rate constant derived from experiments, is related to K_{eq} for reaction (11.25) and k for the rate-limiting step (11.26).

Necessary information: We need the elementary reactions and the rate law with which to compare our predictions.

Strategy: Write the rate law for the elementary rate-limiting step (11.26). Express all concentrations in this rate law in terms of known concentrations in the solution and compare the result to the experimental rate law.

Implementation: The rate law for elementary reaction (11.26) is

$$\frac{\Delta[HOI^-]}{\Delta t} = \frac{\Delta[OI^-]}{\Delta t} = k[HOCl(aq)][I^-(aq)] \tag{11.28}$$

The rate of formation of $OI^-(aq)$ can be equated to the rate of formation of $HOI(aq)$ because reaction (11.27) is fast. $OI^-(aq)$ is produced as fast as $HOI(aq)$ is made.

HOCl(aq) is formed in the equilibrium reaction (11.25), so we write the equilibrium constant expression (in terms of molarities) and solve it for [HOCl(aq)]:

$$K_{eq} = \frac{[HOCl(aq)][OH^-(aq)]}{[OCl^-(aq)][H_2O(l)]}; \quad [HOCl(aq)] = K_{eq} \frac{[OCl^-(aq)][H_2O(l)]}{[OH^-(aq)]}$$

We have kept the concentration of water in these expressions, in order to keep track of all the reactants. Substitute this expression for [HOCl(aq)] into equation (11.28):

continued

$$\frac{\Delta[OI^-]}{\Delta t} = kK_{eq}\frac{[OCl^-(aq)][H_2O(l)]}{[OH^-(aq)]}[I^-(aq)]$$

$$= kK_{eq}[H_2O(l)]\frac{[OCl^-(aq)][I^-(aq)]}{[OH^-(aq)]} \qquad (11.29)$$

The rate law, equation (11.29), has exactly the same form as the experimental rate law, equation (11.24), with $k_{expt} = kK_{eq}[H_2O(l)]$.

Does the answer make sense? The pathway fits both the stoichiometry of the reaction and the rate law. One way to interpret the rate law is to sum all the atoms and ions that appear in the numerator and subtract those in the denominator to give the atoms and ions of the reactants in the rate-limiting step. In this case we have $(ClH_2IO_2^{2-}) - (HO^-) = (ClHIO^-)$, which is correct for the sum of HOCl and I^-, the reactants in the rate-limiting step. Note that our data cannot test whether water is actually involved in the pathway, as shown, because its concentration does not vary in these solutions.

11.32 CHECK THIS

Pathway for $I^-(aq)$-catalyzed decomposition of $H_2O_2(aq)$

Another possible pathway for the $I^-(aq)$-catalyzed decomposition of $H_2O_2(aq)$ is

$$H_2O_2(aq) + H_2O(l) \rightleftharpoons H_3O_2^+(aq) + OH^-(aq) \quad \text{rapid equilibrium, } K_{eq}$$
$$(11.30)$$

$$H_3O_2^+(aq) + I^-(aq) \rightarrow HOI(aq) + H_2O(l) \quad \text{slow, } k \qquad (11.31)$$

$$HOI(aq) + H_2O_2(aq) \rightarrow O_2(g) + I^-(aq) + H^+(aq) + H_2O(l) \quad \text{rapid}$$
$$(11.22)$$

$$H^+(aq) + OH^-(aq) \rightarrow H_2O(l) \quad \text{rapid} \qquad (11.23)$$

(a) Show that the sum of elementary reactions (11.30), (11.31), (11.22), and (11.23) is the overall reaction (11.3) for the decomposition of hydrogen peroxide to produce molecular oxygen and water.
(b) What might the intermediate structure formed in reaction (11.31) look like? Use the intermediate structure shown in reaction (11.21) to guide your thinking.
(c) What does this pathway predict for the rate law for decomposition of $H_2O_2(aq)$? Explain how you get your answer.
(d) How does your predicted rate law from part (c) compare to the experimental rate law, equation (11.14)? If they are different, do you have enough information about the reaction to choose between them? Why or why not? If not, what experiments would you propose that might provide the information necessary to decide which pathway the reaction takes? *Hint:* If you do not vary the concentration of a species to determine its effect on the rate, you cannot exclude the possibility that it takes part in the reaction.

You have now seen how a slow, rate-limiting step that is consistent with the experimental rate law for a reaction can be combined with a series of rapid steps to yield the overall observed stoichiometry of the reaction. When all the steps of a reaction mechanism are summed, intermediate products from one step are reactants in subsequent steps and ultimately cancel out to leave you only the stoichiometric reactants and products. The protonated form of hydrogen peroxide, $H_3O_2^+(aq)$, is in this category in the reaction mechanism in Check This 11.32.

Some essential reactants enter the reaction at an early stage and are formed as products at a later stage, so they cancel out of the overall stoichiometry. The iodide ion, $I^-(aq)$, falls into this category in the $I^-(aq)$-catalyzed decomposition of hydrogen peroxide. Now we can define a **catalyst** as a reactant that is essential to the pathway or mechanism of a reaction and often appears in the rate law, but not in the overall stoichiometry of the reaction. Catalysts do not act in some mysterious way but are part of the mechanism of the catalyzed reaction and behave like any other reactant.

11.33 CHECK THIS

Correlating animated reaction pathways with symbolic reactions

For each of the four proposed pathways (mechanisms) in the *Web Companion*, Chapter 11, Section 11.4.1, describe how the animation illustrates the reaction equations written to symbolize the pathway. Be as specific as possible with your descriptions.

Reaction mechanisms are tentative An important point about our proposed reaction mechanisms is that they are tentative. *We usually can't prove that a reaction mechanism we write is the only one that satisfies all the data*, though we can *reject* proposed mechanisms that don't fit all the data for the reaction of interest. Even though we can't necessarily prove that a particular mechanism is correct or unique, we use it anyway to help predict the course of reactions that haven't been tried. We also use mechanisms as a basis for modifying reactions in directions that could be useful to attain some practical goal, such as producing more of a desired product and less of an undesired by-product. If we find that our predictions don't work, then we have to go back and see where our proposed pathway might need modification, based on the new experimental results, and try again.

Reflection and Projection

Changes in the concentrations of species in a reaction mixture affect the rate of the reaction. We analyzed a few dependencies quantitatively to obtain rate laws, equations that relate the rate of a reaction to the concentrations of species in the reaction mixture. In the examples presented, the rates were directly dependent on the concentration of some of the reactants. These rate laws are first order in their dependence on these reactants. Other reaction orders you may encounter are zeroth order (no dependence on the concentration) and, very occasionally, second order (dependence on the square of the concentration). You have also seen that reaction orders may be negative (rate is inversely dependent on the concentration). Reaction orders and rate laws are determined experimentally. There is no necessary relationship between the stoichiometry of a reaction and the experimental rate law for the reaction.

Assuming that molecules, atoms, and/or ions must come together to react, we introduced the pathway or mechanism of a reaction as a sequence of such elementary reaction events. A reaction mechanism must be consistent with *both* the stoichiometry and the observed rate law for a reaction. Developing plausible reaction mechanisms is one of the hardest and most creative tasks in chemistry. We have tried to show that each of the elementary reactions we proposed is chemically reasonable, but you probably would have had difficulty developing the sequences without help. Figuring out what steps make chemical sense and how to put these together to construct a reaction pathway requires much experience with reactions and with mechanisms that others have constructed. At this stage of your study of chemistry, it's appropriate for you to be able to

- Derive a rate law from experimental rate data.
- Show that a pathway fits an experimental rate law (or other experimental evidence).
- Write a pathway that is parallel to one you already know.

In the next section, we will consider a more general way to analyze first-order reactions and then go on to use the results to investigate what we can learn from the temperature dependence of rates that can help us further characterize reaction pathways.

11.5. First-Order Reactions

Because they are simple to analyze, we have so far focused exclusively on initial rates of reactions. But it isn't difficult to analyze irreversible first-order reactions throughout the entire reaction. The results will help you better understand radioactive decay curves, such as Figure 3.14 in Chapter 3 (Section 3.4), as well as chemical reactions.

Radioactive decay In Chapter 3, Section 3.4, we introduced **radioisotopes**, elemental nuclei that spontaneously emit alpha or beta particles. Isotopes (both radioactive and nonradioactive) are used extensively in chemical and biological studies as well as in medical applications. Many of the radioisotopes used for these purposes decay to other isotopes by emitting a beta particle (electron), for example,

$$^{14}C^{6+} \rightarrow {}^{14}N^{7+} + e^- \tag{11.32}$$

$$^{3}H^{1+} \rightarrow {}^{3}He^{2+} + e^- \tag{11.33}$$

$$^{35}S^{16+} \rightarrow {}^{35}Cl^{17+} + e^- \tag{11.34}$$

All pathways for radioactive decay have in common that the rate law for decay is first order:

$$-\frac{\Delta N}{\Delta t} = k_1 N \qquad \textbf{(11.35)}$$

In equation (11.35), N is the number of unstable nuclei present at a given time, ΔN is the number of nuclei that decay in a short time interval Δt, and k_1 is the first-order rate constant for the reaction. During radioactive decay, N changes with time; but if we choose a short enough time interval, not very many nuclei will decay, and N will be nearly constant. We can rearrange equation (11.35) to

$$-\frac{\Delta N}{N} = k_1 \Delta t \qquad \textbf{(11.36)}$$

For a given time interval, Δt, the right-hand side of equation (11.36) is a constant. This means that the left-hand side also must be a constant. We interpret this relationship to mean that the *fraction* of nuclei that decay in a given time interval is the same, no matter how many nuclei are present. This is a general result *for all first-order reactions: The fraction of reactant that reacts in a given time interval is the same for a particular reaction no matter what the concentration of the reactant.* For example, if 1/1000th of some reactant reacts in the first minute of its reaction, then 1/1000th of the remaining reactant reacts in the next minute and so on throughout the reaction.

Progress of a first-order reaction with time When we analyze related pairs of data points, such as concentration as a function of time in a reaction, we usually try to express the relationship in a way that gives a linear plot of the data. Here we will show an algebraic way to do this for a first-order reaction, using radioactive decay as the example. To make the next few equations easier to write, we will make a substitution:

$$-\frac{\Delta N}{N} = f = k_1 \Delta t$$

The constant, f, is the fraction of the reactant that changes (decays) in time Δt.

Let's call the number of unstable nuclei we start with in our sample (at time zero) N_0. At the end of the first time interval, Δt, the number of nuclei that have decayed is the number we started with, N_0, times the fraction of these that decayed, f:

number of nuclei that decay in 1st time interval = $N_0 \cdot$ f

The number of nuclei left undecayed is the number we started with minus the number that decay:

N_1 = number of nuclei left after 1st time interval = $N_0 - N_0 \cdot$ f = $N_0(1 - $ f$)$

Continuing for another time interval, we get

number of nuclei that decay in 2nd time interval = $N_1 \cdot$ f

N_2 = number of nuclei left after 2nd time interval = $N_1 - N_1 \cdot$ f = $N_1(1 - $ f$)$

$$N_2 = [N_0(1 - \text{f})](1 - \text{f}) = N_0(1 - \text{f})^2 \qquad \textbf{(11.37)}$$

We have specified that Δt and, hence, ΔN must be tiny changes. In the limit of infinitesimal changes, dt and dN, we can use calculus to calculate what happens over a finite reaction time:

$$-\int_{N_0}^{N_t} \frac{dN}{N} = \int_0^t k_1 dt$$

$$-(\ln N_t - \ln N_0) = k_1(t - 0)$$

$$\ln N_t = \ln N_0 - k_1 t$$

or

$$\ln\left(\frac{N_t}{N_0}\right) = -k_1 t$$

If this mathematics is familiar to you, skip the algebraic derivation and start again at the text following equation 11.40.

====== **11.34** CHECK THIS ======

Algebra of exponential change

(a) Explain why the number of nuclei that decay in the 2nd time interval is $N_1 \cdot f$.

(b) Show how equation (11.37) is derived from the preceding equation.

We can generalize this argument to the nth time interval to get

$$N_n = N_0(1 - f)^n \tag{11.38}$$

Equation (11.38) shows how the number of nuclei varies with the number (n) of time intervals that have passed. We can convert equation (11.38) to a linear form by taking the logarithm of both sides:

$$\ln N_n = \ln N_0 + n \ln(1 - f)$$

To make this relationship even simpler, we take advantage of the fact that $\ln(1 - f) \approx -f$, if f is small. (Try it on your calculator for $f = 0.1, 0.01$, and 0.001 to see how small f must be.) Our whole analysis is based on choosing a time interval, Δt, that makes $f (= k_1 \cdot \Delta t)$ small. The approximation gives

$$\ln N_n = \ln N_0 - n \cdot f \tag{11.39}$$

If we follow the nuclear decay reaction for a time t, the number (n) of time intervals (Δt) that elapse is $n = \dfrac{t}{\Delta t}$. The number of nuclei remaining undecayed at time t, N_t, is N_n. Substituting these equivalences into equation (11.39) gives the variation of $\ln N$ with reaction time, t:

$$\ln N_n = \ln N_t = \ln N_0 - n \cdot f = \ln N_0 - \left(\frac{t}{\Delta t}\right)(k_1 \Delta t)$$

$$\ln N_t = \ln N_0 - k_1 t \tag{11.40}$$

Thus, for first-order nuclear decay, the natural logarithm of the number of nuclei remaining, $\ln N_t$, is a linear function of time.

Web Companion

Chapter 3,
Section 3.4.5–9

Review radionuclide counting and half-life in this interactive animated experiment.

① ② ③ ④

Measurement of radioactive decay For a beta-emitting isotope, the number of beta particles emitted is proportional to N_t, the number of nuclei left undecayed at time, t. To determine the number of beta-emitting nuclei that are present in a sample, we use a **scintillation counter** to count how many beta particles the sample is emitting. The sample is put in a liquid that gives off flashes of light when a beta particle hits it and the light flashes (scintillations) are counted by a photodetector. The number of counts per minute, cpm, that are detected is a measure of the amount of a radioisotope in the sample. Figure 11.6 shows plots of the data from a scintillation counter as a function of time for a sample containing ^{32}P, phosphorus-32, a radioisotope that is often used for studying nucleic acids, because it can be incorporated into their sugar-*phosphate* backbones.

Compare Figure 11.6(a) with Figure 3.14. You will see the same exponential decay of the radioactivity in both curves. Figure 3.14 is a generalized curve of percent sample remaining as a function of numbers of half-lives, which is applicable to any first-order decay. Figure 11.6(a) is a specific radioactive decay with sample data, cpm, as a function of time. The horizontal red lines mark the number of counts per minute when the fraction of radioactive nuclei present is 1.00, 0.50, and 0.25, that is, at 0, 1, and 2 half-lives, respectively. Figure 11.6(b) shows the natural logarithm of these same data, ln(cpm), plotted as a function of time. As equation (11.40) predicts, the data fall on a straight line. These data demonstrate how half-lives and first-order rate constants can be measured.

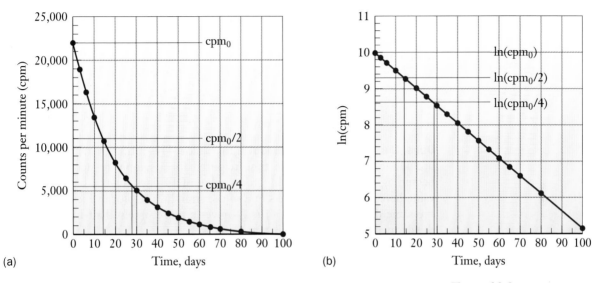

(a)   Time, days   (b)   Time, days

Figure 11.6.

Plots of scintillation counter data as a function of time for a sample containing ^{32}P. The count at time zero was 21,897 cpm. All counts are corrected for background radiation so the counts represent the ^{32}P decay only. The slope of the line in (b) is -4.85×10^{-2} day^{-1}.

11.35 CHECK THIS

Half-life for beta emission from ^{32}P

(a) Use the data in Figure 11.6(a) to find the half-life for beta emission from ^{32}P. Can you get more than one value for comparison and better precision? Show clearly how you get your answers.

(b) Is a stable isotope formed as a product of the beta decay of ^{32}P? Explain how you decide.

Half-life and first-order rate constant The slope of the line in Figure 11.6(b) is $-k_1$, where k_1 is the first-order rate constant for the beta emission reaction. We can use equation (11.40) to see how the half-life for the reaction and the rate constant are related. The time required for a sample to decay from N_0 nuclei to $N_0/2$ nuclei is a half-life, $t_{1/2}$. Substitution into equation (11.40) gives

$$\ln(N_0/2) = \ln N_0 - k_1 t_{1/2}$$

This equation can be rearranged and solved:

$$k_1 t_{1/2} = \ln N_0 - \ln(N_0/2) = \ln\left(\frac{N_0}{N_0/2}\right) = \ln(2) = 0.693 \qquad \textbf{(11.41)}$$

Equation (11.41) shows that the product of the rate constant and half-life for a first-order reaction is a constant equal to ln(2) or 0.693.

11.36 WORKED EXAMPLE

Half-life for beta emission from ^{32}P

Use the slope of the line in Figure 11.6(b) to find the half-life for beta emission from ^{32}P.

Necessary information: We need equation (11.41) and the slope of the line, -4.85×10^{-2} day^{-1}, in Figure 11.6(b).

Strategy: Substitute the known value for k_1 in equation (11.41) and solve for $t_{1/2}$.

Implementation:

$$k_1 t_{1/2} = (4.85 \times 10^{-2} \text{ day}^{-1}) t_{1/2} = 0.693;$$

$$\therefore t_{1/2} = \frac{0.693}{4.85 \times 10^{-2} \text{ day}^{-1}} = 14.3 \text{ days}$$

Does the answer make sense? In Check This 11.35, you read the Figure 11.6(a) graph to get the time required for a sample to decay to half its beginning value. You probably got a value of 14 days for the half-life, so the value here makes sense. The slope of the straight line plot in Figure 11.6(b)—the same data as for Figure 11.6(a)—gives a more precise half-life because all the decay data are used, not just two points.

11.37 CHECK THIS

First-order decay constant and radiocarbon, ^{14}C, dating

(a) The half-life for beta decay of ^{14}C is 5730 years. What is the first-order decay constant for this decay?

(b) A basket woven of twigs was found in an archaeological dig. A sample from the twigs contained 0.213 as much carbon-14 as present-day living plants. Use equation (11.40) and your result from part (a) to find the age of the basket. *Hint:* If necessary, review Chapter 3, Section 3.4, especially Worked Example 3.36 and Check This 3.37.

First-order chemical reactions Many reactions that are first order in a reactant of interest can be analyzed in exactly the same way as radioisotope decays by rewriting equation (11.40) for that reactant in terms of the dimensionless ratio of its molar concentration to the standard concentration (reactant):

$$\ln(\text{reactant})_t = \ln(\text{reactant})_0 - kt \tag{11.42}$$

A plot of the logarithm of the concentration of reactant left unreacted as a function of time will be a straight line with a slope equal to minus the numerical value of k, the first-order rate constant. For example, we can test the data for the yeast-catalyzed decomposition of $H_2O_2(aq)$ from Section 11.2 to see whether the reaction is first order in $[H_2O_2(aq)]$ (or $(H_2O_2(aq))$, which is numerically equal), as we found in Section 11.3 for the $I^-(aq)$-catalyzed decomposition of

$H_2O_2(aq)$. The yeast-catalyzed reaction is likely to depend on the amount of yeast present, but because yeast is a catalyst, we can assume that its concentration does not change during the reaction.

11.38 WORKED EXAMPLE

Rate law for yeast-catalyzed decomposition of $H_2O_2(aq)$

Data for ($H_2O_2(aq)$) as a function of time for yeast-catalyzed decomposition of $H_2O_2(aq)$ are

Time, min	0	1	2	3	4	5
(H_2O_2 (aq))	0.175	0.154	0.136	0.119	0.104	0.091

These data are calculated from the data in Figure 11.2 by using the reaction stoichiometry to convert the amount of oxygen formed to the amount of $H_2O_2(aq)$ remaining unreacted. Use a graphical method to test whether the reaction is first order with respect to ($H_2O_2(aq)$) and, if so, determine the experimental rate constant, k_{expt}, for the reaction.

Necessary information: In addition to the data in the problem, we need equation (11.42).

Strategy: Assume the reaction is first order in ($H_2O_2(aq)$) and plot $\ln(H_2O_2(aq))$ vs. t to see if the data give a straight line. If so, use the slope to find k_{expt} for the reaction.

Implementation: The data for the plot are

Time, min	$\ln(H_2O_2$ (aq))
0	−1.743
1	−1.871
2	−1.994
3	−2.125
4	−2.259
5	−2.399

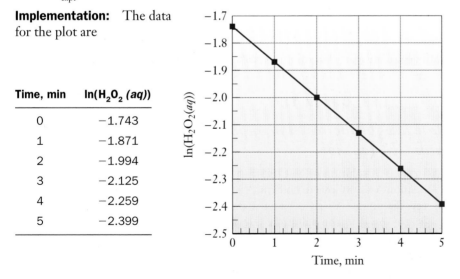

The plot is a straight line with a slope -1.31×10^{-1} min^{-1}. Thus, the reaction is first order in ($H_2O_2(aq)$) (or [$H_2O_2(aq)$]) and $k_{expt} = 1.31 \times 10^{-1}$ min$^{-1} = 2.18 \times 10^{-3}$ sec^{-1}.

Does the answer make sense? The hydrogen peroxide decomposition catalyzed by iodide is first order in [$H_2O_2(aq)$], so it is not surprising to find that the reaction catalyzed by yeast is also. Since the rate of the reaction probably

continued

depends on the amount or concentration of yeast in the solution, it's likely that the rate law has the form: rate = k[yeast][$H_2O_2(aq)$]. Thus, $k_{expt} = k$[yeast], but we have no information about [yeast], so cannot estimate the rate constant, k, in the rate law. The best we can do is compare k_{expt} with the constant in equation (11.11), 8.3×10^{-4} s^{-1}, for the iodide-catalyzed reaction. This constant also included the concentration of the catalyst. The values are comparable, so we can have some confidence that we are in the right ballpark.

Analysis of the catalyzed hydrogen peroxide decompositions is relatively easy because the only reactant concentration that is changing is [$H_2O_2(aq)$]. The other reactant concentration in the rate law equation, [catalyst], remains constant throughout the reaction. Sometimes, even when a reactant is actually consumed in a reaction, we can make its initial concentration large enough, compared to the other reactants, that its concentration changes only a negligible amount during the reaction. This technique, making one or more of the reactant concentrations so large that they change only a small amount during the reaction, is called **flooding.**

You have already studied a system flooded with one reactant, the reaction between the colored form of phenolphthalein and hydroxide ion. In the reactions you carried out in Investigations 11.1 and 11.4, you added one drop (about 0.05 mL) of millimolar concentration phenolphthalein solution to several milliliters of molar concentration hydroxide ion. While the concentration of the colored dianion of phenolphthalein decreased during the reaction, the concentration of hydroxide ion remained constant. In Check This 11.24, you used dependence on the constant initial concentrations of hydroxide ion to find the order with respect to [OH$^-$(aq)]. We can take advantage of flooding by hydroxide ion to determine whether the disappearance of the red phenolphthalein dianion is first order, as reaction (11.2) implies.

11.39 CHECK THIS

Rate law for the reaction of phenolphthalein with OH$^-$(aq) ion

(a) In a 1.0 M solution of NaOH, the decay of the color of the phenolphthalein dianion was followed colorimetrically at a wavelength absorbed by the red dianion. The absorbance data here are directly proportional to the [phth^{2-}(aq)] in the solution. Use a graphical method to find the order of the reaction with respect to [phth^{2-}(aq)]. Clearly explain your procedure.

Time, min	0	0.1	0.2	0.3	0.4	0.5	0.6	0.7	0.8
Absorbance	0.467	0.296	0.192	0.128	0.086	0.058	0.039	0.028	0.021

(b) Use your result from part (a) and your conclusion from Check This 11.24 to write a rate law for the reaction of phenolphthalein with hydroxide ion.

continued

Is your rate law consistent with reaction (11.2) being an elementary, rate-determining reaction? Explain why or why not.

(c) If it is possible, use your rate law in part (b) and the data in part (a) to determine the rate constant for this reaction. If it is not possible, explain why not and what information you lack.

Reflection and Projection

In previous sections, we emphasized kinetic analyses based on initial rates of reaction. In this section, we have expanded the scope of kinetic analyses to include data from more than the first short time interval of a reaction. We mathematically manipulated (in the language of the calculus, "integrated") the differential rate law for first-order reactions to find a graphical way to analyze rate data that would yield straight line plots with slopes related to the rate constants for the reactions. For a first-order reaction, a plot of the logarithm of the first-order reactant species as a function of time is a straight line, if the concentrations of all other species in the reaction are constant. We also related the half-life and rate constant for first-order reactions and applied this relationship to data on radioactive decays, such as radioisotope dating studies. Finally, we introduced the flooding technique that permits us to follow a reaction for a substantial time with the concentrations of most of the relevant reaction species held approximately constant.

The end result of all our approaches is a rate law that we hope will help us understand the pathway (mechanism) for the reaction of interest. We have tried to develop reaction pathways based on elementary reactions that make sense based on the chemistry that you have studied in the previous chapters. The thing that may not make much sense is the great variability in reaction rates. Reactants come together at about the same rate in any reaction (in solution). If, then, reactions occur when reactants come together, why don't all reactions go at about the same rate? The effect of temperature on reaction rate can help us answer this question.

11.6. Temperature and Reaction Rates

11.40 INVESTIGATE THIS

Are UV-sensitive beads affected by temperature?

Do this as a class investigation and work in small groups to discuss and analyze the results. UV-sensitive beads are plastic beads that change color when irradiated with an ultraviolet light (including the sun) and then return to the original color (usually white) when the source of UV light is removed. In Styrofoam® plastic cups, prepare three temperature baths: ice-water, room temperature water, and warm (about 35 °C) water. Irradiate three identical UV-sensitive beads to change their color. Simultaneously, place one colored bead in each temperature bath. Record the temperature of the bath and how long it takes each bead to return to the original color. Repeat to test whether the results are reproducible.

How are UV-sensitive beads affected by temperature?

In Investigate This 11.40, did the time required for the UV-sensitive beads to return to their original color vary with temperature? If so, at what temperature was the rate of return fastest? Slowest? How might you explain your results?

The Arrhenius equation As you found in Investigate This 11.4 and 11.40, reactions have higher rates at higher temperatures. If the products of a reaction don't change, we assume that the rate law has the same form at all temperatures studied. The only factor that can be responsible for the increasing rate with increasing temperature is an increasing rate constant. By the last quarter of the 19th century, the effect of temperature on the rates of many reactions had been studied. Enough data had been obtained to make it possible to show that the temperature dependence (temperature in kelvin) of the rate constants could be expressed in this form:

$$\ln k = (\ln A) - \frac{B}{T} \tag{11.43}$$

In equation (11.43), A and B are constants that are not functions of temperature and are different for each reaction. You will more often see equation (11.43) in its equivalent exponential form:

$$k = A e^{-\left(\frac{B}{T}\right)} \tag{11.44}$$

To convert equation (11.43) to equation (11.44), take the antilogarithm (exponential) of both sides of equation (11.43):

$$e^{\ln k} = k$$

$$e^{(\ln A) - B/T} = e^{\ln A} e^{-B/T} = A e^{-\left(\frac{B}{T}\right)}$$

Equation (11.44) is called the **Arrhenius equation** in honor of Svante Arrhenius (Swedish chemist, 1859–1927), who did much of the analysis that led to this formulation. The units of A are the same as the units of the rate constant, which depends upon the rate law for the reaction. Since an exponent (or power) can have no units, the units of B and T must cancel; therefore the units of B are kelvin, K.

What does the Arrhenius equation tell us?

(a) Show that equations (11.43) and (11.44) give increasing values of k as T increases.

(b) What value of B will give a rate constant at 303 K (30 °C) that is two times larger than the rate constant at 293 K (20 °C)?

Temperature dependence of phenolphthalein-hydroxide reaction

In an investigation like Investigate This 11.4, the times required for the same amount of phenolphthalein color to disappear in a colorimeter at 34.5, 20.4,

continued

and 1.3 °C were 47, 93, and 373 seconds, respectively, in 1.0 M solutions of NaOH. The times were reproducible within about 5–10%. What is the value of B in the Arrhenius equation for reaction 11.2?

Necessary information: The same amount of phenolphthalein dinegative ion reacted in each reaction and the concentration of base was the same in each case. Your results from Check This 11.24 and 11.39 show that the reaction is first order in both $[OH^-(aq)]$ and $[phth^{2-}(aq)]$:

$$-\frac{\Delta[phth^{2-}(aq)]}{\Delta t} = k[phth^{2-}(aq)][OH^-(aq)] \tag{11.45}$$

Strategy: Use a graphical method to obtain B as the slope of a $\ln k$ vs. $1/T$ plot by taking advantage of the fact that the *slopes* of $\ln k$ vs. $1/T$ and $\ln(ak)$ vs. $1/T$ plots are the same, if a is constant. In these reactions, $[OH^-(aq)]$ is the same in each reaction and is so large that it does not vary; the reactions are flooded with $OH^-(aq)$. Thus, equation (11.45) can be treated as a first-order equation in the disappearance of $[phth^{2-}(aq)]$, and equation (11.42) can be used to get ak and hence $\ln(ak)$.

Implementation: For this reaction, equation (11.42) can be written as

$$\ln[phth^{2-}(aq)]_t - \ln[phth^{2-}(aq)]_0 = \ln\left(\frac{[phth^{2-}(aq)]_t}{[phth^{2-}(aq)]_0}\right) = -k[OH^-(aq)]t$$

The reaction time we measure, t, is the time required for the same change in absorbance of each solution. Thus, the ratio of final to initial concentration is the same in all three reactions, so the logarithm of the ratio is also the same: $-k[OH^-(aq)]t = -$(constant). (The constant is negative because the ratio is less than one and the logarithm of a value less than one is negative.) Rearranging, we have

$$\frac{k[OH^-(aq)]}{(\text{constant})} = ak = \frac{1}{t}$$

Substituting our experimental data and doing the calculations gives the values in this table and plotted on the accompanying graph. The bars on each point represent the uncertainty introduced by the uncertainty in the time measurements.

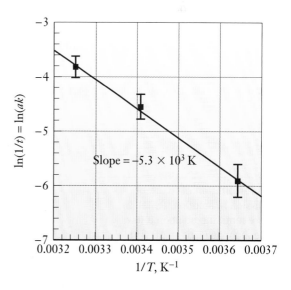

T, K	1/T, K⁻¹	t, s	ak = 1/t, s⁻¹	ln(ak)
274.5	0.00364	373	0.0027	−5.92
293.4	0.00341	93	0.0108	−4.53
307.7	0.00325	47	0.0213	−3.85

The slope of the line on the graph is $-B$, so $B = 5.3 \times 10^3$ K.

Does the answer make sense? Arrhenius found that plots of $\ln k$ vs. $1/T$ were straight lines and our data give a straight-line plot, which makes sense. However, we do not know what B means nor do we have any other values with which to compare our result. Our next task is to discover the meaning of the parameters A and B in the Arrhenius equation.

Temperature dependence of UV-sensitive bead decolorizing reaction

(a) In Investigate This 11.40, you measured the times required for the colored form of UV-sensitive beads to return to their original color at three different temperatures. What is the value of B in the Arrhenius equation for this decolorizing reaction?

(b) How does the value of B for the UV-sensitive bead decolorizing reaction compare to value for the phenolphthalein reaction with hydroxide? Which reaction is more affected by a change in temperature? Explain the reasoning for your answer.

The Arrhenius equation provides a good description of the temperature dependence of reaction rates, but leaves us with three questions: What is the meaning of A? What is the meaning of B? Why does the equation have this particular form? These questions are interrelated, but let's take them up in order.

The frequency factor, A Throughout our discussion of reaction pathways, we have said that reacting species must come together in order to react. For a gas-phase reaction, this means that the species must collide with one another. Such encounters (interactions) between the reactants are quite fleeting, and any reaction has to occur within the short time of a single collision. The situation in solutions, which are of greater interest to us, is quite different.

Reactants in solution are surrounded by solvent molecules and are constantly being jostled about by their nearest neighbors. We can picture the reactant molecules as being surrounded by a cage of solvent molecules, as in Figure 11.7. As the jostling continues, the reactant molecule sooner or later moves to a different solvent cage and thus moves about randomly in the solution. The motion of two reactant molecules with respect to one another is a diffusion process. The number of encounters between reactants in solution depends on how rapidly molecules diffuse in the solution. A is usually called the **frequency factor** because it accounts for the frequency of encounters between molecules in the solution.

Figure 11.7.

Representation of reactant diffusion in solution. Panel (b) represents the solution a few picoseconds (1 ps = 10^{-12} s) later than panel (a).

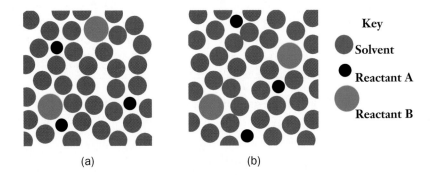

(a)   (b)

Key

● Solvent

● Reactant A

● Reactant B

The size and shape of molecules and ions affect their diffusion through a solvent. However, most species move at about the same speed, so they encounter one another with about the same frequency. For a reaction that takes place on every encounter between two reactants in aqueous solution, the frequency factor, A, is $10^9 – 10^{10}$ $M^{-1} \cdot s^{-1}$. These are said to be **diffusion controlled reactions** because they occur as rapidly as the reactants can diffuse together.

Figure 11.7(a) shows a reactant pair of molecules together in a cage of solvent molecules and Figure 11.7(b) shows that the pair has left the cage without reacting. If reaction does not occur every time a reactant pair is trapped together in a solvent cage, the rate constant for the reaction will be less than the diffusion-controlled rate constant. Reaction might not occur because, in order to react, the reactants have to come together in just the right orientation. This effect will show up as a frequency factor, A, less than the diffusion-controlled limit. Although chemists try to model and calculate these effects theoretically, it is more common to determine the frequency factor experimentally, compare it to the diffusion-controlled limit, and then try to figure out what orientation effects could make the frequency factor less than the limiting value.

> The reaction between hydronium and hydroxide ions is the fastest reaction in aqueous solution, with a rate constant of 1.3×10^{11} $M^{-1} \cdot s^{-1}$. The mechanism of diffusion of these ions is probably different. See Figure 2.26 (Chapter 2, Section 2.12) and the accompanying discussion.

11.45 CONSIDER THIS

How do frequency factors for similar reactions compare?

The frequency factors for reactions (11.46) and (11.47) studied in acetone solutions were 2.5×10^4 and 3.2×10^6 $M^{-1} \cdot s^{-1}$, respectively. How might orientation effects account for the low values compared to the diffusion-limited value? How might you account for the factor of about 100 difference between the two values? Build molecular models of the reactants, if that will help you formulate your explanations.

$$(C_2H_5)_3N\colon + C_2H_5I \rightarrow \left[(C_2H_5)_4N^+\right]I^- \text{ (a salt)} \qquad \textbf{(11.46)}$$

$$\text{\textcircled{N}}\colon + CH_3I \rightarrow \left[\text{\textcircled{N}}^+\!\!-CH_3\right]I^- \text{ (a salt)} \qquad \textbf{(11.47)}$$

For example, when a reaction requires an electrophilic center in one molecule to approach a nucleophilic center in another, the other parts of both molecules can interfere and inhibit their approach simply by being in the way. If two pairs of such reactants are compared and the interfering parts of the molecules are smaller or more compact in one pair, then we would expect the interferences to be smaller and the frequency factor to be larger for that pair. These are the sorts of effects you find in Consider This 11.45.

Activation energy for reaction Another reason a reaction might not occur when the reactants are caged together is that the reactants lack the energy to react. For example, the pathway for reaction (11.47) requires the nucleophilic electron pair on the nitrogen to begin to bond to the electrophilic carbon bonded to the iodine atom. This interaction weakens the carbon–iodine bond and leads to a partial negative charge on the iodine atom and partial positive charge on the nitrogen, as shown for the activated complex in Figure 11.8, the **activation energy diagram** for this reaction.

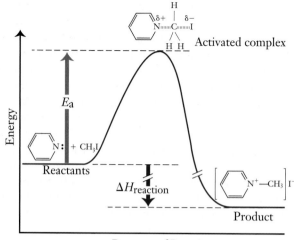

Figure 11.8.

An activation energy diagram for reaction (11.47). In order to emphasize the activation barrier, E_a and $\Delta H_{reaction}$ are not drawn to scale.

For reaction (11.47), $E_a = 58 \text{ kJ·mol}^{-1}$ and $\Delta H_{reaction} = -98 \text{ kJ·mol}^{-1}$. Both values depend somewhat on the polarity of the solvent, because uncharged polar molecules react to give ions, and there is a good deal of solvent reorganization during the reaction.

The **activated complex** is the highest energy species formed as reactants progress toward products in an elementary reaction. The nucleophilic amine and electrophilic iodide have to come together with enough energy to bring about the simultaneous making and breaking of bonds that forms the activated complex shown in Figure 11.8 for this reaction. The **activation energy, E_a,** for a reaction is the *net* energy required for the reactants to form activated complexes, which can then go on to form products (or go back to reactants).

The "progress of reaction" that we are following from left to right on an activation energy diagram is a fuzzy concept. Many different electronic and atomic core rearrangements occur during a reaction, so defining how far the changes have progressed is a complex problem. For our purposes, these details are unimportant. The energies of the reactants, activated complex, and products and a sense of continuous change as the reaction progresses are enough.

11.46 CHECK THIS

Activation energy diagrams

(a) Activation energy diagrams combine both kinetic and thermodynamic information for a reaction. Using the information in Figure 11.8, explain whether reaction (11.47) is exothermic or endothermic.

(b) Sketch the activation energy diagram for a reaction that has the opposite overall thermodynamics from that shown in Figure 11.8.

(c) If reaction (11.47) is reversible, how would the activation energy for the reverse reaction compare to the activation energy shown in Figure 11.8? Explain with a diagram.

(d) Imagine an endothermic reaction that can occur by two pathways, one catalyzed and the other not catalyzed, with two different activation energies. Sketch and label an activation energy diagram that shows the energy as a function of the progress of reaction for both pathways.

Pairs of reactants that come together with too little energy may progress part way toward products (which corresponds to being part way up the barrier in Figure 11.8) but will not be able to form the activated complex and go on to products. While caged by solvent molecules, the pair might gain enough energy from their collisions with solvent molecules to be able to form the activated complex and undergo reaction. If they don't gain enough energy, they ultimately leave the cage unreacted, as shown for the pair in Figure 11.7. Thus, the rate constant for a reaction depends on how many encounters of reactant pairs have an energy E_a or greater. To find the proportion of such encounters, we need to know about energy distributions in molecular systems.

Molecular energy distributions In Section 11.1, when we first looked at the increase in reaction rate with increase in temperature, we said that it is likely that the temperature effect is related to the energy of the reacting system, which increases with temperature. Thus we can guess that B, in the temperature dependent part of the Arrhenius equation, has something to do with energy. Since a minimum energy, E_a, is necessary for reaction to occur, it is likely that B and E_a are related. To understand the relationship, we need to examine the distribution of energy among the molecules in a system at different temperatures shown in Figure 11.9.

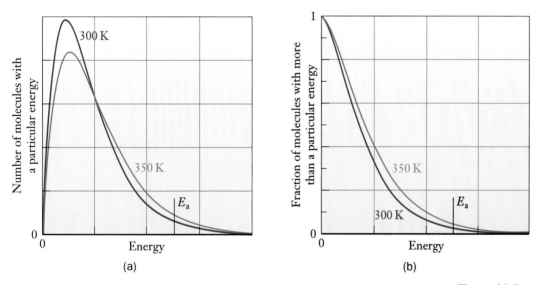

(a)

(b)

Figure 11.9.

Energy distributions for a sample of molecules at two temperatures. Graph (a) is the number of molecules with a given energy. Graph (b) is the fraction of molecules with *more* than a given energy.

For any value on the energy axis in Figure 11.9(a), the value on one of the curves is the number of molecules in the sample that have that energy. Though the two curves have somewhat different shapes, the *total* area under each curve, which is equal to the number of molecules in the sample, is the same for both temperatures. The curves show that the distribution of energies is broader at the higher temperature, that is, there are more molecules with higher energies. Therefore, the average molecular energy, which is related to the energy where the curve is a maximum, is a bit higher at the higher temperature. Figure 11.9 makes the relationship between temperature and energy of the molecules in a system quantitative.

The energy distribution shown by the plots in Figure 11.9 is called the **Maxwell–Boltzmann distribution** because James Clerk Maxwell (Scottish physicist, 1831–1879) and Ludwig Boltzmann (Austrian physicist, 1844–1906) developed the theory describing the distribution. Many kinds of experiments confirm the predictions represented in Figure 11.9. The plots in the figure are for the translational energies of molecules in a gas, but the conclusions we reach are applicable also to the energies of encounters in a solution.

An energy, E_a, is marked on the graphs in Figure 11.9. For the system represented, all encounters with energies higher than this can produce reaction. The area under each curve to the right of E_a gives the number of such encounters. The *fraction* of such encounters is given by the ratio of this area to the total area under the curve and is shown in Figure 11.9(b) and shown enlarged at the high-energy end of the plot in Figure 11.10. What you should observe from all the figures is that the fraction of all encounters with energy E_a or higher is larger for the higher temperature system.

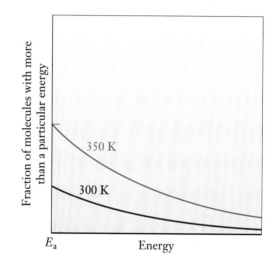

Figure 11.10.

Fraction of encounters with energy greater than E_a from Figure 11.9(b).

━━━━━ **11.47 CONSIDER THIS** ━━━━━

Are there always more encounters with $E > E_a$ at higher temperature?

The position of E_a in Figure 11.9 was chosen arbitrarily at some high energy. Is the observation that "the fraction of all encounters with energy E_a or higher is larger for the higher temperature system" true for any value of E_a? Clearly explain the reasoning for your response.

Our generalization of this important idea is that *the fraction of molecules with energies greater than a given high value, E_a, increases rapidly as the temperature goes up*. The quantitative expression of this relationship, is the **Boltzmann equation:**

$$\text{fraction of molecular encounters with energy } E_a \text{ or greater} = e^{-\left(\frac{E_a}{RT}\right)}$$

$$(11.48)$$

Equation (11.48) is expressed on a molar basis: E_a has units of $J \cdot mol^{-1}$ and R, the gas constant, is $8.314 \, J \cdot mol^{-1} \cdot K^{-1}$.

Accounting for the fraction of molecular encounters that have enough energy to produce reaction, equation (11.48), and the frequency of encounters, A, that have the appropriate orientation for reaction, gives for the rate constant

$$k = Ae^{-\left(\frac{E_a}{RT}\right)} \tag{11.49}$$

Thus, we have identified the meaning of B and the relationship between E_a and B in the Arrhenius equation (11.44): $B = E_a/R$. We can use this relationship to convert the B values we have calculated to activation energies, E_a. We use equation (11.49), the form of the Arrhenius equation you will usually see, to calculate and compare rate constants under different conditions.

11.48 WORKED EXAMPLE

Activation energy for the phenolphthalein-hydroxide reaction

What is the activation energy for the decolorization reaction between phenolphthalein dianion and hydroxide ion, reaction (11.2)?

Necessary information: We need the result from Worked Example 11.43, $B = 5.3 \times 10^3$ K, and the equation $B = E_a/R$.

Strategy: Solve this equation for E_a and substitute the experimental value of B.

Implementation:

$$E_a = B \cdot R = (5.3 \times 10^3 \text{ K})(8.314 \, J \cdot K^{-1} \cdot mol^{-1})$$
$$= 44 \times 10^3 \, J \cdot mol^{-1} = 44 \, kJ \cdot mol^{-1}$$

Does the answer make sense? Most reactions that occur in solution at or near 300 K (room temperature) have activation energies less than $100 \, kJ \cdot mol^{-1}$, so the result for this reaction fits that pattern. At room temperature, there is such a small fraction of encounters with energies of $100 \, kJ \cdot mol^{-1}$ or greater that reactions with higher activation energies cannot get over the energy barrier. Such reactions require higher temperatures and/or catalysts that provide a lower activation energy pathway for reaction.

11.49 CHECK THIS

Activation energy for the UV-sensitive bead decolorizing reaction

Use your results from Check This 11.40 to find the activation energy for the UV-sensitive bead decolorizing reaction.

11.50 WORKED EXAMPLE

Rate constant from Arrhenius equation variables

The activation energy for reaction (11.47) is 58 kJ·mol^{-1}. What is the rate constant, k, for the reaction at 0 °C (273 K) and 50 °C (323 K)?

Necessary information: We need the activation energy, the Arrhenius equation (11.49), and the frequency factor, 3.2×10^6 M^{-1}s^{-1}, from Consider This 11.45.

Strategy: Substitute the known values of T, A, and E_a into equation (11.49) to get k at the desired temperatures.

Implementation:

$$k \text{ (at 273 K)} = (3.2 \times 10^6 \text{ M}^{-1}\text{s}^{-1})e^{-\left[\frac{58 \times 10^3 \text{ J·mol}^{-1}}{(8.314 \text{ J·mol}^{-1}\cdot\text{K}^{-1})(273 \text{ K})}\right]}$$

$$= 2.6 \times 10^{-5} \text{ M}^{-1}\text{s}^{-1}$$

$$k \text{ (at 323 K)} = (3.2 \times 10^6 \text{ M}^{-1}\cdot\text{s}^{-1})e^{-\left[\frac{58 \times 10^3 \text{ J·mol}^{-1}}{(8.314 \text{ J·mol}^{-1}\cdot\text{K}^{-1})(323 \text{ K})}\right]}$$

$$= 1.3 \times 10^{-3} \text{ M}^{-1}\cdot\text{s}^{-1}$$

Does the answer make sense? Rates increase with temperature because the rate constant increases with temperature. Here the higher temperature rate is about 50 times faster.

11.51 CHECK THIS

Rate constant from Arrhenius equation variables

(a) If the activation energy for reaction (11.46) is 50 kJ·mol^{-1}, what is the rate constant, k, for the reaction at 0 °C (273 K) and 50 °C (323 K)?
(b) Is reaction (11.46) or (11.47) faster at 273 K? At 323? Explain your reasoning.

11.52 CHECK THIS

Variation in rates of room temperature reactions

(a) A common rule of thumb is that room temperature reactions approximately double in rate for every 10 °C increase in temperature. [See Consider This 11.42(b).] Does reaction (11.46) or (11.47) or both obey this "rule?" Show how you get your answer.
(b) What does this doubling rule imply about the activation energy of room temperature reactions? Give the reasoning for your answer.

11.7. Light: Another Way to Activate a Reaction

11.53 INVESTIGATE THIS

What factors affect the color of UV-sensitive beads?

Do this as a class investigation and work in small groups to discuss and analyze the results. In Styrofoam® plastic cups, prepare three temperature baths: ice water, room temperature water, and warm (about 35 °C) water. Place an identical UV-sensitive bead in each cup. Place the cups under a source of ultraviolet light so that each bead gets the same amount of illumination. Note and record how dark the color of the bead becomes in each bath. (How can you check to make sure any differences you see are not due to differences between beads?)

11.54 CONSIDER THIS

How does temperature affect the color of UV-sensitive beads?

(a) In Investigate This 11.53, did the darkness or depth of color developed by the UV-sensitive beads change continually over time or reach some constant color in each bath?

(b) Was the depth of color developed by the UV-sensitive beads affected by their temperature? Describe the relationship between the depth of color and the temperature. What explanation can you offer for this relationship?

In Section 11.6, you analyzed the temperature dependence of the decolorizing reaction following illumination of UV-sensitive beads. In Investigate This 11.53, you focused your attention on the color-forming process to see how much color develops under the same amount of illumination at different temperatures. The color development is greater at lower temperatures, which would lead you to conclude that the color-forming reaction cannot be described by the Arrhenius equation for temperature dependence of reactions. The temperature dependence we discussed in Section 11.6 was based on the exponential Boltzmann factor for the distribution of thermal energies in a system. If the energy for a reaction is supplied in a different way, then this factor is not applicable.

The energy for the reaction that produces the color change in the UV-sensitive beads is provided by ultraviolet light. In order to transfer energy to a molecule, a photon of light has to be absorbed by the molecule. The UV-sensitive beads are white (or colorless when melted) when they have not been exposed to ultraviolet light. This means the molecules responsible for the color change do not absorb light in the visible region of the spectrum. They must absorb light in the ultraviolet, which gives them much more energy than is available thermally at room temperature.

11.55 CHECK THIS

Energy of ultraviolet light photons

The lower wavelength (higher energy) limit of the visible region of the spectrum is about 400 nm. What is the energy of a 400 nm photon? How does this energy compare to the activation energies we found for reactions in Section 11.6? *Hint:* The Planck relationship, $E = h \cdot c / \lambda$, from Chapter 4 gives the energy of a single photon. You will need the energy of a mole of photons to compare with molar activation energies.

Rate of photochemical reactions Your result in Check This 11.55 shows that the energy of ultraviolet light photons is enough, over 300 kJ·mol^{-1}, to break chemical bonds. Molecules that absorb a photon of this light will be highly activated and able to undergo chemical reactions. Chemical reactions initiated by absorption of light are called **photochemical reactions.** Since gaining light energy does not depend on encounters among molecules, we would not expect the rates of photochemical reactions to be temperature dependent. What will the rates depend on? Take the reaction of the molecules in the UV-sensitive beads going from their colorless (*L*) to colored (*C*) form as an example:

$$L \xrightarrow{\ h\nu\ } C \tag{11.50}$$

$$\frac{\Delta[C]}{\Delta t} = aI\phi\,[L] \tag{11.51}$$

A compound that is sensitive to UV light and undergoes a reaction like (11.50) is

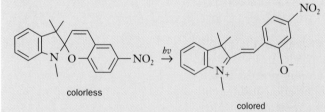

colorless

colored

Compounds like this may be responsible for the color changes in UV-sensitive beads.

The *h*ν over the reaction arrow in equation (11.50) reminds us that the reaction is initiated by light energy. In the rate law, equation (11.51), *I* is the intensity (mol photons·L^{-1}·s^{-1}) of light entering the sample (the bead) and *a* is the fraction of this light absorbed by a unit concentration (1 M) of the *L* molecules. The actual concentration of colorless reactant molecules in the bead is accounted for by their concentration, [*L*]. Thus, the product, *aI*[*L*], is the concentration of activated molecules formed by absorption of light each second. In most photochemical reactions, only a fraction of the activated molecules, ϕ, go on to give product. The rest of the activated molecules lose their energy in various ways and return to the initial form, *L*. The quantity ϕ, the **quantum yield** for the reaction, is a measure of the number of product molecules produced per quantum of light energy absorbed. Quantum yields for elementary reactions like this range from zero (no photochemical reaction) to unity (all photons yield product).

Competing reactions and the steady state The variables in the photochemical rate law, equation (11.51), do not depend on temperature. Thus, the rate of the photochemical reaction, formation of *C* from *L*, does not depend on temperature. Why, then, do you observe a temperature effect on the depth of the color formed by the photochemical reaction? We must account for the reverse thermal reaction:

$$C \rightarrow L \tag{11.52}$$

This is the reaction you observed in Investigate This 11.40 and for which you determined an activation energy in Check This 11.49. The rate law for elementary reaction (11.52) is

$$-\frac{\Delta[C]}{\Delta t} = k[C] \qquad\qquad (11.53)$$

In Investigate This 11.53, you observed that the beads in each bath quickly reached an unchanging depth of color (different at the different temperatures). This means that some constant concentration of the colored form, $[C]$, was present at each temperature. Because reactions (11.50) and (11.52) are going on simultaneously, an unchanging $[C]$ means that a **steady state** has been reached: The rates of formation and decay, equations (11.51) and (11.53), respectively, are equal:

$$\frac{\Delta[C]}{\Delta t} = -\frac{\Delta[C]}{\Delta t}; \quad a I \phi\, [L] = k[C] \qquad\qquad (11.54)$$

When the concentration of a species that forms and decays simultaneously remains constant in a system, a steady state has been reached with the formation and decay reactions going on at the same rate.

We can rearrange equation (11.54) to get the steady-state ratio of the colored to colorless forms of the reactant molecules:

$$\left(\frac{[C]}{[L]}\right)_{\text{steady state}} = \frac{a I \phi}{k} \qquad\qquad (11.55)$$

The numerator on the right of equation (11.55) is a constant, independent of temperature, for a given system, such as the UV-sensitive beads, and a given light intensity. The rate constant, k, in the denominator is temperature dependent.

11.56 CHECK THIS

The effect of temperature on the color of UV-sensitive beads

(a) Does the temperature dependence of k for reaction (11.52), which you found in Check This 11.49, explain the temperature dependence of the steady-state $[C]/[L]$ ratio for the UV-sensitive beads you observed in Investigate This 11.53? Why or why not?

(b) What variable(s) in equation (11.55) could you test to see whether the equation correctly predicts the behavior of the steady-state $[C]/[L]$ ratio?

Light and life Photochemical reactions are essential for life as we know it. All animals on the surface of the earth and most in the seas depend on photosynthesis in microorganisms and plants for food and to maintain an atmosphere of oxygen. The first steps in photosynthesis are reduction–oxidation reactions initiated by the absorption of visible-light photons from sunlight. The outcome of these reactions can be represented as

$$2H_2O(l) \xrightarrow{\;h\nu,\,\text{chloroplasts}\;} O_2(g) + 4H^+(aq) + 4e^- \qquad\qquad (11.56)$$

The notation on the reaction arrow reminds us that the reaction requires light and the chloroplasts in the plant. The electrons are not "free" in the chloroplast, but are captured as reduced forms of molecules like nicotinamide adenine dinucleotide phosphate, NADPH (Check This 10.55, Chapter 10, Section 10.7 and Problem 10.66). These reduced molecules in turn are used to reduce carbon dioxide to glucose as part of a series of reactions that does not involve light.

11.57 CHECK THIS

Visible-light photons and oxidation of water

(a) The standard reduction potential for reduction of oxygen to water (the reverse of reaction (11.56) is 0.816 V at pH 7. What is the standard free energy change for reaction (11.56)? Explain.

(b) The free energy to drive reaction (11.56) is supplied by the energy from visible-light photons, wavelengths in the 400–700 nm range. Is one photon of visible light enough to drive the reaction under standard conditions? Show your reasoning clearly.

Note, once again, how important water is to life. In addition to all the other essential characteristics that we have pointed out, water is the reducing agent required to make the food we eat as well as the source of the oxygen we need to oxidize this food to provide the energy for life.

Reflection and Projection

Most reactions are temperature dependent and increase in rate as the temperature increases. The Arrhenius equation, $k = Ae^{-\left(\frac{E_a}{RT}\right)}$, is the quantitative expression of the temperature dependence of reaction rate constants. The molecular basis of the Arrhenius equation depends on encounters among molecules that lead to reaction. We assume that molecules have to come together in order to react and that their orientation with respect to one another is important in determining whether they react. This orientation requirement leads to Arrhenius frequency factors, A, smaller than the diffusion-controlled encounter frequency.

As reactants interact and change to products, many electron and atomic core rearrangements must occur. These usually require a net input of energy, the activation energy, E_a, in order to form an activated complex, which can proceed to the product(s). To react, pairs of reactant molecules that encounter one another must have energy equal to or greater than E_a. The distribution of energy among molecules at a given temperature is given by the Maxwell–Boltzmann distribution. This distribution leads to the Boltzmann equation, which we use to find the fraction of encounters with energies greater than or equal to E_a. The fraction leads to the exponential term in the Arrhenius equation. The temperature dependence of a reaction rate is a consequence of the activation energy for the reaction.

Another way that reactants can obtain enough energy to form the activated complex is by absorbing light. The energy available from visible and ultraviolet photons is enough to break molecular bonds and can initiate reactions at room temperature that would require heating the system to hundreds or even thousands of degrees to initiate thermally. Photochemical reactions are relatively insensitive to temperature changes, since the energy provided by the photons is usually so much greater than is available thermally.

Reactions that require high temperatures to form the activated complex can often occur under milder conditions, if a catalyst provides a lower energy pathway to the products. Biological reactions must occur under mild conditions, so almost all biochemical reactions are catalyzed by proteins called enzymes, which are the subject of Section 11.10. However, a catalyst cannot make a reaction occur that is not thermodynamically possible. In the next section we will discuss the connections between kinetics and thermodynamics.

11.8. Thermodynamics and Kinetics

11.58 INVESTIGATE THIS

What more can we learn about $H_2O_2(aq)$ decomposition?

Do this as a class investigation and work in small groups to discuss and analyze the results. Measure and record the temperature of 100 mL of 3% aqueous hydrogen peroxide, $H_2O_2(aq)$, solution in a Styrofoam® cup. Add a packet of dried yeast to the solution and use the thermometer to mix the yeast into the solution. Record the temperature and any other observed changes in the system for 3–4 minutes.

11.59 CONSIDER THIS

What can you conclude about $H_2O_2(aq)$ decomposition?

(a) In Investigate This 11.58, you have again decomposed $H_2O_2(aq)$ (as you first did in Investigate This 7.32, Chapter 7, Section 7.7). Is the reaction exothermic or endothermic? Assume that the specific heat of the solution is 4.18 kJ·g^{-1} and that 3% $H_2O_2(aq)$ is 0.88 M. What is $\Delta H_{reaction}$ for the decomposition? Explain your method.

(b) What is the sign of $\Delta S_{reaction}$ for $H_2O_2(aq)$ decomposition, reaction (11.3)? Explain.

(c) Is $\Delta G_{reaction}$ positive or negative for reaction (11.3)? Explain your reasoning.

(d) What can you say about the rate of reaction (11.3) in the absence and presence of yeast? Is this an expected result? What is the basis for your expectation?

Through the first seven chapters of this book, we puzzled over the underlying basis for directionality in chemical reactions. We found that energy (or enthalpy) did not provide the criterion because both exothermic and endothermic reactions are observed to occur. Finally, in Chapter 8, we found that net entropy change (or free energy change), a measure of probability, is the property we can use to predict the direction of change in a system. In Chapter 9, Section 9.1, when we discussed how to identify systems at equilibrium, we considered the example of wood (largely glucose), whose oxidation is greatly favored but which exists unchanged for centuries surrounded by a sea of oxygen. Under the proper conditions, application of a match flame, for example, wood can be oxidized to form carbon dioxide and water as combustion products. Aqueous solutions of hydrogen peroxide provide another example of a reaction, decomposition to oxygen and water, that is highly favored, as you showed in Consider

This 11.59(c). This reaction is so slow, however, that it is almost unobservable over hours or even days. Under proper conditions, for example, the presence of a catalyst such as yeast or any of the others you found in Investigate This 11.6, the reaction proceeds rapidly.

These two examples, and many others like them, show that thermodynamically favored (spontaneous) reactions are not necessarily observed, unless other factors are also favorable. In this chapter, we have examined these other factors that affect the rates of chemical reactions. In order to proceed along the pathway from reactants to products, the reactants have to come together in the appropriate orientation and have sufficient energy to surmount one or more energy barriers along the pathway.

In the combustion of wood, a match flame provides a source of high temperature that gives oxygen enough energy to react with the molecules in the wood to produce the intermediate reactive species that go on to react with one another and more oxygen. These processes produce large amounts of thermal energy, which heats more oxygen to produce more reaction and so on. Thus, once initiated, the oxidation is self-sustaining, which is why one match can create a great deal of combustion. To stop the combustion, you can remove one or the other of the reactants, fuel or oxygen. Or you can remove enough energy, by pouring water on the fire, for example, to halt the self-sustaining initiation processes. The reaction is still thermodynamically favored, but the reactants are not together and/or there is not enough energy to initiate it.

The decomposition of hydrogen peroxide is quite exothermic, as you found in Consider This 11.59(a) and as is shown in Figure 11.11(a). Also shown in the figure is the high activation energy for the uncatalyzed decomposition of hydrogen peroxide. Therefore, even though the decomposition is thermodynamically

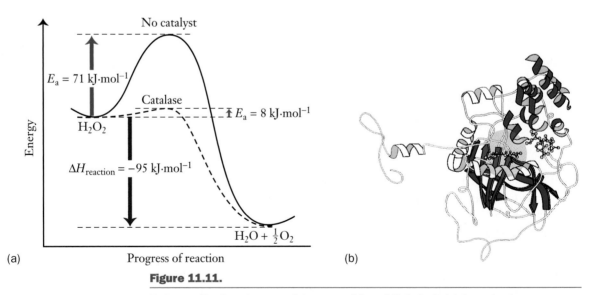

(a)

No catalyst

$E_a = 71$ kJ·mol^{-1}

Catalase

$E_a = 8$ kJ·mol^{-1}

H_2O_2

$\Delta H_{reaction} = -95$ kJ·mol^{-1}

$H_2O + \frac{1}{2}O_2$

Energy

Progress of reaction

(b)

Figure 11.11.

Factors affecting the rate of decomposition of H₂O₂(aq). (a) An activation energy diagram comparing E_a for the uncatalyzed and catalyzed reaction. (b) The heme group in the catalase structure highlighted. The view is from the edge of the flat ring that complexes the iron ion, which interacts with H_2O_2 and helps provide a lower activation energy pathway for the decomposition.

favored (by both enthalpy and entropy), the activation energy makes it very slow at room temperature.

The enzyme catalase (contained in yeast and most other living tissues and organisms) provides an alternative reaction pathway for hydrogen peroxide decomposition. Along this pathway the peroxide molecule interacts within the protein structure with the iron in the heme group (see Chapter 6, Figure 6.10) highlighted in Figure 11.11(b), which lowers the activation energy for decomposition by almost a factor of ten.

11.60 WORKED EXAMPLE

Uncatalyzed decomposition of $H_2O_2(aq)$

The uncatalyzed decomposition of $H_2O_2(aq)$ shows first-order kinetics with a rate of about 10^{-8} M·s^{-1} for a 1 M solution at room temperature, 298 K. What is the first-order rate constant and the Arrhenius frequency factor, A, for this reaction?

Necessary information: We need the first-order rate law, the Arrhenius equation (11.49), and the activation energy, 71 kJ·mol^{-1} from Figure 11.11(a).

Strategy: Use the first-order rate law to find k_1, the first-order rate constant, and then substitute this result and the activation energy into the Arrhenius equation and solve for A.

Implementation: Write and solve the first-order rate law for k_1:

$$\text{rate} = 10^{-8} \text{ M·s}^{-1} = k_1[H_2O_2(aq)] = k_1(1 \text{ M}); \therefore k_1 = 10^{-8} \text{ s}^{-1}$$

The Arrhenius equation is

$$k_1 = 10^{-8} \text{ s}^{-1} = Ae^{-\left[\frac{71,000 \text{ J·mol}^{-1}}{(8.314 \text{ J·mol}^{-1}\cdot\text{K}^{-1})(298 \text{ K})}\right]} = A(3.6 \times 10^{-13})$$

$$A = \frac{10^{-8} \text{ s}^{-1}}{3.6 \times 10^{-13}} = 3 \times 10^{5} \text{ s}^{-1}$$

Does the answer make sense? The Arrhenius frequency factors for first-order reactions are often about 10^{14} s^{-1}, so this value for $H_2O_2(aq)$ decomposition is quite low and in combination with the high activation energy makes the decomposition very slow, as observed. See Check This 11.61(b).

11.61 CHECK THIS

Catalase-catalyzed decomposition of $H_2O_2(aq)$

(a) At 1 M concentrations of $H_2O_2(aq)$ and catalase, the rate of $H_2O_2(aq)$ decomposition is about 10^7 M·s^{-1} at 298 K. What are the second-order rate constant and the Arrhenius frequency factor for this reaction?

(b) The "uncatalyzed," first-order decomposition of $H_2O_2(aq)$ may actually be catalyzed by dust particles and/or trace amounts of ions (or even the walls of the container) that are almost impossible to remove from the solutions. If,

continued

in Worked Example 11.60, the apparent first-order rate constant is actually $k_1 = k_2[\text{trace impurity}]$, what is the second-order Arrhenius frequency factor for [trace impurity] $= 10^{-4}$ M (approximately one part per million impurity)? Does the answer make sense for a second-order reaction in solution? Explain why or why not.

For the examples we have chosen, the combustion of wood (glucose) and the decomposition of hydrogen peroxide, the free energy changes are so large and negative that the reaction at equilibrium (if it can be attained) greatly favors the products. For practical purposes, these reactions can be considered to go in the forward direction only. For other reactions, including many in living systems, the free energy changes are not so large and significant concentrations of both reactants and products are present at equilibrium (if it can be attained). Whether or not equilibrium is attained depends on the rates of the reactions. Note that a catalyst does not change the thermodynamics of a reaction. The enthalpy, entropy, and free energy changes for the overall reaction are the same in the catalyzed and uncatalyzed reaction, so equilibrium in a system is not affected by the presence of a catalyst, except that it is attained more quickly.

A **reversible reaction** takes the same pathway (in reverse) from products to reactants as from reactants to products. If this pathway can be characterized by activation energy diagrams like Figures 11.8 and 11.11(a), which apply to elementary reactions, then the activation energies for the forward, $E_a(\text{forward})$, and reverse, $E_a(\text{reverse})$ reactions can be related through $\Delta H_{\text{reaction}}$:

$$\Delta H_{\text{reaction}} = E_a(\text{forward}) - E_a(\text{reverse}) \qquad \textbf{(11.57)}$$

This relationship, which is demonstrated in Figure 11.12, assumes (as we usually have) that energy and enthalpy changes are numerically equivalent for changes in these systems.

Equation (11.57) provides another connection between the thermodynamics and kinetics of a reaction. In some cases the temperature dependence of the forward and reverse reactions can be studied independently. The differences in activation energies in these cases are consistent with calorimetric measurements of the reaction enthalpies. Now we'll use our knowledge of rates and equilibrium to develop another connection between these two important concepts.

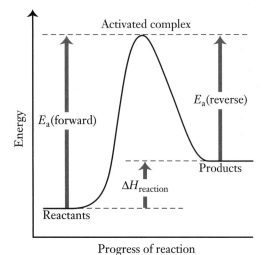

Figure 11.12.

Relationship among energies of activation and enthalpy of reaction.

11.62 CONSIDER THIS

What more can activation energy diagrams tell us?

(a) Is the reaction represented in Figure 11.12 exothermic or endothermic? Explain.

continued

(b) Show that equation (11.57) is valid for this activation energy diagram.
(c) If the temperature is increased, what is the effect on the forward and reverse reactions? Which reaction is more affected? Explain the reasoning for your answers.

When the elementary reaction represented in Figure 11.12 attains equilibrium, the rates of the forward and reverse reactions must be equal because the ratio of products to reactants, the equilibrium constant expression, is unchanging. For simplicity, let the forward and reverse rate laws be $k_f[\text{reactant}]$ and $k_r[\text{product}]$, respectively. Thus, at equilibrium:

$$k_f[\text{reactant}]_{eq} = k_r[\text{product}]_{eq} \qquad \textbf{(11.58)}$$

Equation (11.58) can be rearranged to connect kinetic and thermodynamic constants:

$$\left(\frac{[\text{product}]}{[\text{reactant}]}\right)_{eq} = \frac{k_f}{k_r} = K_{eq} \qquad \textbf{(11.59)}$$

Thus, for elementary reactions, the equilibrium constant can be expressed as the ratio of the forward to the reverse rate constant for the reaction.

In Section 11.6, we found that the increase in a rate constant with increase in temperature depends on the activation energy for the reaction—the larger the activation energy the greater he change in the rate constant for a given change in temperature. Let's apply this finding to the rate constants for the forward and reverse reactions of an elementary reaction system at equilibrium. Assume that, at a given temperature, the system is in equilibrium with the forward and reverse rates equal, as in equation (11.58).

If the temperature is increased, the rate constant for the reaction with the higher activation energy will change more and, therefore, the reaction with higher activation energy will get faster. For an endothermic reaction, as in Figure 11.12, this is the forward reaction, so, as temperature increases, the rate constant ratio will change (primes are the higher temperature values):

$$\frac{k_f{'}}{k_r{'}} > \frac{k_f}{k_r} \qquad \textbf{(11.60)}$$

The new rate constants will mean different forward and reverse reaction rates, unless the product-to-reactant ratio changes.

> The numeric value of the ratio of molar concentrations in equation (11.59) is equivalent to the numeric value of the thermodynamic K_{eq}.

11.63 CHECK THIS

Imbalance of forward and reverse reactions

(a) Show why the changed ratio in equation (11.60) leads to an imbalance of the forward and reverse rate when the temperature is increased for an exothermic system at equilibrium.
(b) Show what the ratio would be and how the argument would go for an exothermic reaction, as in Figure 11.8.

In order to restore equilibrium with equal forward and reverse rates, some more reactant must go to product, so that the rates are again equal, $k_f'[\text{reactant}]_{eq}' = k_r'[\text{product}]_{eq}'$. Comparing equations (11.60) and (11.59), shows that we must have

$$\left(\frac{[\text{product}]}{[\text{reactant}]}\right)_{eq}' > \left(\frac{[\text{product}]}{[\text{reactant}]}\right)_{eq} \quad \text{and} \quad K_{eq}' > K_{eq}$$

Thus, the kinetic analysis predicts that increasing the temperature of an endothermic reaction system at equilibrium will increase the numeric value of the equilibrium constant and increase the ratio of products to reactants.

Le Chatelier's principle predicts that an endothermic reaction at equilibrium will respond to a temperature increase (addition of energy) by going forward to use up some of the added energy and hence to turn more reactant into product. Thus, the kinetic prediction and Le Chatelier's prediction (which is consistent with equilibrium thermodynamics) agree and we find that kinetic and thermodynamic arguments are consistent with one another.

11.9 Outcomes Review

In this final chapter, we tried to bring together much of what you already had learned in order to propose chemically reasonable pathways for chemical reactions. In addition, we introduced new concepts, based on measurements of reaction rates, to help in developing and understanding reaction pathways. We found that rates of reaction can be described in terms of rate laws that involve the concentrations of species in the reaction system and a proportionality constant, the rate constant. The species in the rate law are those that enter the reaction pathway before or during the rate-limiting step for the reaction.

The rate constants for reactions are temperature dependent and their increase with increasing temperature is described by the Arrhenius equation. The form of the Arrhenius equation is a consequence of the necessity for reactants to encounter one another in the right orientation and with enough energy to overcome an activation energy barrier and of the distribution of energy among encounters in the reacting system. Absorption of light by a reactant molecule is a way to initiate reactions that require more energy than is available thermally at room temperature.

Finally, we have to keep in mind that only reactions that are thermodynamically favorable, under the conditions specified, can occur spontaneously. Whether they will actually be observed to occur depends on the kinetics of the reaction pathway(s) available.

Check your understanding of the ideas in this chapter by reviewing these expected outcomes of your study. You should be able to

- Predict the usual direction of the effects of changing concentration, changing temperature, or presence of catalysts on the rate of a reaction [Section 11.1].
- Determine the initial rate of a reaction in units of $M \cdot s^{-1}$ from data for the concentration (or a property that is directly proportional to concentration, such as gas pressure or absorbance) of a reactant or product as a function of time [Section 11.2].
- Use initial rate and concentration data to determine the order of a reaction with respect to the concentrations of species in the solution and write the rate law for the reaction [Section 11.3].
- Explain in words and/or with drawings how the rate of a reaction depends only on the rate of the rate-limiting step in the reaction pathway [Section 11.4].
- Derive the rate law for a reaction whose pathway (mechanism), including knowledge of the rate-limiting step, you are given [Sections 11.4 and 11.5].
- Propose a pathway (mechanism) for a reaction that is similar or analogous to one whose pathway you know [Sections 11.4 and 11.5].
- Use data for concentration (or a property that is directly proportional to concentration) of a reactant or product as a function of time to determine the

rate constant and half-life for a first-order reaction [Section 11.5].

- Use rate constants and/or half-lives for first-order reactions to determine how long a sample has been reacting, given some known initial amount or the amount remaining after a known time [Section 11.5].

- Explain what is meant by flooding a reaction and use rate and concentration data for a reaction that is flooded with respect to a species to find the order of reaction with respect to the concentration of that species and/or find out whether the disappearance of another species is first order [Section 11.5].

- Use the temperature variation of the rate constant (or variables directly proportional to the rate constant, such as rates with the same concentrations of all species) to determine the activation energy for a reaction [Section 11.6].

- Use rate constant and activation energy values to determine the Arrhenius frequency factor for a reaction [Section 11.6].

- Construct and interpret an activation energy diagram, given information about the activation energy and enthalpy change for the reaction [Section 11.6].

- Describe how encounters between reactants in solution differ from collisions between reactants in the gas phase [Section 11.6].

- Describe the origin of the temperature dependence of the rate constant and why the dependence is so strong

when the *average* energy of molecular encounters does not increase so rapidly [Section 11.6].

- Explain how light can initiate reactions that would not otherwise occur at low (room) temperature [Section 11.7].

- Describe how competing photochemical and thermal reactions can lead to a steady-state concentration of a reactive species in a system [Section 11.7].

- Explain why some reactions that are highly favored thermodynamically are not observed to occur [Section 11.8].

- Describe factors that can be changed to provide favorable kinetics for spontaneous reactions that are not otherwise observable [Section 11.8].

- Relate the temperature dependence of an equilibrium constant to the temperature dependences of the forward and reverse rate constants for the reaction [Section 11.8].

- Describe the Michaelis–Menten pathway for enzyme-catalyzed reactions and use kinetic data to characterize these reactions, especially their behavior as a function of substrate and enzyme concentrations [Section 11.10].

- Describe enzyme specificity in terms of functional and substrate specificity and relate specificity to the characteristics of the active sites of enzymes [Section 11.10].

11.10. Extension—Enzymatic Catalysis

11.64 INVESTIGATE THIS

How fast is the reaction of OH⁻(aq) with dissolved CO₂?

(a) Do this as a class investigation and work in small groups to discuss and analyze the results. Add 40 mL of ice-cold carbonated water and 2–3 drops of bromthymol blue indicator solution to a 50-mL erlenmeyer flask containing a magnetic stir bar. Record the color of the solution. Stir the mixture at moderate speed with the magnetic stirrer and quickly add 1 mL of 4 M aqueous sodium hydroxide, NaOH, solution to the stirred mixture. Record the time required for the solution to reach its final color.

(b) Repeat the procedure but add 1 mL of a solution of the enzyme carbonic anhydrase just before the base is added. The amount of enzyme in 1 mL of the enzyme solution is about the same as in one drop (about 50 μL) of your blood.

11.65 CONSIDER THIS

What limits the rate of the reaction of OH⁻(aq)
with dissolved CO₂?

When carbon dioxide dissolves in water to give $CO_2(aq)$, some of it reacts with water to form carbonic acid:

$$CO_2(aq) + H_2O(l) \rightleftharpoons (HO)_2CO(aq) \tag{11.61}$$

Reaction (11.61) is reversible; at equilibrium, most of the CO_2 is present as $CO_2(aq)$. Hydroxide ion, $HO^-(aq)$, added to this solution reacts with the carbonic acid to give the hydrogen carbonate (bicarbonate) ion:

$$(HO)_2CO(aq) + OH^-(aq) \rightarrow H_2O(l) + HOCO_2^-(aq) \quad \text{(fast)} \tag{11.62}$$

(a) Water saturated with $CO_2(g)$ at one atmosphere pressure is approximately 0.1 M in total of $CO_2(aq)$ plus $(HO)_2CO(aq)$. How many moles of $CO_2(aq)$ plus $(HO)_2CO(aq)$ are present in the solutions you used in Investigate This 11.64?

(b) How many moles of $OH^-(aq)$ did you add to each solution in Investigate This 11.64? How do the numbers of moles of $OH^-(aq)$ added compare to the number of moles of $CO_2(aq)$ plus $(HO)_2CO(aq)$?

(c) Do your results suggest that reaction (11.61) in the forward direction is fast or slow? Explain the reasoning for your answer.

(d) What effect does addition of the enzyme have on the reaction? What reaction(s) is the enzyme affecting? Explain how you know.

Enzymes and rate laws Enzymes, like other catalysts, have to interact with the reactants in order to affect the reaction rate. This is just like we've seen for $I^-(aq)$ in reaction (11.21). Therefore, we expect that the rate of an enzyme-catalyzed reaction like reaction (11.61), the hydration of dissolved carbon dioxide to produce carbonic acid, will depend upon the amount of enzyme present. Figure 11.13 shows plots of product formation as a function of time for different concentrations of enzyme in an enzyme-catalyzed reaction.

Figure 11.13.

Product formation *vs.* time in an enzyme-catalyzed reaction. The relative enzyme concentrations for experiments (a), (b), and (c) are 1, 2, and 4, respectively, with all other conditions the same.

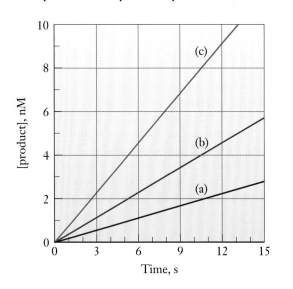

The data in Figure 11.13 can be used to determine the initial rate of the reaction:

$$\text{initial rate} = V_0 = \left(\frac{\Delta(\text{product})}{\Delta t}\right)_{\text{initial}} \qquad \textbf{(11.63)}$$

V_0 is the usual symbol for the initial rate of enzyme-catalyzed reactions. You will also see a lower case v, instead of V_0, used for initial reaction rate (velocity).

11.66 CHECK THIS

Partial rate law for an enzymatic reaction

Use the data from Figure 11.13 to determine the order of the reaction with respect to the enzyme. Write the rate law that relates the initial rate of the reaction, V_0, to the enzyme concentration, [E]. Give your reasoning clearly.

Substrate concentration effects In enzyme-catalyzed reactions, enzymes are almost always present at much lower concentrations than their **substrates,** the reactant compounds whose reactions they catalyze. For reasons that we will discuss below, enzymes are such efficient catalysts that it takes only tiny concentrations to catalyze the reactions necessary to maintain an organism. The consequences of low enzyme concentrations are kinetic results like those in Figure 11.14. Each of the two plots in the figure represents a series of initial rate determinations as the initial concentration of substrate, $[S]_0$ is varied. A different enzyme concentration is used in each series.

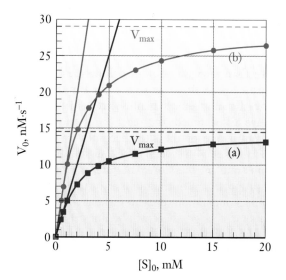

Figure 11.14.

Initial rate, V_0, as a function of initial substrate concentration, $[S]_0$. The enzyme concentration in the (b) series was double that in the (a) series of reactions.

The straight line from the origin through the first few points of each series in Figure 11.14 shows that, at low $[S]_0$, the rate increases linearly with $[S]_0$:

$$V_0 \propto [S]_0 \qquad \textbf{(11.64)}$$

Thus, the reaction is first order with respect to substrate concentration at low $[S]_0$. At high $[S]_0$, the initial rates approach a constant value:

$$V_0 = V_{max} \qquad\qquad\qquad\qquad (11.65)$$

The dashed horizontal lines in Figure 11.14 show the limiting rate, V_{max}, for each series of reactions. (At $[S]_0 = 50$ mM, the initial rates for each series are within 3% of their limiting values.) If a reaction reaches a limiting rate that does not increase with further increase in substrate concentration, we say that the reaction is **saturated,** that is, it is going as fast as it can with the concentration of enzyme present in the system. The data in Figure 11.14 show that the limiting rate, V_{max} (maximum velocity), depends on the concentration of enzyme present in the reaction mixture.

11.67 CONSIDER THIS

What are the initial and maximum rate laws
for an enzyme reaction?

(a) When an enzymatic reaction is saturated, $V_0 = V_{max}$, what is the order of the reaction with respect to the substrate? Explain your reasoning. *Hint:* Review the discussion where reaction order was introduced in Section 11.3.

(b) Use the data from Figure 11.14 to determine how the limiting rate, V_{max}, depends on the concentration of enzyme.

(c) Use your results in part (a) and (b) to write a rate law for the rate of an enzymatic reaction at saturation, that is, the order with respect to enzyme and substrate concentrations.

(d) How does the initial rate of the reaction, V_0, depend on the concentration of enzyme at low substrate concentration? What is the rate law for the rate of an enzymatic reaction at low substrate concentration? That is, show the order with respect to enzyme and substrate concentrations. Explain your reasoning.

Catalysts other than enzymes can also show these saturation effects. We interpret saturation to mean that there are only a limited number of sites for catalysis available to the substrate. When the concentration of substrate is high enough, the sites are all occupied; the reaction of another substrate molecule can't be catalyzed until one of the sites becomes available. The situation is analogous to a supermarket with several checkout stations. When there are only a few shoppers (substrate) in the store, the rate at which shoppers leave with their purchases is proportional to the number of them in the store, since they will find an open checkout station (enzyme) whenever they are finished. When the number of shoppers is large, there will be lines at all checkout stations, and the rate at which shoppers leave will be constant and set by the number of checkout stations available. If more checkout stations are made available, the maximum rate at which shoppers leave will be greater but will still reach saturation if a large enough number of shoppers is in the store.

Michaelis–Menten mechanism for enzyme catalysis Between the two extremes at low and high substrate concentration, the initial-rate data in Figure 11.13 fall on a curve. A mechanism of enzyme action that describes the whole curve quantitatively was proposed in 1913 by two German chemists,

Leonor Michaelis (1875–1949) and Maud Menten (1879–1960), who were studying the rate of hydrolysis of sucrose catalyzed by the yeast enzyme invertase:

$$C_{12}H_{22}O_{11}(aq) + H_2O(l) \xrightarrow{\text{invertase}} C_6H_{12}O_6(aq) + C_6H_{12}O_6(aq)$$

sucrose   glucose   fructose   **(11.66)**

The names of enzymes almost all end with "-ase," as in catal*ase* and invert*ase*, for example.

Enzyme-catalyzed reactions are almost always discussed in terms of the Michaelis–Menten mechanism or some variation of it. Understanding the mechanism is important for understanding the control and direction of reactions in living cells.

Near the beginning of the 20th century several chemists postulated that an enzyme, E, and its substrate, S, formed an enzyme–substrate complex, E-S, prior to reaction. The **Michaelis–Menten mechanism** postulates that this first step is rapidly reversible and that the rate-limiting step in the reaction is the reaction of E-S to produce the reaction product, P. The simplest formulation of the Michaelis–Menten mechanism with all species in aqueous solution is

E + S ⇌ E-S (fast equilibrium with equilibrium constant = K)   **(11.67)**

E-S → E + P (slow, rate-determining step with rate constant = k) **(11.68)**

The mechanism can be shown pictorially as

$$\qquad\qquad\qquad\qquad\qquad\qquad\qquad\qquad\qquad \textbf{(11.69)}$$

Many enzyme reactions are reversible, that is, the enzyme can catalyze the reaction of P to give back S. The Michaelis–Menten mechanism applies in this simple form only to initial reaction rates when almost no P has been formed and we don't have to consider the reverse reaction.

The equilibrium constant expression, in terms of concentrations, for reaction (11.67) at the beginning of the reaction is

$$K = \frac{[E\text{-}S]}{[E][S]_0} \qquad\qquad\qquad\qquad \textbf{(11.70)}$$

The sum of the concentrations of enzyme in both forms, [E] and [E-S], has to equal the total concentration added to the reaction, $[E]_{tot}$:

$$[E]_{tot} = [E] + [E\text{-}S] \qquad\qquad\qquad\qquad \textbf{(11.71)}$$

11.68 CHECK THIS

Concentration of enzyme–substrate complex

Combine equations (11.70) and (11.71) and solve to get [E-S] in terms of K, $[S]_0$, and $[E]_{tot}$.

Because the rate of the reaction is determined by reaction (11.68), the rate-limiting step, we can write the initial rate of formation of product, P, as

$$\frac{\Delta[P]}{\Delta t} = V_0 = k\,[\text{E-S}] \tag{11.72}$$

We want rate laws for observed changes expressed in terms of measurable quantities and we usually can't measure [E-S] experimentally. However, we can substitute your result for [E-S] from Check This 11.68 into (11.72), to get the **Michaelis–Menten equation:**

$$V_0 = k\left(\frac{[\text{E}]_{\text{tot}}[\text{S}]_0}{\left(\dfrac{1}{K}\right) + [\text{S}]_0}\right) \tag{11.73}$$

Limiting cases for the Michaelis–Menten mechanism Equation (11.73) looks a bit complicated, but let's look at the limiting cases, low $[\text{S}]_0$ and high $[\text{S}]_0$. When we run the reactions at lower and lower $[\text{S}]_0$, we reach a point where $[\text{S}]_0 \ll 1/K$. For this low $[\text{S}]_0$ case, $1/K + [\text{S}]_0 \approx 1/K$ and the initial rate equation (11.73) becomes

$$V_0\,(\text{low }[\text{S}]_0) \approx k\left(\frac{[\text{E}]_{\text{tot}}[\text{S}]_0}{\left(\dfrac{1}{K}\right)}\right) = k\,K\,[\text{E}]_{\text{tot}}[\text{S}]_0 \tag{11.74}$$

If the amount of enzyme is held constant, all the factors on the right-hand side of equation (11.74) are constant except $[\text{S}]_0$. The initial rate of reaction increases linearly with $[\text{S}]_0$, as shown by the straight lines through the low $[\text{S}]_0$ points in Figure 11.14. The slopes of those lines are directly proportional to the concentration of enzyme in the reaction, as equation (11.74) predicts and you found in Consider This 11.67(d).

If we run our reactions at very high $[\text{S}]_0$, then $[\text{S}]_0 \gg 1/K$. For this high $[\text{S}]_0$ case, $1/K + [\text{S}]_0 \approx [\text{S}]_0$ and equation (11.73) becomes

$$V_0\,(\text{high }[\text{S}]_0) \approx k\left(\frac{[\text{E}]_{\text{tot}}[\text{S}]_0}{[\text{S}]_0}\right) = k\,[\text{E}]_{\text{tot}} \tag{11.75}$$

For constant amounts of enzyme, both factors on the right-hand side of (11.75) are constant, so the initial rate is a constant (directly proportional to the concentration of enzyme used):

$$V_0\,(\text{high }[\text{S}]_0) = V_{\text{max}} = k\,[\text{E}]_{\text{tot}} \tag{11.76}$$

11.69 CONSIDER THIS

How do you interpret enzyme reaction rate data?

(a) What initial substrate concentration conditions, low $[\text{S}]_0$ or high $[\text{S}]_0$, do you think were used in the enzyme-catalyzed reactions whose results are shown in Figure 11.13? Explain the reasoning for your answer.

(b) Which rate law, equation (11.74) or (11.75), is consistent with the result you got in Check This 11.66? State your reasoning clearly.

Determining and interpreting K Now we have a molecular level interpretation of the experimental results represented by Figures 11.13 and 11.14. The Michaelis–Menten mechanism and rate law, equation (11.73), provide a way to categorize enzyme–substrate interactions by their values of $1/K$. In equation (11.73), substitute V_{max} for $k[E]_{tot}$ [from equation (11.76)] to get an alternative form of the Michaelis–Menten equation:

$$V_0 = \frac{V_{max}[S]_0}{\left(\frac{1}{K}\right) + [S]_0} \tag{11.77}$$

Now consider the case when V_0 has reached half its maximum value, $V_0 = \frac{1}{2}V_{max}$:

$$\frac{1}{2}V_{max} = \frac{V_{max}([S]_0)_{1/2}}{\left(\frac{1}{K}\right) + ([S]_0)_{1/2}} \tag{11.78}$$

$([S]_0)_{1/2}$ reminds you that the conditions are for $V_0 = \frac{1}{2}V_{max}$. Divide through both sides of equation (11.78) by V_{max} and rearrange the result to give

$$\frac{1}{K} = ([S]_0)_{1/2} \tag{11.79}$$

You can determine $1/K$ or K from an experimental value for $([S]_0)_{1/2}$.

11.70 CONSIDER THIS

What is K for the enzyme data in Figure 11.13?

(a) What is K for the enzyme and substrate whose initial rate data are shown in Figure 11.14? Explain how you get your answer.

(b) Are your units for K (expressed in concentrations) appropriate for equation (11.67)? Explain why or why not.

To see how to interpret values of K, return to reaction (11.67) and its equilibrium constant expression (11.70). Reaction (11.67) is the *association* of the enzyme and substrate to give the enzyme–substrate complex. The larger the equilibrium constant, K, is for reaction (11.67), the larger the amount of association. Another way to think about this association is the larger the equilibrium constant, the more tightly the enzyme and substrate are bound in their complex. Thus, a smaller $([S]_0)_{1/2}$ is required to half saturate the reaction, as shown in Figure 11.15. Tight binding is usually a result of good *fit* of the shape and polarity of the substrate with the complementary shape and polarity of the catalytic site of the enzyme, which is discussed below.

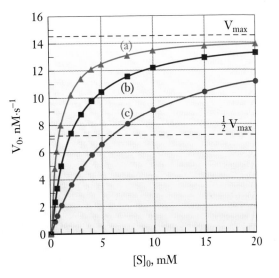

Figure 11.15.

Initial rates *vs.* [S]₀ for enzymes with different Michaelis–Menten *K* values. Enzymes (a), (b), and (c) were reacted with their substrates under conditions that give the same V_{max} values.

11.71 WORKED EXAMPLE

Michaelis–Menten K values

What are the Michaelis–Menten K values for the three enzymes, (a), (b), and (c), for which data are shown in Figure 11.15?

Necessary information: We need equation (11.79) and the data in the figure.

Strategy: Find $([S]_0)_{1/2}$ $\left(= [S]_0 \text{ at } \frac{1}{2}V_{max}\right)$ for each enzyme and then take the inverse to get K.

Implementation: The $\frac{1}{2}V_{max}$ rate is shown by a dashed line in the figure. The corresponding values of $([S]_0)_{1/2}$ are about 0.9 (a little less than 1), 2, and 6 mM for enzymes (a), (b), and (c), respectively. The K values are 1.1, 0.5, and 0.2 mM^{-1}, for enzymes (a), (b), and (c), respectively.

Does the answer make sense? Answer this question in Check This 11.72.

11.72 CHECK THIS

Tightness of substrate binding

Which of the three enzymes represented in Figure 11.15 binds its substrate most tightly? Explain your reasoning.

Enzyme specificity Also related to the fit of enzyme and substrate is the specificity of enzyme catalysis. There are two ways that enzymatic catalysis is specific. **Functional specificity** means that an enzyme usually catalyzes only a specific reaction (or class of reactions). Catalase, for example, catalyzes the decomposition of $H_2O_2(aq)$ to give $O_2(g)$ and H_2O, reaction (11.3), and invertase catalyzes the hydrolysis of sucrose to fructose and glucose, reaction (11.66), but neither enzyme catalyzes other kinds of reactions. Enzymatic catalysis also shows **substrate specificity,** that is, most enzymes can accommodate only one or a very few specific substrates. For example, catalase decomposes only $H_2O_2(aq)$ and urease catalyzes only urea hydrolysis:

$$NH_2CONH_2(aq) + 2H_2O \xrightarrow{\text{urease}} 2NH_4^+(aq) + CO_3^{2-}(aq) \quad \textbf{(11.80)}$$

Other enzymes can be less specific and accommodate several similar substrates but the range is generally fairly limited.

Enzyme active sites The **active site** of an enzyme is the region where the interactions between the enzyme and the substrate catalyze the reaction of the substrate(s). The structure of the active site is the basis for both the functional and substrate specificity. Many detailed structures for enzymes and their active sites are known, mostly by x-ray diffraction analysis. A representation of the structure of catalase is shown as part of the chapter-opening illustration. The details of the structures are complex; the pictures we need are simpler schematic diagrams, such as Figure 11.16, that make clearer the kinds of interactions that are important for functional and substrate specificity.

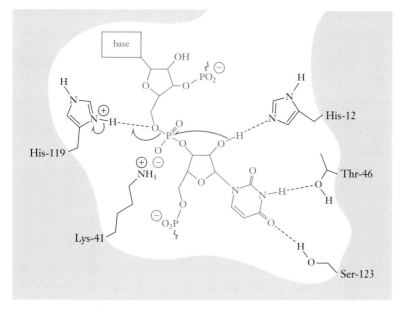

Figure 11.16.

Representation of substrate binding and active site interactions in ribonuclease. The solid blue represents the interior surface of the active site of the enzyme. Five of the amino acid side groups that are essential for binding and catalysis in the active site are shown. The abbreviations stand for histidine, lysine, serine, and threonine and the numbers are the positions of the amino acids along the 124 amino-acid protein chain.

All active sites are in clefts or folds of the overall protein structure, as we have tried to represent in Figure 11.16, not on the surface of the protein. The substrate and other species involved in the reaction have to move into the cleft, in order for the catalyzed reaction to occur. Within the active site cleft are amino acid side groups that interact with functional groups on the reacting substrates. An example is shown at the lower right in Figure 11.16 for the hydrogen bonding of one of the ribonucleic acid, RNA, bases with polar groups in the active site of ribonuclease, an enzyme that cuts up RNA. Other places on the surface of the cleft have nonpolar side groups that interact with nonpolar (hydrophobic) parts of the substrate structure. The combination of these interactions holds the substrate and other reactants in a particular orientation that is favorable for reaction.

To simplify our illustrations, we often represent active-site interactions two-dimensionally, as in Figure 11.16, but remember that the site is three-dimensional and its shape is very important in accommodating substrates. Only certain substrates will have a structure that can be bound by the arrangement of binding positions in the active site.

The arrangement of amino acid side groups is also responsible for catalyzing the reaction, usually by disturbing the electronic distribution of the reactants in ways that make them more susceptible to reaction. For ribonuclease in Figure 11.16 folding of the protein chain brings amino acids from near both ends of the chain together as part of the active site. When pairs of electrons move as shown by the three curved arrows, a P—O bond is broken and the phosphodiester backbone of the ribonucleic acid molecule (in red) is broken. This is the function of the enzyme.

Within or very near the active site, many enzymes have nonprotein groups that help bring about electronic reorganizations favorable to product formation. These nonprotein groups include complexed metal ions (such as zinc ion in carbonic anhydrase), metal-containing groups (such as heme with a bound metal ion in **catalase**), or other relatively small molecules (such as vitamins) that are part of the catalytic pathway.

We sometimes say that an enzyme "recognizes" its substrate(s). Be cautious using terms like this that give cognitive properties to the enzyme. Recognition between enzyme and substrate requires a matching of favorable interactions and is governed by the random motions of the molecules with respect to one another.

Catalase activity and the bombardier beetle Catalase is one of the most active enzymes known. A mole of catalase active sites can decompose about 10^7 mole of $H_2O_2(aq)$ in one second. The ratio of moles of substrate reacted per second to moles of enzyme active sites is called the **turnover number** for the enzyme.

11.73 CHECK THIS

Catalase activity in one millisecond

How many moles of $H_2O_2(aq)$ can be decomposed by one mole of catalase active sites in one millisecond, 10^{-3} s? Explain.

There are many species of bombardier beetle, which vary in the amount of defensive spray they discharge. For the beetle shown in the chapter-opening illustration, about 0.4 μL (1 μL = 10^{-6} L) of 7.4 M $H_2O_2(aq)$ solution (Check This 11.17) reacts for one burst of spray. The number of moles of $H_2O_2(aq)$ that react is $(0.4 \times 10^{-6}$ L·burst$^{-1})(7.4$ M$) = 3 \times 10^{-6}$ mol·burst^{-1}.

11.74 WORKED EXAMPLE

Moles of catalase required by a bombardier beetle

How many moles of catalase active sites are required to decompose 3×10^{-6} mol $H_2O_2(aq)$ in one millisecond (one burst from the beetle)?

Necessary information: We need your result from Check This 11.73; in one millisecond, one mole of enzyme active sites can decompose 10^4 mol of $H_2O_2(aq)$.

Strategy: Use the ratio of moles of active sites to moles of $H_2O_2(aq)$ decomposed in one millisecond (burst) to convert the moles of $H_2O_2(aq)$ decomposed per burst to moles of active sites required.

Implementation:

$$(3 \times 10^{-6} \text{ mol } H_2O_2(aq) \cdot \text{burst}^{-1}) \left[\frac{1 \text{ mol active sites}}{10^4 \text{ mol } H_2O_2(aq) \cdot \text{burst}^{-1}} \right]$$

$$= 3 \times 10^{-10} \text{ mol active sites}$$

Does the answer make sense? The turnover number for the enzyme is large and the amount of substrate, $H_2O_2(aq)$, that reacts is small, so it makes sense that only a small amount of enzyme will be required.

11.75 CHECK THIS

Mass of catalase required by a bombardier beetle

(a) The catalase structure shown in the chapter-opening illustration has one active site in a protein with a molar mass of about 60,000 g. Assuming

continued

catalase is the enzyme in the reaction chamber of the beetle, use the information from Worked Example 11.74 to determine the mass of the protein required by the beetle. Explain.
(b) Is it reasonable for a beetle with a mass of a few grams to contain the mass of protein you calculated in part (a)? Explain the reasoning for your answer.

As you have found throughout this book, chemistry and chemical phenomena are part of everything you do and part of everything around you, including the amazing workings of an insect like the bombardier beetle. We hope you will continue to use the chemistry you have learned to help you better understand the world and its workings.

Chapter 11 Problems

11.1. Pathways of Change

11.1. Suggest two examples of reactions not included in this chapter that are very slow, that require days or years or longer to complete. Suggest two examples of reactions not included in this chapter that are very fast—that are complete in minutes or seconds.

11.2. In Investigate This 11.1 you investigated the time it took the indicator color to fade when phenolphthalein is added to a solution of sodium hydroxide.
(a) Would you predict the color to fade more or less quickly for a 2.0 M solution compared to a 1.0 M solution of sodium hydroxide? Explain your reasoning.
(b) Would you predict the color to fade more or less quickly for a 0.10 M solution compared to a 1.0 M solution of sodium hydroxide? Explain your reasoning.

11.3. Examine the structures from equation 11.1:

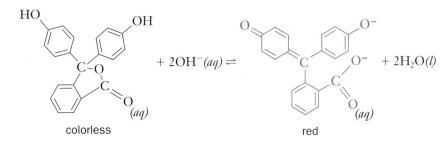

colorless   red

What differences do you seen in the structures before and after adding $OH^-(aq)$? What kind of reaction (see Chapter 6) is this? Explain your choice.

11.4. What do we mean when we talk about the rate of a chemical reaction?

11.5. The bombardier beetle shown in the chapter-opening illustration is tethered to a stiff wire by a blob

of wax on its back. In the stroboscopic study, the tethered beetle is touched with a pair of forceps (the white wedge in each picture) in order to induce it to discharge. The discharge appears as clusters of white dots in three or four successive pictures.
(a) What is the time between discharges? What is the rate of discharge (discharges·s^{-1})? Explain your reasoning.
(b) The normal human ear is sensitive to vibrations (sounds) in the range from about 20 to 20,000 cycles per second. Could the rate of discharge of the beetle be responsible for the sound that accompanies its discharge? Explain.

11.6. In Investigate This 11.1, why does it take longer for the phenolphthalein to go to colorless than it does for the phenolphthalein to turn red? Is this explained by the structures of the two forms of phenolphthalein?

11.7. Consider this reaction beginning with $AgCl(s)$ and then adding the $NH_3(aq)$ solution:

$$AgCl(s) + 2NH_3(aq) \rightleftharpoons Ag(NH_3)_2^+(aq) + Cl^-(aq)$$

(a) What happens to the concentration of ammonia over time?
(b) What happens to the amount of solid silver chloride over time?
(c) What happens to the amount of the silver diammine, $Ag(NH_3)_2^+$, ion over time?

11.8. Chemical reactions often occur faster at higher temperatures. What happens to molecules as their temperature increases? Does this help explain why reactions occur faster at a higher temperature? Explain why or why not.

11.9. Investigate This 11.6 showed that some substances could help catalyze, speed up, the decomposition of hydrogen peroxide, H_2O_2. Investigate This 11.1 showed that reactant concentration could affect the rate of reaction. How could you test to see if a substance, either a reactant or a potential catalyst, changed the rate of a reaction? Describe an experiment that you might conduct.

11.2. Measuring and Expressing Rates of Chemical Change

11.10. Consider a solution reaction that bubbles as the solution goes from clear and colorless to clear blue. Suggest three possible ways you might measure the rate of the process going on in the solution. Explain how your choices are related to the rate.

11.11. Consider the reaction $N_2(g) + 3H_2(g) \rightarrow 2NH_3(g)$.
(a) Sketch a concentration *vs.* time graph for how you predict the concentrations of N_2, H_2, and NH_3 change over time, beginning with a stoichiometric mixture of N_2 and H_2.
(b) Write an expression for the rate of ammonia, NH_3, production in terms of gas pressure, P_{NH_3}, change. Use equation 11.4 to guide you.
(c) Write an expression for the rate of ammonia production in terms of concentration, $[NH_3(g)]$, change.
(d) If N_2 is reacting at a rate of 0.15 M·min^{-1}, what is the rate of ammonia production in these same units?
(e) If N_2 is reacting at a rate of 0.15 M·min^{-1}, what is the rate at which H_2 disappears?
(f) If N_2 is reacting at a rate of 0.15 M·min^{-1}, what is the rate of the reaction written above? Explain.

11.12. Graph these data collected for a reaction that can be symbolized as $A \rightarrow 2B$.

Time, s	[A], mol·L^{-1}
0.0	1.000
10.0	0.833
20.0	0.714
30.0	0.625
40.0	0.555

(a) Calculate the rate of disappearance of **A** at 10.0, 20.0, and 30.0 seconds by approximating the slope of a tangent line at each of these times.
(b) What happens to the rate in part (a) over time? Is this what you might have expected for this reaction? Explain your reasoning.
(c) Calculate the average rate of disappearance of **A** during the time period from 0.0 to 10.0, 10.0 to 20.0, 20.0 to 30.0, and 30.0 to 40.0 seconds.
(d) How do the average rates in part (c) compare to the instantaneous rates you calculated in part (a)? Is there any pattern to the comparisons? Explain.

11.13. Under suitable conditions, the reaction $CH_3OH + H^+ + Cl^- \rightarrow CH_3Cl + H_2O$ occurs in a solution of methanol and hydrochloric acid. The progress of the reaction can be followed by determining the concentration of hydronium ion, $[H^+]$, as a function of reaction time.

Time, min	[H$^+$], M
0	2.12
90	1.95
200	1.74
360	1.54
720	1.19

(a) Graph these data and calculate the average rate of disappearance of H^+ for each time interval.
(b) What do your graph and the average rates in part (a) tell you about the rate of the reaction as a function of time? Explain.
(c) On the same graph as in part (a), plot the concentration of chloromethane, $[CH_3Cl(solvent)]$, as a function of time in this reaction. Explain your reasoning.

11.14. (a) For each of these reactions, write the rate of reaction expressed in terms of the changes in pressure for each reactant and product in the reaction.
   (i) $2NO_2(g) \rightarrow NO(g) + NO_3(g)$
   (ii) $2NOBr(g) \rightarrow 2NO(g) + Br_2(g)$
   (iii) $2N_2O_5(g) \rightarrow 4NO_2(g) + O_2(g)$
(b) For each reaction, what is the relationship of the rate of formation of each product to the rate of decomposition of the reactant? How, if at all, are these relationships connected to the reaction rate expressions you wrote in part (a)? Explain your reasoning.

11.15. 2-Chloro-2-methylpropane, $(CH_3)_3CCl$, reacts in a solution of water and acetone to give 2-methyl-2-propanol, $(CH_3)_3COH$. When a small amount of hydroxide ion, OH^-, is present in the solution, we can write the overall initial reaction as

$$(CH_3)_3CCl + OH^- \rightarrow (CH_3)_3COH + Cl^-$$

Bromphenol blue acid–base indicator is added to the solution and the reaction is followed by timing how long it takes for the indicator to change color (blue to yellow), which shows that all the hydroxide ion has reacted. The results for different initial concentrations of hydroxide ion, $[OH^-]_0$, with all other conditions the same were

Expt #	[OH$^-$]$_0$, M	Time to color change, s
1	0.0025	35
2	0.0030	43
3	0.0015	22

(a) What is the rate of disappearance of the hydroxide ion for each of the experiments? Explain your reasoning.
(b) What is the rate of disappearance of the 2-chloro-2-methylpropane, $(CH_3)_3CCl$, in each of the experiments? What do these results suggest about the dependence of the reaction rate on the concentration of hydroxide ion in the solution? Explain your reasoning.

11.16. (a) If the rate of formation of oxygen from ozone, $2O_3(g) \rightarrow 3O_2(g)$, is 1.8×10^{-3} M·s^{-1} at a certain temperature, what is the rate of decomposition of ozone under these conditions? Explain your reasoning.
(b) What is the reaction rate for this reaction? Explain.

11.17. 🔊 To measure the initial rate of the reaction of iodide ion, $I^-(aq)$, with persulfate ion, $S_2O_8^{2-}(aq)$, a small amount of thiosulfate ion, $S_2O_3^{2-}(aq)$, is added to the reaction mixture. See the "Summary of the Reactions" in the *Web Companion*, Chapter 11, Section 11.2.6, to see why the solution remains clear and colorless until all the thiosulfate has reacted and then suddenly changes color.
(a) This reaction with added thiosulfate is like the changes in the Blue-Bottle reaction, which remains blue for a time and then suddenly changes to colorless, Investigate This 10.61, Chapter 10, Section 10.8. The coupled reactions responsible for the observations on the Blue-Bottle reaction are shown in Figure 10.12. Make a sketch, patterned after Figure 10.12, that shows the coupling of reactions in the iodide-persulfate-thiosulfate system.
(b) What is the coupling species in the iodide-persulfate-thiosulfate system? What reactions are being coupled?
(c) What are the relative rates of the coupled reactions in part (b)? What is the experimental evidence and how do you reason from it to get your response?
(d) During the time the reaction mixture remains colorless, how does the concentration of iodide ion compare with its initial concentration? Explain.
(e) During the time the reaction mixture remains colorless, there is no *net* formation of iodine, $I_2(aq)$. However, the plot shown in the *Web Companion* is labeled with a $\Delta[I_2]$ during this time period, Δt. Why? To what change does this $\Delta[I_2]$ correspond? Explain.
(f) As you consider the points in part (e), do you think the graph might be somewhat misleading? Explain why or why not and, if it's misleading, suggest a way to fix it.

11.18. The reaction $2CO(g) \rightarrow CO_2(g) + C(s)$, which occurs at high temperatures, can be studied by adding $CO(g)$ to a hot reaction vessel and following the decrease in total pressure, P_{total}, of the gases in the constant volume reactor

Time, s	P_{total}, kPa
0	33.2
400	31.6
1000	29.8
1800	27.9

(a) Calculate the average rate of disappearance of $CO(g)$ in kPa·s^{-1} for each time interval. Explain your reasoning. *Hint:* P_{total} is proportional to the number of moles of gas.
(b) Calculate the average rate of appearance of $CO_2(g)$ for each time interval. Explain.

11.3. Reaction Rate Laws

11.19. If a reaction is second order in the concentration of reactant **A** and first order in the concentration of reactant **B,** how will the rate of reaction change if the concentrations of **A** and **B** are both halved? Explain your reasoning.

11.20. For the reaction $CH_3Br(aq) + OH^-(aq) \rightarrow CH_3OH(aq) + Br^-(aq)$, the initial rate of reaction was found to decrease by a factor of two when the concentration of hydroxide ion, $[OH^-(aq)]$, was halved. An increase in the concentration of bromomethane, $[CH_3Br(aq)]$, by a factor of 1.4 increased the rate by a factor of 1.4.
(a) What is the order of this reaction with respect to $[OH^-(aq)]$? Explain your reasoning.
(b) What is the order of this reaction with respect to $[CH_3Br(aq)]$? Explain.
(c) What is the overall order of this reaction? Write the rate law for this reaction. Explain your reasoning.

11.21. Consider the gas-phase reaction $2NO(g) + Cl_2(g) \rightarrow 2NOCl(g)$. Studies show that the initial rate of disappearance of $NO(g)$ is eight times faster when the concentrations of both reactants are doubled and twice as fast when the concentration of chlorine alone is doubled.
(a) What is the order of this reaction with respect to $[Cl_2(g)]$? Explain your reasoning.
(b) What is the order of this reaction with respect to $[NO(g)]$? Explain.
(c) What is the overall order of this reaction? Write the rate law for this reaction. Explain your reasoning.

11.22. The rate of a certain reaction is proportional to the concentration of a reactant and the concentration of a catalyst. During experimental runs on this reaction, the concentration of the catalyst remains constant and the measured initial rates of reaction can be used to determine a rate constant, 8.9×10^{-5} s^{-1}, for the first-order disappearance of the reactant.
(a) If the catalyst concentration is 5.0×10^{-3} M, what is the true rate constant for the overall second order reaction? Explain your reasoning.
(b) What would be the observed initial rate of reaction if the reactant concentration is 0.12 M and the catalyst concentration is 3.5×10^{-3} M? Explain.

11.23. The initial rate of a reaction that is first order in each of two reactants was found to be 7.65×10^{-4} M·s^{-1}, when the initial concentration of each reactant was 0.050 M. What is the numeric value of the rate constant for this reaction? Explain your solution.

11.24. The reaction $2NO(g) + O_2(g) \rightarrow 2NO_2(g)$ was studied by following the initial rate of disappearance of nitric oxide, $-(\Delta NO(g)/\Delta t)_0$, for different initial reactant concentrations.

Expt #	$[NO(g)]_0$, M	$[O_2(g)]_0$, M	$-(\Delta NO(g)/\Delta t)_0$, M·s^{-1}
1	0.0125	0.0183	0.0202
2	0.0250	0.0183	0.0803
3	0.0125	0.0370	0.0409

(a) What is the order of the reaction with respect to $[NO(g)]$? Explain your reasoning.
(b) What is the order of the reaction with respect to $[O_2(g)]_0$? Explain your reasoning.
(c) What is the reaction rate (in M·s^{-1}) for this reaction in each experiment? Explain.
(d) Write the rate law for the reaction and determine the numerical value of the reaction rate constant. Explain your approach and the units of the rate constant.
(e) If the reaction were carried out with $[NO(g)]_0 = [O_2(g)]_0 = 0.0200$ M, what should be the observed rate of disappearance of nitric oxide? Explain.

11.25. In Problem 11.15 some experimental details and data were given for a study of this net reaction: $(CH_3)_3CCl + OH^- \rightarrow (CH_3)_3COH + Cl^-$. Here are more data from this study for experiments in which the initial concentration of hydroxide ion, $[OH^-]_0$, was the same, 0.0025 M, and the initial concentration of 2-chloro-2-methylpropane, $[(CH_3)_3CCl]_0$, was varied.

Expt #	$[(CH_3)_3CCl]_0$, M	Time to color change, s
1	0.015	35
4	0.030	18

(a) What is the rate of disappearance of the hydroxide ion for each of the experiments? Explain your reasoning.
(b) What is the rate of disappearance of the 2-chloro-2-methylpropane, $(CH_3)_3CCl$, in each of the experiments? What do these results suggest about the dependence of the reaction rate on the concentration of 2-chloro-2-methylpropane, $[(CH_3)_3CCl]$, in the solution? Explain your reasoning.
(c) Based on your results from part (b) and from Problem 11.15, write a rate law for this reaction. Explain your reasoning.
(d) What is the numerical value of the rate constant in the rate law you wrote in part (c)? Explain.

11.26. In Check This 11.22, you worked through the *Web Companion*, Chapter 11, Section 11.3.3–6, to find the rate law for this reaction:

$$S_2O_8^{2-}(aq) + 2I^-(aq) \rightarrow I_2(aq) + 2SO_4^{2-}(aq)$$

(a) Can you use the data given on these pages of the *Web Companion* to determine the numerical value for the reaction rate constant in the rate law? If so, what is the value? If not, explain clearly why not.
(b) If you knew that each of the samples in these experiments was initially 0.0015 M in thiosulfate ion, $S_2O_3^{2-}(aq)$, would your answer to part (a) be different? If so, explain specifically how it would be different. If not, explain why not.

11.27. In aqueous solution, permanganate ion, $MnO_4^-(aq)$, reacts with chromium(III) ion, $Cr^{3+}(aq)$: $MnO_4^-(aq) + Cr^{3+}(aq) \rightarrow Mn^{4+}(aq) + CrO_4^{2-}(aq)$. The reaction rate can be studied by measuring the time required for the concentration of $CrO_4^{2-}(aq)$ to increase from zero to 0.020 M. The results for different initial concentrations of the two reactants are

Expt #	Relative $[MnO_4^-(aq)]_0$	Relative $[Cr^{3+}(aq)]_0$	Time to $[CrO_4^{2-}(aq)] = 0.020$ M
1	1	1	23 min
2	2	1	11 min
3	1	0.5	45 min

(a) What type of reaction (see Chapter 6) is this? Explain how you know.
(b) How are the times in the table related to the rates of the reaction? Explain.
(c) What is the reaction order with respect to the concentration of each of the reactants? How do you know?
(d) Write the rate law for the reaction.
(e) In Experiment #3, if the relative $[MnO_4^-(aq)]_0$ had also been changed to 0.5, how long would it have taken for $[CrO_4^{2-}(aq)]$ to reach 0.020 M? Explain your reasoning.

11.28. Ammonia decomposes on a hot tungsten filament: $2NH_3(g) \rightarrow N_2(g) + 3H_2(g)$. The reaction can be followed by measuring the increase in pressure, ΔP, in the system.

Time, s	0	100	200	400	600	800	1000
ΔP, kPa	0	1.46	2.94	5.85	8.81	11.68	14.61

(a) Plot these data and try to deduce the order of reaction with respect to the pressure (or concentration) of ammonia. State the reasoning for your answer.
(b) The initial pressure of ammonia in the system was 26.00 kPa. What is the initial rate of reaction in kPa·s^{-1}? What is the rate constant for the reaction? Explain your reasoning.

11.29. In a series of experiments, the decomposition of hydrogen peroxide, $H_2O_2(aq)$, to water and oxygen gas, $2H_2O_2(aq) \rightarrow 2H_2O(l) + O_2(g)$, was studied using a titration method to determine the concentration of reactant left after 30 minutes of reaction, $[H_2O_2(aq)]_{30}$. The data for various starting concentrations, $[H_2O_2(aq)]_0$ are given in this table.

Expt #	$H_2O_2(aq)]_0$, M	$[H_2O_2(aq)]_{30}$, M
1	0.874	0.812
2	0.356	0.331
3	0.589	0.549

(a) What is the average rate of disappearance of hydrogen peroxide in each of these experiments? Explain.
(b) What is the order of the reaction with respect to $[H_2O_2(aq)]$? Explain your reasoning.
(c) Write the rate law for this reaction. Show how to calculate the rate of reaction for each experiment and the rate constant for the reaction.
(d) If another experiment was done under the same conditions, but starting with 0.234 M hydrogen peroxide, what do you predict would be the concentration of unreacted peroxide after 30 minutes of reaction? Explain.

11.30. In aqueous acidic solution, hydrogen peroxide oxidizes iodide ions to elemental iodine and is, itself, reduced to water. In separate experiments, with all other conditions the same, doubling the initial hydrogen peroxide concentration or doubling the iodide ion concentration caused the initial rate of formation of iodine to double. The rate of formation of iodine was four times higher at pH 1.4 compared to pH 2.0, with all other conditions the same.
(a) Write the balanced equation for this redox reaction.
(b) What are the orders of reaction with respect to $[H_2O_2(aq)]$, $[I^-(aq)]$, and $[H^+(aq)]$? Write the rate law for this reaction. Explain how the rate data lead to these conclusions.
(c) If, under particular conditions, the rate of formation of iodine is 2.5×10^{-3} M·s^{-1}, what is the rate of disappearance of hydrogen peroxide? Of iodide? Explain.
(d) What effect would increasing the pH have on the rate constant for the reaction? Explain your answer.
(e) If an initial reaction solution is diluted by a factor of two by addition of water, what is the effect on the rate constant for the reaction? How is the rate of the reaction affected? Explain the reasoning for your answers.

11.31. The decomposition reaction for azomethane is $CH_3NNCH_3(g) \rightarrow CH_3CH_3(g) + N_2(g)$.
(a) If the initial pressure of azomethane in a reaction vessel is 10.0 kPa, what will be the total pressure in the vessel after one-quarter of the azomethane has decomposed? *Hint:* How will the total number of moles in the flask change if x moles of azomethane decomposes? How is the total pressure in the vessel related to the total number of moles?
(b) Azomethane was introduced into a reaction vessel at 300 °C. The initial pressure of azomethane was 20.0 kPa

and after 2.00 minutes the pressure in the vessel had increased to 21.0 kPa. What fraction of the azomethane had decomposed in this time? What was the initial rate of decomposition (in kPa·s^{-1})? Explain.
(c) In another experiment at 300 °C, the initial pressure of azomethane was 5.00 kPa and the total pressure in the vessel reached 5.25 kPa after 2.00 minutes. What was the initial rate of decomposition of azomethane in this case? How does this rate compare to what you found in part (b)? Show how to use your results to determine the order of the reaction with respect to azomethane. Write the rate law for the reaction and calculate the numeric value of the rate constant. Explain your approach.

11.4. Reaction Pathways or Mechanisms

11.32. In Worked Example 11.31, we stated that "the sum of the series of reactions (11.25), (11.26), and (11.27) is the stoichiometric reaction (11.16)." Show that this statement is true.

11.33. Possible pathways for the reaction of nitrogen dioxide with carbon monoxide are

(i) $NO_2(g) + CO(g) \rightarrow NO(g) + CO_2(g)$ $k_{(i)}$
(ii) $NO_2(g) + NO_2(g) \rightleftharpoons NO(g) + NO_3(g)$
fast equilibrium, $K_{(ii)}$
$NO_3(g) + CO(g) \rightarrow NO_2(g) + CO_2(g)$
slow, $k_{(ii)}$
(iii) $NO_2(g) + NO_2(g) \rightarrow NO(g) + NO_3(g)$
slow, $k_{(iii)}$
$NO_3(g) + CO(g) \rightarrow NO_2(g) + CO_2(g)$ fast

(a) What is the net reaction for each pathway?
(b) Determine the rate law predicted for the reaction by each of the pathways. Explain your reasoning.
(c) What rate experiment(s) would you suggest performing to distinguish among these pathways? What would be the difference(s) in the predicted result(s) from your experiment(s) for each pathway?

11.34. Consider the gas phase reaction of nitric oxide with chlorine to form NOCl:

$$2NO(g) + Cl_2(g) \rightleftharpoons 2NOCl(g)$$

The experimentally determined rate law for this reaction is rate $= k[NO(g)]^2[Cl_2(g)]$. Although this rate law is consistent with a one-step, termolecular (three molecule) reaction, a simultaneous collision among three molecules is not very probable. An alternative pathway has been proposed for which the first step is

$$NO(g) + Cl_2(g) \rightleftharpoons NOCl_2(g) \text{ fast equilibrium, } K$$

(a) Propose a second, rate-limiting step that is consistent with the overall stoichiometry of the reaction.
(b) What overall rate law would be predicted for this two-step pathway? Explain.
(c) How might researchers distinguish experimentally between this alternative pathway and the one-step termolecular pathway?

11.35. In Problems 11.15 and 11.25 some experimental details and data were given for a study of this net reaction: $(CH_3)_3CCl + OH^- \rightarrow (CH_3)_3COH + Cl^-$. A proposed mechanism for this net reaction in a solvent mixture of acetone and water is

$$(CH_3)_3CCl \rightleftharpoons (CH_3)_3C^+ + Cl^-$$
$$\text{slow equilibrium, } K$$

$$(CH_3)_3C^+ + H_2O \rightarrow (CH_3)_3COH_2^+ \qquad \text{fast}$$
$$(CH_3)_3COH_2^+ + OH^- \rightarrow (CH_3)_3COH + H_2O \quad \text{fast}$$

(a) Show that this mechanism gives the observed net reaction in the presence of hydroxide ion in solution.
(b) What is the reaction rate law predicted by this mechanism? Show your reasoning.
(c) Is the rate law you wrote in part (b) consistent with the one you wrote in Problem 11.25(c)? Explain why or why not.

11.36. A possible mechanism for the reaction studied in Problem 11.13 is (the reactants and products are solvated by the mixed solvent, but the solvation is not indicated here)

$$CH_3OH + H^+ \rightleftharpoons CH_3OH_2^+ \qquad \text{fast equilibrium, } K$$
$$CH_3OH_2^+ + Cl^- \rightarrow CH_3Cl + H_2O \qquad \text{slow, } k$$

(a) What is the reaction rate law predicted by this mechanism? Explain your reasoning.
(b) When more data for this reaction, like those in Problem 11.13, were analyzed, the analysis showed that the reaction rate appeared to depend on the square of the hydronium ion concentration, $[H^+]^2$. When the hydrochloric acid, HCl, concentration was doubled, the rate of reaction quadrupled. Is this result consistent with the rate law you wrote in part (a)? If so, show how. If not, show why not.

11.37. Consider these three pathways proposed for the gas phase decomposition of dinitrogen pentoxide, $N_2O_5(g)$:
(i) $N_2O_5(g) \rightleftharpoons NO_2(g) + NO_3(g)$
$$\text{fast equilibrium, } K_{(i)}$$
$$NO_2(g) + NO_3(g) \rightarrow NO(g) + NO_2(g) + O_2(g)$$
$$\text{slow, } k_{(i)}$$
$$NO(g) + NO_3(g) \rightarrow 2NO_2(g) \qquad \text{fast}$$
(ii) $N_2O_5(g) \rightleftharpoons NO_2(g) + NO_3(g)$
$$\text{fast equilibrium, } K_{(ii)}$$
$$NO_3(g) + N_2O_5(g) \rightarrow$$
$$N_2O_4(g) + NO_2(g) + O_2(g) \qquad \text{slow, } k_{(ii)}$$
$$N_2O_4(g) \rightarrow 2NO_2(g) \qquad \text{fast}$$
(iii) $N_2O_5(g) \rightarrow NO_2(g) + NO_3(g) \qquad \text{slow, } k_{(iii)}$
$$NO_3(g) + N_2O_5(g) \rightarrow 3NO_2(g) + O_2(g) \qquad \text{fast}$$
(a) What is the net reaction represented by each of these pathways? Explain your reasoning. *Hint:* Sometimes an elementary reaction has to occur more than once in order to satisfy the stoichiometry of the other elementary reaction(s).
(b) What is the rate law for $N_2O_5(g)$ decomposition predicted by each of these pathways? Explain.

(c) In a series of experiments, the initial rate of disappearance of $N_2O_5(g)$ was found to be a linear function of the initial concentration of $N_2O_5(g)$. What is the experimental rate law suggested by these results? Explain your reasoning.
(d) Does your rate law from part (c) allow you to rule out any of the proposed pathways? If so, which one(s) and why? If not, why not?
(e) What other experiments can you suggest that might help to distinguish among the proposed pathways? What distinguishing results would you expect?

11.38. In a continuing study of the system discussed in Problems 11.15, 11.25, and 11.35, a series of experiments was carried out in which salt, sodium chloride, was dissolved in reaction mixtures that were otherwise identical. The initial concentration of hydroxide ion, $[OH^-]_0$, was 0.0025 M in all these experiments.

Expt #	[NaCl], M	Time to color change, s
1	0	35
5	0.85	35
6	1.71	43
7	2.56	95
8	3.42	268

(a) What is the rate of disappearance of the hydroxide ion for each of the experiments? Explain your reasoning.
(b) What is the rate of disappearance of the 2-chloro-2-methylpropane, $(CH_3)_3CCl$, in each of the experiments? What is the effect of dissolved sodium chloride on the rate of this reaction? Explain your reasoning.
(c) What species in the solution do you think is having the effect you found in part (b)? Show how this effect provides evidence in support of the reaction mechanism in Problem 11.35. Be as specific as possible in your explanation.

11.39. In Sections 11.2., 11.3, and 11.4, we discussed the decomposition of hydrogen peroxide catalyzed by iodide ion in pH 7 solutions. In acidic solutions, as you found in Problem 11.30, the reaction is quite different: $H_2O_2(aq) + 2I^-(aq) + 2H^+(aq) \rightarrow I_2(aq) + 2H_2O(aq)$. The rate law for the reaction is

$$\text{rate} = k[H_2O_2(aq)][I^-(aq)][H^+(aq)]$$

(a) Hydrogen peroxide is a weak Brønsted–Lowry base (like water). In acidic solution, what might be a reasonable reaction for $H_2O_2(aq)$ to undergo? Explain your reasoning. *Hint:* See Check This 11.32.
(b) Write the rate-limiting step for reaction in acidic solution, assuming it is analogous to the rate-limiting step in pH 7 solution. *Hint:* See Check This 11.32.

(c) We said that $E^{o'} = 0.5$ V provides a strong driving force for reaction (11.22) in pH 7 solution:

$$HOI(aq) + H_2O_2(aq) \rightarrow$$
$$O_2(g) + I^-(aq) + H^+(aq) + H_2O(l)$$

What effect will a decrease in pH of the solution have on the cell potential for this reaction? What is the cell potential for this reaction at pH 0? *Hint:* Use Le Chatelier's principle and the Nernst equation.
(d) Use the data in Appendix C to find the cell potential for this reaction at pH 0:

$$HOI(aq) + I^-(aq) + H^+(aq) \rightarrow I_2(aq) + H_2O(l)$$

(e) Although thermodynamically favorable reactions are not necessarily fast, if the reactions in parts (c) and (d) both can proceed, which is more favored at pH 0? Could this comparison explain the difference between the reactions of $H_2O_2(aq)$ with $I^-(aq)$ at pH 0 and pH 7? Show why or why not.
(f) Use the results from the preceding parts to suggest a reaction pathway for the reaction of $H_2O_2(aq)$ with $I^-(aq)$ at pH 0 that fits the rate law and the stoichiometry.

11.40. (a) You can interpret reactions (11.21) and (11.22) as a pair of coupled reactions. What pair of reactants serves to couple the two reactions? Explain.
(b) Make a sketch, patterned after Figures 10.12, 10.14, and 10.15, that shows the coupling of reactions in this $I^-(aq)$-catalyzed decomposition of $H_2O_2(aq)$.
(c) In Chapter 10, Section 10.8, we said that coupling species do not appear in the net reaction for the coupled series of reactions. Is that true in this case as well? Explain.

11.41. Two possible pathways for the gas phase reaction of iodine and hydrogen are

(i) $\quad H_2(g) + I_2(g) \rightleftharpoons 2HI(g)$ $\qquad k_{(i)}$
(ii) $\quad I_2(g) \rightleftharpoons I(g) + I(g)$ $\quad$ fast equilibrium, $K_{(ii)}$
$\qquad H_2(g) + I(g) + I(g) \rightarrow 2HI(g)$ $\quad$ slow, $k_{(ii)}$

(a) Show that both pathways predict the same rate law for the reaction.
(b) Pathway (ii) involves a collision among three particles. Why is this reaction slow? Suggest alternative two-body collisions that would give the same result.

11.42. The net reaction of atmospheric ozone with nitric oxide is

$$O_3(g) + NO(g) \rightarrow NO_2(g) + O_2(g)$$

The experimental rate law for this reaction is

$$-\frac{\Delta[O_3(g)]}{\Delta t} = k[O_3(g)][NO(g)]$$

Among the mechanisms proposed for this reaction are

(i) $\quad O_3(g) + NO(g) \rightarrow O(g) + NO_3(g)$ $\quad$ slow, $k_{(i)}$
$\qquad O(g) + O_3(g) \rightarrow 2O_2(g)$ $\qquad$ fast
$\qquad NO_3(g) + NO(g) \rightarrow 2NO_2(g)$ $\qquad$ fast
(ii) $\quad O_3(g) \rightleftharpoons O(g) + O_2(g)$ $\quad$ fast equilibrium, $K_{(ii)}$
$\qquad NO(g) + O(g) \rightarrow NO_2(g)$ $\qquad$ slow, $k_{(ii)}$

(a) Show that both mechanisms agree with the overall reaction stoichiometry.
(b) What is the rate law predicted for each mechanism? Do these results help you determine whether either of these mechanisms is plausible? Explain why or why not.
(c) Propose another mechanism that would agree with both the reaction stoichiometry and the experimental rate law.

11.5. First-Order Reactions

11.43. In Figure 11.6(b), the horizontal lines showing the fraction of radioactive nuclei present, 1.00, 0.50, and 0.25, are equally spaced on the ln(cpm) axis. What is the spacing? How do you explain this value and why is the spacing equal?

11.44. Phosphorus-32, ^{32}P, is used extensively in molecular biology and biochemical research. To save the expense of radioactive waste disposal, laboratories often store their ^{32}P wastes until the activity has decreased to less than 1% of its starting value and then dispose of it as chemical waste (a safe and appropriate procedure). Suppose you are the laboratory safety officer and assume that you collect the ^{32}P wastes weekly. How many weeks would you have to keep each collection to assure that its activity is less than 1% of the starting value? *Hint:* Use data from Figure 11.6.

11.45. A specimen of bone from an archaeological dig has a carbon-14 radioactivity of 2.93 counts·min^{-1}. If, under the same counting conditions, the activity found in samples of living plants and animals is 12.6 counts·min^{-1}, what is the age of the bone specimen? Explain. *Hint:* Use data from Check This 11.37.

11.46. An antibiotic, **A,** is metabolized in the body by a first-order reaction:

$$-\frac{\Delta[A]}{\Delta t} = k[A]$$

The rate constant, k, depends upon temperature and body mass. At 37 °C, $k = 3.0 \times 10^{-5}$ s^{-1} for a 70-kg person.
(a) How frequently must a 70-kg person take 400-mg pills of the antibiotic, in order to be sure to keep the concentration of antibiotic from falling below 140 mg per 70 kg body mass? You can assume that, upon ingestion, the antibiotic is almost immediately distributed uniformly throughout the body mass. Explain the reasoning for your answer.
(b) At 39 °C, $k = 4.0 \times 10^{-5}$ s^{-1}. If the person in part (a) has a fever of 39 °C, how often must she or he take the antibiotic to maintain a concentration at or above the effective level of 140 mg per 70 kg body mass? Explain.

11.47. These data are for the high temperature (504 °C) decomposition of gaseous dimethyl ether by the net reaction $(CH_3)_2O(g) \rightarrow CH_4(g) + H_2(g) + CO(g)$

Time, sec	0	390	777	1195	3175
Pressure $(CH_3)_2O(g)$, kPa	41.5	35.1	29.8	24.9	10.4

Is the reaction first order, as the net reaction suggests? If so, determine the rate constant for the reaction. If not, explain how you draw this conclusion.

11.48. When heated, cyclopropane, $C_3H_6(g)$, isomerizes to propene, $CH_3CH{=}CH_2(g)$, in a first-order reaction with a rate constant of $6.0 \times 10^{-4}\,s^{-1}$ at 773 K. If cyclopropane at a pressure of 56.7 kPa is placed in a reaction vessel at 773 K, what will be the pressure of propene in the vessel after 15.0 minutes? Explain your reasoning.

11.49. Bacteria in a nutrient medium cause it to become cloudy: The more bacteria, the cloudier the medium. The number (concentration) of bacteria in a nutrient medium is sometimes determined by measuring the absorbance of the medium, which is directly proportional to the number of bacteria present. These are data for a bacterial growth experiment.

Time, min	Absorbance
0	0.053
15	0.095
30	0.167
45	0.301
60	0.533

During the part of their growth called the "log phase," the rate law for bacterial growth is

$$\frac{\Delta(\text{number of bacteria})}{\Delta t} = k(\text{number of bacteria})$$

This equation for the appearance (growth) of bacteria is identical in form to equation (11.35) for the first-order rate of disappearance of a reactant, with the exception that the term on the left here does not have a negative sign.
(a) Why does this equation for growth not have a negative sign?
(b) Use the method outlined in the text (or calculus) to show that the equation above leads to the result: ln(number of bacteria)$_t$ = kt + ln(number of bacteria)$_0$. *Hint:* For small values of f, ln(1 + f) ≈ f.
(c) Plot the data in the table first on an absorbance *vs.* time graph and then in a way that will permit you to determine

the rate constant, k, for the growth. Why is this growth phase called the log phase? Explain your reasoning.
(d) First-order decays or disappearances are characterized by their half-lives. Growth curves are characterized by their doubling times. Show how to use the value of k from part (c) to obtain the doubling time for this bacterial growth.
(e) In another nutrient medium, the doubling time for these bacteria was about 13 minutes. What conclusion(s) can you draw about this medium relative to the first?

11.50. Regulations from the U.S. Public Health Service specify that milk may contain a maximum of 20,000 bacteria per mL when it has been freshly pasteurized. At 5 °C, the usual temperature of a refrigerator compartment, these bacteria are reported to have a doubling time of about 40 hours. After milk is stored in a refrigerator for 10 days, what would we expect the maximum bacteria count to be? Explain your reasoning. *Hint:* See Problem 11.49.

11.51. The decomposition of dinitrogen pentoxide has been studied in the gas phase and in solution as a model system to see how solvents affect reactions. The stoichiometry of the decomposition reaction in carbon tetrachloride, $CCl_4(l)$ (tetrachloromethane), solution is

$$2N_2O_5(CCl_4) \rightarrow 4NO_2(CCl_4) + O_2(g)$$

This is the same net reaction as in the gas phase. These are time and concentration, $[N_2O_5(CCl_4)]$, data for the solution phase reaction at 45 °C.

Time, min	0	184	319	526	867	1198	1877
$[N_2O_5(CCl_4)]$, M	2.33	2.08	1.91	1.67	1.35	1.11	0.72

(a) How many days did this experiment take?
(b) Plot these data on a concentration *vs.* time graph and then in a way that will permit you to determine the numeric value of the rate constant, k, for the reaction, assuming it is first order with respect to $[N_2O_5(CCl_4)]$. Does your plot justify the assumption? Explain your reasoning.
(c) Is your conclusion in part (b) consistent with the gas phase results for this reaction from Problem 11.37? Explain why or why not.
(d) This reaction was followed by measuring the volume of oxygen produced. If the reaction solution volume was 25.0 mL and the evolved gas was measured at a pressure of 95.5 kPa at 45 °C, what volume of gas was measured after the first 184 minutes of reaction? What was the total volume of gas evolved? Explain.

11.52. Show that the time required for a first-order reaction to go to 99.0% completion is twice as long as the time it takes for the reaction to go to 90.0% completion.

11.6. Temperature and Reaction Rates

11.53. (a) If the rate of a reaction at 55 °C is 6.7 times the rate of the reaction at 35 °C, what is the activation energy for the reaction? Explain your reasoning.
(b) At what temperature would the rate of this reaction be twice that at 35 °C? Explain.

11.54. Rate constants for the reaction $2NOCl(g) \rightarrow 2NO(g) + Cl_2(g)$ are 9.4×10^{-6} s^{-1} at 350 K and 6.9×10^{-4} s^{-1} at 400 K.
(a) What is the rate law for this reaction? How do you know?
(b) What is the activation energy for this reaction? Explain your reasoning.
(c) What is the Arrhenius frequency factor for this reaction? Explain.

11.55. Rate constants for the reaction $2N_2O_5 \rightarrow 4NO_2 + O_2$ at two temperatures in the gas phase and in carbon tetrachloride solution are

	25 °C	45 °C
Gas phase	3.38×10^{-5} min^{-1}	43.0×10^{-5} min^{-1}
CCl$_4$ solution	4.69×10^{-5} min^{-1}	Problem 11.51(b)

(a) Determine the activation energy and Arrhenius frequency factor for the gas phase reaction. Explain your reasoning.
(b) Determine the activation energy and Arrhenius frequency factor for the reaction in solution.
(c) In Problem 11.51, we said that this reaction had been studied in the gas phase and in several solvents to see how the rate parameters compare. Does the solvent carbon tetrachloride have a large effect on the rate parameters calculated here? Explain the basis for your conclusion.

11.56. California ground squirrels provoke northern Pacific rattlesnakes to get them to rattle and reveal their temperature (related to rattle frequency), which is an indication of the speed with which they can strike. Cold snakes are more lethargic and their aim is poorer. Use the data from the figure at the right to estimate the activation energy for the rattling process. Explain how you obtain your answer.

11.57. A study has been done to find the optimum temperature to store milk in order to get the best foam for espresso coffee drinks like cappucinos and lattes. Among the conclusions of the study: "Milk should be stored at no more than 40 °F. Every 5 °F increase in storage temperature cuts shelf life by half." What is the activation energy for the process that "cuts the shelf life," that is, spoils the milk? Explain clearly how you obtain your answer.

11.58. Raw (unpasteurized) milk sours in about 9 hours at 20 °C (room temperature), but it takes about 48 hours to sour in a refrigerator at 5 °C.
(a) What is the activation energy for souring of raw milk? Explain your solution.
(b) The milk referred to in Problem 11.57 is presumably pasteurized milk. How does the activation energy for souring of raw milk compare with the activation energy for spoilage of pasteurized milk? What might be the cause of any difference?

11.59. The cyclopentadiene dimer dissociates in the gas phase to cyclopentadiene:

$$\text{dimer} \rightarrow 2 \quad \text{cyclopentadiene}$$

The Arrhenius frequency factor and activation energy for this reaction are 1.3×10^{13} s^{-1} and 146 kJ·mol^{-1}, respectively.
(a) What is the rate constant for this reaction at 200 °C? Explain.
(b) What is the rate of the reaction at 200 °C, if the amounts are stated in pressure units, kPa? Explain.
(c) If a 15.0 kPa sample of the dimer is placed in a 200 °C reaction vessel, how long will it take for 10% of the dimer to dissociate to cyclopentadiene? Explain your reasoning.

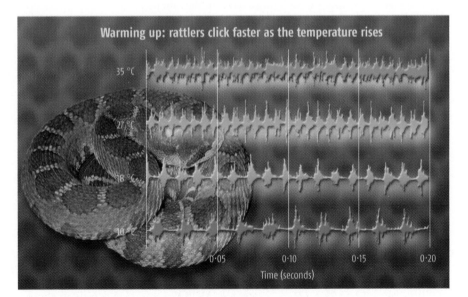

Warming up: rattlers click faster as the temperature rises

35 °C

27 °C

18 °C

10 °C

0·05   0·10   0·15   0·20

Time (seconds)

11.60. The hydrolysis of sucrose, reaction equation (11.66), is also catalyzed by [H$^+$(aq)]. Data for the temperature dependence of the rate constant for this reaction are

Temperature, °C	18	25	32	40	45
k, M^{-1}·s^{-1}	0.0022	0.0054	0.013	0.032	0.055

(a) Use a graphical method to find the activation energy and Arrhenius frequency factor for the acid-catalyzed reaction. Explain your method.
(b) What is the rate law for this reaction? How do you know?
(c) What is the half-life for sucrose at 37 °C, in a solution with [H$^+$(aq)] = 1.0 M? Explain your reasoning.

11.61. Development of a photographic image is controlled by the kinetics of silver halide reduction by the developer. Data for development time for a commercial developer are

Temperature, °C	18	20	21	22	24
Development time, min	10	9	8	7	6

Show how to use these data to estimate the activation energy for the reaction.

11.7. Light: Another Way to Activate a Reaction

11.62. Hydrogen peroxide is always packaged in opaque containers. The container says to keep the hydrogen peroxide out of the light and away from high temperatures.
(a) What effect do you predict light to have on the rate of decomposition? Explain how you reach your conclusion.
(b) What effect do you predict high temperature to have on the rate of decomposition? Explain how you reach your conclusion.
(c) In Investigate This 11.6, you found that many substances catalyze the decomposition of hydrogen peroxide. Would it be a bad idea to dip your finger, or hair, or even a cotton swab into a bottle of hydrogen peroxide from the drugstore? Explain your reasoning.

11.63. Some molecules, **M**, after absorbing a photon that boosts them to an excited state, **M*** (Chapter 4, Section 4.5), emit a photon to return to the ground state. Usually the emission is at a longer wavelength (lower energy) than the absorbed radiation. Excited molecules may also return to the ground state by transferring their energy to other molecules, **Q**:

   (i) **M** + $h\nu_1 \rightarrow$ **M***   rate = (constant)I[**M**]
  (ii) **M*** $\rightarrow$ **M** + $h\nu_2$ rate = k_1[**M***]
 (iii) **M*** + **Q** $\rightarrow$ **M** + **Q***   rate = k_2[**M***][**Q**]

In some cases, **Q*** also loses its energy by emission of light, but in many cases it dissipates the energy in other ways. If a system containing **M** and **Q** molecules is irradiated with a constant intensity, I, of light with a frequency ν_1, a steady state is reached for which the intensity of the emitted radiation, I_e, at frequency ν_2, is a constant.
(a) How does ν_2 compare to ν_1? Explain.
(b) If the intensity of emitted radiation, I_e, is directly proportional to [**M***], what can you say about [**M***] at the steady state? Explain your reasoning.
(c) Derive an expression for the [**M***]/[**M**] ratio at the steady state in terms of the variables in the rate expressions.
(d) How would the intensity of emitted radiation, I_e, vary as [**Q**] varies? Be as specific as you can in your explanation. Explain why molecules like **Q** are often called *quenchers*.

11.8. Thermodynamics and Kinetics

11.64. In Chapter 8, we found that processes that resulted in a net entropy increase were spontaneous, that is, they could occur without any outside influence. Do spontaneous reactions always occur quickly? Explain your response.

11.65. Hydrogen–oxygen mixtures are highly explosive and dangerous. Yet mixtures of hydrogen and oxygen can be kept for long periods of time (perhaps millennia) without reacting. How can you explain this apparent contradiction?

11.66. The dissociation of the dimer of cyclopentadiene shown in Problem 11.59 also proceeds in the reverse direction to form the dimer:

The Arrhenius frequency factor and activation energy for this reaction are 1.3 × 10^6 M^{-1}·s^{-1} and 70. kJ·mol^{-1}, respectively.
(a) Use the information here and in Problem 11.59 to sketch and clearly label an activation energy diagram (energy vs. progress of reaction) for this system.
(b) From your plot in part (a), what is your estimate of the standard enthalpy change, $\Delta H°$, for the above reaction? Explain.
(c) Show how to use bond enthalpies, Table 7.3, Chapter 3, Section 7.7, to estimate $\Delta H°$ for the above reaction. How does the result compare to your estimate in part (b) from the kinetic data? Is the difference, if any, between your estimates what you might expect on the basis of the molecular structures? Explain.
(d) Cyclopentadiene is a volatile liquid, b.p. 40 °C, with a density of about 0.8 kg·L^{-1}. What is the molarity of

the pure liquid? What is the initial rate of the above second-order dimerization reaction in the pure liquid at 25 °C? Explain your method.

(e) About how long will it take for 10% of the cyclopentadiene in the pure liquid to form dimer at 25 °C? Explain your reasoning and any assumptions you make.

(f) Cyclopentadiene is a useful reactant and starting material for many chemical syntheses, but it is not available from chemical suppliers as the pure liquid. How is your result in part (e) related to this lack of availability of the pure liquid?

(g) Chemists who need pure cyclopentadiene purchase the dimer, a viscous liquid, b.p. 170 °C, heat it in a distillation apparatus, collect the volatile cyclopentadiene product, and use it immediately. Use your results from Problem 11.59 and this one to explain how this procedure works kinetically and thermodynamically.

11.67. Which of these statements about elementary reactions are correct and which incorrect? For the ones that are correct, show why they are correct. For the ones that are incorrect, show why they are incorrect and write a corrected statement.

(a) The equilibrium constant for a reaction is the ratio of the reverse to the forward rate at equilibrium.

(b) For a reaction with a very small equilibrium constant, the rate constant for the forward reaction is much smaller than the rate constant for the reverse reaction.

(c) For an endothermic reaction at equilibrium, an increase in temperature decreases both the forward and reverse reaction rates, but they are equal when equilibrium is reattained.

(d) For an exothermic reaction at equilibrium, an increase in temperature increases both the forward and reverse rate constants by the same proportion.

11.68. Consider the elementary reaction, $A \rightarrow B + C$, which is reversible and has been studied going in the reverse direction. The reaction, as written, is endothermic with $\Delta H° = 57$ kJ·mol^{-1}. Species C is colored, so the rate of the reaction can be followed and the equilibrium constant determined colorimetrically. When equal concentrations of B and C, both 0.046 M, were reacted at 25 °C, the initial rate of reaction was 1.12×10^{-5} M·s^{-1}. When equilibrium was attained (no further change in the color of the solution), the concentration of C in the solution was 0.027 M. A series of temperature studies on the rate of the reverse reaction gave an activation energy of 43 kJ·mol^{-1}.

(a) Sketch and label an activation energy diagram for the reaction written above. In particular, show the activation energies and enthalpy of reaction. Explain how you use the data to draw your diagram correctly.

(b) Show how to calculate the equilibrium constant for the reaction written above.

(c) Show how to calculate the rate constant, k_r, for the reverse reaction.

(d) Show how to calculate the rate constant, k_f, for the forward reaction.

(e) Show how to calculate the Arrhenius frequency factors for the forward, A_f, and reverse, A_r, reactions.

(f) If the temperature is increased, will the equilibrium constant increase or decrease? Answer this question using both a thermodynamic argument and a kinetic argument.

(g) Show how to calculate the equilibrium constant at 40 °C for the reaction written above.

11.69. The rate of a first-order reaction decreases with time as the reactant is used up, but the concentration of the reactant should eventually go essentially to zero (or below the limit of detectability). However, for some first-order reactions, the concentration of the reactant does not go to zero, but comes to some constant measurable value.

(a) What is probably happening in the case that the concentration comes to a constant value and does not go to zero?

(b) What experiments would you propose to test your explanation in part (a)? What would you expect the outcomes of the experiments to be if your explanation is correct?

11.70. In your investigations with phenolphthalein and high concentrations of hydroxide ion, [OH$^-$(aq)], Investigate This 11.1 and 11.4, you found that the color of the phenolphthalein dianion, P^{2-}(aq), disappeared from the solution—it all reacted by reaction (11.2) going in the forward direction: P^{2-}(aq) + OH$^-$(aq) $\rightarrow$ POH^{3-}(aq). At lower [OH$^-$(aq)], say 0.10 M, observations on this system are different. Initially, the rate of disappearance of the color is first order, rate = $k_{expt}[P^{2-}$(aq)], where the experimental first-order rate constant is, $k_{expt} = k_f[OH^-$(aq)], just as at the higher [OH$^-$(aq)]. However, as the reaction proceeds, k_{expt} gets smaller (even though [OH$^-$(aq)] is not changing appreciably) and the solution never completely decolorizes but comes to an unchanging light pink color. Compare these observations with the behavior described in Problem 11.69.

(a) How would you characterize the reacting system thermodynamically and kinetically when the solution reaches its unchanging pink color? Be as specific as possible in your explanation.

(b) Assume that reaction (11.2) represents an elementary reaction that can go in both the forward and reverse directions. The rate law for the forward reaction is written above. Write the rate law for the reverse reaction with rate constant k_r.

(c) When [OH$^-$(aq)] = 0.10 M, $k_r \approx \left(\frac{1}{10}\right)k_f[OH^-$(aq)]. Under these conditions, what is the [P^{2-}(aq)]/[POH^{3-}(aq)] ratio when the forward and reverse rates are the same? Show how you get your result.

(d) If [OH$^-$(aq)] = 1.0 M, what is the [P^{2-}(aq)]/[POH^{3-}(aq)] ratio when the forward and reverse rates are the same? Does this result explain

why the solution goes colorless at higher hydroxide concentrations? Explain why or why not.
(e) The reaction, as written, is exothermic. Why? Sketch and label an activation energy diagram for the reaction.
(f) If no other conditions change in the 0.10 M hydroxide solution, will the light pink color get darker or lighter if the temperature of the solution is increased? Explain.

11.10. Extension—Enzymatic Catalysis

11.71. These data are initial rates (1 μmol = 10^{-6} mol) of an enzyme-catalyzed reaction measured with the same amount of enzyme and varying initial concentrations of substrate, S.

$[S]_0$, M	V_0, μmol·min^{-1}
2.0×10^{-2}	60
2.0×10^{-3}	60
2.0×10^{-4}	48
1.5×10^{-4}	45
1.3×10^{-5}	12

(a) What is V_{max} for this reaction? Explain your reasoning.
(b) Why doesn't V_0 continue to increase for $[S]_0$ greater than 2.0×10^{-3} M?
(c) What is the approximate concentration of free enzyme in the reaction with $[S]_0 = 2.0 \times 10^{-2}$ M? Explain.
(d) What is the numeric value of K (with units) in the Michaelis–Menten equation for this reaction? Explain your approach to solving this problem. [Note that K has units, because it is conventionally expressed, equation (11.70), in terms of concentrations, instead of dimensionless concentration ratios.] *Hint:* See equation (11.77).

11.72. Experiments on an enzyme-catalyzed reaction that followed Michaelis–Menten kinetics showed that $K = 1.2 \times 10^5$ M^{-1} and that, with the same amount of enzyme, the initial rate of the reaction, 45 μmol·min^{-1} (1 μmol = 10^{-6} mol), was the same for substrate concentrations of 0.10 M and 0.010 M.
(a) Use the Michaelis–Menten equation to show why the initial reaction rate is the same for both of these substrate concentrations.
(b) Predict the initial rate of reaction for this same amount of enzyme at a substrate concentration of 2.0×10^{-5} M. Explain your reasoning. *Hint:* See equation (11.77).

11.73. Suppose that a mutant enzyme in an organism binds the substrate 10 times more tightly than the native (normal) enzyme, that is, $K_{mutant} = 10K_{native}$.

(a) If nothing else about the catalytic mechanism is different, how will V_{max} for the mutant and native enzymes compare? Explain your reasoning.
(b) For a substrate concentration that is not saturating for either enzyme, how will the initial reaction rates compare? Explain.
(c) On the same graph, sketch and label V_0 *vs.* $[S]_0$ plots for both enzymes.

11.74. The enzyme urease catalyzes the hydrolysis of urea to ammonium and carbonate ions (or bicarbonate under acidic conditions), reaction equation (11.80):

$$NH_2CONH_2(aq) + 2H_2O \xrightarrow{\text{urease}}$$
$$2NH_4^+(aq) + CO_3^{2-}(aq)$$

Initial rate data as a function of substrate, $NH_2CONH_2(aq)$, concentration are

$[NH_2CONH_2(aq)]_0$, M	V_0, M·sec^{-1}
0.00065	0.226
0.00129	0.362
0.00327	0.600
0.00830	0.846
0.0167	0.975
0.0333	1.03

(a) Plot these data on a V_0 *vs.* $[NH_2CONH_2(aq)]_0$ graph and, assuming that the reaction obeys Michaelis–Menten kinetics, estimate V_{max} and K for the reaction.
(b) Show that the Michaelis–Menten equation, in the form given in equation (11.77), can be transformed to this equation by inverting both sides.

$$\frac{1}{V_0} = \left[\frac{\left(\frac{1}{K}\right)}{V_{max}} \right]\left(\frac{1}{[S]_0}\right) + \frac{1}{V_{max}}$$

This is a linear equation for $\frac{1}{V_0}$ plotted as a function of $\frac{1}{[S]_0}$, a "double reciprocal" plot. From the intercept we get $V_{max}\left(= \frac{1}{intercept}\right)$ and the slope divided by the intercept gives $\frac{1}{K}$. This transformed equation is often called the Lineweaver–Burk equation. Plot the data for the urease reaction on a double reciprocal plot and find the numerical values for V_{max} and K.
(c) How do your results from parts (a) and (b) compare?
(d) What initial concentration of urea is required to give an initial rate of reaction that is 99% of V_{max}? Explain.

11.75. The activity of an enzyme is usually defined in terms of the amount of substrate that reacts in a specified time under specified conditions. The unit of activity for a certain bacterial pyrophosphatase is hydrolysis of 10 μmol (1 μmol = 10^{-6} mol) of pyrophosphate in 15 minutes at 37 °C. A sample of the purified enzyme has an activity of 2750 units per milligram of enzyme. Under the test conditions, how many moles of substrate react per second per milligram of enzyme? Explain your reasoning.

11.76. Enzymatic reactions are affected by many factors, including molecules that bind to the enzyme at the site(s) where their substrate(s) must bind to undergo reaction. These molecules are often called *inhibitors* because their usual effect is to slow the enzymatic reaction. The initial rate data here are for an enzymatic reaction in the absence and presence of a constant concentration of an inhibitor.

$[S]_0$, M	V_0, μmol·min^{-1} inhibitor absent	V_0, μmol·min^{-1} inhibitor present
1.0×10^{-4}	28	18
1.5×10^{-4}	36	24
2.0×10^{-4}	43	30
5.0×10^{-4}	63	51
7.5×10^{-4}	74	63

(a) Assuming that the inhibitor binds reversibly to the active site of the enzyme, just as the substrate does, write a series of reactions that includes the inhibitor binding as well as the Michaelis–Menten reactions. Use equation (11.69) as a model and sketch a pictorial representation of your mechanism. Explain, with reference to your mechanism, why the enzymatic reaction is slower with the inhibitor present.
(b) Plot the data with the inhibitor absent and present on the same V_0 vs. $[S]_0$ graph. From your plots, estimate numeric values for V_{max} and K in the absence and presence of a constant amount of inhibitor. Are your results consistent with what you would expect from your mechanism in part (a)? Explain why or why not.
(c) Plot the data on a double-reciprocal graph [see Problem 11.74(b)] and use the equations of the lines to calculate numeric values for V_{max} and K in the absence and presence of a constant amount of inhibitor. How do these results compare with those from part (b)? Do you need to revise any of your explanation from part (b)? If so, how?

11.77. The normal substrate for the enzyme aspartate transcarbamylase is aspartate. Succinate is an inhibitor (see Problem 11.76) of the normal reaction. Based on their structures, does this inhibition make sense? What kinds of interactions in the active site are probably responsible for the binding of succinate in place of aspartate? Explain.

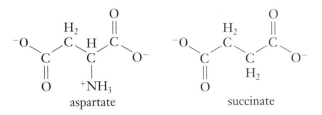

aspartate   succinate

11.78. The rate of almost all reactions, including enzyme-catalyzed reactions, increases with increasing temperature. For enzymatic reactions, however, the rate decreases rapidly beyond a certain temperature (which is dependent on the enzyme under study). How can you account for this behavior? *Hint:* See Chapter 1, Section 1.9.

11.79. In Problem 11.28, you found that the decomposition of ammonia on a hot tungsten wire goes at the same rate even as NH_3 decomposes and its pressure (concentration) decreases. The reaction rate is independent of the amount of ammonia left unreacted (at least for the data we have).
(a) Could this catalytic reaction system be displaying a saturation effect? Discuss this possibility and consider what the sites are that might be saturated, if this is such an effect. What experiment(s) could you do to test your ideas?
(b) Assuming that the results do represent a saturation effect, similar to an enzymatic reaction, sketch a curve showing the initial rate of ammonia decomposition as a function of initial pressure of ammonia from zero to 26 kPa.

General Problems

11.80. The rate law and reaction stoichiometry are important data we can use to propose a reaction pathway, but other information is also helpful. For example, consider the hydrolysis of phosphate esters, such as butyl phosphate (the phosphate and phosphoric acid product transfer protons to water, but these reactions are omitted here):

$$C_4H_9OP(O)(OH)_2 + H_2O \rightarrow C_4H_9OH + (HO)_3PO$$

When this reaction is carried out in water that has been enriched in oxygen-18 (that is, has a higher percentage of O-18 atoms than normal), the oxygen-18 is found in the phosphoric acid product, but not in the alcohol. What does this result suggest about the atoms that interact when the water and ester begin to react? How does this result help to establish a pathway for the reaction? *Hint:* Review Chapter 6, Section 6.7.

11.81. Reaction pathway (i) in Problem 11.37 proposes that dinitrogen pentoxide, O_2NONO_2, reacts by decomposing in a rapid reversible reaction to NO_3 and NO_2, which then can undergo this rate limiting reaction:

$$NO_3 + NO_2 \rightarrow NO_2 + O_2 + NO$$

(a) Write Lewis structures and calculate the formal charges on all the atoms in dinitrogen pentoxide, nitrogen trioxide, and nitrogen dioxide.
(b) Is there electron delocalization that could stabilize any of these molecules? Which one(s)? Could this factor account for any part of the proposed reaction pathway? If so, explain how. If not, explain why not.
(c) What would be a plausible activated complex for the rate limiting reaction written above? Explain why you think it is reasonable and why the reaction might be slow.

11.82. In Problem 11.35, a pathway was proposed for the net reaction:

$$(CH_3)_3CCl + OH^- \rightarrow (CH_3)_3COH + Cl^-$$

This pathway is consistent with the rate law you derived, based on the information in Problems 11.15 and 11.25.

Experiments with $(CH_3)_3CBr$ and $(CH_3)_3CI$ show that they undergo this same reaction with the same rate law, but at different rates.
(a) If the proposed reaction pathway is the same for all three halogen compounds, can it explain the different rates for the reaction? If so, show how. If not, show why not.
(b) Which of the three reactants, the chloro, bromo, or iodo compound, probably reacts most rapidly? Explain the reasoning for your choice. *Hint:* See Table 7.3 (Chapter 7, Section 7.7).
(c) If studies at different temperatures are carried out, which reaction rate will be most affected by the temperature? Explain your reasoning.
(d) In all three reactant systems, there is a side reaction that forms a small amount of 3-methyl propene, $(CH_3)_2C{=}CH_2$. The proposed pathway for this side reaction is a rapid proton transfer (elimination) reaction:

$$(CH_3)_3C^+ + H_2O \rightarrow (CH_3)_2C{=}CH_2 + H_3O^+$$

The ratio of the alkene product to the alcohol product is the same in all three reactant systems. Is this observation consistent with the proposed reaction pathway for these reactions? Explain why or why not.

APPENDIX A

Numerical Answers

CHAPTER 1

Check This 1.7: (a) 7 valence electrons (b) 2 valence electrons (c) 6 valence electrons

Check This 1.43: (a) 358 $J \cdot g^{-1}$ of hexane

Check This 1.46: (a) 0.028 mol H_2O (b) 6.61 mol of sugar (c) 7.39 g of sodium chloride

Check This 1.48: (a) There are 12 mol of carbon atoms, 22 mol of hydrogen atoms in 1 mol of sucrose, and 11 mol of carbon atoms in 1 mol of sucrose. (b) There is 1 mol of Na and 1 mol of Cl atoms in 1 mol of NaCl.

Check This 1.49: moles of methanol = 1.2×10^{-2} mol; moles of hexane = 3.5×10^{-3} mol

Check This 1.52: vaporization energy for water: 1.2 kJ; vaporization energy for methanol: 0.49 kJ; vaporization energy for hexane: 0.12 kJ

Check This 1.58: $\Delta T = 40 \, °C$

End-of-Chapter Problems

1.12. 7320 mL

1.13. 100.0 g

1.16. (a) 1 valence electron, 10 core electrons (b) 7 valence electrons, 28 core electrons (c) 2 valence electrons, 54 core electrons (d) 5 valence electrons, 10 core electrons (e) 6 valence electrons, 10 core electrons

1.17. (a)

Ion	# of protons	# of electrons	# of valence electrons	Core atomic charge	# of core electrons
Na^+	11	10	0	+1	10
K^+	19	18	0	+1	18
Mg^{2+}	12	10	0	+2	8
Ca^{2+}	20	18	0	+2	18
Cl^-	17	18	0	+8	10
Br^-	35	36	0	+8	28

1.21. Mn^{2+} has 25 protons and 23 electrons; Fe^{2+} has 26 protons and 24 electrons; Fe^{3+} has 26 protons and 23 electrons; Cu^{2+} has 29 protons and 27 electrons; Zn^{2+} has 30 protons and 28 electrons.

1.29. (a) 3 (b) 1 (c) 2 (d) 1

1.53. approximately 40–50 °C

1.73. (a) 4.55 kJ (b) 59.8 calories (c) 2.09×10^6 J

1.76. 0.292 $kcal \cdot g^{-1}$

1.77. (a) 46.08 amu (b) 46.08 amu

1.78. (a) 6.02×10^{23} molecules (b) 18.02 g

1.79. (a) 1.721 mol acetone (b) 3.120 mol methanol (c) 2.170 mol dimethyl ether (d) 0.292 mol sucrose

1.80. 3.3×10^{22} molecules of water; 1.9×10^{22} molecules of methanol; 1.0×10^{22} molecules of acetone; 1.3×10^{22} molecules of ethanol; 1.3×10^{22} molecules of dimethyl ether

1.81. 6.02×10^{13} m; about 4.0×10^5% or 4.0×10^3 times farther

1.82. (a) 2 moles of H bonds (b) 0.28 mol (H bonds broken per mole ice melted); 14% H bonds broken

1.89. 310.2 K

1.90. (a) 41.8 J (b) and (c) 104.5 J

1.91. (a) 418 J (b) 1045 J

1.92. 651 J

1.107. (a) for a 70-kg person, 2.72×10^3 mol H_2O; 1.64×10^{27} molecules H_2O (b) 23,230 O atoms (c) 7.06×10^{27} atoms; 1.72×10^{26} N atoms; 2.86×10^2 mol N; 7.54×10^{26} C atoms; 1.25×10^3 mol C

CHAPTER 2

Consider This 2.10: 17

Check This 2.14: (a) 0

Check This 2.15: (a) about six water molecules

Check This 2.26: $-63 \, kJ \cdot mol^{-1}$

Check This 2.41: 1 atom of cobalt, 2 atoms of chlorine, 12 atoms of hydrogen, and 6 atoms of oxygen; Its molar mass is 237.8 $g \cdot mol^{-1}$.

Check This 2.43: 0.15 mol

Check This 2.45: 0.15 M

Check This 2.47: 0.00090 mol

Check This 2.49: 225 g

Check This 2.51: 20 mL

Check This 2.53: 22.5 g

Check This 2.57: 0.00090 mol

Check This 2.59: 0.0014 mol

Check This 2.62: (b) 0.083 g

Check This 2.66: (a) 0.00023 mol excess
(b) 0.019 M $Co^{2+}(aq)$; 0.11 M $Na^+(aq)$; 0.15 M $Cl^-(aq)$

Check This 2.70: $O_2(g)$: 0.0012 M; $CO_2(g)$: 0.0033 M

Check This 2.72: 13.4 M

Check This 2.78: 3.4×10^{-4} M; pH $\approx$ 10.5

End-of-Chapter Problems

2.37. $-1,210$ kJ·mol^{-1}

2.39. (a) 2524 kJ·mol^{-1} (b) 1490 kJ·mol^{-1}
(c) -1034 kJ·mol^{-1}

2.40. (a) 435 kJ·mol^{-1} (b) -452 kJ·mol^{-1}

2.41. 95 kJ·mol^{-1}

2.42. 2957 kJ·mol^{-1}

2.58. (b) about 0.13 mol of $CaCl_2$ (c) about 14 g of $CaCl_2$

2.59. (b) 176.12 g (c) 0.0028 moles (d) 1.7×10^{21} molecules

2.60. (a) 620 g (b) 7.2 g (c) 160 g (d) 3.9 g

2.61. 1.5×10^{20} C atoms

2.62. 4×10^{-10} M DNA

2.63. 4.2×10^{21} $Na^+(aq)$

2.64. (a) 10.8 g (b) 14 g (c) 40 g

2.65. (a) 0.64 M (b) 0.10 M (c) 0.0084 M

2.66. (a) 3.000×10^{-2} M glucose (b) 31.7 mL of glucose

2.67. 0.639 M and 1.000 M solutions, respectively

2.69. 2.3 g NaCl; 0.15 M NaCl

2.70. 0.17 M urea

2.71. 0.10 g (about 100 mg) KH_2PO_4;
3.2×10^{-3} M KH_2PO_4

2.72. 4 mol of C, 14 mol of H, 4 mol of O;
24 mol of atoms; 4 mol of ions

2.73. 1.29 g $Ca_3(PO_4)_2$

2.74. 24.0 mL $SO_3^{2-}(aq)$

2.75. (a) 0.045 mol $Na^+(aq)$, 0.023 mol $SO_4^{2-}(aq)$,
0.018 mol $Ba^{2+}(aq)$, 0.036 mol $Cl^-(aq)$
(b) 0.023 mol $Ba^{2+}(aq)$ (c) 0.018 mol $SO_4^{2-}(aq)$

2.77. (a) 0.01 mol Cl^-; 0.4 g (b) 0.002 mol PO_4^{3-};
0.02 mol Na^+

2.80. (a) 6.4×10^{-4} moles $N_2(g)$
(b) 1.2×10^{-3} moles $O_2(g)$

2.83. approximately 920 g·kg^{-1}

2.86. (a) nitrogen = 0.014 g·(kg water)$^{-1}$;
oxygen = 0.008 g·(kg water)$^{-1}$ (b) 36%

2.91. (a) 2.00 (b) 10.00 (c) 3.30 (d) 7.30

2.92. (a) 1×10^{-4} M (b) 1×10^{-2} M

2.95. (b) 8.0×10^{-2} M $H_3PO_4(aq)$

2.110. (a) Both calcium cation and sulfate anion are about
0.01 m.

2.111. 2.94×10^{-4} M nitrate ion

2.112. (a) 0.613 M

2.114. (a) 1.8×10^{-7}% (b) 1.8×10^{-6}%
(c) 1.8×10^{-4}% (or approximately 0.0002%)

2.115. (a) product is 10^{-14} (b) 10^{-14}

2.118. (a) 119 mL (b) 119 mL

2.119. (b) 12.6 L of vinegar

2.120. (a) 2.25×10^3 mol (b) 1.01×10^3; 57 kg; 2.5 bags;
(c) pH = 5.3

CHAPTER 3

Check This 3.14: $^{206}_{82}$Pb contains 82 p, 124 n, and 82 e^-;
$^{59}_{27}$Co contains 27 p, 32 n, and 27 e^-.

Check This 3.18: $^{206}_{82}$Pb^{2+} has 80 e^- and 124 n.
$^{208}_{82}$Pb^{4+} has 78 e^- and 126 n.

Check This 3.20: (a) 1000 K (b) about 10^9 K

Check This 3.21: (a) 8.5×10^{61} m^3 (b) 1.2×10^{-14} kg·m^3
(c) 1.0×10^{48} kg (d) 5×10^{17} suns (solar masses)

Check This 3.37: 95 days

Check This 3.40: (a) -9.0×10^9 J

Check This 3.42: (a) -1.148×10^{-12} J

Check This 3.43: (b) -4.54×10^{-12} J

Check This 3.45: -7.8×10^{-7} g

Check This 3.47: -2.63×10^{-10} J

Check This 3.48: -7.41×10^8 kJ·mol^{-1}

Check This 3.60: approximately 2.8×10^9 years old

Check This 3.64: 14×10^9 yr

Check This 3.67: (a) 14×10^9 yr

Check This 3.69: $\delta^{13}C = -8.2$

Check This 3.71: (b) 93.3%

End-of-Chapter Problems

3.6. 24%

3.7. The mass of Fe is about 28% the mass of C in the universe.

3.10. The mass of a proton is 1.837×10^3 greater than the mass of the electron.

3.14. 4.18×10^{-23} g

3.16. (a) 28 p; 30 n; 27 e^- (b) 16 p; 16 n; 18 e^- (c) 30 p; 35 n; 28 e^- (d) 17 p; 20 n; 18 e^- (e) 25 p; 30 n; 18 e^- (f) 26 p; 30 n; 24 e^-

3.18. (c) 302 K

3.21. approximately 10^7 K

3.32. about 6400 B.C.

3.33. 0.242

3.34. (b) about 27 days

3.35. (b) 3.9×10^9 yr

3.38. (a) for Ne: -0.129×10^{-11} J·nucleon^{-1} (b) for Si: -0.135×10^{-11} J·nucleon^{-1}

3.39. for Li-6: (a) -5.13×10^{-15} kJ (kg·m^{2}s^{-2}) (b) -8.56×10^{-16} kJ·nucleon^{-1} (c) -5.15×10^8 kJ·mol^{-1}; for Fe-56: (a) -7.83×10^{-14} kJ (b) -1.40×10^{-15} kJ·nucleon^{-1} (c) -8.42×10^8 kJ·mol^{-1}

3.41. (a) $\Delta E = -11300 \times 10^7$ kJ·mol^{-1} (b) $\Delta E = -1610 \times 10^7$ kJ·mol^{-1}

3.42. for Kr-92 + Ba-141: $\Delta E = -1670 \times 10^7$ kJ·mol^{-1}; for Rb-89 + Cs-144: $\Delta E = -1630 \times 10^7$ kJ·mol^{-1}

3.44. -20.6×10^7 kJ

3.45. (a) (i) -6.44×10^{-13} J (ii) -5.21×10^{-13} J

3.46. (a) $\Delta m = 0.00018 \times 10^{-27}$ kg; $\Delta E = 0.98 \times 10^7$ kJ·mol^{-1} (c) $\Delta E = -71.19 \times 10^7$ kJ·mol^{-1} (d) $\Delta E = -70.21 \times 10^7$ kJ·mol^{-1}

3.52. (b) 5.35 µg Ar-40; Ar-40/K-40 mass ratio = 0.107 (c) mass ratio 0.321 (e) about 4.6×10^9 years old

3.53. (b) 102 min

CHAPTER 4

Check This 4.15: (a) 16.3 mm·s^{-1} (b) *amplitude:* top wave: 2 cm; middle wave: 2.5 cm; bottom wave: 1 cm; *wavelength:* top wave: 5 cm; middle wave: 3 cm; bottom wave: 6 cm; *frequency:* top wave: 5 s^{-1}; middle wave:

4 s^{-1}; bottom wave: 3 s^{-1}; *velocity:* top wave: 25 cm·s^{-1} (measured); using eq. (4.2), velocity = 5 cm $\times$ 5 s^{-1} = 25 cm·s^{-1}; top wave satisfies equation (4.2); middle wave: 12 cm·s^{-1} (measured); using eq. (4.2), velocity = 3 cm $\times$ 4 s^{-1} = 12 cm·s^{-1}; middle wave satisfies equation (4.2); bottom wave: 18 cm·s^{-1} (measured); using eq. (4.2), velocity = 6 cm $\times$ 3 s^{-1} = 18 cm·s^{-1}; bottom wave satisfies equation (4.2)

Check This 4.22: (a) 5.83×10^{14} s^{-1} (b) 1.42×10^9 s^{-1}

Check This 4.23: (b) for 101.3 MHz, 2.96 m

Check This 4.28: (c) 4.80×10^{-19} J

Check This 4.33: (a) 3.86×10^{-19} J; 2.32×10^5 J·mol^{-1} (b) 9.42×10^{-25} J (c) for 101.3 MHz, 6.71×10^{-26} J; 4.04×10^{-2} J·mol^{-1}

Check This 4.38: (a) $E_3 - E_2$ for the hydrogen atom = 3.03×10^{-19} J (b) 656 nm

Check This 4.39: (a) 1.06×10^{-19} J; 1.88×10^{-6} m = 1.88 µm

Check This 4.41: (a) 2.43×10^{-11} m (b) 1.04×10^{-34} m (c) 3.97×10^{-10} m = 0.397 nm

Check This 4.42: (a) *top wave:* n = 3 (2 nodes in each wave), wavelength = 0.50 m; *bottom wave:* n = 4 (3 nodes in each wave), wavelength = 0.38 m (b) n = 1, wavelength = 1.5 m

Check This 4.45: (a) 140 J (b) 0.503 kJ·mol^{-1}

Check This 4.51: The atomic radius of Cl is 99 pm, and the atomic radius of C is 79 pm.

Check This 4.67: 8.35×10^{-22} J; 0.503×10^2 kJ·mol^{-1}

End-of-Chapter Problems

4.2. (e) The radius for K is 227 pm, and for Kr the radius is 112 pm.

4.10. (a) 187 seconds

4.11. (a) $\lambda = 2.97$ m (b) $\lambda = 1.2 \times 10^{-8}$ m (c) $\lambda = 1.5 \times 10^{-3}$ m

4.12. (a) $v = 7.5 \times 10^{14}$ Hz (b) $v = 1.2 \times 10^4$ Hz (c) $v = 6.7 \times 10^{11}$ Hz

4.13. (a) 0.255 m·s^{-1}

4.14. $\lambda = 2.62 \times 10^{-2}$ m

4.15. $v = 1.07 \times 10^{15}$ s^{-1}

4.16. 1 m

4.22. $E = 8.316 \times 10^{-28}$ J

4.23. (a) $v = 2.47 \times 10^{19}$ s^{-1}

4.24. (a) $\lambda = 0.600$ m (c) $E = 1.995 \times 10^{-4}$ kJ·mol^{-1}

4.25. (a) 6.69×10^{-26} J·photon^{-1}; 4.03×10^{-2} J·mol^{-1} (b) 4.97×10^{-19} J·photon^{-1}; 2.99×10^5 J·mol^{-1}

4.29. *for potassium:* $v = 3.4 \times 10^{20}$ s^{-1} and $\lambda = 8.9 \times 10^{-13}$ m; *for thorium:* $v = 6.3 \times 10^{20}$ s^{-1} and $\lambda = 4.8 \times 10^{-13}$ m

4.35. sodium, 590 nm; potassium, 410 nm

4.40. $\lambda = 1.5 \times 10^{-34}$ m

4.41. $\lambda = 4.9 \times 10^{-10}$ m

4.43. (a) $\lambda = 6.7 \times 10^{-15}$ m

4.64. (a) 6 (b) 1 (c) 7 (d) 8

4.72. (a) 1 (b) 2 (c) 2 (d) 2

4.73. (a) 3 (b) 1 (c) 0 (d) 5 (e) 4

4.80. (a) $R = 2$ (b) $H = 5248$ kJ·mol^{-1}

4.81. (d) $R = 2/z$ (e) 5,248; 11,808; 32,800 kJ·mol^{-1}, respectively, for He$^+$; Li^{2+}; B^{4+}

CHAPTER 5

Check This 5.28: (a) The observed H—C—C bond angle in ethene is 121.5°. The observed H—C—O bond angle in methanal is 122°. (b) The H—C—C bond angle in ethene and the H—C—O bond angle in methanal are both 125.3°.

End-of-Chapter Problems

5.6. (a) 32 (b) 42

5.16. (a) 1 (b) 16 (c) 8 (d) 6

5.20. (a) 42 (b) 98

5.31. (a) HOCO$_2^-$ has 24, O$_3$ has 18 (c) 3 in HOCO$_2^-$, 2 in O$_3$ (e) 1 in each (f) 1 delocalized in each (g) one single and two 1.5 bonds in HOCO$_2^-$ and two 1.5 bonds in O$_3$

5.33. (a) 24 for nitrate, 18 for nitrite, and 16 for nitronium (c) 3 for nitrate, 2 for nitrite, and 2 for nitronium (e) 1 for nitrate, 1 for nitrite, and 2 for nitronium (f) 0,1 for nitrate, 0,1 for nitrite, and 0,2 for nitronium (g) $1\frac{1}{3}$ for nitrate, $1\frac{1}{2}$ for nitrite, and 2 for nitronium

5.34. (b) 1/3 contribution to the intermediate structure

5.56. (a) 14 (c) 1

5.59. The experimental value is 126 pm.

CHAPTER 6

Check This 6.10: (a)/(b) *test tube #1:* Ca^{2+} = 5.0×10^{-5} mol, C$_2$O$_4^{2-}$ = 4.5×10^{-4} mol, precipitate = 5.0×10^{-5} mol; *test tube #2:* Ca^{2+} = 1.5×10^{-4} mol, C$_2$O$_4^{2}$ = 3.5×10^{-4} mol, precipitate = 1.5×10^{-4} mol; *test tube #3:* Ca^{2+} = C$_2$O$_4^{2-}$ = CaC$_2$O$_4$ = 2.5×10^{-4} mol; *test tube #4:*

Ca^{2+} = 3.5×10^{-4} mol, C$_2$O$_4^{2-}$ = 1.5×10^{-4} mol, precipitate = 1.5×10^{-4} mol

Check This 6.11: (a) 1:2 (b) 2.5, 4.85, 5.95, 6.50, 5.43, and 4.06×10^{-4} mol (c) 20.74×10^{-4} mol (d) Ni^{2+}(aq) = 6.91×10^{-4} mol; C$_4$H$_8$N$_2$O$_2$(alc) = 13.8×10^{-4} mol; total mass = 0.201 g

Check This 6.15: (a) for pH 2.9, [H$_3$O$^+$(aq)] = 1×10^{-3} M

Check This 6.16: 1%

Check This 6.18: 0.1 M

Check This 6.23: (c) for pH 8.6, [OH$^-$(aq)] = 4×10^{-6} M (d) 0.1 M

Check This 6.25: slightly less than 7

Check This 6.49: 22; 20; 22

Check This 6.69: (b) V^{4+}(aq) = 4; H$^-$(aq) = -1; P$_4$(s) = 0; N$_2$(g) = 0; Na$^+$(aq) = 1

Check This 6.73: (b) H = +1; O = -2; Cl = -1 (c) I$^-$ = -1; I$_2$ = 0

Check This 6.79: (a) +2

Check This 6.97: (b) 0.125 M

End-of-Chapter Problems

6.6. 0.583 g

6.7. (a) 0.0683 g (b) 6.18%

6.8. (b) 4.19×10^{-3} (c) 60.9%

6.9. (a) 0.17, 0.33, 0.50, 0.67, and 0.83

6.10. 1:2

6.16. pH = 4–7

6.44. (a) H = 0, N = +1 (b) N = +1, single bonded O = -1, double bonded O = 0 (c) N = 0, H = 0 (d) H = 0, O = 0

6.48. (d) 2 (e) 3

6.52. (a) V = +5, O = -2; H = 0; V = +3, O = -2; H = +1, O = -2 (b) K = 0; Br = 0; K = +1; Br = -1 (c) N = 0; H = 0; N = -3, H = +1

6.53. (a) +6 (b) +4, +6

6.54. (a) -3 (b) +2 (c) +3 (d) +5

6.55. (a) +2 (b) +2 (c) +3 (d) +6 (e) +2

6.57. (c) 2

6.65. (a) MnO$_4^-$, +7; Mn^{2+}, +2; MnO$_2$, +4; MnO$_4^{2-}$, +6

6.66. (a) +6 (c) 65.6%

CHAPTER 7

Check This 7.27: for $\Delta T = -3.5\ °C$, $q_P = 1.6 \times 10^3\ J$

Check This 7.29: $16\ kJ\cdot mol^{-1}$

Check This 7.42: (a) $127\ kJ$

Check This 7.47: (b) $56.32\ kJ\cdot mol^{-1}$

Check This 7.50: (a) $-2801.6\ kJ\cdot mol^{-1}$

Check This 7.51: (a) $-86\ kJ\cdot mol^{-1}$ (b) $-126.3\ kJ\cdot mol^{-1}$

Check This 7.57: $53.7\ kJ$

Check This 7.62: (a) $\Delta E° = 123.8\ kJ$, $\Delta H°_{reaction} = 126.3\ kJ$

Check This 7.69: $29.8\ mL$

Check This 7.71: (a) 0.584 (b) $n_1 = 0.0131\ mol$; $n_2 = 0.0077\ mol$

Consider This 7.74: *Web Companion* $q_V = 1.28\ kJ$, $q_P = 1.57\ kJ$

Consider This 7.76: (a) $0.10\ mol$ (b) $0.12\ mol$ (c) $0.10\ mol$ (d) $0.24\ kJ$

Check This 7.77: *Web Companion* $\Delta H = 15.7\ kJ\cdot mol^{-1}$, $\Delta E = 12.8\ kJ\cdot mol^{-1}$

End-of-Chapter Problems

7.1. (a) 8×10^5 heartbeats (b) 2×10^5 heartbeats

7.13. 25 yards

7.26. $\Delta H = -54\ kJ\cdot mol^{-1}$

7.27. (b) $3\ °C$

7.28. (c) $\Sigma BH_{reactants} = 4725\ kJ$ (d) $-\Sigma BH_{products} = -5956\ kJ$ (e) $\Delta H = -1231\ kJ$

7.30. $\Delta H = +8\ kJ$

7.32. (a) $\Delta H°_{rxn} = -437\ kJ$, $\Delta H°_{rxn} = -563\ kJ$

7.34. (a) $\Delta H°_{rxn} = 0\ kJ$ (b) $\Delta H°_{rxn} = -5606\ kJ$

7.35. (a) $\Delta H°_{rxn} = -732\ kJ$ (b) $\Delta H°_{rxn} = -917\ kJ$

7.36. (a) $-32\ kJ\cdot mol^{-1}$

7.39. $\Delta H°_{rxn} = -133.29\ kJ$

7.41. (a) $-68.0\ kJ$ (d) $-3.7\ kJ$ (e) $-70.2\ kJ$

7.42. (a) $\Delta H° = 4\ kJ$ (b) $\Delta H° = -3\ kJ$

7.43. (a) $\Delta H°_f(NH_3) = -46\ kJ\cdot mol^{-1}$

7.44. (a) reaction 1: $\Delta H°_{rxn} = +131.3\ kJ$; reaction 2: $\Delta H°_{rxn} = -110.5\ kJ$; reaction 3: $\Delta H°_{rxn} = +408.8\ kJ$

7.49. (a) $\Delta H° = -2335\ kJ$ (b) $-1705\ kJ$, 27%

7.51. (a) $4.6 \times 10^3\ J$

7.52. $q_p = 1.5 \times 10^8$, $w = -5 \times 10^7\ J$, $\Delta E = 1.0 \times 10^8\ J$

7.57. (a) $55\ L$ (b) $5.6 \times 10^3\ J$

7.58. work $= +0.62\ kJ$; $\Delta E = -120\ kJ$

7.61. $\Delta H° [NH_4^+(aq)] = -132.2\ kJ\cdot mol^{-1}$ $\Delta H°_{form}[Ca^{2+}(aq)] = -543.9\ kJ\cdot mol^{-1}$

7.67. (a) $76.6\ kPa$

7.68. (a) $3.09 \times 10^{-3}\ mol$ (b) $76.8\ g\cdot mol^{-1}$

7.69. (a) $4.329 \times 10^{-3}\ mol$ (b) $4.329 \times 10^{-3}\ mol$ (c) 27.3%

7.71. (a) $\Delta H = -473\ kJ\cdot mol^{-1}$ (b) $\Delta E = -475\ kJ\cdot mol^{-1}$

7.72. (a) $1 \times 10^{-19}\ N\cdot m$ (or J) (b) $1 \times 10^{-19}\ J$ (c) 100%

7.73. $\Delta H_{hydrolysis} = -34\ kJ\cdot mol^{-1}$

7.74. (a) $\Delta E = 1.8\ J$ (b) $6.2\ atm$ (c) 3 pounds

7.75. $\Delta H° = -1141\ kJ$

CHAPTER 8

Check This 8.16: $0.8\ J\cdot K^{-1}\cdot mol^{-1}$

Check This 8.22: $+4.5\ kJ$

Check This 8.25: $279\ K$

Check This 8.28: $\Delta S° = 536.8\ J\cdot K^{-1}\cdot mol^{-1}$

Check This 8.30: $\Delta H° = -68.0\ kJ\cdot mol^{-1}$; $\Delta G° = -228.0\ kJ\cdot mol^{-1}$

Check This 8.32: $-227.8\ kJ\cdot mol^{-1}$

Check This 8.34: $\Delta H° = 1.5\ kJ\cdot mol^{-1}$; $\Delta S° = -159.4\ J\cdot mol^{-1}\cdot K^{-1}$; $\Delta G° = 49.0\ kJ\cdot mol^{-1}$

Check This 8.45: (a) $0.074\ m$ (b) $4.3 \times 10^2\ g\cdot mol^{-1}$

Check This 8.47: (a) $0.30\ m$

Check This 8.49: $80.6\ °C$

Check This 8.51: $24\ atm$

End-of-Chapter Problems

8.7. (a) 0 (b) 1 (c) 0.02 (d) close to 0 (e) 0 (f) close to 1

8.9. (a) 10

8.10. (a) 15

8.12. (a) 20 (b) $N = 3$: $W = 1$; $N = 6$: $W = 20$; $N = 9$: $W = 84$; $N = 12$: $W = 220$; $N = 15$: $W = 455$; $N = 30$: $W = 4060$; $N = 45$: $W = 14{,}190$; $N = 60$: $W = 34{,}220$ (c) $W = 15{,}504$; $W = 184{,}756$

8.19. (a) 10 (b) $n = 0$: $W = 1$; $n = 1$: $W = 4$; $n = 2$: $W = 10$; $n = 3$: $W = 20$; $n = 4$: $W = 35$; $n = 5$: $W = 56$; $n = 6$: $W = 84$; $n = 7$: $W = 120$; $n = 8$: $W = 165$; $n = 9$: $W = 220$; $n = 10$: $W = 286$

8.37. (a) $\Delta H° = 30.91\ \text{kJ·mol}^{-1}$, $\Delta S° = 93.23\ \text{J/K·mol}^{-1}$
(b) (i) $-19.9\ \text{J·K}^{-1}$ (ii) $-5.52\ \text{J·K}^{-1}$ (iii) $3.11\ \text{J·K}^{-1}$
(c) $58\ °\text{C}$

8.39. (a) $\Delta H°_{l \to g} = 25.94\ \text{kJ·mol}^{-1}$,
$\Delta S°_{l \to g} = 104.0\ \text{J·K}^{-1}\text{·mol}^{-1}$ (b) $249\ \text{K}$

8.40. (a) $38.20\ \text{kJ·mol}^{-1}$ (b) $113.0\ \text{J·K}^{-1}\text{·mol}^{-1}$
(c) $113.0\ \text{J·K}^{-1}\text{·mol}^{-1}$

8.41. (b) $9800\ \text{K}$

8.43. (b) $39\ \text{kJ}$

8.47. (b) $\Delta H° = -802.34\ \text{kJ}$, $\Delta G° = -800.78\ \text{kJ}$,
$\Delta S° = -5.23\ \text{J·K}^{-1}$

8.49. (b) $925\ \text{K}$

8.50. $-4.0\ \text{kJ}$

8.52. (c) $-235\ \text{J·K}^{-1}$

8.54. (a) $\Delta H° = -60.08\ \text{kJ}$, $\Delta S° = -154.3\ \text{J·K}^{-1}$,
$\Delta G° = -14.1\ \text{kJ}$ (b) $-30.04\ \text{kJ·mol}^{-1}$

8.55. *for KCl:* $\Delta H° = 17.21\ \text{kJ·mol}^{-1}$, $\Delta S° =$
$76.4\ \text{J·K}^{-1}\text{·mol}^{-1}$, $\Delta G° = -5.36\ \text{kJ·mol}^{-1}$;
for AgCl: $\Delta H° = 65.49\text{kJ·mol}^{-1}$,
$\Delta S° = 33.0\ \text{J·K}^{-1}\text{·mol}^{-1}$, $\Delta G° = 55.66\ \text{kJ·mol}^{-1}$

8.56. (a) $\Delta G° = -503.3\ \text{kJ}$ (b) $\Delta H° = -631.2\ \text{kJ}$,
$\Delta S° = -429.5\ \text{J·K}^{-1}$, $T = 1470\ \text{K}$

8.57. (a) $\Delta H° = -434.51\ \text{kJ}$, $\Delta S° = 100.6\ \text{J·K}^{-1}$,
$\Delta G° = -505.25\ \text{kJ}$

8.58. (a) $\Delta H° = -1.3\ \text{kJ·mol}^{-1}$, $\Delta S° = 194.4\ \text{J·K}^{-1}$,
$\Delta G° = -59.10\ \text{kJ}$

8.59. (a) $\Delta H° = 131.29\ \text{kJ}$, $\Delta S° = 133.78\ \text{J·K}^{-1}$,
$\Delta G° = 91.40\ \text{kJ}$

8.62. occupy a 1.3 nm cubical box

8.66. (a) 1×10^{10} (b) 2×10^{-14} mol, 3×10^{-9} g

8.75. $-0.272\ °\text{C}$

8.76. (a) $0.703\ \text{m}$ (b) $-3.60\ °\text{C}$ (c) $1.90\ °\text{C}$

8.77. (i) $4.0\ °\text{C}$ (ii) $4.5\ °\text{C}$

8.79. $-1.96\ °\text{C}$

8.80. $-0.141\ °\text{C}$

8.81. $-2.0 \times 10^{-5}\ °\text{C}$

8.82. glucose = 0.31 M, NaCl = 0.16 M

8.83. $3.99 \times 10^4\ \text{g·mol}^{-1}$

8.86. 27 atm

8.95. (a) $W_{\text{solvent}} = 1$, $W_{\text{solute}} = 1$, $W_{\text{total}} = 1$
(b) $W_{\text{solvent}} = 20$, $W_{\text{solute}} = 1$, $W_{\text{total}} = 20$

(c) $W_{\text{solvent}} = 84$, $W_{\text{solute}} = 1$, $W_{\text{total}} = 84$
(d) $W_{\text{solvent}} = 455$, $W_{\text{solute}} = 1$, $W_{\text{total}} = 455$

8.96. (a) total energy = 18 units; 6 units/molecule
(b) total energy = 18 units; average energy per molecule =
6 units/molecule (c) total energy = 18 units; average
energy per molecule = 6 units/molecule

CHAPTER 9

Check This 9.5: $0.0019\ \text{M SCN}^-$

Consider This 9.13: (a) $5.2 \times 10^{-5}\ \text{M}$

Check This 9.15: (a) 1/55 c (c) $1.4 \times 10^{-4}\ \text{M}$

Check This 9.20: 2.88

Check This 9.22: (a) $3.9 \times 10^{-12}\ \text{M}$

Check This 9.28: $pK_b = 4.76$; $K_b = 1.7 \times 10^{-5}$

Consider This 9.29: (b) pOH = 5.27; $(\text{OH}^-(aq)) =$
5.4×10^{-6}

Check This 9.40: pH = 4.85

Check This 9.51: 3.6×10^{-11}

Check This 9.53: (b) $1.8 \times 10^{-5}\ \text{M}$

Check This 9.55: (a) $3.3 \times 10^{-3}\ \text{M}$ (b) $2.2 \times 10^{-6}\ \text{M}$

Check This 9.60: $\Delta G = 2.17\ \text{kJ}$

Check This 9.68: $K_{sp}(273) = 6.25 \times 10^{-5}$;
$K_{sp}(373) = 4.5 \times 10^{-6}$; $\Delta H°_{\text{reaction}} = -22\ \text{kJ·mol}^{-1}$

Check This 9.70: $\Delta S°_{\text{reaction}} = -161\ \text{J·mol}^{-1}\text{·K}^{-1}$,
$\Delta G°_{\text{reaction}} = +26\ \text{kJ·mol}^{-1}$, $K_{sp} = 2.8 \times 10^{-5}$

Check This 9.72: (a) $\Delta H°_{\text{rxn}} = 31.8\ \text{kJ·mol}^{-1}$, $\Delta S°_{\text{rxn}} =$
$99.4\ \text{kJ·mol}^{-1}\text{·K}^{-1}$, $P_{\text{vap}} = 0.418\ \text{bar}$ (= 314 torr)

Consider This 9.73: $2 \times 10^{-12}\ \text{M}$

Consider This 9.74: (a) $\Delta G°'_{\text{rxn}} = -20\ \text{kJ}$, $K' = 3 \times 10^3$
(b) ratio = 3×10^4

Check This 9.75: (a) arterial 100%, venous < 10%

Check This 9.79: $-56.70\ \text{kJ}$

Consider This 9.80: (a) $205.1\ \text{J·mol}^{-1}\text{·K}^{-1}$

Consider This 9.81: (a) $-175.9\ \text{J·mol}^{-1}\text{·K}^{-1}$
(b) $-192.5\ \text{J·mol}^{-1}\text{·K}^{-1}$

Consider This 9.82: (a) $-4.28\ \text{kJ}$ (b) $+0.67\ \text{kJ}$

End-of-Chapter Problems

9.10. 2%

9.11. (b) 0.37

9.12. (a) 1/51

9.13. (a) $P_{HI} = 9.19 \times 10^{-2}$ bar; $P_{I2} = 6.07 \times 10^{-2}$ bar; $P_{H2} = 8.69 \times 10^{-4}$ bar (c) 1.60×10^2

9.14. (b) 5.4×10^{-8}

9.16. (a) pH = 3.02 (b) pH = 2.11

9.17. (b) pH = 7.0 (c) pH = 7.0

9.19. (a) pH = 2.43 (b) pH = 2.59 (c) pH = 8.82 (d) pH = 10.48

9.20. pH = 11.32

9.21. 4.30%

9.22. 6×10^{-4} M

9.23. (b) 2×10^{-5}

9.24. (a) 2.3×10^{-9} (b) pH = 10.94

9.25. (a) pH = 9.73 (b) pH = 8.84

9.26. (a) pH = 2.38, 4% ionization (c) 6% ionization

9.31. pH = 5.54

9.32. [HOAc] = 0.05 M, [OAc$^-$] = 0.20 M

9.33. (a) pH = 3.68 (b) pH = 3.85

9.34. (a) pH = 9.81 (b) pH = 9.61 (c) pH = 10.05

9.35. (a) 0.7 M (b) 3×10^{-7} (c) pH = 10.7 (e) 0.1

9.36. 1.3×10^{-3} M

9.37. (a) 0.0031 L (b) [HOAc*(aq)*] = 0.073 M, [OAc$^-$*(aq)*] = 0.026 M

9.40. (b) 6.07

9.45. for the cation in each solution (a) 1.0×10^{-5} M (b) 3×10^{-17} M (c) 1.2×10^{-4} M (d) 8.0×10^{-3} M

9.46. 0.7 mg

9.47. 1.5×10^{-3} g

9.48. 1.8×10^{-7} mol·L^{-1} $K_{sp} = 3.2 \times 10^{-14}$

9.49. (a) [Fe^{2+}] = 5.8×10^{-6} M, [OH$^-$] = 1.2×10^{-5} M (b) [Fe^{2+}] = 1.3×10^{-14} M, [OH$^-$] = 0.25 M (c) [Fe^{2+}] = 0.25 M, [OH$^-$] = 5.6×10^{-8} M

9.51. 2.5×10^{-4} M

9.52. (a) 2.9×10^{-4} g

9.54. $K_{sp} = 1.1 \times 10^{-18}$

9.55. 5.79×10^{-23}

9.57. pH 7

9.60. $\Delta G° = -7.02$ kJ·mol^{-1}, $K = 17.0$

9.61. -21.1 kJ·mol^{-1}

9.62. (a) -1169.54 kJ; -838 kJ (b) $\Delta S°_{rxn} = -533.64$ J·K^{-1} (c) -1010.52 kJ; -1010.78 kJ (d) 1.97×10^{44}

9.63. (d) $\Delta H°_{rxn} = -127.9$ kJ, $\Delta S°_{rxn} = 74.55$ J·K^{-1}, $\Delta G°_{rxn} = -150.1$ kJ

9.65. (b) -604 kJ·mol^{-1}

9.66. (a) $Q = 0.008$ (d) 0.3 (e) $+3$ kJ·mol^{-1}

9.67. (a) $\Delta G°_{rxn} = -23$ kJ·mol^{-1} (b) 79 (c) $\Delta G_{rxn} = 2$ kJ·mol^{-1}

9.68. $\Delta G° > 0$

9.69. $\Delta G° < 0$

9.72. (b) 180 kJ·mol^{-1}

9.73. (a) $K_{sp} = 1.9 \times 10^{-8}$; solubility = 1.7×10^{-3} M = 0.77 g·L^{-1}

9.74. (a) 303 K: $K_{sp} = 2.36 \times 10^{-4}$; 373 K: $K_{sp} = 1.42 \times 10^{-4}$ (b) $\Delta H°_{reaction} = -7.16$ kJ·mol^{-1}; $\Delta S°_{reaction} = -93.1$ J·K^{-1}·mol^{-1} (c) $\Delta G°_{reaction} = 20.5$ kJ·mol^{-1}; $K_{sp} = 2.55 \times 10^{-4}$

9.75. (a) $\Delta H°_{rxn} = 131.29$ kJ; $\Delta S°_{rxn} = 133.78$ J·K^{-1}; $\Delta G°_{rxn} = 91.40$ kJ (b) 981 K (c) 1

9.76. (b) 410 K (137 °C)

9.77. (a) $\Delta S°_{rxn} = 22.6$ J·K·mol^{-1}, $\Delta H°_{rxn} = -9.82$ kJ·mol^{-1} (b) $\Delta G°_{rxn} = -16.6$ kJ·mol^{-1}

9.78. (a) $\Delta H°_{rxn} = 42.9$ kJ·mol^{-1} (c) $\Delta G°_{rxn} = 6.54$ kJ·mol^{-1}, $K_{eq} = 0.0713$ (d) 0.0713 bar (= 7.13×10^3 Pa = 53.5 torr = 0.0703 atm)

9.79. $\Delta H° = 44.6$ kJ·mol^{-1}; $\Delta S° = 126$ J·mol^{-1}·K^{-1}

9.80. $\Delta H°_{rxn} = 55.9$ kJ·mol^{-1}, $\Delta S°_{rxn} = -80.4$ J·K^{-1}·mol^{-1}, $\Delta G°_{rxn} = 79.9$ kJ·mol^{-1}

9.81. (a) $\Delta H°_{rxn} = 67.6$ kJ·mol^{-1}; $\Delta S°_{rxn} = 156$ J·K^{-1}·mol^{-1} (b) 160 °C

9.84. (a) 0.104 bar (=78 torr) (b) 8.9 kJ

9.85. (b) 7 kJ·mol^{-1} (c) 17 to 1

9.86. (a) $\Delta G°'_{rxn} = 32$ kJ·mol^{-1}, $K = 2.5 \times 10^{-6}$ (b) 4×10^4

9.89. (c) -34 kJ (e) -24 kJ·mol^{-1}

9.91. (a) 8.65 (b) 17.2 (d) $P(N_2O_4) = 0.1995$; $P(NO_2) = 0.152$

9.92. 2.73×10^{-4} mol

9.94. (a) $K(350 °C) = 0.0003$; $K(450 °C) = 0.01$ (b) 1.3×10^2 kJ·mol^{-1} (c) 1.4×10^2 J·K^{-1}·mol^{-1} (d) $T = 770$ K (≈ 500 °C); $K = 0.03$

CHAPTER 10

Check This 10.11: (a) 3.67 (b) 0.853 g Ag; 0.232 g Ni

Check This 10.13: 410 A

Check This 10.25: (a) 0.337 V

Check This 10.30: 2.014 V

Check This 10.42: (a) 0.462 V

Check This 10.47: (c) 0.100 V

Check This 10.52: pH 7: $E_{H^+,H_2} = -0.413$ V; pH 14: $E_{H^+,H_2} = -0.826$ V

Check This 10.55: (a) 0.42 V; -81 kJ·mol^{-1} (b) $K = 1.6 \times 10^{-14}$

Check This 10.60: 0.21 V

Check This 10.64: 0.011 V

Check This 10.67: 0.56 V

Check This 10.69: (a) 0.934 V (O_2, H_2O), -0.88 V (RCOO$^-$, RCHO) (b) 1.81 V

Check This 10.70: (a) about -0.3 V for GOx and 0.8 V for LAC (c) 22%

Check This 10.72: (b) 1.136 V

End-of-Chapter Problems

10.3. (a) 2.2 hr (b) 35.5 g

10.4. (d) 3.5×10^2 kg Mg

10.5. (b) 0.77 g (c) 84%

10.6. (a) 2.585×10^{-3} mol (b) +3

10.7. 52.4 min

10.8. (a) 2.1×10^6 C·month^{-1}, 0.82 amp

10.16. (e) 0.38 mol

10.18. 0.57 V

10.19. (b) 0.46 V

10.21. 2.72 V

10.22. (b) 1.20 V

10.25. (a) 0.61 V (b) 0.91 V (c) 0.23 V (d) 2.30 V

10.28. 1.05 V

10.29. (c) 2.122 V

10.30. (b) 0.736 V (c) 2.35×10^{-3} mol iron

10.31. (b) 0.148 V

10.34. (a) -127.4 kJ·mol^{-1} (b) 15.9 kJ (c) 25.9 g

10.35. (a) 8.282 mol (b) 879 kJ

10.36. (c) 0.22 V, -42 kJ·mol^{-1} (d) 0.000 V and 0.22 V

10.37. (b) 1.085 V

10.38. (a) -1.95×10^3 kJ (b) 2.02 V (c) -0.51 V

10.39. (a) 0.15 V, -87 kJ·mol^{-1} (b) 0.59 V, -3.4×10^2 kJ·mol^{-1} (c) 0.35 V, -68 kJ·mol^{-1} (d) 0.528 V, -101.9 kJ·mol^{-1}

10.40. (c) 0.75 V

10.42. (b) 0.17 V (c) 0.014 M (d) $K = 4.4 \times 10^5$

10.45. (a) 0.100 V (b) pH $= 1.64$

10.46. (b) 1.56 V, $K = 6 \times 10^{52}$ (c) [$Hg_2^{2+}(aq)$] $= 2 \times 10^{-53}$ M; [$Zn^{2+}(aq)$] $= 1.0$ M

10.47. (b) 0.28 V, 54 kJ·mol^{-1} (c) 2.9×10^9 (d) [Sn^{2+}] $= 0.010$ M, [Sn^{4+}] $= 3.4 \times 10^{-14}$

10.48. (a) -0.320 V (b) -108.3 kJ·mol^{-1} (c) -0.282 V, 54.4 kJ·mol^{-1}

10.50. (b) $E° = 1.60$ V; $\Delta G_{rxn} = -463$ kJ·mol^{-1} (c) $K = 2 \times 10^{82}$ (d) -926 kJ

10.51. (a) 2.76 V (b) 0 V (c) 2.0×10^{93}

10.52. (a) 2.122 V (c) 2.06 V (d) 0.19 mol

10.55. (a) $\Delta G° = -26.1$ kJ (b) $K = 3.8 \times 10^4$ (c) -0.802 V

10.56. (a) 0.13 V (b) 1.3×10^{-6} M (c) 0.30 V

10.57. 0.613 V

10.58. (a) 23 kJ·mol^{-1}

10.59. (b) 1.579 V (c) 12 years

10.60. (a) $\Delta G°_i = -276$ kJ·mol^{-1}, $\Delta G°_{ii} = -35$ kJ·mol^{-1}, $\Delta G°_{iii} = 237$ kJ·mol^{-1}, $\Delta G°_{iv} = -74$ kJ·mol^{-1} (b) 0.77 V

10.61. (b) -0.129 V (c) $E(Cl_2, Cl^-) = 1.313$ V, $E(MnO_2, Mn^{2+}) = 1.322$ V, $E = 0.009$ V

10.63. (d) 0.651 M

10.64. (a) 10^{-6} M (b) 0.445 V (c) 0.88 V; $\Delta G = 170$ kJ·mol^{-1}

10.66. (a) 45 kJ·mol^{-1} (b) -5.8 kJ·mol^{-1} (c) 13 kJ·mol^{-1} (d) ratio (i) $\approx 10^{-7}$; coupled reaction ratio ≈ 50

10.67. (a) 1.2×10^{-6} M (b) 1.2×10^{-7}

10.68. (b) $\Delta G°(Ag^+, Ag) = -77.1$ kJ·mol^{-1}; $\Delta G°(AgCl\ K_{sp}) = 56$ kJ·mol^{-1}; $\Delta G°(AgCl, Ag) = -21$ kJ·mol^{-1} (c) 0.22 V

10.69. (a) 5.0×10^{-13} (b) 0.072 V

10.70. (b) $K = 5.3 \times 10^{12}$; (c) 0.047 V

10.71. (b) 3.60

10.73. (a) $E = 0.059$ V (b) $\Delta G = -5.7$ kJ·mol^{-1}

10.74. (a) 222 kJ

10.75. (a) 0.020 (b) $[Cu^{2+}] = 1.7$ M, $[Zn^{2+}] = 0.034$ M (c) 1.7 M (d) 1.1 V

CHAPTER 11

Consider This 11.9: (a) 2.6 kPa, 1.6 kPa

Check This 11.13: (a) 11.8 kPa (b) 1.30×10^{-3} mol (c) 1.32×10^{-3} mol

Check This 11.15: (b) 3.2×10^{-6} mol·s^{-1}

Check This 11.17: 3.7×10^{3} M·s^{-1}

Consider This 11.18: (b) slope $= 2.9 \times 10^{-3}$ s^{-1}

Check This 11.21: 8.3×10^{-3} M^{-1}·s^{-1}

Check This 11.35: (a) 14 days

Check This 11.37: (a) 1.21×10^{-4} years^{-1} (b) $12{,}800$ years

Check This 11.39: (c) 6×10^{-2} M^{-1}·s^{-1}

Consider This 11.42: (b) 6×10^{3} K

Check This 11.51: $k(273$ K$) = 6.8 \times 10^{-6}$ M^{-1}·s^{-1}; $k(373$ K$) = 2.5 \times 10^{-3}$ M^{-1}·s^{-1}

Check This 11.55: 300 kJ·mol^{-1}

Check This 11.57: (a) $\Delta G = 3.15 \times 10^{5}$ J·mol^{-1}

Check This 11.61: (a) $k = 10^{7}$ M^{-1}·s^{-1}; $A = 3 \times 10^{8}$ M^{-1}·s^{-1} (b) $k_2 = 10^{-4}$ M^{-1}·s^{-1}; $A = 3 \times 10^{9}$ M^{-1}·s^{-1}

Consider This 11.65: (a) 0.004 mol (b) 0.004 mol

Consider This 11.70: (a) 500 M^{-1}

Check This 11.73: 10^{4} mol

Check This 11.75: 1.8×10^{-5} g

End-of-Chapter Problems

11.5. (a) 0.00175 s, ≈ 570 discharge·s^{-1}

11.11. (d) 0.30 M·min^{-1} (e) 0.45 M·min^{-1} (f) 0.30 M·min^{-1}

11.12. (a) rate$(10.0$ s$) = -0.013$ M·s^{-1}, rate$(20.0$ s$) = -0.010$ M·s^{-1}, rate$(30.0$ s$) = -0.0075$ M·s^{-1} (c) rate$(0.0$ to 10.0 s$) = -0.0167$ M·s^{-1}, rate$(10.0$ to 20.0 s$) = -0.0119$ M·s^{-1}, rate$(20.0$ to 30.0 s$) = -0.0089$ M·s^{-1}, rate$(30.0$ to 40.0 s$) = -0.0070$ M·s^{-1}

11.13. (a) rate$(0$ to 90 min$) = -0.0019$ M·min^{-1}, rate$(90$ to 200 min$): = -0.0019$ M·min^{-1}, rate$(200$ to 360 min$) = -0.0013$ M·min^{-1}, rate$(360$ to 720 min$) = -0.00097$ M·min^{-1}

11.15. (a) rate$(1) = -7.1 \times 10^{-5}$ M·s^{-1}, rate$(2) = -7.0 \times 10^{-5}$ M·s^{-1}, rate$(3) = -6.8 \times 10^{-5}$ M·s^{-1} (b) -7.0×10^{-5} M·s^{-1} (average)

11.16. (a) -1.2×10^{-3} M·s^{-1} (b) 6×10^{-4} M·s^{-1}

11.18. (a) rate $(0$ to 400 s$) = 8.0 \times 10^{-3}$ kPa·s^{-1}, rate $(400$ to 1000 s$) = 6.0 \times 10^{-3}$ kPa·s^{-1}, rate $(1000$ to 1800 s$) = 4.8 \times 10^{-3}$ kPa·s^{-1} (b) 4.0×10^{-3} kPa·s^{-1}, 3.0×10^{-3} kPa·s^{-1}, and 2.4×10^{-3} kPa·s^{-1}

11.22. (a) 1.8×10^{-2} M^{-1}·s^{-1} (b) 7.5×10^{-6} M·s^{-1}

11.23. 3.1×10^{-3} M^{-1}·s^{-1}

11.24. (d) 3.53×10^{3} M^{-2}·s^{-1} (e) rate $= 0.0282$ M·s^{-1}, rate of disappearance $= 0.0564$ M·s^{-1}

11.25. (a) rate$(1) = -7.1 \times 10^{-5}$ M·s^{-1}, rate$(4) = -14 \times 10^{-5}$ M·s^{-1} (d) 4.7×10^{-3} s^{-1}

11.27. (e) 92 min

11.28. (b) rate $= 1.46 \times 10^{-2}$ kPa·s^{-1}, rate constant $= 1.46 \times 10^{-2}$ kPa·s^{-1}

11.29. (a) rate$_1 = -0.0021$ M·min^{-1}, rate$_2 = -0.00083$ M·min^{-1}, rate$_3 = -0.0013$ M·min^{-1} (c) rate$_1 = 0.0011$ M·min^{-1}, rate$_2 = 0.0004$ M·min^{-1}, rate$_3 = 0.0007$ M·min^{-1}, $k = 0.0013$ min^{-1} (d) 0.216 M

11.30. (c) -2.5×10^{-3} M·s^{-1}; -5.0×10^{-3} M·s^{-1} (e) decreased by a factor of eight

11.31. (a) 12.5 kPa (b) $1/20$, rate $= 8.3 \times 10^{-3}$ kPa·s^{-1} (c) 2.1×10^{-3} s^{-1}

11.38. (a) rates in M·s^{-1} are 7.1×10^{-5}, 7.1×10^{-5}, 5.8×10^{-5}, 2.6×10^{-5}, 0.93×10^{-5} (b) same as in (a)

11.39. (c) 0.3 V (d) 0.894 V

11.43. $\ln 2$

11.44. 14 weeks

11.45. 11.9×10^{3} yr

11.46. (a) 14 hr (b) ≈ 10 hr

11.47. 4.38×10^{-4} s^{-1}

11.48. 33 kPa

11.49. (c) 0.0385 min^{-1} (d) 18 min

11.50. 1.3×10^{6} bacteria·mL^{-1}

11.51. (a) about 1.3 days (b) 6.25×10^{-4} min^{-1} (d) 85.8 mL, 551 mL

11.53. (a) 80. kJ·mol^{-1} (b) 315 K

11.54. (b) 1.0 × 10^2 kJ·mol^{-1} (c) A = 7.9 × 10^9 s^{-1}

11.55. (a) E_a = 100. kJ·mol^{-1}, A = 1.15 × 10^{13} min^{-1}
(b) E_a = 102 kJ·mol^{-1}, A = 3.58 × 10^{13} min^{-1}

11.56. 38 kJ·mol^{-1}

11.57. 160 kJ·mol^{-1}

11.58. (a) 76 kJ·mol^{-1}

11.59. (a) 9.8 × 10^{-4} s^{-1} (b) 9.8 × 10^{-4} kPa·s^{-1} (c) 108 s

11.60. (a) E_a = 91.5 kJ·mol^{-1}, A = 5.9 × 10^{13} M^{-1}·s^{-1}
(c) 30 s

11.61. 64 kJ·mol^{-1}

11.66. (b) −76 kJ·mol^{-1} (c) −148 kJ·mol^{-1}
(d) 12 M, rate = 1.0 × 10^{-4} M·s^{-1} (e) 3.7 hr

11.68. (a) E_a(forward) = 100 kJ·mol^{-1} (b) 0.038
(c) 5.3 × 10^{-3} M^{-1}·s^{-1} (d) 2.0 × 10^{-4} s^{-1}
(e) A_f = 6.8 × 10^{13} s^{-1}, A_r = 1.8 × 10^5 M^{-1}·s^{-1}
(g) 0.11

11.70. (c) 1 to 10 (d) 1 to 100

11.71. (a) V$_{max}$ = 60 μmol·min^{-1} (c) ≈0 (d) 2.0 × 10^4 M^{-1}

11.72. (b) 32 μmol·min^{-1}

11.74. (a) V$_{max}$, = 1.1 M·s^{-1}, 1/K = 2.5 × 10^{-3} M
(b) V$_{max}$ = 1.087 M·s^{-1}, 1/K = 2.5 × 10^{-3}
(d) 0.25 M

11.75. 31 μmol·mg^{-1}·s^{-1}

11.76. (c) inhibitor absent: V$_{max}$ = 96 mmol·min^{-1},
K = 4.2 × 10^3 M^{-1}; inhibitor present:
V$_{max}$ = 96 mmol·min^{-1}, K = 2.3 × 10^3 M^{-1}

APPENDIX B

Thermodynamic Data at 298 K (25 °C)

Substance	$\Delta H°_f$, kJ·mol^{-1}	$S°$, J·K^{-1}·mol^{-1}	$\Delta G°_f$, kJ·mol^{-1}
Aluminum—Al(s)	0	28.33	0
Al^{3+}(aq)	−524.7	−321.7	−481.2
Al$_2$O$_3$(s)	−1675.7	50.92	−1582.3
Barium—Ba(s)	0	62.8	0
Ba^{2+}(aq)	−537.64	9.6	−560.77
BaCO$_3$(s)	−1216.3	112.1	−1137.6
BaSO$_4$(s)	−1465.2	132.2	−1353.1
Bromine—Br$_2$(l)	0	152.23	0
Br(g)	111.88	175.02	82.40
Br$_2$(g)	30.91	245.46	3.11
Br$^-$(aq)	−121.55	82.4	−103.96
HBr(g)	−36.40	198.70	−53.45
Calcium—Ca(s)	0	41.42	0
Ca^{2+}(aq)	−542.83	−53.1	−553.58
Ca(OH)$_2$(s)	−986.09	83.39	−898.49
CaCO$_3$(s), calcite	−1206.9	92.9	−1128.8
CaCO$_3$(s), aragonite	−1207.1	88.7	−1127.8
CaC$_2$(s)	−59.8	69.96	−64.9
CaF$_2$(s)	−1219.6	68.87	−1167.3
CaCl$_2$(s)	−795.8	104.6	−748.1
CaBr$_2$(s)	−682.8	130.	−663.6
CaSO$_4$(s)	−1434.11	106.7	−1321.79
Carbon—(also see pages A15–A16)			
C(s), graphite	0	5.740	0
C(s), diamond	1.895	2.377	2.900
C(g)	716.68	158.10	671.26
CO(g)	−110.53	197.67	−137.17
CO$_2$(g)	−393.51	213.74	−394.36
CO$_3^{2+}$(aq)	−677.14	−56.9	−527.81
Chlorine—Cl$_2$(g)	0	223.07	0
Cl(g)	121.68	165.20	105.68
Cl$^-$(aq)	−167.16	56.5	−131.23
HCl(g)	−92.31	186.91	−95.30

Substance	ΔH°_f, kJ·mol^{-1}	S°, J·K^{-1}·mol^{-1}	ΔG°_f, kJ·mol^{-1}
Copper—Cu(s)	0	33.15	0
Cu$^+$(aq)	71.67	40.6	49.98
Cu^{2+}(aq)	64.77	−99.6	65.49
CuSO$_4$(s)	−771.36	109	−661.8
CuSO$_4$·5H$_2$O(s)	−2279.7	300.4	−1879.7
Fluorine—F$_2$(g)	0	202.78	0
F(g)	78.99	158.75	61.91
F$^-$(aq)	−332.63	−13.8	−278.79
HF(g)	−271.1	173.78	−273.2
Hydrogen—H$_2$(g)	0	130.68	0
H(g)	217.97	114.71	203.25
H$^+$(aq)	0 (assigned)	0 (assigned)	0 (assigned)
H$_2$O(l)	−285.83	69.91	−237.13
H$_2$O(g)	−241.82	188.83	−228.57
H$_2$O$_2$(l)	−187.78	109.6	−120.35
H$_2$O$_2$(aq)	−191.17	143.9	−134.03
Deuterium—D$_2$(g)	0	144.96	0
D$_2$O(l)	−294.60	75.94	−243.44
D$_2$O(g)	−249.20	198.34	−234.54
Iodine—I$_2$(s)	0	116.14	0
I(g)	106.84	180.79	70.25
I$_2$(g)	62.44	260.69	19.33
I$^-$(aq)	−55.19	111.3	−51.57
HI(g)	26.48	206.59	1.70
Iron—Fe(s)	0	27.28	0
Fe^{2+}(aq)	−89.1	−137.7	−78.90
Fe^{3+}(aq)	−48.5	−315.9	−4.7
Fe$_3$O$_4$(s), magnetite	−1118.4	146.4	−1015.4
Fe$_2$O$_3$(s), hematite	−824.2	87.40	−742.2
FeS(s)	−100.0	60.29	−100.4
Lead—Pb(s)	0	64.81	0
Pb^{2+}(aq)	1.7	21.3	−24.43
PbO(s)	−217.86	69.45	−188.49
PbO$_2$(s)	−277.4	76.57	−217.33
PbSO$_4$(s)	−919.94	148.57	−813.14
PbCl$_2$(s)	−359.2	136.4	−313.97
PbBr$_2$(s)	−287.7	161.5	−261.92
PbI$_2$(s)	−175.1	177.0	−171.69

Substance	$\Delta H°_f$, kJ·mol^{-1}	$S°$, J·K^{-1}·mol^{-1}	$\Delta G°_f$, kJ·mol^{-1}
Lithium—Li(s)	0	28.0	0
Li$^+$(aq)	-278.46	14.2	-293.8
Li$_2$O(s)	-595.8		
LiOH(s)	-487.2	50.2	-443.9
LiCl(s)	-408.8		
LiBr(s)	-350.3		
Magnesium—Mg(s)	0	32.68	0
Mg^{2+}(aq)	-466.85	-138.1	-454.8
MgO(s)	-601.70	26.94	-569.43
Mg(OH)$_2$(s)	-924.66	63.1	-833.75
MgCO$_4$(s)	-1095.8	65.7	-1021.1
MgSO$_4$(s)	-1278.2	91.6	-1173.6
MgCl$_2$(s)	-641.8	89.5	-592.3
MgBr$_2$(s)	-524.3	117.2	-503.8
Mercury—Hg(l)	0	76.02	0
Hg(g)	61.32	174.96	31.82
Hg^{2+}(aq)			-164.38
HgO(s)	-90.83	70.29	-58.54
Hg$_2$Cl$_2$(s), calomel	-265.22	192.5	-210.75
HgCl$_2$(s)	-230.1		
Nitrogen—N$_2$(g)	0	191.61	0
N(g)	472.70	153.19	455.58
N$_2$O(g)	82.05	219.85	104.20
NO(g)	90.25	210.76	86.55
NO$_2$(g)	33.18	240.06	51.31
N$_2$O$_4$(g)	9.16	304.29	97.89
NO$_3^-$(aq)	-206.57	146.4	-110.5
NH$_3$(g)	-46.11	192.45	-16.45
NH$_3$(aq)	-80.29	111.3	-26.50
NH$_4^+$(aq)	-132.51	113.4	-79.31
N$_2$H$_4$(l)	50.63	121.21	149.34
NH$_4$Cl(s)	-314.43	94.6	-202.87
NH$_4$NO$_3$(s)	-365.56	151.08	-183.87
Oxygen—O$_2$(g)	0	205.14	0
O(g)	249.4	160.95	230.1
O$_3$(g)	142.7	238.93	163.2
OH$^-$(aq)	-229.99	-10.75	-157.24

Substance	$\Delta H°_f$, kJ·mol^{-1}	$S°$, J·K^{-1}·mol^{-1}	$\Delta G°_f$, kJ·mol^{-1}
Phosphorus—P(s), white	0	41.09	0
$P_4(g)$	58.91	279.98	24.44
$P_4O_{10}(s)$	−2984.0	228.86	−2697.0
$H_3PO_4(aq)$ or $(HO)_3PO(aq)$	−1288.34	158.2	−1142.54
$H_2PO_4^-(aq)$ or $(HO)_2PO_2^-(aq)$	−1302.48	89.1	−1135.1
$HPO_4^{2-}(aq)$ or $HOPO_2^{2-}(aq)$	−1298.7	−35.98	−1094.1
$PO_4^{3-}(aq)$	−1284.07	−217.57	−1025.59
Potassium—K(s)	0	64.18	0
$K^+(aq)$	−252.38	102.5	−283.27
$KOH(s)$	−424.76	78.9	−379.08
$KCl(s)$	−436.75	82.59	−409.14
$KBr(s)$	−393.80	95.90	−380.66
Silver—Ag(s)	0	42.55	0
$Ag^+(aq)$	105.58	72.68	77.11
$AgCl(s)$	−127.07	96.2	−109.79
$AgBr(s)$	−100.37	107.1	−96.90
$AgI(s)$	−61.84	115.5	−66.19
$AgNO_3(s)$	−123.1	140.9	−32.2
Sodium—Na(s)	0	51.21	0
$Na^+(aq)$	−240.12	59.0	−261.91
$NaOH(s)$	−425.61	64.46	−379.49
$NaCl(s)$	−411.15	72.13	−384.14
$NaBr(s)$	−361.06	86.82	−348.98
$NaI(s)$	−287.78	98.53	−286.06
$Na_2CO_3(s)$	−1130.9	135.98	−1047.67
$NaHCO_3(s)$	−947.68	102.09	−851.86
$Na_2SO_4(s)$	−1384.49	149.49	−1266.8
Sulfur—S(s), rhombic	0	31.80	0
S(s), monoclinic	0.33	32.6	0.1
$SO_2(g)$	−296.83	248.22	−300.19
$SO_3(g)$	−395.72	256.76	−371.06
$HSO_4^-(aq)$ or $HOSO_3^-(aq)$	−887.34	131.8	−755.91
$SO_4^{2-}(aq)$	−909.27	20.1	−744.53
$H_2S(g)$	−20.63	205.79	−33.56
$H_2S(aq)$	−39.7	121	−27.83
Zinc—Zn(s)	0	41.63	0
$Zn^{2+}(aq)$	−153.89	−112.1	−147.06
$ZnO(s)$	−348.28	43.64	−318.30
$ZnS(s)$	−202.9	57.7	−198.3
$ZnCl_2(s)$	−415.89	108.37	−369.26
$ZnSO_4(s)$	−978.6	124.7	−871.6

Compounds of Carbon

Name	Formula(state)	$\Delta H°_f$, kJ·mol^{-1}	$S°$, J·K^{-1}·mol^{-1}	$\Delta G°_f$, kJ·mol^{-1}
Hydrocarbons				
methane	$CH_4(g)$	−74.81	186.26	−50.72
ethane	$C_2H_6(g)$	−84.68	229.60	−32.82
propane	$C_3H_8(g)$	−103.85	270.2	−23.49
butane	$C_4H_{10}(g)$	−126.15	310.1	−17.03
pentane	$C_5H_{12}(g)$	−146.44	349	−8.20
ethene	$C_2H_4(g)$	52.26	219.56	68.15
propene	$C_3H_6(g)$	20.42	266.6	62.78
1-butene	$C_4H_8(g)$	1.17	307.4	
1-pentene	$C_5H_{10}(g)$	−20.92	347.6	
ethyne	$C_2H_2(g)$	226.73	200.94	209.20
cyclopropane	$C_3H_6(g)$	53.30	237.4	104.45
cyclobutane	$C_4H_8(g)$	26.65	265.4	
cyclopentane	$C_5H_{10}(g)$	−77.24	292.9	
cyclohexane	$C_6H_{12}(g)$	−123.1	298.2	
	$C_6H_{12}(l)$	−156.4	204.4	26.7
benzene	$C_6H_6(g)$	82.9	269.31	129.72
	$C_6H_6(l)$	49.0	173.3	124.3
Alcohols				
methanol	$CH_3OH(g)$	−200.66	239.81	−161.96
	$CH_3OH(l)$	−238.86	126.8	−166.27
	$CH_3OH(aq)$			−175.23
ethanol	$C_2H_5OH(g)$	−235.10	282.70	−168.49
	$C_2H_5OH(l)$	−277.69	160.7	−174.78
	$C_2H_5OH(aq)$			−180.92
glycerol	$C_3H_5(OH)_3(l)$	−668.6	204.47	−477.06
	$C_3H_5(OH)_3(aq)$			−488.52
Aldehydes and Ketones				
methanal (formaldehyde)	$HCHO(g)$	−115.90	218.78	−109.91
	$HCHO(aq)$			−130.5
ethanal (acetaldehyde)	$CH_3CHO(g)$	−166.36	264.22	−133.30
	$CH_3CHO(l)$	−192.30	160.2	−128.12
	$CH_3CHO(aq)$			−139.24
propanone (acetone)	$(CH_3)_2CO(l)$	−248.1	200.4	−155.39
	$(CH_3)_2CO(aq)$			−161.00

Name	Formula(state)	$\Delta H°_f$, kJ·mol^{-1}	$S°$, J·K^{-1}·mol^{-1}	$\Delta G°_f$, kJ·mol^{-1}
Carboxylic Acids/Ions				
methanoic (formic)	HCOOH(l)	−424.72	128.95	−361.35
ethanoic (acetic)	CH$_3$COOH(l)	−484.5	159.8	−389.9
	CH$_3$COOH(aq)	−485.76	86.6	−396.46
ethanoate (acetate) ion	CH$_3$COO$^-$(aq)			−372.334
oxalic	HOOCCOOH(s)	−827.2	120	−697.9
succinic	HOOCCH$_2$CH$_2$COOH(s)	−940.90	175.7	−747.43
	HOOCCH$_2$CH$_2$COOH(aq)			−746.22
succinate ion	$^-$OOCCH$_2$CH$_2$COO$^-$(aq)			−690.23
pyruvic	CH$_3$COCOOH(l)	−584.5	179.5	−463.38
pyruvate ion	CH$_3$COCOO$^-$(aq)			−474.33
lactic	CH$_3$CHOHCOOH(s)	−694.08	142.26	−522.92
lactate ion	CH$_3$CHOHCOO$^-$(aq)			−517.812
Nitrogen Containing				
hydrogen cyanide	HCN(g)	135.1	201.78	124.7
	HCN(l)	108.87	112.84	124.97
	HCN(aq)	107.1	124.7	119.7
cyanide ion	CN$^-$(aq)	151.0	118.0	165.7
urea	(NH$_2$)$_2$CO(s)	−333.19	104.6	−197.15
	(NH$_2$)$_2$CO(aq)	−319.2	173.85	−203.84
Amino Acids	**side group**			
glycine	—H(s)	−537.2	103.51	−377.69
	—H(aq)			−379.9
alanine	—CH$_3$(s)	−562.7	129.20	−370.24
	—CH$_3$(aq)			−371.71
leucine	—CH$_2$CH(CH$_3$)$_2$(s)	−646.8	211.79	−357.06
	—CH$_2$CH(CH$_3$)$_2$(aq)			−353.09
cysteine	—CH$_2$SH(s)	−533.9	169.9	−343.97
	—CH$_2$SH(aq)			−340.33
aspartic acid	—CH$_2$COOH(s)	−973.37	170.12	−730.23
	—CH$_2$COOH(aq)			−719.98
aspartate ion	—CH$_2$COO$^-$(aq)			−698.69
Sugars				
glucose	C$_6$H$_{12}$O$_6$(s)	−1274.4	212.1	−910.52
	C$_6$H$_{12}$O$_6$(aq)			−917.47
fructose	C$_6$H$_{12}$O$_6$(s)	−1272		
	C$_6$H$_{12}$O$_6$(aq)			
galactose	C$_6$H$_{12}$O$_6$(s)	−1285.37	205.4	−919.43
	C$_6$H$_{12}$O$_6$(aq)			−924.58
sucrose	C$_{12}$H$_{22}$O$_{11}$(s)	−2222.1	360.2	−1544.65
	C$_{12}$H$_{22}$O$_{11}$(aq)			−1551.76
lactose	C$_{12}$H$_{22}$O$_{11}$(s)	−2236.72	386.2	−1566.99
	C$_{12}$H$_{22}$O$_{11}$(aq)			−1569.92

APPENDIX C

Standard Reduction Potentials: $E°$ and $E°'$ (pH 7)

Half-cell contents	Half-cell reaction	$E°$, V	$E°'$(pH 7), V
Strongest oxidizing agent			
$F_2|F^-$	$F_2 + 2e^- \rightleftharpoons 2F^-$	2.87	
H_4XeO_6, XeO_3	$H_4XeO_6 + 2H^+ + 2e^- \rightleftharpoons XeO_3 + 3H_2O$	2.38	
Co^{3+}, Co^{2+}	$Co^{3+} + e^- \rightleftharpoons Co^{2+}$	1.92	
H_2O_2, H_2O	$H_2O_2 + 2H^+ + 2e- \rightleftharpoons 2H_2O$	1.763	
PbO_2, $SO_4^{2-}|PbSO_4$	$PbO_2 + SO_4^{2-} + 4H^+ + 2e^- \rightleftharpoons PbSO_4 + 2H_2O$	1.658	
$HClO|Cl_2$	$2HClO + 2H^+ + 2e^- \rightleftharpoons Cl_2 + 2H_2O$	1.630	
MnO_4^-, Mn^{2+}	$MnO_4^- + 8H^+ + 5e^- \rightleftharpoons Mn^{2+} + 4H_2O$	1.507	
$HOI|I_2$	$2HOI + 2H^+ + 2e^- \rightleftharpoons I_2 + 2H_2O$	1.430	
$Cr_2O_7^{2-}$, Cr^{3+}	$Cr_2O_7^{2-} + 14H^+ + 6e^- \rightleftharpoons 2Cr^{3+} + 7H_2O$	1.36	
$Cl_2|Cl^-$	$Cl_2 + 2e^- \rightleftharpoons 2Cl^-$	1.359	
chlph(II)$^+$|chlph(II)	chlph(II)$^+ + e^- \rightleftharpoons$ chlph(II)		0.9
$MnO_2|Mn^{2+}$	$MnO_2 + 4H^+ + 2e^- \rightleftharpoons Mn^{2+} + 2H_2O$	1.230	
$O_2|H_2O$	$O_2 + 4H^+ + 4e^- \rightleftharpoons 2H_2O$	1.229	0.816
$IO_3^-|I_2$	$2IO_3^- + 12H^+ + 10e^- \rightleftharpoons I_2 + 6H_2O$	1.210	
$Br_2|Br^-$	$Br_2 + 2e^- \rightleftharpoons 2Br^-$	1.087	
$Fe(o\text{-}phen)_3^{3+}$, $Fe(o\text{-}phen)_3^{2+}$	$Fe(o\text{-}phen)_3^{3+} + e^- \rightleftharpoons Fe(o\text{-}phen)_3^{2+}$	1.06	
$NO_3^-|NO$	$NO_3^- + 4H^+ + 3e^- \rightleftharpoons NO + 2H_2O$	0.955	
$Hg^{2+}|Hg$	$Hg^{2+} + 2e^- \rightleftharpoons Hg$	0.908	
$Ag^+|Ag$	$Ag^+ + e^- \rightleftharpoons Ag$	0.799	
$Hg_2^{2+}|Hg$	$Hg_2^{2+} + 2e^- \rightleftharpoons 2Hg$	0.796	
Fe^{3+}, Fe^{2+}	$Fe^{3+} + e^- \rightleftharpoons Fe^{2+}$	0.771	
cytochrome a_3	Fe(III) $+ e^- \rightleftharpoons$ Fe(II)		0.55
chlph(I)$^+$|chlph(I)	chlph(I)$^+ + e^- \rightleftharpoons$ chlph(I)		0.4
Qu, QuH$_2$ (quinhydrone)	$Qu + 2H^+ + 2e^- \rightleftharpoons QuH_2$	0.699	0.286
$O_2|H_2O_2$	$O_2 + 2H^+ + 2e^- \rightleftharpoons H_2O_2$	0.69	0.295
methylene blue	$MB(ox) + 2H^+ + 2e^- \rightleftharpoons MB(red)$	0.54	0.01
$I_2|I^-$	$I_2 + 2e^- \rightleftharpoons 2I^-$	0.5355	
$NiO_2|Ni(OH)_2$	$NiO_2 + 2H_2O + 2e^- \rightleftharpoons Ni(OH)_2 + 2OH^-$	0.49	
$O_2|OH^-$	$O_2 + 2H_2O + 4e^- \rightleftharpoons 4HO^-$	0.403	
vitamin C (ascorbic acid)	dehydroascorbate $+ 2H^+ + 2e^- \rightleftharpoons$ ascorbate $+ H_2O$	0.390	
$Fe(CN)_6^{3-}$, $Fe(CN)_6^{4-}$	$Fe(CN)_6^{3-} + e^- \rightleftharpoons Fe(CN)_6^{4-}$	0.36	
$Cu^{2+}|Cu$	$Cu^{2+} + 2e^- \rightleftharpoons Cu$	0.337	
cytochrome a	Fe(III) $+ e^- \rightleftharpoons$ Fe(II)		0.290
$Hg_2Cl_2|Hg$	$Hg_2Cl_2 + 2e^- \rightleftharpoons 2Hg + 2Cl^-$	0.268	
$AgCl|Ag$	$AgCl + e^- \rightleftharpoons Ag + Cl^-$	0.222	
cytochrome c	Fe(III) $+ e^- \rightleftharpoons$ Fe(II)		0.254
hemoglobin	Fe(III) $+ e^- \rightleftharpoons$ Fe(II)		0.17
Cu^{2+}, Cu^+	$Cu^{2+} + e^- \rightleftharpoons Cu^+$	0.161	

Half-cell contents	Half-cell reaction	$E°$, V	$E°'$(pH 7), V
$S\|H_2S$	$S + 2H^+ + 2e^- \rightleftharpoons H_2S$	0.144	
Sn^{4+}, Sn^{2+}	$Sn^{4+} + 2e^- \rightleftharpoons Sn^{2+}$	0.139	
UQ, UQH_2 (ubiquinone)	$UQ + 2H^+ + 2e^- \rightleftharpoons UQH_2$		0.10
cytochrome b	$Fe(III) + e^- \rightleftharpoons Fe(II)$		0.077
myoglobin	$Fe(III) + e^- \rightleftharpoons Fe(II)$		0.046
fumarate, succinate	$^-O_2CCH{=}CHCO_2^- + 2H^+ + 2e^- \rightleftharpoons {}^-O_2CCH_2CH_2CO_2^-$		0.031
$HgO\|Hg$, OH^-	$HgO + H_2O + 2e^- \rightleftharpoons Hg + 2OH^-$	0.098	
$S_4O_6^{2-}$, $S_2O_3^{2-}$	$S_4O_6^{2-} + 2e^- \rightleftharpoons 2S_2O_3^{2-}$	0.09	
(**SHE**) $H_3O^+\|H_2$	$2H_3O^+ + 2e^- \rightleftharpoons H_2 + 2H_2O$	0.000	
$Fe^{3+}\|Fe$	$Fe^{3+} + 3e^- \rightleftharpoons Fe$	−0.04	
$Pb^{2+}\|Pb$	$Pb^{2+} + 2e^- \rightleftharpoons Pb$	−0.13	
$Sn^{2+}\|Sn$	$Sn^{2+} + 2e^- \rightleftharpoons Sn$	−0.14	
pyruvate, ethanol	$CH_3COCO_2^- + 3H^+ + 2e^- \rightleftharpoons CH_3CH_2OH + CO_2$		−0.14
FAD, $FADH_2$	$FAD + 2H^+ + 2e^- \rightleftharpoons FADH_2$		−0.18
pyruvate, lactate	$CH_3COCO_2^- + 2H^+ + 2e^- \rightleftharpoons CH_3CHOHCO_2^-$		−0.19
acetaldehyde, ethanol	$CH_3CHO + 2H^+ + 2e^- \rightleftharpoons CH_3CH_2OH$		−0.20
$Ni^{2+}\|Ni$	$Ni^{2+} + 2e^- \rightleftharpoons Ni$	−0.23	
$S\|H_2S$	$S + 2H^+ + 2e^- \rightleftharpoons H_2S$		−0.23
1,3-DPG, G3P, P_i	$^{2-}O_3POCH_2CHOHCOOPO_3^{2-} + 2H^+ + 2e^- \rightleftharpoons$ $^{2-}O_3POCH_2CHOHCHO + HOPO_3^{2-}$		−0.29
$PbSO_4\|Pb$	$PbSO_4 + 2e^- \rightleftharpoons Pb + SO_4^{2-}$	−0.356	
$Cd^{2+}\|Cd$	$Cd^{2+} + 2e^- \rightleftharpoons Cd$	−0.40	
$Fe^{2+}\|Fe$	$Fe^{2+} + 2e^- \rightleftharpoons Fe$	−0.41	
NAD^+, NADH	$NAD^+ + H^+ + 2e^- \rightleftharpoons NADH$	−0.105	−0.32
$NADP^+$, NADPH	$NADP^+ + H^+ + 2e^- \rightleftharpoons NADPH$	−0.105	−0.32
$H_3O^+\|H_2$	$2H_3O^+ + 2e^- \rightleftharpoons H_2 + 2H_2O$		−0.414
ferredoxin	$Fe(III) + e^- \rightleftharpoons Fe(II)$		−0.43
gluconate, glucose	$C_5H_{11}O_5CO_2^- + 3H^+ + 2e^- \rightleftharpoons C_5H_{11}O_5CHO + H_2O$		−0.44
3PG, G3P	$^{2-}O_3POCH_2CHOHCO_2^- + 3H^+ + 2e^- \rightleftharpoons$ $^{2-}O_3POCH_2CHOHCHO + H_2O$		−0.55
acetate, acetaldehyde	$CH_3CO_2^- + 3H^+ + 2e^- \rightleftharpoons CH_3CHO + H_2O$		−0.60
$Cr^{3+}\|Cr$	$Cr^{3+} + 3e^- \rightleftharpoons Cr$	−0.74	
$Zn^{2+}\|Zn$	$Zn^{2+} + 2e^- \rightleftharpoons Zn$	−0.763	
$Cd(OH)_2\|Cd$	$Cd(OH)_2 + 2e^- \rightleftharpoons Cd + 2OH^-$	−0.81	
$H_2O\|H_2$	$2H_2O + 2e^- \rightleftharpoons H_2 + 2OH^-$	−0.8281	
$Cr^{2+}\|Cr$	$Cr^{2+} + 2e^- \rightleftharpoons Cr$	−0.89	
$Zn(OH)_2\|Zn$, OH^-	$Zn(OH)_2 + 2e^- \rightleftharpoons Zn + 2OH^-$	−1.25	
$Al^{3+}\|Al$	$Al^{3+} + 3e^- \rightleftharpoons Al$	−1.66	
$Al(OH)_3\|Al$	$Al(OH)_3 + 3e^- \rightleftharpoons Al + 3OH^-$	−2.33	
$Mg^{2+}\|Mg$	$Mg^{2+} + 2e^- \rightleftharpoons Mg$	−2.38	
$Na^+\|Na$	$Na^+ + e^- \rightleftharpoons Na$	−2.71	
$Ca^{2+}\|Ca$	$Ca^{2+} + 2e^- \rightleftharpoons Ca$	−2.76	
$K^+\|K$	$K^+ + e^- \rightleftharpoons K$	−2.92	
$Li^+\|Li$	$Li^+ + e^- \rightleftharpoons Li$	−3.045	

Strongest reducing agent

Photo Credits

Chapter 1

xxxii Arno Gasteiger/PhotoNewZealand.com; **2** (top) J.A. Bell, (bottom) Jones and Atkins, *Chemical Principles* CD-ROM (W. H. Freeman and Company); **3** Richard Megna/Fundamental Photographs; **5** J.A. Bell; **6** (all) J.A. Bell and I.D. Eubanks; **8** David Madison/Bruce Coleman Inc.; **32** (e) Kristian Hilsen/StoneGetty Images; **44** J.A. Bell; **51** Courtesy of NASA/JPL/Caltech.

Chapter 2

72 Craig Aurness/CORBIS, Wothe/Premium Stock/Picture-Quest; **74** J.A. Bell and I.D. Eubanks; **78** I.D. Eubanks; **95** (left) Dr. E. Walker/Photo Researchers, Inc., (right) Susumu Nishinaga/Photo Researchers Inc.; **103** (both) J.A. Bell; **104** J.A. Bell and I.D. Eubanks; **112** J.A. Bell; **131** J.A. Bell; **139** J.A. Bell, art from CD-ROM for Purves et al., *Life: The Science of Biology*, 6th ed. (Sinauer Associates, Inc., 2001); **142** Richard T. Nowitz/CORBIS.

Chapter 3

154 European Southern Observatory; **156** (top) David Parker/Science Photo Library/Photo Researchers, Inc., (bottom) Richard Megna/Fundamental Photographs; **158** © 1994 Wabash Instrument Corp., Fundamental Photographs, NYC; **159** spectra are from http://imagine.gsfc.nasa.gov/docs/science/try_12/spectra.html; **170** Tom Stewart/CORBIS; **175** Wellcome Department of Cognitive Neurology/Science Photo Library/Photo Researchers, Inc.; **176** (top) Courtesy of Oak Ridge Associated Universities, (bottom) J.A. Bell; **180** David Parker/Photo Researchers Inc.; **189** (top) McQuarrie and Rock, *General Chemistry* (W. H. Freeman and Company, 1984); **199** European Southern Observatory; **206** Herron, Kukla, Schrader, Erickson, and DiSpezio, *Heath Chemistry* (D.C. Heath and Co., 1987).

Chapter 4

212 PeriodicTable.com, Theodor Benfry, and Gary Katz; **215** (both) Richard Megna/Fundamental Photographs; **218** Wabash Instrument Corporation; **223** Pauling, *General Chemistry* (W. H. Freeman and Company, 1970); **224** Pauling, *General Chemistry* (W. H. Freeman and Company, 1970); **226** Loren Winters/Visuals Unlimited; **232** (all) M.A. Scharberg and S. E. Branz; **237** © 1994 Wabash Instrument Corp. Fundamental Photographs, NYC; **242** PSSC Matter Waves, Education Development Center, Cambridge, MA as published in Bailar, Moeller, et al., *Chemistry* (Academic Press, 1984); **258** Theodor Benfey, http://140.198.18.108/periodic/ spiraltable.html; **262** Fletcher and Rossing, *The Physics of Musical Instruments* (Springer, 1998); **278** PeriodicTable.com; **279** Science Photo Library/Photo Researchers, Inc.

Chapter 5

280 *Science*, Vol. 289, 2000. © 2003 American Association for the Advancement of Science; **282** © 2003 Richard Megna/Fundamental Photographs, NYC; **301** © 2003 Richard Megna/Fundamental Photographs, NYC; **306** Fletcher and Rossing, *The Physics of Musical Instruments* (Springer, 1998).

Chapter 6

352 © 2003 Richard Megna/Fundamental Photographs, NYC, Fibers Division, Monsanto Chemical Company, Courtesy of Tina Weatherby Carvalho, BEMF; **361** M.A. Scharberg and S. E. Branz; **386** (both) J.A. Bell; **388** (background) W. H. Freeman and Company; **400** Peticolas/Megna/Fundamental Photographs, NYC; **402** Ken Karp; **423** (both) M.A. Scharberg and S. E. Branz.

Chapter 7

436 © 2003 Richard Megna/Fundamental Photographs, NYC; **437** George Stephenson's steam locomotive, *The Rocket*; **440** J.A. Bell; **442** M.A. Scharberg and S.E. Branz; **443** J.A. Bell; **446** (center) James L. Amos/CORBIS, (left inset) Tom Bean, (right inset) George H. H. Huey/CORBIS; **447** J.A. Bell; **489** (left) George Stephenson's steam locomotive, *The Rocket*; **497** R. Silberman.

Chapter 8

512 Courtesy of J. D. Robertson as appeared in Lodish, et al., *Molecular Cell Biology*, 4th edition (W.H. Freeman and Company, 2000); **514** M.A. Scharberg; **543** (all) J.A. Bell; **552** (both) Charles Fisher, Ph.D., Biology Dept., Pennsylvania State University; **570** I.D. Eubanks.

Chapter 9

586 © Susumu Nishinaga/Photo Researchers, Inc., © Jackie Lewin, Royal Free Hospital/Photo Researchers, Inc.; **591** J.A. Bell; **592** J.A. Bell; **594** J.A. Bell; **623** 383;PDB ID:ILXA; Raetz, R., H. & Roderick, S. L. (1995), *Science*; **643** Biophoto Associates; **650** (both) J.A. Bell.

Chapter 10

664 (a) Don W. Fawcett & Keith Porter/Photo Researchers, Inc., (b) From *Science*, 1998, *281*, 64-71; corresponding authors are Bing K. Jap, Lawrence Berkeley Lab and So Iwata, Dept. of Biochemistry, Uppsala U, Sweden, (c) From an article by Chen, Barton, Binyamin, Gao, Zhang, Kim, and Heller (the principal author) in *J. Am. Chem. Soc.* 2001, *123*, 8630-8631. Professor Adam Heller, Department of Chemical Engineering and the Texas Materials Institute, The University of Texas at Austin, Austin, TX 78712, and has given permission to use this photomicrograph.; **673** Alcan, Inc.; **675** Charles E. Rotkin/CORBIS; **692** http://www.ectechnic.co.uk/HOME.HTML; **695** J.A. Bell; **721** From an article by Chen, Barton, Binyamin, Gao, Zhang, Kim, and Heller (the principal author) in *J. Am. Chem. Soc.* 2001, *123*, 8630-8631. Professor Adam Heller, Department of Chemical Engineering and the Texas Materials Institute, The University of Texas at Austin, Austin, TX 78712, and has given permission to use this photomicrograph.

Chapter 11

742 (top and middle) Courtesy of Thomas Eisner, Cornell University, (bottom) E.C. 1.11.1.6; Rossmann, M.G., Proc Natl Acad Sci U S A 82 pp. 1604 (1985); **749** J.A. Bell; **792** E.C. 1.11.1.6; Rossmann, M.G., Proc Natl Acad Sci U S A 82 pp. 1604 (1985); **792** E.C. 1.11.1.6, Rossmann, M.G., Proc Natl Acad Sci U S A 82 pp. 1604 (1985); **815** From "Sssss is for danger," *New Scientist*, 11 Dec. 1999 (#2216), p. 31.

INDEX

Page numbers in bold denote term definitions.

THE ELEMENTS

Element	Symbol	Atomic number	Relative atomic mass	Element	Symbol	Atomic number	Relative atomic mass
Actinium	Ac	89	(227)	Mendelevium	Md	101	(258)
Aluminum	Al	13	26.98	Mercury	Hg	80	200.59
Americium	Am	95	(243)	Molybdenum	Mo	42	95.94
Antimony	Sb	51	121.75	Neodymium	Nd	60	144.24
Argon	Ar	18	39.95	Neon	Ne	10	20.18
Arsenic	As	33	74.92	Neptunium	Np	93	(237)
Astatine	At	85	(210)	Nickel	Ni	28	58.71
Barium	Ba	56	137.34	Niobium	Nb	41	92.91
Berkelium	Bk	97	(247)	Nitrogen	N	7	14.01
Beryllium	Be	4	9.01	Nobelium	No	102	(255)
Bismuth	Bi	83	208.98	Osmium	Os	76	190.2
Bohrium	Bh	107	(262)	Oxygen	O	8	16.00
Boron	B	5	10.81	Palladium	Pd	46	106.4
Bromine	Br	35	79.91	Phosphorus	P	15	30.97
Cadmium	Cd	48	112.40	Platinum	Pt	78	195.09
Calcium	Ca	20	40.08	Plutonium	Pu	94	(244)
Californium	Cf	98	(251)	Polonium	Po	84	(209)
Carbon	C	6	12.01	Potassium	K	19	39.10
Cerium	Ce	58	140.12	Praseodymium	Pr	59	140.91
Cesium	Cs	55	132.91	Promethium	Pm	61	(145)
Chlorine	Cl	17	35.45	Protactinium	Pa	91	(231)
Chromium	Cr	24	52.00	Radium	Ra	88	(226)
Cobalt	Co	27	58.93	Radon	Rn	86	(222)
Copper	Cu	29	63.54	Rhenium	Re	75	186.2
Curium	Cm	96	(247)	Rhodium	Rh	45	102.91
Dubnium	Db	105	(262)	Rubidium	Rb	37	85.47
Dysprosium	Dy	66	162.50	Ruthenium	Ru	44	101.07
Einsteinium	Es	99	(254)	Rutherfordium	Rf	104	(261)
Erbium	Er	68	167.26	Samarium	Sm	62	150.35
Europium	Eu	63	151.96	Scandium	Sc	21	44.96
Fermium	Fm	100	(257)	Seaborgium	Sg	106	(263)
Fluorine	F	9	19.00	Selenium	Se	34	78.96
Francium	Fr	87	(223)	Silicon	Si	14	28.09
Gadolinium	Gd	64	157.25	Silver	Ag	47	107.87
Gallium	Ga	31	69.72	Sodium	Na	11	22.99
Germanium	Ge	32	72.59	Strontium	Sr	38	87.62
Gold	Au	79	196.97	Sulfur	S	16	32.06
Hafnium	Hf	72	178.49	Tantalum	Ta	73	180.95
Hassium	Hs	108	(265)	Technetium	Tc	43	(99)
Helium	He	2	4.00	Tellurium	Te	52	127.60
Holmium	Ho	67	164.93	Terbium	Tb	65	158.92
Hydrogen	H	1	1.0079	Thallium	Tl	81	204.37
Indium	In	49	114.82	Thorium	Th	90	232.0
Iodine	I	53	126.90	Thulium	Tm	69	168.93
Iridium	Ir	77	192.2	Tin	Sn	50	118.69
Iron	Fe	26	55.85	Titanium	Ti	22	47.88
Krypton	Kr	36	83.80	Tungsten	W	74	183.85
Lanthanum	La	57	138.91	Uranium	U	92	238.03
Lawrencium	Lr	103	(262)	Vanadium	V	23	50.94
Lead	Pb	82	207.19	Xenon	Xe	54	131.30
Lithium	Li	3	6.94	Ytterbium	Yb	70	173.04
Lutetium	Lu	71	174.97	Yttrium	Y	39	88.91
Magnesium	Mg	12	24.31	Zinc	Zn	30	65.37
Manganese	Mn	25	54.94	Zirconium	Zr	40	91.22
Meitnerium	Mt	109	(266)				